List of the Elements with Their Atomic Symbols and Atomic Weights

Name	Symbol	Atomic Number	Atomic Weight	Name	Symbol	Atomic Number	Atomic Weight
Actinium	Ac	89	227.028	Mendelevium	Md	101	(258)
Aluminum	Al	13	26.9815	Mercury	Hg	80	200.59
Americium	Am	95	(243)	Molybdenum	Mo	42	95.94
Antimony	Sb	51	121.76	Neodymium	Nd	60	144.24
Argon	Ar	18	39.948	Neon	Ne	10	20.1797
Arsenic	As	33	74.9216	Neptunium	Np	93	237.048
Astatine	At	85	(210)	Nickel	Ni	28	58.693
Barium	Ba	56	137.327	Niobium	Nb	41	92.9064
Berkelium	Bk	97	(247)	Nitrogen	N	7	14.0067
Beryllium	Be	4	9.01218	Nobelium	No	102	(259)
Bismuth	Bi	83	208.980	Osmium	Os	76	190.23
Bohrium	Bh	107	(262)	Oxygen	O	8	15.9994
Boron	B	5	10.811	Palladium	Pd	46	106.42
Bromine	Br	35	79.904	Phosphorus	P	15	30.9738
Cadmium	Cd	48	112.411	Platinum	Pt	78	195.08
Calcium	Ca	20	40.078	Plutonium	Pu	94	(244)
Californium	Cf	98	(251)	Polonium	Po	84	(209)
Carbon	C	6	12.011	Potassium	K	19	39.0983
Cerium	Ce	58	140.115	Praseodymium	Pr	59	140.908
Cesium	Cs	55	132.905	Promethium	Pm	61	(145)
Chlorine	Cl	17	35.4527	Protactinium	Pa	91	231.036
Chromium	Cr	24	51.9961	Radium	Ra	88	226.025
Cobalt	Co	27	58.9332	Radon	Rn	86	(222)
Copper	Cu	29	63.546	Rhenium	Re	75	186.207
Curium	Cm	96	(247)	Rhodium	Rh	45	102.906
Dubnium	Db	105	(262)	Rubidium	Rb	37	85.4678
Dysprosium	Dy	66	162.50	Ruthenium	Ru	44	101.07
Einsteinium	Es	99	(252)	Rutherfordium	Rf	104	(261)
Erbium	Er	68	167.26	Samarium	Sm	62	150.36
Europium	Eu	63	151.965	Scandium	Sc	21	44.9559
Fermium	Fm	100	(257)	Seaborgium	Sg	106	(263)
Fluorine	F	9	18.9984	Selenium	Se	34	78.96
Francium	Fr	87	(223)	Silicon	Si	14	28.0855
Gadolinium	Gd	64	157.25	Silver	Ag	47	107.868
Gallium	Ga	31	69.723	Sodium	Na	11	22.9898
Germanium	Ge	32	72.61	Strontium	Sr	38	87.62
Gold	Au	79	196.967	Sulfur	S	16	32.066
Hafnium	Hf	72	178.49	Tantalum	Ta	73	180.948
Hassium	Hs	108	(265)	Technetium	Tc	43	(98)
Helium	He	2	4.00260	Tellurium	Te	52	127.60
Holmium	Ho	67	164.930	Terbium	Tb	65	158.925
Hydrogen	H	1	1.00794	Thallium	Tl	81	204.383
Indium	In	49	114.818	Thorium	Th	90	232.038
Iodine	I	53	126.904	Thulium	Tm	69	168.934
Iridium	Ir	77	192.22	Tin	Sn	50	118.710
Iron	Fe	26	55.847	Titanium	Ti	22	47.88
Krypton	Kr	36	83.80	Tungsten	W	74	183.84
Lanthanum	La	57	138.906	Uranium	U	92	238.029
Lawrencium	Lr	103	(260)	Vanadium	V	23	50.9415
Lead	Pb	82	207.2	Xenon	Xe	54	131.29
Lithium	Li	3	6.941	Ytterbium	Yb	70	173.04
Lutetium	Lu	71	174.967	Yttrium	Y	39	88.9059
Magnesium	Mg	12	24.3050	Zinc	Zn	30	65.39
Manganese	Mn	25	54.9381	Zirconium	Zr	40	91.224
Meitnerium	Mt	109	(266)				

Media for Improved Performance

The course in general, organic, and biological chemistry for allied health majors has undergone some dramatic technological shifts in recent years:

- More classrooms and instructors each year have equipment for incorporating computer-aided presentation and interaction in the lecture.

- Online assessment is proving to be a strong force for success for students, and it greatly aids instructors in tracking the progress of their students.

- Interactive simulations and dynamic animations are often used to make a point where a written or spoken explanation sometimes isn't enough.

- Integration of the digital resources with the book into a coherent, combined program is often replacing the notion of a book accompanied by print and media ancillaries.

How the Media Update Edition Works

Prepare your lecture

The *Instructor Resource Center (IRC)* on CD/DVD contains lecture outlines from the Instructor Resource Manual, pre-built PowerPoint lecture presentations, a searchable library of all images from the textbook, and all interactive animations and movies for you to incorporate in your lecture.

Prepare your students - Help students prepare for the lecture by assigning a pre-lecture reading comprehension assignment using *Companion Website with Grade Tracker*.

Instructor

Prepare

Student

Focus

Use the practice questions on the *Companion Website* to check whether you understand the basic concepts before the lecture and use the Companion Website's links to the *Activebook* for questions where you are weak.

Catch up — The *Math Toolkit* will help you brush up on basic math skills, so that you can focus on understanding the chemistry.

Illustrate concepts in class

Using the art-only version, or adapting our pre-built lecture presentations from the pre-built lecture PowerPoint presentations on the IRC on CD/DVD, you can take advantage of the high-quality art and lecture guidelines from the book. Use our dynamic animations of microscopic processes, our video clips of difficult demonstrations, our 3D molecular models, and the "what-if" power of our simulations to explain concepts engagingly.

Interact in class - Our support for *Classroom Response Systems (CRS)*, including hardware discounts and "clicker" questions, lets you poll students in class and collect their responses quickly and anonymously. You can instantly determine what misconceptions are tripping up your students, and you will engage them as they discuss the answers to questions carefully designed for response systems.

Assign homework

Use *Companion Website with Grade Tracker* to assign homework online that is automatically graded and stores student scores in an easy-to-use grade book. Question banks include multiple-choice questions and interactive eMedia questions. **Create exams** — Our *TestGen* test-creation software lets you choose questions from deep question banks and format them flexibly for printed exams.

Guide student learning outside of class

There are practice problem sets in *Companion Website with Grade Tracker* that your students can work repeatedly, getting instantaneous feedback on each attempt until they establish mastery. You may elect to track student progress or allow them to work on their own.

Prepare for lab

Do laboratory-based exercises — The Virtual ChemLab Workbook is a realistic, simulated laboratory environment where students can get a feel for what to expect in a real wet lab, or conduct experiments that are not included in your department's laboratory program.

Choose a lab sequence — any of our general chemistry *Lab Manuals* have a wide selection of carefully designed laboratory experiments to follow your syllabus, or use our *Catalyst* system (*www.prenhall.com/catalyst*) to select exactly the experiments you will use from any of our lab manuals and create a customized lab manual for your course.

Lecture   Practice & Assess   Experiment

Follow the lecture

All of the interactive eMedia assets your instructor uses in the classroom are also available on the Companion Website—excellent for student review.

Participate in lecture — *Classroom Response Systems* let you "click" answers to questions posed in class, letting your instructor know on the fly whether you have misunderstood a concept or are confused. Though you might discuss the questions with fellow students, your response is anonymous and not shared with your peers.

Practice

Test your understanding by working through practice problems that provide instant feedback on your correct and incorrect answers. Repeat these assignments until you are comfortable with the material. *The Companion Website* provides practice questions, and your instructor may elect to assign those quizzes for credit in your class.

Prepare for lab

Optimize your valuable time in the laboratory by knowing what to expect when you arrive. Use the *Virtual ChemLab* activities to try out an experiment identical or similar to the experiment in the *Lab Manual*.

Prentice Hall is well poised to help instructors and students take advantage of these gains.

The McMurry/Castellion textbook has always provided a deep suite of digital resources both for the student and for the instructor. However, as instructors and students have become accustomed to using digital resources in their classes, they have also given us feedback about what is especially important and effective for them as well as suggestions on where we should direct our efforts going forward. We have listened to this advice, and in response, we have rethought not only the content delivered but also the structure of how it is delivered in McMurry/Castellion: **Fundamentals of General, Organic, and Biological Chemistry, 4e, Media Update Edition**:

Instructors have asked for

- greater support for digital presentation (PowerPoint™, for example) in the classroom, including in-class questions tailored for use with electronic *Classroom Response Systems* ("clickers") like those provided by InterWrite/PRS and HITT.

- a tracked, graded, online assessment environment that gives students instant feedback on their answers and stores detailed results in an instructor-accessible gradebook.

- assessment centered on the interactive simulations and animations that are used by the instructor in class.

- activities that show the practical application of the chemical principles that students learn in class.

Students have asked for

- practice questions with immediate feedback and stored results that they can use to gauge their mastery

- links to reading material on areas where they need to do more work.

- a way to access all the online resources regardless of what computer they are using or the speed of their internet connection.

All of these solutions, and more, are part of the **Media Update Edition**

The resources that ship with every copy of the text includes:

- An access code to the *Companion Website with Grade Tracker*. The online site features, among other reference resources, graded practice questions that are stored in a gradebook. The students get instant feedback on their answers, and are linked to section-specific reading in an online version of the textbook. Students can make use of Grade Tracker even without the instructor's participation, but can "join" their instructor's class if the instructor sets one up later.

- The *Virtual ChemLab Workbook and CD*. This CD contains software that simulates a real laboratory environment, providing nearly a true hands-on example of chemical principles in action. Guided activities are downloadable from the *Companion Website*.

Special for Instructors

- An *Instructor's Resource Center* on CD/DVD including image galleries from the book, prepared PowerPoint presentations, questions suited for Classroom Response Systems, and TestGen test generation software along with Word™ files of the testbank for the **Media Update Edition**.

- A *Companion Website with GradeTracker* for instructors. This enables the instructor to participate in the GradeTracker at any point during the course by permitting access to the site and allowing the creation of a private "class" for the students that the students already using GradeTracker can join. All results stored so far for each student automatically flow into the instructor's GradeTracker for that class. Instructor's can register for access to this site through *http://www.prenhall.com/tro*.

Fundamentals
of General, Organic,
and Biological Chemistry
Media Update Edition

FOURTH EDITION

John McMurry
Cornell University

Mary E. Castellion
Norwalk, Connecticut

Prentice Hall

Upper Saddle River, New Jersey 07458

Senior Editor: Kent Porter Hamann
Development Editor: Jeanne Allison
Production Editor: Donna F. Young
Executive Marketing Manager: Steve Sartori
Project Manager: Kristen Kaiser
Media Editor: Michael J. Richards
Editorial Assistants: Nancy Bauer and Jackie Howard
Photo Researcher: Diane Austin
Interior Designer: Anne Flanagan
Manufacturing Buyer: Alan Fischer
Director of Creative Services: Paul Belfanti
Senior Managing Editor, Audio & Visual Assets and
    Production: Patricia Burns
Art Development Manager: Jay McElroy
Art Studio: Artworks
        Production Manager: Sean Hogan
        Assistant Production Manager: Ronda Whitson
        Manager, Art Production Technologies: Matt Haas
        Art Project Coordinator: Connie Long
        Illustrators: Royce Copenheaver, Daniel Knopsnyder,
            Mark Landis and Stacy Smith
        Art Quality Assurance: Timothy Nugyen, Pamela Taylor

Art Editor: Tom Benfatti
Art Director: Jon Boylan
Cover Designer and Assistant to Art Director: John Christiana
Executive Managing Editor: Kathleen Schiaparelli
Vice President and Editor in Chief, Development:
    Ray Mullaney
Manager of Electronic Composition: Allyson Grasser
Assistant Manager of Electronic Composition: William Johnson
Electronic Page Makeup: Jackie Ambrosius
Managing Editor, Science Media: Nicole M. Jackson
Assistant Managing Editor, Supplements: Becca Richter
Copy Editor: Luana Richards
Proofreader: Sandra Price
Director, Image Resource Center: Melinda Reo
Manager, Rights and Permissions: Zina Arabia
Interior Image Specialist: Beth Boyd-Brenzel
Cover Image Specialist: Karen Sanatar
Image Permission Coordinator: Rennie Rieger
Cover Image: Michele Burgess/Stock Boston
Other image credits appear in the rearmatter.

© 2006, 2003, 1999, 1996, 1992 by Pearson Education, Inc.
Pearson Prentice Hall
Pearson Education, Inc.
Upper Saddle River, New Jersey 07458

Pearson Prentice Hall™ is a trademark of Pearson Education, Inc.

Printed in the United States of America
10 9 8 7 6 5 4 3 2 1

ISBN 0-13-148684-5

Pearson Education LTD., *London*
Pearson Education Australia PTY, Limited, *Sydney*
Pearson Education Singapore, Pte. Ltd
Pearson Education North Asia Ltd, *Hong Kong*
Pearson Education Canada, Ltd., *Toronto*
Pearson Educación de Mexico, S.A. de C.V.
Pearson Education—Japan, *Tokyo*
Pearson Education Malaysia, Pte. Ltd

Brief Contents

Contents

5 Molecular Compounds   98

6 Chemical Reactions: Classification and Mass Relationships   130

7 Chemical Reactions: Energy, Rates, and Equilibrium   164

24 Lipids   690

25 Lipid Metabolism   718

26 Nucleic Acids and Protein Synthesis   738

27 Genomics   768

28 Protein and Amino Acid Metabolism 788

29 Body Fluids 804

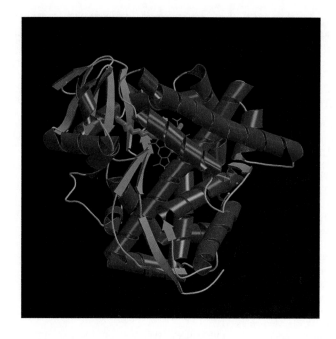

Appendixes A–1

Glossary A–8

Answers to Selected Problems A–15

Photo Credits A–49

Index A–51

Applications and Connections

Applications

Connections

Preface

This textbook is designed to put chemistry in the context of the allied health sciences as well as the other disciplines in which an understanding of the fundamentals of chemistry and of living things is valuable. The coverage in this fourth edition has been developed with considerable input from the many instructors and students who have used our previous three editions.

To teach chemistry all the way from "What is an atom?" to "How do we get energy from glucose?" is a challenge. Throughout our general chemistry and organic chemistry coverage, the focus is on concepts fundamental to the chemistry of living things, and illustrations are drawn from this chemistry wherever practical. In our biochemistry coverage we strive to meet the further challenge of providing a widely useful overview. Our goal is to provide enough detail for thorough understanding while avoiding so much detail that the students are overwhelmed.

The material covered is ample for a thorough, two-term introduction to general, organic, and biological chemistry. This edition retains our unique and well-received integrated biological chemistry sequence (described in greater detail below), which allows for the highest possible degree of flexibility. By varying the topics covered and the time devoted to them, the focus of the course can be adjusted to meet the needs of each instructor's student population and the duration of the course.

Students in this course have their sights set well beyond academic concerns and the laboratory bench. They want to know why: Why must I study the gas laws? Why are molecules important to me as a nurse, soil scientist, or nutritionist? To engage these students in the subject matter, we write in a clear and direct style. We also provide straightforward diagrams, modern molecular models, and thoroughly explained solved problems.

Organization

General Chemistry: Chapters 1–11   The introduction to elements, atoms, the periodic table, and the quantitative nature of chemistry (Chapters 1–3) is followed by chapters that individually highlight the nature of ionic and molecular compounds (Chapters 4 and 5). The next two chapters discuss chemical reactions and their stoichiometry, energies, rates, and equilibria (Chapters 6 and 7). Topics relevant to the chemistry of life follow: Gases, Liquids, and Solids (Chapter 8); Solutions (Chapter 9); and Acids and Bases (Chapter 10). Nuclear Chemistry (Chapter 11) closes the general chemistry sequence.

Organic Chemistry: Chapters 12–17   These chapters concisely focus on what students must know in order to get on with biochemistry. The introduction to hydrocarbons (Chapters 12 and 13) includes the basics of nomenclature, which is thereafter kept to a minimum. Discussion of functional groups with single bonds to oxygen, sulfur, or a halogen (Chapter 14) is followed by a short chapter on amines, which are so important to the chemistry of living things and drugs (Chapter 15). After introducing aldehydes and ketones (Chapter 16), the chemistry of carboxylic acids and their derivatives (including amides) is covered (Chapter 17), with a focus on similarities among the derivatives.

Biological Chemistry: Chapters 18–29   Rather than proceed through the complexities of protein, carbohydrate, lipid, and nucleic acid structure before getting to the roles of these compounds in the body, structure and function are integrated in this text. Protein structure (Chapter 18) is followed by enzyme and coenzyme chemistry (Chapter 19). After that we cover the function of hormones and neurotransmitters, and the action of drugs (Chapter 20). With enzymes introduced, the central pathways and themes of biochemical energy production can be described (Chapter 21). If the time you have available to cover biochemistry is limited, stop with Chapter 21 and your students will have an excellent preparation in the essentials of metabolism. The following chapters cover carbohydrate chemistry (Chapters 22 and 23), then lipid chemistry (Chapters 24 and 25). Next we discuss nucleic acids and protein synthesis (Chapter 26).

The most significant addition of content in this edition is the *new Chapter 27 on Genomics*. The mapping of the human genome has opened the door to a revolution in our understanding of health and disease. Our chapter on genomics is designed to give students an appreciation of this ongoing revolution and sufficient background to understand its future impact on their chosen fields of endeavor.

The last two chapters cover protein and amino acid metabolism (Chapter 28) and provide an overview of the chemistry of body fluids (Chapter 29). In this text, nutrition is not treated as a separate subject, but is integrated with the discussion of each type of biomolecule.

Focus on Learning

Worked Examples   Most Worked Examples, both quantitative and not quantitative, now include a *new Analysis section that precedes the Solution*. The Analysis lays out the approach to solving a problem of the given type. In numerical worked examples, a Ballpark Check that follows the Solution replaces the Ballpark Solution of the previous edition. In this location, it can be used to validate the answer obtained.

New Key Concept Problems are integrated throughout the chapters to focus attention on the use of essential concepts, as do the *Understanding Key Concepts* problems at the end of each chapter. Understanding Key Concepts problems are designed to test students' mastery of the core principles developed in the chapter. Students thus have an opportunity to ask "Did I get it?" before they proceed.

Problems   The problems within the chapters, for which brief answers are given in an appendix, cover every skill and topic to be understood. One or more problems, many of which are *new* to this edition, follow each Worked Example and others stand alone at the ends of sections.

More color-keyed, labeled equations   It is entirely too easy to skip looking at a chemical equation while reading. We have extensively used color to call attention to the aspects of chemical equations and structures under discussion, a continuing feature of this book that has been judged very helpful.

Molecular models   Additional computer-generated molecular models have been introduced and the use of *electrostatic-potential maps for molecular models is new to this edition*.

Key Words   Every key term is boldfaced on its first use, fully defined in the margin adjacent to that use, and listed at the end of the chapter. These are the terms students must understand to get on with the subject at hand. Definitions of all Key Words are collected in the Glossary.

End-of-Chapter Summaries   Here, the answers to the questions posed at the beginning of the chapter provide a summary of what is covered in that chapter. Where appropriate, the types of chemical reactions in a chapter are also summarized.

Focus on Relevancy

For those studying chemistry for the first and perhaps the only time, chemistry may not initially appear exciting. These students benefit from increased emphasis on the relevance and applications of chemistry. In this book, we carefully balance the need for solid science with the need for interesting examples and connections.

- **Applications** are both integrated into and set off from the text. They cover a wide variety of topics, *many new to this edition* and all updated where necessary. Each boxed application provides sufficient information for reasonable understanding and ends with a cross-reference to related end-of-chapter problems that can be assigned by the instructor and solved by the student to determine comprehension of the boxed material. Some favorite Applications retained from previous editions include Chemical Information, Protein Analysis by Electrophoresis, the Biochemistry of Running, and Tooth Decay. *Topics new to this edition* include Prions, Biocatalysis, Plant Hormones, Polysaccharides (their non-food applications), and DNA Fingerprinting.

- **Connections** essays, introduced in the previous edition, focus on practicing professionals who explain how they use chemistry in their daily work. The Connections from the previous edition (physical therapist, respiratory therapist, nuclear medicine technician, sensory evaluation manager, food technologist, dietician, veterinary lab technician) have been retained and new interviews have been added (art conservator, resource conservationist, forensic scientist). This feature is designed to remind students that the chemistry they are learning will benefit them down the road.

Focus on Making Connections Among General, Organic, and Biological Chemistry

This can be a difficult course to teach. Much of what students are interested in lies in the last part of the course, but the material they need to understand the biochemistry is found in the first two-thirds. It is easy to lose sight of the connections among general, organic, and biological chemistry so we have added a special *new* feature, *Concepts to Review*, to call attention to these connections. From Chapter 4 on, the Concepts to Review section at the beginning of the chapter lists topics covered in earlier chapters that form the basis for what is discussed in the current chapter.

We have also retained the successful concept link icons and Looking Ahead notes.

- **Concept link icons** (○○○) are used extensively to indicate places where previously covered material is relevant to the discussion at hand. These links provide for cross-references and also serve to highlight important chemical themes as they are revisited.

- **Looking Ahead notes** call attention to connections between just-covered material and discussions in forthcoming chapters. These notes are designed to illustrate to the students why what they are learning will be useful in what lies ahead.

Making It Easier to Teach: Supplements for Instructors

Instructor's Guide (0-13-045520-2) by Daniel M. Sullivan, University of Nebraska, features lecture outlines with presentation suggestions, teaching tips, suggested in-class demonstrations, and topics for classroom discussion.

Test Item File (0-13-047710-9) by Barbara Mowery, Thomas Nelson Community College, has been revised to reflect the revision in this text book and contains 10% new questions in a bank of over 1400 multiple-choice questions.

TestGen-EQ (0-13-047711-7) The computerized version of the Test Item File is available on a dual-platform CD-ROM. The software available with this database allows you create and tailor exams to your specific needs.

Transparency Pack (0-13-047708-7) Over 200 full-color transparencies chosen from the text put principles into visual perspective and save you time while you are preparing your lectures.

Instructor's Resource CD (0-13-047700-1) An Instructor CD-ROM that contains almost all of the art from the text. Using the included MediaPortfolio software, instructors can browse for figures and other media elements by thumbnail and description as well as search by key word or title. This CD also contains two pre-built PowerPoint Presentations for each chapter as well as PDF files of each image.

Making It Easier to Learn: Supplements for Students

Study Guide and Full Solutions Manual (0-13-047706-0) and **Study Guide and Selected Solutions Manual (0-13-047707-9)**, both by Susan McMurry. The selected version provides solutions only to those problems that have a short answer in the text's Selected Answer appendix (problems numbered red in the text). Both versions explain in detail how the answers to in-text and end-of-chapter problems are obtained. They also contain chapter summaries, study hints, and self-tests for each chapter.

Companion Website for Fundamentals of General, Organic, and Biological Chemistry, Fourth Edition Built to complement the text as part of an integrated course package, the easy-to-use Companion Website features the following content for each chapter: a summary, a gallery of animated objects that demonstrate the ideas within that chapter, a gallery of 3-D molecules that students can manipulate, practice quizzes, a math tutorial covering skills needed to succeed in this course, and links to websites of career interest and other relevance to students in this course. Students can access the site at http://chemistry. prenhall.com/mcmurrygob.

For the Laboratory

Exploring Chemistry: Laboratory Experiments in General, Organic and Biological Chemistry (0-13-047714-1) by Julie R. Peller of Indiana University. Written specifically to accompany *Fundamentals of General, Organic, and Biological Chemistry*, this manual contains 33 fresh and accessible experiments specifically for GOB students.

Annotated Instructor's Manual (0-13-047715-X) by Libbie S. Pelter to **Exploring Chemistry: Laboratory Experiments in GOB** provides the instructor with goals, special instructions, chemical and equipment lists, plus answers for each experiment.

Catalyst: The Prentice Hall Custom Laboratory Program for Chemistry (0-536-67450-7). This CD-ROM allows you to custom-build a chemistry lab manual that matches your content needs and course organization. You can either write your own labs using the Lab Authoring Kit tool, or select from the hundreds of labs available on the CD. This CD also allows you to add your own course notes, syllabi, or other materials.

Acknowledgments

The creation of a modern textbook requires the efforts of many individuals in addition to the authors. We want to extend heartfelt thanks to everyone who participated in making this book and to those who continue to participate in spreading the word about its value.

Kent Porter Hamann, Senior Editor, stood at the helm of our Prentice Hall team. Our special appreciation goes to Ray Mullaney, Editor in Chief of Development, who stepped in to provide assistance and support at critical moments. Once again, Donna Young has shone as one of the best production editors in the business. Donna has masterfully fielded all questions, whether profound or trivial, held together the efforts of a diverse team, and been a pleasure to work with. In addition, our thanks to Heather Scott for the attractive design of this book and to Patricia Burns, Connie Long, and the illustrators at Artworks who rendered the art. Our thanks also go to Kristen Kaiser, who worked so ably on the accompanying student and instructor materials, and to Michael Richards for his work with the accompanying media.

We want to extend personal thanks for the efforts of two individuals who provided valuable assistance in reviewing the material and writing problems: Karen Wiechelman, University of Louisiana; and Kim Waldron, Regis University, Denver. Finally, numerous users of the third edition and others who teach this course have provided constructive and helpful suggestions by reviewing our manuscript for this edition. For their most valuable contributions to the fourth edition, we thank the following individuals:

David Ball, *Cleveland State University*
Boyd Beck, *Snow College*
Mark Benvenuto, *University of Detroit Mercy*
David Blackburn, *University of Minnesota*
Ronald Bost, *North Central Texas College*
Teresa Brown, *Rochester Community and Technical College*
G. Lynn Carlson, *University of Wisconsin—Parkside*
Jens Cavallius, *Case Western Reserve University*
Jeannie T. B. Collins, *University of Southern Indiana*
Bernadette R. Corbett, *Metropolitan Community College*
Mary Dempsey, *University of Minnesota*
Allan A.Gahr, *Gordon College*
Shelley Gaudia, *Lane Community College*
Elliott Goldstein, *Arizona State University*
Judith Iriarte-Gross, *Middle Tennessee State University*
T.G. Jackson, *University of South Alabama*
Martha Kurtz, *Central Washington University*

Richard H. Langley, *Stephen F. Austin State University*
Larry McGahey, *The College of St. Scholastica*
Melvin Merken, *Worcester State College*
Mark E. Ott, *Jackson Community College*
Darryl K. Reach, *University of Arkansas at Little Rock*
David Reinhold, *Western Michigan University*
Theresa Salerno, *Minnesota State University—Mankato*
Michael Serra, *Youngstown State University*
Zhihra Shen, *Utah State University*
Angela Sherman, *College of Notre Dame—Maryland*
Steven Sincoff, *Butte College*
Steven M Socol, *McHenry County College*
Bobby Stanton, *University of Georgia*
Martha Teeter, *Boston College*
Ellen Verdel, *Georgia State University*
Maria Vogt, *Bloomfield College*
Mona Wahby, *Macomb Community College*

The authors invite students and instructors to comment on any aspect of this text and will be pleased to respond. Comments on Chapers 1–14 should be directed to John McMurry, the lead author for these chapters, and comments on Chapters 15–29 should be directed to Mary E. Castellion, the lead author for these chapters.

John McMurry
jem24@cornell.edu

Mary E. Castellion
chemedit@snet.net

A Student's Guide to Using this Text

In designing this text, we have made every effort to provide you, the student, with a set of tools that can make your study of chemistry more efficient and rewarding. As you use the textbook, keep in mind that all the elements on each page—text, figures, molecular structures and models, equations, and the various learning aids described below—are designed to work together. Let them work for you. As with all tools, these will work best if you first learn what each is for and how it should be used. The following four-page "user's manual" will introduce you to the special features of this book and how you can take advantage of them to get the most from the time and effort you devote to studying chemistry.

QUANTITATIVE AND CONCEPTUAL PROBLEM-SOLVING

WORKED EXAMPLES

To succeed in this course you'll have to be able to solve problems. *Worked Examples* appearing frequently throughout the text show you how to (1) analyze different types of problems and devise a sound solution strategy, (2) implement the strategy in working out the problem step-by-step, (3) and validate the answer you obtain. ▶

ANALYSIS ▶

Most Worked Examples, both quantitative and not quantitative, now include a **new** *Analysis* section that precedes the Solution. The Analysis lays out the approach to solving a problem of the given type.

BALLPARK CHECK ▶

Many of the Worked Examples culminate with a *Ballpark Check* that provides you with a quick way of judging whether the answer you've calculated in numerical Worked Examples is reasonable.

KEY CONCEPT PROBLEMS ▶

New *Key Concept Problems* are now integrated within the chapter either appearing at the end of a Worked Example or after the discussion of an important concept. These *Key Concept Problems* focus your attention on essential concepts immediately after their discussion and can provide a way for you to immediately test your understanding of them.

■ WORKED EXAMPLE 6.11

The element boron is produced commercially by the reaction of boric oxide with magnesium at high temperature:

$$B_2O_3(l) + 3\,Mg(s) \rightarrow 2\,B(s) + 3\,MgO(s)$$

What is the percent yield if 675 g of boron is obtained from the reaction of 2350 g of boric oxide? The molecular weight of boric oxide is 69.6 amu.

ANALYSIS To calculate a percent yield, we first have to calculate a theoretical yield and divide that into the actual yield. The theoretical yield is calculated by the same mass-to-mole and mole-to mass conversions used for all stoichiometry problems, as discussed in the previous section.

SOLUTION The theoretical yield of boron from the reaction of 2350 g of boric oxide is

$$2350\ \text{g } B_2O_3 \times \frac{1\ \text{mol } B_2O_3}{69.6\ \text{g } B_2O_3} \times \frac{2\ \text{mol } B}{1\ \text{mol } B_2O_3} \times \frac{10.8\ \text{g } B}{1\ \text{mol } B} = 729\ \text{g } B$$

The percent yield is the actual yield divided by the theoretical yield and multiplied by 100%:

$$\text{Percent yield of } B = \frac{675\ \text{g } B}{729\ \text{g } B} \times 100\% = 92.6\%$$

BALLPARK CHECK The starting 2350 g of B_2O_3 is about 2400/70, or 35 mol. Each mole of B_2O_3 is converted into 2 mol of B (molar mass = 10.8 g/mol), so the theoretical yield should be approximately 35 * 2 * 10 g, or 700 g. Because the theoretical and actual yield are both close to 700 g, the percent yield should be close to 100%.

■ PROBLEM 6.16

What is the theoretical yield of ethyl chloride in the reaction of 19.4 g of ethylene with hydrogen chloride? What is the percent yield if 25.5 g of ethyl chloride is actually formed? (For ethylene, MW = 28.0 amu; for ethyl chloride, MW = 64.5 amu.)

$$H_2C = CH_2 + HCl \rightarrow CH_3CH_2Cl$$

■ KEY CONCEPT PROBLEM 6.9

What is the molecular weight of cytosine, a component of DNA (deoxyribonucleic acid)? (Gray = C, blue = N, red = O, ivory = H.)

Cytosine

◀ SOLUTION

The *Solution* shows you how to apply the appropriate problem-solving strategy and guides you through the steps to follow in obtaining the answer.

◀ PROBLEMS

You can confirm your grasp of the material and hone your problem-solving skills by working the numerous practice *Problems* that cover every skill and topic presented. These *Problems*, many of them new to this edition, follow the worked examples or stand alone at the end of a section. (Brief answers to all these within-chapter Problems can be found at the back of the book.)

STUDENT-FRIENDLY PEDAGOGY

OVERVIEW AND CHAPTER GOALS ▶

Before you start a journey, it's good to know where you're going. Each chapter opens with an outline and introduction, in order to give you an overview of the chapter's contents and how this chapter is related to previous ones. Then, a set of specific goals identifies the important themes and topics of the chapter, highlighting the information that you will need to learn and helping you to structure your study of the material.

---

1. **What are acids and bases?**
   The goal: Be able to recognize acids and bases and write equations for common acid–base reactions.

2. **What is the influence of acid and base strengths on their reactions?**
   The goal: Be able to interpret acid strength using acid dissociation constants K_a and predict the favored direction of acid–base equilibria.

3. **What is the ion-product constant for water?**
   The goal: Be able to write the equation for this constant and use it to find the concentration of H_3O^+ or OH^-.

4. **What is the pH scale for measuring acidity?**
   The goal: Be able to explain the pH scale and find pH from the H_3O^+ concentration.

5. **What is a buffer?**
   The goal: Be able to explain how a buffer maintains pH and how the bicarbonate buffer functions in the body.

6. **How is the acid or base concentration of a solution determined?**
   The goal: Be able to explain how a titration procedure works and use the results of a titration to calculate acid or base concentration in a solution.

---

10 Acids and Bases

The color of these hydrangeas, whether red or blue, depends on the acidity of the soil they're grown in.

CONCEPTS TO REVIEW ▶

At the beginning of each chapter after Chapter 4 is a **new** list of *Concepts to Review*, which are topics from previous chapters that form the basis of the present discussion.

CONCEPTS TO REVIEW

Acids, Bases, and Neutralization Reactions (Sections 4.12 and 6.10)
Reversible Reactions and Chemical Equilibrium (Section 7.6)
Equilibrium Equations and Equilibrium Constants (Section 7.7)
Units of Concentration; Molarity (Section 9.7)
Ion Equivalents (Section 9.10)

CONCEPT LINKS ▶

Concept Links indicate where concepts in the text build on material from earlier chapters. This chain link icon provides a quick visual reminder that new material being discussed relates to a concept introduced previously. The page or section reference cited helps you find the relevant discussion from the earlier section so you can review it if necessary.

The reaction of ammonia with water is a reversible process (Section 7.6) whose equilibrium strongly favors unreacted ammonia. (⟨∞⟩ p. 000) Nevertheless, *some* OH^- ions are produced, so NH_3 is a base.

LOOKING AHEAD ▶

It's always helpful to know *why* you're studying a particular topic or learning a particular concept. Sometimes the real significance of the material being presented doesn't become fully apparent until later, when you encounter it again in a different context. These brief forward looks provide you with a preview of how the material under discussion will prove useful in a later chapter.

⟨∞⟩ LOOKING AHEAD

We'll see in Chapter 28 that the regulation of blood pH by the bicarbonate buffer system is particularly important in preventing the medical conditions called *acidosis* and *alkalosis*. Acidosis, which results when an increase in CO_2 causes blood pH to drop below 7.35, can be brought on by the difficulty in breathing that accompanies asthma and emphysema. Alkalosis, the condition that results when blood pH rises above 7.45, arises either from heavy breathing (hyperventilation) that removes large amounts of CO_2 from the blood or from conditions that raise the concentration of bicarbonate ions. ⟨∞⟩

Chemical species such as B and BH^+, or HA and A^-, whose formulas differ by only one H^+, are called **conjugate acid–base pairs**. Thus, the anion A^- is the **conjugate base** of the acid HA, and HA is the **conjugate acid** of the base A^-. Similarly, B is the conjugate base of the acid BH^+, and BH^+ is the conjugate acid of the base B. To give some examples, acetic acid and acetate ion, the hydronium ion and water, and the ammonium ion and ammonia all make conjugate acid–base pairs.

$$\text{Conjugate acids} \left\{ \begin{array}{l} CH_3COH \rightleftarrows H^+ + CH_3CO^- \\ H_3O^+ \rightleftarrows H^+ + H_2O \\ NH_4^+ \rightleftarrows H^+ + NH_3 \end{array} \right\} \text{Conjugate bases}$$

◀ MARGINAL DEFINITIONS

You'll be learning many new words in this course. To help you master the vocabulary quickly, every important term appears in **boldface** where it's first used and is accompanied by a full definition alongside it in the margin.

Conjugate acid–base pair Two substances whose formulas differ by only a hydrogen ion, H^+.

Conjugate base The substance formed by loss of H^+ from an acid.

Conjugate acid The substance formed by addition of H^+ to a base.

RELEVANCE

APPLICATIONS ▶

These boxed *Applications* essays, all updated where necessary, show how what you learn in this text relates to a variety of interesting subjects. The topics discussed are drawn from everyday life, clinical practice, health and nutrition, ecology, biotechnology, and chemical research. Each boxed application provides sufficient information for reasonable understanding and ends with a cross-reference to related end-of-chapter problems that can be solved if the student has understood the boxed material.

Application Osteoporosis

Bone consists primarily of two components, one mineral and one organic. About 70% of bone is the ionic compound *hydroxyapatite*, $Ca_{10}(PO_4)_6(OH)_2$, called the *trabecular*, or spongy, bone. This mineral component is intermingled in a complex matrix with about 30% by mass of fibers of the protein *collagen*, called the *cortical*, or compact, bone. Hydroxyapatite gives bone its hardness and strength, while collagen fibers add flexibility and resistance to breaking.

Total bone mass in the body increases from birth until reaching a maximum in the mid 30s. By the early 40s, however, an age-related decline in bone mass begins to occur in both sexes. Should this thinning of bones become too great and the bones become too porous and brittle, a clinical condition called *osteoporosis* can result. Osteoporosis is, in fact, the most common of all bone diseases, affecting approximately 25 million people in the U.S. Approximately 1.5 million bone fractures each year are caused by osteoporosis at an estimated health-care cost of $14 billion.

Although both sexes are affected by osteoporosis, the condition is particularly common in post-menopausal women, who undergo cortical bone loss at a rate of 2–3% per year over and above that of the normal age-related

loss. The cumulative lifetime bone loss, in fact, may approach 40–50% in women versus 20–30% in men. It has been estimated that half of all women over age 50 will have an osteoporosis-related bone fracture at some point in their life. Other risk factors in addition to sex include being thin, being sedentary, having a family history of osteoporosis, smoking, and having a diet low in calcium.

No cure yet exists for osteoporosis, but treatment for its prevention and management includes estrogen replacement therapy for post-menopausal women as well as several approved medications called *bisphosphonates* to prevent further bone loss. Calcium supplements are also recommended, as is appropriate weight-bearing exercise. In addition, treatment with sodium fluoride is under active investigation and shows considerable promise. Fluoride ion reacts with hydroxyapatite to give *fluorapatite*, in which OH^- ions are replaced by F^-, increasing both bone strength and density.

$$Ca_{10}(PO_4)_6(OH)_2 + 2\ F^- \ \rightleftharpoons\ Ca_{10}(PO_4)_6F_2$$
Hydroxyapatite   Fluorapatite

See Additional Problems 4.86 and 4.87 at the end of the chapter.

Connection Art Conservator

"There's a lot of chemistry involved in art conservation," says Neil Cockerline, painting conservator. "I get calls frequently from members of the general public asking me, 'What can I use to clean my painting?' It's difficult to explain to them that we have hundreds of chemical combinations that we choose from. What may be appropriate for one painting might be completely inappropriate for another painting."

"Many things can happen in the lifespan of an oil painting, for example, that will affect the chemistry that exists within the structure. For cleaning some oil paintings," continues Cockerline, "you can use very strong solvents. For others, however, you could dissolve the oil paint if you just used distilled water. It all depends on the original materials that were used, the chemistry of those materials, and the aging process."

This simple example illustrates how important it is for art conservators to know their chemistry. When dealing with a work of art that's irreplaceable and worth thousands (or even millions) of dollars, it's critical to pick the right solvent and the right cleaning agent.

Conservators can also use x-rays to uncover secrets. Paints that include heavy-metal pigments show up readily under x-ray examination. The white pigment used most commonly up until the twentieth century, for example, was "lead white," made from a mixture of lead carbonate and lead hydroxide. It's not uncommon today for conservators to discover by x-ray that there is an earlier painting hidden underneath the painting that is visible to the naked eye.

In the photo, Cockerline stands with just such a painting. The original painting (on the right-hand side of the wooden panel) depicts an austere seventeenth century Dutch merchant. At a later date, another artist painted over most of the original work, giving the subject a royal costume while leaving the original face untouched. With that simple change of clothing, the Dutch merchant had been transformed into King Henry VIII of England. After discovering the underlying painting, conservators carefully stripped away the right half of the later painting. The Dutch merchant, at least half of him, was restored to his original glory—another piece of history and art revealed by chemistry.

◀ CONNECTIONS

These interviews with practicing professionals in a variety of fields show how they use chemistry in their daily work. This feature is designed to remind you that the chemistry you are learning can lay the groundwork for many diverse and interesting career paths.

VISUALIZATION

MOLECULAR MODELS ▶

To help students visualize chemical reactions that can't be seen because they take place on the microscopic level, many computer-generated molecular models and electrostatic potential maps are used.

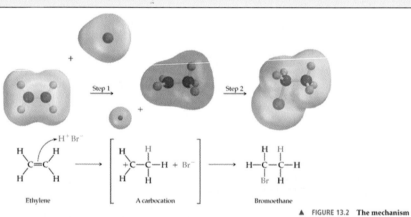

In the first step, the alkene reacts with H^+ from the acid HBr. The carbon–carbon double bond partially breaks, and two electrons move from the double bond to form a new single bond between one of the carbons and the incoming hydrogen. The other double-bond carbon, having had electrons removed from it, now has only six electrons in its outer shell and bears a positive charge. Unlike sodium ion $1Na^+2$ and other metal cations, which are unreactive and easily isolated in salts like NaCl, carbon cations, or *carbocations*, are highly reactive. As soon as the carbocation is formed by reaction of an alkene with H^+, it immediately

▲ **FIGURE 13.2 The mechanism of the addition of HBr to an alkene.** The reaction takes place in two steps and involves a carbocation intermediate. In the first step, two electrons move from the C C double bond to form a C H bond. In the second step, Br^- uses two electrons to form a bond to the positively charged carbon.

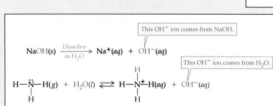

◀ COLOR AND LABELING EQUATIONS

We have extensively used color to call attention to the aspects of chemical equations and structures under discussion. We have also greatly increased the number of explanatory labels set in balloons that focus on the important details in chemical equations and structures.

END-OF-CHAPTER REVIEW

SUMMARY: REVISTING ▶ THE CHAPTER GOALS

The chapter *Summary* mirrors the goals that open the chapter. Each of the questions posed at the start of the chapter is answered by a summary of the essential information needed to attain the corresponding goal. Where appropriate, the types of chemical reactions in a chapter are also summarized.

Summary: Revisiting the Chapter Goals

1. What are acids and bases? According to the *Brønsted–Lowry definition*, an acid is a substance that donates a hydrogen ion (a proton, H^+) and a base is a substance that accepts a hydrogen ion. Thus, the generalized reaction of an acid with a base involves the reversible transfer of a proton:

$$B: + H-A \rightleftharpoons A:^- + H-B^+$$

In aqueous solution, water acts as a base and accepts a proton from an acid to yield a *hydronium ion*, H_3O^+. Reaction of an acid with a metal hydroxide, such as KOH, yields water and a salt; reaction with bicarbonate ion (HCO_3^-) or carbonate ion (CO_3^{2-}) yields water, a salt, and CO_2 gas; and reaction with ammonia yields an ammonium salt.

Key Words

Acid dissociation constant K_a, p. 274
Acid–base indicator, p. 279
Amphoteric, p. 268
Brønsted–Lowry acid, p. 265
Brønsted–Lowry base, p. 266
Buffer, p. 280
Conjugate acid, p. 267
Conjugate acid–base pair, p. 267
Conjugate base, p. 267
Dissociation, p. 271
Hydronium ion, p. 263

KEY WORDS ▲

In the *Key Words* all of the chapter's boldface terms are listed in alphabetical order alongside the summary. Each *Key Word* is cross-referenced to the page where it appears in the text. Definitions of all key words are collected in the Glossary.

UNDERSTANDING KEY ▶ CONCEPTS

Understanding Key Concepts Problems are designed to test your mastery of the core principles developed in the chapter. These end-of-chapter problems will help you review the major ideas in the chapter and essentially form a bridge between the chapter summary and the additional problems that follow. Use these problems to test your grasp of the chapter's main concepts. (You can check your results at the back of the book, where the answers to all Understanding Key Concepts questions are given.)

▪ Understanding Key Concepts

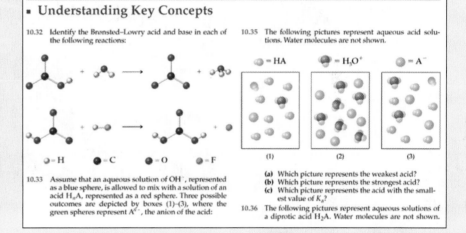

10.32 Identify the Brønsted–Lowry acid and base in each of the following reactions:

○ = H   ● = C   ● = O   ● = F

10.33 Assume that an aqueous solution of OH^-, represented as a blue sphere, is allowed to mix with a solution of an acid H_nA, represented as a red sphere. Three possible outcomes are depicted by boxes (1)–(3), where the green spheres represent A^{n-}, the anion of the acid:

10.35 The following pictures represent aqueous acid solutions. Water molecules are not shown.

○ = HA   ● = H_3O^+ ● = A^-

(1)   (2)   (3)

(a) Which picture represents the weakest acid?
(b) Which picture represents the strongest acid?
(c) Which picture represents the acid with the smallest value of K_a?

10.36 The following pictures represent aqueous solutions of a diprotic acid H_2A. Water molecules are not shown.

ADDITIONAL PROBLEMS ▼

The end-of-chapter *Additional Problems* are divided into sections by topic. The problems are presented in pairs. Even-numbered problems (numbered in red) are answered in the back of the book so that you can check your solution. Each even-numbered problem is accompanied by an odd-numbered problem that deals with the same topic or requires similar skills, affording you additional opportunity for practice.

▪ Additional Problems

Acids and Bases

10.38 What happens when a strong acid such as HBr is dissolved in water?

10.39 What happens when a weak acid such as CH_3CO_2H is dissolved in water?

10.40 What happens when a strong base such as KOH is dissolved in water?

10.41 What happens when a weak base such as NH_3 is dissolved in water?

10.42 What is the difference between a monoprotic acid and a diprotic acid? Give an example of each.

10.43 What is the difference between H^+ and H_3O^+?

10.44 Which of the following are strong acids? Look at Table 10.1 if necessary.

APPLICATION PROBLEMS ▶

Two final sections of exercises end each chapter. If your instructor has asked you to study the *Application boxes*, you can use the Applications Problems to test your understanding of that material.

Applications

10.90 An over-the-counter antacid has $NaAl(OH)_2CO_3$ as the active ingredient. [*Ulcers and Antacids*]
   (a) Write a balanced equation for the reaction of this compound with HCl.
   (b) How many grams of this antacid are required to neutralize 15.0 mL of 0.0955 M HCl?

10.91 Which body fluid is most acidic? Which is most basic? [*pH of Body Fluids*]

10.92 Rain typically has a pH of about 5.6. What is the $[H_3O^+]$ in rain? [*Acid Rain*]

10.93 Acid rain with a pH as low as 1.5 has been recorded in West Virginia. [*Acid Rain*]
   (a) What is the $[H_3O^+]$ in this acid rain?
   (b) How many grams of HNO_3 must be dissolved to make 25 L of solution with a pH of 1.5?

GENERAL QUESTIONS ▶ PROBLEMS

The *General Questions and Problems* give you an opportunity to integrate and synthesize the material you have learned. They are cumulative, pulling together topics from various parts of the chapter and even from previous chapters.

General Questions and Problems

10.94 Alka-Seltzer, a drugstore antacid, contains a mixture of $NaHCO_3$, aspirin, and citric acid, $C_6H_5O_7H_3$. Why does Alka-Seltzer foam and bubble when dissolved in water? Which ingredient is the antacid?

10.95 How many milliliters of 0.50 M NaOH solution are required to titrate 40.0 mL of a 0.10 M H_2SO_4 solution to an end point?

10.96 Which solution contains more acid, 50 mL of a 0.20 N HCl solution or 50 mL of a 0.20 N acetic acid solution? Which has a higher hydronium ion concentration? Which has a lower pH?

10.97 A 0.010 M solution of aspirin has pH 3.3. Is aspirin a strong or a weak acid?

10.98 A 0.15 M solution of HCl is used to titrate 30.0 mL of a $Ca(OH)_2$ solution of unknown concentration. If 140 mL of HCl is required, what is the concentration (in molarity) of the $Ca(OH)_2$ solution?

10.99 Which of the following combinations produces an effective buffer solution?
   (a) NaF and HF
   (b) $HClO_4$ and $NaClO_4$
   (c) NH_4Cl and NH_3
   (d) KBr and HBr

1 Matter and Life

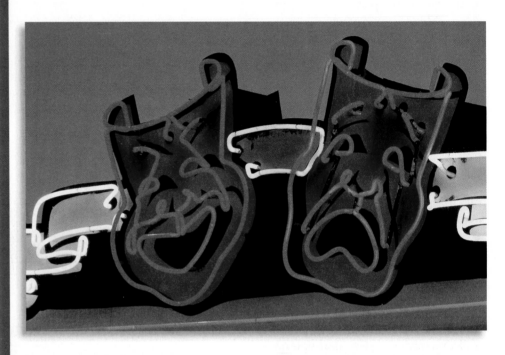

Neon lights contain (surprise!) neon, a gaseous chemical element found on the right side of the periodic table.

Look around you. Everything you see, touch, taste, and smell is made of chemicals. Many of these chemicals—those in rocks, trees, and your own body—occur naturally, but many others are synthetic. The plastics, fibers, and many of the medicines that are so important a part of modern life do not occur in nature but have been created in the chemical laboratory.

Just as everything you see is made of chemicals, many of the natural changes you see taking place around you are the result of *chemical reactions*—the change of one chemical into another. The flowering of a plant in the spring, the color change of a leaf in the fall, and the growth and aging of a human body are all the results of chemical reactions. To understand these and all other natural processes, you must have a basic understanding of chemistry.

As you might expect, the chemistry of living organisms is complex and it's not possible to jump right into it. Thus, the general plan of this book is to increase gradually in complexity, beginning in the first eleven chapters with a grounding in the scientific fundamentals that govern all of chemistry, moving in the next six chapters to look at the nature of the carbon-containing substances, or *organic chemicals*, that compose all living things, and then in the final twelve chapters applying what we've learned to biological chemistry.

We'll begin in this chapter by looking at the following topics:

1. **What is matter?**
   The goal: Be able to discuss the properties of matter and describe the three states of matter.
2. **How is matter classified?**
   The goal: Be able to distinguish between mixtures and pure substances, and between elements and compounds.
3. **What kinds of properties does matter have?**
   The goal: Be able to distinguish between chemical and physical properties.
4. **How are chemical elements represented?**
   The goal: Be able to name and give the symbols of common elements.

1.1 Chemistry: The Central Science

Chemistry is often referred to as "the central science" because it is crucial to all other sciences. In fact, as more and more is learned, the historical dividing lines between chemistry, biology, and physics are fading. Figure 1.1 diagrams the relationship of chemistry and biological chemistry to some other fields of study that deal with living organisms. Regardless of which discipline you're most interested in, the study of chemistry builds the necessary foundation.

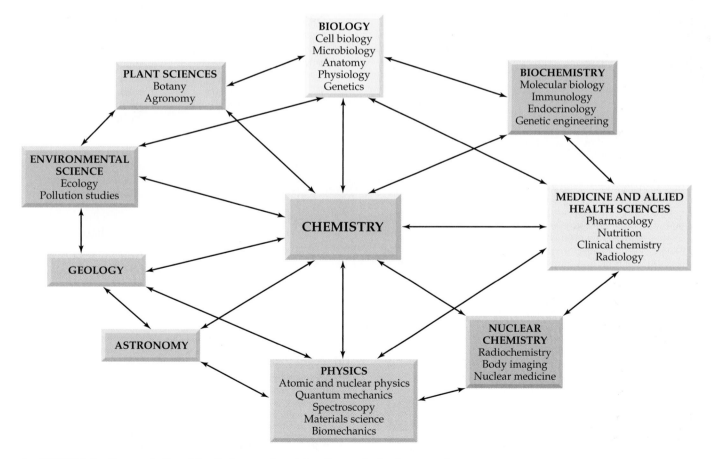

▲ FIGURE 1.1   Some relationships between chemistry, the central science, and other health-related disciplines.

Chemistry   The study of the nature, properties, and transformations of matter.

Matter   The physical material that makes up the universe; anything that has mass and occupies space.

Property   A characteristic useful for identifying a substance or object.

Physical change   A change that does not affect the chemical makeup of a substance or object.

Chemical change   A change in the chemical makeup of a substance.

▲ **FIGURE 1.2**   Samples of the pure substances water, sugar, and baking soda.

Chemistry is the study of matter—its nature, properties, and transformations. **Matter**, in turn, is a catchall word used to describe anything physically real—anything you can see, touch, taste, or smell. In more scientific terms, matter is anything that has mass and volume.

How might we describe different kinds of matter more specifically? Any characteristic that can be used to describe or identify something is called a **property**; size, color, and temperature are all familiar examples. Less familiar properties include *chemical composition*, which describes what matter is made of, and *chemical reactivity*, which describes how matter behaves. Rather than focus on the properties themselves, however, it's often more useful to think about *changes* in properties. Changes are of two types: *physical* and *chemical*. A **physical change** is one that does not alter the chemical makeup of a substance, whereas a **chemical change** is one that *does* alter a substance's chemical makeup. The melting of solid ice to give liquid water, for instance, is a physical change because the water changes only in form but not in chemical makeup. The rusting of an iron bicycle left in the rain, however, is a chemical change because iron combines with oxygen and moisture from the air to give the new substance rust.

Figure 1.2 shows several familiar substances—water, table sugar (sucrose), and baking soda (sodium bicarbonate)— and Table 1.1 lists their composition and some properties. Note in Table 1.1 that the changes occurring when sugar and baking soda are heated are chemical changes because new substances are produced.

TABLE 1.1   Some Properties of Water, Sugar, and Baking Soda

Water	Sugar (Sucrose)	Baking Soda (Sodium Bicarbonate)
Physical properties		
Colorless liquid	White crystals	White powder
Odorless	Odorless	Odorless
Melting point: 0°C	Begins to decompose at 160°C, turning black and giving off water.	Decomposes at 270°C, giving off water and carbon dioxide.
Boiling point: 100°C	—	—
Chemical properties		
Composition:*	Composition:*	Composition:*
11.2% hydrogen	6.4% hydrogen	27.4% sodium
88.8% oxygen	42.1% carbon	1.2% hydrogen
	51.5% oxygen	14.3% carbon
	57.3% oxygen	
Does not burn.	Burns in air.	Does not burn.

*Compositions are given by mass percent.

■ **PROBLEM 1.1**

Which of the following are made of chemicals?
(a) Hairspray   (b) A goldfish
(c) Paint   (d) A watermelon

■ **PROBLEM 1.2**

Identify each of the following as a physical change or a chemical change:
(a) A metal surface being ground   (b) Fruit ripening
(c) Wood burning   (d) A puddle evaporating

1.2 States of Matter

Matter exists in three forms: solid, liquid, and gas. A **solid** has a definite volume and a definite shape that doesn't change regardless of the container it's placed in. A **liquid**, by contrast, has a definite volume but an indefinite shape. The volume of a liquid doesn't change when it's poured into a different container, but its shape does. A **gas** is different still, having neither a definite volume nor a definite shape. A gas expands to fill the volume and take the shape of any container it's placed in (Figure 1.3).

Solid   A substance that has a definite shape and volume.

Liquid   A substance that has a definite volume but that changes shape to fill its container.

Gas   A substance that has neither a definite volume nor a definite shape.

(a) Ice: A solid has a definite volume and a definite shape independent of its container.

(b) Water: A liquid has a definite volume but a variable shape that depends on its container.

(c) Steam: A gas has both variable volume and shape that depend on its container.

▲ **FIGURE 1.3**   **The three states of matter—solid, liquid, and gas.**

Many substances, such as water, can exist in all three phases, or **states of matter**—the solid state, the liquid state, and the gaseous state—depending on the temperature. The conversion of a substance from one state to another is known as a **change of state** and is a common occurrence. The melting of a solid, the freezing or boiling of a liquid, and the condensing of a gas to a liquid are familiar to everyone.

State of matter   The physical state of a substance as a solid, liquid, or gas.

Change of state   The conversion of a substance from one state to another—for example, from liquid to gas.

■ **PROBLEM 1.3**

Formaldehyde is a disinfectant, a preservative, and a raw material for plastics manufacture. Its melting point is −92°C and its boiling point is −21°C. Is formaldehyde a gas, a liquid, or a solid at room temperature (25°C)? (The symbol °C means degrees Celsius.)

■ **PROBLEM 1.4**

Acetic acid, which gives the sour taste to vinegar, has a melting point of 16.6°C and a boiling point of 118°C. Does a bottle of acetic acid contain a solid or a liquid on a chilly morning with the window open and the laboratory at 10°C?

1.3 Classification of Matter

The first question a chemist asks about an unknown substance is whether it is a pure substance or a mixture. Every sample of matter is one or the other. Water and sugar alone are pure substances, but stirring some sugar into a glass of water creates a *mixture*.

Pure substance A substance that has a uniform chemical composition throughout.

Mixture A blend of two or more substances, each of which retains its chemical identity.

Chemical compound A pure substance that can be broken down into simpler substances by chemical reactions.

Reactant A starting substance that undergoes change during a chemical reaction.

Product A substance formed as the result of a chemical reaction.

Chemical reaction A process in which the identity and composition of one or more substances are changed.

Element A fundamental substance that can't be broken down chemically into any simpler substance.

▲ Sugar and water individually are pure substances, but a solution of sugar in water is a mixture.

🌐 **Classification of Matter, Mixtures and Compounds**

What is the difference between a pure substance and a mixture? One difference is that a **pure substance** is uniform in its chemical composition and its properties all the way down to the microscopic level. Every sample of water, sugar, or baking soda, regardless of source, has the composition and properties listed in Table 1.1. A **mixture**, however, can vary in both composition and properties, depending on how it is made. The amount of sugar dissolved in a glass of water will determine the sweetness, boiling point, and other properties of the mixture. Note that you often can't distinguish between a pure substance and a mixture just by looking. The sugar–water mixture *looks* just like pure water but differs on a microscopic level.

Another difference between a pure substance and a mixture is that the components of a mixture can be separated without changing their chemical identities. Water can be separated from a sugar–water mixture, for example, by boiling the mixture to drive off the steam and then condensing the steam to recover the pure water. Pure sugar is left behind in the container.

Pure substances are themselves classified into two groups: those that can undergo a chemical breakdown to yield simpler substances and those that cannot. Any pure material that *can* be broken down into simpler substances by a chemical change is called a **chemical compound**. Water, sugar, baking soda, and millions of other substances are examples. Water, for example, can be chemically changed by passing an electric current through it to produce hydrogen and oxygen. In writing this chemical change, the **reactant** (water) is written on the left, the **products** (hydrogen and oxygen) are written on the right, and an arrow connects the two parts to indicate a chemical change, or **chemical reaction**. The conditions necessary to bring about the reaction are written above and below the arrow.

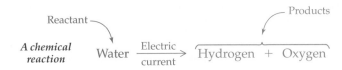

Unlike water, such substances as hydrogen, oxygen, aluminum, gold, and sulfur *cannot* be broken down chemically to yield simpler substances. A pure substance that can't be broken down chemically is called an **element**. At present, 114 elements are known, and all the millions of other substances in the universe are derived from them.

The classification of matter into mixtures, pure compounds, and elements is summarized in Figure 1.4.

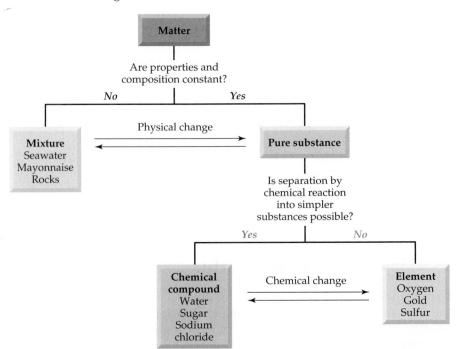

FIGURE 1.4   A scheme for the ▶ classification of matter.

■ **PROBLEM 1.5**

Classify each of the following as a mixture or a pure substance:

(a) Concrete   **(b)** The helium in a balloon

(c) A lead weight   **(d)** Wood

■ **PROBLEM 1.6**

Classify each of the following as a physical change or a chemical change:

(a) Separating a solid from a liquid by filtration

(b) Producing carbon dioxide gas and solid lime by heating limestone

(c) Mixing alcohol and water

1.4 An Example of a Chemical Reaction

Let's take a quick look at an example of a chemical reaction to reinforce some of the ideas discussed in the previous section. The element *nickel* is a hard, shiny metal, and the compound *hydrogen chloride* is a colorless gas that dissolves in water to give a solution called *hydrochloric acid*. When pieces of nickel are added to hydrochloric acid in a test tube, the nickel is slowly eaten away, the colorless solution turns green, and a gas bubbles out of the test tube. Clearly, a chemical reaction is taking place.

Overall, the reaction of nickel with hydrochloric acid can be either written in words or represented in a shorthand notation using symbols, as shown below in brackets. We'll explain the meaning of these symbols in the next section.

Reactants ⟶   Products ⟵

$\overbrace{\text{Nickel} + \text{Hydrochloric acid}} \longrightarrow \overbrace{\text{Nickel chloride} + \text{Hydrogen}}$

$$[\text{Ni} + 2\,\text{HCl} \longrightarrow \text{NiCl}_2 + \text{H}_2]$$

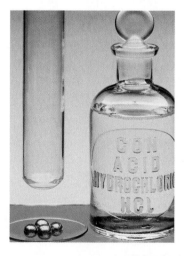

◄ **The reactants** The flat dish contains pieces of nickel, an element that is a typical, lustrous metal. The bottle contains hydrochloric acid, a solution of the chemical compound hydrogen chloride in water. These reactants are about to be combined in the test tube.

◄ **The reaction** As the chemical reaction occurs, the colorless solution turns green when insoluble nickel metal slowly changes into soluble nickel chloride. Bubbles of the element hydrogen are produced and rise slowly through the green solution.

◄ **The product** Hydrogen gas can be collected as it bubbles from the solution. Removal of water from the solution leaves behind the other product, a solid green chemical compound known as nickel chloride.

Application Chemicals, Toxicity, and Risk

Life is not risk-free—we all take many risks each day. We may decide to ride a bike rather than drive, even though the likelihood per mile of being killed on a bicycle is ten times greater than in a car. We may decide to smoke cigarettes, even though smoking kills more than 400,000 people each year in the U.S. Making judgments that affect our health is something we do every day without thinking about it.

What about risks from "chemicals"? News reports sometimes make it seem that our food is covered with pesticides and filled with dangerous additives, that our land is polluted by toxic waste dumps, and that our medicines are unsafe. How bad are the risks from chemicals, and how are the risks evaluated?

First, it's important to realize that *everything*, including your own body, is made of chemicals. There is no such thing as a "chemical-free" food, cosmetic, or anything else. Second, it's important to realize that there is no meaningful distinction between a "natural" substance and a "synthetic" one. Many naturally occurring substances are extraordinarily toxic—strychnine, for example—and many synthetic substances are harmless—polyethylene, for example.

Risk evaluation of chemicals is carried out by exposing test animals, usually mice or rats, to a chemical and then monitoring for signs of harm. To limit the expense and time needed for testing, the amounts administered are hundreds or thousands of times larger than those a person might normally encounter. Even with toxicity established in test animals, the risk to humans is still hard to assess. If a substance is harmful to rats, is it necessarily harmful to humans? How can a large dose for a small animal be translated into a small dose for a large human? All substances are toxic to some organisms to some extent, and the difference between help and harm is often a matter of degree. Vitamin A, for example, is necessary for vision, yet it can cause cancer at high doses. Arsenic trioxide is the most classic of poisons, yet recent work has shown it to be effective at inducing remissions in some types of leukemia. It is sold for drug use under the name Trisenox.

▲ Is this a deadly poison or a treatment for leukemia?

How we evaluate risk is strongly influenced by familiarity. The presence of chloroform in municipal water supplies at a barely detectable level of 0.000 000 01% has caused concern in many cities, yet chloroform has a lower acute toxicity than aspirin, a much more familiar substance. Many foods contain natural ingredients far more toxic than synthetic food additives or pesticide residues, but the risks are ignored because the foods are familiar.

All decisions involve tradeoffs. Does the benefit of a pesticide that will increase the availability of food outweigh a possible health risk to 1 person in 1 million? Do the beneficial effects of a new drug outweigh a potentially dangerous side effect in 0.001% of users? The answers are not always obvious, but it is the responsibility of legislators and well-informed citizens to keep their responses on a factual level.

See Additional Problem 1.52 at the end of the chapter.

1.5 Chemical Elements and Symbols

One hundred fourteen chemical elements are known today. Some are certainly familiar to you—oxygen, helium, iron, aluminum, copper, and gold, for example—but many others are probably unfamiliar—rhenium, niobium, thulium, and promethium. Rather than write out the full names of elements, chemists use a shorthand notation in which elements are referred to by one- or two-letter symbols. The names and symbols of some common elements are listed in Table 1.2, and a complete alphabetical list is given inside the front cover of this book.

TABLE 1.2	**Names and Symbols for Some Common Elements**

Elements with Symbols Based on Modern Names						**Elements with Symbols Based on Latin Names**	
Al	Aluminum	Co	Cobalt	N	Nitrogen	Cu	Copper (*cuprum*)
Ar	Argon	F	Fluorine	O	Oxygen	Au	Gold (*aurum*)
Ba	Barium	He	Helium	P	Phosphorus	Fe	Iron (*ferrum*)
Bi	Bismuth	H	Hydrogen	Pt	Platinum	Pb	Lead (*plumbum*)
B	Boron	I	Iodine	Rn	Radon	Hg	Mercury (*hydrargyrum*)
Br	Bromine	Li	Lithium	Si	Silicon	K	Potassium (*kalium*)
Ca	Calcium	Mg	Magnesium	S	Sulfur	Ag	Silver (*argentum*)
C	Carbon	Mn	Manganese	Ti	Titanium	Na	Sodium (*natrium*)
Cl	Chlorine	Ni	Nickel	Zn	Zinc	Sn	Tin (*stannum*)

Note that all two-letter symbols have only their first letter capitalized, while the second letter is always lowercase. The symbols of most common elements are the first one or two letters of the elements' modern names, such as H (hydrogen) and Al (aluminum). Pay special attention, however, to the elements grouped in the right-hand column in Table 1.2. The symbols for these elements are derived from Latin names, such as Na for sodium, once known as *natrium*. The only way to learn these symbols is to memorize them; fortunately they are few in number.

Only 90 of the 114 elements occur naturally; the remainder have been produced artificially by chemists and physicists. Each element has its own distinctive properties, and just about all of the first 95 elements have been put to use in some way that takes advantage of those properties. As indicated in Table 1.3, which shows the approximate elemental composition of the earth's crust and the human body, the naturally occurring elements are not equally abundant. Oxygen and silicon together account for 75% of the mass in the earth's crust; oxygen, carbon, and hydrogen account for nearly all the mass of a human body.

TABLE 1.3	**Elemental Composition of the Earth's Crust and the Human Body***

Earth's Crust		**Human Body**	
Oxygen	46.1%	Oxygen	61%
Silicon	28.2%	Carbon	23%
Aluminum	8.2%	Hydrogen	10%
Iron	5.6%	Nitrogen	2.6%
Calcium	4.1%	Calcium	1.4%
Sodium	2.4%	Phosphorus	1.1%
Magnesium	2.3%	Sulfur	0.20%
Potassium	2.1%	Potassium	0.20%
Titanium	0.57%	Sodium	0.14%
Hydrogen	0.14%	Chlorine	0.12%

*Mass percent values are given.

Symbols are combined to produce **chemical formulas**, which show by subscripts how many *atoms* (the smallest fundamental units) of different elements are combined in a given chemical compound. For example, the formula H_2O represents water, which contains 2 hydrogen atoms combined with 1 oxygen atom. Similarly, the formula CH_4 represents methane (natural gas), and the formula $C_{12}H_{22}O_{11}$ represents table sugar (sucrose). When no subscript is given for an element, as for carbon in the formula CH_4, a subscript of "1" is understood.

Chemical formula   A notation for a chemical compound using element symbols and subscripts to show how many atoms of each element are present.

H_2O CH_4 $C_{12}H_{22}O_{11}$

2 H atoms   1 C atom   12 C atoms
1 O atom   4 H atoms   22 H atoms
                                11 O atoms

■ **PROBLEM 1.7**

Look at the alphabetical list inside the front cover, and find the symbols for the following elements:

(a) Uranium, the fuel in nuclear reactors

(b) Titanium, the skin of jet fighters

(c) Tungsten, the filament in light bulbs

■ **PROBLEM 1.8**

What elements do the following symbols represent?

(a) Na **(b)** Ca **(c)** Pd

(d) K **(e)** Sr **(f)** Sn

■ **PROBLEM 1.9**

Identify the elements represented in each of the following chemical formulas, and tell the number of atoms of each element:

(a) NH_3 (ammonia)

(b) $NaHCO_3$ (sodium bicarbonate)

(c) C_8H_{18} (octane, a component of gasoline)

(d) $C_6H_8O_6$ (vitamin C)

Periodic table A tabular format listing all known elements.

1.6 Elements and the Periodic Table

The symbols of the 114 known elements are normally presented in a tabular format called the **periodic table**, as shown in Figure 1.5 and the inside front cover of this book. We'll have much more to say about the periodic table and how it's numbered in Sections 3.4–3.8, but will note for now that it is the most important

▼ **FIGURE 1.5 The periodic table of the elements.** Metals appear on the left, nonmetals on the right, and metalloids in a zigzag band between metals and nonmetals. The numbering system is explained in Section 3.4.

1A 1	2A 2	3B 3	4B 4	5B 5	6B 6	7B 7	8	8B 9	10	1B 11	2B 12	3A 13	4A 14	5A 15	6A 16	7A 17	8A 18
1 **H** 1.00794																	2 **He** 4.00260
3 **Li** 6.941	4 **Be** 9.01218											5 **B** 10.81	6 **C** 12.011	7 **N** 14.0067	8 **O** 15.9994	9 **F** 18.9984	10 **Ne** 20.1797
11 **Na** 22.98977	12 **Mg** 24.305											13 **Al** 26.98154	14 **Si** 28.0855	15 **P** 30.9738	16 **S** 32.066	17 **Cl** 35.4527	18 **Ar** 39948
19 **K** 39.0983	20 **Ca** 40.078	21 **Sc** 44.9559	22 **Ti** 47.88	23 **V** 50.9415	24 **Cr** 51.996	25 **Mn** 54.9380	26 **Fe** 55.847	27 **Co** 58.9332	28 **Ni** 58.69	29 **Cu** 63.546	30 **Zn** 65.39	31 **Ga** 69.72	32 **Ge** 72.61	33 **As** 74.9216	34 **Se** 78.96	35 **Br** 79.904	36 **Kr** 83.80
37 **Rb** 85.4678	38 **Sr** 87.62	39 **Y** 88.9059	40 **Zr** 91.224	41 **Nb** 92.9064	42 **Mo** 95.94	43 **Tc** (98)	44 **Ru** 101.07	45 **Rh** 102.9055	46 **Pd** 106.42	47 **Ag** 107.8682	48 **Cd** 112.41	49 **In** 114.82	50 **Sn** 118.710	51 **Sb** 121.757	52 **Te** 127.60	53 **I** 126.9045	54 **Xe** 131.29
55 **Cs** 132.9054	56 **Ba** 137.33	57 ***La** 138.9055	72 **Hf** 178.49	73 **Ta** 180.9479	74 **W** 183.85	75 **Re** 186.207	76 **Os** 190.2	77 **Ir** 192.22	78 **Pt** 195.08	79 **Au** 196.9665	80 **Hg** 200.59	81 **Ti** 204.383	82 **Pb** 207.2	83 **Bi** 208.9804	84 **Po** (209)	85 **At** (210)	86 **Rn** (222)
87 **Fr** (223)	88 **Ra** 226.0254	89 **†Ac** 227.0278	104 **Rf** (261)	105 **Db** (262)	106 **Sg** (266)	107 **Bh** (264)	108 **Hs** (269)	109 **Mt** (268)	110 (271)	111 (272)	112 (277)		114		116		

58 **Ce** 140.12	59 **Pr** 140.9077	60 **Nd** 144.24	61 **PM** (145)	62 **Sm** 150.36	63 **Eu** 151.965	64 **Gd** 157.25	65 **Tb** 158.9254	66 **Dy** 162.50	67 **Ho** 164.9304	68 **Er** 167.26	69 **Tm** 168.9342	70 **Yb** 173.04	71 **Lu** 174.967
90 **Th** 232.0381	91 **Pa** 231.0399	92 **U** 238.0289	93 **Np** 237.048	94 **Pu** (244)	95 **Am** (243)	96 **Cm** (247)	97 **Bk** (247)	98 **Cf** (251)	99 **Es** (252)	100 **Fm** (257)	101 **Md** (258)	102 **No** (259)	103 **Lr** (262)

Metals Metalloids Nonmetals

organizing principle in chemistry. An enormous amount of information is embedded in the periodic table, information that gives chemists the ability to explain known chemical behavior of elements and to predict new behavior. The elements can be roughly divided into three groups: *metals*, *nonmetals*, and *metalloids* (sometimes called *semimetals*).

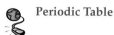

Periodic Table

Ninety of the 114 elements are metals—aluminum, gold, copper, and zinc, for example. **Metals** are solid at room temperature (except for mercury), usually have a lustrous appearance when freshly cut, are good conductors of heat and electricity, and are malleable rather than brittle. That is, a metal can be pounded into a different shape rather than shattering when struck. Note that metals occur on the left side of the periodic table.

Metal A malleable element with a lustrous appearance that is a good conductor of heat and electricity.

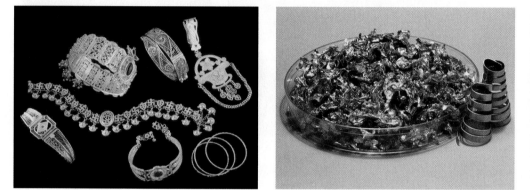

▲ **Metals: Gold, zinc, and copper.** (a) Known for its beauty, gold is very unreactive and is used primarily in jewelry and in electronic components. (b) Zinc, an essential trace element in our diets, has industrial uses ranging from the manufacture of brass, to roofing materials, to batteries. (c) Copper is widely used in electrical wiring, in water pipes, and in coins.

Seventeen elements are **nonmetals**. All are poor conductors of heat and electricity, eleven are gases at room temperature, five are solids, and one is a liquid. Oxygen and nitrogen, for example, are gases present in air; sulfur is a solid found in large underground deposits. Bromine is the only liquid nonmetal. Note that nonmetals occur on the right side of the periodic table.

Nonmetal An element that is a poor conductor of heat and electricity.

▲ **Nonmetals: Nitrogen, sulfur, and bromine.** (a) Nitrogen and (b) sulfur are essential to all living things. Pure nitrogen, which constitutes almost 80% of air, is a gas at room temperature and doesn't condense to a liquid until it is cooled to −328°F. Sulfur, a yellow solid, is found in large underground deposits in Texas and Louisiana. (c) Bromine is a corrosive, dark red liquid.

Metalloid   An element whose properties are intermediate between those of a metal and a nonmetal.

Only seven elements are **metalloids**, so-named because their properties are intermediate between those of metals and nonmetals. Boron, silicon, and arsenic are examples. Note that metalloids occur in a zigzag band between metals on the left and nonmetals on the right side of the periodic table.

Metalloids: Boron and silicon. ▶
(a) Boron is a strong, hard metalloid used in making the composite materials found in military aircraft. (b) Silicon is well known for its use in making computer chips.

TABLE 1.4   Elements Essential for Human Life*

Element	Symbol	Function
Carbon	C	These four elements are present in all living organisms
Hydrogen	H	
Oxygen	O	
Nitrogen	N	
Arsenic	As	May affect cell growth and heart function
Boron	B	Aids in the use of Ca, P, and Mg
Calcium*	Ca	Necessary for growth of teeth and bones
Chlorine*	Cl	Necessary for maintaining salt balance in body fluids
Chromium	Cr	Aids in carbohydrate metabolism
Cobalt	Co	Component of vitamin B_{12}
Copper	Cu	Necessary to maintain blood chemistry
Fluorine	F	Aids in the development of teeth and bones
Iodine	I	Necessary for thyroid function
Iron	Fe	Necessary for oxygen-carrying ability of blood
Magnesium*	Mg	Necessary for bones, teeth, and muscle and nerve action
Manganese	Mn	Necessary for carbohydrate metabolism and bone formation
Molybdenum	Mo	Component of enzymes necessary for metabolism
Nickel	Ni	Aids in the use of Fe and Cu
Phosphorus*	P	Necessary for growth of bones and teeth; present in DNA/RNA
Potassium*	K	Component of body fluids; necessary for nerve action
Selenium	Se	Aids vitamin E action and fat metabolism
Silicon	Si	Helps form connective tissue and bone
Sodium*	Na	Component of body fluids; necessary for nerve and muscle action
Sulfur*	S	Component of proteins; necessary for blood clotting
Zinc	Zn	Necessary for growth, healing, and overall health

*C, H, O, and N are present in all foods. Other elements listed vary in their distribution in different foods. Those marked with an asterisk are *macronutrients*, essential in the diet at more than 100 mg/day; the rest, other than C, H, O, and N, are *micronutrients*, essential at 15 mg or less per day.

Application Mercury and Mercury Poisoning

Mercury, the only metallic element that is liquid at room temperature, has fascinated people for millennia. Egyptian kings were buried in their pyramids along with containers of mercury, alchemists during the Middle Ages used mercury to dissolve gold, and Spanish galleons carried loads of mercury to the New World in the 1600s for use in gold and silver mining. Even its symbol, Hg, from the Latin *hydrargyrum* meaning "liquid silver," hints at mercury's uniqueness.

Much of the recent interest in mercury has concerned its toxicity, but there are some surprises. For example, Hg_2Cl_2 (called *calomel*) is nontoxic and has a long history of medical use as a laxative, yet $HgCl_2$ is also used as a fungicide and rat poison. Dental amalgam, a solid alloy of approximately 50% elemental mercury, 35% silver, 13% tin, and 1% copper has been used safely by dentists for many years to fill tooth cavities, yet exposure to elemental mercury *vapor* for long periods leads to mood swings, headaches, tremors, and loss of hair and teeth. It's thought, in fact, that Abraham Lincoln may have suffered for years from mercury poisoning as a result of taking mercury-laden "blue mass" pills to combat depression.

Why is mercury toxic in some forms but not in others? It turns out that the toxicity of mercury and its compounds is related to solubility. Only soluble mercury compounds are toxic, because they can be transported through the bloodstream to all parts of the body where they react with different enzymes and interfere with various biological processes. Elemental mercury and insoluble mercury compounds become toxic only when converted into soluble compounds, reactions that are extremely slow in the body. Calomel, for example, passes through the body long before it is converted into any soluble compounds. Mercury alloys are safe for dental use because mercury does not evaporate from the alloys and it neither reacts with nor dissolves in saliva. Mercury vapor, however, remains in the lungs when breathed, until it is slowly converted into soluble compounds.

▲ Recent evidence suggests that Abraham Lincoln may have suffered from mercury poisoning because of the pills he took to combat depression.

Of particular concern with regard to mercury toxicity is the environmental danger posed by pollution from both natural and industrial sources. Microorganisms present in lakes and streams are able to convert many mercury-containing wastes into a soluble and highly toxic compound called *methylmercury*. Methylmercury is concentrated to high levels in fish, particularly in shark and swordfish, which are then hazardous when eaten. Although the commercial fishing catch is now monitored carefully, more than 50 deaths from eating contaminated fish were recorded in Minimata, Japan, during the 1950s before the cause of the problem was realized.

See Additional Problem 1.53 at the end of the chapter.

Those elements essential for human life are listed in Table 1.4. In addition to the well-known elements carbon, hydrogen, oxygen, and nitrogen, less familiar elements such as molybdenum and selenium are also important.

◯◯◯ LOOKING AHEAD

The elements listed in Table 1.4 aren't present in our bodies in their free forms, of course. Instead, they are combined into many thousands of different chemical compounds. We'll talk about some compounds formed by metals in Chapter 4 and compounds formed by nonmetals in Chapter 5. ◯◯◯

■ **PROBLEM 1.10**

Look at the periodic table inside the front cover, and locate the following elements:

(a) Cr, chromium   **(b)** Na, sodium   **(c)** P, phosphorus   **(d)** Rn, radon

■ **PROBLEM 1.11**

The seven metalloids are boron (B), silicon (Si), germanium (Ge), arsenic (As), antimony (Sb), tellurium (Te), and astatine (At). Locate them in the periodic table, and tell where they appear with respect to metals and nonmetals.

Key Words

Change of state, p. 3

Chemical change, p. 2

Chemical compound, p. 4

Chemical formula, p. 7

Chemical reaction, p. 4

Chemistry, p. 2

Element, p. 4

Gas, p. 3

Liquid, p. 3

Matter, p. 2

Metal, p. 9

Metalloid, p. 10

Mixture, p. 4

Nonmetal, p. 9

Periodic table, p. 8

Physical change, p. 2

Pure substance, p. 4

Product, p. 4

Property, p. 2

Reactant, p. 4

Solid, p. 3

State of matter, p. 3

Summary: Revisiting the Chapter Goals

1. **What is matter?** *Matter* is anything that has mass and occupies volume—that is, anything physically real. Matter can be classified by its physical state as *solid*, *liquid*, or *gas*. A solid has a definite volume and shape, a liquid has a definite volume but indefinite shape, and a gas has neither a definite volume nor shape.

2. **How is matter classified?** A substance can be characterized as being either *pure* or a *mixture*. A pure substance is uniform in its composition and properties, but a mixture can vary in both composition and properties, depending on how it was made. Every pure substance is either an *element* or a *chemical compound*. Elements are fundamental substances that can't be chemically changed into anything simpler. A chemical compound, by contrast, can be broken down by chemical change into simpler substances.

3. **What kinds of properties does matter have?** A *property* is any characteristic that can be used to describe or identify something. Focusing on changes in properties, a *physical change* of matter is one that does not change the chemical makeup of a substance, whereas a *chemical change* is one that does involve a change in chemical makeup.

4. **How are chemical elements represented?** Elements are represented by one- or two-letter symbols, such as H for hydrogen, Ca for calcium, Al for aluminum, and so on. Most symbols are the first one or two letters of the element name, but some symbols are derived from Latin names—Na (sodium), for example. The 114 known elements are commonly organized into a form called the *periodic table*. Most (90) elements are *metals*, 17 are *nonmetals*, and 7 are *metalloids*.

■ Understanding Key Concepts

The problems in this section are intended as a bridge between the Chapter Summary and the Additional Problems that follow. Primarily visual in nature, they are designed to help you test your grasp of the chapter's most important principles before attempting to solve quantitative problems. Answers to all Key Concept problems are at the end of the book following the appendixes.

1.12 Six of the elements at the far right of the periodic table are gases at room temperature. Identify them using the periodic table inside the front cover of this book.

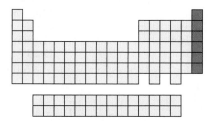

1.13 The so-called "coinage metals" are located near the middle of the periodic table. Identify them using the periodic table inside the front cover of this book.

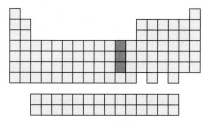

1.14 Identify the three elements indicated on the following periodic table, and tell which is a metal, which is a nonmetal, and which is a metalloid.

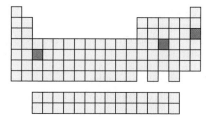

1.15 The radioactive element indicated on the following periodic table is used in smoke detectors. Identify it, and tell whether it is a metal, a nonmetal, or a metalloid.

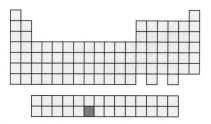

■ Additional Problems

These exercises are divided into sections by topic. Each section begins with review and conceptual questions, followed by numerical problems of varying levels of difficulty. The problems are presented in pairs, with each even-numbered problem followed by an odd-numbered one requiring similar skills. The final section consists of unpaired General Questions and Problems that draw on various parts of the chapter and may even require the use of concepts from previous chapters. All even-numbered problems are answered at the end of the book following the appendixes.

Chemistry and the Properties of Matter

1.16 What is chemistry?

1.17 Which of the following is composed of chemicals?
- **(a)** A rosebush
- **(b)** The substances that give roses their fragrance
- **(c)** The earth in which the bush grows

1.18 Which of the following is a physical change and which is a chemical change?
- **(a)** Boiling water
- **(b)** Decomposing water by passing an electric current through it
- **(c)** Dissolving sugar in water
- **(d)** Exploding of potassium metal when placed in water
- **(e)** Breaking of glass

1.19 Which of the following is a physical change and which is a chemical change?
- **(a)** Steam condensing
- **(b)** Milk souring
- **(c)** Ignition of matches
- **(d)** Breaking of a dinner plate
- **(e)** Nickel sticking to a magnet
- **(f)** Exploding of nitroglycerin

States and Classification of Matter

1.20 Name and describe the three states of matter.

1.21 Name two changes of state, and describe what causes each to occur.

1.22 Sulfur dioxide (SO_2) is a compound produced when sulfur burns in air. It has a melting point of $-72.7°C$ and a boiling point of $-10°C$. In what state does SO_2 exist at room temperature ($25°C$)? (The symbol $°C$ means degrees Celsius.)

1.23 Benzyl salicylate, which melts at $24°C$ and boils at $320°C$, is a chemical compound sometimes used as a sunscreen. In what state is it found at:

(a) $14°C$?   (b) $60°C$?

1.24 Classify each of the following as a mixture or a pure substance:

(a) Pea soup
(b) Seawater
(c) The contents of a propane tank (C_3H_8)
(d) Urine
(e) Lead
(f) A multivitamin tablet

1.25 Classify each of the following as a mixture or a pure substance:

(a) Blood
(b) Silicon
(c) Dishwashing liquid
(d) Toothpaste
(e) Gold
(f) Gaseous ammonia (NH_3)

1.26 What is the difference between an element and a compound?

1.27 What is the difference between a compound and a mixture?

1.28 Classify each of the following as an element, a compound, or a mixture:

(a) Aluminum foil   (b) Table salt
(c) Water   (d) Air
(e) A banana   (f) Notebook paper

1.29 Which of the following terms, (i) mixture, (ii) solid, (iii) liquid, (iv) gas, (v) chemical element, (vi) chemical compound, applies to each of the following substances at room temperature?

(a) Motor oil   (b) Copper
(c) Carbon dioxide   (d) Nitrogen
(e) Sodium bicarbonate

1.30 Hydrogen peroxide, often used in solutions to cleanse cuts and scrapes, breaks down to yield water and oxygen:

$$\text{Hydrogen peroxide} \longrightarrow \text{Water} + \text{Oxygen}$$

(a) Identify the reactants and products.
(b) Which of the substances are chemical compounds, and which are elements?

1.31 When sodium metal is placed in water, the following change occurs:

$$\text{Sodium} + \text{Water} \longrightarrow \text{Hydrogen} + \text{Sodium hydroxide}$$

(a) Identify the reactants and products.
(b) Which of the substances are elements, and which are chemical compounds?

Elements and Their Symbols

1.32 How many elements are presently known? About how many occur naturally?

1.33 Where in the periodic table are the metallic elements found? The nonmetallic elements? The metalloid elements?

1.34 Describe the general properties of metals, nonmetals, and metalloids.

1.35 What is the most abundant element in the earth's crust? In the human body? List the name and symbol for each.

1.36 What are the symbols for the following elements?

(a) Gadolinium (used in color TV screens)
(b) Germanium (used in semiconductors)
(c) Technetium (used in biomedical imaging)
(d) Arsenic (used in pesticides)
(e) Cadmium (used in rechargeable batteries)

1.37 What are the symbols for the following elements?

(a) Zinc   (b) Mercury
(c) Barium   (d) Gold
(e) Silicon   (f) Carbon
(g) Sodium   (h) Lead

1.38 Give the names corresponding to the following symbols:

(a) N   (b) K
(c) Cl   (d) Ca
(e) P   (f) Mn

1.39 Give the names corresponding to the following symbols:

(a) Te   (b) Re
(c) Be   (d) Ar
(e) Pu   (f) Mg

1.40 The symbol CO stands for carbon monoxide, a chemical compound, but the symbol Co stands for cobalt, an element. Explain how you can tell them apart.

1.41 What is wrong with the following statements? Correct them.

(a) The symbol for bromine is BR.
(b) The symbol for manganese is Mg.
(c) The symbol for carbon is Ca.
(d) The symbol for potassium is po.

1.42 What is wrong with the following statements? Correct them.

(a) Water has the formula H2O.
(b) Water is composed of hydrogen and nitrogen.

1.43 What is wrong with the following statements? Correct them.

(a) "Fools gold" is a mixture of copper and sulfur with the formula CoS.
(b) "Laughing gas" is a compound of nitrogen and oxygen with the formula NiO.

1.44 Name the elements combined in the chemical compounds represented by the following formulas:

(a) $MgSO_4$ (b) $FeBr_2$
(c) CoP (d) AsH_3
(e) $CaCr_2O_7$

1.45 How many atoms of what elements are represented by the following formulas?

(a) Propane (LP gas), C_3H_8
(b) Sulfuric acid, H_2SO_4
(c) Aspirin, $C_9H_8O_4$

1.46 The amino acid glycine has the formula $C_2H_5NO_2$. What elements are present in glycine? What is the total number of atoms represented by the formula?

1.47 Ribose, an essential part of ribonucleic acid (RNA), has the formula $C_5H_{10}O_5$. What is the total number of atoms represented by the formula?

1.48 What is the formula for ibuprofen: 13 carbons, 18 hydrogens, and 2 oxygens?

1.49 What is the formula for penicillin V: 16 carbons, 18 hydrogens, 2 nitrogens, 5 oxygens, and 1 sulfur?

1.50 Which of the following two elements is a metal, and which is a nonmetal?

(a) Osmium, a hard, shiny, very dense solid that conducts electricity

(b) Xenon, a colorless, odorless gas

1.51 Which of the following elements is likely to be a metal and which a metalloid?

(a) Tantalum, a hard, shiny solid that conducts electricity

(b) Germanium, a brittle, gray solid that conducts electricity poorly

Applications

1.52 Many over-the-counter medications, beverages, and other products may have unacceptably high risks for certain classes of people. Read the labels on common products such as aspirin, artificially sweetened beverages, alcohol, and cigarettes, and note the types of people for whom each product is of special risk. [*Chemicals, Toxicity, and Risk*]

1.53 Why is Hg_2Cl_2 harmless when swallowed, yet elemental mercury is toxic when breathed? [*Mercury and Mercury Poisoning*]

General Questions and Problems

1.54 Distinguish between the following:

(a) Physical changes and chemical changes

(b) Melting point and boiling point

(c) Reactants and products

(d) Metals and nonmetals

1.55 Are the following statements true or false? If false, explain why.

(a) The combination of sodium and chlorine to produce sodium chloride is a chemical reaction.

(b) The addition of heat to solid sodium chloride until it melts is a chemical reaction.

(c) The formula for a chemical compound that contains lead and oxygen is LiO.

(d) By stirring together salt and pepper we can create a new chemical compound to be used for seasoning food.

1.56 Which of the following are chemical compounds, and which are elements?

(a) H_2O_2 (b) Mo

(c) C   (d) NO

(e) $NaHCO_3$

1.57 A white solid with a melting point of 730°C is melted. When electricity is passed through the resultant liquid, a brown gas and a molten metal are produced. Neither the metal nor the gas can be broken down into anything simpler by chemical means. Classify each—the white solid, the molten metal, and the brown gas—as a mixture, a compound, or an element.

1.58 As a clear red liquid sits at room temperature, evaporation occurs until only a red solid remains. Is the liquid an element, a compound, or a mixture?

1.59 Describe how you could physically separate a mixture of iron filings, table salt, and white sand.

1.60 Small amounts of the following elements in our diets are essential for good health. What is the chemical symbol for each?

(a) Iron   (b) Copper

(c) Cobalt   (d) Molybdenum

(e) Chromium   (f) Fluorine

(g) Sulfur

1.61 Consider the as yet undiscovered elements with atomic numbers 115, 117, and 119. Is element 115 likely to be a metal or a nonmetal? What about element 117? Element 119?

2

Measurements in Chemistry

How accurate are measurements we make? Most references still list the height of Mt. Everest as 29,028 ft, but new measurements show it to be 29,035 ft.

Measurements are essential in chemistry and all its applications. Whenever you use chemical compounds—whether you are adding salt to your cooking, putting antifreeze in your car radiator, or choosing a dosage of medicine—quantities of substances must be measured. A mistake in measuring the quantities could ruin the dinner, damage the engine, or harm the patient.

We'll deal in this chapter with the following questions about measurement:

1. How are measurements made, and what units are used?

The goal: Be able to name and use the metric and SI units of measure for mass, length, volume, and temperature.

2. How good are the reported measurements?

The goal: Be able to interpret the number of significant figures in a measurement and round off numbers in calculations involving measurements.

3. How are large and small numbers best represented?

The goal: Be able to interpret prefixes for units of measure and express numbers in scientific notation.

4. How can a quantity be converted from one unit of measure to another?

The goal: Be able to convert quantities from one unit to another using conversion factors.

5. What techniques are used to solve problems?

The goal: Be able to analyze a problem, use the factor-label method to solve the problem, and carry out a ballpark check of the solution.

6. What are energy, heat, specific heat, density, and specific gravity?

The goal: Be able to define these quantities and use them in calculations.

2.1 Physical Quantities

Height, volume, temperature, and other physical properties that can be measured are called **physical quantities** and are described by both a number and a **unit** of defined size:

61.2 kilograms

Physical quantity A physical property that can be measured.

Unit A defined quantity used as a standard of measurement.

The number alone isn't much good without a unit. If you asked how much blood an accident victim had lost, the answer "three" wouldn't tell you much. Three drops? Three milliliters? Three pints? Three liters? (By the way, an adult human has only 5–6 liters of blood.)

Any physical quantity can be measured in many different units. For example, a person's height might be measured in inches, feet, yards, centimeters, or many other units. To avoid confusion, scientists from around the world have agreed on a system of standard units, called by the French name *Système International d'Unités* (International System of Units), abbreviated *SI*. **SI units** for some common physical quantities are given in Table 2.1. Mass is measured in *kilograms* (kg), length is measured in *meters* (m), volume is measured in *cubic meters* (m^3), temperature is measured in *kelvins* (K), and time is measured in *seconds* (s, not sec).

SI units Units of measurement defined by the International System of Units.

TABLE 2.1   Some SI and Metric Units and Their Equivalents

Quantity	SI Unit (Symbol)	Metric Unit (Symbol)	Equivalents
Mass	Kilogram (kg)	Gram (g)	1 kg = 1000 g = 2.205 lb
Length	Meter (m)	Meter (m)	1 m = 3.280 ft
Volume	Cubic meter (m^3)	Liter (L)	1 m^3 = 1000 L = 264.2 gal
Temperature	Kelvin (K)	Celsius degree (°C)	1 K = 1°C 1°C = 1.8°F
Time	Second (s)	Second (s)	—

SI units are closely related to the more familiar *metric units* used in all industrialized nations of the world except the United States. If you compare the SI and metric units shown in Table 2.1, you'll find that the basic metric unit of mass is the *gram* (g) rather than the kilogram (1 g = 1/1000 kg), the metric unit of volume is the *liter* (L) rather than the cubic meter (1 L = 1/1000 m^3), and the metric unit of temperature is the *Celsius degree* (°C) rather than the kelvin (1 K = 1°C). The meter is the unit of length and the second is the unit of time in both systems. Although SI units are now preferred in scientific research, metric units are still used in some fields. You'll probably find yourself working with both.

In addition to the units listed in Table 2.1, many other widely used units are derived from them. For instance, units of *meters per second* (m/s) are often used for *speed*—the distance covered in a given time. Similarly, units of *grams per cubic centimeter* (g/cm^3) are often used for *density*—the mass of substance in a given volume. We'll see other such derived units in future chapters.

One problem with any system of measure is that the sizes of the units often turn out to be inconveniently large or small for the problem at hand. A biologist describing the diameter of a red blood cell (0.000 006 m) would find the meter to be an inconveniently large unit, but an astronomer measuring the average distance from the earth to the sun (150,000,000,000 m) would find the meter to be inconveniently small. For this reason, metric and SI units can be modified by prefixes to refer to either smaller or larger quantities. For instance, the SI unit for mass—the kilogram—differs by the prefix *kilo-* from the metric unit gram. *Kilo-* indicates that a kilogram is 1000 times as large as a gram:

$$1 \text{ kg} = (1000)(1 \text{ g}) = 1000 \text{ g}$$

Small quantities of active ingredients in medications are often reported in *milligrams* (mg). The prefix *milli-* shows that the unit gram has been divided by 1000, which is the same as multiplying by 0.001:

$$1 \text{ mg} = \left(\frac{1}{1000}\right)(1 \text{ g}) = (0.001)(1 \text{ g}) = 0.001 \text{ g}$$

A list of prefixes is given in Table 2.2, with the most common ones displayed in color. Note that the exponents are multiples of 3 for *mega-*(10^6), *kilo-*(10^3), *milli-*(10^{-3}), *micro-*(10^{-6}), *nano-*(10^{-9}), and *pico-*(10^{-12}). (The use of exponents is reviewed in Section 2.5.) The prefixes *centi-*, meaning 1/100, and *deci-*, meaning 1/10, indicate exponents that are not multiples of 3. *Centi-* is seen most often in the length unit *centimeter* (1 cm = 0.01 m), and *deci-* is used most often in clinical chemistry, where the concentrations of blood components are given in milligrams per deciliter (1 dL = 0.1 L). Note also in Table 2.2 that numbers having five or more digits to the right of the decimal point are shown with thin spaces every three digits for convenience—0.000 001, for example. This manner of writing numbers is becoming more common and will be used throughout this book.

▲ These bacteria have a length of about 0.000 000 5 meter.

TABLE 2.2 Some Prefixes for Multiples of Metric and SI Units

Prefix	Symbol	Base Unit Multiplied by*	Example
mega	M	$1,000,000 = 10^6$	1 megameter (Mm) = 10^6 m
kilo	k	$1,000 = 10^3$	1 kilogram (kg) = 10^3 g
hecto	h	$100 = 10^2$	1 hectogram (hg) = 100 g
deka	da	$10 = 10^1$	1 dekaliter (daL) = 10 L
deci	d	$0.1 = 10^{-1}$	1 deciliter (dL) = 0.1 L
centi	c	$0.01 = 10^{-2}$	1 centimeter (cm) = 0.01 m
milli	m	$0.001 = 10^{-3}$	1 milligram (mg) = 0.001 g
micro	μ	$0.000\ 001 = 10^{-6}$	1 micrometer (μm) = 10^{-6} m
nano	n	$0.000\ 000\ 001 = 10^{-9}$	1 nanogram (ng) = 10^{-9} g
pico	p	$0.000\ 000\ 000\ 001 = 10^{-12}$	1 picogram (pg) = 10^{-12} g

*The scientific notation method of writing large and small numbers (for example, 10^6 for 1,000,000) is explained in Section 2.5.

■ **PROBLEM 2.1**

Give the full name of each of the following units:

(a) mL **(b)** kg **(c)** cm **(d)** km **(e)** μg

■ **PROBLEM 2.2**

Write the symbol for each of the following units:

(a) Liter **(b)** Microliter **(c)** Nanometer **(d)** Megameter

■ **PROBLEM 2.3**

Express each of the following quantities in terms of the basic unit (for example, 1 mL = 0.001 L):

(a) 1 nm **(b)** 1 dg **(c)** 1 km **(d)** 1 mL **(e)** 1 ng

2.2 Measuring Mass

The terms "mass" and "weight," though often used interchangeably, really have quite different meanings. **Mass** is a measure of the amount of matter in an object, whereas **weight** is a measure of the gravitational pull that the earth, moon, or other large body exerts on an object. Clearly, the amount of matter in an object does not depend on location. Whether you're standing on the earth or standing on the moon, the mass of your body is the same. On the other hand, the weight of an object *does* depend on location. Your weight on earth might be 140 lb, but it would only be 23 lb on the moon because the pull of gravity there is only about one-sixth as great.

At the same location, two objects with identical masses have identical weights; that is, gravity pulls equally on both. Thus, the *mass* of an object can be determined by comparing the *weight* of the object to the weight of a known reference standard. Much of the confusion between mass and weight is simply due to a language problem: We speak of "weighing" when we really mean that we are measuring mass by comparing two weights. Figure 2.1 shows two types of balances used for measuring mass in the laboratory.

Mass A measure of the amount of matter in an object.

Weight A measure of the gravitational force that the earth or other large body exerts on an object.

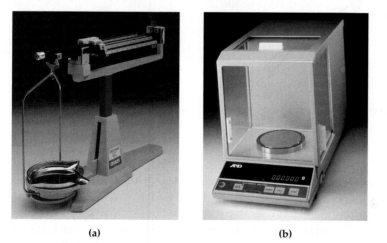

(a) **(b)**

▲ **FIGURE 2.1** (a) The single-pan balance has a sliding counterweight that is adjusted until the weight of the object on the pan is just balanced. (b) A modern electronic balance.

▲ This pile of 400 pennies has a mass of about 1 kg.

One kilogram, the SI unit for mass, is equal to 2.205 lb—too large a quantity for many purposes in chemistry and medicine. Thus, smaller units of mass such as the gram, milligram (mg), and microgram (μg), are more commonly used. Table 2.3 shows the relationships between metric and common units for mass.

TABLE 2.3   Units of Mass

Unit	Equivalent	Unit	Equivalent
1 kilogram (kg)	= 1000 grams = 2.205 pounds	1 ton	= 2000 pounds = 907.03 kilograms
1 gram (g)	= 0.001 kilogram = 1000 milligrams = 0.035 27 ounce	1 pound (lb)	= 16 ounces = 0.454 kilogram = 454 grams
1 milligram (mg)	= 0.001 gram = 1000 micrograms	1 ounce (oz)	= 0.028 35 kilogram = 28.35 grams = 28,350 milligrams
1 microgram (μg)	= 0.000 001 gram = 0.001 milligram		

2.3 Measuring Length and Volume

The meter is the standard measure of length, or distance, in both the SI and metric systems. One meter is 39.37 inches (about 10% longer than a yard), a length that is much too large for most measurements in chemistry and medicine. Other, more commonly used measures of length are the *centimeter* (cm; 1/100 m) and the *millimeter* (mm; 1/1000 m). One centimeter is a bit less than half an inch— 0.3937 inch to be exact. A millimeter, in turn, is 0.039 37 inch, or about the thickness of a dime. Table 2.4 lists the relationships of these units.

TABLE 2.4   Units of Length

Unit	Equivalent	Unit	Equivalent
1 kilometer (km)	= 1000 meters = 0.6214 mile	1 mile (mi)	= 1.609 kilometers = 1609 meters
1 meter (m)	= 100 centimeters = 1000 millimeters = 1.0936 yards = 39.37 inches	1 yard (yd)	= 0.9144 meter = 91.44 centimeters
		1 foot (ft)	= 0.3048 meter = 30.48 centimeters
1 centimeter (cm)	= 0.01 meter = 10 millimeters = 0.3937 inch	1 inch (in.)	= 2.54 centimeters = 25.4 millimeters
1 millimeter (mm)	= 0.001 meter = 0.1 centimeter		

Volume is the amount of space occupied by an object. The SI unit for volume— the cubic meter, m^3—is so large that the liter ($1 \text{ L} = 0.001 \text{ m}^3 = 1 \text{ dm}^3$) is much more commonly used in chemistry and medicine. One liter has the volume of a cube 10 cm (1 dm) on edge and is a bit larger than one U.S. quart. Each liter is further divided into 1000 *milliliters* (mL), with 1 mL being the size of a cube 1 cm on edge, or 1 cm^3. In fact, the milliliter is often called a *cubic centimeter* (cm^3, or cc) in medical work. Figure 2.2 shows the divisions of a cubic meter, and Table 2.5 shows the relationships among units of volume.

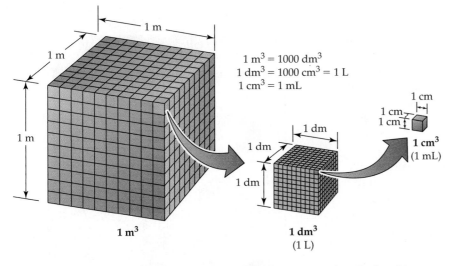

▲ **FIGURE 2.2** A cubic meter is the volume of a cube 1 m on edge. Each cubic meter contains 1000 cubic decimeters (liters), and each cubic decimeter contains 1000 cubic centimeters (milliliters). Thus, there are 1000 mL in a liter and 1000 L in a cubic meter.

TABLE 2.5 Units of Volume

Unit	Equivalent	Unit	Equivalent
1 cubic meter (m^3)	= 1000 liters	1 gallon (gal)	= 3.7856 liters
	= 264.2 gallons	1 quart (qt)	= 0.9464 liter
1 liter (L)	= 0.001 cubic meter		= 946.4 milliliters
	= 1000 milliliters	1 fluid ounce (fl oz)	= 29.57 milliliters
	= 1.057 quarts		
1 deciliter (dL)	= 0.1 liter		
	= 100 milliliters		
1 milliliter (mL)	= 0.001 liter		
	= 1000 microliters		
1 microliter (μL)	= 0.001 milliliter		

▲ The tennis ball weighs 54.07 g on this common laboratory balance, which is capable of determining mass to about 0.01 g.

Uncertainty in Measurement

2.4 Measurement and Significant Figures

How much does a tennis ball weigh? If you put a tennis ball on an ordinary bathroom scale, the scale would probably register 0 lb (or 0 kg if you have a metric scale). If you placed the same tennis ball on a common laboratory balance, however, you might get a reading of 54.07 g. Trying again by placing the ball on an expensive analytical balance like those found in clinical and research laboratories, you might find a weight of 54.071 38 g. Clearly, the precision of your answer depends on the equipment used for the measurement.

Every experimental measurement, no matter how precise, has a degree of uncertainty to it because there is always a limit to the number of digits that can be determined. An analytical balance, for example, might reach its limit in measuring mass to the fifth decimal place, and weighing the tennis ball several times might produce slightly different readings, such as 54.071 39 g, 54.071 38 g, and 54.071 37 g. Also, different people making the same measurement might come up with slightly different answers. How, for instance, would you record the volume of the liquid shown in Figure 2.3?

To indicate the precision of a measurement, the value recorded should use all the digits known with certainty, plus one additional estimated digit that is usually considered uncertain by plus or minus 1 (written as ±1). The total number of digits used to express such a measurement is called the number of **significant figures**. Thus, the quantity 54.07 g has four significant figures

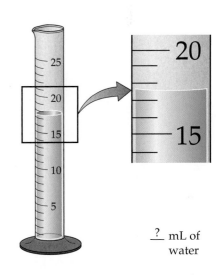

▲ **FIGURE 2.3** What is the volume of liquid in this graduated cylinder?

Significant figures The number of meaningful digits used to express a value.

Application Apothecary Units

Chances are that the bottle of aspirin in your medicine cabinet gives the mass of each tablet in the unit *grains* (gr), where 1 gr = 64.8 mg. Known as an *apothecary unit*, the grain is part of an early system of measurement that was once used by all physicians and pharmacists. The system includes our customary gallons, quarts, and pints, plus some less familiar units that are still used in specialized applications.

The basis for liquid measurement in the apothecary system is the *minim*, a unit originally equal to the volume of one drop of water but now defined as 0.062 mL. Sixty minims exactly equal one *fluidram* (roughly one teaspoonful), and one fluid ounce exactly equals eight fluidrams. The fluidram is still occasionally used for fluid volume in medications and expensive perfumes.

$$1 \text{ minim} = 0.062 \text{ mL}$$
$$1 \text{ fluidram} = 60 \text{ minim} = 3.72 \text{ mL}$$

See Additional Problems 2.74 and 2.75 at the end of the chapter.

▲ The medications dispensed in this medieval pharmacy were probably measured in drams and grains.

(5, 4, 0, and 7), and the quantity 54.071 38 g has seven significant figures. *Remember*: All but one of the significant figures are known with certainty; the last significant figure is only an estimate accurate to ±1.

Uncertain digit

54.07 g   A mass between 54.06 g and 54.08 g (±0.01 g)

Uncertain digit

54.071 38 g   A mass between 54.071 37 g and 54.071 39 g (±0.000 01 g)

Deciding the number of significant figures in a given measurement is usually simple, but can be troublesome when zeros are involved. Depending on the circumstances, a zero might be significant or might be just a space-filler to locate the decimal point. For example, how many significant figures does each of the following measurements have?

94.072 g   Five significant figures (9, 4, 0, 7, 2)

0.0834 cm   Three significant figures (8, 3, 4)

0.029 07 mL Four significant figures (2, 9, 0, 7)

138.200 m   Six significant figures (1, 3, 8, 2, 0, 0)

23,000 kg   *Anywhere* from two (2, 3) to five (2, 3, 0, 0, 0) significant figures

The following rules are helpful for determining the number of significant figures when zeros are present:

Rule 1. Zeros in the middle of a number are like any other digit; they are always significant. Thus, 94.072 g has five significant figures.

Rule 2. Zeros at the beginning of a number are not significant; they act only to locate the decimal point. Thus, 0.0834 cm has three significant figures, and 0.029 07 mL has four.

Rule 3. Zeros at the end of a number and *after* the decimal point are significant. It's assumed that these zeros would not be shown unless they were significant. Thus, 138.200 m has six significant figures. If the value were known to only four significant figures, we would write 138.2 m.

Rule 4. Zeros at the end of a number and *before* an implied decimal point may or may not be significant. We can't tell whether they are part of the measurement or whether they act only to locate the unwritten but implied decimal point. Thus, 23,000 kg may have two, three, four, or five significant figures.

Often, however, a little common sense is useful. A temperature reading of 20°C probably has two significant figures rather than one, because one significant figure would imply a temperature anywhere from 10°C to 30°C and would be of little use. Similarly, a volume given as 300 mL probably has three significant figures. On the other hand, a figure of 150,000,000 km for the distance between the earth and the sun probably has only two or three significant figures because the distance is variable. We'll see a better way to deal with this problem in the next section.

One final point about significant figures: Some numbers, such as those obtained when counting objects and those that are part of a definition, are *exact* and effectively have an unlimited number of significant figures. Thus, a class might have *exactly* 32 students (not 31.9, 32.0, or 32.1), and 1 foot is defined to have *exactly* 12 inches.

▲ There are exactly 32 students in this class, not 31.9 or 32.1.

■ **WORKED EXAMPLE 2.1**

How many significant figures does each of the following measurements have?
(a) 2730.78 m **(b)** 0.0076 mL **(c)** 3400 kg **(d)** 3400.0 m²

SOLUTION
(a) Six (rule 1) **(b)** Two (rule 2)
(c) Two, three, or four (rule 4) **(d)** Five (rule 3)

■ **PROBLEM 2.4**

How many significant figures does each of the following measurements have?
(a) 3.45 m **(b)** 0.1400 kg **(c)** 10.003 L **(d)** 35 cents

■ **KEY CONCEPT PROBLEM 2.5**

What is the temperature reading on the following Celsius thermometer? How many significant figures do you have in your answer?

2.5 Scientific Notation

Rather than write very large or very small numbers in their entirety, it's more convenient to express them using *scientific notation*. A number is written in **scientific notation** as the product of a number between 1 and 10, times the number 10 raised to a power. Thus, 215 is written in scientific notation as 2.15×10^2:

$$215 = 2.15 \times 100 = 2.15 \times (10 \times 10) = 2.15 \times 10^2$$

Scientific notation A number expressed as the product of a number between 1 and 10, times the number 10 raised to a power.

Notice that in this case, where the number is *larger* than 1, the decimal point has been moved *to the left* until it follows the first digit. The exponent on the 10 tells how many places we had to move the decimal point to position it just after the first digit:

$$2\underset{\smile}{15}. \ = \ 2.15 \ \times \ 10^2$$

Decimal point is moved two places to the left, so exponent is 2.

To express a number *smaller* than 1 in scientific notation, we have to move the decimal point *to the right* until it follows the first digit. The number of places moved is the negative exponent of 10. For example, the number 0.002 15 can be rewritten as 2.15×10^{-3}:

$$0.002\ 15 \ = \ 2.15 \ \times \ \frac{1}{1000} \ = \ 2.15 \ \times \ \frac{1}{10 \ \times \ 10 \ \times \ 10} \ = \ 2.15 \ \times \ \frac{1}{10^3} \ = \ 2.15 \ \times \ 10^{-3}$$

$$0.00\underset{\smile}{2}\,15 \ = \ 2.15 \ \times \ 10^{-3}$$

Decimal point is moved three places to the right, so exponent is −3.

To convert a number written in scientific notation to standard notation, the process is reversed. For a number with a *positive* exponent, the decimal point is moved to the *right* a number of places equal to the exponent:

$$3.7962 \ \times \ 10^4 \ = \ 37{,}9\underset{\smile}{62}$$

Positive exponent of 4, so decimal point is moved to the right four places.

For a number with a *negative* exponent, the decimal point is moved to the *left* a number of places equal to the exponent:

$$1.56 \ \times \ 10^{-8} \ = \ 0.000\ 000\ 0\underset{\smile}{15\ 6}$$

Negative exponent of −8, so decimal point is moved to the left eight places.

Scientific notation is particularly helpful for indicating how many significant figures are present in a number that has zeros at the end but to the left of a decimal point. If we read, for instance, that the distance from the earth to the sun is 150,000,000 km, we don't really know how many significant figures are indicated. Some of the zeros might be significant, or they might merely act to locate the decimal point. Using scientific notation, however, we can indicate how many of the zeros are significant. Rewriting 150,000,000 as 1.5×10^8 indicates two significant figures, whereas writing it as 1.500×10^8 indicates four significant figures.

Scientific notation is not ordinarily used for numbers that are easily written, such as 10 or 175, although it's sometimes helpful in doing arithmetic. Rules for doing arithmetic with numbers written in scientific notation are reviewed in Appendix A.

▲ How many molecules are in this 1 g pile of table sugar?

■ **WORKED EXAMPLE 2.2**

There are 1,760,000,000,000,000,000,000 molecules of sucrose (table sugar) in 1 g. Use scientific notation to express this number with four significant figures.

SOLUTION The first four digits—1, 7, 6, and 0—are significant, meaning that only the first of the 19 zeros is significant. Because we have to move the decimal point 21 places to the left to put it after the first significant digit, the answer is 1.760×10^{21}.

■ **WORKED EXAMPLE 2.3**

Use scientific notation to indicate the diameter of a sodium atom: 0.000 000 000 388 m.

SOLUTION There are only three significant figures, because zeros at the beginning of a number are not significant. We have to move the decimal point 10 places to the right to place it after the first digit, so the answer is 3.88×10^{-10} m.

■ WORKED EXAMPLE 2.4

A clinical laboratory found that a blood sample contained 0.0026 g of phosphorus and 0.000 101 g of iron.
(a) Give these quantities in scientific notation.
(b) Give these quantities in the units normally used to report them—milligrams for phosphorus and micrograms for iron.

SOLUTION
(a)

$$0.0026 \text{ g phosphorus } = 2.6 \times 10^{-3} \text{ g phosphorus}$$

$$0.000\,101 \text{ g iron } = 1.01 \times 10^{-4} \text{ g iron}$$

(b) We know that $1 \text{ mg} = 1 \times 10^{-3}$ g, where the exponent is -3. Expressing the amount of phosphorus in milligrams is straightforward because the amount in grams $(2.6 \times 10^{-3} \text{ g})$ already has an exponent of -3. Thus, $2.6 \times 10^{-3} \text{ g} = 2.6$ mg of phosphorus.

$$(2.6 \times 10^{-3} \text{g}) \left(\frac{1 \text{ mg}}{1 \times 10^{-3} \text{g}} \right) = 2.6 \text{ mg}$$

We know that $1 \text{ }\mu\text{g} = 1 \times 10^{-6}$ g, where the exponent is -6. Expressing the amount of iron in micrograms thus requires that we restate the amount in grams so that the exponent is -6. We can do this by moving the decimal point six places to the right:

$$0.000\,101 \text{ g iron } = 101 \times 10^{-6} \text{ g iron } = 101 \text{ }\mu\text{g iron}$$

■ PROBLEM 2.6

Convert the following values to scientific notation:
(a) 58 g   **(b)** 46,792 m   **(c)** 0.000 672 0 cm   **(d)** 345.3 kg

■ PROBLEM 2.7

Convert the following values from scientific notation to standard notation:
(a) 4.885×10^4 mg   **(b)** 8.3×10^{-6} m   **(c)** 4.00×10^{-2} m

■ PROBLEM 2.8

Rewrite the following numbers in scientific notation as indicated:
(a) 600,000 with two significant figures
(b) 1300 with four significant figures
(c) 794,200,000,000 with four significant figures

■ PROBLEM 2.9

Ordinary table salt, or sodium chloride, is made up of small particles called *ions*, which we'll discuss in Chapter 4. If the distance between a sodium ion and a chloride ion is 0.000 000 000 278 m, what is the distance in scientific notation? How many picometers (pm) is this?

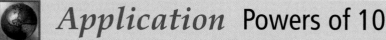
It's not easy to grasp the enormous differences in size represented by powers of 10 (scientific notation), but the following pictures might help. Think of an ordinary household pin. Because the pin is only about 3 cm long, you can see little detail. Imagine, though, that the pin is magnified in size by about 10^2. The image, now 100 times larger, would look something like photo (a).

Now imagine that the pin is magnified by about 10^4 as in photo (b). That small change in exponent, from 2 to 4, makes an extraordinary difference in your view. You see not only the bluntness of the tip (it's not as sharp as you thought) but also that the pin is covered by tiny bacteria about 3×10^{-4} cm in length. Increasing the magnification further, to about 5×10^6, gives you the picture in photo (c). Now you can see the internal details of a bacterium about to undergo cell division. The cell wall, only 70 nm thick, is clearly visible.

The change in magnification from 1 to 10^6 may not seem great, but it represents a million-fold increase—enough to open a whole new world. Powers of 10 are powerful indeed.

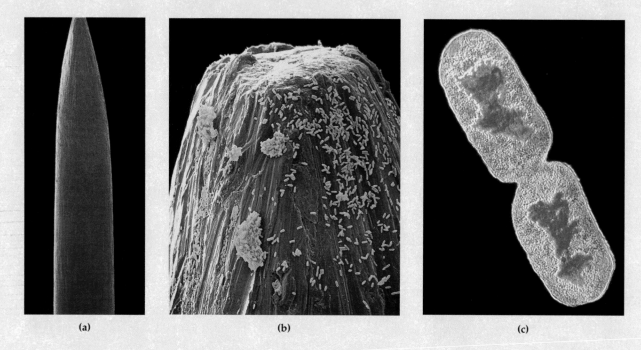

(a)   (b)   (c)

▲ Bacteria on a household pin at different magnifications. The individual pictures are explained in the text.

▲ Calculators often display more digits than are justified by the precision of the data.

2.6 Rounding Off Numbers

It often happens, particularly when doing arithmetic on a pocket calculator, that a quantity appears to have more significant figures than are really justified. For example, you might calculate the gas mileage of your car by finding that it takes 11.70 gallons of gasoline to drive 278 miles:

$$\text{Mileage} = \frac{\text{Miles}}{\text{Gallons}} = \frac{278 \text{ mi}}{11.70 \text{ gal}} = 23.760\ 684 \text{ mi/gal (mpg)}$$

Although the answer on a pocket calculator has eight digits, your measurement is really not as precise as it appears. In fact, as we'll see below, your answer is good to only three significant figures and should be **rounded off** to 23.8 mi/gal.

How do you decide how many digits to keep? The full answer to this question is a bit complex and involves a mathematical treatment called *error analysis*, but for many purposes, a simplified procedure using just two rules is sufficient:

Rounding off   A procedure used for deleting nonsignificant figures.

Significant Figures

Rule 1. In carrying out a multiplication or division, the answer can't have more significant figures than either of the original numbers. This is just a commonsense rule if you think about it. After all, if you don't know the number of miles you drove to better than three significant figures (278 could mean 277, 278, or 279), you certainly can't calculate your mileage to more than the same number of significant figures.

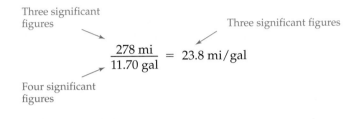

Three significant figures

Three significant figures

$$\frac{278 \text{ mi}}{11.70 \text{ gal}} = 23.8 \text{ mi/gal}$$

Four significant figures

Rule 2. In carrying out an addition or subtraction, the answer can't have more digits after the decimal point than either of the original numbers. For example, if you have 3.18 L of water and you add 0.013 15 L more, you now have 3.19 L. Again, this rule is just common sense. If you don't know the volume you started with past the second decimal place (it could be 3.17, 3.18, or 3.19), you can't know the total of the combined volumes past the same decimal place.

Volume of water at start ⟶   3.18? ?? L   ⟵ Two digits after decimal point
Volume of water addded ⟶ + 0.013 15 L   ⟵ Five digits after decimal point
Total volume of water ⟶   3.19? ?? L   ⟵ Two digits after decimal point

If a calculation has several steps, it's generally best to round off at the end after all the steps have been carried out, keeping the number of significant figures determined by the least accurate number in your calculations. Once you decide how many digits to retain for your answer, the rules for rounding off numbers are straightforward:

Rule 1. If the first digit you remove is 4 or less, drop it and all following digits. Thus, 2.4271 becomes 2.4 when rounded off to two significant figures because the first of the dropped digits (a 2) is 4 or less.

Rule 2. If the first digit you remove is 5 or greater, round the number up by adding a 1 to the digit to the left of the one you drop. Thus, 4.5832 becomes 4.6 when rounded off to two significant figures because the first of the dropped digits (an 8) is 5 or greater.

■ WORKED EXAMPLE 2.5

Suppose that you weigh 124 lb before dinner. How much will you weigh after dinner if you eat 1.884 lb of food?

SOLUTION   Your after-dinner weight is found by adding your original weight to the weight of the food consumed:

$$124 \text{ lb} + 1.884 \text{ lb} = 125.884 \text{ lb (Unrounded)}$$

Because the value of your original weight has no significant figures after the decimal point, your after-dinner weight also must have no significant figures after the decimal point. Thus, 125.884 lb must be rounded off to 126 lb.

■ **WORKED EXAMPLE 2.6**

To make currant jelly, 13.75 cups of sugar was added to 18 cups of currant juice. How much sugar was added per cup of juice?

SOLUTION   The quantity of sugar must be divided by the quantity of juice:

$$\frac{13.75 \text{ cups sugar}}{18 \text{ cups juice}} = 0.763\,888\,89 \frac{\text{cup sugar}}{\text{cup juice}} \quad (\text{Unrounded})$$

The number of significant figures in the answer is limited to two by the quantity 18 cups in the calculation and must be rounded to 0.76 cup of sugar per cup of juice.

■ **PROBLEM 2.10**

Round off each of the following quantities to the indicated number of significant figures:
(a) 2.304 g (three significant figures)
(b) 188.3784 mL (five significant figures)
(c) 0.008 87 L (one significant figure)
(d) 1.000 39 kg (four significant figures)

■ **PROBLEM 2.11**

Carry out the following calculations, rounding each result to the correct number of significant figures:
(a) 4.87 mL + 46.0 mL
(b) 3.4 × 0.023 g
(c) 19.333 m − 7.4 m
(d) 55 mg − 4.671 mg + 0.894 mg
(e) 62,911 ÷ 611

Factor-label method   A problem-solving procedure in which equations are set up so that unwanted units cancel and only the desired units remain.

Conversion factor   An expression of the numerical relationship between two units.

2.7 Problem Solving: Converting a Quantity from One Unit to Another

Many activities in the laboratory and in medicine—measuring, weighing, preparing solutions, and so forth—require converting a quantity from one unit to another. For example: "These pills contain 1.3 grains of aspirin, but I need 200 mg. Is one pill enough?" Converting between units isn't mysterious; we all do it every day. If you run 9 laps around a 400 meter track, for instance, you have to convert between the distance unit "lap" and the distance unit "meter" to find that you have run 3600 m (9 laps times 400 m/lap). If you want to find how many miles that is, you have to convert again to find that 3600 m = 2.237 mi.

The simplest way to carry out calculations involving different units is to use the **factor-label method**. In this method, a quantity in one unit is converted into an equivalent quantity in a different unit by using a **conversion factor** that expresses the relationship between units:

Starting quantity × Conversion factor = Equivalent quantity

▲ These runners have to convert from laps to meters to find out how far they have run.

As an example, we said in Section 2.3 that 1 km = 0.6214 mi. Writing this relationship as a fraction restates it in the form of a conversion factor, either kilometers per mile or miles per kilometer.

Since 1 km = 0.6214 mi, then:

Conversion factors between kilometers and miles

$$\frac{1 \text{ km}}{0.6214 \text{ mi}} = 1 \quad \text{or} \quad \frac{0.6214 \text{ mi}}{1 \text{ km}} = 1$$

Note that this and all other conversion factors are numerically equal to 1 because the value of the quantity above the division line (the numerator) is equal in value to the quantity below the division line (the denominator). Thus, multiplying by a conversion factor is equivalent to multiplying by 1 and so does not change the value of the quantity being multiplied:

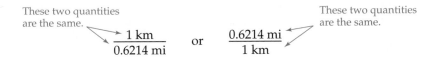

These two quantities are the same. $\frac{1 \text{ km}}{0.6214 \text{ mi}}$ or $\frac{0.6214 \text{ mi}}{1 \text{ km}}$ These two quantities are the same.

The key to the factor-label method of problem solving is that units are treated like numbers and can thus be multiplied and divided (though not added or subtracted) just as numbers can. When solving a problem, the idea is to set up an equation so that all unwanted units cancel, leaving only the desired units. Usually, it's best to start by writing what you know and then manipulating that known quantity. For example, if you know there are 26.22 mi in a marathon and want to find how many kilometers that is, you could write the distance in miles and multiply by the conversion factor in kilometers per mile. The unit "mi" cancels because it appears both above and below the division line, leaving "km" as the only remaining unit.

$$26.22 \text{ mi} \times \frac{1 \text{ km}}{0.6214 \text{ mi}} = 42.20 \text{ km}$$

Starting quantity   Conversion factor   Equivalent quantity

The factor-label method gives the right answer only if the equation is set up so that the unwanted unit (or units) cancel. If the equation is set up in any other way, the units won't cancel and you won't get the right answer. Thus, if you selected the incorrect conversion factor (miles per kilometer) for the above problem, you would end up with an incorrect answer expressed in meaningless units:

Incorrect $26.22 \text{ mi} \times \dfrac{0.6214 \text{ mi}}{1 \text{ km}} = 16.29 \dfrac{\text{mi}^2}{\text{km}}$ Incorrect

■ **WORKED EXAMPLE 2.7**

Write conversion factors for the following pairs of units (use Tables 2.3–2.5):
(a) Deciliters and milliliters   **(b)** Pounds and grams

SOLUTION

(a) Since 1 L = 10 dL = 1000 mL, then 1 dL = 100 mL. The conversion factors are

$$\frac{1 \text{ dL}}{100 \text{ mL}} \quad \text{and} \quad \frac{100 \text{ mL}}{1 \text{ dL}}$$

(b) $\dfrac{1 \text{ lb}}{454 \text{ g}}$ and $\dfrac{454 \text{ g}}{1 \text{ lb}}$

■ **WORKED EXAMPLE 2.8**

(a) Convert 0.75 lb to grams.   **(b)** Convert 0.50 qt to deciliters.

SOLUTION

(a) Set up an equation using the conversion factor from Worked Example 2.7(b) so that the "lb" units cancel and "g" remains:

$$0.75 \text{ lb} \times \frac{454 \text{ g}}{1 \text{ lb}} = 340 \text{ g}$$

(b) In this, as in many problems, it's convenient to use more than one conversion factor. As long as the unwanted units cancel correctly, two or more conversion factors can be strung together in the same calculation. In this case, we can convert first between quarts and milliliters, and then between milliliters and deciliters:

$$0.50 \text{ qt} \times \frac{946.4 \text{ mL}}{1 \text{ qt}} \times \frac{1 \text{ dL}}{100 \text{ mL}} = 4.7 \text{ dL}$$

■ **PROBLEM 2.12**

Write conversion factors for the following pairs of units:

(a) Liters and milliliters   **(b)** Grams and ounces   **(c)** Liters and quarts

■ **PROBLEM 2.13**

Carry out the following conversions:

(a) 16.0 oz = ? g   **(b)** 2500 mL = ? L   **(c)** 99.0 L = ? qt

■ **PROBLEM 2.14**

Convert 0.840 qt to milliliters in a single calculation using more than one conversion factor.

■ **PROBLEM 2.15**

One international nautical mile is defined as exactly 6076.1155 ft, and a speed of 1 knot is defined as one international nautical mile per hour. What is the speed in meters per second of a boat traveling at a speed of 14.3 knots?

2.8 Problem Solving: Estimating Answers

The main drawback to using the factor-label method is that it's possible to get an answer without really understanding what you're doing. It's therefore best after solving a problem to think through a rough estimate, or "ballpark" solution, as a check on your work. If your ballpark check isn't close to the detailed solution, there's a misunderstanding somewhere, and you should think the problem through again. If, for example, you came up with the answer 5.3 cm^3 when calculating the volume of a human cell, you should realize that such an answer couldn't possibly be right. Cells are too tiny to be distinguished with the naked eye, but a volume of 5.3 cm^3 is about the size of a walnut. The Worked Exam-

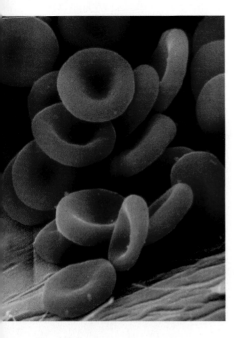

▲ What is the volume of a red blood cell?

ples at the end of this section show how to estimate the answers to simple unit-conversion problems.

The factor-label method and the use of ballpark checks are techniques that will help you solve problems of many kinds, not just unit conversions. Problems sometimes seem complicated, but you can usually sort out the complications by analyzing the problem properly:

- Identify the information given, including units.
- Identify the information needed in the answer, including units.
- Find a relationship between the known information and unknown answer, and plan a series of steps for getting from one to the other.
- Solve the problem.
- Make a rough estimate to be sure your calculated answer is reasonable.

Worked Examples 2.10 and 2.11 illustrate how to use analysis and ballpark checks as an aid in dosage calculations.

■ WORKED EXAMPLE 2.9

A child is 21.5 in. long at birth. How long is this in centimeters?

ANALYSIS The known information is the length of the child in inches. The unknown is the length in centimeters. From Table 2.4, we know that 1 in. = 2.54 cm.

SOLUTION Write the length in inches, and set up an equation with the proper conversion factor so that the inch units cancel. Don't forget to round your answer to the proper number of significant figures.

$$21.5 \text{ in.} \times \frac{2.54 \text{ cm}}{1 \text{ in.}} = 54.6 \text{ cm} \quad (\text{Rounded off from } 54.61)$$

BALLPARK CHECK Since 1 in. = 2.54 cm, it takes about 2.5 times more centimeters than inches to measure the same length. Thus, 21.5 in. is about 2.5 times 20 cm, or approximately 50 cm. The ballpark check agrees with the solution.

■ WORKED EXAMPLE 2.10

A patient requires injection of 0.012 g of a pain killer available as a 15 mg/mL solution. How many milliliters should be administered?

ANALYSIS The known information is the mass of the medication in grams and the concentration of the solution in milligrams per milliliter, which acts as a conversion factor. The unknown is the number of milliliters of solution needed.

SOLUTION

$$(0.012 \text{ g}) \left(\frac{1 \text{ mL}}{15 \text{ mg}} \right) \left(\frac{1000 \text{ mg}}{1 \text{ g}} \right) = 0.80 \text{ mL}$$

▲ How many milliliters should be injected?

BALLPARK CHECK Look for an easy way to round off the numbers in the problem so that you can mentally check the math. You might notice that 0.012 g/15 mg is close to 0.015/15, or 0.001, times 1000 mg/g, giving a ballpark check of 1 mL, which is close to the detailed solution.

■ **WORKED EXAMPLE 2.11**

Administration of digitalis to control atrial fibrillation in heart patients must be carefully regulated because even a modest overdose can be fatal. To take differences between patients into account, dosages are sometimes prescribed in micrograms per kilogram of body weight ($\mu g/kg$). Thus, two people may differ greatly in weight, but both will receive the proper dosage. At a dosage of 20 $\mu g/kg$ body weight, how many milligrams of digitalis should a 160 lb patient receive?

ANALYSIS The known information is (1) the patient's weight in pounds and (2) the required dosage in micrograms per kilogram of body weight. The needed information is the dose in milligrams. Two unit conversions are therefore required—from known pounds of body weight to kilograms and from known micrograms to milligrams. We can solve the problem by using the conversion factors 1 kg = 2.205 lb and 1 mg = 1000 μg, so that all units except milligrams cancel.

SOLUTION

$$160 \text{ lb} \times \frac{1 \text{ kg}}{2.205 \text{ lb}} \times \frac{20 \ \mu g \text{ digitalis}}{1 \text{ kg}} \times \frac{1 \text{ mg}}{1000 \ \mu g} = \begin{array}{l} 1.5 \text{ mg digitalis} \\ \text{(Rounded off)} \end{array}$$

BALLPARK CHECK Since a kilogram is roughly equal to 2 lb, a 160 lb patient has a mass of about 80 kg. At a dosage of 20 $\mu g/kg$, an 80 kg patient should receive 80 × 20 μg, or about 1600 μg of digitalis. Since 1 mg = 1000 μg, it will take only one-thousandth as many milligrams as micrograms. Our rough estimate is therefore about 1.6 mg of digitalis, so the ballpark check agrees with the detailed solution.

■ **PROBLEM 2.16**

(a) How many kilograms does a 7.5 lb infant weigh?

(b) How many milliliters are in a 4.0 fl oz bottle of cough medicine?

■ **PROBLEM 2.17**

Calculate the dosage in milligrams per kilogram body weight for a 135 lb adult who takes two aspirin tablets containing 0.324 g of aspirin each. Calculate the dosage for a 40 lb child who also takes two aspirin tablets.

Temperature The measure of how hot or cold an object is.

▲ Summer temperatures in Death Valley, California, reach over 130°F, but entrants in the annual Badwater to Mt. Whitney 135 mile race run right through it.

2.9 Measuring Temperature

Temperature, the measure of how hot or cold an object is, is commonly reported either in Fahrenheit (°F) or Celsius (°C) units. The SI unit for reporting temperature, however, is the *kelvin* (K). (Note that we say only "kelvin," not "kelvin degree.")

The kelvin and the Celsius degree are the same size—both are 1/100 of the interval between the freezing point of water and the boiling point of water at atmospheric pressure. The only difference between the Kelvin and Celsius temperature scales is that they have different zero points. The Celsius scale assigns a value of 0°C to the freezing point of water, but the Kelvin scale assigns a value of 0 K to the coldest possible temperature, sometimes called *absolute zero*, which is equal to −273.15°C. Thus, 0 K = −273.15°C, and +273.15 K = 0°C. For example, a warm spring day with a temperature of 25°C has a Kelvin tem-

Application Obesity: A Large Problem

In 1962, 12.8% of the U.S. population was considered by the National Institutes of Health to be clinically obese. In 1999, the percentage was 27%. Even children and adolescents are gaining too much weight: Approximately 14% of boys and girls under age 19 are overweight. It looks like a lot of people have been eating too many french fries in the past few decades.

Obesity is defined by reference to *body mass index* (BMI), which is equal to a person's weight in kilograms divided by the square of his or her height in meters. Alternatively, a person's weight in pounds divided by the square of the height in inches is multiplied by 703 to give the BMI. For instance, someone with a weight of 147 lb (66.7 kg) and height of 5′7′′ (67 inches; 1.70 m) has a BMI of 23.

$$BMI = \frac{\text{weight (kg)}}{[\text{height (m)}]^2} \quad \text{or} \quad \frac{\text{weight (lb)}}{[\text{height (in.)}]^2} \times 703$$

A BMI of 25 or above is considered overweight, and a BMI of 30 or above is obese. By these standards, approximately 61% of the U.S. population is overweight.

Health professionals are concerned by the rapid rise in obesity because of the link between BMI and death rates. Many reports have documented the correlation between health and BMI, including a recent study on over 1 million adults. The lowest death risk from any cause, including cancer and heart disease, is associated with a BMI between 22 and 24. Risk increases steadily as BMI increases, more than doubling for a BMI above 29.

Why are more and more people getting larger and larger? Researchers strongly suspect that the obvious answer is correct: a combination of too much food and too little activity. Who among us can resist the urge to "super-size" the fast food or to take an elevator rather than walk up one flight of stairs? Although genetics plays an important role in an individual's ability to put on or take off weight, only disciplined eating and more regular exercise can ultimately reduce the trend of recent decades.

See Additional Problems 2.76 and 2.77 at the end of the chapter.

Weight (lb)

Height	110	115	120	125	130	135	140	145	150	155	160	165	170	175	180	185	190	195	200
5′0′′	21	22	23	24	25	26	27	28	29	30	31	32	33	34	35	36	37	38	39
5′2′′	20	21	22	23	24	25	26	27	27	28	29	30	31	32	33	34	35	36	37
5′4′′	19	20	21	21	22	23	24	25	26	27	27	28	29	30	31	32	33	33	34
5′6′′	18	19	19	20	21	22	23	23	24	25	26	27	27	28	29	30	31	31	32
5′8′′	17	17	18	19	20	21	21	22	23	24	24	25	26	27	27	28	29	30	30
5′10′′	16	17	17	18	19	19	20	21	22	22	23	24	24	25	26	27	27	28	29
6′0′′	15	16	16	17	18	18	19	20	20	21	22	22	23	24	24	25	26	26	27
6′2′′	14	15	15	16	17	17	18	19	19	20	21	21	22	22	23	24	24	25	26
6′4′′	13	14	15	15	16	16	17	18	18	19	19	20	21	21	22	23	23	24	24

Body Mass Index (numbers in boxes)

perature of 25 + 273.15 = 298 K (for most purposes, rounding off to 273 is sufficient):

$$\text{Temperature in K} = \text{Temperature in °C} + 273.15$$

$$\text{Temperature in °C} = \text{Temperature in K} - 273.15$$

For practical applications in medicine and clinical chemistry, the Fahrenheit and Celsius scales are used almost exclusively. The Fahrenheit scale defines the freezing point of water as 32°F and the boiling point of water as 212°F, while 0°C and 100°C are the freezing and boiling points of water on the Celsius scale. Thus, it takes 180 Fahrenheit degrees to cover the same range encompassed by only 100 Celsius degrees, and a Celsius degree is therefore exactly 180/100 = 9/5 = 1.8 times as large as a Fahrenheit degree. Figure 2.4 gives a comparison of all three scales.

Temperature Conversion

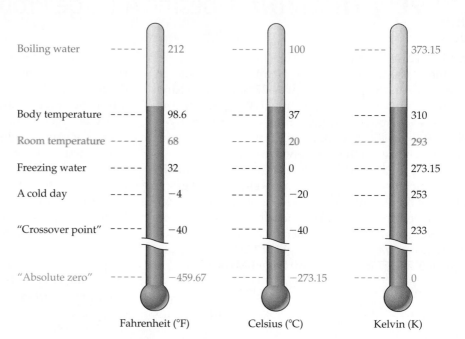

▲ **FIGURE 2.4 A comparison of the Fahrenheit, Celsius, and Kelvin temperature scales.** One Fahrenheit degree is 5/9 the size of a kelvin or Celsius degree.

Converting between the Fahrenheit and Celsius scales is similar to converting between different units of length or volume, but is a bit more complex because two corrections need to be made—one to adjust for the difference in degree size and one to adjust for the different zero points. The size correction is made by using the relationship 1°C = (9/5)°F and 1°F = (5/9)°C. The zero-point correction is made by remembering that the freezing point is higher by 32 on the Fahrenheit scale than on the Celsius scale. Thus, if you want to convert from Celsius to Fahrenheit, you do a size adjustment (multiply °C by 9/5) and then a zero-point adjustment (add 32); if you want to convert from Fahrenheit to Celsius, you find out how many Fahrenheit degrees there are above freezing (by subtracting 32) and then do a size adjustment (multiply by 5/9). The following formulas show the conversion methods:

Celsius to Fahrenheit:

$$°F = \left(\frac{9°F}{5°C} \times °C \right) + 32°F$$

Fahrenheit to Celsius:

$$°C = \frac{5°C}{9°F} \times (°F - 32°F)$$

■ **WORKED EXAMPLE 2.12**

A body temperature above 107°F can be fatal. What does 107°F correspond to on the Celsius scale?

ANALYSIS The known information is the temperature in °F. The unknown is the temperature in °C. The conversion formulas are given in the text.

SOLUTION According to the equation given in the text:

$$°C = \frac{5°C}{9°F} \times (107°F - 32°F) = 41.7°C \quad \text{(Rounded off from 41.666 667°C)}$$

BALLPARK CHECK   A temperature of 107°F is 75 Fahrenheit degrees above freezing (32°F). Because Celsius degrees are a bit less than twice as large as Fahrenheit degrees (1°C = 1.8°F), it takes a bit more than half as many of them, or 40°C, to span the same range. Freezing on the Celsius scale is 0°C, so our rough answer is that 107°F equals about 40°C. This answer seems reasonable if you remember that normal body temperature is 37°C.

■ **PROBLEM 2.18**

The highest land temperature ever recorded was 136°F in Al Aziziyah, Libya, on September 13, 1922. What is this temperature on the Celsius scale?

■ **PROBLEM 2.19**

The use of mercury thermometers is limited by the fact that mercury freezes at −38.9°C. What temperature does this correspond to on the Fahrenheit scale? On the Kelvin scale?

2.10 Energy and Heat

All chemical reactions are accompanied by a change in **energy**, which is defined in scientific terms as *the capacity to do work or supply heat*. There are two fundamental and interconvertible kinds of energy: *potential* and *kinetic*. **Potential energy** is stored energy. The water in a reservoir behind a dam, an automobile poised to coast downhill, and a coiled spring have potential energy waiting to be released. **Kinetic energy**, by contrast, is the energy of motion. When the water falls over the dam and turns a turbine, when the car rolls downhill, or when the spring uncoils and makes the hands on a clock move, the potential energy in each is converted to kinetic energy. Of course, once all the potential energy has been converted, nothing further occurs. The water at the bottom of the dam, the car at the bottom of the hill, and the uncoiled spring no longer have potential energy and thus undergo no further change.

In chemical reactions, the potential energy stored in chemical compounds is often converted into **heat** (Figure 2.5)— the kinetic energy of the moving particles that make up the compound. Because the reaction products have less potential energy than the reactants, we say that the products are *more stable* than the reactants. The term "stable" is used in chemistry to describe a substance that has little remaining potential energy and consequently little tendency to undergo further change.

Energy is measured in SI units by the unit *joule* (J; pronounced "jool"), but the metric *calorie* (cal) is still widely used in medicine. One calorie is the amount of heat necessary to raise the temperature of 1 g of water by 1°C. A *kilocalorie* (kcal), often called a *large calorie (Cal)* or *food calorie* by nutritionists, equals 1000 cal.

$$1000 \text{ cal} = 1 \text{ kcal} \qquad 1000 \text{ J} = 1 \text{ kJ}$$

$$1 \text{ cal} = 4.184 \text{ J} \qquad 1 \text{ kcal} = 4.184 \text{ kJ}$$

Not all substances have their temperatures raised to the same extent when equal amounts of heat energy are added. One calorie raises the temperature of 1 g of water by 1°C but raises the temperature of 1 g of iron by 10°C. The amount of heat needed to raise the temperature of 1 g of a substance by 1°C is called the **specific heat** of the substance. It is measured in units of cal/(g · °C).

$$\text{Specific heat} = \frac{\text{Calories}}{\text{Grams} \times °C}$$

Energy   The capacity to do work or supply heat.

Potential energy   Stored energy.

Kinetic energy   The energy of an object in motion.

Heat   The kinetic energy of moving chemical particles.

Specific heat   The amount of heat that will raise the temperature of 1 g of a substance by 1°C.

▲ **FIGURE 2.5   The reaction of aluminum with bromine releases kinetic energy in the form of heat.** When the reaction is complete, the products undergo no further change.

TABLE 2.6 Specific Heats of Some Common Substances	
Substance	**Specific Heat [cal/(g·°C)]**
Ethanol	0.59
Gold	0.031
Iron	0.106
Mercury	0.033
Sodium	0.293
Water	1.00

Specific heats vary greatly from one substance to another, as shown in Table 2.6. The specific heat of water, 1.00 cal/(g·°C), is higher than that of most other substances, which means that a large transfer of heat is required to change the temperature of a given amount of water by a given number of degrees. One consequence is that the human body, which is about 60% water, is able to maintain a steady internal temperature under changing outside conditions.

Knowing the mass and specific heat of a substance makes it possible to calculate how much heat must be added or removed to accomplish a given temperature change, as shown in Worked Example 2.13.

$$\text{Heat (cal)} = \text{Mass(g)} \times \text{Temperature change(°C)} \times \text{Specific heat}\left(\frac{\text{cal}}{\text{g} \cdot \text{°C}}\right)$$

■ WORKED EXAMPLE 2.13

Taking a bath might use about 95 kg of water. How much energy (in calories) is needed to heat the water from a cold 15°C to a warm 40°C?

ANALYSIS   The known information is the amount of water being heated (95 kg) and the amount of the temperature change 40°C − 15°C = 25°C. The unknown is the total amount of energy needed. Specific heat [1 cal/(g·°C)] acts as a conversion factor.

SOLUTION   Set up an equation so that unwanted units cancel. Two conversion factors are needed.

$$25\text{°C} \times \frac{1000 \text{ g}}{\text{kg}} \times \frac{1.0 \text{ cal}}{(\text{g} \cdot \text{°C})} \times 95 \text{ kg} = 2,400,000 \text{ cal} = 2.4 \times 10^6 \text{ cal}$$

BALLPARK CHECK   The water is being heated 25°C (from 15°C to 40°C), and it therefore takes 25 cal to heat each gram. Since 95 kg of water is 95,000 g, or about 100,000 g, it takes about 25 × 100,000 cal, or 2,500,000 cal, to heat the entire tub of water. The ballpark check agrees with the detailed solution.

■ PROBLEM 2.20

Assuming that Coca-Cola has the same specific heat as water, how much energy in calories is removed when 350 g of Coke (about the contents of one 12 oz can) is cooled from room temperature (25°C) to refrigerator temperature (3°C)?

■ PROBLEM 2.21

What is the specific heat of aluminum if it takes 161 cal to raise the temperature of a 75 g bar by 10.0°C?

Density   The physical property that relates the mass of an object to its volume; mass per unit volume.

▲   Which weighs more, the pillow or the brass weight? In fact, they have a similar mass but the brass weight has a higher *density* because its volume is smaller.

2.11 Density

One further physical quantity that we'll take up in this chapter is **density**, which relates the mass of an object to its volume. Density is usually expressed in units of grams per cubic centimeter (g/cm³) for solids and grams per milliliter (g/mL) for liquids. Thus, if we know the density of a substance, we know both the mass of a given volume and the volume of a given mass. The densities of some common materials are listed in Table 2.7.

$$\text{Density} = \frac{\text{Mass (g)}}{\text{Volume (mL or cm}^3)}$$

TABLE 2.7   Densities of Some Common Materials at 25°C

Substance	Density*	Substance	Density*
Ice (0°C)	0.917	Human fat	0.94
Water (3.98°C)	1.0000	Cork	0.22–0.26
Gold	19.3	Table sugar	1.59
Helium	0.000 194	Balsa wood	0.12
Air	0.001 185	Earth	5.54
Urine	1.003–1.030	Blood plasma	1.027

*Densities are in g/cm^3 for solids and g/mL for liquids and gases.

Most substances change their volume by expanding or contracting when heated or cooled, and densities are therefore temperature-dependent. For example, at 3.98°C, a 1 mL container holds exactly 1.0000 g of water (density = 1.0000 g/mL). As the temperature rises, however, the volume occupied by the water expands so that only 0.9584 g fits in the 1 mL container at 100°C (density = 0.9584 g/mL). When reporting a density, the temperature must also be specified.

Although most substances contract when cooled and expand when heated, water behaves differently. Water contracts when cooled from 100°C to 3.98°C, but below this temperature it begins to *expand* again. The density of liquid water is at its maximum of 1.0000 g/mL at 3.98°C, but decreases to 0.999 87 g/mL at 0°C. When freezing occurs, the density drops still further to a value of 0.917 g/cm^3 for ice at 0°C. Since a less dense substance will float on top of a more dense fluid, ice and any other substance with a density less than that of water will float. Conversely, any substance with a density greater than that of water will sink.

Knowing the density of a liquid is useful because it's often easier to measure a liquid's volume rather than its mass. Suppose, for example, that you need 1.50 g of ethanol. Rather than use a dropper to weigh out exactly the right amount, it would be much easier to look up the density of ethanol (0.7893 g/mL at 20°C) and measure the correct volume (1.90 mL) with a syringe or graduated cylinder. Thus, density acts as a conversion factor between mass (g) and volume (mL).

▲ Ice floats because its density is less than that of water.

$$1.50 \text{ g ethanol} \times \frac{1 \text{ mL ethanol}}{0.7893 \text{ g ethanol}} = 1.90 \text{ mL ethanol}$$

■ **WORKED EXAMPLE 2.14**

What volume of isopropyl alcohol (rubbing alcohol) would you use if you needed 25.0 g? The density of isopropyl alcohol is 0.7855 g/mL at 20°C.

ANALYSIS   The known information is the mass of isopropyl alcohol (25.0 g) and the density (0.7855 g/mL), which acts as a conversion factor between mass and volume. The unknown is the volume of isopropyl alcohol.

SOLUTION   Set up an equation using density as the conversion factor so that the unwanted units cancel:

$$25.0 \text{ g isopropyl alcohol} \times \frac{1 \text{ mL isopropyl alcohol}}{0.7855 \text{ g isopropyl alcohol}} = 31.8 \text{ mL alcohol}$$

BALLPARK CHECK   Since 1 mL of isopropyl alcohol weighs a bit less than 1 g, it will take a bit more than 25 mL to get 25 g—perhaps about 30 mL. The ballpark check agrees with the detailed solution.

Application Measuring Body Fat

Much as we might complain about it, none of us would survive long without the layer of adipose tissue (body fat) lying just beneath our skin. Not only does this fat layer act as a shock absorber and as a thermal insulator to maintain body temperature, it also serves as a long-term energy storehouse. A typical adult body contains about 50% muscles and other cellular material, 24% blood and other fluids, 7% bone, and 19% body fat. Overweight sedentary individuals have a higher fat percentage, while some world-class athletes have as little as 3% body fat.

An individual's percentage of body fat is most easily measured by the skinfold thickness method. The skin at several locations on the arm, shoulder, and waist is pinched, and the thickness of the fat layer beneath the skin is measured with calipers. Comparing the measured results to those in a standard table gives an estimation of percentage body fat.

As an alternative to skinfold measurement, a more accurate assessment of body fat can be made by underwater immersion. In this method, a person climbs into a pool of water, exhales air from the lungs to sink below the surface, and is weighed while totally immersed. The person's underwater body weight is less than the land weight because water gives the body buoyancy. The higher the percentage of body fat, the more buoyant the person and

▲ A person's percentage body fat can be estimated by measuring the thickness of the fat layer under the skin.

the greater the difference between land weight and underwater body weight. Checking the observed buoyancy on a standard table then gives an estimation of body fat. For instance, a person who weighs 60.0 kg on land and 2.9 kg while immersed has about 20% body fat.

See Additional Problems 2.78 and 2.79 at the end of the chapter.

■ **PROBLEM 2.22**
Which of the solids whose densities are given in Table 2.7 will float on water, and which will sink?

■ **PROBLEM 2.23**
Chloroform, once used as an anesthetic agent, has a density of 1.474 g/mL. What volume would you use if you needed 12.37 g?

■ **PROBLEM 2.24**
A glass stopper weighing 16.8 g has a volume of 7.60 cm^3. What is the density of the glass?

2.12 Specific Gravity

For many purposes, ranging from winemaking to medicine, it's more convenient to use *specific gravity* than density. The **specific gravity** (sp gr) of a substance (usually a liquid) is simply the density of the substance divided by the density of water at the same temperature. Because all units cancel, specific gravity is unitless:

Specific gravity The density of a substance divided by the density of water at the same temperature.

$$\text{Specific gravity} = \frac{\text{Density of substance (g/mL)}}{\text{Density of water at the same temperature (g/mL)}}$$

At normal temperatures, the density of water is very close to 1 g/mL. Thus, the specific gravity of a substance is numerically equal to its density and is used in the same way.

The specific gravity of a liquid can be measured using an instrument called a *hydrometer*, which consists of a weighted bulb on the end of a calibrated glass tube, as shown in Figure 2.6. The depth to which the hydrometer sinks when placed in a fluid indicates the fluid's specific gravity: The lower the bulb sinks, the lower the specific gravity of the fluid.

Water that contains dissolved particles can have a specific gravity either higher or lower than 1.00. In winemaking, for instance, the amount of fermentation taking place is gauged by observing the change in specific gravity on going from grape juice, which contains 20% dissolved sugar and has a specific gravity of 1.082, to dry wine, which contains 12% alcohol and has a specific gravity of 0.984. (Pure alcohol has a specific gravity of 0.789.)

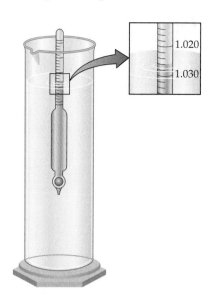

▲ **FIGURE 2.6 A hydrometer for measuring specific gravity.** The instrument has a weighted bulb at the end of a calibrated glass tube. The depth to which the hydrometer sinks in a fluid indicates the fluid's specific gravity.

▲ The amount of fermentation that has taken place in the wine can be measured with a hydrometer.

In medicine, a hydrometer called a *urinometer* is used to indicate the amount of solids dissolved in urine. Although the specific gravity of normal urine is about 1.003–1.030, conditions such as diabetes mellitus or a high fever cause an abnormally high urine specific gravity, indicating either excessive elimination of solids or decreased elimination of water. Abnormally low specific gravity is found in individuals using diuretics—drugs that increase water elimination.

■ **PROBLEM 2.25**

The sulfuric acid solution in an automobile battery typically has a specific gravity of about 1.27. Is battery acid more dense or less dense than pure water

Key Words

Conversion factor, p. 28

Density, p. 36

Factor-label method, p. 28

Energy, p. 35

Heat, p. 35

Kinetic energy, p. 35

Mass, p. 19

Physical quantity, p. 7

Potential energy, p. 35

Rounding off, p. 26

Scientific notation, p. 23

SI units, p. 17

Significant figures, p. 21

Specific gravity, p. 38

Specific heat, p. 35

Temperature, p. 32

Unit, p. 17

Weight, p. 19

Summary: Revisiting the Chapter Goals

1. How are measurements made, and what units are used?

A property that can be measured is called a *physical quantity* and is described by both a number and a label, or *unit*. The preferred units are either those of the International System of Units (*SI units*) or the *metric system*. Mass, the amount of matter an object contains, is measured in *kilograms* (kg) or *grams* (g). Length is measured in *meters* (m). Volume is measured in *cubic meters* (m^3) in the SI system and in *liters* (L) or *milliliters* (mL) in the metric system. Temperature is measured in *kelvins* (K) in the SI system and in *degrees Celsius* (°C) in the metric system.

2. How good are the reported measurements?

When measuring physical quantities or using them in calculations, it's important to indicate the exactness of the measurement by *rounding off* the final answer using the correct number of *significant figures*. All but one of the significant figures in a number is known with certainty; the final digit is estimated to ±1.

3. How are large and small numbers best represented?

Measurements of small and large quantities are usually written in *scientific notation* as the product of a number between 1 and 10, times a power of 10. Numbers greater than 10 have a positive exponent, and numbers less than 1 have a negative exponent. For example, $3562 = 3.562 \times 10^3$, and $0.003\ 91 = 3.91 \times 10^{-3}$.

4. How can a quantity be converted from one unit of measure to another?

A measurement in one unit can be converted to another unit by multiplying by a *conversion factor* that expresses the exact relationship between the units.

5. What techniques are used to solve problems?

Problems are best solved by applying the *factor-label method*, in which units can be multiplied and divided just as numbers can. The idea is to set up an equation so that all unwanted units cancel, leaving only the desired units. Usually it's best to start by identifying the known and needed information, then decide how to convert the known information to the answer, and finally check to make sure the answer is reasonable.

6. What are energy, heat, specific heat, density, and specific gravity?

Energy—the capacity to supply heat or to do work—is of two interconvertible kinds. *Potential energy* is stored energy, and *kinetic energy is the energy of motion. Heat* is the kinetic energy of moving particles in a chemical substance, while *temperature* is a measure of how hot or cold an object is. The *specific heat* of a substance is the amount of heat necessary to raise the temperature of 1 g of the substance by 1°C. Water has an unusually high specific heat, which helps our bodies to maintain an even temperature.

Density, the physical property that relates mass to volume, is expressed in units of grams per milliliter (g/mL) for a liquid or grams per cubic centimeter (g/cm^3) for a solid. The *specific gravity* of a liquid is the density of the liquid divided by the density of water at the same temperature. Because the density of water is approximately 1 g/mL, specific gravity and density have the same numerical value.

■ Understanding Key Concepts

2.26 How many milliliters of water does the graduated cylinder in (a) contain, and how tall in centimeters is the paperclip in (b)? How many significant figures do you have in each answer?

(a)   (b)

2.27 (a) What is the specific gravity of the following solution?
   (b) How many significant figures does your answer have?
   (c) Is the solution more dense or less dense than water?

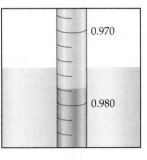

2.28 Assume that you have two graduated cylinders, one with a capacity of 5 mL (a) and the other with a capacity of 50 mL (b). Draw a line in each showing how much liquid you would add if you needed to measure 2.64 mL of water. Which cylinder do you think is more accurate? Explain.

2.29 Assume that identical hydrometers are placed in ethanol (sp gr 0.7893) and in chloroform (sp gr 1.4832). In which liquid will the hydrometer float higher? Explain.

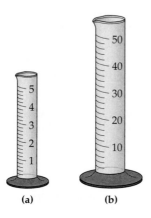

(a)　　　(b)

▪ Additional Problems

Definitions and Units

2.30 What is the difference between a physical quantity and a number?

2.31 What is the difference between mass and weight?

2.32 What are the units used in the SI system to measure mass, volume, length, and temperature?

2.33 What are the units used in the metric system to measure mass, volume, length, and temperature?

2.34 What is the difference between a cubic decimeter (SI) and a liter (metric)?

2.35 What is the difference between a kelvin (SI) and a Celsius degree (metric)?

2.36 Give the full name of each of the following units:

  (a) cL   **(b)** dm   **(c)** mm   **(d)** nL
  (e) mg   **(f)** m^3 **(g)** cc

2.37 Write the symbol for each of the following units:

  (a) Picogram **(b)** Centimeter **(c)** Deciliter
  (d) Microliter **(e)** Milliliter

2.38 How many picograms are in 1 mg? In 35 ng?

2.39 How many microliters are in 1 L? In 20 mL?

Scientific Notation and Significant Figures

2.40 Express the following numbers in scientific notation:

  (a) 9457
  (b) 0.000 07
  (c) 20,000,000,000 (four significant figures)
  (d) 0.012 345
  (e) 652.38

2.41 Convert the following numbers from scientific notation to standard notation:

  (a) 4.87×10^3
  (b) 5.501×10^6
  (c) 2.540×10^{-3}
  (d) 3.68×10^4

2.42 How many significant figures does each of the following numbers have?

  (a) 237,401   **(b)** 0.300
  (c) 3.01   **(d)** 244.4
  (e) 50,000   **(f)** 660

2.43 How many significant figures are in each of the following quantities?

  (a) Distance from New York City to Wellington, New Zealand, 14,397 km
  (b) Average body temperature of a crocodile, 25.6°C
  (c) Melting point of gold, 1064°C
  (d) Diameter of an influenza virus, 0.000 01 mm
  (e) Radius of a phosphorus atom, 0.110 nm

2.44 The diameter of the earth at the equator is 7926.381 mi.

  (a) Round off the earth's diameter to four significant figures, to two significant figures, and to six significant figures.
  (b) Express the earth's diameter in scientific notation.

2.45 Round off each of the numbers in Problem 2.42 to two significant figures, and express them in scientific notation.

2.46 Carry out the following calculations, express each answer to the correct number of significant figures, and include units in the answers

  (a) 9.02 g + 3.1 g
  (b) 88.80 cm + 7.391 cm
  (c) 362 mL − 99.5 mL
  (d) 12.4 mg + 6.378 mg + 2.089 mg

2.47 Carry out the following calculations, express the answers to the correct numbers of significant figures, and include units in the answers.

  (a) 8.08 m × 5.320 m
  (b) 11.1 mi × 24 mi
  (c) 1.350×10^3 mi ÷ 7.311 hr
  (d) 6.80 cm × $(1.00 \times 10^2$ cm$)$ × 2.00 cm

Unit Conversions and Problem Solving

2.48 Carry out the following conversions:
   (a) 3.614 mg to cg
   (b) 12.0 kL to ML
   (c) 14.4 μm to mm
   (d) 6.03×10^{-6} cg to ng

2.49 Carry out the following conversions. Consult Tables 2.3–2.5 as needed.
   (a) 56.4 mi to km and to Mm
   (b) 2.0 L to qt and to fl oz
   (c) 7 ft 2.0 in. to cm and to m
   (d) 1.35 lb to kg and to dg

2.50 Express the following quantities in more convenient units by using SI unit prefixes:
   (a) 9.78×10^4 g
   (b) 1.33×10^{-4} L
   (c) 0.000 000 000 46 g

2.51 What SI unit prefixes correspond to each of the following multipliers?
   (a) 10^3
   (b) 10^{-3}
   (c) 10^6
   (d) 10^{-6}

2.52 The speed limit in Canada is 100 km/hr.
   (a) How many miles per hour is this?
   (b) How many feet per second?

2.53 The speed limit on many secondary roads in the United States is 55 mi/hr.
   (a) How many kilometers per hour is this?
   (b) How many meters per second?

2.54 The diameter of a red blood cell is 6×10^{-6} m. How many red blood cells are needed to make a line 1 in. long?

2.55 The Sears Tower in Chicago has an approximate floor area of 418,000 m^2. How many square feet of floor is this?

2.56 A normal value for blood cholesterol is 200 mg/dL. If a normal adult has a total blood volume of 5 L, how much total cholesterol is present?

2.57 A six-pack of 12 oz soft drinks sells for $2.59, and a 2.0 L bottle of the same soft drink sells for $1.59. What is the cost in dollars per liter for each?

2.58 The white blood cell concentration in normal blood is approximately 12,000 cells/mm^3. How many white blood cells does a normal adult with 5 L of blood have? Express the answer in scientific notation.

2.59 The recommended daily dose of calcium for an 18 year old male is 1200 mg. If 1.0 cup of whole milk contains 290 mg of calcium and milk is his only calcium source, how much milk should an 18 year old male drink each day?

Energy, Heat, and Temperature

2.60 What is the difference between potential energy and kinetic energy?

2.61 What is the difference between heat and temperature?

2.62 What is the temperature of the normal human body (98.6°F) in degrees Celsius? In kelvins?

2.63 The boiling point of liquid nitrogen, used in the removal of warts and in other surgical applications, is −195.8°C. What is this temperature in kelvins and in degrees Fahrenheit?

2.64 Diethyl ether, a substance once used as a general anesthetic, has a specific heat of 0.895 cal/(g · °C). How many calories and how many kilocalories of heat are needed to raise the temperature of 30.0 g of diethyl ether from 10.0°C to 30.0°C?

2.65 Calculate the specific heat of copper if it takes 23 cal to heat a 5.0 g sample from 25°C to 75°C.

2.66 Assume that 50.0 cal of heat is applied to a 15 g sample of sulfur at 20°C. What is the final temperature of the sample if the specific heat of sulfur is 0.175 cal/(g · °C)?

2.67 A 150 g sample of mercury and a 150 g sample of iron are at an initial temperature of 25.0°C. If 250 cal of heat is applied to each sample, what is the final temperature of each? (See Table 2.6.)

Density and Specific Gravity

2.68 Aspirin has a density of 1.40 g/cm^3. What is the volume in cubic centimeters of a tablet weighing 250 mg?

2.69 Gaseous hydrogen has a density of 0.0899 g/L at 0°C. How many liters would you need if you wanted 1.0078 g of hydrogen?

2.70 What is the density of lead (in g/cm^3) if a rectangular bar measuring 0.500 cm in height, 1.55 cm in width, and 25.00 cm in length has a mass of 220.9 g?

2.71 What is the density of lithium metal (in g/cm^3) if a cylindrical wire with a diameter of 2.40 mm and a length of 15.0 cm has a mass of 0.3624 g?

2.72 What is the density of isopropyl alcohol (rubbing alcohol) in grams per milliliter if a 5.000 mL sample weighs 3.928 g at room temperature? What is the specific gravity of isopropyl alcohol?

2.73 Ethylene glycol, commonly used as automobile antifreeze, has a specific gravity of 1.1088 at room temperature. What is the volume of 1.00 kg of ethylene glycol? What is the volume of 2.00 lb of ethylene glycol?

Applications

2.74 Write an alternative conversion factor for each of the following unit equivalents from the apothecary system. [*Apothecary Units*]
   (a) 1 gr = 64.8 mg, 1 mg = ? gr
   (b) 1 fl oz = 8 fluidrams, 1 fluidram = ? fl oz
   (c) 1 fluidram = 3.72 mL, 1 mL = ? fluidram
   (d) 1 minim = 0.062 mL, 1 mL = ? minim
   (e) 1 fl oz = 480 minims, 1 minim = ? fl oz

2.75 (a) How many milligrams of aspirin are in a 5 gr tablet?
   (b) How many drams are in a 1.5 fl oz bottle of perfume? Use conversion factors from Problem 2.74. [*Apothecary Units*]

2.76 How is body mass index (BMI) defined? [*Obesity: A Large Problem*]

2.77 What value for body mass index indicates overweight, and what value indicates obesity? [*Obesity: A Large Problem*]

2.78 What purposes does fat serve in the human body? [*Measuring Body Fat*]

2.79 Liposuction is a technique for removing fat deposits from various areas of the body. How many liters of fat would have to be removed to result in a 5.0 lb weight loss? The density of human fat is 0.94 g/mL. [*Measuring Body Fat*]

General Questions and Problems

2.80 Gemstones are weighed in carats, where 1 carat = 200 mg exactly. What is the mass in grams of the Hope diamond, the world's largest blue diamond, at 44.4 carats?

2.81 If you were cooking in an oven calibrated in Celsius degrees, what temperature would you use if the recipe called for 350°F?

2.82 What dosage in grams per kilogram of body weight does a 130 lb woman receive if she takes two 250 mg tablets of penicillin? How many 125 mg tablets should a 40 lb child take to receive the same dosage?

2.83 A clinical report gave the following data from a blood analysis: iron, 39 mg/dL; calcium, 8.3 mg/dL; cholesterol, 224 mg/dL. Express each of these quantities in grams per deciliter, writing the answers in scientific notation.

2.84 The density of air at room temperature is 1.3 g/L at 0°C. What is the mass of the air (a) in grams and (b) in pounds in a room that is 4.0 m long, 3.0 m wide, and 2.5 m high?

2.85 Approximately 75 mL of blood is pumped by a normal human heart at each beat. Assuming an average pulse of 72 beats per minute, how many milliliters of blood are pumped in one day?

2.86 A doctor has ordered that a patient be given 15 g of glucose, which is available in a concentration of 50.00 g glucose/1000.0 mL of solution. What volume of solution should be given to the patient?

2.87 The density of mercury, commonly used in thermometers, is 13.6 g/mL.

  (a) What is the specific gravity of mercury?
  (b) Will an iron ball bearing float or sink in mercury? The density of iron is 7.86 g/mL.
  (c) What is the mass in pounds of 2.00 L of mercury?

2.88 In a typical person, the level of glucose (also known as blood sugar) is about 85 mg/100 mL of blood. If an average body contains about 11 pt of blood, how many grams and how many pounds of glucose are present in the blood?

2.89 A patient is receiving 3000 mL/day of a solution that contains 5 g of dextrose (glucose) per 100 mL of solution. If glucose provides 4 kcal/g of energy, how many kilocalories per day is the patient receiving from the glucose?

2.90 A rough guide to fluid requirements based on body weight is: 100 mL/kg for the first 10 kg of body weight, 50 mL/kg for the next 10 kg, and 20 mL/kg for weight over 20 kg. What volume of fluid per day is needed by a 55 kg woman? Give the answer with two significant figures.

2.91 Chloral hydrate, a sedative and sleep-inducing drug, is available as a solution labeled 10.0 gr/fluidram. What volume in milliliters should be administered to a patient who is meant to receive 7.5 gr per dose? (1 gr = 64.8 mg; 1 fluidram = 3.72 mL)

2.92 When 1.0 tablespoon of butter is burned or used by our body, it releases 100 kcal (100 food calories) of energy. If we could use all the energy provided, how many tablespoons of butter would have to be burned to raise the temperature of 3.00 L of water from 18.0°C to 90.0°C?

2.93 The density of sulfuric acid, H_2SO_4, is 15.28 lb/gal. What is the density of sulfuric acid in grams per milliliter?

2.94 Sulfuric acid (Problem 2.93) is produced in larger amount than any other chemical: 9.93×10^{10} lb in 2000. What is the volume of this amount in liters?

2.95 The caliber of a gun is expressed by measuring the diameter of the gun barrel in hundredths of an inch. A "22" rifle, for example, has a barrel diameter of 0.22 in. What is the barrel diameter of a .22 rifle in millimeters?

2.96 Amounts of substances dissolved in solution are often expressed as mass per unit volume. For example, normal human blood has a cholesterol concentration of about 200 mg/100 mL. Express this concentration in the following units:

  (a) mg/L   **(b)** μg/mL
  (c) g/L   **(d)** ng/μL

2.97 The element gallium (Ga) has the second largest liquid range of any element, melting at 29.8°C and boiling at 2204°C at atmospheric pressure.

  (a) Is gallium a metal, a nonmetal, or a metalloid?
  (b) What is the density of gallium in g/cm^3 at 25°C if a 1.00-inch cube has a mass of 0.2133 lb?

2.98 A sample of water at 293.2 K was heated for 8 min, 25 s so as to give a constant temperature increase of 3.0°F/min. What is the final temperature of the water in degrees Celsius?

2.99 At a certain point, the Celsius and Fahrenheit scales "cross," giving the same numerical value on both. At what temperature does this crossover occur?

2.100 Imagine that you place a piece of cork measuring 1.30 cm × 5.50 cm × 3.00 cm in a pan of water and that on top of the cork you place a small cube of lead measuring 1.15 cm on each edge. The density of cork is 0.235 g/cm^3, and the density of lead is 11.35 g/cm^3. Will the combination of cork plus lead float or sink?

3

Atoms and the Periodic Table

Periodicity—the presence of regularly repeating patterns—is found throughout nature.

Chemistry must be studied on two levels. In the past two chapters we've dealt with chemistry on the large-scale, or *macroscopic*, level, looking at the properties and transformations of matter that we can see and measure. Now we're ready to look at the small-scale, or *microscopic*, level, studying the behavior and properties of individual atoms. Although scientists have long been convinced of their existence, only within the past twenty years have powerful new instruments made it possible to see individual atoms themselves. In this chapter we'll look at modern atomic theory and at how the structure of atoms influences macroscopic properties.

We'll take up the following questions about atoms and atomic theory:

1. What is the modern theory of atomic structure?
The goal: Be able to explain the major assumptions of atomic theory.

2. How do atoms of different elements differ?
The goal: Be able to explain the composition of different atoms according to the number of protons, neutrons, and electrons they contain.

3. **What are isotopes, and what is atomic weight?**
   The goal: Be able to explain what isotopes are and how they affect an element's atomic weight.
4. **How is the periodic table arranged?**
   The goal: Be able to describe how elements are arranged in the periodic table, name the subdivisions of the periodic table, and relate the position of an element in the periodic table to its electron structure.
5. **How are electrons arranged in atoms?**
   The goal: Be able to explain how electrons are distributed in shells and subshells around the nucleus of an atom.

3.1 Atomic Theory

Take a piece of aluminum foil and cut it in two. Then take one of the pieces and cut *it* in two, and so on. Assuming that you have extremely small scissors and extraordinary dexterity, how long can you keep dividing the foil? Is there a limit, or is matter infinitely divisible into ever smaller and smaller pieces? In fact, there is a limit. The smallest and simplest bit that aluminum (or any other element) can be divided into and still be identifiable as aluminum is called an **atom**, a word derived from the Greek *atomos*, meaning "indivisible."

Atom The smallest and simplest particle of an element.

Chemistry is founded on four fundamental assumptions about atoms and matter, which together make up modern *atomic theory*:

- All matter is composed of atoms.
- The atoms of a given element differ from the atoms of all other elements.
- Chemical compounds consist of atoms combined in specific ratios. That is, only whole atoms can combine—one A atom with one B atom, or one A atom with two B atoms, and so on. The enormous diversity in the substances we see around us is based on the vast number of ways that atoms can combine with one another.
- Chemical reactions change only the way that atoms are combined in compounds; the atoms themselves are unchanged.

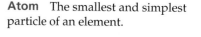

▲ How small can a piece of aluminum foil be cut?

Atoms are extremely small, ranging from about 7.4×10^{-11} m in diameter for a hydrogen atom to 5.24×10^{-10} m for a cesium atom. In mass, atoms vary from 1.67×10^{-24} g for hydrogen to 3.95×10^{-22} g for uranium, one of the heaviest naturally occurring atoms. It's difficult to appreciate just how small atoms are, although it might help if you realize that a fine pencil line is about 3 *million* atoms across, and that even the smallest speck of dust contains about 10^{16} atoms.

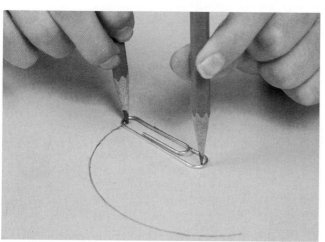

▲ Atoms are so small that this line is about 3 million atoms wide.

Subatomic particles Three kinds of fundamental particles from which atoms are made: protons, neutrons, and electrons.

Proton A positively charged subatomic particle.

Neutron An electrically neutral subatomic particle.

Electron A negatively charged subatomic particle.

Atoms are composed of tiny **subatomic particles** called *protons*, *neutrons*, and *electrons*. A **proton** has a mass of $1.672\,622 \times 10^{-24}$ g and carries a positive $(+)$ electrical charge; a **neutron** has a mass similar to that of a proton $(1.674\,927 \times 10^{-24}$ g) but is electrically neutral; and an **electron** has a mass that is only 1/1836 that of a proton $(9.109\,328 \times 10^{-28}$ g) and carries a negative $(-)$ electrical charge. In fact, electrons are so much lighter than protons and neutrons that their mass is usually ignored. Table 3.1 compares the properties of the three fundamental subatomic particles.

TABLE 3.1   A Comparison of Subatomic Particles

Name	Symbol	Mass (grams)	Mass (amu)	Charge (charge units)
Proton	p	$1.672\,622 \times 10^{-24}$	$1.007\,276$	$+1$
Neutron	n	$1.674\,927 \times 10^{-24}$	$1.008\,665$	0
Electron	e^{-}	$9.109\,328 \times 10^{-28}$	$5.485\,799 \times 10^{-4}$	-1

The masses of atoms and their constituent subatomic particles are so small when measured in grams that it's more convenient to express them on a *relative* mass scale. That is, one atom is assigned a mass, and all others are measured relative to it. The process is like deciding that a golf ball (46.0 g) will be assigned a mass of 1. A baseball (149 g), which is $149/46.0 = 3.24$ times heavier than a golf ball, would then have a mass of about 3.24, a volleyball (270 g) would have a mass of $270/46.0 = 5.87$, and so on.

Atomic mass unit (amu) A convenient unit for describing the mass of an atom; 1 amu = 1/12 the mass of a carbon-12 atom.

The basis for the relative atomic mass scale is an atom of carbon that contains 6 protons and 6 neutrons. Such an atom is assigned a mass of exactly 12 **atomic mass units** (**amu**; also called a *dalton* in honor of the English scientist John Dalton), where 1 amu = $1.660\,539 \times 10^{-24}$ g. Thus, for all practical purposes, both a proton and a neutron have a mass of 1 amu (Table 3.1). Hydrogen atoms are only about one-twelfth as heavy as carbon atoms and have a mass close to 1 amu, magnesium atoms are about twice as heavy as carbon atoms and have a mass close to 24 amu, and so forth.

Nucleus The dense, central core of an atom that contains protons and neutrons.

Subatomic particles are not distributed at random throughout an atom. Rather, the protons and neutrons are packed closely together in a dense core called the **nucleus**. Surrounding the nucleus, the electrons move about rapidly through a large, mostly empty volume of space (Figure 3.1). Measurements show that the diameter of a nucleus is only about 10^{-15} m, while that of the atom itself is about 10^{-10} m. For comparison, if an atom were the size of a large domed stadium, the nucleus would be approximately the size of a small pea in the center of the playing field.

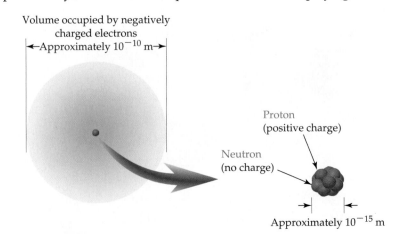

Volume occupied by negatively charged electrons
←Approximately 10^{-10} m→

Proton (positive charge)

Neutron (no charge)

Approximately 10^{-15} m

▲ The relative size of a nucleus in an atom is the same as that of a pea in the middle of this stadium.

▲ **FIGURE 3.1   The structure of an atom.** Protons and neutrons are packed together in the nucleus, while electrons move about in the large surrounding volume. Virtually all the mass of an atom is concentrated in the nucleus.

The structure of the atom is determined by an interplay of different attractive and repulsive forces. Because unlike charges attract one another, the negatively charged electrons are held near the positively charged nucleus. But because like charges repel one another, the negatively charged electrons try to get as far away from one another as possible, accounting for the relatively large volume they occupy. The positively charged protons in the nucleus also repel one another, but are nevertheless held together by a unique attraction called the *nuclear strong force*, which we won't go into.

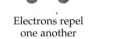
Electrons repel
one another

Protons repel
one another

Protons and electrons
attract one another

■ **WORKED EXAMPLE 3.1**

How many atoms are in a small piece of aluminum foil with a mass of 0.100 g? The mass of an atom of aluminum is 27.0 amu.

ANALYSIS We know the sample mass in grams and the mass of one atom in amu. To find the number of atoms in the sample, two conversions are needed, the first between grams and amu and the second between amu and numbers of atoms. The conversion factor between amu and grams is 1 amu = $1.660\,539 \times 10^{-24}$ g.

SOLUTION Setting up an equation so that the unwanted units cancel gives:

$$(0.100 \text{ g}) \left(\frac{1 \text{ amu}}{1.660\,539 \times 10^{-24} \text{ g}} \right) \left(\frac{1 \text{ Al atom}}{27.0 \text{ amu}} \right) = 2.23 \times 10^{21} \text{ Al atoms}$$

■ **PROBLEM 3.1**

What is the mass in atomic mass units of a nitrogen atom weighing 2.33×10^{-23} g?

■ **PROBLEM 3.2**

What is the mass in grams of 100,000 gold atoms, each weighing 197 amu?

■ **PROBLEM 3.3**

How many atoms are in each of the following?
(a) 1.0 g of hydrogen atoms, each of mass 1.0 amu
(b) 12.0 g of carbon atoms, each of mass 12.0 amu
(c) 23.0 g of sodium atoms, each of mass 23.0 amu

■ **PROBLEM 3.4**

What pattern do you see in your answers to Problem 3.3? (We'll return to this very important pattern in Chapter 6.)

3.2 Elements and Atomic Number

Atoms of different elements differ from one another according to how many protons they contain, a value called the element's **atomic number (Z)**. Thus, if we know the number of protons in an atom, we can identify the element. Any atom with 6 protons, for example, is a carbon atom because carbon has Z = 6.

Atoms are neutral overall and have no net charge because the number of positively charged protons in an atom is the same as the number of negatively charged electrons. Thus, the atomic number also equals the number of electrons in every atom of a given element. Hydrogen, $Z = 1$, has only 1 proton and 1 electron; carbon, $Z = 6$, has 6 protons and 6 electrons; sodium, $Z = 11$, has 11 protons and 11 electrons; and so on up to the newly discovered element with $Z = 116$. In a periodic table, elements are listed in order of increasing atomic number, beginning at the upper left and ending at the lower right.

Atomic number (Z)   The number of protons in atoms of a given element; the number of electrons in atoms of a given element.

The sum of the protons and neutrons in an atom is called the atom's **mass number (A)**. Hydrogen atoms with 1 proton and no neutrons have mass number 1; carbon atoms with 6 protons and 6 neutrons have mass number 12; sodium atoms with 11 protons and 12 neutrons have mass number 23; and so on. Except for hydrogen, atoms generally contain at least as many neutrons as protons, and frequently contain more. There is no simple way to predict how many neutrons a given atom will have.

Mass number (A)   The total number of protons and neutrons in an atom

---

■ WORKED EXAMPLE 3.2

Phosphorus has atomic number $Z = 15$. How many protons, electrons, and neutrons are there in phosphorus atoms that have mass number $A = 31$?

ANALYSIS   The atomic number gives the number of protons, which is the same as the number of electrons, and the mass number gives the total number of protons plus neutrons.

SOLUTION   Phosphorus atoms, with $Z = 15$, have 15 protons and 15 electrons. To find the number of neutrons, subtract the atomic number from the mass number:

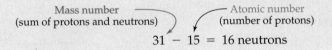

$$\text{Mass number (sum of protons and neutrons)} \quad \text{Atomic number (number of protons)}$$

$$31 - 15 = 16 \text{ neutrons}$$

■ WORKED EXAMPLE 3.3

An atom contains 28 protons and has $A = 60$. Give the number of electrons and neutrons in the atom, and identify the element.

ANALYSIS   The number of protons and the number of electrons are the same and are equal to the atomic number Z, 28 in this case. Subtracting the number of protons (28) from the total number of protons plus neutrons (60) gives the number of neutrons.

SOLUTION   The atom has 28 electrons and $60 - 28 = 32$ neutrons. Looking at the table of elements inside the front cover shows that the element with atomic number 28 is nickel (Ni).

■ PROBLEM 3.5

Use the table of elements inside the front cover to identify those with the following atomic numbers:

(a) $Z = 75$ **(b)** $Z = 3$ **(c)** $Z = 52$

■ **PROBLEM 3.6**

The uranium used in nuclear reactors has $Z = 92$ and $A = 235$. How many protons, neutrons, and electrons are in these uranium atoms?

■ **PROBLEM 3.7**

A certain atom has $A = 52$ and contains 28 neutrons. Identify the element.

3.3 Isotopes and Atomic Weight

All atoms of a given element have the same number of protons—the atomic number Z characteristic of that element—but different atoms of an element can have different numbers of neutrons and therefore different mass numbers. Atoms with identical atomic numbers but different mass numbers are called **isotopes**. Hydrogen, for example, has three isotopes. The most abundant hydrogen isotope, called *protium*, has no neutrons and thus has a mass number of 1. A second hydrogen isotope, called *deuterium*, has one neutron and a mass number of 2; and a third isotope, called *tritium*, has two neutrons and a mass number of 3. Tritium is unstable and does not occur naturally in significant amounts, although it can be made in nuclear reactors (Chapter 11).

Isotopes Atoms with identical atomic numbers but different mass numbers.

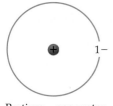 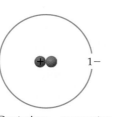 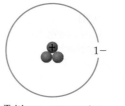

Protium—one proton (⬤) and no neutrons; mass number = 1

Deuterium—one proton (⬤) and one neutron (⬤); mass number = 2

Tritium—one proton (⬤) and two neutrons (⬤); mass number = 3

A specific isotope is represented by showing its mass number as a superscript and its atomic number as a subscript in front of the atomic symbol. Thus, protium is ^1_1H, deuterium is ^2_1H, and tritium is ^3_1H.

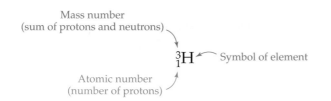

Mass number (sum of protons and neutrons)

^3_1H Symbol of element

Atomic number (number of protons)

Unlike the three isotopes of hydrogen, the isotopes of most elements don't have distinctive names. Instead, they are referred to by giving the mass number after the name of the element. The $^{235}_{92}U$ isotope used in nuclear reactors, for example, is usually referred to as uranium-235, or U-235.

Most naturally occurring elements are mixtures of isotopes. In a large sample of naturally occurring hydrogen atoms, for example, 99.985% have mass number $A = 1$ (protium) and 0.015% have mass number $A = 2$ (deuterium). It's therefore useful to know the *average* mass of the atoms in a large sample, a value called the element's **atomic weight**. For hydrogen, the atomic weight is 1.008 amu. Atomic weights for all elements are given inside the front cover of this book.

To calculate the atomic weight of an element, the individual masses of the naturally occurring isotopes and the percentage of each must be known. Chlorine, for example, occurs on earth as a mixture of 75.77% Cl-35 atoms

Atomic weight The weighted average mass of an element's atoms.

(mass = 34.97 amu) and 24.23% Cl-37 atoms (mass = 36.97 amu). The atomic weight is found by calculating the percentage of the mass contributed by each isotope. For chlorine, the calculation is done in the following way (to four significant figures), giving an atomic weight of 35.46 amu:

$$\begin{aligned}\text{Contribution from } {}^{35}\text{Cl:} & \quad (0.7577)(34.97 \text{ amu}) = 26.50 \text{ amu} \\ \text{Contribution from } {}^{37}\text{Cl:} & \quad (0.2423)(36.97 \text{ amu}) = \underline{8.96 \text{ amu}} \\ & \qquad\qquad\qquad\qquad \text{Atomic weight} = 35.46 \text{ amu}\end{aligned}$$

■ **WORKED EXAMPLE 3.4**

Write the symbol for carbon-12, the isotope of carbon that is the basis for the atomic mass scale.

ANALYSIS   The identity of the atom—carbon—corresponds to atomic number 6, and the mass number of carbon-12 is given in the name: 12.

SOLUTION   The symbol is $^{12}_{6}\text{C}$.

■ **WORKED EXAMPLE 3.5**

Identify element X in the symbol $^{194}_{78}\text{X}$, and give its atomic number, mass number, number of protons, number of electrons, and number of neutrons.

ANALYSIS   The identity of the atom corresponds to the atomic number—78.

SOLUTION   Element X has $Z = 78$, which shows that it is platinum. (Look inside the front cover for a list.) The isotope $^{194}_{78}\text{Pt}$ has a mass number of 194, and we can subtract the atomic number from the mass number to give the number of neutrons. This platinum isotope therefore has 78 protons, 78 electrons, and $194 - 78 = 116$ neutrons.

■ **PROBLEM 3.8**

Chlorine, an element present in common table salt (sodium chloride), has two naturally occurring isotopes, with mass numbers 35 and 37. Write the symbols for both, including their atomic numbers and mass numbers.

■ **PROBLEM 3.9**

Complete each of the following isotope symbols:
(a) $^{11}_{5}\text{?}$ **(b)** $^{56}_{?}\text{Fe}$

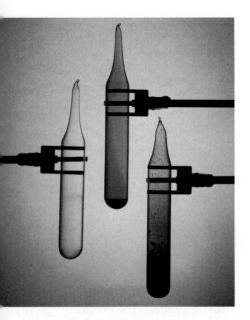

▲ Samples of chlorine, bromine, and iodine, one of Döbereiner's triads of elements with similar chemical properties.

3.4 The Periodic Table

Ten elements have been known since the beginning of recorded history: antimony (Sb), carbon (C), copper (Cu), gold (Au), iron (Fe), lead (Pb), mercury (Hg), silver (Ag), sulfur (S), and tin (Sn). The first "new" element to be found in several thousand years was arsenic (As), discovered in about 1250. In fact, only 24 elements were known up to the time of the American Revolution in 1776.

As the pace of discovery quickened in the late 1700s and early 1800s, chemists began to look for similarities among elements that might make it possible to draw general conclusions. Particularly important was Johann Döbereiner's observation in 1829 that there were several *triads*, or groups of three elements, that appeared to have similar chemical and physical properties. For example, lithium, sodium, and potassium were all known to be silvery metals that react violently

Application **Are Atoms Real?**

Chemistry rests on the belief that matter is composed of the tiny particles we call atoms. Every chemical reaction and every physical law that governs the behavior of matter is explained by chemists in terms of atomic theory. But how do we know that atoms are real and that we aren't just imagining them? The best answer is that we can now actually "see" and manipulate individual atoms through the use of a device called a *scanning tunneling microscope*, or STM. Invented in 1981 by a research team at the IBM Corporation, magnifications of up to 10 million have been achieved, allowing chemists to look directly at atoms themselves. The accompanying photograph shows a computer-enhanced representation of individual iron atoms that have been lined up on a copper surface.

To create an image like that in the photo, a sharp tungsten probe with a tip that is only one or two atoms across is brought near the surface of the sample, and a small voltage is applied. When the tip comes within a few atomic diameters of the sample, a tiny electric current flows from the sample to the probe in a process called *electron tunneling*. The strength of the current flow is extremely sensitive to the distance between sample and probe, varying by as much as a 1000-fold over a distance of one atomic diameter. Passing the probe across the sample while moving it up and down over individual atoms to keep current flow constant gives a two-dimensional map of the probe's path. By then moving the probe back and forth in a series of closely spaced parallel tracks and

▲ Colored scanning tunneling micrograph of the surface of graphite, a form of the element carbon. The hexagonal pattern is related to an identical arrangement of the carbon atoms.

storing the data in a computer, a three-dimensional image of the surface can be constructed.

Most present uses of the scanning tunneling microscope involve studies of surface chemistry, such as the events accompanying the corrosion of metals and the ordering of large molecules in polymers. Work is also underway using STM to determine the structures of complex biological molecules.

See Additional Problems 3.80 and 3.81 at the end of the chapter.

---

with water; chlorine, bromine, and iodine were all known to be colored nonmetals with pungent odors.

Numerous attempts were made in the mid-1800s to account for the similarities among groups of elements, but the great breakthrough came in 1869 when the Russian chemist Dmitri Mendeleev organized the elements into a forerunner of the modern periodic table, introduced previously in Section 1.6 and shown again in Figure 3.2. The table has 114 boxes, each of which tells the symbol, atomic number, and atomic weight of an element:

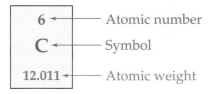

Beginning at the upper left corner of the periodic table, elements are arranged by increasing atomic number into seven horizontal rows, called **periods**, and 18 vertical columns, called **groups**. When organized in this way, *the elements in a given group have similar chemical properties*. Lithium, sodium, potassium, and the other elements in group 1A behave similarly. Chlorine, bromine, iodine, and the other elements in group 7A behave similarly, and so on throughout the table.

Note that different periods (rows) contain different numbers of elements. The first period contains only 2 elements, hydrogen and helium; the second and third

Period One of the 7 horizontal rows of elements in the periodic table.

Group One of the 18 vertical columns of elements in the periodic table.

Main groups · **Transition metal groups** · **Main groups**

Periodic table of the elements (Period 1–7), with group labels 1A–8A (1–18) for main groups and 3B–2B for transition metal groups:

Period	1A/1	2A/2	3B/3	4B/4	5B/5	6B/6	7B/7	8B/8	8B/9	8B/10	1B/11	2B/12	3A/13	4A/14	5A/15	6A/16	7A/17	8A/18
1	1 **H** 1.00794																	2 **He** 4.00260
2	3 **Li** 6.941	4 **Be** 9.01218											5 **B** 10.81	6 **C** 12.011	7 **N** 14.0067	8 **O** 15.9994	9 **F** 18.9984	10 **Ne** 20.1797
3	11 **Na** 22.98977	12 **Mg** 24.305											13 **Al** 26.98154	14 **Si** 28.0855	15 **P** 30.9738	16 **S** 32.066	17 **Cl** 35.4527	18 **Ar** 39948
4	19 **K** 39.0983	20 **Ca** 40.078	21 **Sc** 44.9559	22 **Ti** 47.88	23 **V** 50.9415	24 **Cr** 51.996	25 **Mn** 54.9380	26 **Fe** 55.847	27 **Co** 58.9332	28 **Ni** 58.69	29 **Cu** 63.546	30 **Zn** 65.39	31 **Ga** 69.72	32 **Ge** 72.61	33 **As** 74.9216	34 **Se** 78.96	35 **Br** 79.904	36 **Kr** 83.80
5	37 **Rb** 85.4678	38 **Sr** 87.62	39 **Y** 88.9059	40 **Zr** 91.224	41 **Nb** 92.9064	42 **Mo** 95.94	43 **Tc** (98)	44 **Ru** 101.07	45 **Rh** 102.9055	46 **Pd** 106.42	47 **Ag** 107.8682	48 **Cd** 112.41	49 **In** 114.82	50 **Sn** 118.710	51 **Sb** 121.757	52 **Te** 127.60	53 **I** 126.9045	54 **Xe** 131.29
6	55 **Cs** 132.9054	56 **Ba** 137.33	57 *****La** 138.9055	72 **Hf** 178.49	73 **Ta** 180.9479	74 **W** 183.85	75 **Re** 186.207	76 **Os** 190.2	77 **Ir** 192.22	78 **Pt** 195.08	79 **Au** 196.9665	80 **Hg** 200.59	81 **Tl** 204.383	82 **Pb** 207.2	83 **Bi** 208.9804	84 **Po** (209)	85 **At** (210)	86 **Rn** (222)
7	87 **Fr** (223)	88 **Ra** 226.0254	89 **†Ac** 227.0278	104 **Rf** (261)	105 **Db** (262)	106 **Sg** (266)	107 **Bh** (264)	108 **Hs** (269)	109 **Mt** (268)	110 (271)	111 (272)	112 (277)	114		116			

Lanthanides

58 **Ce** 140.12	59 **Pr** 140.9077	60 **Nd** 144.24	61 **PM** (145)	62 **Sm** 150.36	63 **Eu** 151.965	64 **Gd** 157.25	65 **Tb** 158.9254	66 **Dy** 162.50	67 **Ho** 164.9304	68 **Er** 167.26	69 **Tm** 168.9342	70 **Yb** 173.04	71 **Lu** 174.967

Actinides

90 **Th** 232.0381	91 **Pa** 231.0399	92 **U** 238.0289	93 **Np** 237.048	94 **Pu** (244)	95 **Am** (243)	96 **Cm** (247)	97 **Bk** (247)	98 **Cf** (251)	99 **Es** (252)	100 **Fm** (257)	101 **Md** (258)	102 **No** (259)	103 **Lr** (262)

Metals · Metalloids · Nonmetals

▲ **FIGURE 3.2 The periodic table of the elements.** Each element is identified by a one- or two-letter symbol and is characterized by an *atomic number*. The table begins with hydrogen (H, atomic number 1) in the upper left-hand corner and continues to the yet unnamed element with atomic number 116. The 14 elements following lanthanum (La, atomic number 57) and the 14 elements following actinium (Ac, atomic number 89) are pulled out and shown below the others.

Elements are organized into 18 vertical columns, or *groups*, and 7 horizontal rows, or *periods*. The two groups on the left and the six on the right are the *main groups*; the ten in the middle are the *transition metal groups*. The 14 elements following lanthanum are the *lanthanides*, and the 14 elements following actinium are the *actinides*; together these are known as the *inner transition metals*. Two systems for numbering the groups are explained in the text.

Those elements (except hydrogen) on the left-hand side of the zigzag line running from boron (B) to astatine (At) are *metals*, those elements to the right of the line are *nonmetals*, and those elements abutting the line are *metalloids*.

Main group element An element in one of the two groups on the left or the six groups on the right of the periodic table.

Transition metal element An element in one of the 10 smaller groups near the middle of the periodic table.

Inner transition metal element An element in one of the 14 groups shown separately at the bottom of the periodic table.

periods each contain 8 elements; the fourth and fifth periods each contain 18; the sixth period contains 32; and the seventh period (incomplete as yet) contains 27 elements. Note also that the 14 elements following lanthanum (the *lanthanides*) and the 14 following actinium (the *actinides*) are pulled out and shown below the others.

Groups are numbered in two ways, both shown in Figure 3.2. The two large groups on the far left and the six on the far right are called the **main groups** and are numbered 1A through 8A. The 10 smaller groups in the middle of the table are called the **transition metal groups** and are numbered 1B through 8B. Alternatively, all 18 groups are numbered sequentially from 1 to 18. The 14 groups shown separately at the bottom of the table are called the **inner transition metal groups** and are not numbered.

■ **PROBLEM 3.10**

Locate aluminum in the periodic table, and give its group number and period number.

■ **PROBLEM 3.11**

Identify the group 1B element in period 5 and the group 2A element in period 4.

■ **PROBLEM 3.12**

There are five elements in group 5A of the periodic table. Identify them, and give the period of each.

3.5 Some Characteristics of Different Groups

To see why the periodic table has the name it does, look at the graph of atomic radius versus atomic number in Figure 3.3. The graph shows an obvious *periodicity*, or repeating, rise-and-fall pattern. Beginning on the left with atomic number 1 (hydrogen), the sizes of the atoms increase to a maximum at atomic number 3 (lithium), then decrease to a minimum, then increase again to a maximum at atomic number 11 (sodium), then decrease, and so on. It turns out that the maxima occur for atoms of group 1A elements—Li, Na, K, Rb, Cs, and Fr—and the minima occur for atoms of the group 7A elements.

▲ Sodium, an alkali metal, reacts violently with water to yield hydrogen gas and an alkaline (basic) solution.

 Reactivity of Group 1A Metals; Periodic Trends: Atomic Radii

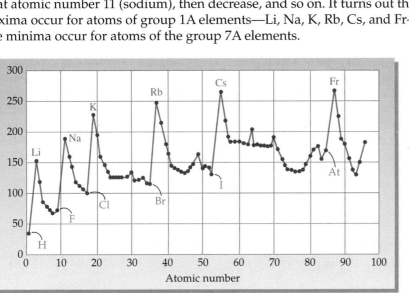

▲ **FIGURE 3.3 A graph of atomic radius in picometers (pm) versus atomic number shows a periodic rise-and-fall pattern.**   The maxima occur for atoms of group 1A elements (Li, Na, K, Rb, Cs, Fr); the minima occur for atoms of the group 7A elements. Accurate data are not available for the group 8A elements.

Alkali metal   An element in group 1A of the periodic table.

Alkaline earth metal   An element in group 2A of the periodic table.

There's nothing unique about the periodicity of atomic radii shown in Figure 3.3. Many other physical and chemical properties can be plotted in a similar way with similar results. In fact, the various elements in a given group of the periodic table usually show remarkable similarities in their properties. Look at the following four groups, for example:

• **Group 1A—Alkali metals:** Lithium (Li), sodium (Na), potassium (K), rubidium (Rb), cesium (Cs), and francium (Fr) are shiny, soft, low-melting metals. All react rapidly (often violently) with water to form products that are highly alkaline, or basic—hence the name *alkali metals*. Because of their high reactivity, the alkali metals are never found in nature in the pure state but only in combination with other elements.

• **Group 2A—Alkaline earth metals:** Beryllium (Be), magnesium (Mg), calcium (Ca), strontium (Sr), barium (Ba), and radium (Ra) are also lustrous, silvery metals, but are less reactive than their neighbors in group 1A. Like the alkali metals, the alkaline earths are never found in nature in the pure state.

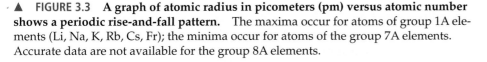

▲ Magnesium, an alkaline earth metal, burns in air.

▲ Chlorine, a halogen, is a toxic, corrosive green gas.

Halogen An element in group 7A of the periodic table.

Noble gas An element in group 8A of the periodic table.

▲ Neon, one of the noble gases, is used in neon lights, signs, and works of art.

- **Group 7A—Halogens:** Fluorine (F), chlorine (Cl), bromine (Br), iodine (I), and astatine (At) are colorful and corrosive nonmetals. All are found in nature only in combination with other elements, such as with sodium in table salt (sodium chloride, NaCl). In fact, the group name *halogen* is taken from the Greek word *hals*, meaning salt.

- **Group 8A—Noble gases:** Helium (He), neon (Ne), argon (Ar), krypton (Kr), xenon (Xe), and radon (Rn) are colorless gases of very low chemical reactivity. Helium, neon, and argon don't combine with any other elements; krypton and xenon combine with very few.

Although the resemblances aren't as pronounced as they are within a single group, *neighboring* elements often behave similarly as well. Thus, as noted in Section 1.6 and indicated in Figure 3.2, the periodic table can be divided into three major classes of elements—*metals*, *nonmetals*, and *metalloids* (metal-like). Metals, the largest category of elements, are found on the left side of the periodic table, bounded on the right by a zigzag line running from boron (B) at the top to astatine (At) at the bottom. Nonmetals are found on the right side of the periodic table, and seven of the elements adjacent to the zigzag boundary between metals and nonmetals are metalloids.

∞ LOOKING AHEAD

Carbon, the element on which life is based, is a group 4A nonmetal near the top right of the periodic table. Clustered near carbon are other elements often found in living organisms, including oxygen, nitrogen, phosphorus, and sulfur. We'll look at the subject of *organic chemistry*—the chemistry of carbon compounds—in Chapters 12–17, and move on to *biochemistry*—the chemistry of living things—in Chapters 18–29. ∞

■ **PROBLEM 3.13**

Identify the following elements as metals, nonmetals, or metalloids:

(a) Ti   (b) Te   (c) Se   (d) Sc   (e) At   (f) Ar

■ **PROBLEM 3.14**

Locate (a) krypton, (b) strontium, (c) nitrogen, and (d) cobalt in the periodic table. Indicate which of the following categories apply to each: (i) metal, (ii) nonmetal, (iii) transition element, (iv) main group element, (v) noble gas.

■ **KEY CONCEPT PROBLEM 3.15**

Identify the element shown in red on the following periodic table, tell its group number, its period number, and whether it is a metal, nonmetal, or metalloid.

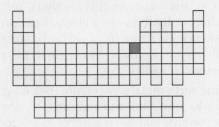

Application The Origin of Chemical Elements

Astronomers believe that the universe began some 15 billion years ago in an extraordinary moment they call the "big bang." Initially, the temperature must have been inconceivably high, but after 1 second, it had dropped to about 10^{10} K and subatomic particles began to form: protons, neutrons, and electrons. After 3 minutes, the temperature had dropped to 10^9 K, and protons began fusing with neutrons to form helium nuclei, $^4_2$He. Matter remained in this form for many millions of years until the expanding universe had cooled to about 10,000 K and electrons were then able to bind to protons and to helium nuclei, forming stable hydrogen and helium atoms.

The attractive force of gravity acting on regions of higher-than-average density slowly produced massive local concentrations of matter and ultimately formed billions of galaxies, each with many billions of stars. As the gas clouds of hydrogen and helium condensed under gravitational attraction and stars formed, their temperatures reached 10^7 K, and their densities reached 100 g/cm^3. Protons and neutrons again fused to yield helium nuclei, generating vast amounts of heat and light.

Most of these early stars probably burned out after a few billion years, but a few were so massive that, as their nuclear fuel diminished, gravitational attraction caused a rapid contraction leading to still higher core temperatures and higher densities—up to 5×10^8 K and 5×10^5 g/cm^3. Under such extreme conditions, larger nuclei were formed, including carbon, oxygen, silicon, magnesium, and iron. Ultimately, the stars underwent a

▲ The stars in the Milky Way galaxy condensed from gas clouds under gravitational attraction.

gravitational collapse resulting in the synthesis of still heavier elements and an explosion visible throughout the universe as a *supernova*.

Matter from exploding supernovas was blown throughout the galaxy, forming a new generation of stars and planets. Our own sun and solar system formed about 4.5 billion years ago from matter released by former supernovas. Except for hydrogen and helium, all the atoms in our bodies and our entire solar system were created more than 5 billion years ago in exploding stars. We and our world are made from the ashes of dying stars.

See Additional Problems 3.82 and 3.83 at the end of this chapter.

3.6 Electronic Structure of Atoms

Why does the periodic table have the shape it does, with periods of different length? Why are periodic variations observed in atomic radii and in so many other characteristics of the elements? And why do elements in a given group of the periodic table show similar chemical behavior? These questions occupied the thoughts of chemists for more than 50 years after Mendeleev, and it was not until well into the 1920s that the answers were established. Today, we know that *the properties of the elements are determined by the arrangement of electrons in their atoms.*

According to the now accepted *quantum mechanical model* of atomic structure, electrons are not perfectly free to move about in an atom. Instead, each electron is restricted to moving about in only a certain region of space within the atom, depending on the amount of energy the electron has. Different electrons have different amounts of energy and thus occupy different regions within the atom. Furthermore, the energies of electrons are *quantized*, or restricted to having only certain values.

To understand the idea of quantization, think about the difference between stairs and a ramp. A ramp is *not* quantized because it changes height continuously. Stairs, by contrast, *are* quantized because they change height only by a

▲ Stairs are *quantized* because they change height in discrete amounts. A ramp, by contrast, is not quantized because it changes height continuously.

Shell (electron) A grouping of electrons in an atom according to energy.

Subshell (electron) A grouping of electrons in a shell according to the shape of the region of space they occupy.

Orbital A region of space within an atom where an electron in a given subshell can be found.

fixed amount. You can climb one stair or two stairs, but you can't climb 1.5 stairs. In the same way, the energy values available to electrons in an atom change only in steps rather than continuously.

Just as a person can be found by giving his or her address within a state, an electron can be found by giving its "address" within an atom. Furthermore, just as a person's address is composed of several successively narrower categories—city, street, and house number—an electron's address is also composed of successively narrower categories—*shell, subshell,* and *orbital.*

The electrons in an atom are grouped around the nucleus into **shells**, roughly like the layers in an onion, according to the energy the electrons contain. The farther a shell is from the nucleus, the larger it is, the more electrons it can hold, and the higher the energies of those electrons. The first shell (the one nearest the nucleus) can hold only 2 electrons, the second shell can hold 8, the third shell can hold 18, and the fourth shell can hold 32 electrons.

Shell number:	1	2	3	4
Electron capacity:	2	8	18	32

Within shells, electrons are further grouped into **subshells** of four different types, identified in order of increasing energy by the letters *s, p, d,* and *f.* The first shell has only an *s* subshell; the second shell has an *s* and a *p* subshell; the third shell has an *s,* a *p,* and a *d* subshell; and the fourth shell has an *s,* a *p,* a *d,* and an *f* subshell. Of the four types, we'll be concerned mainly with *s* and *p* subshells because most of the elements found in living organisms use only these. A specific subshell is symbolized by writing the number of the shell, followed by the letter for the subshell. For example, the designation 3*p* refers to the *p* subshell in the third shell.

Shell number:	1	2	3	4
Subshell designation:	*s*	*s , p*	*s , p , d*	*s , p , d , f*

Note that the number of subshells in a given shell is equal to the shell number. For example, shell number 3 has 3 subshells.

Finally, within each subshell, electrons are grouped into **orbitals**, regions of space within an atom where the specific electrons are most likely to be found. There are different numbers of orbitals within the different kinds of subshells. A given *s* subshell has only 1 orbital, a *p* subshell has 3 orbitals, a *d* subshell has 5 orbitals, and an *f* subshell has 7 orbitals. Each orbital can hold only two electrons, which differ in a property known as *spin.* If one electron in an orbital has a clockwise spin, the other electron in the same orbital must have a counterclockwise spin.

Shell number:	1	2		3			4			
Subshell designation:	*s*	*s ,*	*p*	*s ,*	*p ,*	*d*	*s ,*	*p ,*	*d ,*	*f*
Number of orbitals:	1	1	3	1	3	5	1	3	5	7

Different orbitals have different shapes. Orbitals in *s* subshells are spherical regions centered about the nucleus, while orbitals in *p* subshells are roughly dumbbell-shaped regions (Figure 3.4). As shown in Figure 3.4(b), the three *p* orbitals in a given subshell are oriented at right angles to one another.

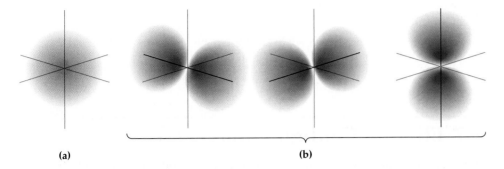

(a) **(b)**

▲ **FIGURE 3.4 The shapes of s and p orbitals.** (a) The s orbitals are spherical, and (b) the p orbitals are dumbbell-shaped. The three p orbitals in a given subshell are oriented at right angles to one another. Each orbital can hold only two electrons.

The overall electron distribution within an atom is summarized in Table 3.2 and in the following list:

- The first shell holds only 2 electrons. They have different spins and are in a single 1s orbital.
- The second shell holds 8 electrons. Two are in a 2s orbital, and 6 are in the three different 2p orbitals.
- The third shell holds 18 electrons. Two are in a 3s orbital, 6 are in three 3p orbitals, and 10 are in five 3d orbitals.
- The fourth shell holds 32 electrons. Two are in a 4s orbital, 6 are in three 4p orbitals, 10 are in five 4d orbitals, and 14 are in seven 4f orbitals.

TABLE 3.2 Electron Distribution in Atoms

Shell number:	1	2		3			4			
Subshell designation:	s	s,	p	s,	p,	d	s,	p,	d,	f
Number of orbitals:	1	1	3	1	3	5	1	3	5	7
Number of electrons:	2	2	6	2	6	10	2	6	10	14
Total electron capacity:	2	8		18			32			

■ **WORKED EXAMPLE 3.6**

How many electrons are present in an atom that has its first and second shells filled and has 4 electrons in its third shell? Name the element.

SOLUTION The first shell of an atom holds 2 electrons in its 1s orbital, and the second shell holds 8 electrons (2 in a 2s orbital and 6 in three 2p orbitals). Thus, the atom has a total of $2 + 8 + 4 = 14$ electrons and must be silicon (Si).

■ **PROBLEM 3.16**

What is the maximum number of electrons that can occupy the following subshells?

(a) 3p subshell **(b)** 2s subshell **(c)** 2p subshell

■ **PROBLEM 3.17**

How many electrons are present in an atom in which the 1s, 2s, and 2p subshells are filled? Name the element.

3.7 Electron Configurations

Electron configuration The specific arrangement of electrons in an atom's shells and subshells.

The exact arrangement of electrons in an atom's shells and subshells is called the atom's **electron configuration** and can be predicted by applying three rules:

Rule 1. **Electrons occupy the lowest-energy orbitals available, beginning with $1s$ and continuing in the order shown in Figure 3.5.** Within each shell, the orbital energies increase in the order s, p, d, f. The overall ordering is complicated, however, by the fact that some "crossover" of energies occurs between orbitals in different shells above the $3p$ level. The $4s$ orbital is lower in energy than the $3d$ orbitals, for example, and is therefore filled first.

Rule 2. **Each orbital can hold only two electrons, which must be of opposite spin.**

Rule 3. **Two or more orbitals with the same energy—the three p orbitals or the five d orbitals in a given shell, for example—are each half filled by one electron before any one orbital is completely filled by addition of the second electron.**

Electron configurations of the first 20 elements are shown in Table 3.3. Notice that the number of electrons in each subshell is indicated by a superscript. For example, the notation $1s^2\,2s^2\,2p^6\,3s^2$ for magnesium means that magnesium atoms have 2 electrons in the first shell, 8 electrons in the second shell, and 2 electrons in the third shell.

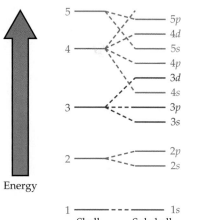

▲ **FIGURE 3.5** **Order of orbital energy levels.** Above the $3p$ level, there is some crossover of energies among orbitals in different shells.

Electron Configurations

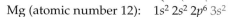

8 electrons in second shell

2 electrons in first shell

2 electrons in third shell

Mg (atomic number 12): $1s^2\,2s^2\,2p^6\,3s^2$

TABLE 3.3 **Electron Configurations of the First 20 Elements**

	Element	Atomic Number	Electron Configuration		Element	Atomic Number	Electron Configuration
H	Hydrogen	1	$1s^1$	Na	Sodium	11	$1s^2\,2s^2\,2p^6\,3s^1$
He	Helium	2	$1s^2$	Mg	Magnesium	12	$1s^2\,2s^2\,2p^6\,3s^2$
Li	Lithium	3	$1s^2\,2s^1$	Al	Aluminum	13	$1s^2\,2s^2\,2p^6\,3s^2\,3p^1$
Be	Beryllium	4	$1s^2\,2s^2$	Si	Silicon	14	$1s^2\,2s^2\,2p^6\,3s^2\,3p^2$
B	Boron	5	$1s^2\,2s^2\,2p^1$	P	Phosphorus	15	$1s^2\,2s^2\,2p^6\,3s^2\,3p^3$
C	Carbon	6	$1s^2\,2s^2\,2p^2$	S	Sulfur	16	$1s^2\,2s^2\,2p^6\,3s^2\,3p^4$
N	Nitrogen	7	$1s^2\,2s^2\,2p^3$	Cl	Chlorine	17	$1s^2\,2s^2\,2p^6\,3s^2\,3p^5$
O	Oxygen	8	$1s^2\,2s^2\,2p^4$	Ar	Argon	18	$1s^2\,2s^2\,2p^6\,3s^2\,3p^6$
F	Fluorine	9	$1s^2\,2s^2\,2p^5$	K	Potassium	19	$1s^2\,2s^2\,2p^6\,3s^2\,3p^6\,4s^1$
Ne	Neon	10	$1s^2\,2s^2\,2p^6$	Ca	Calcium	20	$1s^2\,2s^2\,2p^6\,3s^2\,3p^6\,4s^2$

As you read through the following electron configurations, check the atomic number and the location of each element in the periodic table (Figure 3.2). See if you can detect the relationship between electron configuration and position in the table.

- **Hydrogen($Z = 1$):** The single electron in a hydrogen atom is in the lowest-energy, $1s$ level. The configuration can be represented in either of two ways:

$$\textbf{H} \quad 1s^1 \quad \text{or} \quad \frac{\uparrow}{1s^1}$$

In the written representation, the superscript in the notation $1s^1$ means that the $1s$ orbital is occupied by one electron. In the graphic representation, the $1s$ orbital is indicated by a line and the single electron in this orbital is shown by an up arrow ($\uparrow$). A single electron in an orbital is often referred to as being *unpaired*.

- **Helium($Z = 2$):** The two electrons in helium are both in the lowest-energy, $1s$ orbital, and their spins are *paired*, as represented by up and down arrows ($\uparrow\downarrow$):

$$\textbf{He} \quad 1s^2 \quad \text{or} \quad \frac{\uparrow\downarrow}{1s^2}$$

- **Lithium($Z = 3$):** With the first shell full, the second shell begins to fill. The third electron goes into the $2s$ orbital:

$$\textbf{Li} \quad 1s^2\,2s^1 \quad \text{or} \quad \frac{\uparrow\downarrow}{1s^2} \; \frac{\uparrow}{2s^1}$$

- **Beryllium($Z = 4$):** An electron next pairs up to fill the $2s$ orbital:

$$\textbf{Be} \quad 1s^2\,2s^2 \quad \text{or} \quad \frac{\uparrow\downarrow}{1s^2} \; \frac{\uparrow\downarrow}{2s^2}$$

- **Boron($Z = 5$), Carbon($Z = 6$), Nitrogen($Z = 7$):** The next three electrons enter the three $2p$ orbitals, one at a time. Note that representing the configurations with lines and arrows gives more information than the alternative written notation because the filling and pairing of electrons in individual orbitals within the p subshell is shown.

$$\textbf{B} \quad 1s^2\,2s^2\,2p^1 \quad \text{or} \quad \frac{\uparrow\downarrow}{1s^2} \; \frac{\uparrow\downarrow}{2s^2} \; \underbrace{\frac{\uparrow}{} \; \frac{}{} \; \frac{}{}}_{2p^1}$$

$$\textbf{C} \quad 1s^2\,2s^2\,2p^2 \quad \text{or} \quad \frac{\uparrow\downarrow}{1s^2} \; \frac{\uparrow\downarrow}{2s^2} \; \underbrace{\frac{\uparrow}{} \; \frac{\uparrow}{} \; \frac{}{}}_{2p^2}$$

$$\textbf{N} \quad 1s^2\,2s^2\,2p^3 \quad \text{or} \quad \frac{\uparrow\downarrow}{1s^2} \; \frac{\uparrow\downarrow}{2s^2} \; \underbrace{\frac{\uparrow}{} \; \frac{\uparrow}{} \; \frac{\uparrow}{}}_{2p^3}$$

- **Oxygen($Z = 8$), Fluorine($Z = 9$), Neon($Z = 10$):** Electrons now pair up one by one to fill the three $2p$ orbitals and fully occupy the second shell:

$$\textbf{O} \quad 1s^2\,2s^2\,2p^4 \quad \text{or} \quad \frac{\uparrow\downarrow}{1s^2} \; \frac{\uparrow\downarrow}{2s^2} \; \underbrace{\frac{\uparrow\downarrow}{} \; \frac{\uparrow}{} \; \frac{\uparrow}{}}_{2p^4}$$

$$\textbf{F} \quad 1s^2\,2s^2\,2p^5 \quad \text{or} \quad \frac{\uparrow\downarrow}{1s^2} \; \frac{\uparrow\downarrow}{2s^2} \; \underbrace{\frac{\uparrow\downarrow}{} \; \frac{\uparrow\downarrow}{} \; \frac{\uparrow}{}}_{2p^5}$$

$$\textbf{Ne} \quad 1s^2\,2s^2\,2p^6 \quad \text{or} \quad \frac{\uparrow\downarrow}{1s^2} \; \frac{\uparrow\downarrow}{2s^2} \; \underbrace{\frac{\uparrow\downarrow}{} \; \frac{\uparrow\downarrow}{} \; \frac{\uparrow\downarrow}{}}_{2p^6}$$

- **Sodium to Calcium ($Z = 11–20$):** The same pattern seen for lithium through neon is seen again for sodium ($Z = 11$) through argon ($Z = 18$) as the $3s$ and $3p$ subshells fill up. After argon, however, the first crossover in subshell energies occurs. As indicated in Figure 3.5, the $4s$ subshell is lower in energy than the $3d$ subshell and is filled first. Potassium ($Z = 19$) and calcium ($Z = 20$) therefore have the following electron configurations:

$$\textbf{K } 1s^2\, 2s^2\, 2p^6\, 3s^2\, 3p^6\, 4s^1 \qquad \textbf{Ca } 1s^2\, 2s^2\, 2p^6\, 3s^2\, 3p^6\, 4s^2$$

■ **WORKED EXAMPLE 3.7**

Show how the electron configuration of magnesium can be assigned.

ANALYSIS Magnesium, $Z = 12$, has 12 electrons to be placed in specific orbitals. Assignments are made by putting 2 electrons in each orbital, according to the order shown in Figure 3.5.

- The first 2 electrons are placed in the $1s$ orbital ($1s^2$).
- The next 2 electrons are placed in the $2s$ orbital ($2s^2$).
- The next 6 electrons are placed in the three available $2p$ orbitals ($2p^6$).
- The remaining 2 electrons are both put in the $3s$ orbital ($3s^2$).

SOLUTION Magnesium has the configuration $1s^2\, 2s^2\, 2p^6\, 3s^2$.

■ **WORKED EXAMPLE 3.8**

Write the electron configuration of phosphorus, $Z = 15$, using arrows to show how the electrons in each orbital are paired.

ANALYSIS Phosphorus has 15 electrons, which occupy orbitals according to the order shown in Figure 3.5.

- The first 2 are paired and fill the first shell ($1s^2$).
- The next 8 fill the second shell ($2s^2\, 2p^6$). All electrons are paired.
- The remaining 5 electrons enter the third shell, where 2 fill the $3s$ orbital ($3s^2$) and 3 occupy the $3p$ subshell, one in each of the three p orbitals.

SOLUTION

$$\text{P} \quad \underset{1s^2}{\uparrow\downarrow} \quad \underset{2s^2}{\uparrow\downarrow} \quad \underset{2p^6}{\underbrace{\uparrow\downarrow\ \uparrow\downarrow\ \uparrow\downarrow}} \quad \underset{3s^2}{\uparrow\downarrow} \quad \underset{3p^3}{\underbrace{\uparrow\ \uparrow\ \uparrow}}$$

■ **PROBLEM 3.19**

Write electron configurations for the following elements. (You can check your answers in Table 3.3.)

(a) C   (b) Na   (c) Cl   (d) Ca

■ **PROBLEM 3.20**

Write electron configurations for the elements with atomic numbers 14 and 36.

■ **PROBLEM 3.21**

For an atom containing 33 electrons, identify the incompletely filled subshell, and show the paired and/or unpaired electrons in this subshell using up and down arrows.

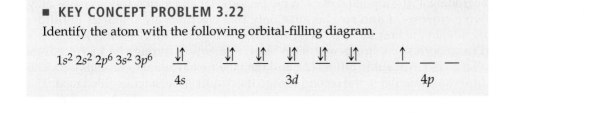

■ **KEY CONCEPT PROBLEM 3.22**

Identify the atom with the following orbital-filling diagram.

$1s^2\ 2s^2\ 2p^6\ 3s^2\ 3p^6$

3.8 Electron Configurations and the Periodic Table

How is an atom's electron configuration related to its chemical behavior, and why do elements with similar behavior occur in the same group of the periodic table? As shown in Figure 3.6, the periodic table can be divided into four regions, or *blocks*, of elements *according to the electron shells and subshells occupied by the subshell filled last.*

- The main group 1A and 2A elements on the left side of the table (plus He) are called the **s-block elements** because an *s* subshell is filled last in these elements.

- The main group 3A–8A elements on the right side of the table (except He) are the **p-block elements** because a *p* subshell is filled last in these elements.

- The transition metals in the middle of the table are the **d-block elements** because a *d* subshell is filled last in these elements.

- The inner transition metals detached at the bottom of the table are the **f-block elements** because an *f* subshell is filled last in these elements.

s-Block element A main group element that results from the filling of an *s* orbital.

p-Block element A main group element that results from the filling of *p* orbitals.

d-Block element A transition metal element that results from the filling of *d* orbitals.

f-Block element An inner transition metal element that results from the filling of *f* orbitals.

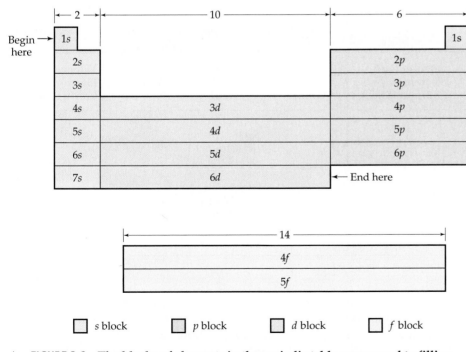

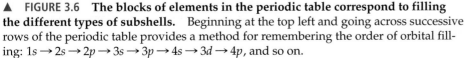

▲ **FIGURE 3.6** **The blocks of elements in the periodic table correspond to filling the different types of subshells.** Beginning at the top left and going across successive rows of the periodic table provides a method for remembering the order of orbital filling: $1s \rightarrow 2s \rightarrow 2p \rightarrow 3s \rightarrow 3p \rightarrow 4s \rightarrow 3d \rightarrow 4p$, and so on.

Thinking of the periodic table as outlined in Figure 3.6 provides a simple way to remember the order of orbital filling shown previously in Figure 3.5.

Beginning at the top left corner of the periodic table, the first row contains only two elements (H and He) because only two electrons are required to fill the s orbital in the first shell, $1s^2$. The second row begins with two s-block elements (Li and Be) and continues with six p-block elements (B through Ne), so electrons fill the next available s orbital ($2s$) and then the first available p orbitals ($2p$). The third row is similar to the second row, so the $3s$ and $3p$ orbitals are filled next. The fourth row again starts with two s-block elements (K and Ca) but is then followed by ten d-block elements (Sc through Zn) and six p-block elements (Ga through Kr). Thus, the order of orbital filling is $4s$ followed by the first available d orbitals ($3d$) followed by $4p$. Continuing through successive rows of the periodic table gives the entire filling order, identical to that shown previously in Figure 3.5.

$$1s \rightarrow 2s \rightarrow 2p \rightarrow 3s \rightarrow 3p \rightarrow 4s \rightarrow 3d \rightarrow 4p \rightarrow 5s \rightarrow$$

$$4d \rightarrow 5p \rightarrow 6s \rightarrow 4f \rightarrow 5d \rightarrow 6p \rightarrow 7s \rightarrow 5f \rightarrow 6d$$

Valence shell The outermost electron shell of an atom.

Valence electron An electron in the outermost, or valence, shell of an atom.

But why do the elements in a given group of the periodic table have similar properties? The answer emerges when you look at Table 3.4, which gives electron configurations for elements in the main groups 1A, 2A, 7A, and 8A. Focusing only on the electrons in the outermost shell, or **valence shell**, *elements in the same group of the periodic table have similar electron configurations in their valence electron shells.* The group 1A elements, for example, all have one **valence electron**, ns^1 (where n represents the number of the outer shell: $n = 2$ for Li; $n = 3$ for Na; $n = 4$ for K; and so on). The group 2A elements have two valence electrons (ns^2); the group 7A elements have seven valence electrons ($ns^2\,np^5$); and the group 8A elements (except He) have eight valence electrons ($ns^2\,np^6$). You might also notice that the group numbers from 1A through 8A give the numbers of valence electrons for the elements in each main group.

TABLE 3.4   Valence-Shell Electron Configurations for Group 1A, 2A, 7A, and 8A Elements

Group	Element	Atomic Number	Valence-Shell Electron Configuration	Group	Element	Atomic Number	Valence-Shell Electron Configuration
1A	Li (Lithium)	3	$2s^1$	7A	F (Fluorine)	9	$2s^2\,2p^5$
	Na (Sodium)	11	$3s^1$		Cl (Chlorine)	17	$3s^2\,3p^5$
	K (Potassium)	19	$4s^1$		Br (Bromine)	35	$4s^2\,4p^5$
	Rb (Rubidium)	37	$5s^1$		I (Iodine)	53	$5s^2\,5p^5$
	Cs (Cesium)	55	$6s^1$				
2A	Be (Beryllium)	4	$2s^2$	8A	He (Helium)	2	$1s^2$
	Mg (Magnesium)	12	$3s^2$		Ne (Neon)	10	$2s^2\,2p^6$
	Ca (Calcium)	20	$4s^2$		Ar (Argon)	18	$3s^2\,3p^6$
	Sr (Strontium)	38	$5s^2$		Kr (Krypton)	36	$4s^2\,4p^6$
	Ba (Barium)	56	$6s^2$		Xe (Xenon)	54	$5s^2\,5p^6$

What is true for the main group elements is also true for the other groups in the periodic table: Atoms within a given group have the same number of valence electrons and have similar electron configurations. *Because the valence electrons are the most loosely held, they are the most important in determining an element's properties.* Similar electron configurations thus explain why the elements in a given group of the periodic table have similar chemical behavior.

■ WORKED EXAMPLE 3.9

Using n to represent the number of the outer shell, write a general outer-shell configuration for the elements in group 6A.

ANALYSIS The elements in group 6A have 6 valence electrons. In each element, the first two of these electrons are in the outer s subshell, giving ns^2, and the next four electrons are in the outer p subshell, giving np^4.

SOLUTION For group 6A, the general outer-shell configuration is $ns^2 np^4$.

■ WORKED EXAMPLE 3.10

How many electrons are in a tin atom? Give the number of electrons in each shell. How many valence electrons are there in a tin atom? Write the outer-shell configuration for tin.

SOLUTION Checking the periodic table shows that tin has atomic number 50 and is in group 4A. The number of electrons in each shell is

Shell number:	1	2	3	4	5
Number of electrons	2	8	18	18	4

As expected from the group number, tin has four valence electrons. They are in the $5s$ and $5p$ subshells (Figure 3.5) and have the configuration $5s^2 5p^2$.

■ PROBLEM 3.23

Identify the group in which all elements have the outer-shell configuration ns^2.

■ PROBLEM 3.24

For chlorine, identify the group number, give the number of electrons in each occupied shell, and write the outer-shell configuration.

■ KEY CONCEPT PROBLEM 3.25

Identify the group number, and write the general outer-shell configuration (for example, ns^1 for group 1A elements) for the elements indicated in red in the following periodic table.

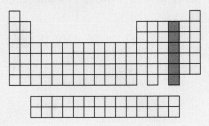

Application Atoms and Light

W hat we see as *light* is really a wave of energy moving through space. The shorter the length of the wave (the *wavelength*), the higher the energy; the longer the wavelength, the lower the energy.

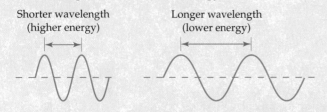

Shorter wavelength
(higher energy)

Longer wavelength
(lower energy)

Visible light has wavelengths in the range 400–600 nm, but that's just one small part of the overall *electromagnetic spectrum*, shown in the accompanying figure. Although we can't see the other wavelengths of electromagnetic energy, we use them for many purposes and their names may be familiar to you: gamma rays, X rays, ultraviolet (UV) rays, infrared (IR) rays, microwaves, and radio waves, for example.

What happens when a beam of electromagnetic energy collides with an atom? If the amount of energy is just right, an electron can be kicked up from its usual energy level to a higher one. An electrical discharge or heat can have the same effect. Many practical applications, from neon lights to fireworks, result.

An atom with its electrons in their usual, lowest-energy locations is said to be in its *ground state*. With one of its electrons promoted to a higher energy, an atom is said to be *excited*. The excited state doesn't last long, though, because the electron quickly drops back to its more stable, ground-state energy level, releasing its extra energy in the process. If the released energy falls in the range of visible light, we can see the result.

In "neon" lights, noble gas atoms are excited by an electric discharge, giving rise to a variety of colors that depend on the gas—red from neon, white from krypton, and blue from argon. Similarly, mercury or sodium atoms

▲ The brilliant colors of fireworks are due to the release of energy from excited atoms as electrons fall from higher to lower energy levels.

excited by electrical energy are responsible for the intense bluish or yellowish light, respectively, provided by some street lamps. In the same manner, metal atoms excited by heat are responsible for the spectacular colors of fireworks—red from strontium, green from barium, and blue from copper, for example.

The concentration of certain metals in body fluids is measured by sensitive instruments relying on the same principle of electron excitation that we see in fireworks. These instruments determine the intensity of the flame color produced by lithium (red), sodium (yellow), and potassium (violet), yielding the concentrations of these metals given in most clinical lab reports. Calcium, magnesium, copper, and zinc concentrations are also found by measuring the energies of excited atoms.

See Additional Problems 3.84 and 3.85 at the end of the chapter.

The electromagnetic spectrum consists of a continuous range of wavelengths, with the familiar visible region accounting for only a small portion near the middle of the range.

Flame Tests for Metals

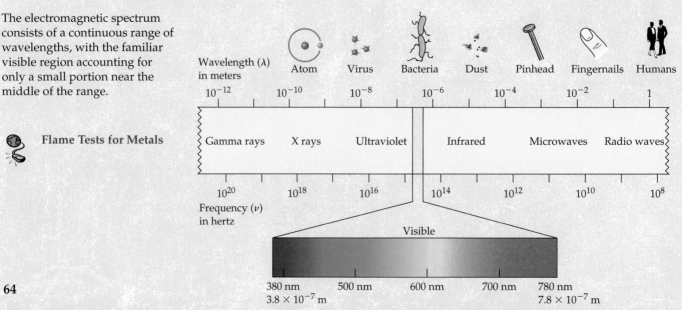

"There's a lot of chemistry involved in art conservation," says Neil Cockerline, painting conservator. "I get calls frequently from members of the general public asking me, 'What can I use to clean my painting?' It's difficult to explain to them that we have hundreds of chemical combinations that we choose from. What may be appropriate for one painting might be completely inappropriate for another painting."

"Many things can happen in the lifespan of an oil painting, for example, that will affect the chemistry that exists within the structure. For cleaning some oil paintings," continues Cockerline, "you can use very strong solvents. For others, however, you could dissolve the oil paint if you just used distilled water. It all depends on the original materials that were used, the chemistry of those materials, and the aging process."

This simple example illustrates how important it is for art conservators to know their chemistry. When dealing with a work of art that's irreplaceable and worth thousands (or even millions) of dollars, it's critical to pick the right solvent and the right cleaning agent.

But art conservation involves much more than just cleaning paintings. Cockerline and his colleagues at the Upper Midwest Conservation Association (UMCA) provide art conservation services for over 150 institutions in a five-state region. They work with art museums, historical societies, private collectors, and even police departments. Some UMCA conservators specialize in paintings, while others specialize in works of art on paper (such as manuscripts, photographs, and books) or in art objects (such as ceramics, sculpture, and ethnographic artifacts).

When conservators begin a restoration project, they learn as much as they can about the work of art and its history. Much like a medical doctor making a diagnosis, an art conservator uses a combination of sophisticated tools and practical experience. "A lot of our work is forensic in nature, trying to determine what has happened to a painting during the course of its life," says Cockerline. A painting is like a multi-layered sandwich, and each layer has a story to tell. Chemistry helps reveal those stories.

Cockerline has spent over two decades working in the conservation field, and one of his favorite analytical techniques is the use of ultraviolet light, more commonly called "black light." According to Cockerline, "Ultraviolet light tells you a lot about the surface of a painting. If you think about posters viewed under a black light, they can look very different than under natural light. That's true with paintings as well. You can determine a lot about things like retouching or varnishes. Different materials fluoresce in characteristic ways. Natural-resin varnishes such as damar and mastic fluoresce a very characteristic yellow-green. Synthetic-resin varnishes will generally fluoresce a dead white."

Conservators can also use x-rays to uncover secrets. Paints that include heavy-metal pigments show up readily under x-ray examination. The white pigment used most commonly up until the twentieth century, for example, was "lead white," made from a mixture of lead carbonate and lead hydroxide. It's not uncommon today for conservators to discover by x-ray that there is an earlier painting hidden underneath the painting that is visible to the naked eye.

In the photo, Cockerline stands with just such a painting. The original painting (on the right-hand side of the wooden panel) depicts an austere seventeenth century Dutch merchant. At a later date, another artist painted over most of the original work, giving the subject a royal costume while leaving the original face untouched. With that simple change of clothing, the Dutch merchant had been transformed into King Henry VIII of England. After discovering the underlying painting, conservators carefully stripped away the right half of the later painting. The Dutch merchant, at least half of him, was restored to his original glory-another piece of history and art revealed by chemistry.

As these examples illustrate, art conservators bring to their jobs an extensive knowledge of chemistry and physics. On a daily basis, they apply their theoretical and practical knowledge of solvents, adhesives, soaps, textiles, polymers, ceramics, metals, and other materials. In addition, conservators must have a thorough grounding in archaeology, anthropology, art history, and conservation ethics. "It's the perfect marriage between art and science," concludes Cockerline.

Key Words

Summary: Revisiting the Chapter Goals

1. What is the modern theory of atomic structure?   All matter is composed of *atoms*. An atom is the smallest and simplest unit into which a sample of an element can be divided while maintaining the properties of the element. Atoms are made up of subatomic particles called *protons, neutrons,* and *electrons.* Protons have a positive electrical charge, neutrons are electrically neutral, and electrons have a negative electrical charge. The protons and neutrons in an atom are present in a dense, positively charged central region called the *nucleus.* Electrons are situated a relatively large distance away from the nucleus, leaving most of the atom as empty space.

2. How do atoms of different elements differ?   Elements differ according to the number of protons their atoms contain, a value called the element's *atomic number* (Z). All atoms of a given element have the same number of protons and an equal number of electrons. The number of neutrons in an atom is not predictable but is generally as great or greater than the number of protons. The total number of protons plus neutrons in an atom is called the atom's *mass number* (A).

3. What are isotopes, and what is atomic weight?   Atoms with identical numbers of protons and electrons but different numbers of neutrons are called *isotopes.* The atomic weight of an element is the weighted average mass of atoms of the element's naturally occurring isotopes measured in *atomic mass units* (amu).

4. How is the periodic table arranged?   Elements are organized into the *periodic table,* consisting of 7 rows, or *periods,* and 18 columns, or *groups.* The two groups on the left of the table and the six groups on the right are called the *main group elements.* The ten groups in the middle are the *transition metal groups,* and the 14 groups pulled out and displayed below the main part of the table are called the *inner transition metal groups.* Within a given group in the table, elements have the same number of valence electrons in their outermost shell and similar electron configurations.

5. How are electrons arranged in atoms?   The electrons surrounding an atom are grouped into layers, or *shells.* Within each shell, electrons are grouped into *subshells,* and within each subshell into *orbitals*—regions of space in which electrons are most likely to be found. The *s* orbitals are spherical, and the *p* orbitals are dumbbell-shaped.

Each orbital and each shell can hold a specific number of electrons. The first shell can hold 2 electrons in an *s* orbital ($1s^2$); the second shell can hold 8 electrons in one *s* and three *p* orbitals ($2s^2\,2p^6$); the third shell can hold 18 electrons in one *s*, three *p*, and five *d* orbitals ($3s^2\,3p^6\,3d^{10}$); and so on. The *electron configuration* of an element is predicted by assigning the element's electrons into orbitals, beginning with the lowest-energy orbital.

▪ Understanding Key Concepts

3.26 Which of the following drawings represents an *s* orbital, and which represents a *p* orbital?

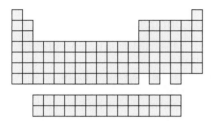

(a)   (b)

3.27 Where on the following outline of a periodic table do the indicated elements or groups of elements appear?

(a) Alkali metals
(b) Halogens
(c) Alkaline earth metals
(d) Transition metals
(e) Hydrogen
(f) Helium

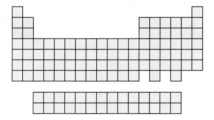

3.28 Approximately where on the following outline of a periodic table does the dividing line between metals and nonmetals fall?

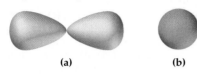

3.29 Is the element marked in red on the following periodic table likely to be a gas, a liquid, or a solid? What is the atomic number of the element in blue? Name at least one other element that is likely to be chemically similar to the element in green.

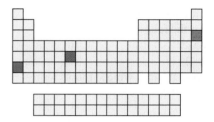

3.30 Where on the blank outline of the periodic table do elements that meet the following descriptions appear?

(a) Elements with the valence-shell electron configuration $ns^2\ np^5$

(b) An element whose third shell contains two *p* electrons

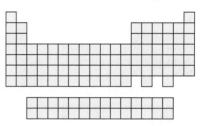

3.31 What atom has the following orbital-filling diagram?

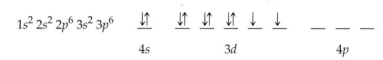

$1s^2\ 2s^2\ 2p^6\ 3s^2\ 3p^6$ ⇅   ⇅ ⇅ ⇅ ↓ ↓   __ __ __

4*s*   3*d*   4*p*

▪ Additional Problems

Atomic Theory and the Composition of Atoms

3.32 What are the four fundamental assumptions about atoms and matter that make up modern atomic theory?

3.33 How do atoms of different elements differ?

3.34 Find the mass in grams of one atom of each of the following elements:

(a) Bi, atomic weight 208.9804 amu
(b) Xe, atomic weight 131.29 amu
(c) He, atomic weight 4.0026 amu

3.35 Find the mass in atomic mass units of each of the following:

(a) 1 O atom, with a mass of 2.66×10^{-23} g
(b) 1 Br atom, with a mass of 1.31×10^{-22} g

3.36 What is the mass in grams of 6.022×10^{23} N atoms of mass 14.01 amu?

3.37 What is the mass in grams of 6.022×10^{23} C atoms of mass 12.00 amu?

3.38 How many O atoms of mass 15.99 amu are in 15.99 g of oxygen?

3.39 How many Na atoms of mass 22.99 amu are in 22.99 g of sodium?

3.40 What are the names of the three subatomic particles? What are their approximate masses in atomic mass units, and what electrical charge does each have?

3.41 Where within an atom are the three types of subatomic particles located?

3.42 Identify each of the following atoms:
(a) Contains 19 protons
(b) Contains 50 protons
(c) Has $Z = 30$

3.43 Identify each of the following atoms:
(a) Contains 15 electrons
(b) Contains 41 protons
(c) Has $Z = 27$

3.44 Give the number of neutrons in each of the naturally occurring isotopes of argon: argon-36, argon-38, argon-40.

3.45 Give the number of protons, neutrons, and electrons in each of the following isotopes:
(a) Al-27
(b) $^{28}_{14}Si$
(c) B-11
(d) $^{115}_{47}Ag$

3.46 Which of the following symbols represent isotopes of the same element?
(a) $^{19}_{9}X$
(b) $^{19}_{10}X$
(c) $^{21}_{9}X$
(d) $^{21}_{12}X$

3.47 Complete each of the following isotope symbols:
(a) $^{206}_{84}?$
(b) $^{224}_{?}Ra$
(c) $^{197}_{?}Au$
(d) $^{84}_{36}?$

3.48 Name the isotope represented by each symbol in Problem 3.46.

3.49 Give the number of neutrons in each isotope listed in Problem 3.47.

3.50 Write symbols for each of the following isotopes:
(a) Its atoms contain 6 protons and 8 neutrons.
(b) Its atoms have mass number 39 and contain 19 protons.
(c) Its atoms have mass number 20 and contain 10 electrons.

3.51 Write symbols for each of the following isotopes:
(a) Its atoms contain 10 electrons and 10 neutrons.
(b) Its atoms have $A = 51$ and $Z = 23$.
(c) Its atoms have $A = 56$ and contain 26 electrons.

3.52 There are three naturally occurring isotopes of carbon, with mass numbers of 12, 13, and 14. How many neutrons does each have? Write the symbol for each isotope, indicating both atomic number and mass number.

3.53 The isotope of iodine with mass number 131 is often used in medicine as a radioactive tracer. Write the symbol for this isotope, indicating both mass number and atomic number.

3.54 Naturally occurring copper is a mixture of 69.17% ^{63}Cu with a mass of 62.93 amu and 30.83% ^{65}Cu with a mass of 64.93 amu. What is the atomic weight of copper?

3.55 Naturally occurring silver is a mixture of 51.84% ^{107}Ag with a mass of 106.91 amu and 48.16% ^{109}Ag with a mass of 108.90 amu. What is the atomic weight of silver?

The Periodic Table

3.56 Why does the third period in the periodic table contain eight elements?

3.57 Why does the fourth period in the periodic table contain 18 elements?

3.58 Americium, atomic number 95, is used in household smoke detectors. What is the symbol for americium? Is americium a metal, a nonmetal, or a metalloid?

3.59 Antimony, $Z = 51$, is alloyed with lead for use in automobile batteries. What is the symbol for antimony? Is antimony a metal, a nonmetal, or a metalloid?

3.60 Answer the following questions for the elements from scandium through zinc:
(a) Are they metals or nonmetals?
(b) To what general class of elements do they belong?
(c) What subshell is being filled by electrons in these elements?

3.61 For (a) calcium, (b) palladium, (c) carbon, and (d) radon, choose which of the following terms apply: (i) metal, (ii) nonmetal, (iii) transition element, (iv) main group element, (v) noble gas, (vi) alkali metal, (vii) alkaline earth metal.

3.62 Name an element in the periodic table that you would expect to be chemically similar to sulfur.

3.63 Name an element in the periodic table that you would expect to be chemically similar to aluminum.

3.64 What elements in addition to lithium make up the alkali metal family?

3.65 What elements in addition to fluorine make up the halogen family?

Electron Configurations

3.66 What is the maximum number of electrons that can go into an orbital?

3.67 What are the shapes and locations within an atom of s and p orbitals?

3.68 What is the maximum number of electrons that can go into the first shell? The second shell? The third shell?

3.69 What is the total number of orbitals in the third shell? The fourth shell?

3.70 How many electrons are present in an atom with its $1s$, $2s$, and $2p$ subshells filled? What is the element?

3.71 How many electrons are present in an atom with its $1s$, $2s$, $2p$, and $3s$ subshells filled and with two electrons in the $3p$ subshell? What is the element?

3.72 Use arrows to show electron pairing in the valence p subshell of:
(a) Sulfur
(b) Bromine
(c) Silicon

3.73 Without looking back in the text, write the electron configurations for the following:
(a) Magnesium, $Z = 12$ (b) Sulfur, $Z = 16$
(c) Neon, $Z = 10$ (d) Cadmium, $Z = 48$

3.74 How many electrons does the element with $Z = 20$ have in its outer shell?

3.75 How many valence electrons do group 4A elements have? Explain.

3.76 Identify the outermost subshell occupied by electrons in beryllium and arsenic atoms.

3.77 What group in the periodic table has the outer-shell configuration $ns^2\ np^4$?

3.78 Give the number of valence electrons in atoms of each of the following elements:
(a) Kr
(b) C
(c) Ca
(d) K
(e) B
(f) Cl

3.79 Using n for the number of the outer shell, write a general outer-shell configuration for the elements in group 7A and in group 1A.

Applications

3.80 What is the advantage of using a scanning tunneling microscope compared to a normal light microscope? [*Are Atoms Real?*]

3.81 What is the main use of scanning tunneling microscopes? [*Are Atoms Real?*]

3.82 What are the first two elements that are made in stars? [*The Origin of Chemical Elements*]

3.83 How are elements heavier than iron made? [*The Origin of Chemical Elements*]

3.84 Which type of electromagnetic energy in each of the following pairs is higher in energy? [*Atoms and Light*]

   (a) Infrared, ultraviolet
   (b) Gamma waves, microwaves
   (c) Visible light, X rays

3.85 Why do you suppose ultraviolet rays from the sun are more damaging to the skin than visible light? [*Atoms and Light*]

General Questions and Problems

3.86 What elements in addition to helium make up the noble gas family?

3.87 Hydrogen is placed in group 1A on many periodic charts, even though it is not an alkali metal. On other periodic charts, though, hydrogen is included with group 7A even though it is not a halogen. Explain.

3.88 Tellurium ($Z = 52$) has a *lower* atomic number than iodine ($Z = 53$), yet it has a *higher* atomic weight (127.60 amu for Te versus 126.90 amu for I). How is this possible?

3.89 What is the atomic number of the yet undiscovered element directly below francium (Fr) in the periodic table?

3.90 Give the number of electrons in each shell for lead.

3.91 Identify the highest-energy occupied subshell in atoms of the following:

   (a) Argon   **(b)** Magnesium
   (c) Technetium **(d)** Iron

3.92 What is the atomic weight of naturally occurring bromine, which contains 50.69% Br-79 of mass 78.92 amu and 49.31% Br-81 of mass 80.91 amu?

3.93 Naturally occurring magnesium consists of three isotopes: 78.99% ^{24}Mg with a mass of 23.99 amu, 10.00% ^{25}Mg with a mass of 24.99 amu, and 11.01% ^{26}Mg with a mass of 25.98 amu. Calculate the atomic weight of magnesium.

3.94 If you had one atom of hydrogen and one atom of carbon, which would weigh more? Explain.

3.95 If you had a pile of 10^{23} hydrogen atoms and another pile of 10^{23} carbon atoms, which of the two piles would weigh more? (See Problem 3.94.)

3.96 If your pile of hydrogen atoms in Problem 3.95 weighed about 1 gram, how much would your pile of carbon atoms weigh?

3.97 Based on your answer to Problem 3.96, how much would you expect a pile of 10^{23} sodium atoms to weigh?

3.98 An unidentified element is found to have an electron configuration by shell of 2 8 18 8 2. To what group and period does this element belong? Is the element a metal or a nonmetal? How many protons does an atom of the element have? What is the name of the element?

3.99 Titanium, atomic number 22, is used in building jet fighter planes because of its combination of high strength and light weight. If titanium has an electron configuration by shell of 2 8 10 2, in what orbital are the outer-shell electrons?

3.100 Zirconium, atomic number 40, is directly beneath titanium (Problem 3.99) in the periodic table. What electron configuration by shell would you expect zirconium to have? Is zirconium a metal or a nonmetal?

3.101 A blood sample is found to contain 8.6 mg/dL of Ca. How many atoms of Ca are present in 8.6 mg? The atomic weight of Ca is 40.08 amu.

3.102 What is wrong with each of the following electron configurations?

   (a) Ni $1s^2\, 2s^2\, 2p^6\, 3s^2\, 3p^6\, 3d^{10}$
   (b) N $1s^2\, 2p^5$
   (c) Si $1s^2\, 2s^2\, 2p\ \underline{\uparrow\downarrow}\ \underline{\ }\ \underline{\ }$
   (d) Mg $1s^2\, 2s^2\, 2p^6\, 3s\ \underline{\uparrow\uparrow}$

3.103 Not all elements follow exactly the electron-filling order described in Figure 3.6. Atoms of which elements are represented by the following electron configurations?

   (a) $1s^2\, 2s^2\, 2p^6\, 3s^2\, 3p^6\, 3d^{10}\, 4s^1$
   (b) $1s^2\, 2s^2\, 2p^6\, 3s^2\, 3p^6\, 3d^{10}\, 4s^2\, 4p^6\, 4d^5\, 5s^1$

3.104 What orbital is filled last in the yet undiscovered element 118?

4

Ionic Compounds

Called tufa towers, these strange forms growing along the shore of California's Mono lake are made of calcium carbonate, $CaCO_3$, a simple ionic compound.

There are more than 19 million known chemical compounds, ranging in size from small *diatomic* (two-atom) substances like carbon monoxide, CO, to deoxyribonucleic acid (DNA), which can contain several *billion* atoms linked together in a precise way. Clearly, there must be some force that holds atoms together in compounds; otherwise, the atoms would simply drift apart and no compounds could exist. The forces that hold atoms together are called *chemical bonds* and are of two major types: *ionic bonds* and *covalent bonds*. In this chapter, we'll look at ionic bonds and at the substances formed by them. In the next chapter, we'll look at covalent bonds.

All chemical bonds result from the electrical attraction between opposite charges—between positively charged nuclei and negatively charged electrons. As a result, the way in which different elements form bonds is related to their different electron configurations.

We'll answer the following questions in this chapter:

1. **What is an ion, what is an ionic bond, and what are the general characteristics of ionic compounds?**
The goal: Be able to describe ions and ionic bonds, and give the general properties of compounds that contain ionic bonds.

2. **What is the octet rule, and how does it apply to ions?**
The goal: Be able to state the octet rule and use it to predict the electron configurations of ions of main group elements.

3. **What is the relationship between an element's position in the periodic table and the formation of its ion?**
The goal: Be able to predict what ions are likely to be formed by atoms of a given element.

4. **What determines the chemical formula of an ionic compound?**
The goal: Be able to write formulas for ionic compounds, given the identities of the ions.

5. **How are ionic compounds named?**
The goal: Be able to name an ionic compound from its formula or give the formula of a compound from its name.

6. **What are acids and bases?**
The goal: Be able to recognize common acids and bases.

CONCEPTS TO REVIEW

The Periodic Table (Sections 3.4 and 3.5)
Electron Configurations (Sections 3.7 and 3.8)

4.1 Ions

A general rule noted by early chemists is that metals, on the left side of the periodic table, tend to form compounds with nonmetals, on the right side of the table. The alkali metals of group 1A, for instance, react with the halogens of group 7A to form a variety of compounds. Sodium chloride (table salt), formed by the reaction of sodium with chlorine, is a familiar example. The names and chemical formulas of some other compounds containing elements from groups 1A and 7A include:

Potassium iodide, KI — Added to table salt to provide iodide ion needed by the thyroid gland

Sodium fluoride, NaF — Added to many municipal water supplies to provide fluoride ion for the prevention of tooth decay

Sodium iodide, NaI — Used in laboratory scintillation counters to detect radiation

Both the compositions and the properties of these alkali metal–halogen compounds are similar. For instance, the two elements always combine in a 1:1 ratio: one alkali metal atom for every halogen atom. Each compound has a high melting point (all are over 500°C); each is a stable, white, crystalline solid; and each is soluble in water. Furthermore, the water solution of each compound conducts electricity, a property that gives a clue to the kind of chemical bond holding the atoms together.

Electricity can flow only through a medium containing charged particles that are free to move. The electrical conductivity of metals, for example, results

▲ A solution of sodium chloride in water conducts electricity, allowing the bulb to light.

from the movement of negatively charged electrons through the metal. But what charged particles might be present in the water solutions of alkali metal–halogen compounds? To answer this question, think about the composition of atoms. Atoms are electrically neutral because they contain equal numbers of protons and electrons. By gaining or losing one or more electrons, however, an atom can be converted into a charged particle called an **ion**.

The *loss* of one or more electrons from a neutral atom gives a *positively* charged ion called a **cation** (**cat**-ion). As we saw in Section 3.8, sodium and other alkali metal atoms have a single electron in their valence shell and an electron configuration symbolized as ns^1. By losing this electron, an alkali metal is converted to a cation.

Ion An electrically charged atom or group of atoms.

Cation A positively charged ion.

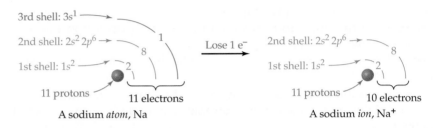

Conversely, the *gain* of one or more electrons by a neutral atom gives a *negatively* charged ion called an **anion** (**an**-ion). Chlorine and other halogen atoms have $ns^2\,np^5$ electrons and can easily gain an additional electron to fill out their valence subshell, thereby forming anions.

Anion A negatively charged ion.

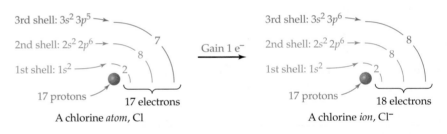

The symbol for a cation is written by adding the positive charge as a superscript to the symbol for the element; an anion symbol is written by adding the negative charge as a superscript. If one electron is lost or gained, the charge is $+1$ or -1 but the number 1 is omitted in the notation, as in Na^+ and Cl^-. If two or more electrons are lost or gained, however, the charge is ± 2 or greater and the number *is* used, as in Ca^{2+} and N^{3-}.

■ **PROBLEM 4.1**

Magnesium atoms lose two electrons when they react. Write the symbol of the ion that results. Is it a cation or an anion?

■ **PROBLEM 4.2**

Oxygen atoms gain two electrons when they react. Write the symbol of the ion that results. Is it a cation or an anion?

Periodic Trends: Ionization Energies

Ionization energy The energy required to remove one electron from a single atom in the gaseous state.

4.2 Periodic Properties and Ion Formation

The ease with which an atom loses an electron to form a positively charged cation is measured by a property called the atom's **ionization energy**, defined as the energy required to remove one electron from a single atom in the gaseous state. Conversely, the ease with which an atom *gains* an electron to form a negatively charged anion is measured by a property called **electron affinity**, defined as the energy released on adding an electron to a single atom in the gaseous state.

Ionization energy
(energy is added)

$$\text{Atom} + \text{Energy} \longrightarrow \text{Cation} + \text{Electron}$$

Electron affinity
(energy is released)

$$\text{Atom} + \text{Electron} \longrightarrow \text{Anion} + \text{Energy}$$

Electron affinity The energy released on adding an electron to a single atom in the gaseous state.

 Periodic Trends: Electron Affinity

The relative magnitudes of ionization energies and electron affinities for elements in the first four rows of the periodic table are shown in Figure 4.1. Because ionization energy measures the amount of energy that must be *added* to pull an electron away from a neutral atom, the small values shown in Figure 4.1 for alkali metals (Li, Na, K) and other elements on the left side of the periodic table mean that these elements lose an electron easily. Conversely, the large values shown for halogens (F, Cl, Br) and noble gases (He, Ne, Ar, Kr) on the right side of the periodic table mean that these elements do not lose an electron easily. Electron affinities, however, measure the amount of energy *released* when an atom gains an electron. Although electron affinities are small compared to ionization energies, the halogens nevertheless have the largest values and therefore gain an electron most easily, while metals have the smallest values and do not gain an electron easily:

Alkali metal   Small ionization energy—electron easily lost
Small electron affinity—electron not easily gained
Net result: Cation formation is favored.

Halogen   Large ionization energy—electron not easily lost
Large electron affinity—electron easily gained
Net result: Anion formation is favored.

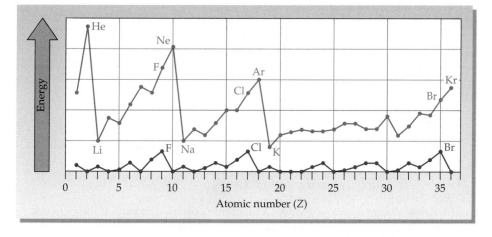

▲ **FIGURE 4.1   Relative ionization energies (red) and electron affinities (blue) for elements in the first four rows of the periodic table.**   Those elements having a value of zero for electron affinity do not accept an electron. Note that the alkali metals (Li, Na, K) have the lowest ionization energies and lose an electron most easily, whereas the halogens (F, Cl, Br) have the highest electron affinities and gain an electron most easily. The noble gases (He, Ne, Ar, Kr) neither gain nor lose an electron easily.

You might also note in Figure 4.1 that main group elements near the *middle* of the periodic table—boron ($Z = 5$), carbon ($Z = 6$, group 4A), and nitrogen ($Z = 7$, group 5A)— neither lose nor gain electrons easily and thus do not form ions easily. We'll see in the next chapter that these elements tend not to form ionic bonds but to form covalent bonds instead.

The result of the tendency for an alkali metal such as sodium to lose an electron and for a halogen such as chlorine to gain an electron is that the two elements react with each other by transfer of an electron from the metal to the halogen (Figure 4.2). The product that results—sodium chloride (NaCl)— is electrically neutral because the positive charge of each Na^+ ion is balanced by the negative charge of each Cl^- ion.

▲ **FIGURE 4.2**   (a) Chlorine is a toxic green gas, sodium is a reactive metal, and sodium chloride is a harmless white solid. (b) Sodium metal burns with an intense yellow flame when immersed in chlorine gas, yielding white sodium chloride "smoke."

■ **WORKED EXAMPLE 4.1**

Look at the periodic trends in Figure 4.1, and predict where the ionization energy of rubidium is likely to fall on the chart.

ANALYSIS   Identify the group number of rubidium (Group 1A), and find where other members of the group appear in Figure 4.1.

SOLUTION   Rubidium (Rb) is the alkali metal below potassium (K) in the periodic table. Since the alkali metals Li, Na, and K all have ionization energies near the bottom of the chart, the ionization energy of rubidium is probably similar.

■ **WORKED EXAMPLE 4.2**

Which element is likely to lose an electron more easily, Mg or S?

ANALYSIS   Identify the group numbers of the elements, and find where members of those groups appear in Figure 4.1.

SOLUTION   Magnesium, a group 2A element is on the left side of the periodic table, has a relatively low ionization energy, and loses an electron easily. Sulfur, a group 6A element, is on the right side of the table, has a higher ionization energy, and loses an electron less easily.

■ **PROBLEM 4.3**

Look at the periodic trends in Figure 4.1, and predict approximately where the ionization energy of xenon is likely to fall.

■ **PROBLEM 4.4**

Which element in each of the following pairs is likely to lose an electron more easily?

(a) Be or B   **(b)** Ca or Co   **(c)** Sc or Se

■ **PROBLEM 4.5**

Which element in each of the following pairs is likely to gain an electron more easily?

(a) H or He **(b)** S or Si **(c)** Na or Mg

4.3 Ionic Bonds

When sodium reacts with chlorine, the product is sodium chloride, a compound completely unlike either of the elements from which it is formed. Sodium is a soft, silvery metal that reacts violently with water, and chlorine is a corrosive, poisonous, green gas. When chemically combined, however, they produce our familiar table salt containing Na^+ ions and Cl^- ions (Figure 4.2). Because opposite electrical charges attract each other, the positive Na^+ ion and negative Cl^- ion are said to be held together by an **ionic bond**.

Ionic bond The electrical attractions between ions of opposite charge in a crystal.

When a vast number of sodium atoms transfer electrons to an equally vast number of chlorine atoms, a visible crystal of sodium chloride results. In this crystal, equal numbers of Na^+ and Cl^- ions are packed together in a regular arrangement. Each positively charged Na^+ ion is surrounded by six negatively charged Cl^- ions, and each Cl^- ion is surrounded by six Na^+ ions (Figure 4.3). This packing arrangement allows each ion to be stabilized by the attraction of unlike charges on its six nearest-neighbor ions, while being as far as possible from ions of like charge.

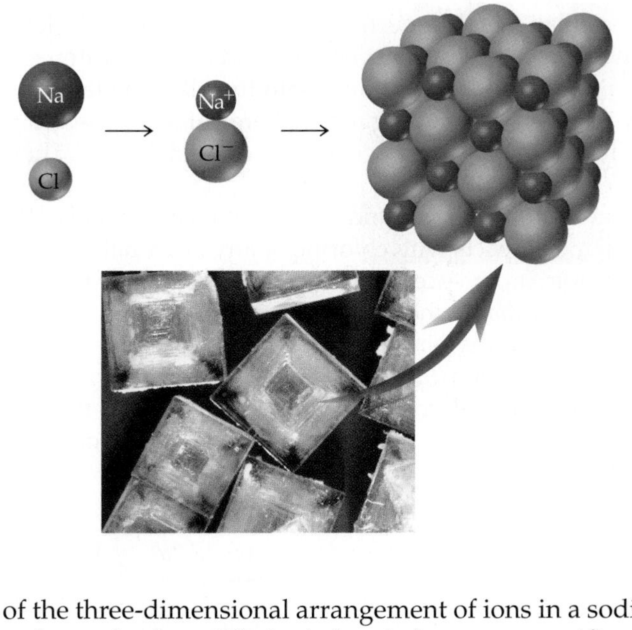

◀ **FIGURE 4.3 The arrangement of Na^+ and Cl^- ions in a sodium chloride crystal.** Each positively charged Na^+ ion is surrounded by six negatively charged Cl^- ions, and each Cl^- ion is surrounded by six Na^+ ions. The crystal is held together by ionic bonds—the attraction between oppositely charged ions.

Because of the three-dimensional arrangement of ions in a sodium chloride crystal, we can't speak of specific ionic bonds between specific pairs of ions. Rather, there are many ions attracted by ionic bonds to their nearest neighbors. We therefore speak of the whole NaCl crystal as being an **ionic solid** and of such compounds as being **ionic compounds**. The same is true of all compounds composed of ions.

Ionic solid A crystalline solid held together by ionic bonds.

Ionic compound A compound that contains ionic bonds.

4.4 Some Properties of Ionic Compounds

Like sodium chloride, ionic compounds are usually crystalline solids. Different ions vary in size and charge, however, and therefore are packed together in crystals in different ways. The ions in each compound settle into a pattern that efficiently fills space and maximizes ionic bonding.

▲ The melting point of sodium chloride is 801°C.

Most Common Ions for Elements

Because the ions in an ionic solid are held rigidly in place by attraction to their neighbors, they can't move about. Once an ionic solid is dissolved in water, however, the ions can move freely, thereby accounting for the electrical conductivity of these compounds in solution.

The high melting points and boiling points observed for ionic compounds are also accounted for by ionic bonding. The attractive force between oppositely charged particles is extremely strong, and the ions need to gain a large amount of energy by being heated to high temperatures for them to loosen their grip on one another. Sodium chloride, for example, melts at 801°C and boils at 1413°C; potassium iodide melts at 681°C and boils at 1330°C.

Despite the strength of ionic bonds, ionic solids shatter if struck sharply. A blow disrupts the orderly arrangement of cations and anions, and the electrical repulsion between ions of like charge that have been pushed together helps to split apart the crystal.

Ionic compounds dissolve in water if the attraction between water and the ions overcomes the attraction of the ions for one another. Compounds like sodium chloride are very soluble and can be dissolved to make solutions of high concentration. Don't be misled, however, by the ease with which sodium chloride and other familiar ionic compounds dissolve in water. Many other ionic compounds are not water-soluble, because water is unable to overcome the ionic forces in many crystals.

4.5 Ions and the Octet Rule

We've seen that alkali metal atoms have a single valence-shell electron, ns^1; halogen atoms have 7 valence electrons, $ns^2\,np^5$; and noble gas atoms have 8 valence electrons, $ns^2\,np^6$. Of the three groups, both the alkali metals and the halogens are extremely reactive, undergoing many chemical reactions and forming many compounds. The noble gases, however, are quite different: They are the least reactive of all elements.

Now look at sodium chloride and similar ionic compounds. When sodium or any other alkali metal reacts with chlorine or any other halogen, the metal transfers an electron from its valence shell to the valence shell of the halogen. Sodium thereby changes its valence-shell electron configuration from $2s^2\,2p^6\,3s^1$ in the atom to $2s^2\,2p^6\,(3s^0)$ in the Na^+ ion, and chlorine changes from $3s^2\,3p^5$ in the atom to $3s^2\,3p^6$ in the Cl^- ion. *In so doing, both sodium and chlorine gain noble gas electron configurations, with 8 valence electrons.* The Na^+ ion has the electron configuration of neon, and the Cl^- ion has the electron configuration of argon.

$$
\begin{array}{ccccccc}
\text{Na} & + & \text{Cl} & \longrightarrow & \text{Na}^+ & + & \text{Cl}^- \\
1s^2\,2s^2\,2p^6\,3s^1 & & 1s^2\,2s^2\,2p^6\,3s^2\,3p^5 & & \underbrace{1s^2\,2s^2\,2p^6}_{\substack{\text{Neon}\\\text{configuration}}} & & \underbrace{1s^2\,2s^2\,2p^6\,3s^2\,3p^6}_{\substack{\text{Argon}\\\text{configuration}}}
\end{array}
$$

Evidently there is something special about having 8 valence electrons (filled s and p subshells) that leads to stability and lack of chemical reactivity. In fact, observations of a great many chemical compounds have shown that main group elements frequently combine in such a way that each winds up with 8 valence electrons, a so-called *electron octet*. This conclusion is summarized in a statement called the **octet rule**:

> **Octet rule** Main group elements tend to undergo reactions that leave them with 8 valence electrons.

Put another way, main group *metals* react so that they attain an electron configuration like that of the noble gas just *before* them in the periodic table, and

Application Minerals and Gems

If you are wearing a sapphire, ruby, emerald, or zircon ring, you have a crystalline ionic compound on your finger. The gem came from the earth's crust, the source of most of our chemical raw materials. These gemstones are *minerals*, which to a geologist means that they are naturally occurring, crystalline chemical compounds. (To a nutritionist, by contrast, a "mineral" is one of the metal ions essential to human health.)

Sapphire and ruby are forms of the mineral *corundum*, or aluminum oxide (Al_2O_3). The blue of sapphire is due to traces of iron and titanium ions also present in the crystal, and the red of ruby is due to traces of chromium ions. Many minerals are *silicates*, meaning

that their anions contain silicon and oxygen combined in a variety of ways. Zircon is zirconium silicate ($ZrSiO_4$), and emerald is a form of the mineral beryl, $Be_3Al_2(Si_6O_{18})$, which is composed of Al^{3+}, Be^{2+}, and silicate rings.

Many minerals not sufficiently beautiful for use in jewelry are valuable as sources for the exotic metals so essential to our industrial civilization. Extraction of the metals from these minerals requires energy and a series of chemical reactions to convert the metal ions into free elements—the reverse of ion formation from atoms.

See Additional Problem 4.80 at the end of the chapter.

(a) (b) (c)

▲ Many minerals are silicates, compounds with polyatomic anions of silicon and oxygen. Their colors are determined largely by the cations they contain. (a) A crystal of tourmaline, a complex silicate of boron, aluminum, and other elements. (b) Dioptase from Namibia and (c) epidote from Pakistan.

reactive main group *nonmetals* react so that they attain an electron configuration like that of the noble gas just *after* them in the periodic table. In both cases, the product ions have filled *s* and *p* subshells in their valence electron shell.

■ WORKED EXAMPLE 4.3

Write the electron configuration of magnesium ($Z = 12$). Show how many electrons a magnesium atom must lose to form an ion with a valence octet, and write the configuration of the ion. Explain the reason for the ion's charge, and write the ion's symbol.

ANALYSIS Write the electron configuration of magnesium as described in Section 3.7 and count the number of electrons in the outermost shell.

SOLUTION Magnesium has the electron configuration $1s^2\,2s^2\,2p^6\,3s^2$. Since the second shell contains an octet of electrons ($2s^2\,2p^6$) while the third shell is only partially filled $(3s^2)$, magnesium can achieve a valence-shell octet by losing the two electrons in the $3s$ subshell. The result is formation of a doubly charged cation, Mg^{2+} with the neon configuration:

$$Mg^{2+} \qquad 1s^2 \, 2s^2 \, 2p^6 \quad (\text{Neon configuration})$$

A neutral magnesium atom has 12 protons and 12 electrons. With the loss of 2 electrons, there is an excess of 2 protons, accounting for the $+2$ charge of the ion, Mg^{2+}.

■ **WORKED EXAMPLE 4.4**

How many electrons must a nitrogen atom, $Z = 7$, gain to attain a noble gas configuration? Write the symbol for the ion formed.

ANALYSIS Write the electron configuration of nitrogen, and see how many more electrons are needed to reach a noble gas configuration.

SOLUTION Nitrogen, a group 5A element, has the electron configuration $1s^2 \, 2s^2 \, 2p^3$. The second shell contains 5 electrons $(2s^2 \, 2p^3)$ and needs 3 more to reach an octet. The result is formation of a triply charged anion, N^{3-}, with the neon configuration:

$$N^{3-} \qquad 1s^2 \, 2s^2 \, 2p^6 \quad (\text{Neon configuration})$$

■ **PROBLEM 4.6**

Write the electron configuration of potassium, $Z = 19$, and show how a potassium atom can attain a noble gas configuration.

■ **PROBLEM 4.7**

How many electrons must an aluminum atom, $Z = 13$, lose to attain a noble gas configuration? Write the symbol for the ion formed.

■ **PROBLEM 4.8**

An oxygen atom, $Z = 8$, has 6 valence electrons. How can an oxygen atom most easily attain a noble gas configuration? Write the electron configuration of the oxygen ion, and identify the noble gas that has the same configuration.

4.6 Electron-Dot Symbols

Electron-dot symbol An atomic symbol with dots placed around it to indicate the number of valence electrons.

Electron-Dot Symbols for Some Main Group Elements

Valence electrons play such an important role in the behavior of atoms that it's useful to have a method for including them with atomic symbols. In an **electron-dot symbol**, dots are placed around the atomic symbol to indicate the number of valence electrons present. A group 1A atom such as sodium, for example, has a single dot, a group 2A atom such as magnesium has two dots, a group 3A atom such as boron has three dots, and so on.

Table 4.1 gives electron-dot symbols for atoms of the first few elements in each main group. As shown, the dots are distributed around the four sides of the element symbol, singly at first and then in pairs. Note that helium differs from other noble gases in having only two valence electrons rather than eight. Nevertheless, helium is considered a member of group 8A because its properties resemble those of the other noble gases and because its highest occupied subshell is filled $(1s^2)$.

| | | | | | | | Noble |
| TABLE 4.1 | | Electron-Dot Symbols for Some Main Group Elements | | | | | |

1A	2A	3A	4A	5A	6A	7A	Noble Gases
H·							He:
Li·	·Be·	·Ḃ·	·Ċ·	·N̈·	·Ö:	·F̈:	:N̈e:
Na·	·Mg·	·Ȧl·	·S̈i·	·P̈:	·S̈:	·C̈l:	:Är:
K·	·Ca·	·Ġa·	·Ġe·	·Äs:	·S̈e:	·B̈r:	:K̈r:

■ **WORKED EXAMPLE 4.5**

Write the electron-dot symbol for any element X in group 5A.

ANALYSIS The group number, 5A, indicates 5 valence electrons. Distribute them around the four sides of the element symbol, first singly and then in pairs.

SOLUTION

$$·\ddot{X}:\qquad \text{(5 electrons)}$$

■ **PROBLEM 4.9**

Write the electron-dot symbol for any element X in group 3A.

■ **PROBLEM 4.10**

Write electron-dot symbols for radon, lead, xenon, and radium.

■ **PROBLEM 4.11**

Write symbols, both with and without electron dots, for the ions formed by the following processes:

(a) Gain of two electrons by selenium

(b) Loss of two electrons by barium

(c) Gain of one electron by bromine

4.7 Ions of Some Common Elements

The periodic table is the key to understanding and remembering which elements form ions and which do not. As shown in Figure 4.4, atoms of elements in the same group tend to form ions of the same charge. The metals of groups 1A and 2A, for example, form only +1 and +2 ions, respectively. The ions of these elements all have noble gas configurations as a result of electron loss from their outermost s subshells. (Note in the following equations that the electrons being lost are shown as products.)

$$\text{Group 1A:} \quad M· \rightarrow M^+ + e^-$$
$$(M = Li,\ Na,\ K,\ Rb,\ or\ Cs)$$

$$\text{Group 2A:} \quad M: \rightarrow M^{2+} + 2e^-$$
$$(M = Be,\ Mg,\ Ca,\ Sr,\ Ba,\ or\ Ra)$$

Four of these ions, Na^+, K^+, Mg^{2+}, and Ca^{2+}, are present in body fluids, where they play extremely important roles in biochemical processes.

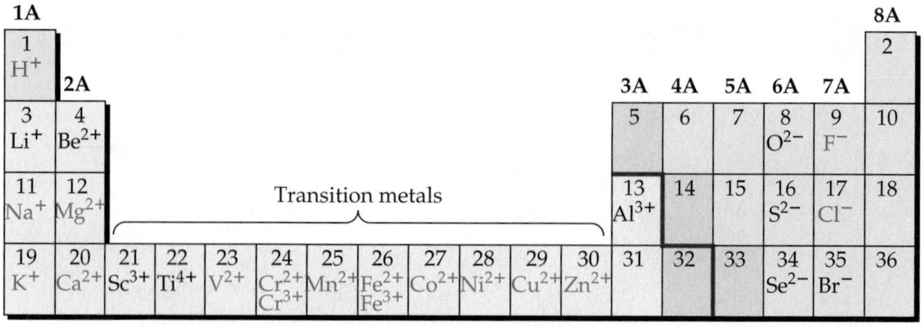

**FIGURE 4.4 Common ions ▶
formed by elements in the first four
periods.** Ions important in biological chemistry are shown in red.

The only group 3A element commonly encountered in ionic compounds is aluminum, which forms Al^{3+} by loss of three electrons from its outermost s and p subshells. Aluminum is not thought to be an essential element in the human diet, although it is known to be present in some organisms.

The first three elements in groups 4A (C, Si, Ge) and 5A (N, P, As) don't ordinarily form either cations or anions because too much energy is required either to remove or to add a sufficient number of electrons. The bonding of these elements is largely covalent and will be described in the next chapter. Carbon, in particular, is the key element on which life is based. Together with hydrogen, nitrogen, phosphorus, and oxygen, carbon is present in all the essential biological compounds that we'll be describing throughout the latter half of this book.

The group 6A elements oxygen and sulfur form large numbers of compounds, some of which are ionic and some of which are covalent. Their ions have noble gas configurations, achieved by gaining two electrons:

$$\text{Group 6A:} \qquad \cdot\ddot{\text{O}}\cdot \; + \; 2\,e^- \; \longrightarrow \; :\!\ddot{\text{O}}\!:^{2-}$$

$$\cdot\ddot{\text{S}}\cdot \; + \; 2\,e^- \; \longrightarrow \; :\!\ddot{\text{S}}\!:^{2-}$$

The halogens are present in many compounds as -1 ions, formed by gaining one electron:

$$\text{Group 7A:} \qquad \cdot\ddot{\text{X}}\!: \; + \; e^- \; \longrightarrow \; :\!\ddot{\text{X}}\!:^{-}$$
$$(X = F, Cl, Br, I)$$

Transition metals lose electrons to form cations, some of which are present in the body. The charges of transition metal cations are not as predictable as those of main group elements, however, because many transition metal atoms can lose one or more d electrons in addition to losing valence s electrons. For example, iron ($\ldots 3s^2\,3p^6\,3d^6\,4s^2$) forms Fe^{2+} by losing two electrons from the $4s$ subshell and also forms Fe^{3+} by losing an additional electron from the $3d$ subshell. Looking at the iron configuration shows why the octet rule is limited to main group elements: Transition metal cations generally don't have valence-shell octets because they would have had to lose *all* their valence-shell d electrons.

Important Points about Ion Formation and the Periodic Table:

- Metals form cations by losing one or more electrons; for example, Na^+, Ca^{2+}, and Fe^{3+}.

- Reactive nonmetals form anions by gaining one or more electrons; for example, Cl^- and O^{2-}.

- Group 1A and 2A metals form +1 and +2 ions, respectively; for example, Li^+ and Mg^{2+}.

- Group 6A nonmetals oxygen and sulfur form the anions O^{2-} and S^{2-}.

- Group 7A elements (the halogens) form -1 ions; for example, F^- and Cl^-.

- Group 8A elements (the noble gases) are unreactive.

- Transition metals form cations, often of more than one type; for example, Fe^{2+} and Fe^{3+}.

■ **WORKED EXAMPLE 4.6**

Which of the following ions is likely to form?
(a) S^{3-} (b) Si^{2+} (c) Sr^{2+}

ANALYSIS   Count the number of valence electrons in each ion. Only main group ions with a valence octet of electrons are likely to form.

SOLUTION
(a) Sulfur is in group 6A, has 6 valence electrons, and needs only two more to reach an octet. Gaining two electrons gives an S^{2-} ion with a noble gas configuration, but gaining three electrons does not. The S^{3-} ion is therefore unlikely to form.
(b) Silicon is a nonmetal in group 4A. Like carbon, it does not form ions because it would have to gain or lose too many (4) electrons to reach a noble gas electron configuration. The Si^{2+} ion does not have an octet and will not form.
(c) Strontium is a metal in group 2A, has only 2 outer-shell electrons, and can therefore lose both to reach a noble gas configuration. The Sr^{2+} ion has an octet forms easily.

■ **PROBLEM 4.12**

Is molybdenum more likely to form a cation or an anion?

■ **PROBLEM 4.13**

Which of the following elements can form more than one cation?
(a) Strontium   (b) Chromium   (c) Bromine

■ **PROBLEM 4.14**

Which of the following ions are likely to form?
(a) Al^{2+} (b) O^- (c) Se^{2-}

4.8 Naming Ions

Main group metal cations in groups 1A, 2A, and 3A are named by identifying the metal, followed by the word "ion," as in the following examples:

$$K^+ \qquad Mg^{2+} \qquad Al^{3+}$$
Potassium ion   Magnesium ion   Aluminum ion

It's sometimes a little confusing to use the same name for both a metal and its ion, and you may occasionally have to stop and think about what is being referred to. For example, it's common practice in nutrition and health-related fields to talk about sodium or potassium in the bloodstream. Because both sodium and potassium *metals* react violently with water, however, they can't possibly be present in blood. The references are to the dissolved sodium and potassium *ions*.

For transition metals such as iron or chromium, which can form more than one type of cation, a method is needed to indicate which ion is being referred to. Two systems are used. The first is an old system that gives the ion with the smaller charge the word ending *-ous* and the ion with the larger charge the ending *-ic*.

Application Salt

I f you're like most people, you feel a little guilty about reaching for the salt shaker at mealtime. The notion that high salt intake and high blood pressure go hand in hand is surely among the most highly publicized pieces of nutritional lore ever to appear.

Salt has not always been held in such disrepute. Historically, salt has been prized since the earliest recorded times as a seasoning and a food preservative. Words and phrases in many languages reflect the importance of salt as a life-giving and life-sustaining substance. We refer to a kind and generous person as "the salt of the earth," for instance, and we speak of being "worth one's salt." In Roman times, soldiers were paid in salt; the English word "salary" is derived from the Latin word for paying salt wages (*salarium*).

Salt is perhaps the easiest of all minerals to obtain and purify. The simplest method, used for thousands of years throughout the world in coastal climates where sunshine is abundant and rainfall is scarce, is to evaporate seawater. Though the exact amount varies depending on the source, seawater contains an average of about 3.5% by mass of dissolved substances, most of which is sodium chloride. It has been estimated that evaporation of all the world's oceans would yield approximately *4.5 million cubic miles* of NaCl.

Only about 10% of current world salt production comes from evaporation of seawater. Most salt is obtained by mining the vast deposits of *halite*, or *rock salt*, formed by evaporation of ancient inland seas. These salt beds vary in thickness up to hundreds of meters and vary in depth from a few meters to thousands of meters below the earth's surface. Salt mining has gone on for at least 3400 years, and the Wieliczka mine in Galicia, Poland, has been worked continuously from A.D. 1000 to the present.

Let's get back now to the dinner table. What about the link between dietary salt intake and high blood pressure? There's no doubt that most people in industrialized nations have a relatively high salt intake, and there's no doubt that high blood pressure among industrialized populations is on the rise. How closely, though, are the two observations related?

In a study called the DASH-Sodium study published in 2001, a strong correlation was found between a change

▲ These chandeliers were carved out of salt by miners at the Wieliczka mine. *(Photo courtesy of James S. Aber, Emporia State University)*

in salt intake and a change in blood pressure. When volunteers cut back their salt intake from 8.3 g per day—roughly what Americans typically consume—to 3.8 g per day, significant drops in blood pressure were found. The largest reduction in blood pressure was seen in people already diagnosed with hypertension, but subjects with normal blood pressure also lowered their readings by several percent.

What should an individual do? The best answer, as in so many things, is to use moderation and common sense. People with hypertension should make a strong effort to lower their sodium intake; others might be well advised to choose unsalted snacks, use less salt in preparing food, and read nutrition labels for sodium content

See Additional Problems 4.81 and 4.82 at the end of this chapter.

The second is a newer system in which the charge on the ion is given as a Roman numeral in parentheses right after the metal name. For example:

	Cr^{2+}	Cr^{3+}
Old name:	Chrom*ous* ion	Chrom*ic* ion
New name:	Chromium(II) ion	Chromium(III) ion

We'll generally emphasize the new system in this book, but it's important to understand both because the old system is often found on labels of commercially supplied chemicals. The small differences between the names in either system illustrate the importance of reading a name very carefully before using a chemical. There are significant differences between compounds consisting of the same two elements but having different charges on the cation. In treating iron-deficiency anemia, for example, iron(II) compounds are preferable because the body absorbs them considerably better than iron(III) compounds.

The names of some common transition metal cations are listed in Table 4.2. Notice that the old names of the copper, iron, and tin ions are derived from their Latin names (*cuprum*, *ferrum*, and *stannum*).

 Naming Anions; Naming Cations

TABLE 4.2	Names of Some Transition Metal Cations		
Element	**Symbol**	**Old Name**	**New Name**
Chromium	Cr^{2+}	Chromous	Chromium(II)
	Cr^{3+}	Chromic	Chromium(III)
Copper	Cu^+	Cuprous	Copper(I)
	Cu^{2+}	Cupric	Copper(II)
Iron	Fe^{2+}	Ferrous	Iron(II)
	Fe^{3+}	Ferric	Iron(III)
Mercury	$*Hg_2^{2+}$	Mercurous	Mercury(I)
	Hg^{2+}	Mercuric	Mercury(II)
Tin	Sn^{2+}	Stannous	Tin(II)
	Sn^{4+}	Stannic	Tin(IV)

*This cation is composed of two mercury atoms, each of which has an average charge of +1.

Anions are named by replacing the ending of the element name with *-ide*, followed by the word "ion" (Table 4.3). For example, the anion formed by fluor*ine* is the fluor*ide* ion, and the anion formed by sul*fur* is the sul*fide* ion.

TABLE 4.3	Names of Some Common Anions	
Element	**Symbol**	**Name**
Bromine	Br^-	Bromide ion
Chlorine	Cl^-	Chloride ion
Fluorine	F^-	Fluoride ion
Iodine	I^-	Iodide ion
Oxygen	O^{2-}	Oxide ion
Sulfur	S^{2-}	Sulfide ion

■ **PROBLEM 4.15**

Name the following ions:

(a) Cu^{2+} **(b)** F^- **(c)** Mg^{2+} **(d)** S^{2-}

■ **PROBLEM 4.16**

Write the symbols for the following ions:

(a) Silver(I) ion   **(b)** Iron(II) ion   **(c)** Cuprous ion   **(d)** Telluride ion

■ **PROBLEM 4.17**

Ringer's solution, which is used intravenously to adjust ion concentrations in body fluids, contains the ions of sodium, potassium, calcium, and chlorine. Give the names and symbols of these ions.

4.9 Polyatomic Ions

Polyatomic ion An ion that is composed of more than one atom.

Polyatomic Ion
Nomenclature

Ions that are composed of more than one atom are called **polyatomic ions**. Most polyatomic ions contain oxygen and another element, and their chemical formulas show by subscripts how many of each type of atom are combined. Sulfate ion, for example, is composed of one sulfur atom and four oxygen atoms, and has a −2 charge: SO_4^{2-}. The atoms in a polyatomic ion are held together by covalent bonds, of the sort discussed in the next chapter, and the entire group of atoms acts as a single unit. A polyatomic ion is charged because it contains a total number of electrons different from the total number of protons in the combined atoms.

The most common polyatomic ions are listed in Table 4.4. Note that the ammonium ion, NH_4^+, is the only cation; all the others are anions. These ions are encountered so frequently in chemistry, biology, and medicine that there's no alternative but to memorize their names and formulas. Fortunately, there aren't many of them.

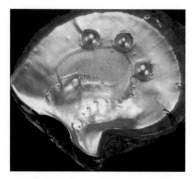

▲ Pearls are made of the ionic compound calcium carbonate.

TABLE 4.4   Some Common Polyatomic Ions

Name	Formula	Name	Formula
Ammonium ion	NH_4^+	Nitrite ion	NO_2^-
Acetate ion	$CH_3CO_2^-$	Permanganate ion	MnO_4^-
Carbonate ion	CO_3^{2-}	Phosphate ion	PO_4^{3-}
Hydrogen carbonate ion (bicarbonate ion)	HCO_3^-	Hydrogen phosphate ion	HPO_4^{2-}
		Dihydrogen phosphate ion	$H_2PO_4^-$
Dichromate ion	$Cr_2O_7^{2-}$	Sulfate ion	SO_4^{2-}
Cyanide ion	CN^-	Hydrogen sulfate ion (bisulfate ion)	HSO_4^-
Hydroxide ion	OH^-		
Hypochlorite ion	OCl^-	Sulfite ion	SO_3^{2-}
Nitrate ion	NO_3^-		

Note in Table 4.4 that several pairs of ions—CO_3^{2-} and HCO_3^-, for example—are related by the presence or absence of a hydrogen ion, H^+. In such instances, the ion with the hydrogen is sometimes named using the prefix $bi-$. Thus, CO_3^{2-} is the carbonate ion and HCO_3^- is the bicarbonate ion; similarly, SO_4^{2-} is the sulfate ion and HSO_4^- is the bisulfate ion.

■ **PROBLEM 4.18**
Name the following ions:
(a) NO_3^- (b) CN^- (c) OH^- (d) HPO_4^{2-}

■ **PROBLEM 4.19**
Write the formulas of the following ions:
(a) Bicarbonate ion   (b) Ammonium ion
(c) Phosphate ion   (d) Permanganate ion

4.10 Formulas of Ionic Compounds

Since all chemical compounds are neutral, it's relatively easy to figure out the formulas of ionic compounds. Once the ions are identified, all that's needed is to decide how many ions of each type give a total charge of zero. Thus, the chemical formula of an ionic compound tells the ratio of anions and cations.

Application Biologically Important Ions

The human body requires many different ions for proper functioning. Several of these ions, such as Ca^{2+}, Mg^{2+}, and HPO_4^{2-}, are used as structural materials in bones and teeth in addition to having other essential functions. Although 99% of Ca^{2+} is contained in bones and teeth, small amounts in body fluids play a vital role in transmission of nerve impulses. Other ions, including essential transition metal ions such as Fe^{2+}, are required for specific chemical reactions in the body. And still others, such as K^+, Na^+, and Cl^-, are present in fluids throughout the body.

Solutions containing ions must have overall neutrality, and several polyatomic anions, especially HCO_3^- and HPO_4^{2-}, are present in body fluids where they help balance the cation charges. Some of the most important ions and their functions are shown in the accompanying table.

See Additional Problems 4.83, 4.84, and 4.85 at the end of the chapter.

Some Biologically Important Ions

Ion	Location	Function	Dietary Source
Ca^{2+}	Outside cell; 99% of Ca^{2+} is in bones and teeth as $Ca_3(PO_4)_2$ and $CaCO_3$	Bone and tooth structure; necessary for blood clotting, muscle contraction, and transmission of nerve impulses	Milk, whole grains leafy vegetables
Fe^{2+}	Blood hemoglobin	Transports oxygen from lungs to cells	Liver, red meat, leafy green vegetables
K^+	Fluids inside cells	Maintains ion concentrations in cells; regulates insulin release and heartbeat	Milk, oranges, bananas, meat
Na^+	Fluids outside cells	Protects against fluid loss; necessary for muscle contraction and transmission of nerve impulses	Table salt, seafood
Mg^{2+}	Fluids inside cells; bone	Present in many enzymes; needed for energy generation and muscle contraction	Leafy green plants seafood, nuts
Cl^-	Fluids outside cells; gastric juice	Maintains fluid balance in cells; helps transfer CO_2 from blood to lungs	Table salt, seafood
HCO_3^-	Fluids outside cells	Controls acid–base balance in blood	By-product of food metabolism
HPO_4^{2-}	Fluids inside cells; bones and teeth	Controls acid–base balance in cells	Fish, poultry, milk

If the ions have the same charge, only one of each ion is needed:

$$K^+ \quad \text{and} \quad F^- \quad \text{form} \quad KF$$

$$Ca^{2+} \quad \text{and} \quad SO_4^{2-} \quad \text{form} \quad CaSO_4$$

If the ions have different charges, however, there must be unequal numbers of anions and cations to have a net charge of zero. When potassium and oxygen combine, for example, it takes two K^+ ions to balance the -2 charge of the O^{2-} ion:

$$2\,K^+ \quad \text{and} \quad O^{2-} \quad \text{form} \quad K_2O$$

The situation is reversed for the Ca^{2+} ion and the Cl^- ion; that is, there is one cation for every two anions:

$$Ca^{2+} \quad \text{and} \quad 2\,Cl^- \quad \text{form} \quad CaCl_2$$

It sometimes helps when writing the formulas for an ionic compound to remember that, when the two ions have different charges, the number of one ion is equal to the charge on the other ion. In magnesium phosphate, for example, the charge on the magnesium ion is $+2$ and the charge on the polyatomic phosphate ion is -3. Thus, there must be three magnesium ions with a total charge of $3 \times (+2) = +6$ and two phosphate ions with a total charge of $2 \times (-3) = -6$ for overall neutrality:

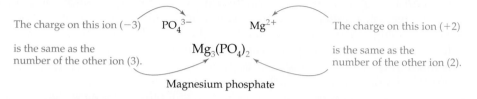

The charge on this ion (-3) PO_4^{3-} Mg^{2+} The charge on this ion $(+2)$

is the same as the number of the other ion (3). $Mg_3(PO_4)_2$ is the same as the number of the other ion (2).

Magnesium phosphate

The formula of an ionic compound shows the lowest possible ratio of atoms in the compound and is thus known as a *simplest formula*. Because there is no such thing as a single neutral *particle* of an ionic compound, however, we use the term **formula unit** to identify the smallest possible neutral *unit* (Figure 4.5). For NaCl, the formula unit is one Na^+ ion and one Cl^- ion; for K_2SO_4, the formula unit is two K^+ ions and one SO_4^{2-} ion; for CaF_2, the formula unit is one Ca^{2+} ion and two F^- ions; and so on.

Formula unit The formula that identifies the smallest neutral unit of a compound.

FIGURE 4.5 Formula units of ▶ ionic compounds. The sum of charges on the ions in a formula unit equals zero

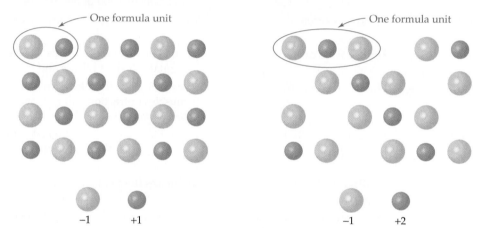

One formula unit   One formula unit

-1 $+1$ -1 $+2$

Once the numbers and kinds of ions in a compound are known, the formula is written using the following rules:

- List the cation first and the anion second; for example, NaCl rather than ClNa.
- Don't write the charges of the ions; for example, KF rather than K^+F^-.
- Use parentheses around a polyatomic ion formula if it has a subscript; for example, $Al_2(SO_4)_3$ rather than Al_2SO_{43}.

■ **WORKED EXAMPLE 4.7**

Write the formula for the compound formed by calcium ions and nitrate ions.

SOLUTION   The two ions are Ca^{2+} and NO_3^-. Two nitrate ions, each with a -1 charge, will balance the $+2$ charge of the calcium ion.

$$Ca^{2+} \qquad \text{Charge} = 1 \times (+2) = +2$$
$$2\,NO_3^- \qquad \text{Charge} = 2 \times (-1) = -2$$

Since there are two nitrate ions, the nitrate formula must be enclosed in parentheses:

$$Ca(NO_3)_2 \qquad \text{Calcium nitrate}$$

■ **PROBLEM 4.20**

Write the formulas for the ionic compounds formed by silver(I) ion with each of the following:

(a) Iodide ion   **(b)** Oxide ion   **(c)** Phosphate ion

■ **PROBLEM 4.21**

Write the formulas for the ionic compounds formed by sulfate ion with each of the following:

(a) Sodium ion   **(b)** Iron(II) ion   **(c)** Chromium(III) ion

■ **PROBLEM 4.22**

The ionic compound containing ammonium ion and carbonate ion gives off the odor of ammonia, a property put to use in smelling salts for reviving someone who has fainted. Write the formula for this compound.

■ **PROBLEM 4.23**

An *astringent* is a compound that causes proteins in blood, sweat, and other body fluids to coagulate, a property put to use in deodorants. Two safe and effective astringents are the ionic compounds of aluminum with sulfate ion and with acetate ion. Write the formulas of both.

■ **KEY CONCEPT PROBLEM 4.24**

Two ionic compounds are represented on the following drawing—red with red and blue with blue. Give a likely formula for each compound.

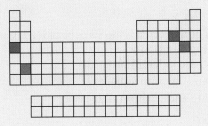

4.11 Naming Ionic Compounds

Ionic compounds are named by citing first the cation and then the anion, with a space between words. To name compounds of transition metals, which often form more than one ion, either the old (-ous, -ic) or the new (Roman numerals) system described in Section 4.8 must be used. Thus, $FeCl_2$ is called iron(II) chloride (or ferrous chloride), and $FeCl_3$ is called iron(III) chloride (or ferric chloride). Ions of elements that form only one type of ion, such as Na^+ and Ca^{2+}, do not need Roman numerals. Table 4.5 lists some common ionic compounds and their uses.

Ionic Compounds

■ **WORKED EXAMPLE 4.8**

Magnesium carbonate is used as an ingredient in Bufferin tablets. Write its formula.

SOLUTION   Look at the cation and the anion parts of the name separately: Magnesium, a group 2A element, forms the doubly positive Mg^{2+} cation; carbonate anion is doubly negative, CO_3^{2-}. Because magnesium carbonate must be neutral, its formula is $MgCO_3$.

TABLE 4.5 Some Common Ionic Compounds and Their Applications

Chemical Name (Common Name)	Formula	Applications
Ammonium carbonate	$(NH_4)_2CO_3$	Smelling salts
Calcium hydroxide (hydrated lime)	$Ca(OH)_2$	Mortar, plaster, whitewash
Calcium oxide (lime)	CaO	Lawn treatment, industrial chemical
Lithium carbonate ("lithium")	Li_2CO_3	Treatment of manic depression
Magnesium hydroxide (milk of magnesia)	$Mg(OH)_2$	Antacid
Magnesium sulfate (Epsom salts)	$MgSO_4$	Laxative, anticonvulsant
Potassium permanganate	$KMnO_4$	Antiseptic, disinfectant*
Potassium nitrate (saltpeter)	KNO_3	Fireworks, matches, and desensitizer for teeth
Silver nitrate	$AgNO_3$	Antiseptic, germicide
Sodium bicarbonate (baking soda)	$NaHCO_3$	Baking powder, antacid, mouthwash, deodorizer
Sodium hypochlorite	$NaOCl$	Disinfectant; active ingredient in household bleach
Zinc oxide	ZnO	Skin protection, in calamine lotion

*An antiseptic kills or inhibits growth of harmful microorganisms on the skin or in the body; a disinfectant kills harmful microorganisms but is generally not used on living tissue.

■ **WORKED EXAMPLE 4.9**

Sodium and calcium both form a wide variety of ionic compounds. Write formulas for the following compounds:
(a) Sodium bromide and calcium bromide
(b) Sodium sulfide and calcium sulfide
(c) Sodium phosphate and calcium phosphate

SOLUTION
(a) $NaBr$ and $CaBr_2$ **(b)** Na_2S and CaS **(c)** Na_3PO_4 and $Ca_3(PO_4)_2$

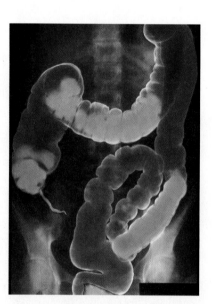

▲ A barium sulfate "cocktail" is given to patients prior to an X ray to help visualize the gastrointestinal tract.

■ **WORKED EXAMPLE 4.10**

Name the following compounds, using Roman numerals to indicate the charges on the cations where necessary:
(a) KF **(b)** $MgCl_2$ **(c)** $AuCl_3$ **(d)** Fe_2O_3

SOLUTION
(a) Potassium fluoride. No Roman numeral is necessary because a group 1A metal forms only one cation.
(b) Magnesium chloride. No Roman numeral is necessary because magnesium (group 2A) forms only Mg^{2+}.
(c) Gold(III) chloride. The Roman numeral is necessary to specify the +3 charge on gold, a transition element that can form other ions.
(d) Iron(III) oxide. Because the three oxide anions (O^{2-}) have a total negative charge of −6, the two iron cations must have a total charge of +6. Thus, each is Fe(III).

■ **PROBLEM 4.25**

Barium sulfate is an ionic compound swallowed by patients before having an X ray of their gastrointestinal tract. Write its formula.

■ **PROBLEM 4.26**

The compound Ag$_2$S is responsible for much of the tarnish found on silverware. Name this compound, and give the charge on the silver ion.

■ **PROBLEM 4.27**

Name the following compounds:

(a) CuO   (b) Ca(CN)$_2$ (c) NaNO$_3$

(d) Cu$_2$SO$_4$ (e) Li$_3$PO$_4$ (f) NH$_4$Cl

■ **PROBLEM 4.28**

Write formulas for the following compounds:

(a) Barium hydroxide   (b) Copper(II) carbonate

(c) Magnesium bicarbonate   (d) Cuprous fluoride

(e) Ferric sulfate   (f) Ferrous nitrate

4.12 H$^+$ and OH$^-$ Ions: An Introduction to Acids and Bases

Two of the most important ions we'll be discussing in the remainder of this book are the hydrogen cation (H$^+$) and the hydroxide anion (OH$^-$). Since a hydrogen *atom* contains one proton and one electron, a hydrogen *cation* is simply a proton. A hydroxide anion, by contrast, is a polyatomic ion in which an oxygen atom is covalently bonded to a hydrogen atom. Although much of Chapter 10 is devoted to the chemistry of H$^+$ and OH$^-$ ions, it's worthwhile taking a preliminary look now.

The importance of the H$^+$ cation and the OH$^-$ anion is that they are fundamental to the concepts of *acids* and *bases*. In fact, one definition of an **acid** is a substance that provides H$^+$ ions when dissolved in water, and one definition of a **base** is a substance that provides OH$^-$ ions when dissolved in water.

    Acid   A substance that provides H$^+$ ions in water; for example, **H**Cl, **H**NO$_3$, **H$_2$**SO$_4$, **H$_3$**PO$_4$

    Base   A substance that provides OH$^-$ ions in water; for example, Na**OH**, K**OH**, Ba(**OH**)$_2$

Hydrochloric acid (HCl), nitric acid (HNO$_3$), sulfuric acid (H$_2$SO$_4$), and phosphoric acid (H$_3$PO$_4$) are among the most common acids. When any of these substances is dissolved in water, H$^+$ ions are formed along with the corresponding anion (Table 4.6).

TABLE 4.6   **Some Common Acids and Ions Derived from Them**

Acids		Ions	
Acetic acid	CH$_3$COOH	Acetate ion	*CH$_3$COO$^-$
Carbonic acid	H$_2$CO$_3$	Hydrogen carbonate ion (bicarbonate ion)	HCO$_3^-$
Hydrochloric acid	HCl	Chloride ion	Cl$^-$
Nitric acid	HNO$_3$	Nitrate ion	NO$_3^-$
Nitrous acid	HNO$_2$	Nitrite ion	NO$_2^-$
Phosphoric acid	H$_3$PO$_4$	Dihydrogen phosphate ion	H$_2$PO$_4^-$
		Hydrogen phosphate ion	HPO$_4^{2-}$
		Phosphate ion	PO$_4^{3-}$
Sulfuric acid	H$_2$SO$_4$	Hydrogen sulfate ion	HSO$_4^-$
		Sulfate ion	SO$_4^{2-}$

*Sometimes written C$_2$H$_3$O$_2^-$ or as CH$_3$CO$_2^-$

Different acids can provide different numbers of H^+ ions. Hydrochloric acid, for instance, provides one H^+ ion; sulfuric acid can provide two H^+ ions; and phosphoric acid can provide three H^+ ions.

Hydrochloric acid: $\qquad HCl \xrightarrow{\text{Dissolve in H}_2\text{O}} H^+ + Cl^-$

Sulfuric acid: $\qquad H_2SO_4 \xrightarrow{\text{Dissolve in H}_2\text{O}} H^+ + HSO_4^-$

$\qquad\qquad\qquad\quad HSO_4^- \xrightarrow{\text{Dissolve in H}_2\text{O}} H^+ + SO_4^{2-}$

Phosphoric acid: $\qquad H_3PO_4 \xrightarrow{\text{Dissolve in H}_2\text{O}} H^+ + H_2PO_4^-$

$\qquad\qquad\qquad\quad H_2PO_4^- \xrightarrow{\text{Dissolve in H}_2\text{O}} H^+ + HPO_4^{2-}$

$\qquad\qquad\qquad\quad HPO_4^{2-} \xrightarrow{\text{Dissolve in H}_2\text{O}} H^+ + PO_4^{3-}$

Sodium hydroxide (NaOH; also known as *lye* or *caustic soda*), potassium hydroxide (KOH; also known as *caustic potash*), and barium hydroxide [Ba(OH)$_2$] are examples of bases. When any of these compounds dissolves in water, OH^- anions go into solution along with the corresponding metal cation. Sodium hydroxide and potassium hydroxide provide one OH^- ion each; barium hydroxide provides two OH^- ions, as indicated by its formula, $Ba(OH)_2$.

Sodium hydroxide: $\qquad NaOH \xrightarrow{\text{Dissolve in H}_2\text{O}} Na^+ + OH^-$

Potassium hydroxide: $\qquad KOH \xrightarrow{\text{Dissolve in H}_2\text{O}} K^+ + OH^-$

Barium hydroxide: $\qquad Ba(OH)_2 \xrightarrow{\text{Dissolve in H}_2\text{O}} Ba^{2+} + 2\,OH^-$

■ **PROBLEM 4.29**

Which of the following compounds are acids, and which are bases? Explain.

(a) HF (b) $Ca(OH)_2$ (c) LiOH (d) HCN

■ **KEY CONCEPT PROBLEM 4.30**

One of the following pictures represents a solution of HCl, and one represents a solution of H_2SO_4. Which is which?

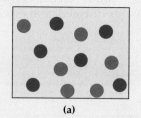

(a)

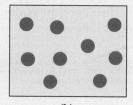

(b)

Application Osteoporosis

Bone consists primarily of two components, one mineral and one organic. About 70% of bone is the ionic compound *hydroxyapatite*, $Ca_{10}(PO_4)_6(OH)_2$, called the *trabecular*, or spongy, bone. This mineral component is intermingled in a complex matrix with about 30% by mass of fibers of the protein *collagen*, called the *cortical*, or compact, bone. Hydroxyapatite gives bone its hardness and strength, while collagen fibers add flexibility and resistance to breaking.

Total bone mass in the body increases from birth until reaching a maximum in the mid 30s. By the early 40s, however, an age-related decline in bone mass begins to occur in both sexes. Should this thinning of bones become too great and the bones become too porous and brittle, a clinical condition called *osteoporosis* can result. Osteoporosis is, in fact, the most common of all bone diseases, affecting approximately 25 million people in the U.S. Approximately 1.5 million bone fractures each year are caused by osteoporosis at an estimated health-care cost of $14 billion.

Although both sexes are affected by osteoporosis, the condition is particularly common in post-menopausal women, who undergo cortical bone loss at a rate of 2–3% per year over and above that of the normal age-related loss. The cumulative lifetime bone loss, in fact, may approach 40–50% in women versus 20–30% in men. It has been estimated that half of all women over age 50 will have an osteoporosis-related bone fracture at some point in their life. Other risk factors in addition to sex include being thin, being sedentary, having a family history of osteoporosis, smoking, and having a diet low in calcium.

No cure yet exists for osteoporosis, but treatment for its prevention and management includes estrogen replacement therapy for post-menopausal women as well as several approved medications called *bisphosphonates* to prevent further bone loss. Calcium supplements are also recommended, as is appropriate weight-bearing exercise. In addition, treatment with sodium fluoride is under active investigation and shows considerable promise. Fluoride ion reacts with hydroxyapatite to give *fluorapatite*, in which OH^- ions are replaced by F^-, increasing both bone strength and density.

$$Ca_{10}(PO_4)_6(OH)_2 + 2\ F^- \longrightarrow Ca_{10}(PO_4)_6F_2$$

Hydroxyapatite   Fluorapatite

See Additional Problems 4.86 and 4.87 at the end of the chapter.

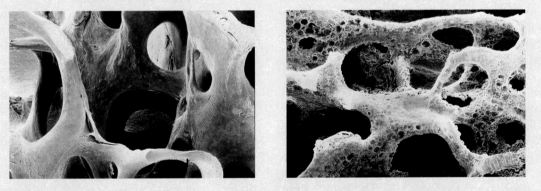

▲ Normal bone is strong and dense; a bone affected by osteoporosis is weak and spongy in appearance.

My name is Judith Fitzpatrick. I'm a physical thera-pist. And here's one of the reasons I love what I do. One of the first rotations that I did when I was starting out was in neonatal pediatrics, which totally terrified me because the preemies are so tiny. On my very first day in pediatrics I got an emergency call from the neonatal Intensive Care Unit. They had a little preemie, maybe as big as a person's hand, with a collapsed lung from a mucus plug that they couldn't get out. The doctor greeted me at the door and said, "We've got to get this baby's chest cleared out and we're depending on you to do that."

I was petrified, because I had never really dealt with little ones—I'd been working with adults for the whole time. The techniques of respiratory physical therapy are designed to mechanically clear airways so that you get better oxygen perfusion. You use clapping, vibration, and suction—you actually put a catheter into the trachea and suction out mucus secretions, eliminating blockages so that the patient can breathe better. And you can do that with anybody from eight weeks premature to 100 years old. But with adults you can use both hands and clap very hard on their chests to make them cough. With the babies you have to tap with one index finger very, very lightly over the chest wall. Sometimes you use an electric toothbrush with a rubber stopper on the side to vibrate the chest to loosen the secretions. And the catheters are so thin—I was terrified that I might perforate his trachea. But I worked with him for maybe 20 minutes or so, con-stantly vibrating, suctioning, and then delivering oxygen so that his oxygen levels wouldn't get too low as I suc-tioned air out of his lungs. Finally the doctor came over and said, "Okay, we're going to bring in the X ray equip-ment now to take a chest X ray and see how it is." About an hour or so later I called back down to the ICU to see how he was doing. The doctor himself got on the phone and said, "I really want to thank you. The baby's lung has expanded and you did a great job."

I've been in practice for over 20 years, and to this day I feel really very validated by that tiny patient!

My original degree was in biology, and I started out in physiology research, studying the medical effects of certain liver enzymes. After a year of doing the day-to-day, nitty gritty research work I felt that I wanted to be in a more personal field. I was spending all my time in the laboratory doing experiments, and I was very isolated. So I decided to go into physical therapy. I chose physical therapy primarily because I really wanted to do some-thing that was very personal and one on one.

My science background had already prepared me with all of the prerequisites for physical therapy school, so I got into a certificate program—a 12 month intensive program of concentrated physical therapy courses, with internships to add hands-on experience. The physical therapy courses involved forms of massage and applica-tion of electrical and heat modalities that are used in the treatment of a variety of conditions. We studied mea-surement techniques; evaluation of muscle function; mus-cle strength; joint range of motion; and kinesiology, which really is a combination of anatomy and physics as applied to movement.

Physiology was really what intrigued me. In physi-cal therapy, you're dealing with patients who are in pain—you're dealing with musculoskeletal systems that are dysfunctional—and unless you understand the phys-iological basis for what you are doing you're really noth-ing more than a technician. And that understanding comes from your basic knowledge of science, in which chemistry is fundamental.

Chemistry is the key to what happens physiologically in any system. Take the physiology of musculoskeletal performance, for example. You need to understand fatigue and how it is related to lactic acid metabolism so that when you're exercising a patient you know how far to go. You have to be very aware of fatigue and its metabolic consequences so that you don't reach a point where you have lactic acid buildup in the muscles, which can cause pain [Chapter 23].

An understanding of chemistry is particularly impor-tant when you use iontophoresis. This technique involves placing electrically charged drug molecules under an electrode of the same charge. Iontophoresis uses the prin-ciple that like charges repel one another to actually drive the medication across the skin, into the local circulatory system around the affected area. This technique enhances the passage of the drug through the skin and allows a deeper penetration than you would get with simple pas-sive transdermal administration, such as a drug-infused skin patch. You can actively transfer larger molecules this way—iontophoresis can deliver molecules that are 10 times larger than those you can use with the passive transdermal systems.

Another advantage of iontophoresis is that it deliv-ers a drug that hasn't already gone through the gastric system, so it's not broken down and metabolized in any way. You avoid some of the side effects that can be caused by metabolites of the drug. Also, you're administering a small dose in a very gradual way, so you don't have the physical trauma that occurs when you inject a large amount of fluid into a constricted area or a tendon. The after-effects can be much less painful and uncomfortable for the patient.

However, there can also be problems with ion-tophoresis. This is where the understanding of chemistry comes in. You have to know not only what medications will give you the treatment effect you need, but also whether they ionize or not. This is important, because if the medications don't ionize you're not going to be able to use iontophoresis for that condition. Once you know that the medication ionizes, you need to ascertain whether it's the positive or the negative ion that is therapeutically active, because that determines which electrode you place the drug under.

Then, because you're using a direct current, there is a buildup of charge at each of those electrodes. At one electrode you are producing hydrogen ions, which create an acid medium. At the other electrode you have a buildup of hydroxyl ions, which create a basic medium [Chapter 10]. You must use caution about where you put the electrodes and how the skin is prepared to reduce skin resistance beforehand. You should avoid potential problem sites that are susceptible to irritation, such as places that are particularly bony or that might have a pimple or a cut. If you're not cautious then your patient will suffer, because, he or she can have local skin irritation or can even develop an electrical burn. So you need to have an understanding of what's happening at your electrodes as well as what the medications are doing.

There are many lesser-known medications that can be administered in this way—not just the familiar steroids or anti-inflammatory medications that are common [Chapter 20], but other things that are used for other purposes. For example, you can administer an anesthetic like xylocaine via iontophoresis to obtain an analgesic effect for an area, or iodine compounds to soften scar tissue. You can also treat calcium deposits: calcific tendinitis or calcific bursitis. The calcium deposits in muscle are actually calcium carbonate, an insoluble salt that creates a hard mass. The physical therapist can use vinegar, which is essentially an acetic acid solution, to dissolve these deposits. The calcium carbonate reacts with the acetate ion of the acetic acid to form calcium acetate salt, which is soluble. The acetate ion displaces the carbonate ion, and you end up with a soluble compound [Chapter 9].

Right now I'm working on a medical condition, lymphedema, that involves the physiology of the circulatory system and fluid balances. Lymphedema is characterized by an accumulation in the tissues of fluid with a high concentration of dissolved protein. It's also associated with a deficient lymphatic flow of some kind. To understand what is happening in this condition you have to understand osmosis and diffusion across semipermeable membranes, which is really basic chemistry [Chapter 9]. How you treat it, both mechanically and pharmacologically, also involves chemistry. For instance, diuretics, which cause the body to excrete more water, may work in the short term with lymphedema. But because the site of action of diuretics is the kidney,

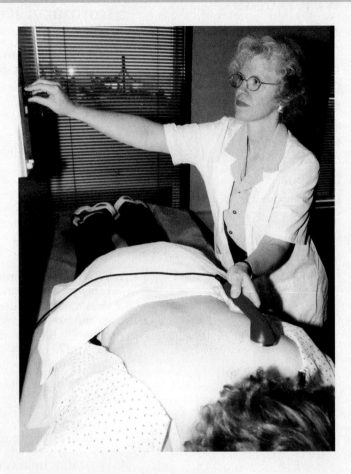

diuretic therapy doesn't actually address what's happening with the protein concentration in the tissue fluids. That high concentration of protein continues to draw more fluid from the vessels, and the deficient lymphatic system can't cope with it [Chapter 29]. And so the long-term effect of a diuretic on a lymphedema patient may actually be harmful instead of helpful.

People are beginning to realize that the current treatment of choice for lymphedema employs a mechanical approach—counter pressure and a type of massage that stimulates the lymphatic vessels. And even though I'm not thinking chemistry when I administer this type of therapy, in my preparation—learning about lymphedema and learning about why what I'm doing works—I rely on my chemistry. And I use it when I'm discussing a case with another medical professional. Any time you want to use a specific treatment, you must be able to explain to a physician why you're making your choice, and I think that's possible only if you really understand what's happening. Knowing the underlying chemistry helps me to do this.

Key Words

Acid, p. 89

Anion, p. 72

Base, p. 76

Cation, p. 72

Electron affinity, p. 73

Electron-dot symbol, p. 78

Formula unit, p. 86

Ion, p. 72

Ionic bond, p. 75

Ionic compound, p. 75

Ionic solid, p. 75

Ionization energy, p. 72

Octet rule, p. 76

Polyatomic ion, p. 84

Summary: Revisiting the Chapter Goals

1. **What is an ion, what is an ionic bond, and what are the general characteristics of ionic compounds?** Atoms are converted into *cations* by the loss of one or more electrons and into *anions* by the gain of one or more electrons. Ionic compounds are composed of cations and anions held together by *ionic bonds*, which result from the attraction between opposite electrical charges. Ionic compounds conduct electricity when dissolved in water, and they are generally crystalline solids with high melting points and high boiling points.

2. **What is the octet rule, and how does it apply to ions?** A valence-shell electron configuration of 8 electrons in filled s and p subshells leads to stability and lack of reactivity, as typified by the noble gases in group 8A. According to the *octet rule*, atoms of main group elements tend to form ions in which they have gained or lost the appropriate number of electrons to reach a noble gas configuration.

3. **What is the relationship between an element's position in the periodic table and the formation of its ion?** Periodic variations in *ionization energy*, the amount of energy that must be supplied to remove an electron from an atom, show that metals lose electrons more easily than nonmetals. As a result, metals usually form cations. Similar periodic variations in *electron affinity*, the amount of energy released on adding an electron to an atom, show that reactive nonmetals gain electrons more easily than metals. As a result, reactive nonmetals usually form anions. Metals from groups 1A and 2A form +1 and +2 cations, respectively. Transition metals form cations without an octet configuration and often have more than one possible charge. Oxygen and sulfur (group 6A) form −2 anions, and halogens (group 7A) form −1 anions. The nonmetals of groups 4A and 5A do not commonly form ions.

4. **What determines the chemical formula of an ionic compound?** Ionic compounds contain appropriate numbers of ions to maintain overall neutrality, thereby providing a means of determining their chemical formulas.

5. **How are ionic compounds named?** Cations have the same name as the metal they are derived from; monatomic anions have the name ending *-ide*. For metals that form more than one ion, a Roman numeral equal to the charge on the ion is added to the name of the cation. Alternatively, the ending *-ous* is added to the name of the cation with the lesser charge and the ending *-ic* is added to the name of the cation with the greater charge. To name an ionic compound, the cation name is given first, with the charge of the metal ion indicated if necessary, and the anion name is given second.

6. **What are acids and bases?** The hydrogen ion (H^+) and the hydroxide ion (OH^-) are among the most important ions in chemistry because they are fundamental to the idea of acids and bases. According to one common definition, an *acid* is a substance that yields H^+ ions when dissolved in water, and a *base* is a substance that yields OH^- ions when dissolved in water.

▪ Understanding Key Concepts

4.31 Where on the blank outline of the periodic table are the following elements found?

(a) Elements that commonly form only one type of cation

(b) Elements that commonly form anions

(c) Elements that can form more than one type of cation

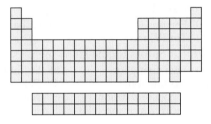

4.32 Where on the blank outline of the periodic table are the following elements found?

(a) Elements that commonly form + 2 ions

(b) Elements that commonly form −2 ions

(c) An element that forms a +3 ion

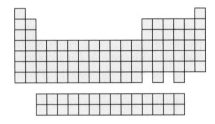

4.33 Which of the following drawings represents an Na atom, which a Ca^{2+} ion, and which an F^- ion?

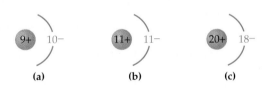

4.34 One of the following drawings represents an Na atom, and one represents an Na^+ ion. Tell which is which, and explain why there is a difference in size.

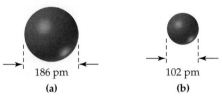

4.35 One of the following drawings represents a Cl atom, and one represents a Cl^- ion. Tell which is which, and explain why there is a difference in size.

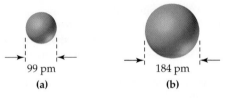

4.36 Three ionic compounds are represented on the following drawing—red with red, blue with blue, and green with green. Give a likely formula for each compound, and tell the name of each.

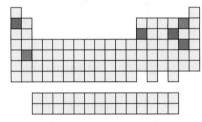

4.37 Each of the pictures (a)–(c) represents one of the following ionic compounds: PbO, $ZnBr_2$, $CrCl_3$. Which is which?

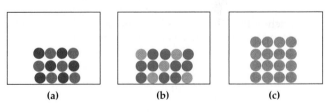

▪ Additional Problems

Ions and Ionic Bonding

4.38 Write electron-dot symbols for the following atoms:

(a) Beryllium   **(b)** Neon

(c) Strontium   **(d)** Aluminum

4.39 Write electron-dot symbols for the following atoms:

(a) Phosphorus   **(b)** Sulfur

(c) Chlorine   **(d)** Barium

4.40 Write equations for loss or gain of electrons by atoms that result in formation of the following ions from the corresponding atoms:

(a) Ca^{2+} **(b)** Au^+

(c) F^- **(d)** Cr^{3+}

4.41 Write symbols for the ions formed by the following:

(a) Gain of 3 electrons by nitrogen

(b) Loss of 1 electron by lithium

(c) Loss of 2 electrons by titanium

(d) Loss of 3 electrons by gold

4.42 Tell whether each of the following statements about ions is true or false:

(a) A cation is formed by addition of one or more electrons to an atom.

(b) Group 4A elements tend to lose 4 electrons to yield ions with a +4 charge.

(c) Group 4A elements tend to gain 4 electrons to yield ions with a −4 charge.

(d) The individual atoms in a polyatomic ion are held together by covalent bonds.

4.43 Tell whether each of the following statements about ionic solids is true or false:

    (a) Ions are randomly arranged in ionic solids.
    (b) All ions are the same size in ionic solids.
    (c) Ionic solids can often be shattered by a sharp blow.
    (d) Ionic solids have low boiling points.

Ions and the Octet Rule

4.44 What is the *octet rule*?

4.45 Why don't H and He obey the octet rule?

4.46 What is the charge of an ion that contains 34 protons and 36 electrons?

4.47 What is the charge of an ion that contains 11 protons and 10 electrons?

4.48 Identify the element X in the following ions, and tell which noble gas has the same electron configuration.

    (a) X^{2+}, a cation with 36 electrons
    (b) X^-, an anion with 36 electrons

4.49 Element Z forms an ion Z^{3+}, which contains 24 protons. What is the identity of Z, how many electrons does Z^{3+} have?

4.50 Write the electron configuration for each of the following ions:

    (a) Rb^+ **(b)** Br^- **(c)** S^{2-}
    (d) Ba^{2+} **(e)** Al^{3+}

4.51 Each of the following ions has a noble gas configuration. Identify the noble gas.

    (a) Ca^{2+} **(b)** Li^+ **(c)** O^{2-}
    (d) F^- **(e)** Mg^{2+}

Periodic Properties and Ion Formation

4.52 Looking only at the periodic table, tell which member of each of the following pairs of atoms has the larger ionization energy and thus loses an electron less easily:

    (a) Li and O   **(b)** Li and Cs
    (c) K and Zn   **(d)** Mg and N

4.53 Looking only at the periodic table, tell which member of each of the following pairs of atoms has the larger electron affinity and thus gains an electron more easily:

    (a) Na and O   **(b)** Cs and Br
    (c) Ca and N

4.54 Which of the following ions are likely to form? Explain.

    (a) Li^{2+} **(b)** K^-
    (c) Mn^{3+} **(d)** Zn^{4+}
    (e) Ne^+

4.55 Which of the following elements are likely to form more than one cation?

    (a) Magnesium   **(b)** Silicon
    (c) Manganese

4.56 Write the electron configurations of Cr^{2+} and Cr^{3+}.

4.57 Write the electron configurations of Co, Co^{2+}, and Co^{3+}.

4.58 Would you expect the ionization energy of Li^+ to be less than, greater than, or the same as the ionization energy of Li? Explain.

4.59. **(a)** Write equations for the loss of an electron by a K atom and the gain of an electron by a K^+ ion.
    (b) What is the relationship between the equations?
    (c) What is the relationship between the ionization energy of a K atom and the electron affinity of a K^+ ion?

Symbols, Formulas, and Names for Ions

4.60 Name the following ions:

    (a) S^{2-} **(b)** Sn^{2+} **(c)** Sr^{2+}
    (d) Mg^{2+} **(e)** Au^+

4.61 Name the following ions in both the old and the new systems:

    (a) Cu^+ **(b)** Fe^{3+} **(c)** Hg^{2+}

4.62 Write symbols for the following ions:

    (a) Selenide ion   **(b)** Oxide ion
    (c) Silver(I) ion

4.63 Write symbols for the following ions:

    (a) Ferrous ion   **(b)** Cobalt(II) ion
    (c) Tin(IV) ion

4.64 Write formulas for the following ions:

    (a) Hydroxide ion   **(b)** Sulfate ion
    (c) Acetate ion   **(d)** Permanganate ion
    (e) Hypochlorite ion   **(f)** Nitrate ion
    (g) Bicarbonate ion

4.65 Name the following ions:

    (a) NH_4^+ **(b)** CN^-
    (c) CO_3^{2-} **(d)** HSO_4^-

Names and Formulas for Ionic Compounds

4.66 Write formulas for the compounds formed by the nitrate ion with the following cations:

    (a) Aluminum   **(b)** Silver
    (c) Zinc   **(d)** Barium

4.67 Write formulas for the compounds formed by the carbonate ion with the following cations:

    (a) Strontium   **(b)** Fe(III)
    (c) Ammonium   **(d)** Sn(IV)

4.68 Write the formula for each of the following substances:

    (a) Sodium bicarbonate (baking soda)
    (b) Potassium nitrate (a backache remedy)
    (c) Calcium carbonate (an antacid)

4.69 Write the formula for each of the following compounds:

    (a) Calcium hypochlorite, used as a swimming pool disinfectant
    (b) Copper(II) sulfate, used to kill algae in swimming pools
    (c) Sodium phosphate, used in detergents to enhance cleaning action

4.70 Complete the following table by writing in the formula of the compound formed by each pair of ions:

	S^{2-}	Cl^-	PO_4^{3-}	CO_3^{2-}
Copper(II)	CuS			
Ca^{2+}				
NH_4^+				
Ferric ion				

4.71 Complete the following table by writing in the formula of the compound formed by each pair of ions:

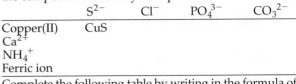

	O^{2-}	HSO_4^-	HPO_4^{2-}	$Cr_2O_7^{2-}$
Na^+	Na_2O			
Zn^{2+}				
NH_4^+				
Ferrous ion				

4.72 Name the following substances:
  (a) $MgCO_3$ **(b)** $Ca(CH_3CO_2)_2$
  (c) $AgCN$ **(d)** $Na_2Cr_2O_7$

4.73 Name the following substances:
  (a) $Fe(OH)_3$
  (b) KNO_2
  (c) $NaOCl$
  (d) $Ba(MnO_4)_2$

4.74 Which of the following structures is most likely to be correct for calcium phosphate?
  (a) Ca_2PO_4
  (b) $CaPO_4$
  (c) $Ca_2(PO_4)_3$
  (d) $Ca_3(PO_4)_2$

4.75 Fill in the missing information to give formulas for the following compounds:
  (a) $Na_?SO_4$ **(b)** $Ba_?(PO_4)_?$ **(c)** $Ga_?(SO_4)_?$

Acids and Bases

4.76 What is the difference between an acid and a base?

4.77 Identify each of the following substances as either an acid or a base:
  (a) HNO_2
  (b) HCN
  (c) $Ca(OH)_2$
  (d) CH_3CO_2H

4.78 Write equations to show how the substances listed in Problem 4.77 give ions when dissolved in water.

4.79 Name the anions that result when the acids in Problem 4.77 are dissolved in water.

Applications

4.80 What is the difference between a geologist's and a nutritionist's definition of a mineral? [*Minerals and Gems*]

4.81 How is salt obtained? [*Salt*]

4.82 What is the effect of normal dietary salt on most people? [*Salt*]

4.83 Where are most of the calcium ions found in the body? [*Biologically Important Ions*]

4.84 While excess sodium ion is considered hazardous, a certain amount is necessary for normal body functions. What is the purpose of sodium in the body? [*Biologically Important Ions*]

4.85 Before a person is allowed to donate blood, a drop of the blood is tested to be sure that it contains a sufficient amount of iron (men, 41 μg/dL; women, 38 μg/dL). Why is this required? [*Biologically Important Ions*]

4.86 Name each ion in hydroxyapatite, $Ca_{10}(PO_4)_6(OH)_2$, give its charge, and show that the formula represents a neutral compound. [*Osteoporosis*]

4.87 Sodium fluoride reacts with hydroxyapatite to give fluorapatite. What is the formula of fluorapatite? [*Osteoporosis*]

General Questions and Problems

4.88 Explain why the hydride ion, H^-, has a noble gas configuration.

4.89 The H^- ion (Problem 4.88) is stable, but the Li^- ion is not. Explain.

4.90 Many compounds containing a metal and a nonmetal are named by the Roman numeral system described in Section 4.8, even though they aren't ionic. Write the chemical formula for each of the following such compounds:
  (a) Chromium(VI) oxide
  (b) Vanadium(V) chloride
  (c) Manganese(IV) oxide
  (d) Molybdenum(IV) sulfide

4.91 The arsenate ion has the formula AsO_4^{3-}. Write the formula of lead(II) arsenate, used as an insecticide.

4.92 One commercially available calcium supplement contains calcium gluconate, a compound that is also used as an anticaking agent in instant coffee.
  (a) If this compound contains one calcium ion for every two gluconate ions, what is the charge on a gluconate ion?
  (b) What is the ratio of iron ions to gluconate ions in iron(III) gluconate, a commercial iron supplement?

4.93 The names given for the following compounds are incorrect. Write the correct name for each compound.
  (a) Cu_3PO_4, copper(III) phosphate
  (b) Na_2SO_4, sodium sulfide
  (c) MnO_2, manganese(II) oxide
  (d) $AuCl_3$, gold chloride
  (e) $Pb(CO_3)_2$, lead(II) acetate
  (f) Ni_2S_3, nickel(II) sulfide

4.94 The formulas given for the following compounds are incorrect. Write the correct formula for each compound.
  (a) Cobalt(II) cyanide, $CoCN_2$
  (b) Uranium(VI) oxide, UO_6
  (c) Tin(II) sulfate, $Ti(SO_4)_2$
  (d) Potassium phosphate, K_2PO_4
  (e) Calcium phosphide, CaP
  (f) Lithium hydrogen sulfate, Li_2HSO_4

4.95 How many protons, electrons, and neutrons are in each of the following ions?
  (a) $^{16}O^{2-}$
  (b) $^{89}Y^{3+}$
  (c) $^{133}Cs^+$
  (d) $^{81}Br^-$

4.96 Element X reacts with element Y to give a product containing X^{3+} ions and Y^{2-} ions.
  (a) Is element X likely to be a metal or a nonmetal?
  (b) Is element Y likely to be a metal or a nonmetal?
  (c) What is the formula of the product?
  (d) What groups of the periodic table are elements X and Y likely to be in?

5

Molecular Compounds

In chemistry, as in architecture, shape is crucially important.

W e saw in the previous chapter that ionic compounds are crystalline solids composed of positively and negatively charged ions. But what about the substances that aren't ionic? With the exception of table salt (NaCl), baking soda ($NaHCO_3$), lime for the garden (CaO), and a few others, most of the materials around us are *not* crystalline, brittle, high-melting ionic solids. We're much more likely to encounter gases like those in air, liquids like water, low-melting solids like butter, or flexible solids like plastics. All these materials are composed of *molecules* rather than ions, all contain *covalent* bonds rather than ionic bonds, and all consist primarily of nonmetal atoms rather than metals.

Among the questions we'll answer in this chapter about molecules and covalent bonds are these:

1. What is a covalent bond?
The goal: Be able to describe the nature of covalent bonds and how they are formed.

2. How does the octet rule apply to covalent bond formation?
The goal: Be able to use the octet rule to predict the numbers of covalent bonds formed by common main group elements.

3. How are molecular compounds represented?
The goal: Be able to interpret molecular formulas and draw Lewis structures for molecules.

4. What is the influence of valence-shell electrons on molecular shape?
The goal: Be able to use Lewis structures to predict molecular geometry.

5. When are bonds and molecules polar?
The goal: Be able to use electronegativity and molecular geometry to predict bond and molecular polarity.

6. What are the major differences between ionic and molecular compounds?
The goal: Be able to compare the structures, compositions, and properties of ionic and molecular compounds.

CONCEPTS TO REVIEW

The Periodic Table (Sections 3.4 and 3.5)
Electron Configurations (Sections 3.7 and 3.8)
The Octet Rule (Section 4.5)
Electron-Dot Symbols (Section 4.6)

5.1 Covalent Bonds

How can we describe the bonding in carbon dioxide, water, polyethylene, and the many millions of non-ionic compounds that make up our bodies and much of the world around us? Simply put, the bonds in such compounds are formed by the *sharing* of electrons between atoms rather than by the complete transfer of electrons from one atom to another. The resultant shared-electron bond is called a **covalent bond**, and the group of atoms held together by covalent bonds is called a **molecule**. A single molecule of water, for example, contains two hydrogen atoms and one oxygen atom covalently bonded to one another in a way we might visualize as shown below:

Covalent bond A bond formed by sharing electrons between atoms.

Molecule A group of atoms held together by covalent bonds.

 +  +

Two hydrogen atoms   One oxygen atom   One water molecule (H_2O)

According to the *octet rule* (Section 4.5), main group elements tend to undergo reactions that leave them with eight valence electrons (or two for hydrogen) and a noble gas electron configuration. (∞, p. 76) Although metals and reactive nonmetals reach an electron octet by gaining or losing an appropriate number of electrons, the nonmetals near the middle of the periodic table reach an electron octet by *sharing* an appropriate number of electrons.

As an example of how covalent bond formation occurs, let's look first at the bond between two hydrogen atoms in a hydrogen molecule, H_2. Recall that a hydrogen *atom* consists of a positively charged nucleus and a single, negatively charged $1s$ valence electron. When two hydrogen atoms come together, electrical interactions occur. Some of these interactions are repulsive—the two positively charged nuclei repel each other and the two negatively charged electrons repel each other. Other interactions, however, are attractive—each nucleus attracts both electrons and each electron attracts both nuclei (Figure 5.1). Because the attractive forces are stronger than the repulsive forces, a covalent bond is formed and the hydrogen atoms stay together.

FIGURE 5.1 A covalent H—H ▶ bond is the net result of attractive and repulsive forces. The nucleus–electron attractions (blue arrows) are greater than the nucleus–nucleus and electron–electron repulsions (red arrows), resulting in a net attractive force that holds the atoms together to form an H_2 molecule.

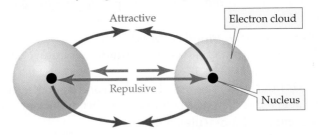

In essence, the electrons act as a kind of "glue" to bind the two nuclei together into an H_2 molecule. Both nuclei are simultaneously attracted to the same electrons and are therefore held together, much as two tug-of-war teams pulling on the same rope are held together.

The two teams are joined ▶ together because both are holding onto the same rope. In a similar way, two atoms are bonded together when both hold onto the same electrons.

Covalent bond formation in the H—H molecule can be visualized by imagining that the spherical $1s$ orbitals from the two individual atoms blend together and *overlap* to give an egg-shaped region in the H_2 molecule. The two electrons in the H—H covalent bond occupy the central region between the nuclei, giving both atoms a share in two valence electrons and the $1s^2$ electron configuration of the noble gas helium. For simplicity, the shared pair of electrons in a covalent bond is often represented as a line between atoms. Thus, the symbols H—H, H:H, and H_2 all represent a hydrogen molecule.

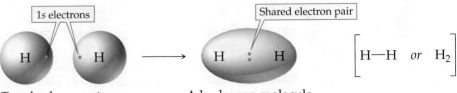

As you might imagine, the magnitudes of the various attractive and repulsive forces between nuclei and electrons in a covalent bond depend on how close the atoms are to each other. If the atoms are too far apart, the attractive forces are small and no bond exists. If the atoms are too close, the repulsive interaction between nuclei is so strong that it pushes the atoms apart. Thus, there is an optimum point where net attractive forces are maximized and where the H_2 molecule is most stable. This optimum distance between nuclei is called the **bond length** and is 74 pm (7.4×10^{-11} m) in the H_2 molecule. On a graph of energy versus internuclear distance, the bond length corresponds to the minimum-energy, most stable arrangement (Figure 5.2).

Bond length The optimum distance between nuclei in a covalent bond.

◀ **FIGURE 5.2** A graph of potential energy versus internuclear distance for the H_2 molecule. If the hydrogen atoms are too far apart, attractions are weak and no bonding occurs. If the atoms are too close, strong repulsions occur. When the atoms are optimally separated, the energy is at a minimum. The distance between nuclei at this minimum-energy point is called the *bond length*.

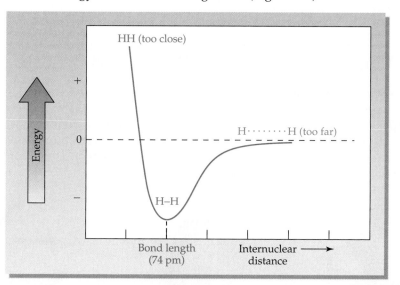

H_2 Bond Formation

As another example of covalent bond formation, look at the chlorine molecule, Cl_2. An individual chlorine atom has seven valence electrons and the valence-shell electron configuration $3s^2 3p^5$. Two of the three $3p$ orbitals are filled by two electrons each, while the third $3p$ orbital holds only one electron. When two chlorine atoms approach each other, the unpaired $3p$ electrons are shared by both atoms in a covalent bond. Each chlorine atom in the resultant Cl_2 molecule now "owns" six outer-shell electrons and "shares" two more, giving each a valence-shell octet like that of the noble gas argon. Such bond formation can be pictured as the overlap of $3p$ orbitals containing the single electrons, with resultant formation of a region of high electron density between the nuclei:

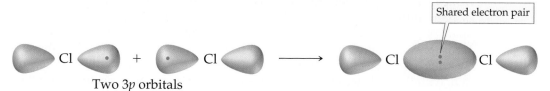

Shared electron pair

Two 3p orbitals

In addition to H_2 and Cl_2, five other elements always exist as *diatomic* (two-atom) molecules (Figure 5.3): Nitrogen (N_2) and oxygen (O_2) are colorless,

◀ **FIGURE 5.3** Diatomic elements in the periodic table.

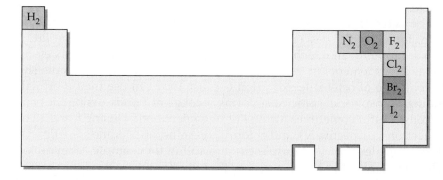

odorless, nontoxic gases present in air; fluorine (F_2) is a pale yellow, highly reactive gas; bromine (Br_2) is a dark red, toxic liquid; and iodine (I_2) is a violet, crystalline solid.

■ PROBLEM 5.1

Draw the iodine molecule using electron-dot symbols, and indicate the shared electron pair. What noble gas configuration do the iodine atoms have in an I_2 molecule?

5.2 Covalent Bonds and the Periodic Table

Molecular compound   A compound that consists of molecules rather than ions.

Covalent bonds can form between unlike atoms as well as between like atoms, making possible a vast number of **molecular compounds**. Water molecules, for example, consist of two hydrogen atoms joined by covalent bonds to an oxygen atom, H_2O; ammonia molecules consist of three hydrogen atoms covalently bonded to a nitrogen atom, NH_3; and methane molecules consist of four hydrogen atoms covalently bonded to a carbon atom, CH_4.

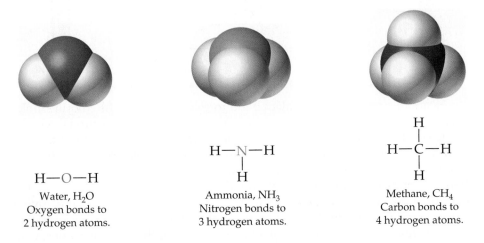

H—O—H

Water, H_2O
Oxygen bonds to
2 hydrogen atoms.

H—N—H
    |
    H

Ammonia, NH_3
Nitrogen bonds to
3 hydrogen atoms.

    H
    |
H—C—H
    |
    H

Methane, CH_4
Carbon bonds to
4 hydrogen atoms.

Note that in all these examples, each atom shares enough electrons to achieve a noble gas configuration: two electrons for hydrogen, and octets for oxygen, nitrogen, and carbon. Hydrogen, with one valence electron, needs a share of one more electron to achieve the helium configuration ($1s^2$) and thus forms one covalent bond. Oxygen, with six valence electrons, needs a share of two more electrons to reach an octet and thus forms two covalent bonds. Nitrogen, with five valence electrons, needs a share of three more electrons and thus forms three covalent bonds. Carbon, with four valence electrons, needs a share of four more electrons and thus forms four covalent bonds. Figure 5.4 summarizes the number of covalent bonds typically formed by common main group elements.

The octet "rule" is a useful guideline, but it has numerous exceptions. Boron, for example, has only three valence electrons it can share and thus forms compounds in which it has only three covalent bonds and six electrons, such as BF_3. Elements in the third row or below in the periodic table show a different type of octet rule exception because they can use their d orbitals for bonding. Phosphorus sometimes forms 5 covalent bonds (using 10 bonding electrons), sulfur sometimes forms 4 or 6 covalent bonds (using 8 and 12 bonding electrons, respectively), and chlorine, bromine, and iodine sometimes form 3, 5, or 7 covalent bonds. Phosphorus and sulfur, for example, form molecules such as PCl_5, SF_4, and SF_6.

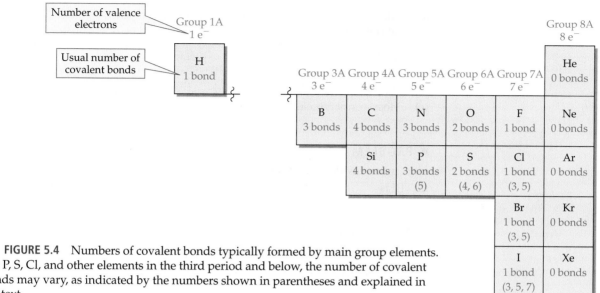

Number of valence electrons → Group 1A 1 e⁻								Group 8A 8 e⁻

Usual number of covalent bonds →

| H
1 bond | | | | | | | | He
0 bonds |

		Group 3A 3 e⁻	Group 4A 4 e⁻	Group 5A 5 e⁻	Group 6A 6 e⁻	Group 7A 7 e⁻	

B 3 bonds	C 4 bonds	N 3 bonds	O 2 bonds	F 1 bond	Ne 0 bonds
	Si 4 bonds	P 3 bonds (5)	S 2 bonds (4, 6)	Cl 1 bond (3, 5)	Ar 0 bonds
				Br 1 bond (3, 5)	Kr 0 bonds
				I 1 bond (3, 5, 7)	Xe 0 bonds

▲ **FIGURE 5.4** Numbers of covalent bonds typically formed by main group elements. For P, S, Cl, and other elements in the third period and below, the number of covalent bonds may vary, as indicated by the numbers shown in parentheses and explained in the text.

BF₃
Boron trifluoride
(6 valence electrons on B)

PCl₅
Phosphorus pentachloride
(10 valence electrons on P)

SF₆
Sulfur hexafluoride
(12 valence electrons on S)

■ **WORKED EXAMPLE 5.1**

Look at Figure 5.4 and tell whether the following molecules are likely to exist:

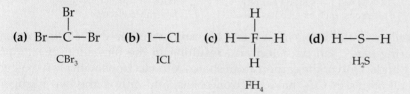

(a) $Br-C-Br$ with Br above and Br below the C — CBr_3 (b) $I-Cl$ — ICl (c) $H-F-H$ with H above and H below — FH_4 (d) $H-S-H$ — H_2S

ANALYSIS Count the number of covalent bonds formed by each element and see if the numbers correspond to those shown in Figure 5.4.

SOLUTION
(a) No. Carbon needs four covalent bonds but has only three in CBr_3.
(b) Yes. Both iodine and chlorine have one covalent bond in ICl.
(c) No. Fluorine can't form more than one covalent bond because it is in the second period and does not have valence d orbitals to use for bonding.
(d) Yes. Sulfur, which is in group 6A like oxygen, often forms two covalent bonds.

■ **WORKED EXAMPLE 5.2**

What are likely formulas for the following molecules?
(a) $SiH_2Cl_?$ (b) $AlBr_?$ (c) $PBr_?$

ANALYSIS The numbers of covalent bonds formed by each element should be those shown in Figure 5.4.

SOLUTION
(a) Silicon typically forms 4 bonds: SiH_2Cl_2
(b) Aluminum typically forms 3 bonds: $AlBr_3$
(c) Phosphorus typically forms 3 bonds: PBr_3

■ **PROBLEM 5.2**
How many covalent bonds are formed by each atom in the following molecules?
(a) PH_3 (b) H_2Se (c) HCl   (d) SiF_4

■ **PROBLEM 5.3**
Lead forms both ionic and molecular compounds. Using Figure 5.4, predict whether a molecular compound containing lead and chlorine is more likely to be $PbCl_4$ or $PbCl_5$.

■ **PROBLEM 5.4**
What are likely formulas for the following molecules?
(a) $CH_2Cl_?$ (b) $BH_?$ (c) $NI_?$ (d) $SiCl_?$

5.3 Multiple Covalent Bonds

The bonding in some molecules can't be explained by the sharing of only two electrons between atoms. For example, the carbon and oxygen atoms in carbon dioxide (CO_2) and the nitrogen atoms in the N_2 molecule can't have electron octets if only two electrons are shared.

$$\cdot \ddot{O} \cdot + \cdot \dot{C} \cdot + \cdot \ddot{O} \cdot$$

$$\cdot \ddot{O} : \dot{C} : \ddot{O} \cdot$$

UNSTABLE—Carbon has only 6 electrons; each oxygen has only 7.

$$\cdot \ddot{N} \cdot + \cdot \ddot{N} \cdot$$

$$\cdot \ddot{N} : \ddot{N} \cdot$$

UNSTABLE—Each nitrogen has only 6 electrons.

The only way the atoms in CO_2 and N_2 can have outer-shell electron octets is by sharing *more* than two electrons, resulting in the formation of *multiple* covalent bonds. Only if the carbon atom shares four electrons with each oxygen atom do all atoms in CO_2 have electron octets, and only if the two nitrogen atoms share six electrons do both have electron octets. A bond formed by sharing two electrons (one pair) is a **single bond**, a bond formed by sharing four electrons (two pairs) is a **double bond**, and a bond formed by sharing six electrons (three pairs) is a **triple bond**. Just as a single bond is represented by a single line between atoms, a double bond is represented by two lines between atoms and a triple bond is represented by three lines.

Single bond A covalent bond formed by sharing one electron pair.

Double bond A covalent bond formed by sharing two electron pairs.

Triple bond A covalent bond formed by sharing three electron pairs.

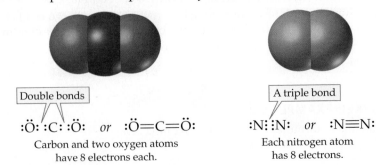

Double bonds

$:\ddot{O}::C::\ddot{O}:$ *or* $:\ddot{O}=C=\ddot{O}:$

Carbon and two oxygen atoms have 8 electrons each.

A triple bond

$:N:::N:$ *or* $:N\equiv N:$

Each nitrogen atom has 8 electrons.

Carbon, nitrogen, and oxygen are the elements most often present in multiple bonds. Carbon and nitrogen form both double and triple bonds; oxygen forms double bonds. Multiple covalent bonding is particularly common in *organic* molecules, which contain the element carbon. For example, ethylene, a simple compound used commercially to induce ripening in fruit, has the formula C_2H_4. The only way for the two carbon atoms to have octets is for them to share four electrons in a carbon—carbon double bond:

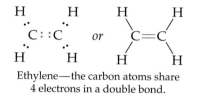   *or*

Ethylene—the carbon atoms share
4 electrons in a double bond.

As another example, acetylene, the gas used in welding, has the formula C_2H_2. Thus, the two acetylene carbons must share six electrons in a carbon—carbon triple bond:

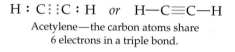   *or*

Acetylene—the carbon atoms share
6 electrons in a triple bond.

▲ Acetylene is frequently used for welding metal because it burns with such a hot flame.

Note that in compounds with multiple bonds like ethylene and acetylene, each carbon atom still forms a total of four covalent bonds.

■ **WORKED EXAMPLE 5.3**

The compound 1-butene contains a multiple bond. In the following representation, however, only the connections between atoms are shown; the multiple bond is not specifically indicated. Identify the position of the multiple bond.

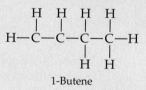

1-Butene

ANALYSIS Look for two adjacent atoms that appear to have fewer than the typical number of covalent bonds, and connect those atoms by a double or triple bond.

SOLUTION

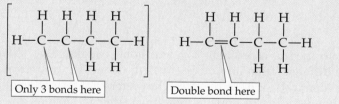

■ **PROBLEM 5.5**

Acetic acid, the organic constituent of vinegar, can be drawn using electron-dot symbols as shown below. How many outer-shell electrons are associated with each atom? Draw the structure using lines rather than dots to indicate covalent bonds.

■ **PROBLEM 5.6**

Identify the position of the double bond in methyl ethyl ketone, a common industrial solvent with the following connections among atoms.

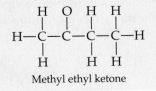

Methyl ethyl ketone

5.4 Coordinate Covalent Bonds

In the covalent bonds we've seen thus far, the shared electrons have come from different atoms. That is, the bonds have resulted from the overlap of two singly occupied valence orbitals, one from each atom. Sometimes, though, a bond is formed by the overlap of a filled orbital on one atom with a vacant orbital on another atom so that both electrons come from the *same* atom. The bond that results is called a **coordinate covalent bond**.

Coordinate covalent bond The covalent bond that forms when both electrons are donated by the same atom.

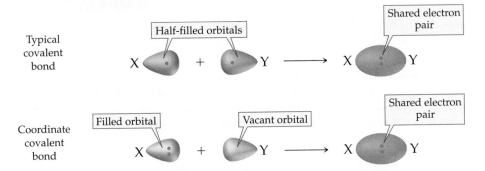

The ammonium ion, NH_4^+, is an example of a species with a coordinate covalent bond. When ammonia reacts in water solution with a hydrogen ion, H^+, the nitrogen atom donates two electrons from a filled valence orbital to form a coordinate covalent bond to the hydrogen ion, which has a vacant $1s$ orbital.

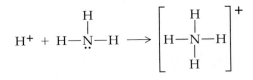

Molecular formula A formula that shows the numbers and kinds of atoms in one molecule of a compound.

Structural formula A molecular representation that shows the connections among atoms by using lines to represent covalent bonds.

Lewis structure A molecular representation that shows both the connections among atoms and the locations of lone-pair valence electrons.

Lone pair A pair of electrons that is not used for bonding.

Once formed, a coordinate covalent bond contains two shared electrons and is no different from any other covalent bond. All four covalent bonds in NH_4^+ are identical, for example. Note, however, that formation of a coordinate covalent bond results in unusual bonding patterns, such as an N atom with 4 covalent bonds rather than the usual 3.

∞∞ **LOOKING AHEAD**

An entire class of substances is based on the ability of transition metals to form coordinate covalent bonds with nonmetals. Called *coordination compounds*, many of these substances have important roles in living organisms. For example, toxic metals can be removed from the bloodstream by forming water-soluble coordination compounds. As another example, we'll see in Chapter 19 that essential metal ions are held in enzyme molecules by coordinate covalent bonds. ∞∞

Application — Carbon Monoxide: A Surprising Molecule

Carbon monoxide is a killer; everyone knows that. It's to blame for an estimated 3500 accidental deaths and suicides each year in the United States, and it is the number 1 cause of all deaths by poisoning. What most people don't know, however, is that our bodies can't function *without* carbon monoxide. A startling discovery made in 1992 has shown that carbon monoxide is a key chemical messenger in the body, used by cells to regulate critical metabolic processes.

The toxicity of carbon monoxide in moderate concentration is due to its ability to bind to hemoglobin molecules in the blood, thereby preventing the hemoglobin from carrying oxygen to tissues. At the same time, though, low concentrations of carbon monoxide are actually *produced* inside the cells of tissues throughout the body. These CO molecules can diffuse from one cell to another, where they stimulate production of a substance called *guanylyl cyclase*. Guanylyl cyclase, in turn, controls the production of another substance called *cyclic GMP*, which regulates many cellular functions.

Levels of carbon monoxide production are particularly high in certain regions of the brain, including those that respond to odors and those that control long-term memory. Although the details of how long-term memories are stored are not clear, it appears that a special kind of cell in the brain's hippocampus is signaled by transfer of a molecular messenger from a neighboring cell. The receiving cell responds back to the signaling cell by releasing carbon monoxide, which causes still more messenger molecules to be sent. After several rounds of this back-and-forth communication, the receiving cell undergoes some sort of change that serves as a memory.

Strong evidence to support this scenario comes from laboratory experiments with rat brains. When CO production is blocked, long-term memories are no longer stored, and those memories that previously existed are erased. When CO production is stimulated, however, memories are again laid down. It's too early for any practical uses to have come from these preliminary experiments, but the time may come when we will be taking "memory pills" to purposely increase the amount of carbon monoxide in our brains.

▲ Carbon monoxide adds to the danger of polluted air, but also functions as a chemical messenger in our bodies.

See Additional Problems 5.86 and 5.87 at the end of the chapter.

5.5 Molecular Formulas and Lewis Structures

Lone pair A pair of electrons that is not used for bonding.

Formulas such as H_2O, NH_3, and CH_4, which show the numbers and kinds of atoms in one molecule of a compound, are called **molecular formulas**. Though important, molecular formulas are limited in their use because they don't provide any information about how the atoms in a given molecule are connected.

Much more useful are **structural formulas**, which use lines between atoms to show how atoms are connected, and **Lewis structures**, which show both the connections among atoms and the placement of unshared valence electrons. In a water molecule, for instance, the oxygen atom shares two electron pairs in covalent bonds with two hydrogen atoms and has two other pairs of valence electrons that are not shared in bonds. Such unshared pairs of valence electrons are called **lone pairs**. In an ammonia molecule, three electron pairs are used in bonding and there is one lone pair. In methane, all four electron pairs are bonding.

▲ Ammonia is used as a fertilizer to supply nitrogen to growing plants.

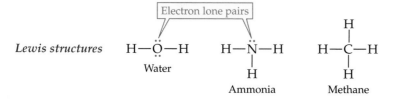

Lewis structures

Electron lone pairs

$$H-\ddot{O}-H$$
Water

$$H-\ddot{N}-H \atop H$$
Ammonia

$$H-\underset{H}{\overset{H}{C}}-H$$
Methane

107

Note how a molecular formula differs from an ionic formula, described previously in Section 4.10. A *molecular* formula gives the number of atoms that are combined in one molecule of a compound, while an *ionic* formula gives only a ratio of ions (Figure 5.5). The formula C_2H_4 for ethylene, for example, says that every ethylene molecule consists of two carbon atoms and four hydrogen atoms. The formula NaCl for sodium chloride, however, says only that there are equal numbers of Na^+ and Cl^- ions in the crystal; the formula says nothing about how the ions interact with one another.

FIGURE 5.5 The distinction ▶ between molecular and ionic compounds. In molecular compounds, the smallest particle is a molecule. In ionic compounds, the smallest particle is an ion.

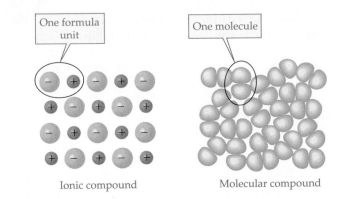

Ionic compound   Molecular compound

5.6 Drawing Lewis Structures

 Drawing Lewis Structures

To draw a Lewis structure, you first have to know the connections among atoms. Sometimes the connections are obvious. Water, for example, can only be H—O—H because only oxygen can be in the middle and form two covalent bonds. Other times, you'll have to be told how the atoms are connected.

Two approaches can be used for drawing Lewis structures once the connections are known. The first is particularly useful for organic molecules like those found in living organisms because common bonding patterns are followed by the atoms. The second approach is a more general, stepwise procedure that works for all molecules.

Lewis Structures for Molecules Containing C, N, O, X (Halogen), and H

As summarized previously in Figure 5.4, carbon, nitrogen, oxygen, halogen, and hydrogen atoms usually maintain consistent bonding patterns:

- H forms one covalent bond.
- C forms four covalent bonds and often bonds to other carbon atoms.
- N forms three covalent bonds and has one lone pair of electrons.
- O forms two covalent bonds and has two lone pairs of electrons.
- Halogens (X = F, Cl, Br, I) form one covalent bond and have three lone pairs of electrons.

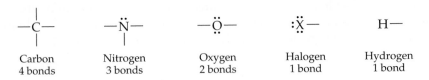

| Carbon 4 bonds | Nitrogen 3 bonds | Oxygen 2 bonds | Halogen 1 bond | Hydrogen 1 bond |

Relying on these common bonding patterns simplifies writing Lewis structures. In ethane (C_2H_6), a constituent of natural gas, for example, three of the four covalent bonds of each carbon atom are used in bonds to hydrogen, and the fourth is a carbon—carbon bond. There is no other arrangement in which all eight atoms can have their usual bonding patterns. In acetaldehyde (C_2H_4O), a substance used in manufacturing perfumes, dyes, and plastics, one carbon has three bonds to hydrogen, while the other has one bond to hydrogen and a double bond to oxygen.

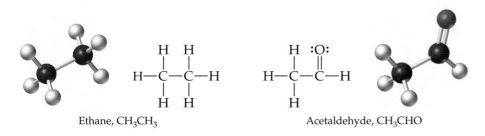

Ethane, CH_3CH_3 Acetaldehyde, CH_3CHO

Because Lewis structures are awkward for larger organic molecules, ethane is more frequently written as a **condensed structure** in which bonds are not specifically shown. In its condensed form, ethane is CH_3CH_3, meaning that each carbon atom has three hydrogen atoms bonded to it (CH_3) and the two CH_3 units are bonded to each other. In the same way, acetaldehyde can be written as CH_3CHO. Note that neither the lone-pair electrons nor the $C=O$ double bond in acetaldehyde are shown explicitly. You'll get a lot more practice with such condensed structures in later chapters.

(Many of the computer-generated pictures we'll be using from now on will be *ball-and-stick models* rather than the space-filling models used previously. Space-filling models are more realistic, but ball-and-stick models do a better job of showing connections and molecular geometry. All models, regardless of type, use a consistent color code in which C is dark gray or black, H is white or ivory, O is red, N is blue, S is yellow, P is dark blue, F is light green, Cl is greenish yellow, Br is brownish red, and I is purple.)

Condensed structure A molecular representation in which bonds are understood by the order in which they are written rather than specifically shown.

Space-filling

Ball-and-stick

■ **WORKED EXAMPLE 5.4**

Draw a Lewis structure for the toxic gas hydrogen cyanide, HCN. The atoms are connected in the order shown.

ANALYSIS With the atoms connected in the order $H—C—N$, the only way that carbon and nitrogen can have their usual numbers of covalent bonds, 4 for C and 3 for N, is for them to be joined by a triple bond.

SOLUTION Including the lone pair on the triply bonded nitrogen atom gives the structure

$$H—C≡N:$$

We can check the structure by noting that all 10 valence electrons (1 e⁻ from H, 4 e⁻ from C, and 5 e⁻ from N) have been used in four covalent bonds and one lone pair.

■ **WORKED EXAMPLE 5.5**

Draw a Lewis structure for vinyl chloride, C_2H_3Cl, a substance used in making poly(vinyl chloride), or PVC, plastic.

ANALYSIS Since H and Cl form only one bond each, the carbon atoms must be bonded to each other. With only four more atoms available, the carbon atoms can't have four covalent bonds each unless they are joined by a double bond.

SOLUTION Putting in the double bond, adding the other three H atoms and one Cl atom, and placing three lone pairs on the chlorine atom gives

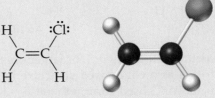

All 18 valence electrons are accounted for in six covalent bonds and three lone pairs.

■ **PROBLEM 5.7**

Methylamine, CH_5N, is responsible for the characteristic odor of decaying fish. Draw a Lewis structure of methylamine.

■ **PROBLEM 5.8**

Draw a Lewis structure of propane (C_3H_8), used to heat homes in rural areas.

■ **PROBLEM 5.9**

Formaldehyde, a substance used as a preservative for biological specimens, has the formula CH_2O. Draw the structural formula of formaldehyde.

■ **PROBLEM 5.10**

Add lone pairs where appropriate to the following structures:

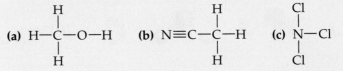

■ **KEY CONCEPT PROBLEM 5.11**

The following molecular model is a representation of methyl methacrylate, a starting material used to prepare Lucite plastic. Only the connections between atoms are shown; multiple bonds are not specifically indicated.

(a) What is the molecular formula of methyl methacrylate?

(b) Indicate the positions of the multiple bonds and lone pairs in methyl methacrylate.

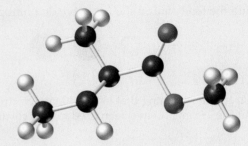

A General Method for Drawing Lewis Structures

A Lewis structure can be drawn for any molecule or ion by following a five-step procedure. Take PCl_3, for example, a substance in which three chlorine atoms surround the central phosphorus atom.

Step 1: Find the total number of valence electrons of all atoms in the molecule or ion. For a polyatomic ion, add one electron for each negative charge or subtract one electron for each positive charge. In PCl_3, for example, phosphorus (group 5A) has 5 valence electrons, and chlorine (group 7A) has 7 valence electrons, giving a total of 26:

$$P + (3 \times Cl) = PCl_3$$

$$5 \text{ e}^- + (3 \times 7 \text{ e}^-) = 26 \text{ e}^-$$

In OH^-, the total is 8 electrons (6 from oxygen, 1 from hydrogen, plus 1 for the negative charge). In NH_4^+, the total is 8 (5 from nitrogen, 1 from each of 4 hydrogens, minus 1 for the positive charge).

Step 2: Draw a line between each pair of connected atoms to represent the two electrons in a covalent bond. Remember that elements in the second row of the periodic table form the number of bonds discussed earlier in this section, while elements in the third row or lower can use more than 8 electrons and form more than the "usual" number of bonds (Figure 5.4). A particularly common pattern is that an atom in the third row or lower occurs as the central atom in a cluster. In PCl_3, for example, the phosphorus atom is in the center with the three chlorine atoms bonded to it:

$$\begin{array}{c} Cl \\ | \\ Cl-P-Cl \end{array}$$

Step 3: Add lone pairs so that each peripheral atom (except H) connected to the central atom gets an octet. In PCl_3, each Cl atom needs three lone pairs:

Step 4: Place all remaining electrons in lone pairs on the central atom. In PCl_3, we have used 24 of the 26 available electrons—6 electrons in three single bonds and 18 electrons in the three lone pairs on each chlorine atom. This leaves 2 electrons for one lone pair on phosphorus:

Step 5: If the central atom does not yet have an octet after all electrons have been assigned, take a lone pair from a neighboring atom and form a multiple bond to the central atom. In PCl_3, each atom has an octet, all 26 available electrons have been used, and the Lewis structure is finished. Worked Example 5.6 shows how to deal with a case where this fifth step is needed.

■ **WORKED EXAMPLE 5.6**

Draw a Lewis structure for sulfur dioxide, SO_2. The connections are O—S—O.

ANALYSIS   Follow the procedure outlined in the text.

SOLUTION

Step 1: The total number of valence electrons in SO_2 is 18, 6 from each atom:

$$S + (2 \times O) = SO_2$$

$$6 \text{ e}^- + (2 \times 6 \text{ e}^-) = 18 \text{ e}^-$$

Step 2: O—S—O   Two covalent bonds use 4 e$^-$.

Step 3: :Ö—S—Ö: Adding three lone pairs to each oxygen to give each an octet uses an additional 12 e$^-$.

Step 4: :Ö—S̈—Ö: The remaining 2 e$^-$ are placed on sulfur, but sulfur still does not have an octet.

Step 5: :Ö—S̈=Ö: Moving one lone pair from a neighboring oxygen atom to form a double bond with the central sulfur atom gives sulfur an octet. (It doesn't matter which side the S=O bond is written on.)

■ **PROBLEM 5.12**

Draw Lewis structures for the following:

(a) Phosgene, $COCl_2$, a poisonous gas
(b) Hypochlorite ion, OCl^-, present in many swimming pool chemicals
(c) Hydrogen peroxide, H_2O_2
(d) Sulfur dichloride, SCl_2

■ **PROBLEM 5.13**

Draw a Lewis structure for nitric acid, HNO_3. The nitrogen atom is in the center and the hydrogen atom is bonded to an oxygen atom.

5.7 The Shapes of Molecules

Look back at the computer-generated drawings of molecules in the previous section and you'll find that the molecules are shown with specific shapes. Acetylene is *linear*, water is *bent*, ammonia is *pyramid-shaped*, methane is *tetrahedral*, and ethylene is flat, or *planar*. What determines such shapes? Why, for example, are the three atoms in water connected at an angle of 104.5° rather than in a straight line? Like so many properties, molecular shapes are related to the numbers and locations of the valence electrons around atoms.

Molecular shapes can be predicted by noting how many bonds and electron pairs surround individual atoms and applying what is called the **valence-shell electron-pair repulsion (VSEPR) model**. The basic idea of the VSEPR model is that the constantly moving valence electrons in bonds and lone pairs make up negatively charged clouds of electrons, which electrically repel one another. The clouds therefore tend to keep as far apart as possible, causing molecules to assume specific shapes. There are three steps to applying the VSEPR model:

Valence-shell electron-pair repulsion (VSEPR) model   A method for predicting molecular shape by noting how many electron charge clouds surround atoms and assuming that the clouds orient as far away from one another as possible.

 VSEPR

Step 1: Draw a Lewis structure of the molecule, and identify the atom whose geometry is of interest.   In a simple molecule like PCl_3 or CO_2, this is usually the central atom.

Step 2: Count the number of electron charge clouds surrounding the atom of interest.   The number of charge clouds is simply the total number of lone pairs plus connections to other atoms. It doesn't matter whether a connection is a single bond or a multiple bond because we're interested only in the *number* of charge clouds, not in how many electrons each cloud contains. The carbon atom in carbon dioxide, for instance, has two double bonds to oxygen (O=C=O), and thus has two charge clouds.

Step 3: Predict molecular shape by assuming that the charge clouds orient in space so that they are as far away from one another as possible. How they achieve this favorable orientation depends on their number, as summarized in Table 5.1.

TABLE 5.1 Molecular Geometry Around Atoms with 2, 3, and 4 Charge Clouds

Number of Bonds	Number of Lone Pairs	Number of Charge Clouds	Molecular Geometry	Example
2	0	2	Linear	O=C=O
3	0	3	Planar triangular	$\begin{matrix}H\\H\end{matrix}$>C=O
2	1		Bent	$\begin{matrix}O\\O\end{matrix}$>S:
4	0	4	Tetrahedral	H—C—H with H top and bottom
3	1		Pyramidal	H—N—H with H
2	2		Bent	H—O—H

Carbon Dioxide; Sulfur Dioxide; Sulfur Trioxide; Nitrogen Trifluoride; Carbon Tetrachloride

If there are only two charge clouds, as occurs on the central atom of CO_2 (two double bonds) and HCN (one single bond and one triple bond), the clouds are farthest apart when they point in opposite directions. Thus, both HCN and CO_2 are linear molecules, with **bond angles** of 180°.

Bond angle The angle formed by three adjacent atoms in a molecule.

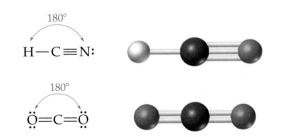

H—C≡N: 180°

These molecules are linear, with bond angles of 180°.

Ö=C=Ö 180°

When there are three charge clouds, as occurs on the central atom in formaldehyde (two single bonds and one double bond) and SO_2 (one single bond, one double bond, and one lone pair), the clouds can be farthest apart if

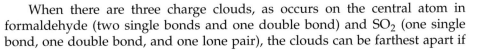

they lie in a plane and point to the corners of an equilateral triangle. Thus, a formaldehyde molecule is planar triangular, with all bond angles near 120°. In the same way, an SO_2 molecule has a planar triangular arrangement of its three electron clouds, but one point of the triangle is occupied by a lone pair. The relationship of the three atoms themselves is therefore bent rather than linear, with an O—S—O bond angle of approximately 120°.

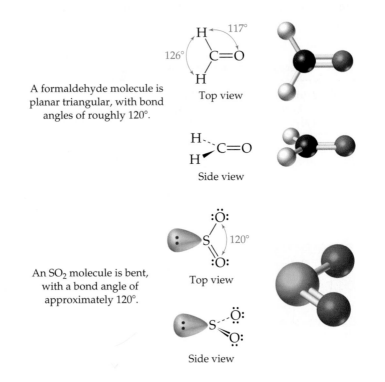

A formaldehyde molecule is planar triangular, with bond angles of roughly 120°.

An SO_2 molecule is bent, with a bond angle of approximately 120°.

When there are four charge clouds, as occurs on the central atom in CH_4 (four single bonds), NH_3 (three single bonds and one lone pair), and H_2O (two single bonds and two lone pairs), the clouds can be farthest apart when they extend to the corners of a *regular tetrahedron*. As illustrated in Figure 5.6, a **regular tetrahedron** is a geometric solid whose four identical faces are equilateral triangles. The central atom is at the center of the tetrahedron, the charge clouds point to the corners, and the angle between lines drawn from the center to any two corners is 109.5°.

Regular tetrahedron A geometric figure with four identical triangular faces.

FIGURE 5.6 The tetrahedral ▶ geometry of an atom surrounded by four charge clouds. The atom is located at the center of the regular tetrahedron, and the four charge clouds point toward the corners. The bond angle between the center and any two corners is 109.5°.

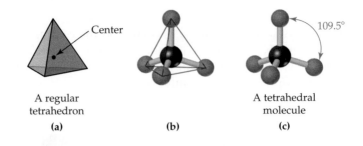

A regular tetrahedron

(a)

A tetrahedral molecule

(b)   (c)

Because valence-shell electron octets are so common, a great many molecules have geometries based on the tetrahedron. In methane (CH_4), for example, the carbon atom has tetrahedral geometry with H—C—H bond angles of exactly 109.5°. In ammonia (NH_3), the nitrogen atom has a tetrahedral arrangement of its four charge clouds, but one corner of the tetrahedron is occupied by a lone pair, resulting in an overall pyramidal shape for the molecule. Similarly, water, which has two corners of the tetrahedron occupied by lone pairs, has an overall bent shape.

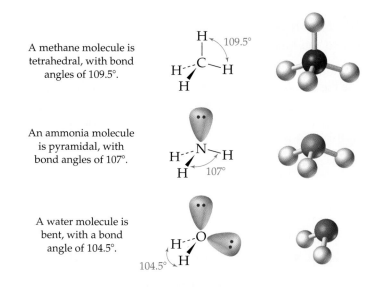

A methane molecule is tetrahedral, with bond angles of 109.5°.

An ammonia molecule is pyramidal, with bond angles of 107°.

A water molecule is bent, with a bond angle of 104.5°.

Note how the three-dimensional shapes of tetrahedral molecules like methane, ammonia, and water are shown: Solid lines are assumed to be in the plane of the paper; a dashed line recedes behind the plane of the paper away from the viewer; and a dark wedged line protrudes out of the paper toward the viewer. This standard method for showing three-dimensionality will be used throughout the rest of the book. Note also that the H—N—H bond angle in ammonia (107°) and the H—O—H bond angle in water (104.5°) are close to, but not exactly equal to, the ideal 109.5° tetrahedral value. The angles are diminished somewhat from their ideal value because the lone-pair charge clouds repel other electron clouds strongly and compress the rest of the molecule.

The geometry around atoms in larger molecules also derives from the shapes shown in Table 5.1. For example, each of the two carbon atoms in ethylene ($H_2C=CH_2$) has three charge clouds, giving rise to planar triangular geometry. It turns out that the molecule as a whole is also planar, with H—C—C and H—C—H bond angles of approximately 120°.

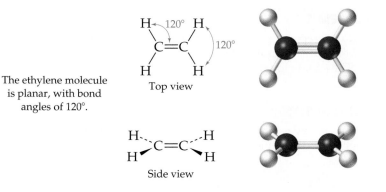

The ethylene molecule is planar, with bond angles of 120°.

Top view

Side view

Carbon atoms bonded to four other atoms are each at the center of a tetrahedron, as shown below for ethane, $H_3C—CH_3$.

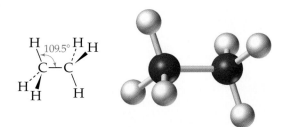

The ethane molecule has tetrahedral carbon atoms, with bond angles of 109.5°.

■ WORKED EXAMPLE 5.7

What shape do you expect for the hydronium ion, H_3O^+?

ANALYSIS   Count the number of charge clouds around the central oxygen atom, and imagine the clouds orienting as far away from one another as possible.

SOLUTION   The Lewis structure for the hydronium ion shows that the oxygen atom has four charge clouds (three single bonds and one lone pair). The hydronium ion is therefore pyramidal with bond angles of approximately 109.5°

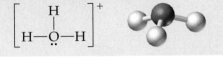

■ WORKED EXAMPLE 5.8

Predict the geometry of an acetaldehyde molecule, CH_3CHO.

SOLUTION   The Lewis structure of acetaldehyde shows that the CH_3 carbon has four charge clouds (four single bonds) and the CHO carbon atom has three charge clouds (two single bonds, one double bond). Table 5.1 indicates that the CH_3 carbon is tetrahedral, while the CHO carbon is planar triangular.

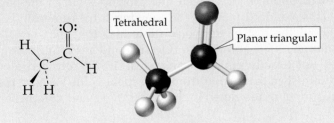

■ PROBLEM 5.14

Predict shapes for the organic molecules chloroform, $CHCl_3$, and dichloro-ethylene, $Cl_2C=CH_2$.

■ PROBLEM 5.15

Electron-pair repulsion influences the shapes of polyatomic *ions* in the same way it influences neutral molecules. What is the shape of the ammonium ion, NH_4^+? The sulfate ion, SO_4^{2-}?

■ PROBLEM 5.16

Hydrogen selenide (H_2Se) resembles hydrogen sulfide (H_2S) in that both compounds have terrible odors and are poisonous. What are their shapes?

■ KEY CONCEPT PROBLEM 5.17

Draw a structure corresponding to the following molecular model of the amino acid methionine, and describe the geometry around the indicated atoms.

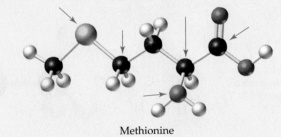

Methionine

How big can a molecule be? The answer is very, *very* big. The really big molecules in our bodies and in many items we buy are all *polymers*. Like a string of beads, a polymer is formed of many repeating units connected in a long chain. Each "bead" in the chain comes from a simple molecule that has formed chemical bonds at both ends, linking it to other molecules. The repeating units can be the same:

$$-a-a-a-a-a-a-a-a-a-a-a-a-$$

or they can be different. If different, they can be connected in an ordered pattern:

$$-a-b-a-b-a-b-a-b-a-b-a-b-$$

or in a random pattern:

$$-a-b-b-a-b-a-a-a-b-a-b-b-$$

Furthermore, the polymer chains can have branches, and the branches can have either the same repeating unit as the main chain or a different one:

```
-a-a-a-a-a-a-a-a-a-a        or      -a-a-a-a-a-a-a-a-a-a
   |        |        |                   |        |        |
   a        a        a                   b        b        b
   |        |        |                   |        |        |
   a        a        a                   b        b        b
   |        |        |                   |        |        |
   a        a        a                   b        b        b
   |        |        |                   |        |        |
```

Still other possible variations include complex, three-dimensional networks of "cross-linked" chains. The rubber used in tires, for example, contains polymer chains connected by cross-linking atoms of sulfur to impart greater rigidity.

We all use synthetic polymers every day—we usually call them "plastics." Common synthetic polymers are made by connecting up to several hundred thousand smaller molecules together, producing giant polymer molecules with masses up to several million atomic mass units. Polyethylene, for example, is made by combining as many as 50,000 ethylene molecules ($H_2C{=}CH_2$) to give a polymer with repeating $-CH_2CH_2-$ units:

$$\text{Many } H_2C{=}CH_2 \longrightarrow -CH_2CH_2CH_2CH_2CH_2CH_2-$$
$$\qquad\quad \text{Ethylene} \qquad\qquad\qquad\qquad \text{Polyethylene}$$

The product is used in such items as chairs, toys, drainpipes, milk bottles, and packaging films.

Nature began to exploit the extraordinary variety of polymer properties long before humans did. In fact, despite great progress in recent years, there is still much to be learned about the polymers in living things. Carbohydrates and proteins are polymers, as are the giant molecules of deoxyribonucleic acid (DNA) that govern the reproduction of viruses, bacteria, plants, and all living creatures. Nature's polymer molecules, though, are larger and more complex than any that chemists have yet created. We'll see the structures of these natural polymers in later chapters.

See Additional Problems 5.88 and 5.89 at the end of the chapter.

▲ These colorful tacks, like thousands of other items, are made of polymers.

5.8 Polar Covalent Bonds and Electronegativity

Electrons in a covalent bond occupy the region between the bonded atoms. If the atoms are identical, as in H_2 and Cl_2, the electrons are attracted equally to both atoms and are shared equally. If the atoms are *not* identical, however, as in HCl, the bonding electrons may be attracted more strongly by one atom than by the other and may thus be shared unequally. Such bonds are said to be

Polar covalent bond A bond in which the electrons are attracted more strongly by one atom than by the other.

polar covalent bonds. In hydrogen chloride, for example, electrons spend more time near the chlorine atom than near the hydrogen. Although the molecule as a whole is neutral, the chlorine is more negative than the hydrogen, resulting in *partial* charges on the atoms. These partial charges are represented by placing a $\delta-$ (Greek lowercase *delta*) on the more negative atom and a $\delta+$ on the more positive atom.

A particularly helpful way of visualizing this unequal distribution of bonding electrons is to look at what is called an *electrostatic potential map*, which uses color to portray the calculated electron distribution in a molecule. In HCl, for example, the electron-poor hydrogen is blue, while the electron-rich chlorine is reddish-yellow.

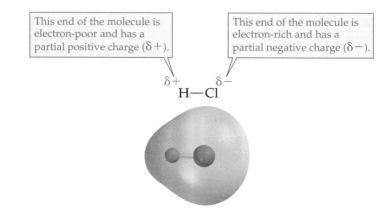

This end of the molecule is electron-poor and has a partial positive charge ($\delta+$).

This end of the molecule is electron-rich and has a partial negative charge ($\delta-$).

$$\overset{\delta+}{\text{H}}-\overset{\delta-}{\text{Cl}}$$

Electronegativity The ability of an atom to attract electrons in a covalent bond.

The ability of an atom to attract electrons in a covalent bond is called the atom's **electronegativity**. Fluorine, the most electronegative element, is assigned a value of 4, and less electronegative atoms are assigned lower values, as shown in Figure 5.7. Metallic elements on the left side of the periodic table attract electrons only weakly and have lower electronegativities, while the halogens and other reactive nonmetal elements on the upper right side of the table attract electrons strongly and have higher electronegativities. Note also in Figure 5.6 that electronegativity generally decreases going down the periodic table within a group.

FIGURE 5.7 Electronegativities of main group elements and several transition metal elements. Reactive nonmetals at the top right of the periodic table are most electronegative, and metals at the lower left are least electronegative. The noble gases are not assigned values.

 Periodic Trends: Electronegativity

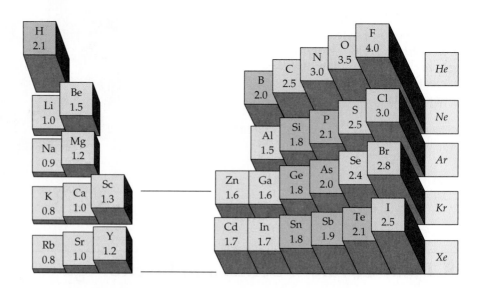

Comparing the electronegativities of bonded atoms makes it possible to compare the polarities of bonds and to predict the occurrence of ionic bonding. Both oxygen (electronegativity 3.5) and nitrogen (3.0), for instance, are more electronegative than carbon (2.5). As a result, both C—O and C—N bonds are polar, with carbon at the positive end. The larger difference in electronegativity values shows that the C—O bond is the more polar of the two.

Less polar ⌐   ⌐More polar

$\delta^+ C \underset{\smile}{\overset{\frown}{-}} N \delta^-$ $\delta^+ C \underset{\smile}{\overset{\frown}{-}} O \delta^-$

Electronegativity   Electronegativity
difference: difference:
$3.0 - 2.5 = 0.5$ $3.5 - 2.5 = 1.0$

As a rule of thumb, electronegativity differences up to 1.9 indicate increasingly polar covalent bonds, and differences of 2 or more indicate substantially ionic bonds. The electronegativity differences show, for example, that the bond between carbon and fluorine is highly polar covalent, the bond between sodium and chlorine is largely ionic, and the bond between rubidium and fluorine is almost completely ionic:

$$\delta^+ C - F \delta^- \quad Na^+ Cl^- \quad Rb^+ F^-$$

Electronegativity
difference: 1.5   2.1   3.2

Note, though, that there is no sharp dividing line between covalent and ionic bonds; most bonds fall somewhere between two extremes.

⊂⊃⊂⊃ LOOKING AHEAD

The values given in Figure 5.7 indicate that carbon and hydrogen have similar electronegativities. As a result, C—H bonds are nonpolar. We'll see in Chapters 12–25 how this fact helps explain the properties of organic and biological compounds, all of which have carbon and hydrogen as their principal constituents.   ⊂⊃⊂⊃

■ **PROBLEM 5.18**

Arrange the elements commonly bonded to carbon in organic compounds, H, N, O, P, and S, in order of increasing electronegativity.

■ **PROBLEM 5.19**

Use electronegativity differences to classify bonds between the following pairs of atoms as ionic, nonpolar covalent, or polar covalent:

(a) I and Cl   (b) Li and O   (c) Br and Br   (d) P and Br

■ **PROBLEM 5.20**

Use the symbols δ^+ and δ^- to identify the location of the partial charges on the polar covalent bonds formed between the following:

(a) Fluorine and sulfur
(b) Phosphorus and oxygen
(c) Arsenic and chlorine

5.9 Polar Molecules

Just as individual bonds can be polar, entire *molecules* can be polar if electrons are attracted more strongly to one part of the molecule than to another. Molecular polarity is due to the sum of all individual bond polarities and lone-pair contributions in the molecule and is often represented by an arrow pointing in the direction that electrons are displaced. The arrow is pointed at the negative end and is crossed at the positive end to resemble a plus sign, $(\delta+)\mapsto(\delta-)$.

Molecular polarity depends on the shape of the molecule as well as the presence of polar covalent bonds and lone pairs. In water, for example, electrons are displaced away from the less electronegative hydrogen atoms toward the more electronegative oxygen atom so that the net polarity points between the two O—H bonds. In chloromethane, CH_3Cl, electrons are attracted from the carbon/hydrogen part of the molecule toward the electronegative chlorine atom so that the net polarity points along the C—Cl bond. Electrostatic potential maps show these polarities clearly, with electron-poor regions in blue and electron-rich regions in red.

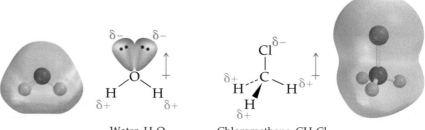

Water, H_2O Chloromethane, CH_3Cl

Furthermore, just because a molecule has polar covalent bonds doesn't mean that the molecule is necessarily polar overall. Carbon dioxide (CO_2) and tetrachloromethane (CCl_4) molecules, for instance, have no net polarity because their symmetrical shapes cause the individual C=O and C—Cl bond polarities to cancel.

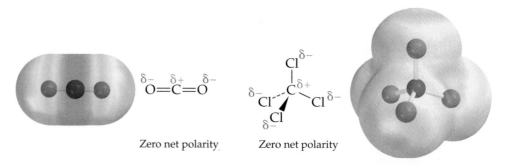

Zero net polarity   Zero net polarity

Polarity has a dramatic effect on the physical properties of molecules, particularly on melting points, boiling points, and solubilities. We'll see numerous examples of such effects in subsequent chapters.

■ WORKED EXAMPLE 5.9

Look at the structures of (a) hydrogen cyanide (HCN) and (b) vinyl chloride (H_2C=CHCl) described in Section 5.7 and Worked Example 5.5, decide whether the molecules are polar, and show the direction of net polarity in each.

ANALYSIS Draw a Lewis structure for each molecule to find its shape, and identify any polar bonds using the electronegativity values in Figure 5.7. Then decide on net polarity by adding the individual contributions.

SOLUTION

(a) The carbon atom in hydrogen cyanide has two charge clouds, making HCN a linear molecule. The C—H bond is relatively nonpolar, but the C≡N bonding electrons are pulled toward the electronegative nitrogen atom. In addition, a lone pair protrudes from nitrogen. Thus, the molecule has a net polarity.

(b) Vinyl chloride, like ethylene, is a planar molecule. The C—H and C=C bonds are nonpolar, but the C—Cl bonding electrons are displaced toward the electronegative chlorine. Thus, the molecule has a net polarity.

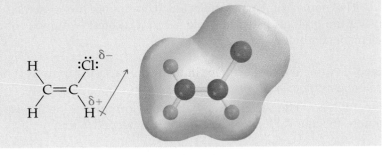

■ **PROBLEM 5.21**

Look at the molecular shape of formaldehyde (CH_2O) described in Section 5.7, decide whether the molecule is polar, and show the direction of net polarity.

■ **PROBLEM 5.22**

Draw a Lewis structure for dimethyl ether (CH_3OCH_3), predict its shape, and tell whether the molecule is polar.

■ **KEY CONCEPT PROBLEM 5.23**

Look at the following electrostatic potential map of methyllithium, and identify the direction of net polarity in the molecule. Explain this polarity based on electronegativity values.

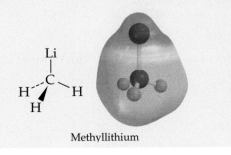

Methyllithium

5.10 Naming Binary Molecular Compounds

When two different elements combine, they form what is called a **binary compound**. The formulas of binary molecular compounds are usually written with the less electronegative element first. Thus, metals are always written

Binary compound A compound formed by combination of two different elements.

before nonmetals, and a nonmetal farther left on the periodic table generally comes before a nonmetal farther right. For example:

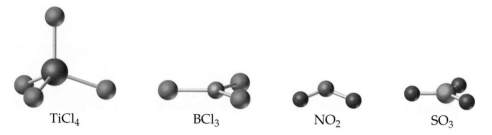

TiCl₄   BCl₃   NO₂   SO₃

TABLE 5.2 **Numerical Prefixes Used in Chemical Names**	
Number	Prefix
1	mono-
2	di-
3	tri-
4	tetra-
5	penta-
6	hexa-
7	hepta-
8	octa-
9	nona-
10	deca-

The names of binary molecular compounds are assigned in two steps, using the prefixes listed in Table 5.2 to indicate the number of atoms of each element combined.

Step 1: Name the first element in the formula, using a prefix if needed to indicate the number of atoms.

Step 2: Name the second element in the formula, using an *-ide* ending as for anions (Section 4.8), along with a prefix if needed. (∞, p. 83)

The prefix *mono-*, meaning one, is omitted except where needed to distinguish between two different compounds with the same elements. For example, the two oxides of carbon are named carbon *mon*oxide for CO and carbon *di*oxide for CO₂. (Note that we say *mon*oxide rather than *mono*oxide.) Some other examples are

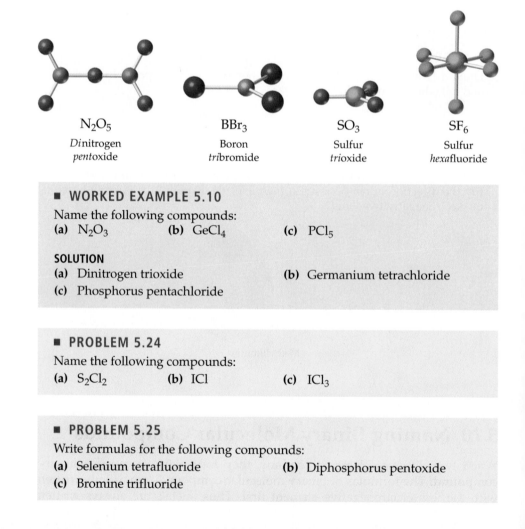

N₂O₅   BBr₃   SO₃   SF₆

*Di*nitrogen *pent*oxide   Boron *tri*bromide   Sulfur *tri*oxide   Sulfur *hexa*fluoride

---

■ **WORKED EXAMPLE 5.10**

Name the following compounds:
(a) N₂O₃   (b) GeCl₄   (c) PCl₅

SOLUTION
(a) Dinitrogen trioxide   (b) Germanium tetrachloride
(c) Phosphorus pentachloride

■ **PROBLEM 5.24**

Name the following compounds:
(a) S₂Cl₂   (b) ICl   (c) ICl₃

■ **PROBLEM 5.25**

Write formulas for the following compounds:
(a) Selenium tetrafluoride   (b) Diphosphorus pentoxide
(c) Bromine trifluoride

Damascenone, by Any Other Name, Would Smell as Sweet

What's in a name? According to Shakespeare's *Romeo and Juliet*, a rose by any other name would smell as sweet. Chemical names, however, often provoke less favorable responses: "It's unpronounceable." "It's too complicated." "It must be something bad."

Regarding pronunciation, chemical names are usually pronounced using every possible syllable. *Phenylpropanolamine*, for instance, a substance used in over-the-counter decongestants, is spoken with seven syllables: phen-yl-pro-pa-**nol**-a-mine.

Regarding complexity, the reason is obvious once you realize that there are more than 19 *million* known chemical compounds: The full name of a chemical compound has to include enough information to tell chemists the composition and structure of the compound. It's as if every person on earth had to have his or her own unique name that described height, hair color, and other identifying characteristics.

But what about chemicals being bad? These days, it seems that a different chemical gets into the news every week, often in a story describing some threat to health or the environment. The unfortunate result is that people sometimes conclude that everything with a chemical name is unnatural and dangerous. Neither is true, though. Acetaldehyde, for instance, is present naturally in most tart, ripe fruits and is often added in small amounts to artificial flavorings. When *pure*, however, acetaldehyde is also a flammable gas that is toxic and explosive in high concentrations.

Similar comparisons of desirable and harmful properties can be made for almost all chemicals, including water, sugar, and salt. The properties of a substance and

▲ Contains the following: β-damascenone, β-ionone, citronellol, geraniol, nerol, eugenol, methyl eugenol, β-phenylethyl alcohol, farnesol, linalool, terpineol, rose oxide, carvone, and many other natural substances.

the conditions surrounding its use must be evaluated before judgments are made. And damascenone, by the way, is the chemical largely responsible for the wonderful odor of roses.

See Additional Problems 5.90 and 5.91 at the end of the chapter.

5.11 Characteristics of Molecular Compounds

We saw in Section 4.4 that ionic compounds have high melting and boiling points because the attractive forces between oppositely charged ions are so strong that the ions are held tightly together. (∞, p. 75) *Molecules*, however, are neutral, so there is no strong electrical attraction between different molecules to hold them together. There are, however, several weaker forces between molecules, called *intermolecular forces*, that we'll look at in more detail in Chapter 8.

When intermolecular forces are very weak, molecules of a substance are so weakly attracted to one another that the substance is a gas at ordinary temperatures. If the forces are somewhat stronger, the molecules are pulled together into a liquid; and if the forces are still stronger, the substance becomes a molecular solid. Even so, the melting points and boiling points of molecular solids are usually lower than those of ionic solids.

In addition to having lower melting points and boiling points, molecular compounds differ from ionic compounds in other ways. Most molecular compounds are insoluble in water, for instance, because they have little attraction to strongly polar water molecules. In addition, they do not conduct electricity when melted because they have no charged particles. Table 5.3 provides a comparison of the properties of ionic and molecular compounds.

TABLE 5.3 A Comparison of Ionic and Molecular Compounds	
Ionic Compounds	**Molecular Compounds**
Smallest components are ions	Smallest components are molecules
Usually composed of metals combined with nonmetals	Usually composed of nonmetals with nonmetals
Crystalline solids	Gases, liquids, or low-melting solids
High melting points	Low melting points
High boiling points (above 700°C)	Low boiling points
Conduct electricity when molten or dissolved in water	Do not conduct electricity
Many are water-soluble	Few are water-soluble
Not soluble in organic liquids	Many are soluble in organic liquids

Key Words

Binary compound, p. 121

Bond angle, p. 113

Bond length, p. 101

Condensed structure, p. 109

Coordinate covalent bond, p. 106

Covalent bond, p. 99

Double bond, p. 104

Electronegativity, p. 118

Lewis structure, p. 106

Lone pair, p. 107

Molecular compound, p. 102

Molecular formula, p. 106

Molecule, p. 99

Polar covalent bond, p. 118

Regular tetrahedron, p. 114

Single bond, p. 104

Structural formula, p. 106

Triple bond, p. 104

Valence-shell electron-pair repulsion (VSEPR) model, p. 112

Summary: Revisiting the Chapter Goals

1. **What is a covalent bond?** A *covalent bond* is formed by the sharing of electrons between atoms rather than by the complete transfer of electrons from one atom to another. Atoms that share two electrons are joined by a *single bond* (such as C—C), atoms that share four electrons are joined by a *double bond* (such as C=C), and atoms that share six electrons are joined by a *triple bond* (such as C≡C). The group of atoms held together by covalent bonds is called a *molecule*.

   Electron sharing typically occurs when a singly occupied valence orbital on one atom *overlaps* a singly occupied valence orbital on another atom. The two electrons occupy both overlapping orbitals and belong to both atoms, thereby bonding the atoms together. Alternatively, electron sharing can occur when a filled orbital containing an unshared, *lone pair* of electrons on one atom overlaps a vacant orbital on another atom to form a *coordinate covalent bond*.

2. **How does the octet rule apply to covalent bond formation?** Depending on the number of valence electrons, different atoms form different numbers of covalent bonds. In general, an atom shares enough electrons to reach a noble gas configuration. Hydrogen, for instance, forms one covalent bond because it needs to share one more electron to achieve the helium configuration ($1s^2$). Carbon and other group 4A elements form four covalent bonds because they need to share four more electrons to reach an octet. In the same way, nitrogen and other group 5A elements form three covalent bonds, oxygen and other group 6A elements form two covalent bonds, and halogens (group 7A elements) form one covalent bond.

3. **How are molecular compounds represented?** Formulas such as H_2O, NH_3, and CH_4, which show the numbers and kinds of atoms in a molecule, are called *molecular formulas*. More useful are *Lewis structures*, which show how atoms are connected in molecules. Covalent bonds are indicated as lines between atoms, and valence electron lone pairs are shown as dots. Lewis structures are drawn by counting the total number of valence electrons in a molecule or polyatomic ion and then placing bonding pairs and shared pairs so that all electrons are accounted for.

4. **What is the influence of valence-shell electrons on molecular shape?** Molecules have specific shapes that depend on the number of electron

charge clouds (bonds and lone pairs) surrounding the various atoms. These shapes can often be predicted using the *valence-shell electron-pair repulsion* (*VSEPR*) model. Atoms with two electron charge clouds adopt linear geometry, atoms with three charge clouds adopt planar triangular geometry, and atoms with four charge clouds adopt tetrahedral geometry.

5. **When are bonds and molecules polar?** Bonds between atoms are *polar covalent* if the bonding electrons are not shared equally between the atoms. The ability of an atom to attract electrons in a covalent bond is the atom's *electronegativity* and is highest for reactive nonmetal elements on the upper right of the periodic table and lowest for metals on the lower left. Comparing electronegativities allows a prediction of whether a given bond is covalent, polar covalent, or ionic. Just as individual bonds can be polar, entire molecules can be polar if electrons are attracted more strongly to one part of the molecule than to another. Molecular polarity is due to the sum of all individual bond polarities and lone-pair contributions in the molecule.

6. **What are the major differences between ionic and molecular compounds?** *Molecular compounds* can be gases, liquids, or low-melting solids. They usually have lower melting points and boiling points than ionic compounds, many are water insoluble, and they do not conduct electricity when melted or dissolved.

▪ Understanding Key Concepts

5.26 Which of the following drawings is more likely to represent an ionic compound and which a covalent compound?

(a)   (b)

5.27 If yellow spheres represent sulfur atoms and red spheres represent oxygen atoms, which of the following drawings depicts a collection of sulfur dioxide molecules?

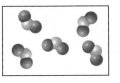 

(a)   (b)

 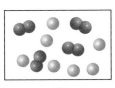

(c)   (d)

5.28 What is the geometry around the central atom in each of the following molecular models? (There are no "hidden" atoms; all atoms in each model are visible.)

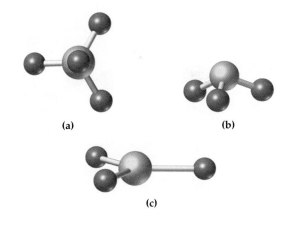

(a)   (b)

(c)

5.29 Three of the following molecular models have a tetrahedral central atom, and one does not. Which is the odd one?

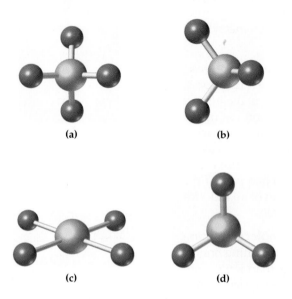

(a)

(b)

(c)

(d)

5.30 The following ball-and-stick molecular model is a representation of acetaminophen, the active ingredient in such over-the-counter headache remedies as Tylenol. The lines indicate only the connections between atoms, not whether the bonds are single, double, or triple. (Red = O, gray = C, blue = N, ivory = H.)

(a) What is the molecular formula of acetaminophen?
(b) Indicate the positions of the multiple bonds in acetaminophen.
(c) What is the geometry around each carbon and each nitrogen?

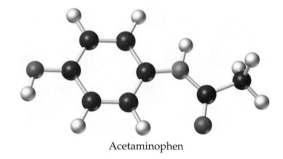

Acetaminophen

5.31 Vitamin C (ascorbic acid) has the following connections among atoms. Complete the Lewis electron-dot structure for vitamin C, showing the positions of any multiple bonds and lone pairs of electrons.

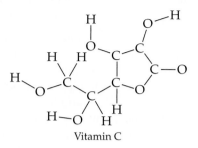

Vitamin C

5.32 The following ball-and-stick molecular model is a representation of thalidomide, a drug that causes terrible birth defects when taken by expectant mothers but that has been approved use against leprosy. The lines indicate only the connections between atoms, not whether the bonds are single, double, or triple (red = O, gray = C, blue = N, ivory = H).

(a) What is the molecular formula of thalidomide?
(b) Indicate the positions of the multiple bonds in thalidomide.
(c) What is the geometry around each carbon and each nitrogen?

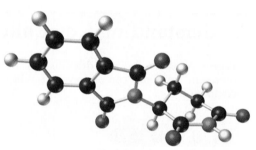

Thalidomide

5.33 Show the position of any electron lone pairs, and indicate the electron-rich and electron-poor regions in acetamide.

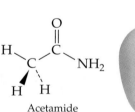

Acetamide

▪ Additional Problems

Covalent Bonds

5.34 What is a covalent bond, and how does it differ from an ionic bond?

5.35 Describe the changes in potential energy that occur as two atoms approach each other to form a covalent bond.

5.36 Which of the following elements would you expect to form (i) diatomic molecules, (ii) mainly covalent bonds, (iii) mainly ionic bonds, (iv) both covalent and ionic bonds? (More than one answer may apply.)

(a) Oxygen   **(b)** Potassium
(c) Phosphorus   **(d)** Iodine
(e) Hydrogen   **(f)** Cesium

5.37 Which of the following pairs of atoms form covalent bonds?
(a) Calcium and fluorine
(b) Carbon and fluorine
(c) Magnesium and fluorine
(d) Silicon and fluorine
(e) Lithium and fluorine

5.38 Look up tellurium ($Z = 52$) in the periodic table and predict how many covalent bonds it is likely to form. Explain.

5.39 Germanium (atomic number 32) is an element used in the manufacture of transistors. Judging from its position in the periodic table, how many covalent bonds does it usually form?

5.40 Which of the following contains a coordinate covalent bond?
(a) $PbCl_2$ **(b)** $Cu(NH_3)_4^{2+}$ **(c)** NH_4^+

5.41 Which of the following contains a coordinate covalent bond?
(a) H_2O **(b)** $HOClO$ **(c)** H_3O^+

5.42 Tin forms both an ionic compound and a covalent compound with chlorine. The ionic compound is $SnCl_2$. Is the covalent compound more likely to be $SnCl_3$, $SnCl_4$, or $SnCl_5$? Explain.

5.43 A compound of gallium with chlorine has a melting point of 77°C and a boiling point of 201°C. Is the compound ionic or covalent? What is a likely formula?

5.44 Nitrous oxide, N_2O, has the following structure. Which bond in N_2O is a coordinate covalent bond?

$$:N\equiv N-\ddot{\underset{\cdot\cdot}{O}}:$$

Nitrous oxide

5.45 Thionyl chloride, $SOCl_2$, has the following structure. Which bond in $SOCl_2$ is a coordinate covalent bond?

Thionyl chloride

Structural Formulas

5.46 Distinguish between the following:
(a) A molecular formula and a structural formula
(b) A structural formula and a condensed structure
(c) A lone pair and a shared pair of electrons

5.47 Assume that you are given samples of two white, crystalline compounds, one of them ionic and one covalent. Describe how you might tell which is which.

5.48 Give the total number of valence electrons in the following molecules:
(a) N_2 **(b)** CO
(c) CH_3CH_2CHO **(d)** OF_2

5.49 Add lone pairs where appropriate to each of the following structures:

(a) $C\equiv O$ **(b)** CH_3SH

(c) $H-\overset{H}{\underset{}{\overset{|}{O^+}}}-H$ **(d)** $H_3C-\overset{H}{\underset{}{\overset{|}{N}}}-CH_3$

5.50 If a research paper appeared reporting the structure of a new molecule with formula C_2H_8, most chemists would be highly skeptical. Why?

5.51 Which of the following three possible structural formulas for $C_3H_6O_2$ is correct? Explain.

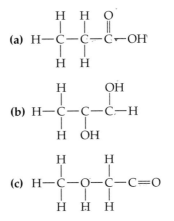

5.52 Convert the following Lewis structures into structural formulas in which lines replace the bonding electrons. Include the lone pairs.

(a) $H:\ddot{\underset{\cdot\cdot}{O}}:\ddot{N}::\ddot{\underset{\cdot\cdot}{O}}:$ **(b)** $H:\overset{H}{\underset{H}{\ddot{C}}}:C:::N:$ **(c)** $H:\ddot{\underset{\cdot\cdot}{F}}:$

5.53 Convert the following Lewis structure for the nitrate ion into a line structure that includes the lone pairs. Why does the nitrate ion have a -1 charge?

5.54 Convert the following structural formulas into condensed structures.

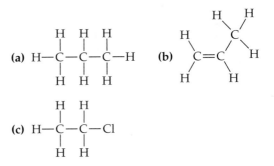

5.55 Acetic acid is the major organic constituent of vinegar. Convert the following structural formula of acetic acid into a condensed structure.

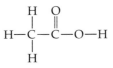

Drawing Lewis Structures

5.56 Draw a Lewis structure for each of the following molecules:
(a) SiF_4 **(b)** $AlCl_3$ **(c)** CF_2Cl_2
(d) SO_3 **(e)** BBr_3 **(f)** NF_3

5.57 Draw a Lewis structure for each of the following molecules:
  (a) Nitrous acid, HNO_2 (H is bonded to an O)
  (b) Ozone, O_3
  (c) Acetaldehyde, CH_3CHO

5.58 Ethanol, or "grain alcohol," has the formula C_2H_6O and contains an O—H bond. Propose a structure for ethanol that is consistent with common bonding patterns.

5.59 Dimethyl ether has the same molecular formula as ethanol (Problem 5.58) but very different properties. Propose a structure for dimethyl ether in which the oxygen is bonded to two carbons.

5.60 Hydrazine, a substance used to make rocket fuel, has the formula N_2H_4. Propose a structure for hydrazine.

5.61 Tetrachloroethylene, C_2Cl_4, is used commercially as a dry cleaning solvent. Propose a structure for tetrachloroethylene based on the common bonding patterns expected in organic molecules. What kind of carbon–carbon bond is present?

5.62 Draw a Lewis structure for carbon disulfide, CS_2, a foul-smelling liquid used as a solvent for fats. What kind of carbon–sulfur bonds are present?

5.63 Draw a Lewis structure for hydroxylamine, NH_2OH.

5.64 The cyanide ion, CN^-, contains a triple bond. Draw a Lewis structure for the cyanide ion and show why it has a negative charge.

5.65 Draw a Lewis structure for each of the following polyatomic ions:
  (a) Formate, HCO_2^- (b) Carbonate, CO_3^{2-}
  (c) Sulfite, SO_3^{2-} (d) Thiocyanate, SCN^-

Molecular Geometry

5.66 Predict the geometry and bond angles around atom A for molecules with the general formulas AB_3 and AB_2E, where B represents another atom and E represents an electron pair.

5.67 Predict the geometry and bond angles around atom A for molecules with the general formulas AB_4, AB_3E, and AB_2E_2, where B represents another atom and E represents an electron pair.

5.68 Sketch the three-dimensional shape of each of the following molecules:
  (a) Chloroform, $CHCl_3$
  (b) Hydrogen sulfide, H_2S
  (c) Ozone, O_3
  (d) Carbon disulfide, CS_2
  (e) Nitrous acid, HNO_2

5.69 Predict the three-dimensional shape of each of the following molecules (see Problem 5.56):
  (a) SiF_4 (b) CF_2Cl_2 (c) SO_3
  (d) BBr_3 (e) NF_3

5.70 Predict the geometry around each carbon atom in the amino acid alanine.

Alanine

5.71 Predict the geometry around each carbon atom in vinyl acetate, a precursor of the poly(vinyl alcohol) polymer used in automobile safety glass.

Vinyl acetate

Polarity of Bonds and Molecules

5.72 Where in the periodic table are the most electronegative elements found, and where are the least electronegative elements found?

5.73 Predict the electronegativity of the yet undiscovered element with $Z = 119$.

5.74 Look at the periodic table, and then order the following elements according to increasing electronegativity: Li, K, Br, C, Cl.

5.75 Look at the periodic table, and then order the following elements according to decreasing electronegativity: C, Ca, Cs, Cl, Cu.

5.76 Which of the following bonds are polar? Identify the negative and positive ends of each bond by using $\delta-$ and $\delta+$.
  (a) I—Br   (b) O—H   (c) C—F
  (d) N—C   (e) C—C

5.77 Which of the following bonds are polar? Identify the negative and positive ends of each bond by using $\delta-$ and $\delta+$.
  (a) O—Br   (b) N—H   (c) P—O
  (d) C—S   (e) C—Li

5.78 Based on electronegativity differences, would you expect bonds between the following pairs of atoms to be largely ionic or largely covalent?
  (a) Na and F   (b) C and Cl
  (c) N and H   (d) Be and Br

5.79 Arrange the following molecules in order of the increasing polarity of their bonds:
  (a) HCl   (b) PH_3
  (c) H_2O (d) CF_4

5.80 Is the planar triangular molecule boron trichloride (BCl_3) polar? Explain.

5.81 Decide whether or not each of the compounds listed in Problem 5.79 is polar, and show the direction of polarity.

Names and Formulas of Molecular Compounds

5.82 Name the following binary compounds:
  (a) PI_3 (b) $AsCl_3$ (c) P_4S_3
  (d) Al_2F_6 (e) NI_3 (f) IF_7

5.83 Name the following compounds:
  (a) $SiCl_4$ (b) NaH
  (c) SbF_5 (d) OsO_4

5.84 Write formulas for the following compounds:
  (a) Nitrogen dioxide
  (b) Sulfur hexafluoride
  (c) Bromine pentaiodide
  (d) Dinitrogen trioxide
  (e) Dinitrogen tetroxide
  (f) Arsenic pentachloride

5.85 Write formulas for the following compounds:
   (a) Selenium dioxide
   (b) Xenon tetroxide
   (c) Dinitrogen pentasulfide
   (d) Triphosphorus tetraselenide

Applications

5.86 What functions does carbon monoxide serve in the brain? [*Carbon Monoxide: A Surprising Molecule*]

5.87 Why is carbon monoxide toxic in high concentration? [*Carbon Monoxide: A Surprising Molecule*]

5.88 How is a polymer formed? [*Very BIG Molecules*]

5.89 Do any polymers exist in nature? Explain. [*Very BIG Molecules*]

5.90 Why are many chemical names so complex? [*Damascenone, by Any Other Name*]

5.91 Can you tell from the name whether a chemical is natural or synthetic? [*Damascenone, by Any Other Name*]

General Questions and Problems

5.92 The discovery in the 1960s that xenon and fluorine react to form a molecular compound was a surprise to most chemists, because it had been thought that noble gases could not form any bonds.
   (a) Why was it thought that noble gases could not form bonds?
   (b) Draw a Lewis structure of XeF_4.

5.93 Acetone, a common solvent used in some nail polish removers, has the molecular formula C_3H_6O and contains a carbon—oxygen double bond.
   (a) Propose two Lewis structures for acetone.
   (b) What is the geometry around the carbon atoms in each of the structures?
   (c) Which of the bonds in each structure are polar?

5.94 The following formulas are unlikely to be correct. What is wrong with each?
   (a) CCl_3 **(b)** N_2H_5
   (c) H_3S **(d)** C_2OS

5.95 Which of the compounds (a) through (d) contain one or more of the following: (i) ionic bonds, (ii) covalent bonds, (iii) coordinate covalent bonds?
   (a) $CaCl_2$ **(b)** $Mg(NO_3)_2$
   (c) BF_4^- **(d)** CBr_4

5.96 The phosphonium ion (PH_4^+) is formed by reaction of phosphine (PH_3) with an acid.
   (a) Draw the Lewis structure of the phosphonium ion.
   (b) Predict its molecular geometry.
   (c) Describe how a fourth hydrogen can be added to PH_3.
   (d) Explain why the ion has a +1 charge.

5.97 Name the following compounds. Be sure to determine whether the compound is ionic or covalent so that you use the proper rules.
   (a) $CaCl_2$ **(b)** $TeCl_2$ **(c)** BF_3
   (d) $MgSO_4$ **(e)** K_2O **(f)** FeF_3
   (g) PF_3

5.98 Draw a Lewis structure of chloral hydrate, known in detective novels as "knockout drops." Indicate all lone pairs.

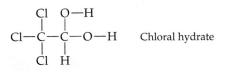

Chloral hydrate

5.99 The dichromate ion, $Cr_2O_7^{2-}$, has neither Cr—Cr nor O—O bonds. Write a Lewis structure.

5.100 Oxalic acid, $H_2C_2O_4$, is a poisonous substance found in uncooked spinach leaves. If oxalic acid has a C—C single bond, draw its Lewis structure.

5.101 Identify the elements in the fourth row of the periodic table that form the following compounds.

   (a) $\ddot{O}=\ddot{X}=\ddot{O}$ **(b)** $\ddot{\underset{..}{F}}\diagdown X \diagup \ddot{\underset{..}{F}}$

5.102 Write Lewis structures for molecules with the following connections, showing the positions of any multiple bonds and lone pairs of electrons.

   (a)
   ```
        O      H
        |      |
   Cl—C—O—C—H
               |
               H
   ```

   (b)
   ```
        H
        |
   H—C—C—C—H
        |
        H
   ```

6

Chemical Reactions
Classification and Mass Relationships

Prehistoric people made these objects by the reduction of iron ore with charcoal.

A log burns in the fireplace, an oyster makes a pearl, a seed grows into a plant—these and almost all other changes you see taking place around you are the result of *chemical reactions*. The study of how and why chemical reactions happen is a major part of chemistry, providing information that is both fascinating and practical. In this chapter we'll begin to look at chemical reactions, starting with a discussion of how to represent them in writing. Next we'll describe the mass relationships among substances involved in chemical reactions, and then we'll introduce a few easily recognized classes of chemical reactions.

Among the questions we'll answer are the following:

1. **How are chemical reactions written?**
   The goal: Given the identities of reactants and products, be able to write a balanced chemical equation or net ionic equation.

2. **What is the mole, and why is it useful in chemistry?**
   The goal: Be able to explain the meaning and uses of the mole and Avogadro's number.

3. **How are molar quantities and mass quantities related?**
   The goal: Be able to convert between molar and mass quantities of an element or compound.

4. **What is the percent yield of a reaction?**
   The goal: Be able to take the amount of product actually formed in a reaction, calculate the amount that could form theoretically, and express the results as a percent yield.

5. **How are chemical reactions of ionic compounds classified?**
   The goal: Be able to recognize precipitation, acid–base neutralization, and redox reactions.

6. **What are oxidation numbers, and how are they used?**
   The goal: Be able to assign oxidation numbers to atoms in compounds and identify the substances oxidized and reduced in a given reaction.

CONCEPTS TO REVIEW

Problem Solving: Converting a Quantity from One Unit to Another (Section 2.7)
Periodic Properties and Ion Formation (Section 4.2)
H^+ and OH^- Ions: An Introduction to Acids and Bases (Section 4.12)

6.1 Chemical Equations

Let's look at a typical example of a chemical reaction. When sodium bicarbonate is heated in the range 50–100°C, sodium carbonate, water, and carbon dioxide are produced. In words, we might write the reaction as

$$\text{Sodium bicarbonate} \xrightarrow{\text{Heat}} \text{Sodium carbonate + Water + Carbon dioxide}$$

Replacing the chemical names with formulas converts the word description of this reaction into a **chemical equation**:

$$\underbrace{2\,NaHCO_3}_{\text{Reactant}} \xrightarrow{\text{Heat}} \underbrace{Na_2CO_3 + H_2O + CO_2}_{\text{Products}}$$

Look at how this equation is written. The **reactants** are written on the left, the **products** are written on the right, and an arrow is placed between them to indicate a chemical change. Conditions necessary for the reaction to occur—heat in this particular instance—are often specified above the arrow.

Why is the number 2 placed before $NaHCO_3$ in the equation? The 2 is necessary because of a fundamental law of nature called the *law of conservation of mass*:

Law of conservation of mass   Matter is neither created nor destroyed in chemical reactions.

Chemical equation   An expression in which symbols and formulas are used to represent a chemical reaction.

Reactant   A substance that undergoes change in a chemical reaction and is written on the left side of the reaction arrow in a chemical equation.

Product   A substance that is formed in a chemical reaction and is written on the right side of the reaction arrow in a chemical equation.

Law of conservation of mass   Matter can be neither created nor destroyed in any physical or chemical change.

131

Balanced equation A chemical equation in which the numbers and kinds of atoms are the same on both sides of the reaction arrow.

Coefficient A number placed in front of a formula to balance a chemical equation.

The bonds between atoms in the reactants are rearranged to form new compounds in chemical reactions, but none of the atoms disappear and no new ones are formed. As a consequence, chemical equations must be **balanced**, meaning that *the numbers and kinds of atoms must be the same on both sides of the reaction arrow.*

The numbers placed in front of formulas to balance equations are called **coefficients**, and they multiply all the atoms in a formula. Thus, the symbol "2 $NaHCO_3$" indicates two units of sodium bicarbonate, which contain 2 Na atoms, 2 H atoms, 2 C atoms, and 6 O atoms ($2 \times 3 = 6$, the coefficient times the subscript for O). Count the numbers of atoms on the right side of the equation to convince yourself that it is indeed balanced.

The substances that take part in chemical reactions may be solids, liquids, or gases, or they may be dissolved in a solvent. Ionic compounds, in particular, frequently undergo reaction in *aqueous solution*—that is, dissolved in water. Sometimes this information is added to an equation by placing the appropriate symbols after the formulas:

(s)	(l)	(g)	(aq)
Solid	Liquid	Gas	Aqueous solution

Thus, the decomposition of solid sodium bicarbonate can be written as

$$2\,NaHCO_3(s) \xrightarrow{\text{Heat}} Na_2CO_3(s) + H_2O(l) + CO_2(g)$$

■ WORKED EXAMPLE 6.1

Interpret in words the following equation for the reaction used in extracting lead metal from its ores. Show that the equation is balanced.

$$2\,PbS(s) + 3\,O_2(g) \longrightarrow 2\,PbO(s) + 2\,SO_2(g)$$

SOLUTION The equation can be read as, "Solid lead(II) sulfide plus gaseous oxygen yields solid lead(II) oxide plus gaseous sulfur dioxide."

To show that the equation is balanced, count the atoms of each element on each side of the arrow:

On the left:	2 Pb	2 S	$(3 \times 2)\,O = 6\,O$
On the right:	2 Pb	2 S	$2\,O + (2 \times 2)\,O = 6\,O$

From 2 PbO   From 2 SO_2

The numbers of atoms of each element are the same in the reactants and products, so the equation is balanced.

■ PROBLEM 6.1

Interpret each of the following equations in words:
(a) $CoCl_2(s) + 2\,HF(g) \rightarrow CoF_2(s) + 2\,HCl(g)$
(b) $Pb(NO_3)_2(aq) + 2\,KI(aq) \rightarrow PbI_2(s) + 2\,KNO_3(aq)$

■ PROBLEM 6.2

Which of the following equations are balanced?
(a) $HCl + KOH \rightarrow H_2O + KCl$
(b) $CH_4 + Cl_2 \rightarrow CH_2Cl_2 + HCl$
(c) $H_2O + MgO \rightarrow Mg(OH)_2$
(d) $Al(OH)_3 + H_3PO_4 \rightarrow AlPO_4 + 2\,H_2O$

6.2 Balancing Chemical Equations

Balancing chemical equations can often be done using a mixture of common sense and trial-and-error. There are four steps:

Balancing Chemical Equations

Step 1: Write an unbalanced equation, using the correct formulas for all reactants and products.   For example, hydrogen and oxygen must be written as H_2 and O_2, rather than as H and O, since we know that both elements exist as diatomic molecules. Remember that *the subscripts in chemical formulas can't be changed in balancing an equation because doing so would change the identity of the substances in the reaction.*

Step 2: Add appropriate coefficients to balance the numbers of atoms of each element.   It helps to begin with elements that appear in only one formula on each side of the equation, which usually means leaving oxygen and hydrogen until last. For example, in the reaction of sulfuric acid with sodium hydroxide to give sodium sulfate and water, we might balance first for sodium. We could do this by adding a coefficient of 2 for NaOH:

$$H_2SO_4 + NaOH \longrightarrow Na_2SO_4 + H_2O \quad \text{(Unbalanced)}$$

$$H_2SO_4 + 2\,NaOH \longrightarrow Na_2SO_4 + H_2O \quad \text{(Balanced for Na)}$$

Add this coefficient . . .   . . . to balance these 2 Na.

If a polyatomic ion appears on both sides of an equation, it is treated as a single unit. For example, the sulfate ion (SO_4^{2-}) in our example is balanced because there is one on the left and one on the right:

$$H_2SO_4 + 2\,NaOH \longrightarrow Na_2SO_4 + H_2O \quad \text{(Balanced for Na and sulfate)}$$

One sulfate here . . .   . . . and one here.

At this point, the equation can be balanced for H and O by adding a coefficient of 2 for H_2O. Then we're finished.

$$H_2SO_4 + 2\,NaOH \longrightarrow Na_2SO_4 + 2\,H_2O \quad \text{(Completely balanced)}$$

4 H and 2 O here.   4 H and 2 O here.

Step 3: Check the equation to make sure the numbers and kinds of atoms on both sides of the equation are the same.

Step 4: Make sure the coefficients are reduced to their lowest whole-number values.   For example, the equation

$$2\,H_2SO_4 + 4\,NaOH \longrightarrow 2\,Na_2SO_4 + 4\,H_2O$$

is balanced, but can be simplified by dividing all coefficients by 2:

$$H_2SO_4 + 2\,NaOH \longrightarrow Na_2SO_4 + 2\,H_2O$$

■ **WORKED EXAMPLE 6.2**

Natural gas (methane, CH_4) burns in oxygen to yield water and carbon dioxide (CO_2). Write a balanced equation for the reaction.

SOLUTION

Step 1: Write the unbalanced equation, using correct formulas for all substances:

$$CH_4 + O_2 \longrightarrow CO_2 + H_2O \quad \text{(Unbalanced)}$$

Step 2: Since carbon appears in one formula on each side of the arrow, let's begin with that element. In fact, there is only one carbon atom in each formula, so the equation is already balanced for that element. Next, note that there are four hydrogen atoms on the left (in CH_4) and only two on the right (in H_2O). Placing a coefficient of 2 before H_2O gives the same number of hydrogen atoms on both sides:

$$CH_4 + O_2 \longrightarrow CO_2 + 2H_2O \quad \text{(Balanced for C and H)}$$

Finally, look at the number of oxygen atoms. There are two on the left (in O_2) but four on the right (two in CO_2 and one in each H_2O). If we place a 2 before the O_2, the number of oxygen atoms will be the same on both sides while the numbers of other elements remain unchanged:

$$CH_4 + 2O_2 \longrightarrow CO_2 + 2H_2O \quad \text{(Balanced for C, H, and O)}$$

Step 3: Check to be sure the numbers of atoms on both sides are the same.

On the left: 1 C   4 H $(2 \times 2)\,O = 4\,O$

On the right: 1 C $(2 \times 2)\,H = 4\,H$ $2\,O + 2\,O = 4\,O$

From CO_2 From $2\,H_2O$

Step 4: Make sure the coefficients are reduced to their lowest whole-number values. In this case, the answer is already correct.

▲ The oxygen in emergency breathing masks comes from heating sodium chlorate.

■ WORKED EXAMPLE 6.3

Sodium chlorate ($NaClO_3$) decomposes when heated to yield sodium chloride and oxygen, a reaction used to provide oxygen for the emergency breathing masks in airliners. Write a balanced equation for this reaction.

SOLUTION   The unbalanced equation is

$$NaClO_3 \longrightarrow NaCl + O_2$$

The equation is already balanced for Na and Cl but is unbalanced for O. To balance O, it's necessary to put a coefficient of 2 before $NaClO_3$ and a coefficient of 3 before O_2, giving 6 O atoms on both sides of the equation:

$$2\,NaClO_3 \longrightarrow NaCl + 3\,O_2 \quad \text{(Balanced for O)}$$

Unfortunately, when we balanced the equation for O, we *unbalanced* it for Na and Cl. Thus, we have to rebalance for Na and Cl by placing a coefficient of 2 before NaCl:

$$2\,NaClO_3 \longrightarrow 2\,NaCl + 3\,O_2 \quad \text{(Balanced for Na, Cl, and O)}$$

Checking gives

On the left: 2 Na 2 Cl $(2 \times 3)\,O = 6\,O$

On the right: 2 Na 2 Cl $(3 \times 2)\,O = 6\,O$

■ PROBLEM 6.3

Ozone (O_3) is formed in the earth's upper atmosphere by the action of solar radiation on oxygen molecules (O_2). Write a balanced equation for the formation of ozone from oxygen.

■ **PROBLEM 6.4**

Balance the following equations:

(a) $Ca(OH)_2 + HCl \rightarrow CaCl_2 + H_2O$

(b) $Al + O_2 \rightarrow Al_2O_3$

(c) $CH_3CH_3 + O_2 \rightarrow CO_2 + H_2O$

(d) $AgNO_3 + MgCl_2 \rightarrow AgCl + Mg(NO_3)_2$

■ **KEY CONCEPT PROBLEM 6.5**

The following diagram represents the reaction of A (red spheres) with B_2 (blue spheres). Write a balanced equation for the reaction.

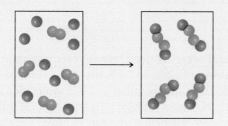

6.3 Avogadro's Number and the Mole

Imagine a laboratory experiment—perhaps the reaction of ethylene (C_2H_4) with hydrogen chloride (HCl) to prepare ethyl chloride (C_2H_5Cl), a colorless, low-boiling liquid used by doctors and athletic trainers as a spray-on anesthetic.

How, though, can you be sure you have a 1 to 1 ratio of reactant molecules in your reaction flask? Since it's impossible to count the right number of molecules, you must weigh them instead. (This is a common method for dealing with all kinds of small objects: nails, nuts, and grains of rice are all weighed rather than counted.) But the weighing approach leads to another problem. How many molecules are there in one gram of ethylene, hydrogen chloride, or any other substance? The answer depends on the identity of the substance because different molecules have different masses.

To determine how many molecules of a given substance are in a certain mass, it's first helpful to define a quantity called *molecular weight*. Just as the *atomic weight* of an element is the average mass of the element's *atoms* (Section 3.3), the **molecular weight (MW)** of a molecule is the average mass of a substance's *molecules*. (∞, p. 49) Numerically, a substance's molecular weight (or **formula weight** for an ionic compound) is equal to the sum of the atomic weights for all the atoms in the molecule or formula unit.

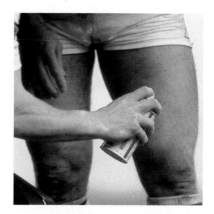

▲ Ethyl chloride is often used as a spray-on anesthetic for athletic injuries.

Molecular weight The sum of the atomic weights of the atoms in a molecule.

Formula weight The sum of the atomic weights of the atoms in one formula unit of any compound.

Molecular weight	The sum of atomic weights of all atoms in a molecule
Formula weight	The sum of atomic weights of all atoms in one formula unit of any compound, whether molecular or ionic

For example, the molecular weight of ethylene (C_2H_4) is 28.0 amu, the molecular weight of HCl is 36.5 amu, and the molecular weight of ethyl chloride (C_2H_5Cl) is 64.5 amu. (The actual values are known more precisely but are rounded off here for convenience.)

For ethylene, C_2H_4:

Atomic weight of 2 C $= 2 \times 12.0$ amu $= 24.0$ amu
Atomic weight of 4 H $= 4 \times 1.0$ amu $= 4.0$ amu

MW of C_2H_4 $\qquad\qquad\qquad\qquad = 28.0$ amu

For hydrogen chloride, HCl:

Atomic weight of H $= 1.0$ amu
Atomic weight of Cl $= 35.5$ amu

MW of HCl $\qquad\qquad = 36.5$ amu

For ethyl chloride, C_2H_5Cl::

Atomic weight of 2 C $= 2 \times 12.0$ amu $= 24.0$ amu

Atomic weight of 5 H $= 5 \times 1.0$ amu $= 5.0$ amu

Atomic weight of Cl $\qquad\qquad\qquad = 35.5$ amu

MW of C_2H_5Cl $\qquad\qquad\qquad = 64.5$ amu

How are molecular weights used? Since the mass ratio of *one* ethylene molecule to *one* HCl molecule is 28.0 to 36.5, the mass ratio of *any* given number of ethylene molecules to the same number of HCl molecules is also 28.0 to 36.5. In other words, a 28.0 to 36.5 *mass* ratio of ethylene and HCl always guarantees a 1 to 1 *number* ratio. *Samples of different substances always contain the same number of molecules or formula units whenever their mass ratio is the same as their molecular or formula weight ratio* (Figure 6.1).

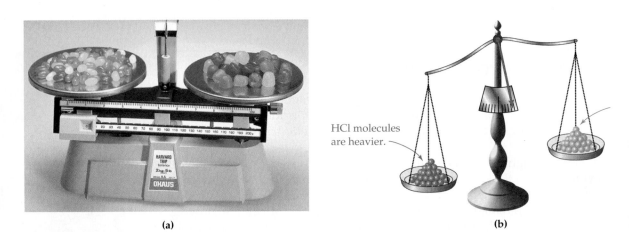

(a)   (b)

HCl molecules
are heavier.

▲ **FIGURE 6.1**   (a) Because one gumdrop weighs more than one jellybean, you can't get equal numbers by taking equal weights. The same is true for atoms or molecules of different substances. (b) Equal numbers of ethylene and HCl molecules always have a mass ratio equal to the ratio of their molecular weights, 28.0 to 36.5.

A particularly convenient way to use this mass/number relationship for molecules is to measure amounts in grams that are numerically equal to molecular weights. If, for instance, you were to carry out your experiment with 28.0 g of ethylene and 36.5 g of HCl, you could be certain that you would have a 1 to 1 ratio of reactant molecules.

When referring to the vast numbers of molecules or formula units that take part in a visible chemical reaction, it's convenient to use a counting unit called a **mole**, abbreviated *mol*. One mole of any substance is the amount whose mass in grams—its *molar mass*—is numerically equal to its molecular or formula weight. One mole of ethylene has a mass of 28.0 g, one mole of HCl has a mass of 36.5 g, and one mole of ethyl chloride has a mass of 64.5 g.

> **Mole**   The amount of a substance whose mass in grams is numerically equal to its molecular or formula weight.

Just how many molecules are there in a mole? Experiments show that one mole of any substance contains 6.022×10^{23} formula units, a value called **Avogadro's number** (abbreviated N_A) after the Italian scientist who first recognized the importance of the mass/number relationship in molecules. Avogadro's

Avogadro's number (N_A)
The number of units in a mole; 6.022×10^{23}.

number of formula units of any substance—that is, one mole—has a mass in grams numerically equal to the molecular weight of the substance.

$1 \text{ mol HCl} = 6.022 \times 10^{23} \text{ HCl molecules} = 36.5 \text{ g HCl}$

$1 \text{ mol C}_2\text{H}_4 = 6.022 \times 10^{23} \text{ C}_2\text{H}_4 \text{ molecules} = 28.0 \text{ g C}_2\text{H}_4$

$1 \text{ mol C}_2\text{H}_5\text{Cl} = 6.022 \times 10^{23} \text{ C}_2\text{H}_5\text{Cl molecules} = 64.5 \text{ g C}_2\text{H}_5\text{Cl}$

◀ These samples of water, sulfur, table sugar, mercury, and copper each contain 1 mole. Do they weigh the same?

How big is Avogadro's number? Our minds can't really conceive of the magnitude of a number like 6.022×10^{23}, but the following comparisons might give you a sense of scale:

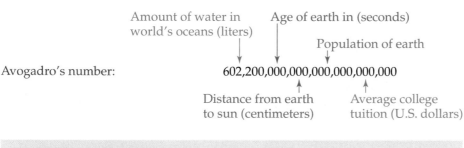

Amount of water in world's oceans (liters)   Age of earth in (seconds)

Population of earth

Avogadro's number: 602,200,000,000,000,000,000,000

Distance from earth to sun (centimeters)   Average college tuition (U.S. dollars)

■ **WORKED EXAMPLE 6.4**

Pseudoephedrine hydrochloride ($C_{10}H_{16}ClNO$) is a nasal decongestant commonly found in cold medications. How many molecules of pseudoephedrine hydrochloride are in a 30.0 mg tablet?

ANALYSIS We are given a mass and need to convert to number of molecules. The conversion factor can be found by calculating the molecular weight of pseudoephedrine hydrochloride and realizing that that amount in grams contains Avogadro's number of molecules (6.022×10^{23}).

SOLUTION The molecular weight of pseudoephedrine is found by summing the atomic weights of all atoms in the molecule:

$$MW = (10\,\cancel{C})\left(12.0\,\frac{amu}{\cancel{C}}\right) + (16\,\cancel{H})\left(1.0\,\frac{amu}{\cancel{H}}\right) + (1\,\cancel{Cl})\left(35.5\,\frac{amu}{\cancel{Cl}}\right)$$

$$+ (1\,\cancel{N})\left(14.0\,\frac{amu}{\cancel{N}}\right) + (1\,\cancel{O})\left(16.0\,\frac{amu}{\cancel{O}}\right) = 201.5 \text{ amu}$$

Since the molecular weight of pseudoephedrine hydrochloride is 201.5 amu, 201.5 g contains 6.022×10^{23} molecules. Using this mass/number relationship as a conversion factor, we find that 30.0 mg of pseudoephedrine hydrochloride contains

$$(30.0\,\cancel{\text{mg}} \text{ pseudoephedrine hydrochloride})\left(\frac{1\,\cancel{g}}{1000\,\cancel{\text{mg}}}\right)\left(\frac{6.022 \times 10^{23} \text{ molecules}}{201.5\,\cancel{g}}\right)$$

$$= 8.97 \times 10^{19} \text{ molecules pseudoephedrine hydrochloride}$$

■ **WORKED EXAMPLE 6.5**

A tiny pencil mark just visible to the naked eye contains about 3×10^{17} atoms of carbon. What is the mass of this pencil mark in grams?

ANALYSIS We are given a number of atoms and need to convert to mass. The conversion factor can be obtained by realizing that the atomic weight of carbon in grams contains Avogadro's number of atoms (6.022×10^{23}).

SOLUTION The atomic weight of carbon is 12.01 amu, so 12.01 g of carbon contains 6.022×10^{23} atoms. Using this relationship as a conversion factor gives

$$(3 \times 10^{17} \text{ atoms of C}) \left(\frac{12.01 \text{ g}}{6.022 \times 10^{23} \text{ atoms}} \right) = 6 \times 10^{-6} \text{ g of carbon}$$

■ **PROBLEM 6.6**

Calculate the molecular weight of each of the following substances:
(a) Ibuprofen, $C_{13}H_{18}O_2$
(b) Phenobarbital, $C_{12}H_{12}N_2O_3$

■ **PROBLEM 6.7**

How many molecules of ascorbic acid (vitamin C, $C_6H_8O_6$) are in a 500 mg tablet?

■ **PROBLEM 6.8**

What is the mass in grams of 5.0×10^{20} molecules of aspirin ($C_9H_8O_4$)?

■ **KEY CONCEPT PROBLEM 6.9**

What is the molecular weight of cytosine, a component of DNA (deoxyribonucleic acid)? (Gray = C, blue = N, red = O, ivory = H.)

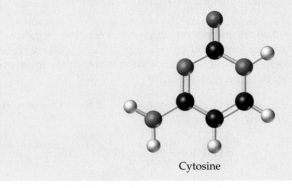

Cytosine

6.4 Gram–Mole Conversions

The mass in grams of one mole of any substance (that is, Avogadro's number of molecules or formula units) is called the **molar mass** of the substance.

Molar mass The mass in grams of one mole of a substance, numerically equal to molecular weight.

> Molar mass = Mass of 1 mole of substance
>
> = Mass of 6.022×10^{23} molecules (formula units) of substance
>
> = Molecular (formula) weight of substance in grams

Application

Did Ben Franklin Have Avogadro's Number? A Ballpark Calculation

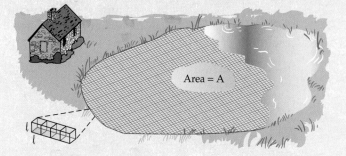

▲ What did these two have in common?

At length being at Clapham, where there is on the common a large pond . . . I fetched out a cruet of oil and dropped a little of it on the water. I saw it spread itself with surprising swiftness upon the surface. The oil, though not more than a teaspoonful, produced an instant calm over a space several yards square which spread amazingly and extended itself gradually . . . making all that quarter of the pond, perhaps half an acre, as smooth as a looking glass. *Excerpt from a letter of Benjamin Franklin to William Brownrigg, 1773.*

Benjamin Franklin, author and renowned statesman, was also an inventor and a scientist. Every schoolchild knows of Franklin's experiment with a kite and a key, demonstrating that lightning is electricity. Less well known is that his measurement of the extent to which oil spreads on water makes possible a simple estimate of molecular size and Avogadro's number.

The calculation goes like this: Avogadro's number is the number of molecules in a mole. So, if we can estimate both the number of molecules and the number of moles in Franklin's teaspoon of oil, we can calculate Avogadro's number. Let's start by calculating the number of molecules in the oil.

1. The volume (V) of oil Franklin used was 1 tsp = 4.9 cm^3, and the area (A) covered by the oil was 1/2 acre = 2.0 × 10^7 cm^2. Let's assume that the oil molecules are tiny cubes that pack closely together and form a layer only one molecule thick. As shown in the accompanying figure, the volume of the oil is equal to the surface area of the layer times the length (l) of the side of one molecule: $V = A \times l$. Rearranging this equation to find the length then gives us an estimate of molecular size:

$$l = \frac{V}{A} = \frac{4.9 \text{ cm}^3}{2.0 \times 10^7 \text{ cm}^2} = 2.5 \times 10^{-7} \text{ cm}$$

2. The area of the oil layer is the area of the side of one molecule (l^2) times the number of molecules (N) of oil: $A = l^2 \times N$. Rearranging this equation gives us the number of molecules:

$$N = \frac{A}{l^2} = \frac{2 \times 10^7 \text{ cm}^2}{(2.5 \times 10^{-7} \text{ cm})^2} = 3.2 \times 10^{20} \text{ molecules}$$

3. To calculate the number of moles, we first need to know the mass (M) of the oil. This could have been determined by weighing, but Franklin neglected to do so. Let's therefore estimate the mass by multiplying the volume (V) of the oil by the density (D) of a typical oil, 0.95 g/cm^3. [Since oil floats on water, it's not surprising that the density of oil is a bit less than the density of water (1.00 g/cm^3).]

$$M = V \times D = 4.9 \text{ cm}^3 \times 0.95 \frac{\text{g}}{\text{cm}^3} = 4.7 \text{ g}$$

4. We now have to make one final assumption about the molecular weight of the oil before completing the calculation. Assuming that a typical oil has MW = 200, then the mass of one mole of oil is 200 g. Dividing the mass of the oil (M) by the mass of one mole gives the number of moles of oil:

$$\text{Moles of oil} = \frac{4.7 \text{ g}}{200 \text{ g/mol}} = 0.024 \text{ mol}$$

5. Finally, the number of molecules per mole—Avogadro's number—can be obtained:

$$\text{Avogadro's number} = \frac{3.2 \times 10^{20} \text{ molecules}}{0.024 \text{ mol}}$$

$$= 1.3 \times 10^{22}$$

The calculation is not very accurate, of course, but Ben wasn't really intending for us to calculate Avogadro's number when he made a rough estimate of how much his oil spread out. Nevertheless, the result isn't too bad for such a simple experiment.

See Additional Problem 6.86 at the end of the chapter.

In effect, molar mass serves as a conversion factor between numbers of moles and mass. If you know how many moles you have, you can calculate their mass; if you know the mass of a sample, you can calculate the number of moles. Suppose, for example, we need to know how much 0.25 mol of water weighs. The molecular weight of H_2O is $(2 \times 1.0 \, amu) + 16.0 \, amu = 18.0 \, amu$, so the molar mass of water is 18.0 g. Thus, the conversion factor between moles of water and mass of water is 18.0 g/mol:

Molar mass used as conversion factor

$$0.25 \, \text{mol } H_2O \times \frac{18.0 \text{ g } H_2O}{1 \text{ mol } H_2O} = 4.5 \text{ g } H_2O$$

Alternatively, suppose we need to know how many moles of water are in 27 g of water. The conversion factor is 1 mol/18.0 g:

Molar mass used as conversion factor

$$27 \, \text{g } H_2O \times \frac{1 \text{ mol } H_2O}{18.0 \text{ g } H_2O} = 1.5 \text{ mol } H_2O$$

Worked Examples 6.6 and 6.7 give more practice in gram–mole conversions.

■ **WORKED EXAMPLE 6.6**

The nonprescription pain relievers Advil and Nuprin contain ibuprofen ($C_{13}H_{18}O_2$), whose molecular weight is 206.3 amu (Problem 6.6a). If the tablets in a bottle of pain reliever contain 0.082 mol of ibuprofen, what is the number of grams of ibuprofen in the bottle?

ANALYSIS We are given a number of moles and asked to find the mass. Molar mass is the conversion factor between the two.

SOLUTION Since the molar mass of ibuprofen is 206.3 g, we have

$$0.082 \, \text{mol } C_{13}H_{18}O_2 \times \frac{206.3 \text{ g } C_{13}H_{18}O_2}{1 \text{ mol } C_{13}H_{18}O_2} = 17 \text{ g } C_{13}H_{18}O_2$$

BALLPARK CHECK One mole of ibuprofen has a mass of about 200 g, so 0.08 mol will have a mass of about $0.08 \times 200 \text{ g} = 16 \text{ g}$.

■ **WORKED EXAMPLE 6.7**

The maximum dose of sodium hydrogen phosphate (Na_2HPO_4, MW = 142.0 amu) that should be taken in one day for use as a laxative is 3.8 g. How many moles of sodium hydrogen phosphate, how many moles of Na^+ ions, and how many total moles of ions are in this dose?

ANALYSIS Molar mass is the conversion factor between mass and number of moles. The chemical formula Na_2HPO_4 shows that each formula unit contains 2 Na^+ ions and 1 HPO_4^{2-} ion.

SOLUTION The number of moles of Na_2HPO_4 is

$$3.8 \, \text{g } Na_2HPO_4 \times \frac{1 \text{ mol } Na_2HPO_4}{142 \text{ g } Na_2HPO_4} = 0.027 \text{ mol } Na_2HPO_4$$

Since 1 mol of Na_2HPO_4 contains 2 mol of Na^+ ions and 1 mol of HPO_4^{2-} ions, we can multiply these values by the number of moles in the sample:

$$\frac{2 \text{ mol } Na^+}{1 \text{ mol } Na_2HPO_4} \times 0.027 \text{ mol } Na_2HPO_4 = 0.054 \text{ mol } Na^+$$

$$\frac{3 \text{ mol ions}}{1 \text{ mol } Na_2HPO_4} \times 0.027 \text{ mol } Na_2HPO_4 = 0.081 \text{ mol ions}$$

■ **PROBLEM 6.10**

How many moles of ethyl alcohol, C_2H_6O, are in a 10.0 g sample? How many grams are in a 0.10 mol sample of ethyl alcohol?

■ **PROBLEM 6.11**

Which weighs more, 5.00 g or 0.0225 mol of acetaminophen ($C_8H_9NO_2$)?

6.5 Mole Relationships and Chemical Equations

The coefficients in a balanced chemical equation tell how many *molecules*, and thus how many *moles*, of each reactant are needed and how many molecules, and thus moles, of each product are formed. You can then use molar mass to calculate reactant and product masses. If, for example, you saw the following balanced equation for the industrial synthesis of ammonia, you would know that 3 mol of H_2 (3 mol $\times$ 2.0 g/mol = 6.0 g) are required for reaction with 1 mol of N_2 (28.0 g) to yield 2 mol of NH_3 (2 mol $\times$ 17.0 g/mol = 34.0 g).

This number of moles of hydrogen . . . . . . reacts with this number of moles of nitrogen . . . to yield this number of moles of ammonia.

$$3 H_2 \ + \ 1 N_2 \longrightarrow 2 NH_3$$

The coefficients can be put in the form of *mole ratios*, which act as conversion factors when setting up factor-label calculations. In the ammonia synthesis, for example, the mole ratio of H_2 to N_2 is 3:1, the mole ratio of H_2 to NH_3 is 3:2, and the mole ratio of N_2 to NH_3 is 1:2.

$$\frac{3 \text{ mol } H_2}{1 \text{ mol } N_2} \quad \frac{3 \text{ mol } H_2}{2 \text{ mol } NH_3} \quad \frac{1 \text{ mol } N_2}{2 \text{ mol } NH_3}$$

Worked Example 6.8 shows an example of how to set up and use mole ratios.

■ **WORKED EXAMPLE 6.8**

Rusting involves the reaction of iron with oxygen to form iron(III) oxide, Fe_2O_3:

$$4 Fe(s) + 3 O_2(g) \rightarrow 2 Fe_2O_3(s)$$

(a) What are the mole ratios of the product to each reactant and of the reactants to each other?

(b) How many moles of iron(III) oxide are formed by the complete oxidation of 6.2 mol of iron?

ANALYSIS AND SOLUTION

(a) The coefficients of a balanced equation represent the mole ratios:

$$\frac{2 \text{ mol Fe}_2\text{O}_3}{4 \text{ mol Fe}} \qquad \frac{2 \text{ mol Fe}_2\text{O}_3}{3 \text{ mol O}_2} \qquad \frac{4 \text{ mol Fe}}{3 \text{ mol O}_2}$$

(b) To find how many moles of Fe_2O_3 are formed, write down the known information—6.2 mol of iron—and select the mole ratio that will allow quantities to cancel, leaving the desired quantity:

$$6.20 \text{ mol Fe} \times \frac{2 \text{ mol Fe}_2\text{O}_3}{4 \text{ mol Fe}} = 3.10 \text{ mol Fe}_2\text{O}_3$$

Note that mole ratios are exact numbers and therefore don't limit the number of significant figures in the result of a calculation.

■ **PROBLEM 6.12**

(a) Balance the following equation, and tell how many moles of nickel will react with 9.81 mol of hydrochloric acid.

$$\text{Ni}(s) + \text{HCl}(aq) \longrightarrow \text{NiCl}_2(aq) + \text{H}_2(g)$$

(b) How many moles of $NiCl_2$ can be formed in the reaction of 6.00 mol of Ni and 12.0 mol of HCl?

■ **PROBLEM 6.13**

Plants convert carbon dioxide and water to glucose ($C_6H_{12}O_6$) and oxygen in the process of photosynthesis. Write a balanced equation for this reaction, and determine how many moles of CO_2 are required to produce 15.0 mol of glucose.

6.6 Mass Relationships and Chemical Equations

Mole ratios make it possible to calculate the *molar* amounts of reactants and products, but actual amounts of substances used in the laboratory are weighed out in grams. Thus, we need to be able to carry out three kinds of conversions when doing chemical arithmetic:

• **Mole-to-mole conversions** are carried out using *mole ratios* as conversion factors. Worked Example 6.8 at the end of the previous section is an example of this kind of calculation.

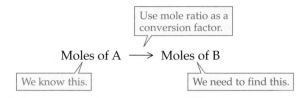

• **Mole-to-mass and mass-to-mole conversions** are carried out using *molar mass* as a conversion factor. Worked Examples 6.6 and 6.7 at the end of Section 6.4 are examples of this kind of calculation.

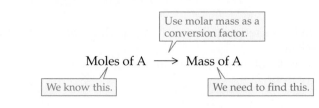

- **Mass-to-mass conversions** are frequently needed, but can't be carried out directly. If you know the mass of substance A and need to find the mass of substance B, you must first convert the mass of A into moles of A, then carry out a mole-to-mole conversion to find moles of B, and then convert moles of B into the mass of B (Figure 6.2).

Overall, there are four steps for determining mass relationships among reactants and products:

Step 1: Write the balanced chemical equation.

Step 2: Choose molar masses and mole ratios to convert the known information into the needed information.

Step 3: Set up the factor-label expression, and calculate the answer.

Step 4: Estimate or check the answer using a ballpark solution.

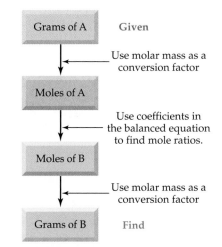

▲ **FIGURE 6.2** A summary of conversions between moles and grams for substances in a chemical reaction. The numbers of moles tell how many molecules of each substance are needed, as given by the coefficients in the balanced equation; the numbers of grams tell what mass of each substance is needed.

  Stoichiometry Calculation

■ **WORKED EXAMPLE 6.9**

In the atmosphere, nitrogen dioxide reacts with water to produce NO and nitric acid, which contributes to pollution by acid rain:

$$3\,NO_2(g) + H_2O(l) \longrightarrow 2\,HNO_3(aq) + NO(g)$$

How many grams of HNO_3 are produced for every 1.0 mol of NO_2 that reacts? The molecular weight of HNO_3 is 63.0 amu.

ANALYSIS We are given the number of moles of a reactant and are asked to find the mass of a product. Problems of this sort always require working in moles and then converting to mass, as outlined in Figure 6.2.

SOLUTION

Step 1: The balanced chemical equation is given above.

Step 2: We first need a mole-to-mole conversion to find the number of moles of product, and then we need a mole-to-mass conversion to find the mass of product. The mole-to-mole calculation is done using the mole ratio of NO_2 to HNO_3 as a conversion factor, and the mole-to-mass calculation is done using the molar mass of HNO_3 (63.0 g/mol) as a conversion factor.

Step 3:

$$1.0\ \text{mol NO}_2 \times \frac{2\ \text{mol HNO}_3}{3\ \text{mol NO}_2} \times \frac{63\ \text{g HNO}_3}{1\ \text{mol HNO}_3} = 42\ \text{g HNO}_3$$

Step 4: BALLPARK CHECK The molar mass of nitric acid is approximately 60 g/mol, and the coefficients in the balanced equation say that 2 mol of HNO_3 are formed for each 3 mol of NO_2 that undergoes reaction. Thus, 1 mol of NO_2 should give about 2/3 mol HNO_3, or 2/3 mol × 60 g/mol = 40 g.

■ **WORKED EXAMPLE 6.10**

The following reaction produced 0.022 g of calcium oxalate (CaC_2O_4). What mass of calcium chloride was used as reactant? (The molar mass of CaC_2O_4 is 128.1 g, and the molar mass of $CaCl_2$ is 111.0 g.)

$$CaCl_2(aq) + Na_2C_2O_4(aq) \longrightarrow CaC_2O_4(s) + 2\,NaCl(aq)$$

ANALYSIS Both the known information and that to be found **are** masses, so this is a mass-to-mass conversion problem. The mass of CaC_2O_4 is first converted into moles, a mole ratio is used to find moles of $CaCl_2$, **and** the number of moles of $CaCl_2$ is converted into mass.

SOLUTION

Step 1: The balanced equation is given above.

Step 2: Convert the mass of CaC_2O_4 into moles, use a mole ratio to find moles of $CaCl_2$, and convert the number of moles of $CaCl_2$ to mass.

Step 3:

$$0.022 \text{ g } CaC_2O_4 \times \frac{1 \text{ mol } CaC_2O_4}{128.1 \text{ g } CaC_2O_4} \times \frac{1 \text{ mol } CaCl_2}{1 \text{ mol } CaC_2O_4} \times \frac{111.0 \text{ g } CaCl_2}{1 \text{ mol } CaCl_2} = 0.019 \text{ g } CaCl_2$$

Step 4: Ballpark Check The balanced equation says that 1 mol of CaC_2O_4 is formed for each mole of $CaCl_2$ that reacts. Because the formula weights of the two substances are similar, it will take about 0.02 g of $CaCl_2$ to form 0.02 g of CaC_2O_4.

■ **PROBLEM 6.14**

Hydrogen fluoride is one of the few substances that reacts with glass (silicon dioxide, SiO_2).

$$4 \, HF(g) + SiO_2(s) \longrightarrow SiF_4(g) + 2 \, H_2O(l)$$

(a) How many moles of HF will react completely with 9.90 mol of SiO_2?
(b) What mass of water (in grams) will be produced by the reaction of 23.0 g of SiO_2?

■ **PROBLEM 6.15**

The tungsten metal used for filaments in light bulbs is made by reaction of tungsten trioxide with hydrogen:

$$WO_3(s) + 3 \, H_2(g) \longrightarrow W(s) + 3 \, H_2O(g)$$

How many grams of tungsten trioxide and how many grams of hydrogen must you start with to prepare 5.00 g of tungsten? For WO_3, MW = 231.8 amu.

▲ HF is one of the few substances that reacts with glass.

6.7 Percent Yield

All the calculations we've done in the last several sections have assumed that 100% of the reactants are converted to products. Only rarely is this the case in practice, though. More frequently, a majority of the reactant molecules behave as written but other processes, called *side reactions*, also occur. In addition, some of the product may be lost in handling. As a result, the amount of product actually formed—the reaction's **yield**—is somewhat less than the amount predicted by theory (the *theoretical yield*). The amount of product actually obtained in a reaction is usually expressed as a **percent yield**:

Yield The amount of product actually formed in a reaction.

Percent yield The percent of the theoretical yield actually obtained from a chemical reaction.

$$\text{Percent yield} = \frac{\text{Actual yield}}{\text{Theoretical yield}} \times 100\%$$

A reaction's actual yield is found by weighing the amount of product obtained. The theoretical yield is found by a mass-to-mass calculation like those illustrated in the preceding section. Worked Example 6.11 shows how to calculate a percent yield.

■ WORKED EXAMPLE 6.11

The element boron is produced commercially by the reaction of boric oxide with magnesium at high temperature:

$$B_2O_3(l) + 3\,Mg(s) \longrightarrow 2\,B(s) + 3\,MgO(s)$$

What is the percent yield if 675 g of boron is obtained from the reaction of 2350 g of boric oxide? The molecular weight of boric oxide is 69.6 amu.

ANALYSIS   To calculate a percent yield, we first have to calculate a theoretical yield and divide that into the actual yield. The theoretical yield is calculated by the same mass-to-mole and mole-to mass conversions used for all stoichiometry problems, as discussed in the previous section.

SOLUTION   The theoretical yield of boron from the reaction of 2350 g of boric oxide is

$$2350\text{ g B}_2\text{O}_3 \times \frac{1\text{ mol B}_2\text{O}_3}{69.6\text{ g B}_2\text{O}_3} \times \frac{2\text{ mol B}}{1\text{ mol B}_2\text{O}_3} \times \frac{10.8\text{ g B}}{1\text{ mol B}} = 729\text{ g B}$$

The percent yield is the actual yield divided by the theoretical yield and multiplied by 100%:

$$\text{Percent yield of B} = \frac{675\text{ g B}}{729\text{ g B}} \times 100\% = 92.6\%$$

BALLPARK CHECK   The starting 2350 g of B_2O_3 is about 2400/70, or 35 mol. Each mole of B_2O_3 is converted into 2 mol of B (molar mass = 10.8 g/mol), so the theoretical yield should be approximately $35 \times 2 \times 10$ g, or 700 g. Because the theoretical and actual yield are both close to 700 g, the percent yield should be close to 100%.

■ WORKED EXAMPLE 6.12

The reaction of ethylene with water to give ethyl alcohol (CH_3CH_2OH) occurs in 78.5% yield. How many grams of ethyl alcohol are formed by reaction of 25.0 g of ethylene? (For ethylene, MW = 28.0 amu; for ethyl alcohol, MW = 46.0 amu.)

$$H_2C{=}CH_2 + H_2O \longrightarrow CH_3CH_2OH$$

ANALYSIS   Treat this as a typical stoichiometry problem to find the amount of ethyl alcohol that can theoretically be formed from 25.0 g of ethylene, and then multiply the answer by 78.5% to find the amount actually formed.

SOLUTION   The theoretical yield of ethyl alcohol is

$$25.0\text{ g ethylene} \times \frac{1\text{ mol ethylene}}{28.0\text{ g ethylene}} \times \frac{1\text{ mol ethyl alc.}}{1\text{ mol ethylene}} \times \frac{46.0\text{ g ethyl alc.}}{1\text{ mol ethyl alc.}}$$

$$= 41.1\text{ g ethyl alcohol}$$

and the amount actually formed is

$$41.1\text{ g ethyl alc.} \times 0.785 = 32.3\text{ g ethyl alcohol}$$

BALLPARK CHECK   The 25.0 g of ethylene is a bit less than 1 mol, so a bit less than 0.78 mol of ethyl alcohol will form—perhaps about 3/4 mol, or $3/4 \times 46$ g = 34 g.

■ PROBLEM 6.16

What is the theoretical yield of ethyl chloride in the reaction of 19.4 g of ethylene with hydrogen chloride? What is the percent yield if 25.5 g of ethyl chloride is actually formed? (For ethylene, MW = 28.0 amu; for ethyl chloride, MW = 64.5 amu.)

$$H_2C{=}CH_2 + HCl \longrightarrow CH_3CH_2Cl$$

■ PROBLEM 6.17

The reaction of ethylene oxide with water to give ethylene glycol (automobile antifreeze) occurs in 96.0% yield. How many grams of ethylene glycol are formed by reaction of 35.0 g of ethylene oxide? (For ethylene oxide, MW = 44.0 amu; for ethylene glycol, MW = 62.0 amu.)

$$\underset{\text{Ethylene oxide}}{H_2C{-}CH_2} \overset{O}{} \quad + \quad H_2O \quad \longrightarrow \quad \underset{\text{Ethylene glycol}}{HOCH_2CH_2OH}$$

▲ Reaction of aqueous $Pb(NO_3)_2$ with aqueous KI gives a yellow precipitate of PbI_2.

Precipitate An insoluble solid that forms in solution during a chemical reaction.

Salt An ionic compound formed from reaction of an acid with a base.

6.8 Classes of Chemical Reactions

One of the best ways to understand any subject is to look for patterns that help to categorize large amounts of information. When learning about chemical reactions, for instance, it's helpful to group the reactions of ionic compounds into three general classes: *precipitation reactions, acid–base neutralization reactions,* and *oxidation–reduction reactions.* This is not the only possible way of categorizing reactions nor does the list include all possibilities, but it's useful nonetheless. Let's look briefly at an example of each of the three reaction classes before studying them in more detail in subsequent sections.

- **Precipitation reactions** are processes in which an insoluble solid called a **precipitate** forms when reactants are combined in aqueous solution. Most precipitations take place when the anions and cations of two ionic compounds change partners. For example, an aqueous solution of lead(II) nitrate reacts with an aqueous solution of potassium iodide to yield an aqueous solution of potassium nitrate plus an insoluble yellow precipitate of lead iodide:

$$Pb(NO_3)_2(aq) + 2\,KI(aq) \longrightarrow 2\,KNO_3(aq) + PbI_2(s)$$

- **Acid–base neutralization reactions** are processes in which an acid reacts with a base to yield water plus an ionic compound called a **salt**. We'll look at both acids and bases in more detail in Chapter 10, but you might recall for the moment that we previously defined acids as compounds that produce H^+ ions and bases as compounds that produce OH^- ions when dissolved in water (Section 4.12). (∞, p. 89) Thus, a neutralization reaction removes H^+ and OH^- ions from solution and yields neutral H_2O. The reaction between hydrochloric acid and sodium hydroxide is a typical example:

$$HCl(aq) + NaOH(aq) \longrightarrow H_2O(l) + NaCl(aq)$$

Note that in this reaction, the "salt" produced is sodium chloride, or common table salt. In a general sense, however, *any* ionic compound produced in an acid–base reaction is also called a salt.

• **Oxidation–reduction reactions,** or **redox reactions,** are processes in which one or more electrons are transferred between reaction partners (atoms, molecules, or ions). As a result of this transfer, the charges on atoms in the various reactants change. When metallic magnesium reacts with iodine vapor, for instance, a magnesium atom gives an electron to each of two iodine atoms, forming an Mg^{2+} ion and two I^- ions. The charge on the magnesium changes from 0 to +2, and the charge on each iodine changes from 0 to −1:

$$Mg(s) + I_2(g) \longrightarrow MgI_2(s)$$

■ **PROBLEM 6.18**

Classify each of the following processes as a precipitation, acid–base neutralization, or redox reaction.
(a) $AgNO_3(aq) + KCl(aq) \rightarrow AgCl(s) + KNO_3(aq)$
(b) $2\,Al(s) + 3\,Br_2(l) \rightarrow 2\,AlBr_3(s)$
(c) $Ca(OH)_2(aq) + 2\,HNO_3(aq) \rightarrow 2\,H_2O(l) + Ca(NO_3)_2(aq)$

6.9 Precipitation Reactions and Solubility Guidelines

Let's now look in more detail at precipitation reactions. To predict whether a precipitation reaction will occur on mixing aqueous solutions of two ionic compounds, you must know the **solubilities** of the potential products—how much of each compound will dissolve in a given amount of solvent at a given temperature. If a substance has a low solubility in water, then it is likely to precipitate from an aqueous solution. If a substance has a high solubility in water, then no precipitate will form.

Solubility is a complex matter, and it's not always possible to make correct predictions. As a rule of thumb, though, the following solubility guidelines for ionic compounds are useful.

Rule 1 **A compound is probably soluble if it contains one of the following** *cations*:
 • Group 1A cation: Li^+, Na^+, K^+, Rb^+, Cs^+
 • Ammonium ion: NH_4^+

Rule 2 **A compound is probably soluble if it contains one of the following** *anions*:
 • Halide: Cl^-, Br^-, I^-
 except Ag^+, Hg_2^{2+}, and Pb^{2+} compounds
 • Nitrate (NO_3^-), perchlorate (ClO_4^-), acetate $(CH_3CO_2^-)$, sulfate (SO_4^{2-})
 except Ba^{2+}, Hg_2^{2+}, and Pb^{2+} sulfates

If a compound does *not* contain at least one of the ions listed above, it is probably *not* soluble. Thus, Na_2CO_3 is soluble because it contains a group 1A cation, and $CaCl_2$ is soluble because it contains a halide anion. The compound $CaCO_3$, however, is *insoluble* because it contains none of the ions listed above. The guidelines are given in a different form in Table 6.1.

Let's try a problem. What will happen if aqueous solutions of sodium nitrate ($NaNO_3$) and potassium sulfate (K_2SO_4) are mixed? To answer this question, look at the guidelines to find the solubilities of the two possible products, Na_2SO_4 and KNO_3. Because both have group 1A cations (Na^+ and K^+), both are water-soluble and no precipitation will occur. If aqueous solutions of silver nitrate

Solubility The amount of a compound that will dissolve in a given amount of solvent at a given temperature.

 Precipitation Reactions

▲ Reaction of aqueous $AgNO_3$ with aqueous Na_2CO_3 gives a white precipitate of Ag_2CO_3.

TABLE 6.1 General Solubility Guidelines for Ionic Compounds in Water	
Soluble	**Exceptions**
Ammonium compounds (NH_4^+)	None
Lithium compounds (Li^+)	None
Sodium compounds (Na^+)	None
Potassium compounds (K^+)	None
Nitrates (NO_3^-)	None
Perchlorates (ClO_4^-)	None
Acetates ($CH_3CO_2^-$)	None
Chlorides (Cl^-)	
Bromides (Br^-)	Ag^+, Hg_2^{2+}, and Pb^{2+} compounds
Iodides (I^-)	
Sulfates (SO_4^{2-})	Ba^{2+}, Hg_2^{2+}, and Pb^{2+} compounds

($AgNO_3$) and sodium carbonate (Na_2CO_3) are mixed, however, the guidelines predict that a precipitate of insoluble silver carbonate (Ag_2CO_3) will form.

$$2\,AgNO_3(aq) + Na_2CO_3(aq) \longrightarrow Ag_2CO_3(s) + 2\,NaNO_3(aq)$$

■ **WORKED EXAMPLE 6.13**

Will a precipitation reaction occur when aqueous solutions of $CdCl_2$ and $(NH_4)_2S$ are mixed?

SOLUTION Identify the two potential products, and predict the solubility of each using the guidelines in the text. In the present instance, $CdCl_2$ and $(NH_4)_2S$ might give CdS and NH_4Cl. Since the guidelines predict that CdS is insoluble, a precipitation reaction will occur:

$$CdCl_2(aq) + (NH_4)_2S(aq) \longrightarrow CdS(s) + 2\,NH_4Cl(aq)$$

■ **PROBLEM 6.19**

Predict the solubility of the following compounds:
(a) $CdCO_3$ **(b)** Na_2S **(c)** $PbSO_4$ **(d)** $(NH_4)_3PO_4$ **(e)** Hg_2Cl_2

■ **PROBLEM 6.20**

Predict whether a precipitation reaction will occur in each of the following situations:
(a) $NiCl_2(aq) + (NH_4)_2S(aq) \rightarrow$ **(b)** $AgNO_3(aq) + CaBr_2(aq) \rightarrow$

6.10 Acids, Bases, and Neutralization Reactions

Neutralization reaction The reaction of an acid with a base.

When acids and bases are mixed in the right proportion, both acidic and basic properties disappear because of a **neutralization reaction.** The most common kind of neutralization reaction occurs between an acid that we can generalize as HA, and a metal hydroxide we can generalize as MOH, to yield water and a salt. The H^+ ion from the acid combines with the OH^- ion from the base to give neutral H_2O, while the anion from the acid (A^-) combines with the cation from the base (M^+) to give the salt:

 Dissolution of $Mg(OH)_2$ by Acid

A neutralization reaction: $HA(aq) + MOH(aq) \longrightarrow H_2O(l) + MA(aq)$
  Acid   Base   Water   A salt

Application — Gout and Kidney Stones: Problems in Solubility

One of the major pathways in the body for the breakdown of the nucleic acids DNA and RNA is by conversion to a substance called *uric acid*, $C_5H_4N_4O_3$, so named because it was first isolated in 1776 from urine. Most people excrete about 0.5 g of uric acid every day in the form of sodium urate, the salt that results from an acid–base reaction of uric acid. Unfortunately, the amount of sodium urate that dissolves in water (or urine) is fairly low—only about 0.07 mg/mL at the normal body temperature of 37°C. When too much sodium urate is produced or mechanisms for its elimination fail, its concentration in blood and urine rises, and the excess sometimes precipitates in the joints and kidneys.

Gout is a disorder of nucleic acid metabolism that primarily affects middle-aged men (only 5% of gout patients are women). It is characterized by an increased sodium urate concentration in blood, leading to the deposit of sodium urate crystals in soft tissue around the joints, particularly in the hands and at the base of the big toe. Deposits of the sharp, needle-like crystals cause an extremely painful inflammation that can lead ultimately to arthritis and even to bone destruction.

Just as increased sodium urate concentration in blood can lead to gout, increased concentration in urine can result in the formation of one kind of *kidney stones*, small crystals that precipitate in the kidney. Although often quite small, kidney stones cause excruciating pain when they pass through the ureter, the duct that carries

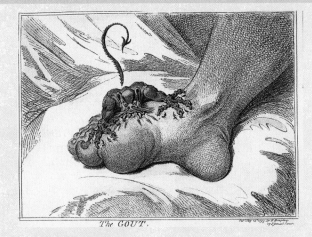

▲ This old cartoon leaves little doubt about how painful gout can be.

urine from the kidney to the bladder. In some cases, complete blockage of the ureter occurs.

Treatment of excessive sodium urate production involves both dietary modification and drug therapy. Foods such as liver, sardines, and asparagus should be avoided, and drugs such as allopurinol can be taken to lower production of sodium urate. Allopurinol functions by inhibiting the action of an enzyme called *xanthine oxidase*, thereby blocking a step in nucleic acid metabolism.

See Additional Problem 6.87 at the end of the chapter.

The reaction of hydrochloric acid with potassium hydroxide to produce potassium chloride is an example:

$$HCl(aq) + KOH(aq) \longrightarrow H_2O(l) + KCl(aq)$$

Another kind of neutralization reaction is that between an acid and a carbonate (or bicarbonate) to yield water, a salt, and carbon dioxide. Hydrochloric acid reacts with potassium carbonate, for example, to give H_2O, KCl, and CO_2:

$$2\,HCl(aq) + K_2CO_3(aq) \longrightarrow H_2O(l) + 2\,KCl(aq) + CO_2(g)$$

The reaction occurs because the carbonate ion (CO_3^{2-}) reacts initially with H^+ to yield H_2CO_3, which is unstable and immediately decomposes to give CO_2 plus H_2O.

We'll defer a more complete discussion of carbonates as bases until Chapter 10, but note for now that they yield OH^- ions when dissolved in water just as KOH and other bases do.

$$K_2CO_3(s) + H_2O \xrightarrow{\text{Dissolve in water}} 2\,K^+(aq) + HCO_3^-(aq) + OH^-(aq)$$

∞∞ **LOOKING AHEAD**

Acids and bases are enormously important in biological chemistry. We'll see in Chapter 18, for instance, how acids and bases affect the structure and properties of proteins. ∞∞

■ **WORKED EXAMPLE 6.14**

Write an equation for the neutralization reaction of aqueous HBr and aqueous $Ba(OH)_2$.

SOLUTION The reaction of HBr with $Ba(OH)_2$ involves the combination of a proton (H^+) from the acid with OH^- from the base to yield water and a salt ($BaBr_2$).

$$2\,HBr(aq) + Ba(OH)_2(aq) \longrightarrow 2\,H_2O(l) + BaBr_2(aq)$$

■ **PROBLEM 6.21**

Write and balance equations for each of the following acid–base neutralization reactions:
(a) $CsOH(aq) + H_2SO_4(aq) \rightarrow$
(b) $Ca(OH)_2(aq) + CH_3CO_2H(aq) \rightarrow$
(c) $NaHCO_3(aq) + HBr(aq) \rightarrow$

6.11 Net Ionic Equations

In the equations we've been writing up to this point, all the substances involved in reactions have been written using their full formulas. In the precipitation reaction of lead(II) nitrate with potassium iodide mentioned in Section 6.8, for example, only the parenthetical (aq) indicated that the reaction actually takes place in aqueous solution, and nowhere was it explicitly indicated that ions are involved:

$$Pb(NO_3)_2(aq) + 2\,KI(aq) \longrightarrow 2\,KNO_3(aq) + PbI_2(s)$$

In fact, though, lead(II) nitrate, potassium iodide, and potassium nitrate dissolve in water to yield solutions of ions. Thus, it's more accurate to write the reaction as an **ionic equation**, in which all the ions are explicitly shown:

Ionic equation An equation in which ions are explicitly shown.

An ionic equation: $Pb^{2+}(aq) + 2\,NO_3^-(aq) + 2\,K^+(aq) + 2\,I^-(aq)$
$$\longrightarrow 2\,K^+(aq) + 2\,NO_3^-(aq) + PbI_2(s)$$

A look at this ionic equation shows that the NO_3^- and K^+ ions undergo no change during the reaction. They appear on both sides of the reaction arrow and act merely as **spectator ions**, which are present but play no role. The actual reaction, when stripped to its essentials, can be described more simply by writing a **net ionic equation**, which includes only the ions that undergo change and ignores all spectator ions.

Spectator ion An ion that appears unchanged on both sides of a reaction arrow.

Net ionic equation An equation that does not include spectator ions.

Ionic equation: $Pb^{2+}(aq) + 2\,\cancel{NO_3^-}(aq) + 2\,\cancel{K^+}(aq) + 2\,I^-(aq) \longrightarrow$
$$2\,\cancel{K^+}(aq) + 2\,\cancel{NO_3^-}(aq) + PbI_2(s)$$

Writing a Net Ionic Equation

Net ionic equation: $Pb^{2+}(aq) + 2\,I^-(aq) \rightarrow PbI_2(s)$

Note that a net ionic equation, like all chemical equations, must be balanced both for atoms and for charge, with all coefficients reduced to their lowest whole numbers. Note also that all compounds that do *not* give ions in solution—all insoluble compounds and all molecular compounds—are represented by their full formulas.

■ WORKED EXAMPLE 6.15

Write net ionic equations for the following reactions:
(a) $AgNO_3(aq) + ZnCl_2(aq) \rightarrow$ (b) $HCl(aq) + Ca(OH)_2(aq) \rightarrow$

SOLUTION

(a) The solubility guidelines discussed in Section 6.9 predict that a precipitate of insoluble AgCl will form when aqueous solutions of Ag^+ and Cl^- are mixed. Writing all the ions separately gives an ionic equation, and eliminating spectator ions Zn^{2+} and NO_3^- gives the net ionic equation.

Ionic equation: $2\,Ag^+(aq) + 2\,\cancel{NO_3^-(aq)} + \cancel{Zn^{2+}(aq)} + 2\,Cl^-(aq) \longrightarrow$
$$2\,AgCl(s) + Zn^{2+}(aq) + 2\,NO_3^-(aq)$$

Net ionic equation: $2\,Ag^+(aq) + 2\,Cl^-(aq) \longrightarrow 2\,AgCl(s)$

The coefficients can all be divided by 2 to give

Net ionic equation: $Ag^+(aq) + Cl^-(aq) \longrightarrow AgCl(s)$

A check shows that the equation is balanced for atoms and charge (zero on each side).

(b) Allowing the acid HCl to react with the base $Ca(OH)_2$ will lead to a neutralization reaction. Writing the ions separately, and remembering to write a complete formula for water, gives an ionic equation. Then eliminating the spectator ions and dividing the coefficients by 2 gives the net ionic equation.

Ionic equation: $2\,H^+(aq) + 2\,Cl^-(aq) + Ca^{2+}(aq) + 2\,OH^-(aq) \longrightarrow$
$$2\,H_2O(l) + \cancel{Ca^{2+}(aq)} + \cancel{2\,Cl^-(aq)}$$

Net ionic equation: $H^+(aq) + OH^-(aq) \longrightarrow H_2O(l)$

A check shows that atoms and charges are the same on both sides of the equation.

■ PROBLEM 6.22

Write net ionic equations for the following reactions:
(a) $2\,Li(s) + Pb(NO_3)_2(aq) \rightarrow 2\,LiNO_3(aq) + Pb(s)$
(b) $2\,KOH(aq) + H_2SO_4(aq) \rightarrow K_2SO_4(aq) + 2\,H_2O(l)$
(c) $CuS(s) + 2\,HCl(aq) \rightarrow CuCl_2(aq) + H_2S(g)$

6.12 Redox Reactions

Redox reactions, the third and final category of reactions we'll discuss, is more complex than the other two. Look, for instance, at the following examples and see if you can tell what they have in common. Copper metal reacts with aqueous silver nitrate to form silver metal and aqueous copper(II) nitrate; iron rusts in air to form iron(III) oxide; the zinc metal container on the outside of a battery

Oxidation–reduction (redox) reaction A reaction in which electrons are transferred from one atom to another.

▲ The copper wire reacts with aqueous Ag^+ ion and becomes coated with metallic silver. At the same time, copper(II) ions go into solutions, producing the blue color.

Oxidation The loss of one or more electrons by an atom.

Reduction The gain of one or more electrons by an atom.

 Formation of Silver Crystals

Reducing agent A reactant that causes a reduction by giving up electrons to another reactant.

reacts with manganese dioxide and ammonium chloride inside the battery to generate electricity and give aqueous zinc chloride plus manganese(III) oxide. Although these and many thousands of other reactions appear unrelated, all are examples of **oxidation–reduction, or redox, reactions**.

$$Cu(s) + 2\,AgNO_3(aq) \longrightarrow 2\,Ag(s) + Cu(NO_3)_2(aq)$$

$$2\,Fe(s) + 3\,O_2(g) \longrightarrow Fe_2O_3(s)$$

$$Zn(s) + 2\,MnO_2(s) + 2\,NH_4Cl(s) \longrightarrow$$
$$ZnCl_2(aq) + Mn_2O_3(s) + 2\,NH_3(aq) + H_2O(l)$$

Historically, the word *oxidation* referred to the combination of an element with oxygen to yield an oxide, and the word *reduction* referred to the removal of oxygen from an oxide to yield the element. Today, though, the words have taken on a much broader meaning. An **oxidation** is now defined as the loss of one or more electrons by an atom, and a **reduction** is the gain of one or more electrons. Thus, an oxidation–reduction reaction, or redox reaction, is one in which *electrons are transferred from one atom to another.*

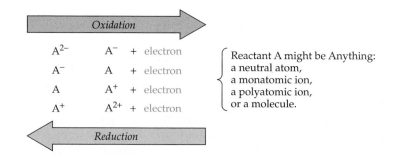

Take the reaction of copper with aqueous Ag^+ as an example. Copper metal gives an electron to each of two Ag^+ ions, forming Cu^{2+} and silver metal. Copper is oxidized in the process, and Ag^+ is reduced. You can follow the transfer of the electrons by noting that the charge on the copper increases from 0 to +2 when it loses two electrons, while the charge on Ag^+ decreases from +1 to 0 when it gains an electron.

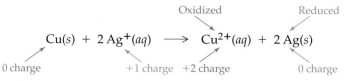

Similarly, in the reaction of aqueous iodide ion with bromine, iodide ion gives an electron to bromine, forming iodine and bromide ion. Iodide ion is oxidized as its charge increases from −1 to 0, and bromine is reduced as its charge decreases from 0 to −1.

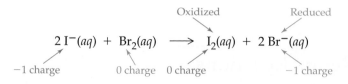

As these examples show, oxidation and reduction always occur together. Whenever one substance loses an electron (is oxidized), another substance must gain that electron (be reduced). The substance that gives up an electron and causes a reduction—the copper atom in the reaction of Cu with Ag^+ and the iodide ion in the reaction of I^- with Br_2—is called a **reducing agent**. The substance that gains an electron and causes an oxidation—the silver ion in the reaction of Cu with Ag^+ and the bromine molecule in the reaction of I^- with Br_2—is called an **oxidizing agent**. The charge on the reducing agent increases during the reaction, while the charge on the oxidizing agent decreases.

Reducing agent	Loses one or more electrons
	Causes reduction
	Undergoes oxidation
	Becomes more positive (or less negative)
Oxidizing agent	Gains one or more electrons
	Causes oxidation
	Undergoes reduction
	Becomes more negative (or less positive)

Oxidizing agent A reactant that causes an oxidation by taking electrons from another reactant.

Among the simplest of redox processes is the reaction of an element, usually a metal, with an aqueous cation to yield a different element and a different ion. Iron metal reacts with aqueous copper(II) ion, for example, to give iron(II) ion and copper metal. Similarly, magnesium metal reacts with aqueous acid to yield magnesium ion and hydrogen gas. In both cases, the reactant element (Fe or Mg) is oxidized, and the reactant ion (Cu^{2+} or H^+) is reduced.

$$Fe(s) + Cu^{2+}(aq) \longrightarrow Fe^{2+}(aq) + Cu(s)$$
$$Mg(s) + 2\,H^+(aq) \longrightarrow Mg^{2+}(aq) + H_2(g)$$

The reaction of a metal with water or aqueous acid (H^+) to release H_2 gas is a particularly important process. The alkali metals and alkaline earth metals on the left side of the periodic table are the most powerful reducing agents (electron donors), so powerful that they even react with pure water, in which the concentration of H^+ is very low. By contrast, metals toward the middle of the periodic table, such as iron and chromium, react only with aqueous acids but not with water. Those metals near the bottom right of the periodic table, such as platinum and gold, react with neither aqueous acid nor water. At the other extreme, the reactive nonmetals at the top right of the periodic table are extremely weak reducing agents and instead are powerful oxidizing agents (electron acceptors).

Redox reactions involve almost every element in the periodic table, and they occur in a vast number of processes throughout nature, biology, and industry. Here are just a few examples:

▲ Magnesium metal reacts with aqueous acid to give hydrogen gas and Mg^{2+} ion.

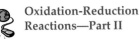
Oxidation-Reduction Reactions—Part II

- **Corrosion** is the deterioration of a metal by oxidation, such as the rusting of iron in moist air. The economic consequences of rusting are enormous: It has been estimated that up to one-fourth of the iron produced in the United States is used to replace bridges, buildings, and other structures that have been destroyed by corrosion. (The raised dot in the formula $Fe_2O_3 \cdot H_2O$ for rust indicates that one water molecule is associated with each Fe_2O_3 in an undefined way.)

$$4\,Fe(s) + 3\,O_2(g) \xrightarrow{H_2O} 2\,Fe_2O_3 \cdot H_2O(s)$$
$$\text{Rust}$$

- **Combustion** is the burning of a fuel by oxidation with oxygen in air. Gasoline, fuel oil, natural gas, wood, paper, and other organic substances of carbon and hydrogen are the most common fuels. Even some metals, though, will burn in air; magnesium and calcium are examples.

$$CH_4(g) + 2\,O_2(g) \longrightarrow CO_2(g) + 2\,H_2O(l)$$
Methane
(natural gas)

$$2\,Mg(s) + O_2(g) \longrightarrow 2\,MgO(s)$$

- **Respiration** is the process of breathing and using oxygen for the many biological redox reactions that provide the energy needed by living organisms. We'll see in Chapters 21–22 that energy is released from food molecules slowly and in complex, multistep pathways, but the overall result of respiration is similar to that of a combustion reaction. For example, the simple sugar glucose ($C_6H_{12}O_6$) reacts with O_2 to give CO_2 and H_2O according to the following equation:

$$C_6H_{12}O_6 + 6\,O_2 \longrightarrow 6\,CO_2 + 6\,H_2O + \text{energy}$$

Glucose
(a carbohydrate)

- **Bleaching** makes use of redox reactions to decolorize or lighten colored materials. Dark hair is bleached to turn it blond, clothes are bleached to remove stains, wood pulp is bleached to make white paper, and so on. The exact oxidizing agent used depends on the situation: hydrogen peroxide (H_2O_2) is used for hair, sodium hypochlorite (NaOCl) is used for clothes, and elemental chlorine is used for wood pulp, but the principle is always the same. In all cases, colored organic materials are destroyed by reaction with strong oxidizing agents.

- **Metallurgy,** the science of extracting and purifying metals from their ores, makes use of numerous redox processes. Worldwide, approximately 800 million tons of iron is produced each year by reduction of the mineral hematite, Fe_2O_3, with carbon monoxide.

$$Fe_2O_3(s) + 3\,CO(g) \longrightarrow 2\,Fe(s) + 3\,CO_2(g)$$

■ **PROBLEM 6.23**

Identify the oxidized reactant, the reduced reactant, the oxidizing agent, and the reducing agent in each of the following reactions:
(a) $Fe(s) + Cu^{2+}(aq) \rightarrow Fe^{2+}(aq) + Cu(s)$
(b) $Mg(s) + Cl_2(g) \rightarrow MgCl_2(s)$
(c) $2\,Al(s) + Cr_2O_3(s) \rightarrow 2\,Cr(s) + Al_2O_3(s)$

■ **PROBLEM 6.24**

Potassium, a silvery metal, reacts with bromine, a corrosive, reddish liquid, to yield potassium bromide, a white solid. Write the balanced equation, and identify the oxidizing and reducing agents.

Reactions with Oxygen

▲ Sulfur burns in air to yield SO_2. Is this a redox reaction?

6.13 Recognizing Redox Reactions

How can you tell when a redox reaction is taking place? When ions are involved, it's simply a matter of determining whether there is a change in the charges. When molecular substances are involved, though, it isn't as obvious. Is the combining of sulfur with oxygen a redox reaction? If so, which partner is the oxidizing agent and which is the reducing agent?

$$S(s) + O_2(g) \rightarrow SO_2(g)$$

Because oxygen is more electronegative than sulfur (Section 5.8), the oxygen atoms in SO_2 attract the electrons in the S—O bonds more strongly than sulfur does, giving the oxygen atoms a larger share of the electrons than sulfur. (∞, p. 118) By extending the ideas of oxidation and reduction to an increase

Application Batteries

Imagine life without batteries: no cars (they couldn't be started very easily without an electrical outlet nearby), no heart pacemakers, no flashlights, no hearing aids, no portable computers, radios, cellular phones, or thousands of other things. Modern society couldn't exist without batteries.

▲ How could society function without batteries?

Although they come in many types and sizes, all batteries are based on redox reactions. In a typical redox reaction carried out in the laboratory—say, the reaction of zinc metal with Ag^+ to yield Zn^{2+} and silver metal—the reactants are simply mixed in a flask and electrons are transferred by direct contact between the reactants. In a battery, however, the two reactants are kept in separate compartments and the electrons are transferred through a wire running between them.

The common household battery used for flashlights and radios is the *dry-cell*, developed in 1866. One reactant is a can of zinc metal, and the other is a paste of solid manganese dioxide. A graphite rod sticks into the MnO_2 paste to provide electrical contact, and a moist paste of ammonium chloride separates the two reactants. When the zinc can and the graphite rod are connected by a wire, zinc sends electrons flowing through the wire toward the MnO_2 in a redox reaction. The resultant electrical current can then be used to light a bulb or power a radio. The accompanying figure shows a cutaway view of a dry-cell battery.

$$Zn(s) + 2\,MnO_2(s) + 2\,NH_4Cl(s) \longrightarrow$$
$$ZnCl_2(aq) + Mn_2O_3(s) + 2\,NH_3(aq) + H_2O(l)$$

Closely related to the dry-cell battery is the familiar *alkaline* battery, in which the ammonium chloride paste is replaced by an alkaline, or basic, paste of NaOH or KOH. The alkaline battery has a longer life than the standard dry-cell battery because the zinc container corrodes less easily under basic conditions. The redox reaction is

$$Zn(s) + 2\,MnO_2(s) \longrightarrow ZnO(aq) + Mn_2O_3(s)$$

The batteries used in implanted medical devices such as pacemakers must be small, corrosion resistant, reliable, and able to last up to 10 years. Nearly all pacemakers being implanted today—about 750,000 each year—use titanium-encased lithium-iodine batteries, whose redox reaction is

$$2\,Li(s) + I_2(s) \longrightarrow 2\,LiI(aq)$$

See Additional Problem 6.88 at the end of the chapter.

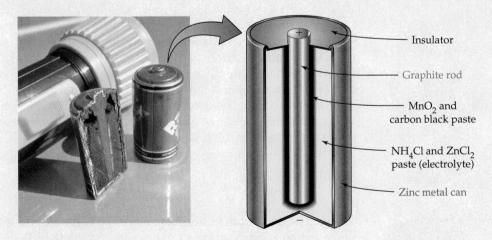

▲ A dry-cell battery. The cutaway view shows the two reactants that make up the redox reaction.

155

or decrease in electron *sharing* instead of complete electron *transfer*, we can say that the sulfur atom is oxidized in its reaction with oxygen because it loses a share in some electrons, while the oxygen atoms are reduced because they gain a share in some electrons.

A formal system has been devised for keeping track of changes in electron sharing, and thus for determining whether atoms are oxidized or reduced in reactions. We assign to each atom in a substance a value called an **oxidation number** (or *oxidation state*), which indicates whether the atom is neutral, electron-rich, or electron-poor. By comparing the oxidation number of an atom before and after reaction, we can tell whether the atom has gained or lost shares in electrons. Note that *oxidation numbers don't necessarily imply ionic charges*. They are simply a convenient device for keeping track of electrons in redox reactions.

The rules for assigning oxidation numbers are straightforward:

Oxidation number A number that indicates whether an atom is neutral, electron-rich, or electron-poor.

    Assignment of Oxidation Numbers

- **An atom in its elemental state has an oxidation number of 0.**

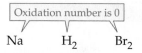

- **A monatomic ion has an oxidation number equal to its charge.**

- **In a molecular compound, an atom usually has the same oxidation number it would have if it were a monatomic ion.** Recall from Chapters 4 and 5 that the less electronegative elements (hydrogen and metals) on the left side of the periodic table tend to form cations, and the more electronegative elements (oxygen, nitrogen, and the halogens) near the top right of the periodic table tend to form anions. (∞, p. 79) Hydrogen and metals therefore have positive oxidation numbers in most compounds, while reactive nonmetals generally have negative oxidation numbers. Hydrogen is usually +1, oxygen is usually −2, nitrogen is usually −3, and halogens are usually −1:

$$
\overset{+1}{\text{H}}-\overset{-1}{\text{Cl}} \qquad \overset{+1}{\text{H}}-\overset{-2}{\text{O}}-\overset{+1}{\text{H}} \qquad \overset{+1}{\text{H}}-\overset{-3}{\underset{\underset{\overset{|}{\text{H}}\,\leftarrow\,+1}{}}{\text{N}}}-\overset{+1}{\text{H}}
$$

For compounds with more than one nonmetal element, such as SO_2, NO, or CO_2, the more electronegative element—oxygen in these examples—has a negative oxidation number and the less electronegative element has a positive oxidation number. Thus, in answer to the question posed at the beginning of the section, the combining of sulfur with oxygen to form SO_2 is a redox reaction because the oxidation number of sulfur increases from 0 to +4 while that of oxygen decreases from 0 to −2.

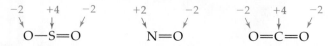

- **The sum of the oxidation numbers in a neutral compound is 0.** With this rule, the oxidation number of any atom in a compound can be found if the oxidation numbers of the other atoms are known. In the SO_2 example just mentioned, each of the two O atoms has an oxidation number of −2, so the S atom must have an oxidation number of +4. In HNO_3, the H atom has an oxidation number of +1 and the strongly electronegative O atom has an oxidation number of -2, so the N atom must have an oxidation number of $+5$. In a polyatomic ion, the sum of the oxidation numbers equals the charge on the ion.

$$\overset{+1}{H}-\overset{-2}{O}-\overset{+5}{N}=\overset{-2}{O} \qquad \text{Total} = 1 + 5 + 3(-2) = 0$$
$$\underset{-2}{|}$$
$$O$$

Worked Examples 6.16 and 6.17 show further instances of assigning and using oxidation numbers.

■ WORKED EXAMPLE 6.16

What is the oxidation number of the titanium atom in $TiCl_4$? Name the compound using a Roman numeral (Section 4.11).

SOLUTION Chlorine, a reactive nonmetal, is more electronegative than titanium and has an oxidation number of -1. Because there are four chlorine atoms in $TiCl_4$, the oxidation number of titanium must be $+4$. The compound is named titanium(IV) chloride. Note that the Roman numeral IV in the name of this molecular compound refers to the oxidation number $+4$ rather than to a true ionic charge.

■ WORKED EXAMPLE 6.17

Use oxidation numbers to show that the production of iron metal from its ore (Fe_2O_3) by reaction with charcoal (C) is a redox reaction. Which reactant has been oxidized, and which has been reduced? Which reactant is the oxidizing agent, and which is the reducing agent?

$$2\,Fe_2O_3(s) + 3\,C(s) \longrightarrow 4\,Fe(s) + 3\,CO_2(g)$$

SOLUTION The idea is to assign oxidation numbers to both reactants and products, and see if there has been a change. In the production of iron from Fe_2O_3, the oxidation number of Fe changes from $+3$ to 0, and the oxidation number of C changes from 0 to $+4$. Iron has thus been reduced (decrease in oxidation number), and carbon ion has been oxidized (increase in oxidation number). Oxygen is neither oxidized nor reduced because its oxidation number does not change. Carbon is the reducing agent, and Fe_2O_3 is the oxidizing agent.

$$\overset{+3}{2\,Fe_2}\overset{-2}{O_3} + \overset{0}{3\,C} \longrightarrow \overset{0}{4\,Fe} + \overset{+4\ -2}{3\,CO_2}$$

■ PROBLEM 6.25

What are the oxidation numbers of the metal atoms in the following compounds? Name each, using the oxidation number as a Roman numeral.
(a) VCl_3 (b) $SnCl_4$ (c) CrO_3 (d) $Cu(NO_3)_2$ (e) $NiSO_4$

■ PROBLEM 6.26

Assign an oxidation number to each atom in the following substances:
(a) N_2O_4 (b) $HClO_4$ (c) MnO_4^- (d) H_2SO_4

Application Photography: A Series of Redox Reactions

Photography is so common that most people never give a moment's thought to how remarkable the process really is. Ordinary black-and-white photographic film consists of a celluloid strip that has been coated with a gelatin emulsion containing very tiny crystals, or "grains," of a silver halide, usually AgBr. There is a considerable amount of art as well as science to making the film, and the recipes used by major manufacturers are well-guarded secrets. When exposed to light, the surfaces of the AgBr grains turn dark because of a light-induced redox reaction in which Br^- transfers an electron to Ag^+, producing atoms of elemental silver. (The Br_2 produced in the reaction reacts with the gelatin emulsion.) Those areas of the film exposed to the brightest light have the largest number of silver atoms, and those areas exposed to the least light have the smallest number.

$$2\,AgBr \xrightarrow{\text{Light}} 2\,Ag + Br_2$$

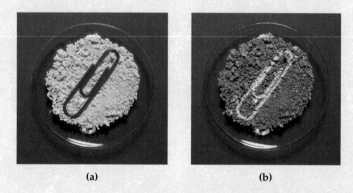

| (a) | (b) |

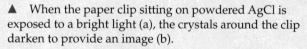

▲ When the paper clip sitting on powdered AgCl is exposed to a bright light (a), the crystals around the clip darken to provide an image (b).

Perhaps surprisingly in view of what a finished photograph looks like, only a few hundred out of many trillions of Ag^+ ions in each grain are reduced to Ag atoms, and the latent image produced on the film is still invisible at this point. The key to silver halide photography is the *developing* process, in which the latent image is amplified.

By mechanisms still not understood in detail, the presence of a relatively tiny number of Ag atoms on the surface of an AgBr grain sensitizes the remaining Ag^+ ions in the grain toward further reduction when the film is exposed to the organic reducing agent hydroquinone. Those grains that have been exposed to the strongest light—and thus have more Ag atoms—reduce and darken quickly, while those grains with fewer Ag atoms reduce and darken more slowly. By carefully monitoring the amount of time allowed for reduction of the AgBr grains with hydroquinone, it's possible to amplify the latent image on the exposed film and make it visible. At the desired point, further developing is stopped by transferring the film from the basic hydroquinone solution to a weakly acidic one.

Once the image is fully formed, the film is *fixed* by washing away the remaining unreduced AgBr so that the film is no longer sensitive to light. Although pure AgBr is insoluble in water, it is made soluble by reaction with a solution of sodium thiosulfate, $Na_2S_2O_3$, called *hypo* by photographers.

$$AgBr(s) + 2\,S_2O_3^{2-}(aq) \longrightarrow$$
$$Ag(S_2O_3)_2^{3-}(aq) + Br^-(aq)$$

At this point, the film contains a negative image formed by a layer of black, finely divided silver metal, a layer that is denser and darker in those areas exposed to the most light but lighter in those areas exposed to the least light. To convert this negative image into the final printed photograph, the entire photographic procedure is repeated a second time. Light is passed through the negative image onto special photographic paper that is coated with the same kind of gelatin–AgBr emulsion used on the original film. Developing the photographic paper with hydroquinone and fixing the image with sodium thiosulfate reverses the negative image, and a final, positive image is produced. The whole process from film to print is carried out billions of times and consumes over 3 million pounds of silver each year.

See Additional Problem 6.89 at the end of the chapter.

■ **PROBLEM 6.27**

Show by using oxidation numbers whether the following are redox reactions:

(a) $Na_2S(aq) + NiCl_2(aq) \rightarrow 2\,NaCl(aq) + NiS(s)$

(b) $2\,Na(s) + 2\,H_2O(l) \rightarrow 2\,NaOH(aq) + H_2(g)$

(c) $C(s) + O_2(g) \rightarrow CO_2(g)$

(d) $CuO(s) + 2\,HCl(aq) \rightarrow CuCl_2(aq) + H_2O(l)$

(e) $2\,MnO_4^-(aq) + 5\,SO_2(g) + 2\,H_2O(l) \rightarrow$
$$2\,Mn^{2+}(aq) + 5\,SO_4^{2-}(aq) + 4\,H^+(aq)$$

Summary: Revisiting the Chapter Goals

1. **How are chemical reactions written?** Chemical equations must be *balanced*; that is, the numbers and kinds of atoms must be the same in the reactants and the products. To balance an equation, *coefficients* are placed before formulas but the formulas themselves can't be changed.

2. **What is the mole, and why is it useful in chemistry?** A *mole* refers to *Avogadro's number* (6.022×10^{23}) of formula units of a substance. One mole of any substance has a mass (a *molar mass*) equal to the molecular or formula weight of the substance in grams. Because equal numbers of moles contain equal numbers of formula units, molar masses act as conversion factors between numbers of molecules and masses in grams.

3. **How are molar quantities and mass quantities related?** The coefficients in a balanced chemical equation represent the numbers of moles of reactants and products in a reaction. Thus, the ratios of coefficients act as *mole ratios* that relate amounts of reactants and/or products. By using molar masses and mole ratios in factor-label calculations, unknown masses or molar amounts can be found from known masses or molar amounts.

4. **What is the percent yield of a reaction?** The *yield* of a reaction is the amount of product obtained. The *percent yield* is the amount of product obtained divided by the amount theoretically possible and multiplied by 100%.

5. **How are chemical reactions of ionic compounds classified?** Three common types of reactions of ionic compounds can be identified: *Precipitation reactions* are processes in which an insoluble solid called a *precipitate* is formed. Most precipitations take place when the anions and cations of two ionic compounds change partners. Solubility guidelines for ionic compounds are used to predict when precipitation will occur.

Acid–base neutralization reactions are processes in which an acid reacts with a base to yield water plus an ionic compound called a *salt*. Since acids produce H^+ ions and bases produce OH^- ions when dissolved in water, a neutralization reaction removes H^+ and OH^- ions from solution and yields neutral H_2O.

Oxidation–reduction (redox) reactions are processes in which one or more electrons are transferred between reaction partners. An *oxidation* is defined as the loss of one or more electrons by an atom, and a *reduction* is the gain of one or more electrons. An *oxidizing agent* causes the oxidation of another reactant by accepting electrons, and a *reducing agent* causes the reduction of another reactant by donating electrons.

6. **What are oxidation numbers, and how are they used?** *Oxidation numbers* are assigned to atoms in reactants and products to provide a measure of whether an atom is neutral, electron-rich, or electron-poor. By comparing the oxidation number of an atom before and after reaction, we can tell whether the atom has gained or lost shares in electrons and thus whether a redox reaction has occurred.

Key Words

■ Understanding Key Concepts

6.28 Assume that the mixture of substances in drawing (a) undergoes a reaction. Which of the drawings (b)–(d) represents a product mixture consistent with the law of mass conservation?

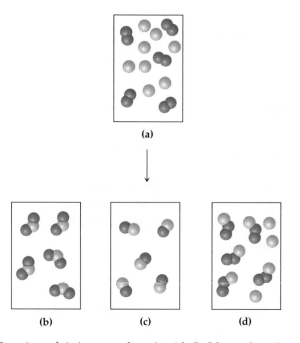

(a)

(b)   **(c)**   **(d)**

6.29 Reaction of A (green spheres) with B (blue spheres) is shown schematically in the following diagram:

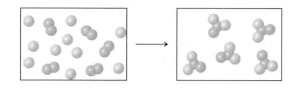

Which equation best describes the reaction?

(a) $A_2 + 2 B \rightarrow A_2B_2$
(a) $10 A + 5 B_2 \rightarrow 5 A_2B_2$
(a) $2 A + B_2 \rightarrow A_2B_2$
(a) $5 A + 5 B_2 \rightarrow 5 A_2B_2$

6.30 If blue spheres represent nitrogen atoms and red spheres represent oxygen atoms, which box represents reactants and which represents products for the reaction $2 NO(g) + O_2(g) \rightarrow 2 NO_2(g)$?

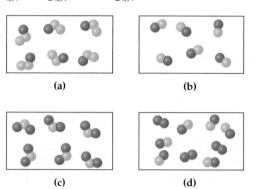

(a)   **(b)**

(c)   **(d)**

6.31 Methionine, an amino acid used by organisms to make proteins, can be represented by the following ball-and-stick molecular model. Write the formula for methionine, and give its molecular weight (red = O, gray = C, blue = N, yellow = S, ivory = H).

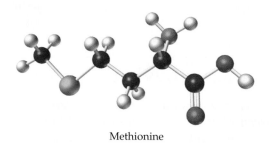

Methionine

6.32 The following diagram represents the reaction of A_2 (red spheres) with B_2 (blue spheres):

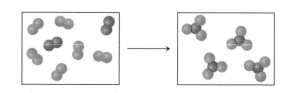

(a) Write a balanced equation for the reaction.
(b) How many moles of product can be made from 1.0 mol of A_2? From 1.0 mol of B_2?

6.33 Assume that an aqueous solution of a cation (represented as a red sphere) is allowed to mix with a solution of an anion (represented as a yellow sphere). Three possible outcomes are represented by boxes (1)–(3):

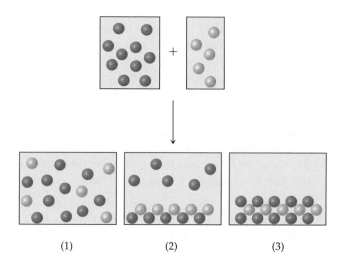

(1)   **(2)**   **(3)**

Which outcome corresponds to each of the following reactions?

(a) $2 Na^+(aq) + CO_3^{2-}(aq) \rightarrow$
(b) $Ba^{2+}(aq) + CrO_4^{2-}(aq) \rightarrow$
(c) $2 Ag^+(aq) + SO_3^{2-}(aq) \rightarrow$

6.34 Assume that an aqueous solution of a cation (represented as a blue sphere) is allowed to mix with a solution of an anion (represented as a green sphere) and that the following result is obtained:

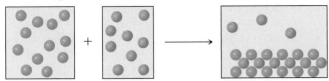

Which combinations of cation and anion, chosen from the following lists, are compatible with the observed results? Explain.

Cations: Na^+, Ca^{2+}, Ag^+, Ni^{2+}

Anions: Cl^-, CO_3^{2-}, CrO_4^{2-}, NO_3^-

6.35 The following drawing represents the reaction of ethylene oxide with water to give ethylene glycol, a compound used as automobile antifreeze. What mass in grams of ethylene oxide is needed to react with 9.0 g of water, and what mass in grams of ethylene glycol is formed?

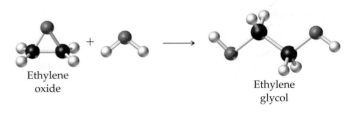

Ethylene oxide

Ethylene glycol

Additional Problems

Balancing Chemical Equations

6.36 What is meant by the term "balanced equation"?

6.37 Why is it not possible to balance an equation by changing the subscript on a substance, say from H_2O to H_2O_2?

6.38 Write balanced equations for the following reactions:
  (a) Gaseous sulfur dioxide reacts with water to form aqueous sulfurous acid (H_2SO_3).
  (b) Liquid bromine reacts with solid potassium metal to form solid potassium bromide.
  (c) Gaseous propane (C_3H_8) burns in oxygen to form gaseous carbon dioxide and water vapor.

6.39 Balance the following equation for the synthesis of hydrazine, N_2H_4, a substance used as rocket fuel.

$$NH_3(g) + Cl_2(g) \rightarrow N_2H_4(l) + NH_4Cl(s)$$

6.40 Which of the following equations are balanced? Balance those that need it.
  (a) $2 C_2H_6(g) + 5 O_2(g) \rightarrow 2 CO_2(g) + 6 H_2O(l)$
  (b) $3 Ca(OH)_2(aq) + 2 H_3PO_4(aq) \rightarrow$
  $$Ca_3(PO_4)_2(aq) + 6 H_2O(l)$$
  (c) $Mg(s) + O_2(g) \rightarrow 2 MgO(s)$
  (d) $K(s) + H_2O(l) \rightarrow KOH(aq) + H_2(g)$

6.41 Balance the following equations:
  (a) $Hg(NO_3)_2(aq) + LiI(aq) \rightarrow LiNO_3(aq) + HgI_2(s)$
  (b) $I_2(s) + Cl_2(g) \rightarrow ICl_5(s)$
  (c) $Al(s) + O_2(g) \rightarrow Al_2O_3(s)$
  (d) $CuSO_4(aq) + AgNO_3(aq) \rightarrow$
  $$Ag_2SO_4(s) + Cu(NO_3)_2(aq)$$
  (e) $Mn(NO_3)_3(aq) + Na_2S(aq) \rightarrow$
  $$Mn_2S_3(s) + NaNO_3(aq)$$
  (f) $NO_2(g) + O_2(g) \rightarrow N_2O_5(g)$
  (g) $P_4O_{10}(s) + H_2O(l) \rightarrow H_3PO_4(aq)$

6.42 Write a balanced equation for the reaction of aqueous sodium bicarbonate ($NaHCO_3$) with aqueous sulfuric acid (H_2SO_4) to yield CO_2, Na_2SO_4, and H_2O.

6.43 When organic compounds are burned, they react with oxygen to form CO_2 and H_2O. Write balanced equations for the combustion of the following:
  (a) C_4H_{10} (butane, used in lighters)
  (b) C_2H_6O (ethyl alcohol, used in gasohol and as race car fuel)
  (c) C_8H_{18} (octane, a component of gasoline)

Molar Masses and Moles

6.44 What is a mole of a substance? How many molecules are in one mole of a molecular compound?

6.45 What is the difference between molecular weight and formula weight? Between molecular weight and molar mass?

6.46 How many Ca^{2+} ions are in a mole of $CaCl_2$? How many Cl^- ions?

6.47 How many moles of ions are in 1.75 mol of K_2SO_4?

6.48 How many calcium atoms are in 16.2 g of calcium?

6.49 What is the mass in grams of 2.68×10^{22} atoms of uranium?

6.50 Caffeine has the formula $C_8H_{10}N_4O_2$. If an average cup of coffee contains approximately 125 mg of caffeine, how many moles of caffeine are in one cup?

6.51 How many moles of aspirin, $C_9H_8O_4$, are in a 500 mg tablet?

6.52 What is the molar mass of diazepam (Valium), $C_{16}H_{13}ClN_2O$?

6.53 Calculate the molar masses of the following substances:
  (a) Benzene, C_6H_6
  (b) Sodium bicarbonate, $NaHCO_3$
  (c) Chloroform, $CHCl_3$
  (d) Penicillin V, $C_{16}H_{18}N_2O_5S$

6.54 How many moles are present in 5.00 g samples of each of the compounds listed in Problem 6.53?

6.55 Iron(II) sulfate, $FeSO_4$, is used in the clinical treatment of iron-deficiency anemia. What is the molar mass of $FeSO_4$? How many moles of $FeSO_4$ are in a standard 300 mg tablet?

6.56 What is the mass in grams of 0.0015 mol of aspirin, $C_9H_8O_4$?

6.57 How many grams are present in 0.050 mol samples of each of the compounds listed in Problem 6.53?

Mole and Mass Relationships from Chemical Equations

6.58 At elevated temperatures in an automobile engine, N_2 and O_2 can react to yield NO, an important cause of air pollution.
  (a) Write a balanced equation for the reaction.
  (b) How many moles of N_2 are needed to react with 7.50 mol of O_2?
  (c) How many moles of NO can be formed when 3.81 mol of N_2 reacts?
  (d) How many moles of O_2 must react to produce 0.250 mol of NO?

6.59 Ethyl acetate reacts with H_2 in the presence of a catalyst to yield ethyl alcohol:

$$C_4H_8O_2(l) + H_2(g) \rightarrow C_2H_6O(l)$$

(a) Write a balanced equation for the reaction.
(b) How many moles of ethyl alcohol are produced by reaction of 1.5 mol of ethyl acetate?
(c) How many grams of ethyl alcohol are produced by reaction of 1.5 mol of ethyl acetate with H_2?
(d) How many grams of ethyl alcohol are produced by reaction of 12.0 g of ethyl acetate with H_2?
(e) How many grams of H_2 are needed to react with 12.0 g of ethyl acetate?

6.60 Ammonia, NH_3, is prepared for use as a fertilizer by reacting N_2 with H_2.

(a) Write a balanced equation for the reaction.
(b) How many moles of N_2 are needed for reaction to make 16.0 g of NH_3?
(c) How many grams of H_2 are needed to react with 75.0 g of N_2?

6.61 Iron metal can be produced from the mineral hematite, Fe_2O_3, by reaction with carbon:

$$Fe_2O_3(s) + C(s) \rightarrow Fe(s) + CO_2(g)$$

(a) Balance the equation.
(b) How many moles of carbon are needed to react with 525 g of hematite?
(c) How many grams of carbon are needed to react with 525 g of hematite?

6.62 An alternative method for preparing pure iron from Fe_2O_3 (Problem 6.61) is by reaction with carbon monoxide:

$$Fe_2O_3(s) + CO(g) \rightarrow Fe(s) + CO_2(g)$$

(a) Balance the equation.
(b) How many grams of CO are needed to react with 3.02 g of Fe_2O_3?
(c) How many grams of CO are needed to react with 1.68 mol of Fe_2O_3?

6.63 Magnesium metal burns in oxygen to form magnesium oxide, MgO.

(a) Write a balanced equation for the reaction.
(b) How many grams of oxygen are needed to react with 25.0 g of Mg? How many grams of MgO will result?
(c) How many grams of Mg are needed to react with 25.0 g of O_2? How many grams of MgO will result?

6.64 Titanium metal is obtained from the mineral rutile, TiO_2. How many kilograms of rutile are needed to produce 95 kg of Ti?

6.65 In the preparation of iron from hematite (Problem 6.61), how many kilograms of iron are present in 105 kg of hematite?

Percent Yield

6.66 Once made by heating wood in the absence of air, methanol (CH_3OH) is now made by reacting carbon monoxide and hydrogen at high pressure.

$$CO(g) + 2H_2(g) \rightarrow CH_3OH(l)$$

(a) How many grams of CH_3OH can be made from 10.0 g of CO if it all reacts?
(b) If 9.55 g of CH_3OH was recovered when the amounts in part (a) were used, what was the percent yield?

6.67 We saw in Problem 6.61 that iron metal can be made by heating iron ore with carbon according to the (unbalanced) equation:

$$Fe_2O_3(s) + C(s) \rightarrow Fe(s) + CO_2(g)$$

(a) How many kilograms of iron can be obtained from the reaction of 75.0 kg of Fe_2O_3?
(b) If 51.3 kg of iron was obtained from the reaction in part (a), what was the percent yield?

6.68 Dichloromethane, CH_2Cl_2, the solvent used to decaffeinate coffee beans, is prepared by reaction of CH_4 with Cl_2.

(a) Write the balanced equation. (HCl is also formed.)
(b) How many grams of Cl_2 are needed to react with 50.0 g of CH_4?
(c) How many grams of dichloromethane are formed from 50.0 g of CH_4 if the reaction occurs in 76% yield?

6.69 Cisplatin $[Pt(NH_3)_2Cl_2]$, a compound used in cancer treatment, is prepared by reaction of ammonia with potassium tetrachloroplatinate:

$$K_2PtCl_4 + 2NH_3 \rightarrow 2KCl + Pt(NH_3)_2Cl_2$$

(a) How many grams of NH_3 are needed to react with 55.8 g of K_2PtCl_4?
(b) How many grams of cisplatin are formed from 55.8 g of K_2PtCl_4 if the reaction takes place in 95% yield?

Types of Chemical Reactions

6.70 Write balanced ionic equations and net ionic equations for the following reactions:

(a) Aqueous hydrofluoric acid is neutralized by aqueous calcium hydroxide.
(b) Aqueous magnesium hydroxide is neutralized by aqueous nitric acid.

6.71 Write balanced ionic equations and net ionic equations for the following reactions:

(a) A precipitate of barium sulfate forms when aqueous solutions of barium nitrate and potassium sulfate are mixed.
(b) Zinc ion and hydrogen gas form when zinc metal reacts with aqueous sulfuric acid.

6.72 Which of the following substances are likely to be soluble in water?

(a) ZnS   (b) $Au_2(CO_3)_3$
(c) $PbCl_2$ (d) MnO_2

6.73 Which of the following substances are likely to be soluble in water?

(a) Ag_2O (b) $Ba(NO_3)_2$
(c) $SnCO_3$ (d) Al_2S_3

6.74 Use the solubility guidelines in Section 6.9 to predict whether a precipitation reaction will occur when aqueous solutions of the following substances are mixed.

(a) NaOH + $HClO_4$ (b) $FeCl_2$ + KOH
(c) $(NH_4)_2SO_4$ + $NiCl_2$

6.75 Use the solubility guidelines in Section 6.9 to predict whether precipitation reactions will occur between the listed pairs of reactants. Write balanced equations for those reactions that should occur.

(a) NaBr and $Hg_2(NO_3)_2$
(b) $CuCl_2$ and K_2SO_4
(c) $LiNO_3$ and $Ca(CH_3CO_2)_2$
(d) $(NH_4)_2CO_3$ and $CaCl_2$
(e) KOH and $MnBr_2$
(f) Na_2S and $Al(NO_3)_3$

6.76 Write net ionic equations for the following reactions:

(a) $Mg(s) + CuCl_2(aq) \rightarrow MgCl_2(aq) + Cu(s)$
(b) $2KCl(aq) + Pb(NO_3)_2(aq) \rightarrow$

$$PbCl_2(s) + 2KNO_3(aq)$$

(c) $2Cr(NO_3)_3(aq) + 3Na_2S(aq) \rightarrow$

$$Cr_2S_3(s) + 6NaNO_3(aq)$$

6.77 Write net ionic equations for the following reactions:

(a) $2\,AuCl_3(aq) + 3\,Sn(s) \rightarrow 3\,SnCl_2(aq) + 2\,Au(s)$

(b) $2\,NaI(aq) + Br_2(l) \rightarrow 2\,NaBr(aq) + I_2(s)$

(c) $2\,AgNO_3(aq) + Fe(s) \rightarrow Fe(NO_3)_2(aq) + 2\,Ag(s)$

Redox Reactions and Oxidation Numbers

6.78 Where in the periodic table are the best reducing agents found? The best oxidizing agents?

6.79 Where in the periodic table are the most easily reduced elements found? The most easily oxidized?

6.80 Tell in each of the following instances whether the substance gains electrons or loses electrons in a redox reaction:

(a) An oxidizing agent

(b) A reducing agent

(c) A substance undergoing oxidation

(d) A substance undergoing reduction

6.81 Tell for each of the following substances whether the oxidation number increases or decreases in a redox reaction:

(a) An oxidizing agent

(b) A reducing agent

(c) A substance undergoing oxidation

(d) A substance undergoing reduction

6.82 Assign an oxidation number to each element in the following compounds:

(a) NO_2 (b) SO_3 (c) $COCl_2$ (d) CH_2Cl_2

6.83 Assign an oxidation number to the metal in each of the following compounds:

(a) $CoCl_3$ (b) $FeSO_4$ (c) UO_3

(d) CuF_2 (e) TiO_2 (f) SnS

6.84 Which element is oxidized and which is reduced in each of the following reactions?

(a) $Si(s) + 2\,Cl_2(g) \rightarrow SiCl_4(l)$

(b) $Cl_2(g) + 2\,NaBr(aq) \rightarrow Br_2(aq) + 2\,NaCl(aq)$

(c) $SbCl_3(s) + Cl_2(g) \rightarrow SbCl_5(s)$

6.85 Which element is oxidized and which is reduced in each of the following reactions?

(a) $2\,SO_2(g) + O_2(g) \rightarrow 2\,SO_3(g)$

(b) $2\,Na(s) + Cl_2(g) \rightarrow 2\,NaCl(s)$

(c) $CuCl_2(aq) + Zn(s) \rightarrow ZnCl_2(aq) + Cu(s)$

(d) $2\,NaCl(aq) + F_2(g) \rightarrow 2\,NaF(aq) + Cl_2(g)$

Applications

6.86 What do you think are some of the errors involved in calculating Avogadro's number by spreading oil on a pond? [*Did Ben Franklin Have Avogadro's Number?*]

6.87 Sodium urate, the principal constituent of some kidney stones, has the formula $NaC_5H_3N_4O_3$. In aqueous solution, the solubility of sodium urate is only 0.067 g/L. How many moles of sodium urate dissolve in 1.00 L of water? [*Gout and Kidney Stones*]

6.88 What is the reducing agent in a typical dry-cell battery? [*Batteries*]

6.89 Explain how the negative image on photographic film is converted to a positive image when printed. [*Photography: A Series of Redox Reactions*]

General Questions and Problems

6.90 Zinc metal reacts with hydrochloric acid (HCl) according to the equation:

$$Zn(s) + 2\,HCl(aq) \rightarrow ZnCl_2(aq) + H_2(g)$$

(a) How many grams of hydrogen would be produced if 15.0 g of zinc reacted?

(b) Is this a redox reaction? If so, tell what is reduced, what is oxidized, what is the reducing agent, and what is the oxidizing agent.

6.91 Lithium oxide is used aboard the space shuttle to remove water from the atmosphere according to the equation

$$Li_2O(s) + H_2O(g) \rightarrow 2\,LiOH(s)$$

(a) How many grams of Li_2O must be carried on board to remove 80.0 kg of water?

(b) Is this a redox reaction? Why or why not?

6.92 Balance the following equations:

(a) The thermite reaction, used in welding:

$$Al(s) + Fe_2O_3(s) \rightarrow Al_2O_3(l) + Fe(l)$$

(b) The explosion of ammonium nitrate:

$$NH_4NO_3(s) \rightarrow N_2(g) + O_2(g) + H_2O(g)$$

6.93 Batrachotoxin, $C_{31}H_{42}N_2O_6$, an active component of South American arrow poison, is so toxic that $0.05\ \mu g$ can kill a person. How many molecules is this?

6.94 Look at the solubility guidelines in Section 6.9 and predict whether a precipitate will form when $CuCl_2(aq)$ and $Na_3PO_4(aq)$ are mixed. If so, write both the balanced equation and the net ionic equation for the process.

6.95 When table sugar (sucrose, $C_{12}H_{22}O_{11}$) is heated, it decomposes to form C and H_2O.

(a) Write a balanced equation for the process.

(b) How many grams of carbon are formed by the breakdown of 60.0 g of sucrose?

(c) How many grams of water are formed when 6.50 g of carbon are formed?

6.96 Although Cu is not sufficiently active to react with acids, it can be dissolved by concentrated nitric acid, which functions as an oxidizing agent according to the following equation:

$$Cu(s) + 4\,HNO_3(aq) \rightarrow$$
$$Cu(NO_3)_2(aq) + 2\,NO_2(g) + 2\,H_2O(l)$$

(a) Write the net ionic equation for this process.

(b) Are 35.0 g of HNO_3 sufficient to dissolve 5.00 g of copper?

6.97 The net ionic equation for the Breathalyzer test used to indicate alcohol concentration in the body is

$$16\,H^+(aq) + 2\,Cr_2O_7^{2-}(aq) + 3\,C_2H_6O(aq) \rightarrow$$
$$3\,C_2H_4O_2(aq) + 4\,Cr^{3+}(aq) + 11\,H_2O(l)$$

(a) How many grams of $K_2Cr_2O_7$ must be used to consume 1.50 g of C_2H_6O?

(b) How many grams of $C_2H_4O_2$ can be produced from 80.0 g of C_2H_6O?

6.98 Ethyl alcohol is formed by enzyme action on sugars and starches during fermentation:

$$C_6H_{12}O_6 \rightarrow 2\,CO_2 + 2\,C_2H_6O$$

If the density of ethyl alcohol is 0.789 g/mL, how many quarts can be produced by the fermentation of 100.0 lb of sugar?

6.99 Balance the following equations:

(a) $Al(OH)_3(aq) + HNO_3(aq) \rightarrow$
$$Al(NO_3)_3(aq) + H_2O(l)$$

(b) $AgNO_3(aq) + FeCl_3(aq) \rightarrow$
$$AgCl(s) + Fe(NO_3)_3(aq)$$

(c) $(NH_4)_2Cr_2O_7(s) \rightarrow Cr_2O_3(s) + H_2O(g) + N_2(g)$

(d) $Mn_2(CO_3)_3(s) \rightarrow Mn_2O_3(s) + CO_2(g)$

7 Chemical Reactions

Energy, Rates, and Equilibrium

A chemical reaction, like skiing, takes place spontaneously only if it goes downhill in energy. The reverse process, going uphill in energy, does not occur spontaneously.

There are many questions we still haven't answered about reactions. Why, for instance, do reactions occur? Certainly, the fact that a balanced equation can be written isn't enough. We can write a balanced equation for the reaction of gold with water, for example, but the reaction doesn't occur in practice, so your gold jewelry is safe in the shower.

Balanced, but does not occur $2\ Au(s)\ +\ 3\ H_2O(l) \rightarrow Au_2O_3(s)\ +\ 3\ H_2(g)$

To describe reactions more completely, we need to know the answers to several fundamental questions. Is energy released or absorbed when a reaction occurs? Is a given reaction fast or slow? Does a reaction continue until all reactants are converted to products or is there a point beyond which no additional product forms?

We'll answer these and the following questions about reactions in this chapter:

1. **What energy changes take place during reactions?**
   The goal: Be able to explain the factors that influence energy changes in chemical reactions.

2. **What is "free energy," and what is the criterion for spontaneity in chemistry?**
   The goal: Be able to define enthalpy, entropy, and free-energy changes, and explain how the values of these quantities affect chemical reactions.

3. **What determines the rate of a chemical reaction?**
   The goal: Be able to explain activation energy and other factors that determine reaction rate.

4. **What is chemical equilibrium?**
   The goal: Be able to describe what occurs in a reaction at equilibrium and write the equilibrium equation for a given reaction.

5. **What is Le Châtelier's principle?**
   The goal: Be able to state Le Châtelier's principle and use it to predict the effect on reactions of changes in temperature, pressure, and concentration.

CONCEPTS TO REVIEW

Energy and Heat (Section 2.10)
Chemical Equations (Section 6.1)

7.1 Heat Changes During Chemical Reactions

Why does chlorine react so easily with many elements and compounds while nitrogen does not? What difference between Cl_2 molecules and N_2 molecules accounts for their different reactivities? The answer is that the nitrogen–nitrogen triple bond is much *stronger* than the chlorine–chlorine single bond and can't be broken as easily in chemical reactions.

The strength of a covalent bond is measured by its **bond dissociation energy**, defined as the amount of energy that must be supplied to break the bond and separate the atoms in an isolated gaseous molecule. The triple bond in N_2, for example, has a bond dissociation energy of 226 kcal/mol, while the single bond in chlorine has a bond dissociation energy of only 58 kcal/mol:

Bond dissociation energy The amount of energy that must be supplied to break a bond and separate the atoms in an isolated gaseous molecule.

$$:N:::N: \xrightarrow{226\ kcal/mol} :\overset{\cdot}{N}\cdot\ +\ \cdot\overset{\cdot}{N}: \qquad N_2 \text{ bond dissociation energy } =\ 226\ kcal/mol$$

$$:\overset{\cdot\cdot}{\underset{\cdot\cdot}{Cl}}:\overset{\cdot\cdot}{\underset{\cdot\cdot}{Cl}}: \xrightarrow{58\ kcal/mol} :\overset{\cdot\cdot}{\underset{\cdot\cdot}{Cl}}\cdot\ +\ \cdot\overset{\cdot\cdot}{\underset{\cdot\cdot}{Cl}}: \qquad Cl_2 \text{ bond dissociation energy } =\ 58\ kcal/mol$$

A chemical change like bond breaking that absorbs heat is said to be **endothermic**, from the Greek words *endon* (within) and *therme* (heat), meaning that *heat* is put *in*. The reverse of bond breaking is bond formation, a process that *releases* heat and is described as **exothermic**, from the Greek *exo* (outside), meaning that heat goes *out*. The amount of energy released in forming a bond is numerically the same as that absorbed in breaking it. When Cl_2 molecules are

Endothermic A process or reaction that absorbs heat and has a positive ΔH.

Exothermic A process or reaction that releases heat and has a negative ΔH.

pulled apart into atoms, 58 kcal/mol of heat is absorbed; when chlorine atoms combine to give Cl_2, 58 kcal/mol of heat is released. Similarly, when nitrogen atoms combine to give N_2, 226 kcal/mol of heat is released.

$$:\dot{N}\cdot \;+\; \cdot\dot{N}: \;\longrightarrow\; :N:::N: \;+\; 226\text{ kcal/mol heat released}$$

$$:\ddot{\underset{..}{C}l}\cdot \;+\; \cdot\ddot{\underset{..}{C}l}: \;\longrightarrow\; :\ddot{\underset{..}{C}l}:\ddot{\underset{..}{C}l}: \;+\; 58\text{ kcal/mol heat released}$$

The same energy relationships that govern bond breaking and bond formation apply to every physical or chemical change. That is, the amount of heat transferred during a change in one direction is numerically equal to the amount of heat transferred during the change in the opposite direction. Only the *direction* of the heat transfer is different. This relationship reflects a fundamental law of nature called the *law of conservation of energy*:

Law of conservation of energy   Energy can be neither created nor destroyed in any physical or chemical change.

If more energy could be released by an exothermic reaction than was consumed in its reverse, the law would be violated and we could "manufacture" energy out of nowhere by cycling back and forth between forward and reverse reactions—a clear impossibility.

In every chemical reaction, some bonds are broken and some are formed. The difference between the energy absorbed in breaking bonds and that released in forming bonds is called the **heat of reaction** and is a quantity that can be measured. Heats of reaction that are measured when a reaction is held at constant pressure are represented by the abbreviation ΔH, where Δ (the Greek capital letter delta) is a general symbol used to indicate "a change in," and H is a quantity called *enthalpy*. Thus, the value of ΔH represents the **enthalpy change** that occurs during a reaction. The terms *enthalpy change* and *heat of reaction* are often used interchangeably, but we'll generally use the latter term in this book.

Heat of reaction ΔH = Energy of bonds formed in products minus
(Enthalpy change)   energy of bonds broken in reactants

Thermite

▲ The so-called "thermite" reaction of aluminum with iron(III) oxide is so strongly exothermic that it will melt iron.

7.2 Exothermic and Endothermic Reactions

When the total strength of the bonds formed in the products is *greater* than the total strength of the bonds broken in the reactants, energy is released and a reaction is exothermic. All combustion reactions are exothermic; for example, burning 1 mol of methane releases 213 kcal of heat. The heat released in an exothermic reaction can be thought of as a reaction product, and ΔH is assigned a *negative* value because heat *leaves*.

An exothermic reaction—negative ΔH

Heat is a product.

$$CH_4(g) \;+\; 2\,O_2(g) \;\longrightarrow\; CO_2(g) \;+\; 2\,H_2O(l) \;+\; 213\text{ kcal}$$

or

$$CH_4(g) \;+\; 2\,O_2(g) \;\longrightarrow\; CO_2(g) \;+\; 2\,H_2O(l) \qquad \Delta H = -213\text{ kcal/mol}$$

Note that ΔH is given in units of kilocalories per mole, where "per mole" means the reaction of *molar amounts of products and reactants as represented by the coefficients of the balanced equation*. Thus, the value $\Delta H = -213$ kcal/mol refers to the amount of heat released when 1 mol (16.0 g) of methane reacts with 2 mol of O_2 to give 1 mol of CO_2 gas and 2 mol of liquid H_2O. If we were to double the amount of methane from 1 mol to 2 mol, the amount of heat released would also double.

The quantities of heat released in the combustion of several fuels, including natural gas (which is primarily methane), are compared in Table 7.1. The values are given in kilocalories per gram to make comparisons easier. You can see from the table why there is interest in the potential of hydrogen as a fuel.

When the total energy of the bonds formed in the products is *less* than the total energy of the bonds broken in the reactants, energy is absorbed and a reaction is endothermic. The combination of nitrogen and oxygen to give nitrogen(II) oxide, a gas present in automobile exhaust, is such a reaction. The heat added in an endothermic reaction is like a reactant, and ΔH is assigned a *positive* value because heat is *added*.

An endothermic reaction—positive ΔH

Heat is a reactant.

$$N_2(g) + O_2(g) + 43 \text{ kcal} \longrightarrow 2 \text{ NO}(g)$$

or

$$N_2(g) + O_2(g) \longrightarrow 2 \text{ NO}(g) \qquad \Delta H = +43 \text{ kcal/mol}$$

Important points about heat transfers and chemical reactions

- An exothermic reaction releases heat to the surroundings; ΔH is negative.
- An endothermic reaction absorbs heat from the surroundings; ΔH is positive.
- The reverse of an exothermic reaction is endothermic.
- The reverse of an endothermic reaction is exothermic.
- The amount of heat absorbed or released in the reverse of a reaction is equal to that released or absorbed in the forward reaction, but ΔH has the opposite sign.

Worked Example 7.1 shows how to calculate the amount of heat absorbed or released for reaction of a given amount of reactant. All that's needed is the balanced equation and its accompanying ΔH. Mole ratios and molar masses are used for finding masses of reactants or products, as discussed in Sections 6.5 and 6.6.

■ WORKED EXAMPLE 7.1

Methane undergoes combustion with O_2 according to the following equation:

$$CH_4(g) + 2 \text{ O}_2(g) \longrightarrow CO_2(g) + 2 \text{ H}_2O(l) \quad \Delta H = -213 \text{ kcal/mol}$$

(a) How much heat is released during the combustion of 0.35 mol of methane (MW = 16.0 amu)?

(b) How much heat is released during the combustion of 8.00 g of methane?

(c) How much heat is released when 2.50 mol of O_2 reacts completely with methane?

ANALYSIS The value of ΔH given for the reaction, −213 kcal, is the amount of heat released when one mole of methane reacts with O_2. We need to find the amount of heat released when amounts other than one mole react.

SOLUTION

(a) Finding the amount of heat released by combustion of 0.35 mol of methane can be done with a factor-label calculation:

$$0.35 \text{ mol CH}_4 \times \frac{-213 \text{ kcal}}{1 \text{ mol CH}_4} = -75 \text{ kcal}$$

TABLE 7.1 Energy Values of Some Common Fuels

Fuel	Energy Value (kcal/g)
Wood (pine)	4.3
Ethyl alcohol	7.1
Coal (anthracite)	7.4
Crude oil (Texas)	10.5
Gasoline	11.5
Natural gas	11.7
Hydrogen	34.0

 Reduction of CuO

▲ The flares atop these oil wells are caused by burning methane gas.

(b) Going from a given mass of methane to the amount of heat released in a reaction requires finding the number of moles of methane by including molar mass in the calculation:

$$8.00 \text{ g } \cancel{CH_4} \times \frac{1 \text{ mol } \cancel{CH_4}}{16.0 \text{ g } \cancel{CH_4}} \times \frac{-213 \text{ kcal}}{1 \text{ mol } \cancel{CH_4}} = -107 \text{ kcal}$$

(c) Finding the amount of heat released by combustion of 2.5 mol of oxygen can be done with a factor-label calculation:

$$2.50 \cancel{\text{ mol } O_2} \times \frac{-213 \text{ kcal}}{2 \cancel{\text{ mol } O_2}} = -266 \text{ kcal}$$

BALLPARK CHECK
(a) Since 213 kcal is released for each mole of methane that reacts, 0.35 mol of methane should release about 1/3 of 213 kcal, or about 70 kcal.
(b) One mole of methane (MW = 16.0 amu) has a mass of 16 g, so 8 g of methane is 1/2 mole. Thus, about 1/2 of 213 kcal, or 106 kcal, is released from combustion of 8 g.
(c) The balanced equation shows that 213 kcal is released for each 2 mol of oxygen that reacts. Thus, 2.5 mol of oxygen should release a bit more than 213 kcal, perhaps about 250 kcal.

■ **PROBLEM 7.1**

In photosynthesis, green plants convert carbon dioxide and water into glucose $(C_6H_{12}O_6)$ according to the following equation:

$$6 \text{ CO}_2(g) + 6 \text{ H}_2O(l) + 678 \text{ kcal} \longrightarrow C_6H_{12}O_6(aq) + 6 \text{ O}_2(g)$$

(a) Is the reaction exothermic or endothermic?
(b) What is the value of ΔH for the reaction?
(c) Write the equation for the reverse of the reaction, including heat as a reactant or product.

■ **PROBLEM 7.2**

The following equation shows the conversion of aluminum oxide (from the ore bauxite) to aluminum:

$$2 \text{ Al}_2O_3(s) \longrightarrow 4 \text{ Al}(s) + 3 \text{ O}_2(g) \quad \Delta H = +801 \text{ kcal/mol}$$

(a) Is the reaction exothermic or endothermic?
(b) How many kilocalories are required to produce 1.00 mol of aluminum?
(c) How many kilocalories are required to produce 10.0 g of aluminum?

■ **PROBLEM 7.3**

How much heat is absorbed during production of 145 g of NO by the combination of nitrogen and oxygen?

$$N_2(g) + O_2(g) \longrightarrow 2 \text{ NO}(g) \quad \Delta H = +43 \text{ kcal/mol}$$

Application Energy from Food

Any serious effort to lose weight usually leads to studying the caloric values of foods. Have you ever wondered where the numbers quoted on food labels come from?

Food is "burned" in the body to yield H_2O, CO_2, and energy, just as natural gas is burned in furnaces to yield the same products. In fact, the "caloric value" of a food is just the heat of reaction for complete combustion of the food (minus a small correction factor). The value is the same whether the food is burned in the body or in the laboratory. One gram of protein releases 4 kcal, 1 g of table sugar (a carbohydrate) releases 4 kcal, and 1 g of fat releases 9 kcal (see Table).

Caloric Values of Some Foods

Substance, Sample Size	Caloric Value (kcal)
Protein, 1 g	4
Carbohydrate, 1 g	4
Fat, 1 g	9
Alcohol, 1 g	7.1
Cola drink, 12 fl oz (369 g)	160
Apple, one medium (138 g)	80
Iceberg lettuce, 1 cup shredded (55 g)	5
White bread, 1 slice (25 g)	65
Hamburger patty, 3 oz (85 g)	245
Pizza, 1 slice (120 g)	290
Vanilla ice cream, 1 cup (133 g)	270

The caloric value of a food is usually given in "Calories" (note the capital C), where 1 Cal = 1000 cal = 1 kcal. To determine these values experimentally, a carefully dried and weighed food sample is placed together with oxygen into an instrument called a *calorimeter*, the food is ignited, the temperature change is measured, and the amount of heat given off is calculated from the temperature change. In the calorimeter, the heat from the food is released very quickly and the temperature

▲ Eating this dessert would give your body 550 Calories. Burning the dessert in a calorimeter would release 550 kcal as heat.

rises dramatically. Clearly, though, something a bit different goes on when food is burned in the body, otherwise we'd burst into flames after a meal!

It's a fundamental principle of chemistry that the total heat released or absorbed in going from reactants to products is the same, no matter how many reactions are involved. The body applies this principle by withdrawing energy from food a bit at a time in a long series of interconnected reactions rather than all at once in a single reaction. These and other reactions continually taking place in the body—called the body's *metabolism*—will be examined in later chapters.

See Additional Problem 7.70 at the end of the chapter.

7.3 Why Do Chemical Reactions Occur? Free Energy

There is a natural tendency for events leading to lower energy to occur spontaneously. Water falls downhill, for instance, releasing its stored (potential) energy and reaching a lower-energy, more stable position. Similarly, a wound-up spring uncoils when set free. Applying this lesson to chemistry, the obvious conclusion is that exothermic processes—those that release heat energy—should be spontaneous. At the same time, endothermic processes, which absorb heat energy, should be nonspontaneous. Often, these conclusions are correct, but not always. Many, but not all, exothermic processes take place spontaneously, and many, but not all, endothermic processes are nonspontaneous.

▲ There is a natural tendency for events leading to lower energy to occur spontaneously. Thus, water always flows *down* a waterfall, not up.

Spontaneous process A process or reaction that, once started, proceeds on its own without any external influence.

Before exploring the situation further, it's important to understand what the word "spontaneous" means in chemistry, for it's not quite the same as in everyday language. A **spontaneous process** is one that, once started, proceeds on its own without any external influence. The change does not necessarily happen quickly, like a spring suddenly uncoiling or a car coasting downhill. It can also happen slowly, like the gradual rusting away of an abandoned bicycle. A *nonspontaneous process*, by contrast, takes place only in the presence of a continuous external influence: Energy must be continually expended to rewind a spring or push a car uphill. The reverse of a spontaneous process is always nonspontaneous.

As an example of a process that takes place spontaneously yet absorbs heat, think about what happens when you take an ice cube out of the refrigerator. The ice spontaneously melts to give liquid water above 0°C, even though it *absorbs* heat energy from the surroundings. What this and other spontaneous processes that absorb heat energy have in common is *an increase in molecular disorder or randomness*. When the solid ice melts, the H_2O molecules are no longer locked in position but are now free to move around randomly in the liquid water.

Entropy (S) A measure of the amount of molecular disorder in a system.

The amount of disorder in a system is called the system's **entropy**, symbolized S and expressed in units of calories per mole kelvin [cal/(mol·K)]. The greater the disorder or randomness of the particles in a substance or mixture, the larger the value of S (Figure 7.1). Gases have more disorder and higher entropy than liquids because particles in the gas move around more freely than particles in the liquid. Similarly, liquids have higher entropy than solids. In chemical reactions, entropy increases when, for example, a gas is produced from a solid or when 2 moles of reactants split into 4 moles of products.

Lower entropy
Solid ⟶ Liquid ⟶ Gas
Fewer moles of reactants ⟶ More moles of products
Higher entropy

FIGURE 7.1 The mixture on the right has more disorder and a higher entropy than the mixture on the left, and therefore has a higher value of S. The value of the entropy change, ΔS, for converting the mixture on the left to that on the right is positive because entropy increases.

Stir
Entropy increases (positive ΔS)

The entropy *change* for a process, ΔS, has a *positive* value if disorder increases, because disorder is added. The melting of ice to give water is an example. Conversely, ΔS has a *negative* value if disorder decreases. The freezing of water to give ice is an example.

It thus appears that two factors determine the spontaneity of a chemical or physical change: the release or absorption of heat, ΔH, and the increase or decrease in entropy, ΔS. *To decide whether a process is spontaneous, both the enthalpy change and the entropy change must be taken into account.*

When ΔH and ΔS are both favorable, a process is spontaneous; when both are unfavorable, a process is nonspontaneous. Clearly, however, the two factors don't have to operate in the same direction. It's possible for a process to be *unfavored* by enthalpy (absorb heat and have a positive ΔH) yet be *favored* by entropy (increase in disorder and have a positive ΔS). The melting of an ice cube above 0°C, for which $\Delta H = +6.01$ kJ/mol and $\Delta S = +22.0$ J/(mol·K), is just such a process. To take both heat of reaction (ΔH) and the change in disorder (ΔS) into account when determining the spontaneity of a process, a quantity called the **free-energy change** (ΔG) is needed:

Free-energy change

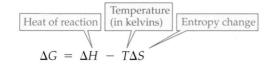

$$\Delta G = \Delta H - T\Delta S$$

The value of the free-energy change ΔG determines spontaneity. A negative value for ΔG means that free energy is released and the reaction or process is spontaneous. Such events are said to be **exergonic**. A positive value for ΔG means that free energy must be added and the process is nonspontaneous. Such events are said to be **endergonic**.

Exergonic A spontaneous reaction or process that releases free energy and has a negative ΔG.

Endergonic A nonspontaneous reaction or process that absorbs free energy and has a positive ΔG.

Spontaneous process ΔG is negative; free energy is released; process is exergonic.

Nonspontaneous process ΔG is positive; free energy is added; process is endergonic.

 Formation of Sodium Chloride

The equation for the free-energy change shows that spontaneity also depends on temperature (T). At low temperatures, the value of $T\Delta S$ is often small so that ΔH is the dominant factor. At a high enough temperature, however, the value of $T\Delta S$ can become larger than ΔH. Thus, an endothermic process that is nonspontaneous at low temperature can become spontaneous at a higher temperature. An example is the industrial synthesis of hydrogen by reaction of carbon with water:

$$\Delta H = +31.3 \text{ kcal/mol (Unfavorable)}$$

$$C(s) + H_2O(l) \longrightarrow CO(g) + H_2(g)$$

$$\Delta S = +32 \text{ cal/(mol·K) (Favorable)}$$

The reaction has an unfavorable (positive) ΔH term but a favorable (positive) ΔS term because disorder increases when a solid and a liquid are converted into two gases. No reaction occurs if carbon and water are mixed together at 25°C (298 K) because the unfavorable ΔH is larger than the favorable $T\Delta S$. Above about 700°C (973 K), however, the favorable $T\Delta S$ becomes larger than the unfavorable ΔH, so the reaction becomes spontaneous.

Important points about spontaneity and free energy

- A spontaneous process, once begun, proceeds without any external assistance and is exergonic; that is, it has a negative value of ΔG.

- A nonspontaneous process requires continuous external influence and is endergonic; that is, it has a positive value of ΔG.

- The value of ΔG for the reverse of a reaction is numerically equal to the value of ΔG for the forward reaction, but has the opposite sign.

- Some nonspontaneous processes become spontaneous with an increase in temperature.

⊂◯◯ **LOOKING AHEAD**

We'll see in later chapters that a knowledge of free-energy changes is especially important for understanding how metabolic reactions work. Living organisms can't raise their temperatures to convert nonspontaneous reactions into spontaneous reactions, so they must resort to other strategies, which we'll explore in Chapter 21. ◯◯◯

■ **WORKED EXAMPLE 7.2**

Does entropy increase or decrease in the following processes?
(a) Smoke from a cigarette disperses throughout a room rather than remaining in a cloud over the smoker's head.
(b) Water boils, changing from liquid to vapor.
(c) A chemical reaction occurs: $3 H_2(g) + N_2(g) \rightarrow 2 NH_3(g)$

ANALYSIS Entropy is a measure of molecular disorder, Entropy increases when the products are more disordered than the reactants; entropy decreases when the products are less disordered than the reactants.

SOLUTION
(a) Entropy increases because smoke particles are more disordered in the larger volume.
(b) Entropy increases because H_2O molecules have more freedom and disorder in the gas than in the liquid.
(c) Entropy decreases because 4 mol of reactant gas particles become 2 mol of product gas particles, with a consequent decrease in freedom and disorder.

■ **WORKED EXAMPLE 7.3**

The industrial method for synthesizing hydrogen by reaction of carbon with water has $\Delta H = +31.3$ kcal/mol and $\Delta S = +32$ cal/(mol · K). What is the value of ΔG for the reaction at 27°C (300 K)? Is the reaction spontaneous or nonspontaneous at this temperature?

$$C(s) + H_2O(l) \longrightarrow CO(g) + H_2(g)$$

SOLUTION Use the free-energy equation to determine the value of ΔG. (Remember that ΔS has units of *calories* per mol degree.)

$$\Delta G = \Delta H - T\Delta S$$

$$= +31.3 \frac{\text{kcal}}{\text{mol}} - (300\text{ K})\left(+32\frac{\text{cal}}{\text{mol} \cdot \text{K}}\right)\left(\frac{1\text{ kcal}}{1000\text{ cal}}\right)$$

$$= +21.7 \frac{\text{kcal}}{\text{mol}}$$

Because ΔG is positive, the reaction is nonspontaneous at 300 K.

■ **PROBLEM 7.4**

Does entropy increase or decrease in the following processes?
(a) After raking your leaves into a neat pile, a breeze blows them all over your lawn.
(b) Gasoline fumes escape as the fuel is pumped into your car.
(c) $Mg(s) + Cl_2(g) \rightarrow MgCl_2(s)$

■ **PROBLEM 7.5**

Lime (CaO) is prepared by the decomposition of limestone ($CaCO_3$).

$$CaCO_3(s) \longrightarrow CaO(s) + CO_2(g) \quad \Delta G = +31 \text{ kcal/mol at } 25°C$$

(a) Does the reaction occur spontaneously at 25°C?

(b) Does entropy increase or decrease in this reaction?

(c) Would you expect the reaction to be spontaneous at higher temperatures?

■ **PROBLEM 7.6**

The melting of solid ice to give liquid water has $\Delta H = +1.44$ kcal/mol and $\Delta S = +5.26$ cal/(mol · K).

(a) What is the value of ΔG for the melting process at $-10°C$ (263 K)? Is melting spontaneous or nonspontaneous at this temperature?

(b) What is the value of ΔG for the melting process at 0°C (273 K)? Is melting spontaneous or nonspontaneous at this temperature?

(c) What is the value of ΔG for the melting process at $+10°C$ (283 K)? Is melting spontaneous or nonspontaneous at this temperature?

■ **KEY CONCEPT PROBLEM 7.7**

The following drawing portrays a reaction of the type $A(s) \rightarrow B(s) + C(g)$, where the different colored spheres represent different molecular structures. Assume that the reaction has $\Delta H = -23.5$ kcal/mol.

(a) What is the sign of ΔS for the reaction?

(b) Is the reaction likely to be spontaneous at all temperatures, nonspontaneous at all temperatures, or spontaneous at some but nonspontaneous at others?

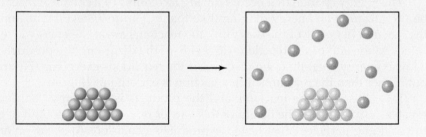

7.4 How Do Chemical Reactions Occur? Reaction Rates

Just because a chemical reaction has a favorable free-energy change doesn't mean it occurs rapidly. The value of ΔG tells us only whether a reaction *can* occur; it says nothing about how *fast* the reaction will occur or about the details of the molecular changes that take place during the reaction. It's now time to look into these other matters.

For a chemical reaction to occur, reactant particles must collide, some chemical bonds have to break, and new bonds have to form. Not all collisions lead to products, however. One requirement for a productive collision is that the colliding molecules must approach with the right orientation so that the atoms about to form new bonds can connect. In the reaction of ozone (O_3) with nitric oxide (NO) to give oxygen (O_2) and nitrogen dioxide (NO_2), for example, the two reactants must collide so that the nitrogen atom of NO strikes a terminal oxygen atom of O_3 (Figure 7.2).

FIGURE 7.2 For a collision ▶ between NO and O_3 molecules to give O_2 and NO_2, the molecules must collide so that the right atoms come into contact. No bond can form if the molecules collide with the wrong orientation.

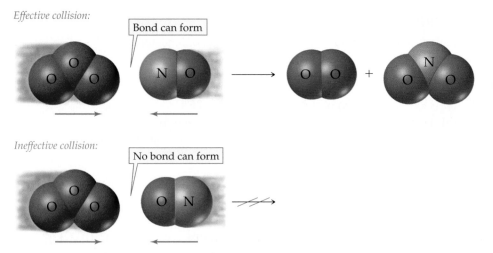

Effective collision:

Bond can form

Ineffective collision:

No bond can form

▲ Matches are unreactive at room temperature but burst into flames when struck. The frictional heat produced on striking provides enough energy to start the combustion

▼ **FIGURE 7.3** **Reaction energy diagrams are used to show energy changes during a chemical reaction.** A reaction begins on the left and proceeds to the right. (a) In an exergonic reaction, the product energy level is lower than that of reactants. (b) In an endergonic reaction, the situation is reversed. The height of the barrier between reactant and product energy levels is the activation energy, E_{act}. The difference between reactant and product energy levels is the free-energy change, ΔG.

Another requirement for a reaction to occur is that the collision must take place with enough energy to break the appropriate bonds in the reactant. If the reactant particles are moving slowly, a collision might be too gentle to overcome the repulsion between electrons in the different reactants, and the particles will simply bounce apart. Only if the collision is sufficiently energetic will a reaction ensue.

For this reason, many reactions with a favorable free-energy change don't occur at room temperature. To get such a reaction started, energy (heat) must be added. The heat causes the reactant particles to move faster, thereby increasing both the frequency and the force of the collisions. We all know that matches burn, for instance, but we also know that they don't burst into flame until struck. The heat of friction provides enough energy for a few molecules to react. Once started, the reaction sustains itself as the energy released by reacting molecules gives other molecules enough energy to react.

The energy change that occurs during the course of a chemical reaction can be visualized in an energy diagram like that in Figure 7.3. At the beginning of the reaction (left side of the diagram), the reactants are at the energy level indicated. At the end of the reaction (right side of the diagram), the products are at a lower energy level than the reactants if the reaction is exergonic (Figure 7.3a) but higher than the reactants if the reaction is endergonic (Figure 7.3b).

Lying between the reactants and the products is an energy "barrier" that must be surmounted. The height of this barrier represents the amount of energy the colliding particles must have for productive collisions to occur, an amount called the **activation energy,** E_{act}, of the reaction. The size of the activation energy determines the **reaction rate**, or how fast the reaction occurs. The lower the activation energy, the greater the number of productive collisions in a given amount of time, and the faster the reaction. Conversely, the higher the activation energy, the lower the number of productive collisions, and the slower the reaction.

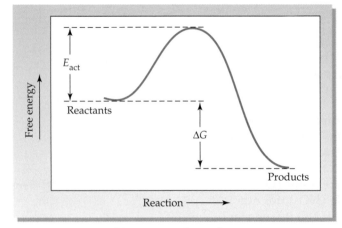

(a) An exergonic reaction

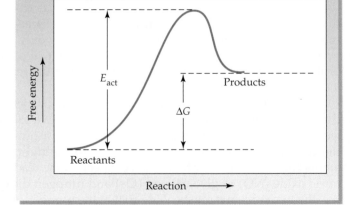

(b) An endergonic reaction

Note that the size of the activation energy and the size of the free-energy change are unrelated. A reaction with a large E_{act} will take place very slowly even if it has a large negative ΔG. Every reaction is different; each has its own characteristic activation energy and free-energy change.

Activation energy (E_{act}) The amount of energy necessary for reactants to surmount the energy barrier to reaction; determines reaction rate.

Reaction rate A measure of how rapidly a reaction occurs; determined by E_{act}.

■ **WORKED EXAMPLE 7.4**

Draw an energy diagram for a reaction that is very fast but has only a small negative free-energy change.

ANALYSIS   A very fast reaction is one that has a small E_{act}. A reaction that has a small negative free-energy change is one that is favorable but has only a small energy difference between starting materials and products.

SOLUTION

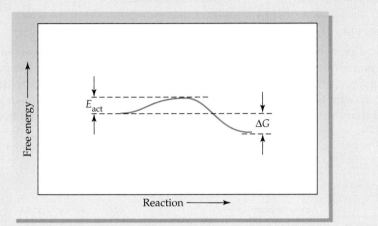

■ **PROBLEM 7.8**

Draw an energy diagram for a reaction that is very slow but highly favorable.

■ **PROBLEM 7.9**

Draw an energy diagram for a reaction that is slightly unfavorable.

7.5 Effects of Temperature, Concentration, and Catalysts on Reaction Rates

Several things can be done to help reactants over an activation energy barrier and thereby speed up a reaction. Let's look at some possibilities.

Temperature

One possibility for increasing a reaction rate is to add energy to the reactants by raising the temperature. With more energy in the system, the reactants move faster, so the frequency of collisions increases. Furthermore, the force with which collisions occur increases, making them more likely to overcome the activation barrier. As a rule of thumb, a 10°C rise in temperature causes a reaction rate to double.

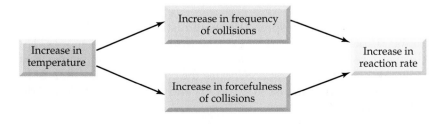

Concentration

A second possibility for speeding up a reaction is to increase the concentrations of the reactants. With reactants more crowded together, collisions become more frequent and reactions more likely. Flammable materials burn more rapidly in pure oxygen than in air, for instance, because the concentration of O_2 molecules is higher (air is approximately 21% oxygen). Hospitals must therefore take extraordinary precautions to be sure that no flames are used near patients receiving oxygen. Although different reactions respond differently to concentration changes, doubling or tripling a reactant concentration often doubles or triples the reaction rate.

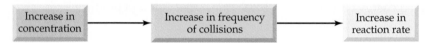

Catalysts

A third possibility for speeding up a reaction is to add a **catalyst**—a substance that accelerates a chemical reaction but is itself unchanged in the process. For example, such metals as nickel, palladium, and platinum catalyze the addition of hydrogen to the carbon–carbon double bonds in vegetable oils to yield semi-solid margarine. Without the metal catalyst, the reaction does not occur.

Catalyst A substance that speeds up the rate of a chemical reaction but is itself unchanged.

 Effect of a Catalyst on the Rate of a Reaction

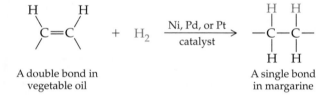

A double bond in vegetable oil   A single bond in margarine

A catalyst doesn't affect the energy level of either reactants or products. Rather, it increases a reaction rate by letting a reaction take place through an alternative pathway with a lower energy barrier. In a reaction energy diagram, the catalyzed reaction has a lower activation energy (Figure 7.4).

FIGURE 7.4 A reaction energy diagram for a reaction in the presence (green curve) and absence (red curve) of a catalyst. The catalyzed reaction has a lower E_{act} because it uses an alternative pathway with a lower energy barrier. The free-energy change ΔG is unaffected by the presence of a catalyst.

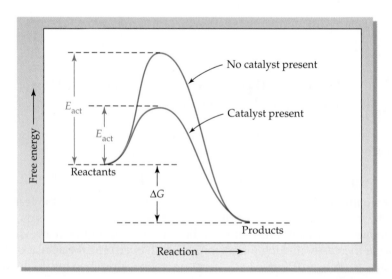

In addition to their widespread use in industry, we also rely on catalysts to reduce the air pollution created by exhaust from automobile engines. The catalytic converters in most automobiles are tubes packed with catalysts of two types (Figure 7.5). One catalyst accelerates the complete combustion of hydrocarbons and CO in the exhaust to give CO_2 and H_2O, while the other decomposes NO to N_2 and O_2.

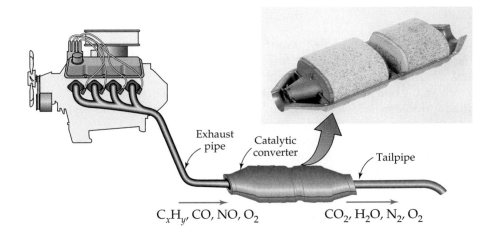

◀ **FIGURE 7.5** A catalytic converter. The exhaust gases from an automobile pass through a two-stage catalytic converter. In one stage, carbon monoxide and unburned hydrocarbons are converted to CO_2 and H_2O. In another stage, NO is converted to N_2 and O_2.

C_xH_y, CO, NO, O_2 → Exhaust pipe → Catalytic converter → Tailpipe → CO_2, H_2O, N_2, O_2

Table 7.2 summarizes the effects of changing conditions on reaction rates.

TABLE 7.2	Effects of Changes in Reaction Conditions on Rates
Change	**Effect**
Concentration	Increase in reactant concentration increases rate. Decrease in reactant concentration decreases rate.
Temperature	Increase in T increases rate. Decrease in T decreases rate.
Catalyst added	Increases reaction rate.

⧉ LOOKING AHEAD

The thousands of biochemical reactions continually taking place in our bodies are catalyzed by large protein molecules called *enzymes*. Since almost every reaction is catalyzed by its own specific enzyme, the study of enzyme structure, activity, and control is a central part of biochemistry. We'll look more closely at enzymes and how they work in Chapters 19 and 20. ⧉

■ **PROBLEM 7.10**

Ammonia is synthesized industrially by reaction of nitrogen and hydrogen in the presence of an iron catalyst according to the equation $3 H_2(g) + N_2(g) \rightarrow 2 NH_3(g)$. What effect will the following changes have on the reaction rate?

(a) The temperature is raised from 600 K to 700 K.

(b) The iron catalyst is removed.

(c) The concentration of H_2 gas is halved.

7.6 Reversible Reactions and Chemical Equilibrium

Many chemical reactions result in the virtually complete conversion of reactants into products. When sodium metal reacts with chlorine gas, for example, both are entirely consumed. The sodium chloride product is so much more stable than the reactants that, once started, the reaction keeps going until it is complete.

What happens, though, when the reactants and products are of approximately equal stability? This is the case, for example, in the reaction of acetic

Application Regulation of Body Temperature

Maintaining normal body temperature is crucial. If the body's thermostat is unable to maintain a temperature of 37°C, the rates of the many thousands of chemical reactions that take place constantly in the body will change accordingly, with potentially disastrous consequences.

If, for example, a skater fell through the ice of a frozen lake, *hypothermia* could soon result. Hypothermia is a dangerous state that occurs when the body is unable to generate enough heat to maintain normal temperature. All chemical reactions in the body slow down because of the lower temperature, energy production drops, and death can result. Slowing the body's reactions can also be used to advantage, however. During open-heart surgery, the heart is stopped and maintained at about 15°C, while the body, which receives oxygenated blood from an external pump, is cooled to 25–32°C.

Conversely, a marathon runner on a hot, humid day might become overheated, and *hyperthermia* could result. Hyperthermia, also called *heat stroke*, is an uncon-

trolled rise in temperature as the result of the body's inability to lose sufficient heat. Chemical reactions in the body are accelerated at higher temperatures, the heart struggles to pump blood faster to supply increased oxygen, and brain damage can result if the body temperature rises above 41°C.

Body temperature is maintained both by the thyroid gland and by the hypothalamus region of the brain, which act together to regulate metabolic rate. When the body's environment changes, temperature receptors in the skin, spinal cord, and abdomen send signals to the hypothalamus, which contains both heat-sensitive and cold-sensitive neurons.

Stimulation of the heat-sensitive neurons on a hot day causes a variety of effects: Impulses are sent to stimulate the sweat glands, dilate the blood vessels of the skin, decrease muscular activity, and reduce metabolic rate. Sweating cools the body through evaporation; approximately 540 cal is removed by evaporation of 1.0 g of sweat. Dilated blood vessels cool the body by allowing more blood to flow close to the surface of the skin where heat is removed by contact with air. Decreased muscular activity and a reduced metabolic rate cool the body by lowering internal heat production.

Stimulation of the cold-sensitive neurons on a cold day also causes a variety of effects: The hormone epinephrine is released to stimulate metabolic rate; peripheral blood vessels contract to decrease blood flow to the skin and prevent heat loss; and muscular contractions increase to produce more heat, resulting in shivering and "goosebumps."

One further comment: Drinking alcohol to warm up on a cold day actually has the opposite effect. Alcohol causes blood vessels to dilate, resulting in a warm feeling as blood flow to the skin increases. Although the warmth feels good temporarily, body temperature ultimately drops as heat is lost through the skin at an increased rate.

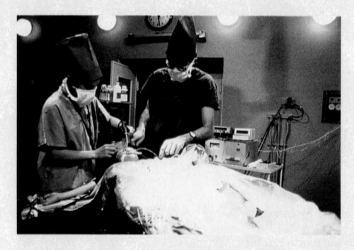

▲ The body is cooled to 25–32°C by immersion in ice prior to open-heart surgery to slow down metabolism.

See Problems 7.71 and 7.72 at the end of the chapter

---

acid (the main organic constituent of vinegar) with ethyl alcohol to yield ethyl acetate, a solvent used in nail-polish remover and glue.

$$\underset{\text{Acetic acid}}{CH_3\overset{\overset{\text{O}}{\|}}{C}OH} + \underset{\text{Ethyl alcohol}}{HOCH_2CH_3} \underset{\text{Or this direction?}}{\overset{\text{This direction?}}{\rightleftarrows}} \underset{\text{Ethyl acetate}}{CH_3\overset{\overset{\text{O}}{\|}}{C}OCH_2CH_3} + \underset{\text{Water}}{H_2O}$$

Reversible reaction A reaction that can go in either direction, from products to reactants or reactants to products.

Imagine the situation if you mix acetic acid and ethyl alcohol. The two would begin to form ethyl acetate and water. But as soon as ethyl acetate and water formed, *they* would begin to go back to acetic acid and ethyl alcohol. Such a reaction, which can easily go in either direction, is said to be **reversible** and is indicated by a double arrow ($\rightleftharpoons$) in equations. The reaction read from left to right as written is referred to as the *forward reaction*, and the reaction from right to left is referred to as the *reverse reaction*.

Now suppose you mix some ethyl acetate and water. The same thing would occur: As soon as small quantities of acetic acid and ethyl alcohol form, the reaction in the other direction would begin to take place. No matter which pair of reactants is mixed together, both reactions occur until ultimately the concentrations of reactants and products reach constant values and undergo no further change. At this point, the reaction vessel contains all four substances—acetic acid, ethyl acetate, ethyl alcohol, and water—and the reaction is said to be in a state of **chemical equilibrium**.

Since the reactant and product concentrations undergo no further change once equilibrium is reached, you might conclude that the forward and reverse reactions have stopped. That's not the case, however. The forward reaction takes place rapidly at the beginning of the reaction but then slows down as reactant concentrations decrease. At the same time, the reverse reaction takes place slowly at the beginning but then speeds up as product concentrations increase (Figure 7.6). Ultimately, the forward and reverse rates become equal and change no further.

Chemical equilibrium A state in which the rates of forward and reverse reactions are the same.

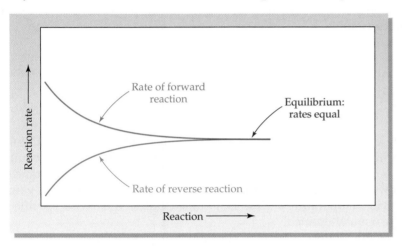

◄ **FIGURE 7.6 Reaction rates in an equilibrium reaction.** The forward rate is large initially but decreases as the concentrations of reactants drop. The reverse rate is small initially but increases as the concentrations of products increase. At equilibrium, the forward and reverse reaction rates are equal.

Chemical equilibrium is an active, dynamic condition. All substances present are continuously being made and unmade at the same rate, so their concentrations are constant at equilibrium. As an analogy, you might think of two floors of a building connected by up and down escalators. If the number of people moving up is the same as the number of people moving down, the numbers of people on each floor remain constant. *Individual people* are continuously changing from one floor to the other, but the *total populations* of the two floors are in equilibrium.

Note that it's not necessary for the concentrations of reactants and products at equilibrium to be equal (just as it's not necessary for the numbers of people on two floors connected by escalators to be equal). Equilibrium can be reached at any point between pure products and pure reactants. The extent to which the forward or reverse reaction is favored over the other is a characteristic property of a given reaction under given conditions.

7.7 Equilibrium Equations and Equilibrium Constants

Let's look at the details of a specific equilibrium reaction. Suppose that you allow various mixtures of sulfur dioxide and oxygen to come to equilibrium with sulfur trioxide and then measure the concentrations of all three gases in the mixtures.

$$2\,SO_2(g) + O_2(g) \rightleftharpoons 2\,SO_3(g)$$

▲ When the number of people moving up is the same as the number of people moving down, the number of people on each floor remains constant, and the two populations are in equilibrium.

No matter what the original concentrations were, and no matter what concentrations remain at equilibrium, a constant numerical value of 429 will be obtained if the equilibrium concentrations are substituted into the following expression. (The use of square brackets is a general notation for indicating that the concentration of the enclosed substance is measured in units of moles per liter.)

$$\frac{[SO_3]^2}{[SO_2]^2[O_2]} = 429 \quad \text{(at a constant temperature of 727°C)}$$

Numerous experiments like that just described have led to a general equation that is valid for any reaction. Let's consider a general, reversible reaction:

$$a\,A + b\,B + \cdots \rightleftharpoons m\,M + n\,N + \cdots$$

where A, B, ... are the reactants; M, N, ... are the products; and $a, b, \ldots, m, n, \ldots$ are the coefficients in the balanced equation. At equilibrium, the composition of the reaction mixture obeys the following *equilibrium equation*, where K is the **equilibrium constant**.

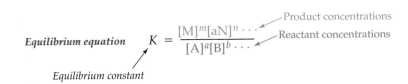

Equilibrium equation $K = \dfrac{[M]^m[aN]^n \cdots}{[A]^a[B]^b \cdots}$ Product concentrations
Reactant concentrations

Equilibrium constant

The equilibrium constant K is the number obtained by multiplying the equilibrium concentrations of the products and dividing by the equilibrium concentrations of the reactants, with the concentration of each substance raised to a power equal to its coefficient in the balanced equation. The value of K varies with temperature—25°C is assumed unless otherwise specified—and units are usually omitted.

The value of the equilibrium constant indicates the position of a reaction at equilibrium. If the forward reaction is favored, the product term $[M]^m[N]^n$ is larger than the reactant term $[A]^a[B]^b$, and the value of K is larger than 1. If instead the reverse reaction is favored, $[M]^m[N]^n$ is smaller than $[A]^a[B]^b$ at equilibrium, and the value of K is smaller than 1.

For a reaction such as the combination of hydrogen and oxygen to form water vapor, the equilibrium constant is enormous (3.1×10^{81}), showing how greatly the formation of water is favored. Equilibrium is effectively nonexistent for such reactions, and the reaction is described as *going to completion*.

On the other hand, the equilibrium constant is very small for a reaction such as the combination of nitrogen and oxygen at 25°C to give NO (4.7×10^{-31}), showing what we know from observation—that N_2 and O_2 in the air don't combine noticeably at room temperature:

▲ When ignited, hydrogen gas reacts completely with oxygen to yield water.

 Formation of Water

$$N_2(g) + O_2(g) \rightleftharpoons 2\,NO(g) \quad K = \frac{[NO]^2}{[N_2][O_2]} = 4.7 \times 10^{-31}$$

When K is close to 1, say between 10^3 and 10^{-3}, significant amounts of both reactants and products are present at equilibrium. An example is the reaction of acetic acid with ethyl alcohol to give ethyl acetate (Section 7.6). For this reaction, $K = 3.4$.

$$CH_3CO_2H + CH_3CH_2OH \rightleftharpoons CH_3CO_2CH_2CH_3 + H_2O$$

$$K = \frac{[CH_3CO_2CH_2CH_3][H_2O]}{[CH_3CO_2H][CH_3CH_2OH]} = 3.4$$

We can summarize the meaning of equilibrium constants in the following way:

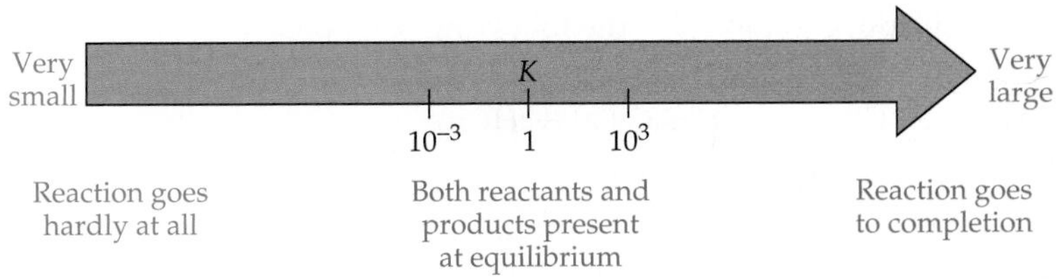

K **very much larger than 1**	Only products are present at equilibrium; reaction goes essentially to completion.
K **between 1 and 1000**	More products than reactants are present at equilibrium.
K **between 1 and 0.001**	More reactants than products are present at equilibrium.
K very much smaller than 1	Only reactants are present at equilibrium; essentially no reaction occurs.

■ **WORKED EXAMPLE 7.5**

The first step in the industrial synthesis of hydrogen is the reaction of steam with methane to give carbon monoxide and hydrogen. Write the equilibrium equation for the reaction.

$$H_2O(g) + CH_4(g) \rightleftharpoons CO(g) + 3\ H_2(g)$$

ANALYSIS The equilibrium constant K is the number obtained by multiplying the equilibrium concentrations of the products (CO and H_2) and dividing by the equilibrium concentrations of the reactants (H_2O and CH_4), with the concentration of each substance raised to the power of its coefficient in the balanced equation.

SOLUTION

$$K = \frac{[CO][H_2]^3}{[H_2O][CH_4]}$$

■ **WORKED EXAMPLE 7.6**

In the reaction of Cl_2 with PCl_3, the concentrations of reactants and products were determined experimentally at equilibrium and found to be 7.2 mol/L for PCl_3, 7.2 mol/L for Cl_2, and 0.050 mol/L for PCl_5.

$$PCl_3(g) + Cl_2(g) \rightleftharpoons PCl_5(g)$$

Write the equilibrium equation, and calculate the equilibrium constant for the reaction. Is the forward or the reverse reaction favored?

ANALYSIS
All the coefficients in the balanced equation are 1, so the equilibrium constant equals the concentration of the product, PCl_5, divided by the product of the concentrations of the two reactants, PCl_3 and Cl_2. Insert the values given for each concentration, and calculate the value of K.

SOLUTION

$$K = \frac{[PCl_5]}{[PCl_3][Cl_2]} = \frac{0.050}{(7.2)(7.2)} = 9.6 \times 10^{-4}$$

The value of K is less than 1, so the reverse reaction is favored. Note that units are omitted.

■ **PROBLEM 7.11**

Write equilibrium equations for the following reactions:

(a) $N_2O_4(g) \rightleftarrows 2\, NO_2(g)$

(b) $CH_4(g) + Cl_2(g) \rightleftarrows CH_3Cl(g) + HCl(g)$

(c) $2\, BrF_5(g) \rightleftarrows Br_2(g) + 5\, F_2(g)$

■ **PROBLEM 7.12**

Do the following reactions favor reactants or products at equilibrium?

(a) Sucrose(aq) + H_2O(l) $\rightleftarrows$ Glucose(aq) + Fructose(aq) $K = 1.4 \times 10^5$

(b) $NH_3(aq)$ + H_2O(l) $\rightleftarrows$ NH_4^+(aq) + OH^-(aq) $K = 1.6 \times 10^{-5}$

(c) $Fe_2O_3(s)$ + $3\, CO(g)$ $\rightleftarrows$ $2\, Fe(s)$ + $3\, CO_2(g)$ K (at 727°C) = 24.2

■ **PROBLEM 7.13**

For the reaction $H_2(g) + I_2(g) \rightleftharpoons 2\, HI(g)$, equilibrium concentrations at 25°C are $[H_2] = 0.0250$ mol/L, $[I_2] = 0.0869$ mol/L, and $[HI] = 0.251$ mol/L. What is the value of K at 25°C?

■ **KEY CONCEPT PROBLEM 7.14**

The following pictures represent two similar reactions that have achieved equilibrium:

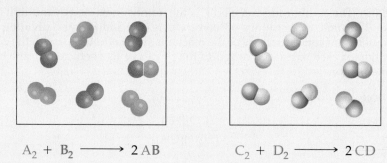

$A_2 + B_2 \longrightarrow 2\, AB$ $C_2 + D_2 \longrightarrow 2\, CD$

Which reaction has the larger equilibrium constant, and which has the smaller equilibrium constant?

7.8 Le Châtelier's Principle: The Effect of Changing Conditions on Equilibria

LeChâtelier's Principle

The effect on a chemical equilibrium of a change in reaction conditions can be predicted by a general rule called *Le Châtelier's principle*:

> **Le Châtelier's principle**   When a stress is applied to a system at equilibrium, the equilibrium shifts to relieve the stress.

The word "stress" in this context means any change in concentration, pressure, volume, or temperature that disturbs the original equilibrium and causes the rates of the forward and reverse reactions to become temporarily unequal.

We saw in Section 7.5 that reaction rates are affected by changes in temperature and concentration, and by addition of a catalyst. But what about equilibria? Are they similarly affected? The answer is that changes in concentration, temperature, and pressure *do* affect equilibria, but that addition of a catalyst does not (except to reduce the time it takes to reach equilibrium). The change that a catalyst causes in E_{act} affects forward and reverse reactions equally so that equilibrium concentrations are the same in both the presence and the absence of the catalyst.

Effect of Changes in Concentration

Let's look at the effect of a concentration change by considering the reaction of CO with H_2 to form CH_3OH (methanol). Once equilibrium has been reached, the concentrations of the reactants and product are constant, and the forward and reverse reaction rates are equal.

$$CO(g) + 2\ H_2(g) \rightleftharpoons CH_3OH(g)$$

What will happen if the concentration of CO is increased? To relieve the "stress" of added CO, according to Le Châtelier's principle, the extra CO must be used up. In other words, the rate of the forward reaction must increase to consume CO. You can think of the CO added on the left as "pushing" the equilibrium to the right:

$$[CO \longrightarrow]$$
$$CO(g) + 2\ H_2(g) \rightleftharpoons CH_3OH(g)$$

Of course, as soon as more CH_3OH forms, the reverse reaction will also speed up and some product will be converted back to CO and H_2. Ultimately, the forward and reverse reaction rates will adjust until they are again equal and equilibrium is reestablished. At this new equilibrium state, the value of $[H_2]$ will be lower, because more H_2 has reacted with the added CO, and the value of $[CH_3OH]$ will be higher. The changes offset each other, however, so the value of the equilibrium constant K remains constant.

$$CO(g) + 2\ H_2(g) \rightleftharpoons CH_3OH(g)$$

If this increases . . .   . . . then this decreases . . .   . . . and this increases . . .

. . . but this remains constant. $K = \dfrac{[CH_3OH]}{[CO]\,[H_2]^2}$

What will happen if CH_3OH is added to the reaction at equilibrium? Some of the methanol will react to yield CO and H_2, making the values of $[CO]$, $[H_2]$, and $[CH_3OH]$ higher when equilibrium is reestablished. Again, though, the value of K is unchanged.

If this increases . . .

$$CO(g) + 2\ H_2(g) \rightleftharpoons CH_3OH(g)$$

. . . then this increases . . .   . . . and this increases . . .

. . . but this remains constant. $K = \dfrac{[CH_3OH]}{[CO]\,[H_2]^2}$

Finally, what will happen if a reactant is continuously supplied or a product is continuously removed? Because the concentrations are continuously changing, equilibrium can never be reached. As a result, it's sometimes possible to force a reaction to produce large quantities of a desirable product even when the equilibrium constant is unfavorable. Take the reaction of acetic acid with ethanol to yield ethyl acetate, for example. As discussed in the previous section, the equilibrium constant K for this reaction is 3.4, meaning that substantial amounts of reactants and products are both present at equilibrium. If, however, the ethyl acetate is removed as soon as it's formed, the production of more and more product is forced to occur, in accord with Le Châtelier's principle.

> Continuously removing this product from the reaction forces more of it to be produced.

$$\underset{\text{Acetic acid}}{CH_3\overset{O}{\overset{\|}{C}}OH} + \underset{\text{Ethyl alcohol}}{CH_3CH_2OH} \rightleftharpoons \underset{\text{Ethyl acetate}}{CH_3\overset{O}{\overset{\|}{C}}OCH_2CH_3} + H_2O$$

Metabolic reactions sometimes take advantage of this effect, with one reaction prevented from reaching equilibrium by the continuous consumption of its product in a further reaction.

Effect of Changes in Temperature and Pressure

 Temperature Dependence of Equilibrium

We said in Section 7.1 that the reverse of an exothermic reaction is always endothermic. Equilibrium reactions are therefore exothermic in one direction and endothermic in the other. Le Châtelier's principle predicts that an increase in temperature will cause an equilibrium to shift in favor of the endothermic reaction so the additional heat is absorbed. Conversely, a decrease in temperature will cause an equilibrium to shift in favor of the exothermic reaction so additional heat is released. In other words, you can think of heat as a reactant or product whose increase or decrease stresses an equilibrium just as a change in reactant or product concentration does.

Endothermic reaction (heat is absorbed)   Favored by increase in temperature

Exothermic reaction (heat is released)   Favored by decrease in temperature

In the exothermic reaction of N_2 with H_2 to form NH_3, for example, raising the temperature favors the reverse reaction, which absorbs the heat:

$$[\longleftarrow \text{Heat}]$$
$$N_2(g) + 3\,H_2(g) \rightleftharpoons 2\,NH_3(g) + \text{Heat}$$

What about changing the pressure? Pressure influences an equilibrium only if one or more of the substances involved is a gas. As predicted by Le Châtelier's principle, increasing the pressure in such a reaction shifts the equilibrium in the direction that decreases the number of molecules in the gas phase and thus decreases the pressure. For the ammonia synthesis, increasing the pressure favors the forward reaction because 4 moles of gas is converted to 2 moles of gas.

$$[\text{Pressure} \longrightarrow]$$
$$\underbrace{N_2(g) + 3\,H_2(g)}_{\text{4 moles of gas}} \rightleftharpoons \underbrace{2\,NH_3(g)}_{\text{2 moles of gas}}$$

The effects of changing reaction conditions on equilibria are summarized in Table 7.3.

TABLE 7.3	Effects of Changes in Reaction Conditions on Equilibria
Change	**Effect**
Concentration	Increase in reactant concentration or decrease in product concentration favors forward reaction.
	Increase in product concentration or decrease in reactant concentration favors reverse reaction.
Temperature	Increase in T favors endothermic reaction.
	Decrease in T favors exothermic reaction.
Pressure	Increase in pressure favors side with fewer moles of gas.
	Decrease in pressure favors side with more moles of gas.
Catalyst added	Equilibrium reached more quickly; value of K unchanged.

⬡⬡⬡ LOOKING AHEAD

We'll see in Chapter 21 how Le Châtelier's principle is exploited to keep chemical "traffic" moving through the body's metabolic pathways. It often happens that one reaction in a series is prevented from reaching equilibrium because its product is continuously consumed in another reaction.   ⬡⬡⬡

■ **WORKED EXAMPLE 7.7**

Nitrogen reacts with oxygen to give NO:

$$N_2(g) + O_2(g) \rightleftharpoons 2\,NO(g) \quad \Delta H = +43\ kcal/mol$$

Explain the effects on reactant and product concentrations of the following changes:
(a) Increasing temperature
(b) Increasing the concentration of NO
(c) Adding a catalyst

SOLUTION
(a) The reaction is endothermic (positive ΔH), so increasing the temperature will favor the forward reaction. The concentration of NO will be higher at equilibrium.
(b) Increasing the concentration of NO, a product, will favor the reverse reaction. At equilibrium, the concentrations of both N_2 and O_2, as well as that of NO, will be higher.
(c) A catalyst will accelerate the rate at which equilibrium is reached, but the concentrations at equilibrium will be unchanged.

■ **PROBLEM 7.15**

Will the yield of SO_3 at equilibrium be favored by a higher or lower pressure? By a higher or lower temperature?

$$2\,SO_2(g) + O_2(g) \rightleftharpoons 2\,SO_3(g) \quad \Delta H = -47\ kcal/mol$$

■ **PROBLEM 7.16**

What effect will each of the listed changes have on the position of the equilibrium in the reaction of carbon with hydrogen?

$$C(s) + 2\,H_2(g) \rightleftharpoons CH_4(g) \quad \Delta H = -18\ kcal/mol$$

(a) Increasing temperature
(b) Increasing pressure
(c) Allowing CH_4 to escape continuously from the reaction vessel

Application   Nitrogen Fixation

All plants and animals need nitrogen—it's present in all proteins and nucleic acids, and it's the fourth most abundant element in the human body. Because the triple bond in the N_2 molecule is so strong, however, plants and animals can't use the free element directly. It's up to nature and the fertilizer industry to convert N_2, which makes up 78% of the atmosphere, into usable nitrogen compounds in a process called *nitrogen fixation*. Plants can use ammonia (NH_3), nitrates (NO_3^-), urea (H_2NCONH_2), and other simple, water-soluble compounds as their sources of nitrogen. Animals then get their nitrogen by eating plants or other plant-eating animals (see Figure).

Most industrial fertilizer production begins with the *Haber process*, the reaction between nitrogen and hydrogen to give ammonia:

The Haber process
$$N_2(g) + 3\ H_2(g) \longrightarrow 2\ NH_3(g)$$

Nature, of course, was fixing nitrogen long before the fertilizer industry got started. Microorganisms, such as blue-green algae and the bacteria that live in the roots of alfalfa, beans, peas, and other leguminous plants, fix nitrogen from the atmosphere by converting N_2 to ammonia.

Fixed nitrogen is also created by lightning, whose released energy causes the endothermic combination of N_2 and O_2 in the atmosphere to form NO. Further reaction with O_2 or O_3 ultimately produces water-soluble nitrates that plants can use. Lightning strikes somewhere on earth an average of 100 times each second, producing almost half of the world's fixed nitrogen supply.

See Additional Problem 7.73 at the end of the chapter.

The nitrogen cycle, showing ▶ how N_2 in the atmosphere is fixed, used by plants and animals, and then returned to the atmosphere.

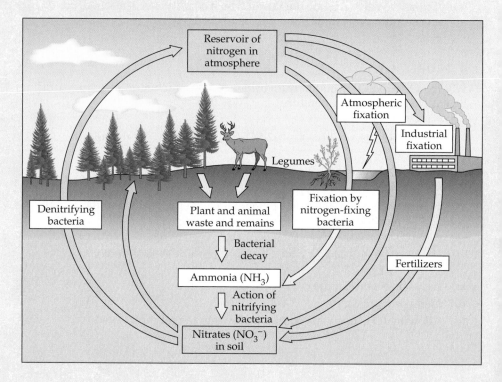

Summary: Revisiting the Chapter Goals

1. **What energy changes take place during reactions?** The strength of a covalent bond is measured by its *bond dissociation energy*, the amount of energy that must be supplied to break the bond in an isolated gaseous molecule. For any reaction, the heat released or absorbed by changes in bonding is called the *heat of reaction*, or *enthalpy change* (ΔH). If the total strength of the bonds formed in a reaction is greater than the total strength of the bonds broken, then heat is released (negative ΔH) and the reaction is said to be *exothermic*. If the total strength of the bonds formed in a reaction is less than the total strength of the bonds broken, then heat is absorbed (positive ΔH) and the reaction is said to be *endothermic*.

2. **What is "free energy," and what is the criterion for spontaneity in chemistry?** *Spontaneous reactions* are those that, once started, continue without external influence; nonspontaneous reactions require a continuous external influence. Spontaneity depends on two factors, the amount of heat absorbed or released in a reaction (ΔH) and the *entropy change* (ΔS), which measures the change in molecular disorder in a reaction. Spontaneous reactions are favored by a release of heat (negative ΔH) and an increase in disorder (positive ΔS). The *free-energy change* (ΔG) takes both factors into account, according to the equation $\Delta G = \Delta H - T\Delta S$. A negative value for ΔG indicates spontaneity, and a positive value for ΔG indicates nonspontaneity.

3. **What determines the rate of a chemical reaction?** A chemical reaction occurs when reactant particles collide with proper orientation and sufficient energy. The exact amount of collision energy necessary is called the *activation energy* (E_{act}). A high activation energy results in a slow reaction because few collisions occur with sufficient force, whereas a low activation energy results in a fast reaction. Reaction rates can be increased by raising the temperature, by raising the concentrations of reactants, or by adding a *catalyst*, which accelerates a reaction without itself undergoing any change.

4. **What is chemical equilibrium?** A reaction that can occur in either the forward or reverse direction is *reversible* and will ultimately reach a state of *chemical equilibrium*. At equilibrium, the forward and reverse reactions occur at the same rate, and the concentrations of reactants and products are constant. Every reversible reaction has a characteristic *equilibrium constant* (K), given by an *equilibrium equation*.

For the reaction: $aA + bB + \cdots \rightleftharpoons mM + nN + \cdots$

$$K = \frac{[M]^m[N]^n \cdots}{[A]^a[B]^b \cdots}$$

Product concentrations raised to powers equal to coefficients

Reactant concentrations raised to powers equal to coefficients

If K is larger than 1, the forward reaction is favored; if K is less than 1, the reverse reaction is favored.

5. **What is Le Châtelier's principle?** *Le Châtelier's principle* states that when a stress is applied to a system in equilibrium, the equilibrium shifts so that the stress is relieved. Applying this principle allows prediction of the effects of changes in temperature, pressure, and concentration.

Key Words

Activation energy (E_{act}), p. 175
Bond dissociation energy, p. 165
Catalyst, p. 176
Chemical equilibrium, p. 179
Endergonic, p. 171
Endothermic, p. 165
Enthalpy change (ΔH), p. 166
Entropy (S), p. 170
Equilibrium constant (K), p. 180
Exergonic, p. 171
Exothermic, p. 165
Free-energy change (ΔG), p. 171
Heat of reaction, p. 166
Law of conservation of energy, p. 166
Le Châtelier's principle, p. 182
Reaction rate, p. 175
Reversible reaction, p. 178
Spontaneous process, p. 170

▪ Understanding Key Concepts

7.17 What are the signs of ΔH, ΔS, and ΔG for the spontaneous conversion of a crystalline solid into a gas? Explain.

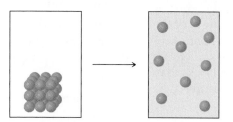

7.18 What are the signs of ΔH, ΔS, and ΔG for the spontaneous condensation of a vapor to a liquid? Explain.

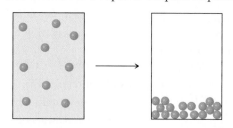

7.19 Consider the following spontaneous reaction of A_2 molecules (red) and B_2 molecules (blue):

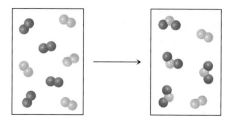

(a) Write a balanced equation for the reaction.
(b) What are the signs of ΔH, ΔS, and ΔG for the reaction? Explain.

7.20 Two curves are shown in the following energy diagram:

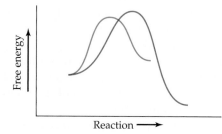

(a) Which curve represents the faster reaction, and which the slower?
(b) Which curve represents the spontaneous reaction, and which the nonspontaneous?

7.21 Two curves are shown in the following energy diagram. Which curve represents the catalyzed reaction, and which the uncatalyzed?

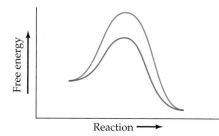

7.22 Draw energy diagrams for the following situations:
(a) A slow reaction with a large negative ΔG
(b) A fast reaction with a small positive ΔG

7.23 The following drawing portrays a reaction of the type $A(s) \rightarrow B(g) + C(g)$, where the different colored spheres represent different molecular structures. Assume that the reaction has $\Delta H = +9.1$ kcal/mol.

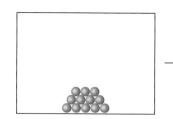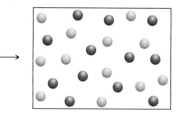

(a) What is the sign of ΔS for the reaction?
(b) Is the reaction likely to be spontaneous at all temperatures, nonspontaneous at all temperatures, or spontaneous at some but nonspontaneous at others?

■ Additional Problems

Enthalpy and Heat of Reaction

7.24 Is the total enthalpy (H) of the reactants for an endothermic reaction greater than or less than the total enthalpy of the products?

7.25 What is meant by the term "heat of reaction"? What other name is given to the heat of a reaction?

7.26 The vaporization of Br_2 from the liquid to the gas state requires 7.4 kcal/mol.
  (a) What is the sign of ΔH for this process?
  (b) How many kilocalories are needed to vaporize 6.5 mol of Br_2?
  (c) How many kilocalories are needed to evaporate 75 g of Br_2?

7.27 Converting liquid water to solid ice releases 1.44 kcal/mol.
  (a) What is the sign of ΔH for this process?
  (b) How many kilocalories are released by freezing 3.4 mol of H_2O?
  (c) How many kilocalories are released by freezing 25 g of H_2O?
  (d) How many kilocalories are absorbed by melting 1 mol of ice?

7.28 Glucose, also known as "blood sugar," has the formula $C_6H_{12}O_6$.
  (a) Write the equation for the combustion of glucose with O_2 to give CO_2 and H_2O.
  (b) If 3.8 kcal is released by combustion of each gram of glucose, how many kilocalories are released by the combustion of 1.50 mol of glucose?
  (c) What is the minimum amount of energy a plant must absorb to produce 15.0 g of glucose?

7.29 During the combustion of 5.00 g of ethanol, C_2H_5OH, 35.5 kcal is released.
  (a) Write a balanced equation for the combustion reaction.
  (b) What is the sign of ΔH for this reaction?
  (c) How much energy is released by the combustion of 1.00 mol of C_2H_5OH?
  (d) How many grams and how many moles of C_2H_5OH must be burned to release 450.0 kcal?
  (e) How many kilocalories are released by the combustion of 10.0 g of C_2H_5OH?
  (f) How many grams of C_2H_5OH must be burned to raise the temperature of 500.0 mL of water from 20.0°C to 100.0°C?

Entropy and Free Energy

7.30 Which of the following processes results in an increase in disorder of the system?
  (a) A drop of ink spreading out when it is placed in water
  (b) Steam condensing into drops on windows
  (c) Constructing a building from loose bricks

7.31 Which of the following processes results in a decrease in disorder of the system?
  (a) Assembling a jigsaw puzzle
  (b) $I_2(s) + 3\ F_2(g) \rightarrow 2\ IF_3(g)$
  (c) A precipitate forming when two solutions are mixed

7.32 What is meant by a "spontaneous" process?

7.33 How is the sign of the free-energy change related to the spontaneity of a process?

7.34 What are the two factors that affect the spontaneity of a reaction?

7.35 What is the difference between an exothermic reaction and an exergonic reaction?

7.36 Why are most spontaneous reactions exothermic?

7.37 Is it possible for a reaction to be nonspontaneous yet exothermic? Explain.

7.38 For the reaction $NaCl(s) \xrightarrow{\text{Water}} Na^+(aq) + Cl^-(aq)$,
$$\Delta H = +1.00\ \text{kcal/mol}.$$
  (a) Is this process endothermic or exothermic?
  (b) Does entropy increase or decrease in this process?
  (c) Table salt (NaCl) readily dissolves in water. Explain, based on your answers to parts (a) and (b).

7.39 For the reaction $2\ Hg(l) + O_2(g) \rightarrow 2\ HgO(s)$,
$$\Delta H = -43\ \text{kcal/mol}.$$
  (a) Does entropy increase or decrease in this process?
  (b) Is this process spontaneous?

7.40 The reaction of gaseous H_2 and liquid Br_2 to give gaseous HBr has $\Delta H = -17.4$ kcal/mol and $\Delta S = 27.2$ cal/(mol·K).
  (a) Write the balanced equation for this reaction.
  (b) Does entropy increase or decrease in this process?
  (c) Is this process spontaneous at all temperatures? Explain.
  (d) What is the value of ΔG for the reaction at 300 K?

7.41 The following reaction is used in the industrial synthesis of PVC polymer:
$$Cl_2(g) + H_2C{=}CH_2(g) \longrightarrow$$
$$ClCH_2CH_2Cl(l) \quad \Delta H = -52\ \text{kcal/mol}$$
  (a) Is ΔS positive or negative for this process?
  (b) Is this process spontaneous at all temperatures? Explain.

Rates of Chemical Reactions

7.42 What is the activation energy of a reaction?

7.43 Which reaction is faster, one with $E_{act} = +10$ kcal/mol or one with $E_{act} = +5$ kcal/mol? Explain.

7.44 Draw energy diagrams for exergonic reactions that meet the following descriptions:
  (a) A slow reaction that has a small free-energy change
  (b) A fast reaction that has a large free-energy change

7.45 Draw an energy diagram for a reaction whose products have the same free energies as its reactants. What is free-energy change in this case?

7.46 Give two reasons why increasing temperature increases the rate of a reaction.

7.47 Why does increasing concentration generally increase the rate of a reaction?

7.48 What is a catalyst, and what effect does it have on the activation energy of a reaction?

7.49 If a catalyst changes the activation energy of a forward reaction from 28.0 kcal/mol to 23.0 kcal/mol, what effect does it have on the reverse reaction?

7.50 For the reaction $C(s,\ \text{diamond}) \rightarrow C(s,\ \text{graphite})$, $\Delta G = -0.693$ kcal/mol at 25°C.
  (a) According to this information, do diamonds spontaneously turn into graphite?
  (b) In light of your answer to part (a), why can diamonds be kept unchanged for thousands of years?

7.51 For the reaction $2\,H_2(g) + 2\,C(s) \rightarrow H_2C{=}CH_2(g)$, $\Delta G = +16.3$ kcal/mol at 25°C.

   (a) Is the reaction of hydrogen with carbon to yield ethylene spontaneous at 25°C?

   (b) Is it possible to find a catalyst for the reaction of hydrogen with carbon at 25°C?

Chemical Equilibria

7.52 What is meant by the term "chemical equilibrium"? Must amounts of reactants and products be equal at equilibrium?

7.53 Why don't catalysts alter the amounts of reactants and products present at equilibrium?

7.54 Write the equilibrium equations for the following reactions:

   (a) $2\,CO(g) + O_2(g) \rightleftharpoons 2\,CO_2(g)$

   (b) $C_2H_6(g) + 2\,Cl_2(g) \rightleftharpoons C_2H_4Cl_2(g) + 2\,HCl(g)$

   (c) $HF(aq) + H_2O(l) \rightleftharpoons H_3O^+(aq) + F^-(aq)$

   (d) $3\,O_2(g) \rightleftharpoons 2\,O_3(g)$

7.55 Write the equilibrium equations for the following reactions, and tell whether reactants or products are favored in each case.

   (a) $S_2(g) + 2\,H_2(g) \rightleftharpoons 2\,H_2S(g)$ $K = 2.8 \times 10^{-21}$

   (b) $CO(g) + 2\,H_2(g) \rightleftharpoons CH_3OH(g)$ $K = 10.5$

   (c) $Br_2(g) + Cl_2(g) \rightleftharpoons 2\,BrCl(g)$ $K = 58.0$

   (d) $I_2(g) \rightleftharpoons 2\,I(g)$ $K = 6.8 \times 10^{-3}$

7.56 For the reaction $N_2O_4(g) \rightleftharpoons 2\,NO_2(g)$, the equilibrium concentrations at 25°C are $[NO_2] = 0.025$ mol/L and $[N_2O_4] = 0.0869$ mol/L. What is the value of K at 25°C?

7.57 Fluorine can react with oxygen to yield oxygen difluoride:

$$2\,F_2(g) + O_2(g) \rightleftharpoons 2\,OF_2(g)$$

What is the value of K if the following concentrations are found at equilibrium? $[O_2] = 0.200$ mol/L, $[F_2] = 0.0100$ mol/L, $[OF_2] = 0.0633$ mol/L

7.58 Use your answer from Problem 7.56 to calculate the following:

   (a) $[N_2O_4]$ at equilibrium when $[NO_2] = 0.12$ mol/L

   (b) $[NO_2]$ at equilibrium when $[N_2O_4] = 0.12$ mol/L

7.59 Use your answer from Problem 7.57 to calculate the following:

   (a) $[O_2]$ at equilibrium when $[OF_2] = 0.22$ mol/L and $[F_2] = 0.0300$ mol/L

   (b) $[OF_2]$ at equilibrium when $[F_2] = 0.080$ mol/L and $[O_2] = 0.650$ mol/L

7.60 Would you expect to find relatively more reactants or more products for the reaction in Problem 7.56 if the pressure were raised?

7.61 Would you expect to find relatively more reactants or more products for the reaction in Problem 7.57 if the pressure were lowered?

Le Châtelier's Principle

7.62 Oxygen can be converted into ozone by the action of lightning or electric sparks:

$$3\,O_2(g) \rightleftharpoons 2\,O_3(g)$$

For this reaction, $\Delta H = +68$ kcal/mol and $K = 2.68 \times 10^{-29}$ at 25°C.

   (a) Is the reaction exothermic or endothermic?

   (b) Are the reactants or the products favored at equilibrium?

   (c) Explain the effect on the equilibrium of:

   (1) Increasing pressure

   (2) Increasing the concentration of $O_2(g)$

   (3) Increasing the concentration of $O_3(g)$

   (4) Adding a catalyst

   (5) Increasing the temperature

7.63 Hydrogen chloride can be made from the reaction of chlorine and hydrogen:

$$Cl_2(g) + H_2(g) \rightarrow 2\,HCl(g)$$

For this reaction, $K = 2.6 \times 10^{33}$ and $\Delta H = -44$ kcal/mol at 25°C.

   (a) Is the reaction endothermic or exothermic?

   (b) Are the reactants or the products favored at equilibrium?

   (c) Explain the effect on the equilibrium of:

   (1) Increasing pressure

   (2) Increasing the concentration of $HCl(g)$

   (3) Increasing the concentration of $Cl_2(g)$

   (4) Decreasing the concentration of $H_2(g)$

   (5) Adding a catalyst

7.64 When each of the following equilibria is disturbed by increasing the pressure, does the concentration of reaction products increase, decrease, or remain the same?

   (a) $2\,CO_2(g) \rightleftharpoons 2\,CO(g) + O_2(g)$

   (b) $N_2(g) + O_2(g) \rightleftharpoons 2\,NO(g)$

   (c) $Si(s) + 2\,Cl_2(g) \rightleftharpoons SiCl_4(g)$

7.65 For each of the following equilibria, use Le Châtelier's principle to predict the direction of the reaction when the pressure is increased.

   (a) $C(s) + H_2O(g) \rightleftharpoons CO(g) + H_2(g)$

   (b) $2\,H_2(g) + O_2(g) \rightleftharpoons 2\,H_2O(g)$

   (c) $2\,Fe(s) + 3\,H_2O(g) \rightleftharpoons Fe_2O_3(s) + 3\,H_2(g)$

7.66 The reaction $CO(g) + H_2O(g) \rightleftharpoons CO_2(g) + H_2(g)$ has $\Delta H = -9.8$ kcal/mol. Does the amount of H_2 in an equilibrium mixture increase or decrease when the temperature is increased?

7.67 The reaction $3\,O_2(g) \rightleftharpoons 2\,O_3(g)$ has $\Delta H = +68$ kcal/mol. Does the equilibrium constant for the reaction increase or decrease when the temperature increases?

7.68 The reaction $H_2(g) + I_2(g) \rightleftharpoons 2\,HI(g)$ has $\Delta H = -2.2$ kcal/mol. Will the equilibrium concentration of HI increase or decrease when:

   (a) I_2 is added?

   (b) H_2 is removed?

   (c) A catalyst is added?

   (d) The temperature is increased?

7.69 The reaction $Fe^{3+}(aq) + Cl^-(aq) \rightleftharpoons FeCl^{2+}(aq)$ is endothermic. How will the equilibrium concentration of $FeCl^{2+}$ change when:

   (a) $Fe(NO_3)_3$ is added?

   (b) Cl^- is precipitated by addition of $AgNO_3$?

   (c) The temperature is increased?

   (d) A catalyst is added?

Applications

7.70 Which provides more energy, 1 g of carbohydrate or 1 g of fat? [*Energy from Food*]

7.71 What body organs help to regulate body temperature? [*Regulation of Body Temperature*]

7.72 What is the purpose of blood vessel dilation? [*Regulation of Body Temperature*]

7.73 What does it mean to "fix" nitrogen, and what natural processes accomplish nitrogen fixation? [*Nitrogen Fixation*]

General Questions and Problems

7.74 For the production of ammonia from its elements, $\Delta H = -22$ kcal/mol.

(a) Is this process endothermic or exothermic?

(b) How many kilocalories are involved in the production of 0.500 mol of NH_3?

7.75 Magnetite, an iron ore with formula Fe_3O_4, can be reduced by treatment with hydrogen to yield iron metal and water vapor.

(a) Write the balanced equation.

(b) This process requires 36 kcal for every 1.00 mol of Fe_3O_4 reduced. How much energy (in kilocalories) is required to produce 75 g of iron?

(c) How many grams of hydrogen are needed to produce 75 g of iron?

(d) This reaction has $K = 2.3 \times 10^{-18}$. Are the reactants or the products favored?

7.76 Hemoglobin (Hb) reacts reversibly with O_2 to form HbO_2, a substance that transfers oxygen to tissues:

$$Hb(aq) + O_2(aq) \rightleftharpoons HbO_2(aq)$$

Carbon monoxide (CO) is attracted to Hb 140 times more strongly than O_2 and establishes another equilibrium:

$$Hb(aq) + CO(aq) \rightleftharpoons Hb(CO)(aq)$$

(a) Explain, using Le Châtelier's principle, why inhalation of CO can cause weakening and eventual death.

(b) Still another equilibrium is established when both O_2 and CO are present:

$$Hb(CO)(aq) + O_2(aq) \rightleftharpoons HbO_2(aq) + CO(aq)$$

Explain, using Le Châtelier's principle, why pure oxygen is often administered to victims of CO poisoning.

7.77 Many hospitals administer glucose intravenously to patients. If 3.8 kcal is provided by each gram of glucose, how many grams must be administered to maintain a person's normal basal metabolic needs of about 1700 kcal/day?

7.78 For the evaporation of water, $H_2O(l) \rightarrow H_2O(g)$, at 100°C, $\Delta H = +9.72$ kcal/mol.

(a) How many kilocalories are needed to vaporize 10.0 g of $H_2O(l)$?

(b) How many kilocalories are released when 10.0 g of $H_2O(g)$ is condensed?

7.79 Ammonia reacts slowly in air to produce nitrogen monoxide and water vapor:

$$NH_3(g) + O_2(g) \rightleftharpoons NO(g) + H_2O(g) + \text{Heat}$$

(a) Balance the equation.

(b) Write the equilibrium equation.

(c) Explain the effect on the equilibrium of:
(1) Raising the pressure
(2) Adding $NO(g)$
(3) Decreasing the concentration of NH_3
(4) Lowering the temperature

7.80 Methanol, CH_3OH, is used as race car fuel.

(a) Write the balanced equation for the combustion of methanol.

(b) $\Delta H = -174$ kcal/mol methanol for the process. How many kilocalories are released by burning 50.0 g of methanol?

7.81 Sketch an energy diagram for a system in which the forward reaction has $E_{act} = +25$ kcal/mol and the reverse reaction has $E_{act} = +35$ kcal/mol.

(a) Is the forward process endergonic or exergonic?

(b) What is the value of ΔG for the reaction?

7.82 Aluminum metal reacts with chlorine with a spectacular display of sparks.

$$2\ Al(s) + 3\ Cl_2(g) \rightarrow 2\ AlCl_3(s)\quad \Delta H = -336.6\ \text{kcal/mol}$$

How much heat (in kilocalories) is released on reaction of 5.00 g of Al?

7.83 How much heat (in kilocalories) is evolved or absorbed in the reaction of 1.00 g of Na with H_2O? Is the reaction exothermic or endothermic?

$$2\ Na(s) + 2\ H_2O(l) \rightarrow$$
$$2\ NaOH(aq) + H_2(g)\quad \Delta H = -88.0\ \text{kcal/mol}$$

8

Gases, Liquids, and Solids

CONTENTS

The three states of matter—solid ice, liquid water, and gaseous steam—are all apparent during this winter eruption of Old Faithful in Yellowstone National Park.

The past seven chapters have dealt with matter at the atomic level. We've seen that all matter is composed of atoms, ions, or molecules; that these particles are in constant motion; that atoms combine to make compounds using chemical bonds; and that physical and chemical changes are accompanied by the release or absorption of energy. Now it's time to look at a different aspect of matter, concentrating not on the properties and small-scale behavior of individual atoms but on the properties and large-scale behavior of visible amounts of matter.

The major questions we'll answer in this chapter are the following:

1. **What is the kinetic–molecular theory of gases?**
   The goal: Be able to state the assumptions of the kinetic–molecular theory of gases and use these assumptions to explain the behavior of gases.

2. **What are the gas laws?**
   The goal: Be able to use Boyle's law, Charles's law, Gay-Lussac's law, and Avogadro's law to explain the effect on gases of a change in pressure, volume, or temperature.

3. **What is the ideal gas law?**
   The goal: Be able to use the ideal gas law to find the pressure, volume, temperature, or molar amount of a gas sample.

4. **What is partial pressure?**
   The goal: Be able to define partial pressure and use Dalton's law of partial pressures.

5. **What are the major intermolecular forces?**
   The goal: Be able to explain dipole–dipole forces, London dispersion forces, and hydrogen bonding, and recognize which of these forces affect a given molecule.

6. **What are the various kinds of solids, and how do they differ?**
   The goal: Be able to recognize the different kinds of solids and describe their characteristics.

7. **What occurs during a change of state?**
   The goal: Be able to apply the concepts of heat change, equilibrium, and vapor pressure to changes of state.

CONCEPTS TO REVIEW

Ionic Bonds (Section 4.3)
Polar Covalent Bonds and Polar Molecules (Sections 5.8 and 5.9)
Enthalpy, Entropy, and Free Energy (Sections 7.1-7.3)

8.1 States of Matter and Their Changes

Matter can exist in any of three phases, or *states*—solid, liquid, and gas. Which of the three states exists under a given set of conditions depends on the strength of the attractive forces among the particles in the substance. In a gas, the attractive forces between particles are so weak that the particles move about freely, are far apart, and have almost no influence on one another. In a liquid, the attractive forces are stronger, pulling the particles close together but still allowing them considerable freedom to move about. In a solid, the attractive forces are so strong that the atoms, molecules, or ions are held in a specific arrangement and can only wiggle around in place (Figure 8.1).

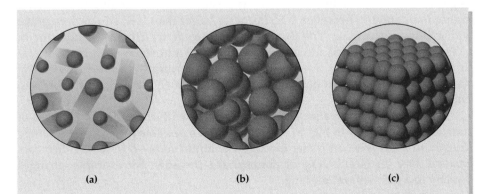

(a)   (b)   (c)

◄ **FIGURE 8.1   A molecular comparison of gases, liquids, and solids.**   (a) In a gas, the particles feel little attraction for one another and are free to move about randomly. (b) In a liquid, the particles are held close together by attractive forces but are free to slide over one another. (c) In a solid, the particles are rigidly held in a specific arrangement.

Change of state The change of a substance from one state of matter (gas, liquid, or solid) to another.

The transformation of a substance from one state to another is called a *phase change*, or a **change of state**. Every change of state is reversible and, like all chemical and physical processes, is characterized by a free-energy change, ΔG. A change of state that is spontaneous in one direction (exergonic, negative ΔG) is nonspontaneous in the other direction (endergonic, positive ΔG). As always, the free-energy change ΔG has both an enthalpy term ΔH and a temperature-dependent entropy term ΔS, according to the equation $\Delta G = \Delta H - T\Delta S$. (You might want to reread Section 7.3 to brush up on these ideas.) (∞ , p. 169)

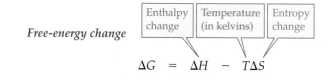

The enthalpy change ΔH is a measure of the heat absorbed or released during a given change of state. In the melting of a solid to a liquid, for example, heat is absorbed and ΔH is positive (endothermic). In the reverse process—the freezing of a liquid to a solid—heat is released and ΔH is negative (exothermic). Look at the change between ice and water for instance:

Melting: $H_2O(s) \longrightarrow H_2O(l)$ $\Delta H = +1.44 \text{ kcal/mol}$

Freezing: $H_2O(l) \longrightarrow H_2O(s)$ $\Delta H = -1.44 \text{ kcal/mol}$

The entropy change ΔS is a measure of the change in molecular disorder or freedom that occurs during a process. In the melting of a solid to a liquid, for example, disorder increases because particles gain freedom of motion, so ΔS is positive. In the reverse process—the freezing of a liquid to a solid—disorder decreases as particles are locked into position, so ΔS is negative. Look at the change between ice and water:

Melting: $H_2O(s) \longrightarrow H_2O(l)$ $\Delta S = +5.26 \text{ cal/(mol} \cdot \text{K)}$

Freezing: $H_2O(l) \longrightarrow H_2O(s)$ $\Delta S = -5.26 \text{ cal/(mol} \cdot \text{K)}$

As with all processes that are unfavored by one term in the free-energy equation yet favored by the other, the sign of ΔG depends on the temperature (Section 7.3). The melting of ice, for instance, is unfavored by a positive ΔH yet is favored by a positive ΔS. Thus, at a low temperature, the unfavorable ΔH is larger than the favorable $T\Delta S$, so ΔG is positive and no melting occurs. At a higher temperature, however, $T\Delta S$ becomes larger than ΔH, so ΔG is negative and melting *does* occur. The exact temperature at which the changeover in behavior occurs is called the **melting point (mp)** and represents the temperature at which solid and liquid coexist in equilibrium. In the corresponding change from a liquid to a gas, the two states are in equilibrium at the **boiling point (bp)**.

Melting point (mp) The temperature at which solid and liquid are in equilibrium.

Boiling point (bp) The temperature at which liquid and gas are in equilibrium.

The names and enthalpy changes associated with the different changes of state are summarized in Figure 8.2. Note that it's possible for a solid to change directly to a gas without going through the liquid state—a process called *sublimation*. Dry ice (solid CO_2) at atmospheric pressure, for example, changes directly to a gas without melting.

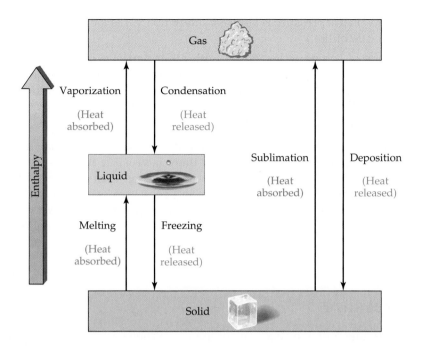

◀ FIGURE 8.2 **Changes of state.** The changes are endothermic from bottom to top and exothermic from top to bottom. Solid and liquid states are in equilibrium at the melting point; liquid and gas states are in equilibrium at the boiling point.

■ **PROBLEM 8.1**

The change of state from liquid H_2O to gaseous H_2O has $\Delta H = +9.72$ kcal/mol and $\Delta S = +26.1$ cal/(mol · K).

(a) Is the change from liquid to gaseous H_2O favored or disfavored by ΔH? by ΔS?

(b) What is the value of ΔG for the change from liquid to gaseous H_2O at 373 K?

(c) What are the values of ΔH and ΔS for the change from gaseous to liquid H_2O?

8.2 Gases and the Kinetic–Molecular Theory

Gases behave quite differently from liquids and solids. Gases, for instance, have low densities and are easily compressed to a smaller volume when placed under pressure, a property that allows them to be stored in large tanks. Liquids and solids, by contrast, are much more dense and much less compressible. Furthermore, gases undergo a far larger expansion or contraction when their temperature is changed than do liquids and solids.

The behavior of gases can be explained by a group of assumptions known as the **kinetic–molecular theory of gases.** We'll see in the next several sections how the following assumptions account for the observable properties of gases:

- **A gas consists of many particles, either atoms or molecules, moving about at random with no attractive forces between them.** Because of this random motion, different gases mix together quickly.

- **The amount of space occupied by the gas particles themselves is much smaller than the amount of space between particles.** Most of the volume taken up by a gas is empty space, accounting for the ease of compression and low densities of gases.

- **The average kinetic energy of gas particles is proportional to the Kelvin temperature.** Thus, gas particles have more kinetic energy and move faster as the temperature increases. (In fact, gas particles move much faster than you might suspect. The average speed of a helium atom at room temperature

Kinetic–molecular theory of gases A group of assumptions that explain the behavior of gases.

 Kinetic Energy in a Gas

and atmospheric pressure is approximately 1.36 km/s, or 3000 mi/hr, nearly that of a rifle bullet.)

- **Collisions of gas particles, either with other particles or with the wall of their container, are elastic; that is, the total kinetic energy of the particles is constant. The pressure of a gas against the walls of its container is the result of collisions** of the gas particles with the walls. The more collisions and the more forceful each collision, the higher the pressure.

Ideal gas A gas that obeys all the assumptions of the kinetic–molecular theory.

A gas that obeys all the assumptions of the kinetic–molecular theory is called an **ideal gas.** In practice, though, there is no such thing as a perfectly ideal gas. All gases behave somewhat differently than predicted by the kinetic–molecular theory when, at very high pressures or very low temperatures, their particles get closer together and interactions between particles become significant. As a rule, however, most real gases display nearly ideal behavior under normal conditions.

8.3 Pressure

We're all familiar with the effects of air pressure. When you fly in an airplane, the change in air pressure against your eardrums as the plane climbs or descends can cause a painful "popping." When you pump up a bicycle tire, you increase the pressure of air against the inside walls of the tire until the tire feels hard.

Pressure The force per unit area pushing against a surface.

In scientific terms, **pressure (P)** is defined as a force (F) per unit area (A) pushing against a surface; that is, $P = F/A$. In the bicycle tire, for example, the pressure you feel is the force of air molecules colliding with the inside walls of the tire. The units you probably use for tire pressure are pounds per square inch (psi), where 1 psi is equal to the pressure exerted by a 1 pound object resting on a 1 square inch surface.

We on earth are under pressure from the atmosphere, the blanket of air pressing down on us (Figure 8.3). Atmospheric pressure is not constant, however; it varies slightly from day to day depending on the weather, and it also varies with altitude. Air pressure is about 14.7 psi at sea level but only about 4.7 psi on the summit of Mt. Everest.

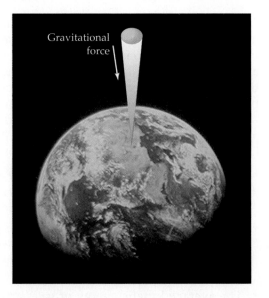

FIGURE 8.3   Atmospheric ▶ pressure.   A column of air weighing 14.7 lb presses down on each square inch of the earth's surface at sea level, resulting in what we call atmospheric pressure.

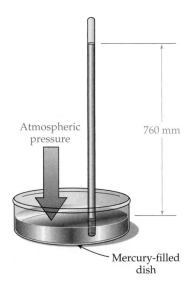

▲   **FIGURE 8.4   Measuring atmospheric pressure.**   A mercury barometer is used to measure atmospheric pressure by determining the height of a mercury column in a sealed glass tube. The downward pressure of the mercury in the column is exactly balanced by the outside atmospheric pressure, which presses down on the mercury in the dish and pushes it up into the column.

One of the most commonly used units of pressure is the *millimeter of mercury,* abbreviated *mm Hg* and often called a *torr* (after the Italian physicist Evangelista Torricelli). This unusual unit dates back to the early 1600s when Torricelli made the first mercury *barometer.* As shown in Figure 8.4, a barometer consists of a long, thin tube that is sealed at one end, filled with mercury, and then inverted into a dish of mercury. Some mercury runs from the tube into the dish until the downward pressure of the mercury in the column is exactly balanced by the outside atmospheric pressure, which presses down on the mercury in the dish and pushes it up into the column. The height of the mercury column varies depending on the altitude and weather conditions, but standard atmospheric pressure at sea level is defined to be exactly 760 mm.

Gas pressure inside a container is often measured using an open-end *manometer*, a simple instrument similar in principle to the mercury barometer. As shown in Figure 8.5, an open-end manometer consists of a U-tube filled with mercury, with one end connected to a gas-filled container and the other end open to the atmosphere. The difference between the heights of the mercury levels in the two arms of the U-tube indicates the difference between the pressure of the gas in the container and the pressure of the atmosphere. If the gas pressure inside the container is less than atmospheric, the mercury level is higher in the arm connected to the container (Figure 8.5a). If the gas pressure inside the container is greater than atmospheric, the mercury level is higher in the arm open to the atmosphere (Figure 8.5b).

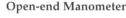
Open-end Manometer

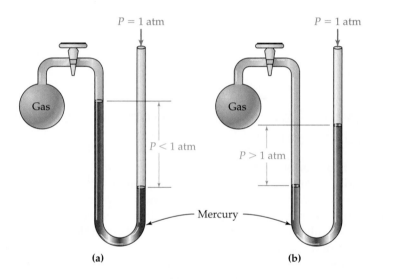

▲ **FIGURE 8.5 Open-end manometers for measuring pressure in a gas-filled bulb.** (a) When the pressure in the gas-filled container is lower than atmospheric, the mercury level is higher in the arm open to the container. (b) When the pressure in the container is higher than atmospheric, the mercury level is higher in the arm open to the atmosphere.

Pressure is given in the SI system by a unit named the *pascal* (Pa), where 1 Pa = 0.007 500 mm Hg (or 1 mm Hg = 133.32 Pa). Measurements in pascals are becoming more common, and many clinical laboratories have made the switchover. Higher pressures are often still given in *atmospheres* (atm), where 1 atm = 760 mm Hg exactly.

Pressure units: 1 atm = 760 mm Hg = 14.7 psi = 101,325 Pa

1 mm Hg = 1 torr = 133.32 Pa

■ **PROBLEM 8.2**

The air pressure outside a jet airliner flying at 35,000 ft is about 220 mm Hg. How many atmospheres is this? How many pounds per square inch? How many pascals?

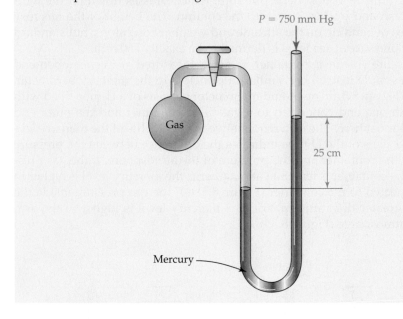

■ **KEY CONCEPT PROBLEM 8.3**

What is the pressure of the gas inside the following manometer (in mm Hg) if outside pressure is 750 mm Hg??

$P = 750$ mm Hg

25 cm

Gas

Mercury

8.4 Boyle's Law: The Relation Between Volume and Pressure

The physical behavior of all gases is much the same, regardless of identity. Helium and chlorine, for example, are completely different in their *chemical* behavior, but are very similar in many of their physical properties. Observations of many different gases by scientists in the 1700s led to the formulation of what are now called the **gas laws**, which make it possible to predict the influence of pressure (P), volume (V), temperature (T), and molar amount (n) on any gas or mixture of gases. We'll begin by looking at *Boyle's law*, which describes the relation between volume and pressure.

Imagine that you have a sample of gas inside a cylinder that has a movable plunger at one end (Figure 8.6). What would happen if you were to double the pressure on the gas by pushing the plunger down, while keeping the temperature constant? Since the gas particles would be forced closer together, the volume of the sample would decrease.

Gas laws A series of laws that predict the influence of pressure (P), volume (V), and temperature (T) on any gas or mixture of gases.

FIGURE 8.6 Boyle's law. The ▶ volume of a gas decreases proportionately as its pressure increases. If the pressure of a gas sample is doubled, the volume is halved.

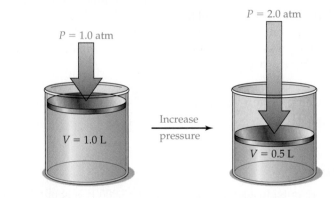

$P = 1.0$ atm

$P = 2.0$ atm

$V = 1.0$ L

Increase pressure

$V = 0.5$ L

According to **Boyle's law**, the volume of a fixed amount of gas at a constant temperature is inversely proportional to its pressure, meaning that volume and

Application Blood Pressure

Having your blood pressure measured is a quick and easy way to get an indication of the state of your circulatory system. Although blood pressure varies with age, a normal adult male has a reading near 120/80 mm Hg, and a normal adult female has a reading near 110/70 mm Hg. Abnormally high values signal an increased risk of heart attack and stroke.

Pressure varies greatly in different types of blood vessels. Usually, though, measurements are carried out on arteries in the upper arm as the heart goes through a full cardiac cycle. *Systolic pressure* is the maximum pressure developed in the artery just after contraction, as the heart forces the maximum amount of blood into the artery. *Diastolic pressure* is the minimum pressure that occurs at the end of the heart cycle.

Blood pressure is most often measured by a *sphygmomanometer*, a device consisting of a squeeze bulb, a flexible cuff, and a mercury manometer. The cuff is placed around the upper arm over the brachial artery and inflated by the squeeze bulb to about 200 mm Hg pressure, an amount great enough to squeeze the artery shut and prevent blood flow. Air is then slowly released from the cuff, and pressure drops. As cuff pressure reaches the systolic pressure, blood spurts through the artery, creating a turbulent tapping sound that can be heard through a stethoscope. The pressure registered on the manometer at the moment the first sounds are heard is the systolic blood pressure.

Sounds continue until the pressure in the cuff becomes low enough to allow diastolic blood flow. At

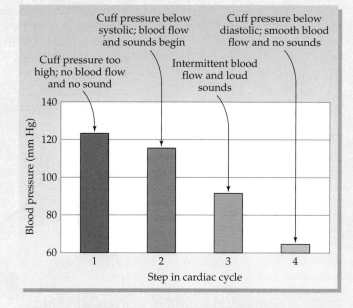

▲ The sequence of events during blood pressure measurement, including the sounds heard.

this point, blood flow becomes smooth, no sounds are heard, and a diastolic blood pressure reading is recorded on the manometer. Readings are usually recorded as systolic/diastolic, for example, 120/80. The accompanying Figure shows the sequence of events during measurement.

See Additional Problem 8.96 at the end of the chapter.

---

pressure change in opposite directions. As pressure goes up, volume goes down; as pressure goes down, volume goes up (Figure 8.7).

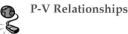 **P-V Relationships**

Boyle's law The volume of a gas is inversely proportional to its pressure for a fixed amount of gas at a constant temperature. That is, P times V is constant when the amount of gas n and the temperature T are kept constant. (The symbol $\propto$ means "is proportional to," and k denotes a constant value.)

$$\text{Volume } (V) \propto \frac{1}{\text{Pressure } (P)}$$

$$\text{or} \quad PV = k \text{ (A constant value)}$$

Because $P \times V$ is a constant value for a fixed amount of gas at a constant temperature, the starting pressure (P_1) times the starting volume (V_1) must equal the final pressure (P_2) times the final volume (V_2). Thus, Boyle's law can be used to find the final pressure or volume when the starting pressure or volume is changed.

$$\text{Since} \quad P_1 V_1 = k \quad \text{and} \quad P_2 V_2 = k$$

$$\text{then} \quad P_1 V_1 = P_2 V_2$$

$$\text{so} \quad P_2 = \frac{P_1 V_1}{V_2} \quad \text{and} \quad V_2 = \frac{P_1 V_1}{P_2}$$

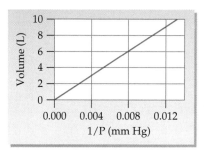

▲ **FIGURE 8.7 Boyle's law.**
Volume is inversely proportional to pressure for a fixed amount of gas at a constant temperature. As 1/pressure goes up, the volume also goes up.

199

As an example of Boyle's law behavior, think about what happens every time you breathe. Between breaths, the pressure inside the lungs is equal to atmospheric pressure. When inhalation takes place, the diaphragm and rib cage expand, increasing the volume of the lungs and thereby decreasing the pressure inside them (Figure 8.8). Air must then move into the lungs to equalize their pressure with that of the atmosphere. When exhalation takes place, the diaphragm and rib cage contract, decreasing the volume of the lungs and increasing pressure inside them. Now gases move out of the lungs until pressure is again equalized with the atmosphere.

FIGURE 8.8 Boyle's law in ▶ breathing. During inhalation, the diaphragm moves down and the rib cage moves up and out, thus increasing lung volume, decreasing pressure, and drawing in air. During exhalation, lung volume decreases, pressure increases, and air moves out.

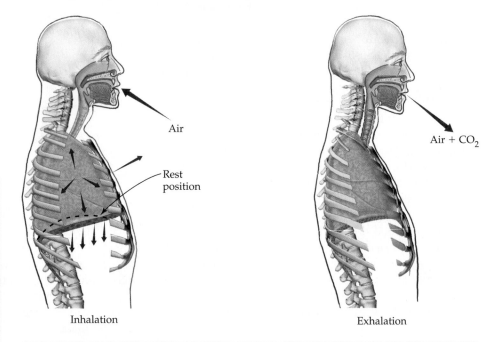

Inhalation

Exhalation

▲ Squeezing this balloon decreases its volume and increases the inside pressure. If squeezed hard enough, it will break.

■ WORKED EXAMPLE 8.1

In a typical automobile engine, the fuel/air mixture in a cylinder is compressed from 1.0 atm to 9.5 atm. If the uncompressed volume of the cylinder is 750 mL, what is the volume when fully compressed?

ANALYSIS This is a Boyle's-law problem because the volume and pressure in the cylinder change while the amount of gas and the temperature remain constant. According to Boyle's law, the pressure of the gas times its volume is constant:

$$P_1V_1 = P_2V_2$$

Of the four variables in this equation, we know that $P_1 = 1.0$ atm, $V_1 = 750$ mL, and $P_2 = 9.5$ atm, and we need to find V_2.

SOLUTION Solving the above equation for V_2 and substituting in the known values gives

$$V_2 = \frac{P_1V_1}{P_2} = \frac{(1.0 \text{ atm})(750 \text{ mL})}{9.5 \text{ atm}} = 79 \text{ mL}$$

BALLPARK CHECK Since the pressure *increases* approximately tenfold (from 1.0 atm to 9.5 atm), the volume must *decrease* to approximately one-tenth, from 750 mL to about 75 mL.

■ PROBLEM 8.4

An oxygen cylinder used for breathing has a volume of 5.0 L at 90 atm pressure. What volume would the same amount of oxygen have at the same temperature if the pressure were 1.0 atm?

8.5 Charles's Law: The Relation Between Volume and Temperature

Imagine that you again have a sample of gas inside a cylinder with a plunger at one end. What would happen if you were to double the sample's Kelvin temperature while letting the plunger move freely to keep the pressure constant? The gas particles would move with twice as much energy and would collide twice as forcefully with the walls. To maintain a constant pressure, the volume of the gas in the cylinder must double (Figure 8.9).

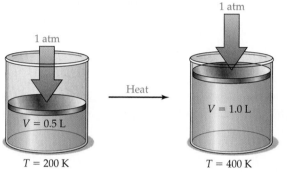

1 atm

1 atm

Heat

$V = 0.5$ L

$V = 1.0$ L

$T = 200$ K

$T = 400$ K

◀ **FIGURE 8.9 Charles's law.**
The volume of a gas is directly proportionately to its Kelvin temperature at constant n and P. If the Kelvin temperature is doubled, the volume is doubled.

According to **Charles's law**, the volume of a fixed amount of gas at a constant pressure is directly proportional to its Kelvin temperature. Note the difference between *directly* proportional in Charles's law and *inversely* proportional in Boyle's law. Directly proportional quantities change in the same direction: As temperature goes up or down, volume also goes up or down (Figure 8.10).

Charles's law The volume of a gas is directly proportional to its Kelvin temperature for a fixed amount of gas at a constant pressure. That is, V divided by T is constant when n and P are held constant.

$$V \propto T \quad \text{(In kelvins)}$$

$$\text{or} \quad \frac{V}{T} = k \text{ (A constant value)}$$

$$\text{or} \quad \frac{V_1}{T_1} = \frac{V_2}{T_2}$$

As an example of Charles's law, think of what happens when a hot-air balloon is inflated. Heating causes the air inside to expand and fill the balloon.

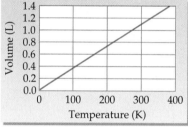

▲ **FIGURE 8.10 Charles's law.**
Volume is directly proportional to the Kelvin temperature for a fixed amount of gas at a constant pressure. As the temperature goes up, the volume also goes up.

◀ The volume of the gas in the balloon increases as it is heated, causing a decrease in density and allowing the balloon to rise.

■ **WORKED EXAMPLE 8.2**

An average adult inhales a volume of 0.50 L of air with each breath. If the air is warmed from room temperature ($20°C = 293$ K) to body temperature ($37°C = 310$ K) while in the lungs, what is the volume of the air when exhaled?

ANALYSIS This is a Charles's-law problem because the volume and temperature of the air change while the amount and pressure remain constant. According to Charles's law, the volume of the gas divided by its temperature in kelvins is constant:

$$\frac{V_1}{T_1} = \frac{V_2}{T_2}$$

Of the four variables in this equation, we know that $V_1 = 0.50$ L, $T_1 = 293$ K, and $T_2 = 310$ K, and we need to find V_2.

SOLUTION Solving for V_2 and carrying out the calculation gives

$$V_2 = \frac{V_1 T_2}{T_1} = \frac{(0.50 \text{ L})(310 \text{ K})}{293 \text{ K}} = 0.53 \text{ L}$$

BALLPARK CHECK Starting from a temperature of about 300°C, a rise of 3°C represents a 1% change. Thus, a temperature rise of 17°C (from 293 K to 310 K) is a roughly 6% change. The volume must also increase by approximately 6%, from 0.50 L to about 0.53 L.

■ **PROBLEM 8.6**

A sample of chlorine gas has a volume of 0.30 L at 273 K and 1 atm pressure. What is its volume at 350 K and 1 atm pressure? At 500°C and 1 atm?

8.6 Gay-Lussac's Law: The Relation Between Pressure and Temperature

Imagine next that you have a fixed amount of gas in a sealed container whose volume remains constant. What would happen if you were to double the temperature (in kelvins)? The gas particles would move with twice as much energy and would collide with the walls of the container with twice as much force. Thus, the pressure in the container would double. According to **Gay-Lussac's law**, the pressure of a fixed amount of gas at constant volume is directly proportional to its Kelvin temperature. As temperature goes up or down, pressure also goes up or down (Figure 8.11).

Gay-Lussac's law   The pressure of a gas is directly proportional to its Kelvin temperature for a fixed amount of gas at a constant volume. That is, P divided by T is constant when n and V are held constant.

$$P \propto T \text{ (In kelvins)}$$

$$\text{or} \quad \frac{P}{T} = k \text{ (A constant value)}$$

$$\text{or} \quad \frac{P_1}{T_1} = \frac{P_2}{T_2}$$

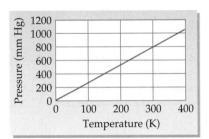

▲ **FIGURE 8.11   Gay-Lussac's law.** Pressure is directly proportional to the temperature in kelvins for a fixed amount of gas at a constant volume. As the temperature goes up, the pressure also goes up.

As an example of Gay-Lussac's law, think of what happens when an aerosol can is thrown into an incinerator. As the can gets hotter, pressure builds up inside and the can explodes (hence the warning statement on aerosol cans).

■ **WORKED EXAMPLE 8.3**

What would the inside pressure become if an aerosol can with an initial pressure of 4.5 atm were heated in a fire from room temperature ($20°C$) to $600°C$?

ANALYSIS   This is a Gay-Lussac's-law problem because the pressure and temperature inside the can change while the amount and the volume remain constant. According to Gay-Lussac's law, the pressure of the gas divided by its temperature in kelvins is constant:

$$\frac{P_1}{T_1} = \frac{P_2}{T_2}$$

Of the four variables in this equation, we know P_1, T_1, and T_2, and we need to find P_2. As always, it's first necessary to convert the given temperatures from degrees Celsius to kelvins ($20°C = 293$ K; $600°C = 873$ K).

SOLUTION   Solving for P_2 and carrying out the calculation gives

$$P_2 = \frac{P_1 T_2}{T_1} = \frac{(4.5 \text{ atm}) (873 \text{ K})}{293 \text{ K}} = 13 \text{ atm}$$

BALLPARK CHECK   Since the kelvin temperature increases approximately threefold (from about 300 K to about 900 K), the pressure must also increase by approximately threefold, from 4.5 atm to about 14 atm.

■ **PROBLEM 8.7**

Driving on a hot day causes tire temperature to rise. What is the pressure inside an automobile tire at $45°C$ if the tire has a pressure of 30 psi at $15°C$? Assume that the volume and amount of air in the tire remain constant.

8.7 The Combined Gas Law

Since PV, V/T, and P/T all have constant values for a fixed amount of gas, these relationships can be merged into a **combined gas law**, which holds true whenever the amount of gas is fixed.

Combined gas law $\dfrac{PV}{T} = k$ (A constant value)

or $\dfrac{P_1 V_1}{T_1} = \dfrac{P_2 V_2}{T_2}$

If any five of the six quantities in this equation are known, the sixth can be calculated. Furthermore, if any of the three variables T, P, or V is constant, that variable drops out of the equation, leaving behind Boyle's law, Charles's law, or Gay-Lussac's law. As a result, *the combined gas law is the only equation you need to remember for a fixed amount of gas*. Worked Example 8.4 gives a sample calculation.

Gas Laws

Since
$$\frac{P_1V_1}{T_1} = \frac{P_2V_2}{T_2}$$

At constant T: $\quad \dfrac{P_1V_1}{T} = \dfrac{P_2V_2}{T} \quad$ gives $\quad P_1V_1 = P_2V_2 \quad$ (Boyle's law)

At constant P: $\quad \dfrac{PV_1}{T_1} = \dfrac{PV_2}{T_2} \quad$ gives $\quad \dfrac{V_1}{T_1} = \dfrac{V_2}{T_2} \quad$ (Charles's law)

At constant V: $\quad \dfrac{P_1V}{T_1} = \dfrac{P_2V}{T_2} \quad$ gives $\quad \dfrac{P_1}{T_1} = \dfrac{P_2}{T_2} \quad$ (Gay-Lussac's law)

■ **WORKED EXAMPLE 8.4**

A 6.3 L sample of helium gas stored at 25°C and 1.0 atm pressure is transferred to a 2.0 L tank and maintained at a pressure of 2.8 atm. What temperature is needed to maintain this pressure?

ANALYSIS This is a combined-gas-law problem because pressure, volume, and temperature change while the amount of helium remains constant. According to the combined gas law, the pressure of the gas times its volume and divided by its temperature in kelvins is constant:

$$\frac{P_1V_1}{T_1} = \frac{P_2V_2}{T_2}$$

Of the six variables in this equation, we know P_1, V_1, T_1, P_2, and V_2, and we need to find T_2. As always, it's first necessary to convert the given temperature from degrees Celsius to kelvins (25°C = 298 K).

SOLUTION Solving the combined gas law equation for T_2 gives

$$T_2 = \frac{P_2V_2T_1}{P_1V_1} = \frac{(2.8 \text{ atm})(2.0 \text{ L})(298 \text{ K})}{(1.0 \text{ atm})(6.3 \text{ L})} = 260 \text{ K } (-13°C)$$

BALLPARK CHECK Since the volume goes down by a factor of about 3 (from 6.3 L to 2.0 L) and the pressure goes up by a factor of about 3 (from 1.0 atm to 2.8 atm), the two changes roughly offset one another and the temperature won't change much.

■ **PROBLEM 8.8**

A weather balloon is filled with helium to a volume of 250 L at 22°C and 745 mm Hg. The balloon ascends to an altitude where the pressure is 570 mm Hg, and the temperature is −64°C. What is the volume of the balloon at this altitude?

■ **PROBLEM 8.9**

Another weather balloon is filled with helium to a volume of 225 L at 18°C and 763 mm Hg. The balloon ascends to an altitude where its volume rises to 1365 L and the temperature is −55°C. What is the pressure at this altitude?

8.8 Avogadro's Law: The Relation Between Volume and Molar Amount

Let's look at one final gas law, which takes changes in amount of gas into account. Imagine that you have two different volumes of a gas at the same temperature and pressure. How many moles does each sample contain? According to **Avogadro's law**, the volume of a gas is directly proportional to its molar amount at a constant pressure and temperature (Figure 8.12). A sample that contains twice the molar amount has twice the volume.

> **Avogadro's law**   The volume of a gas is directly proportional to its molar amount at a constant pressure and temperature. That is, V divided by n is constant when P and T are held constant.
>
> $$\text{Volume } (V) \propto \text{ Number of moles } (n)$$
>
> or $\dfrac{V}{n} = k$ (A constant value; the same for all gases)
>
> or $\dfrac{V_1}{n_1} = \dfrac{V_2}{n_2}$

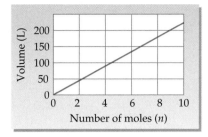

▲ **FIGURE 8.12   Avogadro's law.** Volume is directly proportional to the molar amount, n, at a constant temperature and pressure. As the number of moles goes up, the volume also goes up.

Because the particles in a gas are so tiny compared to the empty space surrounding them, there is no interaction among gas particles and the chemical identity of the particles doesn't matter. Thus, the value of the constant k in the equation $V/n = k$ is the same for all gases, and it's therefore possible to compare the molar amounts of *any* two gases simply by comparing their volumes at the same temperature and pressure.

Notice that the *values* of temperature and pressure don't matter; it's only necessary that T and P for the two gases be the same. To simplify comparisons of gas samples, however, it's convenient to define a set of conditions called **standard temperature and pressure (STP)**:

> **Standard temperature and pressure (STP)**   0°C (273.15 K);   1 atm (760 mm Hg)

At standard temperature and pressure, one mole of any gas (6.02×10^{23} particles) has a volume of 22.4 L, a quantity called the **standard molar volume** (Figure 8.13).

> **Standard Molar volume of any gas at STP**   22.4 L/mol

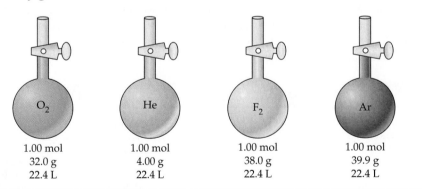

O$_2$	He	F$_2$	Ar
1.00 mol	1.00 mol	1.00 mol	1.00 mol
32.0 g	4.00 g	38.0 g	39.9 g
22.4 L	22.4 L	22.4 L	22.4 L

◄ **FIGURE 8.13   Avogadro's law.** Each of these 22.4 L bulbs contains 1.00 mol of gas at 0°C and 1 atm pressure.

■ **WORKED EXAMPLE 8.5**

Use the standard molar volume of a gas at STP (22.4 L) to find how many moles of air at STP are in a room measuring 4.11 m wide by 5.36 m long by 2.58 m high.

ANALYSIS   We first need to find the volume of the room and then use standard molar volume as a conversion factor to find the number of moles.

SOLUTION   The volume of the room is the product of its three dimensions:

$$\text{Volume} = (4.11 \text{ m})(5.36 \text{ m})(2.58 \text{ m}) = 56.8 \text{ m}^3$$

$$= 56.8 \text{ m}^3 \times \frac{1000 \text{ L}}{1 \text{ m}^3} = 5.68 \times 10^4 \text{ L}$$

The number of moles of gas (air) in the room is

$$5.68 \times 10^4 \; \cancel{L} \times \frac{1 \; mol}{22.4 \; \cancel{L}} = 2.54 \times 10^3 \; mol$$

There are 2.54×10^3 moles of gas (air) in the room.

■ **PROBLEM 8.10**

How many moles of methane gas, CH_4, are in a 1.00×10^5 L storage tank at STP? How many grams of methane is this? How many grams of carbon dioxide gas could the same tank hold?

8.9 The Ideal Gas Law

The relationships among the four variables P, V, T, and n for gases can be combined into a single expression called the **ideal gas law**. If you know the values of any three of the four quantities, you can calculate the value of the fourth.

$$\textbf{Ideal gas law} \quad \frac{PV}{nT} = R \quad \text{(A constant value)}$$

$$\text{or} \quad PV = nRT$$

Gas constant (R) The constant R in the ideal gas law, $PV = nRT$.

The constant R in the ideal gas law (instead of the usual k) is called the **gas constant**. It's value depends on the units chosen for pressure, with the two most common values

$$\text{For } P \text{ in atmospheres:} \quad R = 0.0821 \; \frac{L \cdot atm}{mol \cdot K}$$

$$\text{For } P \text{ in millimeters Hg:} \quad R = 62.4 \; \frac{L \cdot mm \; Hg}{mol \cdot K}$$

In using the ideal gas law, it's important to choose the correct value of R and, if necessary, to convert volume into liters and temperature into kelvins.

Table 8.1 summarizes the various gas laws, and Worked Example 8.6 shows how to use the ideal gas law.

TABLE 8.1   A Summary of the Gas Laws

Gas law		Variables	Constant
Boyle's law	$P_1V_1 = P_2V_2$	P, V	n, T
Charles's law	$V_1/T_1 = V_2/T_2$	V, T	n, P
Gay-Lussac's law	$P_1/T_1 = P_2/T_2$	P, T	n, V
Combined gas law	$P_1V_1/T_1 = P_2V_2/T_2$	P, V, T	n
Avogadro's law	$V_1/n_1 = V_2/n_2$	V, n	P, T
Ideal gas law	$PV = nRT$	P, V, T, n	R

■ **WORKED EXAMPLE 8.6**

How many moles of air are in the lungs of an average person with a total lung capacity of 3.8 L? Assume that the person is at 1.0 atm pressure and has a normal body temperature of 37°C.

ANALYSIS This is an ideal-gas-law problem because it asks for a value of n when P, V, and T are known: $n = PV/RT$. The volume is given in the correct unit of liters, but temperature must be converted to kelvins (37°C = 37 + 273 K = 310 K).

SOLUTION

$$n = \frac{PV}{RT} = \frac{(1.0 \text{ atm})(3.8 \text{ L})}{\left(0.0821 \dfrac{\text{L} \cdot \text{atm}}{\text{mol} \cdot \text{K}}\right)(310 \text{ K})} = 0.15 \text{ mol}$$

There is 0.15 mol of air (that is, 0.15 mol of gas particles) in the lungs of an average person.

■ **WORKED EXAMPLE 8.7**

Methane gas is sold in steel cylinders with a volume of 43.8 L containing 5.54 kg. What is the pressure in atmospheres inside the cylinder at a temperature of 20.0°C (293.15 K)? The molecular weight of methane (CH_4) is 16.0 amu.

ANALYSIS This is an ideal-gas-law problem because it asks for a value of P when V and T are known and information to calculate n is provided: $P = nRT/V$.

SOLUTION First, calculate the number of moles n of methane in the cylinder by using molar mass (16.0 g/mol) as a conversion factor.

$$n = (5.54 \text{ kg methane})\left(\frac{1000 \text{ g}}{1 \text{ kg}}\right)\left(\frac{1 \text{ mol}}{16.0 \text{ g}}\right) = 346 \text{ mol methane}$$

Then use the ideal gas law to calculate the pressure:

$$P = \frac{nRT}{V} = \frac{(346 \text{ mol})\left(0.0821 \dfrac{\text{L} \cdot \text{atm}}{\text{mol} \cdot \text{K}}\right)(293 \text{ K})}{43.8 \text{ L}} = 190 \text{ atm}$$

■ **PROBLEM 8.11**

An aerosol spray deodorant can with a volume of 350 mL contains 3.2 g of propane gas (C_3H_8) as propellant. What is the pressure in the can at 20°C?

■ **PROBLEM 8.12**

A helium gas cylinder of the sort used to fill balloons has a volume of 180 L and a pressure of 2200 psi (150 atm) at 25°C. How many moles of helium are in the tank? How many grams?

■ **KEY CONCEPT PROBLEM 8.13**

Show the approximate level of the movable piston in drawings (a) and (b) after the indicated changes have been made to the initial gas sample.

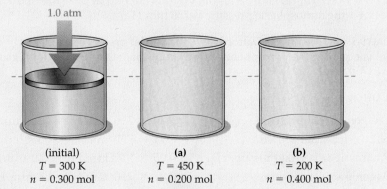

(initial)	(a)	(b)
$T = 300$ K	$T = 450$ K	$T = 200$ K
$n = 0.300$ mol	$n = 0.200$ mol	$n = 0.400$ mol

1.0 atm

8.10 Partial Pressure and Dalton's Law

According to the kinetic–molecular theory, each particle in a gas acts independently of all others because there are no attractive forces between them and they are so far apart. To any individual particle, the chemical identity of its neighbors is irrelevant. Thus, *mixtures* of gases behave the same as pure gases and obey the same laws.

Dry air, for example, is a mixture of about 21% oxygen, 78% nitrogen, and 1% argon by volume, which means that 21% of atmospheric air pressure is caused by O_2 molecules, 78% by N_2 molecules, and 1% by Ar atoms. The contribution of each gas in a mixture to the total pressure of the mixture is called the **partial pressure** of that gas. According to **Dalton's law**, the total pressure exerted by a gas mixture (P_{total}) is the sum of the partial pressures of the components in the mixture:

$$\text{Dalton's law} \quad P_{total} = P_{gas\,1} + P_{gas\,2} + \cdots$$

Partial pressure The contribution of a given gas in a mixture to the total pressure.

In dry air at a total air pressure of 760 mm Hg, the partial pressure caused by the contribution of O_2 is $(0.21) \times (760 \text{ mm Hg}) = 160 \text{ mm Hg}$, the partial pressure of N_2 is $(0.78) \times (760 \text{ mm Hg}) = 593 \text{ mm Hg}$, and that of argon is 7 mm Hg. *The partial pressure exerted by each gas in a mixture is the same pressure that the gas would exert if it were alone.* This makes sense if you think about it: The pressure exerted by each gas depends on the frequency of collisions of its molecules with the walls of the container. But this frequency doesn't change when other gases are present because the different molecules have no influence on one another.

The partial pressure of a specific gas is represented by adding the formula of the gas as a subscript to the symbol for pressure. You might see the partial pressure of oxygen represented as P_{O_2}, for instance. Moist air inside the lungs at 37°C and atmospheric pressure has the following average composition at sea level. Note that P_{total} is equal to atmospheric pressure, 760 mm Hg.

$$
\begin{aligned}
P_{total} &= P_{N_2} + P_{O_2} + P_{CO_2} + P_{H_2O} \\
&= 573 \text{ mm Hg} + 100 \text{ mm Hg} + 40 \text{ mm Hg} + 47 \text{ mm Hg} \\
&= 760 \text{ mm Hg}
\end{aligned}
$$

The composition of air doesn't change appreciably with altitude, but the total pressure decreases rapidly. The partial pressure of oxygen in the air therefore decreases with increasing altitude, and it is this change that leads to difficulty in breathing at high elevations.

▲ At an altitude of 4000 m (about 13,000 ft) the partial pressure of oxygen is only about 97 mm Hg, or 60% of that at sea level.

■ WORKED EXAMPLE 8.8

Humid air on a warm summer day is approximately 20% oxygen, 74% nitrogen, 5% water vapor, and 1% argon. What is the partial pressure of each component if the atmospheric pressure is 750 mm Hg?

ANALYSIS According to Dalton's law, the partial pressure of any gas in a mixture is equal to the percent concentration of the gas times the total gas pressure (750 mm Hg).

SOLUTION

Oxygen partial pressure (P_{O_2}) : $0.20 \times 750 \text{ mm Hg} = 150 \text{ mm Hg}$
Nitrogen partial pressure (P_{N_2}) : $0.74 \times 750 \text{ mm Hg} = 550 \text{ mm Hg}$
Water vapor partial pressure (P_{H_2O}) : $0.05 \times 750 \text{ mm Hg} = 40 \text{ mm Hg}$
Argon partial pressure (P_{Ar}) : $0.01 \times 750 \text{ mm Hg} = 8 \text{ mm Hg}$

CHECK   Note that the sum of the partial pressures must equal the total pressure (within rounding error):

$$(150 + 550 + 40 + 8) \text{ mm Hg} = 748 \text{ mm Hg}$$

■ **PROBLEM 8.14**

Assuming a total pressure of 9.5 atm, what is the partial pressure of each component in the mixture of 98% helium and 2.0% oxygen breathed by deep-sea divers? How does the partial pressure of oxygen in diving gas compare with its partial pressure in normal air?

■ **PROBLEM 8.15**

Determine the percent composition of air in the lungs from the following composition in partial pressures: $P_{N_2} = 573$ mm Hg, $P_{O_2} = 100$ mm Hg, $P_{CO_2} = 40$ mm Hg, $P_{H_2O} = 47$ mm Hg; all at 37°C and 1 atm pressure.

■ **PROBLEM 8.16**

What is the partial pressure of oxygen in the lungs at an altitude where the atmospheric pressure is 685 mm Hg?

■ **KEY CONCEPT PROBLEM 8.17**

Assume that you have a mixture of He (MW = 4 amu) and Xe (MW = 131 amu) at 300 K. Which of the drawings (a)–(c) best represents the mixture (blue = He; green = Xe)?

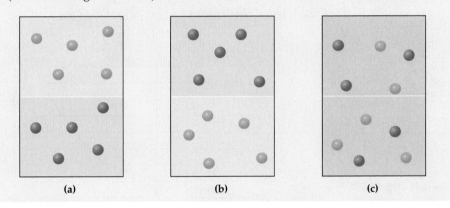

    (a)   (b)   (c)

8.11 Intermolecular Forces

What determines whether a substance is a gas, a liquid, or a solid at a given temperature? Why does rubbing alcohol evaporate much more readily than water? Why do molecular compounds have lower melting points than ionic compounds? To answer these and a great many other such questions, we need to look into the nature of **intermolecular forces**—the forces that act *between different molecules* rather than within an individual molecule.

    In gases, the intermolecular forces are negligible so the gas molecules act independently of one another. In liquids and solids, however, intermolecular forces are strong enough to hold the molecules in close contact. As a general rule, the stronger the intermolecular forces in a substance, the more difficult it is to separate the molecules, and the higher the melting point and boiling point of the substance.

Intermolecular force A force that acts between molecules and hold molecules close to one another in liquids and solids.

Greenhouse Gases and Global Warming

The mantle of gases surrounding the earth is far from the uniform mixture you might expect. Although atmospheric pressure decreases in a regular way at higher altitudes, the profile of temperature versus altitude is much more complex, as shown in the accompanying Figure. Four regions of the atmosphere have been defined based on this temperature curve. The temperature in the *troposphere*, the region nearest the earth's surface, decreases regularly up to about 12 km altitude, where it reaches a minimum value, and then increases in the *stratosphere*, up to about 50 km. In the *mesosphere* (50–85 km), the temperature again decreases, but it then increases in the *thermosphere*. To give you a feeling for these altitudes, passenger jets normally fly near the top of the troposphere at altitudes of 10–12 km, and the world altitude record for aircraft is 37.65 km—roughly in the middle of the stratosphere.

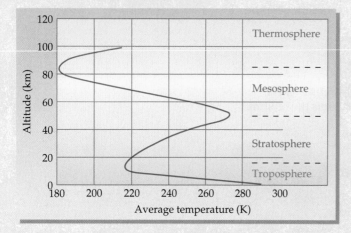

▲ The variation of average temperature with altitude in the earth's atmosphere. Four regions can be defined based on these temperature variations.

Not surprisingly, it's the layer nearest the earth's surface—the troposphere—that is the most easily disturbed by human activities and has the greatest impact on the earth's surface conditions. Among those impacts, the so-called *greenhouse effect* is much in the news today.

The basis for concern about the greenhouse problem is the fear that human activities over the past century may have disturbed the earth's delicate thermal balance. One component of that balance is the radiant energy the earth's surface receives from the sun, some of which is radiated back into space. Although much of this radiation passes out through the atmosphere, some is absorbed by atmospheric gases—particularly water vapor, carbon dioxide, and methane. This absorbed radiation warms the atmosphere and acts to maintain a relatively stable temperature at the earth's surface. Should increasing amounts of radiation be absorbed, however, increased atmospheric heating could result, and global temperatures could rise.

Measurements show that the concentration of atmospheric CO_2 has been rising in the last 150 years, largely because of the increased use of fossil fuels, from an estimated 290 parts per million (ppm) in 1850 to a current level of 370 ppm. Thus, there is concern among many atmospheric scientists that increased absorption of radiant energy and widespread global warming might follow. The final years of the 1990s were, in fact, the warmest of the 20th century, leading some scientists to believe that global warming has already started. The best computer models presently available predict a potential warming by as much as 3°C by the year 2050, an amount that could result in an increased melting of polar ice caps and a resultant rise in ocean levels.

See Additional Problems 8.97 and 8.98 at the end of the chapter.

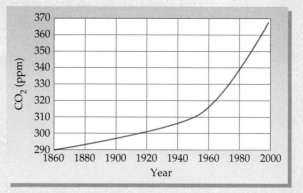

▲ Concentrations of atmospheric CO_2 have increased dramatically in the last 150 years because of increased fossil fuel use. Atmospheric scientists worry that global atmospheric warming may occur as a result.

There are three major types of intermolecular forces: *dipole–dipole, London dispersion*, and *hydrogen bonding*. Let's discuss each in turn.

Dipole–Dipole Forces

Recall from Sections 5.8 and 5.9 that many molecules contain polar covalent bonds and may therefore have a net molecular polarity. (∞, pp. 117, 119) In such cases, the positive and negative ends of different molecules are attracted to one another by what is called a **dipole–dipole force** (Figure 8.14).

Dipole–dipole forces are weak, with strengths on the order of 1 kcal/mol compared to the 70–100 kcal/mol typically found for the strength of a covalent bond. Nevertheless, the effects of dipole–dipole forces are important, as can be seen by looking at the difference in boiling points between polar and nonpolar molecules. Butane, for instance, is a nonpolar molecule with a molecular weight of 58 amu and a boiling point of −0.5°C, while acetone has the same molecular weight yet boils 57°C higher because it is polar. (Recall from Section 5.8 how molecular polarities can be visualized using electrostatic potential maps. ∞∞∞, p. 118)

Dipole–dipole force The attractive force between positive and negative ends of polar molecules.

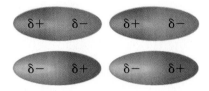

▲ **FIGURE 8.14** The positive and negative ends of polar molecules are attracted to one another by dipole–dipole forces. As a result, polar molecules have higher boiling points than nonpolar molecules of similar size.

London dispersion force The short-lived attractive force due to the constant motion of electrons within molecules.

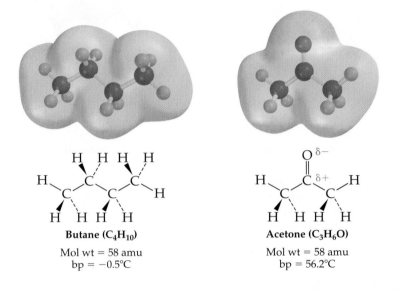

Butane (C$_4$H$_{10}$)
Mol wt = 58 amu
bp = −0.5°C

Acetone (C$_3$H$_6$O)
Mol wt = 58 amu
bp = 56.2°C

London Dispersion Forces

Only polar molecules experience dipole–dipole forces, but all molecules, regardless of structure, experience *London dispersion forces*. **London dispersion forces** are caused by the constant motion of electrons within molecules. Take even a simple nonpolar molecule like Br$_2$, for example. Averaged over time, the distribution of electrons throughout the molecule is uniform, but at any given *instant* there may be more electrons at one end of the molecule than at the other (Figure 8.15). At that instant, the molecule has a short-lived polarity. Electrons in neighboring molecules are attracted to the positive end of the polarized molecule, resulting in a polarization of the neighbor and creation of an attractive London dispersion force that holds the molecules together. As a result, Br$_2$ is a liquid at room temperature rather than a gas.

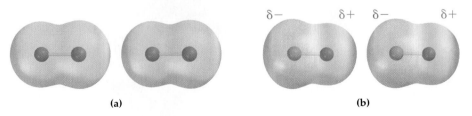

(a)   (b)

◀ **FIGURE 8.15** (a) Averaged over time, the electron distribution in a Br$_2$ molecule is symmetrical. (b) At any given instant, however, the electron distribution may be unsymmetrical, resulting in a temporary polarity that induces a complementary polarity in neighboring molecules.

London dispersion forces are weak—in the range 0.5–2.5 kcal/mol—but they increase with molecular weight and amount of surface contact between molecules. The larger the molecular weight, the more electrons there are moving about and the greater the temporary polarization of a molecule. The larger the amount of surface contact, the greater the close interaction between different molecules.

The effect of surface contact on the magnitude of London dispersion forces can be seen by comparing a roughly spherical molecule with a flatter, more linear one having the same molecular weight. Both 2,2-dimethylpropane and pentane, for instance, have the same formula (C_5H_{12}), but the nearly spherical shape of 2,2-dimethylpropane allows for less surface contact with neighboring molecules than does the more linear shape of pentane (Figure 8.16). As a result, London dispersion forces are smaller for 2,2-dimethylpropane, molecules are held together less tightly, and the boiling point is correspondingly lower: 9.5°C for 2,2-dimethylpropane versus 36°C for pentane.

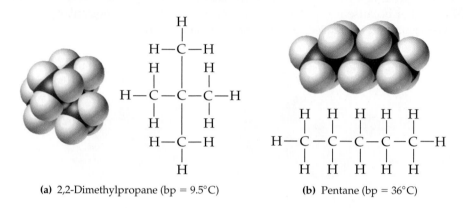

(a) 2,2-Dimethylpropane (bp = 9.5°C)   (b) Pentane (bp = 36°C)

▲ **FIGURE 8.16   London dispersion forces.**   More compact molecules like 2,2-dimethylpropane have smaller surface areas, weaker London dispersion forces, and lower boiling points. By comparison, flatter, less compact molecules like pentane have larger surface areas, stronger London dispersion forces, and higher boiling points.

Hydrogen Bonds

In many ways, hydrogen bonding is responsible for life on earth. It causes water to be a liquid rather than a gas at ordinary temperatures, and it is the primary intermolecular force that holds huge biomolecules in the shapes needed to play their essential roles in biochemistry. Deoxyribonucleic acid (DNA), for instance, contains two long molecular strands coiled around each other and held together by hydrogen bonds (Figure 8.17).

FIGURE 8.17   Molecular models ▶ of deoxyribonucleic acid (DNA). Two long strands are wrapped around each other and held together by hydrogen bonds.

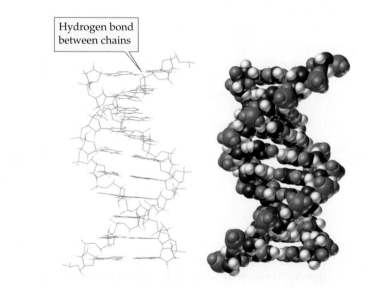

Hydrogen bond between chains

 Hydrogen Bonding

Hydrogen bond The attraction between a hydrogen atom bonded to an electronegative O, N, or F atom and another nearby electronegative O, N, or F atom.

A **hydrogen bond** is an attractive interaction between an unshared electron pair on an electronegative O, N, or F atom and a positively polarized hydrogen atom bonded to another electronegative O, N, or F. For example, hydrogen bonds occur in both water and ammonia:

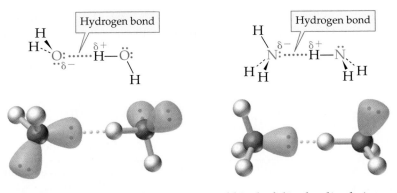

Hydrogen bonding is really just a special kind of dipole–dipole interaction. The O—H, N—H, and F—H bonds are highly polar, with a partial positive charge on the hydrogen and a partial negative charge on the electronegative atom. In addition, the hydrogen atom has no inner-shell electrons to act as a shield around its nucleus, and it has a small size so it can be approached closely. As a result, the dipole–dipole attractions involving positively polarized hydrogens are unusually strong, and hydrogen bonds result. Water, in particular, is able to form a vast three-dimensional network of hydrogen bonds because each H_2O molecule has two hydrogens and two electron pairs (Figure 8.18).

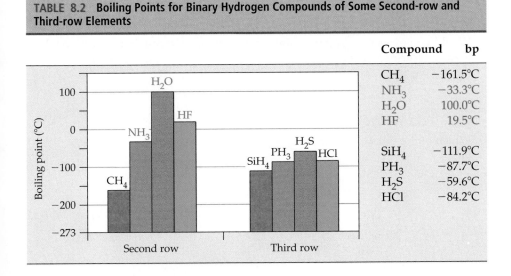

◀ **FIGURE 8.18 Hydrogen bonding in water.** The intermolecular attraction in water is especially strong because each oxygen atom has two lone pairs and two hydrogen atoms, allowing the formation of as many as four hydrogen bonds per molecule. Individual hydrogen bonds are constantly being formed and broken.

Hydrogen bonds can be quite strong, with energies up to 10 kcal/mol. To see the effect of hydrogen bonding, look at Table 8.2, which compares the boiling points of binary hydrogen compounds of second-row elements with their third-row counterparts. Because NH_3, H_2O, and HF molecules are held tightly together by hydrogen bonds, an unusually large amount of energy must be

TABLE 8.2 Boiling Points for Binary Hydrogen Compounds of Some Second-row and Third-row Elements

Compound	bp
CH_4	−161.5°C
NH_3	−33.3°C
H_2O	100.0°C
HF	19.5°C
SiH_4	−111.9°C
PH_3	−87.7°C
H_2S	−59.6°C
HCl	−84.2°C

added to separate them in the boiling process. As a result, the boiling points of NH_3, H_2O, and HF are much higher than the boiling points of their second-row neighbor CH_4 and of related third-row compounds.

A summary and comparison of the various kinds of intramolecular forces is shown in Table 8.3.

Attractive Forces

TABLE 8.3	A Comparison of Intermolecular Forces	
Force	**Strength**	**Characteristics**
Dipole–dipole	Weak (1 kcal/mol)	Occurs between polar molecules
London dispersion	Weak (0.5–2.5 kcal/mol)	Occurs between all molecules; strength depends on size
Hydrogen bond	Moderate (2–10 kcal/mol)	Occurs between molecules with O—H, N—H, and F—H bonds

∞ LOOKING AHEAD

Dipole–dipole forces, London dispersion forces, and hydrogen bonds are traditionally called "intermolecular forces" because of their influence on the properties of molecular compounds. But these same forces can also operate between different parts of a very large molecule. In this context they are often referred to as "noncovalent interactions." We'll see in later chapters how noncovalent interactions determine the shapes of biologically important molecules such as proteins and nucleic acids   ∞

■ WORKED EXAMPLE 8.9

Identify the intermolecular forces that influence the properties of the following compounds:
(a) Methane, CH_4 **(b)** HCl   **(c)** Acetic acid, CH_3CO_2H

SOLUTION
(a) Methane is a nonpolar molecule, so it has only London dispersion forces.
(b) HCl is a polar molecule, so it has both dipole–dipole forces and London dispersion forces.
(c) Acetic acid is a polar molecule with an O—H bond. Thus, it has dipole–dipole forces, London dispersion forces, and hydrogen bonds.

■ PROBLEM 8.18

Would you expect the boiling points to increase or decrease in each of the following series? Explain.
(a) Kr, Ar, Ne   **(b)** Cl_2, Br_2, I_2,

■ PROBLEM 8.19

Which of the following compounds form hydrogen bonds?

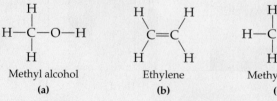

Methyl alcohol	Ethylene	Methylamine
(a)	**(b)**	**(c)**

■ **PROBLEM 8.20**

Identify the intermolecular forces (dipole–dipole, London dispersion, hydrogen bonding) that influence the properties of the following compounds:

(a) Ethane, CH_3CH_3
(b) Ethyl alcohol, CH_3CH_2OH
(c) Ethyl chloride, CH_3CH_2Cl

8.12 Liquids

Molecules are in constant motion in the liquid state, just as they are in gases. If a molecule happens to be near the surface of a liquid, and if it has enough energy, it can break free of the liquid and escape into the gas state, called **vapor**. In an open container, the now gaseous molecule will wander away from the liquid, and the process will continue until all the molecules escape from the container (Figure 8.19a). This, of course, is what happens during *evaporation*. We're all familiar with seeing a puddle of water evaporate after a rainstorm.

Vapor The gas molecules in equilibrium with a liquid.

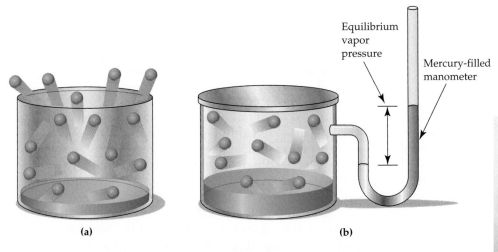

Equilibrium vapor pressure

Mercury-filled manometer

(a)   (b)

▲ **FIGURE 8.19 The transfer of molecules between liquid and gas states.** (a) Molecules can escape from an open container and drift away until the liquid has entirely evaporated. (b) Molecules in a closed container can't escape. Instead, they reach an equilibrium in which the rates of molecules leaving the liquid and returning to the liquid are equal, and the concentration of molecules in the gas state is constant.

If the liquid is in a closed container, the situation is different because the gaseous molecules can't escape. Thus, the random motion of the molecules will occasionally bring them back into the liquid. After the concentration of molecules in the gas state has increased sufficiently, the number of molecules reentering the liquid becomes equal to the number escaping from the liquid (Figure 8.19b). At this point, a dynamic equilibrium exists, exactly as in a chemical reaction at equilibrium. Evaporation and condensation take place at the same rate, and the concentration of vapor in the container is constant as long as the temperature doesn't change.

Once molecules have escaped from the liquid into the gas state, they are subject to all the gas laws previously discussed. In a closed container at equilibrium, for example, the gas molecules make their own contribution to the total pressure of the gas above the liquid according to Dalton's law (Section 8.10). We call this contribution the **vapor pressure** of the liquid.

▲ Because bromine is colored, it's possible to see its gaseous reddish vapor above the liquid.

 Diffusion of Bromine Vapor

Vapor pressure The partial pressure of gas molecules in equilibrium with a liquid.

Vapor pressure depends both on temperature and on the chemical identity of a liquid. As the temperature rises, molecules become more energetic and more likely to escape into the gas state. Thus, vapor pressure rises with increasing temperature until ultimately it becomes equal to the pressure of the atmosphere (Figure 8.20). At this point, bubbles of vapor form under the surface and force their way to the top, giving rise to the violent action observed during a vigorous boil. At an atmospheric pressure of exactly 760 mm Hg, boiling occurs at what is called the **normal boiling point**.

Normal boiling point The boiling point at a pressure of exactly 1 atmosphere.

FIGURE 8.20   A plot showing ▶ **the change of vapor pressure with temperature for ethyl ether, ethyl alcohol, and water.**   At a liquid's boiling point, its vapor pressure is equal to atmospheric pressure. The boiling points commonly reported are those at 760 mm Hg.

   Vapor Pressure vs. Temperature

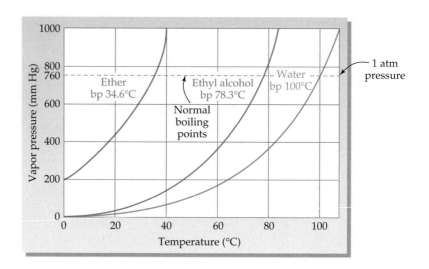

▲   Medical instruments are sterilized in this autoclave by heating them with water at high pressure.

If atmospheric pressure is higher or lower than normal, the boiling point of a liquid changes accordingly. At high altitudes, for example, atmospheric pressure is lower than at sea level, and boiling points are also lower. On top of Mt. Everest (29,035 ft; 8850 m), atmospheric pressure is about 245 mm Hg and the boiling temperature of water is only 71°C. If the atmospheric pressure is higher than normal, the boiling point is also higher. This principle is used in strong vessels known as *autoclaves*, in which water at high pressure is heated to the temperatures needed for sterilizing medical and dental instruments (170°C).

Many familiar properties of liquids can be explained by the intermolecular forces just discussed. We all know, for instance, that some liquids, such as water or gasoline, flow easily when poured, whereas others, such as motor oil or maple syrup, flow sluggishly.

The measure of a liquid's resistance to flow is called its *viscosity*. Not surprisingly, viscosity is related to the ease with which individual molecules move around in the liquid and thus to the intermolecular forces present. Substances such as gasoline, which have small, nonpolar molecules, experience only weak intermolecular forces and have relatively low viscosities, whereas more polar substances such as glycerin [$C_3H_5(OH)_3$] experience stronger intermolecular forces and so have higher viscosities.

Another familiar property of liquids is *surface tension*, the resistance of a liquid to spread out and increase its surface area. The beading-up of water on a newly waxed car and the ability of a water strider to walk on water are both due to surface tension.

Surface tension is caused by the difference between the intermolecular forces experienced by molecules at the surface of the liquid and those experienced by molecules in the interior. Molecules in the interior of a liquid are surrounded and experience maximum intermolecular forces, while molecules at the surface have fewer neighbors and feel weaker forces. Surface molecules are therefore less stable, and the liquid acts to minimize their number by minimizing the surface area (Figure 8.21).

▲   Surface tension allows a water strider to walk on water without penetrating the surface.

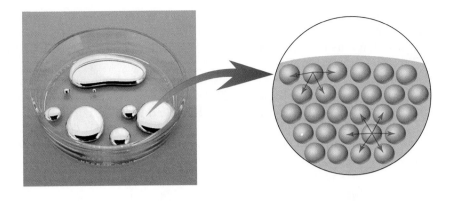

◀ **FIGURE 8.21** Surface tension is caused by the different forces experienced by molecules in the interior of a liquid and those on the surface. Molecules on the surface are less stable because they feel fewer attractive forces, so the liquid acts to minimize their number by minimizing the area of the surface.

8.13 Water: A Unique Liquid

Ours is a world based on water. Water covers nearly 71% of the earth's surface, it accounts for 66% of the mass of an adult human body, and it is needed by all living things. The water in our blood forms the transport system that circulates substances throughout our body, and water is the medium in which all biochemical reactions are carried out. Largely because of its strong hydrogen bonding, water has many properties that are quite different from those of other compounds.

Water has the highest specific heat of any liquid (Section 2.10), giving it the capacity to absorb a large quantity of heat while changing only slightly in temperature. (◯◯◯, p. 35) As a result, large lakes and other bodies of water tend to moderate the air temperature and climate of surrounding areas. Another consequence of the high specific heat of water is that the human body is better able to maintain a steady internal temperature under changing outside conditions.

◀ Large lakes moderate the air temperature of surrounding areas because of their capacity to store a large amount of heat.

In addition to a high specific heat, water has an unusually high *heat of vaporization* (540 cal/g), meaning that it carries away a large amount of heat when it evaporates. You can feel the effect of water evaporation on your wet skin when the wind blows. Even when comfortable, your body is still relying for cooling on the heat carried away from the skin and lungs by evaporating water. The heat generated by the chemical reactions of metabolism is carried by blood to the skin, where water moves through cell walls to the surface and evaporates. When metabolism, and therefore heat generation, speeds up, blood flow increases and capillaries dilate so that heat is brought to the surface faster.

Water is also unique in what happens as it changes from a liquid to a solid. Most substances are more dense as solids than as liquids because molecules are more closely packed in the solid than in the liquid. Water, however, is different. Liquid water has a maximum density of 1.000 g/mL at 3.98°C but then becomes *less* dense as it cools. When it freezes, its density decreases still further to 0.917 g/mL.

As water freezes, each molecule is locked into position by hydrogen bonding to four other water molecules (Figure 8.22). The resulting structure has more open space than does liquid water, accounting for its lower density. As a result, ice floats on liquid water, and lakes and rivers freeze from the top down. If the reverse were true, fish would be killed in winter as they became trapped in ice at the bottom.

FIGURE 8.22   Ice consists of ▶ individual H$_2$O molecules held rigidly together in an ordered manner by hydrogen bonds.

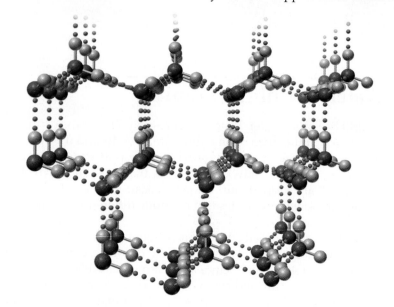

8.14 Solids

It's obvious from a brief look around that most substances are solids rather than liquids or gases. It's also obvious that there are many different kinds of solids. Some, such as iron and aluminum, are hard and metallic; others, such as sugar and table salt, are crystalline and easily broken; and still others, such as rubber and many plastics, are soft and amorphous.

The most fundamental distinction between solids is that some are crystalline and some are amorphous. A **crystalline solid** is one whose particles— whether atoms, ions, or molecules—have an ordered arrangement extending

Crystalline solid A solid whose atoms, molecules, or ions are rigidly held in an ordered arrangement.

Crystalline solids, such as the ▶ minerals shown here, have flat faces and distinct angles. These regular macroscopic features reflect a similarly ordered arrangement of particles at the atomic level.

over a long range. This order on the atomic level is also seen on the visible level, because crystalline solids usually have flat faces and distinct angles.

Crystalline solids can be further categorized as ionic, molecular, covalent network, or metallic. *Ionic solids* are those like sodium chloride, whose constituent particles are ions. A crystal of sodium chloride is composed of alternating Na^+ and Cl^- ions ordered in a regular three-dimensional arrangement and held together by ionic bonds, as discussed in Section 4.3. (p. 75) *Molecular solids* are those like sucrose or ice, whose constituent particles are molecules held together by the intermolecular forces discussed in Sections 8.11 and 8.12. A crystal of ice, for example, is composed of H_2O molecules held together in a regular way by hydrogen bonding (Figure 8.22). *Covalent network solids* are those like diamond (Figure 8.23) or quartz (SiO_2), whose atoms are linked together by covalent bonds into a giant three-dimensional array. In effect, a covalent network solid is one *very* large molecule.

◀ The sand on this beach is silica, SiO_2, a covalent network solid. Each grain of sand is essentially one large molecule.

◀ **FIGURE 8.23** Diamond is a covalent network solid—one very large molecule of carbon atoms linked by covalent bonds.

Metallic solids, such as silver or iron, can be viewed as vast three-dimensional arrays of metal cations immersed in a sea of electrons that are free to move about. This continuous electron sea acts both as a glue to hold the cations together and as a mobile carrier of charge to conduct electricity. Furthermore, the fact that bonding attractions extend uniformly in all directions explains why metals are malleable rather than brittle. When a metal crystal receives a sharp blow, no spatially oriented bonds are broken; instead, the electron sea simply adjusts to the new distribution of cations.

An **amorphous solid**, by contrast with a crystalline solid, is one whose constituent particles are randomly arranged and have no ordered long-range structure. Amorphous solids often result when liquids cool before they can achieve internal order, or when their molecules are large and tangled together, as happens

Amorphous solid A solid whose particles do not have an orderly arrangement.

in many polymers. Glass is an amorphous solid, as are tar, the gemstone opal, and some hard candies. Amorphous solids differ from crystalline solids by softening over a wide temperature range rather than having sharp melting points, and by shattering to give pieces with curved rather than planar faces.

A summary of the different types of solids and their characteristics is given in Table 8.4.

TABLE 8.4   Types of Solids

Substance	Smallest Unit	Interparticle Forces	Properties	Examples
Ionic solid	Ions	Attraction between positive and negative ions	Brittle and hard; high-melting; crystalline	NaCl, KI, $Ca_3(PO_4)_2$
Molecular solid	Molecules	Intermolecular forces	Soft; low- to moderate melting; crystalline	Ice, wax, frozen CO_2, all solid organic compounds
Covalent network	Atoms	Covalent bonds	Very hard; very high-melting; crystalline	Diamond, quartz (SiO_2), tungsten carbide (WC)
Metal or alloy	Metal atoms	Metallic bonding (attraction between metal ions and surrounding mobile electrons)	Lustrous; soft (Na) to hard (Ti); high-melting; crystalline	Elements (Fe, Cu, Sn, ...), bronze (CuSn alloy), amalgams (Hg + other metals)
Amorphous solid	Atoms, ions, or molecules (including polymer molecules)	Any of the above	Noncrystalline; no sharp mp; able to flow (may be very slow); curved edges when shattered	Glasses, tar, some plastics

8.15 Changes of State

What happens when a solid is heated? As more and more energy is added, molecules begin to stretch, bend, and vibrate more vigorously, and atoms or ions wiggle about with more energy. Finally, if enough energy is added and the motions become vigorous enough, the melting point is reached and particles start to break free from one another. Addition of further heat continues the melting process until all particles have broken free and are in the liquid phase. The quantity of heat required to completely melt a substance once it has reached its melting point is called its **heat of fusion.** After melting is complete, further addition of heat causes the temperature of the liquid to rise.

The change of a liquid into a vapor proceeds in the same way as the change of a solid into a liquid. When you first put a pan of water on the stove, all the added heat goes into raising the temperature of the water. Once the boiling point is reached, further absorbed heat goes into freeing molecules from their neighbors as they escape into the gas state. The quantity of heat needed to completely vaporize a liquid once it has reached its boiling point is called its **heat of vaporization.** A liquid that has a low heat of vaporization, like rubbing alcohol (isopropyl alcohol), evaporates rapidly and is said to be *volatile.* If you spill a volatile liquid on your skin, you will feel a cooling effect as it evaporates because it is absorbing heat from your body.

Water, because of its unusually strong intermolecular forces, has a particularly high heat of vaporization. Thus, water evaporates more slowly than many other liquids, takes a long time to boil away, and absorbs more heat in the process. A so-called *heating curve* for H_2O that indicates the temperature and state changes as heat is added is shown in Figure 8.24, and a list of heats of fusion and heats of vaporization for some common substances is given in Table 8.5. (p. 222)

Heat of fusion The quantity of heat required to completely melt a substance once it has reached its melting point.

Heat of vaporization The quantity of heat needed to completely vaporize a liquid once it has reached its boiling point.

Application Biomaterials for Joint Replacement

Freely movable joints in the body, such as those in the shoulder, knee, or hip, are formed by the meeting of two bones. The bony surfaces are not in direct contact, of course; rather, they are covered by cartilage for nearly frictionless motion and are surrounded by a fluid-containing capsule for lubrication. The hip, for instance, is a ball-and-socket joint, formed where the rounded upper end of the femur meets a cup-shaped part of the pelvic bone called the acetabulum.

Unfortunately, joints can wear out or fail, particularly when the cartilage is damaged by injury or diseased by degenerative arthritis. At some point, it may even be necessary to replace the failing joint—an estimated 300,000 joint-replacement surgeries are performed each year in the U.S. Yet even total joint replacement is not without problems, for the average lifetime of an artificial hip joint is only 8–12 years.

The first joint-replacement material, used in 1962, was stainless steel, but slow corrosion in the body led to its abandonment in favor of more resistant titanium or cobalt–chromium alloys. A typical modern hip-replacement joint consists of three parts: a polished metal ball to replace the head of the femur, a titanium alloy stem that is cemented into the shaft of the femur for stability, and a polyethylene cup to replace the hip socket.

Even with these materials, though, abrasion at the ball/cup contact can lead to joint wear. Over time, the constant repetitive movement of the polyethylene socket over the metal ball results in billions of microscopic polyethylene particles being sloughed off into the surrounding fluid. In addition, the cement holding the metal stem to the femur can slowly degrade, releasing other particles. The foreign particle are attacked by the body's immune system, resulting in the release of enzymes that cause the death of adjacent bone cells, loosening of the metal stem, and ultimate failure of the joint.

A potential solution to the problem involves the use of *biomaterials*—new materials that are created specifically for use in biological systems and do not provoke an immune response. It has been found, for instance, that the titanium stem of the artificial joint can bonded to an extremely thin layer of calcium phosphate, $Ca_3(PO_4)_2$, a close relative of hydroxyapatite $Ca_{10}(PO_4)_6(OH)_2$, the primary mineral constituent of bone (see the Chapter 4 Application, *Osteoporosis*). Natural bone then grows into the calcium phosphate, forming a strong natural bond to the stem and making cement unnecessary. Other biomaterials are being designed to replace the polyethylene socket.

See Additional Problem 8.99 at the end of the chapter.

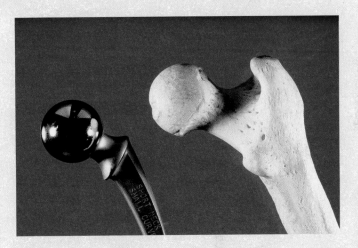

▲ This artificial hip joint is bonded to a thin layer of calcium phosphate to stimulate attachment of natural bone.

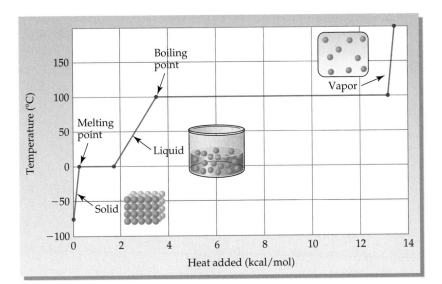

◀ **FIGURE 8.24 A heating curve for water, showing the temperature and state changes that occur when heat is added.** The horizontal lines at 0°C and 100°C represent the heat of fusion and heat of vaporization, respectively.

Changes of State

TABLE 8.5	Melting Points, Boiling Points, Heats of Fusion, and Heats of Vaporization of Some Common Substances			
Substance	Melting Point (°C)	Boiling Point (°C)	Heat of Fusion (cal/g)	Heat of Vaporization (cal/g)
Ammonia	−77.7	−33.4	84.0	327
Butane	−138.4	−0.5	19.2	92.5
Ether	−116	34.6	23.5	85.6
Ethyl alcohol	−117.3	78.5	26.1	200
Isopropyl alcohol	−89.5	82.4	21.4	159
Sodium	97.8	883	14.3	492
Water	0.0	100.0	79.7	540

■ **WORKED EXAMPLE 8.10**

Naphthalene, an organic substance often used in mothballs, has a heat of fusion of 35.7 cal/g and a molar mass of 128.0 g/mol. How much heat in kilocalories is required to melt 0.300 mol of naphthalene?

ANALYSIS The heat of fusion tells how much heat is required to melt 1 g. To find the amount of heat needed to melt 0.300 mol, we need a mole-to-mass conversion.

SOLUTION

$$(0.300 \text{ mol naphthalene}) \left(\frac{128.0 \text{ g}}{1 \text{ mol}} \right) = 38.4 \text{ g naphthalene}$$

Multiplying by the heat of fusion then gives the answer.

$$(38.4 \text{ g naphthalene}) \left(\frac{35.7 \text{ cal}}{1 \text{ g naphthalene}} \right) = 1370 \text{ cal} = 1.37 \text{ kcal}$$

■ **PROBLEM 8.21**

How much heat in kilocalories is required to melt and to boil 1.50 mol of isopropyl alcohol (rubbing alcohol; molar mass = 60.0 g/mol)? The heat of fusion and heat of vaporization of isopropyl alcohol are given in Table 8.5.

Application CO$_2$ as an Environmentally Friendly Solvent

How can CO$_2$ be a solvent? After all, carbon dioxide is a gas, not a liquid, at room temperature. Furthermore, CO$_2$ at atmospheric pressure doesn't become liquid even when cooled. When the temperature drops to $-78°C$ at 1 atm pressure, CO$_2$ goes directly from gas to solid (dry ice) without first becoming liquid. Only when the pressure is raised does liquid CO$_2$ exist. At a room temperature of 22.4°C, a pressure of 60 atm is needed to force gaseous CO$_2$ molecules close enough together so they condense to a liquid. Even as a liquid, though, CO$_2$ is not a particularly good solvent. Only when it enters an unusual and rarely seen state of matter called the *supercritical state* does CO$_2$ become a remarkable solvent.

To understand the supercritical state of matter, think about the liquid and gas states at the molecular level. In the liquid state, molecules are packed closely together, and most of the available volume is taken up by the molecules themselves. In the gas state, molecules are far apart, and most of the available volume is empty space. In the supercritical state, however, the situation is intermediate between liquid and gas. There is *some* space between molecules, but not much. The molecules are too far apart to be truly a liquid, yet they are too close together to be truly a gas. Supercritical CO$_2$ exists when the pressure is above 72.8 atm and the temperature is above 31.2°C—a high enough pressure to force molecules close together and a high enough temperature to allow the molecules to be apart.

Because open spaces already exist between CO$_2$ molecules, it is energetically easy for dissolved molecules to slip in, and supercritical CO$_2$ is therefore an extraordinarily good solvent. Among its many applications, supercritical CO$_2$ is used in the beverage and food processing industries to decaffeinate coffee beans and to obtain spice extracts from vanilla, pepper, cloves, nutmeg and other seeds. In the cosmetics and perfume industry, fragrant oils are extracted from flowers using supercritical CO$_2$. Perhaps the most important future application is the use of carbon dioxide for dry-cleaning clothes, thereby replacing environmentally harmful chlorinated solvents.

The use of supercritical CO$_2$ as a solvent has many benefits, including the fact that it is nontoxic and non-flammable. Most important, though, is that the technology is environmentally friendly. Industrial processes using CO$_2$ are designed as closed systems so that CO$_2$ is recaptured after use and continually recycled. There are no organic solvent vapors released into the atmosphere and no toxic liquids to seep into groundwater supplies, as can occur with current procedures using chlorinated organic solvents. The future looks bright for this new technology.

See Additional Problems 8.100 and 8.101 at the end of the chapter.

▲ The caffeine in these coffee beans can be removed by extraction with supercritical CO$_2$.

The speakers blare the call up and down the hospital hallways: "Code blue! Code blue!"

A flurry of white coats appears, some individuals walking fast, some running, others actually trying not to run ... everyone careening down the hall into one room where a critically ill patient has gone into cardiac and respiratory arrest. John Kuo, a respiratory therapist at a major hospital in New York City, races into the room wheeling the respiratory cart ahead of him.

"I save people's lives every day; that's just part of my job. We have special beepers because we're on the revival team. First the beeper goes off and then we'll hear our code over the PA system—Code Blue, which means Cardiac Emergency. One of the doctors used to tell us 'Don't run to the code. Just walk very fast.' Because if you run to the code and you get out of breath, then you get overheated and light-headed—you can't think as clearly or see what you're doing, and that's not beneficial to the patient. So I try not to run there ... although I often run anyway!

"When we get there, our job is to ventilate the patient and get oxygenated air into the lungs as soon as possible. The patient isn't breathing on his own, so we have to do the breathing for him. You put the mask on the face covering the mouth and nose, and you force the air into the lungs with the hand pump. It's like squeezing a big balloon. This part is just the basic gas laws at work—you squeeze the bag and make its volume smaller, the pressure rises. The higher pressure forces air into the lungs, and they have to expand—increase in volume.

"At that point somebody's checking for the pulse. If there's no pulse the cardiac team starts compressing the heart, doing CPR. They take care of the heart and we take care of the lungs.

"I've been a respiratory therapist for nine years. A respiratory therapist helps keep the patients breathing. That's the essence of what we do, whether it's in a hospital or in a home care setting. We do everything from teaching simple breathing exercises to full-blown life support. But the most important and most stressful job is being part of a cardiac arrest team. The team is made up of doctors, nurses, and respiratory therapists. Each person has a role, and ours is to maintain and keep the airway open and oxygenate the patient. Because once you stop breathing, your heart stops.

"It's also our job to determine the levels of gases in the blood. We take a blood sample, and a machine measures how much CO_2 and oxygen it contains [Chapter 9], along with its acidity [Chapter 10]. We draw arterial blood because we need to find out if the patient is breathing in enough oxygen and breathing out enough CO_2. The blood has to supply enough oxygen to the tissues for the cells to carry on their regular metabolism [Chapter 21]. By the time the blood gets back to the heart by way of the venous system, we can't get an accurate assessment of how well the lungs are doing, because the tissues have already used up much of the oxygen [Chapter 28]. So we get blood from the arterial side. And we soak the sample in ice because you don't want the blood cells to use up the oxygen in their own metabolism. The cold slows down their metabolic processes [Chapter 19].

"The blood gas machine can return a reading in just 10–15 minutes, but for the analysis to be accurate, we must supply certain information. The barometric pressure and the temperature can both affect the readings—the gas laws again—so we need to know these when we calibrate the equipment. For example, the machine assumes normal body temperature, so if the patient is running a fever, we have to punch that information into the machine's computer.

"I think that chemistry really helps people in the health care field. You have to understand the principles that underlie what you do. My teacher used to say, 'The things you learn now probably won't make sense to you now. But as you start doing therapy, eventually something will click and things will start to fall into place in your mind. You will begin to understand why a particular patient responds in a particular way, why some drugs function correctly under certain conditions and not others.' And she was right."

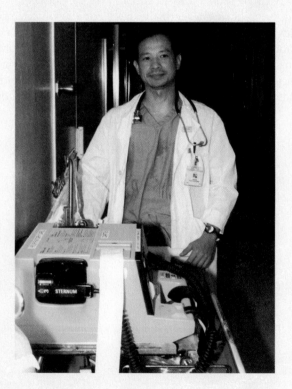

Summary: Revisiting the Chapter Goals

1. **What is the kinetic–molecular theory of gases?** According to the *kinetic–molecular theory of gases*, the physical behavior of gases can be explained by assuming that they consist of particles moving rapidly at random, separated from other particles by great distances, and colliding without loss of energy. Gas *pressure* is the result of molecular collisions with a surface.

2. **What are the gas laws?** *Boyle's law* says that the volume of a fixed amount of gas at constant temperature is inversely proportional to its pressure ($P_1V_1 = P_2V_2$). *Charles's law* says that the volume of a fixed amount of gas at constant pressure is directly proportional to its Kelvin temperature ($V_1/T_1 = V_2/T_2$). *Gay-Lussac's law* says that the pressure of a fixed amount of gas at constant volume is directly proportional to its Kelvin temperature ($P_1/T_1 = P_2/T_2$). Boyle's law, Charles's law, and Gay-Lussac's law together give the *combined gas law* ($P_1V_1/T_1 = P_2V_2/T_2$), which applies to changing conditions for a fixed quantity of gas. *Avogadro's law* says that equal volumes of gases at the same temperature and pressure contain the same number of moles ($V_1/n_1 = V_2/n_2$).

3. **What is the ideal gas law?** The four gas laws together give the *ideal gas law*, $PV = nRT$, which relates the effects of temperature, pressure, volume, and molar amount. At 0°C and 1 atm pressure, called *standard temperature and pressure (STP)*, 1 mol of any gas (6.02×10^{23} molecules) occupies a volume of 22.4 L.

4. **What is partial pressure?** The amount of pressure exerted by an individual gas in a mixture is called the *partial pressure* of the gas. According to *Dalton's law*, the total pressure exerted by the mixture is equal to the sum of the partial pressures of the individual gases.

5. **What are the major intermolecular forces?** There are three major types of *intermolecular forces*, which act to hold molecules near one another in solids and liquids. *Dipole–dipole forces* are the electrical attractions that occur between polar molecules. *London dispersion forces* occur between all molecules as a result of temporary molecular polarities due to unsymmetrical electron distribution. These forces increase in strength with molecular weight and with the surface area of molecules. *Hydrogen bonding*, the strongest of the three intermolecular forces, occurs between a hydrogen atom bonded to O, N, or F and a nearby O, N, or F atom.

6. **What are the various kinds of solids, and how do they differ?** Solids are either crystalline or amorphous. *Crystalline solids* are those whose constituent particles have an ordered arrangement; *amorphous solids* lack internal order and do not have sharp melting points. There are several kinds of crystalline solids: *Ionic solids* are those like sodium chloride, whose constituent particles are ions. *Molecular solids* are those like ice, whose constituent particles are molecules held together by intermolecular forces. *Covalent network solids* are those like diamond, whose atoms are linked together by covalent bonds into a giant three-dimensional array. *Metallic solids*, such as silver or iron, also consist of large arrays of atoms, but their crystals have metallic properties such as electrical conductivity.

7. **What occurs during a change of state?** When a solid is heated, particles begin to move around freely at the *melting point*, and the substance becomes liquid. The amount of heat necessary to melt a given amount of solid at its melting point is its *heat of fusion*. As a liquid is heated, molecules escape from the surface of a liquid until an equilibrium is reached between liquid and gas, resulting in a *vapor pressure* of the liquid. At a liquid's *boiling point*, its vapor pressure equals atmospheric pressure, and the entire liquid is converted into gas. The amount of heat necessary to vaporize a given amount of liquid at its boiling point is called its *heat of vaporization*.

Key Words

Amorphous solid, p. 219

Avogadro's law, p. 205

Boiling point (bp), p. 194

Boyle's law, p. 199

Change of state, p. 194

Charles's law, p. 201

Combined gas law, p. 203

Crystalline solid, p. 218

Dalton's law, p. 208

Dipole–dipole force, p. 211

Gas constant (R), p. 206

Gas laws, p. 198

Gay-Lussac's law, p. 202

Heat of fusion, p. 220

Heat of vaporization, p. 220

Hydrogen bond, p. 212

Ideal gas, p. 196

Ideal gas law, p. 206

Intermolecular force, p. 209

Kinetic–molecular theory of gases, p. 195

London dispersion force, p. 211

Melting point (mp), p. 194

Normal boiling point, p. 216

Partial pressure, p. 208

Pressure (P), p. 196

Standard temperature and pressure (STP), p. 205

Standard molar volume, p. 205

Vapor, p. 215

Vapor pressure, p. 215

▪ Understanding Key Concepts

8.22 Assume that you have a sample of gas in a cylinder with a movable piston, as shown in the following drawing:

Redraw the apparatus to show what the sample will look like after the following changes:

(a) The temperature is increased from 300 K to 450 K at constant pressure.

(b) The pressure is increased from 1 atm to 2 atm at constant temperature.

(c) The temperature is decreased from 300 K to 200 K and the pressure is decreased from 3 atm to 2 atm.

8.23 Assume that you have a sample of gas at 350 K in a sealed container, as represented in part (a). Which of the drawings (b)– (d) represents the gas after the temperature is lowered from 350 K to 150 K?

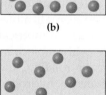

(b)

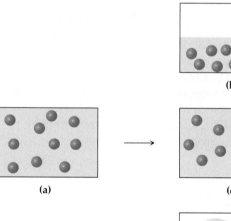

(a)   (c)

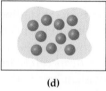

(d)

8.24 Assume that drawing (a) represents a sample of H_2O at 200 K. Which of the drawings (b)–(d) represents what the sample will look like when the temperature is raised to 300 K?

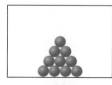

(a)

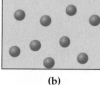

(b)

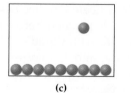

(c)

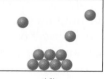

(d)

8.25 Three bulbs, two of which contain different gases and one of which is empty, are connected as shown in the following drawing:

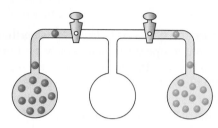

Redraw the apparatus to represent the gases after the stopcocks are opened and the system is allowed to come to equilibrium.

8.26 Redraw the following open-end manometer to show what it would look like when stopcock A is opened.

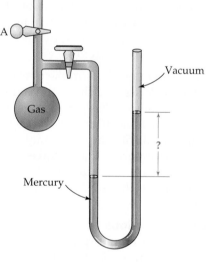

8.27 The following diagram represents the heating curve of a hypothetical substance:

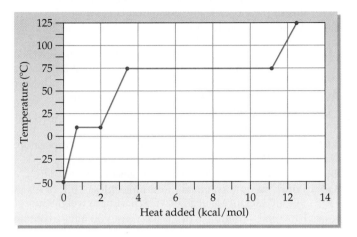

(a) What is the melting point of the substance?
(b) What is the boiling point of the substance?
(c) Approximately what is the heat of fusion for the substance in kcal/mol?
(d) Approximately what is the heat of vaporization for the substance in kcal/mol?

8.28 Show the approximate level of the movable piston in drawings (a), (b), and (c) after the indicated changes have been made to the gas.

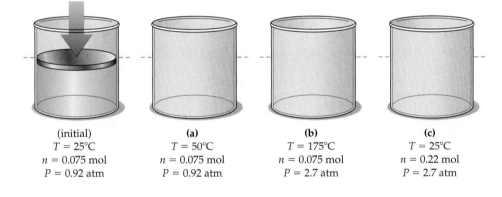

(initial)	(a)	(b)	(c)
$T = 25°C$	$T = 50°C$	$T = 175°C$	$T = 25°C$
$n = 0.075$ mol	$n = 0.075$ mol	$n = 0.075$ mol	$n = 0.22$ mol
$P = 0.92$ atm	$P = 0.92$ atm	$P = 2.7$ atm	$P = 2.7$ atm

8.29 What is the partial pressure of each gas—red, yellow, and green—if the total pressure inside the following container is 600 mm Hg?

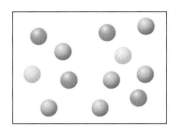

▪ Additional Problems

Gases and Pressure

8.30 How is 1 atm of pressure defined?

8.31 List four common units for measuring pressure.

8.32 What are the four assumptions of the kinetic–molecular theory of gases?

8.33 How does the kinetic–molecular theory of gases explain gas pressure?

8.34 Convert the following values into mm Hg:
(a) Standard pressure
(b) 0.25 atm
(c) 7.5 atm
(d) 28.0 in. Hg
(e) 41.8 Pa

8.35 Atmospheric pressure at the top of Mt. Whitney in California is 440 mm Hg.
(a) How many atmospheres is this?
(b) How many pascals is this?

8.36 What is the pressure (in mm Hg) inside a container of gas connected to a mercury-filled, open-end manometer of the sort shown in Figure 8.5 when the level in the arm connected to the container is 17.6 cm lower than the level in the arm open to the atmosphere, and the atmospheric pressure reading outside the apparatus is 754.3 mm Hg?

8.37 What is the pressure (in atmospheres) inside a container of gas connected to a mercury-filled, open-end manometer of the sort shown in Figure 8.5 when the level in the arm connected to the container is 28.3 cm higher than the level in the arm open to the atmosphere, and the atmospheric pressure reading outside the apparatus is 1.021 atm?

Boyle's Law

8.38 What is Boyle's law, and what variables must be kept constant for the law to hold?

8.39 Does Boyle's law involve a direct proportionality or an inverse proportionality?

8.40 The pressure of gas in a 600.0 mL cylinder is 65.0 mm Hg. To what volume must it be compressed to increase the pressure to 385 mm Hg?

8.41 The volume of a balloon is 3.50 L at 1.00 atm. What is the pressure if the balloon is compressed to 2.00 L?

8.42 Oxygen gas is commonly sold in 50.0 L steel containers at a pressure of 150 atm. What volume would the gas occupy at a pressure of 1.00 atm if its temperature remained unchanged?

8.43 A syringe has a volume of 10.0 cc at 14.7 psi. If the tip is blocked so that air can't escape, what pressure is required to decrease the volume to 2.00 cc?

Charles's Law

8.44 What is Charles's law, and what variables must be kept constant for the law to hold?

8.45 Does Charles's law involve a direct proportionality or an inverse proportionality?

8.46 A hot-air balloon has a volume of 960 L at 18°C. To what temperature must it be heated to raise its volume to 1200 L, assuming the pressure remains constant?

8.47 A hot-air balloon has a volume of 875 L. What is the temperature of the balloon if its volume changes to 955 L when heated to 56°C?

8.48 A gas sample has a volume of 125 mL at 25°C. What is its volume at 37°C?

8.49 A balloon has a volume of 43.0 L at 20°C. What is its volume at −5°C?

Gay-Lussac's Law

8.50 What is Gay-Lussac's law, and what variables must be kept constant for the law to hold?

8.51 Does Gay-Lussac's law involve a direct or an inverse proportionality?

8.52 A glass laboratory flask is filled with gas at 25°C and 0.95 atm pressure, sealed, and then heated to 125°C. What is the pressure inside the flask?

8.53 An aerosol can has an internal pressure of 3.75 atm at 25°C. What temperature is required to raise the pressure to 16.6 atm?

Combined Gas Law

8.54 A gas has a volume of 2.84 L at 1.00 atm and 0°C. At what temperature does it have a volume of 7.50 L at 520 mm Hg?

8.55 A helium balloon has a volume of 3.50 L at 22.0°C and 1.14 atm. What is its volume if the temperature is increased to 30.0°C and the pressure is increased to 1.20 atm?

8.56 When H_2 gas was released by the action of HCl on Zn, the volume of H_2 collected was 55.0 mL at 26°C and 749 mm Hg. What is the volume of the H_2 at 0°C and 1.00 atm pressure (STP)?

8.57 A compressed-air tank carried by scuba divers has a volume of 8.00 L and a pressure of 140 atm at 20°C. What is the volume of air in the tank at 0°C and 1.00 atm pressure (STP)?

8.58 What is the effect on the pressure of a gas if you simultaneously:
  (a) Halve its volume and double its Kelvin temperature?
  (b) Double its volume and halve its Kelvin temperature?

8.59 What is the effect on the volume of a gas if you simultaneously:
  (a) Halve its pressure and double its Kelvin temperature?
  (b) Double its pressure and double its Kelvin temperature?

8.60 A sample of oxygen produced in a laboratory experiment had a volume of 590 mL at a pressure of 775 mm Hg and a temperature of 352 K. What is the volume of this sample at 25°C and 800.0 mm Hg pressure?

8.61 A small cylinder of helium gas used for filling balloons has a volume of 2.30 L and a pressure of 1850 atm at 25°C. How many balloons can you fill if each one has a volume of 1.5 L and a pressure of 1.25 atm at 25°C?

Avogadro's Law and Standard Molar Volume

8.62 Explain Avogadro's law using the kinetic–molecular theory of gases.

8.63 What conditions are defined as standard temperature and pressure (STP)?

8.64 How many liters does a mole of gas occupy at STP?

8.65 Which sample contains more molecules: 1.0 L of O_2 at STP, or 1.0 L of H_2 at STP? Which sample weighs more?

8.66 How many milliliters of Cl_2 gas must you have to obtain 0.20 g at STP?

8.67 Assume that you have 1.75 g of the deadly gas hydrogen cyanide, HCN. What is the volume of the gas at STP?

8.68 A typical room is 4.0 m long, 5.0 m wide, and 2.5 m high. What is the total mass of the oxygen in the room assuming that the gas in the room is at STP and that air contains 21% oxygen and 79% nitrogen?

8.69 What is the total mass of nitrogen in the room described in Problem 8.68?

Ideal Gas Law

8.70 What is the ideal gas law?

8.71 How does the ideal gas law differ from the combined gas law?

8.72 Which sample contains more molecules: 2.0 L of Cl_2 at STP, or 3.0 L of CH_4 at 300 K and 1.5 atm? Which sample weighs more?

8.73 Which sample contains more molecules: 2.0 L of CO_2 at 300 K and 500 mm Hg, or 1.5 L of N_2 at 57°C and 760 mm Hg? Which sample weighs more?

8.74 If 15.0 g of CO_2 gas has a volume of 0.30 L at 310 K, what is its pressure in mm Hg?

8.75 If 20.0 g of N_2 gas has a volume of 4.00 L and a pressure of 6.0 atm, what is its temperature in degrees Celsius?

8.76 If 18.0 g of O_2 gas has a temperature of 350 K and a pressure of 550 mm Hg, what is its volume?

8.77 How many moles of a gas will occupy a volume of 0.55 L at a temperature of 347 K and a pressure of 2.5 atm?

Dalton's Law and Partial Pressure

8.78 What is meant by *partial pressure*?

8.79 What is Dalton's law?

8.80 If the partial pressure of oxygen in air at 1.0 atm is 160 mm Hg, what is the partial pressure on the summit of Mt. Whitney, where atmospheric pressure is 440 mm Hg? Assume that the percent oxygen is the same.

8.81 A special gas mixture used in bacterial growth chambers contains 1.00% by weight CO_2 and 99.0% O_2. What is the partial pressure (in atmospheres) of each gas at a total pressure of 0.977 atm?

Liquids and Intermolecular Forces

8.82 What is the vapor pressure of a liquid?

8.83 What is the value of a liquid's vapor pressure at its normal boiling point?

8.84 What is the effect of pressure on a liquid's boiling point?

8.85 What is a liquid's heat of vaporization?

8.86 What characteristic must a compound have to experience each of the following intermolecular forces?

(a) London dispersion forces
(b) Dipole–dipole forces
(c) Hydrogen bonding

8.87 In which of the following compounds are dipole–dipole attractions the most important intermolecular force?

(a) N_2 (b) HCN
(c) CCl_4 (d) $MgBr_2$
(e) CH_3Cl (f) CH_3CO_2H

8.88 Dimethyl ether (CH_3OCH_3) and ethanol (C_2H_5OH) have the same formula (C_2H_6O), but the boiling point of dimethyl ether is −25°C while that of ethanol is 78°C. Explain.

8.89 Iodine is a solid at room temperature (mp = 113.5°C) while bromine is a liquid (mp = −7°C). Explain in terms of intermolecular forces.

8.90 The heat of vaporization of water is 9.72 kcal/mol.

(a) How much heat (in kilocalories) is required to vaporize 3.00 mol of H_2O?
(b) How much heat (in kilocalories) is released when 255 g of steam condenses?

8.91 The heat of vaporization of octane (C_8H_{18}), an important component of gasoline, is 8.25 kcal/mol. How much heat (in kilocalories) is required to vaporize 100.0 g of octane?

Solids

8.92 What is the difference between an amorphous and a crystalline solid?

8.93 List three kinds of crystalline solids, and give an example of each.

8.94 The heat of fusion of acetic acid, the principal organic component of vinegar, is 45.9 cal/g. How much heat (in kilocalories) is required to melt 1.66 mol of solid acetic acid?

8.95 The heat of fusion of sodium metal is 630 cal/mol. How much heat (in kilocalories) is required to melt 245 g of sodium?

Applications

8.96 What is the difference between a systolic and a diastolic pressure reading? Is a blood pressure of 180/110 within the normal range? [*Blood Pressure*]

8.97 What are the three most important greenhouse gases? [*Greenhouse Gases and Global Warming*]

8.98 How can an increase in atmospheric CO_2 affect the earth's surface temperature? [*Greenhouse Gases and Global Warming*]

8.99 What is a "biomaterial"? [*Biomaterials for Joint Replacement*]

8.100 What is the supercritical state of matter? [*CO_2 as an Environmentally Friendly Solvent*]

8.101 What are the environmental advantages of using supercritical CO_2 in place of chlorinated organic solvents? [*CO_2 as an Environmentally Friendly Solvent*]

General Questions and Problems

8.102 Use the kinetic–molecular theory to explain why gas pressure increases if the temperature is raised and the volume is kept constant.

8.103 Hydrogen and oxygen react according to the equation $2 H_2 + O_2 \rightarrow 2 H_2O$. According to Avogadro's law, how many liters of hydrogen are required to react with 2.5 L of oxygen at STP?

8.104 If 3.0 L of hydrogen and 1.5 L of oxygen at STP react to yield water, how many moles of water are formed? What gas volume does the water have at a temperature of 100°C and 1 atm pressure?

8.105 Approximately 240 mL/min of CO_2 is exhaled by an average adult at rest. Assuming a temperature of 37°C and 1 atm pressure, how many moles of CO_2 is this?

8.106 How many grams of CO_2 are exhaled by an average resting adult in 24 hours (see Problem 8.105)?

8.107 Imagine that you have two identical containers, one containing hydrogen at STP and the other containing oxygen at STP. How could you tell which was which without opening them?

8.108 One mole of any gas has a volume of 22.4 L at STP. What is the molecular weight of each of the following gases, and what are their densities in grams per liter at STP?

(a) CH_4 (b) CO_2 (c) O_2

8.109 Gas pressure outside the space shuttle is approximately 1×10^{-14} mm Hg at a temperature of approximately 1 K. If the gas is almost entirely hydrogen atoms (H, not H_2), what volume of space is occupied by 1 mol of atoms? What is the density of H gas in atoms per liter?

8.110 Ethylene glycol, $C_2H_6O_2$, has one OH bonded to each carbon.

(a) Draw the Lewis dot structure of ethylene glycol.
(b) Draw the Lewis dot structure of chloroethane, C_2H_5Cl.
(c) Chloroethane has a slightly higher molar mass than ethylene glycol, yet has a much lower boiling point (3°C versus 198°C). Explain.

8.111 A gas tank containing N_2 has a volume of 25.0 L and a pressure of 35.0 atm. A second tank, containing O_2, has a volume of 10.0 L and a pressure of 15.0 atm. Both gases are pumped into a third tank with a volume of 15.0 L.

(a) What is the partial pressure of each gas in the new tank?
(b) What is the total pressure in the new tank?

8.112 The *Rankine* temperature scale used in engineering is to the Fahrenheit scale as the Kelvin scale is to the Celsius scale. That is, 1 Rankine degree is the same size as 1 Fahrenheit degree, and 0°R = absolute zero.

(a) What temperature corresponds to the freezing point of water on the Rankine scale?
(b) What is the value of the gas constant R on the Rankine scale in $(L \cdot atm)/(°R \cdot mol)$?

8.113 Isooctane, C_8H_{18}, is the component of gasoline from which the term *octane rating* derives.

(a) Write a balanced equation for the combustion of isooctane to yield CO_2 and H_2O.
(b) Assuming that gasoline is 100% isooctane and that the density of isooctane is 0.792 g/mL, what mass of CO_2 (in kilograms) is produced each year by the annual U.S. gasoline consumption of 4.6×10^{10} L?
(c) What is the volume (in liters) of this CO_2 at STP?

9

Solutions

Deep-sea divers who ascend too quickly can develop the bends, *a condition caused by increased solubility of nitrogen in the blood at high pressure.*

U p to this point, we've been concerned primarily with pure substances, both elements and compounds. In day-to-day life, however, most of the materials we come in contact with are mixtures. Air, for example, is a gaseous mixture containing primarily oxygen and nitrogen; blood is a liquid mixture of many different components; and many rocks are solid mixtures of different minerals.

We'll look closely in this chapter at the characteristics and properties of mixtures, with particular attention to the homogeneous mixtures we call *solutions*.

Among the questions we'll answer are the following:

1. **What are solutions, and what factors affect solubility?**
   The goal: Be able to define the different kinds of mixtures and explain the influence on solubility of solvent and solute structure, temperature, and pressure.

2. **How is the concentration of a solution expressed?**
   The goal: Be able to define, use, and convert between the most common ways of expressing solution concentrations.

3. **How are dilutions carried out?**
   The goal: Be able to calculate the concentration of a solution prepared by dilution and explain how to make a desired dilution.

4. **What is an electrolyte?**
   The goal: Be able to recognize strong and weak electrolytes and nonelectrolytes, and express electrolyte concentrations.

5. **How do solutions differ from pure solvents in their behavior?**
   The goal: Be able to explain vapor pressure lowering, boiling point elevation, and freezing point depression for solutions.

6. **What is osmosis?**
   The goal: Be able to describe osmosis and some of its applications.

CONCEPTS TO REVIEW

Enthalpy Changes (Section 7.1)
Chemical Equilibrium (Section 7.6)
Le Châtelier's Principle (Section 7.8)
Partial Pressure of Gases (Section 8.10)
Intermolecular Forces and Hydrogen Bonds (Section 8.11)
Vapor Pressure (Section 8.12)

9.1 Mixtures and Solutions

As we saw in Section 1.3, a *mixture* is an intimate combination of two or more substances, both of which retain their chemical identities. (⚬⚬⚬, p. 3) Mixtures are often classified as either *heterogeneous* or *homogeneous*, depending on their appearance. **Heterogeneous mixtures** are those in which the mixing is not uniform and which therefore have regions of different composition. Granite and many other rocks, for instance, have a grainy character due to the heterogeneous mixing of different minerals. **Homogeneous mixtures** are those in which the mixing *is* uniform and that therefore have the same composition throughout. Seawater, a homogeneous mixture of soluble ionic compounds in water, is an example.

Homogeneous mixtures can be further classified according to the size of their particles as either *solutions* or *colloids*. **Solutions**, the most important class of homogeneous mixtures, contain particles the size of a typical ion or small molecule—roughly 0.1–2 nm in diameter. **Colloids**, such as milk and fog, are also homogeneous in appearance but contain larger particles than solutions—in the range 2–500 nm diameter.

Liquid solutions, colloids, and heterogeneous mixtures can be distinguished in several ways. For example, liquid solutions are transparent (although they may be colored). Colloids may appear transparent if the particle size is small, but they have a murky or opaque appearance if the particle size is larger. Neither solutions nor colloids separate on standing, and the particles in both are too small to be removed by filtration. Heterogeneous mixtures are murky or opaque, and their particles will slowly settle on prolonged standing. House paint is an example.

Heterogeneous mixture A nonuniform mixture that has regions of different composition.

Homogeneous mixture A uniform mixture that has the same composition throughout.

Solution A homogeneous mixture that contains particles the size of a typical ion or small molecule.

Colloid A homogeneous mixture that contains particles in the range 2–500 nm diameter.

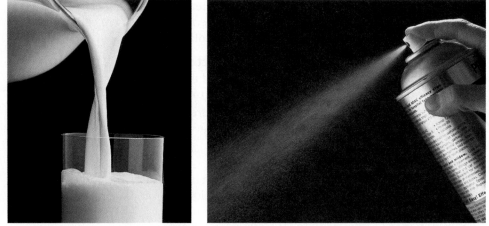

▲ (a) Wine is a solution of dissolved molecules, and (b) milk is a colloid with fine particles that do not separate out on standing. (c) An aerosol spray, by contrast, is a heterogeneous mixture of small particles visible to the naked eye.

Table 9.1 gives some examples of solutions, colloids, and heterogeneous mixtures. It's interesting to note that blood has characteristics of all three. About 45% by volume of blood consists of suspended red and white cells, which settle slowly on standing; the remaining 55% is *plasma*, which contains ions in solution and colloidal protein molecules.

TABLE 9.1   Some Characteristics of Solutions, Colloids, and Heterogeneous Mixtures

Kind of Mixture	Particle Size	Examples	Characteristics
Solution	<2.0 nm	Air, seawater, gasoline, wine	Transparent to light; does not separate on standing; nonfilterable
Colloid	2.0–500 nm	Butter, milk, fog, pearl	Often murky or opaque to light; does not separate on standing; nonfilterable
Heterogeneous	>500 nm	Blood, paint, aerosol sprays	Murky or opaque to light; separates on standing; filterable

Although we usually think of a solid dissolved in a liquid when talking about solutions, solutions actually occur in all three phases of matter (Table 9.2). Metal alloys like 14-karat gold (58% gold with silver and copper) and brass (10–40% zinc with copper), for instance, are solutions of one solid with another. For solutions in which a gas or solid is dissolved in a liquid, the dissolved substance is called the **solute** and the liquid is called the **solvent**. When one liquid is dissolved in another, the minor component is usually considered the solute and the major component is the solvent.

Solute   A substance dissolved in a liquid.

Solvent   The liquid in which another substance is dissolved.

TABLE 9.2   Some Different Kinds of Solutions

Kind of Solution	Example
Gas in gas	Air (O_2, N_2, Ar, and other gases)
Gas in liquid	Seltzer water (CO_2 in water)
Gas in solid	H_2 in palladium metal
Liquid in liquid	Gasoline (mixture of hydrocarbons)
Liquid in solid	Dental amalgam (mercury in silver)
Solid in liquid	Seawater (NaCl and other salts in water)
Solid in solid	Metal alloys such as 14-karat gold (Au, Ag, and Cu)

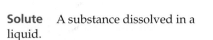

■ **PROBLEM 9.1**

Classify each of the following liquid mixtures as heterogeneous or homogeneous. Further classify each homogeneous mixture as a solution or colloid.

(a) Orange juice   **(b)** Apple juice   **(c)** Hand lotion   **(d)** Tea

---

9.2 The Solution Process

What determines whether a substance is soluble in a given liquid? Solubility depends primarily on the strength of the attractions between solute and solvent particles relative to the strengths of the attractions within the pure substances. Ethyl alcohol is soluble in water, for example, because hydrogen bonding (Section 8.11) is nearly as strong between water and ethyl alcohol molecules as it is between water molecules alone or ethyl alcohol molecules alone. (∞, p. 212)

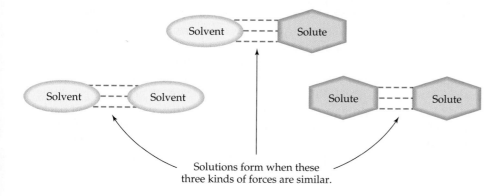

Solutions form when these three kinds of forces are similar.

A good rule of thumb for predicting solubility is that "like dissolves like," meaning that substances with similar intermolecular forces form solutions with one another, while substances with different intermolecular forces do not (Section 8.11). (∞, p. 209) Polar solvents dissolve polar and ionic solutes; nonpolar solvents dissolve nonpolar solutes; and hydrogen-bonding solvents dissolve hydrogen-bonding solutes. Thus, a polar, hydrogen-bonding compound like water dissolves ethyl alcohol and sodium chloride, whereas a nonpolar organic compound like hexane (C_6H_{14}) dissolves other nonpolar organic compounds like fats and oils. Water and oil, however, do not dissolve one another, as summed up by the old saying, "Oil and water don't mix." The intermolecular forces between water molecules are so strong that after an oil–water mixture is shaken, the water layer re-forms, squeezing out the oil molecules.

Water solubility is not limited to ionic compounds and ethyl alcohol. Many polar organic substances, such as sugars, amino acids, and even some proteins, dissolve in water. In addition, small, moderately polar organic molecules such as chloroform ($CHCl_3$) are soluble in water to a limited extent. When mixed with water, a small amount of the organic compound dissolves, but the remainder forms a separate liquid layer. As the number of carbon atoms in organic molecules increases, though, water solubility decreases.

The process of dissolving a solid in a liquid can be visualized as shown in Figure 9.1 for sodium chloride. When NaCl crystals are put in water, ions at the crystal surface come into contact with polar water molecules. Positively charged Na^+ ions are attracted to the negatively polarized oxygen of water, while negatively charged Cl^- ions are attracted to the positively polarized hydrogens. The combined forces of attraction between an ion and several water molecules pull the ion away from the crystal, exposing a fresh surface, until ultimately the crystal dissolves. Once in solution, Na^+ and Cl^- ions are completely surrounded by solvent molecules, a phenomenon called **solvation** (or,

▲   Oil and water don't mix because the two substances have different kinds of intermolecular forces.

Solvation   The clustering of solvent molecules around a dissolved solute molecule or ion.

FIGURE 9.1 **Dissolution of an** ▶
NaCl crystal in water. Polar water
molecules surround the individual
Na^+ and Cl^- ions at an exposed
edge or corner, pulling them from
the crystal surface into solution and
surrounding them. Note how the
negatively polarized oxygens of
water molecules cluster around Na^+
ions and the positively polarized
hydrogens cluster around Cl^- ions.

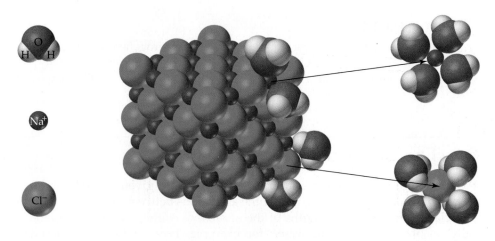

 Dissolution of NaCl in Water

specifically for water, *hydration*). The water molecules form a loose shell around
the ions, stabilizing them by electrical attraction.

Like all chemical and physical changes, the dissolution of a substance in a
solvent has associated with it a heat change, or *enthalpy* change (Section 7.1).
(∞, p. 166) Some substances dissolve exothermically, releasing heat and
warming the resultant solution, while other substances dissolve endothermic-
ally, absorbing heat and cooling the resultant solution. Calcium chloride, for
example, *releases* 19.4 kcal/mol of heat energy when it dissolves in water, but
ammonium nitrate (NH_4NO_3) *absorbs* 6.1 kcal/mol of heat energy. Athletes
and others take advantage of both situations when they use instant hot packs or
cold packs to treat injuries. Both hot and cold packs consist of a pouch of water
and a dry chemical, such as $CaCl_2$ or $MgSO_4$ for hot packs and NH_4NO_3 for
cold packs. Squeezing the pack breaks the pouch and the solid dissolves, either
raising or lowering the temperature.

■ **WORKED EXAMPLE 9.1**

Which of the following pairs of substances would you expect to form solutions?
(a) Benzene (C_6H_6) and toluene (C_7H_8)
(b) Oil and methyl alcohol (CH_3OH)

ANALYSIS Identify the kinds of intermolecular forces in each substance (Section 8.1). Substances with similar intermolecular forces tend to form solutions.

SOLUTION
(a) Benzene and toluene have similar formulas and similar polarity. They
therefore form a solution.
(b) Oil is nonpolar, but methyl alcohol is polar and forms hydrogen bonds.
The two substances are so dissimilar that they do not form a solution.

■ **PROBLEM 9.2**

Which of the following pairs of substances would you expect to form solutions?
(a) CCl_4 and water
(b) Gasoline and $MgSO_4$
(c) Hexane (C_6H_{14}) and heptane (C_7H_{16})
(d) Ethyl alcohol (C_2H_5OH) and heptanol ($C_7H_{15}OH$)

9.3 Solid Hydrates

Some ionic compounds attract water strongly enough to hold onto water molecules even when crystalline, forming what are called *solid hydrates*. For example,
the plaster of Paris used to make decorative objects and casts for broken limbs is

calcium sulfate hemihydrate, $CaSO_4 \cdot \frac{1}{2} H_2O$. The dot between $CaSO_4$ and $\frac{1}{2} H_2O$ in the formula indicates that for every two $CaSO_4$ formula units in the crystal there is also one water molecule present.

$$CaSO_4 \cdot \tfrac{1}{2} H_2O \quad \text{A solid hydrate}$$

After being ground up and mixed with water to make plaster, $CaSO_4 \cdot \frac{1}{2} H_2O$ gradually changes into the crystalline dihydrate $CaSO_4 \cdot 2 H_2O$, known as *gypsum*. During the change, the plaster hardens and expands in volume, causing it to fill a mold or shape itself closely around a broken limb. Table 9.3 lists some other ionic compounds that are handled primarily as hydrates.

▲ Plaster of Paris ($CaSO_4 \cdot \frac{1}{2} H_2O$) slowly turns into gypsum ($CaSO_4 \cdot 2 H_2O$) when added to water. In so doing, the plaster hardens and expands, causing it to fill a mold.

TABLE 9.3   Some Common Solid Hydrates

Formula	Name	Uses
$AlCl_3 \cdot 6 H_2O$	Aluminum trichloride hexahydrate	Antiperspirant
$CaSO_4 \cdot 2 H_2O$	Calcium sulfate dihydrate (gypsum)	Cements, wallboard molds
$CaSO_4 \cdot \frac{1}{2} H_2O$	Calcium sulfate hemihydrate (plaster of paris)	Casts, molds
$CuSO_4 \cdot 5 H_2O$	Copper(II) sulfate pentahydrate (blue vitriol)	Pesticide, germicide topical fungicide
$MgSO_4 \cdot 7 H_2O$	Magnesium sulfate heptahydrate (epsom salts)	Laxative, anticonvulsant
$Na_2B_4O_7 \cdot 10 H_2O$	Sodium tetraborate decahydrate (borax)	Cleaning compounds, fireproofing agent
$Na_2S_2O_3 \cdot 5 H_2O$	Sodium thiosulfate pentahydrate (hypo)	Photographic fixer

Still other ionic compounds attract water so strongly that they pull water vapor from humid air to become hydrated. Compounds that show this behavior, such as calcium chloride ($CaCl_2$), are called **hygroscopic** and are often used as drying agents. You might have noticed a small bag of a hygroscopic compound (probably silica gel, SiO_2) included in the packing material of a new stereo or VCR to keep humidity low during shipping.

Hygroscopic   Having the ability to pull water molecules from the surrounding atmosphere.

■ **PROBLEM 9.3**
Write the formula of sodium sulfate decahydrate, known as Glauber's salt and used as a laxative.

■ **PROBLEM 9.4**
What mass of Glauber's salt must be used to provide 1.00 mol of sodium sulfate?

9.4 Solubility

We saw in Section 9.2 that ethyl alcohol is soluble in water because hydrogen bonding is nearly as strong between water and ethyl alcohol molecules as it is between water molecules alone or ethyl alcohol molecules alone. So similar are the forces in this particular case, in fact, that the two liquids are **miscible**, or mutually soluble in all proportions. Ethyl alcohol will continue to dissolve in water no matter how much is added.

Not all substances continue to dissolve regardless of amount added, however. Imagine, for instance that you are asked to prepare a saline solution (aqueous NaCl). You might measure out some water, add solid NaCl, and stir the

Miscible   Mutually soluble in all proportions.

mixture. Dissolution will occur rapidly at first but will then slow down as more and more NaCl is added. Eventually the dissolution stops because an equilibrium is reached when the numbers of Na^+ and Cl^- ions leaving a crystal and going into solution are equal to the numbers of ions returning from solution to the crystal. At this point, the solution is said to be **saturated**. A maximum of 35.8 g of NaCl will dissolve in 100 mL of water at 20°C. Any amount above this limit will simply sink to the bottom of the container and sit there.

Saturated solution A solution that contains the maximum amount of dissolved solute at equilibrium.

The equilibrium reached by a saturated solution is like the equilibrium reached by a reversible reaction (Section 7.6). (∞, p. 177) Both are dynamic situations in which no *apparent* change occurs because the rates of forward and backward processes are equal. Solute particles leave the solid surface and reenter the solid from solution at the same rate.

$$Solid\ solute \underset{Crystallize}{\overset{Dissolve}{\rightleftharpoons}} Solution$$

Solubility The maximum amount of a substance that will dissolve in a given amount of solvent at a specified temperature.

The maximum amount of a substance that will dissolve in a given amount of a solvent at a given temperature, usually expressed in grams per 100 mL (g/100 mL), is called the substance's **solubility**. Solubility is a characteristic property of a specific solute–solvent combination, and different substances have greatly differing solubilities. Only 9.6 g of sodium hydrogen carbonate will dissolve in 100 mL of water at 20°C, for instance, but 204 g of sucrose will dissolve under the same conditions.

9.5 The Effect of Temperature on Solubility

As anyone who has ever made tea or coffee knows, temperature often has a dramatic effect on solubility. The compounds in tea leaves or coffee beans, for instance, dissolve easily in hot water but not in cold water. The effect of temperature is different for every substance, however, and is usually unpredictable. As shown in Figure 9.2, the solubilities of most molecular and ionic solids increase with increasing temperature, but the solubilities of others (NaCl) are almost unchanged, and the solubilities of still others [$Ce_2(SO_4)_3$] decrease with increasing temperature.

FIGURE 9.2   Solubilities of ▶ some solids in water as a function of temperature. Most substances become more soluble as temperature rises, although the exact relationship is usually complex.

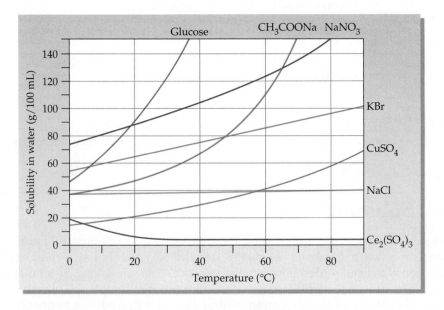

Supersaturated solution A solution that contains more than the maximum amount of dissolved solute; a nonequilibrium situation.

Solids that are more soluble at high temperature than at low temperature can sometimes form what are called **supersaturated** solutions, which contain even more solute than a saturated solution. Suppose, for instance, that a large amount of a substance is dissolved at a high temperature. As the solution cools, the solubility decreases and the excess solute should precipitate to maintain equilibrium. But if the cooling is done very slowly, and if the container stands quietly, crystallization might not occur immediately and a supersaturated solution might result. Such a solution is unstable, however, and precipitation can occur dramatically when a tiny seed crystal is added to initiate crystallization (Figure 9.3).

The influence of temperature on the solubility of gases, unlike solids, *is* predictable: Addition of heat decreases the solubility of most gases (helium is the only common exception). One result of this temperature-dependent decrease in gas solubility can sometimes be noted in a stream or lake near the outflow of warm water from an industrial operation. As water temperature increases, the concentration of dissolved oxygen in the water decreases, killing fish that can't tolerate the lower oxygen levels.

▲ FIGURE 9.3 **A supersaturated solution of sodium acetate in water.** When a tiny seed crystal is added, larger crystals rapidly grow and precipitate from the solution until equilibrium is reached.

■ **PROBLEM 9.5**

Look at the graph of solubility versus temperature in Figure 9.2, and estimate the solubility of KBr in water at 50°C in g/(100 mL).

9.6 The Effect of Pressure on Solubility: Henry's Law

Pressure has virtually no effect on the solubility of a solid or liquid, but it has a strong effect on the solubility of a gas. According to **Henry's law**, the solubility (or concentration) of a gas in a liquid is directly proportional to the partial pressure of the gas over the liquid. (Recall from Section 8.10 that each gas in a mixture exerts a partial pressure independent of other gases present ∞, p. 208) If the partial pressure of the gas doubles, solubility doubles; if the gas pressure is halved, solubility is halved (Figure 9.4).

Henry's law   The solubility (or concentration) of a gas is directly proportional to the partial pressure of the gas if the temperature is constant. That is, concentration (C) divided by pressure (P) is constant when T is constant.

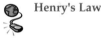 Henry's Law

$$\text{or} \quad \frac{C}{P_{gas}} = k \quad \text{(At a constant temperature)}$$

 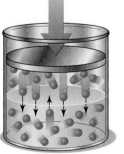 

    (a) Equilibrium   **(b)** Pressure increase   **(c)** Equilibrium restored

▲ **FIGURE 9.4 Henry's law.** The solubility of a gas is directly proportional to its partial pressure. An increase in pressure causes more gas molecules to enter solution until equilibrium is restored between the dissolved and undissolved gas.

▲ The CO_2 gas dissolved under pressure comes out of solution when the bottle is opened and the pressure drops.

Henry's law can be explained using Le Châtelier's principle (Section 7.8), which states that when a system at equilibrium is placed under stress, the equilibrium shifts to relieve that stress. (∞, p. 182) In the case of a saturated solution of a gas in a liquid, an equilibrium exists whereby gas molecules enter and leave the solution at the same rate. When the system is stressed by increasing the pressure of the gas, more gas molecules go into solution to relieve that increase. Conversely, when the pressure of the gas is decreased, more gas molecules come out of solution to relieve the decrease.

$$[\text{Pressure} \longrightarrow]$$
$$\text{Gas} + \text{Solvent} \rightleftharpoons \text{Solution}$$

As an example of Henry's law in action, think about the fizzing that occurs when you open a bottle of soft drink or champagne. The bottle is sealed under greater than 1 atm of CO_2 pressure, causing some of the CO_2 to dissolve. When the bottle is opened, however, CO_2 pressure drops and gas comes fizzing out of solution.

Writing Henry's law in the form $P_{gas} = C/k$ shows that partial pressure can be used to express the concentration of a gas in a solution, a practice especially common in health-related sciences. Table 9.4 gives some typical values and illustrates the convenience of having the same unit for concentration of a gas in both air and blood. Compare the oxygen partial pressures in saturated alveolar air (air in the lungs) and in arterial blood, for instance. The values are almost the same because the gases dissolved in blood come to equilibrium with the same gases in the lungs.

TABLE 9.4   Partial Pressures and Normal Gas Concentrations in Body Fluids

Sample	Partial Pressure (mm Hg)			
	P_{N_2}	P_{O_2}	P_{CO_2}	P_{H_2O}
Inspired air (dry)	597	159	0.3	3.7
Alveolar air (saturated)	573	100	40	47
Expired air (saturated)	569	116	28	47
Arterial blood	573	95	40	
Venous blood	573	40	45	
Peripheral tissues	573	40	45	

If the partial pressure of a gas over a solution changes while the temperature is constant, the new solubility of the gas can be found easily. Because C/P is a constant value at constant temperature, Henry's law can be restated to show how one of the variables changes if the other changes:

$$\frac{C_1}{P_1} = \frac{C_2}{P_2} = k \quad (\text{At a fixed temperature})$$

Worked Example 9.2 gives an illustration of how to use this equation.

■ WORKED EXAMPLE 9.2

At a partial pressure of oxygen in the atmosphere of 159 mm Hg, the solubility of oxygen in blood is 0.44 g/100 mL. What is the solubility of oxygen in blood at 11,000 ft, where the partial pressure of O_2 is 56 mm Hg?

ANALYSIS   According to Henry's law, the solubility of the gas divided by its pressure is constant:

$$\frac{C_1}{P_1} = \frac{C_2}{P_2}$$

Of the four variables in this equation, we know that $P_1 = 159$ mm Hg, $C_1 = 0.44$ g/100 mL, and $P_2 = 56$ mm Hg, and we need to find C_2.

SOLUTION Solving the above equation for C_2 and substituting in the known values gives

$$C_2 = \frac{C_1 P_1}{P_1} = \frac{(0.44 \text{ g}/100 \text{ mL})(56 \text{ mm Hg})}{159 \text{ mm Hg}} = 0.15 \text{ g}/100 \text{ mL}$$

BALLPARK CHECK The pressure drops by a factor of about 3 (from 159 mm Hg to 56 mm Hg), so solubility must also drop by a factor of 3 (from 0.44 g/100 mL to about 0.15 g/100 mL).

■ **PROBLEM 9.6**

At 20°C and a partial pressure of 760 mm Hg, the solubility of CO_2 in water is 0.169 g/100 mL. What is the solubility of CO_2 at 2.5×10^4 mm Hg?

■ **PROBLEM 9.7**

At a total atmospheric pressure of 1.00 atm, the partial pressure of CO_2 in air is approximately 4.0×10^{-4} atm. Using the data in Problem 9.6, what is the solubility of CO_2 in an open bottle of seltzer water at 20°C?

9.7 Units of Concentration

Although we might speak casually of a solution of, say, orange juice as either "dilute" or "concentrated," laboratory work usually requires an exact knowledge of a solution's concentration. As indicated in Table 9.5, there are several common methods for expressing concentration. The units differ, but all the methods describe how much solute is present in a given quantity of solution.

TABLE 9.5 Some Units for Expressing Concentration

Concentration Measure	Solute Measure	Solution Measure
Molarity, M	Moles	Volume (L)
Weight/volume percent, (w/v)%	Weight (g)	Volume (mL)
Volume/volume percent, (v/v)%	Volume*	Volume*
Parts per million, ppm	Parts*	10^6parts*

*Any units can be used as long as they are the same for both solute and solution.

Let's look at each of the four concentration measures listed in Table 9.5 individually, beginning with *molarity*.

Mole/Volume Concentration: Molarity

We saw in Chapter 6 that the various relationships between amounts of reactants and products in chemical reactions are calculated in *moles* (Sections 6.4–6.6). Thus, the most generally useful means of expressing concentration in the laboratory is **molarity (M),** the number of moles of solute dissolved per liter of solution. For example, a solution made by dissolving 1.00 mol (58.5 g) of NaCl in enough water to give 1.00 L of solution has a concentration of 1.00 mol/L, or 1.00 M. The

Application Breathing and Oxygen Transport

Humans, like all animals, need oxygen. When we breathe, the freshly inspired air travels through the bronchial passages and enters the approximately 150 million alveolar sacs of the lungs, where it picks up moisture and mixes with air remaining from the previous breath. Oxygen then diffuses through the delicate walls of the lung alveoli and into arterial blood, which transports it to all body tissues.

Only about 3% of the oxygen in blood is dissolved; the rest is chemically bound to *hemoglobin* molecules, large proteins that contain *heme* groups embedded in them. Each hemoglobin molecule contains four heme groups, and each heme group contains an iron atom that is able to bind one O_2 molecule. Thus, a single hemoglobin molecule can bind four molecules of oxygen. The entire system of oxygen transport and delivery in the body depends on the pickup and release of O_2 by hemoglobin (Hb) according to the following series of equilibria:

$$Hb + O_2 \rightleftharpoons Hb(O_2)$$

$$Hb(O_2) + O_2 \rightleftharpoons Hb(O_2)_2$$

$$Hb(O_2)_2 + O_2 \rightleftharpoons Hb(O_2)_3$$

$$Hb(O_2)_3 + O_2 \rightleftharpoons Hb(O_2)_4$$

The positions of the different equilibria depend on the concentration of O_2 in the various tissues, as measured by partial pressure (P_{O_2}, Table 9.4). In hard-working, oxygen-starved muscles, where P_{O_2} is low, oxygen is released from hemoglobin as the equilibria shift toward the left, according to Le Châtelier's principle. In the lung, where P_{O_2} is high, oxygen is absorbed by hemoglobin as the equilibria shift toward the right.

The amount of oxygen carried by hemoglobin at any given value of P_{O_2} is usually expressed as a percent saturation and can be found from the curve shown in the accompanying Figure. The saturation is 97.5% in the lungs, where $P_{O_2} = 100$ mm Hg, meaning that each hemoglobin is carrying close to its maximum possible amount of 4 O_2 molecules. When $P_{O_2} = 26$ mm Hg, however, the saturation drops to 50%.

What about people who live at high altitudes? In Leadville, Colorado, for example, where the altitude is 10,156 ft, the partial pressure of O_2 in the lungs is only about 68 mm Hg. Hemoglobin is only 90% saturated with O_2 at this pressure, meaning that less oxygen is available for delivery to the tissues. The body copes

▲ At high altitudes, the partial pressure of oxygen in the air is too low to saturate hemoglobin sufficiently. Additional oxygen is therefore needed.

with the situation by producing more hemoglobin molecules, which provide more capacity for O_2 transport and drive the Hb + O_2 equilibria to the right.

See Additional Problem 9.88 at the end of the chapter.

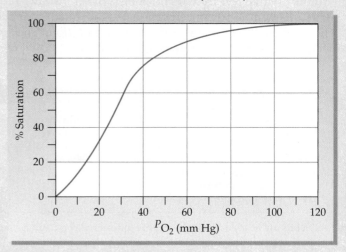

▲ An oxygen-carrying curve for hemoglobin. The percent saturation of the oxygen binding sites on hemoglobin depends on the partial pressure of oxygen (P_{O_2}).

molarity of any solution is found by dividing the number of moles of solute by the number of liters of solution.

$$\text{Molarity (M)} = \frac{\text{Moles of solute}}{\text{Liters of solution}}$$

Note that a solution of a given molarity is prepared by dissolving the solute in enough solvent to give a *final* solution volume of 1.00 L, not by dissolving it in an *initial* volume of 1.00 L. If an initial volume of 1.00 L were used, the final solution volume might be a bit larger than 1.00 L because of the additional volume of the solute. In practice, the appropriate amount of solute is weighed and placed in a *volumetric flask*, as shown in Figure 9.5. Enough solvent is then added to dissolve the solute, and further solvent is added until an accurately calibrated final volume is reached. The solution is then shaken until it is uniformly mixed.

(a)　　　(b)　　　(c)

◀ **FIGURE 9.5 Preparing a solution of known molarity.** (a) A measured number of moles of solute is placed in a volumetric flask. (b) Enough solvent is added to dissolve the solute by swirling. (c) Further solvent is carefully added until the calibration mark on the neck of the flask is reached, and the solution is shaken until uniform.

Solution Formation from a Solid

Molarity can be used as a conversion factor to relate the volume of a solution to the number of moles of solute it contains. If we know the molarity and volume of a solution, we can calculate the number of moles of solute. If we know the number of moles of solute and the molarity of the solution, we can find the solution's volume.

$$\text{Molarity} = \frac{\text{Moles of solute}}{\text{Volume of solution (L)}}$$

$$\frac{\text{Moles of}}{\text{solute}} = \text{Molarity} \times \frac{\text{Volume of}}{\text{solution}} \qquad \frac{\text{Volume}}{\text{of solution}} = \frac{\text{Moles of solute}}{\text{Molarity}}$$

Figure 9.6 gives a flow diagram showing how molarity is used in calculating the quantities of reactants or products in a chemical reaction, and Worked Examples 9.4 and 9.5 show how the calculations are done. Note that Problem 9.10 employs *millimolar* (mM) concentrations, which are useful in healthcare fields for expressing low concentrations such as are often found in body fluids (1 mM = 0.001 M).

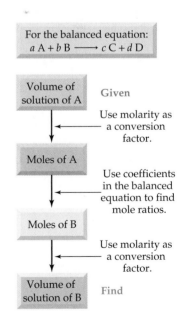

▲ **FIGURE 9.6** A flow diagram summarizing the use of molarity for conversions between solution volume and moles to find quantities of reactants and products for chemical reactions in solution.

■ **WORKED EXAMPLE 9.3**

What is the molarity of a solution made by dissolving 2.355 g of sulfuric acid (H_2SO_4) in water and diluting to a final volume of 50.0 mL? The molar mass of H_2SO_4 is 98.1 g/mol.

ANALYSIS Molarity is defined as moles of solute per liter of solution: $M = mol/L$. Thus, we must first find the number of moles of sulfuric acid by doing a mass-to-mole conversion, and then divide the number of moles by the volume of the solution.

SOLUTION

$$(2.355 \text{ g } H_2SO_4)\left(\frac{1 \text{ mol } H_2SO_4}{98.1 \text{ g } H_2SO_4}\right) = 0.0240 \text{ mol } H_2SO_4$$

$$\left(\frac{0.0240 \text{ mol } H_2SO_4}{50.0 \text{ mL}}\right)\left(\frac{1000 \text{ mL}}{1 \text{ L}}\right) = 0.480 \text{ M}$$

BALLPARK CHECK The molar mass of sulfuric acid is about 100 g/mol, so 2.355 g is roughly 0.025 mol. The volume of the solution is 50.0 mL, or 0.05 L, so we have about 0.025 mol of acid in 0.05 L of solution, or about 0.5 M concentration.

■ WORKED EXAMPLE 9.4

A blood concentration of 0.065 M ethyl alcohol (EtOH) is sufficient to induce a coma. At this concentration, what is the total mass of alcohol (in grams) in an adult male whose total blood volume is 5.6 L? The molar mass of ethyl alcohol is 46.0 g/mol.

ANALYSIS We are given a molarity (0.065 M) and a volume (5.6 L), which will allow us to calculate the number of moles of alcohol in the blood. A mole-to-mass conversion will then give the mass of alcohol.

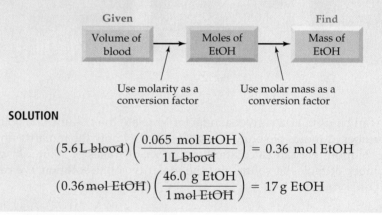

SOLUTION

$$(5.6 \text{ L blood})\left(\frac{0.065 \text{ mol EtOH}}{1 \text{ L blood}}\right) = 0.36 \text{ mol EtOH}$$

$$(0.36 \text{ mol EtOH})\left(\frac{46.0 \text{ g EtOH}}{1 \text{ mol EtOH}}\right) = 17 \text{ g EtOH}$$

■ WORKED EXAMPLE 9.5

Within our stomachs, gastric juice that is about 0.1 M in HCl aids in digestion. How many milliliters of gastric juice will react completely with a 500 mg tablet of magnesium hydroxide (frequently used as an antacid)? The molar mass of $Mg(OH)_2$ is 58.3 g/mol, and the balanced equation is

$$2 \text{ HCl}(aq) + Mg(OH)_2(aq) \longrightarrow MgCl_2(aq) + 2 \text{ H}_2O(l)$$

ANALYSIS We are given a molarity and need to find a volume. To do this, we have to find how many moles of HCl are present. We can do this by first calculating how many moles of $Mg(OH)_2$ are in 500 mg and then checking the coefficients of the balanced equation, which say that 1 mol of $Mg(OH)_2$ reacts with 2 mol of HCl.

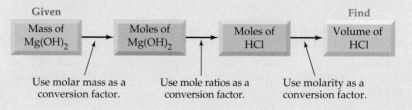

SOLUTION

$$(500 \ \cancel{\text{mg Mg(OH)}_2}) \left(\frac{1 \, \text{g}}{1000 \, \cancel{\text{mg}}} \right) \left(\frac{1 \, \text{mol Mg(OH)}_2}{58.3 \, \cancel{\text{g Mg(OH)}_2}} \right) = 0.008 \, 58 \ \text{mol Mg(OH)}_2$$

$$(0.008 \, 58 \ \cancel{\text{mol Mg(OH)}_2}) \left(\frac{2 \, \cancel{\text{mol HCl}}}{1 \, \cancel{\text{mol Mg(OH)}_2}} \right) \left(\frac{1 \, \text{L HCl}}{0.1 \, \cancel{\text{mol HCl}}} \right) = 0.2 \, \text{L} (200 \, \text{mL})$$

■ **PROBLEM 9.8**

What is the molarity of a solution that contains 50.0 g of vitamin B_1 hydrochloride (molar mass = 337 g) in 160 mL of solution?

■ **PROBLEM 9.9**

How many moles of solute are present in the following solutions?
(a) 125 mL of 0.25 M $NaNO_3$ **(b)** 450 mL of 1.5 M HNO_3

■ **PROBLEM 9.10**

The concentration of cholesterol ($C_{27}H_{46}O$) in blood is approximately 5.0 mM. How many grams of cholesterol are in 250 mL of blood?

■ **PROBLEM 9.11**

What mass (in grams) of calcium carbonate is needed to react completely with 75 mL of 0.10 M HCl according to the following equation?

$$2 \, \text{HCl}(aq) + \text{CaCO}_3(aq) \longrightarrow \text{CaCl}_2(aq) + \text{H}_2\text{O}(l) + \text{CO}_2(g)$$

Weight/Volume Percent Concentration [(w/v)%]

One of the most common methods for expressing percent concentration is to give the number of grams (weight) as a percentage of the number of milliliters (volume) of the final solution—called the **weight/volume percent concentration [(w/v)%]**. Mathematically, (w/v)% concentration is found by taking the number of grams of solute per milliliter of solution and multiplying by 100%.

$$(\text{w/v})\% \ \text{concentration} = \frac{\text{Mass of solute (g)}}{\text{Volume of solution (mL)}} \times 100\%$$

For example, if 15 g of glucose is dissolved in enough water to give 100 mL of solution, the glucose concentration is 15 g/100 mL, or 15% (w/v).

$$\frac{15 \, \text{g glucose}}{100 \, \text{mL solution}} \times 100\% = 15\% \, (\text{w/v})$$

To prepare 100 mL of a specific weight/volume solution, the weighed solute is dissolved in just enough solvent to give a final volume of 100 mL, not in an initial volume of 100 mL solvent. (If the solute were dissolved in 100 mL of *solvent*, the final volume of the *solution* could well be a bit larger than 100 mL, since the volume of the solute would be included.) In practice, solutions are prepared using a volumetric flask, as shown previously in Figure 9.5. Worked Example 9.6 illustrates how weight/volume percent concentration is found from a known mass and volume of solution.

■ WORKED EXAMPLE 9.6

A solution of heparin sodium, an anticoagulant for blood, contains 1.8 g of heparin sodium dissolved to make a final volume of 15 mL of solution. What is the weight/volume percent concentration of this solution?

ANALYSIS Weight/volume percent concentration is defined as the mass of the solution in grams divided by the volume in milliliters and multiplied by 100%.

SOLUTION

$$(w/v)\% \text{ concentration} = \frac{1.8 \text{ g heparin sodium}}{15 \text{ mL}} \times 100\% = 12\% \ (w/v)$$

■ WORKED EXAMPLE 9.7

How many grams of NaCl are needed to prepare 250 mL of a 1.5% (w/v) saline solution?

ANALYSIS We are given a concentration and a volume, and we need to find the mass of solute by rearranging the equation for (w/v)% concentration.

SOLUTION

$$\text{Since} \quad (w/v)\% = \frac{\text{Mass of solute in g}}{\text{Volume of solution in mL}} \times 100\%$$

$$\text{then} \quad \text{Mass of solute in g} = \frac{(\text{Volume of solution in mL})[(w/v)\%]}{100\%}$$

$$= \frac{(250)(1.5\%)}{100\%} = 3.8 \ g \ \text{NaCl}$$

■ WORKED EXAMPLE 9.8

How many milliliters of a 0.75% (w/v) solution of the food preservative sodium benzoate are needed to obtain 45 mg?

ANALYSIS We are given a concentration and a mass, and we need to find the volume of solution by rearranging the equation for (w/v)% concentration. Remember that 45 mg = 0.045 g.

SOLUTION

$$\text{Since} \quad (w/v)\% = \frac{\text{Mass of solute in g}}{\text{Volume of solution in mL}} \times 100\%$$

$$\text{then} \quad \text{Volume of solution in mL} = \frac{(\text{Mass of solute in g})(100\%)}{(w/v)\%}$$

$$= \frac{(0.045 \text{ g})(100\%)}{0.75\%} = 6.0 \ \text{mL}$$

■ PROBLEM 9.12

In clinical lab reports, some concentrations are given in mg/dL. Convert an 8.6 mg/dL concentration of Ca^{2+} to weight/volume percent.

■ PROBLEM 9.13

What is the weight/volume percent concentration of a solution that contains 23 g of potassium iodide in 350 mL of aqueous solution?

■ **PROBLEM 9.14**

How many grams of solute are needed to prepare the following solutions?
(a) 100.0 mL of 12% (w/v) glucose ($C_6H_{12}O_6$)
(b) 75 mL of 2.0% (w/v) KCl

Volume/Volume Percent Concentration [(v/v)%]

The concentration of a solution made by dissolving one liquid in another is often given by expressing the volume of solute as a percentage of the volume of final solution—the **volume/volume percent concentration [(v/v)%]**. Mathematically, the volume of the solute (usually in milliliters) per milliliter of solution is multiplied by 100%.

$$(v/v)\% \text{ concentration} = \frac{\text{Volume of solute (mL)}}{\text{Volume of solution (mL)}} \times 100\%$$

For example, if 10.0 mL of ethyl alcohol is dissolved in enough water to give 100.0 mL of solution, the ethyl alcohol concentration is (10.0 mL/100.0 mL) × 100% = 10.0%(v/v).

■ **WORKED EXAMPLE 9.9**

How many milliliters of methyl alcohol are needed to prepare 75 mL of a 5.0% (v/v) solution?

ANALYSIS We are given a solution volume (75 mL) and a concentration [5.0% (v/v), meaning 5.0 mL solute/100 mL solution]. The concentration acts as a conversion factor for finding the amount of methyl alcohol needed.

SOLUTION

$$(75\,\text{mL solution})\left(\frac{5.0 \text{ mL methyl alcohol}}{100\,\text{mL solution}}\right) = 3.8 \text{ mL methyl alcohol}$$

■ **PROBLEM 9.15**

How would you use a 500.0 mL volumetric flask to prepare a 7.5% (v/v) solution of acetic acid in water?

■ **PROBLEM 9.16**

What volume of solute (in milliliters) is needed to prepare the following solutions?
(a) 100 mL of 22% (v/v) ethyl alcohol
(b) 150 mL of 12% (v/v) acetic acid

Parts per Million (ppm)

When concentrations are very small, as often occurs in dealing with trace amounts of pollutants or contaminants, **parts per million (ppm)** or **parts per billion (ppb)** are used. The "parts" can be in any unit of either mass or volume as long as the units of both solute and solvent are the same.

$$\text{ppm} = \frac{\text{Mass of solute (g)}}{\text{Mass of solution (g)}} \times 10^6 \quad \text{or} \quad \frac{\text{Volume of solute (mL)}}{\text{Volume of solution (mL)}} \times 10^6$$

$$\text{ppb} = \frac{\text{Mass of solute (g)}}{\text{Mass of solution (g)}} \times 10^9 \quad \text{or} \quad \frac{\text{Volume of solute (mL)}}{\text{Volume of solution (mL)}} \times 10^9$$

To take an example, the maximum allowable concentration in air of the organic solvent benzene (C_6H_6) is currently set by government regulation at 1 ppm. A concentration of 1 ppm means that if you were to take a million "parts" of air in any unit—say, mL—then 1 of those parts would be benzene vapor and the other 999,999 parts would be other gases:

$$1 \text{ ppm} = \frac{1 \text{ mL}}{1,000,000 \text{ mL}} \times 10^6$$

■ WORKED EXAMPLE 9.10

The maximum allowable concentration of chloroform, $CHCl_3$, in drinking water is 100 ppb. What is the maximum amount (in grams) of chloroform allowed in a glass of water that contains 400 g (400 mL)?

ANALYSIS We are given a solution amount (400 g) and a concentration (100 ppb). This concentration of 100 ppb means:

$$100 \text{ ppb} = \frac{\text{Mass of solute (g)}}{\text{Mass of solution (g)}} \times 10^9$$

This equation can be rearranged to find the mass of solute.

SOLUTION

$$\text{Mass of solute (g)} = \frac{\text{Mass of solution (g)}}{10^9} \times 100 \text{ ppb}$$

$$= \frac{400 \text{ g}}{10^9} \times 100 \text{ ppb} = 4 \times 10^{-5} \text{ g} \quad (\text{or } 0.04 \text{ mg})$$

■ PROBLEM 9.17

What is the concentration in ppm of sodium fluoride in tap water that has been fluoridated by the addition of 32 mg of NaF for every 20 kg of solution?

■ PROBLEM 9.18

The maximum allowable amounts of lead and copper allowed in drinking water are 0.015 mg/kg for lead and 1.3 mg/kg for copper. Express these values in ppm, and tell the maximum amount of each (in grams) allowed in 100 g of water.

9.8 Dilution

Many solutions, from orange juice to chemical reagents, are stored in high concentrations and then prepared for use by *dilution*—that is, by adding additional solvent to lower the concentration. For example, you might make up 1/2 gal of orange juice by adding water to a canned concentrate. In the same way, you might buy a medicine or chemical reagent in concentrated solution and dilute it before use.

The key fact to remember about dilution is that the amount of *solute* remains constant; only the *volume* is changed by adding more solvent. If, for example, the initial and final concentrations are given in molarity, then we know that the number of moles of solute is the same both before and after dilution, and can be determined by multiplying molarity times volume:

▲ Orange juice concentrate is diluted with water before drinking.

Number of moles = Molarity (mol/L) × Volume (L)

Because the number of moles remains constant, we can set up the following equation, where M_1 and V_1 refer to the solution before dilution, and M_2 and V_2 refer to the solution after dilution:

$$\text{Moles of solute} = M_1V_1 = M_2V_2$$

This equation can be rewritten to solve for M_2, the concentration of the solution after dilution:

$$M_2 = M_1 \times \frac{V_1}{V_2} \qquad \text{where} \qquad \frac{V_1}{V_2} \qquad \text{is a } \textit{dilution factor}$$

The equation shows that the concentration after dilution (M_2) can be found by multiplying the initial concentration (M_1) by a **dilution factor**, which is simply the ratio of the initial and final solution volumes (V_1/V_2). If, for example, the solution volume *increases* by a factor of 5, from 10 mL to 50 mL, then the concentration must *decrease* to 1/5 its initial value because the dilution factor is 10 mL/50 mL, or 1/5. Worked Example 9.11 shows how to use this relationship for calculating dilutions.

The relationship between concentration and volume can also be used to find what volume of initial solution to start with to achieve a given dilution:

$$\text{Since} \qquad M_1V_1 = M_2V_2$$

$$\text{then} \qquad V_1 = V_2 \times \frac{M_2}{M_1}$$

In this case, V_1 is the initial volume that must be diluted to prepare a less concentrated solution with volume V_2. The initial volume is found by multiplying the final volume (V_2) by the ratio of the final and initial concentrations (M_2/M_1). For example, to decrease the concentration of a solution to 1/5 its initial value, the initial volume must be 1/5 the desired final volume. Worked Example 9.12 gives a sample calculation.

> **Dilution factor** The ratio of the initial and final solution volumes (V_1/V_2).

 Solution Formation by Dilution

■ **WORKED EXAMPLE 9.11**

What is the final concentration if 75 mL of a 3.5 M glucose solution is diluted to a volume of 450 mL?

ANALYSIS The number of moles of solute is constant, so

$$M_1V_1 = M_2V_2$$

Of the four variables in this equation, we know the initial concentration M_1 (3.5 M), the initial volume V_1 (75 mL), and the final volume V_2 (450 mL), and we need to find the final concentration M_2.

SOLUTION Solving the above equation for M_2 and substituting in the known values gives

$$M_2 = \frac{M_1V_1}{V_2} = \frac{(3.5 \text{ M glucose})(75\,\text{mL})}{450\,\text{mL}} = 0.58 \text{ M glucose}$$

BALLPARK CHECK The volume increases by a factor of 6, from 75 mL to 450 mL, so the concentration must decrease by a factor of 6, from 3.5 M to about 0.6 M.

■ **WORKED EXAMPLE 9.12**

Aqueous NaOH can be purchased at a concentration of 1.0 M. How would you use this concentrated solution to prepare 750 mL of 0.32 M NaOH?

ANALYSIS   The number of moles of solute is constant, so

$$M_1V_1 = M_2V_2$$

Of the four variables in this equation, we know the initial concentration M_1 (1.0 M), the final volume V_2 (750 mL), and the final concentration M_2 (0.32 M), and we need to find the initial volume V_1.

SOLUTION   Solving the above equation for V_2 and substituting in the known values gives

$$V_1 = \frac{V_2M_2}{M_1} = \frac{(750 \text{ mL})(0.32\,M)}{1.0\,M} = 240 \text{ mL}$$

To prepare the desired solution, take 240 mL of 1.0 M NaOH and dilute it with water to a final volume of 750 mL.

BALLPARK CHECK   The concentration decreases by a factor of about 3, from 1.0 M to 0.30 M, so the final volume must be about 3 times greater than the initial volume. Thus, a final volume of 750 mL corresponds to an initial volume of about 250 mL.

■ **PROBLEM 9.19**

Hydrochloric acid is normally purchased at a concentration of 12.0 M. What is the final concentration if 100.0 mL of 12.0 M HCl is diluted to 500.0 mL?

■ **PROBLEM 9.20**

Aqueous ammonia is commercially available at a concentration of 16.0 M. How much of the concentrated solution would you use to prepare 500.0 mL of a 1.25 M solution?

9.9 Ions in Solution: Electrolytes

Look at Figure 9.7, which shows a light bulb connected to a power source through a circuit that is interrupted by two metal strips dipped into a beaker of liquid. When the strips are dipped into pure water, the bulb remains dark, but when they are dipped into an aqueous NaCl solution, the circuit is closed and the bulb lights. This simple demonstration shows, as mentioned previously in Section 4.1, that ionic compounds in aqueous solution can conduct electricity. (∞, p. 71)

FIGURE 9.7   A simple demon- ▶
stration to show that electricity can
flow through a solution of ions.
(a) With pure water in the beaker, the circuit is incomplete, no electricity flows, and the bulb does not light. (b) With a concentrated NaCl solution in the beaker, the circuit is complete, electricity flows, and the light bulb glows.

(a)   (b)

Substances like NaCl that conduct an electric current when dissolved in water are called **electrolytes**. Conduction occurs because negatively charged Cl^- anions migrate through the solution toward the metal strip connected to the positive terminal of the power source, while positively charged Na^+ cations migrate toward the strip connected to the negative terminal. As you might expect, the ability of a solution to conduct electricity depends on the concentration of ions in solution. Distilled water contains virtually no ions and is nonconducting; ordinary tap water contains low concentrations of dissolved ions (mostly Na^+, K^+, Mg^{2+}, Ca^{2+}, and Cl^-) and is weakly conducting; and a concentrated solution of NaCl is strongly conducting.

Ionic substances like NaCl that ionize completely when dissolved in water are called **strong electrolytes**, while molecular substances like acetic acid (CH_3CO_2H) that are only partially ionized are **weak electrolytes**. Molecular substances like glucose that do not produce ions when dissolved in water are **nonelectrolytes**.

Electrolyte   A substance that produces ions and therefore conducts electricity when dissolved in water.

   Electrolytes and Nonelectrolytes

Strong electrolyte   A substance that ionizes completely when dissolved in water.

Weak electrolyte   A substance that is only partly ionized in water.

Nonelectrolyte   A substance that does not produce ions when dissolved in water.

   Strong and Weak Electrolytes

Strong electrolyte; completely ionized	$NaCl(s) \xrightarrow[\text{in water}]{\text{Dissolve}} Na^+(aq) + Cl^-(aq)$
Weak electrolyte; partly ionized	$CH_3CO_2H(l) \underset{\text{in water}}{\overset{\text{Dissolve}}{\rightleftharpoons}} CH_3CO_2^-(aq) + H^+(aq)$
Nonelectrolyte; not ionized	$Glucose(s) \xrightarrow[\text{in water}]{\text{Dissolve}} Glucose(aq)$

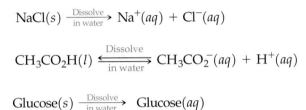

9.10 Electrolytes in Body Fluids: Equivalents and Milliequivalents

What would happen if NaCl and KBr were dissolved in the same solution? Because the cations (K^+ and Na^+) and anions (Cl^- and Br^-) are all mixed together and no reactions occur between them, an identical solution could just as well be made from KCl and NaBr. Thus, we can no longer speak of having an NaCl + KBr solution; we can only speak of having a solution with four different ions in it.

A similar situation exists for blood and other body fluids, which contain many different anions and cations. Since they are all mixed together, we can't "assign" specific cations to specific anions, and we can't talk about specific ionic compounds. Instead, we're interested only in individual ions and in the total numbers of positive and negative charges. To discuss such mixtures, it's useful to use a new term—*equivalents* of ions.

One **equivalent (Eq)** of an ion is an amount equal to the molar mass of the ion divided by the number of charges it has:

$$\text{One equivalent of Ion} = \frac{\text{Molar mass of ion (g)}}{\text{Number of charges on ion}}$$

If the ion has a charge of +1 or −1, one equivalent is simply the formula weight of the ion in grams. Thus, one equivalent of Na^+ is 23 g, and one equivalent of Cl^- is 35.5 g. If the ion has a charge of +2 or −2, however, one equivalent is equal to the ion's formula weight in grams divided by 2. Thus, one equivalent of Mg^{2+} is (24.3 g)/2 = 12.2 g, and one equivalent of CO_3^{2-} is [12.0 g + (3 × 16.0 g)]/2 = 30.0 g.

Because ion concentrations in body fluids are often low, clinical chemists find it more convenient to talk about *milliequivalents* of ions rather than equivalents. One milliequivalent (mEq) of an ion is 1/1000 of an equivalent. For example, the normal concentration of Na^+ in blood is 0.14 Eq/L, or 140 mEq/L.

$$1 \text{ mEq} = 0.001 \text{ Eq} \qquad 1 \text{ Eq} = 1000 \text{ mEq}$$

TABLE 9.6 Concentrations of Major Electrolytes in Blood Plasma

Cation	Concentration (mEq/L)
Na^+	136–145
Ca^{2+}	4.5–6.0
K^+	3.6–5.0
Mg^{2+}	3

Anion	Concentration (mEq/L)
Cl^-	98–106
HCO_3^-	25–29
SO_4^{2-} and HPO_4^{2-}	2

Average concentrations of the major electrolytes in blood plasma are given in Table 9.6. As you might expect, the total milliequivalents of positively and negatively charged electrolytes must be equal to maintain electrical neutrality. Adding the milliequivalents of positive and negative ions in Table 9.6, however, shows a higher concentration of positive ions than negative ions. The difference, called the *anion gap*, is made up by the presence of negatively charged proteins and the anions of organic acids.

■ **WORKED EXAMPLE 9.13**

The normal concentration of Ca^{2+} in blood is 5.0 mEq/L. How many milligrams of Ca^{2+} are in 1.00 L of blood?

ANALYSIS We are given a volume and a concentration in mEq/L, and we need to find an amount in grams. Thus, we need to calculate how many grams are in 1 Eq of Ca^{2+} and then use concentration as a conversion factor between volume and mass.

SOLUTION The molar mass of Ca^{2+} is 40.1 g, but because Ca^{2+} has two charges, we must divide the molar mass by 2:

$$1 \text{ Eq } Ca^{2+} = \frac{40.1 \text{ g } Ca^{2+}}{2} = 20.1 \text{ g } Ca^{2+}$$

Next, convert Eq to mEq and calculate how many milligrams are in 5.0 mEq/L Ca^{2+}:

$$\left(\frac{20.1 \text{ g } Ca^{2+}}{1 \text{ Eq}}\right)\left(\frac{1000 \text{ mg}}{1 \text{ g}}\right)\left(\frac{1 \text{ Eq}}{1000 \text{ mEq}}\right)\left(\frac{5.0 \text{ mEq}}{1.0 \text{ L}}\right) = 100 \text{ mg/L } Ca^{2+}$$

There are 100 mg of Ca^{2+} in 1.0 L of blood.

■ **PROBLEM 9.21**

How many grams are in one equivalent of each of the following ions?
(a) K^+ (b) Br^- (c) Mg^{2+} (d) SO_4^{2-}

■ **PROBLEM 9.22**

How many grams are in 1 mEq of each ion in Problem 9.21?

■ **PROBLEM 9.23**

Look at the data in Table 9.6, and calculate how many milligrams of Mg^{2+} are in 250 mL of blood.

9.11 Properties of Solutions

The properties of solutions are similar in many respects to those of pure solvents, but there are also some interesting and important differences. One such difference is that a solution has a higher boiling point than the pure solvent; another is that a solution has a lower freezing point. Pure water boils at 100.0°C and freezes at 0.0°C, for example, but a 1.0 M solution of NaCl in water boils at 101.0°C and freezes at −3.7°C.

The elevation of boiling point and the lowering of freezing point for a solution as compared with a pure solvent are examples of **colligative properties**—properties that depend on the *concentration* of a dissolved solute but not on its

Colligative property A property of a solution that depends only on the number of dissolved particles, not on their chemical identity.

Application
Electrolytes, Fluid Replacement, and Sports Drinks

Athletes sweat. And the hotter the day, the more intense and longer lasting the activity, the more they sweat. Sweat loss during strenuous exercise on a hot day can amount to as much as 2 liters per hour, and the total sweat loss during a 24 hour endurance run can exceed 16 liters, or approximately 35 pounds.

The composition of sweat is highly variable, not only with the individual but also with the time during the exercise and the athlete's overall conditioning. Typically, however, sweat contains about 30–40 mEq/L Na^+ ion and about 5–10 mEq/L K^+ ion. In addition, there are small amounts of other metal ions such as Mg^{2+}, and there are sufficient Cl^- ions to balance the charge (35–50 mEq/L).

Obviously, all the water and dissolved electrolytes lost by an athlete through sweating must be replaced. Otherwise, dehydration, hyperthermia and heat stroke, dizziness, nausea, muscle cramps, impaired kidney function, and other difficulties ensue. As a rule of thumb, a sweat loss equal to 5% of body weight—about 3.5 liters for a 150 pound person—is the maximum amount that can be safely allowed for a well-conditioned athlete.

▲ Severe dehydration due to excessive sweat loss can lead to serious medical problems for endurance athletes.

Plain water works perfectly well to replace sweat lost during short bouts of activity up to a few hours in length, but a carbohydrate–electrolyte beverage, or "sports drink," is much superior for rehydrating during and after longer activity in which substantial amounts of electrolytes have been lost. Some of the better-known sports drinks are little more than overpriced sugar–water solutions, but others are carefully formulated and highly effective for fluid replacement. Nutritional research has shown that a serious sports drink should meet the following criteria. There are several dry-powder mixes on the market to choose from.

- The drink should contain 6–8% of soluble complex carbohydrates (about 15 grams per 8 ounce serving) and only a small amount of simple sugar for taste. The complex carbohydrates, which usually go by the name "maltodextrin," provide a slow release of glucose into the bloodstream. Not only does the glucose provide a steady source of energy, it also enhances the absorption of water from the stomach.

- The drink should contain electrolytes to replenish those lost in sweat. Approximately 20 mEq/L Na^+, 10 mEq/L K^+, and 4 mEq/L Mg^{2+} are recommended. These amounts correspond to about 100 mg sodium, 100 mg potassium, and 25 mg magnesium per 8 ounce serving.

- The drink should be noncarbonated, since carbonation can cause gastrointestinal upset during exercise, and it should not contain caffeine, which acts as a diuretic.

- The drink should taste good so the athlete will want to drink it. Thirst is a poor indicator of fluid requirements, and most people will drink less than needed unless a beverage is flavored.

In addition to complex carbohydrates, electrolytes, and flavorings, some sports drinks also contain vitamin A (as beta-carotene), vitamin C (ascorbic acid), and selenium, which act as antioxidants to protect cells from damage. Some drinks also contain the amino acid glutamine, which appears to lessen lactic acid buildup in muscles and to help muscles bounce back more quickly after an intense workout.

See Additional Problems 9.89 and 9.90 at the end of the chapter.

chemical identity. Other colligative properties are a lower vapor pressure for a solution compared with the pure solvent and *osmosis*, the migration of solvent molecules through a semipermeable membrane.

Colligative properties

- Vapor pressure is lower for a solution than for a pure solvent.
- Boiling point is higher for a solution than for a pure solvent.
- Freezing point is lower for a solution than for a pure solvent.
- Osmosis occurs when a solution is separated from a pure solvent by a semipermeable membrane.

Vapor Pressure Lowering of Solutions

We said in Section 8.12 that the vapor pressure of a liquid depends on the equilibrium between molecules entering and leaving the liquid surface. (∞, p. 215) If, however, the liquid molecules are mixed with other particles that don't evaporate, then evaporation is more difficult and the vapor pressure of a solution is lower than that of the pure solvent (Figure 9.8). Note that the *identity* of the solute particles is irrelevant; only their concentration matters.

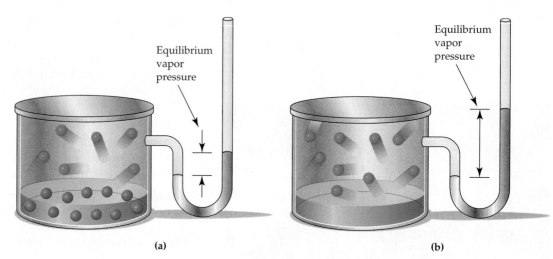

Equilibrium vapor pressure

Equilibrium vapor pressure

(a)   (b)

▲ **FIGURE 9.8 Vapor pressure lowering of solution.** (a) The vapor pressure of a solution is lower than (b) the vapor pressure of the pure solvent because fewer solvent molecules are able to escape from the surface of the solution.

Boiling Point Elevation of Solutions

A consequence of the vapor pressure lowering for a solution is that the boiling point of the solution is higher than that of the pure solvent. Recall from Section 8.12 that boiling occurs when the vapor pressure of a liquid reaches atmospheric pressure. (∞, p. 216) But because the vapor pressure of a solution is lower than that of the pure solvent at a given temperature, the solution must be heated to a higher temperature for its vapor pressure to reach atmospheric pressure. Figure 9.9 shows a close-up plot of vapor pressure versus temperature for pure water and for a 1.0 M NaCl solution. The vapor pressure of pure water reaches atmospheric pressure (760 mm Hg) at 100.0°C, but the vapor pressure of the NaCl solution does not reach the same point until 101.0°C.

For each mole of solute particles added, regardless of chemical identity, the boiling point of 1 kg of water is raised by 0.51°C. The addition of 1 mol of a molecular substance like glucose to 1 kg of water therefore raises the boiling point from 100.0°C to 100.51°C. The addition of 1 mol of NaCl per kilogram of water raises the boiling point by 2 × 0.51°C = 1.02°C, however, because the solution contains 2 mol of solute particles—Na^+ and Cl^- ions.

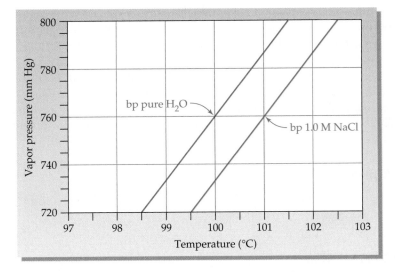

◄ **FIGURE 9.9 A close-up plot of vapor pressure versus temperature for pure water (red curve) and for a 1.0 M NaCl solution (green curve).** Pure water boils at 100.0°C, but the solution does not boil until 101.0°C.

■ **PROBLEM 9.24**

What is the boiling point of a solution of 0.75 mol of KBr in 1.0 kg of water?

■ **PROBLEM 9.25**

When 1.0 mol of HF is dissolved in 1.0 kg of water, the boiling point of the resulting solution is 100.5°C. Is HF a strong electrolyte or a weak electrolyte? Explain.

■ **KEY CONCEPT PROBLEM 9.26**

The following diagram shows plots of vapor pressure versus temperature for a solvent and a solution.

(a) Which curve represents the pure solvent, and which the solution?

(b) What are the approximate boiling points of the pure solvent and the solution?

(c) What is the approximate concentration of the solution in mol/kg if 1 mol of solute particles raise the boiling point of 1 kg of solvent by 3.63°C?

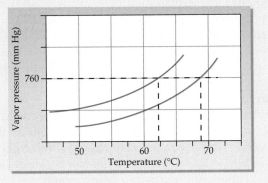

Freezing Point Depression of Solutions

Just as a solution has a lower vapor pressure and consequently higher boiling point than a pure solvent, it also has a lower freezing point. Motorists in cold climates take advantage of this effect when they add "antifreeze" to the water in automobile cooling systems. Antifreeze is a nonvolatile solute, usually ethylene glycol ($HOCH_2CH_2OH$), that is added in sufficient concentration to lower the freezing point below the lowest expected outdoor temperature. In the same way, salt sprinkled on icy roads lowers the freezing point of ice below the road temperature and thus causes ice to melt.

▲ Salt is spread on icy roads to lower the melting point of the ice.

Freezing point depression has much the same cause as vapor pressure lowering. Solute molecules are dispersed between solvent molecules throughout the solution, thereby making it more difficult for solvent molecules to come together and organize into ordered crystals.

For each mole of nonvolatile solute particles, the freezing point of 1 kg of water is lowered by 1.86°C. Thus, addition of 1 mol of antifreeze per kilogram of water lowers the freezing point from 0.00°C to −1.86°C, and addition of 1 mol of NaCl (2 mol of particles) per kilogram lowers the freezing point from 0.00°C to −3.72°C.

■ **PROBLEM 9.27**

What is the freezing point of a solution of 1.0 mol of glucose in 1.0 kg of water?

■ **PROBLEM 9.28**

When 0.5 mol of a certain ionic substance is dissolved in 1.0 kg of water, the freezing point of the resulting solution is −2.8°C. How many ions does the substance give when it dissolves?

9.12 Osmosis and Osmotic Pressure

Certain materials, including those that make up the membranes around living cells, are *semipermeable*. They allow water and other small molecules to pass through, but they block the passage of large solute molecules or ions. When a solution and a pure solvent, or two solutions of different concentration, are separated by such a membrane, solvent molecules pass through the membrane in a process called **osmosis**. Although the passage of solvent through the membrane takes place in both directions, passage from the pure solvent side to the solution side is more favored and occurs more often. As a result, the amount of liquid on the pure solvent side decreases, the amount of liquid on the solution side increases, and the concentration of the solution decreases.

Perhaps the simplest explanation of osmosis is seen by looking at what happens on the molecular level. As shown in Figure 9.10, a solution inside a bulb is separated by a semipermeable membrane from pure solvent in the outer container. Solvent molecules in the container, because of their somewhat higher concentration, approach the membrane a bit more frequently than molecules in the bulb, thereby passing through more often and causing the liquid level in the attached tube to rise.

As the liquid in the tube rises, its increased weight creates an increased pressure that pushes solvent back through the membrane until the rates of forward and reverse passage become equal and the liquid level stops rising. The amount of pressure necessary to achieve this equilibrium is called the **osmotic pressure** of the solution. Osmotic pressures can be extremely high, even for relatively dilute solutions. The osmotic pressure of a 0.15 M NaCl solution at 25°C, for example, is 7.3 atm, a value that will support a difference in water level of approximately 250 ft!

As with other colligative properties, the amount of osmotic pressure depends only on the concentration of solute particles, not on their identity. Thus, it's convenient to use a new unit, *osmolarity* (osmol), to describe the concentration of particles in solution. The **osmolarity** of a solution is equal to the number of moles of dissolved particles (ions or molecules) per liter of solution. A 0.2 M glucose solution, for instance, has an osmolarity of 0.2 osmol, but a 0.2 M solution of NaCl has an osmolarity of 0.4 osmol because it contains 0.2 mol of Na$^+$ ions and 0.2 mol of Cl$^-$ ions.

Osmosis The passage of solvent through a semipermeable membrane separating two solutions of different concentration.

Osmotic pressure The amount of external pressure applied to the more concentrated solution to halt the passage of solvent molecules across a semipermeable membrane.

Osmolarity (osmol) The sum of the molarities of all dissolved particles in a solution.

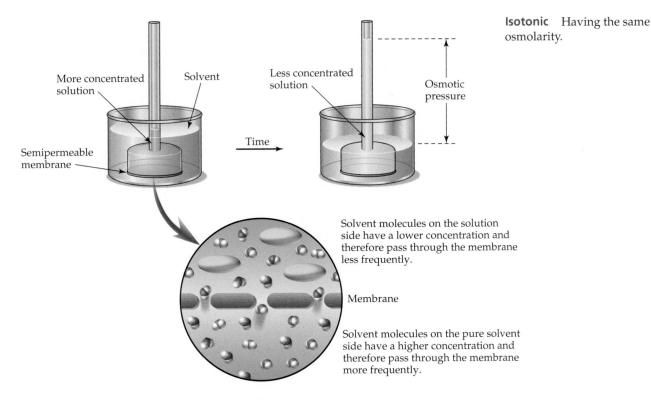

Solvent molecules on the solution side have a lower concentration and therefore pass through the membrane less frequently.

Membrane

Solvent molecules on the pure solvent side have a higher concentration and therefore pass through the membrane more frequently.

Osmosis is particularly important in living organisms because the membranes around cells are semipermeable. The fluids both inside and outside cells must therefore have the same osmolarity to prevent buildup of osmotic pressure and consequent rupture of the cell membrane.

In blood, the plasma surrounding red blood cells has an osmolarity of approximately 0.30 osmol and is said to be **isotonic** with (that is, has the same osmolarity as) the cell contents. If the cells are removed from plasma and placed in 0.15 M NaCl (called *physiological saline solution*), they are unharmed because the osmolarity of the saline solution (0.30 osmol) is the same as that of plasma. If, however, red blood cells are placed in pure water or in any solution with an osmolarity much lower than 0.30 osmol (a *hypotonic* solution), water will pass through the membrane into the cell, causing the cell to swell up and burst, a process called *hemolysis*.

Finally, if red blood cells are placed in a solution having an osmolarity greater than the cell contents (a *hypertonic* solution), water will pass out of the cells into the surrounding solution, causing the cells to shrivel, a process called *crenation*. Figure 9.11 shows red blood cells under all three conditions: isotonic, hypotonic, and hypertonic. Clearly, it's critical that any solution used intravenously be isotonic to prevent red blood cells from being destroyed.

▲ **FIGURE 9.10 The phenomenon of osmosis.** A solution inside the bulb is separated from pure solvent in the outer container by a semipermeable membrane. Solvent molecules in the container have a slightly higher concentration than molecules in the bulb and therefore pass through the membrane more frequently. The liquid in the tube therefore rises until an equilibrium is reached. At equilibrium, the osmotic pressure exerted by the column of liquid in the tube is sufficient to prevent further net passage of solvent.

▼ **FIGURE 9.11** Red blood cells in (a) an isotonic solution are normal in appearance, while the cells in (b) a hypotonic solution are swollen because of water gain, and those in (c) a hypertonic solution are shriveled because of water loss.

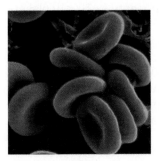

(a)

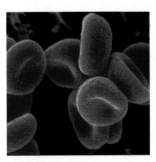

(b)

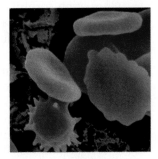

(c)

■ **WORKED EXAMPLE 9.14**

The solution of glucose commonly used intravenously has a concentration of 5.0% (w/v) glucose. What is the osmolarity of this solution? Glucose has MW = 180 amu.

ANALYSIS   Since glucose is a molecular substance that does not give ions in solution, the osmolarity of the solution is the same as the molarity. Recall from Section 9.7 that a solution of 5.0% (w/v) glucose has a concentration of 5.0 g glucose per 100 mL of solution, which is equivalent to 50 g per liter of solution:

$$5.0\%(\text{w/v}) = \frac{5.0\text{ g glucose}}{100\text{ mL solution}} \times 100\%$$

$$\left(\frac{5.0\text{ g glucose}}{100\text{ mL solution}}\right)\left(\frac{1000\text{ mL}}{1\text{ L}}\right) = 50\frac{\text{g glucose}}{\text{L solution}}$$

Finding the molar concentration of glucose requires a mass-to-mole conversion.

SOLUTION

$$\left(\frac{50\text{ g glucose}}{1\text{ L}}\right)\left(\frac{1\text{ mol}}{180\text{ g}}\right) = 0.28\text{ M glucose} = 0.28\text{ osmol}$$

■ **PROBLEM 9.29**

What is the osmolarity of the following solutions?
(a) 0.35 M KBr   (b) 0.15 M glucose + 0.05 M K_2SO_4

■ **PROBLEM 9.30**

What is the osmolarity of a typical oral rehydration solution (ORS) for infants that contains 90 mEq/L Na^+, 20 mEq/L K^+, 110 mEq/L Cl^-, and 2.0% glucose? Glucose has MW = 180 amu.

9.13 Dialysis

Dialysis is similar to osmosis, except that the pores in a dialysis membrane are larger than those in an osmotic membrane so that both solvent molecules and small solute particles can pass through; only large colloidal particles such as proteins can't pass. (The exact dividing line between a "small" molecule and a "large" one is imprecise, and dialysis membranes with a variety of pore sizes are available.) Dialysis membranes include animal bladders, parchment, and cellophane.

Perhaps the most important medical use of dialysis is in artificial kidney machines, where *hemodialysis* is used to cleanse the blood of patients whose kidneys malfunction (Figure 9.12). Blood is diverted from the body and pumped through a long cellophane dialysis tube suspended in an isotonic solution formulated to contain many of the same components as blood plasma. These substances—glucose, NaCl, $NaHCO_3$, and KCl—have the same concentrations in the dialysis solution as they do in blood so that they have no net passage through the membrane.

Small waste materials such as urea pass through the dialysis membrane from the blood to the solution side where they are washed away, but cells, proteins, and other important blood components are prevented by their size from passing through the membrane. In addition, the dialysis fluid concentration can be controlled so that imbalances in electrolytes are corrected. The wash solution is changed every 2 hours, and a typical hemodialysis procedure lasts for 4–7 hours.

Application Timed-Release Medications

There's much more in most medications than medicine. Even something as simple as a generic aspirin tablet contains a binder to keep it from crumbling, a filler to bring it to the right size and help it disintegrate in the stomach, and a lubricant to keep it from sticking to the manufacturing equipment. Timed-release medications are more complex still.

The widespread use of timed-release medication dates from the introduction of Contac decongestant in 1961. The original idea was a simple one: Tiny beads of medicine were encapsulated by coating them with varying thicknesses of a slow-dissolving polymer. Those beads with a thinner coat dissolve and release their medicine more rapidly; those with a thicker coat dissolve more slowly. Combining the right number of beads with the right thicknesses into a single capsule makes possible the gradual release of medication over a predictable time.

The technology of timed-release medications has become much more sophisticated in recent years, and the kinds of medications that can be delivered have become more numerous. Some medicines, for instance, either damage the stomach lining or are destroyed by the highly acidic environment in the stomach but can be delivered safely if given an *enteric coating*. The enteric coating is a polymeric material formulated so that it is stable in acid but reacts and is destroyed when it passes into the more basic environment of the intestines.

One clever new device for timed release of medication through the skin uses the osmotic effect to force a drug from its reservoir. Useful only for drugs that don't dissolve in water, the device is divided into two compartments, one containing medication covered by a perforated membrane and the other containing a hygroscopic material (Section 9.3) covered by a semipermeable membrane. As moisture from the air diffuses through the membrane into the compartment with the hygroscopic material, the buildup of osmotic pressure squeezes the medication out of the other compartment through tiny holes.

See Additional Problem 9.91 at the end of the chapter.

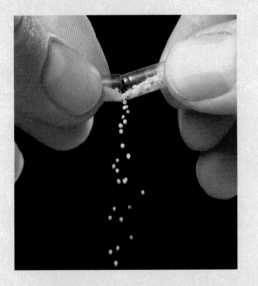

▲ The small beads of medicine are coated with different thicknesses of a slow-dissolving polymer so that they dissolve and release medicine at different times.

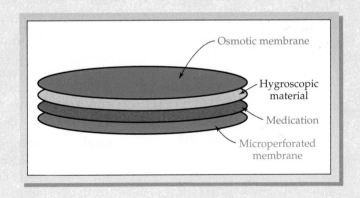

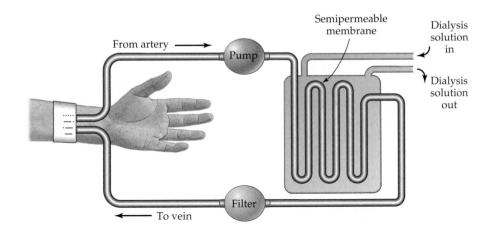

◀ **FIGURE 9.12 Operation of a hemodialysis unit used for purifying blood.** Blood is pumped from an artery through a coiled semipermeable membrane of cellophane. Small waste products pass through the membrane and are washed away by an isotonic dialysis solution.

As noted above, colloidal particles are too large to pass through a semipermeable membrane. Protein molecules, in particular, don't cross semipermeable membranes and thus play an essential role in determining the osmolarity of body fluids. The distribution of water and solutes across the capillary walls that separate blood plasma from the fluid surrounding cells is controlled by the balance between blood pressure and osmotic pressure. The pressure of blood inside the capillary tends to push water out of the plasma (filtration), but the osmotic pressure of colloidal protein molecules tends to draw water into the plasma (reabsorption). The balance between the two processes varies with location in the body. At the arterial end of a capillary, where blood pumped from the heart has a higher pressure, filtration is favored. At the venous end, where blood pressure is lower, reabsorption is favored, causing waste products from metabolism to enter the bloodstream.

Key Words

Summary: Revisiting the Chapter Goals

1. **What are solutions, and what factors affect solubility?** Mixtures are classified as either *heterogeneous*, if the mixing is nonuniform, or *homogeneous*, if the mixing is uniform. *Solutions* are homogeneous mixtures that contain particles the size of ions and molecules (<2.0 nm diameter), whereas larger particles (2.0–500 nm diameter) are present in *colloids*.

   The maximum amount of one substance (the *solute*) that can be dissolved in another (the *solvent*) is called the substance's *solubility*. Substances tend to be mutually soluble when their intermolecular forces are similar. The solubility in water of a solid often increases with temperature, but the solubility of a gas decreases with temperature. Pressure significantly affects gas solubilities, which are directly proportional to their partial pressure over the solution (*Henry's law*).

2. **How is the concentration of a solution expressed?** The concentration of a solution can be expressed in several ways, including molarity, weight/ weight percent composition, weight/volume percent composition, and parts per million. *Molarity*, which expresses concentration as the number of moles of solute per liter of solution, is the most useful method when calculating quantities of reactants or products for reactions in aqueous solution.

3. **How are dilutions carried out?** A dilution is carried out by adding more solvent to an existing solution. Only the amount of solvent changes; the amount of solute remains the same. Thus, the molarity times the volume of the dilute solution is equal to the molarity times the volume of the concentrated solution: $M_1V_1 = M_2V_2$.

4. **What is an electrolyte?** Substances that form ions when dissolved in water and whose water solutions therefore conduct an electric current are called *electrolytes*. Substances that ionize completely in water are *strong electrolytes*, those that ionize partially are *weak electrolytes*, and those that do not ionize are *nonelectrolytes*. Body fluids contain small amounts of many different electrolytes, whose concentrations are expressed in *equivalents*.

5. **How do solutions differ from pure solvents in their behavior?** In comparing a solution to a pure solvent, the solution has a lower vapor pressure at a given temperature, a higher boiling point, and a lower melting point. Called *colligative properties*, these effects depend only on the number of dissolved particles, not on their chemical identity.

6. **What is osmosis?** *Osmosis* occurs when solutions of different concentration are separated by a semipermeable membrane that allows solvent molecules to pass but blocks the passage of solute ions and molecules. Solvent flows from the more dilute side to the more concentrated side until sufficient

osmotic pressure builds up and stops the flow. An effect similar to osmosis occurs when membranes of larger pore size are used. In *dialysis*, the membrane allows the passage of solvent and small dissolved molecules but prevents passage of proteins and larger particles.

▪ Understanding Key Concepts

9.31 Assume that two liquids are separated by a semipermeable membrane. Make a drawing that shows the situation after equilibrium is reached.

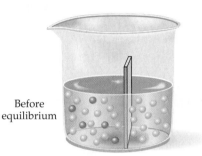

Before
equilibrium

9.32 When 1 mol of HCl is added to 1 kg of water, the boiling point increases by 1.0°C, but when 1 mol of acetic acid, CH₃CO₂H, is added to 1 kg of water, the boiling point increases by only 0.5°C. Explain.

9.33 When 1 mol of HF is added to 1 kg of water, the freezing point decreases by 1.9°C, but when 1 mol of HBr is added, the freezing point decreases by 3.7°C. Which is more highly separated into ions, HF or HBr? Explain.

9.34 The following plot shows the solubilities of two substances as a function of temperature. Which of the substances is a gas, and which is a liquid? Explain.

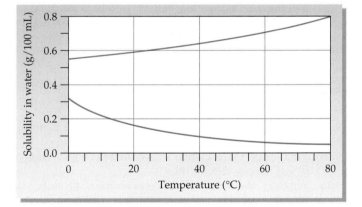

9.35 Assume that you have two full beakers, one containing pure water (blue) and the other containing an equal volume of a 10% (w/v) solution of glucose (green). Which of the drawings (a)–(c) best represents the two beakers after they have stood uncovered for several days and partial evaporation has occurred? Explain.

(a)   (b)   (c)

9.36 A beaker containing 150.0 mL of 0.1 M glucose is represented by (a). Which of the drawings (b)–(d) represents the solution that results when 50.0 mL is withdrawn from (a) and then diluted by a factor of 4?

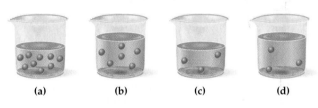

(a)   (b)   (c)   (d)

9.37 The following diagram shows parts of the vapor pressure curves for a solvent and a solution. Which curve is which?

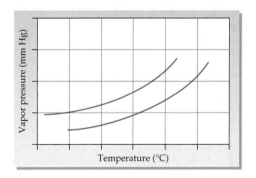

▪ Additional Problems

Solutions and Solubility

9.38 What is the difference between a homogeneous mixture and a heterogenous one?

9.39 How can you tell a solution from a colloid?

9.40 What characteristic of water allows it to dissolve ionic solids?

9.41 Why does water not dissolve motor oil?

9.42 Which of the following are solutions?
  (a) Italian salad dressing
  (b) Rubbing alcohol
  (c) Algae in pond water
  (d) Black coffee

9.43 Which of the following pairs of liquids are likely to be miscible?

(a) H_2SO_4 and H_2O
(b) C_8H_{18} and C_6H_6
(c) C_2H_5OH and H_2O
(d) CS_2 and CCl_4

9.44 The solubility of NH_3 in water at an NH_3 pressure of 760.0 mm Hg is 51.8 g/100 mL. What is the solubility of NH_3 if the partial pressure is reduced to 250.0 mm Hg?

9.45 At 159 mm Hg, the solubility of oxygen in blood is 0.44 g/100 mL. If pure oxygen at 760 mm Hg is administered to a patient, what is the concentration of oxygen in the blood assuming Henry's law is followed?

Concentration and Dilution of Solutions

9.46 Is a solution highly concentrated if it's saturated? Is a solution saturated if it's highly concentrated?

9.47 How is weight/volume percent concentration defined?

9.48 How is molarity defined?

9.49 How is volume/volume percent concentration defined?

9.50 How would you prepare 500.0 mL of a 5.0% (v/v) ethyl alcohol solution?

9.51 A dilute solution of boric acid, $B(OH)_3$, is often used as an eyewash. How would you prepare 500.0 mL of a 0.50% (w/v) boric acid solution?

9.52 Describe how you would prepare 250 mL of a 0.10 M NaCl solution.

9.53 Describe how you would prepare 1.0 L of a 7.5% (w/v) KBr solution.

9.54 What is the weight/volume percent concentration of the following solutions?

(a) 5.0 g KCl in 75 mL of solution
(b) 15 g sucrose in 350 mL of solution

9.55 The concentration of glucose in blood is approximately 90 mg/100 mL. What is the weight/volume percent concentration of glucose?

9.56 How many grams of each substance are needed to prepare the following solutions?

(a) 50.0 mL of 8.0% (w/v) KCl
(b) 200.0 mL of 7.5% (w/v) acetic acid

9.57 Which of the following solutions is more concentrated?

(a) 0.50 M KCl or 5.0% (w/v) KCl
(b) 2.5% (w/v) $NaHSO_4$ or 0.025 M $NaHSO_4$

9.58 If you had only 23 g of KOH remaining in a bottle, how many milliliters of 10.0% (w/v) solution could you prepare? How many milliliters of 0.25 M solution?

9.59 Assuming a blood glucose concentration of 90 mg/ 100 mL and a blood volume of 5.0 L, how many grams of glucose are present in the blood of an average adult?

9.60 The lethal dosage of potassium cyanide in rats is 10 mg KCN/kg body weight. What is this concentration in ppm?

9.61 The maximum concentration set by the U.S. Public Health Service for arsenic in drinking water is 0.050 mg/kg. What is this concentration in ppm and in ppb?

9.62 What is the molarity of the following solutions?

(a) 12.5 g $NaHCO_3$ in 350.0 mL solution
(b) 45.0 g H_2SO_4 in 300.0 mL solution
(c) 30.0 g NaCl dissolved to make 500.0 mL solution

9.63 How many moles of solute are in the following solutions?

(a) 200 mL of 0.30 M acetic acid, CH_3CO_2H
(b) 1.50 L of 0.25 M NaOH
(c) 750 mL of 2.5 M nitric acid, HNO_3

9.64 How many milliliters of a 0.75 M HCl solution do you need to obtain 0.0040 mol of HCl?

9.65 Nalorphine, a relative of morphine, is used to combat withdrawal symptoms in heroin users. How many milliliters of a 0.40% (w/v) solution of nalorphine must be injected to obtain a dose of 1.5 mg?

9.66 A flask containing 450 mL of 0.50 M H_2SO_4 was accidentally knocked to the floor. How many grams of $NaHCO_3$ do you need to put on the spill to neutralize the acid according to the following equation?

$$H_2SO_4(aq) + 2\ NaHCO_3(aq) \rightarrow$$
$$Na_2SO_4(aq) + 2\ H_2O(l) + 2\ CO_2(g)$$

9.67 How many milliliters of 0.0200 M $Na_2S_2O_3$ solution are needed to dissolve 0.450 g of AgBr?

$$AgBr(s) + 2\ Na_2S_2O_3(aq) \rightarrow$$
$$Na_3Ag(S_2O_3)_2(aq) + NaBr(aq)$$

9.68 How much water must you add to 100.0 mL of orange juice concentrate if you want the final juice to be 20.0% of the strength of the original?

9.69 How much water would you add to 100.0 mL of 0.500 M NaOH if you wanted the final concentration to be 0.150 M?

9.70 Concentrated (12.0 M) hydrochloric acid is sold for household and industrial purposes under the name "muriatic acid." How many milliliters of 0.500 M HCl solution can be made from 25.0 mL of 12.0 M HCl solution?

9.71 Dilute solutions of $NaHCO_3$ are sometimes used in treating acid burns. How many milliliters of 0.100 M $NaHCO_3$ solution are needed to prepare 750.0 mL of 0.0500 M $NaHCO_3$ solution?

Electrolytes

9.72 What is an electrolyte?

9.73 Give an example of a strong electrolyte and a nonelectrolyte.

9.74 What does it mean when we say that the concentration of Ca^{2+} in blood is 3.0 mEq/L?

9.75 What is the total anion concentration (in mEq/L) of a solution that contains 5.0 mEq/L Na^+, 12.0 mEq/L Ca^{2+}, and 2.0 mEq/L Li^+?

9.76 Kaochlor, a 10% KCl solution, is an oral electrolyte supplement administered for potassium deficiency. How many milliequivalents of K^+ are in a 30 mL dose?

9.77 A solution of sulfate ion (SO_4^{2-}) is 0.025 M. What is this concentration in Eq/L and in mEq/L?

9.78 Look up the concentration of Cl^- ion in blood (Table 9.6) and calculate how many grams of Cl^- are in 100.0 mL blood.

9.79 Normal blood contains 3 mEq/L of Mg^{2+}. How many milligrams of Mg^{2+} are present in 100.0 mL of blood?

Properties of Solutions

9.80 Which lowers the freezing point of 2.0 kg of water more, 0.20 mol NaOH or 0.20 mol Ba(OH)$_2$? Both compounds are strong electrolytes. Explain.

9.81 Which solution has the higher boiling point, 0.500 M glucose or 0.300 M KCl? Explain.

9.82 Methanol, CH$_3$OH, is sometimes used as an antifreeze for the water in automobile windshield washer fluids. How many grams of methanol must be added to 5.00 kg of water to lower its freezing point to −10.0°C? For each mole of solute, the freezing point of 1 kg of water is lowered 1.86°C.

9.83 What is the boiling point of the solution produced by adding 350 g of ethylene glycol (molar (molar mass = 62.1 g/mol) to 1.5 kg of water? For each mole of nonvolatile solute, the boiling point of 1 kg of water is raised 0.51°C.

Osmosis

9.84 Why does a red blood cell swell up and burst when placed in pure water?

9.85 What does it mean when we say that a 0.15 M NaCl solution is isotonic with blood, whereas distilled water is hypotonic?

9.86 Which of the following solutions has the higher osmolarity?
 (a) 0.25 M KBr or 0.20 M Na$_2$SO$_4$
 (b) 0.30 M NaOH or 3.0% (w/v) NaOH

9.87 Which of the following solutions will give rise to a greater osmotic pressure at equilibrium: 5.00 g of NaCl in 350.0 mL water or 35.0 g of glucose in 400.0 mL water? For NaCl, MW = 58.5 amu; for glucose, MW = 180 amu.

Applications

9.88 How does the body increase oxygen availability at high altitude? [*Breathing and Oxygen Transport*]

9.89 What are the major electrolytes in sweat, and what are their approximate concentrations in mEq/L? [*Electrolytes, Fluid Replacement, and Sports Drinks*]

9.90 Why is a sports drink more effective than plain water for rehydration after extended exercise? [*Electrolytes, Fluid Replacement, and Sports Drinks*]

9.91 How does an enteric coating on a medication work? [*Timed-Release Medications*]

General Questions and Problems

9.92 Uric acid, the principal constituent of some kidney stones, has the formula C$_5$H$_4$N$_4$O$_3$. In aqueous solution, the solubility of uric acid is only 0.067 g/L. Express this concentration in (w/v)%, in ppm, and in molarity.

9.93 Emergency treatment of cardiac arrest victims sometimes involves injection of a calcium chloride solution directly into the heart muscle. How many grams of CaCl$_2$ are administered in an injection of 5.0 mL of a 5.0% (w/v) solution? How many milliequivalents of Ca^{2+} does this correspond to?

9.94 Nitric acid, HNO$_3$, is available commercially at a concentration of 16 M. What volume would you use to prepare 750 mL of a 0.20 M solution?

9.95 How much 16 M nitric acid (see Problem 9.94) would react with 5.50 g of KOH according to the following equation?

$$HNO_3(aq) + KOH(aq) \rightarrow H_2O(l) + KNO_3(aq)$$

9.96 An old bottle of 12.0 M hydrochloric acid has only 25 mL left in it. What is the HCl concentration if enough water is added so that the final volume is 525 mL?

9.97 One test for vitamin C (ascorbic acid, C$_6$H$_8$O$_6$) is based on the reaction of the vitamin with iodine:

$$C_6H_8O_6(aq) + I_2(aq) \rightarrow C_6H_6O_6(aq) + 2\ HI(aq)$$

 (a) If 25.0 mL of a fruit juice requires 13.0 mL of 0.0100 M I$_2$ solution for reaction, what is the molarity of the ascorbic acid in the fruit juice?
 (b) The Food and Drug Administration recommends that 60 mg of ascorbic acid be consumed per day. How many milliliters of the fruit juice in part (a) must a person drink to obtain the recommended dosage?

9.98. *Ringer's solution*, used in the treatment of burns and wounds, is prepared by dissolving 8.6 g of NaCl, 0.30 g of KCl, and 0.33 g of CaCl$_2$ in water and diluting to a volume of 1.00 L. What is the molarity of each of the three components?

9.99 What is the osmolarity of Ringer's solution (see Problem 9.98)? Is it hypotonic, isotonic, or hypertonic with blood plasma (0.30 osmol)?

9.100 In many states, a person with a blood alcohol concentration of 0.080% (v/v) is considered legally drunk. What volume of total alcohol does this concentration represent, assuming a blood volume of 5.0 L?

9.101 Ammonia is very soluble in water (51.8 g/L at 20°C and 760 mm Hg).
 (a) Show how NH$_3$ can hydrogen-bond to water.
 (b) What is the solubility of ammonia in water in moles per liter?

9.102 Cobalt(II) chloride, a blue solid, can absorb water from the air to form cobalt(II) chloride hexahydrate, a pink solid. The equilibrium is so sensitive to moisture in the air that CoCl$_2$ is used as a humidity indicator.
 (a) Write a balanced equation for the equilibrium. Be sure to include water as a reactant to produce the hexahydrate.
 (b) How many grams of water are released by the decomposition of 2.50 g of cobalt(II) chloride hexahydrate?

9.103 How many milliliters of 0.150 M BaCl$_2$ are needed to react completely with 35.0 mL of 0.200 M Na$_2$SO$_4$? How many grams of BaSO$_4$ will be formed?

9.104 Many compounds are only partially dissociated into ions in aqueous solution. Trichloroacetic acid (CCl$_3$CO$_2$H), for instance, is partially dissociated in water according to the equation

$$CCl_3CO_2H(aq) \rightleftharpoons H^+(aq) + CCl_3CO_2^-(aq)$$

What is the percentage of molecules dissociated if the freezing point of a solution prepared by dissolving 0.100 mol of trichloroacetic acid in 100.0 g of water is −2.53°C? For each mole of solute, the freezing point of 1 kg of water is lowered 1.86°C.

10

Acids and Bases

The color of these hydrangeas, whether red or blue, depends on the acidity of the soil they're grown in.

Acids! The word evokes images of dangerous, corrosive liquids that eat away everything they touch. Although a few well-known substances such as sulfuric acid (H_2SO_4) do indeed fit this description, most acids are relatively harmless. In fact, many acids, such as ascorbic acid (vitamin C), are necessary for life.

We've already touched on the subject of acids and bases on several occasions, but it's now time for a more detailed study that will answer the following questions:

1. What are acids and bases?
The goal: Be able to recognize acids and bases and write equations for common acid–base reactions.

2. **What is the influence of acid and base strengths on their reactions?**
   The goal: Be able to interpret acid strength using acid dissociation constants K_a and predict the favored direction of acid–base equilibria.
3. **What is the ion-product constant for water?**
   The goal: Be able to write the equation for this constant and use it to find the concentration of H_3O^+ or OH^-.
4. **What is the pH scale for measuring acidity?**
   The goal: Be able to explain the pH scale and find pH from the H_3O^+ concentration.
5. **What is a buffer?**
   The goal: Be able to explain how a buffer maintains pH and how the bicarbonate buffer functions in the body.
6. **How is the acid or base concentration of a solution determined?**
   The goal: Be able to explain how a titration procedure works and use the results of a titration to calculate acid or base concentration in a solution.

CONCEPTS TO REVIEW

Acids, Bases, and Neutralization Reactions (Sections 4.12 and 6.10)
Reversible Reactions and Chemical Equilibrium (Section 7.6)
Equilibrium Equations and Equilibrium Constants (Section 7.7)
Units of Concentration; Molarity (Section 9.7)
Ion Equivalents (Section 9.10)

10.1 Acids and Bases in Aqueous Solution

Let's take a moment to review what we said about acids and bases in Sections 4.12 and 6.10 before going on to a more systematic study:

- An acid is a substance that gives hydrogen ions, H^+, when dissolved in water.

- A base is a substance that gives hydroxide ions, OH^-, when dissolved in water.

- The neutralization reaction of an acid with a base yields water plus a *salt*, an ionic compound composed of the cation from the base and the anion from the acid.

The above definitions of acids and bases were proposed in 1887 by the Swedish chemist Svante Arrhenius and are useful for many purposes. They aren't, however, completely sufficient. One problem is that the definitions are too limited because they refer only to reactions that take place in aqueous solution. (We'll see shortly how the definitions can be broadened.) Another problem is that the H^+ ion is so reactive it doesn't exist in water. Instead, H^+ reacts with H_2O to give the **hydronium ion,** H_3O^+. When gaseous HCl dissolves in water, for instance, H_3O^+ and Cl^- are formed. As described in Section 5.8, electrostatic potential maps show that the hydrogen of HCl is positively polarized and electron-poor (blue) while the oxygen of water is negatively polarized and electron-rich (red).

Hydronium ion The H_3O^+ ion, formed when an acid reacts with water.

 Introduction to Aqueous Acids

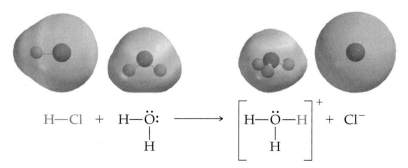

Introduction to Aqueous Bases

Thus, the Arrhenius definition has to be modified to say that an acid yields H_3O^+ in water rather than H^+. In practice, however, the notations H_3O^+ and $H^+(aq)$ are often used interchangeably.

The Arrhenius definition of a base is correct as far as it goes, but it's important to realize that the OH^- ions "produced" by the base can come from either of two sources. Metal hydroxides, such as NaOH, KOH, and $Ba(OH)_2$, are ionic compounds that already contain OH^- ions and merely release those ions when they dissolve in water. Ammonia, however, is not ionic and contains no OH^- ions in its structure. Nonetheless, ammonia is a base because it undergoes a reaction with water when it dissolves, producing NH_4^+ and OH^- ions.

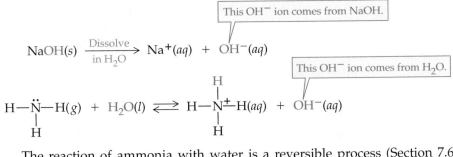

The reaction of ammonia with water is a reversible process (Section 7.6) whose equilibrium strongly favors unreacted ammonia. (∞, p. 177) Nevertheless, *some* OH^- ions are produced, so NH_3 is a base.

10.2 Some Common Acids and Bases

Acids and bases are present in a variety of foods and consumer products. Acids generally have a sour taste, and nearly every sour food contains an acid: Lemons, oranges, and grapefruit contain citric acid, for instance, and sour milk contains lactic acid. Bases aren't so obvious in foods, but most of us have them stored under the kitchen or bathroom sink. Bases are present in many household cleaning agents, from perfumed toilet soap, to ammonia-based window cleaners, to the substance you put down the drain to dissolve hair, grease, and other materials that clog it.

Some of the most common acids and bases are listed below. It's a good idea at this point to learn their names and formulas, because we'll be referring to them often.

▲ Sulfuric acid is used as an electrolyte in automobile batteries.

- **Sulfuric acid, H_2SO_4,** is manufactured in greater quantity than any other industrial chemical. Some 48 millions tons were prepared in the United States in 1999 for use in many hundreds of industrial processes, including the preparation of phosphate fertilizers. Its most common consumer use is as the acid found in automobile batteries. As anyone who has splashed battery acid on their skin or clothing knows, sulfuric acid is highly corrosive and can cause painful burns.

- **Hydrochloric acid, HCl,** or *muriatic acid* as it was historically known, has many industrial applications, including its use in metal cleaning and in the manufacture of high-fructose corn syrup. Aqueous HCl is also present as "stomach acid" in the digestive systems of most mammals.

- **Phosphoric acid, H_3PO_4,** is used in vast quantities in the manufacture of phosphate fertilizers. In addition, it is also used as an additive in foods and toothpastes. The tart taste of many soft drinks is due to the presence of phosphoric acid.

- **Nitric acid, HNO_3,** is a strong oxidizing agent that is used for many purposes, including the manufacture of ammonium nitrate fertilizer and military explosives. When spilled on the skin, it leaves a characteristic yellow coloration because of its reaction with skin proteins.

- **Acetic acid, CH₃CO₂H,** is the primary organic constituent of vinegar. It also occurs in all living cells and is used in many industrial processes such as the preparation of solvents, lacquers, and coatings.
- **Sodium hydroxide, NaOH,** also called *caustic soda* or *lye,* is the most commonly used of all bases. Industrially, it is used in the production of aluminum from its ore, in the production of glass, and in the manufacture of soap from animal fat. Concentrated solutions of NaOH can cause severe burns if allowed to sit on the skin for long. Drain cleaners often contain NaOH because it reacts with the fats and proteins found in grease and hair.
- **Calcium hydroxide, Ca(OH)₂,** or *slaked lime,* is made industrially by treating lime (CaO) with water. It has many applications, including its use in mortars and cements. An aqueous solution of Ca(OH)₂ is often called *limewater.*
- **Magnesium hydroxide, Mg(OH)₂,** or *milk of magnesia,* is an additive in foods, toothpaste, and many over-the-counter medications. Antacids such as Rolaids, Mylanta, and Maalox, for instance, all contain magnesium hydroxide.
- **Ammonia, NH₃,** is used primarily as a fertilizer, but it also has many other industrial applications including the manufacture of pharmaceuticals and explosives. A dilute solution of ammonia is frequently used around the house as a glass cleaner.

▲ Soap is manufactured by the reaction of vegetable oils and animal fats with the bases NaOH and KOH.

10.3 The Brønsted–Lowry Definition of Acids and Bases

The Arrhenius definition of acids and bases discussed in Section 10.1 applies only to reactions that take place in aqueous solution. A far more general definition was proposed in 1923 by the Danish chemist Johannes Brønsted and the English chemist Thomas Lowry. A **Brønsted–Lowry acid** is any substance that is able to give a hydrogen ion, H^+, to another molecule or ion. Since a hydrogen *atom* consists of a proton and an electron, a hydrogen *ion,* H^+, is simply a proton. Thus, we often refer to acids as *proton donors.* The reaction need not occur in water, and a Brønsted–Lowry acid need not give appreciable concentrations of H_3O^+ ions in water.

Different acids can supply different numbers of H^+ ions, as we saw in Section 4.12. (⟳, p. 89) Acids with one proton to donate, such as HCl and HNO_3, are called *monoprotic acids;* H_2SO_4 is a *diprotic acid* because it has two protons; and H_3PO_4 is a *triprotic acid* because it has three protons.

Brønsted–Lowry acid A substance that can donate a hydrogen ion, H^+, to another molecule or ion.

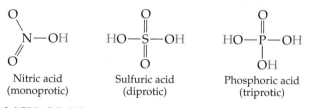

Nitric acid (monoprotic)   Sulfuric acid (diprotic)   Phosphoric acid (triprotic)

Acetic acid (CH_3CO_2H), an example of an organic acid, actually has a total of four hydrogens, but only the one bonded to the electronegative oxygen is positively polarized and therefore acidic. The three hydrogens bonded to carbon are not acidic. Most organic acids are similar in that they contain many hydrogen atoms, but only the one in the —CO_2H group (blue in the electrostatic potential map) is acidic.

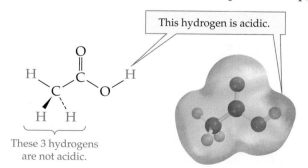

This hydrogen is acidic.

These 3 hydrogens are not acidic.

Just as a Brønsted–Lowry acid is a substance that *donates* H^+ ions, a **Brønsted–Lowry base** is a substance that *accepts* H^+ from an acid. The reaction need not occur in water, and the Brønsted–Lowry base need not give appreciable concentrations of OH^- ions in water. Gaseous NH_3, for example, acts as a base to accept H^+ from gaseous HCl and yield the ionic solid $NH_4^+Cl^-$.

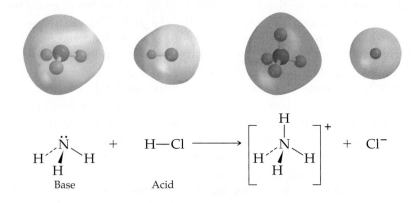

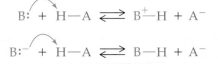

Base   Acid

Putting the acid and base definitions together, *an acid–base reaction is one in which a proton is transferred*. The general reaction between proton-donor acids and proton-acceptor bases can be represented as

$$B: \; + \; H—A \; \rightleftarrows \; B\overset{+}{—}H \; + \; A^-$$

$$B:^- \; + \; H—A \; \rightleftarrows \; B—H \; + \; A^-$$

where the abbreviation HA represents a Brønsted–Lowry acid and B: or B:$^-$ represents a Brønsted–Lowry base. Notice in these acid–base reactions that both electrons in the product B—H bond come from the base, as indicated by the curved arrow flowing from the electron pair of the base to the hydrogen atom of the acid. Thus, the B—H bond that forms is a coordinate covalent bond (Section 5.4). (p. 106) In fact, a base *must* have such a lone pair of electrons; without them, it could not accept H^+ from an acid.

A base can be either neutral (B:) or negatively charged (B:$^-$). If the base is neutral, then the product has a positive charge (BH^+) after H^+ has added. Ammonia is an example:

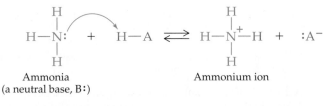

Ammonia   Ammonium ion
(a neutral base, B:)

If the base is negatively charged, then the product is neutral (BH). Hydroxide ion is an example:

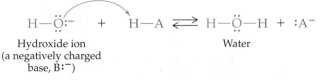

Hydroxide ion   Water
(a negatively charged
base, B:$^-$)

An important consequence of the Brønsted–Lowry definitions is that the *products* of an acid–base reaction are themselves acids and bases. When the acid HA donates a proton to the base B, the product A^- is itself a base because it can accept back a proton. At the same time, BH^+ is itself an acid because it can donate back a proton. That is, acid–base reactions are reversible (Section 7.7), although the equilibrium constant for a given reaction is often large so that the reversibility is not noticeable. (p. 179)

$$\underset{\text{Base}}{B\!:} + \underset{\text{Acid}}{H\text{—}A} \rightleftharpoons \underset{\text{Base}}{:\!A^-} + \underset{\text{Acid}}{\overset{+}{B}\text{—}H}$$

Conjugate acid–base pair

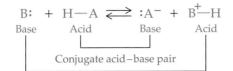

Chemical species such as B and BH^+, or HA and A^-, whose formulas differ by only one H^+, are called **conjugate acid–base pairs**. Thus, the anion A^- is the **conjugate base** of the acid HA, and HA is the **conjugate acid** of the base A^-. Similarly, B is the conjugate base of the acid BH^+, and BH^+ is the conjugate acid of the base B. To give some examples, acetic acid and acetate ion, the hydronium ion and water, and the ammonium ion and ammonia all make conjugate acid–base pairs.

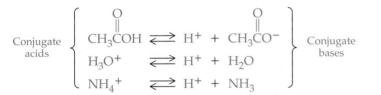

Conjugate acid–base pair Two substances whose formulas differ by only a hydrogen ion, H^+.

Conjugate base The substance formed by loss of H^+ from an acid.

Conjugate acid The substance formed by addition of H^+ to a base.

Conjugate Acids and Bases

Conjugate acids / Conjugate bases

$$CH_3\overset{O}{\overset{\|}{C}}OH \rightleftharpoons H^+ + CH_3\overset{O}{\overset{\|}{C}}O^-$$
$$H_3O^+ \rightleftharpoons H^+ + H_2O$$
$$NH_4^+ \rightleftharpoons H^+ + NH_3$$

■ WORKED EXAMPLE 10.1

Identify each of the following as a Brønsted–Lowry acid or base:

(a) PO_4^{3-} **(b)** $HClO_4$ **(c)** CN^-

ANALYSIS A Brønsted–Lowry acid must have a hydrogen that it can donate as H^+, and a Brønsted–Lowry base must have an atom with a lone pair of electrons that can bond to H^+. Typically, a Brønsted–Lowry base is an anion derived by loss of H^+ from an acid.

SOLUTION

(a) The phosphate anion (PO_4^-) is a Brønsted–Lowry base derived by loss of 3 H^+ ions from phosphoric acid, H_3PO_4.

(b) Perchloric acid ($HClO_4$) is a Brønsted–Lowry acid because it can donate an H^+ ion.

(c) The cyanide ion (CN^-) is a Brønsted–Lowry base derived by removal of H^+ ion from hydrogen cyanide, HCN.

■ WORKED EXAMPLE 10.2

Write formulas for:

(a) The conjugate acid of the cyanide ion, CN^-

(b) The conjugate base of perchloric acid, $HClO_4$

ANALYSIS A conjugate acid is formed by adding H^+ to a base; a conjugate base is formed by removing H^+ from an acid.

SOLUTION

(a) HCN is the conjugate acid of CN^-.

(b) ClO_4^- is the conjugate base of $HClO_4$.

■ PROBLEM 10.1

Which of the following would you expect to be Brønsted–Lowry acids?

(a) HCO_2H **(b)** H_2S **(c)** $SnCl_2$

■ PROBLEM 10.2

Which of the following would you expect to be Brønsted–Lowry bases?

(a) SO_3^{2-} **(b)** Ag^+ **(c)** F^-

■ **PROBLEM 10.3**

Write formulas for:

(a) The conjugate acid of HS^-

(b) The conjugate acid of PO_4^{3-}

(c) The conjugate base of H_2CO_3

(d) The conjugate base of NH_4^+

■ **KEY CONCEPT PROBLEM 10.4**

For the following reaction, identify the Brønsted–Lowry acids, bases, and conjugate acid–base pairs.

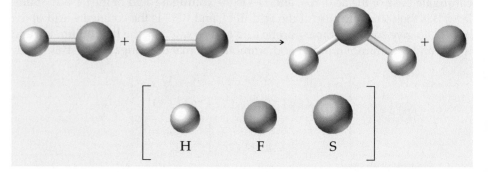

10.4 Water as Both an Acid and a Base

Water is neither an acid nor a base in the Arrhenius sense because it does not contain appreciable concentrations of either H_3O^+ or OH^-. In the Brønsted–Lowry sense, however, water is *both* an acid and a base. When water reacts as a Brønsted–Lowry acid, it *donates* a proton to a base. In its reaction with acetate ion, for example, water donates H^+ to the acetate ion to form acetic acid:

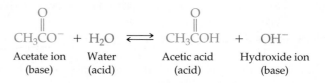

| Acetate ion | Water | Acetic acid | Hydroxide ion |
| (base) | (acid) | (acid) | (base) |

When water reacts as a Brønsted–Lowry base, it *accepts* H^+ from an acid. This, of course, is exactly what happens when an acid such as HCl dissolves in water, as discussed in Section 10.1.

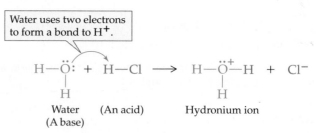

Amphoteric A substance that can react as either an acid or a base.

Substances like water, which can react as either an acid or a base depending on the circumstances, are said to be **amphoteric** (am-pho-**tare**-ic). When water acts as an acid, it donates H^+ and becomes OH^-; when it acts as a base, it accepts H^+ and becomes H_3O^+.

■ **PROBLEM 10.5**

Is water an acid or a base in each of the following reactions?

(a) $H_3PO_4(aq) + H_2O(l) \rightarrow H_2PO_4^-(aq) + H_3O^+(aq)$

(b) $F^-(aq) + H_2O(l) \rightarrow HF(aq) + OH^-(aq)$

(c) $NH_4^+(aq) + H_2O(aq) \rightarrow NH_3(aq) + H_3O^+(aq)$

10.5 Some Common Acid–Base Reactions

Among the most common of the many kinds of Brønsted–Lowry acid–base reactions are those of an acid with hydroxide ion, with bicarbonate or carbonate ion, and with ammonia or a related nitrogen-containing compound. Let's look briefly at each of the three types.

Reaction of Acids with Hydroxide Ion

Acids react with metal hydroxides to yield water and a salt in a neutralization reaction (Section 6.10): (∞, p. 148)

$$HCl(aq) + KOH(aq) \longrightarrow H_2O(l) + KCl(aq)$$
(An acid)   (A base)   (Water)   (A salt)

Such reactions are usually written with a single arrow because their equilibria lie far to the right and they have very large equilibrium constants ($K = 5 \times 10^{15}$; Section 7.7). The net ionic equation (Section 6.11) for all such reactions makes clear why the properties of the acid and base disappear in neutralization reactions: The hydrogen ions and hydroxide ions are used up in the formation of water.

$$H^+(aq) + OH^-(aq) \longrightarrow H_2O(l)$$

■ PROBLEM 10.6

Maalox, an over-the-counter antacid, contains aluminum hydroxide, $Al(OH)_3$, and magnesium hydroxide, $Mg(OH)_2$. Write balanced equations for the reaction of both with stomach acid (HCl).

Reaction of Acids with Bicarbonate and Carbonate Ion

Bicarbonate ion, HCO_3^-, reacts with acid by accepting H^+ to yield carbonic acid, H_2CO_3. Similarly, carbonate ion accepts two protons in its reaction with acid. It turns out, though, that the H_2CO_3 reaction product is unstable, rapidly decomposing to yield carbon dioxide gas and water:

$$H^+(aq) + HCO_3^-(aq) \longrightarrow [H_2CO_3(aq)] \longrightarrow H_2O(l) + CO_2(g)$$
$$2\ H^+(aq) + CO_3^{2-}(aq) \longrightarrow [H_2CO_3(aq)] \longrightarrow H_2O(l) + CO_2(g)$$

Most metal carbonates are insoluble in water—marble, for example, is almost pure calcium carbonate, $CaCO_3$—but they nevertheless react easily with aqueous acid. In fact, geologists often test for carbonate-bearing rocks simply by putting a few drops of aqueous HCl on the rock and watching to see if bubbles of CO_2 form (Figure 10.1).

▲ **FIGURE 10.1** Marble, which is primarily $CaCO_3$, releases bubbles of CO_2 when treated with hydrochloric acid.

■ PROBLEM 10.7

Write balanced equations for the following reactions:
(a) $KHCO_3(aq) + H_2SO_4(aq) \rightarrow ?$ **(b)** $MgCO_3(aq) + HNO_3(aq) \rightarrow ?$

Reaction of Acids with Ammonia

Acids react with ammonia to yield ammonium salts such as ammonium chloride, NH_4Cl, most of which are water-soluble.

$$NH_3(aq) + HCl(aq) \longrightarrow NH_4Cl(aq)$$

Living organisms contain a group of compounds called *amines*, which contain ammonia-like nitrogen atoms bonded to carbon. Amines react with acids

Application Ulcers and Antacids

Peptic ulcers are one of the hazards of fast-paced modern life—nature's way of telling us to slow down. Ulcers occur either in the stomach itself or in the upper part of the small intestine (the duodenum) when the protective mucosal lining is penetrated and gastric juices begin to dissolve the stomach or intestinal wall. The resultant lesion causes considerable pain and can lead to eventual perforation of the wall. So common are they that nearly 10% of the U.S. population will suffer from an ulcer at some point in their life.

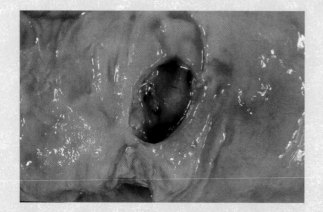

▲ A peptic ulcer, or lesion in the stomach wall.

For many years, it was thought that ulcers were caused by excess stomach acid eating into the stomach lining. Treatment therefore centered on the use of drugs such as cimetidine (Tagamet) and ranitidine (Zantac), which inhibit gastric acid secretion, and on antacids, which lower stomach acidity. As you might expect, all the common over-the-counter antacid remedies are bases—hydroxides, carbonates, or bicarbonates—as shown in the accompanying Table. Since aluminum and calcium salts tend to cause constipation, most formulations also contain magnesium salts, which have a counteracting laxative effect. Sodium bicarbonate ($NaHCO_3$, baking soda) is less suitable than other bases because it tends to enter the bloodstream too rapidly.

Ingredients of Some Antacids

Trade Name	Active Ingredient
Alka-Seltzer	$NaHCO_3 + KHCO_3$
Gaviscon	$Al(OH)_3 + MgCO_3$
Gelusil	$Al(OH)_3 + Mg(OH)_2$
Maalox	$Al(OH)_3 + Mg(OH)_2$
Mylanth	$Al(OH)_3 + Mg(OH)_2$
Rolaids	$CaCO_3 + Mg(OH)_2$
Tums	$CaCO_3$

Although much of the pain and damage associated with an ulcer is due to acid secretion, which increases in times of stress, the actual *cause* of the ulcer is not gastric acid at all. Rather, ulcers are caused by the bacteria *Helicobacter pylori*, which, remarkably enough, is able to live and thrive in the highly acidic environment of the stomach. Antibiotic treatment with a mixture of metronidazole plus erythromycin or clarithromycin has proven to be greater than 90% effective at curing the ulcer and preventing its recurrence.

See Additional Problem 10.90 at the end of the chapter.

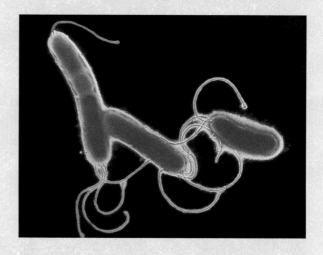

▲ Cells of *Helicobacter pylori*, the bacterium that causes most stomach ulcers.

just as ammonia does, yielding water-soluble salts. Methylamine, for example, an organic compound found in rotting fish, reacts with HCl:

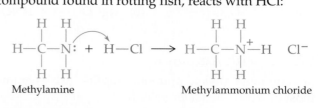

Methylamine Methylammonium chloride

◯◯◯ **LOOKING AHEAD**

We'll see in Chapter 15 that amines occur throughout all living organisms, both plant and animal, as well as in many pharmaceutical agents. Other amines called amino acids form the building blocks from which proteins are made, as we'll see in Chapter 18.

■ **PROBLEM 10.8**

What products would you expect from the reaction of ammonia and sulfuric acid in aqueous solution?

$$NH_3(aq) + H_2SO_4(aq) \rightarrow ?$$

■ **PROBLEM 10.9**

Show how ethylamine ($C_2H_5NH_2$) reacts with hydrochloric acid to form an ethylammonium salt.

10.6 Acid and Base Strength

Acids differ in their ability to give up a proton, as indicated in Table 10.1. The six acids at the top of the table are **strong acids**, meaning that they give up a proton easily and are essentially 100% **dissociated**, or split apart into ions, in water. Those remaining are **weak acids**, meaning that they give up a proton with difficulty and are substantially less than 100% dissociated in water. In a similar way, the bases at the top of the table are **weak bases** because they have little affinity for a proton, and the bases at the bottom of the table are **strong bases** because they grab and hold a proton tightly.

Strong acid An acid that gives up H^+ easily and is essentially 100% dissociated in water.

Dissociation The splitting apart of an acid in water to give H^+ and an anion.

Weak acid An acid that gives up H^+ with difficulty and is less than 100% dissociated in water.

Weak base A base that has only a slight affinity for H^+ and holds it weakly.

Strong base A base that has a high affinity for H^+ and holds it tightly.

TABLE 10.1   Relative Strengths of Acids and Conjugate Bases

		Acid			Conjugate Base		
Increasing acid strength	Strong acids: 100% dissociated	Perchloric acid	$HClO_4$	ClO_4^-	Perchlorate ion	Little or no reaction as bases	**Increasing base strength**
		Sulfuric acid	H_2SO_4	HSO_4^-	Hydrogen sulfate ion		
		Hydriodic acid	HI	I^-	Iodide ion		
		Hydrobromic acid	HBr	Br^-	Bromide ion		
		Hydrochloric acid	HCl	Cl^-	Chloride ion		
		Nitric acid	HNO_3	NO_3^-	Nitrate ion		
		Hydronium ion	H_3O^+	H_2O	**Water**		
	Weak acids	Hydrogen sulfate ion	HSO_4^-	SO_4^{2-}	Sulfate ion	Very weak bases	
		Phosphoric acid	H_3PO_4	$H_2PO_4^-$	Dihydrogen phosphate ion		
		Nitrous acid	HNO_2	NO_2^-	Nitrite ion		
		Hydrofluoric acid	HF	F^-	Fluoride ion		
		Acetic acid	CH_3COOH	CH_3COO^-	Acetate ion		
	Very weak acids	Carbonic acid	H_2CO_3	HCO_3^-	Bicarbonate ion	Weak bases	
		Dihydrogen phosphate ion	$H_2PO_4^-$	HPO_4^{2-}	Hydrogen phosphate ion		
		Ammonium ion	NH_4^+	NH_3	Ammonia		
		Hydrocyanic acid	HCN	CN^-	Cyanide ion		
		Bicarbonate ion	HCO_3^-	CO_3^{2-}	Carbonate ion		
		Hydrogen phosphate ion	HPO_4^{2-}	PO_4^{3-}	Phosphate ion		
		Water	H_2O	OH^-	**Hydroxide ion**	Strong base	

Note that diprotic acids, such as sulfuric acid, undergo two stepwise dissociations in water. The first dissociation yields HSO_4^- and occurs to the extent of nearly 100%, so H_2SO_4 is a strong acid. The second dissociation yields SO_4^{2-} and takes place to a much lesser extent because separation of a positively charged H^+ from the negatively charged HSO_4^- anion is difficult. Thus, HSO_4^- is a weak acid.

$$H_2SO_4(l) \xrightarrow{\;H_2O\;} H_3O^+(aq) + HSO_4^-(aq)$$

$$HSO_4^-(aq) + H_2O(l) \rightleftharpoons H_3O^+(aq) + SO_4^{2-}(aq)$$

Perhaps the most striking feature of Table 10.1 is the inverse relationship between acid strength and base strength. *The stronger the acid, the weaker its conjugate base; the weaker the acid, the stronger its conjugate base.* HCl, for example, is a strong acid, so Cl^- is a very weak base. H_2O, however, is a very weak acid, so OH^- is a strong base.

Why is there an inverse relationship between acid strength and base strength? To answer this question, think about what it means for an acid or base to be strong or weak. A strong acid H—A is one that readily gives up a proton, meaning that its conjugate base A^- has little affinity for the proton. But this is exactly the definition of a weak base—a substance that has little affinity for a proton.

$$H{-}A \;+\; H_2O \;\rightleftharpoons\; H_3O^+ \;+\; A^-$$

If this is a strong acid because it gives up a proton readily . . .

. . . then this is a weak base because it has little affinity for a proton.

In the same way, a weak acid is one that gives up a proton with difficulty, meaning that its conjugate base has a high affinity for the proton. But this is just the definition of a strong base—a substance that has a high affinity for the proton.

$$H{-}A \;+\; H_2O \;\rightleftharpoons\; H_3O^+ \;+\; A^-$$

If this is a weak acid because it gives up a proton with difficulty . . .

. . . then this is a strong base because it has a high affinity for a proton.

Knowing the relative strengths of different acids as shown in Table 10.1 makes it possible to predict the direction of proton-transfer reactions. *An acid–base proton-transfer equilibrium always favors reaction of the stronger acid and formation of the weaker acid.* That is, the proton always leaves the stronger acid (whose weaker conjugate base can't hold the proton) and always ends up in the weaker acid (whose stronger conjugate base holds the proton tightly). Put another way, in a contest for the proton, the stronger base always wins.

Stronger acid + Stronger base $\rightleftharpoons$ Weaker base + Weaker acid

To try out this rule, let's compare the reactions of acetic acid with water and with hydroxide ion. The idea is to write the equation, identify the acid on each side of the arrow, and then decide which acid is stronger and which is weaker. For example, the reaction of acetic acid with water to give acetate ion and hydronium ion is favored in the reverse direction, because acetic acid is a weaker acid than H_3O^+.

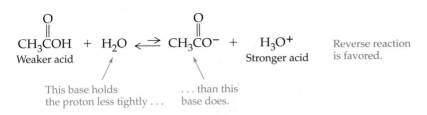

On the other hand, the reaction of acetic acid with hydroxide ion to give acetate ion and water is favored in the forward direction, because acetic acid is a stronger acid than H_2O.

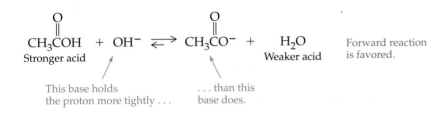

■ **WORKED EXAMPLE 10.3**

Write a balanced equation for the proton-transfer reaction between phosphate ion (PO_4^{3-}) and water, and determine in which direction the equilibrium is favored.

ANALYSIS   Look in Table 10.1 to see the relative acid and base strengths of the species involved in the reaction. The acid–base proton-transfer equilibrium will favor reaction of the stronger acid and formation of the weaker acid.

SOLUTION   Phosphate ion is the anion of a weak acid and is therefore a strong base. Table 10.1 shows that HPO_4^{2-} is a stronger acid than H_2O and OH^- is a stronger base than HPO_4^-, so the reaction is favored in the reverse direction.

$$PO_4^{3-}(aq) + H_2O(l) \xleftrightarrow{} HPO_4^{2-}(aq) + OH^-(aq)$$

Weaker base   Weaker acid   Stronger acid   Stronger base

■ **PROBLEM 10.10**

Use Table 10.1 to identify the stronger acid in each of the following pairs:
(a) H_2O or NH_4^+ **(b)** H_2SO_4 or CH_3CO_2H **(c)** HCN or H_2CO_3

■ **PROBLEM 10.11**

Use Table 10.1 to identify the stronger base in each of the following pairs:
(a) F^- or Br^- **(b)** OH^- or HCO_3^-

■ **PROBLEM 10.12**

Write a balanced equation for the proton-transfer reaction between hydrogen phosphate ion and hydroxide ion. Identify each acid–base pair, and determine in which direction the equilibrium is favored.

■ **KEY CONCEPT PROBLEM 10.13**

Look at the following electrostatic potential map of the amino acid alanine, and identify the most acidic hydrogens in the molecule:

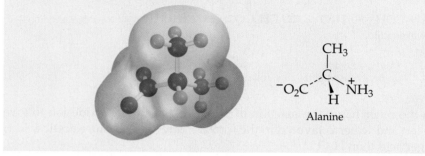

Alanine

10.7 Acid Dissociation Constants

The reaction of a weak acid with water, like any chemical equilibrium, can be described by an equilibrium equation (Section 7.7), where square brackets indicate the concentrations of the enclosed species in molarity (mol/L). (∞, p. 179)

For the reaction $HA(aq) + H_2O(l) \rightleftharpoons H_3O^+(aq) + A^-(aq)$

we have $K = \dfrac{[H_3O^+][A^-]}{[HA][H_2O]}$

Because water is a solvent for the reaction as well as a participant, its concentration is essentially constant and has no effect on the equilibrium. Therefore, we usually put the equilibrium constant K and the water concentration $[H_2O]$ together to make a new constant called the **acid dissociation constant, K_a**. The acid dissociation constant is simply the hydronium ion concentration $[H_3O^+]$ times the conjugate base concentration $[A^-]$ divided by the undissociated acid concentration $[HA]$.

Acid dissociation constant $K_a = K\,[H_2O] = \dfrac{[H_3O^+][A^-]}{[HA]}$

For a strong acid, the H_3O^+ and A^- concentrations are large, so K_a is also large. In fact, the K_a values for strong acids such as HCl are so large that it's difficult and not very useful to measure them. For a weak acid, however, the H_3O^+ and A^- concentrations are small, so K_a is small. Table 10.2 gives some K_a values for some common acids and illustrates several important points:

• Strong acids have K_a values much greater than 1 because dissociation is favored.

• Weak acids have K_a values much less than 1 because dissociation is not favored.

• Donation of each successive H^+ from a polyprotic acid is more difficult than the one before it, so K_a values become successively lower.

• Most organic acids, which contain the $-CO_2H$ group, have K_a values near 10^{-5}.

■ **PROBLEM 10.14**

Benzoic acid has $K_a = 6.5 \times 10^{-5}$, and citric acid has $K_a = 7.2 \times 10^{-4}$. Which of the two is the stronger acid?

TABLE 10.2   Some Acid Dissociation Constants, K_a, at 25°C

Acid	K_a	Acid	K_a
Hydrofluoric acid (HF)	3.5×10^{-4}	*Polyprotic acids*	
Hydrocyanic acid (HCN)	4.9×10^{-10}	Sulfuric acid	
Ammonium ion (NH_4^+)	5.6×10^{-10}	H_2SO_4	Large
		HSO_4^-	1.2×10^{-2}
Organic acids		Phosphoric acid	
Formic acid (HCOOH)	1.8×10^{-4}	H_3PO_4	7.5×10^{-3}
Acetic acid (CH_3COOH)	1.8×10^{-5}	$H_2PO_4^-$	6.2×10^{-8}
Propanoic acid	1.3×10^{-5}	HPO_4^{2-}	2.2×10^{-13}
(CH_3CH_2COOH)		Carbonic acid	
Ascorbic acid (vitamin C)	7.9×10^{-5}	H_2CO_3	4.3×10^{-7}
		HCO_3^-	5.6×10^{-11}

10.8 Dissociation of Water

Like all weak acids, pure water is slightly dissociated into ions. But because each dissociation yields one H_3O^+ ion and one OH^- ion, the concentrations of the two ions are identical. At 25°C, the concentration of each is 1.00×10^{-7} M.

$$H_2O(l) + H_2O(l) \rightleftharpoons H_3O^+(aq) + OH^-(aq)$$

so $$K = \frac{[H_3O^+][OH^-]}{[H_2O][H_2O]}$$

and $$K_a = K[H_2O] = \frac{[H_3O^+][OH^-]}{[H_2O]}$$

where $$[H_3O^+] = [OH^-] = 1.00 \times 10^{-7}\,M \;(\text{At } 25°C)$$

As both a reactant and a solvent, the concentration of water is essentially constant. We can therefore put the acid dissociation constant K_a and the water concentration $[H_2O]$ together to make a new constant called the **ion-product constant for water** (K_w), which is simply the H_3O^+ concentration times the OH^- concentration. At 25°C, $K_w = 1.00 \times 10^{-14}$.

Ion-product constant for water $$K_w = K_a[H_2O] = [H_3O^+][OH^-]$$
$$= (1.00 \times 10^{-7})(1.00 \times 10^{-7})$$
$$= 1.00 \times 10^{-14} \quad (\text{at } 25°C)$$

The importance of the equation $K_w = [H_3O^+][OH^-]$ is that it applies to all aqueous solutions, not just to pure water. Since the product of $[H_3O^+]$ times $[OH^-]$ is always constant for any solution, we can determine the concentration of one species if we know the concentration of the other. If an acid is present in solution, for instance, so that $[H_3O^+]$ is large, then $[OH^-]$ must be small. If a base is present in solution so that $[OH^-]$ is large, then $[H_3O^+]$ must be small. For example, for a 0.10 M HCl solution, we know that $[H_3O^+] = 0.10$ M because HCl is 100% dissociated. Thus, we can calculate that $[OH^-] = 1.0 \times 10^{-13}$ M.

Since $$K_w \times [H_3O^+][OH^-] = 1.00 \times 10^{-14}$$

then $$[OH^-] = \frac{K_w}{[H_3O^+]} = \frac{1.00 \times 10^{-14}}{0.10} = 1.0 \times 10^{-13}\,M$$

Similarly, for a 0.10 M NaOH solution, we know that $[OH^-] = 0.10$ M, so $[H_3O^+] = 1.0 \times 10^{-13}$ M:

$$[H_3O^+] = \frac{K_w}{[OH^-]} = \frac{1.00 \times 10^{-14}}{0.10} = 1.0 \times 10^{-13} \text{ M}$$

Solutions are identified as acidic, neutral, or basic (*alkaline*) according to the value of their H_3O^+ and OH^- concentrations:

Acidic solution: $[H_3O^+] > 10^{-7}$ M $[OH^-] < 10^{-7}$ M

Neutral solution: $[H_3O^+] = 10^{-7}$ M $[OH^-] = 10^{-7}$ M

Basic solution: $[H_3O^+] < 10^{-7}$ M $[OH^-] > 10^{-7}$ M

■ **WORKED EXAMPLE 10.4**

Milk has an H_3O^+ concentration of 4.5×10^{-7} M. What is the value of $[OH^-]$? Is milk acidic, neutral, or basic?

ANALYSIS   The OH^- concentration can be found by dividing K_w by $[H_3O^+]$. An acidic solution has $[H_3O^+] > 10^{-7}$ M, a neutral solution has $[H_3O^+] = 10^{-7}$ M, and a basic solution has $[H_3O^+] < 10^{-7}$ M.

SOLUTION   Milk is slightly acidic because its $[H_3O^+]$ is slightly larger than 1×10^{-7} M.

$$[OH^-] = \frac{K_w}{[H_3O^+]} = \frac{1.00 \times 10^{-14}}{4.5 \times 10^{-7}} = 2.2 \times 10^{-8} \text{ M}$$

■ **PROBLEM 10.15**

Identify the following solutions as either acidic or basic. What is the value of $[OH^-]$ in each?
(a) Beer, $[H_3O^+] = 3.2 \times 10^{-5}$ M
(b) Household ammonia, $[H_3O^+] = 3.1 \times 10^{-12}$ M

10.9 Measuring Acidity in Aqueous Solution: pH

In many fields, from medicine to chemistry to winemaking, it's necessary to know the exact concentration of H_3O^+ or OH^- in a solution. If, for example, the H_3O^+ concentration in blood varies only slightly from a value of 4.0×10^{-8} M, death can result.

Although correct, it's nevertheless awkward to refer to low concentrations of H_3O^+ using molarity. If you were asked which concentration is higher, 9.0×10^{-8} M or 3.5×10^{-7} M, you'd probably have to stop and think for a moment before answering. Fortunately, there's an easier way to express and compare H_3O^+ concentrations—the *pH scale*.

The pH of an aqueous solution is a number, usually between 0 and 14, that indicates the H_3O^+ concentration of the solution. A pH smaller than 7 corresponds to an acidic solution, a pH larger than 7 corresponds to a basic solution, and a pH of exactly 7 corresponds to a neutral solution. The pH scale and pH values of some common substances are shown in Figure 10.2.

Mathematically, the **pH** of a solution is defined as the negative common logarithm of the H_3O^+ concentration:

$$pH = -\log[H_3O^+]$$

pH A measure of the acid strength of a solution; the negative common logarithm of the H_3O^+ concentration.

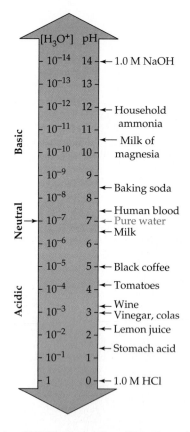

▲ **FIGURE 10.2   The pH scale and the pH values of some common substances.**   A low pH corresponds to a strongly acidic solution, a high pH corresponds to a strongly basic solution, and a pH of 7 corresponds to a neutral solution.

If you've studied logarithms, you might remember that the common logarithm of a number is the power to which 10 must be raised to equal the number. The pH definition can therefore be restated as

$$[H_3O^+] = 10^{-pH}$$

For example, in neutral water at 25°C, where $[H_3O^+] = 1 \times 10^{-7}$ M, the pH is 7; in a strong acid solution where $[H_3O^+] = 1 \times 10^{-1}$ M, the pH is 1; and in a strong base solution where $[H_3O^+] = 1 \times 10^{-14}$ M, the pH is 14.

Acidic solution: pH < 7 $[H_3O^+] > 1 \times 10^{-7}$ M

Neutral solution: pH = 7 $[H_3O^+] = 1 \times 10^{-7}$ M

Basic solution: pH > 7 $[H_3O^+] < 1 \times 10^{-7}$ M

Keep in mind that the pH scale covers an enormous range of acidities because it is a *logarithmic* scale, which involves powers of 10 (Figure 10.3). A change of only 1 pH unit means a tenfold change in $[H_3O^+]$; a change of 2 pH units means a hundredfold change in $[H_3O^+]$; and a change of 12 pH units means a change of 10^{12} (a million million) in $[H_3O^+]$.

It might help give you a feeling for the size of the quantities involved to think of a typical backyard swimming pool, which contains about 100,000 L of water. You would have to add only 0.10 mol of HCl (3.7 g) to lower the pH of the pool from 7.0 (neutral) to 6.0, but you would have to add 10,000 mol of HCl (370 kg!) to lower the pH of the pool from 7.0 to 1.0.

■ **WORKED EXAMPLE 10.5**

The H_3O^+ concentration in coffee is about 1×10^{-5} M. What pH is this?

ANALYSIS   The negative of the exponent of the H_3O^+ concentration gives the pH: $[H_3O^+] = 10^{-pH}$.

SOLUTION   Since the exponent of 1×10^{-5} M is -5, the pH is 5.

■ **WORKED EXAMPLE 10.6**

Lemon juice has a pH of about 2. What $[H_3O^+]$ is this?

ANALYSIS   The $[H_3O^+]$ is 10^{-pH}.

SOLUTION   Since pH = 2, $[H_3O^+] = 1 \times 10^{-2}$ M

■ **WORKED EXAMPLE 10.7**

A cleaning solution was found to have $[OH^-] = 1 \times 10^{-3}$ M. What is the pH?

ANALYSIS   To find pH, the value of $[H_3O^+]$ must first be found by using the equation: $[H_3O^+] = K_w/[OH^-]$.

SOLUTION

$$[H_3O^+] = \frac{K_w}{[OH^-]} = \frac{1.00 \times 10^{-14}}{1 \times 10^{-3}} = 1 \times 10^{-11} \text{ M}$$

The pH is therefore 11.

■ **WORKED EXAMPLE 10.8**

What is the pH of a 0.01 M solution of HCl?

ANALYSIS   To find pH, we must first find the value of $[H_3O^+]$.

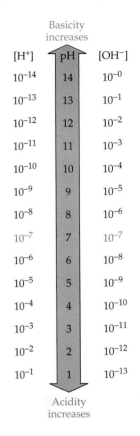

[H⁺]	pH	[OH⁻]
	Basicity increases	
10^{-14}	14	10^{-0}
10^{-13}	13	10^{-1}
10^{-12}	12	10^{-2}
10^{-11}	11	10^{-3}
10^{-10}	10	10^{-4}
10^{-9}	9	10^{-5}
10^{-8}	8	10^{-6}
10^{-7}	7	10^{-7}
10^{-6}	6	10^{-8}
10^{-5}	5	10^{-9}
10^{-4}	4	10^{-10}
10^{-3}	3	10^{-11}
10^{-2}	2	10^{-12}
10^{-1}	1	10^{-13}
	Acidity increases	

▲ **FIGURE 10.3   The relationship of the pH scale to H^+ and OH^- concentrations.**

▲   Adding only a teaspoonful of concentrated (6 M) hydrochloric acid would lower the pH of this pool from 7 to 6. Lowering the pH from 7 to 1 would take 400 gallons.

SOLUTION   Since HCl is a strong acid (Table 10.1), it is 100% dissociated and the H_3O^+ concentration is the same as the HCl concentration. $[H_3O^+] = 0.01$ M, or 1×10^{-2} M, and pH = 2.

■ **PROBLEM 10.16**

Which solution has the higher H_3O^+ concentration, one with pH = 5 or one with pH = 9? Which has the higher OH^- concentration?

■ **PROBLEM 10.17**

Give the pH of solutions with the following concentrations:
(a) $[H_3O^+] = 1 \times 10^{-5}$ M   **(b)** $[OH^-] = 1 \times 10^{-9}$ M

■ **PROBLEM 10.18**

Give the hydronium ion concentrations of solutions with the following values of pH. Which of the solutions is most acidic? Which is most basic?
(a) pH 13   **(b)** pH 3   **(c)** pH 8

■ **PROBLEM 10.19**

What is the pH of a 1×10^{-4} M solution of HNO_3?

10.10 Working with pH

Converting between pH and H_3O^+ concentration is easy when the pH is a whole number, but how do you find the H_3O^+ concentration of blood, which has a pH of 7.4, or the pH of a solution with $[H_3O^+] = 4.6 \times 10^{-3}$ M? Sometimes it's sufficient to make an estimate. The pH of blood (7.4) is between 7 and 8, so the H_3O^+ concentration of blood must be between 10^{-7} and 10^{-8} M. To be exact about finding pH values, though, requires a calculator.

Converting from pH to $[H_3O^+]$ requires finding the *antilogarithm* of the negative pH, which is done on many calculators with an "INV" key and a "log" key. Converting from $[H_3O^+]$ to pH requires finding the logarithm, which is commonly done with a "log" key and an "expo" or "EE" key for entering exponents of 10. Consult your calculator instructions if you're not sure how to use these keys. Remember that the sign of the number given by the calculator must be changed from minus to plus to get the pH.

The H_3O^+ concentration in blood with pH = 7.4 is

$$[H_3O^+] = \text{antilog}(-7.4) = 4 \times 10^{-8}$$

The pH of a solution with $[H_3O^+] = 4.6 \times 10^{-3}$ M is

$$pH = -\log(4.6 \times 10^{-3}) = -(-2.34) = 2.34$$

A note about significant figures: An antilogarithm contains the same number of digits that the original number has to the right of the decimal point. A logarithm contains the same number of digits to the right of the decimal point that the original number has:

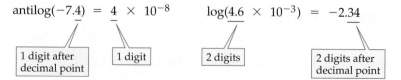

■ **WORKED EXAMPLE 10.9**

Soft drinks usually have a pH of approximately 3.1. What is the $[H_3O^+]$ concentration in a soft drink?

ANALYSIS Converting from a pH value to an $[H_3O^+]$ concentration requires finding an antilogarithm on a calculator.

SOLUTION Entering the negative pH on a calculator (-3.1) and pressing the "INV" and "log" keys gives the answer 7.943×10^{-4}, which must be rounded off to 8×10^{-4} since the pH has only one digit to the right of the decimal point.

■ **WORKED EXAMPLE 10.10**

What is the pH of a 0.0045 M solution of $HClO_4$?

ANALYSIS

Finding pH requires first finding $[H_3O^+]$.

SOLUTION

Since $HClO_4$ is a strong acid (Table 10.1), it is 100% dissociated, and the H_3O^+ concentration is the same as the $HClO_4$ concentration. Thus, $[H_3O^+] =$ 0.0045 M, or 4.5×10^{-3} M. Taking the negative logarithm gives pH = 2.35.

■ **PROBLEM 10.20**

Identify each of the following solutions as acidic or basic, estimate $[H_3O^+]$ values for each, and rank them in order of increasing acidity:

(a) Saliva, pH = 6.5   **(b)** Pancreatic juice, pH = 7.9
(c) Orange juice, pH = 3.7   **(d)** Wine, pH = 3.5

■ **PROBLEM 10.21**

Find the pH of the following solutions:

(a) Seawater with $[H_3O^+] = 5.3 \times 10^{-9}$ M
(b) A urine sample with $[H_3O^+] = 8.9 \times 10^{-6}$ M

■ **PROBLEM 10.22**

What is the pH of a 0.0025 M solution of HCl?

10.11 Laboratory Determination of Acidity

There are several ways to measure the pH of a solution. The simplest but least accurate method is to use an **acid–base indicator**, a dye that changes color depending on the pH of the solution. For example, the well-known dye *litmus* is red below pH 4.8 but blue above pH 7.8; the indicator *phenolphthalein* (fee-nol-**thay**-lein) is colorless below pH 8.2 but red above pH 10; and so on. To make pH determination particularly easy, a mixture of indicators known as *universal indicator* gives approximate pH measurements in the range 2–10 (Figure 10.4a). Also

Acid–base indicator A dye that changes color depending on the pH of a solution.

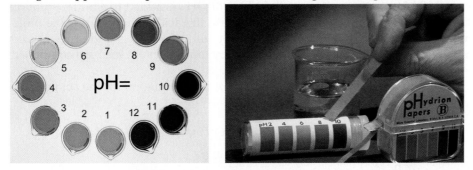

(a)   (b)

◀ **FIGURE 10.4 Finding pH.** (a) The color of universal indicator in solutions of known pH from 1 to 12. (b) Testing pH with a paper strip. Comparing the color of the strip with the code on the package gives the approximate pH.

Application pH of Body Fluids

ach fluid in our bodies must have a pH range suited to its function, as shown in the accompanying Table. The stability of cell membranes, the shapes of huge protein molecules that must be folded in certain ways to function, and the activities of enzymes are all dependent on appropriate H_3O^+ concentrations.

pH of Body Fluids

Fluid	pH
Blood plasma	7.4
Interstitial fluid	7.4
Cytosol	7.0
Saliva	5.8–7.1
Gastric juice	1.6–1.8
Pancreatic juice	7.5–8.8
Intestinal juice	6.3–8.0
Urine	4.6–8.0
Sweat	4.0–6.8

Let's take the body fluids listed in the Table in order. Blood plasma and the interstitial fluid surrounding cells, which together comprise one-third of body water, have a slightly basic pH of 7.4. In fact, one of the functions of blood is to neutralize the acid by-products of cellular metabolism. The fluid within cells, called the *cytosol*, is slightly more acidic than the fluid outside, so a pH differential exists.

The strongly acidic environment in the stomach has three important functions. First, it aids in the digestion of proteins by causing them to denature, or unfold. Second, it kills most of the bacteria we consume along with our food. Third, it converts the enzyme that breaks down proteins from an inactive form to the active form.

When the acidic mixture of partially digested food (*chyme*) leaves the stomach and enters the small intestine, it triggers secretion by the pancreas of an alkaline fluid containing bicarbonate ions. A principal function of this pancreatic juice and other fluids within the intestine is to dilute and neutralize the hydrochloric acid carried along from the stomach.

Urine has a wide normal pH range, depending on the diet and recent activities. It is generally acidic, though, because one important function of urine is to eliminate a quantity of hydrogen ion equal to that produced by the body each day. Without this elimination, the body would soon be overwhelmed by acid.

See Additional Problem 10.91 at the end of the chapter.

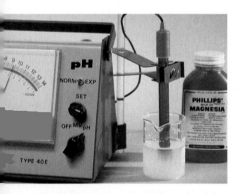

▲ **FIGURE 10.5 Using a pH meter to obtain an accurate reading of pH.** Is milk of magnesia acidic or basic?

Buffer A combination of substances that act together to prevent a drastic change in pH; usually a weak acid and its conjugate base.

available are rolls of "pH paper," which make it possible to determine pH simply by putting a drop of solution on the paper and comparing the color that appears to the color on a calibration chart (Figure 10.4b).

A much more accurate method of determining pH is to use an electronic pH meter like the one shown in Figure 10.5. Electrodes are dipped into the solution, and the pH is read from the meter.

10.12 Buffer Solutions

Much of the body's chemistry depends on maintaining the pH of blood and other fluids within narrow limits. This is accomplished through the use of **buffers**— combinations of substances that act together to prevent a drastic change in pH.

Most buffers are mixtures of a weak acid and a roughly equal concentration of its conjugate base—for example, a solution that contains 0.10 M acetic acid and 0.10 M acetate ion. If a small amount of OH^- is added to a buffer solution, the pH increases, but not by much because the acid component of the buffer neutralizes the added OH^-. If a small amount of H_3O^+ is added to a buffer solution, the pH decreases, but again not by much because the base component of the buffer neutralizes the added H_3O^+.

To see why buffer solutions work, look at the equation for the acid dissociation constant of an acid HA.

For the reaction: $HA(aq) + H_2O(l) \rightleftharpoons A^-(aq) + H_3O^+(aq)$

we have $K_a = \dfrac{[H_3O^+][A^-]}{[HA]}$

Rearranging this equation shows that the value of $[H_3O^+]$, and thus the pH, depends on the ratio of the undissociated acid concentration to the conjugate base concentration, $[HA]/[A^-]$:

$$[H_3O^+] = K_a \frac{[HA]}{[A^-]}$$

In the case of the acetic acid–acetate ion buffer, for instance, we have

$$[H_3O^+] = K_a \frac{[CH_3CO_2H]}{[CH_3CO_2^-]}$$

Initially, the pH of the 0.10 M acetic acid–0.10 M acetate ion buffer solution is 4.74. When acid is added, most is removed by reaction with $CH_3CO_2^-$, so the concentration of CH_3CO_2H increases and the concentration of $CH_3CO_2^-$ decreases. But as long as the changes in $[CH_3CO_2H]$ and $[CH_3CO_2^-]$ are relatively small, the ratio of $[CH_3CO_2H]$ to $[CH_3CO_2^-]$ doesn't change much and there is little change in the pH. When base is added to the buffer, most is removed by reaction with CH_3CO_2H, so the concentration of CH_3CO_2H decreases and the concentration of $CH_3CO_2^-$ increases. Here too, though, as long as the concentration changes are relatively small, there is little change in the pH.

The ability of a buffer solution to resist changes in pH when acid or base is added is illustrated in Figure 10.6. Addition of 0.010 mol of H_3O^+ to 1.0 L of pure water changes the pH from 7 to 2, and addition of 0.010 mol of OH^- changes the pH from 7 to 12. A similar addition of acid to 1.0 L of a 0.10 M acetic acid–0.10 M acetate ion buffer, however, changes the pH only from 4.74 to 4.68, and addition of base changes the pH only from 4.74 to 4.85.

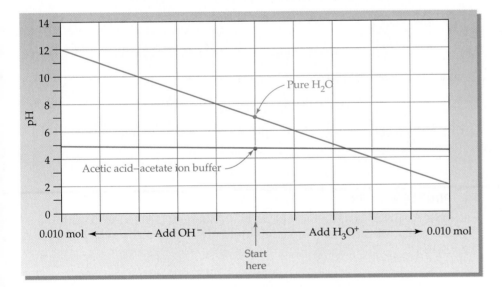

◀ FIGURE 10.6 A comparison of the change in pH when 0.010 mol of acid and 0.010 mol of base are added to 1.0 L of pure water and to 1.0 L of a 0.10 M acetic acid–0.10 M acetate ion buffer. The pH of the water varies between 12 and 2, while the pH of the buffer varies only between 4.85 and 4.68.

The effective pH range of a buffer depends on the K_a of the acid HA and on the relative concentrations of HA and conjugate base A^-. In general, the most effective buffers meet the following conditions:

- The ratio of [HA] to $[A^-]$ should be close to 1, so that neither additional acid nor additional base will change the pH of the solution dramatically.
- The amounts of [HA] and $[A^-]$ in the buffer should be approximately 10 times greater than the amounts of either acid or base you expect to add so that the $[A^-]/[HA]$ ratio does not undergo a large change.

■ WORKED EXAMPLE 10.11

What is the pH of a buffer solution that contains 0.100 M HF and 0.100 M NaF? The K_a of HF is 3.5×10^{-4}.

ANALYSIS Finding the pH of a solution requires finding the $[H_3O^+]$ concentration, which can be calculated from the relationship $[H_3O^+] = K_a [HA]/[A^-]$.

SOLUTION

$$[H_3O^+] = K_a\frac{[HA]}{[A^-]} = (3.5 \times 10^{-4})\frac{[0.100]}{[0.100]} = 3.5 \times 10^{-4} \text{ M}$$

Taking the negative logarithm of $[H_3O^+]$ then gives the pH:

$$pH = -\log(3.5 \times 10^{-4}) = 3.46$$

■ WORKED EXAMPLE 10.12

What is the pH of 1.00 L of the 0.100 M hydrofluoric acid–0.100 M fluoride ion buffer system described in Worked Example 10.11 after 0.020 mol of NaOH is added?

ANALYSIS Initially, the 0.100 M HF–0.100 M NaF buffer has pH = 3.46, as calculated in Worked Example 10.11. When 0.020 mol of NaOH is added to 1.00 L of the buffer, the HF concentration *decreases* from 0.100 M to 0.080 M as a result of an acid–base reaction. At the same time, the F^- concentration *increases* from 0.100 M to 0.120 M because additional F^- is produced by the neutralization:

$$HF(aq) + OH^-(aq) \longrightarrow H_2O(l) + F^-(aq)$$

With the concentrations of HF and F^- known, $[H_3O^+]$ can be calculated.

SOLUTION

$$pH = -\log(2.3 \times 10^{-4}) = 3.6$$

Taking the negative logarithm of $[H_3O^+]$ gives the pH:

$$pH = -\log(2.3 \times 10^{-4}) = 3.64$$

The addition of 0.020 mol of base causes the pH of the buffer to rise only from 3.46 to 3.64.

■ PROBLEM 10.23

What is the pH of 1.00 L of the 0.100 M hydrofluoric acid–0.100 M fluoride ion buffer system described in Worked Example 10.11 after 0.020 mol of HNO_3 is added?

■ PROBLEM 10.24

What is the pH of a buffer system that contains 0.050 M formic acid and 0.060 M sodium formate? The K_a of formic acid is 1.8×10^{-4}.

■ PROBLEM 10.25

A mixture of HCl and NaCl is a very poor buffer system, but a mixture of HF and NaF is a good buffer. Explain, taking into account the relative acid and base strengths of the components in each system.

10.13 Buffers in the Body

The pH of body fluids is maintained by three major buffer systems. Two of these buffers, the carbonic acid–bicarbonate system and the dihydrogen phosphate–hydrogen phosphate system, depend on weak acid–conjugate base interactions exactly like those of the acetate buffer system described in the preceding section.

$$H_2CO_3(aq) + H_2O(l) \rightleftharpoons HCO_3^-(aq) + H_3O^+(aq)$$

$$H_2PO_4^-(aq) + H_2O(l) \rightleftharpoons HPO_4^{2-}(aq) + H_3O^+(aq)$$

The third buffer system depends on the ability of proteins to act as either proton acceptors or proton donors at different pH values.

To illustrate the action of buffers in the body, let's look at the carbonic acid–bicarbonate system, the principal buffer in blood serum and other extracellular fluids. (The hydrogen phosphate system is the major buffer within cells.) Because carbonic acid is unstable and therefore in equilibrium with CO_2 and water, there is an extra step in the bicarbonate buffer mechanism:

$$CO_2(aq) + H_2O(l) \rightleftharpoons H_2CO_3(aq) \rightleftharpoons HCO_3^-(aq) + H_3O^+(aq)$$

As a result, the bicarbonate buffer system is intimately related to the elimination of CO_2, which is continuously produced in cells and transported to the lungs to be exhaled.

Because most CO_2 is present simply as the dissolved gas rather than as H_2CO_3, the acid dissociation constant for carbonic acid in blood can be written using $[CO_2]$:

$$K_a = \frac{[H_3O^+][HCO_3^-]}{[CO_2]}$$

which can be rearranged to

$$[H_3O^+] = K_a \frac{[CO_2]}{[HCO_3^-]}$$

An increase in $[CO_2]$ raises $[H_3O^+]$ and lowers pH.

A decrease in $[CO_2]$ lowers $[H_3O^+]$ and raises pH.

This rearranged equation shows that an increase in $[CO_2]$ makes the ratio of $[CO_2]$ to $[HCO_3^-]$ larger, thereby increasing $[H_3O^+]$ and decreasing the pH; that is, the blood becomes more acidic. Similarly, a decrease in $[CO_2]$ makes the ratio of $[CO_2]$ to $[HCO_3^-]$ smaller, thereby decreasing $[H_3O^+]$ and increasing the pH; that is, the blood becomes more basic. At the normal blood pH of 7.4, the $[CO_2]/[HCO_3^-]$ ratio is about 1 to 20.

The relationships between the bicarbonate buffer system, the lungs, and the kidneys are shown in Figure 10.7. Under normal circumstances the reactions shown in the figure are at equilibrium. Addition of excess acid (red arrows) causes formation of H_2CO_3 and results in lowering of H_3O^+ concentration. Removal of acid (blue arrows) causes formation of more H_3O^+ by dissociation of H_2CO_3. The maintenance of pH by this mechanism is supported by a reserve of bicarbonate ions in body fluids. Such a buffer can accommodate large additions of H_3O^+ before there is a significant change in the pH, a condition that meets the body's needs because excessive production of acid is a more common body condition than excessive loss of acid.

FIGURE 10.7 Relationships of ▶ the bicarbonate buffer system to the lungs and the kidneys. The red and blue arrows show the responses to the stresses of increased or decreased respiratory rate and removal or addition of acid.

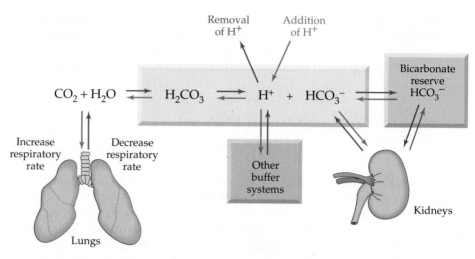

A change in the breathing rate provides a quick further adjustment in the bicarbonate buffer system. When the CO_2 concentration in the blood starts to rise, the breathing rate increases to remove CO_2, thereby decreasing the acid concentration (red arrows in Figure 10.7). When the CO_2 concentration in the blood starts to fall, the breathing rate decreases and acid concentration increases (blue arrows in Figure 10.7).

Additional backup to the bicarbonate buffer system is provided by the kidneys. Each day a quantity of acid equal to that produced in the body is excreted in the urine. In the process, the kidney returns HCO_3^- to the extracellular fluids, where it becomes part of the bicarbonate reserve.

◯◯◯ LOOKING AHEAD

We'll see in Chapter 28 that the regulation of blood pH by the bicarbonate buffer system is particularly important in preventing the medical conditions called *acidosis* and *alkalosis*. Acidosis, which results when an increase in CO_2 causes blood pH to drop below 7.35, can be brought on by the difficulty in breathing that accompanies asthma and emphysema. Alkalosis, the condition that results when blood pH rises above 7.45, arises either from heavy breathing (hyperventilation) that removes large amounts of CO_2 from the blood or from conditions that raise the concentration of bicarbonate ions.   ◯◯◯

10.14 Acid and Base Equivalents

We said in Section 9.10 that it's sometimes useful to think in terms of ion *equivalents* when we're primarily interested in the ion itself rather than the compound that produced the ion. (◯◯◯, p. 244) For similar reasons, it's also useful sometimes to consider acid or base equivalents.

One equivalent (Eq) of an acid is equal to the molar mass of the acid divided by the number of H^+ ions produced per formula unit. Thus, one equivalent of an acid is the weight in grams that can donate one mole of H^+ ions. Similarly, one equivalent of a base is the weight in grams that can produce one mole of OH^- ions:

$$\text{One equivalent of acid} = \frac{\text{Molar mass of acid (g)}}{\text{Number of } H^+ \text{ ions produced}}$$

$$\text{One equivalent of base} = \frac{\text{Molar mass of base (g)}}{\text{Number of } OH^- \text{ ions produced}}$$

One equivalent of the monoprotic acid HCl is 36.5 g, the molar mass of the acid, but one equivalent of the diprotic acid H_2SO_4 is 49.0 g, the molar mass of the acid (98.0 g) divided by 2 since one mole of H_2SO_4 can produce two moles of H^+.

$$\text{One equivalent of } H_2SO_4 = \frac{\text{Molar mass of } H_2SO_4}{2} = \frac{98.0 \text{ g}}{2} = 49.0 \text{ g}$$

Divide by 2 because H_2SO_4 is diprotic.

Using acid–base equivalents has two practical advantages: First, they are convenient when only the acidity or basicity of a solution is of interest rather than the identity of the acid or base. Second, they show quantities that are chemically equivalent in their properties: 36.5 g of HCl and 49.0 g of H_2SO_4 are chemically equivalent quantities because each reacts with one equivalent of base. *One equivalent of any acid neutralizes one equivalent of any base.*

Because acid–base equivalents are so useful, clinical chemists sometimes express acid and base concentrations in *normality* rather than molarity. The **normality (N)** of an acid or base solution is defined as the number of equivalents (or milliequivalents) of acid or base per liter of solution. For example, a solution made by dissolving 1.0 equivalent (49.0 g) of H_2SO_4 in water to give 1.0 L of solution has a concentration of 1.0 Eq/L, or 1.0 N. Similarly, a solution that contains 0.010 Eq/L of acid is 0.010 N and has an acid concentration of 10 mEq/L.

$$\text{Normality (N)} = \frac{\text{Equivalents of acid or base}}{\text{Liters of solution}}$$

The values of molarity (M) and normality (N) are the same for monoprotic acids such as HCl but are not the same for diprotic or triprotic acids. A solution made by diluting 1.0 equivalent (49.0 g, 0.50 mol) of the diprotic acid H_2SO_4 to a volume of 1.0 L has a *normality* of 1.0 Eq/L, or 1.0 N, but a *molarity* of 0.50 M. For any acid or base, normality is always equal to molarity times the number of H^+ or OH^- ions produced per formula unit.

Normality of acid = (Molarity of acid) × (Number of H^+ ions produced)
Normality of base = (Molarity of base) × (Number of OH^- ions produced)

■ **WORKED EXAMPLE 10.13**

How many equivalents are in 3.1 g of the diprotic acid H_2S? H_2S has MW = 34.0 amu.

ANALYSIS   The number of acid or base equivalents is calculated by doing a gram-to-mole conversion using molar mass as the conversion factor and then multiplying by the number of H^+ ions produced.

SOLUTION

$$(3.1 \text{ g } H_2S)\left(\frac{1 \text{ mol } H_2S}{34.0 \text{ g } H_2S}\right)\left(\frac{2 \text{ Eq } H_2S}{1 \text{ mol } H_2S}\right) = 0.18 \text{ Eq } H_2S$$

■ **WORKED EXAMPLE 10.14**

What is the normality of a solution made by diluting 6.5 g of H_2SO_4 to a volume of 200 mL? What is the concentration of this solution in milliequivalents per liter? H_2SO_4 has MW = 98.0 amu.

ANALYSIS   Calculate how many equivalents of H_2SO_4 are in 6.5 g by using the molar mass of the acid, 98.0 g/mol, as a conversion factor, and then determine the normality of the acid:

SOLUTION

$$(6.5 \text{ g } H_2SO_4)\left(\frac{1 \text{ mol } H_2SO_4}{98.0 \text{ g } H_2SO_4}\right)\left(\frac{2 \text{ Eq } H_2SO_4}{1 \text{ mol } H_2SO_4}\right) = 0.13 \text{ Eq } H_2SO_4$$

$$\left(\frac{0.13 \text{ Eq } H_2SO_4}{200 \text{ mL}}\right)\left(\frac{1000 \text{ mL}}{1 \text{ L}}\right) = 0.65 \text{ Eq } H_2SO_4/L = 0.65 \text{ N}$$

The concentration of the sulfuric acid solution is 0.65 N, or 650 mEq/L.

■ **PROBLEM 10.26**

How many equivalents are in each of the following?
(a) 5.0 g HNO_3 **(b)** 12.5 g $Ca(OH)_2$ **(c)** 4.5 g H_3PO_4

■ **PROBLEM 10.27**

What would the normalities of the solutions be if each of the samples in Problem 10.26 were dissolved in water and diluted to a volume of 300.0 mL?

10.15 Titration

Determining the pH of a solution gives the solution's H_3O^+ concentration but not necessarily its total acid concentration. That's because the two aren't the same thing. The H_3O^+ concentration gives only the amount of acid that has dissociated into ions, while total acid concentration gives the sum of dissociated plus undissociated acid. In a 0.10 M solution of acetic acid, for instance, the total acid concentration is 0.10 M, yet the H_3O^+ concentration is only 0.0013 M (pH = 2.89) because acetic acid is a weak acid that is only about 1% dissociated.

The total acid or base concentration of a solution can be found by carrying out a **titration** procedure, as shown in Figure 10.8. Let's assume, for instance, that we have an HCl solution whose acid concentration we want to find. (We could equally well have an NaOH solution whose base concentration we want to find.) We begin by measuring out a known volume of the HCl solution and adding an acid–base indicator. Next, we fill a calibrated glass tube called a *buret* with an NaOH solution of known concentration, and we slowly add the NaOH

Titration A procedure for determining the total acid or base concentration of a solution.

FIGURE 10.8 Titration of an ▶ acid solution of unknown concentration with a base solution of known concentration. (a) A measured volume of the acid solution is placed in the flask along with an indicator. (b) Base of known concentration is then added from a buret until the color change of the indicator shows that neutralization is complete (the *end point*).

 Acid-Base Titration

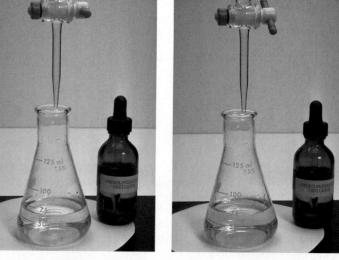

(a)   (b)

to the HCl until neutralization is complete (the *end point*), as signalled by a color change in the indicator.

Reading from the buret gives the volume of the NaOH solution that has reacted with the known volume of HCl. Knowing both the concentration and volume of the NaOH solution then lets us calculate the molar amount of NaOH, and the coefficients in the balanced equation let us find the molar amount of HCl that has been neutralized. Dividing the molar amount of HCl by the volume of the HCl solution gives the concentration. The calculation thus involves mole–volume conversions just like those done previously in Section 9.7. (∞, p. 239) Figure 10.9 shows a flow diagram of the strategy, and Worked Example 10.15 gives an example calculation.

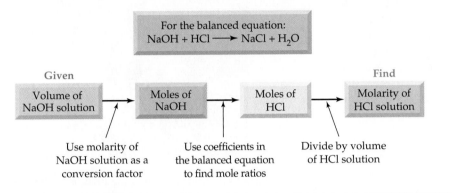

◀ **FIGURE 10.9 A flow diagram for an acid–base titration.** This diagram summarizes the calculations needed to determine the concentration of an HCl solution by titration with an NaOH solution of known concentration. The steps are similar to those shown previously in Figure 9.6.

■ **WORKED EXAMPLE 10.15**

When a 5.00 mL sample of household vinegar (dilute aqueous acetic acid) was titrated, 44.5 mL of 0.100 M NaOH solution was required to reach the end point. What is the acid concentration of vinegar in molarity, normality, and milliequivalents per liter? The neutralization reaction is

$$CH_3CO_2H(aq) + NaOH(aq) \longrightarrow CH_3CO_2^-Na^+(aq) + H_2O(l)$$

ANALYSIS To find the molarity of the vinegar, we need to know the number of moles of acetic acid dissolved in the 5.00 mL sample. Following the flow diagram in Figure 10.9, we first find the number of moles of NaOH, then find the number of moles of acid, and then divide by the volume of the acid solution. Because acetic acid is a monoprotic acid, the normality of the solution is numerically the same as its molarity.

SOLUTION

$$\text{Moles } CH_3CO_2H = (44.5 \text{ mL NaOH})\left(\frac{0.100 \text{ mol NaOH}}{1000 \text{ mL}}\right)\left(\frac{1 \text{ mol } CH_3CO_2H}{1 \text{ mol NaOH}}\right)$$

$$= 0.004\,45 \text{ mol } CH_3CO_2H$$

$$\text{Concentration of } CH_3CO_2H = \frac{0.004\,45 \text{ mol } CH_3CO_2H}{0.005\,00 \text{ L}} = 0.890 \text{ M} = 0.890 \text{ N}$$

The acid concentration is 890 mEq/L.

■ **PROBLEM 10.28**

To determine the concentration of the acid in an old bottle of aqueous HCl whose label had become unreadable, a titration was carried out. What is the HCl concentration if 58.4 mL of 0.250 M NaOH was required to titrate a 20.0 mL sample of the acid?

■ **PROBLEM 10.29**

How many milliliters of 0.150 M NaOH are required to neutralize 50.0 mL of 0.200 M H_2SO_4?

Application Acid Rain

As the water that has evaporated from oceans and lakes condenses into raindrops, it dissolves small quantities of gases from the atmosphere. Under normal conditions, rain is slightly acidic, with a pH close to 5.6, because of dissolved CO_2. In recent decades, however, the acidity of rainwater in many industrialized areas of the world has increased by a factor of over 100, to a pH between 3 and 3.5.

The primary cause of this so-called *acid rain* is industrial and automotive pollution. Each year, large power plants and smelters—particularly those located in developing countries and former Eastern Bloc countries—pour millions of tons of sulfur dioxide (SO_2) gas into the atmosphere, where some is oxidized by air to produce sulfur trioxide (SO_3). Sulfur oxides then dissolve in rain to form dilute sulfurous acid and sulfuric acid:

$$SO_2(g) + H_2O(l) \longrightarrow H_2SO_3(aq)$$

Sulfurous acid

$$SO_3(g) + H_2O(l) \longrightarrow H_2SO_4(aq)$$

Sulfuric acid

▲ This limestone statue adorning the Rheims Cathedral in France has been severely eroded by acid rain.

Nitrogen oxides produced by the high-temperature reaction of N_2 with O_2 in coal-burning plants and in automobile engines make a further contribution to the problem. Nitrogen dioxide (NO_2) dissolves in water to form dilute nitric acid (HNO_3) and nitric oxide (NO):

$$3\,NO_2(g) + H_2O(l) \longrightarrow 2\,HNO_3(aq) + NO(g)$$

Oxides of both sulfur and nitrogen have always been present in the atmosphere, produced by such natural sources as volcanoes and lightning bolts, but their amounts have increased dramatically over the last century because of industrialization.

Many processes in nature require such a fine pH balance that they are dramatically upset by the shift that has occurred in the pH of rain. Thousands of lakes in the Adirondack region of upper New York State and in southeastern Canada have become so acidic that all fish life has disappeared. Massive tree die-offs have occurred throughout central and eastern Europe as acid rain has lowered the pH of the soil and has leached nutrients from leaves. Countless marble statues have been slowly dissolved away as their calcium carbonate has been attacked by acid rain.

$$CaCO_3(s) + 2\,H^+(aq) \longrightarrow$$
$$Ca^{2+}(aq) + H_2O(l) + CO_2(g)$$

Fortunately, acidic emissions from automobiles and power plants have been greatly reduced in recent years, particularly in the United States and Canada. Nitrogen oxide emissions have been lowered by equipping automobiles with catalytic converters (Section 7.5), which catalyze the decomposition of nitrogen oxides to N_2 and O_2. Sulfur dioxide emissions from power plants have been reduced by *scrubbing* combustion products before they are emitted from plant smoke stacks. The process involves addition of an aqueous suspension of lime (CaO) to the combustion chamber and the stack, where it reacts with SO_2 to give calcium sulfite ($CaSO_3$)

Unfortunately, scrubbers are expensive, and the $CaSO_3$, which has no commercial uses, must be disposed of in landfills. Much more work on methods to control acidic emissions remains to be done because the problem will grow more serious as sources of low-sulfur coal are exhausted and power plants are forced to rely on more abundant sources of high-sulfur coal.

See Additional Problems 10.92 and 10.93 at the end of the chapter.

■ **PROBLEM 10.30**

A 21.5 mL sample of a KOH solution of unknown concentration required 16.1 mL of 0.150 M H_2SO_4 solution to reach the end point in a titration. What is the molarity of the KOH solution?

10.16 Acidity and Basicity of Salt Solutions

It's tempting to think of all salt solutions as neutral; after all, they come from the neutralization reaction between an acid and a base. In fact salt solutions can be neutral, acidic or basic, depending on the ions present, because some ions react with water to produce H_3O^+ and some ions react with water to produce OH^-. To predict the acidity of a salt solution, it's convenient to classify salts according to the acid and base from which they would be formed in a neutralization reaction. The classification and some examples are given in Table 10.3.

TABLE 10.3   Acidity and Basicity of Salt Solutions

Acid That Anion Is From	Base That Cation Is From	Solution	Example
Strong	Weak	Acidic	NH_4Cl, NH_4NO_3
Weak	Strong	Basic	$NaHCO_3$, $K(CH_3CO_2)$
Strong	Strong	Neutral	$NaCl$, KBr, $Ca(NO_3)_2$
Weak	Weak	More information needed	

The general rule for predicting the acidity or basicity of a salt solution is that the stronger partner from which the salt was formed dominates. That is, a salt formed from a strong acid and a weak base yields an acidic solution because the strong acid dominates; a salt formed from a weak acid and a strong base yields a basic solution because the base dominates; and a salt formed from a strong acid and a strong base yields a neutral solution because neither acid nor base dominates. Let's look at some examples.

Salt of Strong Acid + Weak Base → Acidic Solution

A salt such as NH_4Cl, which can be formed by reaction of a strong acid (HCl) with a weak base (NH_3), yields an acidic solution. The Cl^- ion does not react with water, but the NH_4^+ ion is a weak acid that gives H_3O^+ ions:

$$NH_4^+(aq) + H_2O(l) \rightleftharpoons NH_3(aq) + H_3O^+(aq)$$

Salt of Weak Acid + Strong Base → Basic Solution

A salt such as sodium bicarbonate, which can be formed by reaction of a weak acid (H_2CO_3) with a strong base (NaOH), yields a basic solution. The Na^+ ion does not react with water, but the HCO_3^- ion is a weak base that gives OH^- ions:

$$HCO_3^-(aq) + H_2O(l) \rightleftharpoons H_2CO_3(aq) + OH^-(aq)$$

Salt of Strong Acid + Strong Base → Neutral Solution

A salt such as NaCl, which can be formed by reaction of a strong acid (HCl) with a strong base (NaOH), yields a neutral solution. Neither the Cl^- ion nor the Na^+ ion react with water.

Salt of Weak Acid + Weak Base

Both cation and anion in this type of salt react with water, so we can't predict whether the resulting solution will be acidic or basic without quantitative information. The ion that reacts to the greater extent with water will govern the pH—it may be either the cation or the anion.

■ **WORKED EXAMPLE 10.16**

Predict whether each of the following salts produces an acidic, basic, or neutral solution:

(a) $BaCl_2$

(b) NaCN

(c) NH_4NO_3

ANALYSIS Look in Table 10.1 to see the classification of acids and bases as strong or weak.

SOLUTION

(a) $BaCl_2$ gives a neutral solution because it is formed from a strong acid (HCl) and a strong base [$Ba(OH)_2$].

(b) NaCN gives a basic solution because it is formed from a weak acid (HCN) and a strong base (NaOH).

(c) NH_4NO_3 gives an acidic solution because it is formed from a strong acid (HNO_3) and a weak base (NH_3).

■ **PROBLEM 10.31**

Predict whether each of the following salts produces an acidic, basic, or neutral solution:

(a) K_2SO_4

(b) Na_2HPO_4

(c) MgF_2

(d) NH_4Br

Summary: Revisiting the Chapter Goals

1. What are acids and bases? According to the *Brønsted–Lowry definition*, an acid is a substance that donates a hydrogen ion (a proton, H^+) and a base is a substance that accepts a hydrogen ion. Thus, the generalized reaction of an acid with a base involves the reversible transfer of a proton:

$$B: + H—A \rightleftharpoons A:^- + H—B^+$$

In aqueous solution, water acts as a base and accepts a proton from an acid to yield a *hydronium ion*, H_3O^+. Reaction of an acid with a metal hydroxide, such as KOH, yields water and a salt; reaction with bicarbonate ion (HCO_3^-) or carbonate ion (CO_3^{2-}) yields water, a salt, and CO_2 gas; and reaction with ammonia yields an ammonium salt.

2. What is the influence of acid and base strengths on their reactions? Different acids and bases differ in their ability to give up or accept a proton. A *strong acid* gives up a proton easily and is 100% *dissociated* in aqueous solution; a *weak acid* gives up a proton with difficulty, is only slightly dissociated in water, and establishes an equilibrium between dissociated and undissociated forms. Similarly, a *strong base* accepts and holds a proton readily, whereas a *weak base* has a low affinity for a proton and establishes an equilibrium in aqueous solution. The two substances that are related by the gain or loss of a proton are called a *conjugate acid–base pair*. The exact strength of an acid is defined by an *acid dissociation constant*, K_a:

For the reaction $HA + H_2O \rightleftharpoons H_3O^+ + A^-$

we have $K_a = \dfrac{[H_3O^+][A^-]}{[HA]}$

A proton-transfer reaction always takes place in the direction that favors formation of the weaker acid.

3. What is the ion-product constant for water? Water is *amphoteric*; that is, it can act as either an acid or a base. Water also dissociates slightly into H_3O^+ ions and OH^- ions, the product of whose concentrations in any aqueous solution is the *ion-product constant for water*, $K_w = [H_3O^+][OH^-] = 1.00 \times 10^{-14}$ at 25°C.

4. What is the pH scale for measuring acidity? The acidity or basicity of an aqueous solution is given by its *pH*, defined as the negative logarithm of the hydronium ion concentration, $[H_3O^+]$. A pH below 7 means an acidic solution; a pH equal to 7 means a neutral solution; and a pH above 7 means a basic solution.

5. What is a buffer? The pH of a solution can be controlled through the use of a *buffer* that acts to remove either added H^+ ions or added OH^- ions. Most buffer solutions consist of roughly equal amounts of a weak acid and its conjugate base. The bicarbonate buffer present in blood and the hydrogen phosphate buffer present in cells are particularly important examples.

6. How is the acid or base concentration of a solution determined? Acid (or base) concentrations are determined in the laboratory by *titration* of a solution of unknown concentration with a base (or acid) solution of known strength until an indicator signals that neutralization is complete.

Key Words

Acid dissociation constant(K_a), p. 274

Acid–base indicator, p. 279

Amphoteric, p. 268

Brønsted–Lowry acid, p. 265

Brønsted–Lowry base, p. 266

Buffer, p. 280

Conjugate acid, p. 267

Conjugate acid–base pair, p. 267

Conjugate base, p. 267

Dissociation, p. 271

Hydronium ion, p. 263

Ion-product constant for water(K_w), p. 275

Normality (N), p. 285

pH, p. 276

Strong acid, p. 271

Strong base, p. 271

Titration, p. 286

Weak acid, p. 271

Weak base, p. 271

■ Understanding Key Concepts

10.32 Identify the Brønsted–Lowry acid and base in each of the following reactions:

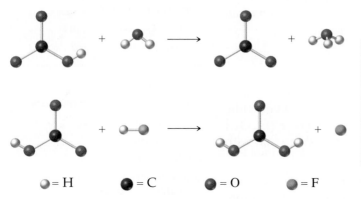

◯ = H   ● = C   ● = O   ● = F

10.33 Assume that an aqueous solution of OH^-, represented as a blue sphere, is allowed to mix with a solution of an acid H_nA, represented as a red sphere. Three possible outcomes are depicted by boxes (1)–(3), where the green spheres represent A^{n-}, the anion of the acid:

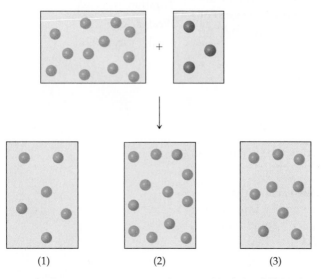

Which outcome corresponds to each of the following reactions?
(a) $HF + OH^- \rightarrow H_2O + F^-$
(b) $H_2SO_3 + 2\,OH^- \rightarrow 2\,H_2O + SO_3^{2-}$
(c) $H_3PO_4 + 3\,OH^- \rightarrow 3\,H_2O + PO_4^{3-}$

10.34 Electrostatic potential maps of acetic acid (CH_3CO_2H) and ethyl alcohol (CH_3CH_2OH) are shown. Identify the most acidic hydrogen in each, and tell which of the two is likely to be the stronger acid.

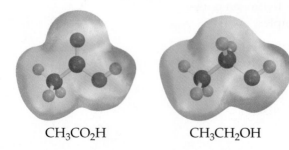

CH_3CO_2H CH_3CH_2OH

10.35 The following pictures represent aqueous acid solutions. Water molecules are not shown.

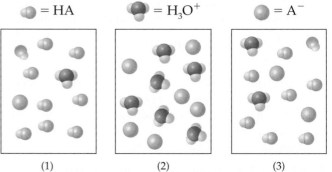

◐ = HA   ● = H_3O^+ ◯ = A^-

(1)   (2)   (3)

(a) Which picture represents the weakest acid?
(b) Which picture represents the strongest acid?
(c) Which picture represents the acid with the smallest value of K_a?

10.36 The following pictures represent aqueous solutions of a diprotic acid H_2A. Water molecules are not shown.

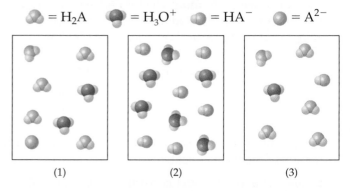

= H_2A = H_3O^+ = HA^- = A^{2-}

(1)   (2)   (3)

(a) Which picture represents a solution of a weak diprotic acid?
(b) Which picture represents an impossible situation?

10.37 Assume that the red spheres in the buret represent H^+ ions, the blue spheres in the flask represent OH^- ions, and you are carrying out a titration of the base with the acid. If the volumes in the buret and the flask are identical and the concentration of the acid in the buret is 1.00 M, what is the concentration of the base in the flask?

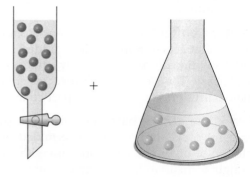

▪ Additional Problems

Acids and Bases

10.38 What happens when a strong acid such as HBr is dissolved in water?

10.39 What happens when a weak acid such as CH_3CO_2H is dissolved in water?

10.40 What happens when a strong base such as KOH is dissolved in water?

10.41 What happens when a weak base such as NH_3 is dissolved in water?

10.42 What is the difference between a monoprotic acid and a diprotic acid? Give an example of each.

10.43 What is the difference between H^+ and H_3O^+?

10.44 Which of the following are strong acids? Look at Table 10.1 if necessary.

(a) $HClO_4$ (b) H_2CO_3
(c) H_3PO_4 (d) NH_4^+
(e) HI   (f) $H_2PO_4^-$

10.45 Which of the following are weak bases? Look at Table 10.1 if necessary.

(a) NH_3 (b) $Ca(OH)_2$
(c) HPO_4^{2-} (d) LiOH
(e) CN^- (f) NH_2^-

Brønsted–Lowry Acids and Bases

10.46 Identify each of the following substances as a Brønsted–Lowry base, a Brønsted–Lowry acid, or neither:

(a) HCN   (b) $CH_3CO_2^-$
(c) $AlCl_3$ (d) H_2CO_3
(e) Mg^{2+} (f) $CH_3NH_3^+$

10.47 Label the Brønsted–Lowry acids and bases in the following equations, and tell which substances are conjugate acid–base pairs.

(a) $CO_3^{2-}(aq) + HCl(aq) \rightarrow HCO_3^-(aq) + Cl^-(aq)$
(b) $H_3PO_4(aq) + NH_3(aq) \rightarrow$
$$H_2PO_4^-(aq) + NH_4^+(aq)$$
(c) $NH_4^+(aq) + CN^-(aq) \rightleftarrows NH_3(aq) + HCN(aq)$
(d) $HBr(aq) + OH^-(aq) \rightarrow H_2O(l) + Br^-(aq)$
(e) $H_2PO_4^-(aq) + N_2H_4(aq) \rightleftarrows$
$$HPO_4^{2-}(aq) + N_2H_5^+(aq)$$

10.48 Write the formulas of the conjugate acids of the following Brønsted–Lowry bases:

(a) $ClCH_2CO_2^-$ (b) C_5H_5N
(c) SeO_4^{2-} (d) $(CH_3)_3N$

10.49 Write the formulas of the conjugate bases of the following Brønsted–Lowry acids:

(a) HCN   (b) $(CH_3)_2NH_2^+$
(c) H_3PO_4 (d) $HSeO_3^-$

10.50 The hydrogen-containing anions of many polyprotic acids are amphoteric. Write equations for HCO_3^- and $H_2PO_4^-$ acting as bases with the strong acid HCl and as acids with the strong base NaOH.

10.51 Write balanced equations for proton-transfer reactions between the listed pairs. Indicate the conjugate pairs, and determine the favored direction for each equilibrium.

(a) HCl and PO_4^{3-} (b) HCN and SO_4^{2-}
(c) $HClO_4$ and NO_2^- (d) CH_3O^- and HF

10.52 Tums, a drugstore remedy for acid indigestion, contains $CaCO_3$. Write an equation for the reaction of Tums with gastric juice (HCl).

10.53 Write balanced equations for the following acid–base reactions:

(a) $LiOH + HNO_3 \rightarrow$
(b) $BaCO_3 + HI \rightarrow$
(c) $H_3PO_4 + KOH \rightarrow$
(d) $Ca(HCO_3)_2 + HCl \rightarrow$
(e) $Ba(OH)_2 + H_2SO_4 \rightarrow$

Acid and Base Strength: K_a and pH

10.54 How is K_a defined? Write the equation for K_a for the generalized acid HA.

10.55 Rearrange the equation you wrote in Problem 10.54 to solve for $[H_3O^+]$ in terms of K_a.

10.56 How is K_w defined, and what is its numerical value at 25°C?

10.57 How is pH defined?

10.58 Write the expression for the K_a of the ammonium ion, NH_4^+, in water.

10.59 Write the expressions for the acid dissociation constants for the three successive dissociations of phosphoric acid, H_3PO_4, in water.

10.60 Find K_a values in Table 10.2, and decide which acid in each of the following pairs is stronger:

(a) HCO_2H or HF
(b) HSO_4^- or HCN
(c) $H_2PO_4^-$ or HPO_4^{2-}
(d) $CH_3CH_2CO_2H$ or CH_3CO_2H

10.61 Which substance in each of the following pairs is the stronger base? Look at Table 10.1 if necessary.

(a) OH^- or PO_4^{3-}
(b) Br^- or NO_2^-
(c) NH_3 or OH^-
(d) CN^- or H_2O
(e) I^- or HPO_4^{2-}

10.62 The electrode of a pH meter was placed in a sample of urine, and a reading of 7.9 was obtained. Is the sample acidic, basic, or neutral?

10.63 A 0.10 M solution of the deadly poison hydrogen cyanide, HCN, has a pH of 5.2. Is HCN acidic? Is it strong or weak?

10.64 Normal gastric juice has a pH of about 2. Assuming that gastric juice is primarily aqueous HCl, what is the HCl concentration?

10.65 Human spinal fluid has a pH of 7.4. Approximately what is the H_3O^+ concentration of spinal fluid?

10.66 What is the approximate pH of a 0.10 M solution of a strong monoprotic acid?

10.67 What is the approximate pH of a 0.10 M solution of a strong base such as KOH?

10.68 Approximately what pH does each of the following H_3O^+ concentrations correspond to?

(a) Fresh egg white: $[H_3O^+] = 2.5 \times 10^{-8}$ M
(b) Apple cider: $[H_3O^+] = 5.0 \times 10^{-4}$ M
(c) Household ammonia: $[H_3O^+] = 2.3 \times 10^{-12}$ M

10.69 What are the H_3O^+ concentrations of solutions with the following pH values?
  (a) pH 4
  (b) pH 11
  (c) pH 0
  (d) pH 1.38
  (e) pH 7.96

10.70 What is the OH^- concentration of each solution in Problem 10.69?

10.71 What is the OH^- concentration for each solution in Problem 10.68? Rank the solutions according to increasing acidity.

Buffers

10.72 What are the two components of a buffer system? How does a buffer work to hold pH nearly constant?

10.73 Which system would you expect to be a better buffer: $HNO_3 + Na^+NO_3^-$, or $CH_3CO_2H + CH_3CO_2^-Na^+$? Explain.

10.74 The pH of a buffer solution containing 0.10 M acetic acid and 0.10 M sodium acetate is 4.74. Write the equations for reaction of this buffer with a small amount of HNO_3 and with a small amount of NaOH.

10.75 Blood serum generally has a pH of 7.4.
  (a) What happens to the blood pH when a person breathes rapidly and lowers the concentration of CO_2 in blood? Write the equation involved.
  (b) What happens to the blood pH when a person has difficulty breathing and the concentration of CO_2 in blood increases? Write the equation involved.

10.76 What is the pH of a buffer system that contains 0.200 M hydrocyanic acid (HCN) and 0.150 M sodium cyanide (NaCN)? The K_a of hydrocyanic acid is 4.9×10^{-10}.

10.77 What is the pH of 1.00 L of the 0.200 M hydrocyanic acid–0.150 M cyanide ion buffer system described in Problem 10.76 after 0.020 mol of HCl is added? After 0.020 mol of NaOH is added?

Concentrations of Acid and Base Solutions

10.78 What does it mean when we talk about acid and base *equivalents*?

10.79 How is normality defined as a means of expressing acid or base concentration?

10.80 How many equivalents are in 500.0 mL of 0.50 M HNO_3? Of 0.50 M H_3PO_4?

10.81 How many equivalents of NaOH are needed to react with 0.035 Eq of the triprotic acid H_3PO_4?

10.82 How many milliliters of 0.0050 N KOH are required to neutralize 25 mL of 0.0050 N H_2SO_4? To neutralize 25 mL of 0.0050 N HCl?

10.83 Because hydrogen cyanide is a gas at room temperature, it is easier to measure by volume than by weight. How many liters of HCN at STP are required to make 250 mL of 0.10 N solution?

10.84 How many equivalents of an acid or base are in each of the following?
  (a) 0.25 mol $Mg(OH)_2$
  (b) 2.5 g $Mg(OH)_2$
  (c) 15 g CH_3CO_2H

10.85 What mass of citric acid (triprotic; $C_6H_5O_7H_3$) is needed to produce 152 mEq?

10.86 What are the molarity and the normality of a solution made by dissolving 5.0 g $Ca(OH)_2$ in enough water to make 400.0 mL of solution?

10.87 What are the molarity and the normality of a solution made by dissolving 25 g of citric acid ($C_6H_5O_7H_3$, a triprotic acid) in enough water to make 750 mL of solution?

10.88 Titration of a 10.0 mL solution of KOH required 15.0 mL of 0.0250 M H_2SO_4 solution. What is the molarity of the KOH solution?

10.89 If 35.0 mL of a 0.100 N acid solution is needed to reach the end point in titration of 21.5 mL of a base solution, what is the normality of the base solution?

Applications

10.90 An over-the-counter antacid has $NaAl(OH)_2CO_3$ as the active ingredient. [*Ulcers and Antacids*]
  (a) Write a balanced equation for the reaction of this compound with HCl.
  (b) How many grams of this antacid are required to neutralize 15.0 mL of 0.0955 M HCl?

10.91 Which body fluid is most acidic? Which is most basic? [*pH of Body Fluids*]

10.92. Rain typically has a pH of about 5.6. What is the $[H_3O^+]$ in rain? [*Acid Rain*]

10.93 Acid rain with a pH as low as 1.5 has been recorded in West Virginia. [*Acid Rain*]
  (a) What is the $[H_3O^+]$ in this acid rain?
  (b) How many grams of HNO_3 must be dissolved to make 25 L of solution with a pH of 1.5?

General Questions and Problems

10.94 Alka-Seltzer, a drugstore antacid, contains a mixture of $NaHCO_3$, aspirin, and citric acid, $C_6H_5O_7H_3$. Why does Alka-Seltzer foam and bubble when dissolved in water? Which ingredient is the antacid?

10.95 How many milliliters of 0.50 M NaOH solution are required to titrate 40.0 mL of a 0.10 M H_2SO_4 solution to an end point?

10.96 Which solution contains more acid, 50 mL of a 0.20 N HCl solution or 50 mL of a 0.20 N acetic acid solution? Which has a higher hydronium ion concentration? Which has a lower pH?

10.97 A 0.010 M solution of aspirin has pH 3.3. Is aspirin a strong or a weak acid?

10.98 A 0.15 M solution of HCl is used to titrate 30.0 mL of a $Ca(OH)_2$ solution of unknown concentration. If 140 mL of HCl is required, what is the concentration (in molarity) of the $Ca(OH)_2$ solution?

10.99 Which of the following combinations produces an effective buffer solution?
   (a) NaF and HF
   (b) $HClO_4$ and $NaClO_4$
   (c) NH_4Cl and NH_3
   (d) KBr and HBr

10.100 One method of analyzing ammonium salts is to treat them with NaOH and then heat the solution to remove the NH_3 gas formed.

$$NH_4^+(aq) + OH^-(aq) \rightarrow NH_3(g) + H_2O(l)$$

   (a) Label the Brønsted–Lowry acid–base pairs.
   (b) If 2.86 L of NH_3 at 60°C and 755 mm Hg is produced by the reaction of NH_4Cl, how many grams of NH_4Cl were in the original sample?

10.101 What is the pH of the solution formed by diluting 25.0 mL of 0.40 M NaOH to a final volume of 2.00 L?

10.102 Sodium oxide, Na_2O, reacts with water to give NaOH.
   (a) Write a balanced equation for the reaction.
   (b) What is the pH of the solution prepared by allowing 1.55 g of Na_2O to react with 500.0 mL of water? Assume that there is no volume change.
   (c) How many milliliters of 0.0100 M HCl are needed to neutralize the NaOH solution prepared in (b)?

11 Nuclear Chemistry

The age of this mummy is 3100 years, as determined by radiocarbon dating.

In all the reactions we've discussed thus far, only the *bonds* between atoms have changed; the chemical identities of atoms themselves have remained unchanged. Anyone who reads the paper or watches television knows, however, that atoms *can* change, often resulting in the conversion of one element into another. Atomic weapons, nuclear energy, and radioactive radon gas in our homes are all topics of societal importance, and all involve *nuclear chemistry—* the study of the properties and reactions of atomic nuclei.

We'll answer the following questions about nuclear chemistry in this chapter:

1. **What is a nuclear reaction, and how are equations for nuclear reactions balanced?**
   The goal: Be able to write and balance equations for nuclear reactions.

2. **What are the different kinds of radioactivity?**
   The goal: Be able to list the characteristics of three common kinds of radiation—α, β, and γ.

3. **How are the rates of nuclear reactions expressed?**
   The goal: Be able to explain half-life and calculate the quantity of a radioisotope remaining after a given number of half-lives.

4. **What is ionizing radiation?**
   The goal: Be able to describe the properties of the different types of ionizing radiation and their potential for harm to living tissue.

5. **How is radioactivity measured?**
   The goal: Be able to describe the common units for measuring radiation.

6. **What is transmutation?**
   The goal: Be able to explain nuclear bombardment and balance equations for nuclear bombardment reactions.

7. **What are nuclear fission and nuclear fusion?**
   The goal: Be able to explain nuclear fission and nuclear fusion.

CONCEPTS TO REVIEW

Atomic Theory (Section 3.1)
Isotopes (Section 3.3)

11.1 Nuclear Reactions

Recall from Section 3.2 that an atom is characterized by its *atomic number*, Z, and its *mass number*, A. (∞, p. 47) The atomic number, written below and to the left of the element symbol, gives the number of protons in the nucleus. The mass number, written above and to the left of the element symbol, gives the total number of **nucleons**, a general term for both protons (p) and neutrons (n). The most common isotope of carbon, for example, has 12 nucleons: 6 protons and 6 neutrons: $^{12}_{6}\text{C}$.

Nucleon A general term for both protons and neutrons.

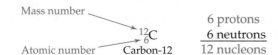

Atoms with identical atomic numbers but different mass numbers are called *isotopes* (Section 3.3), and the nucleus of a specific isotope is called a **nuclide**. (∞, p. 49) Thirteen isotopes of carbon are known—two occur commonly (^{12}C and ^{13}C) and one (^{14}C) is produced in small amounts in the upper atmosphere by the action of neutrons from cosmic rays on ^{14}N. The remaining ten carbon isotopes have been produced artificially. Only the two commonly occurring isotopes are stable indefinitely; the others undergo spontaneous **nuclear reactions**, which change their nuclei. Carbon-14, for example, slowly decomposes to give nitrogen-14 plus an electron, a process we can write as

Nuclide The nucleus of a specific isotope of an element.

Nuclear reaction A reaction that changes an atomic nucleus, usually causing the change of one element into another.

$$^{14}_{6}\text{C} \longrightarrow {}^{14}_{7}\text{N} + {}^{0}_{-1}\text{e}$$

The electron is often written as $_{-1}^{0}\text{e}$, where the superscript 0 indicates that the mass of an electron is essentially zero when compared with that of a proton or neutron, and the subscript -1 indicates that the charge is -1. (The subscript in this instance is not a true atomic number.)

Nuclear reactions, such as the spontaneous decay of ^{14}C, are distinguished from chemical reactions in several ways:

- A *nuclear* reaction involves a change in an atom's nucleus, usually producing a different element. A *chemical* reaction, by contrast, involves only a change in distribution of the outer-shell electrons around the atom and never changes the nucleus itself or produces a different element.

- Different isotopes of an element have essentially the same behavior in chemical reactions but often have completely different behavior in nuclear reactions.

- The rate of a nuclear reaction is unaffected by a change in temperature or pressure or by the addition of a catalyst.

- The nuclear reaction of an atom is essentially the same whether it is in a chemical compound or in uncombined, elemental form.

- The energy change accompanying a nuclear reaction can be up to several million times greater than that accompanying a chemical reaction. The nuclear transformation of 1.0 g of uranium-235 releases 3.4×10^8 kcal, for example, while the chemical combustion of 1.0 g of methane releases only 12 kcal.

11.2 The Discovery and Nature of Radioactivity

The discovery of *radioactivity* dates to the year 1896 when the French physicist Henri Becquerel made a remarkable observation. While investigating the nature of phosphorescence—the luminous glow of some minerals and other substances that remains when the light is suddenly turned off—Becquerel happened to place a sample of a uranium-containing mineral on top of a photographic plate that had been wrapped in black paper and put in a drawer to protect it from sunlight. On developing the plate, Becquerel was surprised to find a silhouette of the mineral. He concluded that the mineral was producing some kind of unknown radiation, which passed through the paper and exposed the photographic plate.

Marie Sklodowska Curie and her husband, Pierre, took up the challenge and began a series of investigations into this new phenomenon, which they termed **radioactivity**. They found that the source of the radioactivity was the element uranium (U) and that two previously unknown elements, which they named polonium (Po) and radium (Ra), were also radioactive. For these achievements, Becquerel and the Curies shared the 1903 Nobel Prize in physics.

Further work on radioactivity by the English scientist Ernest Rutherford established that there were at least two types of radiation, which he named *alpha* (α) and *beta* (β) after the first two letters of the Greek alphabet. Shortly thereafter, a third type of radiation was found and named for the third Greek letter, *gamma* (γ).

Subsequent studies showed that when the three kinds of radiation are passed between two plates with opposite electrical charges, each is affected differently. Alpha radiation bends toward the negative plate and must therefore have a positive charge. Beta radiation, by contrast, bends toward the positive plate and must have a negative charge, while gamma radiation does not bend toward either plate and has no charge (Figure 11.1).

Radioactivity The spontaneous emission of radiation from a nucleus.

Separation of Alpha, Beta, and Gamma Rays

FIGURE 11.1 The effect of an ▶ electric field on α, β, and γ radiation. The radioactive source in the shielded box emits radiation, which passes between the two electrically charged plates. Alpha radiation is deflected toward the negative plate, β radiation is deflected toward the positive plate, and γ radiation is not deflected.

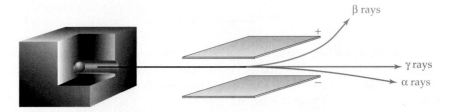

Another difference among the three kinds of radiation soon became apparent when it was discovered that α and β radiations are composed of small particles with a measurable mass, while **γ radiation** consists of high-energy electromagnetic waves and has no mass (see the Application "Atoms and Light" in Chapter 3). (∞, p. 64) Rutherford was able to show that a **β particle** is an electron (e^-) and that an **α particle** is simply a helium nucleus, He^{2+}. (Recall that a helium *atom* consists of two protons, two neutrons, and two electrons. When the two electrons are removed, the remaining helium nucleus, or α particle, has only the two protons and two neutrons.)

Yet a third difference among the three kinds of radiation is their penetrating power. Because of their relatively large mass, α particles move slowly (up to about one-tenth the speed of light) and can be stopped by a few sheets of paper or by the top layer of skin. Beta particles, because they are much lighter, move at up to nine-tenths the speed of light and have about 100 times the penetrating power of α particles. A block of wood or heavy protective clothing is necessary to stop β radiation, which can otherwise penetrate the skin and cause burns and other damage. Gamma rays move at the speed of light (3.00×10^8 m/s) and have about 1000 times the penetrating power of α particles. A lead block several inches thick is needed to stop γ radiation, which can otherwise penetrate and damage the body's internal organs.

The characteristics of the three kinds of radiation are summarized in Table 11.1. Note that an α particle, even though it's an ion with a 2+ charge, is usually written using the symbol $_2^4He$ without the charge. A β particle is usually written $_{-1}^0e$, as noted previously.

γ Radiation Radioactivity consisting of high-energy light waves.

β Particle An electron (e^-), emitted as β radiation.

α Particle A helium nucleus (He^{2+}), emitted as α radiation.

TABLE 11.1 Characteristics of α, β, and γ Radiation

Type of Radiation	Symbol	Charge	Composition	Mass (amu)	Velocity	Relative Penetrating Power
Alpha	α, $_2^4He$	+2	Helium nucleus	4	Up to 10% speed of light	Low (1)
Beta	β, $_{-1}^0e$	−1	Electron	1/1823	Up to 90% speed of light	Medium (100)
Gamma	γ, $_0^0\gamma$	0	High-energy radiation	0	Speed of light (3.00×10^8 m/s)	High (1000)

11.3 Stable and Unstable Isotopes

Every element in the periodic table has at least one radioactive isotope, or **radioisotope**, and more than *3300* radioisotopes are known. Their radioactivity is the result of having unstable nuclei, although the exact causes of this instability aren't fully understood. Radiation is emitted when an unstable radioactive nucleus, or **radionuclide**, spontaneously changes into a more stable one.

For elements in the first few rows of the periodic table, stability is associated with a roughly equal number of neutrons and protons (Figure 11.2). Hydrogen, for example, has stable $_1^1H$ (protium) and $_1^2H$ (deuterium) isotopes, but its $_1^3H$ isotope (tritium) is radioactive. As elements get heavier, the number of neutrons relative to protons in stable nuclei increases. Lead-208, for example, the most abundant stable isotope of lead, has 126 neutrons and 82 protons in its nuclei. Nevertheless, of the 35 known isotopes of lead, only 3 are stable while 32 are radioactive. In fact, there are only 264 stable isotopes among all the elements. All isotopes of elements with atomic numbers higher than that of bismuth (83) are radioactive.

Radioisotope A radioactive isotope.

Radionuclide The nucleus of a radioactive isotope.

FIGURE 11.2   A plot showing ▶ the numbers of neutrons and protons for known isotopes of the first 18 elements.   Stable (nonradioactive) isotopes of these elements have equal or nearly equal numbers of neutrons and protons.

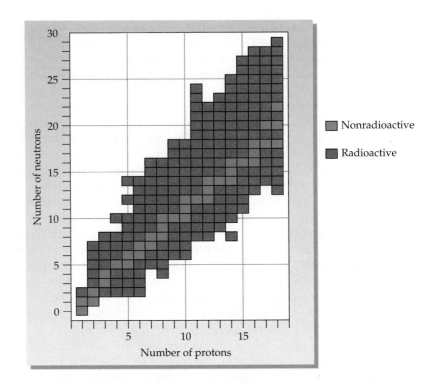

FIGURE 11.2   A plot showing ▶ the numbers of neutrons and protons for known isotopes of the first 18 elements.   Stable (nonradioactive) isotopes of these elements have equal or nearly equal numbers of neutrons and protons.

Most of the more than 3300 known radioisotopes have been made in high-energy particle accelerators by reactions that we'll describe in Section 11.10. Such isotopes are called *artificial radioisotopes* because they aren't found in nature. All isotopes of the transuranium elements (those heavier than uranium) are artificial. The much smaller number of radioactive isotopes found in the earth's crust, such as $^{238}_{92}U$, are called *natural radioisotopes*.

Aside from their radioactivity, different radioisotopes of the same element have the same chemical properties as stable isotopes, which accounts for their great usefulness as *tracers*. A chemical compound tagged with a radioactive atom undergoes exactly the same reactions as its nonradioactive counterpart. The difference is that the tagged compound can be located with a radiation detector and its whereabouts determined, as we'll see in the Application "Body Imaging" later in this chapter.

11.4 Nuclear Decay

Think for a minute about the consequences of α and β radiation. If radioactivity involves the spontaneous emission of a small particle from an unstable atomic nucleus, then the nucleus itself must undergo a change. With that understanding of radioactivity came the startling discovery that atoms of one element can change into atoms of another element, something that had previously been thought impossible. The spontaneous emission of a particle from an unstable nucleus is called **nuclear decay**, or *radioactive decay*, and the resulting change of one element into another is called **transmutation.**

Nuclear decay The spontaneous emission of a particle from an unstable nucleus.

Transmutation The change of one element into another.

Nuclear decay: Radioactive element ⟶ New element + Emitted particle

Let's look at what happens to a nucleus when nuclear decay occurs.

Alpha Emission

When an atom of uranium-238 ($^{238}_{92}U$) emits an α particle, the nucleus loses two protons and two neutrons. Because the number of protons in the nucleus has now changed from 92 to 90, the *identity* of the atom has changed from uranium to thorium. Furthermore, since the total number of nucleons has decreased by 4, uranium-238 has become thorium-234 ($^{234}_{90}$Th) (Figure 11.3).

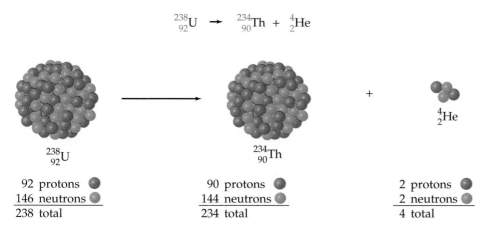

$$^{238}_{92}U \longrightarrow {}^{234}_{90}Th + {}^{4}_{2}He$$

$^{238}_{92}$U	$^{234}_{90}$Th	$^{4}_{2}$He
92 protons	90 protons	2 protons
146 neutrons	144 neutrons	2 neutrons
238 total	234 total	4 total

◀ **FIGURE 11.3 Alpha emission.** Emission of an α particle from an atom of uranium-238 produces an atom of thorium-234.

Note that the equation for a nuclear reaction is not balanced in the usual chemical sense because the kinds of atoms are not the same on both sides of the arrow. Instead, we say that a nuclear equation is balanced when the number of nucleons is the same on both sides of the equation and when the sums of the charges on the nuclei plus any ejected subatomic particles (protons or electrons) are the same on both sides. In the decay of $^{238}_{92}$U to give $^{4}_{2}$He and $^{234}_{90}$Th, for example, there are 238 nucleons and 92 nuclear charges on both sides of the nuclear equation.

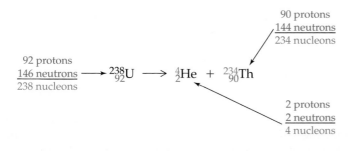

■ WORKED EXAMPLE 11.1

Polonium-208 is one of the α emitters studied by Marie Curie. Write the equation for the α decay of polonium-208, and identify the element formed.

ANALYSIS   Look up the atomic number of polonium (84) in the periodic table, and write the known part of the nuclear equation, using the standard symbol for polonium-208:

$$^{208}_{84}Po \longrightarrow {}^{4}_{2}He + ?$$

Then calculate the mass number and atomic number of the product element, and write the final equation.

SOLUTION   The mass number of the product is 208 − 4 = 204, and the atomic number is 84 − 2 = 82. A look at the periodic table identifies the element with atomic number 82 as lead (Pb).

$$^{208}_{84}Po \longrightarrow {}^{4}_{2}He + {}^{204}_{82}Pb$$

Check your answer by making sure that the mass numbers and atomic numbers on the two sides of the equation are balanced:

Mass numbers: 208 = 4 + 204   Atomic numbers: 84 = 2 + 82

■ **PROBLEM 11.1**

High levels of radioactive radon-222 ($^{222}_{86}\text{Rn}$) have been found in many homes built on radium-containing rock, leading to the possibility of health hazards. What product results from α emission by radon-222?

■ **PROBLEM 11.2**

What isotope of radium (Ra) is converted into radon-222 by α emission?

Beta Emission

Whereas α emission leads to the loss of two protons and two neutrons from the nucleus, β emission involves the *decomposition* of a neutron to yield an electron and a proton. The electron is ejected as a β particle, and the proton is retained by the nucleus. Note that the electrons emitted during β radiation come from the *nucleus* and not from the occupied orbitals surrounding the nucleus.

The net result of β emission is that the atomic number of the atom increases by 1 because there is a new proton. The mass number of the atom remains the same, however, because a neutron has changed into a proton leaving the total number of nucleons unchanged. For example, iodine-131 ($^{131}_{53}\text{I}$), a radioisotope used in detecting thyroid problems, undergoes nuclear decay by β emission to yield xenon-131 ($^{131}_{54}\text{Xe}$):

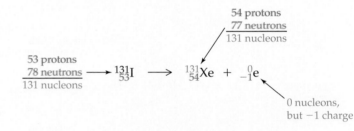

Note that the superscripts (mass numbers) are balanced in this equation because a β particle has a mass near zero, and the subscripts are balanced because a β particle has a charge of -1.

■ **WORKED EXAMPLE 11.2**

Write a balanced nuclear equation for the β decay of chromium-55.

ANALYSIS Write the known part of the nuclear equation:

$$^{55}_{24}\text{Cr} \longrightarrow {}^{\ 0}_{-1}\text{e} + ?$$

Then calculate the mass number and atomic number of the product element, and write the final equation.

SOLUTION The mass number of the product stays at 55, and the atomic number is $24 + 1 = 25$, so the product is manganese-55.

$$^{55}_{24}\text{Cr} \longrightarrow {}^{\ 0}_{-1}\text{e} + {}^{55}_{25}\text{Mn}$$

Check your answer by making sure that the mass numbers and atomic numbers on the two sides of the equation are balanced:

Mass numbers: $55 = 0 + 55$ Atomic numbers: $24 = -1 + 25$

■ **PROBLEM 11.3**

Carbon-14, a β emitter, is a rare isotope used in dating archaeological artifacts. Write a nuclear equation for the decay of carbon-14.

■ **PROBLEM 11.4**

Write nuclear equations for β emission from the following radioisotopes:

(a) $^{3}_{1}H$ (b) $^{210}_{82}Pb$ (c) $^{20}_{9}F$

Gamma Emission

Emission of γ rays, unlike the emission of α and β particles, causes no change in mass or atomic number because γ rays are simply high-energy electromagnetic waves. Although γ emission can occur alone, it usually accompanies α or β emission as a mechanism for the new nucleus that results from a transmutation to get rid of some extra energy.

Since γ emission affects neither mass number nor atomic number, it is often omitted from nuclear equations. Nevertheless, γ rays are of great importance: Their penetrating power makes them by far the most dangerous kind of external radiation for humans and also makes them useful in numerous medical applications. Cobalt-60, for example, is used in cancer therapy as a source of penetrating γ rays that kill cancerous tissue.

$$^{60}_{27}Co \longrightarrow {}^{60}_{28}Ni + {}^{0}_{-1}e + {}^{0}_{0}\gamma$$

Positron Emission

In addition to α, β, and γ radiation, there is another common type of radioactive decay process called *positron emission*, which involves the conversion of a proton in the nucleus into a neutron plus an ejected **positron**, $^{0}_{1}e$ or β^{+}. A positron, which can be thought of as a "positive electron," has the same mass as an electron but a positive charge. The result of positron emission is a decrease in the atomic number of the product nucleus because a proton has changed into a neutron, but no change in the mass number. Potassium-40, for example, undergoes positron emission to yield argon-40, a nuclear reaction important in geology for dating rocks. Note once again that the sum of the two subscripts on the right of the nuclear equation ($18 + 1 = 19$) is equal to the subscript in the $^{40}_{19}K$ nucleus on the left.

Positron A "positive electron," which has the same mass as an electron but a positive charge.

Electron Capture

Electron capture, symbolized E.C., is a process in which the nucleus captures an inner-shell electron from the surrounding electron cloud, thereby converting a proton into a neutron. The mass number of the product nucleus is unchanged, but the atomic number decreases by 1, just as in positron emission. The conversion of mercury-197 into gold-197 is an example:

Electron capture A process in which the nucleus captures an inner-shell electron from the surrounding electron cloud, thereby converting a proton into a neutron.

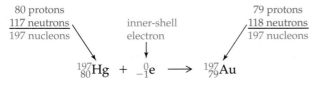

Characteristics of the five different kinds of radioactive decay processes are summarized in Table 11.2.

TABLE 11.2 A Summary of Radioactive Decay Processes				
Process	Symbol	Change in Atomic Number	Change in Mass Number	Change in Neutron Number
α emission	$^4_2$He or α	-2	-4	-2
β emission	$^0_{-1}$e or β^-	$+1$	0	-1
γ emission	$^0_0\gamma$ or γ	0	0	0
Positron emission	$^0_1$e or β^+	-1	0	$+1$
Electron capture	E.C.	-1	0	$+1$

■ **WORKED EXAMPLE 11.3**

Write balanced nuclear equations for each of the following processes:
(a) Electron capture by polonium-204: $^{204}_{84}$Po $+ \, ^0_{-1}$e $\rightarrow$?
(b) Positron emission from xenon-118: $^{118}_{54}$Xe $\rightarrow \, ^0_1$e $+$?

ANALYSIS The key to writing nuclear equations is to make sure that the number of nucleons is the same on both sides of the equation and that the number of charges is the same.

SOLUTION
(a) In electron capture, the mass number is unchanged and the atomic number decreases by 1, giving bismuth-204: $^{204}_{84}$Po $+ \, ^0_{-1}$e $\rightarrow \, ^{204}_{83}$Bi

(b) In positron emission, the mass number is unchanged and the atomic number decreases by 1, giving iodine-118: $^{118}_{54}$Xe $\rightarrow \, ^0_1$e $+ \, ^{118}_{53}$I

■ **PROBLEM 11.5**

Write nuclear equations for positron emission from the following radioisotopes:
(a) $^{38}_{20}$Ca   **(b)** $^{118}_{54}$Xe   **(c)** $^{79}_{37}$Rb

■ **PROBLEM 11.6**

Write nuclear equations for electron capture by the following radioisotopes:
(a) $^{62}_{30}$Zn   **(b)** $^{110}_{50}$Sn   **(c)** $^{81}_{36}$Kr

■ **KEY CONCEPT PROBLEM 11.7**

Identify the isotopes involved, and tell the type of decay process occurring in the following nuclear reaction:

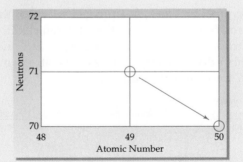

Application Medical Uses of Radioactivity

The origins of nuclear medicine date from 1901 when the French physician Henri Danlos first used radium in the treatment of a tuberculous skin lesion. Since that time, the use of radioactivity has become a crucial part of modern medical care, both diagnostic and therapeutic. Current nuclear techniques can be grouped into three classes: (1) in vivo procedures, (2) radiation therapy, and (3) imaging procedures. We'll describe the first two here and the third one later in this chapter in the Application "Body Imaging."

In Vivo Procedures

In vivo studies—those that take place inside the body—are carried out to assess the functioning of a particular organ or body system. A *radiopharmaceutical* agent is administered, and its path in the body—whether absorbed, excreted, diluted, or concentrated—is determined by analysis of blood or urine samples.

Among the many in vivo procedures utilizing radioactive agents is a simple method for the determination of whole-blood volume by injecting a known quantity of red blood cells labeled with radioactive chromium-51. After a suitable interval to allow the labeled cells to be distributed evenly throughout the body, a blood sample is taken and blood volume is calculated by comparing the concentration of labeled cells in the blood with the quantity of labeled cells injected.

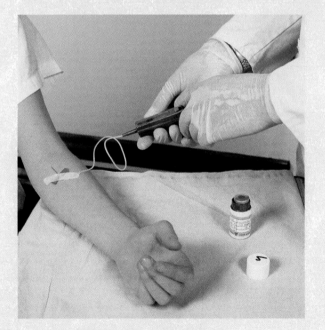

▲ A person's blood volume can be found by injecting a small amount of radioactive chromium-51 and measuring the dilution factor.

Therapeutic Procedures

Therapeutic procedures—those in which radiation is purposely used as a weapon to kill diseased tissue—can involve either external or internal sources of radiation. External radiation therapy for the treatment of cancer is often carried out with γ rays emanating from a cobalt-60 source. The highly radioactive source is shielded by a thick lead container and has a small opening directed toward the site of the tumor. By focusing the radiation beam on the tumor, the tumor can receive the full exposure while the exposure of surrounding parts of the body is minimized. Nevertheless, enough healthy tissue is affected so that most patients treated in this manner suffer the effects of radiation sickness.

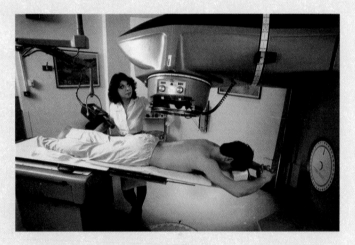

▲ Cobalt-60 is used in cancer therapy as a source of γ rays.

Internal radiation therapy is a much more selective technique than external therapy. In the treatment of thyroid disease, for example, a radioactive substance such as iodine-131 is administered. This powerful β emitter is incorporated into the iodine-containing hormone thyroxine, which concentrates in the thyroid gland. Because β particles penetrate no farther than several millimeters, the localized I-131 produces a high radiation dose that destroys only the surrounding diseased tissue. To treat some tumors, such as those in the female reproductive system, a radioactive source is placed physically close to the tumor for a specific amount of time.

See Additional Problems 11.64 and 11.65 at the end of the chapter.

11.5 Radioactive Half-Life

The rate of radioactive decay varies greatly from one radioisotope to another. Some radioisotopes, such as uranium-238, decay at a barely perceptible rate over billions of years, while others, such as carbon-17, decay within thousandths of a second.

Half-life ($t_{1/2}$) The amount of time required for one-half of a radioactive sample to decay.

Rates of nuclear decay are measured in units of **half-life** ($t_{1/2}$), defined as the amount of time required for one-half of the radioactive sample to decay. For example, the half-life of iodine-131 is 8.021 days. If today you have a certain amount of $^{131}_{53}I$, say 1.000 g, then 8.021 days from now only 0.500 g of $^{131}_{53}I$ will remain because one-half of the sample will have decayed into $^{131}_{54}Xe$. After 8.021 more days (16.063 total) only 0.250 g of $^{131}_{53}I$ will remain; after another 8.021 days (24.084 total) only 0.125 g will remain; and so on. Each passage of a half-life causes the decay of one-half of whatever sample remains. The half-life is the same no matter what the size of the sample, the temperature, or any other external conditions. There is no known way to slow down, speed up, or otherwise change the characteristics of radioactive decay.

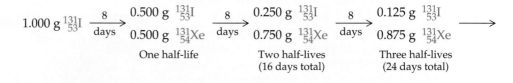

$$1.000 \text{ g } ^{131}_{53}I \xrightarrow[\text{days}]{8} \begin{array}{c} 0.500 \text{ g } ^{131}_{53}I \\ 0.500 \text{ g } ^{131}_{54}Xe \end{array} \xrightarrow[\text{days}]{8} \begin{array}{c} 0.250 \text{ g } ^{131}_{53}I \\ 0.750 \text{ g } ^{131}_{54}Xe \end{array} \xrightarrow[\text{days}]{8} \begin{array}{c} 0.125 \text{ g } ^{131}_{53}I \\ 0.875 \text{ g } ^{131}_{54}Xe \end{array} \longrightarrow$$

One half-life   Two half-lives   Three half-lives
(16 days total)   (24 days total)

The percent of radioisotope remaining after the passage of each half-life is represented by the curve in Figure 11.4.

FIGURE 11.4 The decay of a ▶ radioactive nucleus over time. All nuclear decays follow this curve, whether the half-lives are measured in years, days, minutes, or seconds. That is, 50% of the sample remains after one half-life, 25% after two half-lives, 12.5% after three half-lives, and so on.

Radioactive Decay

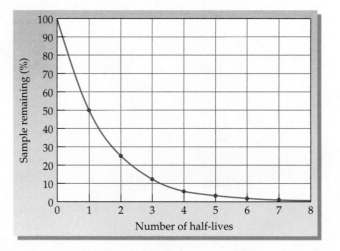

The half-lives of some useful radioisotopes are given in Table 11.3. As you might expect, radioisotopes that are used internally for medical applications have fairly short half-lives so that they decay rapidly and do not remain in the body for prolonged periods.

■ **WORKED EXAMPLE 11.4**

Phosphorus-32, a radioisotope used in leukemia therapy, has a half-life of about 14 days. Approximately what percent of a sample remains after 8 weeks?

ANALYSIS Determine how many half-lives have elapsed, and then multiply the starting amount (100%) by 1/2 for each of the half-lives that has elapsed.

TABLE 11.3   Half-Lives of Some Useful Radioisotopes

Radioisotope	Symbol	Radiation	Half-life	Use
Tritium	$^{3}_{1}H$	β	12.33 years	Biochemical tracer
Carbon-14	$^{14}_{6}C$	β	5730 years	Archaeological dating
Sodium-24	$^{24}_{11}Na$	β	14.959 hours	Examining circulation
Phosphorus-32	$^{32}_{15}P$	β	14.262 days	Leukemia therapy
Potassium-40	$^{40}_{19}K$	β, β^{+}	1.277×10^{9} years	Geological dating
Cobalt-60	$^{60}_{27}Co$	β, γ	5.271 years	Cancer therapy
Arsenic-74	$^{74}_{33}As$	β^{+}	17.77 days	Locating brain tumors
Technetium-99*m* *	$^{99m}_{43}Tc$	γ	6.01 hours	Brain scans
Iodine-131	$^{131}_{53}I$	β	8.021 days	Thyroid therapy
Uranium-235	$^{235}_{92}U$	α, γ	7.038×10^{8} years	Nuclear reactors

*The *m* in technetium-99*m* stands for *metastable*, meaning that the nucleus undergoes γ emission but does not change its mass number or atomic number.

SOLUTION   Since one half-life of $^{32}_{15}P$ is 14 days (2 weeks), 8 weeks represents four half-lives. The amount that remains is

Four half-lives

$$\text{Final percentage} = 100\% \times (\tfrac{1}{2} \times \tfrac{1}{2} \times \tfrac{1}{2} \times \tfrac{1}{2})$$
$$= 100\% \times \tfrac{1}{16} = 6.25\%$$

■ **PROBLEM 11.8**

The half-life of carbon-14, an isotope used in archaeological dating, is 5730 years. What percentage of $^{14}_{6}C$ remains in a sample estimated to be 17,000 years old?

■ **KEY CONCEPT PROBLEM 11.9**

What is the half-life of the radionuclide that shows the following decay curve?

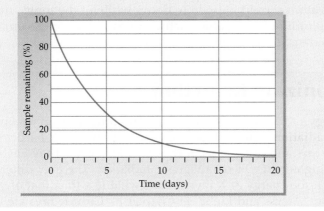

11.6 Radioactive Decay Series

When a radioactive isotope decays, nuclear change occurs and a different element is formed. Often, this newly formed nucleus is stable, but sometimes the product nucleus is itself radioactive and undergoes further decay. In fact, some radioactive nuclei undergo a whole **decay series** of nuclear disintegrations

Decay series   A sequential series of nuclear disintegrations leading from a heavy radioisotope to a non-radioactive product.

before they ultimately reach a nonradioactive product. Uranium-238, for example, undergoes a series of 14 sequential nuclear reactions, ultimately stopping at lead-206 (Figure 11.5).

FIGURE 11.5   The decay series from $^{238}_{92}$U to $^{206}_{82}$Pb. Each isotope except for the last is radioactive and undergoes nuclear decay. The long slanted arrows represent α emissions, and the short horizontal arrows represent β emissions.

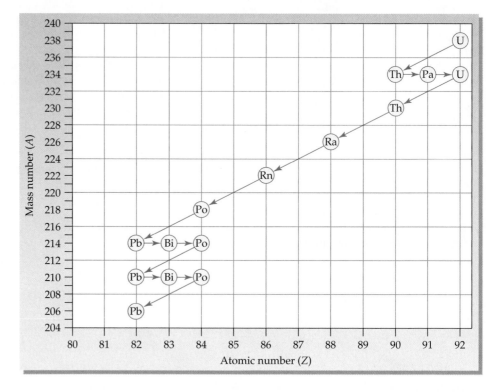

One of the intermediate radionuclides in the U-238 decay series is radon-222, a gas produced by α decay of radium-226 whose half-life is 1600 years. Rocks, soil, and building materials that contain radium are sources of radon-222, which can seep through cracks in basements and get into the air inside homes and other buildings. Radon itself is a gas that passes in and out of the lungs without being incorporated into body tissue. If, however, a radon-222 atom should happen to decay while in the lungs, the solid decay product polonium-218 results. Further decay of the Po-218 emits α particles, which can damage lung tissue.

▲ Radon gas can be detected and measured by this home detection kit.

Ionizing radiation   A general name for high-energy radiation of all kinds.

X rays   Electromagnetic radiation with an energy somewhat less than that of γ rays.

Cosmic rays   A mixture of high-energy particles—primarily of protons and various atomic nuclei—that shower the earth from outer space.

11.7 Ionizing Radiation

High-energy radiation of all kinds is often grouped together under the name **ionizing radiation**. This includes not only α particles, β particles, and γ rays, but also *X rays* and *cosmic rays*. **X rays** are like γ rays: They have no mass and consist of high-energy electromagnetic radiation. The only difference between them is that the energy of X rays is somewhat less than that of γ rays (see the Application "Atoms and Light" in Chapter 3). **Cosmic rays** are not rays at all but are a mixture of high-energy particles that shower the earth from outer space. They consist primarily of protons, along with some α and β particles.

The interaction of any kind of ionizing radiation with a molecule knocks out an electron, converting the atom or molecule into an extremely reactive ion:

$$\text{Molecule} \xrightarrow[\text{radiation}]{\text{ionizing}} \text{Ion} + e^-$$

The reactive ion can react with other molecules nearby, creating still other fragments that can in turn cause further reactions. In this manner, a large dose of ionizing radiation can destroy the delicate balance of chemical reactions in living cells, ultimately causing the death of an organism.

A small dose of ionizing radiation may not cause visible symptoms but can nevertheless be dangerous if it strikes a cell nucleus and damages the genetic machinery inside. The resultant changes might lead to a genetic mutation, to cancer, or to cell death. The nuclei of rapidly dividing cells, such as those in bone marrow, the lymph system, the lining of the intestinal tract, or an embryo, are the most readily damaged. It's because of the susceptibility of rapidly dividing cells to radiation effects that ionizing radiation is able to selectively destroy cancer cells. Some properties of ionizing radiation are summarized in Table 11.4.

TABLE 11.4 Some Properties of Ionizing Radiation

Type of Radiation	Energy Range*	Penetrating Distance in Water **
α	3–9 MeV	0.02–0.04 mm
β	0–3 MeV	0–4 mm
X	100 eV–10 keV	0.01–1 cm
γ	10 keV–10 MeV	1–20 cm

*The energies of subatomic particles are often measured in electron volts (eV): 1 eV = 6.703×10^{-19} cal.
**Distance at which one-half of the radiation is stopped.

The effects of ionizing radiation on the human body vary with the energy of the radiation, its distance from the body, the length of exposure, and the location of the source outside or inside the body. When coming from outside the body, γ rays and X rays are potentially more harmful than α and β particles because they pass through clothing and skin and into the body's cells. Alpha particles are stopped by clothing and skin, and β particles are stopped by wood or several layers of clothing. These types of radiation are much more dangerous when emitted within the body, however, because all their radiation energy is given up to the immediately surrounding tissue. Alpha emitters are especially hazardous internally and are almost never used in medical applications.

Health professionals who work with X rays or other kinds of ionizing radiation protect themselves by surrounding the source with a thick layer of lead or other material. Protection from radiation is also afforded by controlling the distance between the worker and the radiation source because radiation intensity (I) decreases with the square of the distance from the source. The intensities of radiation at two different distances, 1 and 2, are given by the equation

$$\frac{I_1}{I_2} = \frac{d_2{}^2}{d_1{}^2}$$

For example, suppose a source delivers 16 units of radiation at a distance of 1.0 m. Doubling the distance to 2.0 m would decrease the radiation intensity to one-fourth:

$$\frac{16 \text{ units}}{I_2} = \frac{(2 \text{ m})^2}{(1 \text{ m})^2}$$

$$I_2 = 16 \text{ units} \times \frac{1 \text{ m}^2}{4 \text{ m}^2} = 4 \text{ units}$$

■ **WORKED EXAMPLE 11.5**

If a radiation source gives 75 units of radiation at a distance of 2.4 m, at what distance would the source give 25 units of radiation?

ANALYSIS Radiation intensity (I) decreases with the square of the distance (d) from the source according to the equation

$$\frac{I_1}{I_2} = \frac{d_2^2}{d_1^2}$$

Of the four variables in this equation, we know that $I_1 = 75$ units, $I_2 = 25$ units, and $d_1 = 2.4$ m, and we need to find d_2.

SOLUTION Solve the above equation for d_2, and substitute in the known values:

$$d_2^2 = \frac{I_1 d_1^2}{I_2} \quad \text{so} \quad d_2 = \sqrt{\frac{I_1 d_1^2}{I_2}}$$

$$d_2 = \sqrt{\frac{(75 \text{ units})(2.4 \text{ m})^2}{(25 \text{ units})}} = 4.2 \text{ m}$$

■ **PROBLEM 11.10**

A β-emitting radiation source gives 250 units of radiation at a distance of 4.0 m. At what distance does the radiation drop to one-tenth its original value?

▲ This photographic film badge is a common device for monitoring radiation exposure.

▲ Radiation is conveniently detected and measured using this scintillation counter, which electronically counts the flashes produced when radiation strikes a phosphor.

11.8 Detecting Radiation

Small amounts of naturally occurring radiation have always been present, but people have been aware of it only within the past 100 years. The problem is that radiation is invisible. We can't see, hear, smell, touch, or taste radiation, no matter how high the dose. We can, however, detect radiation by taking advantage of its ionizing properties.

The simplest device for detecting exposure to radiation is the photographic film badge worn by people who routinely work with radioactive materials. The film is protected from exposure to light, but any other radiation striking the badge causes the film to fog (remember Becquerel's discovery). At regular intervals, the film is developed and compared with a standard to indicate the radiation exposure.

Perhaps the best-known method for detecting and measuring radiation is the *Geiger counter*, an argon-filled tube containing two electrodes (Figure 11.6). The inner walls of the tube are coated with an electrically conducting material and given a negative charge, while a wire in the center of the tube is given a positive charge. As radiation enters the tube through a thin window, it strikes and ionizes argon atoms, which briefly conduct a tiny electric current between the walls and the center electrode. The passage of the current is detected, amplified, and used to produce a clicking sound or to register on a meter. The more radiation that enters the tube, the more frequent the clicks. Geiger counters are useful for seeking out a radiation source in a large area and for gauging the intensity of emitted radiation.

The most versatile method for measuring radiation in the laboratory is the *scintillation counter*, a device in which a substance called a *phosphor* emits a flash of light when struck by radiation. The number of flashes are counted electronically and converted into an electrical signal.

Application Irradiated Food

The idea of irradiating food to kill harmful bacteria is not a new one; it goes back almost as far as the earliest studies on radiation. Not until the 1940s did serious work get under way, however, when U.S. Army scientists found that irradiation increased the shelf-life of ground beef. Nevertheless, widespread civilian use of the technique has been a long time in coming, spurred on in recent years by outbreaks of food poisoning that have resulted in several deaths.

The principle of food irradiation is simple: Exposure of contaminated food to ionizing radiation—usually γ rays produced by cobalt-60 or cesium-137 destroys the genetic material of any bacteria or other organisms present, thereby killing them. Irradiation will not, however, kill viruses or priors, the cause of mad-cow disease. The food itself undergoes little if any change when irradiated and does not itself become radioactive. The only real argument against food irradiation, in fact, is that it is *too* effective. Knowing that irradiation will kill all harmful organisms, a food processor might be tempted to cut back on normal sanitary practices.

The U.S. Food and Drug Administration, after studying the matter extensively, has declared that food irradiation is safe and that it does not appreciably alter the vitamin or other nutritional content of food. Spices, fruits, pork, and vegetables were approved for irradia-

▲ Irradiating this meat with γ rays will make it much safer to eat by killing any harmful bacteria.

tion in 1986, followed by poultry in 1990 and red meat, particularly ground beef, in 1997. In 2000, approval was extended to whole eggs and sprouting seeds. Should the food industry adopt irradiation of meat as its standard practice, such occurrences as the 1993 Seattle outbreak of *E. coli* poisoning caused by undercooked hamburgers will become a thing of the past.

See Additional Problems 11.66 and 11.67 at the end of the chapter.

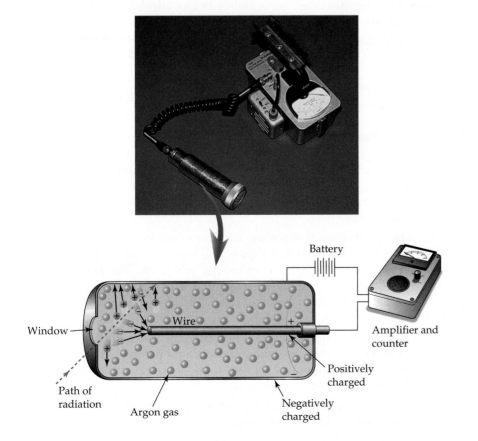

◀ **FIGURE 11.6 A Geiger counter for measuring radiation.** As radiation enters the tube through a thin window, it ionizes argon atoms and produces electrons that conduct a tiny electric current between the walls and the center electrode. The current flow then registers on the meter.

Battery

Window

Wire

Path of radiation

Argon gas

Positively charged

Negatively charged

Amplifier and counter

11.9 Measuring Radiation

Radiation intensity is expressed in different ways, depending on what characteristic of the radiation is measured (Table 11.5). Some units measure the number of nuclear decay events; others measure exposure to radiation or the biological consequences of radiation.

TABLE 11.5	Common Units for Measuring Radiation	
Unit	**Quantity Measured**	**Description**
Curie (Ci)	Decay events	Amount of radiation equal to 3.7×10^{10} disintegrations per second
Roentgen (R)	Ionizing intensity	Amount of radiation producing 2.1×10^9 charges per cubic centimeter of dry air
Rad	Energy absorbed per gram of tissue	1 rad = 1 R
Rem	Tissue damage	Amount of radiation producing the same damage as 1 R of X rays
Sievert (Sv)	Tissue damage	1 Sv = 100 rem

- Curie The *curie* (Ci), the *millicurie* (mCi), and the *microcurie* (μCi) measure the number of radioactive disintegrations occurring each second in a sample. One curie is the decay rate of 1 g of radium, equal to 3.7×10^{10} disintegrations per second; 1 mCi = 0.001 Ci = 3.7×10^7 disintegrations per second; and 1 μCi = 0.000 001 Ci = 3.7×10^4 disintegrations per second.

  The dosage of a radioactive substance administered orally or intravenously is usually given in millicuries. To calculate the size of a dose, it's necessary to determine the decay rate of the isotope solution per milliliter. Because the emitter concentration is constantly decreasing as it decays, the activity must be measured immediately before administration. Suppose, for example, that a solution containing iodine-131 for a thyroid function study is found to have a decay rate of 0.020 mCi/mL and the dose administered is to be 0.050 mCi. The amount of the solution administered must be

$$\frac{0.050 \text{ mCi}}{\text{Dose}} \times \frac{1 \text{ mL } {}^{131}\text{I solution}}{0.020 \text{ mCi}} = 2.5 \text{ mL } {}^{131}\text{I solution/dose}$$

- **Roentgen** The *roentgen* (R) is a unit for measuring the ionizing intensity of γ or X radiation. In other words, the roentgen measures the capacity of the radiation for affecting matter. One roentgen is the amount of radiation that produces 2.1×10^9 units of charge in 1 cm^3 of dry air at atmospheric pressure. Each collision of ionizing radiation with an atom produces one ion, or one unit of charge.

- **Rad** The *rad* (radiation absorbed dose) is a unit for measuring the energy absorbed per gram of material exposed to a radiation source and is defined as the absorption of 1×10^{-5} J of energy per gram. The energy absorbed varies with the type of material irradiated and the type of radiation. For most purposes, though, the roentgen and the rad are so close that they can be considered identical when used for X rays and γ rays: 1 R = 1 rad.

- **Rem** The *rem* (roentgen equivalent for man) measures the amount of tissue damage caused by radiation, taking into account the differences in energy of different types of radiation. One rem is the amount of radiation that produces the same effect as 1 R of X rays.

  Rems are the preferred units for medical purposes because they measure equivalent doses of different kinds of radiation. For example, 1 rad of α radiation causes 20 times more tissue damage than 1 rad of γ rays, but 1 rem of α radiation and 1 rem of γ rays cause the same amount of damage. Thus, the rem takes both ionizing intensity and biological effect into account, while the rad deals only with intensity.

• **SI Units**   In the SI system, the *becquerel* (Bq) is defined as one disintegration per second. The SI unit for energy absorbed is the *gray* (Gy; 1 Gy = 100 rad). For radiation dose, the SI unit is the *sievert* (Sv), which is equal to 100 rem.

The biological consequences of different radiation doses are given in Table 11.6. Although the effects seem fearful, the average radiation dose received annually by most people is only about 0.27 rem. About 80% of this *background radiation* comes from natural sources (rocks and cosmic rays); the remaining 20% comes from medical procedures such as X rays and from consumer products. The amount due to emissions from nuclear power plants and to fallout from testing of nuclear weapons in the 1950s is barely detectable.

TABLE 11.6   Biological Effects of Short-Term Radiation on Humans

Dose (rem)	Biological Effects
0–25	No detectable effects
25–100	Temporary decrease in white blood cell count
100–200	Nausea, vomiting, longer-term decrease in white blood cells
200–300	Vomiting, diarrhea, loss of appetite, listlessness
300–600	Vomiting, diarrhea, hemorrhaging, eventual death in some cases
Above 600	Eventual death in nearly all cases

■ **PROBLEM 11.11**

Estimates of the radiation released during the 1986 Chernobyl nuclear power plant disaster in Russia imply a worldwide increase in background radiation of about 5 mrem. By what percentage will this amount increase the annual dose of most people?

■ **PROBLEM 11.12**

A solution of selenium-75, a radioisotope used in the diagnosis of pancreatic disease, was found just prior to administration to have an activity of 44 μCi/mL. How many milliliters should be administered intravenously for a dose of 175 μCi?

11.10   Artificial Transmutation

Very few of the approximately 3300 known radioisotopes occur naturally. Most are made from stable isotopes by **artificial transmutation**, the change of one atom into another brought about by nuclear bombardment reactions.

When an atom is bombarded with a high-energy particle, such as a proton, neutron, α particle, or even the nucleus of another element, an unstable nucleus is created in the collision. A nuclear change then occurs, and a different element is produced. For example, transmutation of ^{14}N to ^{14}C occurs in the upper atmosphere when neutrons produced by cosmic rays collide with atmospheric nitrogen. In the collision, a neutron dislodges a proton (^{1}H) from the nitrogen nucleus as the neutron and nucleus fuse together:

$$^{14}_{7}\text{N} + {}^{1}_{0}\text{n} \longrightarrow {}^{14}_{6}\text{C} + {}^{1}_{1}\text{H}$$

Artificial transmutation   The change of one atom into another brought about by a nuclear bombardment reaction.

Application Body Imaging

We're all familiar with the appearance of a standard X-ray image, produced when X rays pass through the body and the intensity of the radiation that exits is recorded on film. X-ray imaging is, however, only one of a host of noninvasive imaging techniques that are now in common use.

Among the most widely used imaging techniques are those that give diagnostic information about the health of various parts of the body by analyzing the distribution pattern of a radioactively tagged substance in the body. A radiopharmaceutical agent that is known to concentrate in a specific organ or other part is injected into the body, and its distribution pattern is monitored by an external radiation detector such as a γ-ray camera. Depending on the medical condition, a diseased part might concentrate more of the radiopharmaceutical than normal and thus show up on the film as a radioactive hot spot against a cold background. Alternatively, the diseased part might concentrate less of the radiopharmaceutical than normal and thus show up as a cold spot on a hot background.

Among the radioisotopes most widely used for diagnostic imaging is technetium-99*m*, whose short half-life of only 6 hours minimizes the patient's exposure to radioactivity. Bone scans using this nuclide, such as that shown in the accompanying photograph, are an important tool in the diagnosis of cancer and other conditions.

Several other techniques now used in medical diagnosis are made possible by *tomography*, a technique in which computer processing allows production of images through "slices" of the body. In X-ray tomography, commonly known as *CAT* or *CT* scanning (computerized tomography), the X-ray source and an array of detectors move rapidly in a circle around a patient's body, collecting up to 90,000 readings. CT scans can detect structural abnormalities such as tumors without the use of radioactive materials.

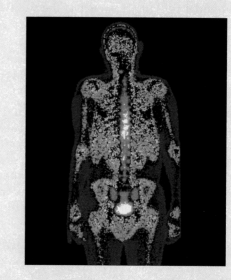

▲ A bone scan carried out with radioactive technetium-99*m*. Color has been added to help the visualization.

Combining tomography with radioisotope imaging gives cross-sectional views of regions that concentrate a radioactive substance. One such technique, *positron emission tomography* (PET), utilizes radioisotopes that emit positrons and ultimately yield γ rays. Oxygen-15, nitrogen-13, carbon-11, and fluorine-18 are commonly used for PET because they can be readily incorporated into many physiologically active compounds. An ^{18}F-labeled glucose derivative, for instance, is useful for imaging brain regions that respond to various stimuli.

The disadvantage of PET scans is that the necessary radioisotopes are so short-lived that they must be produced on-site immediately before use. The cost of PET is therefore high, because a hospital must install and maintain the necessary nuclear facility.

See Additional Problem 11.68 at the end of the chapter.

Artificial transmutation can lead to the synthesis of entirely new elements never before seen on earth. In fact, all the *transuranium elements*—those elements with atomic numbers greater than 92—have been produced by bombardment reactions. For example, plutonium-241 (^{241}Pu) can be made by bombardment of uranium-238 with α particles:

$$^{238}_{92}\text{U} + {}^{4}_{2}\text{He} \rightarrow {}^{241}_{94}\text{Pu} + {}^{1}_{0}\text{n}$$

Plutonium-241 is itself radioactive, with a half-life of 14.35 years, decaying by β emission to yield americium-241. Americium-241, in turn, decays by α emission with a half-life of 432.2 years. (If the name sounds vaguely familiar, it's because americium is used commercially in making smoke detectors.)

$$^{241}_{94}\text{Pu} \rightarrow {}^{241}_{95}\text{Am} + {}^{0}_{-1}\text{e}$$

Note that all the equations just given for artificial transmutations are balanced. The sum of the mass numbers and the sum of the charges are the same on both sides of each equation.

■ WORKED EXAMPLE 11.6

Californium-246 is formed by bombardment of uranium-238 atoms. If four neutrons are also formed, what particle is used for the bombardment?

ANALYSIS   First write an incomplete nuclear equation incorporating the known information:

$$^{238}_{92}\text{U} + \text{?} \longrightarrow {}^{246}_{98}\text{Cf} + 4\,{}^{1}_{0}\text{n}$$

Then find the numbers of nucleons and charges necessary to balance the equation. In this instance, there are 238 nucleons on the left and $246 + 4 = 250$ nucleons on the right, so the bombarding particle must have $250 - 238 = 12$ nucleons. Furthermore, there are 92 nuclear charges on the left and 98 on the right, so the bombarding particle must have $98 - 92 = 6$ protons.

SOLUTION   The missing particle is ${}^{12}_{6}\text{C}$.

$$^{238}_{92}\text{U} + {}^{12}_{6}\text{C} \longrightarrow {}^{246}_{98}\text{Cf} + 4\,{}^{1}_{0}\text{n}$$

▲ Smoke detectors contain a small amount of americium-241. The α particles emitted by Am-241 ionize the air within the detector, causing it to conduct a tiny electric current. When smoke enters the chamber, conductivity drops and an alarm is triggered.

■ PROBLEM 11.13

What isotope results from α decay of the americium-241 in smoke detectors?

■ PROBLEM 11.14

The element berkelium, first prepared at the University of California at Berkeley in 1949, is made by α bombardment of ${}^{241}_{95}\text{Am}$. Two neutrons are also produced during the reaction. What isotope of berkelium results from this transmutation? Write a balanced nuclear equation.

■ PROBLEM 11.15

Write a balanced nuclear equation for the reaction of argon-40 with a proton:

$$^{40}_{18}\text{Ar} + {}^{1}_{1}\text{H} \rightarrow \text{?} + {}^{1}_{0}\text{n}$$

11.11   Nuclear Fission and Nuclear Fusion

We saw in the previous section that particle bombardment of various elements causes artificial transmutation and results in the formation of new, usually heavier elements. Under very special conditions with a very few isotopes, however, different kinds of nuclear events occur. Certain very heavy nuclei can split apart, and certain very light nuclei can fuse together. The two resultant processes—**nuclear fission** for the fragmenting of heavy nuclei and **nuclear fusion** for the joining together of light nuclei—have changed the world since their discovery in the late 1930s and early 1940s.

Nuclear fission   The fragmenting of heavy nuclei.

Nuclear fusion   The joining together of light nuclei.

Nuclear Fission

Uranium-235 is the only naturally occurring isotope that undergoes nuclear fission. When U-235 is bombarded by a stream of relatively slow-moving neutrons, its nucleus splits to give isotopes of other elements. The split can take place in more than 400 different ways, and more than 800 different fission products have been identified. One of the more frequently occurring pathways generates barium-142 and krypton-91, along with two additional neutrons plus the one neutron that initiated the fission:

$$\ce{_0^1n + _{92}^{235}U -> _{56}^{142}Ba + _{36}^{91}Kr + 3\,_0^1n}$$

As indicated by the balanced nuclear equation above, *one* neutron is used to initiate fission of a U-235 nucleus, but *three* neutrons are released. Thus, a nuclear **chain reaction** can be started: 1 neutron initiates one fission that releases 3 neutrons; the 3 neutrons initiate three new fissions that release 9 neutrons; the 9 neutrons initiate nine fissions that release 27 neutrons; and so on at an ever faster pace (Figure 11.7). If the sample size is small, many of the neutrons escape before initiating additional fission events, and the chain reaction stops. If a sufficient amount of ^{235}U is present, however, an amount called the **critical mass**, then the chain reaction becomes self-sustaining. Under high-pressure conditions that confine the ^{235}U to a small volume, the chain reaction can occur so rapidly that a nuclear explosion results. For ^{235}U, the critical mass is about 56 kg, although the amount can be reduced to approximately 15 kg by placing a coating of ^{238}U around the ^{235}U to reflect back some of the escaping neutrons.

Chain reaction A reaction that, once started, is self-sustaining.

Critical mass The minimum amount of radioactive material needed to sustain a nuclear chain reaction.

FIGURE 11.7 A chain reaction. ▶ Each fission event produces additional neutrons that induce more fissions. The rate of the process increases at each stage. Such chain reactions usually lead to the formation of many different fission products in addition to the two indicated.

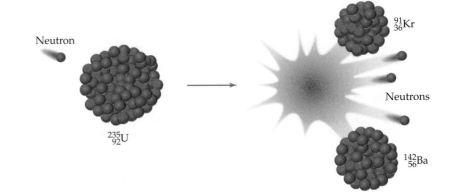

An enormous quantity of heat is released during nuclear fission—the fission of just 1.0 g of uranium-235 produces 3.4×10^8 kcal, for instance. This heat can be used to convert water to steam, which can be harnessed to turn huge generators and produce electric power. The use of nuclear power is much more advanced in some countries than in others, with Lithuania and France leading the way by generating about 86% and 77%, respectively, of their electricity in nuclear plants. In the United States, only about 22% of the electricity is nuclear-generated.

The major objections causing so much public debate about nuclear power plants are twofold: safety and waste disposal. Although a nuclear explosion is not possible under the conditions that exist in a power plant, there is a serious potential radiation hazard should an accident rupture the containment vessel holding the nuclear fuel and release radioactive substances to the environment. Perhaps even more important is the problem posed by disposal of radioactive wastes from nuclear plants. Many of these wastes have such long half-lives that hundreds or even thousands of years must elapse before they will be safe for humans to approach. How to dispose of such hazardous materials safely is an unsolved problem.

▲ Energy from the fission of uranium-235 is used to produce steam and generate electricity in this nuclear power plant.

■ **PROBLEM 11.16**

What other isotope besides tellurium-137 is produced by nuclear fission of uranium-235?

$$\ce{_{92}^{235}U + _0^1n -> _{52}^{137}Te + 2\,_0^1n + ?}$$

Application Archaeological Radiocarbon Dating

Biblical scrolls are found in a cave near the Dead Sea. Are they authentic? A mummy is discovered in an Egyptian tomb. How old is it? The burned bones of a man are dug up near Lubbock, Texas. How long ago did humans live on the North American continent? Using a technique called *radiocarbon dating*, archaeologists can answer these and many other questions. (The Dead Sea Scrolls are 1900 years old and authentic, the mummy is 3100 years old, and the human remains found in Texas are 9900 years old.)

Radiocarbon dating depends on the slow and constant production of radioactive carbon-14 atoms in the upper atmosphere by bombardment of nitrogen atoms with neutrons from cosmic rays. Carbon-14 atoms combine with oxygen to yield $^{14}CO_2$, which slowly mixes with ordinary $^{12}CO_2$ and is then taken up by plants during photosynthesis. When these plants are eaten by animals, carbon-14 enters the food chain and is distributed evenly throughout all living organisms.

As long as a plant or animal is living, a dynamic equilibrium is established in which the organism excretes or exhales the same amount of ^{14}C that it takes in. As a result, the ratio of ^{14}C to ^{12}C in the living organism is the same as that in the atmosphere—about 1 part in 10^{12}. When the plant or animal dies, however, it no longer takes in more ^{14}C. Thus, the $^{14}C/^{12}C$ ratio in the organism slowly decreases as ^{14}C undergoes radioactive decay. At 5730 years (one ^{14}C half-life) after the death of the organism, the $^{14}C/^{12}C$ ratio has decreased by a factor of 2; at 11,460 years after death, the $^{14}C/^{12}C$ ratio has decreased by a factor of 4; and so on.

▲ Radiocarbon dating has determined that these charcoal paintings in the Lascaux cave in France are approximately 15,000 years old.

By measuring the amount of ^{14}C remaining in the traces of any once-living organism, archaeologists can determine how long ago the organism died. Human hair from well-preserved remains, charcoal or wood fragments from once-living trees, and cotton or linen from once-living plants are all useful sources for radiocarbon dating. The accuracy of the technique lessens as a sample gets older and the amount of ^{14}C it contains diminishes, but artifacts with an age of 1000–20,000 years can be dated with reasonable accuracy.

See Additional Problems 11.69, 11.70, and 11.71 at the end of the chapter.

Nuclear Fusion

Just as heavy nuclei such as ^{235}U release energy when they undergo *fission*, very light nuclei such as the isotopes of hydrogen release enormous amounts of energy when they undergo *fusion*. In fact, it's just such a fusion reaction of hydrogen nuclei to produce helium that powers our sun and other stars. Among the processes thought to occur in the sun are those in the following sequence leading to helium-4:

$$^1_1H + {}^2_1H \rightarrow {}^3_2He$$

$$^3_2He + {}^3_2He \rightarrow {}^4_2He + 2\,{}^1_1H$$

$$^3_2He + {}^1_1H \rightarrow {}^4_2He + {}^0_1e$$

Nuclear fusion occurs under the conditions found in stars, where the temperature is on the order of 10^8 K and where nuclei are stripped of all their electrons. On earth, however, the necessary conditions for nuclear fusions have been generated only in the so-called "hydrogen bomb." The dream nevertheless exists that controlled nuclear fusion can provide the ultimate cheap, clean power source. The fuel is deuterium (^2H), available in the oceans in limitless amounts, and there are few radioactive by-products. Providing high enough temperatures and holding material long enough for nuclei to react is a gigantic technical challenge that has yet to be met.

Nuclear medicine. It sounds ominous, evoking images of mushroom clouds, reactors, and meltdowns. But in fact, nuclear medicine is a perfectly safe specialty of radiology, utilizing radioisotopes for diagnosing and treating a wide variety of medical problems.

Ask John T. Green. He's the assistant supervisor of nuclear medicine at a university–affiliated medical center. He's been a nuclear radiation technologist for over 12 years, and he loves his work. The role of his department is to evaluate the function of various body parts and organs. John uses radioisotopes to detect and assess lung, heart, liver, kidney, and other soft tissue abnormalities.

John explains, "We give the isotope to our patients in chemical form, either as intravenous injection or by mouth. There are certain chemical compounds that have an affinity for specific cells, and those are the ones we use to deliver the radioisotopes to the tissues we are interested in."

"We utilize several kinds of isotopes in our hospital. One is gallium-67, a gamma-ray emitter used to study soft tissue abscesses or abnormalities. This isotope is produced in cyclotrons—we get fresh shipments of it weekly. We incorporate it into a compound that tags onto the patient's white blood cells. Gallium-67 has a half-life of 78 hours, so we inject the material intravenously and then have the patient come back 48 hours later for a scan. We scan the whole body with a gamma camera. This is an instrument that detects gamma rays and processes the emissions to produce an image that the radiologist can then study. We look for lymphomas, which are malignancies involving one kind of white blood cell. And because white blood cells migrate to sites of infection, we can find abscesses or other inapparent infections that might explain a fever of unknown origin that the patient is running."

The thyroid has a remarkable ability to take up iodine, which it incorporates into the hormone thyroxine [Chapter 20]. This fact is exploited by John's department, where iodine-131 is used for treatment of thyroid abnormalities or cancer. They also use smaller quantities of the same radioisotope as a diagnostic tool in examining patients who have previously had surgery for thyroid cancer, to search for any indication that the cancer may have spread. Another iodine isotope, iodine-123, is used strictly for diagnostic purposes, to examine size and shape of the thyroid gland itself. The rate at which this isotope is taken up by the gland enables investigators to tell whether a patient's thyroid is overactive, underactive, or normal.

John's department also employs technetium-99*m*, an isotope with many uses. They make up several different tagging kits with technetium, depending on the organ they want to examine. One common target is bone; another is the liver. Thallium scans, by contrast, are strictly for the heart. Their main purpose is to look for ischemic Cardiac tissue—cells that have been injured by being deprived of adequate oxygen, as happens during a heart attack. John explains, "The clinicians examine the patients at different stress levels and at resting levels. They'll have the patient exercise on a treadmill, monitored by the cardiologist as well as by the nuclear medicine technologist. Thallium is then administered and a scan is performed. Once that stress state is determined, the cardiologist wants to compare the results with those from a resting state. So the patient is told to come back maybe three or four hours later and they take another scan to see what has changed.

How did John feel about chemistry in school? "I didn't really give much thought to chemistry when I was in school. I didn't think it was difficult, but I didn't think it was something that I wanted to do for the rest of my life either. I didn't look at it as something that would have much of a role in my future. But now I use it every day, The coolest thing I can think of about this field is that it's constantly changing. There's always a new exam, a new procedure to learn about. And that's kind of special.

Summary: Revisiting the Chapter Goals

1. **What is a nuclear reaction, and how are equations for nuclear reactions balanced?** A *nuclear reaction* is one that changes an atomic nucleus, causing the change of one element into another. Loss of an α particle leads to a new atom whose atomic number is 2 less than that of the starting atom. Loss of a β particle leads to an atom whose atomic number is 1 greater than that of the starting atom:

$$\alpha \text{ emission:} \quad {}^{238}_{92}\text{U} \rightarrow {}^{234}_{90}\text{Th} + {}^{4}_{2}\text{He}$$

$$\beta \text{ emission:} \quad {}^{131}_{53}\text{I} \rightarrow {}^{131}_{54}\text{Xe} + {}^{0}_{-1}\text{e}$$

A nuclear reaction is balanced when the sum of the *nucleons* (protons and neutrons) is the same on both sides of the reaction arrow and when the sum of the charges on the nuclei plus any ejected subatomic particles is the same.

2. **What are the different kinds of radioactivity?** *Radioactivity* is the spontaneous emission of radiation from the nucleus of an unstable atom. The three major kinds of radiation are called *alpha* (α), *beta* (β), and *gamma* (γ). Alpha radiation consists of helium nuclei, small particles containing two protons and two neutrons (^4_2He); β radiation consists of electrons ($^0_{-1}\text{e}$); and γ radiation consists of high-energy light waves. Every element in the periodic table has at least one radioactive isotope, or *radioisotope*.

3. **How are the rates of nuclear reactions expressed?** The rate of a nuclear reaction is expressed in units of *half-life* ($t_{1/2}$), where one half-life is the amount of time necessary for one-half of the radioactive sample to decay.

4. **What is ionizing radiation?** High-energy radiation of all types—α particles, β particles, γ rays, and X rays—is called *ionizing radiation*. When any of these kinds of radiation strikes an atom, it dislodges an electron and gives a reactive ion that can be lethal to living cells. Gamma rays and X rays are the most penetrating and most harmful types of external radiation; α and β particles are the most dangerous types of internal radiation because of their high energy and the resulting damage to surrounding tissue.

5. **How is radioactivity measured?** Radiation intensity is expressed in different ways according to the property being measured. The *curie* (Ci) measures the number of radioactive disintegrations per second in a sample; the *roentgen* (R) measures the ionizing ability of radiation; the *rad* measures the amount of radiation energy absorbed per gram of tissue; and the *rem* measures the amount of tissue damage caused by radiation. Radiation effects become noticeable with a human exposure of 25 rem and become lethal at an exposure above 600 rem.

6. **What is transmutation?** *Transmutation* is the change of one element into another brought about by a nuclear reaction. Most known radioisotopes do not occur naturally but are made by bombardment of an atom with a high-energy particle. In the ensuing collision between particle and atom, a nuclear change occurs and a new element is produced by *artificial transmutation*.

7. **What are nuclear fission and nuclear fusion?** With a very few isotopes, including $^{235}_{92}\text{U}$, the nucleus is split apart by neutron bombardment to give smaller fragments. A large amount of energy is released during this *nuclear fission*, leading to use of the reaction for generating electric power. *Nuclear fusion* results when small nuclei such as those of tritium (^3_1H) and deuterium (^2_1H) combine to give a heavier nucleus.

Key Words

α **Particle**, p. 299

Artificial transmutation, p. 313

β **Particle**, p. 299

Chain reaction, p. 316

Cosmic rays, p. 308

Critical mass, p. 316

Decay series, p. 307

Electron capture, p. 303

γ **Radiation**, p. 299

Half-life ($t_{1/2}$), p. 306

Ionizing radiation, p. 308

Nuclear decay, p. 300

Nuclear fission, p. 315

Nuclear fusion, p. 315

Nuclear reaction, p. 297

Nucleon, p. 297

Nuclide, p. 297

Positron, p. 303

Radioactivity, p. 298

Radioisotope, p. 299

Radionuclide, p. 299

Transmutation, p. 300

X rays, p. 308

▪ Understanding Key Concepts

11.17 Magnesium-28 decays by β emission to give aluminum-28. If yellow spheres represent $^{28}_{12}Mg$ atoms and blue spheres represent $^{28}_{13}Al$ atoms, how many half-lives have passed in the following sample?

11.18 Refer to Figure 11.4 and then make a drawing similar to that in Problem 11.15 representing the decay of a sample of $^{28}_{12}Mg$ after approximately 4 half-lives have passed.

11.19 Write the symbol of the isotope represented by the following drawing. Blue spheres represent neutrons, and red spheres represent protons.

11.20 Shown below is a portion of the decay series for plutonium-241 ($^{241}_{94}Pu$). The series has two kinds of arrows: shorter arrows pointing right and longer arrows pointing left. Which arrow corresponds to an α emission, and which to a β emission? Explain.

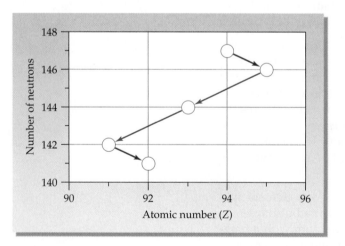

11.21 Identify and write the symbol for each of the nuclides in the decay series shown in Problem 11.20.

11.22 Identify the isotopes involved, and tell the type of decay process occurring in the following nuclear reaction:

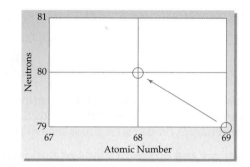

11.23 What is the half-life of the radionuclide that shows the following decay curve?

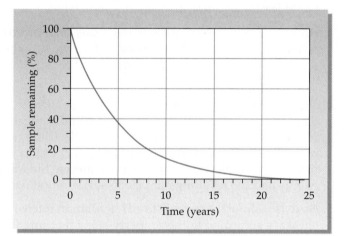

■ Additional Problems

Radioactivity

11.24 What does it mean to say that a substance is radioactive?

11.25 Describe how α radiation, β radiation, γ radiation, positron emission, and electron capture differ.

11.26 List several ways in which a nuclear reaction differs from a chemical reaction.

11.27 What word is used to describe the change of one element into another?

11.28 What symbol is used for an α particle in a nuclear equation?

11.29 What symbol is used for a β particle in a nuclear equation?

11.30 Which kind of radiation, α, β, or γ, has the highest penetrating power, and which has the lowest?

11.31 What happens when ionizing radiation strikes an atom in a chemical compound?

11.32 How does ionizing radiation lead to cell damage?

11.33 What are the main sources of background radiation?

11.34 How can a nucleus emit an electron during β decay when there are no electrons present in the nucleus to begin with?

11.35 What is the difference between an α particle and a helium atom?

Nuclear Decay and Transmutation

11.36 What does it mean to say that a nuclear equation is balanced?

11.37 What are transuranium elements, and how are they made?

11.38 What happens to the mass number and atomic number of an atom that emits an α particle? A β particle?

11.39 What happens to the mass number and atomic number of an atom that emits a γ ray? A positron?

11.40 How does nuclear fission differ from normal radioactive decay?

11.41 What characteristic of uranium-235 fission causes a chain reaction?

11.42 What products result from radioactive decay of the following β emitters?

(a) $^{35}_{16}\text{S}$　　(b) $^{24}_{10}\text{Ne}$　　(c) $^{90}_{38}\text{Sr}$

11.43 What products result from radioactive decay of the following α emitters?

(a) $^{190}_{78}\text{Pt}$　　(b) $^{208}_{87}\text{Fr}$　　(c) $^{245}_{96}\text{Cm}$

11.44 Identify the starting radioisotopes that give the following products:

(a) $? \rightarrow {}^{140}_{56}\text{Ba} + {}^{0}_{-1}\text{e}$　　(b) $? \rightarrow {}^{242}_{94}\text{Pu} + {}^{4}_{2}\text{He}$

11.45 What products are formed in the following nuclear reactions?

(a) $^{109}_{47}\text{Ag} + {}^{4}_{2}\text{He} \rightarrow ?$　　(b) $^{10}_{5}\text{B} + {}^{4}_{2}\text{He} \rightarrow ? + {}^{1}_{0}\text{n}$

11.46 Balance the following equations for the nuclear fission of $^{235}_{92}\text{U}$:

(a) $^{235}_{92}\text{U} + {}^{1}_{0}\text{n} \rightarrow {}^{160}_{62}\text{Sm} + {}^{72}_{30}\text{Z n} + ?\,{}^{1}_{0}\text{n}$

(b) $^{235}_{92}\text{U} + {}^{1}_{0}\text{n} \rightarrow {}^{87}_{35}\text{Br} + ? + 3\,{}^{1}_{0}\text{n}$

11.47 Complete and balance the following nuclear equations:

(a) $^{126}_{50}\text{Sn} \rightarrow {}^{0}_{-1}\text{e} + ?$

(b) $^{210}_{88}\text{Ra} \rightarrow {}^{4}_{2}\text{He} + ?$

(c) $^{76}_{36}\text{Kr} + {}^{0}_{-1}\text{e} \rightarrow ?$

11.48 For centuries, alchemists dreamed of turning base metals into gold. The dream finally became reality when it was shown that mercury-198 can be converted into gold-198 on bombardment by neutrons. What small particle is produced in addition to gold-198? Write a balanced nuclear equation for the reaction.

11.49 Meitnerium-266 ($^{266}_{109}\text{Mt}$) was prepared in 1982 by bombardment of bismuth-209 atoms with iron-58. What other product must also have been formed? Write a balanced nuclear equation for the transformation.

Half-Life

11.50 What does it mean when we say that strontium-90, a waste product of nuclear power plants, has a half-life of 28.8 years?

11.51 What percentage of the original radioactivity remains in a sample after two half-lives have passed?

11.52 Selenium-75, a β emitter with a half-life of 120 days, is used medically for pancreas scans. Approximately how much selenium-75 would remain from a 0.050 g sample that has been stored for 1 year?

11.53 Approximately how long would it take a sample of selenium-75 to lose 99% of its radioactivity? (See Problem 11.52.)

11.54 The half-life of mercury-197 is 64.1 hours. If a patient undergoing a kidney scan is given 5.0 ng of mercury-197, how much will remain after 6 days? After 27 days?

11.55 Gold-198, a β emitter used to treat leukemia, has a half-life of 2.695 days. The standard dosage is about 1.0 mCi/kg body weight.

(a) What is the product of the β emission of gold-198?

(b) How long does it take a 30.0 mCi sample of gold-198 to decay so that only 3.75 mCi remains?

(c) How many millicuries are required by a 70.0 kg adult?

Measuring Radioactivity

11.56 Describe how a Geiger counter works.

11.57 Describe how a film badge works.

11.58 Why are rems the preferred units for measuring the health effects of radiation?

11.59 Approximately what amount (in rems) of short-term exposure to radiation produces noticeable effects in humans?

11.60 Match each unit in the left-hand column with the property being measured in the right-hand column:

(1) curie　　(a) Ionizing intensity of radiation

(2) rem　　(b) Amount of tissue damage

(3) rad　　(c) Number of disintegrations per second

(4) roentgen　　(d) Amount of radiation per gram of tissue

11.61 Sodium-24 is used to study the circulatory system and to treat chronic leukemia. It is administered in the form of saline (NaCl) solution, with a therapeutic dosage of 180 μCi/kg body weight. How many milliliters of a 5.0 mCi/mL solution are needed to treat a 68 kg adult?

11.62 A Se-75 source is producing 300 rem at a distance of 2.0 m. What distance is needed to decrease the intensity of exposure to below 25 rem, the level at which no effects would be detectable?

11.63 If a radiation source has an intensity of 650 rem at 1.0 m, what is its intensity at 50.0 m?

Applications

11.64 Why is radiation therapy used for cancer treatment? [*Medical Uses of Radioactivity*]

11.65 What are the three main classes of techniques used in nuclear medicine? [*Medical Uses of Radioactivity*]

11.66 What is the purpose of food irradiation, and how does it work? [*Irradiated Food*]

11.67 What kind of radiation is used to treat food? [*Irradiated Food*]

11.68 What are the advantages of CT and PET relative to conventional X rays? [*Body Imaging*]

11.69 Why is ^{14}C dating useful only for samples that contain material from objects that were once alive? [*Archaeological Radiocarbon Dating*]

11.70 Why can't ^{14}C dating be used to tell if a piece of leather was made in 1850? [*Archaeological Radiocarbon Dating*]

11.71 Some dried beans with a ^{14}C/^{12}C ratio one-eighth of the current value were found in an old cave. How old are the beans? [*Archeological Radiocarbon Dating*]

General Questions and Problems

11.72 Harmful chemical spills can often be cleaned up by treatment with another chemical. For example, a spill of H_2SO_4 might be neutralized by addition of $NaHCO_3$. Why can't the harmful radioactive wastes from nuclear power plants be cleaned up just as easily?

11.73 Why is a scintillation counter or Geiger counter more useful for determining the existence and source of a new radiation leak than a film badge?

11.74 Technetium-99m, used for brain scans and to monitor heart function, is formed by decay of Mo-99.

   (a) By what type of decay does Mo-99 produce Tc-99m?

   (b) Molybdenum-99 is formed by neutron bombardment of a natural isotope. If one neutron is absorbed and there are no other by-products of this process, from what isotope is Mo-99 formed?

11.75 The half-life of Tc-99m (Problem 11.74) is 6.01 hours. If a sample with an initial activity of 15 μCi is injected into a patient, what is the activity in 24 hours, assuming that none of the sample is excreted?

11.76 Plutonium-238 is an α emitter used to power batteries for heart pacemakers.
(a) What is the product of this emission?
(b) Why is a pacemaker battery enclosed in a metal case before being inserted into the chest cavity?

11.77 Why do dentists place a dense apron over a patient before taking X rays?

11.78 High levels of radioactive fallout after the 1986 accident at the Chernobyl nuclear power plant in Russia resulted in numerous miscarriages and many instances of farm animals born with severe defects. Why are embryos and fetuses particularly susceptible to the effects of radiation?

11.79 What are the main advantages of fusion relative to fission as an energy source? What are the drawbacks?

11.80 Write balanced nuclear equations for the following processes:
(a) α emission of ^{162}Re
(b) β emission of ^{188}W

11.81 Balance the following transmutation reactions:
(a) $^{253}_{99}\text{Es} + ? \rightarrow ^{256}_{101}\text{Md} + ^{1}_{0}\text{n}$
(b) $^{250}_{98}\text{Cf} + ^{11}_{5}\text{B} \rightarrow ? + 4\,^{1}_{0}\text{n}$

11.82 The most abundant isotope of uranium, ^{238}U, does not undergo fission. In a *breeder reactor*, however, a ^{238}U atom captures a neutron and emits two beta particles to make a fissionable isotope of plutonium, which can then be used as fuel in a nuclear reactor. Write a balanced nuclear equation.

11.83 Boron is used in *control rods* for nuclear reactors because it can absorb neutrons and emit alpha particles. Balance the equation:

$$^{10}_{5}\text{B} + ^{1}_{0}\text{n} \rightarrow ? + ^{4}_{2}\text{He}$$

11.84 Thorium-232 decays by a ten-step series, ultimately yielding lead-208. How many alpha particles and how many beta particles are emitted?

11.85 Californium-246 is formed by bombardment of uranium-238 atoms. If four neutrons are formed as by-products, what particle is used for the bombardment?

12

Introduction to Organic Chemistry
Alkanes

The gasoline, kerosene, and other products of this petroleum refinery are primarily mixtures of simple organic compounds called alkanes.

As knowledge of chemistry slowly grew in the 1700s, mysterious differences were noted between compounds obtained from living sources and those obtained from minerals. It was found, for instance, that chemicals from living sources were often liquids or low-melting solids, while chemicals from mineral sources were usually high-melting solids. Furthermore, chemicals from living sources were generally more difficult to purify and work with than those from minerals. To express these differences, the term *organic chemistry* was introduced to mean the study of compounds from living organisms, while *inorganic chemistry* was used to refer to the study of compounds from minerals.

Today we know that there are no fundamental differences between organic and inorganic compounds: The same scientific principles are applicable to both. The only common characteristic of compounds from living sources is that they contain the element carbon. Thus, organic chemistry is now defined as the study of carbon compounds.

Why is carbon special? The answer derives from its position in the periodic table. As a group 4A nonmetal, carbon atoms have the unique ability to form four strong covalent bonds. Furthermore, carbon atoms can bond together, forming long chains and rings. Of all the elements, only carbon is able to form such an immense array of compounds, from methane with one carbon atom to DNA with billions of carbons.

In this and the next five chapters, we'll look at the chemistry of organic compounds, beginning with answers to the following questions:

1. **What are functional groups, and how are they used to classify organic molecules?**
   The goal: Be able to classify organic molecules into families by functional group.
2. **What are isomers?**
   The goal: Be able to recognize and draw constitutional isomers.
3. **How are organic molecules drawn?**
   The goal: Be able to convert between structural formulas and condensed or line structures.
4. **What are alkanes and cycloalkanes, and how are they named?**
   The goal: Be able to name an alkane or cycloalkane from its structure, or write the structure, given the name.
5. **What are the general properties and chemical reactions of alkanes?**
   The goal: Be able to describe the physical properties of alkanes and the products formed in the combustion and halogenation reactions of alkanes.

CONCEPTS TO REVIEW

Covalent Bonds (Sections 5.1 and 5.2)
Multiple Covalent Bonds (Section 5.3)
Drawing Lewis Structures (Section 5.6)
VSEPR and Molecular Shapes (Section 5.7)
Polar Covalent Bonds (Section 5.8)

12.1 The Nature of Organic Molecules

Let's begin a study of **organic chemistry**—the chemistry of carbon compounds—by reviewing what we've seen in earlier chapters about the structures of organic molecules:

Organic chemistry The study of carbon compounds.

- **Carbon is tetravalent; it always forms four bonds** (Section 5.2). In methane, for example, carbon is connected to four hydrogen atoms:

Methane, CH_4

- **Organic molecules have covalent bonds** (Section 5.2). In ethane, for example, the bonds result from the sharing of two electrons, either between C and C atoms or between C and H atoms:

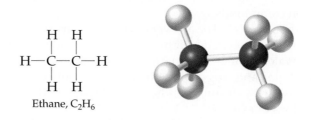

Ethane, C_2H_6

- **Organic molecules contain polar covalent bonds when carbon bonds to an electronegative element on the right side of the periodic table** (Section 5.8). In chloromethane, for example, the electronegative chlorine atom attracts electrons more strongly than carbon, resulting in polarization of the C—Cl bond so that carbon has a partial positive charge, $\delta+$ and chlorine has a partial negative charge, $\delta-$. In electrostatic potential maps (Section 5.8), the chlorine atom is therefore red and the carbon blue.

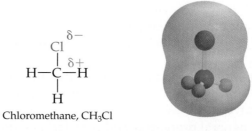

Chloromethane, CH_3Cl

- **Carbon can form multiple covalent bonds by sharing more than two electrons with a neighboring atom** (Section 5.3). In ethylene, for example, the two carbon atoms share four electrons in a double bond; in acetylene, the two carbons share six electrons in a triple bond:

H—C≡C—H
Acetylene, C_2H_2

- **Organic molecules have specific three-dimensional shapes** (Section 5.7). When carbon is bonded to four atoms, as in methane, CH_4 the bonds are oriented toward the four corners of a regular tetrahedron with carbon in the center. Such three-dimensionality is commonly shown using normal lines for bonds in the plane of the page, dashed lines for bonds receding behind the page, and wedged lines for bonds coming out of the page:

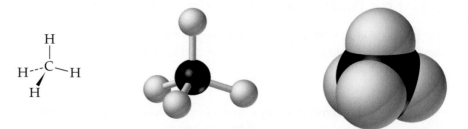

- **Organic molecules often contain hydrogen, nitrogen, and oxygen, in addition to carbon** (Section 5.6). Nitrogen can form single, double, and triple bonds to carbon; oxygen can form single and double bonds:

C—N   C—O   C—H
C=N   C=O
C≡N

Covalent bonding makes organic compounds quite different from the inorganic compounds we've been concentrating on up to this point. For example, inorganic compounds such as NaCl have high melting points and high boiling points because they consist of large numbers of oppositely charged ions held together by strong electrical attractions (Section 4.4). (∞, p. 75) Organic compounds, by contrast, consist of atoms joined by covalent bonds into individual molecules. Because the molecules are attracted to one another only by weak intermolecular forces (Section 8.11), organic compounds generally have lower melting and boiling points than do inorganic salts. (∞, p. 209) Many simple organic compounds are liquids at room temperature, and a few are even gases.

Still other important differences between organic and inorganic compounds include solubility and electrical conductivity. Whereas many inorganic compounds dissolve in water to yield solutions of ions that conduct electricity (Section 9.9), most organic compounds are insoluble in water and do not conduct electricity. (∞, p. 248) Only small polar organic molecules, such as glucose and ethyl alcohol, or large molecules with many polar groups, such as some proteins, dissolve in water. This lack of water solubility for organic compounds has important practical consequences, including the difficulty in removing greasy dirt and in cleaning up environmental oil spills.

◄   Oil spills can be a serious environmental problem because of the lack of solubility of oil in water.

∞ LOOKING AHEAD

The interior of a living cell is largely a water solution containing many hundreds of different compounds. We'll see in later chapters how cells use membranes composed of water-insoluble organic molecules to enclose their watery interiors and to regulate the flow of substances across the cell boundary. ∞

12.2 Families of Organic Molecules: Functional Groups

There are more than 18 *million* organic compounds described in the scientific literature. Each has unique physical properties, such as melting point and boiling point, and each has unique chemical properties. How can we ever understand them all?

The situation isn't as hopeless as it sounds because chemists have learned through experience that organic compounds can be classified into families according to their structural features and that the chemical behavior of the members of a family is often predictable. Instead of 18 million compounds with random chemical reactivity, there are just a few general families of organic compounds whose chemistry falls into simple patterns.

Functional group An atom or group of atoms within a molecule that has a characteristic structure and chemical behavior.

The structural features that allow us to class compounds together are called *functional groups*. A **functional group** is a group of atoms within a larger molecule that has a characteristic structure and chemical behavior. A given functional group tends to undergo the same reactions in every molecule it's a part of. For example, the carbon–carbon double bond is one of the most common functional groups. Ethylene (C_2H_4) the simplest compound with a double bond, undergoes many chemical reactions similar to those of oleic acid $(C_{18}H_{34}O_2)$, a much larger and more complex compound that also contains a double bond. Both, for example, react with hydrogen in the same manner, as shown in Figure 12.1. These identical reactions with hydrogen are typical: *The chemistry of an organic molecule is primarily determined, not by its size and complexity, but by the functional groups it contains.*

FIGURE 12.1 The reactions of ▶ **(a) ethylene and (b) oleic acid with hydrogen.** The carbon–carbon double bond functional group adds two hydrogen atoms in both cases, regardless of the complexity of the rest of the molecule.

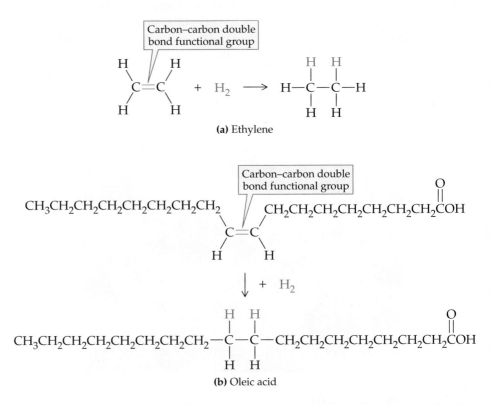

Table 12.1 lists some of the most important families of organic molecules and their distinctive functional groups. Compounds that contain a $C{=}C$ double bond, for instance, are in the *alkene* family; compounds that have an —OH group are in the *alcohol* family; and so on.

Much of the chemistry discussed in this and the next five chapters is the chemistry of the families listed in Table 12.1, so it's best to memorize the names and become familiar with their structures now. Note that they fall into three groups:

Hydrocarbon An organic compound that contains only carbon and hydrogen.

- The first four families in Table 12.1 are **hydrocarbons**, organic compounds that contain only carbon and hydrogen. *Alkanes* have only single bonds and contain no functional groups. As we'll see, the absence of functional groups makes alkanes relatively unreactive. *Alkenes* contain a carbon–carbon double bond functional group; *alkynes* contain a carbon–carbon triple bond functional group; and *aromatic* compounds contain a six-membered ring of carbon atoms with three alternating double bonds.

- The next four families in Table 12.1 have functional groups that contain only single bonds and have a carbon atom bonded to an electronegative atom. *Alkyl halides* have a carbon–halogen bond; *alcohols* have a carbon–oxygen bond; *ethers* have two carbons bonded to the same oxygen; and *amines* have a carbon–nitrogen bond.

TABLE 12.1 Some Important Families of Organic Molecules

Family Name	Functional Group Structure*	Simple Example	Name Ending
Alkane	Contains only C—H and C—C single bonds	CH_3CH_3 Ethane	-ane
Alkene	C=C	$H_2C=CH_2$ Ethylene	-ene
Alkyne	—C≡C—	H—C≡C—H Acetylene (Ethyne)	-yne
Aromatic	(benzene ring)	Benzene	None
Alkyl halide	—C—X (X=F, Cl, Br, I)	CH_3—Cl Methyl chloride	None
Alcohol	—C—O—H	CH_3—OH Methyl alcohol (Methanol)	-ol
Ether	—C—O—C—	CH_3—O—CH_3 Dimethyl ether	None
Amine	—C—N	CH_3—NH_2 Methylamine	-amine
Aldehyde	—C—C—H (C=O)	CH_3—C—H Acetaldehyde (Ethanal)	-al
Ketone	—C—C—C— (C=O)	CH_3—C—CH_3 Acetone	-one
Carboxylic acid	—C—C—OH (C=O)	CH_3—C—OH Acetic acid	-ic acid
Anhydride	—C—C—O—C—C— (C=O, C=O)	CH_3—C—O—C—CH_3 Acetic anhydride	None
Ester	—C—C—O—C— (C=O)	CH_3—C—O—CH_3 Methyl acetate	-ate
Amide	—C—C—NH_2, —C—C—N—H, —C—C—N— (C=O)	CH_3—C—NH_2 Acetamide	-amide

*The bonds whose connections aren t specified are assumed to be attached to carbon or hydrogen atoms in the rest of the molecule.

- The remaining families in Table 12.1 have functional groups that contain a carbon–oxygen double bond: *aldehydes, ketones, carboxylic acids, anhydrides, esters,* and *amides.*

Functional Groups

■ **PROBLEM 12.1**

Locate and identify the functional groups in **(a)** lactic acid, from sour milk; **(b)** methyl methacrylate, used in making Lucite and Plexiglas; and **(c)** phenylalanine, an amino acid found in proteins.

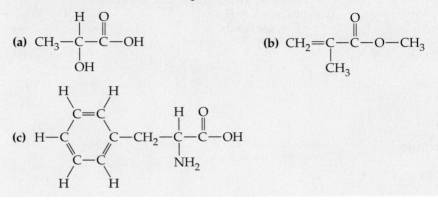

■ **PROBLEM 12.2**

Propose structures for molecules that fit the following descriptions:
(a) C_2H_4O containing an aldehyde functional group
(b) $C_3H_6O_2$ containing a carboxylic acid functional group

12.3 The Structure of Organic Molecules: Alkanes and Their Isomers

Alkane A hydrocarbon that has only single bonds.

Hydrocarbons that contain only single bonds belong to the family of organic molecules called **alkanes**. If we imagine ways that one carbon and four hydrogens can combine, there is only one possibility: methane, CH_4. If we imagine ways that two carbons and six hydrogens can combine, only ethane, CH_3CH_3, is possible. And if we imagine the combination of three carbons with eight hydrogens, only propane, $CH_3CH_2CH_3$, is possible.

Methane, Ethane, Propane

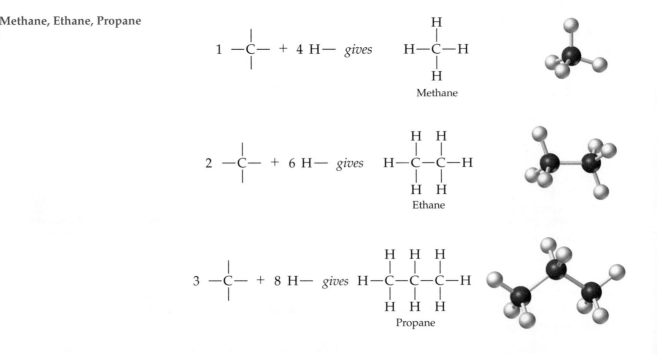

If larger numbers of carbons and hydrogens combine, however, more than one kind of molecule can be formed. There are two ways in which molecules with the formula C_4H_{10} can be formed: The four carbons can either be in a row or have a branched arrangement. Similarly, there are three ways in which molecules with the formula C_5H_{12} can be formed; and so on for larger alkanes. Compounds with all their carbons connected in a continuous line are called **straight-chain alkanes**, while those with a branching connection of carbons are called **branched-chain alkanes**. Note that in a straight-chain alkane, you can draw a line through all the carbon atoms without lifting your pencil from the paper. In a branched-chain alkane, however, you must either lift your pencil from the paper or retrace your steps to draw a line through all the carbons.

Straight-chain alkane An alkane that has all its carbons connected in a row.

Branched-chain alkane An alkane that has a branching connection of carbons.

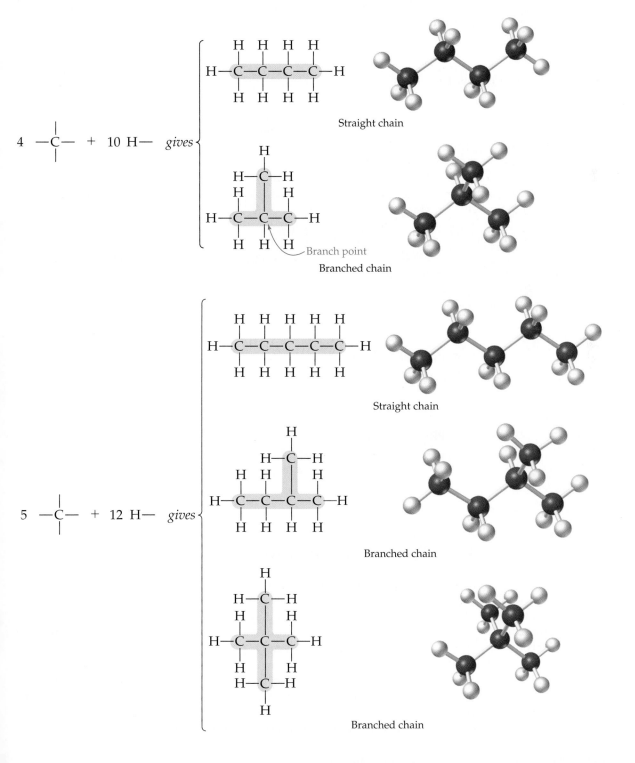

Straight chain

Branched chain

Straight chain

Branched chain

Branched chain

Isomers Compounds with the same molecular formula but different structures.

Constitutional isomers Compounds with the same molecular formula but different connections among their atoms.

Compounds with the same formula but different structures are called **isomers**. More specifically, the two different C_4H_{10} compounds and the three different C_5H_{12} compounds are **constitutional isomers**—compounds with the same molecular formula but with different connections among their constituent atoms. Needless to say, the number of possible alkane isomers grows rapidly as the number of carbon atoms increases.

Different constitutional isomers are completely different chemical compounds. They have different structures, different physical properties such as melting point and boiling point, and potentially different physiological properties. For example, ethyl alcohol and dimethyl ether both have the formula C_2H_6O but ethyl alcohol is a liquid with a boiling point of 78.5°C while dimethyl ether is a gas with a boiling point of −23°C (Table 12.2). Clearly, molecular formulas by themselves aren't very useful in organic chemistry; a knowledge of structure is also necessary.

TABLE 12.2 Some Properties of Ethyl Alcohol and Dimethyl Ether

Name and Molecular Formula	Structure		Boiling Point	Melting Point	Physiological Activity
Ethyl alcohol C_2H_6O			78.5°C	− 117.3°C	Central-nervous-system depressant
Dimethyl ether C_2H_6O			− 23°C	− 138.5°C	Nontoxic; anesthetic at high concentration

■ **PROBLEM 12.3**

Draw the straight-chain isomer with the formula C_7H_{16}.

■ **PROBLEM 12.4**

Draw two branched-chain isomers with the formula C_7H_{16}.

12.4 Drawing Organic Structures

Condensed structure A shorthand way of drawing structures in which C—C and C—H bonds are understood rather than shown.

Drawing structural formulas that show every atom and every bond in a molecule is both time-consuming and awkward, even for relatively small molecules. Much easier is the use of **condensed structures** (Section 5.6), which are simpler but show the essential information about which functional groups are present and how atoms are connected. (∞, p. 108) In condensed structures, C—C and C—H single bonds aren't shown; rather, they're "understood." If a carbon atom has three hydrogens bonded to it, we write CH_3; if the carbon has two hydrogens bonded to it, we write CH_2; and so on. For example, the four-carbon, straight-chain alkane (butane) and its branched-chain isomer (2-methylpropane) can be written as the following condensed structures:

 Condensed Structural Formula

Application Natural versus Synthetic

Prior to the development of organic chemistry as a science, only substances obtained directly from plants and animals were available for treating diseases, dying clothes, cleansing and perfuming our bodies, and so forth. Extracts of the opium poppy, for instance, have been used since the seventeenth century for the relief of pain. The prized purple dye called *Tyrian purple*, obtained from a Middle Eastern mollusk, has been known since antiquity. Oils distilled from bergamot, sweet bay, rose, and lavender, have been employed for centuries in making perfume.

Many of these naturally occurring substances were first used without any knowledge of their chemical composition. As organic chemistry developed, though, chemists learned how to work out the structures of compounds. The disease-curing properties of limes and other citrus fruits, for example, were known for centuries, but the chemical structure of vitamin C, the active ingredient, was not determined until 1933. Today there is a revival of interest in folk remedies, and a large effort is being made to identify medicinally important chemical compounds found in plants.

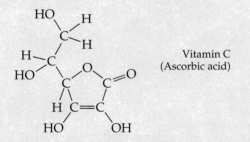

Vitamin C
(Ascorbic acid)

Once a structure is known, organic chemists try to *synthesize* the compound in the laboratory. If the starting materials are inexpensive and the process is simple enough, it may be more economical to manufacture a compound than to isolate it from a natural substance. In the case of vitamin C, a complete synthesis was achieved in 1933, and it is now much cheaper to synthesize it from glucose than to extract it from natural

sources. Worldwide, more than 80 million pounds are synthesized each year.

Some individuals still demand vitamins only from natural sources, assuming that "natural" is somehow better. Although eating an orange is certainly better for you than taking a vitamin C tablet, the difference lies in the many other substances present in food. The vitamin C is exactly the same, whether "natural" or "synthetic," just as the NaCl produced by reacting sodium and chlorine in the laboratory is identical to the NaCl found in the ocean.

See Additional Problem 12.62 at the end of the chapter.

▲ Whether from the laboratory or from food, the vitamin C in these two sources is the same.

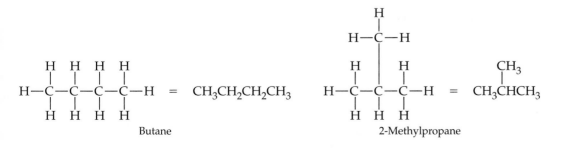

Butane   2-Methylpropane

Note that the horizontal bonds between carbons aren't shown—the CH_3 and CH_2 units are simply placed next to one another—but that the vertical bond in 2-methylpropane *is* shown for clarity. Occasionally, as a further simplification, a

333

row of CH_2 groups is shown in parentheses, with a subscript equal to the number of groups. For example, the six-carbon, straight-chain alkane (hexane) can be written as

$$CH_3CH_2CH_2CH_2CH_2CH_3 = CH_3(CH_2)_4CH_3$$

■ **PROBLEM 12.5**

Draw the following three isomers of C_5H_{12} as condensed structures:

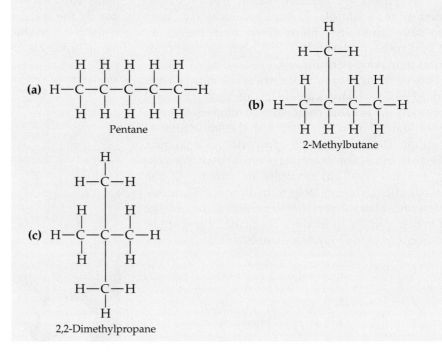

(a) Pentane

(b) 2-Methylbutane

(c) 2,2-Dimethylpropane

12.5 The Shapes of Organic Molecules

Every carbon atom in an alkane has its four bonds pointing toward the four corners of a tetrahedron, but chemists don't usually worry about three-dimensional shapes when writing condensed structures. Condensed structures don't imply any particular three-dimensional shape; they only indicate the connections between atoms without specifying geometry.

In fact, butane has no one single shape because *rotation* takes place around carbon–carbon single bonds. The two parts of a molecule joined by a carbon–carbon single bond are free to spin around the bond, giving rise to an infinite number of possible three-dimensional geometries, or **conformations**. A given butane molecule might be fully extended at one instant but twisted an instant later (Figure 12.2). An actual sample of butane contains a great many molecules that are constantly changing shape. At any given instant, however, most of the molecules have the least crowded, extended conformation shown in Figure 12.2a. The same is true for all other alkanes: At any given instant most molecules are in the least crowded conformation.

Conformation The specific three-dimensional arrangement of atoms in a molecule.

FIGURE 12.2 **Some conformations of butane (there are many others as well).** The least crowded, extended conformation in (a) is the lowest-energy one.

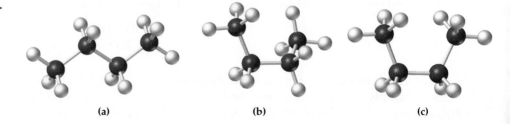

(a)   (b)   (c)

As long as any two structures show identical connections between atoms, they represent identical compounds no matter how the structures are drawn. Sometimes you have to mentally rotate structures to see whether they're the same or different. To see that the following two structures represent the same compound rather than two isomers, picture one of them flipped right-to-left so that the red CH_3 groups are on the same side.

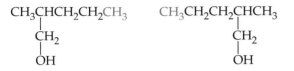

■ **WORKED EXAMPLE 12.1**

The following structures all have the formula C_7H_{16}. Which of them represent the same molecule?

$$\begin{array}{cc} \quad CH_3 & \quad CH_3 \\ \quad | & \quad | \\ \textbf{(a)}\ \ CH_3CHCH_2CH_2CH_2CH_3 & \textbf{(b)}\ \ CH_3CH_2CH_2CH_2CHCH_3 \end{array}$$

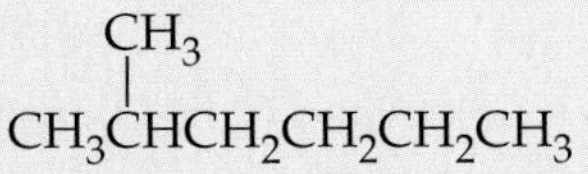

$$\begin{array}{c} \quad\quad CH_3 \\ \quad\quad | \\ \textbf{(c)}\ \ CH_3CH_2CH_2CHCH_2CH_3 \end{array}$$

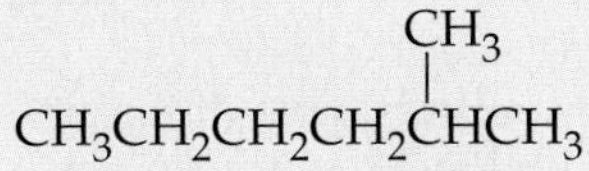

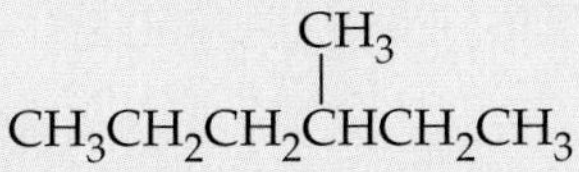

ANALYSIS Pay attention to the *connections* between atoms. Don't get confused by the apparent differences caused by writing a structure right to left versus left to right.

SOLUTION Molecule (a) has a straight chain of six carbons with a —CH_3 branch on the second carbon from the end. Molecule (b) also has a straight chain of six carbons with a —CH_3 branch on the second carbon from the end and is therefore identical to (a). The only difference between (a) and (b) is that one is written "forward" and one is written "backward." Molecule (c), by contrast, has a straight chain of six carbons with a —CH_3 branch on the *third* carbon from the end and is therefore an isomer of (a) and (b).

■ **WORKED EXAMPLE 12.2**

Are the following pairs of compounds the same, isomers, or unrelated?

$$\begin{array}{cc} \quad CH_3 & \quad CH_3 \\ \quad | & \quad | \\ \textbf{(a)}\ \ CH_3CHCH_2CH_2 & CH_3CHCH_2CH_2CH_3 \\ \quad | \\ \quad CH_3 \end{array}$$

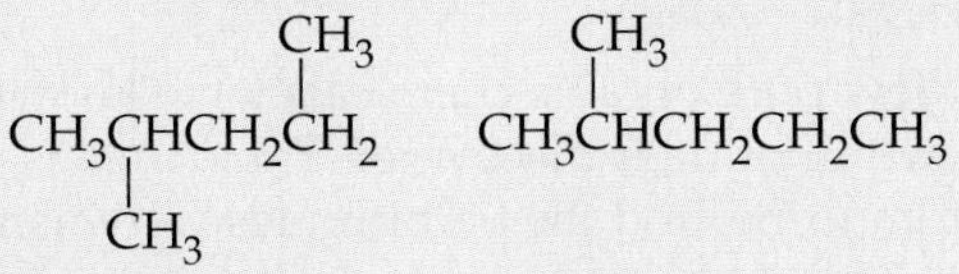

$$\begin{array}{cc} & \quad\quad CH_2CH_3 \\ & \quad\quad | \\ \textbf{(b)}\ \ CH_3CH_2CHCH_3 & CH_3CHCH_2 \\ \quad\quad | & \quad | \\ \quad\quad CH_2CH_3 & CH_3 \end{array}$$

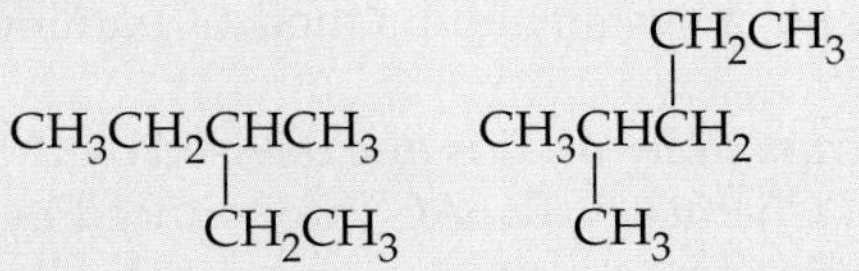

$$\begin{array}{cc} & \quad O \\ & \quad \| \\ \textbf{(c)}\ \ CH_3CH_2OCH_3 & CH_3CH_2CH \end{array}$$

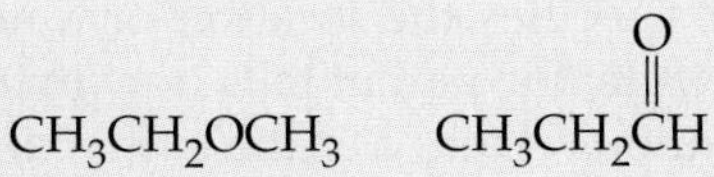

ANALYSIS First compare molecular formulas to see if the compounds are related, and then look at the structures to see if they are the same compound or isomers. Find the longest continuous carbon chain in each, and then compare the locations of the substituents connected to the longest chain.

SOLUTION

(a) Both compounds have the same molecular formula (C_6H_{14}) so they are related. Since the $-CH_3$ group is on the second carbon from the end of a five-carbon chain in both cases, these compounds are identical.

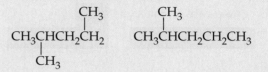

(b) Both compounds have the same molecular formula (C_6H_{14}) and the longest chain in each is five carbon atoms. A comparison shows, however, that the $-CH_3$ group is on the middle carbon atom in one structure and on the second carbon atom in the other. These compounds are isomers.

$$CH_3CH_2CHCH_3 \qquad CH_3CHCH_2$$
$$\overset{|}{C}H_2CH_3 \qquad \overset{|}{C}H_3 \qquad \overset{|}{C}H_2CH_3$$

(c) These compounds have different formulas (C_3H_8O and C_3H_6O), so they are unrelated and are not isomers.

■ **PROBLEM 12.6**

Which of the following structures are identical?

(a)
$$\overset{\overset{\textstyle CH_3}{|}}{CH_2}CH_2\overset{\overset{\textstyle CH_3}{|}}{CH}CH_2CH_3$$

(b)
$$CH_3CH_2CH_2\overset{\overset{\textstyle CH_3}{|}}{\underset{\underset{\textstyle CH_3}{|}}{C}}CH_3$$

(c)
$$CH_3CH_2\overset{\overset{\textstyle CH_3}{|}}{CH}CH_2CH_2CH_3$$

■ **PROBLEM 12.7**

There are five isomers with the formula C_6H_{14}. Draw as many as you can.

12.6 Naming Alkanes

In earlier times, when relatively few pure organic chemicals were known, new compounds were named at the whim of their discoverer. Thus, urea is a crystalline substance first isolated from urine, and the barbiturates were named by their discoverer in honor of his friend Barbara. As more and more compounds became known, however, the need for a systematic method of naming compounds became apparent.

The system of naming (*nomenclature*) now used is one devised by the International Union of Pure and Applied Chemistry, IUPAC (pronounced **eye**-you-pack). In the IUPAC system for organic compounds, a chemical name has three parts: *prefix*, *parent*, and *suffix*. The prefix specifies the location of functional groups and other **substituents** in the molecule; the parent tells how many carbon atoms are present in the longest continuous chain; and the suffix identifies what family the molecule belongs to:

Substituent An atom or group of atoms attached to a parent compound.

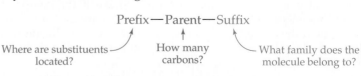

Application Displaying Molecular Shapes

Molecular shapes are critical to the proper functioning of biological molecules. The tiniest difference in shape can cause two compounds to behave differently or to have different physiological effects in the body. It's therefore critical that chemists be able both to determine molecular shapes with great precision and to visualize these shapes in useful ways.

Three-dimensional shapes of molecules are determined by *X-ray crystallography*, a technique that allows us to "see" molecules in a crystal using X-ray waves rather than light waves. The molecular "picture" obtained by X-ray crystallography looks at first like a series of regularly spaced dark spots against a lighter background on a photographic film. After computerized manipulation of the data, however, recognizable

molecules can be drawn. Relatively small molecules like morphine are usually displayed in either a ball-and-stick format (a), which emphasizes the connections among atoms, or in a space-filling format (b), which emphasizes the overall shape.

Enormous biological molecules like enzymes and other proteins are best displayed on computer terminals where their structures can be enlarged, rotated, and otherwise manipulated for the best view. An immunoglobulin molecule, for instance, is so large that little detail can be seen in ball-and-stick or space-filling views. Nevertheless, a cleft inside the molecule is visible if the model is rotated in just the right way.

See Additional Problem 12.63 at the end of the chapter.

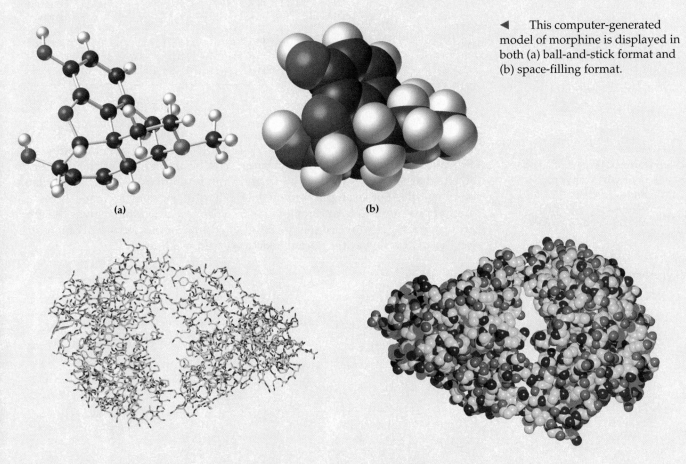

◀ This computer-generated model of morphine is displayed in both (a) ball-and-stick format and (b) space-filling format.

(a) (b)

▲ Computer-generated models of an immunoglobulin, one of the antibodies in blood that protect us from harmful invaders such as bacteria and viruses.

Straight-chain alkanes are named by counting the number of carbon atoms and adding the family suffix -ane. With the exception of the first four compounds—*meth*ane, *eth*ane, *prop*ane, and *but*ane—whose parent names have historical origins, the alkanes are named from Greek numbers according to the number of carbons present (Table 12.3). Thus, *pent*ane is the five-carbon alkane, *hex*ane is the six-carbon alkane, and so on. The first ten alkane names are so common that they should be memorized.

TABLE 12.3	Names of Straight-Chain Alkanes	
No. of Carbons	**Structure**	**Name**
1	CH_4	*Meth*ane
2	CH_3CH_3	*Eth*ane
3	$CH_3CH_2CH_3$	*Prop*ane
4	$CH_3CH_2CH_2CH_3$	*But*ane
5	$CH_3CH_2CH_2CH_2CH_3$	*Pent*ane
6	$CH_3CH_2CH_2CH_2CH_2CH_3$	*Hex*ane
7	$CH_3CH_2CH_2CH_2CH_2CH_2CH_3$	*Hept*ane
8	$CH_3CH_2CH_2CH_2CH_2CH_2CH_2CH_3$	*Oct*ane
9	$CH_3CH_2CH_2CH_2CH_2CH_2CH_2CH_2CH_3$	*Non*ane
10	$CH_3CH_2CH_2CH_2CH_2CH_2CH_2CH_2CH_2CH_3$	*Dec*ane

Alkyl group The part of an alkane that remains when a hydrogen atom is removed.

Methyl group The $-CH_3$ alkyl group.

Ethyl group The $-CH_2CH_3$ alkyl group.

Substituents such as $-CH_3$ and $-CH_2CH_3$ that branch off the main chain are called **alkyl groups**. An alkyl group can be thought of as the part of an alkane that remains when one hydrogen atom is removed to create an available bonding site. For example, removal of a hydrogen from methane gives the **methyl group,** $-CH_3$, and removal of a hydrogen from ethane gives the **ethyl group,** $-CH_2CH_3$. Notice that these alkyl groups are named simply by replacing the -ane ending of the parent alkane with an -yl ending.

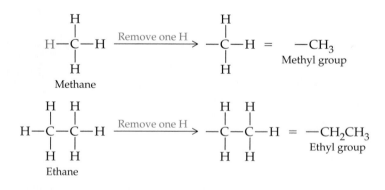

Both methane and ethane have only one "kind" of hydrogen. It doesn't matter which of the four methane hydrogens is removed, so there is only one possible methyl group. Similarly, it doesn't matter which of the six equivalent ethane hydrogens is removed, so only one ethyl group is possible.

The situation is more complex for larger alkanes, which contain more than one kind of hydrogen. Propane, for example, has two different kinds of hydrogens. Removal of any one of the six hydrogens attached to an end carbon yields a straight-chain alkyl group called **propyl**, whereas removal of one of the two hydrogens attached to the central carbon yields a branched-chain alkyl group called **isopropyl**.

Propyl group The straight-chain alkyl group $-CH_2CH_2CH_3$.

Isopropyl group The branched-chain alkyl group $-CH(CH_3)_2$.

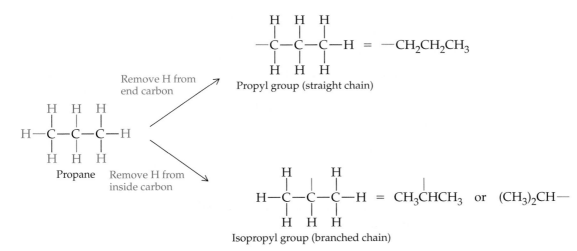

It's important to realize that alkyl groups themselves are not compounds and that the "removal" of a hydrogen from an alkane is just a way of looking at things, not a chemical reaction. Alkyl groups are simply partial structures that help us name compounds. The names of some common alkyl groups are listed in Table 12.4.

TABLE 12.4   Some Common Alkyl Groups*

			CH_3
CH_3-	CH_3CH_2-	$CH_3CH_2CH_2-$	$CH_3\overset{|}{C}H-$
Methyl	Ethyl	*n*-propyl	Isopropyl
		CH_3	
$CH_3CH_2CH_2CH_2-$	$CH_3\overset{|}{C}HCH_2CH_3$	$CH_3\overset{|}{C}HCH_2-$	$CH_3\overset{|}{\underset{|}{C}}CH_3$
n-Butyl	*sec*-Butyl	Isobutyl	CH_3
			tert-Butyl

*The red bond shows the connection to the rest of the molecule.

Notice that four butyl (four-carbon) groups are listed in Table 12.4: butyl, *sec*-butyl, isobutyl, and *tert*-butyl. The prefix *sec*- stands for *secondary*, and the prefix *tert*- stands for *tertiary*, referring to the number of other carbon atoms attached to the branch point. There are four possible substitution patterns, called *primary*, *secondary*, *tertiary*, and *quaternary*. A **primary** carbon atom has one other carbon attached to it, a **secondary** carbon atom has two other carbons attached, a **tertiary** carbon atom has three other carbons attached, and a **quaternary** carbon atom has four other carbons attached.

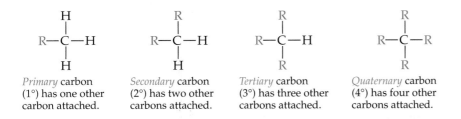

Primary carbon (1°) has one other carbon attached.   *Secondary* carbon (2°) has two other carbons attached.   *Tertiary* carbon (3°) has three other carbons attached.   *Quaternary* carbon (4°) has four other carbons attached.

The symbol **R** is used here and in later chapters as a general abbreviation for any organic substituent. You might think of it as representing the **R**est of the molecule, which we're not bothering to specify. The group R might represent methyl, ethyl, propyl, or any of a vast number of other possibilities. For example, the general-

Classification of Atoms and Bonds in an Organic Molecule

Primary carbon atom   A carbon atom with one other carbon attached to it.

Secondary carbon atom   A carbon atom with two other carbons attached to it.

Tertiary carbon atom   A carbon atom with three other carbons attached to it.

Quaternary carbon atom   A carbon atom with four other carbons attached to it.

ized formula R—OH for an alcohol might refer to CH_3OH, CH_3CH_2OH, or any of a great many others.

Branched-chain alkanes can be named by following four steps:

Step 1: Name the main chain. Find the longest continuous chain of carbons, and name the chain according to the number of carbon atoms it contains. The longest chain may not be immediately obvious because it may not always be written on one line; you may have to "turn corners" to find it.

Nomenclature of Alkanes

$$CH_3-CH_2$$
$$|$$
$$CH_3-CH-CH_2-CH_3$$

Name as a substituted pentane, not as a substituted butane, because the *longest* chain has five carbons.

Step 2: Number the carbon atoms in the main chain, beginning at the end nearer the first branch point:

$$\overset{CH_3}{\underset{|}{\underset{2}{CH}}}$$
$$\underset{1}{CH_3}-\underset{2}{CH}-\underset{3}{CH_2}-\underset{4}{CH_2}-\underset{5}{CH_3}$$

The first (and only) branch occurs at C2 if we start numbering from the left, but would occur at C4 if we started from the right by mistake.

Step 3: Identify the branching substituents, and number each according to its point of attachment to the main chain:

$$\overset{CH_3}{\underset{|}{}}$$
$$\underset{1}{CH_3}-\underset{2}{CH}-\underset{3}{CH_2}-\underset{4}{CH_2}-\underset{5}{CH_3}$$

The main chain is a pentane. There is one —CH_3 substituent group connected to C2 of the chain.

If there are two substituents on the same carbon, assign the same number to both. There must always be as many numbers in the name as there are substituents.

$$CH_2-CH_3$$
$$|$$
$$\underset{1}{CH_3}-\underset{2}{CH_2}-\underset{3}{C}-\underset{4}{CH_2}-\underset{5}{CH_2}-\underset{6}{CH_3}$$
$$|$$
$$CH_3$$

The main chain is a hexane. There are two substituents, a —CH_3 and a —CH_2CH_3, both connected to C3 of the chain.

Step 4: Write the name as a single word, using hyphens to separate the numbers from the different prefixes and commas to separate numbers if necessary. If two or more different substituent groups are present, cite them in alphabetical order. If two or more identical substituents are present, use one of the prefixes *di-*, *tri-*, *tetra-*, and so forth, but don't use these prefixes for alphabetizing purposes.

$$\overset{CH_3}{\underset{|}{}}$$
$$\underset{1}{CH_3}-\underset{2}{CH}-\underset{3}{CH_2}-\underset{4}{CH_2}-\underset{5}{CH_3}$$

2-Methylpentane (a five-carbon main chain with a 2-methyl substituent)

$$CH_2-CH_3$$
$$|$$
$$\underset{1}{CH_3}-\underset{2}{CH_2}-\underset{3}{C}-\underset{4}{CH_2}-\underset{5}{CH_2}-\underset{6}{CH_3}$$
$$|$$
$$CH_3$$

3-Ethyl-3-methylhexane (a six-carbon main chain with 3-ethyl and 3-methyl substituents cited alphabetically)

$$\underset{1}{CH_3}-\underset{2}{CH_2}$$
$$|$$
$$\underset{1}{CH_3}-\underset{3}{C}-\underset{4}{CH_2}-\underset{5}{CH_2}-\underset{6}{CH_3}$$
$$|$$
$$CH_3$$

3,3-Dimethylhexane (a six-carbon main chain with two 3-methyl substitutents)

■ **WORKED EXAMPLE 12.3**

What is the IUPAC name of the following alkane?

$$\overset{CH_3}{\underset{|}{}} \qquad \overset{CH_3}{\underset{|}{}}$$
$$CH_3-CH-CH_2-CH_2-CH-CH_2-CH_3$$

ANALYSIS   Follow the four steps outlined in the text.

Step 1: The longest continuous chain of carbon atoms is seven, so the main chain is a *hept*ane.

Step 2: Number the main chain beginning at the end nearer the first branch:

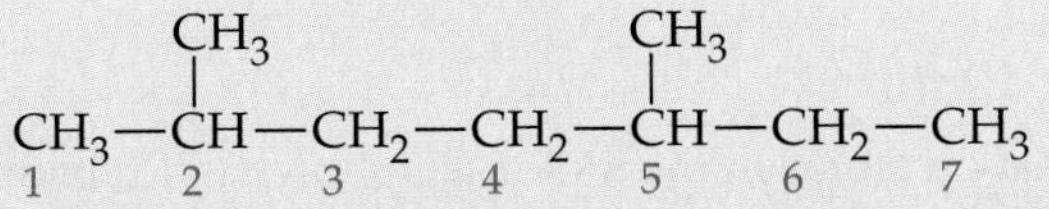

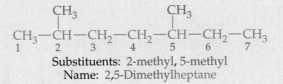

Step 3: Identify and number the substituents (a 2-methyl and a 5-methyl in this case).

Step 4: Write the name as one word, using the prefix *di-* because there are two methyl groups. Separate the two numbers by a comma, and use a hyphen between the numbers and the word.

SOLUTION

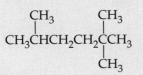

Substituents: 2-methyl, 5-methyl
Name: 2,5-Dimethylheptane

■ WORKED EXAMPLE 12.4

Identify each of the carbon atoms in the following molecule as primary, secondary, tertiary, or quaternary.

$$CH_3CHCH_2CH_2CCH_3$$

with CH_3 groups above the 2nd and 5th carbons and a CH_3 group below the 5th carbon.

ANALYSIS   Look at each carbon atom in the molecule, count the number of other carbon atoms attached, and make the assignment accordingly: primary—one carbon attached; secondary—two carbons attached; tertiary—three carbons attached; quaternary—four carbons attached.

SOLUTION

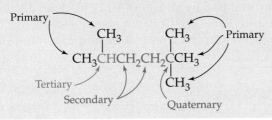

■ PROBLEM 12.8

What are the IUPAC names of the following alkanes?

(a) 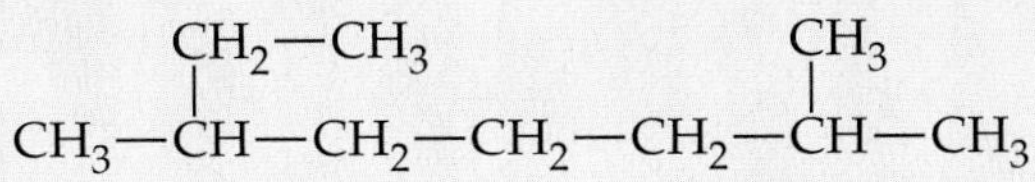 $CH_3-CH-CH_2-CH_2-CH_2-CH-CH_3$
with a CH_2-CH_3 group on the 2nd carbon and a CH_3 group on the 6th carbon.

(b) 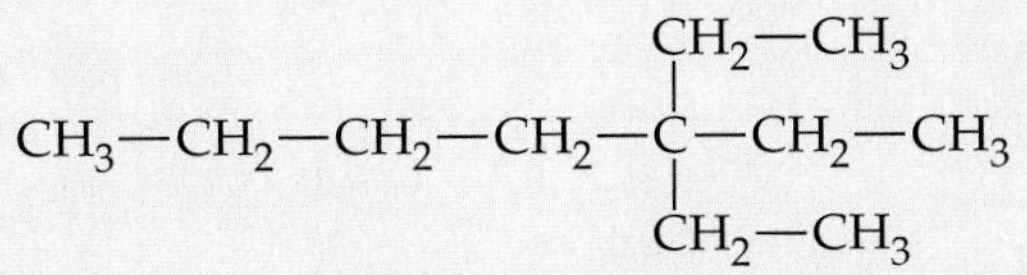 $CH_3-CH_2-CH_2-CH_2-C-CH_2-CH_3$
with a CH_2-CH_3 group above the central carbon and a CH_2-CH_3 group below it.

■ **PROBLEM 12.9**

Draw structures corresponding to the following IUPAC names:

(a) 3-Methylhexane   **(b)** 3,4-Dimethyloctane

(c) 2,2,4-Trimethylpentane

■ **PROBLEM 12.10**

Identify the carbon atoms in the molecules shown in Problem 12.9 as primary, secondary, tertiary, or quaternary.

■ **PROBLEM 12.11**

Draw and name alkanes that meet the following descriptions:

(a) An alkane with a tertiary carbon atom

(b) An alkane that has both a tertiary and a quaternary carbon atom

■ **KEY CONCEPT PROBLEM 12.12**

What are the IUPAC names of the following alkanes?

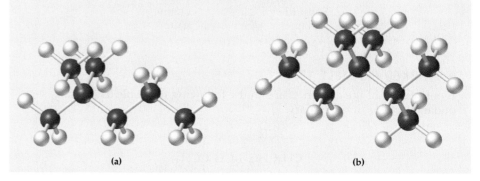

(a)   (b)

12.7 Properties of Alkanes

Alkanes contain only nonpolar C—C and C—H bonds, so the only intermolecular forces influencing them are weak London dispersion forces (Section 8.11). (∞∞, p. 211) The effect of these forces is shown in the regularity with which the melting points and boiling points of straight-chain alkanes increase with molecular size (Figure 12.3). The first four alkanes—methane, ethane, propane, and butane—are gases at room temperature and pressure. Alkanes with 5–15

FIGURE 12.3 The boiling ▶ points and melting points for the C_1–C_{14} straight-chain alkanes increase with molecular size.

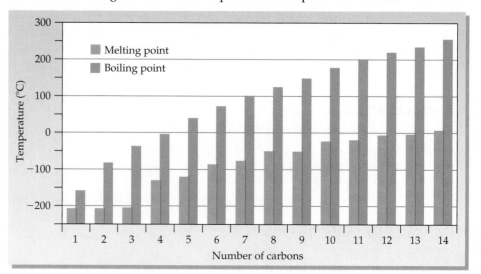

carbon atoms are liquids, and those with 16 or more carbon atoms are generally low-melting, waxy solids.

In keeping with their low polarity, alkanes are insoluble in water but soluble in nonpolar organic solvents, including other alkanes (Section 9.2). (∞∞, p. 233) Because alkanes are generally less dense than water, they float on its surface. Low-molecular-weight alkanes are volatile and must be handled with care because their vapors are flammable. Mixtures of alkane vapors and air can explode when ignited by a single spark.

The physiological effects of alkanes are limited. Methane, ethane, and propane gases are nontoxic, and the danger of inhaling them lies in potential suffocation due to lack of oxygen. Breathing the vapor of higher alkanes in large concentrations can induce loss of consciousness. There is also a danger in breathing droplets of liquid alkanes because they dissolve nonpolar substances in lung tissue and cause pneumonia-like symptoms.

Mineral oil, petroleum jelly, and paraffin wax are mixtures of higher alkanes. All are harmless to tissue and are used in numerous food and medical applications. Mineral oil passes through the body unchanged and is sometimes used as a laxative. Petroleum jelly (sold as Vaseline) softens, lubricates, and protects the skin. Paraffin wax is used in candle making.

▲ The waxy coating that makes these apples so shiny is a mixture of higher-molecular-weight alkanes.

Properties of Alkanes

- Odorless or mild odor; colorless; tasteless; nontoxic
- Nonpolar; insoluble in water but soluble in nonpolar organic solvents; less dense than water
- Flammable
- Not very reactive

12.8 Reactions of Alkanes

Alkanes don't react with acids, bases, or most other common laboratory reagents. Their only major reactions are with oxygen (combustion) and with halogens (halogenation).

Combustion

The reaction of an alkane with oxygen occurs during **combustion** in an engine or furnace (Section 7.2). (∞∞, p. 166) Carbon dioxide and water are the products of complete combustion of any hydrocarbon, and a large amount of heat is released. Some examples were given in Table 7.1.

Combustion A chemical reaction that produces a flame, usually because of burning with oxygen.

$$CH_4(g) + 2\,O_2(g) \longrightarrow CO_2(g) + 2\,H_2O(l) \qquad \Delta H = -213\,\text{kcal/mol}$$

When hydrocarbon combustion is incomplete because of faulty engine or furnace performance, carbon monoxide and carbon-containing soot are among the products. Carbon monoxide is a highly toxic and dangerous substance, especially so because it has no odor and can easily go undetected (See the Application "Carbon Monoxide: A Surprising Molecule" in Chapter 5). Breathing air that contains as little as 2% CO for only one hour can cause respiratory and nervous system damage or death. The supply of oxygen to the brain is cut off by carbon monoxide because it binds strongly to blood hemoglobin at the site where oxygen is normally bound. By contrast with CO, CO_2 is nontoxic and causes no harm, except by suffocation when present in high concentration.

■ **PROBLEM 12.13**

Write a balanced equation for the complete combustion of ethane with oxygen.

Halogenation

The second notable reaction of alkanes is *halogenation*, the replacement of an alkane hydrogen by a chlorine or bromine in a process initiated by heat or light. Complete chlorination of methane, for example, yields carbon tetrachloride:

$$CH_4 + 4\,Cl_2 \xrightarrow{\text{Heat or light}} CCl_4 + 4\,HCl$$

Although the above equation for the reaction of methane with chlorine is balanced, it doesn't fully represent what actually happens. In fact, this reaction, like many organic reactions, yields a mixture of products:

$$CH_4 + Cl_2 \longrightarrow CH_3Cl + HCl$$
$$\xrightarrow{Cl_2} CH_2Cl_2 + HCl$$
$$\xrightarrow{Cl_2} CHCl_3 + HCl$$
$$\xrightarrow{Cl_2} CCl_4 + HCl$$

CH_3Cl, chloromethane
CH_2Cl_2, dichloromethane
$CHCl_3$, chloroform
CCl_4, carbon tetrachloride

When writing the equation for an organic reaction, attention is usually focused on converting a particular reactant into a desired product. Because they are of little interest, any minor by-products are often ignored. Also, inorganic products, such as the HCl formed in the chlorination of methane, are often of little interest. Thus, it's not always necessary to balance the equation for an organic reaction as long as the reactant, the major product, and any necessary reagents and conditions are shown. A chemist who plans to convert methane into bromomethane might therefore write the equation as

$$CH_4 \xrightarrow[\text{Light, heat}]{Br_2} CH_3Br$$

Like many equations for organic reactions, this equation isn't balanced.

■ **PROBLEM 12.14**

Write the structures of all possible products with one and two chlorine atoms that can be formed in the reaction of propane with Cl_2.

12.9 Cycloalkanes

Cycloalkane An alkane that contains a ring of carbon atoms.

The organic compounds described thus far have all been open-chain, or *acyclic*, alkanes. **Cycloalkanes**, which contain rings of carbon atoms, are also well known and are widespread throughout nature. Compounds of all ring sizes from 3 through 30 and beyond have been prepared in the laboratory. The two simplest cycloalkanes—cyclopropane and cyclobutane—contain three and four carbon atoms, respectively:

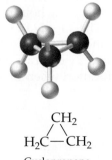

$$\begin{array}{c} CH_2 \\ / \quad \backslash \\ H_2C—CH_2 \end{array}$$

Cyclopropane
(mp −128°C, bp −33°C)

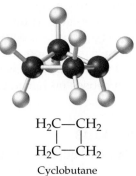

$$\begin{array}{c} H_2C—CH_2 \\ | \quad | \\ H_2C—CH_2 \end{array}$$

Cyclobutane
(mp −50°C, bp −12°C)

Application Petroleum

Many alkanes occur naturally throughout plants and animals, but natural gas and petroleum deposits provide the world's most abundant supply. *Natural gas* consists chiefly of methane, along with smaller amounts of ethane, propane, and butane. *Petroleum* is a complex liquid mixture of hydrocarbons that must be separated, or *refined*, into different *fractions* before it can be used. Petroleum refining begins with distillation to separate the crude oil into three main fractions according to boiling points: straight-run gasoline (bp 30–200°C), kerosene (bp 175–300°C), and gas oil (bp 275–400°C). The residue is then further distilled under reduced pressure to recover lubricating oils, waxes, and asphalt.

Distillation of petroleum is just the beginning of the process for making gasoline. It has long been known that straight-chain alkanes burn less smoothly in engines than do branched-chain alkanes, as measured by a compound's *octane number*. Heptane, a straight-chain alkane and particularly poor fuel, is assigned an octane rating of 0, while 2,2,4-trimethylpentane, a branched-chain alkane commonly known as isooctane, is given a rating of 100. Straight-run gasoline, with its high percentage of unbranched alkanes, is thus a poor fuel. Petroleum chemists, however, have devised methods to remedy the problem. One of these methods, called *catalytic cracking*, involves taking the kerosene fraction (C_{11}–C_{14}) and "cracking" it into smaller C_3–C_5 molecules at high temperature. These small hydrocarbons are then catalytically recombined to yield C_7–C_{10} branched-chain molecules that are perfectly suited for use as automobile fuel.

See Additional Problems 12.64 and 12.65 at the end of the chapter.

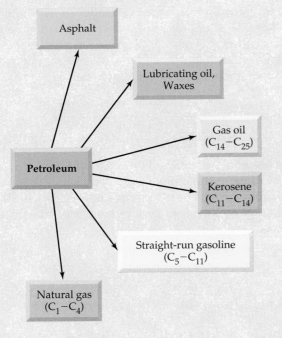

▲ The principal products of petroleum refining.

▲ The crude petroleum from this well is an enormously complex mixture of alkanes and other organic molecules.

Note that the C—C—C bond angles in cyclopropane and cyclobutane rings would be 60° and 90° respectively, if the rings were flat—values that are considerably compressed from the normal 109.5° tetrahedral value. As a result, these compounds are less stable and more reactive than other cycloalkanes. The five-membered (cyclopentane) ring has nearly ideal bond angles, as does the six-membered (cyclohexane) ring. Cyclohexane adopts a puckered, nonplanar shape called a *chair conformation*, in which the carbon atoms have 109° tetrahedral bond angles. Both cyclopentane and cyclohexane rings are therefore stable, and many naturally occurring and biochemically active molecules contain such rings.

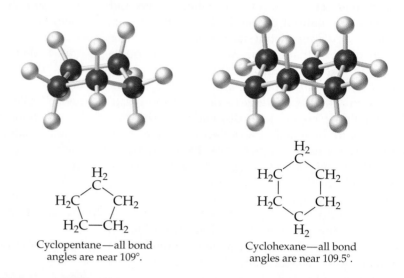

Cyclopentane—all bond angles are near 109°.

Cyclohexane—all bond angles are near 109.5°.

Cyclic and acyclic alkanes are similar in many of their properties. Cyclopropane and cyclobutane are gases, while larger cycloalkanes are liquids or solids. Like alkanes, cycloalkanes are nonpolar, insoluble in water, and flammable. Because of their cyclic structures, however, cycloalkane molecules are more rigid and less flexible than their open-chain counterparts. Rotation is not possible around the carbon–carbon bonds in cycloalkanes without breaking open the ring.

12.10 Drawing and Naming Cycloalkanes

Even condensed structures become awkward when working with large molecules that contain rings. Thus, a more streamlined way of drawing structures is often used in which cycloalkanes are represented simply by *polygons*. A triangle represents cyclopropane, a square represents cyclobutane, a pentagon represents cyclopentane, and so on.

Cyclopropane   Cyclobutane   Cyclopentane   Cyclohexane   Cycloheptane

Notice that carbon and hydrogen atoms aren't shown in these **line structures**. A carbon atom is simply understood to be at every intersection of lines, and the proper number of hydrogen atoms necessary to give each carbon atom four covalent bonds is supplied mentally. Methylcyclohexane, for example, looks like this in a line structure:

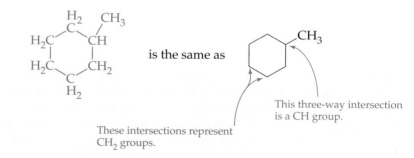

is the same as

This three-way intersection is a CH group.

These intersections represent CH₂ groups.

Cycloalkanes are named by a straightforward extension of the rules for open-chain alkanes. In most cases, only two steps are needed:

Step 1: Use the cycloalkane name as the parent. That is, compounds are named as alkyl-substituted cycloalkanes rather than as cycloalkyl-substituted alkanes. If there is only one substituent on the ring, it's not even necessary to assign a number because all ring positions are identical.

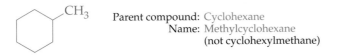

Parent compound: Cyclohexane
Name: Methylcyclohexane
(not cyclohexylmethane)

Step 2: Number the substituents. Start numbering at the group that has alphabetical priority, and proceed around the ring in the direction that gives the second substituent the lower possible number.

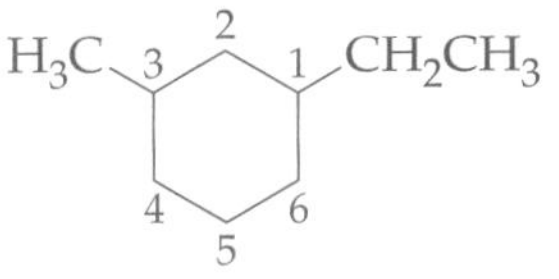

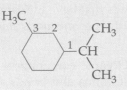

1-Ethyl-3-methylcyclohexane
(not 1-ethyl-5-methylcyclohexane or
1-methyl-3-ethylcyclohexane or
1-methyl-5-ethylcyclohexane)

■ **WORKED EXAMPLE 12.5**

What is the IUPAC name of the following cycloalkane?

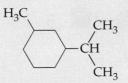

ANALYSIS Follow the two steps outlined in the text.

Step 1: The parent cycloalkane is cyclohexane because it contains six carbons, and there are two substituents—a methyl group and an isopropyl group.

Step 2: Number the compound beginning at the group having alphabetical priority (isopropyl rather than methyl), and proceed around the ring in a direction that gives the second group the lower possible number (3 rather than 5).

SOLUTION

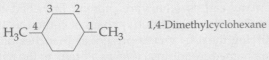

1-Isopropyl-3-methylcyclohexane

■ **WORKED EXAMPLE 12.6**

Draw a line structure for 1,4-dimethylcyclohexane.

ANALYSIS Draw a hexagon to represent a cyclohexane ring, and attach a —CH₃ group at an arbitrary position that becomes C1. Then count around the ring to C4, and attach another —CH₃ group.

SOLUTION Note that the second methyl group is written here as H₃C— because it is attached on the left side of the ring.

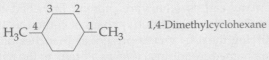

1,4-Dimethylcyclohexane

■ **PROBLEM 12.15**

What are the IUPAC names of the following cycloalkanes?

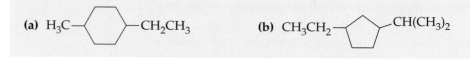

(a) H₃C— ⬡ —CH₂CH₃   (b) CH₃CH₂— ⬠ —CH(CH₃)₂

■ **PROBLEM 12.16**

Draw line structures representing the following IUPAC names:
(a) 1,1-Diethylcyclohexane   **(b)** 1,3,5-Trimethylcycloheptane

■ **KEY CONCEPT PROBLEM 12.17**

What is the IUPAC name of the following cycloalkane?

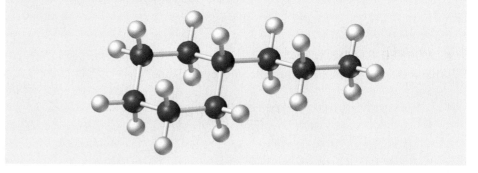

Key Words

Alkane, p. 330

Alkyl group, p. 338

Branched-chain alkane, p. 331

Combustion, p. 343

Condensed structure, p. 332

Conformation, p. 334

Constitutional isomers, p. 332

Cycloalkane, p. 344

Ethyl group, p. 338

Functional group, p. 328

Hydrocarbon, p. 328

Isomers, p. 332

Isopropyl group, p. 338

Line structure, p. 346

Methyl group, p. 338

Propyl group, p. 338

Organic chemistry, p. 325

Primary carbon atom, p. 339

Quaternary carbon atom, p. 339

Secondary carbon atom, p. 339

Straight-chain alkane, p. 331

Substituent, p. 336

Tertiary carbon atom, p. 339

Summary: Revisiting the Chapter Goals

1. **What are functional groups, and how are they used to classify organic molecules?** Organic compounds can be classified into various families according to the functional groups they contain. A *functional group* is a part of a larger molecule and is composed of a group of atoms that has characteristic structure and chemical reactivity. A given functional group undergoes nearly the same chemical reactions in every molecule where it occurs.

2. **What are isomers?** *Isomers* are compounds that have the same formula but different structures. Isomers that differ in their connections among atoms are called *constitutional isomers*.

3. **How are organic molecules drawn?** Organic compounds can be represented by *structural formulas* in which all atoms and bonds are shown, by *condensed structures* in which not all bonds are drawn, or by *line structures* in which the carbon skeleton is represented by lines and the locations of C and H atoms are understood.

4. **What are alkanes and cycloalkanes, and how are they named?** Compounds containing only carbon and hydrogen are called *hydrocarbons*, and hydrocarbons that have only single bonds are called *alkanes*. A *straight-chain alkane* has all its carbons connected in a row, a *branched-chain alkane* has a branching connection of atoms somewhere along its chain, and a *cycloalkane* has a ring of carbon atoms.

Straight-chain alkanes are named by adding the family ending *-ane* to a parent that tells how many carbon atoms are present. Branched-chain alkanes are named by using the longest continuous chain of carbon atoms for the parent and then identifying the *alkyl groups* present as branches off the main chain. Cycloalkanes are named by adding *cyclo-* as a prefix to the name of the alkane.

5. What are the general properties and chemical reactions of alkanes?
Alkanes are generally soluble only in nonpolar organic solvents, have weak intermolecular forces, and are nontoxic. Their principal chemical reactions are *combustion*, a reaction with oxygen to give carbon dioxide and water, and *halogenation*, a reaction in which hydrogen atoms are replaced by chlorine or bromine.

Summary of Reactions

1. Combustion of an alkane with oxygen (Section 12.8):

$$CH_4 + 2\,O_2 \longrightarrow CO_2 + 2\,H_2O$$

2. Halogenation of an alkane to yield an alkyl halide (Section 12.8):

$$CH_4 + Cl_2 \longrightarrow CH_3Cl + HCl$$

▪ Understanding Key Concepts

12.18 Convert each of the following models into a condensed formula (gray = C; ivory = H; red = O):

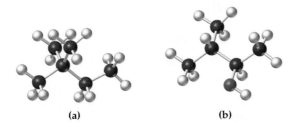

(a)   (b)

12.19 Convert each of the following models into a line drawing (gray = C; ivory = H; red = O; blue = N):

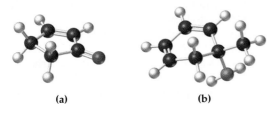

(a)   (b)

12.20 Identify the functional groups in each of the following compounds:

(a)   (b)

12.21 Give systematic names for the following alkanes:

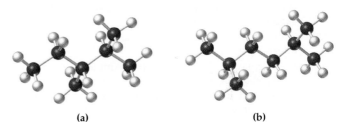

(a)   (b)

12.22 Give systematic names for the following cycloalkanes:

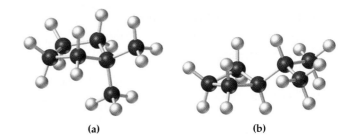

(a)   (b)

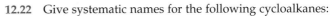

12.23   The following two compounds are isomers, even though both can be named 1,3-dimethylcyclopentane. What is the difference between them?

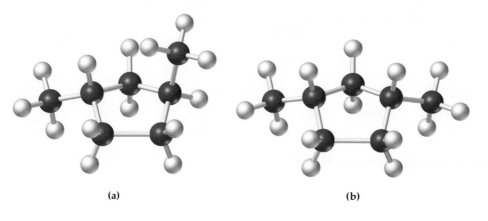

(a)   (b)

▪ Additional Problems

Organic Molecules and Functional Groups

12.24   What characteristics of carbon makes possible the existence of so many different organic compounds?

12.25   What are functional groups, and why are they important?

12.26   Why are most organic compounds nonconducting and insoluble in water?

12.27   If you were given two unlabeled bottles, one containing hexane and one containing water, how could you tell them apart?

12.28   What is meant by the term *polar covalent bond*? Give an example of such a bond.

12.29   Give examples of compounds that are members of the following families:
   (a) Alcohol   **(b)** Amine
   (c) Carboxylic acid   **(d)** Ether

12.30   Locate and identify the functional groups in the following molecules:

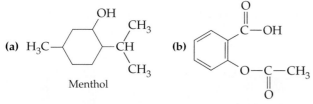

(a) Menthol   (b) Aspirin

12.31   Identify the functional groups in the following molecules:

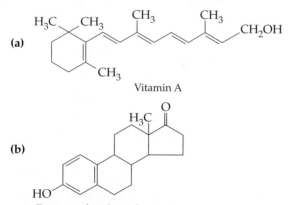

(a) Vitamin A

(b) Estrone, a female sex hormone

12.32   Propose structures for molecules that fit the following descriptions:
   (a) A ketone with the formula $C_5H_{10}O$
   (b) An ester with the formula $C_6H_{12}O_2$
   (c) A compound with the formula $C_2H_5NO_2$ that is both an amine and a carboxylic acid

12.33   Propose structures for molecules that fit the following descriptions:
   (a) An amide with the formula C_4H_9NO
   (b) An aldehyde that has a ring of carbons, $C_6H_{10}O$
   (c) An aromatic compound that is also an ether, $C_8H_{10}O$

Alkanes and Isomers

12.34   What requirement must be met for two compounds to be isomers?

12.35   If one compound has the formula C_5H_{10} and another has the formula C_4H_{10}, are the two compounds isomers? Explain.

12.36   What is the difference between a secondary carbon and a tertiary carbon? Between a primary carbon and a quaternary carbon?

12.37   Why can't a compound have a *quintary* carbon (five R groups attached to C)?

12.38   Give examples of compounds that meet the following descriptions:
   (a) An alkane that has two tertiary carbons
   (b) A cycloalkane that has only secondary carbons

12.39   Give examples of compounds that meet the following descriptions:
   (a) A branched-chain alkane that has only primary and quaternary carbons
   (b) A cycloalkane that has three substituents

12.40   There are three isomers with the formula C_3H_8O. Draw their structures.

12.41   Write condensed structures for each of the following molecular formulas. You may have to use rings and/or multiple bonds in some instances.
   (a) C_2H_7N
   (b) C_4H_8
   (c) C_2H_4O
   (d) CH_2O_2

12.42 How many isomers can you write that fit the following descriptions?
  (a) Alcohols with formula $C_4H_{10}O$
  (b) Amines with formula C_3H_9N
  (c) Ketones with formula $C_5H_{10}O$

12.43 How many isomers can you write that fit the following descriptions?
  (a) Aldehydes with formula $C_5H_{10}O$
  (b) Esters with formula $C_4H_8O_2$
  (c) Carboxylic acids with formula $C_4H_8O_2$

12.44 Which of the following pairs of structures are identical, which are isomers, and which are unrelated?

  (a) $CH_3CH_2CH_3$ and $\underset{\displaystyle CH_2CH_3}{CH_3}$

  (b) $CH_3{-}\underset{\underset{\textstyle H}{|}}{N}{-}CH_3$ and $CH_3CH_2{-}\underset{\underset{\textstyle H}{|}}{N}{-}H$

  (c) $CH_3CH_2CH_2{-}O{-}CH_3$ and
  $CH_3CH_2CH_2{-}\overset{\overset{\textstyle O}{\|}}{C}{-}CH_3$

  (d) $CH_3{-}\overset{\overset{\textstyle O}{\|}}{C}{-}CH_2CH_2CH(CH_3)_2$ and
  $CH_3CH_2{-}\overset{\overset{\textstyle O}{\|}}{C}{-}CH_2CH_2CH_2CH_3$

  (e) $CH_3CH{=}CHCH_2CH_2{-}O{-}H$ and
  $CH_3CH_2\underset{\underset{\textstyle CH_3}{|}}{CH}{-}\overset{\overset{\textstyle O}{\|}}{C}{-}H$

12.45 Which of the structures in each group represent the same compound, and which represent isomers?

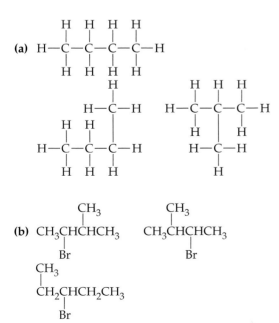

12.46 What is wrong with each of the following structures?
  (a) $CH_3{=}CHCH_2CH_2OH$

  (b) $CH_3CH_2CH{=}\overset{\overset{\textstyle O}{\|}}{C}{-}CH_3$

  (c) $CH_2CH_2CH_2C{\equiv}\underset{\underset{\textstyle CH_3}{|}}{C}CH_3$ (with CH_3 above)

12.47 There are two things wrong with the following structure. What are they?

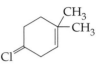

Alkane Nomenclature

12.48 What are the IUPAC names of the following alkanes?

  (a) $CH_3CH_2CH_2CH_2\underset{\underset{\textstyle CH_3}{|}}{\overset{\overset{\textstyle CH_2CH_3}{|}}{CH}}CHCH_2CH_3$

  (b) $CH_3CH_2CH_2\underset{\underset{\textstyle CH_2CH_3}{|}}{\overset{\overset{\textstyle CH_3CHCH_3}{|}}{CH}}CH_2CHCH_3$

  (c) $CH_3\underset{\underset{\textstyle CH_3}{|}}{\overset{\overset{\textstyle CH_3}{|}}{C}}CH_2CH_2CH_2\overset{\overset{\textstyle CH_3}{|}}{C}HCH_3$

  (d) $CH_3CH_2CH_2\underset{\underset{\textstyle CH_3CHCH_3}{|}}{\overset{\overset{\textstyle CH_2CH_2CH_2CH_3}{|}}{C}}CH_3$

  (e) $CH_3\underset{\underset{\textstyle CH_3}{|}}{\overset{\overset{\textstyle CH_3}{|}}{C}}CH_2\underset{\underset{\textstyle CH_3}{|}}{\overset{\overset{\textstyle CH_3}{|}}{C}}CH_3$

  (f) $CH_3CH_2\underset{\underset{\textstyle CH_3CH_2}{|}}{\overset{\overset{\textstyle CH_3CH_2}{|}}{C}}CH_2\overset{\overset{\textstyle CH_3}{|}}{C}H$ (CH_3 at end)

  (g) $CH_3(CH_2)_7\underset{\underset{\textstyle CH_3}{|}}{\overset{\overset{\textstyle CH_3}{|}}{C}}{-}CH_3$

12.49 Give IUPAC names for each of the five isomers with the formula C_6H_{14}.

12.50 Write condensed structures for each of the following compounds:
  (a) 3-Ethylhexane
  (b) 2,2,3-Trimethylpentane
  (c) 3-Ethyl-3,4-dimethylheptane
  (d) 5-Isopropyl-2-methyloctane
  (e) 2,2,6,6-Tetramethyl-4-propylnonane
  (f) 1,1-Dimethylcyclopentane

12.51 Draw structures corresponding to the following IUPAC names:

(a) Cyclooctane
(b) 1,2,3,4-Tetramethylcyclobutane
(c) 4-Ethyl-1,1-dimethylcyclohexane
(d) Ethylcycloheptane
(e) 1,3,5-Triethylcyclohexane
(f) 1,2,4,5-Tetramethylcyclooctane

12.52 Name the following cycloalkanes:

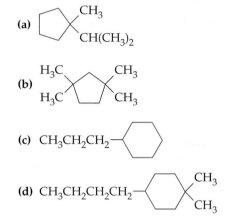

(a)

(b)

(c) CH₃CH₂CH₂—

(d) CH₃CH₂CH₂CH₂—

12.53 Name the following cycloalkanes:

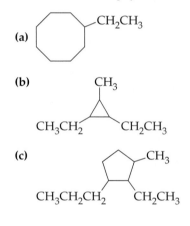

(a)

(b)

(c)

12.54 The following names are incorrect. Tell what is wrong with each, and provide the correct names.

(a)

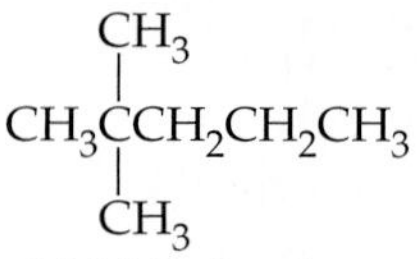

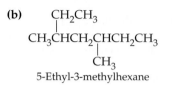

2,2-Methylpentane

(b)

CH₃CHCH₂CHCH₂CH₃

5-Ethyl-3-methylhexane

(c)

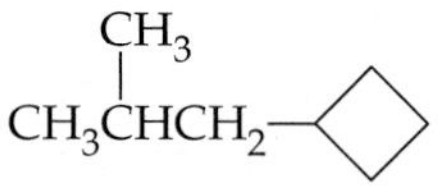

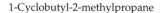

1-Cyclobutyl-2-methylpropane

12.55 The following names are incorrect. Write the structural formula that agrees with the apparent name, and then write the correct name of the compound.

(a) 2-Ethylbutane
(b) 2-Isopropyl-2-methylpentane
(c) 5-Ethyl-1,1-methylcyclopentane
(d) 3-Ethyl-3,5,5-trimethylhexane
(e) 1,2-Dimethyl-4-ethylcyclohexane
(f) 2,4-Diethylpentane
(g) 5,5,6,6-Methyl-7,7-ethyldecane

12.56 Draw structures and give IUPAC names for the nine isomers of C_7H_{16}.

12.57 Draw the structural formulas and name all cyclic isomers with the formula C_5H_{10}.

Reactions of Alkanes

12.58 Propane, commonly known as LP gas, burns in air to yield CO_2 and H_2O. Write a balanced equation for the reaction.

12.59 Write a balanced equation for the combustion of isooctane, C_8H_{16}, a component of gasoline.

12.60 Write the formulas of the three singly chlorinated isomers formed when 2,2-dimethylbutane reacts with Cl_2 in the presence of light.

12.61 Write the formulas of the seven doubly brominated isomers formed when 2,2-dimethylbutane reacts with Br_2 in the presence of light.

Applications

12.62 Why does a synthetically produced version of a compound have exactly the same properties as a "natural" compound? [*Natural Versus Synthetic*]

12.63 Why is it important to know the shape of a molecule? [*Displaying Molecular Shapes*]

12.64 How does petroleum differ from natural gas? [*Petroleum*]

12.65 What types of hydrocarbons burn most efficiently in an automobile engine? [*Petroleum*]

General Questions and Problems

12.66 Identify the functional groups in the following molecules:

(a) Testosterone, a male sex hormone

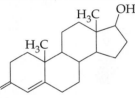

(b) Aspartame, an artificial sweetener

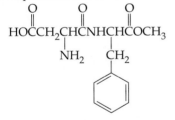

12.67 Label each carbon in Problem 12.66 as primary, secondary, tertiary, or quaternary.

12.68 If someone reported the preparation of a compound with the formula C_3H_9 most chemists would be skeptical. Why?

12.69 Most lipsticks are about 70% castor oil and wax. Why is lipstick more easily removed with petroleum jelly than with water?

12.70 When cyclohexane is exposed to Br_2 in the presence of light, reaction occurs. Write the formulas of:

(a) All possible monobromination products
(b) All possible dibromination products

12.71 Which do you think has a higher boiling point, pentane or neopentane (2,2-dimethylpropane)? Why?

12.72 Propose structures for the following:

(a) An aldehyde, C_4H_8O
(b) A bromo-substituted alkene, $C_6H_{11}Br$
(c) A cycloalkane, C_6H_{12}
(d) A diene (dialkene), C_5H_8

13

Alkenes, Alkynes, and Aromatic Compounds

Flamingos owe their color to alkene pigments in their diet. Without these compounds, their feathers eventually turn white.

I n this and the remaining four chapters on organic chemistry, we'll examine some families of organic compounds whose functional groups give them characteristic properties. Compounds in the three families described in this chapter all contain carbon–carbon multiple bonds. *Alkenes*, such as ethylene, contain a double-bond functional group; *alkynes*, such as acetylene, contain a triple-bond functional group; and *aromatic compounds*, such as benzene, contain a six-membered ring of carbon atoms usually pictured as having three alternating double bonds. All three functional groups are widespread in nature and are found in many biologically important molecules.

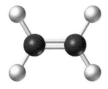
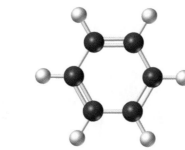

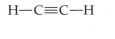

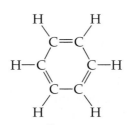

Ethylene	Acetylene	Benzene
(an *alkene*—has a C–C double bond)	(an *alkyne*—has a C–C triple bond)	(an *aromatic compound*— has a six-membered ring with three double bonds)

In this chapter, we'll answer the following questions:

1. What are alkenes, alkynes, and aromatic compounds?
The goal: Be able to recognize the functional groups in these three families of unsaturated organic compounds and give examples of each.

2. How are alkenes, alkynes, and aromatic compounds named?
The goal: Be able to name an alkene, alkyne, or simple aromatic compound from its structure, or write the structure, given the name.

3. What are cis–trans isomers?
The goal: Be able to identify cis–trans isomers of alkenes and predict their occurrence.

4. What are the categories of organic reactions?
The goal: Be able to recognize and describe addition, elimination, substitution, and rearrangement reactions.

5. What are the typical reactions of alkenes, alkynes, and aromatic compounds?
The goal: Be able to predict the products of reactions of alkenes, alkynes, and aromatic compounds.

6. How do organic reactions take place?
The goal: Be able to show how addition reactions occur.

CONCEPTS TO REVIEW

VSEPR and Molecular Shapes (Section 5.7)
The Shapes of Organic Molecules (Section 12.5)
Naming Alkanes (Section 12.6)

Saturated A molecule whose carbon atoms bond to the maximum number of hydrogen atoms.

Unsaturated A molecule that contains a carbon–carbon multiple bond, to which more hydrogen atoms can be added.

Alkene A hydrocarbon that contains a carbon–carbon double bond.

13.1 Alkenes and Alkynes

Alkanes, introduced in Chapter 12, are often referred to as **saturated** because each carbon atom bonds to the maximum number of hydrogen atoms and no more hydrogen atoms can be added. Alkenes and alkynes, however, are said to be **unsaturated** because they contain carbon–carbon multiple bonds to which more hydrogen atoms can be added. **Alkenes** are hydrocarbons that

Alkyne A hydrocarbon that contains a carbon–carbon triple bond.

contain carbon–carbon double bonds, and **alkynes** are hydrocarbons that contain carbon–carbon triple bonds.

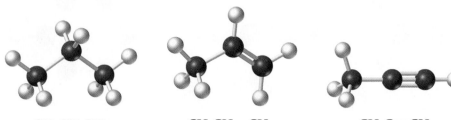

CH₃CH₂CH₃
Propane—an alkane
(*saturated*)

CH₃CH=CH₂
Propene—an alkene
(*unsaturated*)

CH₃C≡CH
Propyne—an alkyne
(*unsaturated*)

▲ Polyethylene, the most widely produced polymer in the world, is made from ethylene, a simple alkene.

Simple alkenes are made in vast quantities in the petroleum industry by thermal "cracking" of the alkanes in petroleum. Alkanes are rapidly heated to high temperatures (750–900°C), causing them to crack apart into reactive fragments that then reunite or rearrange into lower-molecular-weight molecules such as ethylene ($H_2C=CH_2$) and propene ($CH_3CH=CH_2$).

Most of the organic chemicals used in making drugs, explosives, paints, plastics, and pesticides are synthesized by routes that begin with alkenes. More ethylene is produced in the United States each year than any other organic chemical—approximately 28 million tons in 2000, most of it used for making polyethylene.

Curiously, ethylene is also formed in the leaves, flowers, and roots of plants, where it acts as a hormone to control seedling growth, stimulate root formation, and regulate fruit ripening. The ability of ethylene to hasten ripening was discovered indirectly by citrus growers who at one time ripened green oranges in rooms heated with kerosene stoves. When the stoves were replaced with other heaters, the fruit, to their surprise, no longer ripened. Ethylene from incomplete combustion of the kerosene, not heat from the stoves, had caused the ripening. Today, fruit is intentionally exposed to ethylene to continue the ripening process during shipment to market.

13.2 Naming Alkenes and Alkynes

Alkenes and alkynes are named in the IUPAC system by a series of rules similar to those used for alkanes (Section 12.6). (∞∞, p. 336) The parent names indicating the number of carbon atoms in the main chain are the same as those for alkanes, with the -*ene* suffix used in place of -*ane* for alkenes and the -*yne* suffix used for alkynes.

Step 1: Name the parent compound. Find the longest chain containing the double or triple bond, and name the parent compound by adding the suffix -*ene* or -*yne* to the name for the main chain.

CH₃CH₂CH₂CH=CH₂   Name as a *pentene*—a five-carbon chain containing a double bond.

CH₃CH₂CH₂C≡CCH₃   Name as a *hexyne*—a six-carbon chain containing a triple bond.

CH₃CH₂CH₂
⟍
　　C=CHCH₃   Name as a *hexene*—a six-carbon chain containing a triple bond . . .
⟋
CH₃CH₂CH₂

CH₃CH₂CH₂
⟍
　　C=CHCH₃   . . . *not* as a heptene, because the double bond is not contained in the seven-carbon chain.
⟋
CH₃CH₂CH₂

Step 2: Number the carbon atoms in the main chain, beginning at the end nearer the multiple bond. If the multiple bond is an equal distance from both ends, begin numbering at the end nearer the first branch point.

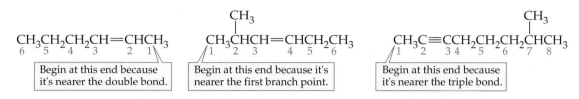

Cyclic alkenes are called **cycloalkenes**. The double-bond carbon atoms in substituted cycloalkenes are numbered 1 and 2 so as to give the first substituent the lower number:

Cycloalkene   A cyclic hydrocarbon that contains a double bond.

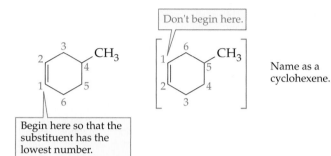

Step 3: Write the full name. Assign numbers to the branching substituents, and list the substituents alphabetically. Use commas to separate numbers and hyphens to separate words from numbers. Indicate the position of the multiple bond in the chain by giving the number of the *first* multiple-bonded carbon. If more than one double bond is present, identify the position of each and use the appropriate name ending *-diene, -triene, -tetraene,* and so forth. For example,

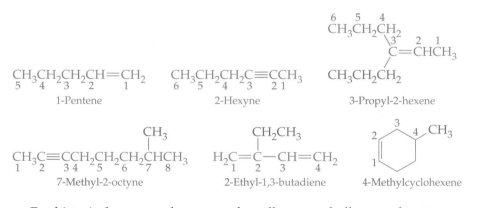

For historical reasons, there are a few alkenes and alkynes whose names don't conform strictly to the rules. For instance, the two-carbon alkene $H_2C=CH_2$ should properly be called *ethene,* but the name *ethylene* has been used for so long that it is now accepted by IUPAC. Similarly, the three-carbon alkene *propene* ($CH_3CH=CH_2$) is usually called *propylene.* The simplest alkyne, $HC\equiv CH$, should be known as *ethyne* but is more often called *acetylene.*

■ **WORKED EXAMPLE 13.1**

What is the IUPAC name of the following alkene?

$$CH_3CH_2CH_2-\overset{\overset{\displaystyle H_3C}{|}}{C}=\overset{\overset{\displaystyle CH_2CH_3}{|}}{C}-CH_3$$

ANALYSIS   Follow the steps described in the text.

Step 1: Find the longest continuous chain containing the double bond. In this case, we have to turn a corner to find that it's a heptene:

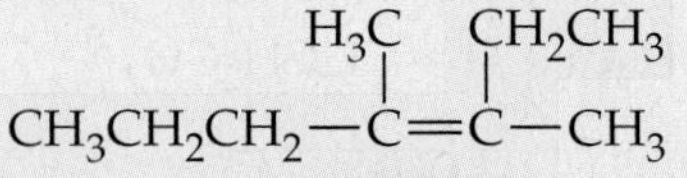

H₃C   CH₂CH₃

CH₃CH₂CH₂—C=C—CH₃   Name as a *heptene.*

Step 2: Number the chain from the end nearer the double bond. The first double-bond carbon is C4 starting from the left-hand end, but C3 starting from the right:

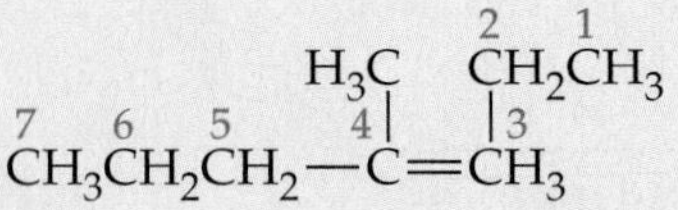

$\overset{2}{\text{C}}\overset{1}{\text{H}_2\text{CH}_3}$

$\underset{7}{\text{CH}_3}\underset{6}{\text{CH}_2}\underset{5}{\text{CH}_2}—\overset{4}{\text{C}}=\overset{3}{\text{CH}_3}$ Name as a substituted *3-heptene.*

Step 3: Identify the substituents, and write the name.

SOLUTION

H₃C   CH₂CH₃   Substituents: 3-methyl, 4-methyl

CH₃CH₂CH₂—C=C—CH₃   Name: 3,4-Dimethyl-3-heptene

■ **WORKED EXAMPLE 13.2**

Draw the structure of 3-ethyl-4-methyl-2-pentene.

ANALYSIS   First, identify the parent name (*pent*) and write the number of carbon atoms it indicates in a straight line. Then, counting from one end, put in the double bond between carbons 2 and 3:

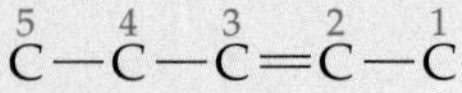

$\overset{5}{\text{C}}—\overset{4}{\text{C}}—\overset{3}{\text{C}}=\overset{2}{\text{C}}—\overset{1}{\text{C}}$ 2-Pentene

Next, add the ethyl and methyl substituents on carbons 3 and 4, and write in the additional hydrogen atoms so that each carbon atom has four bonds.

SOLUTION

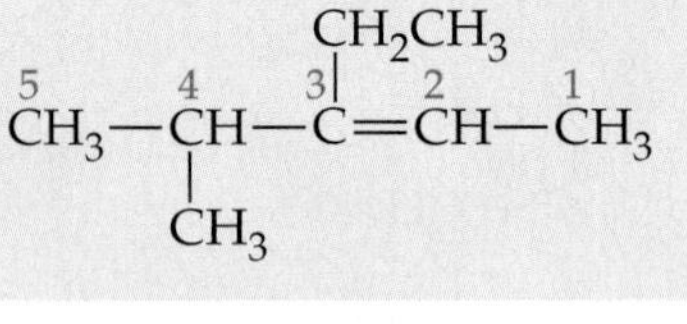

CH₂CH₃

$\underset{5}{\text{CH}_3}—\underset{4}{\text{CH}}—\overset{3}{\text{C}}=\overset{2}{\text{CH}}—\overset{1}{\text{CH}_3}$ 3-Ethyl-4-methyl-2-pentene

CH₃

■ **PROBLEM 13.1**

What are the IUPAC names of the following compounds?

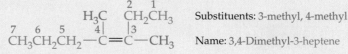

(a) 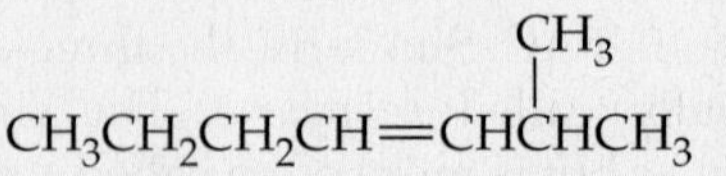 $CH_3CH_2CH_2CH=CHCHCH_3$ (with CH_3 on the CH)   (b) 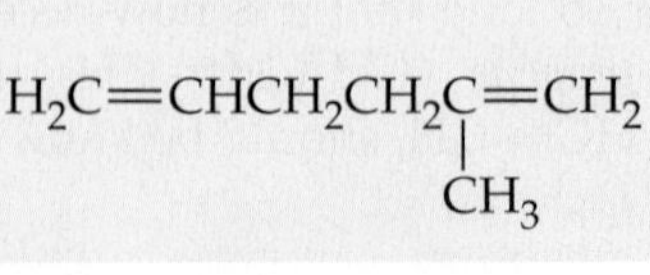 $H_2C=CHCH_2CH_2C=CH_2$ (with CH_3 substituent)

■ **PROBLEM 13.2**

Draw structures corresponding to the following IUPAC names:

(a) 3-Methyl-1-heptene   (b) 4,4-Dimethyl-2-pentyne
(c) 2-Methyl-2-hexene   (d) 3-Ethyl-2,2-dimethyl-3-hexene

■ **KEY CONCEPT PROBLEM 13.3**

What are the IUPAC names of the following alkenes?

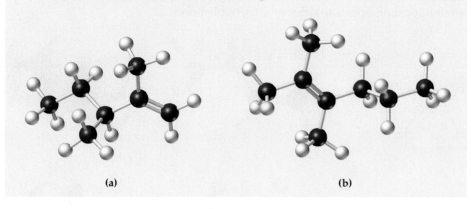

(a)   (b)

13.3 The Structure of Alkenes: Cis–Trans Isomerism

Alkenes and alkynes differ from alkanes in shape because of their multiple bonds. Whereas methane is tetrahedral, ethylene is flat (planar), and acetylene is linear (straight), as predicted by the VSEPR model discussed in Section 5.7. (∞, p. 112)

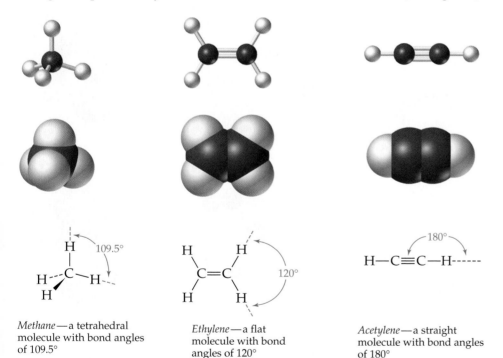

Methane—a tetrahedral molecule with bond angles of 109.5°

Ethylene—a flat molecule with bond angles of 120°

Acetylene—a straight molecule with bond angles of 180°

The two carbons and four attached atoms that make up the double-bond functional group lie in a plane. Unlike the situation in alkanes, where free rotation around the C—C single bond occurs (Section 12.5), there is no rotation around double bonds. (∞, p. 334) As a consequence, a new kind of isomerism is possible for alkenes.

To see this new kind of isomerism, look at the following C_4H_8 compounds. When written as condensed structures, there appear to be three alkene isomers of formula C_4H_8: 1-butene, 2-butene, and 2-methylpropene. In fact, though, there are *four*. 1-Butene and 2-butene are constitutional isomers because their double bonds occur at different positions along the chain, while 2-methylpropene is isomeric with both because it has a different connection of carbon atoms. But because rotation can't occur around carbon–carbon double bonds, *there are two different 2-butenes*. In one isomer, the two groups are close together on the same side of the double bond; in the other isomer, the two methyl groups are far apart on opposite sides of the double bond.

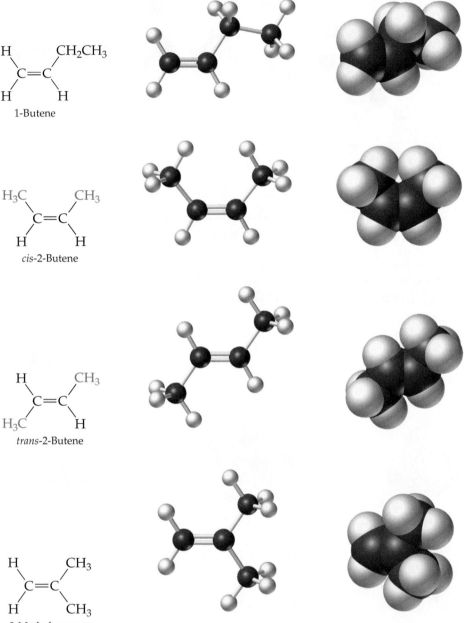

1-Butene

cis-2-Butene

trans-2-Butene

2-Methylpropene

Cis–trans isomers Alkenes that have the same connections between atoms but differ in their three-dimensional structures because of the way that groups are attached to different sides of the double bond.

The two 2-butenes are called **cis–trans isomers**. They have the same formula and connections between atoms but have different three-dimensional structures because of the way that groups are attached to different sides of the double bond. The isomer with its methyl groups on the same side of the double bond is named *cis*-2-butene, and the isomer with its methyl groups on opposite sides of the double bond is named *trans*-2-butene.

Cis–trans isomerism occurs in an alkene whenever each double-bond carbon is bonded to two different substituent groups. If one of the double-bond carbons is attached to two identical groups, cis–trans isomerism is not possible. In 2-methyl-1-butene, for example, cis–trans isomerism is not possible because C1 is bonded to two identical groups (hydrogen atoms). You can convince yourself of this by mentally flipping one of the following two structures top-to-bottom and seeing that it becomes identical to the other structure.

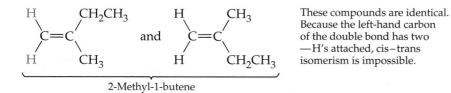

2-Methyl-1-butene

These compounds are identical. Because the left-hand carbon of the double bond has two —H's attached, cis–trans isomerism is impossible.

In 2-pentene, however, the following structures do not become identical when flipped, so cis–trans isomerism does occur.

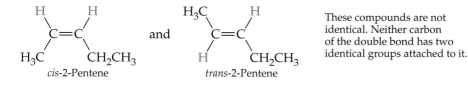

cis-2-Pentene   trans-2-Pentene

These compounds are not identical. Neither carbon of the double bond has two identical groups attached to it.

■ **WORKED EXAMPLE 13.3**

Draw structures for both the cis and trans isomers of 2-hexene.

ANALYSIS First, draw a condensed structure of 2-hexene to see what groups are attached to the double-bond carbons:

$$CH_3CH = CHCH_2CH_2CH_3$$
$$1 2 3 4 5 6$$

Attached to C2: —H and —CH₃
Attached to C3: —H and —CH₂CH₂CH₃

Next, draw two double bonds. Choose one end of each double bond, and attach its groups in the same way to generate two identical partial structures:

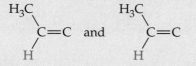

Finally, attach groups to the other end in the two possible ways.

SOLUTION

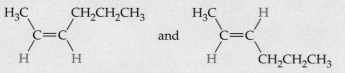

The structure with the two hydrogens on the same side of the double bond is the cis isomer, and that with the two hydrogens on opposite sides is the trans isomer.

■ **PROBLEM 13.4**

Which of the following substances exist as cis–trans isomers? Draw both isomers.
(a) 3-Heptene   **(b)** 2-Methyl-2-hexene   **(c)** 5-Methyl-2-hexene

■ **PROBLEM 13.5**

Draw both cis and trans isomers of 3,4-dimethyl-3-hexene.

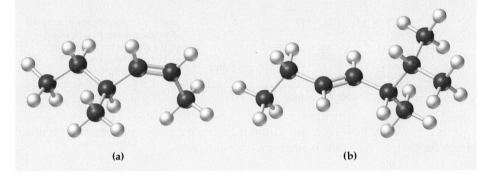

■ **KEY CONCEPT PROBLEM 13.6**

Name the following compounds, including the appropriate *cis-* or *trans-* prefix.

(a)　　　　　(b)

13.4 Properties of Alkenes and Alkynes

Alkenes and alkynes resemble alkanes in many respects. The bonds in alkenes and alkynes are nonpolar, and the physical properties of these compounds are influenced mainly by weak London dispersion forces (Section 8.11). (∞∞, p. 211) Alkenes and alkynes with 1–4 carbon atoms are gases, and boiling points increase with the size of the molecules.

Like alkanes, alkenes and alkynes are insoluble in water, soluble in nonpolar solvents, and less dense than water. They are flammable and nontoxic, although those that are gases present explosion hazards when mixed with air. Unlike alkanes, alkenes are quite reactive because of their double bonds. As we'll see in the next section, alkenes undergo addition of various reagents to their double bonds to yield saturated products.

Properties of Alkenes and Alkynes

* Nonpolar; insoluble in water; soluble in nonpolar organic solvents; less dense than water
* Flammable; nontoxic
* Alkenes display cis–trans isomerism when each double-bond carbon atom has different substituents
* Chemically reactive at the multiple bond

13.5 Kinds of Organic Reactions

Before looking at the chemistry of alkenes, let's first see some general reactivity patterns that make the task of organizing and categorizing organic reactions much simpler. There are four particularly important kinds of organic reactions: *additions, eliminations, substitutions*, and *rearrangements*.

Addition reaction A general reaction type in which a substance X—Y adds to the multiple bond of an unsaturated reactant to yield a saturated product that has only single bonds.

* **Addition Reactions** Additions occur when two reactants add together to form a single product with no atoms "left over." We can generalize the process as:

These two reactants add together . . .　　$A + B \longrightarrow C$　　. . . to give this single product.

Application The Chemistry of Vision

Does eating carrots really improve your vision? Although carrots probably don't do much to help someone who is already on a proper diet, it's nevertheless true that the chemistry of carrots and the chemistry of vision are related. Both involve alkenes.

Carrots, peaches, sweet potatoes, and other yellow vegetables are rich in β-carotene, a purple-orange alkene that provides our main dietary source of vitamin A. The conversion of β-carotene to vitamin A takes place in the mucosal cells of the small intestine, where enzymes cut the molecule in half to yield an alcohol. Vitamin A in excess of the body's immediate needs is stored in the liver, from which it can be transported to the eye. In the eye, vitamin A is converted into a compound called *retinal*, which undergoes cis–trans isomerization of its C11–C12 double bond to produce 11-*cis*-retinal. Reaction with the protein *opsin* then produces the light-sensitive substance *rhodopsin*.

The human eye has two kinds of light-sensitive cells, *rod cells* and *cone cells*. The 3 million or so rod cells are primarily responsible for seeing in dim light, whereas the 100 million cone cells are responsible for seeing in bright light and for the perception of bright colors. When light strikes the rod cells, cis–trans isomerization of the C11–C12 double bond occurs and 11-*trans*-rhodopsin, also called metarhodopsin II, is produced. This cis–trans isomerization is accompanied by a change in molecular geometry, which in turn causes a nerve impulse to be sent to the brain where it is perceived as vision. Metarhodopsin II is then changed back to 11-*cis*-retinal for use in another vision cycle.

See Additional Problems 13.66 and 13.67 at the end of the chapter.

▲ Rod cells in the frog eye.

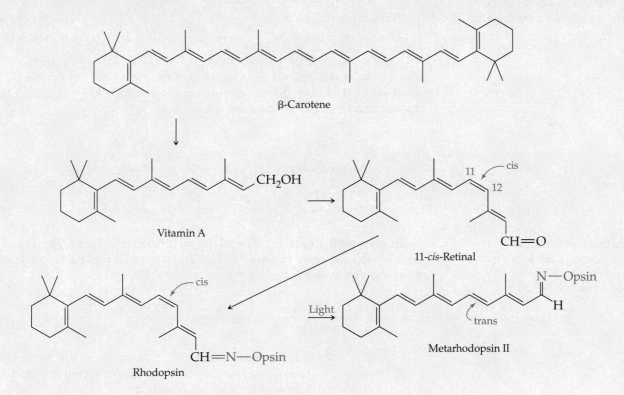

363

As an example of an addition reaction, we'll soon see that alkenes, such as ethylene, react with H_2 to yield alkanes:

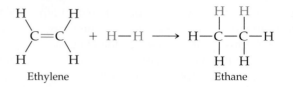

Ethylene   Ethane

Elimination reaction A general reaction type in which a saturated reactant yields an unsaturated product by losing groups from two adjacent carbons.

• **Elimination Reactions** Eliminations are, in a sense, the opposite of addition reactions. Eliminations occur when a single reactant splits into two products, a process we can generalize as:

This one reactant . . . $A \longrightarrow B + C$ . . . splits apart to give these two products.

As an example of an elimination reaction, we'll see in the next chapter that alcohols, such as ethanol, split apart into an alkene and water when treated with an acid catalyst:

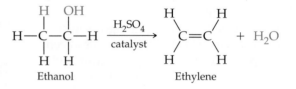

Ethanol   Ethylene

Substitution reaction A general reaction type in which an atom or group of atoms in a molecule is replaced by another atom or group of atoms.

• **Substitution Reactions** Substitutions occur when two reactants exchange parts to give two new products, a process we can generalize as:

These two reactants exchange parts . . . $AB + C \longrightarrow AC + B$ . . . to give these two products.

As an example of a substitution reaction, we saw in Section 12.8 that alkanes, such as methane, react with Cl_2 in the presence of ultraviolet light to yield alkyl chlorides. A —Cl group substitutes for the —H group of the alkane, and two new products result:

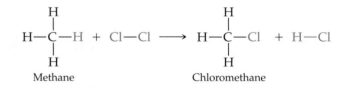

Methane   Chloromethane

Rearrangement reaction A general reaction type in which a molecule undergoes bond reorganization to yield an isomer.

• **Rearrangement Reactions** Rearrangements occur when a single reactant undergoes a reorganization of bonds and atoms to yield a single isomeric product, a process we can generalize as:

This one reactant . . . $A \longrightarrow B$ . . . gives this isomeric product.

We won't be studying any rearrangement reactions in this book, but we might note as an example that *cis*-2-butene can be converted into its isomer *trans*-2-butene by treatment with an acid catalyst:

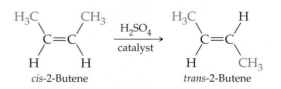

cis-2-Butene   *trans*-2-Butene

■ PROBLEM 13.7

Classify each of the following reactions as an addition, elimination, substitution, or rearrangement:

(a) $CH_3Br + NaOH \rightarrow CH_3OH + NaBr$
(b) $H_2C{=}CH_2 + HCl \rightarrow CH_3CH_2Cl$
(c) $CH_3CH_2Br \rightarrow H_2C{=}CH_2 + HBr$

13.6 Reactions of Alkenes and Alkynes

Most of the reactions of carbon–carbon multiple bonds are *addition reactions*. A generalized reagent we might write as X–Y adds to the multiple bond in the unsaturated reactant to yield a saturated product that has only single bonds. Alkenes and alkynes react similarly in many ways, but we'll look mainly at alkenes in this chapter because they're more common.

Addition Reactions of Alkenes

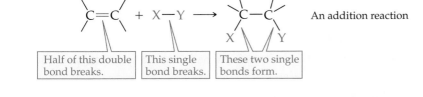

An addition reaction

Addition of H$_2$ to Alkenes and Alkynes: Hydrogenation

Alkenes and alkynes react with hydrogen in the presence of a metal catalyst such as palladium to yield the corresponding alkane product:

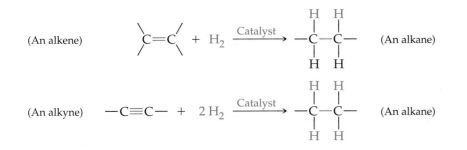

For example,

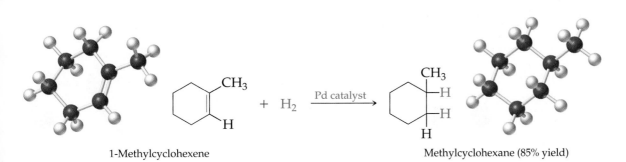

1-Methylcyclohexene   Methylcyclohexane (85% yield)

The addition of hydrogen to an alkene, a process called **hydrogenation**, is used commercially to convert unsaturated vegetable oils, which contain numerous double bonds, to the saturated fats used in margarine and cooking fats. We'll see the structures of these fats and oils in Chapter 24.

Hydrogenation The addition of H$_2$ to a multiple bond to give a saturated product.

■ **WORKED EXAMPLE 13.4**

What product would you obtain from the following reaction?

$$CH_3CH_2CH_2CH{=}CHCH_3 + H_2 \xrightarrow{Pd} ?$$

ANALYSIS   Rewrite the reactant, showing a single bond and two partial bonds in place of the double bond:

$$\underset{|\quad\quad|}{CH_3CH_2CH_2CH{-}CHCH_3}$$

Then, add a hydrogen to each carbon atom of the double bond, and rewrite the product in condensed form:

$$\underset{\underset{H\quad\;\;H}{|\quad\;\;|}}{CH_3CH_2CH_2CH{-}CHCH_3} \quad \text{is the same as} \quad \underset{\text{Hexane}}{CH_3CH_2CH_2CH_2CH_2CH_3}$$

SOLUTION   The reaction is

$$CH_3CH_2CH_2CH = CHCH_3 + H_2 \xrightarrow{Pd} CH_3CH_2CH_2CH_2CH_2CH_3$$

■ **PROBLEM 13.8**

Write the structures of the products from the following hydrogenation reactions:

(a) $CH_3CH_2CH{=}CH_2 + H_2 \xrightarrow{Pd} ?$
(b) *cis*-2-Butene + $H_2 \xrightarrow{Pd} ?$
(c) *trans*-2-Butene + $H_2 \xrightarrow{Pd} ?$

(d) 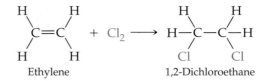 $={CH_2} + H_2 \xrightarrow{Pd} ?$

Addition of Cl_2 and Br_2 to Alkenes: Halogenation

Alkenes react with the halogens Br_2 and Cl_2 to give 1,2-dihaloalkane addition products in a **halogenation** reaction:

Halogenation   The addition of Cl_2 or Br_2 to a multiple bond to give a dihalide product.

Addition of Halogens to Alkenes

$$\underset{\text{Ethylene}}{\overset{\displaystyle H\quad\quad H}{\underset{\displaystyle H\quad\quad H}{C{=}C}}} + Cl_2 \longrightarrow \underset{\text{1,2-Dichloroethane}}{\overset{\displaystyle H\quad\quad H}{\underset{\displaystyle Cl\quad\; Cl}{H{-}C{-}C{-}H}}}$$

For example,

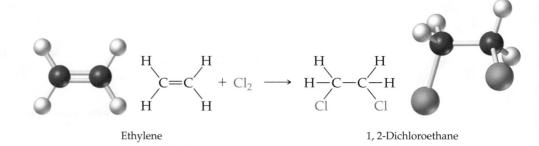

Ethylene   1, 2-Dichloroethane

This reaction, the first step in making poly(vinyl chloride) plastics, is used to manufacture nearly 8 million tons of 1,2-dichloroethane each year in the United States.

Reaction with Br_2 also provides a convenient test for the presence in a molecule of a carbon–carbon double or triple bond (Figure 13.1). A few drops of a reddish-brown solution of Br_2 are added to a sample of an unknown compound. Immediate disappearance of the color as the bromine reacts to form a colorless dibromide reveals the presence of the multiple bond.

(a)   (b)

◄ **FIGURE 13.1 Testing for unsaturation with bromine.** (a) No color change results when the bromine solution is added to hexane (C_6H_{14}). (b) Disappearance of the bromine color when it is added to 1-hexene (C_6H_{12}) indicates the presence of a double bond.

 Testing for Unsaturated Hydrocarbons with Bromine

■ **PROBLEM 13.9**

What products would you expect from the following halogenation reactions?

(a) 2-Methylpropene + Br_2 ⟶ ?   (b) 1-Pentene + Cl_2 ⟶ ?

$$\text{(c)} \quad CH_3CH_2CH{=}\underset{\underset{CH_3}{|}}{\overset{\overset{CH_3}{|}}{C}}CH_2CHCH_3 + Cl_2 \longrightarrow ?$$

Addition of HBr and HCl to Alkenes

Alkenes react with hydrogen bromide (HBr) to yield *alkyl bromides* (R—Br) and with hydrogen chloride (HCl) to yield *alkyl chlorides* (R—Cl), in what are called **hydrohalogenation** reactions:

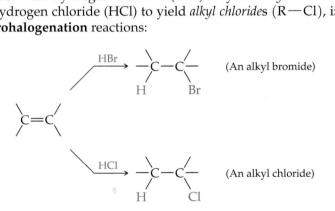

Hydrohalogenation The addition of HCl or HBr to a multiple bond to give an alkyl halide product.

 Addition of HBr to an Alkene

The addition of HBr to 2-methylpropene is an example:

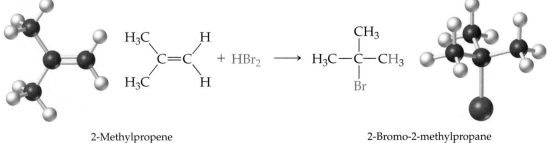

2-Methylpropene   2-Bromo-2-methylpropane

Look carefully at the above example. Only one of the two possible addition products is obtained. 2-Methylpropene *could* add HBr to give 1-bromo-2-methylpropane, but it doesn't; it gives only 2-bromo-2-methylpropane.

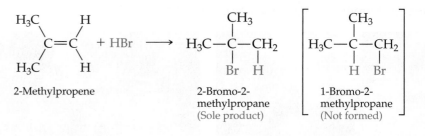

This result is typical of what happens when HBr and HCl add to an alkene in which one of the double-bond carbons has more hydrogens than the other. The results of such additions can be predicted by **Markovnikov's rule**, formulated in 1869 by the Russian chemist Vladimir Markovnikov:

Markovnikov's rule In the addition of HX to an alkene, the H becomes attached to the carbon that already has the most H's, and the X becomes attached to the carbon that has fewer H's.

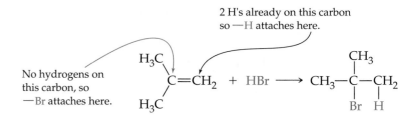

■ **WORKED EXAMPLE 13.5**

What product do you expect from the following reaction?

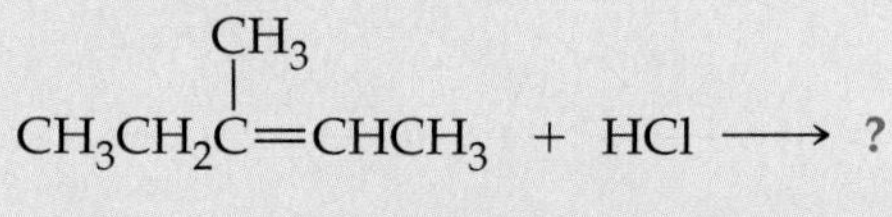

$$CH_3CH_2\overset{\overset{\displaystyle CH_3}{|}}{C}=CHCH_3 \ + \ HCl \longrightarrow ?$$

ANALYSIS The reaction of an alkene with HCl leads to formation of an alkyl chloride addition product according to Markovnikov's rule. To make a prediction, look at the starting alkene and count the number of hydrogens attached to each double-bond carbon. Then write the product by attaching H to the carbon that has more hydrogens and attaching Cl to the carbon that has fewer hydrogens.

SOLUTION

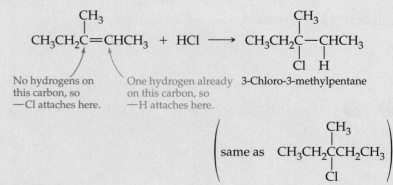

■ **WORKED EXAMPLE 13.6**

From what alkene might 2-chloro-3-methylbutane be made?

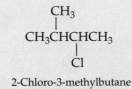

2-Chloro-3-methylbutane

ANALYSIS 2-Chloro-3-methylbutane is an alkyl chloride that might be made by addition of HCl to an alkene. To generate the possible alkene precursors, remove the —Cl group and an —H atom from adjacent carbons, and replace with a double bond.

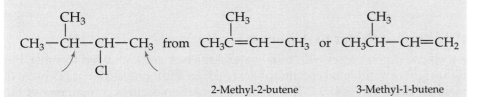

2-Methyl-2-butene   3-Methyl-1-butene

Look at the possible alkene addition reactions to see which is compatible with Markovnikov's rule. In this case, addition to 3-methyl-1-butene is compatible.

SOLUTION

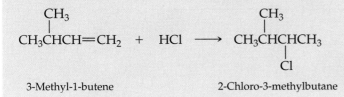

3-Methyl-1-butene   2-Chloro-3-methylbutane

■ **PROBLEM 13.10**

What products do you expect from the following reactions?

(a) Cyclohexene + HBr $\longrightarrow$?

(b) 1-Hexene + HCl $\longrightarrow$?

(c) 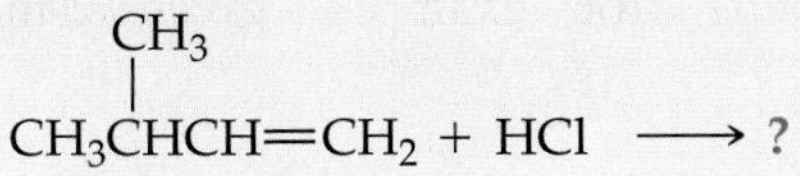 $\underset{\text{CH}_3}{\text{CH}_3\text{CHCH}}{=}\text{CH}_2$ + HCl $\longrightarrow$?

■ **PROBLEM 13.11**

What alkenes are the following alkyl halides likely to have been made from? (Careful, there may be more than one answer.)

(a) 3-Chloro-3-ethylpentane   (b) 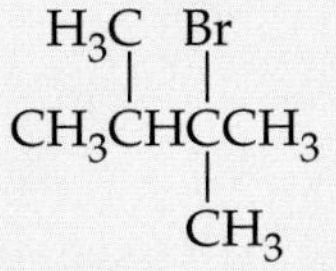 $\underset{\text{CH}_3}{\overset{\text{H}_3\text{C} \quad \text{Br}}{\text{CH}_3\text{CHCCH}_3}}$

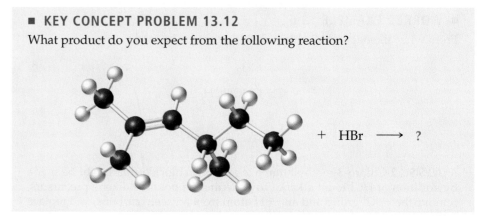

■ **KEY CONCEPT PROBLEM 13.12**

What product do you expect from the following reaction?

$+$ HBr $\longrightarrow$?

Addition of Water to Alkenes: Hydration

An alkene doesn't react with pure water if the two are just mixed together, but if the right experimental conditions are used, an addition reaction takes place to yield an *alcohol* (R—OH). This **hydration** reaction occurs on treatment of the alkene with water in the presence of a strong acid catalyst, such as H_2SO_4. In fact, nearly 100 million gallons of ethyl alcohol (ethanol) are produced each year in the United States by this method.

Hydration The addition of water to a multiple bond to give an alcohol product.

Addition of Water to an Alkene

An alcohol

For example,

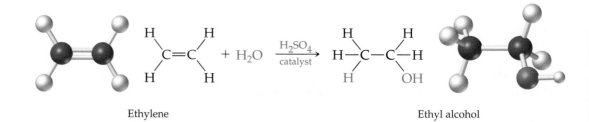

Ethylene

Ethyl alcohol

As with the addition of HBr and HCl, Markovnikov's rule can be used to predict the product when water adds to an unsymmetrically substituted alkene. Hydration of 2-methylpropene, for example, gives 2-methyl-2-propanol:

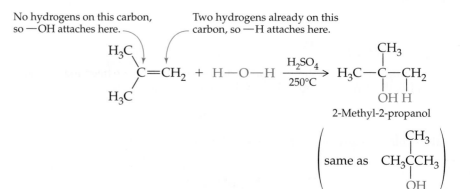

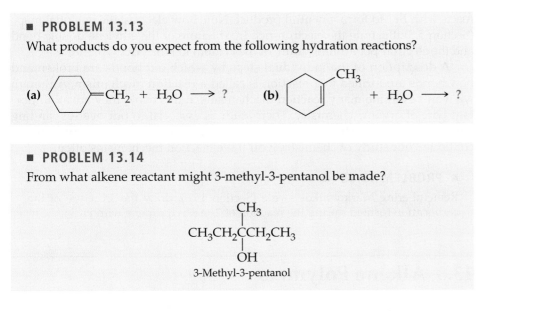

■ **PROBLEM 13.13**

What products do you expect from the following hydration reactions?

(a) [structure: cyclohexane ring with =CH₂] + H₂O ⟶ ?   (b) [structure: cyclohexene ring with CH₃] + H₂O ⟶ ?

■ **PROBLEM 13.14**

From what alkene reactant might 3-methyl-3-pentanol be made?

$$CH_3$$
$$CH_3CH_2CCH_2CH_3$$
$$OH$$

3-Methyl-3-pentanol

13.7 How an Alkene Addition Reaction Occurs

How do alkene addition reactions take place? Do two molecules, say ethylene and HBr, simply collide and immediately form a product molecule of bromoethane, or is the process more complex? Detailed studies have shown that alkene addition reactions take place in two distinct steps, as illustrated in Figure 13.2 for the addition of HBr to ethylene.

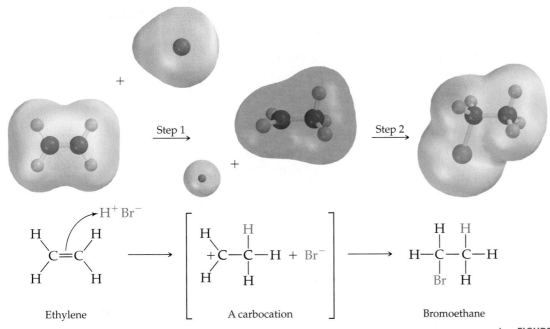

Ethylene   A carbocation   Bromoethane

In the first step, the alkene reacts with H^+ from the acid HBr. The carbon–carbon double bond partially breaks, and two electrons move from the double bond to form a new single bond between one of the carbons and the incoming hydrogen. The other double-bond carbon, having had electrons removed from it, now has only six electrons in its outer shell and bears a positive charge. Unlike sodium ion (Na^+) and other metal cations, which are unreactive and easily isolated in salts like NaCl, carbon cations, or *carbocations*, are highly reactive. As soon as the carbocation is formed by reaction of an alkene with H^+, it immediately

▲ **FIGURE 13.2 The mechanism of the addition of HBr to an alkene.** The reaction takes place in two steps and involves a carbocation intermediate. In the first step, two electrons move from the C=C double bond to form a C—H bond. In the second step, Br^- uses two electrons to form a bond to the positively charged carbon.

Reaction mechanism   A description of the individual steps by which old bonds are broken and new bonds formed are in a reaction.

reacts with Br^- to form a neutral product. Note how electrostatic potential maps (Section 5.8) illustrate the electron-rich (red) nature of the ethylene double bond and the electron-poor (blue) nature of the H atom in HBr. (∞, p. 118)

A description of the individual steps by which old bonds are broken and new bonds are formed in a reaction is called a **reaction mechanism**. Although we won't examine many reaction mechanisms in this book, they are an important part of organic chemistry. Their study is essential to our ever-expanding ability to understand biochemistry and the physiological effects of drugs. If you continue your study of chemistry, you'll see reaction mechanisms often.

> ■ **PROBLEM 13.15**
>
> Remembering Markovnikov's rule (Section 13.6), draw the structure of the carbocation formed during the reaction of 2-methylpropene with HCl.

13.8 Alkene Polymers

Polymer   A large molecule formed by the repetitive bonding together of many smaller molecules.

Monomer   A small molecule that is used to prepare a polymer.

A **polymer** is a large molecule formed by the repetitive bonding together of many smaller molecules called **monomers.** As we'll see in later chapters, biological polymers occur throughout nature. Cellulose and starch are polymers built from sugars, proteins are polymers built from amino acids, and the DNA that makes up our genetic heritage is a polymer built from nucleic acids. Although the basic idea is the same, synthetic polymers are much simpler than biopolymers because the starting monomer units are usually small, simple organic molecules.

Many simple alkenes (often called *vinyl monomers* because the partial structure $H_2C=CH-$ is known as a *vinyl group*) undergo *polymerization* reactions when treated with the proper catalyst. Ethylene yields polyethylene on polymerization, propylene yields polypropylene, and styrene yields polystyrene. The polymer product might have anywhere from a few hundred to a few thousand monomer units incorporated into a long, repeating chain.

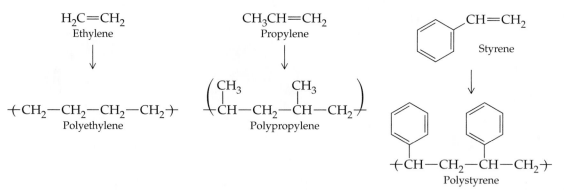

The fundamental reaction in the polymerization of a vinyl monomer resembles the additions to double bonds described in the previous section. The addition to an alkene of a species called an *initiator* yields a reactive intermediate, which in turn adds to a second alkene molecule to produce another reactive intermediate, which adds to a third alkene molecule, and so on. Because the result is continuous addition of one monomer after another to the end of the growing polymer chain, polymers made in this way are called *chain-growth polymers.*

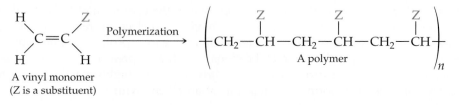

Variations in the substituent group Z attached to the double bond impart different properties to the product, as illustrated by the alkene polymers listed in Table 13.1.

TABLE 13.1 Some Alkene Polymers and Their Uses

Monomer Name	Monomer Structure	Polymer Name	Uses
Ethylene	$H_2C{=}CH_2$	Polyethylene	Packaging, bottles
Propylene	$H_2C{=}CH{-}CH_3$	Polypropylene	Bottles, rope, pails, medical tubing
Vinyl chloride	$H_2C{=}CH{-}Cl$	Poly(vinyl chloride)	Insulation, plastic pipe
Styrene	$H_2C{=}CH{-}\hexagon$	Polystyrene	Foams and molded plastics
Styrene and butadiene	$H_2C{=}CH{-}\hexagon$ and $H_2C{=}CHCH{=}CH_2$	Styrene-butadiene rubber (SBR)	Synthetic rubber for tires
Acrylonitrile	$H_2C{=}CH{-}C{\equiv}N$	Orlon, Acrilan	Fibers, outdoor carpeting
Methyl methacrylate	$H_2C{=}\overset{\displaystyle O}{\underset{\displaystyle CH_3}{C}}COCH_3$	Plexiglas, Lucite	Windows, contact lenses, fiber optics
Tetrafluoroethylene	$F_2C{=}CF_2$	Teflon	Nonstick coatings, bearings, replacement heart valves and blood vessels

The properties of a polymer depend not only on the monomer but also on the average size of the huge molecules in a particular sample and on how extensively they branch. The long molecules in straight-chain polyethylene pack closely together, giving a rigid material called *high-density polyethylene*, which is mainly used in bottles for products such as milk and motor oil. When polyethylene molecules contain many branches, they can't pack together so tightly and instead form a flexible material called *low-density polyethylene*, which is used mainly in packaging materials.

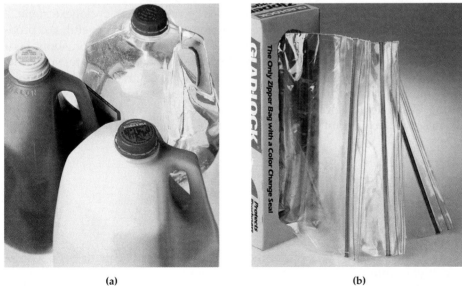

(a) (b)

▲ (a) High-density polyethylene is strong and rigid enough to be used in many kinds of bottles and toys. (b) Low-density polyethylene is moisture-proof but flexible and is used in plastic bags and packaging.

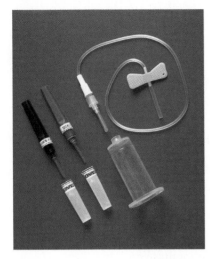

▲ These disposable polypropylene syringes are used once and then discarded.

Polymer technology has come a long way since the development of synthetic rubber, nylon, Plexiglas, and Teflon. The use of polymers has changed the nature of activities from plumbing and carpentry to clothing and auto manufacture. In the health care fields, the use of inexpensive, disposable equipment is now common.

■ **PROBLEM 13.16**

Write the structure of a segment of poly(vinyl acetate), a polymer used for the springy soles in running shoes. The structure of the monomer is

$$H_2C=CHOCCH_3 \quad \text{Vinyl acetate}$$

■ **PROBLEM 13.17**

Write the structures of the monomers used to make the following polymers:

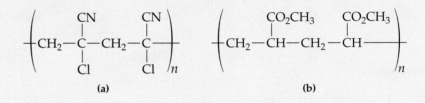

(a)   (b)

13.9 Aromatic Compounds and the Structure of Benzene

▲ The odor of cherries is due to benzaldehyde, an aromatic compound.

Aromatic The class of compounds containing benzene-like rings.

In the early days of organic chemistry, the word *aromatic* was used to describe many fragrant substances from fruits, trees, and other natural sources. It was soon realized, however, that substances grouped as aromatic behave differently from most other organic compounds. Today, the term **aromatic** is used by chemists to refer to the class of compounds containing benzene-like rings.

Benzene, the simplest aromatic compound, is a flat, symmetrical molecule with the molecular formula C_6H_6. It is often represented as cyclohexatriene, a six-membered carbon ring with three double bonds. Though useful, the problem with this representation is that it gives the wrong impression about benzene's chemical reactivity and bonding. Because benzene appears to have three double bonds, you might expect it to react with H_2, Br_2, HCl, and H_2O to give the same kinds of addition products that alkenes do. But this expectation is wrong. Benzene and other aromatic compounds are much less reactive than alkenes and don't normally undergo addition reactions.

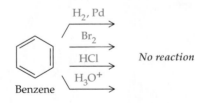

Benzene's relative lack of chemical reactivity is a consequence of its structure. If you were to draw a six-membered ring with alternating single and double bonds, where would you place the double bonds? There are two equivalent possibilities (Figure 13.3b), neither of which is fully correct by itself. Experimental evidence shows that all six carbon–carbon bonds in benzene are identical, so a picture with three double bonds and three single bonds can't be right.

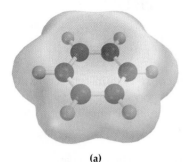

Two equivalent structures, which differ in the position of their double-bond electrons. Neither structure is correct by itself.

(a) (b) (c)

◄ **FIGURE 13.3** **Some representations of benzene.** (a) An electrostatic potential map shows the equivalency of the carbon–carbon bonds. Benzene is usually represented by two equivalent structures in (b) or by the single structure in (c).

The properties of benzene can best be explained by assuming that its true structure is an *average* of the two equivalent conventional structures. The double-bond electrons are not held between specific pairs of atoms but are instead free to move over the entire ring. Each carbon–carbon bond is thus intermediate between a single bond and a double bond. The name **resonance** is given to this phenomenon where the true structure of a molecule is an average among two or more possible conventional structures, and a special double-headed arrow (⟷) is used to show the resonance relationship.

Because the real structure of benzene is intermediate between the two forms shown in Figure 13.3b, it's difficult to represent benzene with the standard conventions using lines for covalent bonds. Thus, we sometimes represent the double bonds as a circle inside the six-membered ring, as shown in Figure 13.3c. Usually, though, we draw the ring with three double bonds, with the understanding that it is an aromatic ring with equivalent bonding all around.

Simple aromatic hydrocarbons like benzene are nonpolar, insoluble in water, volatile, and flammable. Unlike alkanes and alkenes, however, several aromatic hydrocarbons are toxic. Benzene itself has been implicated as a cause of leukemia, and the dimethyl-substituted benzenes are central-nervous-system depressants.

Everything we've said about the structure and stability of the benzene ring also applies to the ring when it has substituents, for example, in compounds such as the bacteriocidal agent hexachlorophene and the flavoring ingredient vanillin:

Resonance The phenomenon where the true structure of a molecule is an average among two or more conventional structures.

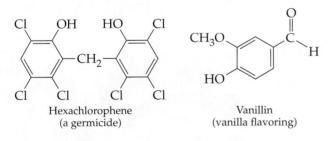

Hexachlorophene
(a germicide)

Vanillin
(vanilla flavoring)

The benzene ring is also present in many biomolecules and retains its characteristic properties in these compounds as well. In addition, "aromaticity" is not limited to rings containing only carbon. Many aromatic compounds, such as pyridine, indole, and adenine, contain nitrogen atoms. These and all compounds that contain a substituted benzene ring, or a similarly stable six-membered ring in which double-bond electrons are equally shared around the ring, are classified as aromatic compounds.

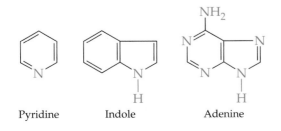

Pyridine Indole Adenine

Application Polycyclic Aromatic Hydrocarbons and Cancer

The definition of the term *aromatic* can be extended beyond simple monocyclic (one-ring) compounds to include *polycyclic* aromatic compounds—substances that have two or more benzene-like rings joined together by a common bond. Naphthalene, familiar for its use in mothballs, is the simplest and best-known polycyclic aromatic compound.

In addition to naphthalene, there are many polycyclic aromatic compounds that are more complex. Benz[*a*]pyrene, for example, contains five benzene-like rings joined together; ordinary graphite (the "lead" in pencils) consists of enormous two-dimensional sheets of benzene-like rings stacked one on top of the other.

Even benzene itself can cause certain types of cancer on prolonged exposure, so breathing the fumes of benzene and other volatile aromatic compounds in the laboratory should be avoided.

See Additional Problems 13.72 and 13.73 at the end of the chapter.

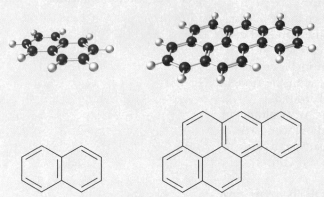

Naphthalene   Benz[*a*]pyrene

A graphite segment

Perhaps the most notorious polycyclic aromatic hydrocarbon is benz[*a*]pyrene, one of the carcinogenic (cancer-causing) substances found in chimney soot, cigarette smoke, and charcoal-broiled meat. Exposure to even a tiny amount is sufficient to induce a skin tumor in susceptible mice.

After benz[*a*]pyrene is taken into the body by eating or inhaling, the body attempts to rid itself of the foreign substance by converting it into a water-soluble metabolite called a *diol epoxide*, which can be excreted. Unfortunately, the diol epoxide metabolite reacts with and binds to cellular DNA, thereby altering the DNA and leading to mutations or cancer.

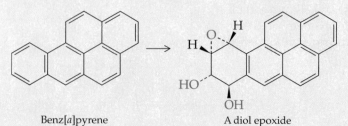

Benz[*a*]pyrene   A diol epoxide

▲ Would you eat this? It contains small amounts of benz[*a*]pyrene.

376

13.10 Naming Aromatic Compounds

Substituted benzenes are named using -*benzene* as the parent. Thus, C_6H_5Br is bromobenzene, $C_6H_5CH_2CH_3$ is ethylbenzene, and so on. No number is needed for monosubstituted benzenes because all the ring positions are identical.

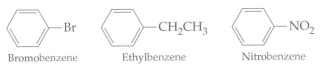

| Bromobenzene | Ethylbenzene | Nitrobenzene |

Disubstituted aromatic compounds are named using one of the prefixes *ortho*-, *meta*-, or *para*-. An *ortho*- or *o*-disubstituted benzene has its two substituents in a 1,2 relationship on the ring; a *meta*- or *m*-disubstituted benzene has its two substituents in a 1,3 relationship; and a *para*- or *p*-disubstituted benzene has its substituents in a 1,4 relationship.

| *ortho*-Dibromobenzene | *meta*-Chloronitrobenzene | *para*-Dimethylbenzene |

Many substituted aromatic compounds have common names in addition to their systematic names. For example, methylbenzene is familiarly known as *toluene*, hydroxybenzene as *phenol*, aminobenzene as *aniline*, and so on, as shown in Table 13.2. Frequently, these common names are used together with *ortho*-, *meta*-, or *para*- prefixes. For example:

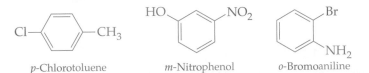

| *p*-Chlorotoluene | *m*-Nitrophenol | *o*-Bromoaniline |

TABLE 13.2 Common Names of Some Aromatic Compounds

Structure	Name	Structure	Name
⬡—CH₃	Toluene	H₃C—⬡—CH₃	*para*-Xylene
⬡—OH	Phenol	⬡—C(=O)—OH	Benzoic acid
⬡—NH₂	Aniline	⬡—C(=O)—H	Benzaldehyde

Occasionally, the benzene ring itself might be considered a substituent group attached to another parent compound. When this happens, the name **phenyl** (pronounced **fen**-nil) is used for the C_6H_5- unit:

Phenyl The C_6H_5- group.

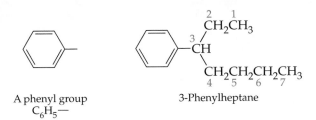

A phenyl group
C_6H_5-

3-Phenylheptane

■ **WORKED EXAMPLE 13.7**

Draw the structure of *m*-chloroethylbenzene.

ANALYSIS *m*-Chloroethylbenzene has a benzene ring with two substituents, chloro and ethyl, in a meta relationship. Draw a benzene ring, and attach one of the substituents, say chloro, to any position:

Now, go to a meta position two carbons away from the chloro-substituted carbon, and attach the second (ethyl) substituent.

SOLUTION

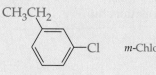

m-Chloroethylbenzene

■ **PROBLEM 13.18**

What are the IUPAC names for the following compounds?

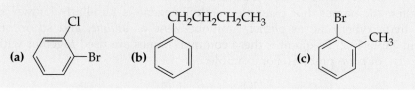

(a)   (b)   (c)

■ **PROBLEM 13.19**

Draw structures corresponding to the following names:

(a) *o*-Dibromobenzene   (b) *p*-Nitrotoluene
(c) *m*-Diethylbenzene   (d) *m*-Isopropylphenol

■ **KEY CONCEPT PROBLEM 13.20**

Name the following compounds (red = O, blue = N, red-brown = Br):

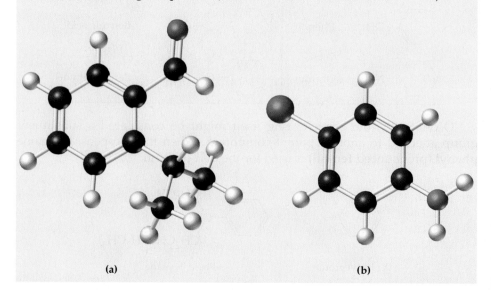

(a)   (b)

13.11 Reactions of Aromatic Compounds

Unlike alkenes, which undergo addition reactions, aromatic compounds usually undergo *substitution* reactions. That is, a group Y substitutes for one of the hydrogen atoms on the aromatic ring without changing the ring itself. It doesn't matter which of the six ring hydrogens in benzene is replaced because all six are equivalent.

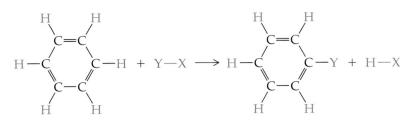

Nitration is the substitution of a *nitro group* ($-NO_2$) for one of the ring hydrogens. The reaction occurs when benzene reacts with nitric acid in the presence of sulfuric acid as catalyst:

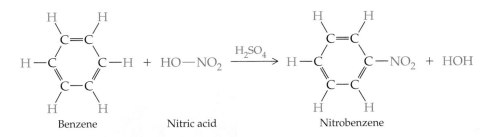

Nitration of aromatic rings is a key step in the synthesis both of explosives like TNT (trinitrotoluene) and of many important pharmaceutical agents. Nitrobenzene itself is the industrial starting material for the preparation of aniline, which is used to make many of the brightly colored dyes in clothing.

Halogenation is the substitution of a halogen atom, usually bromine or chlorine, for one of the ring hydrogens. The reaction occurs when benzene reacts with Br_2 or Cl_2 in the presence of iron as catalyst:

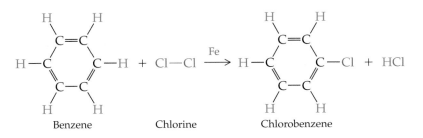

Sulfonation is the substitution of a sulfonic acid group ($-SO_3H$) for one of the ring hydrogens. The reaction occurs when benzene reacts with concentrated sulfuric acid and SO_3:

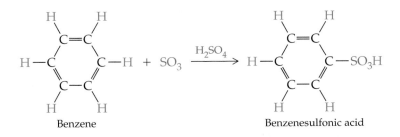

▲ Many dyes used for clothing are derivatives of aminobenzene, or aniline. Aniline itself is made from nitrobenzene.

Nitration The substitution of a nitro group ($-NO_2$) for a hydrogen on an aromatic ring.

Halogenation The substitution of a halogen group ($-X$) for a hydrogen on an aromatic ring.

Sulfonation The substitution of a sulfonic acid group ($-SO_3H$) for a hydrogen on an aromatic ring.

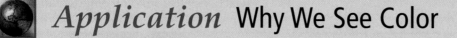
The purple dye mauve, the plant pigment cyanidin, and a host of other organic compounds are brightly colored. What do all these compounds have in common? If you look carefully at each structure, you'll see that each has numerous alternating double and single bonds.

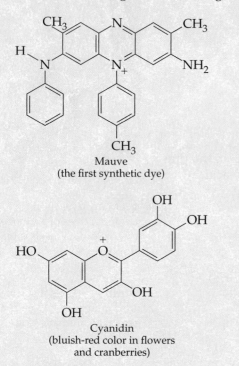

Mauve
(the first synthetic dye)

Cyanidin
(bluish-red color in flowers
and cranberries)

In describing benzene, we said that the double-bond electrons are spread out, or *delocalized*, over the whole molecule. The same phenomenon occurs whenever there are many alternating double and single bonds: The double-bond electrons form a delocalized region of electron density in the molecule that is capable of absorbing light. Organic compounds such as benzene, with small numbers of delocalized electrons, absorb in the ultraviolet region of the electromagnetic spectrum, which our eyes can't detect. Compounds with longer stretches of alternating double and single bonds absorb in the visible region.

The color that we see is complementary to the color that's absorbed; that is, we see what is left of the white light after certain colors have been absorbed. For example, the plant pigment cyanidin absorbs greenish-yellow light and thus appears reddish-blue. The same principle applies to almost all colored organic compounds: They contain large regions of delocalized electrons and absorb some portion of the visible spectrum.

See Additional Problems 13.74 and 13.75 at the end of the chapter.

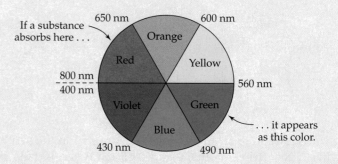

▲ Using an artist's color wheel, it's possible to determine the observed color of a substance by knowing the color of the light absorbed. Observed and absorbed colors are complementary. Thus, if a substance absorbs red light, it has a green color.

Aromatic-ring sulfonation is a key step in the synthesis of such compounds as the sulfa-drug family of antibiotics.

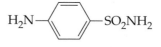

Sulfanilamide—a sulfa antibiotic

■ **PROBLEM 13.21**

Write the products from reaction of the following reagents with *p*-xylene (*p*-dimethylbenzene).

(a) Br_2 and Fe   **(b)** HNO_3 and H_2SO_4 **(c)** SO_3 and H_2SO_4

■ **PROBLEM 13.22**

Reaction of Br_2 and Fe with toluene (methylbenzene) might lead to one or more of *three* substitution products. Show the structure of each.

Summary: Revisiting the Chapter Goals

1. What are alkenes, alkynes, and aromatic compounds? *Alkenes* are hydrocarbons that contain a carbon–carbon double bond, and *alkynes* are hydrocarbons that contain a carbon–carbon triple bond. *Aromatic compounds* contain six-membered, benzene-like rings and are usually written with three double bonds. In fact, though, there is equal bonding between neighboring carbon atoms because the double-bond electrons are symmetrically spread around the entire ring. All three families are said to be *unsaturated* because they have fewer hydrogens than corresponding alkanes.

2. How are alkenes, alkynes, and aromatic compounds named? Alkenes are named using the family ending *-ene*, while alkynes use the family ending *-yne*. Disubstituted benzenes are named with *-benzene* as the parent and with one of the prefixes *ortho-* (1,2 substitution), *meta-* (1,3 substitution), or *para-* (1,4 substitution).

3. What are cis–trans isomers? The lack of rotation around carbon–carbon double bonds leads to *cis–trans isomerism* for disubstituted alkenes. The cis isomer has two hydrogens on the same side of the double bond; the trans isomer has them on opposite sides.

4. What are the categories of organic reactions? The four categories of organic reactions are *addition, elimination, substitution,* and *rearrangement* reactions. Addition reactions occur when two reactants add together to form a single product with no atoms left over. Eliminations occur when a single reactant splits into two products. Substitution reactions occur when two reactants exchange parts to give two new products. Rearrangement reactions occur when a single reactant undergoes a reorganization of bonds and atoms to yield a single isomeric product.

5. What are the typical reactions of alkenes, alkynes, and aromatic compounds? Alkenes and alkynes undergo addition reactions to their multiple bonds. Addition of hydrogen to an alkene (*hydrogenation*) yields an alkane product; addition of Cl_2 or Br_2 (*halogenation*) yields a 1,2-dihaloalkane product; addition of HBr and HCl (*hydrohalogenation*) yields an alkyl halide product; and addition of water (*hydration*) yields an alcohol product. *Markovnikov's rule* predicts that in the addition of HX or H_2O to a double bond, the H becomes attached to the carbon that has more H's, and the X or OH becomes attached to the carbon that has fewer H's.

Aromatic compounds are unusually stable but can be made to undergo substitution reactions, in which one of the ring hydrogens is replaced by another group ($C_6H_6 \rightarrow C_6H_5Y$). Among these substitutions are *nitration* (substitution of $-NO_2$ for $-H$), *halogenation* (substitution of $-Br$ or $-Cl$ for $-H$), and *sulfonation* (substitution of $-SO_3H$ for $-H$).

6. How do organic reactions take place? A description of the individual steps by which old bonds are broken and new bonds are formed in a reaction is called a *reaction mechanism*. The addition reaction of an alkene with HX takes place in two steps. In the first step, the alkene uses two electrons to bond to H^+, giving a positively charged species called a *carbocation*. In the second step, the carbocation reacts with the halide ion to give the final product.

Key Words

Addition reaction, p. 362

Alkene, p. 355

Alkyne, p. 356

Aromatic, p. 374

Cis–trans isomers, p. 360

Cycloalkene, p. 357

Elimination reaction, p. 364

Halogenation (alkene), p. 366

Halogenation (aromatic), p. 379

Hydration, p. 370

Hydrogenation, p. 365

Hydrohalogenation, p. 367

Markovnikov's rule, p. 368

Monomer, p. 372

Nitration, p. 379

Phenyl, p. 377

Polymer, p. 372

Reaction mechanism, p. 372

Rearrangement reaction, p. 364

Resonance, p. 375

Saturated, p. 355

Substitution reaction, p. 364

Sulfonation, p. 379

Unsaturated, p. 355

Summary of Reactions

1. Reactions of alkenes and alkynes (Section 13.6)

(a) Addition of H_2 to yield an alkane (hydrogenation):

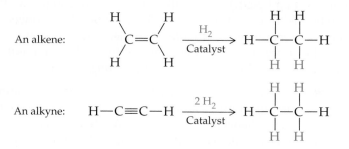

(b) Addition of Cl_2 or Br_2 to yield a dihalide (halogenation):

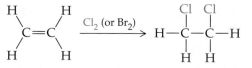

(c) Addition of HCl or HBr to yield an alkyl halide (hydrohalogenation):

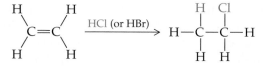

(d) Addition of H_2O to yield an alcohol (hydration):

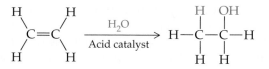

2. Reactions of aromatic compounds (Section 13.11)

(a) Substitution of an —NO_2 group to yield a nitrobenzene (nitration):

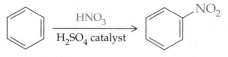

(b) Substitution of a Cl or Br atom to yield a halobenzene (halogenation):

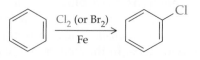

(c) Substitution of an —SO_3H group to yield a benzenesulfonic acid (sulfonation):

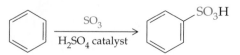

▪ Understanding Key Concepts

13.23 Name the following alkenes, and predict the products of their reaction with (1) HBr, (2) H_2O and an acid catalyst

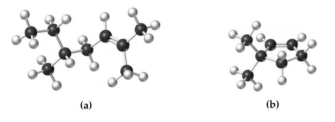

(a)   (b)

13.24 Name the following alkynes:

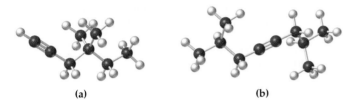

(a)   (b)

13.25 Give IUPAC names for the following substances (red = O, red-brown = Br):

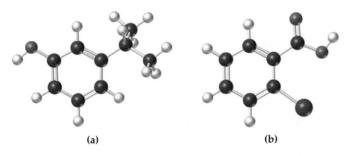

(a)   (b)

13.26 Draw the product from reaction of each of the following substances with (1) Br_2 and iron catalyst, (2) SO_3 and H_2SO_4 catalyst (red = O):

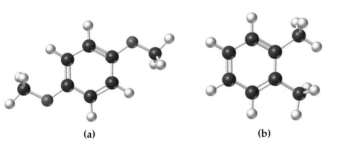

(a)   (b)

13.27 Alkynes undergo hydrogenation to give alkanes, just as alkenes do. Draw and name the products that would result from hydrogenation of the alkynes shown in Problem 13.24.

13.28 We saw in Section 13.9 that benzene can be represented by either of two resonance forms, which differ in the positions of the double bonds in the aromatic ring. Naphthalene, a polycyclic aromatic compound, can be represented by *three* forms with different double bond positions. Draw all three structures, showing the double bonds in each (the following molecular model of naphthalene shows only the connections among atoms).

13.29 The following structure is that of a carbocation intermediate in the reaction of an alkene with HCl. Draw the structure of the alkene reactant.

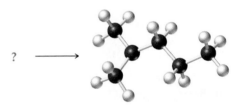

? ⟶

▪ Additional Problems

Naming Alkenes, Alkynes, and Aromatic Compounds

13.30 Why are alkenes, alkynes, and aromatic compounds said to be unsaturated?

13.31 Not all compounds that smell nice are called "aromatic," and not all compounds called "aromatic" smell nice. Explain.

13.32 What family name endings are used for alkenes, alkynes, and substituted benzenes?

13.33 What prefixes are used in naming each of the following?
  (a) A 1,3-disubstituted benzene
  (b) A 1,4-disubstituted benzene

13.34 Write structural formulas for compounds that meet the following descriptions:
  (a) An alkene with five carbons
  (b) An alkyne with four carbons
  (c) A substituted aromatic hydrocarbon with eight carbons

13.35 Write structural formulas for compounds that meet the following descriptions:

   (a) An alkene, C_8H_{16}, that can't have cis–trans isomers

   (b) An aromatic alkene, $C_{10}H_{12}$, that has cis–trans isomers

13.36 What are the IUPAC names of the following compounds?

   (a) $CH_3CH_2CH_2CH=CH_2$

   (b) $CH_3\overset{\overset{\displaystyle CH_3}{|}}{C}HCH_2C\equiv CCH_3$

   (c) $(CH_3)_2C=C(CH_3)_2$

   (d) $CH_3CH=\overset{\overset{\displaystyle CH_3}{|}}{C}-\overset{\underset{\displaystyle CH_2CH_3}{|}}{C}=CH_2$

   (e)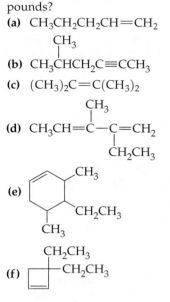

   (f)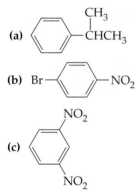

13.37 Give IUPAC names for the following aromatic compounds:

   (a)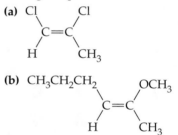

   (b) $Br-\!\!\langle\ \rangle\!\!-NO_2$

   (c)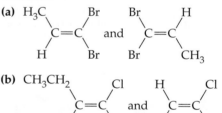

13.38 Draw structures corresponding to the following IUPAC names:

   (a) *cis*-2-Hexene

   (b) 2-Methyl-3-hexene

   (c) 2-Methyl-1,3-butadiene

   (d) *trans*-3-Heptene

   (e) *m*-Bromotoluene

   (f) *o*-Ethylphenol

   (g) *m*-Dipropylbenzene

13.39 Draw structures corresponding to the following names:

   (a) Aniline

   (b) Phenol

   (c) *o*-Xylene

   (d) Toluene

   (e) Benzoic acid

   (f) *p*-Nitrophenol

   (g) *o*-Chloroaniline

   (h) 3,3-Diethyl-6-methyl-4-nonene

13.40 There are three alkynes with the formula C_5H_8. Draw and name them.

13.41 Draw and name all aromatic compounds with the formula C_7H_7Br.

13.42 Excluding cis–trans isomers, there are five alkenes with the formula C_5H_{10}. Draw structures for as many as you can, and give their IUPAC names.

13.43 How many dienes (compounds with two double bonds) are there with the formula C_5H_8? Draw and name as many as you can.

Alkene Cis–Trans Isomers

13.44 What requirement must be met for an alkene to show cis–trans isomerism?

13.45 Why don't alkynes show cis–trans isomerism?

13.46 Which of the alkenes in Problem 13.42 can exist as cis–trans isomers?

13.47 Which of the compounds in Problem 13.43 can exist as cis–trans isomers?

13.48 Draw structures of the following compounds, indicating the cis or trans geometry of the double bond if necessary:

   (a) *cis*-3-Heptene

   (b) *cis*-4-Methyl-2-pentene

   (c) *trans*-2,5-Dimethyl-3-hexene

13.49 Draw structures of the double-bond isomers of the following compounds:

   (a)
$$\underset{H}{\overset{Cl}{}}C=C\underset{CH_3}{\overset{Cl}{}}$$

   (b)
$$\underset{H}{\overset{CH_3CH_2CH_2}{}}C=C\underset{CH_3}{\overset{OCH_3}{}}$$

13.50 Which of the following pairs are isomers, and which are identical?

   (a)
$$\underset{H}{\overset{H_3C}{}}C=C\underset{Br}{\overset{Br}{}} \quad\text{and}\quad \underset{Br}{\overset{Br}{}}C=C\underset{CH_3}{\overset{H}{}}$$

   (b)
$$\underset{Cl}{\overset{CH_3CH_2}{}}C=C\underset{H}{\overset{Cl}{}} \quad\text{and}\quad \underset{Cl}{\overset{H}{}}C=C\underset{CH_2CH_3}{\overset{Cl}{}}$$

13.51 Draw the other cis–trans isomer for each of the following molecules:

   (a)
$$\underset{CH_3\overset{|}{C}H\!\!\underset{CH_3}{}}{\overset{H_3C}{}}C=C\underset{H}{\overset{CH_3}{}}$$

   (b)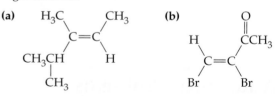

Kinds of Reactions

13.52 What is the difference between a substitution reaction and an addition reaction?

13.53 Give an example of an addition reaction.

13.54 If 2-methyl-2-pentene were somehow converted into 1-hexene, what kind of reaction would that be?

13.55 If bromocyclohexane were somehow converted into cyclohexene, what kind of reaction would that be?

13.56 Identify the type of reaction for each of the following:

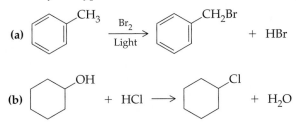

13.57 Identify the type of reaction for each of the following:

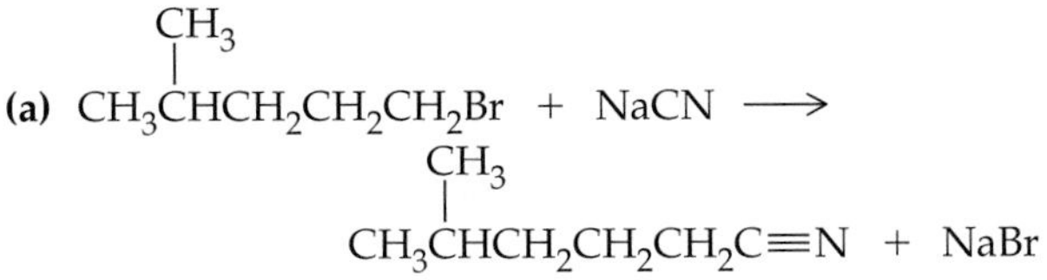

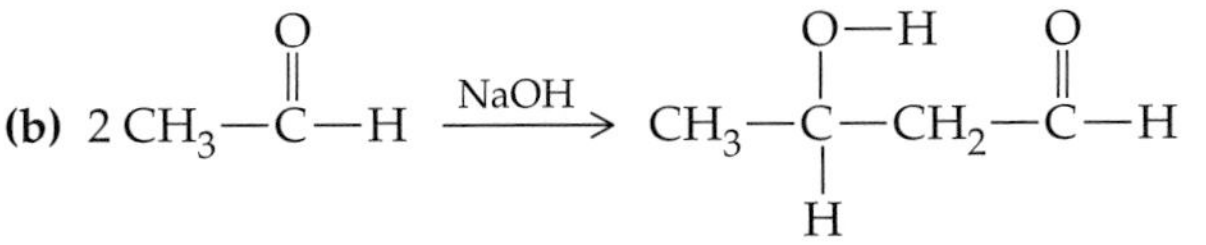

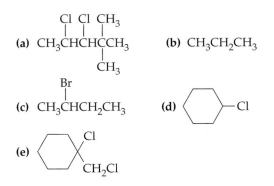

Reactions of Alkenes and Alkynes

13.58 Write equations for the reaction of 2,3-dimethyl-2-butene with the following:

(a) H_2 and Pd catalyst
(b) Br_2
(c) HBr
(d) H_2O and H_2SO_4 catalyst

13.59 Write equations for the reaction of 2-methyl-2-butene with the reagents shown in Problem 13.58.

13.60 What alkene could you use to make the following products? Draw the structure of the alkene, and tell what other reagent is also required for the reaction to occur.

(a)
$$\underset{Cl}{\overset{Cl}{\underset{|}{CH_3CHCHCCH_3}}}\overset{CH_3}{\underset{CH_3}{\overset{|}{C}}}$$

(b) $CH_3CH_2CH_3$

(c)
$$\underset{Br}{\overset{Br}{\underset{|}{CH_3CHCH_2CH_3}}}$$

(d) cyclohexane—Cl

(e) cyclohexane with Cl and CH_2Cl substituents

13.61 2,2-Dibromo-3-methylpentane can be prepared by an addition reaction of excess HBr with an alkyne. Draw the structure of the alkyne, name it, and write the reaction.

13.62 Draw the carbocation formed as an intermediate when HCl adds to styrene (phenylethylene).

13.63 4-Methyl-1-pentyne reacts with HBr in a 1:1 molar ratio to yield an addition product, $C_6H_{11}Br$. Draw the structures of two possible products. Assuming that Markovnikov's rule is followed, predict which of the two structures you drew is formed, and draw the carbocation involved as an intermediate.

13.64 Polyvinylpyrrolidone (PVP) is often used in hair sprays to hold hair in place. Draw a few units of the PVP polymer. The vinylpyrrolidone monomer unit has the structure

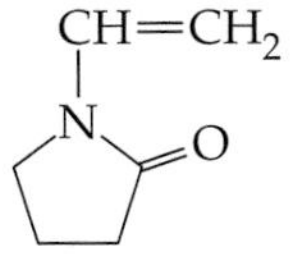

13.65 Saran, used as a plastic wrap for foods, is a polymer with the following structure. What is the monomer unit of Saran?

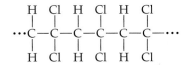

Reactions of Aromatic Compounds

13.66 Benzene reacts with only one of the following four reagents under ordinary conditions. Which of the four is it, and what is the structure of the product?

(a) H_2 and Pd catalyst
(b) Br_2 and iron catalyst
(c) HBr
(d) H_2O and H_2SO_4 catalyst

13.67 Write equations for the reaction of *p*-dichlorobenzene with the following:

(a) Br_2 and Fe catalyst
(b) HNO_3 and H_2SO_4 catalyst
(c) H_2SO_4 and SO_3
(d) Cl_2 and Fe catalyst

13.68 Benzene and other aromatic compounds don't normally react with hydrogen in the presence of a palladium catalyst. If very high pressures (200 atm) and high temperatures are used, however, benzene will add three molecules of H_2 to give an addition product. What is a likely structure for the product?

13.69 The explosive trinitrotoluene, or TNT, is made by carrying out three successive nitration reactions on toluene. If these nitrations take place in the ortho and para positions relative to the methyl group, what is the structure of TNT?

Applications

13.70 What is the difference in the purpose of the rod cells and the cone cells in the eye? [*The Chemistry of Vision*]

13.71 Describe the isomerization that occurs when light strikes the rhodopsin in the eye. [*The Chemistry of Vision*]

13.72 What is a polycyclic aromatic hydrocarbon? [*Polycyclic Aromatic Hydrocarbons and Cancer*]

13.73 How does benz[*a*]pyrene cause cancer? [*Polycyclic Aromatic Hydrocarbons and Cancer*]

13.74 Naphthalene is a white solid. Does it absorb light in the visible or in the ultraviolet range? [*Why We See Color*]

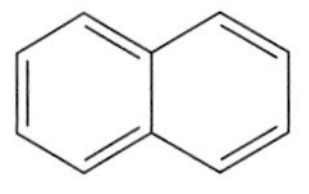

Naphthalene

13.75 Tetrabromofluorescein is a purple dye often used in lipsticks. If the dye is purple, what color does it absorb? [*Why We See Color*]

General Questions and Problems

13.76 Why do you suppose small-ring cycloalkenes like cyclohexene don't exist as cis–trans isomers, whereas large-ring cycloalkenes like cyclodecene *do* show isomerism?

13.77 Salicylic acid (*o*-hydroxybenzoic acid) is used as starting material to prepare aspirin. Draw the structure of salicylic acid.

13.78 "Super glue" is an alkene polymer of the following monomer unit. Draw a representative segment of the structure of super glue.

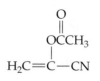

13.79 The following names are incorrect by IUPAC rules. Draw the structures represented by the following names, and write correct names.
(a) 2-Methyl-4-hexene
(b) 5,5-Dimethyl-3-hexyne
(c) 2-Butyl-1-propene
(d) 1,5-Dibromobenzene
(e) 1,2-Dimethyl-3-cyclohexene
(f) 3-Methyl-2,4-pentadiene

13.80 Assume that you have two unlabeled bottles, one with cyclohexane and one with cyclohexene. How could you tell them apart by carrying out chemical reactions?

13.81 Assume you have two unlabeled bottles, one with cyclohexene and one with benzene. How could you tell them apart by carrying out chemical reactions?

13.82 The compound *p*-dichlorobenzene has been used as an insecticide. Draw its structure.

13.83 Menthene, a compound found in mint plants, has the formula $C_{10}H_{18}$ and the IUPAC name 1-isopropyl-4-methylcyclohexene. What is the structure of menthene?

13.84 Cinnamaldehyde, the pleasant-smelling substance found in cinnamon oil, has the following structure. What product would you expect to obtain from reaction of cinnamaldehyde with hydrogen and a palladium catalyst?

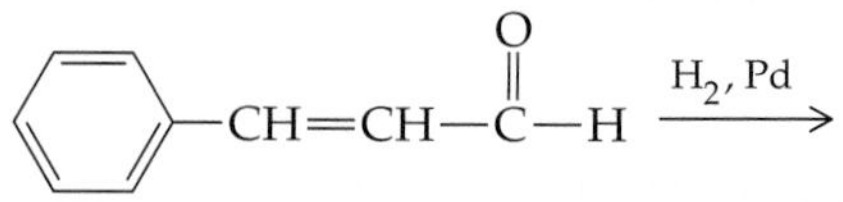

$$\text{benzene}-CH=CH-\overset{\overset{\textstyle O}{\|}}{C}-H \xrightarrow{H_2,\,Pd}$$

13.85 Predict the products of the following reactions:

(a) $CH_3CH_2CH=CH\overset{\overset{\textstyle CH_3}{|}}{C}HCH_3 \xrightarrow{H_2,\,Pd}$?

(b) 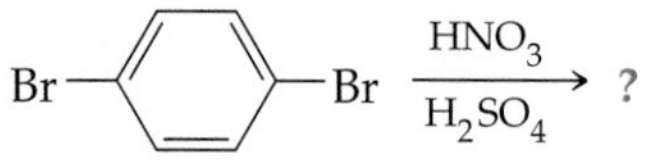 $Br-\text{benzene}-Br \xrightarrow[H_2SO_4]{HNO_3}$?

(c) 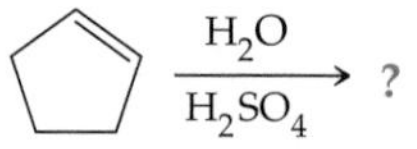 cyclopentene $\xrightarrow[H_2SO_4]{H_2O}$?

(d) $CH_3\overset{\overset{\textstyle CH_3CHCH_3}{|}}{C}HCH_2CH=CH_2 \xrightarrow{HBr}$?

(e) $CH_3C\equiv CCH_2CH_3 \xrightarrow{H_2,\,Pd}$?

(f) 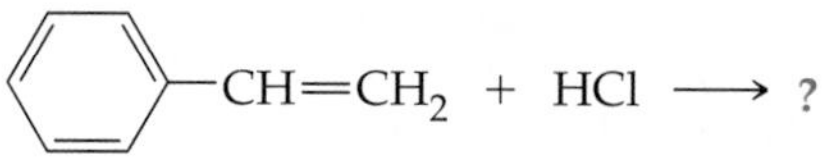 $\text{benzene}-CH=CH_2 + HCl \longrightarrow$?

13.86 Two products are possible when 2-pentene is treated with HBr. Write the structures of the possible products, and explain why they are made in about equal amounts.

13.87 Benzene is a liquid at room temperature, while naphthalene, another aromatic compound (Problem 13.74), is a solid. Account for this difference in physical properties.

13.88 Ocimene, a compound isolated from the herb basil, has the IUPAC name 3,7-dimethyl-1,3-6-octatriene.

   (a) Draw its structure.

   (b) Draw the structure of the compound formed if enough HBr is added to react with all the double bonds in ocimene.

13.89 Describe how you could prepare the following compound from an alkene. Draw the formula of the alkene, name it, and list the inorganic reactants needed for the conversion.

$$\begin{array}{c} \quad\ \text{HO}\quad\ \text{CH}_3 \\ \quad\ | \quad\quad | \\ \text{CH}_3\text{CH}_2-\text{C}-\text{C}-\text{CH}_3 \\ \quad\ | \quad\quad | \\ \quad\ \text{H}_3\text{C}\quad \text{CH}_3 \end{array}$$

13.90 Which of the following compounds are capable of cis–trans isomerism?

$$\begin{array}{c} \quad\quad\ \text{CH}_3 \\ \quad\quad\ | \end{array}$$
(a) $CH_3CHCH=CHCH_3$

$$\begin{array}{c} \quad\quad\quad\ \text{CH}=\text{CH}_2 \\ \quad\quad\quad\ | \end{array}$$
(b) $CH_3CH_2CHCH_3$

$$\begin{array}{c} \quad\quad\quad\ \text{Cl} \\ \quad\quad\quad\ | \end{array}$$
(c) $CH_3CH=CHCHCH_2CH_3$

14

Some Compounds with Oxygen, Sulfur, or a Halogen

On April 7, 1853, Queen Victoria of England gave birth to her eighth child while anesthetized by chloroform, changing forever the practice of obstetrics.

The past two chapters have dealt primarily with hydrocarbons, but most organic compounds contain other elements in addition to carbon and hydrogen. In this chapter we'll concentrate on functional groups that have single bonds to the electronegative atoms oxygen, sulfur, and the halogens.

Alcohols and *phenols*, for example, are compounds that have an organic part bonded to an —OH group. Alcohols are used widely in industry and are among the most abundant of all naturally occurring molecules. *Ethers* have two organic groups bonded to the same oxygen and are particularly useful as solvents. *Thiols*, which contain an organic part bonded to an —SH group, are widespread in living organisms, and *alkyl halides*, which contain an organic part bonded to a halogen, are valuable in many industrial processes.

Questions we'll answer in this chapter include the following:

1. **What are the distinguishing features of alcohols, phenols, ethers, thiols, and alkyl halides?**
   The goal: Be able to describe the structures and uses of compounds with these functional groups.
2. **How are alcohols, phenols, ethers, thiols, and alkyl halides named?**
   The goal: Be able to give systematic names for the simple members of these families and write their structures, given the names.
3. **What are the general properties of alcohols, phenols, and ethers?**
   The goal: Be able to describe such properties as polarity, hydrogen bonding, and water solubility.
4. **Why are alcohols and phenols weak acids?**
   The goal: Be able to explain why alcohols and phenols are acids.
5. **What are the main chemical reactions of alcohols and thiols?**
   The goal: Be able to describe and predict the products of the dehydration of alcohols and of the oxidation of alcohols and thiols.

CONCEPTS TO REVIEW

Oxidation and Reduction (Section 6.12)
Hydrogen Bonds (Section 8.11)
Acid Dissociation Constants (Sections 10.7–10.8)
Naming Alkanes (Section 12.6)

14.1 Alcohols, Phenols, and Ethers

An **alcohol** is a compound that has an –OH group (a *hydroxyl group*) bonded to a saturated, alkane-like carbon atom; a **phenol** has an –OH group bonded directly to an aromatic, benzene-like ring; and an **ether** has an oxygen atom bonded to two organic groups. Compounds in all three families can be thought of as organic relatives of water in which one or both of the H_2O hydrogens has been replaced by an organic substituent. For example:

Alcohol A compound that has an —OH group bonded to a saturated, alkane-like carbon atom, R—OH.

Phenol A compound that has an —OH group bonded directly to an aromatic, benzene-like ring, Ar—OH.

Ether A compound that has an oxygen atom bonded to two organic groups, R—O—R.

CH_3CH_2OH
Ethyl alcohol

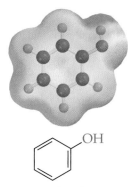

OH

Phenol

$CH_3CH_2OCH_2CH_3$
Diethyl ether

The structural similarity between alcohols and water also leads to similarities in many of their physical properties. For example, compare the boiling points of ethyl alcohol, dimethyl ether, propane, and water:

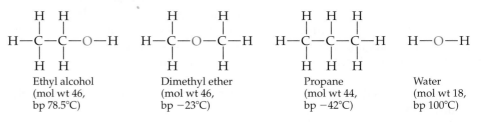

Ethyl alcohol (mol wt 46, bp 78.5°C)   Dimethyl ether (mol wt 46, bp −23°C)   Propane (mol wt 44, bp −42°C)   Water (mol wt 18, bp 100°C)

Ethyl alcohol, dimethyl ether, and propane have similar molecular weights, yet ethyl alcohol boils more than 100° higher than the other two. In fact, the boiling point of ethyl alcohol is close to that of water. Why should this be?

We said in Section 8.11 that the high boiling point of water is due to hydrogen bonding—the attraction between a lone pair of electrons on the electronegative oxygen in one molecule with the positively polarized —OH hydrogen on another molecule. (∞, p. 212) This attraction holds molecules together and prevents their easy escape into the vapor phase. In a similar manner, hydrogen bonds form between alcohol (or phenol) molecules (Figure 14.1). Alkanes and ethers don't have hydroxyl groups, however, and can't form hydrogen bonds. As a result, they have lower boiling points. Ethers, in fact, resemble alkanes in many of their properties.

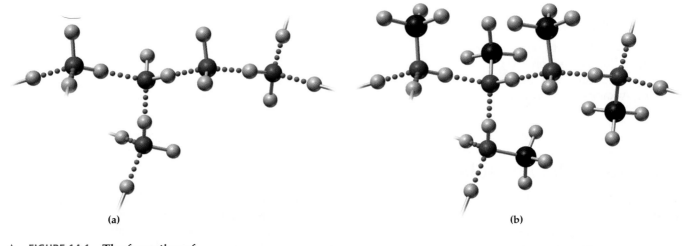

(a)   (b)

▲ **FIGURE 14.1 The formation of hydrogen bonds in water (a) and in alcohols (b).** Because of the hydrogen bonds, the easy escape of molecules into the vapor phase is prevented, resulting in high boiling points.

■ **PROBLEM 14.1**

Identify each of the following compounds as an alcohol, a phenol, or an ether:

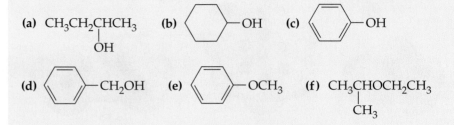

■ **PROBLEM 14.2**

What is the difference between a hydroxyl group and hydroxide ion?

14.2 Some Common Alcohols

Simple alcohols are among the most commonly encountered of all organic chemicals. They are useful as solvents, antifreeze agents, and disinfectants, and they are involved in the metabolic processes of all living organisms.

Methyl Alcohol (CH₃OH, Methanol)

Methyl alcohol, the simplest member of the alcohol family, is commonly known as *wood alcohol* because it was once prepared by heating wood in the absence of air. Today it is made in large quantities by reaction of carbon monoxide with hydrogen. Methanol is used industrially as a starting material for preparing formaldehyde ($H_2C{=}O$) and methyl *tert*-butyl ether (MTBE), an octane booster added to gasoline.

Methyl alcohol
(CH_3OH, methanol)

$$CO(g) + 2\,H_2(g) \xrightarrow[\text{High pressure, 250°C}]{\text{Cu catalyst}} CH_3OH(l)$$

Methyl alcohol is colorless, miscible with water, and toxic to humans when ingested or inhaled. It causes blindness in low doses (about 15 mL for an adult) and death in larger amounts (100–250 mL).

Ethyl Alcohol (CH₃CH₂OH, Ethanol)

Ethyl alcohol is one of the oldest known pure organic chemicals; its production by fermentation of grain and sugar goes back many thousands of years. Sometimes called *grain alcohol*, ethyl alcohol is the "alcohol" present in all table wines (10–13%), beers (3–5%), and distilled liquors (35–90%). During fermentation, starches or complex sugars are converted to simple sugars ($C_6H_{12}O_6$), which are then converted to ethyl alcohol:

Ethyl alcohol
(CH_3CH_2OH, ethanol)

$$C_6H_{12}O_6 \xrightarrow{\text{Yeast enzymes}} 2\,CH_3CH_2OH + 2\,CO_2$$

◀ Grapes are turned into wine by fermentation of their sugar in this vat.

The maximum alcohol concentration produced by fermentation alone is about 14% by volume, but higher alcohol concentrations can be produced by distillation of the fermentation product or by addition of a distilled product such as brandy. Alcohol for nonbeverage use is often *denatured* by addition of an unpleasant tasting substance such as methyl alcohol, camphor, or kerosene. The denatured alcohol that results is exempt from the tax applied to the sale of consumable alcohol.

Industrially, most ethyl alcohol is made by hydration of ethylene (Section 13.6). (∞, p. 370) Distillation yields a 95% ethyl alcohol plus 5% water mixture, and subsequent removal of the water gives 100% ethyl alcohol, known as *absolute alcohol*.

In some states, a blend of ethyl alcohol and gasoline called *gasohol* is commercially available. Gasohol is a desirable fuel because it produces fewer air pollutants than gasoline and because ethyl alcohol can be produced from corn and other starchy plants rather than from dwindling fossil fuel reserves.

Isopropyl Alcohol [(CH$_3$)$_2$CHOH]

Isopropyl alcohol, often called *rubbing alcohol*, is used as a 70% mixture with water for rubdowns because it cools the skin through evaporation and causes pores to close. It's also used as a solvent for medicines, as a sterilant for instruments, and as a skin cleanser before drawing blood or giving injections. In fact, the "medicinal" odor we associate with doctors' offices is often that of isopropyl alcohol. Although not as toxic as methyl alcohol, isopropyl alcohol is much more toxic than ethyl alcohol.

Ethylene Glycol (HOCH$_2$CH$_2$OH)

Ethylene glycol, a dialcohol, is a colorless liquid that is miscible with water and insoluble in nonpolar solvents. Its two major uses are as antifreeze and as a starting material for making polyester films and fibers, such as Dacron. In humans, ethylene glycol is a central-nervous-system depressant; a lethal dose for an adult is about 100 mL.

Glycerol [HOCH$_2$CH(OH)CH$_2$OH]

Like ethylene glycol, the *trialcohol* glycerol is a colorless liquid that is miscible with water. Unlike ethylene glycol, it's not toxic but has a sweet taste that makes it useful in candy and prepared foods. Often called *glycerin*, glycerol is also used in cosmetics and tobacco as a moisturizer, in plastics manufacture, in antifreeze and shock-absorber fluids, and as a solvent. We'll see in Chapter 24 that the glycerol molecule also provides the structural backbone of animal fats and vegetable oils.

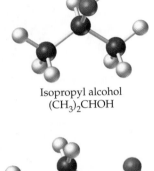

Isopropyl alcohol
(CH$_3$)$_2$CHOH

Ethylene glycol
HOCH$_2$CH$_2$OH

Glycerol
HOCH$_2$CH(OH)CH$_2$OH

 Nomenclature of Alcohols

14.3 Naming Alcohols

Alcohols with one —OH group are often referred to by common names that identify the alkyl group and then add the word "alcohol." Thus, the two-carbon alcohol is ethyl alcohol, the three-carbon alcohol is propyl alcohol, and so on. In the IUPAC system, alcohols are named using the *-ol* ending for the parent compound:

Step 1: Name the parent compound. Find the longest chain that has the hydroxyl substituent attached, and name the chain by replacing the *-e* ending of the corresponding alkane with *-ol*:

$$\underset{\text{CH}_3\text{CHCH}_2\text{CHCH}_2\text{CH}_3}{\overset{\text{CH}_3 \qquad \text{OH}}{|\qquad\quad|}}$$

Name as a *hexanol*—
a six-carbon chain
containing a hydroxyl group.

If the compound is a cyclic alcohol, add the *-ol* ending to the name of the parent cycloalkane. For example:

Cyclopentanol

Step 2: Number the carbon atoms in the main chain. Begin at the end nearer the hydroxyl group, ignoring the location of other substituents:

▲ Ethylene glycol is used as automobile antifreeze.

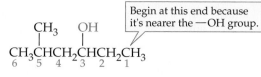

In a cyclic alcohol, begin with the carbon that bears the —OH group and proceed in a direction that gives the other substituents the lowest possible numbers.

Step 3: Write the name, placing the number that locates the hydroxyl group immediately before the parent compound name.   Number all other substituents according to their positions, and list them alphabetically. Note that in a cyclic alcohol, it's not necessary to use the number "1" to specify the location of the —OH group.

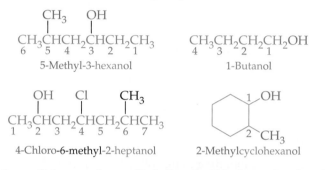

Dialcohols, or *diols*, are often called **glycols**. Ethylene glycol is the simplest glycol; propylene glycol is often used as a solvent for medicines that need to be inhaled or rubbed onto the skin. Numbering starts from the end closer to an —OH group, and the *–diol* name ending is used.

Glycol   A dialcohol, or diol.

$$HOCH_2CH_2OH \qquad CH_3\overset{\displaystyle OH}{\overset{|}{C}}HCH_2OH$$

Ethylene glycol
(1,2-Ethanediol)

Propylene glycol
(1,2-Propanediol)

As was true of alkyl groups (Section 12.6), alcohols are classified as primary, secondary, or tertiary according to the number of carbon substituents bonded to the hydroxyl-bearing carbon. (⬭, p. 339) Alcohols with one substituent are said to be *primary*, those with two substituents are *secondary*, and those with three substituents are *tertiary*. The substituent groups need not be the same, so we'll use the representations R, R′, and R″ to indicate different substituent groups.

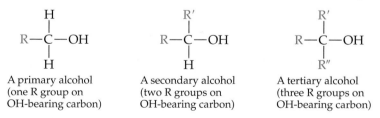

A primary alcohol
(one R group on
OH-bearing carbon)

A secondary alcohol
(two R groups on
OH-bearing carbon)

A tertiary alcohol
(three R groups on
OH-bearing carbon)

■ **WORKED EXAMPLE 14.1**

Give the systematic name of the following alcohol, and classify it as primary, secondary, or tertiary:

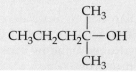

ANALYSIS First, identify the longest carbon chain, and number the carbon atoms beginning at the end nearer the —OH group. The longest chain attached to the —OH has five carbon atoms.

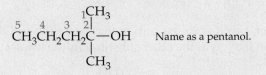

Name as a pentanol.

Next, identify and number the hydroxyl group and the substituents. Finally, write the name of the compound.

SOLUTION

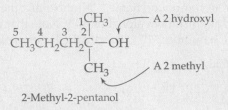

2-Methyl-2-pentanol

Since the —OH group is bonded to a carbon atom that has three alkyl substituents, this is a tertiary alcohol.

■ **PROBLEM 14.3**

Draw structures corresponding to the following names:
(a) 3-Methyl-3-pentanol   (b) Cyclohexanol
(c) 2-Methyl-4-heptanol   (d) 2-Butanol
(e) 3-Chloro-1-propanol

■ **PROBLEM 14.4**

Give systematic names for the following compounds:

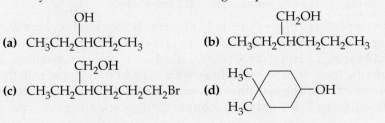

■ **PROBLEM 14.5**

Identify each of the alcohols in Problems 14.3 and 14.4 as primary, secondary, or tertiary.

14.4 Properties of Alcohols

Alcohols are much more polar than hydrocarbons because of the electronegative oxygen atom that withdraws electrons from the neighboring atoms. Because of this polarity, hydrogen bonding (Section 8.11) occurs and has a strong influence on alcohol properties. (∞, p. 212)

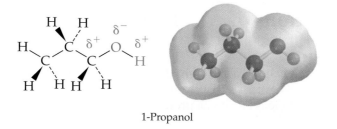

1-Propanol

Straight-chain alcohols with up to 12 carbon atoms are liquids, and each boils at a considerably higher temperature than the related alkane. Alcohols with a small organic part, such as methanol and ethanol, are like water in their solubility behavior. Methanol and ethanol are miscible with water, with which they can form hydrogen bonds, and they can dissolve small amounts of many salts. Nevertheless, both are also miscible with many organic solvents. Alcohols with a larger organic part, such as 1-heptanol, are much more like alkanes and less like water. 1-Heptanol is nearly insoluble in water and can't dissolve salts but does dissolve alkanes.

$$CH_3-OH \qquad\qquad CH_3CH_2CH_2CH_2CH_2CH_2CH_2-OH$$

Methanol: has a small organic part and is therefore water-like.

1-Heptanol: has a large organic part and is therefore alkane-like.

Alcohols with two or more —OH groups can form more than one hydrogen bond. They are therefore higher boiling and more water-soluble than similar alcohols with one —OH group. Compare 1-butanol and 1,4-butanediol for example:

$CH_3CH_2CH_2CH_2OH$ $\left\{ \begin{array}{l} \text{bp 117°C,} \\ \text{water solubility} \\ \text{of 7 g/100 ml.} \end{array} \right.$
1-Butanol

$HOCH_2CH_2CH_2CH_2OH$ $\left\{ \begin{array}{l} \text{Added —OH} \\ \text{raises bp to} \\ \text{230°C and gives} \\ \text{miscibility with} \\ \text{water} \end{array} \right.$
1,4-Butanediol

■ **PROBLEM 14.6**

Which of the following compounds has the highest boiling point?
(a) $CH_3CH_2CH_2OH$ (b) $CH_3CH_2OCH_3$ (c) $CH_3CH_2CH_3$

■ **PROBLEM 14.7**

Which of the following compounds is most soluble in water?

(a) $CH_3(CH_2)_{10}CH_2OH$ (b) $CH_3CH_2\underset{\underset{\displaystyle OH}{|}}{CH}CH_3$ (c) $CH_3CH_2OCH_3$

14.5 Reactions of Alcohols

Dehydration

Dehydration The loss of water from an alcohol to yield an alkene.

Alcohols undergo loss of water (**dehydration**) on treatment with a strong acid catalyst. The —OH group is lost from one carbon, and an —H is lost from an adjacent carbon to yield an alkene product:

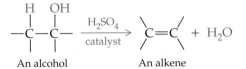

For example:

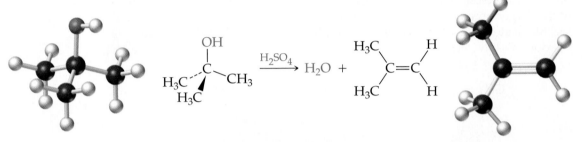

tert-Butyl alcohol   2-Methylpropene

When more than one alkene can result from dehydration of an alcohol, a mixture of products is usually formed. A good rule of thumb is that the major product is the one that has the greater number of alkyl groups attached to the double-bond carbons. For example, dehydration of 2-butanol leads to a mixture containing 80% 2-butene and only 20% 1-butene:

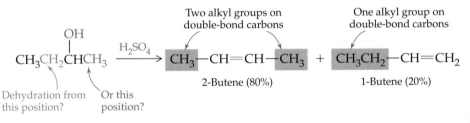

■ **WORKED EXAMPLE 14.2**

What products would you expect from the following dehydration reaction? Which product will be major, and which minor?

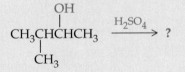

ANALYSIS   Find the hydrogens on carbons next to the OH-bearing carbon, and rewrite the structure to emphasize these hydrogens:

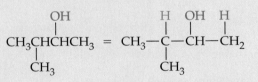

Then, remove the possible combinations of —H and —OH, drawing a double bond where —H and —OH have come from:

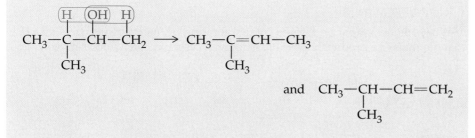

and CH_3—CH—CH=CH_2
              |
              CH_3

Finally, determine which of the alkenes has the larger number of alkyl substituents on its double-bond carbons and is therefore the major product.

SOLUTION

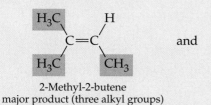

2-Methyl-2-butene	and	3-Methyl-1-butene
major product (three alkyl groups)		minor product (one alkyl group)

■ **WORKED EXAMPLE 14.3**

What alcohol(s) would yield 4-methyl-2-hexene on dehydration?

$$CH_3$$
$$|$$
$$CH_3CH_2CHCH\!=\!CHCH_3$$

4-Methyl-2-hexene

ANALYSIS   The double bond in the alkene is formed by removing —H and —OH from adjacent carbons of the starting alcohol. There are two possibilities for this removal, depending on which carbon is bonded to —OH and which is bonded to —H.

SOLUTION

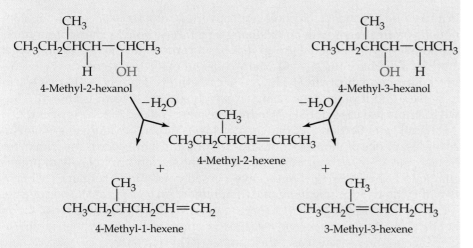

Dehydration of 4-methyl-2-hexanol would yield 4-methyl-2-hexene as major product, along with 4-methyl-1-hexene. Dehydration of 4-methyl-3-hexanol would give 4-methyl-2-hexene as minor product, along with 3-methyl-3-hexene as major product.

■ **PROBLEM 14.8**

What alkenes might be formed by dehydration of the following alcohols? If more than one product is possible in a given case, indicate which is major.

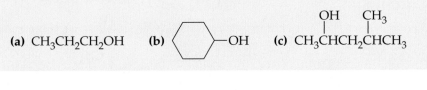

(a) $CH_3CH_2CH_2OH$ **(b)** ⬡—OH   **(c)** $CH_3\overset{\underset{\displaystyle |}{OH}}{C}HCH_2\overset{\underset{\displaystyle |}{CH_3}}{C}HCH_3$

■ **PROBLEM 14.9**

What alcohols would yield the following alkenes on dehydration?

(a) $(CH_3)_2C=C(CH_3)_2$ **(b)** $CH_3CH_2CH=CH_2$

■ **KEY CONCEPT PROBLEM 14.10**

What alkene(s) might be formed by dehydration of the following alcohol (red = O)?

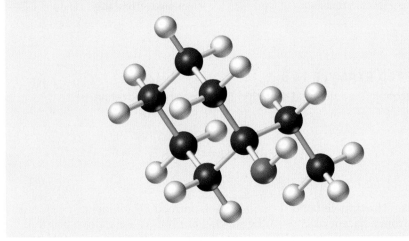

Oxidation

Carbonyl group   The C=O functional group.

Primary and secondary alcohols are converted into *carbonyl*-containing compounds on treatment with an oxidizing agent. A **carbonyl group** (pronounced car-bo-**neel**) is a functional group that has a carbon atom joined to an oxygen atom by a double bond, C=O. Many different oxidizing agents can be used—potassium permanganate ($KMnO_4$) or potassium dichromate ($K_2Cr_2O_7$), for example—and it often doesn't matter which specific reagent is chosen. Thus, we'll simply use the symbol [O] to indicate a generalized oxidizing agent.

Recall from Section 6.12 that an *oxidation* is defined in inorganic chemistry as the loss of one or more electrons by an atom, and a *reduction* is the gain of one or more electrons. (∞, p. 152) Because of the size and complexity of organic compounds, however, the words have a broader meaning in organic chemistry than they do with simple inorganic compounds. An *organic oxidation* is one that increases the number of C—O bonds and/or decreases the number of C—H bonds. Conversely, an *organic reduction* is one that decreases the number of C—O bonds and/or increases the number of C—H bonds.

Changes in Oxidation State

In the oxidation of an alcohol, two hydrogen atoms are removed from the alcohol and are converted into water during the reaction by the oxidizing agent [O]. One hydrogen comes from the —OH group, and the other hydrogen from the carbon atom bonded to the —OH group. In the process, a new C—O bond is formed and a C—H bond is broken.

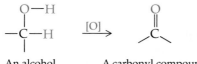

An alcohol   A carbonyl compound

Different kinds of carbonyl-containing products are formed, depending on the structure of the starting alcohol and on the reaction conditions. Primary alcohols (RCH_2OH) are converted either into *aldehydes* ($RCH{=}O$) if carefully controlled conditions are used, or into *carboxylic acids* (RCO_2H) if an excess of oxidant is used:

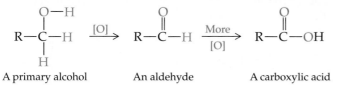

A primary alcohol   An aldehyde   A carboxylic acid

For example:

$$CH_3CH_2CH_2CH_2OH \xrightarrow{[O]} CH_3CH_2CH_2\overset{\displaystyle O}{\overset{\|}{C}}H \xrightarrow[\text{[O]}]{\text{More}} CH_3CH_2CH_2\overset{\displaystyle O}{\overset{\|}{C}}{-}OH$$

1-Butanol   Butanal   Butanoic acid

Secondary alcohols (R_2CHOH) are converted into *ketones* ($R_2C{=}O$) on treatment with oxidizing agents:

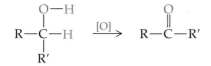

A secondary alcohol   A ketone

For example:

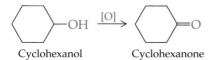

Cyclohexanol   Cyclohexanone

Tertiary alcohols don't normally react with oxidizing agents because they don't have a hydrogen on the carbon atom to which the —OH group is bonded:

$$R{-}\overset{\displaystyle O{-}H}{\underset{\displaystyle R'}{\overset{|}{\underset{|}{C}}}}{-}R'' \xrightarrow{[O]} \quad \textit{No reaction}$$

A tertiary alcohol

⊙⊙⊙ LOOKING AHEAD

We'll see in Chapter 23 that alcohol oxidations are critically important steps in many biological processes. When lactic acid builds up in tired, overworked muscles, for example, the liver removes it by oxidizing it to pyruvic acid. Our bodies, of course, don't use $K_2Cr_2O_7$ or $KMnO_4$ for the oxidation; instead, they use specialized, highly selective enzymes to carry out their chemistry. Regardless of the details, though, the net chemical transformation is the same whether carried out in a laboratory flask or in a living cell.   ⊙⊙⊙

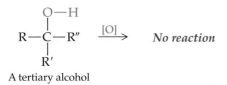

Lactic acid   Pyruvic acid

■ **WORKED EXAMPLE 14.4**

What is the product of the following oxidation reaction?

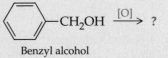

Benzyl alcohol

ANALYSIS The starting material is a primary alcohol, so it will be converted first to an aldehyde and then to a carboxylic acid. To find the structures of these products, first rewrite the starting alcohol to emphasize the hydrogen atoms on the hydroxyl-bearing carbon:

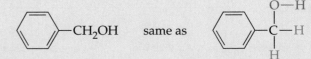

Next, remove two hydrogens, one from the —OH group and one from the hydroxyl-bearing carbon. In their place, make a C=O double bond. This is the aldehyde product that will form initially. Finally, convert the aldehyde to a carboxylic acid by removing the hydrogen from the —CH=O group and replacing it with an —OH group.

SOLUTION

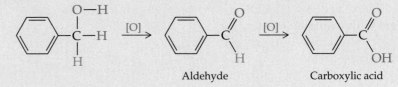

Aldehyde   Carboxylic acid

■ **PROBLEM 14.11**

What products would you expect from oxidation of the following alcohols?

(a) $CH_3CH_2CH_2OH$ (b) $CH_3CHCH_2CH_2CH_3$ with OH

(c)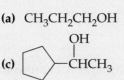

■ **PROBLEM 14.12**

From what alcohols might the following carbonyl-containing products have been made?

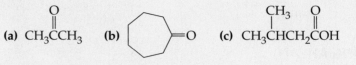

■ **KEY CONCEPT PROBLEM 14.13**

From what alcohols might the following carbonyl-containing products have been made (red = O, red-brown = Br)?

(a)   (b)

Application Ethyl Alcohol as a Drug and a Poison

Ethyl alcohol is classified for medical purposes as a central-nervous-system (CNS) depressant. Its direct effects (being "drunk") resemble the response to anesthetics and are quite predictable. At a blood alcohol concentration of 80–300 mg/dL, motor coordination and pain perception are affected, accompanied by loss of balance, slurred speech, and amnesia. At a concentration of 300–400 mg/dL, there may be nausea and loss of consciousness. Further increases in blood alcohol levels cause progressive loss of protective reflexes in stages like those of surgical anesthesia. Above 600 mg/dL of blood alcohol, spontaneous respiration and cardiovascular regulation are affected, ultimately resulting in death.

The passage of ethyl alcohol through the body begins with its absorption in the stomach and small intestine, followed by rapid distribution to all body fluids and organs. In the pituitary gland, alcohol inhibits the production of a hormone that regulates urine flow, causing increased urine production and dehydration. In the stomach, ethyl alcohol stimulates production of acid. Throughout the body, it causes blood vessels to dilate, resulting in flushing of the skin and a sensation of warmth as blood moves into capillaries beneath the surface. The result, though, is not a warming of the body but an increased loss of heat at the surface, making alcoholic beverages a poor choice in cold weather.

Alcohol metabolism occurs mainly in the liver and proceeds by oxidation in two steps, first to acetaldehyde and then to acetic acid. One of the hydrogen atoms lost in the oxidation at each stage binds to the biochemical oxidizing agent NAD$^+$ (nicotinamide adenine dinucleotide), and the other leaves as a hydrogen ion, H$^+$. When continuously present in the bodies of chronic alcoholics, alcohol and acetaldehyde are toxic, leading to devastating physical and metabolic deterioration. The liver usually suffers the worst damage because it is the major site of alcohol metabolism.

▲ The Breathalyzer test measures blood alcohol concentration.

The quick and uniform distribution of ethyl alcohol in body fluids, the ease with which it crosses lung membranes, and its ready oxidizability provide the basis for tests of blood alcohol concentration. The Breathalyzer test measures alcohol concentration in expired air by the color change that occurs when the bright yellow-orange oxidizing agent potassium dichromate ($K_2Cr_2O_7$) is reduced to blue-green chromium(III). The color change can be interpreted by instruments to give an accurate measure of alcohol concentration in the blood. As an alternative method, the Intoxilyzer test uses a beam of infrared light to measure blood alcohol levels. In many states, driving with a blood alcohol level above 0.10% (100 mg/dL) is illegal, and more than 50% of states have lowered the legal limit to 0.08%.

See Additional Problems 14. 54–14.57 at the end of the chapter.

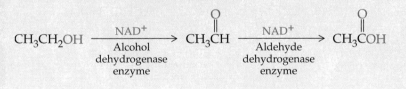

14.6 Phenols

The word *phenol* is the name both of a specific compound (hydroxybenzene, C_6H_5OH) and of a family of compounds. Phenol itself, formerly called carbolic acid, is a medical antiseptic first used by Joseph Lister in 1867. Lister showed that the occurrence of postoperative infection dramatically decreased when phenol was used to cleanse the operating room and the patient's skin. Because phenol numbs the skin, it also became popular in topical drugs for pain and itching and in treating sore throats.

The medical use of phenol is now restricted because it can cause severe skin burns and has been found to be toxic, both by ingestion and by absorption through the skin. The once common use of phenol for treating diaper rash is especially hazardous because phenol is more readily absorbed through a rash. Only solutions containing less than 1.5% phenol or lozenges containing a maximum of 50 mg of phenol are now allowed in nonprescription drugs. Many mouthwashes and throat lozenges contain alkyl-substituted phenols such as thymol as active ingredients for pain relief.

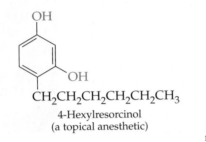

4-Hexylresorcinol
(a topical anesthetic)

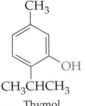

Thymol
(a topical anesthetic; occurs naturally in the herb thyme)

Alkyl-substituted phenols such as the cresols (methylphenols) are common as *disinfectants* in hospitals and elsewhere. By contrast with an antiseptic, which safely kills microorganisms on living tissue, a disinfectant should be used only on inanimate objects. The germicidal properties of phenols can be partially explained by their ability to disrupt the permeability of cell walls of microorganisms.

Phenols are usually named with the ending -*phenol* rather than -*benzene*. For example:

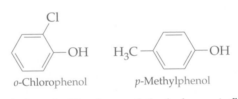

o-Chlorophenol   *p*-Methylphenol

The properties of phenols, like those of alcohols, are influenced by the presence of the electronegative oxygen atom and by hydrogen bonding. Most phenols are water-soluble to some degree and have higher melting and boiling points than similarly substituted alkylbenzenes.

Biomolecules containing a hydroxyl-substituted benzene ring include the amino acid tyrosine as well as many other compounds.

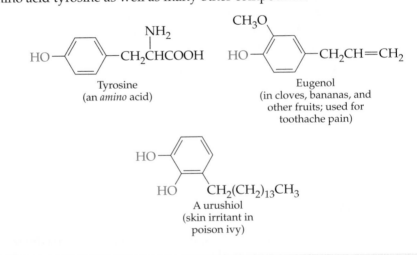

Tyrosine
(an *amino* acid)

Eugenol
(in cloves, bananas, and other fruits; used for toothache pain)

A urushiol
(skin irritant in poison ivy)

▲ Careful! The urushiol in this poison ivy plant can cause severe skin rash.

■ **PROBLEM 14.14**

Draw structures for the following:

(a) *m*-Bromophenol   (b) *p*-Ethylphenol

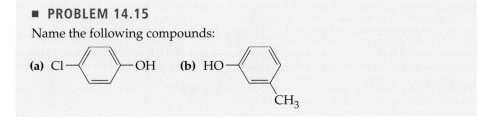

■ **PROBLEM 14.15**

Name the following compounds:

(a) Cl——OH (b) HO—

14.7 Acidity of Alcohols and Phenols

Alcohols and phenols are weakly acidic because of the positively polarized OH hydrogen. They dissociate slightly in aqueous solution and establish equilibria between neutral and anionic forms:

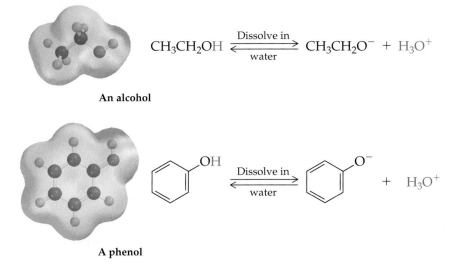

An alcohol

A phenol

Methanol and ethanol are about as acidic as water itself (Section 10.8), with K_a values near 10^{-15}. (∞, p. 275) In fact, they are so little dissociated in water that their aqueous solutions are neutral (pH 7). Thus, an **alkoxide ion** (RO^-), or anion of an alcohol, is as strong a base as hydroxide ion, OH^-. An alkoxide ion is produced by reaction of an alkali metal with an alcohol, just as hydroxide ion is produced by reaction of an alkali metal with water. For example:

Alkoxide ion The anion resulting from deprotonation of an alcohol, RO^-.

$$2 H_2O + 2 Na \longrightarrow 2 Na^+ OH^- + H_2$$

$$2 CH_3OH + 2 Na \longrightarrow 2 Na^+ {}^-OCH_3 + H_2$$

Methanol   Sodium methoxide

In contrast to alcohols, phenols are considerably more acidic than water. Phenol itself, for example, has $K_a = 1.0 \times 10^{-10}$, meaning that phenols react with dilute aqueous sodium hydroxide to give an anion.

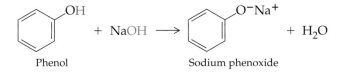

Phenol   Sodium phenoxide

Application Phenols as Antioxidants

If you occasionally read the labels on food packages, the names "butylated hydroxytoluene" and "butylated hydroxyanisole," or their abbreviations BHT and BHA, are probably familiar. You can find them on most cereal, cookie, and cracker boxes. Both compounds are substituted phenols.

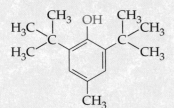

Butylated hydroxytoluene
(BHT)

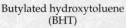

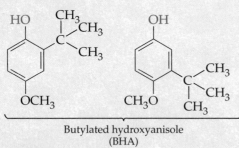

Butylated hydroxyanisole
(BHA)

Foods that contain unsaturated fats—those having carbon–carbon double bonds—become rancid when oxygen from the air reacts with their double bonds, producing ketones and other oxygen-containing substances with bad smells and tastes. The chemistry of oxidative rancidity is complex but involves the formation of reactive substances that contain unpaired electrons and are known as *free radicals*. Each free radical reacts further with O₂ in a *chain reaction* that ultimately leads to the destruction of a great many fat molecules.

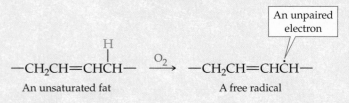

BHT and BHA prevent oxidation by donating a hydrogen atom from their —OH group to the free radical as soon as it forms, thereby converting the radical back to the starting fat and interrupting the destructive chain reaction. In the process, the BHA or BHT is converted into a stable and unreactive free radical, which causes no damage. Vitamin E, a natural antioxidant within the body, acts similarly.

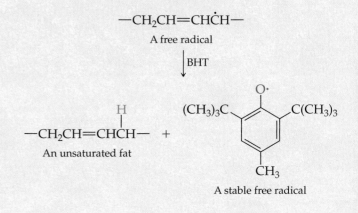

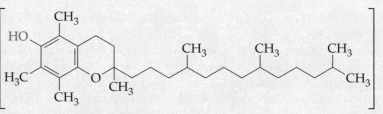

Vitamin E (a naturally occurring antioxidant)

The formation of free radicals is suspected of playing a role in both cancer and the normal aging of living tissue. Although there is no conclusive evidence, antioxidants may well be effective in slowing the progress of both conditions.

See Additional Problems 14.58 and 14.59 at the end of the chapter.

▲ Most baked goods contain the antioxidants BHA and BHT to help preserve freshness.

14.8 Ethers

Ethers—compounds with two organic groups bonded to the same oxygen atom—are named by identifying the two organic groups and adding the word *ether*. (The compound frequently referred to simply as "ether" is actually diethyl ether.)

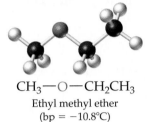

$$CH_3—O—CH_3$$
Dimethyl ether
(bp = −24.5°C)

$$CH_3—O—CH_2CH_3$$
Ethyl methyl ether
(bp = −10.8°C)

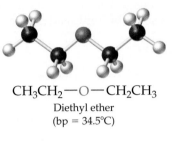

$$CH_3CH_2—O—CH_2CH_3$$
Diethyl ether
(bp = 34.5°C)

Compounds containing the oxygen atom in a ring are classified as cyclic ethers and are often given common names.

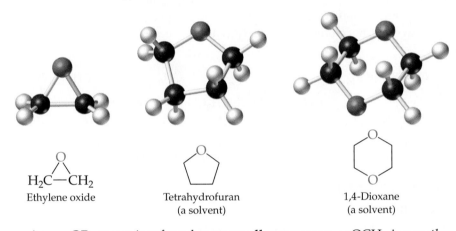

Ethylene oxide

Tetrahydrofuran
(a solvent)

1,4-Dioxane
(a solvent)

An —OR group is referred to as an **alkoxy group**; —OCH$_3$ is a *methoxy* group, —OCH$_2$CH$_3$ is an *ethoxy* group, and so on. These names are used when the ether functional group is present in a compound that also has other functional groups. For example:

Alkoxy group   An —OR group.

$$CH_3CH_2OCH_2CH_2OH$$

2-Ethoxyethanol

o-Methoxyphenol

Nomenclature of Ethers

Though they contain polar C—O bonds, ethers lack the —OH group of water and alcohols, and thus do not form hydrogen bonds to one another. Simple ethers are therefore higher boiling than comparable alkanes but lower boiling than alcohols. The ether oxygen can hydrogen-bond with water, causing dimethyl ether to be water-soluble and diethyl ether to be partially miscible with water. As with alcohols, ethers with larger organic groups are often insoluble in water.

Ethers are alkane-like in many of their properties and don't react with most acids, bases, or other reagents. Ethers do, however, react readily with oxygen, and the simple ethers are highly flammable. On standing in air, many ethers form explosive *peroxides*, compounds that contain an O—O bond. Thus, ethers must be handled with care and stored in the absence of oxygen.

Diethyl ether, the best-known ether, is now used primarily as a solvent but was for many years a popular anesthetic. Its value as an inhalation anesthetic was discovered in the 1840s, and it was a mainstay of the operating room until the 1940s. Although it acts quickly and is very effective, ether is far from ideal

▲ The maturation of this silkworm moth is controlled by a hormone that contains a three-membered ether ring.

as an anesthetic because it has a long recovery time and it often induces nausea. Moreover, its effectiveness is strongly counterbalanced by its hazards. Diethyl ether is a highly volatile, flammable liquid whose vapor forms explosive mixtures with air.

Diethyl ether has now been replaced by safer, less flammable anesthetics such as enflurane and isoflurane (See the following Application "Inhaled Anesthetics"). Both compounds were products of an intensive effort during the 1960s during which more than 400 halogenated ethers were synthesized in a search for improved anesthetics.

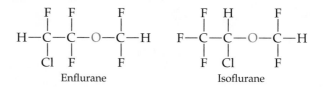

Ethers are found throughout the plant and animal kingdoms. Some are present in plant oils and are used in perfumes; others have a variety of biological roles. Juvenile hormone, for example, is a cyclic ether that helps govern the growth of the silkworm moth. The three-membered ether ring (an *epoxide* ring) in the juvenile hormone is unusually reactive because of strained 60° bond angles.

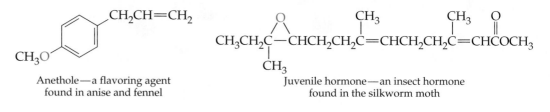

Anethole—a flavoring agent found in anise and fennel

Juvenile hormone—an insect hormone found in the silkworm moth

---

■ **PROBLEM 14.16**

Name the following compounds:

(a) $CH_3OCH_2CH_2CH_3$ (b) 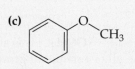   (c)

14.9 Thiols and Disulfides

Thiol A compound that contains an —SH group, R—SH.

Sulfur is just below oxygen in group 6A of the periodic table, and many oxygen-containing compounds have sulfur analogs. For example, **thiols** (R—SH), also called *mercaptans*, are sulfur analogs of alcohols. The systematic name of a thiol is formed by adding *-thiol* to the parent hydrocarbon name. Otherwise, thiols are named in the same way as alcohols.

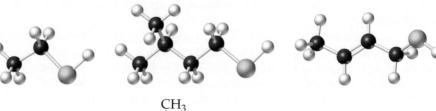

CH_3CH_2SH
Ethanethiol

$CH_3CHCH_2CH_2SH$
3-Methyl-1-butanethiol

$CH_3CH{=}CHCH_2SH$
2-Butene-1-thiol

Application Inhaled Anesthetics

William Morton's demonstration in 1846 of ether-induced anesthesia during dental surgery ranks as one of the most important medical breakthroughs of all time. Before that date, all surgery had been carried out with the patient fully conscious. Use of chloroform ($CHCl_3$) as an anesthetic quickly followed Morton's work, popularized by Queen Victoria of England, who in 1853 gave birth to a child while anesthetized by chloroform.

Hundreds of substances in addition to ether and chloroform have subsequently been shown to act as inhaled anesthetics. Halothane, enflurane, isoflurane, and methoxyflurane are at present the most commonly used agents in hospital operating rooms. All four are potent at relatively low doses, are nontoxic, and are nonflammable, an important safety feature.

The resultant changes in the fluidity and shape of the membranes apparently decrease the ability of sodium ions to pass into the nerve cells, thereby blocking the firing of nerve impulses.

Depth of anesthesia is determined by the concentration of anesthetic agent that reaches the brain. Brain concentration, in turn, depends on the solubility and transport of the anesthetic agent in the bloodstream and on its partial pressure in inhaled air. Anesthetic potency is usually expressed as a *minimum alveolar concentration* (MAC), defined as the concentration of anesthetic in inhaled air that results in anesthesia in 50% of patients. As shown in the following Table, nitrous oxide, N_2O, is the least potent of the common anesthetics and methoxyflurane is the most potent agent; a partial pressure of only 1.2 mm Hg is sufficient to anesthetize 50% of patients.

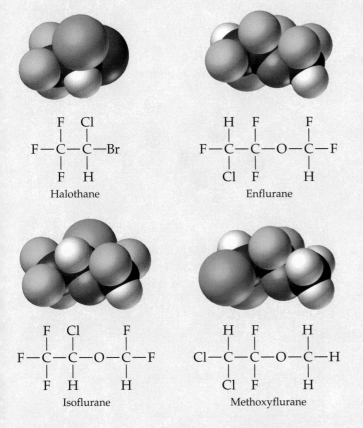

Halothane

Enflurane

Isoflurane

Methoxyflurane

Relative Potency of Inhaled Anesthetics

Anesthetic	MAC (%)	MAC (Partial pressure, mm Hg)
Nitrous oxide		>760
Enflurane	1.7	13
Isoflurane	1.4	11
Halothane	0.75	5.7
Methoxyflurane	0.16	1.2

See Additional Problems 14.60 and 14.61 at the end of the chapter.

Despite their importance, surprisingly little is known about how inhaled anesthetics work in the body. Remarkably, the potency of different inhaled anesthetics correlates well with their solubility in olive oil, leading many scientists to believe that anesthetics act by dissolving in the fatty membranes surrounding nerve cells.

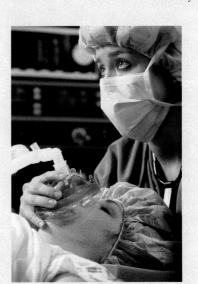

▲ Inhaled anesthetic agents are used to prepare patients for surgery.

▲ Skunks repel predators by releasing several thiols with appalling odors.

Disulfide A compound that contains a sulfur–sulfur bond, RS—SR.

The most outstanding characteristic of thiols is their appalling odor. Skunk scent is caused by two of the simple thiols shown above, 3-methyl-1-butanethiol and 2-butene-1-thiol. Thiols are also in the air whenever garlic and onions are being sliced, or when there's a natural gas leak. Natural gas itself is odorless, but a low concentration of methanethiol (CH_3SH) is added as a safety measure to make leak detection easy.

Thiols react with mild oxidizing agents such as Br_2 in water to yield **disulfides**, RS—SR. Two thiols join together in this reaction, the hydrogen from each is lost, and a bond forms between the two sulfurs:

$$RSH + HSR \xrightarrow{[O]} RSSR$$
Two thiol molecules   A disulfide

For example:

$$H_3C-S-H + H-S-CH_3 \xrightarrow{[O]} CH_3-S-S-CH_3 + H_2O$$
Methanethiol   Dimethyl disulfide

The reverse reaction occurs when a disulfide is treated with a reducing agent, represented by [H]:

$$RSSR \xrightarrow{[H]} RSH + RSH$$

Thiols are important biologically because they occur as a functional group in the amino acid cysteine, which is part of many proteins:

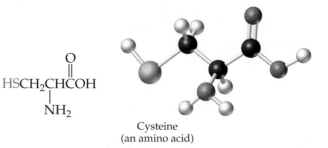

$$\underset{\underset{NH_2}{|}}{HSCH_2CHCOH} \overset{O}{\underset{}{\|}}$$

Cysteine
(an amino acid)

The easy formation of S—S bonds between two cysteines helps pull large protein molecules into the shapes they need to function. Hair protein, for example, is unusually rich in —S—S— and —SH groups. When hair is "permed," some disulfide bonds are broken and others are then formed. As a result, the hair proteins are held in a different shape (Figure 14.2).

FIGURE 14.2 Chemistry can ▶ curl your hair. A permanent wave results when disulfide bridges are formed between —SH groups in hair protein molecules.

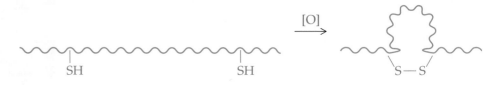

■ **PROBLEM 14.17**
What disulfides would you obtain from oxidation of the following thiols?
(a) $CH_3CH_2CH_2SH$ **(b)** 3-Methyl-1-butanethiol (skunk scent)

14.10 Halogen-Containing Compounds

Alkyl halide A compound that has an alkyl group bonded to a halogen atom, R—X.

The simplest halogen-containing compounds are the **alkyl halides**, RX, where R is an alkyl group and X is a halogen. Their common names are formed by giving the name of the alkyl group followed by the halogen name with an *-ide* ending. The compound CH_3Br, for example, is commonly called *methyl bromide*.

Systematic names are obtained by considering the halogen atom as a substituent on a parent alkane. The parent alkane is named in the usual way by selecting the longest continuous chain (Section 12.6) and numbering from the end nearer the first substituent, either alkyl or halogen. (∞, p. 336) The *halo*-substituent name is then given as a prefix, just as if it were an alkyl group. A few common halogenated compounds, such as chloroform ($CHCl_3$), are also known by nonsystematic names.

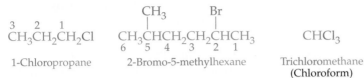

$\overset{3}{CH_3}\overset{2}{CH_2}\overset{1}{CH_2}Cl$	$CH_3\overset{6}{CH}CH_2CH_2CHCH_3$	$CHCl_3$
1-Chloropropane	2-Bromo-5-methylhexane	Trichloromethane (Chloroform)

Halogenated organic compounds have a variety of medical and industrial uses. Ethyl chloride is used as a topical anesthetic because it cools the skin through rapid evaporation; halothane is an important anesthetic. Chloroform was once employed as an anesthetic and as a solvent for cough syrups and other medicines, but is now considered too toxic for such uses. Bromotrifluoromethane, CF_3Br, is useful for extinguishing fires in aircraft and electronic equipment because it is nonflammable and nontoxic, and it evaporates without a trace.

Although a large number of halogen-containing organic compounds are found in nature, especially in marine organisms, few are significant in human biochemistry. One exception is thyroxine, an iodine-containing hormone secreted by the thyroid gland. A deficiency of iodine in the human diet leads to a low thyroxine level, which causes a swelling of the thyroid gland called *goiter*. To ensure adequate iodine in the diet of people who live far from an ocean, potassium iodide is sometimes added to table salt.

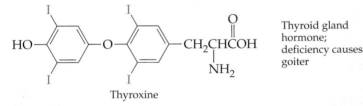

Thyroid gland hormone; deficiency causes goiter

Thyroxine

▲ This coral has many halogen-containing organic compounds.

Halogenated compounds are also used widely in industry and agriculture. Dichloromethane (CH_2Cl_2, methylene chloride), trichloromethane ($CHCl_3$, chloroform), and trichloroethylene ($Cl_2C{=}CHCl$) are used as solvents and degreasing agents, although their use is diminishing as less polluting alternatives become available. Because these substances are excellent solvents for the greases in skin, continued exposure often causes dermatitis.

Halogenated herbicides such as 2,4-D and fungicides such as Captan have resulted in vastly increased crop yields in recent decades, and the widespread application of chlorinated insecticides such as DDT is largely responsible for the progress made toward worldwide control of malaria and typhus. Despite their enormous benefits, chlorinated pesticides present problems because they persist in the environment and are not broken down rapidly. They remain in the fatty tissues of organisms and accumulate up the food chain as larger organisms consume smaller ones. Eventually, the concentration in some animals becomes high enough to cause harm. In an effort to maintain a balance between the value of halogenated pesticides and the harm they can do, the use of many has been restricted, and others have been banned altogether.

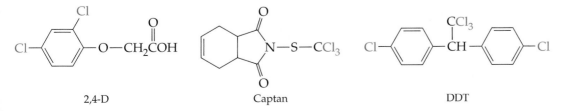

2,4-D	Captan	DDT

Application

Chlorofluorocarbons and the Ozone Hole

Newspaper stories about a "hole" in the ozone layer have appeared with regularity in recent years. What began as speculation about potential problems is now accepted as fact: Up to 65% of the stratospheric ozone over the South Pole disappears in the polar spring when the so-called *ozone hole* develops, before returning to near normal levels in the autumn. More recently, a similar though less dramatic decrease in polar ozone has been found over the North Pole.

Although toxic to all life forms at high concentrations, ozone (O_3) is critically important in the upper atmosphere because it acts as a shield to protect the earth from intense solar radiation. If the global ozone layer were depleted, more solar radiation would reach the earth, causing an increase in the incidence of skin cancer and eye cataracts.

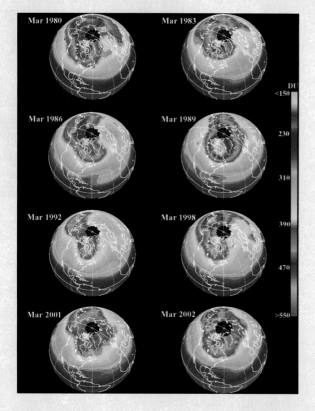

▲ False-color images of the ozone hole over the Arctic region for the period from 1980 through early 2002. Ozone concentrations are given in Dobson Units (DU) according to the color code indicated.

Diminished growth of oceanic plankton and stunting of the growth of terrestrial crops are also predicted.

The principal cause of ozone depletion is the presence in the atmosphere of a group of alkyl halides called *chlorofluorocarbons*, or CFCs, familiar to many as the Freons manufactured by the DuPont Company. The chlorofluorocarbons are simple alkyl halides in which all the hydrogens of an alkane have been replaced by chlorine and fluorine. Fluorotrichloromethane (CCl_3F, Freon 11) and dichlorodifluoromethane (CCl_2F_2, Freon 12) are two of the most common CFCs in industrial use.

Because they are inexpensive and stable, yet not toxic, flammable, or corrosive, CFCs are ideal as propellants in aerosol cans, as refrigerants, as solvents, and as fire extinguishing agents. In addition, they are used for blowing bubbles into foamed plastic insulation. Unfortunately, the stability that makes CFCs so useful also causes them to persist in the environment. Molecules released at ground level slowly find their way into the upper atmosphere, where they undergo a complex series of reactions that ultimately result in ozone destruction. When ultraviolet (UV) light strikes a CFC molecule, a C—Cl bond breaks, producing a reactive chlorine atom, Cl. The chlorine atom then reacts with ozone to yield oxygen and ClO.

$$CFCl_3 \xrightarrow{\text{UV light}} CFCl_2 + Cl$$

$$Cl + O_3 \longrightarrow ClO + O_2$$

Recognition of the problem led the U.S. government in 1980 to ban the use of CFCs as aerosol propellants, although they are still used as refrigerants. Worldwide efforts to reduce CFC use began with an international agreement reached in 1987, and a total ban on the industrial production and atmospheric release (though not continued use) of CFCs took effect in 1996. The ban has not been wholly successful, however, because of a substantial black market that has developed, particularly in Russia and China where up to $300 million per year of illegal CFCs are manufactured. Even with these stringent efforts, amounts of CFCs in the stratosphere will continue to rise until the early 2000s and will not return to acceptable levels until the middle of the century.

See Additional Problems 14.62 and 14.63 at the end of the chapter.

CFCs and Stratospheric Ozone

■ **PROBLEM 14.18**

Give systematic names for the following alkyl halides:

(a) 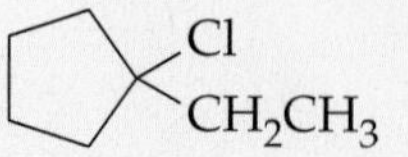 [structure: cyclopentane ring with Cl and CH₂CH₃ substituents]

(b) $CH_3CH_2CHCH_2CHCH_2CH_3$ with CH_3 and Br substituents

"To be real honest, chemistry kind of scared me when I was introduced to the subject in junior high and high school. I really was most interested in biology," says John Crellin, Resource Conservationist for the U.S. Department of Agriculture (USDA). "I didn't have the big picture in mind, and I wasn't able to see beyond the drudgeries of learning new kinds of math and memorizing formulas. But as I studied biology side by side with my chemistry class, I began to understand that it's nearly impossible to study one without the other. Chemistry became more fun and more 'real' to me when I understood how important chemistry is in everyday life. I began to see how things relate in all of the sciences."

Trained as an agronomist and a soil scientist, John has worked in the Midwest for USDA's Natural Resources Conservation Service (NRCS) for the past 14 years. NRCS, formerly called the Soil Conservation Service, was established by the federal government as a national response to the Dust Bowl catastrophe of the mid-1930s. Today, NRCS is the USDA's lead conservation agency, providing technical assistance to farmers, ranchers, and other land users. "Soil and water quantity and quality are our main focus," says John. "And we achieve this by helping people to improve our land's health and productivity by restoring wetlands, forestland and wildlife resources and employing sound cropland management decisions."

"In my current job, I work primarily with agricultural landowners and small rural landowners. I'm responsible for technical planning when it comes to soil and water conservation, especially for farmers who receive some sort of financial help from USDA or state programs. I develop conservation plans for highly erodible lands, for wetland restorations, and for land that qualifies for the Conservation Reserve Program. I spend a lot of time in the field, interviewing landowners to determine their needs, goals, and economic capabilities. I also make field visits to collect data, assess land limitations, and determine the most appropriate conservation practices for that site."

Two of the important subjects where John applies his chemical knowledge are pesticides and nutrient management.

The appropriate use of pesticides (including insecticides, herbicides, and fungicides) is an important, but sometimes misunderstood, part of today's agriculture. "In my job I need a basic understanding of how pesticides behave when applied to the soil," says John. "I need to understand how they affect the chemistry of the targets they are meant to suppress. I need to understand if they will have an adverse effect on non-target life, such as beneficial insects, plants, humans, or other animals. Do the chemicals 'leach,' which means do they move down into the soil profile? Do the chemicals change into other forms when they are in soil? How long do the chemicals persist before they are degraded by microbes or diluted through the soil profile?"

John finds that most of the farmers who work a lot with agricultural chemicals understand the appropriate uses and the limitations of pesticides. Many of these farmers recently began planting new, genetically engineered varieties of corn and soybeans (see "Biotechnology: Applications of DNA," pp. 734–36). With these new varieties, farmers are able to use low tillage agriculture, thereby minimizing soil erosion while at the same time using fewer, safer, and more cost-effective pesticides.

In some cases, however, John has to educate landowners. He often works with small landowners and suburbanites who are moving out into the country on smaller plots of land, such as two to five acres. If the owners want to put the land into grassland as part of a conservation plan, it is often necessary to use a herbicide, such as "Roundup," while preparing the site for seeding. (One of the most widely used herbicides, Roundup contains a chemical compound called glyphosate that inhibits the production of an enzyme essential for plant growth.)

"Some people are totally opposed to any kind of pesticide. The name scares them," says John. "I have to help them understand the basics of various pesticides and how chemicals might react under different conditions. Does the chemical leach into the groundwater? Does it degrade quickly? How selective is it with regards to plant or insect control?"

Another agricultural and environmental subject of increasing importance in the last few years is "nutrient management." In order to increase the productivity of their soil, farmers often add nutrients in the form of supplementary fertilizers, such as commercial fertilizer, animal manure, or a sludge byproduct from a manufacturing process. Among the nutrients necessary for plant growth are the primary macronutrients (nitrogen, phosphorus, and potassium), the secondary macronutrients (calcium, magnesium, and sulfur), and the micronutrients (including boron, chlorine, copper, iron, manganese, molybdenum, and zinc).

As a soil scientist, John helps landowners understand how these nutrients interact with the soil. The interactions depend to a great extent on the physical, chemical, and biological characteristics of that particular soil sample. "There's a lot of chemistry that takes place underground. For example, the amount of clay or organic matter is important, because those materials have a lot of cation exchange sites that are able to hold beneficial nutrients in reserve." Other chemical factors that are important to consider when making nutrient management decisions include soil pH and whether the nutrients are in a chemical form that makes them available to plants (see "Nitrogen Fixation," p. 180).

Accurate nutrient management recommendations are important, because problems can result from either inadequate or excessive fertilization. Too little fertilizer leads to poor plant growth, but too much fertilizer can also reduce plant growth and quality. In addition, excessive applications of fertilizer, including manure, can be harmful to the environment.

"Sometimes people want to add more fertilizer than they really need," observes John. "They think that a little bit extra will help. But with today's very small profit margins on farms in the Midwest, farmers need to make every penny count. Having a good understanding of soil chemistry and fertilizer requirements can save someone from spending extra money that will not help in producing a profitable crop yield. And it also helps protect the environment."

Key Words

Alcohol, p. 389

Alkoxide ion, p. 403

Alkoxy group, p. 405

Alkyl halide, p. 408

Carbonyl group, p. 398

Dehydration, p. 396

Disulfide, p. 408

Ether, p. 389

Glycol, p. 393

Phenol, p. 389

Thiol, p. 406

Summary: Revisiting the Chapter Goals

1. **What are the distinguishing features of alcohols, phenols, ethers, thiols, and alkyl halides?** An *alcohol* has an —OH group (a *hydroxyl* group) bonded to a saturated, alkane-like carbon atom; a *phenol* has an —OH group bonded directly to an aromatic ring; and an *ether* has an oxygen atom bonded to two organic groups. *Thiols* are sulfur analogs of alcohols, R—SH. *Alkyl halides* contain a halogen atom bonded to an alkyl group, R—X.

   The —OH group is present in many biochemically active molecules. Phenols are notable for their use as disinfectants and antiseptics; ethers are used primarily as solvents. Thiols are found in proteins. Halogenated compounds are rare in human biochemistry but are widely used in industry as solvents and in agriculture as herbicides, fungicides, and insecticides.

2. **How are alcohols, phenols, ethers, thiols, and alkyl halides named?** Alcohols are named using the *-ol* ending, and phenols are named using the *-phenol* ending. Ethers are named by identifying the two organic groups attached to oxygen, followed by the word *ether*. Thiols use the name ending *-thiol*, and alkyl halides are named as halo-substituted alkanes.

3. **What are the general properties of alcohols, phenols, and ethers?** Both alcohols and phenols are like water in their ability to form hydrogen bonds. As the size of the organic part increases, alcohols become less soluble in water. Ethers do not hydrogen-bond and are more alkane-like in their properties.

4. **Why are alcohols and phenols weak acids?** Like water, alcohols and phenols are weak acids that can donate H^+ from their —OH group to a strong base. Alcohols are similar to water in acidity; phenols are more acidic than water and will react with aqueous NaOH.

5. What are the main chemical reactions of alcohols and thiols? Alcohols undergo loss of water (*dehydration*) to yield alkenes when treated with a strong acid, and they undergo *oxidation* to yield compounds that contain a *carbonyl group* ($C=O$). Primary alcohols (RCH_2OH) are oxidized to yield either aldehydes (RCHO) or carboxylic acids (RCO_2H); secondary alcohols (R_2CHOH) are oxidized to yield ketones ($R_2C=O$); and tertiary alcohols are not oxidized. Thiols react with mild oxidizing agents to yield *disulfides* (RSSR), a reaction of importance in protein chemistry. Disulfides can be reduced back to thiols.

Summary of Reactions

1. Reactions of alcohols (Section 14.5)

(a) Loss of H_2O to yield an alkene (dehydration):

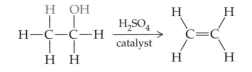

(b) Oxidation to yield a carbonyl compound:

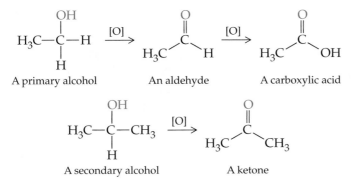

2. Reactions of thiols (Section 14.10); oxidation to yield a disulfide:

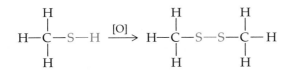

▪ Understanding Key Concepts

14.19 Give IUPAC names for the following compounds (gray = C, red = O, ivory = H):

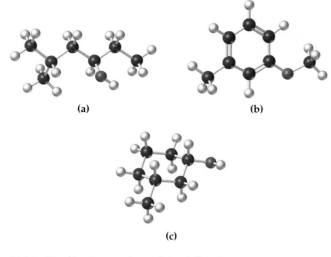

(a)   (b)

(c)

14.20 Predict the product of the following reaction:

$\xrightarrow{H_2SO_4}$

14.21 Predict the product of the following reaction:

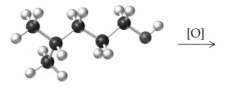

$\xrightarrow{[O]}$

14.22 Predict the product of the following reaction (gray = C, yellow = S, ivory = H):

$\xrightarrow{[O]}$

14.23 What alcohols might the following carbonyl compounds have been made from (reddish-brown = Br)?

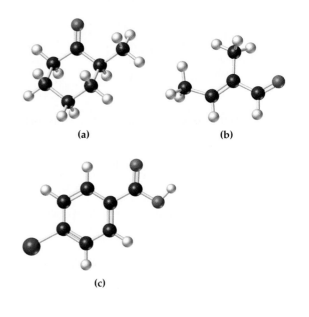

(a)   (b)

(c)

▪ Additional Problems

Alcohols, Ethers, and Phenols

14.24 How do alcohols, ethers, and phenols differ structurally?

14.25 What is the structural difference between primary, secondary, and tertiary alcohols?

14.26 Why do alcohols have higher boiling points than ethers of the same molecular weight?

14.27 Which is the stronger acid, ethanol or phenol?

14.28 The steroidal compound prednisone is often used to treat poison ivy and poison oak inflammations. Identify the functional groups present in prednisone.

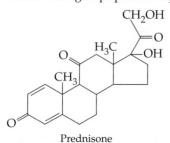

Prednisone

14.29 Vitamin E has the structure shown on page 404. Identify the functional group to which each oxygen belongs.

14.30 Give systematic names for the following alcohols:

(a)
$$CH_3CH_2\overset{\overset{\displaystyle CH_2OH}{|}}{C}HCH_2CH_2CH_3$$

(b) $(CH_3)_2CHCH_2CH_2OH$

(c)
$$HOCH_2CH_2\underset{\underset{\displaystyle OH}{|}}{C}HCH_2OH$$

(d)
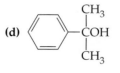

(e)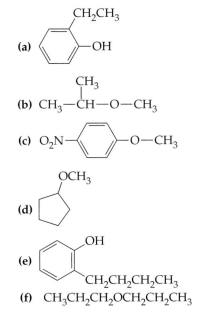

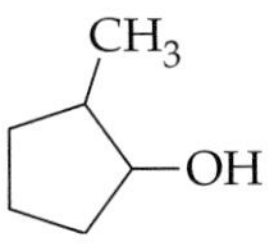

(f) 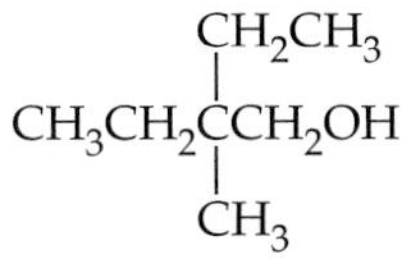 CH₃CH₂CCH₂OH

(with CH₂CH₃ above and CH₃ below the central carbon)

14.31 Give systematic names for the following compounds:

(a)

(b) CH₃—CH—O—CH₃ (with CH₃ above)

(c) O₂N— —O—CH₃

(d)

(e)
CH₂CH₂CH₂CH₃

(f) CH₃CH₂CH₂OCH₂CH₂CH₃

14.32 Draw structures corresponding to the following names:
(a) 2,4-Dimethyl-2-pentanol
(b) 2,2-Dimethylcyclohexanol
(c) 5,5-Diethyl-1-heptanol
(d) 3-Ethyl-3-hexanol
(e) 2,3,7-Trimethylcyclooctanol
(f) 3,3-Diethyl-1,6-hexanediol

14.33 Draw structures corresponding to the following names:
(a) Ethyl phenyl ether
(b) *o*-Dihydroxybenzene (catechol)
(c) *p*-Bromophenyl *tert*-butyl ether
(d) *p*-Nitrophenol
(e) 2,4-Diethoxy-3-methylpentane
(f) 4-Methoxy-3-methyl-1-pentene

14.34 Identify each of the alcohols named in Problem 14.30 as primary, secondary, or tertiary.

14.35 Locate the alcohol functional groups in prednisone (Problem 14.28), and identify each as primary, secondary, or tertiary.

14.36 Arrange the following six-carbon compounds in order of their expected boiling points, and explain your ranking:
(a) Hexane
(b) 1-Hexanol
(c) Dipropyl ether

14.37 Glucose is much more soluble in water than 1-hexanol even though both contain six carbons. Explain.

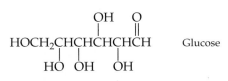

Reactions of Alcohols

14.38 What type of product is formed on oxidation of a secondary alcohol?

14.39 What structural feature prevents tertiary alcohols from undergoing oxidation reactions?

14.40 What product(s) can form on oxidation of a primary alcohol?

14.41 What type of product is formed on reaction of an alcohol with Na metal?

14.42 Assume that you have samples of the following two compounds, both with formula C₇H₈O. Both compounds dissolve in ether, but only one of the two dissolves in aqueous NaOH. How could you use this information to distinguish between them?

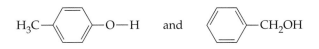

14.43 Assume that you have samples of the following two compounds, both with formula C₇H₁₄O. What simple chemical reaction will allow you to distinguish between them? Explain.

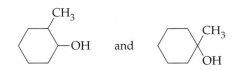

14.44 The following alkenes can be prepared by dehydration of an appropriate alcohol. Show the structure of the alcohol in each case. If the alkene can arise from dehydration of more than one alcohol, show all possibilities.

(a) 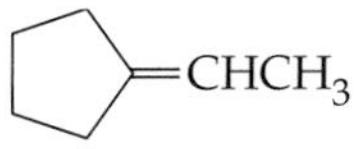 =CHCH₃

(b) 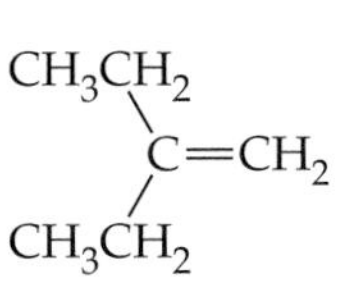
CH₃CH₂
   C=CH₂
CH₃CH₂

(c) 3-Hexene

(d) 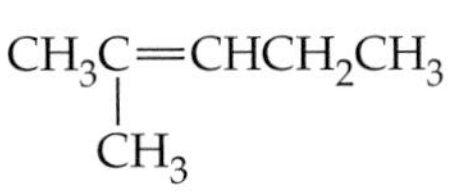 CH₃C=CHCH₂CH₃ (with CH₃ below)

(e) 1,3-Butadiene

(f) 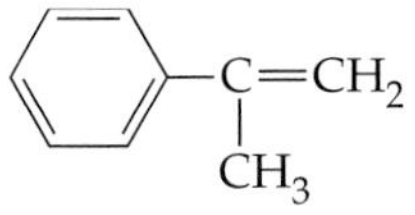 —C=CH₂ (with CH₃ below)

14.45 What alkenes might be formed by dehydration of the following alcohols? If more than one product is possible, indicate which you expect to be major.

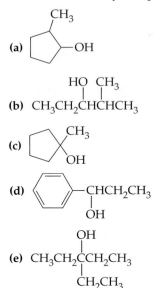

(b) HO CH₃
CH₃CH₂CHCHCH₃

(d) [benzene ring]—CHCH₂CH₃
                            |
                           OH

(e) OH
CH₃CH₂CCH₂CH₃
      |
     CH₂CH₃

14.46 What carbonyl-containing products would you obtain from oxidation of the following alcohols? If no reaction occurs, write "NR."

(a) [benzene ring]—CHCH₃
                       |
                      OH

(b) CH₃CHCH₂OH
          |
         CH₃

(c) 3-Methyl-3-pentanol

(d)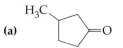
[cyclobutane]—OH

(e) H₃C OH
CH₃CH₂CHCCH₃
          |
         CH₃

(f) [benzene ring]—CHCH₂CH₃
                       |
                      OH

14.47 What alcohols would you oxidize to obtain the following carbonyl compounds?

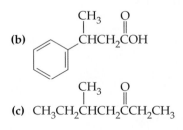

(b) CH₃ O
[benzene ring]—CHCH₂COH

(c) CH₃   O
CH₃CH₂CHCH₂CCH₂CH₃

Thiols and Disulfides

14.48 What is the most noticeable characteristic of thiols?

14.49 What is the structural relationship between a thiol and an alcohol?

14.50 The amino acid cysteine forms a disulfide when oxidized. What is the structure of the disulfide?

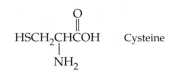

O
‖
HSCH₂CHCOH   Cysteine
        |
       NH₂

14.51 Oxidation of a dithiol such as 1,4-pentanedithiol forms a cyclic disulfide. Draw the structure of the cyclic disulfide.

SH
|
CH₃CHCH₂CH₂CH₂SH
1,4-Pentanedithiol

14.52 The boiling point of propanol is 97°C, much higher than that of either ethanethiol (37°C) or chloroethane (13°C) even though all three compounds have similar molecular weights. Explain.

14.53 Propanol is very soluble in water, while ethanethiol and chloroethane are only slightly soluble. Explain.

Applications

14.54 Is ethanol a stimulant or a depressant? [*Ethyl Alcohol as a Drug and a Poison*]

14.55 At what blood alcohol concentration does speech begin to be slurred? What is the approximate lethal concentration of ethyl alcohol in the blood? [*Ethyl Alcohol as a Drug and a Poison*]

14.56 Cirrhosis of the liver is a common disease of alcoholics. Why is the liver particularly affected by alcohol consumption? [*Ethyl Alcohol as a Drug and a Poison*]

14.57 Describe the basis of the Breathalyzer test for alcohol concentration. [*Ethyl Alcohol as a Drug and a Poison*]

14.58 What is a free radical? [*Phenols as Antioxidants*]

14.59 What vitamin appears to be a phenolic antioxidant? [*Phenols as Antioxidants*]

14.60 What substance was used as the first general anesthetic? [*Inhaled Anesthetics*]

14.61 How is "minimum alveolar concentration" for an anesthetic defined? [*Inhaled Anesthetics*]

14.62 Ozone is considered to be an air pollutant at the earth's surface. Why is it beneficial in the upper atmosphere? [*Chlorofluorocarbons and the Ozone Hole*]

14.63 Chlorofluorocarbons (CFCs) are still widely used as coolants in refrigerators and air-conditioners, but states have legislated methods for disposal of CFC-containing appliances. Why? [*Chlorofluorocarbons and the Ozone Hole*]

General Questions and Problems

14.64 Name all ether and alcohol isomers with formula C₄H₁₀O, and write their structural formulas.

14.65 Thyroxine (Section 14.11) is synthesized in the body by reaction of thyronine with iodine. Write the reaction, and tell what kind of process is occurring.

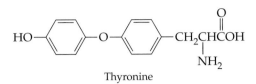

Thyronine

14.66 Neither 1-nonanol nor decane is water-soluble. Explain.

14.67 Phenols undergo the same kind of substitution reactions that other aromatic compounds do (Section 13.11). Formulate the reaction of *p*-methylphenol with Br_2 to give a mixture of two substitution products.

14.68 What is the difference between an antiseptic and a disinfectant?

14.69 Write the formulas and IUPAC names for the following common alcohols:
  (a) Rubbing alcohol
  (b) Wood alcohol
  (c) Grain alcohol
  (d) Diol used as antifreeze

14.70 Name the following compounds:

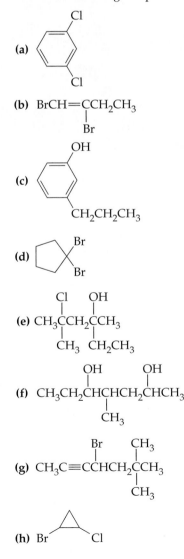

14.71 Complete the following reactions:

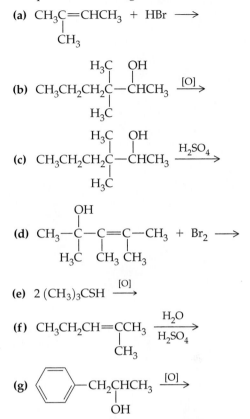

14.72 The odor of roses is due to geraniol.

$$CH_3C{=}CHCH_2CH_2C{=}CHCH_2OH$$
$$\quad\ \ CH_3 \qquad\qquad CH_3$$
Geraniol

  (a) What is the systematic name of geraniol?
  (b) When geraniol is oxidized, the aldehyde citral, one of the compounds responsible for lemon scent, is formed. Write the structure of citral.

14.73 Concentrated ethanol solutions can be used to kill microorganisms. At low concentrations, however, such as in some wines, the microorganisms can survive and cause oxidation of the alcohol. What is the structure of the acid formed?

14.74 "Flaming" desserts, such as cherries jubilee, use the ethanol in brandy or other distilled spirits as the flame carrier. Write the equation for the combustion of ethanol.

14.75 We said in Chapter 13 that H_2SO_4 catalyzes the addition of water to alkenes to form alcohols. In this chapter, however, we saw that H_2SO_4 is also used to dehydrate alcohols to make alkenes. Why do you think that sulfuric acid can serve two purposes—aiding in both hydration and dehydration?

15

Amines

CONTENTS

Caffeine, which we consume in coffee, sodas, chocolate, and over-the-counter medications is, like many other nitrogen-containing compounds, physiologically active.

Among the classes of organic compounds, we have now looked at hydrocarbons and those with single bonds joining carbon to oxygen, sulfur, or halogens. This chapter is devoted to amines, which have single bonds between carbon and nitrogen. The amines are present in so many kinds of essential biomolecules and important pharmaceutical agents that they are worthy of a chapter to themselves.

Consider that DNA, the blueprint for our heredity, relies on cyclic amines to carry the code that directs the synthesis of all of our proteins (Section 26.10), or that many of the biomolecules that carry chemical messages from place to place in our bodies are amines (the hormones and neurotransmitters, Chapter 20). Histamine, for example, the compound that initiates hay fever and other allergic reactions, is an amine. Therefore, many of the drugs that we use to mimic or to control the activity of histamine, the antihistamines present in cold and allergy medications, are amines. These are but a few examples of the biochemical roles played by amines.

In this chapter we'll consider the following questions:

1. What are the different types of amines?
The goal: Be able to recognize primary, secondary, tertiary, and heterocyclic amines, as well as quaternary ammonium ions.

2. How are amines named?
The goal: Be able to name simple amines and write their structures, given the names.

3. **What are the general properties of amines?**
   The goal: Be able to describe amine properties such as hydrogen bonding, solubility, boiling point, and basicity.
4. **How do amines react with water and acids?**
   The goal: Be able to predict the products of the acid–base reactions of amines and ammonium ions.
5. **What are alkaloids?**
   The goal: Be able to describe the sources of alkaloids, name some examples, and tell how their properties are typical of amines.

CONCEPTS TO REVIEW

Lewis structures (Section 5.6)
Acids and bases (Sections 6.10, 10.1, 10.2)
Organic molecules (Sections 12.1, 12.2)

15.1 Amines

Amines contain one or more organic groups bonded to nitrogen in compounds with the general formulas RNH_2, R_2NH, and R_3N. In the same way that alcohols and ethers can be thought of as organic derivatives of water, amines are organic derivatives of ammonia. They are classified as *primary, secondary,* or *tertiary* according to how many organic groups are bonded to the nitrogen atom. The organic groups (represented below by colored rectangles) may be large or small, and they may be the same or different. The ball-and-stick models show amines in which all the R groups are ethyl groups.

Amine A compound that has one or more organic groups bonded to nitrogen; primary, RNH_2; secondary, R_2NH; or tertiary, R_3N.

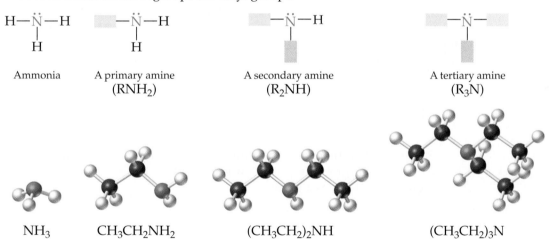

Ammonia   A primary amine A secondary amine A tertiary amine
               (RNH_2) (R_2NH) (R_3N)

NH_3 $CH_3CH_2NH_2$ $(CH_3CH_2)_2NH$ $(CH_3CH_2)_3N$

Note that each amine nitrogen atom has a lone pair of electrons. The lone pair is not always written, but it is always there. When a fourth group bonds to the nitrogen through this lone pair, the product is a **quaternary ammonium ion**, which has a positive charge and forms ionic compounds with anions [for example, $(CH_3CH_2)_4N^+\,Cl^-$].

Quaternary ammonium ion A positive ion with four organic groups bonded to the nitrogen atom.

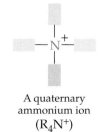

A quaternary
ammonium ion
(R_4N^+)

The groups bonded to an amine nitrogen atom may be alkyl or aryl groups and may or may not have other functional groups. For example:

CH_3NH_2

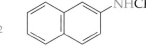

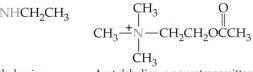

Methylamine
(a primary alkyl amine)

Aniline
(a primary
aromatic amine)

N-Ethylnaphthylamine
(a secondary aromatic amine)

Acetylcholine, a neurotransmitter
(a quaternary ammonium ion)

Primary alkyl amines (RNH_2) are named simply by identifying the alkyl group attached to nitrogen and adding the suffix -*amine* to the alkyl group name.

Ethylamine   Isopropylamine   Cyclohexylamine

Secondary (R_2NH) and tertiary (R_3N) amines with two or three identical groups are named by adding the appropriate prefix, *di-* or *tri-*, to the alkyl group name.

$$CH_3CH_2CH_2\underset{\underset{H}{|}}{N}CH_2CH_2CH_3 \qquad CH_3CH_2\underset{\underset{CH_2CH_3}{|}}{N}CH_2CH_3$$

Dipropylamine   Triethylamine

When the R groups are different in secondary or tertiary amines, the compounds are named as *N*-substituted derivatives of a primary amine. The parent compound is chosen as the primary amine that contains the largest of the R groups, and the other groups are considered to be *N*-substituents (*N* because they're attached directly to nitrogen). The following compounds, for example, are named as propylamines because the propyl group in each is the largest alkyl group:

$$CH_3CH_2\underset{\underset{H}{|}}{N}CH_2CH_2CH_3 \qquad CH_3\underset{\underset{CH_3}{|}}{N}CH_2CH_2CH_3$$

N-Ethylpropylamine   *N,N*-Dimethylpropylamine

Amino group   The $-NH_2$ functional group.

The simplest aromatic amine is known by the common name aniline. The $-NH_2$ functional group is an **amino group**, and when this group is a substituent, *amino-* is used as a prefix.

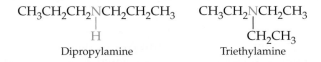

Aniline   *N*-Methylaniline   3-Aminopropanoic acid

The amino acids that make up all proteins have the general structure shown below—a primary amino group bonded to the carbon atom next to a *carboxylic acid group* ($-COOH$), with a side chain on the same carbon:

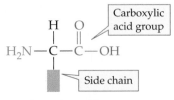

An amino acid

Adenine, a nitrogen-containing cyclic compound , is one of the four amines that provide the code in our DNA.

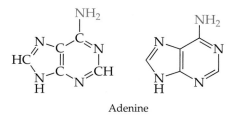

Adenine

In cyclic structures, the C's and the H's bonded to them are usually omitted, as shown above on the right. The H's bonded to N's are always included.

⊂∞⊃ LOOKING AHEAD

All amino acids contain both the amino functional group, $-NH_2$, and the carboxylic acid functional group, $-COOH$ (in addition to whatever functional groups are part of the side chain). The chemistry of the carboxylic acid group is discussed in Chapter 17. The amino acids and their combination to form proteins are covered in Chapter 18.   ⊂∞⊃

▪ WORKED EXAMPLE 15.1

Would you expect amines to be bases in aqueous solution? Explain the answer and write the equation for the reaction of ethylamine with water.

ANALYSIS   Like ammonia, amines have a lone pair of electrons on the nitrogen atom. Because ammonia is a base that reacts with water to accept a hydrogen ion that becomes bonded to the lone pair, it is reasonable to expect that amines are bases that react in the same manner.

SOLUTION   The reactions as bases for ammonia and ethylamine are

$$NH_3 + H_2O \rightleftharpoons NH_4^+ + OH^-$$

$$CH_3CH_2NH_2 + H_2O \rightleftharpoons CH_3CH_2NH_3^+ + OH^-$$

▪ WORKED EXAMPLE 15.2

Write the structure of *N,N*-diethylbutylamine and identify it as a primary, secondary, or tertiary amine.

ANALYSIS   The name shows that butylamine, the four-carbon alkyl amine, is the parent compound. The "*N,N*-diethyl" part of the name shows that two ethyl groups are bonded to the amino nitrogen.

SOLUTION   The structure shows that three alkyl groups are bonded to the N atom. This is a tertiary amine.

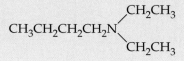

▪ WORKED EXAMPLE 15.3

Name the compound below. Is it a primary, secondary, or tertiary amine?

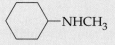

Application Chemical Information

Suppose you're reading an article in a magazine about caffeine in soft drinks and you decide you'd like to know something about the chemical called "caffeine." Where would you look for information?

The first place might be a chemical handbook, where you would find the structure and an entry like the one shown at the bottom of this page from *Lange's Handbook of Chemistry*. This entry includes some physical properties and a reference to Beilstein's *Handbook of Organic Chemistry*, a publication that would give more chemical information (but it's not so easily available). A second standard chemical handbook is the *CRC Handbook of Chemistry and Physics* (CRC Press, Boca Raton, FL).

With your curiosity not yet satisfied, you could look next in another generally available and reliable source of chemical information, *The Merck Index: An Encyclopedia of Chemicals and Drugs*, first published in 1889 as a list of products of the Merck pharmaceutical company. The index has since grown way beyond its original intended use to a standard reference work giving information on the preparation, properties, and uses of over 10,000 chemical compounds. It has a strong medical emphasis and, where appropriate, lists toxicity data, therapeutic uses in both human and veterinary medicine, and the physiological effects and cautions associated with hazardous chemicals. In some libraries, you may find *The Merck Index* available on CD-ROM or online via the Internet.

The entry for caffeine in *The Merck Index*, reproduced on the facing page, begins with a list of alternative chemical names and a capitalized entry that's the name of a medication containing caffeine. One of the joys of using *The Merck Index* is that every one of the alternative names of every substance in the book, including the drug names, appears in the index, so no matter what name you come across, you can discover exactly what substance it refers to.

Next comes information about the molecular formula and sources of the compound, followed by a list of patents and journal references to articles about caffeine, including an account of its first synthesis in 1895. After the structure is a paragraph about the physical properties of caffeine, followed by a listing of the properties of some important derivatives of caffeine. The final lines give the therapeutic uses of caffeine, which is a central-nervous-system (CNS) stimulant.

If you knew only the formula for an unknown molecule that happened to be caffeine ($C_8H_{10}N_4O_2$), you could also search in the molecular formula index of *The Merck Index*. In a current edition, you would find two molecules with that formula—caffeine and a bronchodilator with a structure similar to caffeine.

Having learned from *The Merck Index* that caffeine is present in No-Doz, you can next turn to *The Physicians Desk Reference* (the *PDR*) to learn about the use of caffeine in drugs. While intended primarily for physicians, the *PDR* is readily available in book stores and libraries. It also may be available on CD-ROM. For each drug, it contains the full product labeling information provided by the manufacturers. You can go directly to the entry for No-Doz via the product name index and learn more about the effects and cautions accompanying the intended use of this medication. By using the Generic and Chemical Name Index in the *PDR*, you can also discover that caffeine is present in 27 other medications ranging from nonprescription analgesics like Anacin to a potent prescription drug for migraine headaches that combines caffeine with belladonna and ergotamine (both alkaloids, Section 15.6) and sodium pentobarbital.

Name	Formula	Formula weight	Beilstein reference	Density	Refractive index	Melting point	Boiling point	Flash point	Solubility in 100 parts solvent
Caffeine		194.19	26,461	1.23_4^{18}		238	subl 178		2.1 aq; 1.5 alc; 18 chl; 0.19 eth; 1 bz

Caffeine information from *Lange's Handbook of Chemistry*. Reproduced from *Dean/Lange's Handbook of Chemistry*, 13th Ed. (1987), Entry C1, p. 7–194, with permission of McGraw-Hill, Inc.

ANALYSIS The cyclohexyl group is the largest alkyl group bonded to N, so the compound is named as a cyclohexylamine. One methyl group is bonded to the nitrogen, and this is indicated in the name by the prefix *N*.

SOLUTION The name is *N*-methylcyclohexylamine. The compound has two groups bonded to N and is therefore a secondary amine.

You might think also of searching on the Internet for information about caffeine. A search for "caffeine" and "chemistry" together at the time this book was written turned up about 5500 hits using one well-known search engine and over 15,000 hits using another. Starting with the reference books described here, however, remains the most efficient way to find the molecular structure and other fundamentals.

1635. Caffeine. *3,7-Dihydro-1,3,7-trimethyl-1 H-purine-2,6-dione*; 1,3,7-trimethylxanthine; 1,3,7-trimethyl-2,6-dioxopurine; coffeine; thein; guaranine; methyltheobromine; No-Doz. $C_8H_{10}N_4O_2$; mol wt 194.19. C 49.48%, H 5.19%, N 28.85%, O 16.48%. Occurs in tea, coffee, maté leaves; also in guarana paste and cola nuts: Shuman, **U.S.** pat. **2,508,545** (1950 to General Foods). Obtained as a by-product from the manuf of caffeine-free coffee: Barch, **U.S.** pat. **2,817,588** (1957 to Standard Brands); Nutting, **U.S.** pat. **2,802,739** (1957 to Hill Bros. Coffee); Adler, Earle, U.S. pat. **2,933,395** (1960 to General Foods). Crystal structure: Sutor, *Acta Cryst.* **11**, 453 (1958). Synthesis: Fischer, Arch, *Ber.* **28**, 2473, 3135 (1895); . . . .

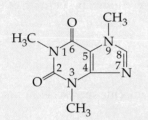

Hexagonal prisms by sublimation, mp 238°. Sublimes 178°. Fast sublimation is obtained at 160–165° under 1 mm press. at 5 mm distance. d $^{18}_{4}$ 1.23. pH of 1% soln 6.9. Aq solns of caffeine salts dissociate quickly. Absorption spectrum: Hartley, *J. Chem. Soc.* **87**, 1802 (1905). One gram dissolves in 46 ml water, 5.5 ml water at 80°, 1.5 ml boiling water, 66 ml alcohol, 22 ml alcohol at 60°, 50 ml acetone, 5.5 ml chloroform, 530 ml ether, 100 ml benzene, 22 ml boiling benzene. Freely sol in pyrrole; in tetrahydrofuran contg about 4% water; also sol in ethyl acetate; slightly in petr ether. Soly in water is increased by alkali benzoates, cinnamates, citrates or salicylates. LD_{50} orally in mice, hamsters, rats, rabbits (mg/kg): 127, 230, 355, 246 (males); 137, 249, 247, 224 (females) (Palm).

Monohydrate, felted needles, contg 8.5% H_2O. Efflorescent in air; complete dehydration takes place at 80°. . . . .

THERAP CAT: CNS stimulant.

THERAP CAT (VET): Has been used as a cardiac and respiratory stimulant and as a diuretic.

NO DOZ® Tablets
[*no'doz*]

COMPOSITION

Each tablet contains 100 mg. Caffeine. Other Ingredients: Cornstarch, Flavors, Mannitol, Microcrystalline Cellulose, Stearic Acid, Sucrose.

INDICATIONS

Helps restore mental alertness or wakefulness when experiencing fatigue or drowsiness.

DOSAGE AND ADMINISTRATION

Adults and children 12 years of age and over: One or two tablets not more often than every 3 to 4 hours.

CAUTION

Do not take without consulting physician if under medical care. No stimulant should be substituted for normal sleep in activities requiring physical alertness.

WARNING

For occasional use only. Not intended for use as a substitute for sleep. If fatigue or drowsiness persists or continues to recur, consult a doctor. The recommended dose of this product contains about as much caffeine as a cup of coffee. Limit the use of caffeine-containing medications, foods, or beverages while taking this product because too much caffeine may cause nervousness, irritability, sleeplessness and, occasionally, rapid heart beat. Do not give to children under 12 years of age. **KEEP THIS AND ALL MEDICINES OUT OF THE REACH OF CHILDREN. IN CASE OF ACCIDENTAL OVERDOSE, SEEK PROFESSIONAL ASSISTANCE OR CONTACT A POISON CONTROL CENTER IMMEDIATELY. As with any drug, if you are pregnant or nursing a baby, seek the advice of a health professional before using this product**.

OVERDOSE

Typical of caffeine.

HOW SUPPLIED

NO DOZ® is supplied as:

A circular white tablet with "NoDoz" debossed on one side. . . . .

Caffeine information from The Merck Index. *Reproduced from* The Merck Index, *11th Ed. (1989), S. Budavari, M.J. O'Neil, A. Smith, P.E. Heckelman, Eds., by permission of the copyright owner, Merck & Co., Inc., Rahway, N.J., U.S.A.,* © *Merck & Co., Inc., 1989.*

Entry on No-Doz® from The Physicians Desk Reference. Reproduced from The Physicians Desk Reference with permission of Medical Economics Company, Inc., Oradell, N.J.

See Additional Problem 15.43 at the end of the chapter.

■ **PROBLEM 15.1**

Identify the following compounds as primary, secondary, or tertiary amines.

(a) $CH_3CH_2CH_2NH_2$

(b) $CH_3CH_2NHCHCH_3$

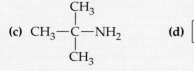

(d)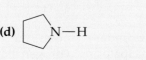

(e)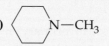

■ **PROBLEM 15.2**

What are the names of these amines?

(a) $CH_3CH_2CH_2NH_2$

(b)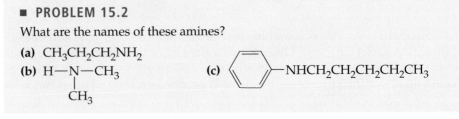

(c)

■ **PROBLEM 15.3**

Draw structures corresponding to the following names:

(a) Butylamine

(b) *N*-Methylethylamine

(c) *N,N*-Dimethylaniline

(d) 2-Aminobutanol

■ **KEY CONCEPT PROBLEM 15.4**

Draw the structure of tetramethylammonium ion, count its valence electrons, and explain why the ion has a positive charge.

■ **KEY CONCEPT PROBLEM 15.5**

Draw the condensed formula of the molecule shown below and name it.

15.2 Properties of Amines

The lone electron pair on the nitrogen in amines, like the lone electron pair in ammonia, causes amines to be weak bases by forming a bond with an H^+ ion from an acid or water (Section 15.4).

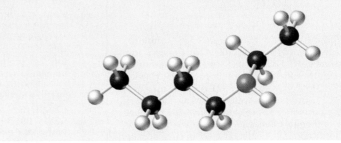

▲ **FIGURE 15.1 Hydrogen bonding of a secondary amine.**
Hydrogen bonding between $(CH_3CH_2)_2NH$ molecules is shown by red dots.

Also, in primary and secondary amines, the unshared electron pair on the nitrogen atom can participate in hydrogen bonds to hydrogen atoms from water or other primary or secondary amines (Figure 15.1).

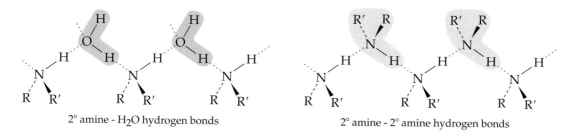

2° amine - H₂O hydrogen bonds   2° amine - 2° amine hydrogen bonds

Because of the hydrogen bonding, primary and secondary amines have higher boiling points than alkanes of similar size. (∞, p. 342) They are, however, lower boiling than alcohols of similar size because nitrogen atoms are less electronegative than oxygen atoms and so form weaker hydrogen bonds.

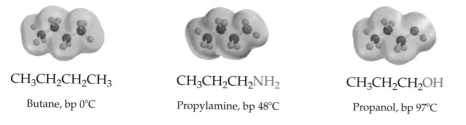

$CH_3CH_2CH_2CH_3$ $CH_3CH_2CH_2NH_2$ $CH_3CH_2CH_2OH$

Butane, bp 0°C   Propylamine, bp 48°C   Propanol, bp 97°C

In fact, mono-, di-, and trimethylamine and ethylamine are gases at room temperature. Other common amines with molecules of moderate size are liquids (Table 15.1).

TABLE 15.1 Boiling Points of Some Simple Amines

Structure	Name	Boiling Point (°C)	Structure	Name	Boiling Point (°C)
NH_3	Ammonia	−33.3	**Secondary amines**		
Primary amines			$(CH_3)_2NH$	Dimethylamine	7.4
CH_3NH_2	Methylamine	−6.3	$(CH_3CH_2)_2NH$	Diethylamine	56.3
$CH_3CH_2NH_2$	Ethylamine	16.6	$((CH_3)_2CH)_2NH$	Diisopropylamine	84
$(CH_3)_3CNH_2$	*tert*-Butylamine	44.4	**Tertiary amines**		
⬡—NH_2	Aniline	184.1	$(CH_3)_3N$	Trimethylamine	3
			$(CH_3CH_2)_3N$	Triethylamine	89.3
			⬡N (pyridine)	Pyridine	115

Tertiary amine molecules have no hydrogen atoms attached to nitrogen. They therefore cannot hydrogen-bond with each other and are lower boiling than other amines or alcohols of similar molecular weight. Compare the boiling point of trimethylamine (3°C) with those of propylamine (48°C) and the other compounds shown in Table 15.1.

All amines, however, can hydrogen-bond to water molecules through the lone electron pair on their nitrogen atoms. As a result, amines with up to about six carbon atoms are quite soluble in water.

Many volatile amines have strong odors. Some smell like ammonia and others like stale fish or decaying meat. The protein in flesh contains amine groups, and the smaller, volatile amines produced during decay and protein breakdown are responsible for the odor of rotten meat.

Another significant property of amines is that many are physiologically active. The simpler amines are irritating to the skin, eyes, and mucous membranes and are toxic by ingestion. Some of the more complex amines from

▲ A good test for fresh fish. If the fishy odor is strong, the fish is not fresh.

plants (Section 15.6) are very poisonous. On the other hand, all living things contain a wide variety of amines, and many useful drugs are amines.

Properties of Amines

- Primary and secondary amines are hydrogen-bonded and thus are higher boiling than alkanes, but lower boiling than alcohols.
- Tertiary amines are lower boiling than secondary or primary amines because hydrogen bonding is not possible.
- The simplest amines are gases; other common amines are liquids.
- Volatile amines have unpleasant odors.
- Simple amines are water-soluble because of hydrogen bonding.
- Amines are weak Brønsted–Lowry bases (Section 15.4).
- Many amines are physiologically active, and many are toxic (see the Application "Toxicology," p. 435).

■ **PROBLEM 15.6**

Arrange the following compounds in order of increasing boiling point. Explain why you placed them in that order.

(a) CH_3—$\overset{\overset{\displaystyle CH_3}{|}}{N}$—$CH_2CH_3$ (b) $CH_3CH_2CH_2CH_2OH$ (c) $CH_3CH_2CH_2CH_2NH_2$

■ **PROBLEM 15.7**

Draw the structures of (a) ethylamine and (b) trimethylamine. Use dashed lines to show where they would form hydrogen bonds to water molecules.

15.3 Heterocyclic Nitrogen Compounds

Heterocycle A ring that contains nitrogen or some other atom in addition to carbon.

In many nitrogen-containing compounds, the nitrogen atom is in a ring with carbon atoms. Such rings that contain atoms other than carbon are known as **heterocycles**. Heterocyclic nitrogen compounds may be nonaromatic or aromatic. Piperidine, for example, is a saturated heterocyclic amine with a six-membered ring, and pyridine is an aromatic heterocyclic amine which, like other aromatic compounds, is often represented on paper as a ring with alternating double and single bonds.

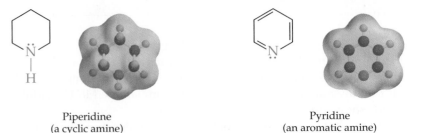

Piperidine
(a cyclic amine)

Pyridine
(an aromatic amine)

The names and structures of several heterocyclic nitrogen compounds are given in Table 15.2. You need not memorize these names and structures, but you should realize that such rings are very common in natural compounds from plants and animals. For example, nicotine, from tobacco leaves, contains one pyridine ring and one pyrrolidine ring; quinine, an antimalarial drug isolated from the bark of the South American *Cinchona* tree, contains a quinoline ring system plus a nitrogen ring with a two-carbon bridge across it. The amino acid tryptophan contains an indole ring system in addition to its amino group.

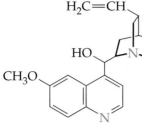

Nicotine
from tobacco
(an insecticide; an
active ingredient
in cigarette smoke)

Quinine
from the *Cinchona* Tree
(an antimalarial drug)

Tryptophan
(an amino acid)

⧐ LOOKING AHEAD

Hydrogen bonding between the hydrogen atoms of amines and the oxygen or nitrogen atoms of other groups helps to determine the shape of many biomolecules. Such attractions contribute to the complex shapes into which large protein molecules are folded (Section 18.8). Hydrogen bonding of amine groups also plays a crucial role in the helical structure of the molecule that carries hereditary information—deoxyribonucleic acid, DNA (Section 26.4).   ⧐

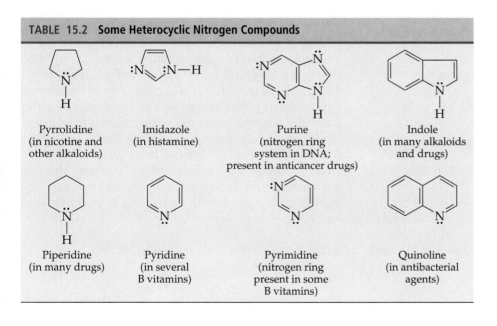

TABLE 15.2   **Some Heterocyclic Nitrogen Compounds**

Pyrrolidine
(in nicotine and
other alkaloids)

Imidazole
(in histamine)

Purine
(nitrogen ring
system in DNA;
present in anticancer drugs)

Indole
(in many alkaloids
and drugs)

Piperidine
(in many drugs)

Pyridine
(in several
B vitamins)

Pyrimidine
(nitrogen ring
present in some
B vitamins)

Quinoline
(in antibacterial
agents)

▪ PROBLEM 15.8

Consult Table 15.1 and identify:

(a) Two amines that are gases at room temperature

(b) A heterocyclic amine

(c) A compound with an amine group on an aromatic ring

▪ PROBLEM 15.9

Consult Table 15.2 and write the molecular formulas for piperidine and purine.

Application NO: A Small Molecule with Big Responsibilities

Imagine a molecule that can lower blood pressure, kill invading bacteria, and enhance memory. Imagine also that this is not a large biomolecule with many functional groups, but a molecule with just two atoms. What a major surprise it has been to discover that nitric oxide (NO) is just such a molecule. Even more surprising for a molecule with such a diversity of biochemical functions, pure nitric oxide is a gas at room and body temperatures.

A close look at NO shows that it has 11 valence electrons (5 from N and 6 from O). The atoms are joined by a double bond, $\cdot \ddot{N}{=}\ddot{O}\!:$, which gives O an electron octet and leaves N with 7 valence electrons. The unpaired electron makes the molecule a *free radical* and therefore very reactive—it quickly grabs onto another electron in whatever way it can. In fact, the lifetime of free NO in the body is around 100 milliseconds.

Nitric oxide is synthesized in the linings of blood vessels and elsewhere from oxygen and the amino acid arginine. In blood vessels, NO activates reactions in smooth muscle cells that cause dilation and a resulting decrease in blood pressure. This discovery explains the action of drugs long used to treat angina, the pain experienced during exertion by individuals with partially blocked blood vessels. Nitroglycerin and other drugs of this type release NO.

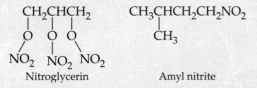

Nitroglycerin   Amyl nitrite

Alfred Nobel, who became wealthy as the developer of dynamite, suffered from angina shortly before his death. He wrote to a friend, "Isn't it the irony of fate that I have been prescribed nitroglycerin, to be taken internally. They call it Trinitrin, so as not to scare the chemist and the public." Nobel refused to take Trinitrin.

▲ **A dynamite explosion at a gold mining site.** Nitroglycerin is the active ingredient in dynamite, which made Alfred Nobel a very wealthy man.

In a further extension of the irony, the Nobel Prize in Physiology or Medicine was awarded in 1998 to three individuals who had studied the role of NO in physiology (Ferid Murad, Robert F. Furchgott, and Lous J. Ignarro). One of them had discovered that it is the NO released by nitroglycerin that relieves the pain of angina.

As you might expect, the discoveries about the role of NO in human health and disease have opened the door to possible therapeutic applications. Compounds that release NO are being considered for inhibition of the growth of cancerous cells, protection of vessel walls after surgery, promotion of wound healing, and killing parasites. Another potential application is the incorporation of an NO-releasing chemical into a coating for implanted medical devices with the goal of preventing bacterial infection at the implant site. These applications are an amazing group for a molecule that was once known mainly for its role in helping to generate smog and acid rain.

See Additional Problem 15.44 at the end of the chapter.

■ PROBLEM 15.10

Which of the following compounds are heterocyclic nitrogen compounds?

(a) [imidazole ring]—$CH_2CH_2NH_3$

(b) [benzene ring]—NH_2

(c) HO—[benzene ring]—$CH_2CHCO_2^-$
  $^+NH_3$

(d) HO—[indole ring with $CH_2CH_2{}^+NH_3$]

15.4 Basicity of Amines

Aqueous solutions of amines are weakly basic because of formation of the OH^- ion in equilibria like the following between the amines and their **ammonium ions**.

$$CH_3CH_2NH_2 + H_2O \rightleftharpoons CH_3CH_2NH_3^+ + OH^-$$

$$(CH_3CH_2)_2NH + H_2O \rightleftharpoons (CH_3CH_2)_2NH_2^+ + OH^-$$

$$(CH_3CH_2)_3N + H_2O \rightleftharpoons (CH_3CH_2)_3NH^+ + OH^-$$

Ammonium ion   A positive ion formed by addition of hydrogen to ammonia or an amine (may be primary, secondary, or tertiary).

Notice that in the reverse of these reactions, ammonium ions can react as acids in the presence of bases to regenerate the amines.

Ammonium ions are also formed in the reactions of amines with the hydronium ion in acidic solutions.

$$CH_3CH_2NH_2 + H_3O^+ \rightleftharpoons CH_3CH_2NH_3^+ + H_2O$$

$$(CH_3CH_2)_2NH + H_3O^+ \rightleftharpoons (CH_3CH_2)_2NH_2^+ + H_2O$$

$$(CH_3CH_2)_3N + H_3O^+ \rightleftharpoons (CH_3CH_2)_3NH^+ + H_2O$$

The positive ions formed by addition of H^+ to alkylamines are named by replacing the ending *-amine* by the ending *-ammonium*. To name the ions of heterocyclic amines, the amine name is modified by replacing the *-e* with *-ium*. For example:

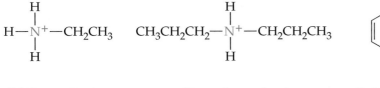

Ethylammonium ion   Dipropylammonium ion   Pyridinium ion
(from ethylamine)   (from dipropylamine)   (from pyridine)

In the aqueous environment of blood and other body fluids, amines are present as ammonium ions and are written as such in the context of biochemistry. For example, histamine and serotonin (both neurotransmitters; Sections 20.8, 20.9) are represented as follows:

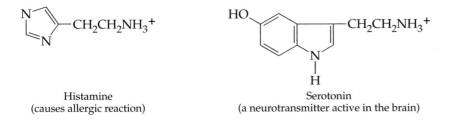

Histamine   Serotonin
(causes allergic reaction)   (a neurotransmitter active in the brain)

In general, nonaromatic amines (such as $CH_3CH_2NH_2$) are slightly stronger bases than ammonia, and aromatic amines (such as $C_6H_5NH_2$ or pyridine, Table 15.2) are weaker bases than ammonia:

Basicity: Nonaromatic amines > Ammonia > Aromatic amines

■ **PROBLEM 15.11**

Write an equation for the acid–base equilibrium of dimethylamine and water. Label both acids and both bases.

The chemical reactions that keep us alive and functioning occur in the aqueous solutions known as "body fluids"—blood, digestive juices, urine, and the fluid inside cells. But for organic compounds of all classes, water solubility decreases as the hydrocarbon-like portions of the molecules become larger and molecular weight increases. How does the body manage its reactions in water solution, especially when large and complex biomolecules are involved?

Acidic and basic functional groups are part of many biomolecules. At the pH of body fluids, many of these groups are ionized and thus are water-soluble. (∞, p. 233) The most common ionized functional groups in biomolecules are carboxylate groups (from carboxylic acids, —COOH, discussed in Section 17.3), phosphate groups (as well as diphosphates and triphosphates, discussed in Section 17.8), and ammonium groups.

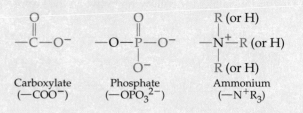

Carboxylate (—COO⁻) Phosphate (—OPO₃²⁻) Ammonium (—N⁺R₃)

For example, nicotinamide adenine dinucleotide, an important biochemical oxidizing agent (Section 21.7), has three charges as it exists in body fluids: two negative charges from a diphosphate and one positive charge from a quaternary amine.

In the biochemistry that lies ahead, you will see that most of the major biochemical pathways occur in the aqueous medium of the cytosol inside cells. It would be disastrous if these pathways could be shut down by diffusion of reactants out of the cells. Such diffusion would require passage of intermediates through the nonpolar

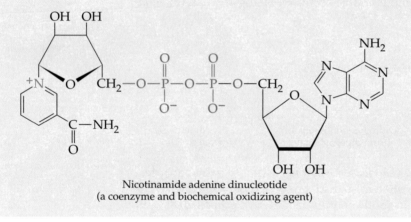

Nicotinamide adenine dinucleotide
(a coenzyme and biochemical oxidizing agent)

■ **PROBLEM 15.12**

Complete the following equations:

(a) $CH_3CH_2\underset{\underset{CH_3}{|}}{CH}NH_2$ + HBr(*aq*) ⟶ ?

(b) ⬡—NH₂ + HCl(*aq*) ⟶ ?

(c) $CH_3CH_2NH_2$ + CH_3COOH(*aq*) ⟶ ?

(d) $CH_3NH_3{}^+Cl^-$ + NaOH(*aq*) ⟶ ?

■ **PROBLEM 15.13**

Name the organic ions produced in reactions (a)–(c) in Problem 15.12.

medium of the cell walls. Diffusion does not occur because the intermediates are ionized within the cytosol.

Drugs too must be soluble in body fluids and must be transported in fluids from their entry point in the body to their site of action. Many drugs are weak acids or bases and therefore are present as their ions in body fluids. For example:

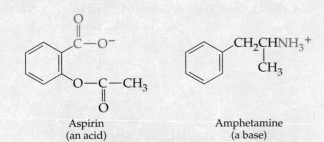

Aspirin
(an acid)

Amphetamine
(a base)

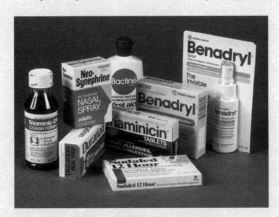

▲ **Over-the-counter ammonium salts**. An ammonium salt is the active ingredient in each of these over-the-counter medications.

The extent of ionization of a drug helps determine how it is distributed in the body, because ions are less likely to cross cell membranes than uncharged molecules. Weak acids like aspirin are less ionized in the acidic environment in the stomach and are absorbed readily there. Weak bases are not significantly absorbed in the stomach but are better absorbed in the more basic environment in the small intestine.

Many drugs must be delivered to the body in their more water-soluble forms as salts, in what is sometimes known as the "solubility switch." Converting such amines as phenylephrine (the decongestant in Neo-Synephrine) to ammonium hydrochlorides increases their solubility to the point where delivery in solution is possible.

Phenylephrine hydrochloride
(a decongestant)

See Additional Problems 15.45–47 at the end of the chapter.

■ **PROBLEM 15.14**

Which is the stronger base?

(a) Ammonia or ethylamine

(b) Triethylamine or pyridine

■ **PROBLEM 15.15**

Write the formulas for the ammonium ions formed by the following amines:

Epinephrine
(a biochemical messenger)

Amphetamine
(a central-nervous-system
stimulant and abused drug)

431

15.5 Amine Salts

Ammonium salt An ionic compound composed of an ammonium cation and an anion; an amine salt.

An **ammonium salt** (also known as an *amine salt*) is composed of a cation and an anion and is named by combining the ion names. For example, in methylammonium chloride ($CH_3NH_3^+Cl^-$), the methylammonium ion, $CH_3NH_3^+$, is the cation and the chloride ion is the anion.

Ammonium salts are generally odorless, white, crystalline solids that are much more water-soluble than neutral amines because they're ionic. For example:

$$CH_3CH_2CH_2CH_2-\underset{\underset{CH_2CH_2CH_2CH_3}{|}}{N}-CH_2CH_2CH_2CH_3 \ + \ HCl(aq) \ \rightleftharpoons \ CH_3CH_2CH_2CH_2-\overset{\overset{H}{|}}{\underset{\underset{CH_2CH_2CH_2CH_3}{|}}{N^+}}-CH_2CH_2CH_2CH_3 \ Cl^-(aq)$$

Tributylamine   Hydrochloric   Tributylammonium chloride

(water-insoluble)   acid   (water-soluble)

In an older system, amine salt formulas were written and named by combining the structures and names of the amine and the acid. In this system, methylammonium chloride is written $CH_3NH_2 \cdot HCl$ and named methylamine hydrochloride. You'll often see this system used with drugs that are amine salts. For example, diphenhydramine is one of a family of antihistamines available in over-the-counter medications. Antihistamines of this family are oily liquids, so they are converted to amine salts for formulation into medications.

$$(C_6H_5)_2CHOCH_2CH_2N(CH_3)_2 \cdot HCl \quad \text{or} \quad (C_6H_5)_2CHOCH_2CH_2CH_2NH(CH_3)_2{}^+Cl^-$$

Diphenhydramine hydrochloride (Benadryl), an antihistamine

If a free amine is needed, it is easily regenerated from an amine salt by treatment with a base:

$$CH_3NH_3^+ Cl^-(aq) + NaOH(aq) \longrightarrow CH_3NH_2(aq) + NaCl(aq) + H_2O(l)$$

Quaternary ammonium salt An ionic compound composed of a quaternary ammonium ion and an anion.

Quaternary ammonium ions have four organic groups bonded to the nitrogen atom. With no H atom that can be removed by a base and no lone pair that can bond to H^+, they are not bases and their structure in solution is unaffected by changes in pH. Their salts are known as **quaternary ammonium salts**. One commonly encountered quaternary ammonium salt has the following structure, where R represents a range of C_8 to C_{18} alkyl groups:

$$\text{C}_6\text{H}_5-CH_2-\overset{\overset{CH_3}{|}}{\underset{\underset{CH_3}{|}}{N^+}}-R \ Cl^- \qquad R = -C_8H_{17} \ \text{to} \ -C_{18}H_{37}$$

Benzalkonium chloride
(an antiseptic and disinfectant)

The compound has antimicrobial properties and also acts as a detergent. As a dilute solution, it is used in surgical scrubs and for sterile storage of instruments, but concentrated solutions are harmful to body tissues.

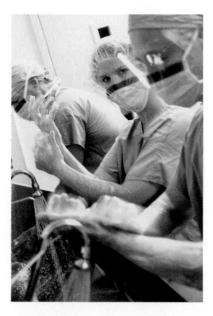

▲ **The preoperative surgical scrub.** Several quaternary ammonium compounds have germicidal properties that suit them for this use.

■ **PROBLEM 15.16**

Write the structures of the following compounds:

(a) Hexyldimethylammonium chloride

(b) Ethylammonium bromide

■ **PROBLEM 15.17**

Identify each compound in Problem 15.16 as the salt of a primary, secondary, or tertiary amine.

■ **PROBLEM 15.18**

Write an equation for the formation of the free amine from butylammonium chloride by reaction with aqueous OH^-.

■ **PROBLEM 15.19**

Compare the structure of Benadryl (p. 432) with the general antihistamine structure:

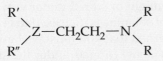

Does Benadryl have that general structure? Write a comparison of the structures.

■ **PROBLEM 15.20**

Write the structure of benzylamine hydrochloride in two different ways, and name the hydrochloride as an ammonium salt.

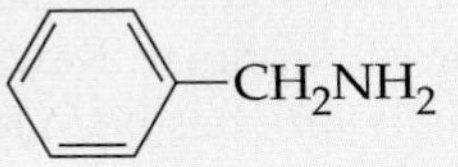

$\quad$ —CH_2NH_2 Benzylamine

15.6 Amines in Plants: Alkaloids

The roots, leaves, and fruits of flowering plants are a rich source of nitrogen compounds. These compounds, once called "vegetable alkali" because their water solutions are basic, are now referred to as **alkaloids**.

The molecular structures of many thousands of alkaloids have been determined. Most are bitter-tasting, physiologically active, and toxic to human beings and other animals in sufficiently high doses. Many of us are familiar with the physiological activity of two alkaloids—caffeine (p. 423) and nicotine (p. 427), which are stimulants. Quinine (p. 427) is used as a standard for bitterness: Even a 1×10^{-6} M solution tastes bitter. For a long time, quinine was the only drug available for treating malaria (caused by a parasitic protozoan).

The bitterness and poisonous nature of alkaloids probably evolved to protect plants from being devoured by animals. The three poisonous compounds described here—coniine, atropine, and solanine—illustrate some more of the many types of alkaloid structures:

Alkaloid A naturally occurring nitrogen-containing compound isolated from a plant; usually basic, bitter, and poisonous.

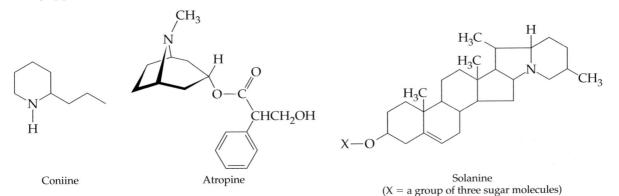

Coniine   Atropine   Solanine
(X = a group of three sugar molecules)

▲ A potato that has turned green because of exposure to sunlight. Before it is eaten, this potato must be peeled to remove all of the green chlorophyll so that the poisonous alkaloid solanine is also removed.

▲ All parts of the poppy, including poppy seeds, contain morphine. Eating poppy seeds can introduce enough morphine into your body fluids to show up in a laboratory blood test.

- **Coniine** is extracted from poison hemlock (*Conium maculatum*). It was the poison Socrates used to end his life after being convicted of corrupting Greek youth with philosophical discussions.

- **Atropine** is the toxic substance in the herb known as *deadly nightshade* or *belladonna (Atropa belladonna)*. In Meyerbeer's opera, *L'Africaine*, the heroine sings of the peaceful death this plant can bring before committing suicide over her lost love. Like many other alkaloids, atropine acts on the central nervous system, a property sometimes applied in medications (in appropriately low dosage!) to reduce cramping of the digestive tract.

- **Solanine**, an even more potent poison than atropine, is found in potatoes and tomatoes, both of which belong to the same botanical family as the deadly nightshade (*Solanaceae*). The tiny amount of solanine in properly stored potatoes only contributes to their characteristic flavor. But when potatoes are exposed to sunlight or stored under very cold or very warm conditions, the production of solanine is increased to levels that can be dangerous. If you've ever been warned that you must peel green potatoes deeply, this is the reason. The alkaloids are formed under the skin. They aren't destroyed during cooking, but can be removed by peeling. Sunlight fortunately also stimulates the formation of chlorophyll under the skin. The green color of the chlorophyll provides a warning, and by peeling away all of the green color, you have most likely removed the excess solanine. Potato sprouts also contain solanine and should be cut out before potatoes are cooked.

Some alkaloids are notable not as poisons but as pain relievers (*analgesics*), as sleep inducers, and for the creation of euphoria. Raw opium, a paste derived from the opium poppy (*Papaver somniferum*), has been known for these properties since ancient times. About 20 alkaloids are present in the poppy, including morphine and codeine. The free alkaloids are oily liquids, not very soluble in water. The medicinal use of morphine for pain was enhanced in the sixteenth century when the German physician Paracelsus extracted opium into brandy to produce *laudanum*, essentially a solution of morphine in alcohol. A similar extract (10% opium by weight in alcohol) is still sometimes prescribed for diarrhea, as is *paregoric*, a more dilute solution of opium combined with anise oil, glycerin, benzoic acid, and camphor. Heroin, another close relative of morphine, does not occur naturally but is easily synthesized. Within the body, removal of the $CH_3C=O$ groups converts heroin back to morphine.

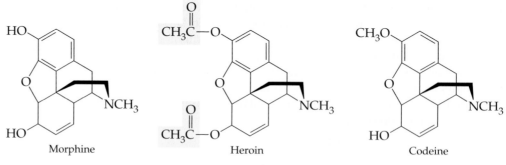

Morphine   Heroin   Codeine

Application Toxicology

Toxicology is the science devoted to poisons—their identification, their effects, their modes of action, and methods of protecting against them or counteracting their effects. It is a science with many subspecialties. *Clinical* toxicology is concerned with the treatment of individuals harmed by toxic agents. *Forensic* toxicology deals with the effects of toxic agents as they relate to criminal cases, most notably in drug abuse or intentional poisonings. A third branch of toxicology—*environmental* toxicology—is concerned with toxic substances purposefully or accidentally introduced into our surroundings. Such environmental pollutants might be harmful to humans, to other animals, or to plants. Yet another important branch of toxicology focuses on the beneficial uses of poisons—for example, pesticides in agriculture or chemotherapeutic agents to kill cancer cells.

Some toxic substances have the general effect of harming any kind of tissue. Strong acids such as sulfuric acid and strong bases such as sodium hydroxide (lye) "burn" by destroying all cells they contact.

Many more poisons have a molecular structure that allows them to interact with a specific biomolecule. Several of the most poisonous substances known are *neurotoxins*, which bind to proteins that form ion channels in nerve cell membranes. Blocking the channels prevents transmission of nerve impulses, causing paralysis and death by suffocation. Tetrodotoxin, a poison of this type, is produced by the puffer fish. Despite this, puffer fish is considered a delicacy in Japan, where the risk is minimized by allowing only well-trained and certified chefs to remove the toxin-containing parts of the fish.

▲ A puffer fish—a highly poisonous creature—and a skilled chef preparing a puffer fish by cutting out the toxic parts.

A number of poisons act at receptors that normally bind acetylcholine, a neurotransmitter (a molecule that carries chemical messages between nerve cells). For example, muscarine, the active ingredient in the highly poisonous mushroom *Amanita muscaria*, duplicates the action of acetylcholine, overstimulating certain of its receptors. In this case (though not always), you can easily see the structural similarity between the natural chemical messenger and its poisonous mimic:

$$CH_3-\overset{\overset{\displaystyle O}{\|}}{C}-O-CH_2-CH_2-\overset{+}{N}(CH_3)_3$$
Acetylcholine (ACh)
(a neurotransmitter)

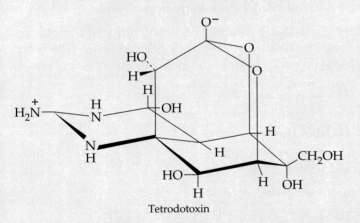

Tetrodotoxin

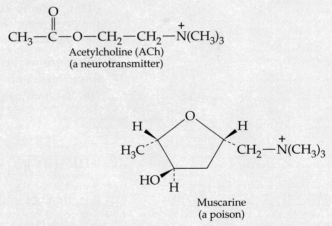

Muscarine
(a poison)

Chemical change in the body can be caused by poisons, medications, or natural chemical messengers (hormones or neurotransmitters, discussed in Chapter 20). Change is initiated when a messenger molecule encounters a structurally compatible receptor on the surface of a cell or an enzyme that catalyzes a necessary reaction.

Ideally, a thorough understanding of any poison includes knowing how it acts at the molecular level and knowing a molecular mechanism by which its effects can be reversed. Such knowledge is a major goal of toxicology.

See Additional Problem 15.48 at the end of the chapter.

Key Words

Summary: Revisiting the Chapter Goals

1. **What are the different types of amines?** *Amines* are classified as *primary, secondary,* or *tertiary,* depending on whether they have one, two, or three organic groups bonded to nitrogen. These amines can all add hydrogen to form *ammonium ions,* which have four bonds to the nitrogen, which bears a single positive charge. Ions with four organic groups bonded to nitrogen are known as *quaternary ammonium ions.* In *heterocyclic amines,* the nitrogen of the amine group is bonded to two carbon atoms that are part of a ring.

2. **How are amines named?** Primary amine names have *-amine* added to the alkyl group name, and secondary and tertiary amines with identical R groups have *di-* and *tri-* prefixes. When the R groups are different, amines are named as *N-substituted derivatives* of the amine with the largest R group. Ions derived from amines are named by replacing *-amine* in the name with *-ammonium.* The $—NH_2$ group as a substituent is called an *amino group.*

3. **What are the general properties of amines?** Amines have an unshared electron pair on nitrogen that is available to accept a proton or for hydrogen bonding. Primary and secondary amine molecules hydrogen-bond to each other, but tertiary amine molecules cannot do so. Thus, the general order of boiling points for molecules of comparable size is

Hydrocarbons $<$ Tertiary amines $<$ Primary and secondary amines $<$ Alcohols

All amines can, however, hydrogen-bond to other kinds of molecules, and for this reason small amine molecules are water-soluble. Many amines are physiologically active. Volatile amines have strong, unpleasant odors.

4. **How do amines react with water and acids?** Amines are weak bases and establish equilibria with water by adding H^+ to form ammonium ions (RNH_3^+, $R_2NH_2^+$, R_3NH^+) and hydroxide ions (OH^-). They react with acids to form ammonium salts. Ammonium ions react as acids (proton donors) in the presence of a base. *Quaternary ammonium ions* (R_4N^+) have no lone electron pair and are not bases, nor can they form hydrogen bonds.

5. **What are alkaloids?** *Alkaloids* are naturally occurring nitrogen compounds found in plants. They are all bases, most with a bitter taste. Like other amines, many are physiologically active, notably as poisons or analgesics.

Summary of Reactions

1. Reactions of amines (Section 15.4)

(a) Acid–base reaction with water:

$$CH_3CH_2NH_2 + H_2O \rightleftharpoons CH_3CH_2NH_3^+ + OH^-$$

(b) Acid–base reaction with a strong acid to yield an amine salt:

$$CH_3CH_2NH_2 + HCl \longrightarrow CH_3CH_2NH_3^+Cl^-$$

2. Reaction of ammonium ion (Section 15.4) or amine salt (Section 15.5)

Acid–base reaction of primary, secondary, or tertiary amine salt (or ion) with a base to regenerate the amine:

$$CH_3CH_2NH_3^+Cl^- + NaOH \longrightarrow CH_3CH_2NH_2 + NaCl + H_2O$$

■ Understanding Key Concepts

15.21

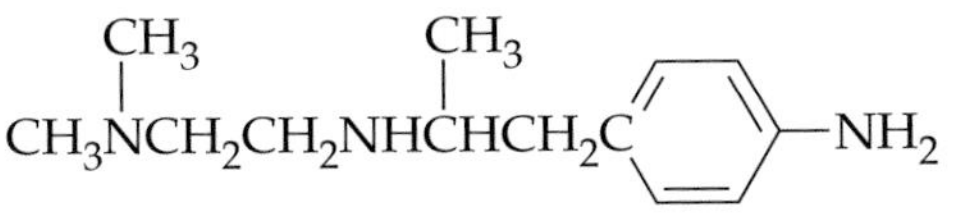

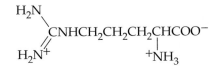

(a) For the compound above, identify each nitrogen as part of a primary, secondary, or tertiary amine.
(b) Which amine group(s) would be able to participate in hydrogen bonding?

15.22 The structure of the amino acid arginine is shown below.

$$H_2N \diagdown$$
$$C{=}NHCH_2CH_2CH_2CHCOO^-$$
$$H_2N^+ \diagup {}^+NH_3$$

(a) Which amine groups would be able to participate in hydrogen bonding?
(b) Is arginine likely to be water-soluble? Explain.

15.23 Draw structures to illustrate hydrogen bonding (similar to those on p. 425) between the following compounds.

(a) $CH_3{-}NH{-}CH_2{-}CH_2{-}CH_3$ and
$CH_3{-}NH{-}CH_2{-}CH_2{-}CH_3$

(b) 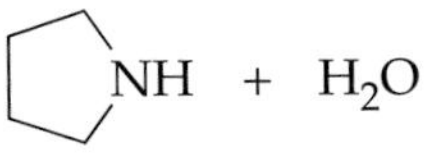 ⬠NH + H_2O

(c) $NH_3{-}CH_2{-}CH_2{-}CH_3$ and ⬠NH

15.24 Explain what bonds must be made or broken and where the electrons go when the hydrogen bonded water between the two amines shown at the top of p. 425 reacts to form an amine, ammonium ion, and OH^-.

15.25 Arrange the following compounds in the order of decreasing base strength.

$$NH_3 \qquad (CH_3)_2NH \qquad \text{⬡}{-}NH_2$$

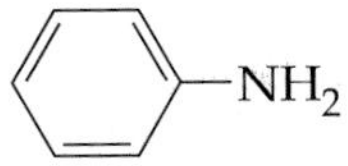

15.26 Complete each of the following equations:

(a) 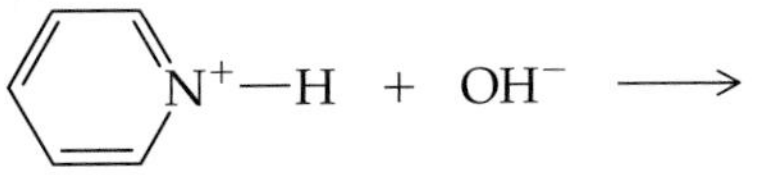 $\text{⬡}N^+{-}H + OH^- \longrightarrow$

(b) 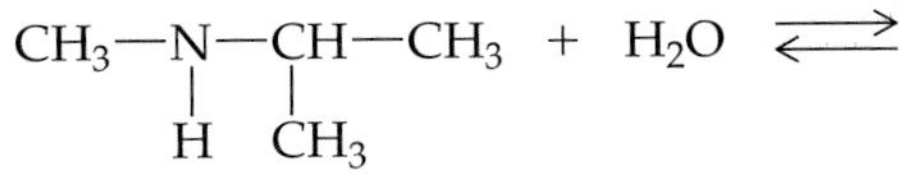 $CH_3{-}\underset{\underset{H}{|}}{N}{-}\underset{\underset{CH_3}{|}}{CH}{-}CH_3 + H_2O \rightleftharpoons$

(c) $(CH_3)_4N^+ + OH^- \longrightarrow$

(d) 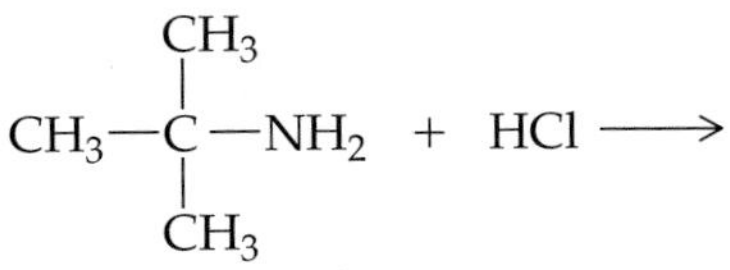 $CH_3{-}\underset{\underset{CH_3}{|}}{\overset{\overset{CH_3}{|}}{C}}{-}NH_2 + HCl \longrightarrow$

■ Additional Problems

Amines and Ammonium Salts

15.27 Draw the structures corresponding to the following names:
(a) Propylamine
(b) Diethylamine
(c) *N*-Methylpropylamine

15.28 Draw the structures corresponding to the following names:
(a) *N*-Methylbutylamine
(b) *N*-Ethylcyclopentylamine
(c) *m*-Propylaniline

15.29 Name the following amines, and identify them as primary, secondary, or tertiary:

(a) $CH_3CH_2CH_2CH_2NH_2$ (b) 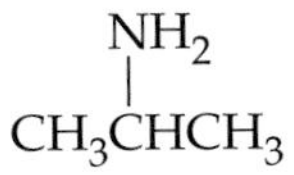 $CH_3\underset{\underset{NH_2}{|}}{CH}CH_3$

15.30 Name the following amines, and identify them as primary, secondary, or tertiary:

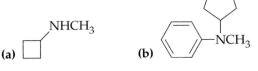

(a)   (b)

15.31 Is water a weaker or stronger base than ammonia?

15.32 Which is a stronger base, diethyl ether or diethylamine?

15.33 Give names for structures, and structures for names, of the following ammonium salts. Indicate whether each is the ammonium salt of a primary, secondary, or tertiary amine.

(a) $CH_3NH_3^+ Cl^-$ (b) 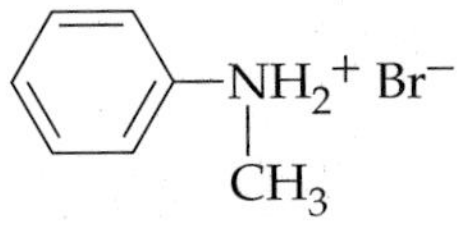 $\text{⬡}{-}\underset{\underset{CH_3}{|}}{N}H_2^+ Br^-$

(c) *N*-Propylbutylammonium bromide

15.34 Give names for structures, and structures for names, of the following ammonium salts. Indicate whether each is the ammonium salt of a primary, secondary, or tertiary amine.

(a)

(b) Anilinium chloride

(c) *N*-Ethyl-*N*-isopropylhexylammonium chloride

15.35 The compound L-dopa is used medically for its potent activity against Parkinson's disease, a chronic disease of the central nervous system. Identify the functional groups present in L-dopa.

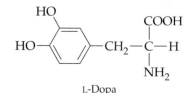

L-Dopa

15.36 Identify the functional groups in cocaine.

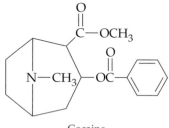

Cocaine

15.37 Most illicit cocaine is actually cocaine hydrochloride—the product of the reaction of cocaine (Problem 15.36) with HCl. Show the structure of cocaine hydrochloride.

15.38 When quinine (an antimalarial drug, p. 427) reacts with HCl, which nitrogen is first to form the ammonium salt? Show the structure of quinine hydrochloride.

Reactions of Amines

15.39 Complete the following equations:

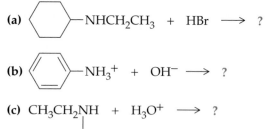

15.40 Complete the following equations:

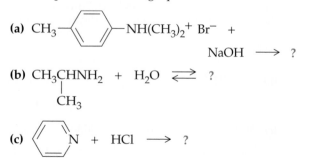

15.41 Many hair conditioners contain an ammonium salt such as the following to help prevent "fly-away" hair. Will this salt react with acids or bases? Why or why not?

$$CH_3(CH_2)_{15} \quad CH_3$$
$$\overset{+}{N} \qquad Cl^-$$
$$CH_3(CH_2)_{15} \quad CH_3$$

15.42 Choline has the following structure. Do you think that this substance reacts with aqueous hydrochloric acid? If so, what is the product? If not, why not?

$$HO-CH_2-CH_2-\overset{+}{N}(CH_3)_3$$

Applications

15.43 The *Handbook of Chemistry and Physics* indicates that adenine has a solubility of 0.09 g/100 mL of cold water, is slightly soluble in alcohol, and is insoluble in chloroform and ether. If you wished to extract caffeine, but not adenine, from ground coffee beans, what would be your solvent of choice? Why? [*Chemical Information*]

15.44 In the last 10 years or so, there has been a lot of interest in NO (nitric oxide). What are five functions that have been attributed to NO? [*NO: A Small Molecule with Big Responsibilities*]

15.45 Which of the following drug preparations are more readily absorbed? [*Organic Compounds in Body Fluids and the "Solubility Switch"*]

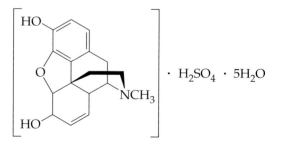

Morphine sulfate

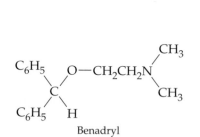

Benadryl

Thorazine hydrochloride

15.46 Promazine, a potent antipsychotic tranquilizer, is administered as the hydrochloride salt. Write the formula of the salt (there is only one HCl in the salt). [*Organic Compounds in Body Fluids and the "Solubility Switch"*]

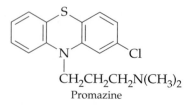

Promazine

15.47 Turn to the citric acid cycle (Figure 10.9, Section 10.8) and list the names of the intermediates in the cycle. Why are they not listed as acids? [*Organic Compounds in Body Fluids and the "Solubility Switch"*]

15.48 (a) Should you decide to pursue toxicology, what are three major areas in which you might work?

(b) As you study a new toxin, what are three questions that need to be answered so that you can better understand it, and hopefully develop an antidote for the toxin? [*Toxicology*]

General Questions and Problems

15.49 1-Propylamine, 1-propanol, acetic acid, and butane have about the same molar masses. Which would you expect to have the (a) highest boiling point, (b) lowest boiling point, (c) least solubility in water, and (d) least chemical reactivity? Explain.

15.50 Explain why decylamine is much less soluble in water than ethylamine.

15.51 Propose structures for amines that fit these descriptions:

(a) A secondary amine with formula $C_5H_{13}N$

(b) A tertiary amine with formula $C_6H_{13}N$

15.52 *para*-Aminobenzoic acid (PABA) is a common ingredient in sunscreens. Draw the structure of PABA.

15.53 PABA (Problem 15.52) is used by certain bacteria as a starting material from which folic acid (a necessary vitamin) is made. Sulfa drugs such as sodium sulfanilamide work because they resemble PABA. The bacteria try to metabolize the sulfa drug, fail to do so, and die due to lack of folic acid.

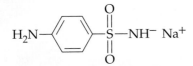

(a) Describe how this structure is similar to that of PABA.

(b) Why do you think the sodium salt, rather than the neutral compound, is used as the drug?

15.54 Acyclovir is an antiviral drug used to treat herpes infections. It has the following structure:

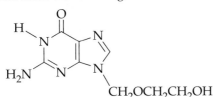

(a) What heterocyclic base (Table 15.2) is the parent of this compound?

(b) Label the other functional groups present.

15.55 Which is the stronger base, trimethylamine or pyridine? In which direction will the following reaction proceed?

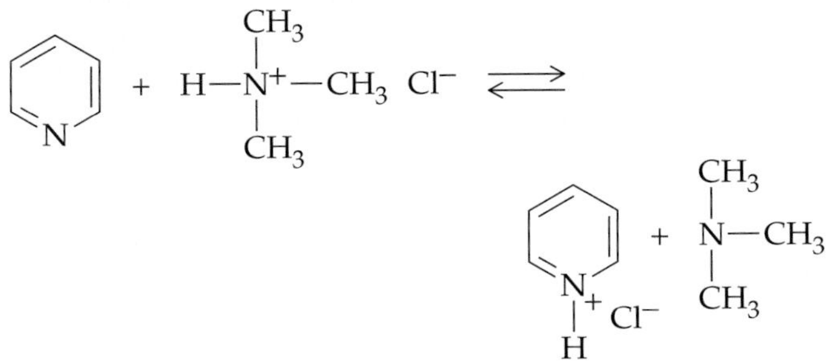

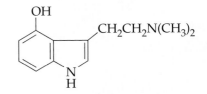

15.56 How do amines differ from analogous alcohols in (a) odor, (b) basicity, and (c) boiling point?

15.57 What two undesirable characteristics are often associated with alkaloids?

15.58 Name the following compounds:

(a) 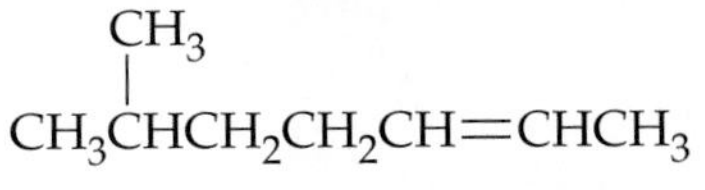 $\overset{\overset{\displaystyle CH_3}{|}}{CH_3CHCH_2CH_2CH}=CHCH_3$

(b) 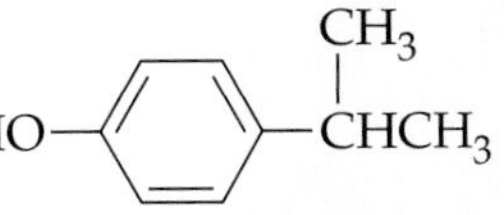 HO—⟨benzene ring⟩—$\overset{\overset{\displaystyle CH_3}{|}}{CHCH_3}$

(c) $(CH_3CH_2CH_2CH_2)_2NH$

15.59 Complete the following equations:

(a) 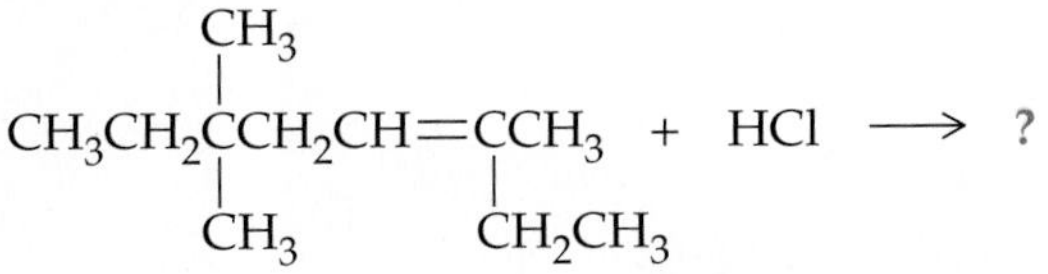 $CH_3CH_2\overset{\overset{\displaystyle CH_3}{|}}{\underset{\underset{\displaystyle CH_3}{|}}{C}}CH_2CH=\overset{\overset{\displaystyle }{}}{\underset{\underset{\displaystyle CH_2CH_3}{|}}{C}}CH_3$ + HCl ⟶ ?

(b) 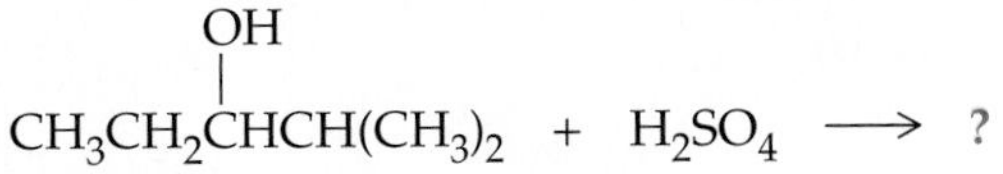 $CH_3CH_2\overset{\overset{\displaystyle OH}{|}}{CH}CH(CH_3)_2$ + H_2SO_4 ⟶ ?

(c) $2\ CH_3CH_2SH \overset{[O]}{\longrightarrow}$?

(d) 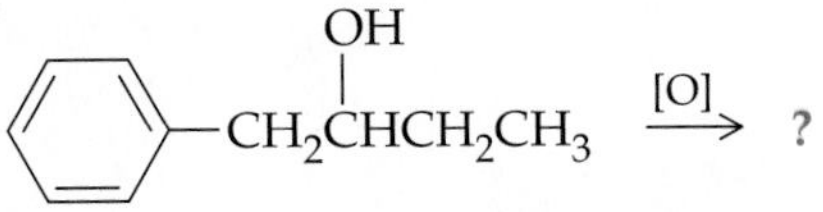 ⟨benzene ring⟩—$CH_2\overset{\overset{\displaystyle OH}{|}}{CH}CH_2CH_3 \overset{[O]}{\longrightarrow}$?

(e) $(CH_3)_3N$ + H_2O ⟶ ?

(f) $(CH_3)_3N$ + HCl ⟶ ?

(g) $(CH_3)_3NH^+$ + OH^- ⟶ ?

15.60 Hexylamine and triethylamine have the same molar mass. The boiling point of hexylamine is 129°C, while that of triethylamine is only 89°C. Explain these observations.

15.61 Lemon juice, which contains citric acid, is traditionally recommended for removing the odor associated with cleaning fish. What functional group is responsible for a "fishy" odor, and why does lemon juice work to remove the odor?

15.62 Psilocin is a hallucinogenic compound that is isolated from mushrooms and has the structure shown below. What heterocyclic base (Table 15.2) is the parent of this compound?

OH
⟨indole ring structure⟩ $CH_2CH_2N(CH_3)_2$
N
H

15.63 Why is aniline not considered to be a heterocyclic nitrogen compound?

15.64 Benzene and pyridine are both single-ring, aromatic compounds. Benzene is a neutral compound that is insoluble in water. Pyridine, with a similar molar mass, is basic and completely miscible with water. Explain these phenomena.

15.65 Name the organic reactants in Problems 15.39 and 15.40.

16 Aldehydes and Ketones

The volatile chemicals in the fragrance of a rose are being measured by delicate sensors. Without doubt, many aldehydes and ketones will be detected.

In this and the next chapter, we move on from compounds with only single covalent bonds in their functional groups to several families of compounds that contain a *carbonyl group*. The carbonyl group has a carbon atom and an oxygen atom connected by a double bond, C=O. *Aldehydes* and *ketones*, the subject of this chapter, are the two simplest families of carbonyl compounds. In aldehydes, the carbonyl group is bonded to one carbon atom and a hydrogen atom, so that the—CHO group falls at one end of a molecule. In ketones, the carbonyl group is bonded to two carbon atoms and thus falls in the middle of a molecule.

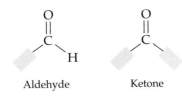

Aldehyde   Ketone

Where might you encounter aldehydes or ketones? The aromas of many flowers and plants derive largely from natural aldehydes, and we take advantage of this fact in many ways. Cookies baking in the oven fill the air with the aroma of cinnamon, vanilla, or almond—all natural flavors and aromas from aldehydes. Have you ever burned a citronella candle to repel mosquitoes? Or enjoyed sniffing a fresh rose? These too are the aromas of aldehydes. Among ketones, jasmone from the jasmine flower and muscone from the male musk deer are vital to the complex formulas of expensive perfumes. Aldehyde and ketone functional groups also play essential roles in the biomolecules that will be our focus later in this book.

In this chapter, we'll answer the following questions:

1. **What is the carbonyl group?**
   The goal: Be able to recognize the carbonyl group and describe its polarity and shape.
2. **How are ketones and aldehydes named?**
   The goal: Be able to name the simple members of these families and write their structures, given the names.
3. **What are the general properties of aldehydes and ketones?**
   The goal: Be able to describe such properties as polarity, hydrogen bonding, and water solubility.
4. **What are some of the significant occurrences and applications of aldehydes and ketones?**
   The goal: Be able to specify where aldehydes and ketones are found, list their major applications, and discuss some important members of each family.
5. **What are the results of the oxidation and reduction of aldehydes and ketones?**
   The goal: Be able to describe and predict the products of the oxidation and reduction of aldehydes and ketones.
6. **What are hemiacetals and acetals, and how are they formed?**
   The goal: Be able to recognize hemiacetals and acetals, describe the conditions under which they are formed, and predict the products of hemiacetal and acetal formation and acetal hydrolysis.

CONCEPTS TO REVIEW

Electronegativity and molecular polarity (Sections 5.8, 5.9)
Hydrogen bonding (Section 8.11)
Naming alkanes (Section 12.6)

16.1 The Carbonyl Group

Carbonyl compounds are distinguished by the presence of a **carbonyl group** (C=O) and are classified according to what is bonded to the carbonyl carbon, as illustrated in Table 16.1.

Carbonyl compound Any compound that contains a carbonyl group C=O.

Carbonyl group A functional group that has a carbon atom joined to an oxygen atom by a double bond, $\diagdown$C=O.

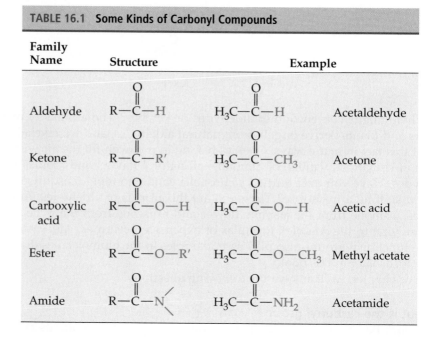

TABLE 16.1   Some Kinds of Carbonyl Compounds

Family Name	Structure	Example	
Aldehyde	R—C(=O)—H	H$_3$C—C(=O)—H	Acetaldehyde
Ketone	R—C(=O)—R′	H$_3$C—C(=O)—CH$_3$	Acetone
Carboxylic acid	R—C(=O)—O—H	H$_3$C—C(=O)—O—H	Acetic acid
Ester	R—C(=O)—O—R′	H$_3$C—C(=O)—O—CH$_3$	Methyl acetate
Amide	R—C(=O)—N	H$_3$C—C(=O)—NH$_2$	Acetamide

Since oxygen attracts electrons more strongly than carbon (it's more electro-negative; Section 5.8), carbonyl groups are strongly polarized, with a partial positive charge on the carbon atom and a partial negative charge on the oxygen atom. (∞, p. 119) As described in this and the next chapter, the polarity of the carbonyl group contributes to its reactivity.

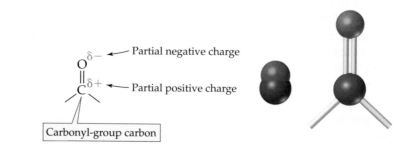

Another property common to all carbonyl groups is planarity. The carbonyl carbon atom is surrounded by three regions of electron density, and the bond angles between the three substituents on carbon are 120°, or close to it.

120° angles, in a planar triangle

Aldehyde   A compound that has a carbonyl group bonded to one carbon and one hydrogen, RCHO.

Ketone   A compound that has a carbonyl group bonded to two carbons in organic groups that can be the same or different, R$_2$C=O, RCOR′.

It's useful to divide carbonyl compounds into two groups based on their chemical properties. In one group are **aldehydes** and **ketones**, which have similar properties because their carbonyl groups are bonded to atoms that don't attract electrons strongly—carbon and hydrogen. In the second group are *carboxylic acids*, *esters*, and *amides*. The carbonyl-group carbon in these compounds is bonded to an oxygen or nitrogen atom, which *does* attract electrons strongly. This second group of carbonyl-containing compounds is discussed in Chapter 17.

Because of the planar, triangular arrangement of atoms around the carbonyl group, there are various ways of representing carbonyl compound structures on paper. Sometimes the bonds of the carbonyl carbon are written at 120° angles to remind us that such angles are present in the molecules. The structures in Table 16.1, which emphasize the location of the double bond, do not fit well on a single line of type, so the simplified formulas shown below are often used for aldehydes and ketones:

Aldehydes

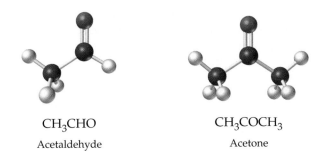

O
‖
R—C—H   RCHO

Ketones

O
‖
R—C—R′   RCOR′ or R₂C=O

For example:

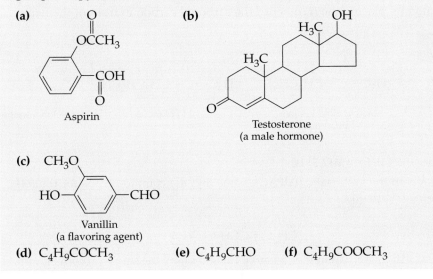

CH₃CHO

Acetaldehyde

CH₃COCH₃

Acetone

The aldehyde group, you will notice, can only be connected to one carbon atom and therefore is always at the end of a carbon chain. The ketone group, by contrast, must be connected to two groups, and thus always occurs within a carbon chain.

■ **PROBLEM 16.1**

Which of the following molecules contain aldehyde or ketone functional groups? Copy the formulas and circle these functional groups.

(a)

O
‖
OCCH₃

—COH
‖
O

Aspirin

(b)

OH

H₃C

H₃C

O

Testosterone
(a male hormone)

(c) CH₃O

HO—

—CHO

Vanillin
(a flavoring agent)

(d) C₄H₉COCH₃   (e) C₄H₉CHO   (f) C₄H₉COOCH₃

■ **PROBLEM 16.2**

Draw the structures of compounds (d) and (e) in Problem 16.1 to show all individual atoms and all covalent bonds.

∞ **LOOKING AHEAD**

Aldehyde or ketone groups are present in biomolecules with a wide range of functions, from the steroid hormones that regulate sexual function (Section 20.5), to the bases that are essential to nucleic acids and the genetic code (Section 26.2). Most distinctively, the structure and reactions of aldehydes and ketones are fundamental to the chemistry of carbohydrates, those in our diet and those that provide energy and structure to our bodies (Chapters 22, 23).   ∞

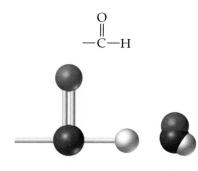

16.2 Naming Aldehydes and Ketones

The simplest aldehydes are known by their common names, which end in *-aldehyde*, for example, formaldehyde, acetaldehyde, and benzaldehyde. To name aldehydes systematically in the IUPAC system, the final *-e* of the name of the alkane with the same number of carbons is replaced by *-al*. The three-carbon aldehyde derived from propane is named systematically as propanal, the four-carbon aldehyde as butanal, and so on. When substituents are present, the chain is numbered beginning with 1 for the carbon atom at the —CHO end, as illustrated below for 3-methylbutanal.

Aldehydes

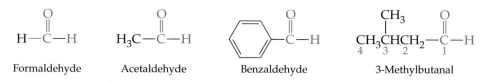

Formaldehyde Acetaldehyde Benzaldehyde 3-Methylbutanal

The simplest ketone is almost always referred to as acetone. Some ketones are best known by common names that give the names of the two alkyl groups bonded to the carbonyl carbon followed by the word *ketone*—for example, methyl ethyl ketone, shown below. Ketones are named systematically by replacing the final *-e* of the corresponding alkane name with *-one* (pronounced *own*). The numbering of the alkane chain begins at the end nearest the carbonyl group. As shown here for 2-pentanone, the location of the carbonyl group is indicated by placing the number of the carbonyl carbon in front of the name.

Ketones

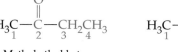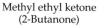

Acetone Methyl ethyl ketone (2-Butanone) 2-Pentanone Cyclohexanone

■ **WORKED EXAMPLE 16.1**

Give both the systematic (IUPAC), name and the common name for the following compound:

$$CH_3CH_2CCH_2CH_2CH_3$$

ANALYSIS The compound is a ketone, as shown by the single carbonyl group bonded to two alkyl groups. The IUPAC system identifies and numbers carbon chain to indicate where the carbonyl group is located.

$$CH_3CH_2CCH_2CH_2CH_3$$
1 2 3 4 5 6

The common name relies on naming the two alkyl groups.

SOLUTION The systematic name is 3-hexanone. The common name is ethyl propyl ketone.

■ **PROBLEM 16.3**

Draw structures corresponding to the following names:
(a) Hexanal
(b) Methyl phenyl ketone
(c) 4-Methyl-2-hexanone
(d) Isobutyl methyl ketone

Application Chemical Warfare Among the Insects

Life in the insect world is a jungle. Predators abound, just waiting to make a meal of any insect that happens along. To survive, insects have evolved extraordinarily effective means of chemical protection. Take the humble millipede *Apheloria corrugata*, for example. When attacked by ants, the millipede protects itself by discharging benzaldehyde cyanohydrin.

Cyanohydrins [RCH(OH)C≡N] are formed by addition of the toxic gas HCN (hydrogen cyanide) to ketones or aldehydes. Like the reaction of a ketone or aldehyde with an alcohol to yield a hemiacetal (Section 16.7), the reaction with HCN to yield a cyanohydrin is reversible. Thus, the benzaldehyde cyanohydrin secreted by the millipede can decompose to yield benzaldehyde and HCN. The millipede is protected when this decomposition reaction releases deadly cyanide gas, a remarkably clever and very effective kind of chemical warfare.

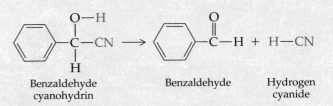

Benzaldehyde cyanohydrin → Benzaldehyde + Hydrogen cyanide

The potent chemical weapon of the bombardier beetle is benzoquinone, the simplest member of a class of compounds that are cyclohexadienediones (cyclohexene rings with two double bonds and two carbonyl groups). When threatened, the bombardier beetle initiates the enzyme-catalyzed oxidation of a dihydroxybenzene by hydrogen peroxide. A cloud of irritating benzoquinone vapor shoots out of the beetle's defensive organ at up to 100°C, with such force that it sounds like a pistol shot.

See Additional Problems 16.48–49 at the end of the chapter.

▲ **A bombardier beetle defending itself**. The beetle is spraying boiling hot benzoquinone at a predator. The benzoquinone is produced in a fraction of a second by the oxidation of dihydroxybenzene.

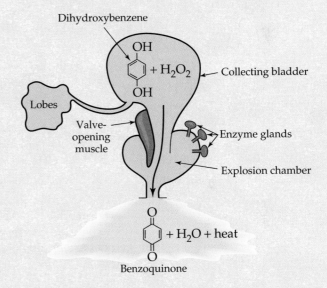

▲ **Defensive organ of the bombardier beetle.**

■ **PROBLEM 16.4**

Give systematic, IUPAC names for the following compounds:

(a) $CH_3CH_2CH_2CH_2\overset{\displaystyle O}{\overset{\displaystyle \|}{C}}H$

(b) $CH_3CH_2\overset{\displaystyle O}{\overset{\displaystyle \|}{C}}CH_2CH_3$

(c) $CH_3CH_2\overset{\displaystyle CH_3}{\overset{\displaystyle |}{C}}HCH_2CH_2\overset{\displaystyle O}{\overset{\displaystyle \|}{C}}H$

(d) Dipropyl ketone

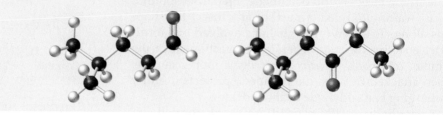

Which of the these two molecules is a ketone and which is an aldehyde? Write the condensed formulas for both of them.

16.3 Properties of Aldehydes and Ketones

The polarity of the carbonyl group makes aldehydes and ketones moderately polar compounds. As a result, they are higher boiling than alkanes with similar molecular weights. Because they have no hydrogen atoms bonded to oxygen or nitrogen, however, their molecules don't hydrogen-bond with each other, and this makes aldehydes and ketones lower boiling than alcohols. In a series of compounds with similar molecular weights, the alkane is lowest boiling, the alcohol is highest boiling, and the aldehyde and ketone fall in between.

$CH_3CH_2CH_2CH_3$

Butane, bp 0°C

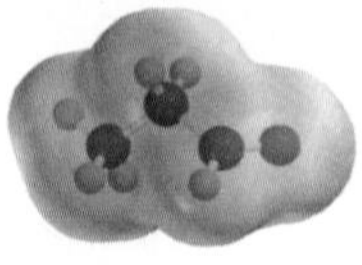

$$CH_3CH_2\overset{\displaystyle O}{\overset{\|}{C}}H$$

Propanal, bp 50°C

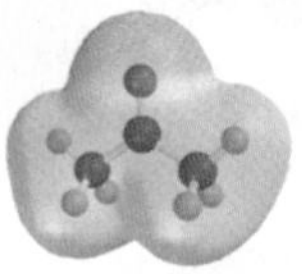

$$CH_3\overset{\displaystyle O}{\overset{\|}{C}}CH_3$$

Acetone, bp 56°C

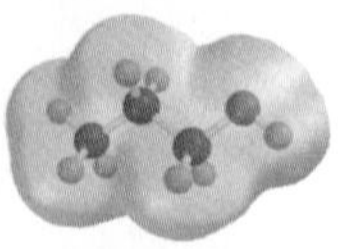

$CH_3CH_2CH_2OH$

Propanol, bp 97°C

Formaldehyde (HCHO), the simplest aldehyde, is a gas, and acetaldehyde (CH_3CHO) boils close to room temperature. The other simple aldehydes and ketones are liquids (Table 16.2), and those with more than 12 carbon atoms are solids.

TABLE 16.2 Physical Properties of Some Simple Aldehydes and Ketones

Structure	Name	Boiling Point (°C)	Water Solubility (g/100 mL H_2O)
HCHO	Formaldehyde	−21	55
CH_3CHO	Acetaldehyde	21	Soluble
CH_3CH_2CHO	Propanal	49	16
$CH_3CH_2CH_2CHO$	Butanal	76	7
$CH_3CH_2CH_2CH_2CHO$	Pentanal	103	1
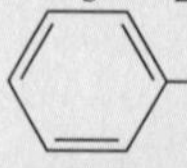 ⬡—CHO	Benzaldehyde	178	0.3
CH_3COCH_3	Acetone	56	Soluble
$CH_3CH_2COCH_3$	2-Butanone	80	26
$CH_3CH_2CH_2COCH_3$	2-Pentanone	102	6
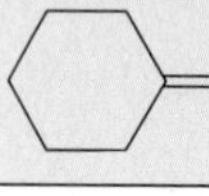 ⬡=O	Cyclohexanone	156	2

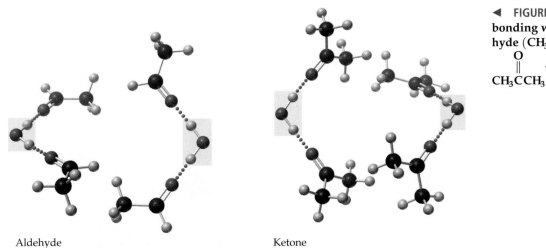

Aldehyde   Ketone

Aldehydes and ketones are soluble in common organic solvents, and those with fewer than five carbon atoms are also soluble in water because they're able to hydrogen-bond with water molecules (Figure 16.1).

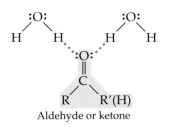

Aldehyde or ketone

▲ A perfume that might contain citronellal or civetone.

Simple ketones are excellent solvents because they dissolve both polar and nonpolar compounds. With increasing numbers of carbon atoms, aldehydes and ketones become more alkane-like and less water-soluble.

The structures of a few naturally occurring aldehydes and ketones with distinctive odors are shown below. Citronellal (used in citronella candles) is one of about a dozen compounds with similar structures that contribute to the aroma of oils extracted from geraniums, roses, citronella (a tropical plant), and lemon grass. All are used in soaps, cosmetics, and perfumes.

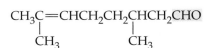

Citronellal
(insect repellant, also used
in perfumes; from citronella
and lemon grass oils)

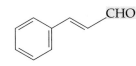

Cinnamaldehyde
(cinnamon flavor in
foods, drugs; from
cinnamon bark)

Camphor
(cools the skin in
liniment and itch remedies,
but is not safe at over 2.5%
concentration; from
the camphor tree)

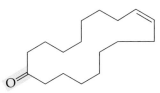

Civetone
(musky odor in perfumes;
from the scent gland of
the civet cat)

Camphor has been used in remedies for thousands of years but has been judged ineffective in most situations by the U.S. Food and Drug Administration. Although camphor does cool the skin, it appears to have little if any therapeutic value, despite its "medicinal" smell. It is, however, used effectively to repel moths.

The lower-boiling aldehydes and ketones are flammable and form explosive mixtures with air. The simple ketones have generally low toxicity, while the simple aldehydes, especially formaldehyde, are toxic.

Application Vanilla: What Kind Is Best?

You're standing at the flavoring shelf in the supermarket, thinking about the lovely pound cake you're going to bake. Which should you buy, "artificial vanilla" or "natural vanilla flavoring"? Or maybe those very expensive real vanilla beans? Chemical logic tells you that an "artificial" flavoring compound—that is, one made synthetically from other chemicals—should have exactly the same taste and aroma as the identical compound obtained from a plant. That is precisely correct. Pure vanillin, the principal flavoring ingredient in vanilla extract, has the same structure and properties, whatever its source.

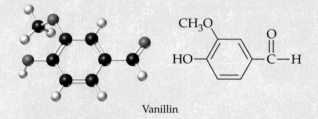

Vanillin

▲ Vanilla beans at a local market.

But the other contents of the two bottles are not the same, and herein lies the difference. Synthetic vanillin is made from waste lignin, a by-product of wood pulp that contains mostly aromatic rings. The "artificial" vanilla contains no flavoring agents other than this synthetic vanillin. The natural flavoring is extracted from vanilla beans that have been dried in the sun. There is about 1 g of vanillin in 400 g of vanilla beans. Some flavorful components from the other 399 g also come over into the extract. The resulting richer flavor is, like that of all natural flavorings and aromas, a blend of the contributions from many chemical compounds—sometimes thousands.

See Additional Problem 16.50 at the end of the chapter.

Properties of Aldehydes and Ketones

- No hydrogen bonding occurs between aldehyde or ketone molecules, but these molecules are polar.
- Aldehydes and ketones are lower boiling than alcohols, but higher boiling than alkanes.
- Common aldehydes and ketones are liquids.
- Simple aldehydes and ketones are water-soluble due to hydrogen bonding with water molecules, and ketones are good solvents.
- Volatile aldehyde and ketone vapors are flammable.
- Many aldehydes and ketones have distinctive odors.
- Simple ketones are less toxic than simple aldehydes.

■ **KEY CONCEPT PROBLEM 16.6**

For the compound shown below, choose which of each pair of properties is most likely applicable.

$$\text{CH}_3\text{CH}_2\overset{\displaystyle \overset{\text{O}}{\|}}{\text{C}}\text{CH}_2\text{CH}_3$$

(a) Polar or nonpolar

(b) Flammable or nonflammable

(c) Solid or liquid

(d) A bp of 250°C or a bp of 100°C

16.4 Some Common Aldehydes and Ketones

Formaldehyde (HCHO): Toxic but Useful

At room temperature, pure formaldehyde is a colorless gas with a pungent, suffocating odor. Low concentrations in the air (0.1–1.1 ppm) can cause eye, throat, and bronchial irritation, and higher concentrations can trigger asthma attacks. Skin contact can produce dermatitis. Because formaldehyde is formed during incomplete combustion of hydrocarbon fuels, it's partly responsible for the irritation caused by smog-laden air. Formaldehyde is very toxic by ingestion, and as a product of the biochemical breakdown of methanol it can cause serious kidney damage, coma, and death.

Formaldehyde is commonly sold as a 37% (weight/weight percent, meaning 37 g of formaldehyde in 100 g of solution) aqueous solution under the name *formalin*. It kills viruses, fungi, and bacteria by reaction with the $-NH_2$ groups in proteins, allowing for its use in disinfecting and sterilizing equipment. It is too harsh for use on the skin. On standing, formaldehyde polymerizes into a solid known as *paraformaldehyde*. At one time, paraformaldehyde candles were burned to disinfect hospital rooms that had been occupied by patients with contagious diseases.

In the chemical industry, the major use of formaldehyde is in production of polymers with applications such as adhesives for binding plywood, foam insulation for buildings, textile finishes, and hard and durable manufactured objects such as telephone parts. The first completely synthetic and commercially successful plastic was a polymer of phenol and formaldehyde known as *Bakelite*, once widely used for such items as pot handles, fountain pens, and cameras. Urea–formaldehyde polymers are now more widely used than Bakelite. Once the final polymerization of such materials is finished, no further melting and reshaping is possible because of the cross-linked, three-dimensional structure. In the general formulas below, the red wavy lines indicate bonds to the rest of the polymer; the CH_2 groups are from the formaldehyde molecules.

▲ **A shark preserved in formaldehyde.** This shark and its tank were part of an exhibit in a major art museum.

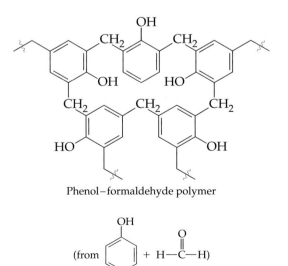

Phenol–formaldehyde polymer

Urea–formaldehyde polymer

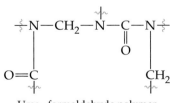

Formaldehyde polymers, especially when new, release formaldehyde into the air. Because of concern over the toxicity and possible carcinogenicity of formaldehyde from polymeric materials, their use in some household applications has been limited.

Products made of Bakelite, the ▶ first commercially successful polymer. Bakelite is resistant to electricity, heat, and chemical deterioration, and is easily molded. It was widely used in electrical equipment, heat-resistant handles, fountain pens, cameras, and even jewelry.

Acetaldehyde (CH_3CHO): Sweet Smelling but Narcotic

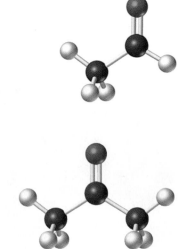

Acetaldehyde is a sweet-smelling, flammable liquid present in ripe fruits, notably apples. It is less toxic than formaldehyde, and small amounts are produced in the normal breakdown of carbohydrates. Acetaldehyde is, however, a general narcotic, and large doses can cause respiratory failure. Chronic exposure produces symptoms like those of alcoholism. (See Application "Ethyl Alcohol as a Drug and a Poison," Chapter 14.)

Acetone (CH_3COCH_3): A Super Solvent

Acetone, a liquid at room temperature, is perhaps the most widely used of all organic solvents. It can dissolve most organic compounds and is also miscible with water. In paint stores, acetone is sold for general-purpose cleanup work, and it is the solvent in many varnishes, lacquers, and nail polish removers.

Acetone is highly volatile and is a serious fire and explosion hazard when allowed to evaporate in a closed space. No chronic health risk has been associated with acetone exposure. Unfortunately, however, it is one of a large group of readily available products that include volatile solvents (some others are benzene, chloroform, model-airplane glue, and nail polish remover) that are inhaled to produce alcohol-like intoxication, sometimes with serious outcomes.

When the biochemical breakdown of fats and carbohydrates to yield energy is out of balance (for example, in starvation or diabetes mellitus), acetone is produced in the liver, a condition that in severe cases leaves the odor of acetone on a patient's breath (◯◯◯ Section 25.7).

◯◯◯ LOOKING AHEAD

In the biochemistry chapters that lie ahead, you'll find that all of the simplest sugars—the monosaccharides (Section 22.4)—contain either an aldehyde group or a ketone group. Glucose, the six-carbon sugar shown below, plays a major role in metabolism as the primary fuel molecule for energy generation (Section 23.2)   ◯◯◯

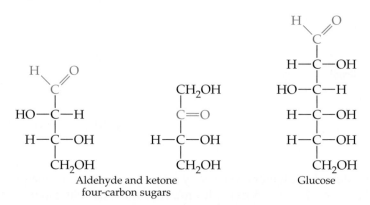

Aldehyde and ketone four-carbon sugars

Glucose

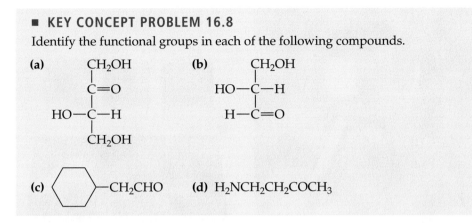

■ **KEY CONCEPT PROBLEM 16.8**
Identify the functional groups in each of the following compounds.

16.5 Oxidation of Aldehydes

Alcohols can be oxidized to aldehydes or ketones (∞, p. 398), and aldehydes can be further oxidized to carboxylic acids. In aldehyde oxidation, the hydrogen bonded to the carbonyl carbon is replaced by an —OH group. Ketones, because they don't have this hydrogen, do not react with oxidizing agents (except with those strong enough to destroy the molecule).

For example:

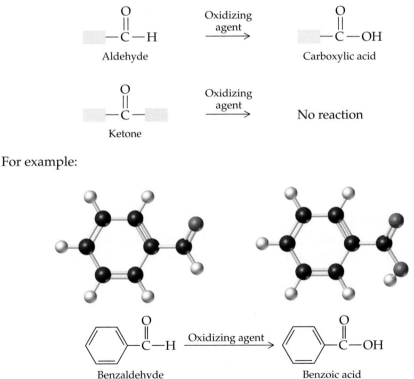

Benzaldehyde   Benzoic acid

Among the mild oxidizing agents that cause the conversion from aldehyde to carboxylic acid, one is oxygen in the air. To prevent air oxidation, aldehydes are often stored under nitrogen gas.

Because ketones cannot be oxidized, treatment with a mild oxidizing agent can be used as a test to distinguish between aldehydes and ketones. *Tollens' reagent* is the most visually appealing oxidizing agent for aldehydes. Treatment of an aldehyde with this reagent, in which the Ag$^+$ ion in [Ag(NH$_3$)$_2$]$^+$ is the oxidizing agent, rapidly yields the carboxylic acid anion and metallic silver. If the reaction is done in a clean glass container, metallic silver deposits on the

FIGURE 16.2 **The Tollens' and** ▶
Benedict's tests for aldehydes.
(left) In the Tollens' test, colorless
silver ion (Ag^+) is reduced to
metallic silver. (right) In the Benedict's test for aldehyde-containing
sugars, the blue copper(II) ion
(Cu^{2+}) is reduced to copper(I) in
brick-red copper(I) oxide (Cu_2O).
In the photo, the Cu^{2+} containing
reagent solution is on the left. A few
drops of glucose produce the greenish color in the center. A large quantity of glucose produces the
brick-red precipitate on the right. In
both tests, the aldehyde is oxidized
to the carboxylic acid.

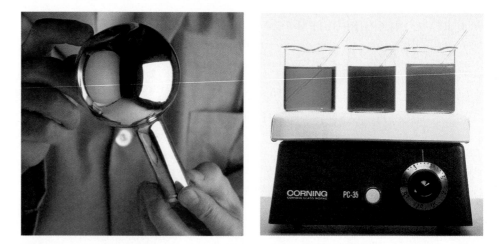

inner walls, producing a beautiful shiny mirror (Figure 16.2, left). Before modern instrumental methods were available, chemists had to rely on such visible
chemical changes to identify chemical compounds.

Tollens' test

$$RCHO + Ag(NH_3)_2^+ \xrightarrow{NH_3,\ H_2O} RCOO^- + NH_4^+ + Ag$$

Tollens' reagent   Silver
(colorless)   mirror

A test with another mild oxidizing agent, known as *Benedict's reagent*, also
relies on reduction of a metal to produce visible evidence for the presence of
aldehydes. The reagent solution contains blue copper(II) ion that is reduced to
give a precipitate of red copper (I) oxide in the reaction with an aldehyde
(Figure 16.2, right). Benedict's reagent has been used extensively as a test for
sugars in the urine (although more specific tests are now preferred, as discussed in Application "Glucose in Blood and Urine," Chapter 23).

Benedict's test

$$RCHO + Cu^{2+} \xrightarrow{Buffer} RCOO^- + Cu_2O$$

Blue   Brick-red
in solution   solid

■ **PROBLEM 16.9**
Draw structures of the products you would obtain by treating the following
compounds with Tollens' reagent. If no reaction occurs, write "NR."

$$\overset{\displaystyle CH_3}{\underset{\displaystyle |}{}}$$

(a) $CH_3CHCH_2CH_2CH_2CHO$ **(b)** 2,2-Dimethylpentanal
(c) 2-Methyl-3-pentanone

16.6 Reduction of Aldehydes and Ketones

The reduction of a carbonyl group is the addition of hydrogen across the double bond to produce an —OH group, a reaction that is the reverse of the oxidation of an alcohol:

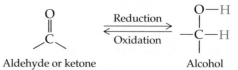

Aldehyde or ketone   Alcohol

Aldehydes are reduced to primary alcohols, and ketones are reduced to secondary alcohols:

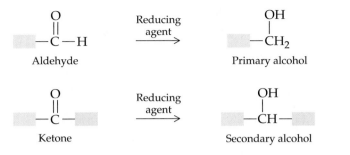

These reductions occur by formation of a bond to the carbonyl carbon atom by a hydride ion (H:⁻) accompanied by bonding of a hydrogen ion (H⁺) to the carbonyl oxygen atom. The reductions make good sense when you think about the polarity of the carbonyl group. The carbonyl-group carbon has a partial positive charge because electrons are drawn away by the electronegative oxygen atom, so the negatively charged hydride ion is drawn to this carbon atom. Because the oxygen atom has a partial negative charge, the positively charged hydrogen atom is attracted there.

$$O^{\delta-} \longleftarrow H^+ \quad \text{attracted here}$$
$$C^{\delta+} \longleftarrow :H^- \quad \text{attracted here}$$

Note that a hydride ion (H:⁻) has a lone pair of valence electrons. Both electrons are used to form a covalent bond to the carbonyl carbon. This change leaves a negative charge on the carbonyl oxygen. Aqueous acid is then added, H⁺ bonds to the oxygen, and a neutral alcohol results. Thus, the two new hydrogen atoms in the alcohol product come from different sources.

Reduction of an aldehyde

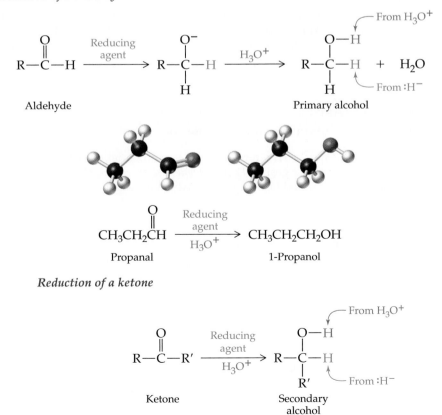

Reduction of a ketone

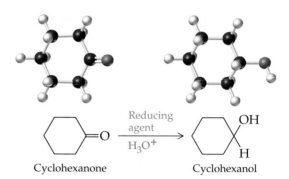

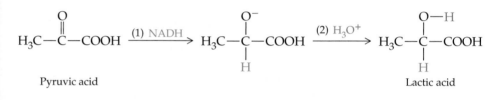

Cyclohexanone → Cyclohexanol

Reducing agent / H_3O^+

In biological systems, the reducing agent for a carbonyl group is often the coenzyme nicotinamide adenine dinucleotide, which cycles between reacting as a reducing agent (NADH) and an oxidizing agent (NAD^+) by the loss and gain of a hydride ion ($H:^-$). The biochemical reduction of pyruvic acid, a ketone-containing acid that plays a pivotal role in energy production, utilizes NADH (∞∞, Section 21.4). The reaction occurs in active skeletal muscles. Vigorous exercise causes a buildup of the reduction product, lactic acid, leading to a tired, flat feeling and sometimes to muscle cramps.

▲ Benzaldehyde is one of the hundreds of chemicals that contribute to the complex flavor of raspberries.

$$H_3C-\overset{O}{\underset{}{C}}-COOH \xrightarrow{\text{(1) NADH}} H_3C-\overset{O^-}{\underset{H}{C}}-COOH \xrightarrow{\text{(2) } H_3O^+} H_3C-\overset{O-H}{\underset{H}{C}}-COOH$$

Pyruvic acid   Lactic acid

∞∞ LOOKING AHEAD

The reduction of aldehydes and ketones to alcohols is an important reaction in living cells, and NADH is commonly the source of the hydride ion. It donates $H:^-$ to an aldehyde or ketone to yield an anion that then picks up H^+ from surrounding aqueous fluids. The major role of NADH as a biochemical reducing agent is introduced in Section 21.7. ∞∞

■ **WORKED EXAMPLE 16.2**

What product would you obtain by reduction of benzaldehyde?

SOLUTION First, draw the structure of the starting material, showing the double bond in the carbonyl group. Then rewrite the structure showing only a single bond between C and O, along with partial bonds to both C and O:

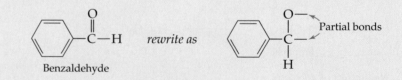

Benzaldehyde   *rewrite as*   Partial bonds

Finally, attach hydrogen atoms to the two partial bonds and rewrite the product:

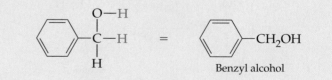

Benzyl alcohol

Application Is It Poisonous, or Isn't It?

In its broadest meaning, the term "drug" refers to any chemical agent other than food that affects living organisms. More commonly, the term is applied to substances that prevent or treat disease, or to addictive substances. By contrast, a "poison" or "toxic substance" is any chemical agent that harms living organisms. (See Application "Toxicology" in Chapter 15.) How can we tell one from the other?

Strange as it may seem, the categories of "drug" and "poison" are not mutually exclusive. Often a substance that in low concentrations can cure disease or alleviate symptoms can also cause injury or death when taken in larger amounts. The sixteenth century German physician Paracelsus understood that most substances can't be absolutely categorized as "safe" or "toxic." He expressed this insight in a famous phrase: "The dose makes the poison."

The term *dose* refers to the amount of a substance that enters the body at a single time. A small dose of a substance might be a life-saving medication, while a large dose of the same substance might be poisonous. Vitamin A, for example, is essential to our health, and our daily diet should include about 1 mg, but a single dose larger than 200 mg can cause nausea, fatigue, and other unpleasant symptoms and is classified as a toxic dose.

One standard method for evaluating toxicity, in use since the 1920s, is the LD_{50}, or lethal dose, test. The LD_{50} is a measure of the toxicity of a single dose, known as *acute* toxicity. (By contrast, *chronic* toxicity is the result of less-than-acute exposure over a long period of time.) A substance is fed in varying doses to laboratory animals, frequently rats or mice. The result of the test is reported as the LD_{50}, the dose that kills 50% of the animals in a uniform population, say all male rats with a similar genetic makeup being fed an identical diet. Because LD_{50} values are listed in standard chemical references, such as *The Merck Index* (see Application "Chemical Information," Chapter 15), they are the most easily found toxicity data.

By comparing LD_{50} values, relative toxicities can be evaluated. The LD_{50} values in the table, reported as the dose in grams per kilogram of body weight of the test animal, show that in this group of compounds formaldehyde is the most toxic and acetone the least toxic by ingestion.

	LD_{50}
Formaldehyde, 37% aqueous solution	800 mg/kg
Acetaldehyde	1.9 g/kg
Butyraldehyde	5.8 g/kg
Acetone	8.4 g/kg

To put these values in perspective, compare them with the LD_{50} of 2×10^{-4} mg/kg for the extremely toxic batrachotoxin, an arrow poison extracted from frogs in South America, or the LD_{50} of 0.23 mg/kg for muscarine, the poisonous chemical in *Amanita muscaria* mushrooms.

▲ *Amanita muscaria*, a highly toxic mushroom. To pick natural mushrooms, one must know very well the difference between poisonous and nonpoisonous varieties.

For therapeutic use in humans, a compound must show a comfortably wide margin between the dose that produces the desired effect and the dose that produces an acute toxic effect. Aspirin, for example, has an LD_{50} of 1.75 g/kg in rats. An average therapeutic dose for humans is two aspirin tablets, which contain only 650 mg, or 0.01 g/kg for an average 65 kg person—1/175 of the LD_{50}, if we assume that the results for rats can be extrapolated to human beings. This is a fairly comfortable safety margin, but not an extraordinarily generous one (overdoses of aspirin, accidental or deliberate, still cause a significant number of fatalities each year).

The LD_{50} test is controversial, it has many drawbacks and many advantages. Among the drawbacks are the need to sacrifice large numbers of animals; the wide variation of results with animal characteristics such as species, sex, age, and diet; and the difficulty of extrapolating results from test animals to humans. (We obviously cannot conduct LD_{50} tests in humans.) The advantages include the relative speed of the test; the information it can provide on the cause of toxicity (obtained via post mortem examination of the animals); and the value of the test as a first approximation of hazards to those exposed.

The alternative to animal testing of toxicity is in vitro testing—testing not carried out in living animals, but rather in laboratory glassware. (The literal meaning of "in vitro" is "in glass.") Most such tests rely on observing the effects of chemicals on cultured living cells.

See Additional Problems 16.51–53 at the end of the chapter.

■ **PROBLEM 16.10**

What product would you obtain from reduction of the following ketones and aldehydes?

(a) $(CH_3)_2CHCHO$ **(b)** *m*-Chlorobenzaldehyde   **(c)** Cyclopentanone

---

■ **PROBLEM 16.11**

What ketones or aldehydes might be reduced to yield the following alcohols?

(a)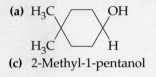

(b) $HOCH_2CH_2CH_2CHCH_3$
　　　　　　　　　　$\underset{\displaystyle CH_3}{|}$

(c) 2-Methyl-1-pentanol

16.7 Addition of Alcohols: Hemiacetals and Acetals

Hemiacetal Formation

Addition reaction, aldehydes and ketones Addition of an alcohol or other compound to the carbon–oxygen double bond to give a carbon–oxygen single bond.

Hemiacetal A compound with both an alcohol-like —OH group and an ether-like —OR group bonded to the same carbon atom.

Aldehydes and ketones undergo **addition reactions** in which the carbon–oxygen double bond is converted to a single bond. The products of addition reactions with alcohols are known as *hemiacetals* and *acetals* **Hemiacetals** have both an alcohol-like —OH group and an ether-like —OR group bonded to what was the carbonyl carbon atom. The H from the alcohol bonds to the carbonyl-group oxygen, and the OR from the alcohol bonds to the carbonyl-group carbon.

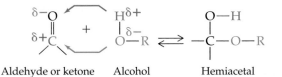

Aldehyde or ketone　　Alcohol　　　　Hemiacetal

- **The negatively polarized alcohol oxygen atom adds to the positively polarized carbonyl carbon** (similar to what happens in reduction of the carbonyl group) Almost all carbonyl-group reactions follow this same polarity pattern.
- **The reaction is reversible**. Hemiacetals rapidly revert back to aldehydes or ketones by loss of alcohol and establish an equilibrium with the aldehyde or ketone.

　　Ethanol (CH_3CH_2OH) forms hemiacetals with acetaldehyde and acetone as follows:

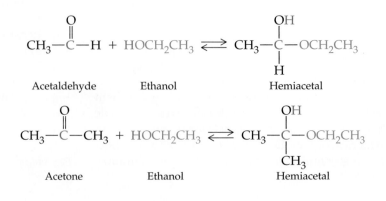

In practice, hemiacetals are often too unstable to be isolated. When equilibrium is reached, very little hemiacetal is present. A major exception occurs when the —OH and —CHO functional groups that react are part of the *same* molecule. The resulting cyclic hemiacetal is more stable than a noncyclic hemiacetal. Because of their greater stability, most simple sugars exist mainly in the cyclic hemiacetal form, as shown below for glucose, rather than in the open-chain form shown on p. 450.

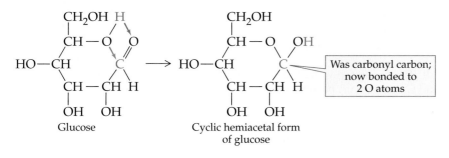

HO—CH   Was carbonyl carbon; now bonded to 2 O atoms

Glucose   Cyclic hemiacetal form of glucose

The cyclic form of glucose is customarily written as

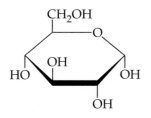

Acetal Formation

If a small amount of acid catalyst is added to the reaction of an alcohol with an aldehyde or ketone, the initially formed hemiacetal is converted into an acetal. An **acetal** is a compound that has *two* ether-like —OR groups bonded to what was the carbonyl carbon atom.

Acetal A compound that has two ether-like —OR groups bonded to the same carbon atom.

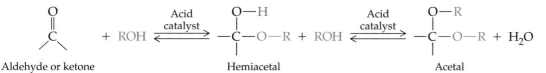

Aldehyde or ketone   Hemiacetal   Acetal

For example:

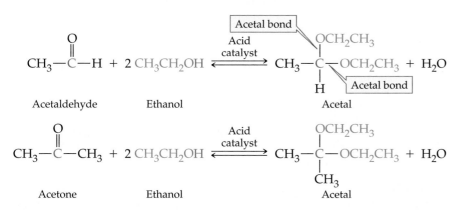

(In old nomenclature, a distinction was made by calling the compounds from aldehydes "hemiacetals," and those from ketones "hemiketals"; thus, you may see "hemiketal" or "ketal" in some publications.)

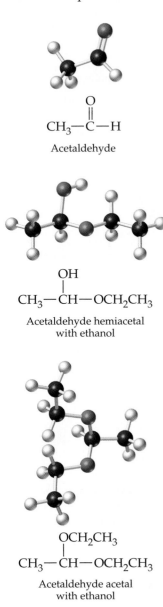

O
||
CH$_3$—C—H

Acetaldehyde

OH
|
CH$_3$—CH—OCH$_2$CH$_3$

Acetaldehyde hemiacetal
with ethanol

OCH$_2$CH$_3$
|
CH$_3$—CH—OCH$_2$CH$_3$

Acetaldehyde acetal
with ethanol

▲ **FIGURE 16.3 Acetaldehyde and its hemiacetal and acetal.** This hemiacetal and acetal are formed by reactions with ethanol.

The hemiacetal and acetal formed by reaction of acetaldehyde with ethanol are shown in Figure 16.3.

■ **WORKED EXAMPLE 16.3**

Write the structure of the intermediate hemiacetal and the acetal final product formed in the following reaction:

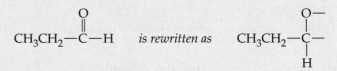

$$CH_3CH_2CH + 2\,CH_3OH \xrightarrow[\text{catalyst}]{\text{Acid}} ?$$

SOLUTION First, rewrite the structure showing only a single bond between C and O, along with partial bonds to both C and O:

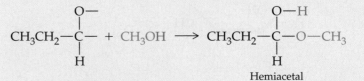

O   O—
|| |
CH$_3$CH$_2$—C—H   *is rewritten as*   CH$_3$CH$_2$—C—
                                          |
                                          H

Next, add one molecule of the alcohol (CH$_3$OH in this case) by attaching —H to the oxygen partial bond and —OCH$_3$ to the carbon partial bond. This yields the hemiacetal intermediate:

O—   O—H
| |
CH$_3$CH$_2$—C— + CH$_3$OH ⟶ CH$_3$CH$_2$—C—O—CH$_3$
| |
H   H

Hemiacetal

Finally, replace the —OH group of the hemiacetal with an —OCH$_3$ from a second molecule of alcohol. This yields the acetal product and water:

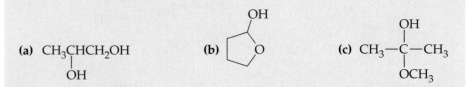

O—H   O—CH$_3$
| |
CH$_3$CH$_2$—C—O—CH$_3$ + CH$_3$OH ⟶ CH$_3$CH$_2$—C—O—CH$_3$ + H$_2$O
| |
H   H

Acetal

■ **WORKED EXAMPLE 16.4**

Which of the following compounds are hemiacetals?

(a) CH$_3$CHCH$_2$OH
          |
          OH

(b) [cyclic structure with OH and O]

(c) CH$_3$—C—CH$_3$
          |
with OH above and OCH$_3$ below the central C

ANALYSIS To identify a hemiacetal, look for a carbon atom attached to two oxygen atoms, one in an —OH group and one in an —OR group. Note that the O of the —OR group can be part of a ring.

SOLUTION Compound (a) contains two O atoms, but they are bonded to different C atoms; it is not a hemiacetal. Compound (b) has one ring C atom bonded to two oxygen atoms, one in the substituent —OH group and one bonded to the rest of the ring, which is the R group; it is a cyclic hemiacetal. Compound (c) also contains a C atom bonded to one —OH group and one —OR group, so it too is a hemiacetal.

■ **WORKED EXAMPLE 16.5**

Which of the following compounds are acetals?

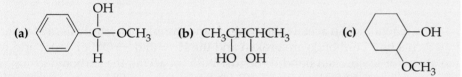

(a) CH₃CHOCH₂CH₃   (b) CH₃C—OCH₃   (c)
    |
    OCH₂CH₃

ANALYSIS As in identifying hemiacetals, look for a carbon atom that has single bonds to two oxygen atoms, but in this case both of them in —OR groups.

SOLUTION In (a), the central carbon atom is bonded to —CH₃, —H, and two —OCH₂CH₃ groups, so the compound is an acetal. Compound (b) does have a carbon atom bonded to two oxygen atoms, but one of the bonds is a double bond rather than a single bond, so this is not an acetal. Compound (c) has an oxygen atom in a ring, making it also part of an —OR group, where R is the ring. Since one of the carbons connected to the O in the ring is also connected to an —OCH₂CH₃ group, compound (c) is an acetal.

■ **PROBLEM 16.12**

Which of the following compounds are hemiacetals?

(a)
        OH
        |
    —C—OCH₃
        |
        H
    (b) CH₃CHCHCH₃
          | |
        HO OH
    (c)
    —OH
    OCH₃

■ **PROBLEM 16.13**

Draw the structures of the hemiacetals formed in these reactions:

(a) CH₃CH₂CH₂CHO + CH₃CH₂OH ⟶ ?

           O
           ‖
(b) CH₃CH₂CCH₂CH(CH₃)₂ + CH₃OH ⟶ ?

■ **PROBLEM 16.14**

Draw the structure of each acetal final product formed in the reactions shown in Problem 16.13.

■ **PROBLEM 16.15**

Which of the following compounds are hemiacetals or acetals?

(a)
        OCH₃
        |
  CH₃—C—CH₃
        |
        OCH₃
    (b) HOCHCH₂CH₂CH₃
          |
        OCH₂CH₃

        O
        ‖
(c) CH₃CCH₂OH
    (d)
    —OCH₂CH₃

■ **PROBLEM 16.16**

Of the compounds in Problem 16.15 that are acetals or hemiacetals, which were formed from aldehydes and which were formed from ketones? Explain what indicates the difference.

Acetal Hydrolysis

Because acetal formation is an equilibrium reaction, the extent to which the reaction proceeds in either direction can be controlled by changing the reaction conditions. Therefore, the aldehyde or ketone from which an acetal was formed can be regenerated by reversing the reaction. To do this requires an acid catalyst and a large quantity of water (a product of acetal formation) to drive the reaction back towards the aldehyde or ketone (Le Châtelier's principle). (∞ , p. 182)

$$
\underset{\text{Acetal}}{
\begin{array}{c} O-R \\ | \\ -C-O-R \\ | \end{array}} + H-OH \;\xrightleftharpoons[]{\substack{\text{Acid} \\ \text{catalyst}}}\; \underset{\substack{\text{Aldehyde or} \\ \text{ketone}}}{
\begin{array}{c} O \\ \| \\ \diagup C \diagdown \end{array}} + R-OH + R-O-H
$$

For example:

$$
\underset{\underset{CH_3}{|}}{\overset{\overset{OCH_3}{|}}{CH_3-C-OCH_3}} + H_2O \;\xrightleftharpoons[]{\substack{\text{Acid} \\ \text{catalyst}}}\; CH_3-\overset{\overset{O}{\|}}{C}-CH_3 + 2\,CH_3OH
$$

Hydrolysis A reaction in which a bond or bonds are broken and the H— and —OH of water add to the atoms of the broken bond or bonds.

The reaction of an acetal with water is an example of **hydrolysis**, a reaction in which a bond or bonds are broken and the H— and —OH of water add to the atoms of the broken bond or bonds. With an acetal, the carbonyl group is regenerated as the bonds to R and OR are broken, the —OH from water bonds to the R group, and the —H from water bonds to the —OR group. The result is two molecules of the alcohol, R—O—H.

∞ LOOKING AHEAD

Consider for a moment that biochemical reactions take place in an environment where water molecules are always available, along with enzyme catalysts precisely suited to the necessary reactions. In this environment, it is not surprising that hydrolysis reactions play an important role. During digestion, hydrolysis breaks bonds in carbohydrates (Section 23.1), triacylglycerols (Section 25.1), and proteins (Section 28.1). ∞

■ **WORKED EXAMPLE 16.6**

Write the structure of the aldehyde or ketone that would be formed by hydrolysis of the following acetal:

$$(CH_3)_2CHCH_2CH(OCH_2CH_3)_2 + H_2O \xrightarrow{H_3O^+} \;?$$

SOLUTION The products will be the aldehyde or ketone and the alcohol from which the acetal could have been formed. The method described will lead to the answer if it is not apparent. First, rewrite the starting acetal so that the two C—O acetal bonds are more evident:

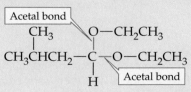

(continued p. 462)

Connection Sensory Evaluation Manager

Did you know you could taste a color? Teresa Pendergast does. She's in charge of sensory evaluation at a leading flavorings company, and it's her job to see that the foods that manufacturers ask them to flavor actually taste the way they're supposed to. That grape lollipop had better taste like a bunch of purple grapes—orange soda ought to be full of orange juiciness.

But tastes are funny. Make a red lollipop, but instead of flavoring it strawberry or cherry, flavor it lime. Then color another lollipop green but flavor it cherry. People will automatically say that the green one is lime and the red one is cherry, even though that's not what they're tasting at all. Terry runs sensory evaluation panels, and that interesting psychological twist makes her work a little more difficult. When she's setting up the individualized booths for each of her tasting panelists, she often must use blue, red, or yellow lights to mask subtle differences in the appearance of the foods. "If you're trying to compare two products, there might be a shade difference in color that you can't really help for some reason," she explains. "You need to camouflage that so the taster's choice is not influenced by visual cues."

Perhaps the manufacturer of a particular ranch style salad dressing would like to make a fat-free version of that dressing. It's not as easy as it sounds, because when the fat is removed from a substance both the taste and the texture change. Terry says, "When they've taken out all the fats and substituted gums or other starches, our job is to use flavors to compensate—to make it taste like their full-fat product."

The first step is to evaluate the flavor of the original ranch dressing. In order to do this, Terry has trained a panel of employees to discriminate the different components of the target food's flavor profile. She explains, "One of the major areas that I work in is called descriptive analysis. What that entails is training people to taste a product and describe what they're tasting—to actually recreate the flavor profile in words. How do you describe that ranch dressing you're eating? It's a complex mixture of ingredients. For instance, someone might taste mayonnaise, buttermilk, garlic, onion, sourness or sweetness. They might also taste something else. In a lot of these fat-free products tasters tell us they detect a "cardboard" note. So what we do is actually take cardboard, wet it, and have them smell it. And we ask, 'Is this the smell you mean?'"

Terry's group works to identify references and define attributes for all the components of the ranch dressing. Buttermilk might be used as a reference for the taste of sour. For other components the chemists actually make up solutions. For sweetness they use a 7% sucrose solution; for salt, a 0.1% solution in water. The panelists taste these and use them as benchmarks.

Taking chemistry in school was important for Terry because she often runs into it at work. Different chemicals are major components of specific flavors. For grape flavor, a main ingredient is methyl anthranilate; for cherry it is benzaldehyde; for banana, it is amyl acetate. The chief constituent of spearmint is *l*-carvone. When their suppliers bring in raw materials, they talk in chemical terms, and it helps to have that background knowledge. Chemical knowledge also comes in handy because Terry's group works very closely with flavorists, who are chemists. According to Terry, "When we speak with the chemists, we talk in technical terms. We might refer to a flavor as tasting aldehydic, or phenolic. Another common flavor reference is *cis*-3-hexen-1-ol, which has a 'green,' or leafy quality—it's actually derived from the strawberry leaf. When you use these terms to describe the flavor or product they're trying to match, it gives them a better idea of what they have to do to make their flavor come closer to the target."

Next, remove the acetal —R and —OR groups, which are —CH_2CH_3 and —OCH_2CH_3 in this case. Convert these groups to alcohol molecules by adding the —OH and —H from H_2O. This change leaves the C and O of the acetal with bonds to be completed.

$$CH_3CHCH_2-\underset{\underset{H}{|}}{\overset{\overset{CH_3}{|}}{C}}-\overset{O}{\underset{}{|}} \quad + \quad CH_3CH_2O-H + CH_3CH_2-OH$$

Finally, convert the —O on the acetal carbon to a carbonyl group and rewrite the structure of the product. In this example, the product is an aldehyde.

$$CH_3CHCH_2-\overset{\overset{CH_3}{|}}{\underset{\underset{H}{|}}{C}}-\overset{O}{\underset{}{|}} \longrightarrow CH_3CHCH_2-\overset{\overset{CH_3}{|}}{\overset{O}{\overset{||}{C}}}-H \quad \text{or} \quad (CH_3)_2CHCH_2CHO$$
3-Methylbutanal

■ **PROBLEM 16.17**

What aldehydes or ketones result from the following acetal hydrolysis reactions? What alcohol is formed in each case?

(a) $\text{⟨benzene ring⟩}-CH_2C(OCH_3)_2CH_2CH_3 \xrightarrow{H_3O^+} ?$

(b) $CH_3CH_2CH_2OCHOCH_2CH_2CH_3 \xrightarrow{H_3O^+} ?$
$\qquad\qquad\qquad\quad |$
$\qquad\qquad\quad CH_2CH_3$

(c) $CH_3CH_2CH_2OCH_2OCH_2CH_2CH_3 \xrightarrow{H_3O^+} ?$

Key Words

Acetal, p. 457

Addition reaction, aldehydes and ketones, p. 456

Aldehyde, p. 442

Carbonyl compound, p. 441

Carbonyl group, p. 441

Hemiacetal, p. 456

Hydrolysis, p. 460

Ketone, p. 440

Summary: Revisiting the Chapter Goals

1. **What is the carbonyl group?** The *carbonyl group* is a carbon atom connected by a double bond to an oxygen atom, $C=O$. Because of the electronegativity of oxygen, the $C=O$ group is polar, with a partial negative charge on oxygen and a partial positive charge on carbon. The oxygen and the two substituents on the carbonyl-group carbon atom form a planar triangle.

2. **How are ketones and aldehydes named?** The simplest *aldehydes* and *ketones* are known by common names (formaldehyde, acetaldehyde, benzaldehyde, acetone). Aldehydes are named systematically by replacing the final *-e* in an alkane name with *-al* and when necessary numbering the chain starting with 1 at the —CHO group. Ketones are named systematically by replacing the final *-e* in an alkane name with *-one* and numbering starting with 1 at the end nearer the $C=O$ group. The location of the carbonyl group is indicated by placing the number of its carbon before the name. Some common names of ketones identify each alkyl group separately.

3. **What are the general properties of aldehydes and ketones?** Aldehyde and ketone molecules are moderately polar, do not hydrogen-bond with

each other, but can hydrogen-bond with water molecules. The smaller ones are water-soluble, and the ketones are excellent solvents. In comparable series of compounds, aldehydes and ketones are higher boiling than alkanes but lower boiling than alcohols. Many aldehydes and ketones have distinctive, pleasant odors.

4. **What are some of the significant occurrences and applications of aldehydes and ketones?** Aldehydes and ketones are present in many plants, where they contribute to their aromas. Such natural aldehydes and ketones are widely used in perfumes and flavorings. Formaldehyde (an irritating and toxic substance) is used in polymers, is present in smog-laden air, and is produced biochemically from ingested methanol. Acetone is a widely used solvent and is a by-product of food breakdown during diabetes and starvation. Simple sugars are aldehydes or ketones.

5. **What are the results of the oxidation and reduction of aldehydes and ketones?** Mild oxidizing agents (Tollens' and Benedict's reagents) convert aldehydes to carboxylic acids but have no effect on ketones. With reducing agents, hydride ion $(H:^-)$ adds to the C of the $C=O$ group in an aldehyde or ketone and hydrogen ion (H^+) adds to the O to produce primary or secondary alcohols, respectively.

6. **What are hemiacetals and acetals, and how are they formed?** Aldehydes and ketones establish equilibria with alcohols to form hemiacetals or acetals. The relatively unstable *hemiacetals*, which have an —OH and an —OR on what was the carbonyl carbon, result from addition of one alcohol molecule. The more stable *acetals*, which have two —OR groups on the carbonyl carbon, form by addition of a second alcohol molecule to a hemiacetal. The aldehyde or ketone can be regenerated from an acetal by treatment with an acid catalyst and a large quantity of water, which is an example of a *hydrolysis* reaction.

Summary of Reactions

1. Reactions of aldehydes

(a) Oxidation to yield a carboxylic acid (Section 16.5):

$$CH_3CH_2\overset{\overset{\displaystyle O}{\|}}{C}H \xrightarrow{[O]} CH_3CH_2\overset{\overset{\displaystyle O}{\|}}{C}OH$$

(b) Reduction to yield a primary alcohol (Section 16.6):

$$CH_3CH_2\overset{\overset{\displaystyle O}{\|}}{C}H \xrightarrow{[H]} CH_3CH_2CH_2OH$$

(c) Addition of alcohol to yield a hemiacetal or acetal (Section 16.7):

$$CH_3\overset{\overset{\displaystyle O}{\|}}{C}H + CH_3CH_2OH \longrightarrow CH_3\overset{\overset{\displaystyle H}{|}}{\underset{\underset{\displaystyle OH}{|}}{C}}OCH_2CH_3$$

$$CH_3\overset{\overset{\displaystyle H}{|}}{\underset{\underset{\displaystyle OH}{|}}{C}}OCH_2CH_3 + CH_3CH_2OH \longrightarrow CH_3\overset{\overset{\displaystyle H}{|}}{\underset{\underset{\displaystyle OCH_2CH_3}{|}}{C}}OCH_2CH_3 + H_2O$$

2. Reactions of ketones

a. Reduction to yield a secondary alcohol (Section 16.6):

$$\underset{\substack{\| \\ O}}{CH_3CCH_3} \xrightarrow{[H]} \underset{\substack{| \\ OH}}{CH_3CHCH_3}$$

b. Addition of an alcohol to yield a hemiacetal or acetal (Section 16.7):

$$\underset{\substack{\| \\ O}}{CH_3CCH_3} + CH_3CH_2OH \longrightarrow \underset{\substack{| \\ OH}}{\overset{\overset{\displaystyle CH_3}{|}}{CH_3C}}-OCH_2CH_3$$

$$\underset{\substack{| \\ OH}}{\overset{\overset{\displaystyle CH_3}{|}}{CH_3C}}-OCH_2CH_3 + CH_3CH_2OH \longrightarrow \underset{\substack{| \\ OCH_2CH_3}}{\overset{\overset{\displaystyle CH_3}{|}}{CH_3COCH_2CH_3}} + H_2O$$

3. Reaction of acetals (Section 16.7)

Hydrolysis to regenerate an aldehyde or ketone:

$$\underset{\substack{| \\ OCH_2CH_3}}{CH_3CHOCH_2CH_3} \xrightarrow[H_2O]{H^+} \underset{\substack{\| \\ O}}{CH_3CH} + 2\,CH_3CH_2OH$$

▪ Understanding Key Concepts

16.18 The carbonyl group can be reduced by addition of a hydride ion (H^-) and a proton (H^+). Removal of H^- and H^+ from an alcohol results in a carbonyl group.

$$\underset{\diagup \diagdown}{\overset{\overset{\displaystyle O}{\|}}{C}} + H^- + H^+ \rightleftharpoons \underset{\diagup \diagdown}{\overset{\overset{\displaystyle O-H}{|}}{C}}-H$$

(a) To which atom of the carbonyl is the hydride ion added, and why?

(b) In the reaction above, indicate which reaction arrow represents reduction and which represents oxidation.

16.19 A fundamental difference between aldehydes and ketones is that one can be oxidized to carboxylic acids, but the other cannot. Which is which? What is an example of a test to differentiate aldehydes from ketones?

16.20 In the diagram below, indicate with dashed lines where hydrogen bonds would form. Explain why you chose these atoms to hydrogen bond.

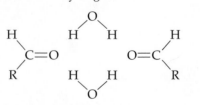

16.21 (a) Describe what happens in the reaction of an aldehyde with an alcohol. What is necessary for this reaction to occur?

(b) Copy the structures below and use lines to show where new bonds are formed. Cross out bonds that no longer exist as the aldehyde and alcohol react to form a hemiacetal.

$$R-\underset{\substack{| \\ H}}{\overset{\overset{\displaystyle O}{\|}}{C}} \qquad \overset{H}{\underset{\displaystyle O-R'}{|}}$$

16.22 Glucose is the major sugar in mammalian blood. We often see it represented as the "free aldehyde" and the cyclic hemiacetal forms shown below. Explain why using the cyclic structural formula is preferred for the hemiacetal.

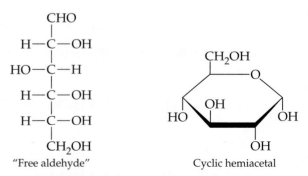

"Free aldehyde"   Cyclic hemiacetal

16.23 Describe the two types of addition reactions that aldehydes and ketones undergo with alcohols.

▪ Additional Problems

Aldehydes and Ketones

16.24 Draw a structure for a compound that meets each of these descriptions:
- (a) A ketone, C_3H_6O
- (b) An aldehyde with 6 carbons
- (c) A ketoaldehyde, $C_4H_6O_2$
- (d) A hydroxyketone, $C_3H_6O_2$

16.25 Draw a structure for a compound that meets each of these descriptions:
- (a) A ketone, C_5H_8O
- (b) An aldehyde with 8 carbons
- (c) A ketoaldehyde, $C_6H_{10}O_2$
- (d) A hydroxyketone, $C_5H_8O_2$

16.26 Indicate which compounds contain aldehyde or ketone carbonyl groups.

(a) $CH_3CH_2\overset{\displaystyle H}{\underset{}{C}}{=}O$

(b) $O{=}CCH_2CH_2CHCH_3$
   with NH_2 and CH_3 substituents

(c) $CH_3CH_2{-}O{-}CH{=}CH_2$

(d) $CH_3CH_2C(OCH_3)_3$

(e) $CH_3CHCOOH$
   with CH_3 substituent

(f) $CH_3COCH_2CH_2OH$

16.27 Indicate which compounds contain aldehyde or ketone carbonyl groups.

(a) CH_3CH_2CHO

(b) $(CH_3)_2C(OH)CH_2CH_2CH_3$

(c) $CH_3{-}\langle\text{benzene}\rangle{-}CONH_2$

(d) $CH_3CHCH_2CHCH_3$
   with OH and OCH_3 substituents

(e) $CH_3CH_2COCH_2CH_3$

(f) CH_3COOCH_3

16.28 Draw structures corresponding to the following aldehyde and ketone names:
- (a) Heptanal
- (b) 4,4-Dimethylpentanal
- (c) *o*-Chlorobenzaldehyde
- (d) 3-Heptanone
- (e) 2,4-Dimethyl-3-pentanone
- (f) *m*-Nitroacetophenone

16.29 Draw structures corresponding to the following aldehyde and ketone names:
- (a) 3-Hydroxy-2,2,4-trimethylpentanal
- (b) 4-Ethyl-2-isopropylhexanal
- (c) *p*-Bromobenzaldehyde
- (d) Cyclohexanone
- (e) 1,1,1-Trichloro-2-butanone
- (f) 2-Methyl-3-hexanone

16.30 Give systematic names for the following aldehydes and ketones:

(a) CH_3CH_2CHCHO
   with CH_3 substituent

(b) $CH_3CH_2CH_2CHCH_3$
   with CHO substituent

(c) $(CH_3)_3CCHO$

(d) $CH_3\overset{\displaystyle O}{\overset{\|}{C}}CH_2CH_3$

(e) $CH_3\overset{\displaystyle O}{\overset{\|}{C}}CH_2CH_2CHCH_3$
   with CH_3 substituent

16.31 Give systematic names for the following aldehydes and ketones:

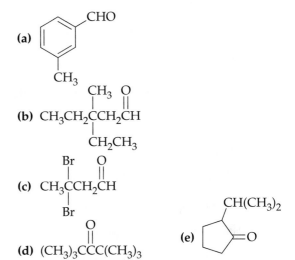

(a) (benzene ring with CHO and CH₃ substituents)

(b) $CH_3CH_2\overset{\displaystyle CH_3}{\underset{\displaystyle CH_2CH_3}{C}}CH_2\overset{\displaystyle O}{\overset{\|}{C}}H$

(c) $CH_3\overset{\displaystyle Br}{\underset{\displaystyle Br}{C}}CH_2\overset{\displaystyle O}{\overset{\|}{C}}H$

(d) $(CH_3)_3C\overset{\displaystyle O}{\overset{\|}{C}}C(CH_3)_3$

(e) (cyclopentanone ring with $CH(CH_3)_2$ substituent and $={=}O$)

16.32 The following names are incorrect. What is wrong with each?
- (a) 1-Butanone   (b) 4-Butanone   (c) 3-Butanone

16.33 The following names are incorrect. What is wrong with each?
- (a) Cyclohexanal   (b) 2-Butanal   (c) 1-Pentanone

Reactions of Aldehydes and Ketones

16.34 What kind of compound is produced when an aldehyde reacts with an alcohol in a 1:1 ratio? Does this reaction also form a second product?

16.35 What kind of compound is produced when an aldehyde reacts with an alcohol in a 1:2 ratio in the presence of an acid catalyst? Does this reaction also form a second product?

16.36 Draw the structures of the products formed when the following compounds react with Tollens' reagent. With a reducing agent.
- (a) Cyclopentanone
- (b) Hexanal

(c) $CH_3CH_2CH_2CHCH_2CH_3$
   with CHO substituent

16.37 Draw the structures of the products formed when the following compounds react: (1) with Tollens' reagent; (2) with a reducing agent.

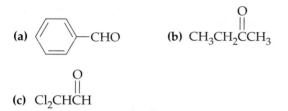

(a) ⟨benzene⟩—CHO   **(b)** $CH_3CH_2\overset{O}{\overset{||}{C}}CH_3$

(c) $Cl_2CH\overset{O}{\overset{||}{C}}H$

16.38 Draw the structures of the aldehydes and primary alcohols that might be oxidized to yield each of the following carboxylic acids:

(a) H_3C—⟨benzene⟩—COOH

(b) $CH_3CH_2\overset{COOH}{\overset{|}{C}}HCH_2\overset{CH_3}{\overset{|}{C}}HCH_3$

(c) $CH_3CH{=}CHCOOH$

16.39 Draw the structures of the aldehydes and primary alcohols that might be oxidized to yield each of the following carboxylic acids:

(a) ⟨benzene⟩—COOH, OH   **(b)** CH_3COOH

(c) $CH_3COCOOH$

16.40 Write the structures of the hemiacetals that result from reactions (a) and (b). Write the structures of the hydrolysis products of the compounds in (c) and (d).

(a) 2-Butanone + 1-Propanol ⟶ ?

(b) Butanal + Isopropyl alcohol ⟶ ?

(c) $CH_3CH_2CH_2\overset{O-CH_2CH_3}{\overset{|}{C}}H-O-CH_3$ ⟶ ?

(d)
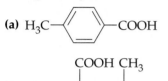
⟶ ?

16.41 Write the structures of the hemiacetals that result from reactions (a) and (b). Write the structures of the hydrolysis products of the compounds in (c) and (d).

(a) Acetone + Ethanol ⟶ ?

(b) Hexanal + 2-Butanol ⟶ ?

(c)

⟶ ?

(d)
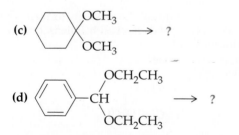
⟶ ?

16.42 Cyclic hemiacetals sometimes form if an alcohol group in one part of a molecule adds to a carbonyl group elsewhere in the same molecule. What is the structure of the open-chain hydroxy aldehyde from which the following hemiacetal might form?

A cyclic hemiacetal

16.43 Galactose exists largely in the cyclic hemiacetal form shown. Draw the structure of galactose in its open-chain hydroxy aldehyde form.

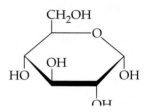

16.44 What products result from hydrolysis of the following cyclic acetal?

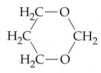

16.45 Acetals are usually made by reaction of an aldehyde or ketone with two molecules of a monoalcohol. If an aldehyde or ketone reacts with *one* molecule of a dialcohol, however, a cyclic acetal results. Draw the structure of the cyclic acetal formed in the following reaction:

⟨cyclohexane⟩=O + HO—CH_2CH_2—OH ⟶ ?

16.46 Aldosterone is a key steroid involved in controlling the sodium–potassium balance in the body. Identify the functional groups in aldosterone.

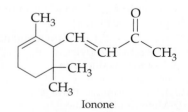

16.47 Ionone is a compound that is responsible for the odor of violets. Identify the functional groups in ionone.

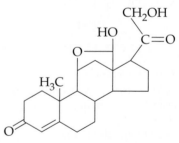

Ionone

Applications

16.48 In oxidation–reduction reactions there must be a reduction associated with each oxidation and vice versa. When enzymes within the bombardier beetle's defensive organ catalyze oxidation of *para*-dihydroxybenzene, what is reduced in the reaction? Write the complete oxidation–reduction reaction and be sure to balance the equation. [*Chemical Warfare Among the Insects*]

16.49 HCN is quite toxic. How do you suppose that a millipede can use this weapon without killing itself? [*Chemical Warfare Among the Insects*]

16.50 For both natural vanilla extracts and "artificial" vanilla, the vanillin is extracted from the source with a solvent. Look at the structure of vanillin in the application, predict whether water, ethanol, or diethyl ether would be the best solvent for good extraction, and explain your choice. Remember, the solvent also must be suitable for human consumption. Check a vanilla extract bottle to see what was used for the extraction. [*Vanilla: What Kind Is Best?*]

16.51 Both HCN and benzaldehyde have the aroma of almonds. If a liquid had a faint almond odor, how would you test it to determine whether the aldehyde or the acid was present? [*Is It Poisonous, or Isn't It?*]

16.52 **(a)** What is an advantage of *in vitro* acute toxicity testing?
(b) Why is there still considerable uncertainty associated with *in vitro* acute toxicity testing? (*Hint*: Since you are testing only one certain cell type, what might be happening in other cell types?) [*Is It Poisonous, or Isn't It?*]

16.53 Which of three compounds with the LD 50's listed below is the most toxic and which is the least toxic? **(a)** 23 g/kg **(b)** 23 mg/kg **(c)** 18 mg/kg [*Is It Poisonous, or Isn't It?*]

General Questions and Problems

16.54 The following compound is used in the fragrance industry. Name it.

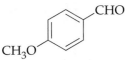

16.55 Can the following alcohol, $(CH_3)_3COH$, be formed by the reduction of an aldehyde or ketone? Why or why not?

16.56 Many flavorings and perfumes are partially based on fragrant aldehydes and ketones. Why do you think the portion of the odor due to the ketone is more stable than that due to the aldehyde?

16.57 One problem with burning some plastics is the release of formaldehyde. What are some of the physiological effects of exposure to formaldehyde?

16.58 Chloral hydrate, a potent sedative and component in "knockout" drops, is formed by reacting trichloroacetaldehyde with water in a reaction analogous to hemiacetal formation. Draw the formula of chloral hydrate.

16.59 Name the following compounds:

(a) $CH_3CH_2\overset{\displaystyle O}{\overset{\displaystyle \|}{C}}CH(CH_3)_2$ **(b)** $CH_2{=}CHCH_2CH_2CH{=}CH_2$

(c)

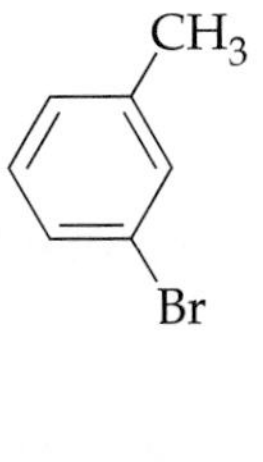

(d) $(CH_3)_3CCHC\overset{\displaystyle O}{\overset{\displaystyle \|}{C}}CH_2CH_3$ with CH_3 substituent

16.60 Name the following compounds:

(a)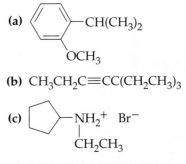

(b) $CH_3CH_2C{\equiv}CC(CH_2CH_3)_3$

(c) cyclopentyl$-NH_2^+$ Br^- with CH_2CH_3

(d) $(CH_3CH_2)_2N(CH_2)_5CH_3$

16.61 Draw the structural formulas of the following compounds:
(a) *o*-Nitrobenzaldehyde
(b) 2,3-Dichlorocyclopentanone
(c) Butyl methyl ether
(d) 2,4-Dimethyl-3-pentanol

16.62 Draw the structural formulas of the following compounds:
(a) 2,3,3-Triiodobutanal
(b) 1,1,3-Tribromoacetone
(c) 4-Amino-4-methyl-2-pentanone

16.63 Complete the following equations:

(a) $CH_3CH{=}C(CH_3)_2 + H_2 \xrightarrow{Pd} ?$

(b) $CH_3CH_2\overset{\displaystyle CH_2OH}{\underset{\displaystyle CH_2CH_3}{C}}CH_2CH_3 \xrightarrow{[O]} ?$

(c) $CH_3\overset{\displaystyle O}{\overset{\displaystyle \|}{C}}CH_2CH_2CH_3 \xrightarrow[H_3O^+]{\text{Reducing agent}} ?$

(d) 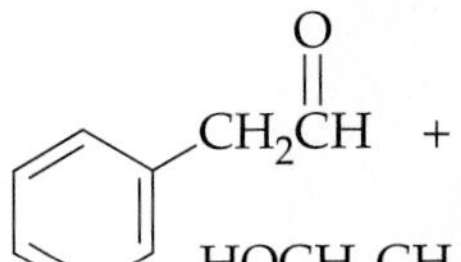 benzene$-CH_2\overset{\displaystyle O}{\overset{\displaystyle \|}{C}}H + HOCH_2CH_2CH_3 \longrightarrow \underset{\text{(Hemiacetal)}}{?}$

16.64 Complete the following equations:

(a) 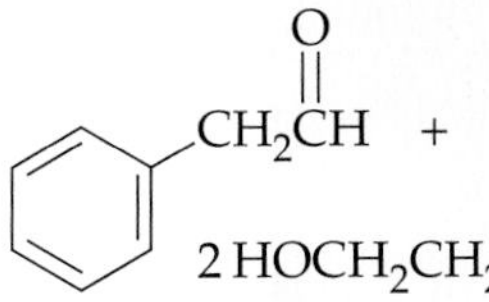 benzene$-CH_2\overset{\displaystyle O}{\overset{\displaystyle \|}{C}}H + 2\,HOCH_2CH_2CH_3 \longrightarrow \underset{\text{(Acetal)}}{?}$

(b) $CH_3CH{=}\overset{\displaystyle CH_2CH_3}{\underset{}{C}}CH_2CH_2CH_3 + HCl \longrightarrow ?$

(c) 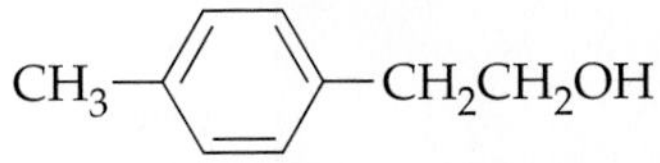 $CH_3{-}$benzene$-CH_2CH_2OH \xrightarrow{H_2SO_4} ?$

16.65 How could you differentiate between 3-hexanol and hexanal using a simple chemical test?

16.66 The liquids 1-butanol, 1-butylamine, and butanal have similar molar masses. Assign the observed boiling points of 78°C, 75°C, and 117°C to these compounds and explain your choices.

16.67 2-Butanone has a solubility of 26 g/100 mL of H_2O, but 2-heptanone, which is found in clove and cinnamon bark oils, is only very slightly soluble in water. Explain the difference in solubility of these two ketones.

17

Carboxylic Acids and Their Derivatives

The active ingredient in aspirin is acetylsalicylic acid, one of our most useful carboxylic acid derivatives.

The simplest carbonyl-containing functional groups, the aldehydes and ketones, have been described in Chapter 16. In this chapter, we move on to carbonyl compounds that are *carboxylic acids* or *derivatives of carboxylic acids*—the *esters* and *amides*. Also, because of their major role in biochemistry, esters of phosphoric acid are introduced here.

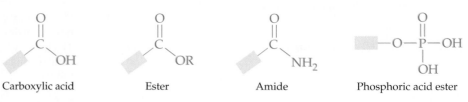

Carboxylic acid   Ester   Amide   Phosphoric acid ester

Carboxylic acids are easily converted to esters and amides, and the esters and amides are easily converted back to the carboxylic acids. These properties make carboxylic acids, amides, and esters useful in chemical synthesis, whether in biochemistry or in the chemical industry. The reactivity of carboxylic acids, for example, makes possible the formation of polymers— both the synthetic kind and the proteins that keep us alive.

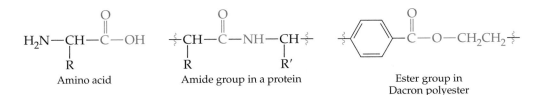

Amino acid Amide group in a protein Ester group in Dacron polyester

The study of these compounds will prepare you for Chapters 18 and 19, where we will explore how proteins are put together and what their roles are in living things.

In this chapter we'll answer the following questions:

1. **What are the general structures and properties of carboxylic acids and their derivatives?**
   The goal: Be able to describe and compare the structures, reactions, hydrogen bonding, water solubility, boiling points, and acidity or basicity of carboxylic acids, esters, and amides.

2. **How are carboxylic acids, esters, and amides named?**
   The goal: Be able to name the simple members of these families and write their structures, given the names.

3. **What are some occurrences and applications of significant carboxylic acids, esters, and amides?**
   The goal: Be able to identify the general occurrence and some important members of each family.

4. **How are esters and amides synthesized from carboxylic acids and converted back to carboxylic acids?**
   The goal: Be able to describe and predict the products of the ester- and amide-forming reactions of carboxylic acids and the hydrolysis of esters and amides.

5. **What are the organic phosphoric acid derivatives?**
   The goal: Be able to recognize and write the structures of phosphate esters and their ionized forms.

CONCEPTS TO REVIEW

Acid–base chemistry (Sections 10.1–10.6)
Naming alkanes (Section 12.6)

17.1 Carboxylic Acids and Their Derivatives: Properties and Names

Carboxylic acids have an —OH group bonded to the carbonyl carbon atom. In their derivatives, the —OH group is replaced by other groups. **Esters** have an —OR group bonded to the carbonyl carbon atom. **Amides** have an —NH₂, —NHR, or NR₂ group bonded to the carbonyl carbon atom.

Note that carboxylic acids, esters, and amides have in common a carbonyl carbon atom bonded either to an oxygen or to a nitrogen. The resulting polarity of their functional groups and the structural similarities of carboxylic acids, esters, and amides account for many similarities in the properties of these compounds.

Carboxylic acid A compound that has a carbonyl group bonded to a carbon atom and an —OH group, RCOOH.

Ester A compound that has a carbonyl group bonded to a carbon atom and an —OR group, RCOOR'.

Amide A compound that has a carbonyl group bonded to a carbon atom and a nitrogen atom group, RCONR'₂, where the R' groups may be alkyl groups or hydrogen atoms.

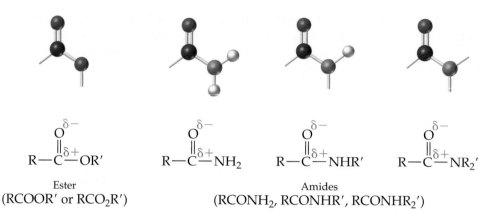

Carboxylic acid
(RCOOH or RCO₂H)

Ester
(RCOOR′ or RCO₂R′)

Amides
(RCONH₂, RCONHR′, RCONHR₂′)

For example, since carboxylic acids and their derivatives all contain polar functional groups, all are higher boiling than comparable alkanes.

Carboxylic acids and their derivatives also have in common their participation in **carbonyl-group substitution reactions**, in which a group we represent as —Z replaces (substitutes for) the group bonded to the carbonyl carbon atom.

Carbonyl-group substitution reaction   A reaction in which a new group replaces (substitutes for) a group attached to a carbonyl-group carbon in an acyl group.

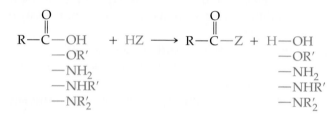

For example, esters are made by such reactions:

$$CH_3-\overset{O}{\underset{\|}{C}}-OH + H-OCH_2CH_3 \rightleftharpoons CH_3-\overset{O}{\underset{\|}{C}}-OCH_2CH_3 + H-OH$$

Acetic acid   Ethanol   Ethyl acetate   Water
(a carboxylic acid)   (an ester)

And esters can be converted back to carboxylic acids by such reactions:

$$CH_3-\overset{O}{\underset{\|}{C}}-OCH_2CH_3 + H-OH \rightleftharpoons CH_3-\overset{O}{\underset{\|}{C}}-OH + H-OCH_2CH_3$$

Ethyl acetate   Water   Acetic acid   Ethanol

Acyl group   An RC=O group.

The parts of molecules in which an alkyl group is bonded to the carbonyl carbon atom are known as **acyl groups**. Note that in carbonyl-group substitution reactions, it is the group bonded to the acyl group that changes.

Acyl groups

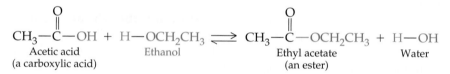

■ **PROBLEM 17.1**

Identify each of the following formulas as that of a carboxylic acid, an amide, an ester, or none of these.

(a) $CH_3\overset{O}{\underset{\|}{C}}NH_2$ (b) CH_3OCH_3 (c) CH_3COOH (d) $CH_3COOCH_2CH_3$

(e) CH_3COCH_3 (f) $CH_3CH_2CONHCH_3$ (g) $CH_3CH_2NH_2$ (h) $CH_3CH_2\overset{O}{\underset{\|}{C}}NH_2$

Carboxylic Acids

The most significant property of carboxylic acids is their behavior as acids. They surrender the hydrogen of the **carboxyl group**, COOH, to bases and establish acid–base equilibria in aqueous solution (a property further discussed in Section 17.3). The common carboxylic acids share the concentration-dependent corrosive properties of all acids but are not generally hazardous to human health in other ways.

Like alcohols, carboxylic acids form hydrogen bonds with each other so that even formic acid (HCOOH), the lightest and simplest carboxylic acid, is a liquid at room temperature with a boiling point of 101°C.

Carboxyl group The —COOH functional group.

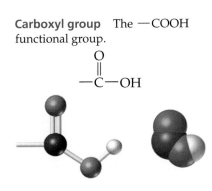

Carboxylic acids pair up by hydrogen bonding, as illustrated for formic acid.

Acids with saturated straight-chain R groups of up to nine carbon atoms are volatile liquids with strong, sharp odors, and those with up to four carbons are water-soluble. Acids with larger saturated R groups are waxy, odorless solids; their water solubility falls off as the size of the hydrophobic, alkane-like portion increases relative to the size of the water-soluble —COOH portion.

Carboxylic acids are named systematically (in the IUPAC system) by replacing the final -e of the corresponding alkane name with -oic acid. The three-carbon acid is propanoic acid; the straight-chain four-carbon acid is butanoic acid; and so on. If alkyl substituents or other functional groups are present, the chain is numbered beginning at the —COOH end, as in 3-methylbutanoic acid and 2-hydroxypropanoic acid, which is better known as lactic acid, the acid present in sour milk.

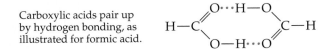

Propanoic acid   3-Methylbutanoic acid   Lactic acid (2-Hydroxypropanoic acid)

Because carboxylic acids were among the first organic compounds to be isolated and purified, the simple ones all have common names that are used more often than the systematic names. Recognizing the common acid names given in Table 17.1 is important, as they provide the basis for the common names of derivatives of these acids. When using these names, the carbon atoms next to the —COOH group are identified by Greek letters α, β, and so on, rather than numbers. In this system, for example, the three-carbon acid is *propionic acid*, and the second C=O group (a *keto* group) next to the —COOH group is an α-keto group.

α-Ketopropionic acid
(Pyruvic acid,
a key biochemical
intermediate)

α-Aminopropionic acid
(Alanine)

In alanine, as in all common amino acids, the –NH₂ group is on the alpha carbon atom (the C next to –COOH).

Acetyl group A CH₃C=O group.

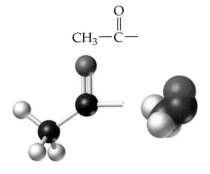

The acyl group that remains when a carboxylic acid loses its —OH is named by replacing the -ic at the end of the acid name with -oyl. The acyl group from acetic acid is traditionally called an **acetyl group**, however.

     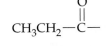

Acetyl group   Propanoyl group   Benzoyl group

TABLE 17.1 Physical Properties of Some Carboxylic Acids

Structure	Common Name	Melting Point (°C)	Boiling Point (°C)
HCOOH	Formic	8	101
CH_3COOH	Acetic	17	118
CH_3CH_2COOH	Propionic	−22	141
$CH_3CH_2CH_2COOH$	Butyric	−4	163
$CH_3CH_2CH_2CH_2COOH$	Valeric	−34	185
$CH_3(CH_2)_{16}COOH$	Stearic	70	383
HOOCCOOH	Oxalic	190	Decomposes
$HOOCCH_2COOH$	Malonic	135	Decomposes
$HOOCCH_2CH_2COOH$	Succinic	188	Decomposes
$HOOCCH_2CH_2CH_2COOH$	Glutaric	98	Decomposes
$H_2C{=}CHCOOH$	Acrylic	13	141
$CH_3CH{=}CHCOOH$	Crotonic	72	185
⬡—COOH	Benzoic	122	249
⬡(COOH)(OH)	Salicylic	159	Decomposes

Dicarboxylic acids, which contain two —COOH groups, are named systematically by adding the ending *-dioic acid* to the alkane name (the *-e* is retained). Here also, the simple dicarboxylic acids are usually known by common names. Oxalic acid (IUPAC name, ethanedioic acid) is found in plants of the genus *Oxalis*, which includes rhubarb and spinach. You will encounter succinic acid, glutaric acid, and several other dicarboxylic acids when we come to the generation of biochemical energy and the citric acid cycle (Section 21.8).

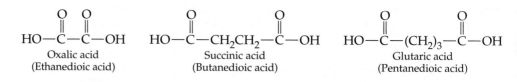

Oxalic acid
(Ethanedioic acid)

Succinic acid
(Butanedioic acid)

Glutaric acid
(Pentanedioic acid)

Unsaturated acids are named systematically in the IUPAC system with the ending *-enoic*. For example, the simplest unsaturated acid, $H_2C{=}CH_2COOH$, is named propenoic acid. It is, however, best known as acrylic acid, which is a raw material for acrylic polymers.

⫘ LOOKING AHEAD

Biochemistry is dependent on the continual breakdown of food molecules. Frequently this process requires transfer of acetyl groups from one molecule to another. Acetyl-group transfer occurs, for example, at the beginning of the citric acid cycle, which is central to production of life-sustaining energy (Section 21.8). ⫘

▪ WORKED EXAMPLE 17.1

Give the systematic and common names for the following compound:

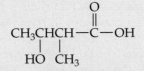

SOLUTION For the systematic name of a carboxylic acid, first identify the longest chain containing the —COOH group and number it starting with the carboxyl-group carbon:

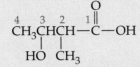

The parent compound is the four-carbon acid, butanoic acid. It has a methyl group on carbon 2 and a hydroxy group on carbon 3, giving the name 3-hydroxy-2-methylbutanoic acid.

The common name for the four-carbon acid is butyric acid (Table 17.1), and the substituents are located by Greek letters rather than numbers. The common name of this acid is β-hydroxy-α-methylbutyric acid.

▪ PROBLEM 17.2

Draw the structures of the following acids:
(a) 3-Methylhexanoic acid   **(b)** *o*-Nitrobenzoic acid

▪ PROBLEM 17.3

Write the complete structural formula of malonic acid (refer to Table 17.1), showing all bonds.

▪ PROBLEM 17.4

Identify the acid that will be formed by addition of hydrogen to the double bond in acrylic acid (refer to Table 17.1).

Esters

When the —OH of the carboxyl group is converted to the —OR of an ester group (—COOR), the ability of the molecules to hydrogen-bond with each other is lost. Simple esters are therefore lower boiling than the acids from which they are derived.

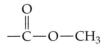

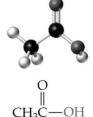
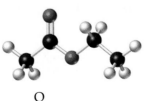
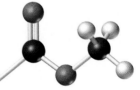

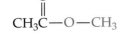

Acetic acid, bp 118°C   Methyl ester, bp 57°C   Ethyl ester, bp 77°C

The simple esters are colorless, volatile liquids with pleasant odors, and many of them contribute to the natural fragrance of flowers and ripe fruits. The vapors of volatile esters can be irritating and have a narcotic effect. When ingested, individual esters vary widely in their acute toxic effects. The lower-molecular-weight esters are somewhat soluble in water and are quite flammable. Esters are neither acids nor bases in aqueous solution.

Ester names consist of two words. The first is the name of the alkyl group R in the ester group —COOR. The second is the name of the parent acid, with the family-name ending *-ic acid* replaced by *-ate*. (Note that the order of the two parts of the name is the reverse of the order in which ester condensed formulas are usually written.)

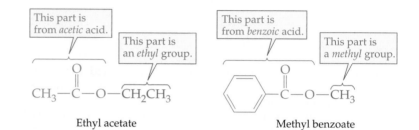

Ethyl acetate   Methyl benzoate

▲ Methyl butyrate can simulate the taste of these beautiful fresh apples.

Both common and systematic names are derived in this manner. An ester of the straight-chain, four-carbon carboxylic acid, for example, can be named systematically as a butanoate (from butanoic acid) or by a common name as a butyrate (from butyric acid):

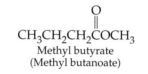

$$CH_3CH_2CH_2COCH_3$$
Methyl butyrate
(Methyl butanoate)

An ester used as a food flavoring to give the taste of apples

■ **WORKED EXAMPLE 17.2**

What is the structure of butyl acetate?

SOLUTION The two-word name consisting of an alkyl group name followed by an acid name with an *-ate* ending shows that the compound is an ester. The name "acetate" shows that the RCOO part of the molecule is from acetic acid (CH_3COOH). The "butyl" part of the name shows that a butyl group has replaced H in the carboxyl group. Therefore the structure is

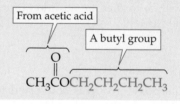

■ **WORKED EXAMPLE 17.3**

What is the name of the compound shown?

$$CH_3(CH_2)_{16}COCH_2CH_2CH_3$$

SOLUTION The compound has the general formula RCOOR', so it is an ester. The acid part of the molecule (RCOO—) is from stearic acid (see Table 17.1). The R' group has three C atoms and is therefore a propyl group. The compound is propyl stearate.

■ **PROBLEM 17.5**

Draw the structures of
(a) Propyl benzoate   **(b)** Ethyl propanoate   **(c)** Methyl acrylate
(See Table 17.1.)

■ **PROBLEM 17.6**

Which of the following compounds has the highest boiling point and which has the lowest boiling point? Explain your answer.
(a) CH_3OCH_3 **(b)** CH_3COOH **(c)** $CH_3CH_2CH_3$

■ **PROBLEM 17.7**

In each of the following pairs of compounds, which would you expect to be more soluble in water? Why?

(a) $C_8H_{17}COOH$ or $CH_3CH_2CH_2COOH$

(b) $CH_3CHCOOH$ or $CH_3CH_2COOCHCH_3$
$\quad\quad\;\; |$ $\quad\quad\quad\quad\quad\quad\quad |$
$\quad\quad CH_3$ $\quad\quad\quad\quad\quad\quad CH_3$

Amides

Amides may contain an —NH_2 group or may have one or two R′ groups bonded to the nitrogen. *Unsubstituted amides* ($RCONH_2$) can form three strong hydrogen bonds to other amide molecules and are thus higher melting and higher boiling than the acids from which they are derived.

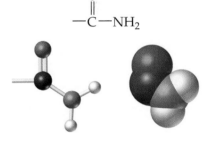

Hydrogen bonding in $\overset{\displaystyle O}{\overset{\displaystyle \|}{RCNH_2}}$

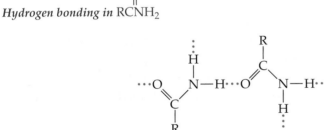

Except for the simplest one (formamide, $HCONH_2$, a liquid), the low-molecular-weight unsubstituted amides are solids that are soluble in both water, with which they form hydrogen bonds, and organic solvents. *Monosubstituted amides* ($RCONHR′$) can also form hydrogen bonds to each other, but *disubstituted amides* ($RCONR′_2$) cannot do so and are therefore lower boiling.

$\overset{O}{\overset{\|}{CH_3COH}}$	$\overset{O}{\overset{\|}{CH_3CNH_2}}$	$\overset{O}{\overset{\|}{CH_3CNHCH_3}}$	$\overset{O}{\overset{\|}{CH_3CN(CH_3)_2}}$
Acetic acid (bp 118°C)	Acetamide (bp 222°C)	N-Methylacetamide (bp 206°C)	N,N-Dimethylacetamide (bp 165°C)

Note the distinction between amines (Chapter 15) and amides. The nitrogen atom is bonded to a carbonyl-group carbon in an amide, but not in an amine:

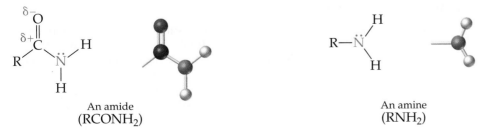

An amide   An amine
($RCONH_2$)   (RNH_2)

The positive end of the carbonyl group attracts the unshared pair of electrons on nitrogen strongly enough to prevent it from acting as a base by accepting a hydrogen atom. As a result, amides are not basic like amines.

Amides with an unsubstituted —NH_2 group are named by replacing the *-ic acid* or *-oic acid* of the corresponding carboxylic acid name with *-amide*. For example, the two-carbon amide derived from acetic acid is called acetamide. If the nitrogen atom of the amide has alkyl substituents on it, the compound is named by first specifying the alkyl group and then identifying the amide name. The alkyl substituents are preceded by the letter *N* to identify them as being attached directly to nitrogen.

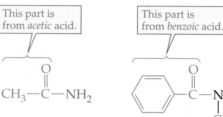
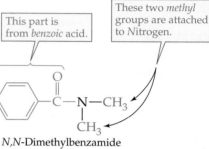

Acetamide   N,N-Dimethylbenzamide

To review, some derivatives of acetic acid are shown below:

Methyl acetate   Acetamide   N-Methylacetamide   N,N-Dimethylacetamide   N-Ethyl-N-Methylacetamide

Properties of Carboxylic Acids, Esters, and Amides

- All undergo carbonyl-group substitution reactions.
- Esters and amides can be made from acids and converted back to acids.
- Hydrogen bonding is strong in pure acids and unsubstituted or monosubstituted amides; disubstituted amides and esters do not hydrogen-bond.
- Simpler acids and esters are liquids; all unsubstituted amides except formamide are solids.
- Carboxylic acids give acidic aqueous solutions. Esters and amides are neither acids nor bases.
- Small esters are somewhat water-soluble, and small amides are water-soluble.
- Volatile acids have strong, sharp odors; volatile esters have pleasant, fruity odors.

⊂⊃⊃ LOOKING AHEAD

In later chapters you will see that the fundamental bonding connections in proteins are amide bonds (Section 18.2) and those in oils and fats are ester bonds (Section 25.2).   ⊂⊃⊃

■ **WORKED EXAMPLE 17.4**

Pairs of acid molecules react via carbonyl-group substitution to form compounds known as *acid anhydrides*. (*Anhydride* means "without water.") Relate the reactants below to the substitution reaction pattern (on p. 470) and complete the equation.

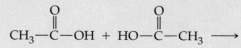

SOLUTION   The reaction here fits the substitution reaction pattern as follows:

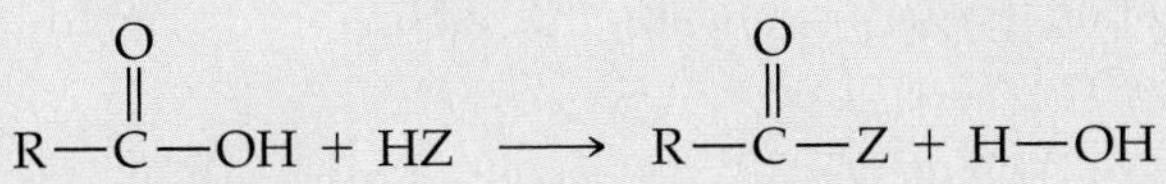

$$R-\overset{\displaystyle O}{\overset{\|}{C}}-OH + HZ \longrightarrow R-\overset{\displaystyle O}{\overset{\|}{C}}-Z + H-OH$$

with Z equal to

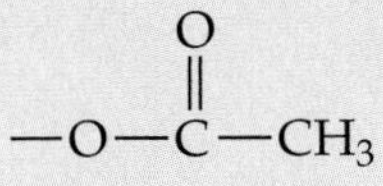

$$-O-\overset{\displaystyle O}{\overset{\|}{C}}-CH_3$$

The reaction is

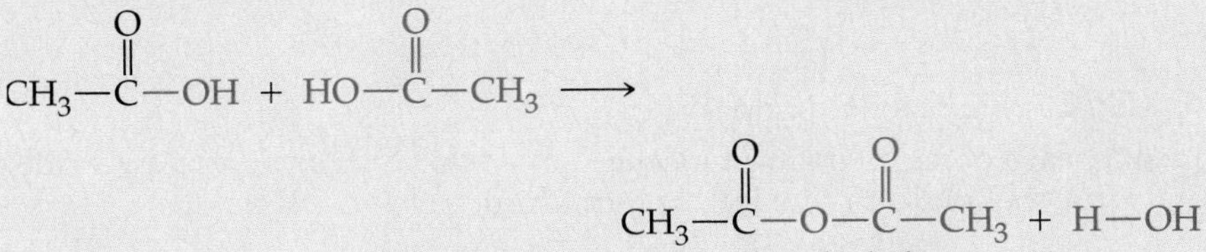

$$CH_3-\overset{\displaystyle O}{\overset{\|}{C}}-OH + HO-\overset{\displaystyle O}{\overset{\|}{C}}-CH_3 \longrightarrow$$

$$CH_3-\overset{\displaystyle O}{\overset{\|}{C}}-O-\overset{\displaystyle O}{\overset{\|}{C}}-CH_3 + H-OH$$
Acetic anhydride

(Carboxylic acid anhydrides are of little interest in biochemistry. They easily react with water to give back acids in the reverse of the reaction described here and are not found as biomolecules.)

■ **PROBLEM 17.8**

Write condensed structures for propionic acid and an ester and unsubstituted, *N*-monosubstituted, and *N,N*-disubstituted amides derived from propionic acid. Name the derivatives.

■ **PROBLEM 17.9**

What are the names of the following compounds?

(a) 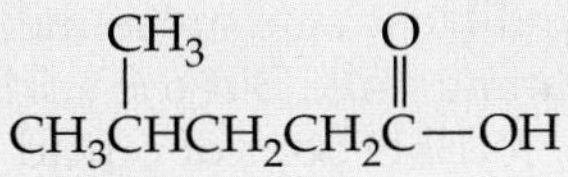 $CH_3\overset{\overset{\textstyle CH_3}{|}}{C}HCH_2CH_2\overset{\overset{\textstyle O}{\|}}{C}-OH$

(b) 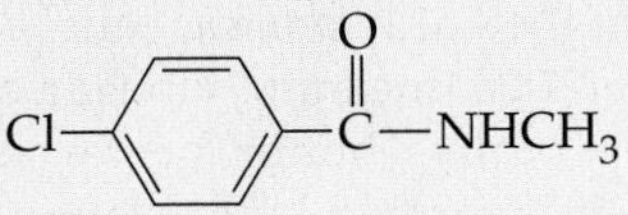 $Cl-\langle\text{benzene ring}\rangle-\overset{\overset{\textstyle O}{\|}}{C}-NHCH_3$

■ **PROBLEM 17.10**

Draw structures corresponding to these names:
(a) 4-Methylpentanamide   (b) *N*-Methylbutanamide

■ **PROBLEM 17.11**

Match partial structures (a)–(d) with the following classes of compounds: (i) *α*-amino carboxylic acid, (ii) monosubstituted amide, (iii) methyl ester, (iv) carboxylic acid.

(a) 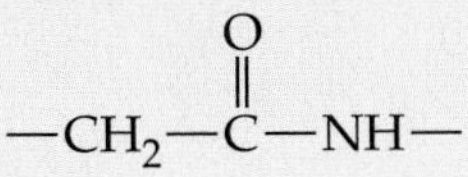 $-CH_2-\overset{\overset{\textstyle O}{\|}}{C}-NH-$

(b) $-\overset{\overset{\textstyle O}{\|}}{\underset{\underset{\textstyle NH_2}{|}}{C}}H-\overset{\overset{\textstyle O}{\|}}{C}-OH$

(c) $-CH_2-\overset{\overset{\textstyle O}{\|}}{C}-OH$

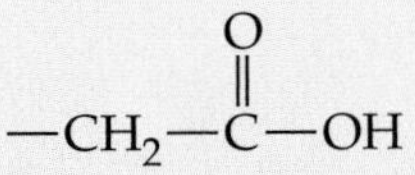

(d) 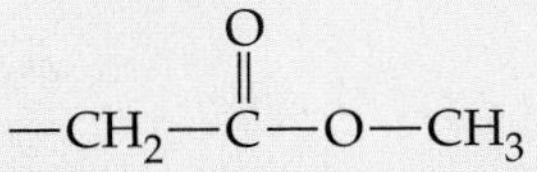 $-CH_2-\overset{\overset{\textstyle O}{\|}}{C}-O-CH_3$

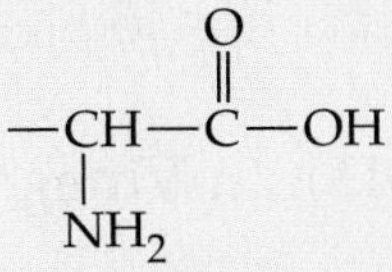

■ **PROBLEM 17.12**

Classify each of compounds (a)–(f) as one of the following: (i) amide, (ii) ester, (iii) carboxylic acid.

(a) CH_3COOCH_3

(b) $RCONHR$

(c) C_6H_5COOH

(d)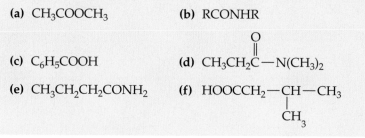

(e) $CH_3CH_2CH_2CONH_2$

(f) $HOOCCH_2—CH—CH_3$
                        $|$
                        CH_3

■ **KEY CONCEPT PROBLEM 17.13**

Identify each of the following molecules as an ester, carboxylic acid, or amide, and write the condensed molecular formula for each.

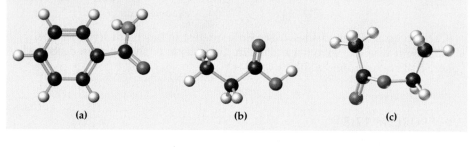

(a)   (b)   (c)

▲ If left out of the refrigerator too long, the butter will develop the terrible odor of butyric acid.

17.2 Some Common Carboxylic Acids

Carboxylic acids occur throughout the plant and animal kingdoms, and the common names of many come from the plants or animals in which they were first identified. Formic acid is known as the chemical that puts the sting in ant bites (from the Latin *formica*, "ant"). Acetic acid and lactic acid are named from the Latin *acetum*, meaning vinegar, and *lactis*, meaning milk. Butyric acid (from the Latin *butyrum*, "butter") is responsible for the terrible odor of rancid butter (in which it is formed by oxidation of butyraldehyde). Caproic acid (from the Latin *caper*, "goat") was first isolated from the skin of goats (which also have a distinctive odor). The esters of long-chain carboxylic acids such as stearic acid are components of all animal fats and vegetable oils.

Some naturally occurring carboxylic acids

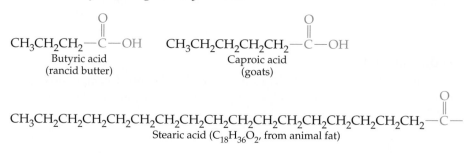

$$CH_3CH_2CH_2—\overset{\overset{\textstyle O}{\|}}{C}—OH$$
Butyric acid
(rancid butter)

$$CH_3CH_2CH_2CH_2CH_2—\overset{\overset{\textstyle O}{\|}}{C}—OH$$
Caproic acid
(goats)

$$CH_3CH_2CH_2CH_2CH_2CH_2CH_2CH_2CH_2CH_2CH_2CH_2CH_2CH_2CH_2CH_2CH_2—\overset{\overset{\textstyle O}{\|}}{C}—OH$$
Stearic acid ($C_{18}H_{36}O_2$, from animal fat)

Acetic Acid (CH_3COOH): In Vinegar

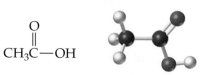

$$CH_3\overset{\overset{\textstyle O}{\|}}{C}—OH$$

Everyone recognizes the sour taste of acetic acid, the most common carboxylic acid, because vinegar is a solution of 4– 8% acetic acid in water (with various flavoring agents). When fermentation of grapes, apples, and other fruits proceeds in the presence of ample oxygen, oxidation goes beyond the formation of ethanol to the formation of acetic acid (∞, Section 14.5). Aqueous acetic acid solutions are also common laboratory reagents.

In concentrations over 50%, acetic acid is corrosive and can damage the skin, eyes, nose, and mouth. There is no pain when the concentrated acid is spilled on unbroken skin, but painful blisters form 30 minutes or more later. Pure acetic acid is known as *glacial acetic acid* because with just a slight amount of cooling below room temperature the liquid forms icy-looking crystals (at 17°C). Acetic acid is a reactant in many industrial processes and is sometimes used as a solvent, especially for other carboxylic acids. As a food additive, it is used to adjust acidity.

Citric Acid: In Citrus Fruits and Blood

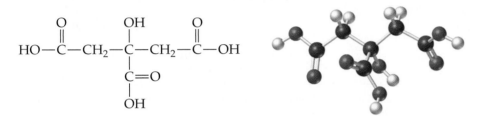

Citric acid is produced by almost all plants and animals during metabolism, and its normal concentration in human blood is about 2 mg/100 mL. Citrus fruits owe their tartness to citric acid; for example, lemon juice contains 4– 8% and orange juice about 1% citric acid. Pure citric acid is a white, crystalline solid (mp 153°C) that is very soluble in water. Citric acid and its salts are extensively used in pharmaceuticals, foods, and cosmetics. They serve to buffer pH in shampoos and hair-setting lotions; or add tartness to candies and soft drinks; or react with bicarbonate ion to produce the fizz in AlkaSeltzer.

∞ LOOKING AHEAD

Citric acid lends its name to the *citric acid cycle*, part of the major biochemical pathway that leads directly to the generation of energy. Citric acid is the product of the first reaction of the eight-reaction cycle, which is presented in Section 21.8. From one to three carboxyl groups are present in each of the reactants in the cycle. ∞

17.3 Acidity of Carboxylic Acids

Carboxylic acids are weak acids that establish equilibria in aqueous solution with **carboxylate ions**, $RCOO^-$. The carboxylate ions are named by replacing the *-ic* ending in the carboxylic acid name with *-ate* (giving the same names and endings used in naming esters). At pH 7.4 in body fluids, carboxylic acids exist mainly as their carboxylate ions.

Carboxylate anion The anion that results from ionization of a carboxylic acid, $RCOO^-$.

$$CH_3\overset{O}{\underset{\|}{C}}-OH + H_2O \rightleftharpoons CH_3\overset{O}{\underset{\|}{C}}-O^- + H_3O^+$$
Acetic acid Acetate ion

$$CH_3\overset{O}{\underset{\|}{C}}-\overset{O}{\underset{\|}{C}}-OH + H_2O \rightleftharpoons CH_3\overset{O}{\underset{\|}{C}}-\overset{O}{\underset{\|}{C}}-O^- + H_3O^+$$
Pyruvic acid Pyruvate ion

Many carboxylic acids have about the same acid strength as acetic acid, as shown by the acid ionization constants in Table 17.2. (∞, p. 274) There are some exceptions, though. Trichloroacetic acid, used in preparing microscope slides and in precipitating protein from body fluids, is a strong acid that must be handled with the same respect as sulfuric acid.

Name	Structure	Acid Dissociation Constant (K_a)
Trichloroacetic acid	Cl_3CCOOH	2.3×10^{-1}
Chloroacetic acid	$ClCH_2COOH$	1.4×10^{-3}
Formic acid	$HCOOH$	1.8×10^{-4}
Acetic acid	CH_3COOH	1.8×10^{-5}
Propanoic acid	CH_3CH_2COOH	1.3×10^{-5}
Hexanoic acid	$CH_3(CH_2)_4COOH$	1.3×10^{-5}
Benzoic acid	C_6H_5COOH	6.5×10^{-5}
Acrylic acid	$H_2C{=}CHCOOH$	5.6×10^{-5}
Oxalic acid	$HOOCCOOH$	5.4×10^{-2}
	$^-OOCCOOH$	5.2×10^{-5}
Glutaric acid	$HOOC(CH_2)_3COOH$	4.5×10^{-5}
	$^-OOC(CH_2)_3COOH$	3.8×10^{-6}

TABLE 17.2 Carboxylic Acid Dissociation Constants*

*The acid dissociation constant K_a is the equilibrium constant for the ionization of an acid; the smaller its value, the weaker the acid:

$$RCOOH + H_2O \rightleftharpoons RCOO^- + H_3O^+ \qquad K_a = \frac{[RCOO^-][H_3O^+]}{[RCOOH]}$$

Carboxylic acid salt An ionic compound containing a carboxylic acid anion and a cation.

Carboxylic acids undergo neutralization reactions with bases as do other acids. With strong bases such as sodium hydroxide, a carboxylic acid reacts to give water and a **carboxylic acid salt**, as shown here for the formation of sodium acetate. (Note that, like all such aqueous acid-base reactions, this reaction proceeds much more favorably in the forward direction than in the reverse direction, and is thus written with a single arrow.) As for all salts, a carboxylic acid salt is named with cation and anion names.

 Nomenclature of Carboxylic Acids and Their Derivatives

$$\underset{\substack{\text{Acetic acid}\\\text{(a weak acid)}}}{CH_3{-}\overset{\overset{\displaystyle O}{\|}}{C}{-}O{-}H(aq)} + \underset{\substack{\text{Sodium}\\\text{hydroxide}}}{Na^+\ OH^-(aq)} \longrightarrow \underset{\text{Sodium acetate}}{CH_3{-}\overset{\overset{\displaystyle O}{\|}}{C}{-}O^-\ Na^+(aq)} + H_2O$$

The sodium and potassium salts of carboxylic acids are ionic solids that are usually far more soluble in water than the carboxylic acids themselves and are used when the solubility of an acid must be increased. For example, benzoic acid has a water solubility of only 3.4 g/L at 25°C, whereas for sodium benzoate the water solubility is 550 g/L. The formation of carboxylic acid salts, like the formation of amine salts, is useful in creating water-soluble derivatives of drugs. (See Application "Organic Compounds in Body Fluids and the 'Solubility Switch,'" Chapter 15.)

■ **WORKED EXAMPLE 17.5**

Write the structural formulas of trichloroacetic acid and acetic acid and explain why trichloroacetic acid is the much stronger acid of the two.

SOLUTION

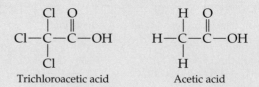

Trichloroacetic acid   Acetic acid

Application Acids for the Skin

A strongly acid solution can damage the skin as seriously as a flame can. Nevertheless, as administered by physicians such solutions have found a place in the treatment of a variety of skin conditions. And weakly acidic solutions used in skin treatments fall at the borderline of the distinction between prescription drugs and cosmetics.

Trichloroacetic acid, a strong acid (see Table 17.2 in Section 17.3), is used for chemical peeling of the skin (a treatment for eczema or psoriasis), for removal of acne scars, and sometimes for removing wrinkles. In what is the equivalent of a first- or second-degree burn, a surface layer of the skin is destroyed by reaction with the acid. The depth of removal (and the length of the healing period) varies with the strength of the acid and how long it is left on the skin. As the result of healing, old skin is replaced by a new, smoother skin surface.

A number of naturally occurring α-hydroxy acids that are weak acids (similar to acetic acid) provide less aggressive skin treatments.

▲ A skin-care product relying on the benefits of an alpha-hydroxy acid.

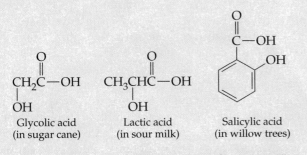

| Glycolic acid (in sugar cane) | Lactic acid (in sour milk) | Salicylic acid (in willow trees) |

Glycolic acid is used by physicians for a "mini" peel that can be done quickly and from which the skin returns to a normal appearance in just a few hours. In higher concentrations, glycolic acid is used for "spot" removal of precancerous lesions or unsightly brown, thickened skin (*keratoses*).

The effectiveness of α-hydroxy acids for conditions that produce dry, flaking, itchy skin has brought them into the cosmetics market. In recent years, "alpha-hydroxys" have been highly promoted for removing wrinkles, moisturizing skin, and generally making us more beautiful. The alpha-hydroxys have a further promotional advantage because they are all "nature's own" chemicals. For example, lactic acid, which at 12% concentration is a prescription drug, is present in many over-the-counter lotions and creams at lower concentration.

While the use by adults of cosmetics containing 10% or less of alpha-hydroxy acid is considered safe, one general caution has been noted. Use of alpha-hydroxys may increase the sensitivity of the skin to ultraviolet radiation from the sun. Thus, an individual using alpha-hydroxys should be diligent about applying a sunscreen of at least SPF 15 before going outside.

See Additional Problems 17.76–77 at the end of the chapter.

The structural difference is the replacement of three hydrogen atoms on the alpha carbon by three chlorine atoms. The chlorines are much more electronegative than the hydrogen and therefore draw electrons away from the rest of the molecule in trichloroacetic acid. (○○○, p. 117) The result is that the hydrogen atom in the —COOH group in trichloroacetic acid is held less strongly and is much more easily removed than the corresponding hydrogen atom in acetic acid, making trichloroacetic acid a much stronger acid.

■ **PROBLEM 17.14**

Write the products of the following reactions:

(a) $CH_3CH_2CH_2COOH + KOH \rightarrow ?$

(b) 2-Methylpentanoic acid $+ Ba(OH)_2 \rightarrow ?$

Application Acid Salts As Food Additives

Run your eyes over the lists of ingredients on the labels of soft drinks, cookies, cakes, dried sauce mixes, or preserved meats. Chances are you'll find the names of some acid salts.

Scanning the label on a package of strawberry jam-filled cookies, for example, turns up *sodium benzoate, sodium propionate, potassium sorbate,* and *sodium citrate.* What's the purpose of all these food additives? The first three are preservatives. Sodium benzoate prevents the growth of microorganisms, especially in acidic foods. Used in many soft drinks, it's one of the most common food additives. Sodium propionate, also very common, prevents the growth of mold in baked goods. Potassium sorbate, the salt of an unsaturated acid that occurs naturally in many plants (sorbic acid, $CH_3CH\!\!=\!\!CHCH\!\!=\!\!CHCOOH$), is a good mold and fungus inhibitor. The fourth ingredient is the trisodium salt of citric acid. In the cookies, it's combined with citric acid to buffer the acidity of the strawberry jam.

A package of dehydrated cream sauce with bacon bits and noodles includes a group of less familiar acid salts. Disodium guanylate, disodium inosinate, and monosodium glutamate are salts of acids that occur naturally in meats. As food additives, these salts serve as flavor enhancers by imparting a "meaty" flavor and are sometimes found in packaged foods that should taste "meaty" but don't contain much meat.

Many foods also contain two other additives that are salts, not of carboxylic acids, but of compounds with acidic hydroxyl groups: sodium ascorbate and sodium erythorbate. Ascorbic acid is vitamin C, and erythorbic acid is an isomer of vitamin C. Both salts act as antioxidants (Section 21.10) and help to maintain the color of cured meat such as bacon, though only the ascorbate has value as a vitamin.

▲ Certain acid salts can keep this from happening to your bread so quickly.

All the additives mentioned are salts of acids that occur naturally in plants and animals, and all have been approved by the U.S. Food and Drug Administration (FDA). Some additives are essential— without preservatives certain foods would harbor disease-causing microorganisms. Other additives make convenience foods possible or make food more appealing by enhancing flavor, color, or consistency. It could be argued that such additives are not essential, although that doesn't necessarily mean they are harmful. Do we really need dry powders that turn into cream sauce when water, butter, and milk are added? Each of us must decide this for ourselves.

See Additional Problems 17.78– 79 at the end of the chapter.

Common food additives

Sodium benzoate

Sodium ascorbate
(an antioxidant)

Monosodium glutamate
(flavor enhancer)

■ PROBLEM 17.15
Write the formulas of calcium formate and sodium acrylate (refer to Table 17.1).

■ PROBLEM 17.16
Suppose that sodium acetate and disodium glutarate are dissolved in water. Write the formulas of the ions present in the solution (refer to Table 17.1).

17.4 Reactions of Carboxylic Acids: Ester and Amide Formation

The reactions of alcohols and amines with carboxylic acids follow the same pattern—both result in substitution of other groups for the —OH of the acid and formation of water as a by-product. With alcohols the —OH of the acid is replaced by the —OR of the alcohol. With amines the —OH of the acid is replaced by the —NR_2 of the amine.

Ester formation

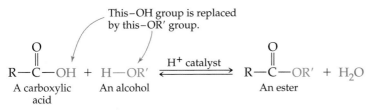

Amide formation

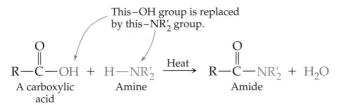

Esterification

In the laboratory, ester formation, known as **esterification**, is carried out by warming a carboxylic acid with an alcohol in the presence of a strong-acid catalyst such as sulfuric acid. For example:

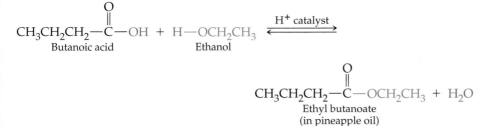

Esterification reactions are reversible and often reach equilibrium with substantial amounts of both reactants and products present. Ester formation is favored either by using a large excess of the alcohol or by continuously removing one of the products (for example, by distilling off a low-boiling ester). Both techniques are applications of Le Châtelier's principle. (◐◑, p. 183)

> ■ **WORKED EXAMPLE 17.6**
>
> The flavor ingredient in oil of wintergreen is an ester that can be made by reaction of *o*-hydroxybenzoic acid (salicylic acid) with methanol. What is its structure?
>
> $$\text{(structure)} + CH_3OH \xrightarrow{H^+} ?$$

Esterification The reaction between an alcohol and a carboxylic acid to yield an ester plus water.

▲ Pineapples are among the many fruits with flavors derived from esters.

SOLUTION   First, write the two reaction partners so that the —COOH group of the acid and the —OH group of the alcohol are facing each others:

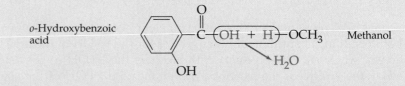

Next, remove —OH from the acid and —H from the alcohol to form water and then join the two resulting organic fragments with a single bond. The product is the ester:

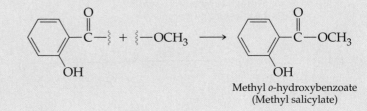

Methyl *o*-hydroxybenzoate
(Methyl salicylate)

■ PROBLEM 17.17

Raspberry oil contains an ester that can be made by reaction of formic acid with 2-methyl-1-propanol. What is its structure?

$$HCOOH + (CH_3)_2CHCH_2OH \rightarrow ?$$

■ PROBLEM 17.18

What carboxylic acid and what alcohol are needed to make each of the following esters?

(a)

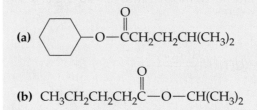

(b) $CH_3CH_2CH_2CH_2\overset{\text{O}}{\overset{\|}{C}}{-}O{-}CH(CH_3)_2$

Amide Formation

Unsubstituted amides are formed by the reaction of carboxylic acids with ammonia, for example:

Acetic acid   Acetamide

Substituted amides are produced in reactions between primary or secondary amines and carboxylic acids, for example:

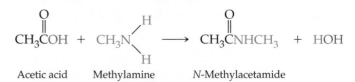

Acetic acid   Methylamine   *N*-Methylacetamide

The amide formation reactions must be heated to proceed as shown. In each case, the overall reaction is formation of an amide accompanied by formation of water by the —OH group of the acid and an —H atom from ammonia or an amine.

∞∞∞ LOOKING AHEAD

Proteins are constructed of long chains of amino acids held together by amide bonds.

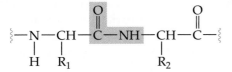

The biochemical synthesis of proteins, described in Section 26.10, is a strictly controlled process in which amino acids with different R groups must be assembled in exactly the correct order. ∞∞∞

■ WORKED EXAMPLE 17.7

The mosquito and tick repellent DEET (diethyltoluamide) can be prepared by reaction of diethylamine with *p*-methylbenzoic acid (toluic acid). What is the structure of DEET?

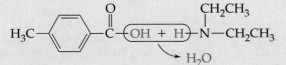

SOLUTION First, rewrite the equation so that the —OH of the acid and the —H of the amine are facing each other:

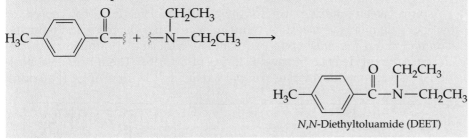

Next, remove the —OH from the acid and the —H from the nitrogen atom of the amine to form water. Then join the two resulting fragments together to form the amide product.

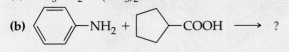

N,N-Diethyltoluamide (DEET)

■ PROBLEM 17.19

Draw structures of the amides that can be made from the reactants below:

(a) $CH_3NH_2 + (CH_3)_2CHCOOH \longrightarrow$?

(b)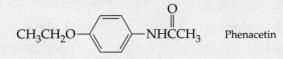

■ PROBLEM 17.20

What carboxylic acid and what amine would you use if you wanted to prepare phenacetin, once used in headache remedies but now banned because of its potential for causing kidney damage?

$$CH_3CH_2O- \hspace{-0.5em}\bigcirc\hspace{-0.5em} -NHCCH_3 \quad \text{Phenacetin}$$

17.5 Aspirin and Other Over-the-Counter Carboxylic Acid Derivatives

Esters and amides have many uses in medicine, in industry, and in living systems. Their importance in the industrial production of polyesters and polyamides is explored in Section 17.7. Here we discuss a few of the most familiar over-the-counter medications that are carboxylic acid derivatives.

Aspirin and Other Salicylic Acid Derivatives

Aspirin is a white, crystalline solid (mp 135°C) that's a member of the group of drugs known as salicylates: esters of salicyclic acid. Chemically, aspirin is acetylsalicylic acid, an ester formed between acetic acid and the —OH group of salicylic acid.

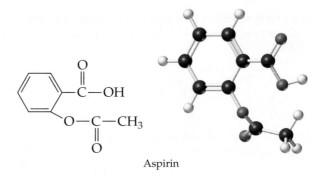

Aspirin

As early as the fifth century BC, Hippocrates knew that chewing the bark of a willow tree would relieve pain. By the 1800s, salicylic acid had been identified as the active ingredient produced in the body from willow bark, and many chemical modifications of salicylic acid were investigated. Outstanding success came with aspirin. Aspirin is now over 100 years old as a widely available medication, having been brought to market in 1899 in Germany by Frederich Bayer and Company (Figure 17.1).

Aspirin has an amazing array of therapeutic actions. It is best known for providing pain relief (an *analgesic*), reducing fever (an *antipyretic*), and reducing inflammation (an *anti-inflammatory*). As an analgesic and anti-inflammatory, it is the first line of defense in several forms of arthritis. There is even a beef-flavored aspirin on the market for pain relief in arthritic dogs (Palaprin).

FIGURE 17.1 Pharmaceutical ad from the 1900s. The ad points out that aspirin was a substitute for salicylates. Substitutes have since been found for some of the other pharmaceuticals listed, as well.

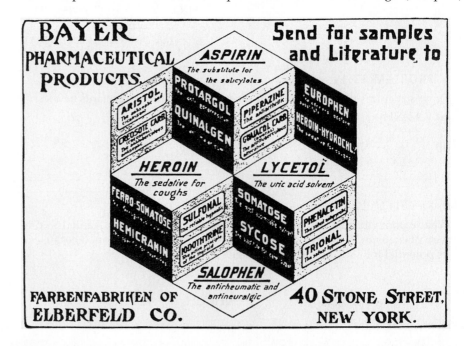

(Because there is a greater risk of toxicity in dogs, the doggy aspirin is available only through veterinarians and must be used with careful supervision.)

For a drug that has been in use so long and in such large quantity (29 billion tablets per year in the United States), it's amazing that discoveries about the physiological effects of aspirin are still being made. In recent years, aspirin has been found to inhibit the clumping of blood platelets and thereby to protect against heart attacks caused by blood clots. Small regular doses of aspirin (for example, one baby aspirin, which is 85 mg, per day as compared to 325 mg in the usual tablet) are being recommended for some individuals at risk for heart attack. (See Section 24.9 for the chemical action of aspirin.)

In 1982 the relationship was discovered between *Reye's syndrome*, a rare children's liver disorder, and the use of aspirin to treat influenza or chickenpox. It is now strongly recommended that aspirin not be given to a child whose fever might be caused by these diseases.

The principal undesirable side effects of aspirin are gastric bleeding and gastrointestinal distress. Aspirin is only slightly ionized in the acidic environment of the stomach but causes trouble once inside the stomach lining, where its ionization produces ions that do not easily exit across cell membranes. Aspirin inhibits the action of an enzyme necessary for coagulation of blood cells. For several days after taking two aspirin tablets, the time it takes for bleeding to stop is doubled. Regular use of aspirin as an anti-inflammatory can cause the loss of 3– 8 mL of blood a day. The usual blood loss of a few milliliters per tablet is not harmful, but more extensive bleeding can occur in susceptible individuals.

Numerous efforts have been made to produce a modification of aspirin that would retain all of its beneficial effects while eliminating the negative ones, but no salicylate has yet proved as effective. The sodium salt of aspirin causes less bleeding, but has no action against blood clotting. Conversion of the carboxylic acid group to an amide gives a medication with very little gastric irritation but no anti-inflammatory activity and unreliable benefits.

Methyl salicylate, or oil of wintergreen (Worked Example 17.6), is too poisonous to be of any value as an oral medication. Therapeutically it is useful as a *counterirritant*, a substance that relieves internal pain by stimulating nerve endings in the skin. Methyl salicylate is one of the active ingredients in liniments such as Ben Gay and Heet.

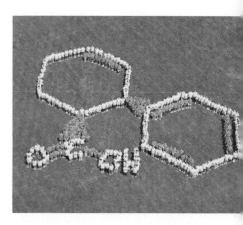

▲ The 18-millionth chemical compound. In June 1998, 300 people formed this structure to celebrate the registration by *Chemical Abstracts* of its 18-millionth compound, a carboxylic acid with potential as an anti-inflammatory and pain-relieving medication. By 2001 over 20 million compounds had been registered.

Acetaminophen

An alternative to aspirin for pain relief is acetaminophen (best known by the trade name Tylenol), an amide that also contains a hydroxyl group:

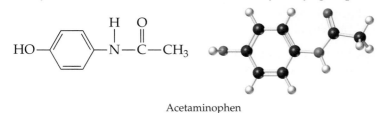

Acetaminophen

Like aspirin, acetaminophen reduces fever, but unlike aspirin it is not an anti-inflammatory agent. The major advantage of acetaminophen as compared to aspirin is that because it is not a carboxylic acid it does not induce internal bleeding. For this reason, it is the most valuable simple pain reliever for individuals prone to bleeding or recovering from surgery or wounds. Overdoses of acetaminophen can cause kidney and liver damage, however.

Benzocaine and Lidocaine

Benzocaine is a local anesthetic used in many over-the-counter *topical* preparations (those applied to the skin surface) for such conditions as cold sores, poison ivy, sore throats, and hemorrhoids. Like other local anesthetics, it works by blocking the transmission of impulses by sensory nerves. Benzocaine is one of a

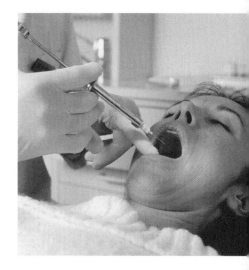

▲ Lidocaine is commonly used as a dental pain killer.

family of structurally related local anesthetics that includes lidocaine (Xylocaine), which is most commonly administered by injection to prevent pain during dental work. Because benzocaine is less soluble than lidocaine, it cannot be used in this manner.

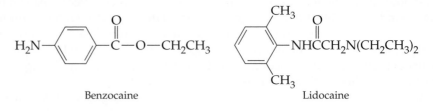

Benzocaine   Lidocaine

■ **PROBLEM 17.21**

Salsalate is another salicylate used as an aspirin alternative for those who are hypersensitive to aspirin. Draw the structure of salicylic acid and the structure of salsalate, which is an ester formed by the reaction of two molecules of salicylic acid.

■ **KEY CONCEPT PROBLEM 17.22**

Examine the structures of aspirin, acetaminophen, benzocaine, and lidocaine. For each of these compounds, indicate whether it is acidic, basic, or neither.

17.6 Hydrolysis of Esters and Amides

Recall that in hydrolysis a bond or bonds are broken and the H— and —OH of water add to the atoms that had been in the broken bond. Esters and amides undergo hydrolysis to give back carboxylic acids plus alcohols or amines in reactions that follow the carbonyl-group substitution pattern (⚬⚬⚬, p. 470).

For esters, the net effect of hydrolysis is substitution of —OH for —OR:

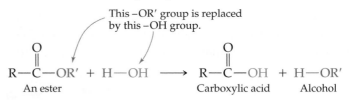

For amides, the net effect of hydrolysis is substitution of —OH for —NH₂ or the substituted amide nitrogen:

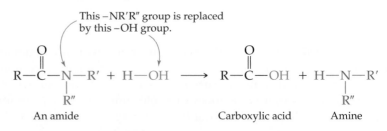

Ester Hydrolysis

Both acids and bases can cause ester hydrolysis. Acid-catalyzed hydrolysis is simply the reverse of the esterification. An ester is treated with water in the presence of a strong acid such as sulfuric acid, and hydrolysis takes place. An excess of water pushes the equilibrium to the right. For example:

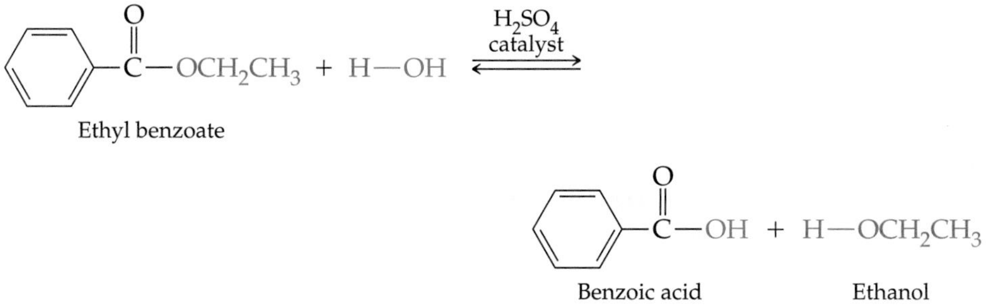

Ethyl benzoate

Benzoic acid   Ethanol

Ester hydrolysis by reaction with a base such as NaOH or KOH is known as **saponification** (after the Latin word *sapo*, "soap"). The product of saponification is a carboxylate anion rather than a free carboxylic acid. (The use of saponification in making soap is discussed in Section 24.4.)

Saponification The reaction of an ester with aqueous hydroxide ion to yield an alcohol and the metal salt of a carboxylic acid.

Methyl butanoate

Sodium butanoate   Methanol

■ WORKED EXAMPLE 17.8

What product would you obtain from acid-catalyzed hydrolysis of ethyl formate, a flavor constituent of rum?

Ethyl formate

SOLUTION The name of an ester gives a good indication of the names of the two products. Thus, ethyl formate yields ethyl alcohol and formic acid. To find the product structures in a more systematic way, write the structure of the ester and locate the bond between the carbonyl-group carbon and the — OR ′ group:

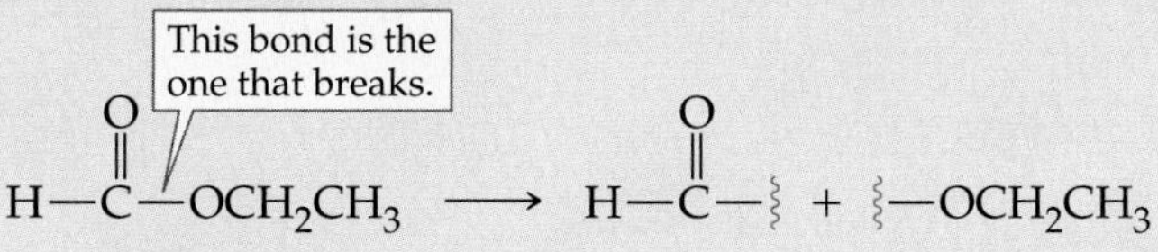

This bond is the one that breaks.

Next, carry out a hydrolysis reaction on paper. First form the carboxylic acid product by connecting an — OH to the carbonyl-group carbon. Then add an — H to the — OCH$_2$CH$_3$ group to form the alcohol product.

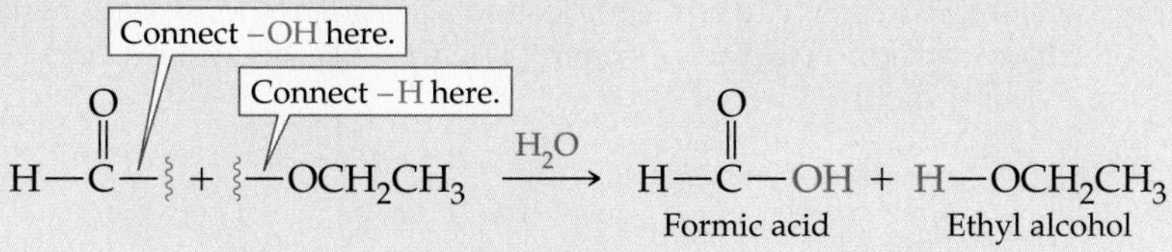

Connect –OH here.

Connect –H here.

Formic acid   Ethyl alcohol

■ PROBLEM 17.23

If a bottle of aspirin tablets has the aroma of vinegar, it is time to discard those tablets. Explain why, and include a chemical equation in the explanation.

■ **PROBLEM 17.24**

What products would you obtain from acid-catalyzed hydrolysis of the following esters?

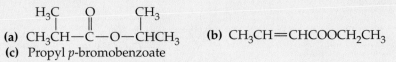

(a) CH₃CH—C—O—CHCH₃   **(b)** CH₃CH=CHCOOCH₂CH₃

(c) Propyl *p*-bromobenzoate

Amide Hydrolysis

Amides are stable in water but undergo hydrolysis with heating in the presence of acids or bases. The products are the carboxylic acid and amine from which the amide could have been synthesized.

$$\underset{\substack{\|\\O}}{RC}-NHR + H-OH \longrightarrow \underset{\substack{\|\\O}}{RC}-OH + RN\underset{H}{\overset{H}{\diagdown}}$$

The products are the ions formed by the acid or the amine in the presence of bases or acids, respectively. For example, in the hydrolysis of *N*-methylacetamide:

$$\underset{\substack{\|\\O}}{CH_3C}-NHCH_3 + H_3O^+ \longrightarrow \underset{\substack{\|\\O}}{CH_3C}OH + CH_3NH_3^+$$

$$\underset{\substack{\|\\O}}{CH_3C}-NHCH_3 + OH^- \longrightarrow \underset{\substack{\|\\O}}{CH_3C}-O^- + CH_3NH_2$$

∞ LOOKING AHEAD

In Chapter 28 you will see that the cleavage of amide bonds by hydrolysis is the key process that occurs in the stomach during digestion of proteins.   ∞

■ **WORKED EXAMPLE 17.9**

What are the carboxylic acid and amine produced by the hydrolysis of *N*-ethylbutanamide?

$$CH_3CH_2CH_2-\underset{\substack{\|\\O}}{C}-NHCH_2CH_3 + H_2O \longrightarrow ?$$

N-Ethylbutanamide

SOLUTION   First, look at the name of the starting amide. Often, the name of the amide indicates the names of the two products. Thus, *N*-ethylbutanamide will yield ethylamine and butanoic acid. To be more systematic about finding the product structures, write the amide and locate the bond between the carbonyl-group carbon and the nitrogen. Then break this amide bond and write the two fragments:

This amide bond is the one that breaks.

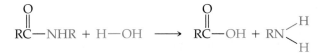

CH₃CH₂CH₂C—NHCH₂CH₃ → CH₃CH₂CH₂C—⟩ + ⟨—NHCH₂CH₃

Next, carry out a hydrolysis reaction on paper and form the products by connecting an —OH to the carbonyl-group carbon and an —H to the nitrogen:

Connect –OH here.

Connect –H here.

$$CH_3CH_2CH_2C\overset{O}{\underset{}{\|}}\!\!-\!\!\{ \ + \ \}\!\!-\!\!NHCH_2CH_3 \xrightarrow{H_2O}$$

$$CH_3CH_2CH_2\overset{O}{\underset{}{\overset{\|}{C}}}\!\!-\!\!OH \ + \ H\!\!-\!\!NHCH_2CH_3$$

Butanoic acid   Ethylamine

■ PROBLEM 17.25

What carboxylic acids and amines result from hydrolysis of the following amides?

(a) $CH_3CH\!\!=\!\!CH\overset{O}{\underset{}{\overset{\|}{C}}}NHCH_3$ **(b)** *N,N*-Diethyl-*p*-chlorobenzamide

---

17.7 Polyamides and Polyesters

Imagine what would happen if a molecule with *two* carboxylic acid groups were to react with a molecule that has *two* amino groups. Amide formation could join the two molecules together, but further reactions could then link more and more molecules together until a giant chain resulted. This is exactly what happens when certain kinds of synthetic polymers are made.

Nylons are *polyamides* produced by reaction of diamines with diacids. One type of nylon (nylon 66) is made by heating adipic acid (hexanedioic acid) with hexamethylenediamine (1,6-hexanediamine) at 280°C.

Synthesis of Nylon 610

$$n \ HOOC\!\!-\!\!(CH_2)_4\!\!-\!\!COOH$$
Adipic acid

$$+$$

$$n \ H_2N\!\!-\!\!(CH_2)_6\!\!-\!\!NH_2$$
Hexamethylenediamine

$$\xrightarrow[-H_2O]{280°} \left[\overset{O}{\underset{}{\overset{\|}{C}}}\!\!-\!\!(CH_2)_4\!\!-\!\!\overset{O}{\underset{}{\overset{\|}{C}}}\!\!-\!\!NH\!\!-\!\!(CH_2)_6\!\!-\!\!NH\right]_n$$
Nylon 66, a polyamide
(repeating unit)

The polymer molecules are composed of thousands of the repeating unit, shown above enclosed in square brackets. (You'll see in the next chapter that proteins are also polyamides; they differ from nylon in having 20 different amino acids in the polymer chain, rather than identical repeating units.)

The properties of nylon suit it for many kinds of applications. High impact strength, abrasion resistance, and a naturally slippery surface make nylon an excellent material for bearings and gears. It can be formed into very strong fibers, making it valuable for a range of applications from nylon stockings, to clothing, to mountaineering ropes and carpets. Sutures and replacement arteries are also fabricated from nylon, which is resistant to deterioration in body fluids.

Just as diacids and diamines react to yield polyamides, diacids and dialcohols react to yield *polyesters*. The most widely used polyester is made by the reaction of terephthalic acid (1,4-benzenedicarboxylic acid) with ethylene glycol:

▲ Nylon being pulled from the interface between adipic acid and hexamethylenediamine.

$$n \ HO\!\!-\!\!\overset{O}{\underset{}{\overset{\|}{C}}}\!\!-\!\!\!\!\raisebox{-2pt}{⬡}\!\!\!\!-\!\!\overset{O}{\underset{}{\overset{\|}{C}}}\!\!-\!\!OH \ + \ n \ HO\!\!-\!\!CH_2\!\!-\!\!CH_2\!\!-\!\!OH \xrightarrow{-H_2O}$$

Terephthalic acid   Ethylene glycol

$$\left[\overset{O}{\underset{}{\overset{\|}{C}}}\!\!-\!\!\!\!\raisebox{-2pt}{⬡}\!\!\!\!-\!\!\overset{O}{\underset{}{\overset{\|}{C}}}\!\!-\!\!O\!\!-\!\!CH_2\!\!-\!\!CH_2\!\!-\!\!O\right]_n$$

Poly(ethylene terephthalate), a polyester
(repeating unit)

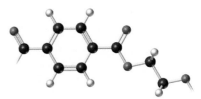

Repeating unit of polyester

Application Kevlar: A Life-Saving Polymer

You're a police officer about to take part in capturing a dangerous, well-armed criminal. Your temperament and training will, you know, protect you. But there's another protection you are very grateful for—your bulletproof vest. It is filled with fibers made of Kevlar, an amazing polymer that protects police officers, firefighters, bicycle riders, lumberjacks, and others engaged in hazardous activities.

Kevlar is a polyamide created at DuPont by chemist Stephanie L. Kwolek and introduced to commercial applications in the 1970s. Today there is an ever-expanding list of applications for Kevlar. In its various forms it is five times stronger than steel, almost half as dense as fiberglass, highly resistant to damage by chemicals, dimensionally stable, very difficult to cut or break, a poor electrical conductor, and flame-resistant (if ignited, it self-extinguishes).

Like nylon, Kevlar is produced by the reaction of a dicarboxylic acid with a diamine, and because it contains aromatic rings, it is classified as a *polyaramide*.

▲ Kevlar chaps offer protection against a straying chain saw.

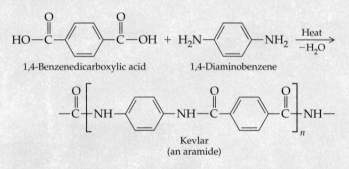

1,4-Benzenedicarboxylic acid   1,4-Diaminobenzene

Kevlar
(an aramide)

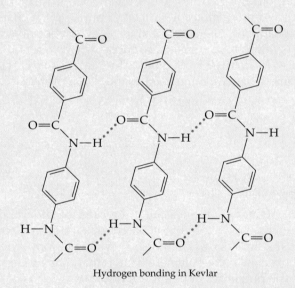

Hydrogen bonding in Kevlar

The great strength of Kevlar results from the way that uniformly arranged hydrogen bonding holds the polymer chains together, as shown in the structural diagram at the right.

Some Uses of Kevlar

- Bulletproof vests
- Heat-protective apparel
- Cut-resistant gloves and other apparel
- Helmets for firefighters and bicycle riders
- Automotive and industrial hoses
- Structural composites for boats and aircraft
- Emergency tow lines for boats
- Brake linings and other friction-resistant applications

In 1995, Kwolek was inducted into the National Inventors Hall of Fame. Her achievements in designing a tough polymer are thus recognized alongside those of inventors such as Louis Pasteur, Alexander Graham Bell, Thomas Edison, and Henry Ford.

See Additional Problem 17.80 at the end of the chapter.

We know this polyester best in clothing fiber, where it has the trade name Dacron. Under the name Mylar it is used in plastic film and recording tape. Its chemical name, poly(ethylene terephthalate) or PET, is usually applied when it is used in clear, flexible soft-drink bottles.

■ **KEY CONCEPT PROBLEM 17.26**

Give the structure of the repeating units in the polymers that would be formed in the reactions of the following compounds.

(a) n HOCCH$_2$CH$_2$COH + n HOCH$_2$CH$_2$OH

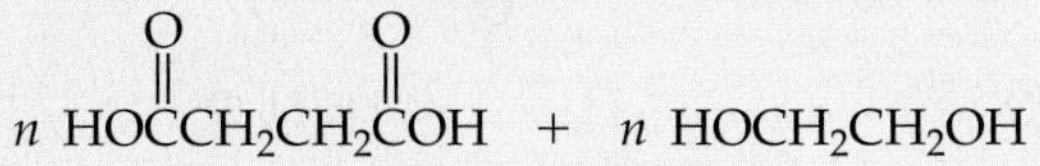

(b) n HOC—⟨ ⟩—COH + n H$_2$NCH$_2$CH$_2$NH$_2$

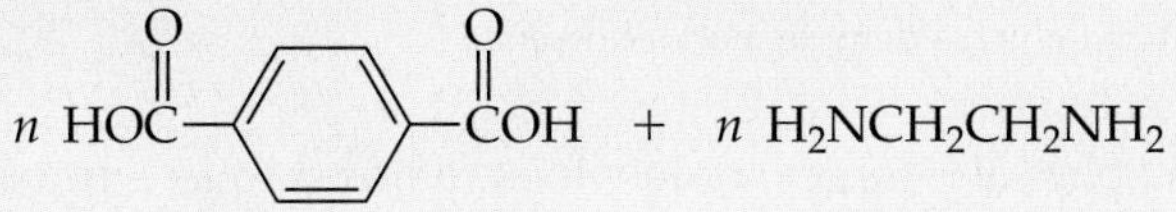

17.8 Phosphoric Acid Derivatives

Phosphoric acid is an inorganic acid with three ionizable hydrogen atoms (red), allowing it to form three different anions:

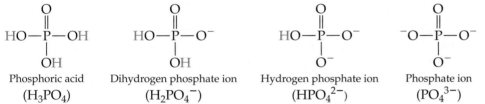

Phosphoric acid
(H$_3$PO$_4$)   Dihydrogen phosphate ion
(H$_2$PO$_4^-$)   Hydrogen phosphate ion
(HPO$_4^{2-}$)   Phosphate ion
(PO$_4^{3-}$)

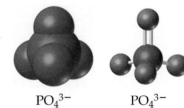

PO$_4^{3-}$ PO$_4^{3-}$

Phosphoric acid reacts with alcohols to form **phosphate esters**. It may be esterified at one, two, or all three of its —OH groups by reaction with an alcohol. Reaction with one molecule of methanol gives the monoester:

Phosphate ester A compound formed by reaction of an alcohol with phosphoric acid; may be a monoester, ROPO$_3$H$_2$; a diester, (RO)$_2$PO$_3$H; or a triester, (RO)$_3$PO; also may be a di- or triphosphate.

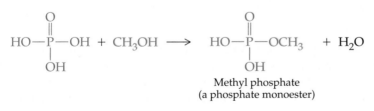

Methyl phosphate
(a phosphate monoester)

The corresponding diester and triester are

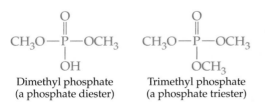

Dimethyl phosphate
(a phosphate diester)   Trimethyl phosphate
(a phosphate triester)

Phosphate monoesters and diesters are acids because they still contain acidic hydrogen atoms. Thus, in neutral or alkaline solutions, including most body fluids, they are present as ions. In biochemical formulas and equations, the phosphate groups are therefore usually written in their ionized forms. For example, you will most often see the formula for glyceraldehyde monophosphate, a key intermediate in the metabolism of glucose (Section 23.2) written as an ion in one of the two ways shown below:

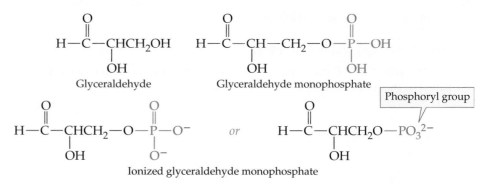

Glyceraldehyde

Glyceraldehyde monophosphate

Phosphoryl group

Ionized glyceraldehyde monophosphate

Phosphoryl group The $-PO_3^{2-}$ group in organic phosphates.

The $-PO_3^{2-}$ group as part of a larger molecule is referred to as a **phosphoryl group**.

Two molecules of phosphoric acid lose water between them to form a phosphoric acid anhydride. The resulting acid (*pyrophosphoric acid*, or *diphosphoric acid*) reacts with yet another phosphoric acid molecule to give *triphosphoric acid*.

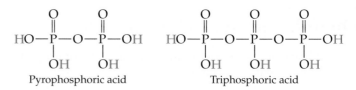

Pyrophosphoric acid

Triphosphoric acid

These acids, too, can form esters, which are known as diphosphates and triphosphates. In the following two methyl phosphate esters, written in their ionized forms, note the difference between the $C-O-P$ ester linkage and the $P-O-P$ phosphoric anhydride linkages:

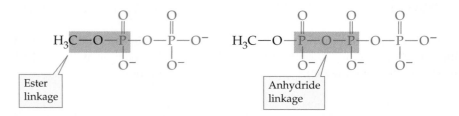

Ester linkage

Anhydride linkage

Phosphorylation Transfer of a phosphoryl group, $-PO_3^2$, between organic molecules.

Transfer of a phosphoryl group from one molecule to another is known as **phosphorylation**. In biochemical reactions, the phosphoryl groups are often provided by a triphosphate (adenosine triphosphate, ATP), which is converted to a diphosphate (adenosine diphosphate, ADP) in a reaction accompanied by the release of energy. The addition and removal of phosphoryl groups is a common mechanism for regulating the activity of biomolecules (⊂∞ Section 19.9).

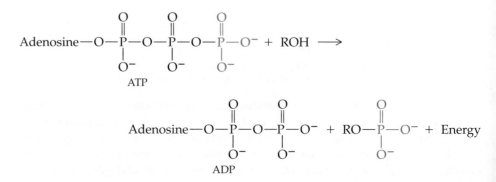

ATP

ADP

Organic Phosphates

- Organic phosphates contain —C—O—P— linkages; those with one, two, or three R groups have the general formulas $ROPO_3H_2$, $(RO)_2PO_2H$, and $(RO)_3PO$.

- Organic phosphates with one or two R groups (monoesters, $ROPO_3^{2-}$, or diesters, $(RO)_2PO_2^-$) are acids and are ionized in body fluids.

- The diphosphate and triphosphate groups, which are important in biomolecules, contain one or two P—O—P anhydride linkages, respectively.

- Phosphorylation is the transfer of a phosphoryl group ($-PO_3^{2-}$) from one molecule to another.

■ PROBLEM 17.27

Write the formula for the phosphate monoester formed from butyl alcohol in both its nonionized and ionized forms.

■ PROBLEM 17.28

Identify the functional group in each of the following compounds and give the structures of the products of hydrolysis of these compounds.

(a) $CH_3\overset{\overset{\displaystyle O}{\|}}{C}NH_2$ (b) $CH_3CH_2OPO_3^{2-}$ (c) $CH_3CH_2\overset{\overset{\displaystyle O}{\|}}{C}OCH_3$

■ PROBLEM 17.29

In the structure of acetyl coenzyme A drawn below, identify a phosphate monoester group, a phosphorus anhydride linkage, two amide groups, and the acetyl group.

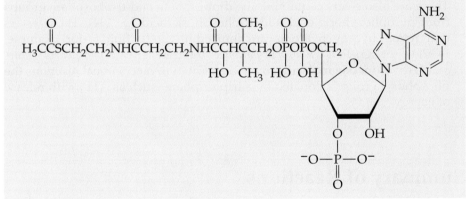

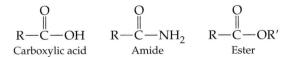

Summary: Revisiting the Chapter Goals

1. What are the general structures and properties of carboxylic acids and their derivatives? *Carboxylic acids*, *amides*, and *esters* have the general structures

$$R-\overset{\overset{\displaystyle O}{\|}}{C}-OH \qquad R-\overset{\overset{\displaystyle O}{\|}}{C}-NH_2 \qquad R-\overset{\overset{\displaystyle O}{\|}}{C}-OR'$$

Carboxylic acid   Amide   Ester

They undergo *carbonyl-group substitution reactions*. Most carboxylic acids are weak acids (a few are strong acids), but esters and amides are neither acids nor bases. Acids and unsubstituted or monosubstituted amides hydrogen-bond with each other, while ester and disubstituted amide molecules do not

Key Words

Acetyl group, p. 471

Acyl group, p. 470

Amide, p. 469

Carbonyl-group substitution reaction, p. 470

Carboxyl group, p. 471

Carboxylate anion, p. 479

Carboxylic acid, p. 469

Carboxylic acid salt, p. 480

Ester, p. 469

Esterification, p. 483

Phosphate ester, p. 493

Phosphoryl group, p. 494

Phosphorylation, p. 494

Saponification, p. 489

do so. Simple acids and esters are liquids; all amides except formamide are solids. The simpler compounds of all three classes are water-soluble or partially water-soluble.

2. **How are carboxylic acids, esters, and amides named?** Many carboxylic acids are best known by their common names (Table 17.1), and these names are the basis for the common names of esters and amides. Esters are named with two words: The first is the name of the alkyl group that has replaced the —H in —COOH, and the second is the name of the parent acid with *-ic acid* replaced by *-ate* (for example, methyl acetate). For amides, the ending *-amide* is used, and where there are R groups on the N, these are named first, preceded by *N* (for example, *N*-methylacetamide).

3. **What are some occurrences and applications of significant carboxylic acids, esters, and amides?** Natural carboxylic acids and esters are common; the acids have bad odors, while esters contribute to the pleasant odors of fruits and flowers. Acetic acid and citric acid occur in vinegar and citrus fruits, respectively. Aspirin and other salicylates are esters; acetaminophen (Tylenol) is an amide; benzocaine is representative of a family of amides that are local anesthetics. Proteins and nylon are polymers containing amide bonds. Fats and oils are esters, as are polyesters such as Dacron.

4. **How are esters and amides synthesized from carboxylic acids and converted back to carboxylic acids?** In ester formation, the —OH of a carboxylic acid group is replaced by the — OR group of an alcohol. In amide formation, the —OH group of a carboxylic acid is replaced by —NH_2 from ammonia or —NHR or —NHR_2 from an amine. Hydrolysis with acids or bases adds —H and —OH to the atoms from the broken bond to restore the carboxylic acid and the alcohol, ammonia, or amine.

5. **What are the organic phosphoric acid derivatives?** Phosphoric acid forms mono-, di-, and triesters: $ROPO_3H_2$, $(RO)_2PO_2H$, and $(RO)_3PO$. There are also esters containing the diphosphate and triphosphate groups from pyrophosphoric acid and triphosphoric acid (p. 494). Those esters that retain hydrogen atoms are ionized in body fluids—for example, $ROPO_3^{2-}$, $(RO)_2PO_2^-$. *Phosphorylation* is the transfer of a *phosphoryl group*, —PO_3^{2-}, from one molecule to another. Often in biochemical reactions the phosphoryl group is donated by a triphosphate (such as ATP) with release of energy.

Summary of Reactions

1. Reactions of carboxylic acids

(a) Acid–base reaction with water (Section 17.3):

$$\underset{\text{CH}_3\overset{\displaystyle O}{\overset{\|}{\text{C}}}\text{OH}}{} + \text{H}_2\text{O} \rightleftharpoons \underset{\text{CH}_3\overset{\displaystyle O}{\overset{\|}{\text{C}}}\text{O}^-}{} + \text{H}_3\text{O}^+$$

(b) Acid–base reaction with a strong base to yield a carboxylic acid salt (Section 17.3):

$$\underset{\text{CH}_3\overset{\displaystyle O}{\overset{\|}{\text{C}}}\text{OH}}{} + \text{NaOH}(aq) \longrightarrow \underset{\text{CH}_3\overset{\displaystyle O}{\overset{\|}{\text{C}}}\text{O}^-\ \text{Na}^+}{} + \text{H}_2\text{O}$$

(c) Substitution with an alcohol to yield an ester (Section 17.4):

$$\underset{\substack{\|\\ \text{CH}_3\text{COH}}}{\overset{\text{O}}{}} + \text{CH}_3\text{OH} \xrightarrow{\text{H}^+} \underset{\substack{\|\\ \text{CH}_3\text{COCH}_3}}{\overset{\text{O}}{}} + \text{H}_2\text{O}$$

(d) Substitution with an amine to yield an amide (Section 17.4):

$$\underset{\substack{\|\\ \text{CH}_3\text{COH}}}{\overset{\text{O}}{}} + \text{CH}_3\text{NH}_2 \longrightarrow \underset{\substack{\|\\ \text{CH}_3\text{CNHCH}_3}}{\overset{\text{O}}{}} + \text{H}_2\text{O}$$

2. Reactions of esters

(a) Hydrolysis to yield an acid and an alcohol (Section 17.6):

$$\underset{\substack{\|\\ \text{CH}_3\text{COCH}_3}}{\overset{\text{O}}{}} \xrightarrow[\text{H}_2\text{O}]{\text{H}^+} \underset{\substack{\|\\ \text{CH}_3\text{COH}}}{\overset{\text{O}}{}} + \text{CH}_3\text{OH}$$

(b) Hydrolysis with a strong base to yield a carboxylate anion and an alcohol (saponification; Section 17.6):

$$\underset{\substack{\|\\ \text{CH}_3\text{CH}_2\text{CH}_2\text{CH}_2\text{CH}_2\text{COCH}_3}}{\overset{\text{O}}{}} + \text{NaOH}(aq) \xrightarrow{\text{H}_2\text{O}}$$

$$\underset{\substack{\|\\ \text{CH}_3\text{CH}_2\text{CH}_2\text{CH}_2\text{CH}_2\text{CO}^-}}{\overset{\text{O}}{}} \text{Na}^+ + \text{CH}_3\text{OH}$$

3. Reactions of amides

(a) Hydrolysis to yield an acid and an amine (Section 17.6):

$$\underset{\substack{\|\\ \text{CH}_3\text{CNHCH}_3}}{\overset{\text{O}}{}} \xrightarrow[\text{H}_2\text{O}]{\text{H}^+ \text{ or OH}^-} \underset{\substack{\|\\ \text{CH}_3\text{COH}}}{\overset{\text{O}}{}} + \text{CH}_3\text{NH}_2$$

4. Phosphate reactions (Section 17.8)

(a) Phosphate ester formation

$$\text{HO}-\overset{\overset{\text{O}}{\|}}{\underset{\underset{\text{OH}}{|}}{\text{P}}}-\text{OH} + \text{CH}_3\text{OH} \longrightarrow \text{HO}-\overset{\overset{\text{O}}{\|}}{\underset{\underset{\text{OH}}{|}}{\text{P}}}-\text{OCH}_3 + \text{H}_2\text{O}$$

(b) Phosphorylation

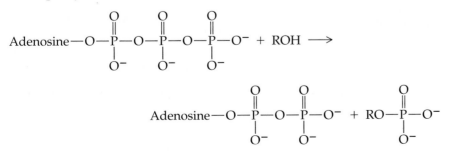

■ Understanding Key Concepts

17.30 Muscle cells deficient in oxygen will reduce pyruvate (an intermediate in metabolism) to lactate at a cellular pH of approximately 7.4:

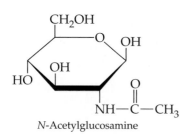

(a) Why do we say pyruvate and lactate, rather than pyruvic acid and lactic acid?

(b) Alter the above structures to create pyruvic acid and lactic acid.

(c) Show hydrogen bonding of water to both pyruvate and lactate. Would you expect a major difference in water solubility of lactate and pyruvate? Explain.

17.31 *N*-Acetylglucosamine is an important component on the surfaces of cells.

(a) Under what chemical conditions might the acetyl group be removed, changing the nature of the cell-surface components?

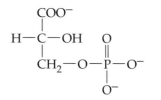

N-Acetylglucosamine

(b) Draw the structures of the products of acid hydrolysis.

17.32 One phosphorylated form of glycerate is 3-phospho-glycerate (a metabolic intermediate):

$$
\begin{array}{c}
\text{COO}^- \\
| \\
\text{H}-\text{C}-\text{OH} \qquad \text{O} \\
| \qquad\qquad \| \\
\text{CH}_2-\text{O}-\text{P}-\text{O}^- \\
| \\
\text{O}^-
\end{array}
$$

(a) Identify the type of linkage between glycerate and phosphate.

(b) 1,3-Bisphosphoglycerate (two phosphates on glycerate) has an anhydride linkage between the carbonyl at C1 of glycerate and phosphate. Draw the structure of 1,3-bisphosphoglycerate (another metabolic intermediate).

17.33 The names of the first four dicarboxylic acids can be remembered by using the first letter of each word of the saying "*Oh, my such good …*" to remind us of *oxalate, malonate, succinate,* and *glutarate* (as these acids would occur at physiological pH). Write the structures of these four dicarboxylate anions.

17.34

$$
\begin{array}{c}
\text{NH}_2 \\
| \\
\text{HOOC}-\text{CH}-\text{CH}_2-\text{CH}_2-\text{OH}
\end{array}
$$

(a) Draw the cyclic ester resulting from cyclization of the compound above. Cyclic esters are called *lactones*. A common name for this lactone is homoserine lactone.

(b) If two molecules of the linear compound above reacted to form an ester, what would be the structure of the ester product?

(c) If two molecules of the linear compound above reacted to form an amide, what would be the structure of the amide product?

17.35. (a) Draw the structures of the following compounds and use dashed lines to indicate where they can form hydrogen bonds to other molecules of the same kind: (i) acetic acid, (ii) methyl acetate, (iii) acetamide.

(b) Arrange these compounds in the order of increasing boiling points and explain the order.

17.36 For each of the following compounds, give the systematic name and indicate whether the compound is an acid, a base, or neither.

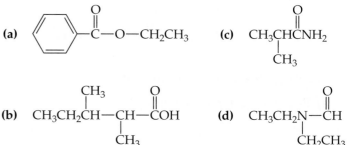

17.37 Volicitin, in the "spit" from beet armyworms, causes corn plants to produce volatile compounds. Draw the hydrolysis products from volicitin that match the common names given below.

(a) Glutamic acid (α-aminoglutaric acid)
(b) Ammonia
(c) 17-Hydroxylinolenic acid

$$
\begin{array}{c}
\qquad\qquad\qquad\qquad\qquad \text{O} \quad\; \text{H} \quad\; \text{COOH} \qquad\qquad\quad \text{O} \\
\qquad\qquad\qquad\qquad\qquad \| \qquad | \qquad | \qquad\qquad\qquad\;\; \| \\
\text{CH}_2-\text{CH}{=}\text{CH}-(\text{CH}_2)_7-\text{C}-\text{N}-\text{CH}-\text{CH}_2-\text{CH}_2-\text{C}-\text{NH}_2 \\
| \\
\text{CH}{=}\text{CH}-\text{CH}_2-\text{CH}{=}\text{CH}-\text{CH}-\text{CH}_3 \\
\qquad\qquad\qquad\qquad\qquad\qquad\qquad | \\
\qquad\qquad\qquad\qquad\qquad\qquad\quad \text{OH}
\end{array}
$$

Volicitin

Additional Problems

Carboxylic Acids

17.38 Write the equation for the ionization of benzoic acid in water.

17.39 Assume that you have a sample of propanoic acid dissolved in water.
   (a) Draw the structure of the major species present in the water solution.
   (b) Now assume that aqueous HCl is added to the propanoic acid solution until pH 2 is reached. Draw the structure of the major species present.
   (c) Finally, assume that aqueous NaOH is added to the propanoic acid solution until pH 12 is reached. Draw the structure of the major species present.

17.40 There are four carboxylic acids with the formula $C_5H_{10}O_2$. Draw and name each one.

17.41 Draw and name three different carboxylic acids with the formula $C_7H_{14}O_2$.

17.42 Give systematic names for the following carboxylic acids:

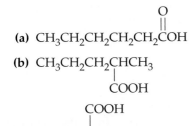

17.43 Give systematic names for the following carboxylic acids:

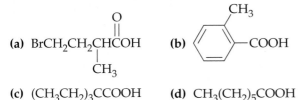

17.44 Give systematic names for the following carboxylic acid salts:

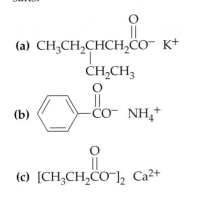

17.45 Give systematic names and common names for the following carboxylic acid salts:

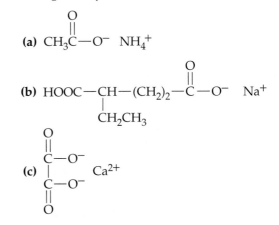

17.46 Draw structures corresponding to these names:
   (a) 3,4-Dimethylpentanoic acid
   (b) Triphenylacetic acid
   (c) *m*-Ethylbenzoic acid
   (d) Methylammonium butanoate

17.47 Draw structures corresponding to these names:
   (a) 2,2-Dichlorobutanoic acid
   (b) 3-Hydroxyhexanoic acid
   (c) 3,3-Dimethyl-4-phenylpentanoic acid

17.48 Malic acid, a dicarboxylic acid found in apples, has the systematic name hydroxybutanedioic acid. Draw its structure.

17.49 Fumaric acid is a metabolic intermediate that has the systematic name *trans*-2-butenedioic acid. Draw its structure.

17.50 What is the formula for the disodium salt of malic acid (see Problem 17.48)?

17.51 Aluminum acetate is used as an antiseptic ingredient in some skin-rash ointments. Draw its structure.

Esters and Amides

17.52 Draw and name compounds that meet these descriptions:
   (a) Three different amides with the formula $C_5H_{11}NO$
   (b) Three different esters with the formula $C_6H_{12}O_2$

17.53 Draw and name compounds that meet these descriptions:
   (a) Three different amides with the formula $C_6H_{13}NO$
   (b) Three different esters with the formula $C_5H_{10}O_2$

17.54 Give systematic names for structures, and structures for names:

(a) $CH_3COCH_2CH_2CHCH_3$ with O above first C and CH_3 on the CH

(b) $CH_3CHCH_2CH_2COCH_3$ with CH_3 and O

   (c) Cyclohexyl acetate

   (d) Phenyl *o*-hydroxybenzoate

17.55 Give systematic names for structures, and structures for names:

(a) Methyl pentanoate

(b) Isopropyl 2-methylbutanoate

(c) $(CH_3)_3CCOCH_2CH_3$

(d)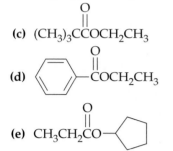

(e) $CH_3CH_2CO\!\!-\!\!\bigcirc$

17.56 Draw the structures of the carboxylic acids and alcohols you would use to prepare each of the esters in Problem 17.54.

17.57 Draw the structures of the carboxylic acids and alcohols you would use to prepare each of the esters in Problem 17.55.

17.58 Give systematic names for structures, and structures for names:

(a)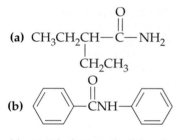

(b)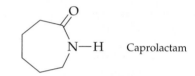

(c) *N*-Ethyl-*N*-methylbenzamide

(d) 2,3-Dibromohexanamide

17.59 Give systematic names for structures, and structures for names:

(a) 3-Methylpentanamide

(b) *N*-Phenylacetamide

(c) $\overset{\overset{\displaystyle O}{\|}}{H}CN(CH_3)_2$

(d) $CH_3CH_2\overset{\overset{\displaystyle O}{\|}}{C}NH\overset{\overset{\displaystyle CH_3}{|}}{C}HCH_3$

17.60 Show how you would prepare each of the amides in Problem 17.58 from the appropriate carboxylic acid and amine.

17.61 What compounds are produced from hydrolysis of each of the amides in Problem 17.59?

Reactions of Carboxylic Acids and Their Derivatives

17.62 Procaine, a local anesthetic whose hydrochloride is Novocain, has the following structure. Identify the functional groups present, and show the structures of the alcohol and carboxylic acids you would use to prepare procaine.

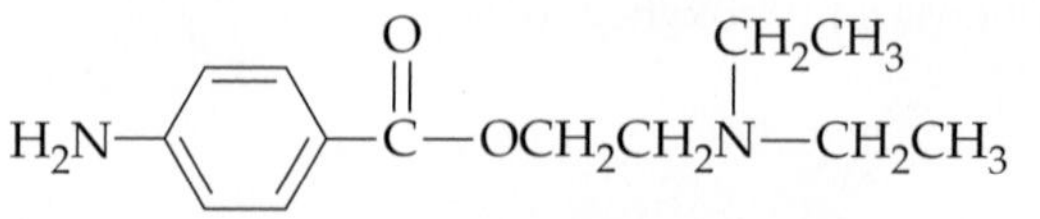

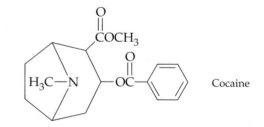

17.63 Lidocaine (Xylocaine) is a local anesthetic closely related to procaine. Identify the functional groups present in lidocaine, and show how you might prepare it from a carboxylic acid and an amine.

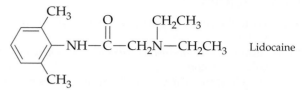

17.64 Lactones are cyclic esters in which the carboxylic acid part and the alcohol part are connected to form a ring. What product(s) would you expect to obtain from hydrolysis of butyrolactone?

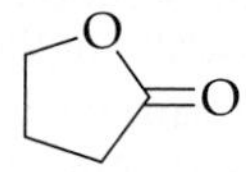

Butyrolactone (a cyclic ester)

17.65 A *lactam* is a cyclic amide in which the carboxylic acid part and the amine part are connected. Draw the structure of the product(s) from hydrolysis of caprolactam, an industrial precursor of nylon.

Caprolactam

17.66 Cocaine, an alkaloid isolated from the leaves of the South American coca plant, *Erythroxylon coca*, has the structure indicated. Identify the functional groups present, and give the structures of the products you would obtain from hydrolysis of cocaine.

Cocaine

17.67 Household soap is a mixture of the sodium or potassium salts of long-chain carboxylic acids that arise from saponification of animal fat.

(a) Identify the functional groups present in the fat molecule shown in the reaction below.

(b) Draw the structures of soap molecules produced in the following reaction:

$$CH_2\!\!-\!\!O\!\!-\!\!CO(CH_2)_{14}CH_3$$
$$CH\!\!-\!\!O\!\!-\!\!CO(CH_2)_7CH\!\!=\!\!CH(CH_2)_7CH_3 \xrightarrow{3\ KOH} ?$$
$$CH_2\!\!-\!\!O\!\!-\!\!CO(CH_2)_{16}CH_3$$
A fat

Polyesters and Polyamides

17.68 Baked-on paints used for automobiles and many appliances are often based on *alkyds*, such as can be made from terephthalic acid (p. 491) and glycerol (below). Sketch a section of the resultant polyester polymer. Note that the glycerol can be esterified at any of the three alcohol groups, providing *cross-linking* to form a very strong surface.

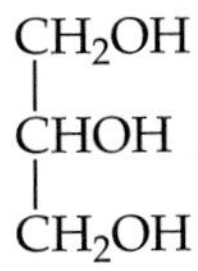

$$CH_2OH$$
$$|$$
$$CHOH$$
$$|$$
$$CH_2OH$$

Glycerol

17.69 A simple polyamide can be made from ethylenediamine (below) and oxalic acid (p. 472). Draw a few units of the polymer formed.

$$NH_2—CH_2—CH_2—NH_2$$

Ethylenediamine

Phosphate Esters and Anhydrides

17.70 The following phosphate ester is an important intermediate in carbohydrate metabolism. What products would result from hydrolysis of this ester?

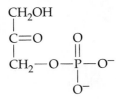

17.71 In the compound below:

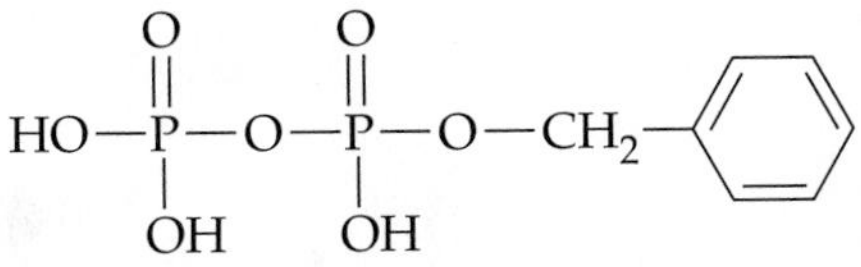

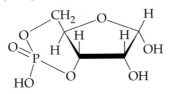

 (a) Identify the ester linkage.
 (b) Identify the anhydride linkage.
 (c) Show the complete acid hydrolysis products.

17.72 The metabolic intermediate acetyl phosphate is a mixed anhydride formed from acetic acid and phosphoric acid. What is the structure of acetyl phosphate?

17.73 Acetyl phosphate (see Problem 17.72) has what is called "high phosphoryl-group transfer potential." Write a reaction where there is phosphoryl-group transfer from acetyl phosphate to ethanol to make a phosphate ester.

17.74 For the structure below, give a general description of the linkage between the phosphate and ribose. (*Hint:* It starts with cyclic phosphate....)

17.75 Differentiate between a phosphate diester and a diphosphate. Give an example of each.

Applications

17.76 Name the two compounds below, and explain what chemical properties account for how they are used in skin treatment. [*Acids for the Skin*]

$$CCl_3COOH \qquad CH_3CHCOOH$$
$$\qquad\qquad\qquad\qquad |$$
$$\qquad\qquad\qquad\qquad OH$$

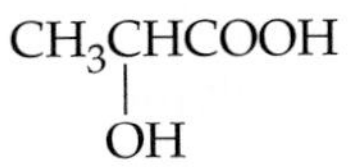

17.77 Which is an α-hydroxy acid, succinic acid or malic acid, and which is the stronger acid? [*Acids for the Skin*]

$$^-OOC—CH_2—CH_2—COO^- \quad \text{Succinic acid}$$

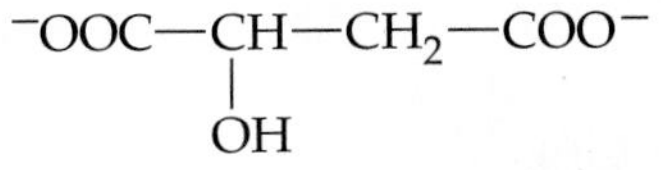

$$^-OOC—CH—CH_2—COO^- \quad \text{Malic acid}$$
$$\qquad\qquad |$$
$$\qquad\qquad OH$$

17.78 Write the structure for the trisodium salt of citric acid and then show the reaction of trisodium citrate with hydrogen ions as it buffers against acidity from fruit juices. [*Acid Salts as Food Additives*]

17.79 Against what type of microorganisms are sodium benzoate and potassium sorbate particularly effective? [*Acid Salts as Food Additives*]

17.80 Kevlar appears to be nearly indestructible; however, there are still groups of chemicals to which it is not resistant. Based on the chemistry of Kevlar, name one group of chemicals that Kevlar is susceptible to, and explain why it is not resistant. [*Kevlar: A Life-Saving Polymer*]

General Questions and Problems

17.81 Three amide isomers, *N,N*-dimethylformamide, *N*-methylacetamide, and propanamide, have respective boiling points of 153°C, 202°C, and 213°C. Explain these boiling points in light of the structural formulas.

17.82 Salol, the phenyl ester of salicylic acid, is used as an intestinal antiseptic. Draw the structure of phenyl salicylate.

17.83 Propanamide and methyl acetate have about the same molar mass, both are quite soluble in water, and yet the boiling point of propanamide is 213°C while that of methyl acetate is 57°C. Explain.

17.84 Mention at least two simple chemical tests by which you could distinguish between benzaldehyde and benzoic acid.

17.85 Write the formula of the triester formed from glycerol and stearic acid (Table 17.1).

17.86 Name these compounds.

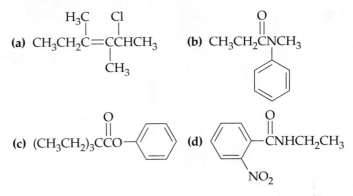

18

Amino Acids and Proteins

CONTENTS

When they grow up, these piglets will provide useful protein of several kinds.

The word *protein* is a familiar one. Taken from the Greek *proteios*, meaning "primary," "protein" is an apt description for the biological molecules that are of primary importance to all living organisms. Approximately 50% of your body's dry weight is protein. Some proteins, such as the collagen in connective tissue, serve a structural purpose. Others direct responses to internal and external conditions. And still other proteins defend the body against foreign invaders. Most importantly, as enzymes, proteins catalyze almost every chemical reaction that occurs in your body. Because of their importance and the role they play in all biochemical functions, we have chosen to discuss proteins, which are polymers of amino acids, in this first chapter devoted to biochemistry.

In this chapter, we'll look at the following questions about amino acids and proteins:

1. **What are the structural features of amino acids?**
   The goal: Be able to describe and recognize amino acid structures and illustrate how they are connected in proteins.

2. **What are the properties of amino acids?**
   The goal: Be able to describe how the properties of amino acids depend on their side chains and how their ionic charges vary with pH.

3. **Why do amino acids have "handedness"?**
   The goal: Be able to explain what is responsible for handedness and recognize simple molecules that display this property.

4. **What is the primary structure of a protein and what conventions are used for drawing and naming primary structures?**
   The goal: Be able to define protein primary structure, explain how primary structures are represented, and draw a simple protein structure, given its amino acid sequence.

5. **What types of interactions determine the overall shapes of proteins?**
   The goal: Be able to describe and recognize disulfide bonds, hydrogen bonding along the protein backbone, and noncovalent interactions between amino acid side chains in proteins.

6. **What are the secondary and tertiary structures of proteins?**
   The goal: Be able to define these structures and the attractive forces that determine their nature, describe the α-helix and β-sheet, and distinguish between fibrous and globular proteins.

7. **What is quaternary protein structure?**
   The goal: Be able to define quaternary structure and give examples of proteins with quaternary structure.

8. **What chemical properties do proteins have?**
   The goal: Be able to describe protein hydrolysis and denaturation, and give some examples of agents that cause denaturation.

CONCEPTS TO REVIEW

Acid–base properties (Sections 6.10, 10.9, 17.3)
Hydrolysis reactions (Section 17.6)
Intermolecular forces (Section 8.11)
Polymers (Application, p 117; Sections 13.8, 17.7)

18.1 An Introduction to Biochemistry

Biochemistry—now we're ready to investigate the chemical basis of life. Medical professionals are faced with biochemistry every day because all diseases are associated with abnormalities in biochemistry. Nutritionists evaluate our dietary needs based on our biochemistry. And the pharmaceutical industry designs molecules that mimic or alter the action of biomolecules. The ultimate goal of biochemistry is to understand the structures of biomolecules and the relationship between their structures and functions.

Biochemistry is becoming the common ground for the life sciences. Microbiology, botany, zoology, immunology, pathology, physiology, toxicology, neuroscience, cell biology—in all these fields, answers to fundamental questions are being found at the molecular level.

The principal classes of biomolecules are *proteins, carbohydrates, lipids,* and *nucleic acids.* Some biomolecules are small and have only a few functional groups. Others are huge and their biochemistry is governed by the interactions

▲ **A medical conference.** Biochemistry may help to solve the problem under discussion.

Monomers and Polymers Part 1

of large numbers of functional groups. Proteins, the subject of this chapter; nucleic acids (Chapters 26, 27); and large carbohydrates (Section 22.9) are all polymers, some containing hundreds, thousands, or even millions of repeating units.

Biochemical reactions must continuously break down food molecules, generate and store energy, build up new biomolecules, and eliminate waste. Each biomolecule has its own role to play in these processes. But despite the huge size of some biomolecules and the complexity of their interactions, their functional groups and chemical reactions are no different from those of simpler organic molecules. *All the principles of chemistry introduced thus far apply to biochemistry*. Of the functional groups introduced in previous chapters, those listed in Table 18.1 are of greatest importance in biomolecules.

TABLE 18.1   Functional Groups of Importance in Biochemical Molecules

Functional Group	Structure	Type of Biomolecule
Amino group	$-NH_3^+$, $-NH_2$	Amino acids and proteins (Sections 18.3, 18.7)
Hydroxyl group	$-OH$	Monosaccharides (carbohydrates) and glycerol: a component of triacylglycerols (lipids) (Sections 22.4, 24.2)
Carbonyl group	$\overset{\text{O}}{\overset{\|}{-\text{C}-}}$	Monosaccharides (carbohydrates); in acetyl group (CH_3CO) used to transfer carbon atoms during catabolism (Sections 22.4, 21.4, 21.8)
Carboxyl group	$\overset{\text{O}}{\overset{\|}{-\text{C}-\text{OH}}}$, $\overset{\text{O}}{\overset{\|}{-\text{C}-\text{O}^-}}$	Amino acids, proteins, and fatty acids (lipids) (Sections 18.3, 18.7, 24.2)
Amide group	$\overset{\text{O}}{\overset{\|}{-\text{C}-\text{N}-}}$	Links amino acids in proteins; formed by reaction of amino group and carboxyl group (Section 18.7)
Carboxylic acid ester	$\overset{\text{O}}{\overset{\|}{-\text{C}-\text{O}-\text{R}}}$	Triacylglycerols (and other lipids); formed by reaction of carboxyl group and hydroxyl group (Section 24.2)
Phosphates, mono-, di-, tri-	$-\overset{\|}{\underset{\|}{\text{C}}}-\text{O}-\overset{\text{O}}{\overset{\|}{\underset{\underset{\text{O}^-}{\|}}{\text{P}}}}-\text{O}^-$	ATP and many metabolism intermediates (Sections 17.8, 21.5, and throughout metabolism sections)
	$-\overset{\|}{\underset{\|}{\text{C}}}-\text{O}-\overset{\text{O}}{\overset{\|}{\underset{\underset{\text{O}^-}{\|}}{\text{P}}}}-\text{O}-\overset{\text{O}}{\overset{\|}{\underset{\underset{\text{O}^-}{\|}}{\text{P}}}}-\text{O}^-$	
	$-\overset{\|}{\underset{\|}{\text{C}}}-\text{O}-\overset{\text{O}}{\overset{\|}{\underset{\underset{\text{O}^-}{\|}}{\text{P}}}}-\text{O}-\overset{\text{O}}{\overset{\|}{\underset{\underset{\text{O}^-}{\|}}{\text{P}}}}-\text{O}-\overset{\text{O}}{\overset{\|}{\underset{\underset{\text{O}^-}{\|}}{\text{P}}}}-\text{O}^-$	
Hemiacetal group	$-\overset{\|}{\underset{\underset{\text{OR}}{\|}}{\text{C}}}-\text{OH}$	Cyclic forms of monosaccharides; formed by a reaction of carbonyl group with hydroxyl group (Sections 16.7, 22.4)
Acetal group	$-\overset{\|}{\underset{\underset{\text{OR}}{\|}}{\text{C}}}-\text{OR}$	Connects monosaccharides in disaccharides and larger carbohydrates; formed by reaction of carbonyl group with hydroxyl group (Sections 16.7, 22.7, 22.9)

○○○ **LOOKING AHEAD**

The focus in the rest of this book is on human biochemistry and the essential structure–function relationships of biomolecules. In this and the next two chapters, we examine the structure of proteins and the roles of proteins and other molecules in controlling biochemical reactions. Next, we take an overview of metabolism and the production of energy (Chapter 21). Then we discuss the structure and function of carbohydrates (Chapters 22 and 23), the structure and function of lipids (Chapters 24 and 25), the role of nucleic acids in protein synthesis and heredity (Chapters 26, 27), the metabolism of proteins (Chapter 28), and the chemistry of body fluids (Chapter 29). ○○○

18.2 Protein Structure and Function: An Overview

Proteins are polymers of **amino acids**. Every amino acid in a protein contains an amine group ($-NH_2$), a carboxyl group (COOH), and an R group, all bonded to a central carbon atom. The amino acids in proteins are **α-amino acids**—the amine group in each is connected to the carbon atom "*alpha* to" (next to) the carboxylic acid group. The R groups may be hydrocarbons, or they may contain a functional group.

Protein A large biological molecule made of many amino acids linked together through amide (peptide) bonds.

Amino acid A molecule that contains both an amino group and a carboxylic acid functional group.

Alpha- (α-) amino acid An amino acid in which the amino group is bonded to the carbon atom next to the —COOH group.

$-NH_2$ and $-R$ are bonded to carbon next to $-$COOH

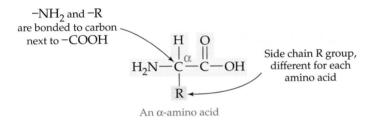

Side chain R group, different for each amino acid

An α-amino acid

Two or more amino acids can link together by forming amide bonds (○○○, p. 484), which are known as **peptide bonds** when they occur in proteins. A *dipeptide* results from the formation of a peptide bond between the $-NH_2$ group of one amino acid and the $-$COOH group of a second amino acid. For example, valine and cysteine are connected in a dipeptide as follows:

Peptide bond An amide bond that links two amino acids together.

🌐 **Monomers and Polymers Part 3**

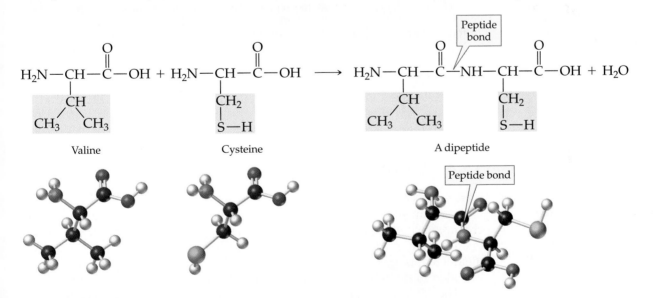

Valine   Cysteine   A dipeptide

A *tripeptide* results from linkage of three amino acids via two peptide bonds, and so on. Any number of amino acids can link together to form a linear chainlike polymer—a *polypeptide*.

Proteins have four levels of structure, each of which is explored later in this chapter.

- *Primary structure* is the sequence of amino acids in a protein chain (Section 18.7).
- *Secondary structure* is the regular and repeating spatial organization of neighboring segments of protein chains (Section 18.9).
- *Tertiary structure* is the overall shape of a protein molecule (Section 18.10) produced by regions of secondary structure combined with the overall bending and folding of the protein chain.
- *Quaternary structure* refers to the overall structure of proteins composed of more than one polypeptide chain (Section 18.11).

What roles do proteins play in living things? No doubt, you are aware that a hamburger is produced from muscle protein and that we depend on our own muscle proteins for every move we make. But this is only one of many essential roles of proteins. They provide *structure* and *support* to tissues and organs throughout our bodies. As *hormones* and *enzymes*, they control all aspects of metabolism. In body fluids, water-soluble proteins pick up other molecules for *storage* or *transport*. And the proteins of the immune system provide *protection* against invaders. To accomplish their biological functions, which are summarized in Table 18.2, some proteins must be tough and fibrous, whereas others must be globular and soluble in body fluids. The overall shape of a protein molecule, as you will see often in the following chapters, is essential to the role of that protein in our metabolism.

TABLE 18.2	**Classification of Proteins by Function**	
Type	**Function**	**Example**
Enzymes	Catalysts	*Amylase*—begins digestion of carbohydrates by hydrolysis
Hormones	Regulate body functions by carrying messages to receptors	*Insulin*—facilitates use of glucose for energy generation
Storage proteins	Make essential substances available when needed	*Myoglobin*—stores oxygen in muscles
Transport proteins	Carry substances through body fluids	*Serum albumin*—carries fatty acids in blood
Structural proteins	Provide mechanical shape and support	*Collagen*—provides structure to tendons and cartilage
Protective proteins	Defend the body against foreign matter	*Immunoglobulin*—aids in destruction of invading bacteria
Contractile proteins	Do mechanical work	*Myosin and actin*—govern muscle movement

18.3 Amino Acids

Side chain (amino acid)   The group bonded to the carbon next to the carboxyl group in an amino acid; different in different amino acids.

Nature uses the 20 α-amino acids listed in Table 18.3 to build the proteins in all living organisms; 19 of them differ only in the identity of the R group, or **side chain**, attached to the α carbon. The remaining amino acid (proline) is a secondary amine whose nitrogen and α carbon atoms are joined in a five-membered ring. Each amino acid has a three-letter shorthand code that is included in the table—for example, Ala (alanine), Gly (glycine), Pro (proline).

The 20 protein amino acids are classified as neutral, acidic, or basic, depending on the nature of their side chains. The 15 neutral amino acids are further divided into those with nonpolar side chains and those with polar functional groups such as amide or hydroxyl groups in their side chains. As we

TABLE 18.3   The 20 Protein Amino Acids with their Abbreviations and Isoelectric Points. The structures are written here in their fully ionized forms. These ions and the isoelectric points given in parantheses are explained in Section 18.4.

Nonpolar Side Chains

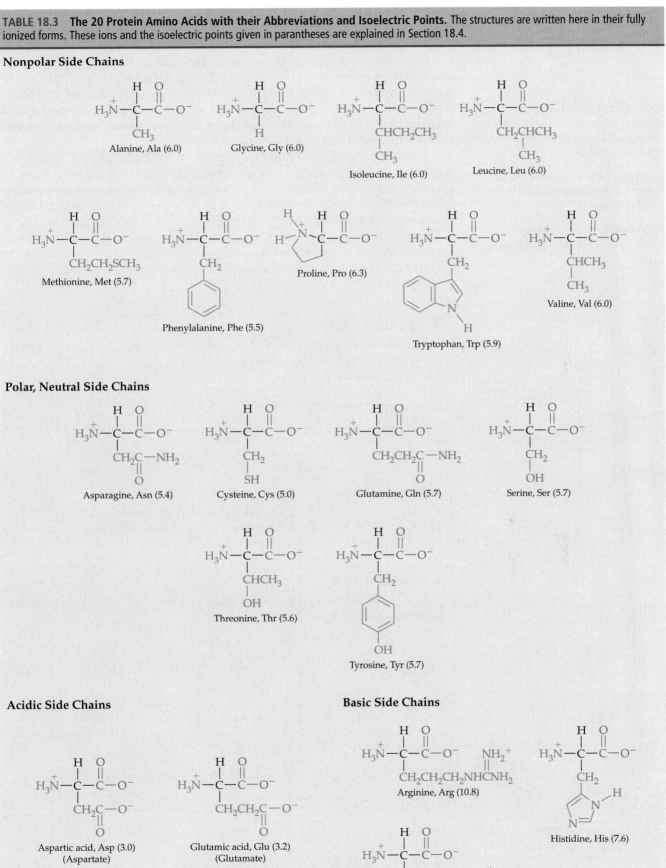

Alanine, Ala (6.0)

Glycine, Gly (6.0)

Isoleucine, Ile (6.0)

Leucine, Leu (6.0)

Methionine, Met (5.7)

Phenylalanine, Phe (5.5)

Proline, Pro (6.3)

Tryptophan, Trp (5.9)

Valine, Val (6.0)

Polar, Neutral Side Chains

Asparagine, Asn (5.4)

Cysteine, Cys (5.0)

Glutamine, Gln (5.7)

Serine, Ser (5.7)

Threonine, Thr (5.6)

Tyrosine, Tyr (5.7)

Acidic Side Chains

Aspartic acid, Asp (3.0)
(Aspartate)

Glutamic acid, Glu (3.2)
(Glutamate)

Basic Side Chains

Arginine, Arg (10.8)

Histidine, His (7.6)

Lysine, Lys (9.7)

Noncovalent forces   Forces of attraction other than covalent bonds that can act between molecules or within molecules.

Hydrophobic   Water-fearing; a hydrophobic substance does not dissolve in water.

Hydrophilic   Water-loving; a hydrophilic substance dissolves in water.

explore the structure and function of proteins, you will see that it is the sequence of amino acids in a protein and the chemical nature of their side chains that enable proteins to perform their varied functions.

Intermolecular forces are of central importance in determining the shapes and functions of proteins. (∞ Section 8.11) In the context of biochemistry it is more meaningful to refer to all interactions other than covalent bonding as **noncovalent forces**. Noncovalent forces act between different molecules or between different parts of the same large molecule, which is often the case in proteins (Section 18.8).

The nonpolar side chains are described as **hydrophobic** (water-fearing)—they are not attracted to water molecules. To avoid aqueous body fluids, they gather into clusters that provide a water-free environment, often a pocket within a large protein molecule. The polar, acidic, and basic side chains are **hydrophilic** (water-loving)—they *are* attracted to polar water molecules. They interact with water molecules much as water molecules interact with one another. Attractions between water molecules and hydrophilic groups on the surface of folded proteins imparts water solubility to the proteins. (∞ Section 9.2)

■ **PROBLEM 18.1**

Name the common amino acids that contain an aromatic ring, contain sulfur, are alcohols, and have alkyl-group side chains.

■ **PROBLEM 18.2**

Draw alanine showing the tetrahedral geometry of its α carbon.

■ **PROBLEM 18.3**

Choose one amino acid with a nonpolar side chain and one with a polar side chain; draw the two dipeptides formed by these two amino acids.

■ **PROBLEM 18.4**

Which of the following compounds is an α-amino acid and which is not an α-amino acid?

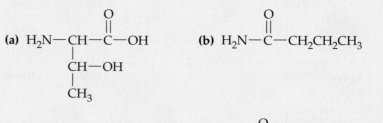

■ **PROBLEM 18.5**

Which of the following pairs of amino acids can form hydrogen bonds between their side-chain groups?
(a) Phe, Thr   **(b)** Asn, Ser   **(c)** Thr, Tyr   **(d)** Gly, Trp

■ **PROBLEM 18.6**

In the ball-and-stick model of valine near the beginning of Section 18.2, identify the carboxyl group, the amino group, and the R group.

18.4 Acid–Base Properties of Amino Acids

Amino acids contain both an acidic group, —COOH, and a basic group, —NH$_2$. As you might expect, these two groups can undergo an intramolecular acid–base reaction. The result is transfer of the hydrogen from the —COOH group to the —NH$_2$ group to form a *dipolar* ion, an ion that has one positive charge and one negative charge and is thus electrically neutral. Dipolar ions are known as **zwitterions** (from the German *zwitter*, "hybrid"). The zwitterion form of threonine is shown here and those of the other amino acids in proteins are given in Table 18.3.

Zwitterion A neutral dipolar ion that has one + charge and one − charge.

 Proteins and Amino Acids

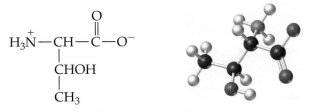

Threonine—zwitterion

Because they're zwitterions, amino acids have many of the physical properties we associate with salts.(◖◗◗ Section 4.4) Pure amino acids can form crystals, have high melting points, and are soluble in water but not in hydrocarbon solvents.

In acidic solution (low pH), amino acid zwitterions accept protons on their basic —COO$^-$ groups to leave only the positively charged NH$_3^+$ groups. In basic solution (high pH), amino acid zwitterions *lose* protons from their acidic —NH$_3^+$ groups to leave only the negatively charged COO$^-$ groups.

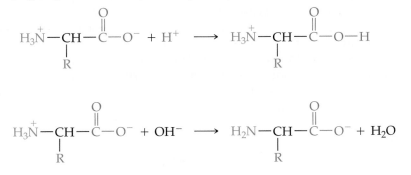

Amino acids are never present in the completely nonionized form in either the solid state or aqueous solution. The charge of an amino acid molecule at any given moment depends on the particular amino acid and the pH of the medium. The pH at which the net positive and negative charges are evenly balanced is the amino acid's **isoelectric point (pI)**. At this point, the overall charge of all the amino acids in a sample is zero. (The mathematical relationship between isoelectric points and acid dissociation constants is discussed in Appendix C.)

The two amino acids with acidic side chains, aspartic acid and glutamic acid, have isoelectric points at more acidic (lower) pH values than those with neutral side chains. Since the side chain —COOH groups of these compounds are substantially ionized at physiological pH of 7.4, these amino acids are usually referred to as *aspartate* and *glutamate*, the names of the anions formed when the —COOH groups in the side chains are ionized. (Recall that the same convention is used, for example, for sulfate ion from sulfuric acid or nitrate ion from nitric acid.)

Isoelectric point (pI) The pH at which a sample of an amino acid has equal numbers of + and − charges.

Application Nutrition in Health and Disease

To a professional, nutrition is more about the chemical components of what we eat than the flavor, form, and texture of our food. No matter what the recipe, once inside our bodies, it's only the quantity and fate of the proteins, carbohydrates, fats, minerals, and vitamins that matters.

A massive reevaluation of what we eat has been going on in the United States. Increasingly, fruits and vegetables are "in" and red meat is "out." The ongoing changes are in response to experimental evidence about the relationships between body chemistry and health. Fruits and vegetables contain a growing catalog of "phytochemicals" (which just means chemicals from plants) that appear to counteract cancer, heart disease, and a host of other conditions. And meat—especially the fat and cholesterol that naturally occur in meat—most certainly plays a role in the development of heart disease.

As a guide to a healthy diet, the U.S. Department of Agriculture, after lengthy study and consultation, developed the *Food Guide Pyramid* in 1992. The dietary guidelines outlined in the Pyramid were reaffirmed in the latest edition of the Dietary Guidelines for Americans, 5e, 2000. Foods that should be eaten in smaller amounts are at the top of the pyramid, and foods that should be eaten in larger amounts are at the bottom. the principal sources of protein are at the second level, in the milk and meat groups. (In later chapters, we'll return to the pyramid as we discuss the role of fats and carbohydrates in our diets.)

The pyramid provides guidelines for average healthy adults who eat a typical American diet. Guidelines for infants, the elderly, and persons with diseases must be tailored to specific needs. Designing a diet for such individuals depends on understanding body chemistry, food chemistry, and the distinctive characteristics of the individuals to be fed. In specifying a diet for a medical patient, a nutritionist must know the patient's medical history, blood and urine levels of key chemicals, medications, physical dimensions (underweight or obese?), energy requirements, and allergies (if any). In the past decade, modified versions of the Food Guide Pyramid for various ethnic and cultural groups, for young children, for the elderly, for vegetarians, and individuals looking for "healthy weight" dietary guidelines have developed. (Examples of these alternate pyramids can be seen at www.mayo.edu/news/pyramid.jpg; www.usda.gov/cnpp/KidsPyra/; and www.nal.usda.gov/fnic)

To give just one example of the implications of patient assessment, consider a person taking an amine antidepressant such as phenelzine (Nardil). Phenelzine acts by inhibiting the enzyme that removes amino groups from amino acids during their normal metabolism (Section 27.3). Some foods, especially those that are aged, fermented, or decayed, contain tyramine, an amine produced from the protein amino acid tyrosine. Notice the structural similarities among these compounds.

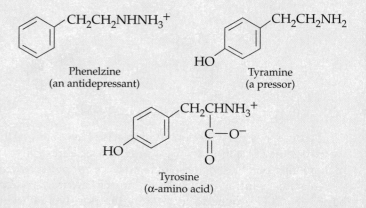

Phenelzine
(an antidepressant)

Tyramine
(a pressor)

Tyrosine
(α-amino acid)

Tyramine is a *pressor*—it constricts blood vessels and elevates blood pressure, which can cause irregular heartbeat, severe headache, and in serious cases, intracranial hemorrhage and cardiac failure. Ordinarily, tyramine is inactivated by removal of its amino group by the same enzymes that act on the protein amino acids. In the presence of phenelzine, however, this does not occur. A person taking phenelzine must avoid foods that contain high levels of tyramine. There is also some indication (though not proven) that tyramine in foods triggers migraine headaches.

Foods Containing Relatively High Levels of Tyramine

Cheese

Smoked fish

Meat that is not fresh

Liver

Dry sausage

Sauerkraut

Yeast extracts

Beer and ale

Chianti and vermouth wines

▲ **The Food Guide Pyramid from the U.S. Department of Agriculture.**

See Additional Problem 18.72 at the end of the chapter.

■ **WORKED EXAMPLE 18.1**

Consider the structures of phenylalanine and serine in Table 18.3. Which of these two acids has a hydrophobic side chain and which has a hydrophilic side chain?

ANALYSIS Identify the side chains. The side chain in phenylalanine is a hydrocarbon. The side chain in serine contains a hydroxyl group.

SOLUTION The hydrocarbon side chain in phenylalanine is a hydrocarbon, which is nonpolar and hydrophobic. The hydroxyl group in the side chain of serine is polar and is therefore hydrophilic.

■ **WORKED EXAMPLE 18.2**

Look up the zwitterion structure of valine in Table 18.3. Draw valine as it would be found at low pH and at high pH.

ANALYSIS At low pH, which is acidic, basic groups may gain H^+ and at high pH, which is basic, acidic groups may lose H^+. In the zwitterion form of an amino acid, the $-COO^-$ group is basic and the $-NH_3^+$ is acidic.

SOLUTION Valine has an alkyl group side chain that is unaffected by pH. At low pH, which is acidic, valine adds a hydrogen ion to its carboxyl group to give the structure on the left below. At high pH, which is basic, valine loses a hydrogen ion from its acidic $-NH_3^+$ group to give the structure on the right below.

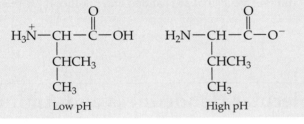

■ **PROBLEM 18.7**

Draw the structure of glutamic acid at low pH and at high pH.

■ **PROBLEM 18.8**

Use the definitions of acids and bases as proton donors and proton acceptors to explain which functional group in the zwitterion form of an amino acid is an acid and which is a base. (See zwitterion of threonine, p. 509)

Chiral Having right- or left-handedness; able to have two different mirror-image forms.

18.5 Handedness

Are you right-handed or left-handed? Although you may not often think about it, handedness affects almost everything you do. It also affects the biochemical activity of molecules.

Anyone who plays softball knows that the last available glove always fits the wrong hand. This happens because your hands aren't identical. Rather, they're mirror images. When you hold your left hand up to a mirror, the image you see looks like your right hand (Figure 18.1). Try it. In another way to describe this condition, note that the mirror images of your hand cannot be superimposed on each other; one won't completely fit on top of the other. Objects that have handedness in this manner are said to be **chiral** (pronounced **ky**-ral, from the Greek *cheir*, meaning "hand").

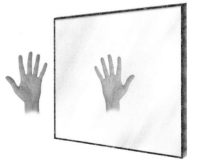

▲ **FIGURE 18.1 The meaning of *mirror image.*** If you hold your left hand up to a mirror, the image you see looks like your right hand.

▲ **FIGURE 18.2 The meaning of superimposable.** It is easy to mentally place the chair on top of its mirror image.

Achiral The opposite of chiral; having no right- or left-handedness and no nonsuperimposable mirror images

Not all objects are handed, of course. There's no such thing as a right-handed tennis ball or a left-handed coffee mug. When a tennis ball or a coffee mug is held up to a mirror, the image reflected is identical to the ball or mug itself. Objects like the coffee mug that lack handedness are said to be nonchiral, or **achiral**. Their mirror images are superimposable. Take a minute to convince yourself of this by studying the chair in Figure 18.2.

■ **PROBLEM 18.9**
Which of the following objects are chiral?

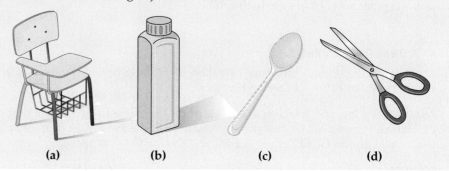

(a)　　　　(b)　　　　(c)　　　　(d)

■ **PROBLEM 18.10**
List three common objects that are handed and another three that aren't.

🌐 Chirality

18.6 Molecular Handedness and Amino Acids

Just as certain objects are chiral, certain molecules are also chiral. Alanine and propane provide a comparison between chiral and achiral molecules.

Alanine, a chiral molecule 　　　　　　　　　　　　　　*Propane, an achiral molecule*

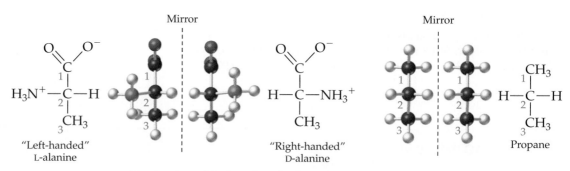

Alanine is a chiral molecule. Its mirror images cannot be superimposed on each other. As a result, alanine exists in two different forms that are mirror images of each other: a "right-handed" form known as D-alanine and a "left-handed" form known as L-alanine. Propane, by contrast, is an achiral molecule. The molecule and its mirror image are identical and it has no left- and right-handed isomers.

What is it that determines why some molecules are chiral but others aren't? Can we predict chirality from structural formulas? Recall from Section 5.7 that carbon forms four bonds oriented to the four corners of an imaginary tetrahedron. (∞, pp. 114–115) The formulas for alanine and propane are drawn below in a manner that emphasizes the four groups bonded to the central carbon atom. In alanine, this carbon is connected to *four different groups*: a —COO⁻ group, an —H atom, an —NH₃⁺ group, and a —CH₃ group:

$$
\begin{array}{ll}
& 1.\,-COO^- \\
& 2.\,-H \\
& 3.\,-NH_3^+ \\
& 4.\,-CH_3
\end{array} \Big\} \text{Different}
$$

Alanine (chiral)

$$
\begin{array}{ll}
1.\,-CH_3 \\
2.\,-CH_3
\end{array}\Big\}\text{Identical}
\quad
\begin{array}{ll}
3.\,-H \\
4.\,-H
\end{array}\Big\}\text{Identical}
$$

Propane (achiral)

Such a carbon atom is referred to as a **chiral carbon atom**, or *chirality center*. The presence of one chiral carbon atom always produces a chiral molecule that exists in mirror image forms. Thus, alanine is chiral. In propane the central carbon atom is bonded to two pairs of identical groups, and the two other carbon atoms are each bonded to three hydrogen atoms. The propane molecule has no chiral carbon atoms and is therefore achiral. (If a molecule has two or more chiral carbon atoms, it may or may not be chiral, depending on its overall shape.)

The two mirror-image forms of a chiral molecule like alanine are called **enantiomers or optical isomers** ("optical" because of their effect on polarized light; we'll discuss this in Section 22.2). The mirror-image relationship of the enantiomers of a compound with four different groups on one carbon atom is illustrated in Figure 18.3.

Like other isomers, enantiomers have the same formula but different arrangements of their atoms. More specifically, enantiomers are one kind of **stereoisomer**, compounds that have the same formula and atoms with the same connections but different spatial arrangements. (Cis–trans isomers are stereoisomers, too.) (⚬⚬⚬ Section 13.3) Pairs of enantiomers have many of the same physical properties. Both enantiomers of alanine, for example, have the same melting point, the same solubility in water, the same isoelectric point, and the same density. But pairs of enantiomers always differ in their effect on polarized light and in how they react with other molecules that are also chiral. Most importantly, pairs of enantiomers often differ in their biological activity, odors, tastes, or activity as drugs. For example, the very different natural flavors of spearmint and caraway seeds are attributed to the two enantiomers shown below:

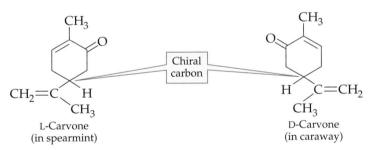

L-Carvone (in spearmint)

D-Carvone (in caraway)

What about the amino acids listed in Table 18.3? Are any of them chiral? Of the 20 common amino acids, 19 are chiral because they have four different groups bonded to their α carbons, —H, —NH₂, —COOH, and —R (the side chain). Only glycine, H_2NCH_2COOH, is achiral; its α carbon is bonded to two hydrogen atoms. Even though the naturally occurring chiral α-amino acids have pairs of enantiomers, nature uses only a single isomer of each for making proteins. For historical reasons (as you will see in Section 22.2), the naturally occurring isomers are all classified as left-handed, or L-amino acids.

The artificial sweetener aspartame (sold as Equal or NutraSweet) provides another excellent illustration of the delicate nature of the structure–function relationship and its role in biochemistry. Aspartame is the methyl ester of a dipeptide made from aspartate and phenylalanine in which both amino acids

Chiral carbon atom (chirality center) A carbon atom bonded to four different groups.

Enantiomers, optical isomers The two mirror-image forms of a chiral molecule.

Stereoisomers Isomers that have the same molecular and structural formulas, but different spatial arrangements of their atoms.

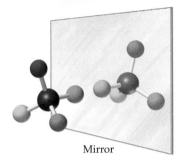

Mirror

▲ **FIGURE 18.3 A chiral molecule.** The central atom is bonded to four different groups; the molecule is therefore chiral.

▲ Spearmint leaves and caraway seeds. The very different flavors of these food seasonings are imparted by a pair of enantiomers, which interact in different ways with our taste buds.

▲ This artificial sweetener would not taste sweet if either of the two amino acids in its molecular structure were D rather than L isomers.

have the naturally occurring "left-handed" or L chirality. By contrast, the methyl esters of aspartate and phenylalanine dipeptides that have two D isomers or one D isomer combined with one L isomer are bitter.

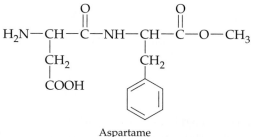

Aspartame
(methyl ester of aspartylphenylalanine)

⊙⊙⊙ LOOKING AHEAD

Amino acids, as you have seen, are chiral. Chirality is an important property of another major class of biomolecules. The individual sugar units in all carbohydrates are chiral, a topic addressed in Sections 22.2 and 22.3.

■ **WORKED EXAMPLE 18.3**

Lactic acid can be isolated from sour milk. Is lactic acid chiral?

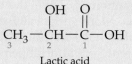

Lactic acid

ANALYSIS A molecule is chiral if it contains one C atom bonded to four different groups. Identify any C atoms that meet this condition.

SOLUTION To find out if lactic acid is chiral, list the groups attached to each carbon:

Groups on carbon 1	Groups on carbon 2	Groups on carbon 3
1. —OH	1. —COOH	1. —CH(OH)COOH
2. =O	2. —OH	2. —H
3. —CH(OH)CH$_3$	3. —H	3. —H
	4. —CH$_3$	4. —H

Next, look at the lists to see if any carbon is attached to four different groups. Of the three carbons, carbon 2 has four different groups, and lactic acid is therefore chiral.

■ **PROBLEM 18.11**

2-Aminopropane is an achiral molecule, but 2-aminobutane is chiral. Explain.

■ **PROBLEM 18.12**

Which of the following molecules are chiral?
(a) 3-Chloropentane   **(b)** 2-Chloropentane   **(c)** CH$_3$CHCH$_2$CHCH$_2$CH$_3$
|
CH$_3$ CH$_3$

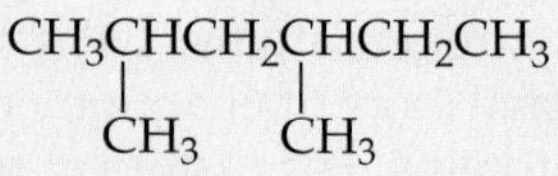

■ **PROBLEM 18.13**

Two of the 20 common amino acids have two chiral carbon atoms in their structures. Identify these amino acids and their chiral carbon atoms.

■ **KEY CONCEPT PROBLEM 18.14**

There are two isomers that have the formula C_2H_4BrCl. Draw them and identify any chiral carbons.

18.7 Primary Protein Structure

The **primary structure** of a protein is the sequence in which its amino acids are lined up and connected by peptide bonds. Along the *backbone* of the protein is a chain of alternating peptide bonds and α carbons. The amino acid side chains (R_1, R_2, ...) are substituents along the backbone, where they are bonded to the α carbons.

Primary protein structure The sequence in which amino acids are linked by peptide bonds in a protein.

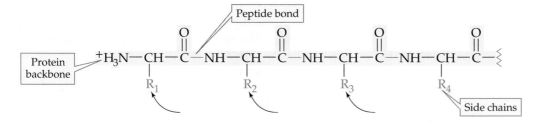

The carbon and nitrogen atoms along the backbone lie in a zigzag arrangement, with tetrahedral bonding around the α carbons. The electrons of each carbonyl-group double bond are shared to a considerable extent with the adjacent C—N bond. This sharing makes the C—N bond sufficiently like a double bond that there is no rotation around it. (⚬⚬⚬ Section 13.3) The result is that the carbonyl group, the —NH group bonded to it, and the two adjacent α carbons form a rigid, planar unit. The side-chain groups extend to opposite sides of the chain. Whether in a dipeptide or a huge polymer chain, these peptide units are planar.

Planar units along a protein chain

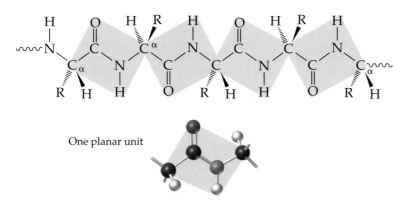

One planar unit

A pair of amino acids—for example, alanine and serine—can be combined to form two different dipeptides. The alanine —COO⁻ can react with the serine —NH$_3^+$:

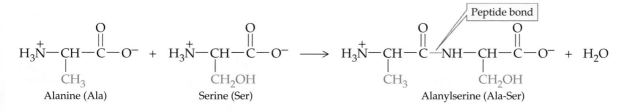

Or the serine —COO⁻ can react with the alanine —NH₃⁺:

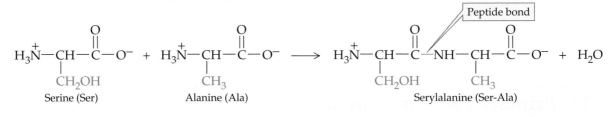

Serine (Ser)   +   Alanine (Ala)   ⟶   Serylalanine (Ser-Ala)   +   H₂O

Amino-terminal (N-terminal) amino acid The amino acid with the free —NH₃⁺ group at the end of a protein.

Carboxyl-terminal (C-terminal) amino acid The amino acid with the free —COO⁻ group at the end of a protein.

Residue (amino acid) An amino acid unit in a polypeptide.

By convention, peptides and proteins are always written with the **amino-terminal (N-terminal) amino acid** (the one with the free —NH₃⁺ group) on the left, and the **carboxyl-terminal (C-terminal) amino acid** (the one with the free —COO⁻ group) on the right. The individual amino acids joined in the chain are referred to as **residues**.

A peptide is named by citing the amino acid residues in order, starting at the N-terminal acid and ending with the C-terminal acid. All residue names except the C-terminal one have the *-yl* ending instead of *-ine*, as in alanylserine (abbreviated Ala-Ser) or serylalanine (Ser-Ala).

The primary structure of a protein is the result of the amino acids being lined up one by one to form peptide bonds in precisely the correct order. Consider that there are six ways in which three different amino acids can be joined, more than 40,000 ways in which eight amino acids can be joined, and more than 360,000 ways in which ten amino acids can be joined. Despite the rapid increase in the number of possible combinations with the number of amino acid residues present, only the one correct isomer can do the job. For example, human *angiotensin II* must have its eight amino acids arranged in exactly the correct order:

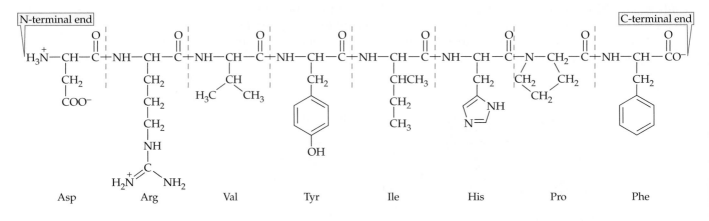

Asp   Arg   Val   Tyr   Ile   His   Pro   Phe

If not, this hormone will not participate as it should in regulating blood pressure.

So crucial is primary structure to function—no matter how big a protein—that the change of only one amino acid can sometimes drastically alter a protein's biological properties. Sickle-cell anemia is the best-known example of the potentially devastating result of amino acid substitution. It is a hereditary disease caused by a genetic defect that replaces one amino acid (glutamate, Glu) in each of two polypeptide chains of the hemoglobin molecule with another (valine, Val).

Sickle-cell anemia is named for the "sickle" shape of affected red blood cells. (A sickle is a tool with a curved blade and a short handle that is used to cut vegetation.) The sickling of the cells and the resultant painful, debilitating, and potentially fatal disease is entirely the result of the single amino acid substitution. The change replaces a hydrophilic, carboxylic acid–containing side chain with a hydrophobic, neutral hydrocarbon side chain and alters the shape of hemoglobin. (The effect of the change in charge on electrophoresis is illustrated in the Application, "Protein Analysis by Electrophoresis," p. 523)

Hemoglobin is the molecule that carries oxygen in the blood and releases it where it is needed. Sickling takes place in cells carrying the sickle-cell form of hemoglobin that has released oxygen. In this state, a hydrophobic pocket is exposed on the surface of the hemoglobin. The hydrophobic valine side chain is drawn into this pocket, and as this combination takes place in successive hemoglobins, insoluble, fibrous chains are formed. The stiff fibers force red blood cells into the sickled shape.

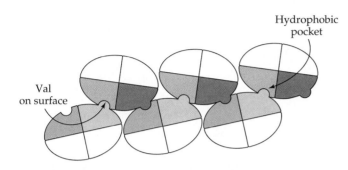

▲ Normal (left) and sickled red blood cells (right)   .

Normal deoxyhemoglobin does not form such fibers because the $—COO^-$ side chain in glutamate is too hydrophilic to enter the pocket. Also, the pocket is not available in oxygen-carrying hemoglobin because of a change in shape that occurs when the molecule picks up oxygen.

Sickled red blood cells are fragile, and because they are inflexible, they tend to collect and block capillaries, causing inflammation and pain, and possibly blocking blood flow in a manner that can damage major organs. Also, they have a short lifespan, which causes afflicted individuals to become severely anemic.

The percentage of individuals carrying the genetic trait for sickle-cell anemia is highest among people in ethnic groups with origins in tropical regions where malaria is prevalent. The ancestors of these individuals survived because if they were infected with malaria it was not fatal. Malaria-carrying parasites enter red blood cells and reproduce there. In a person with the sickle-cell trait, the cells respond by sickling and the parasites cannot multiply. As a result, the genetic trait for sickle-cell anemia is carried forward in the surviving population.

∞∞∞ LOOKING AHEAD

More than any other kind of biomolecule, proteins are in control of our biochemistry. Are you wondering how each of our thousands of proteins is produced with all their amino acids lined up in the correct order? The information necessary to do this is stored in DNA, and the remarkable machinery that does the job resides in the nuclei of our cells. Chapter 26 provides the details of how protein synthesis is accomplished.   ∞∞∞

■ WORKED EXAMPLE 18.4

Draw the structure of the dipeptide Ala-Gly.

ANALYSIS You need the names and structures of the two amino acids. Since alanine is named first, it is amino-terminal and glycine is carboxyl-terminal. Ala-Gly must have a peptide bond between the alanine $—COO^-$ and the glycine–NH.

SOLUTION The structures of alanine and glycine, and the structure of the Ala-Gly dipeptide are

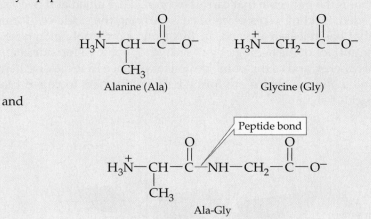

Alanine (Ala)   Glycine (Gly)

and

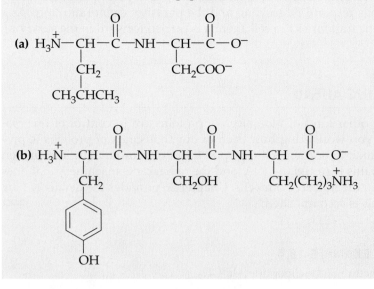

Ala-Gly

■ **PROBLEM 18.15**

(a) Use the three-letter shorthand notations to name the isomeric tripeptides that could be made from serine, tyrosine, and glycine.

(b) Draw the complete structure of the tripeptides that have glycine as the amino-terminal amino acid.

■ **PROBLEM 18.16**

Using three-letter abbreviations, show the six tripeptides that contain valine, histidine, and alanine.

■ **PROBLEM 18.17**

Identify the amino acids in the following dipeptide and tripeptide, and write the abbreviated forms of the peptide names.

■ **PROBLEM 18.18**

Copy the structure of the tripeptide in Problem 18.17b and circle the two planar regions along the backbone.

Application Proteins in the Diet

From a biochemical viewpoint, what are our protein requirements? Proteins are a necessary part of the daily diet because our bodies don't store proteins as they do carbohydrates and fats. Children need large amounts of protein for proper growth, and adults need protein to replace what is lost each day by normal biochemical reactions. Furthermore, 9 of the 20 amino acids are not synthesized by adult humans and must be obtained in the diet. These are known as the *essential amino acids* (histidine, isoleucine, leucine, lysine, methionine, phenylalanine, threonine, tryptophan, valine).

The total recommended daily amount of protein for an adult, which is the *minimum* required for good health, is 0.8 g per kilogram of body weight. For a 70 kg (154 lb) male, this is 56 g, and for a 55 kg (121 lb) female, it's 44 g. For reference, a McDonald's Big Mac contains 25 g of protein (along with 34 g of fat). The average protein intake in the United States is about 110 g/day, well above what most of us need.

Not all foods are equally good sources of protein. A *complete* protein source provides the nine essential amino acids in sufficient amount to meet our minimum daily needs. Most meat and dairy products meet this requirement, but many vegetable sources such as wheat and corn don't.

Vegetarians must be certain to adopt a diet that includes all of the essential amino acids, which means consuming a variety of foods. In some regions, food combinations that automatically provide *complementary* proteins (proteins that together supply all of the essential amino acids) are traditional, for example, rice and lentils in India, corn tortillas and beans in Mexico, and rice and black-eyed peas in the southern United States. The grains are low in lysine and threonine, while the legumes (lentils, beans, and peas) supply these amino acids, but are low in methionine and tryptophan, which are present in grains.

Protein is the major source of nitrogen in the diet. A healthy adult is normally at nitrogen equilibrium, meaning that the amount of nitrogen taken in each day is equal to the amount excreted. Infants and children, pregnant women, those recovering from starvation, and those with healing wounds are usually in *positive nitrogen balance*—they're excreting less nitrogen than they consume, a condition to be expected when new tissue is growing. The reverse condition, *negative nitrogen balance*, occurs when more nitrogen is excreted than consumed. This happens when protein intake is inadequate, during starvation, and in a number of

▲ Complementary protein food combination: beans and rice.

pathologic conditions including malignancies, malabsorption syndromes, and kidney disease.

Health and nutrition professionals group all disorders caused by inadequate protein intake as *protein-energy malnutrition* (PEM). Children, because of their higher protein needs, suffer most from this kind of malnutrition. The problem is rampant where meat and milk are in short supply and where the dietary staples are vegetables or grains. An individual is malnourished to some degree if *any* of the essential amino acids are deficient in their diet.

Protein deficiency alone is rare, however, and its symptoms are usually accompanied by those of vitamin deficiencies, infectious diseases, and starvation. At one end of the spectrum is *kwashiorkor*, in which protein is deficient although caloric intake may be adequate. Children with kwashiorkor have edema (swelling due to water retention), have an enlarged liver, and are underdeveloped. The word "kwashiorkor" is from the language of Ghana and is translated as "the sickness the older child gets when the next child is born." The onset of kwashiorkor comes when weaning from mother's milk results in conversion to a high-carbohydrate, low-protein diet. At the other end of the spectrum is *marasmus*, which is the result of starvation. As distinguished from kwashiorkor, marasmus in children is identified with severe muscle wasting, below-normal stature, and poor response to treatment.

See Additional Problems 18.73–76 at the end of the chapter.

■ **KEY CONCEPT PROBLEM 18.19**

Endoproteases are enzymes that cut proteins at specific points within their sequences. Chymotrypsin is an endoprotease that cuts on the C-terminal side of aromatic amino acids. Identify in Table 18.3 the three amino acids that have aromatic side chains. Now determine the number of pieces that would result from the treatment of vasopressin, which has the structure shown below, with chymotrypsin. Write out the sequences of these fragments using three-letter designators for each amino acid.

Asp-Tyr-Phe-Glu-Asn-Cys-Pro-Lys-Gly

18.8 Shape-Determining Interactions in Proteins

Without interactions between atoms in amino acid side chains or along the backbone, protein chains would twist about randomly in body fluids like spaghetti strands in boiling water. The essential structure–function relationship for each protein depends on the polypeptide chain being held in its necessary shape by these interactions. Before we look at the secondary, tertiary, and quaternary structures of proteins, it will be helpful to understand the kinds of interactions that determine the shapes of protein molecules.

Hydrogen Bonds Along the Backbone

Hydrogen bonds form when a hydrogen atom bonded to a highly electronegative atom is attracted to another highly electronegative atom that has an unshared electron pair. The hydrogens in the —NH— groups and the oxygens in the —C=O groups along protein backbones meet these conditions:

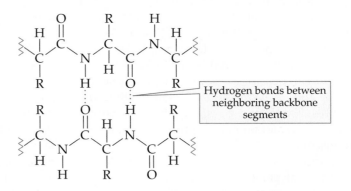

Hydrogen bonds between neighboring backbone segments

This type of hydrogen bonding creates pleated sheet and helical secondary structures, as described in Section 18.9 and as illustrated in the imaginary protein in Figure 18.4.

Hydrogen Bonds of R Groups with Each Other or with Backbone Atoms

Some amino acid side chains contain atoms that can form hydrogen bonds. Side-chain hydrogen bonds can connect different parts of a protein molecule, sometimes nearby and sometimes far apart along the chain. In the protein in Figure 18.4, hydrogen bonds between side chains have created folds in in two places. Often hydrogen-bonding side chains are present on the surface of a folded protein, where they can hydrogen-bond with surrounding water molecules.

Ionic Attractions Between R Groups (Salt Bridges)

Where there are ionized acidic and basic side chains, the attraction between their positive and negative charges creates what are sometimes known as *salt*

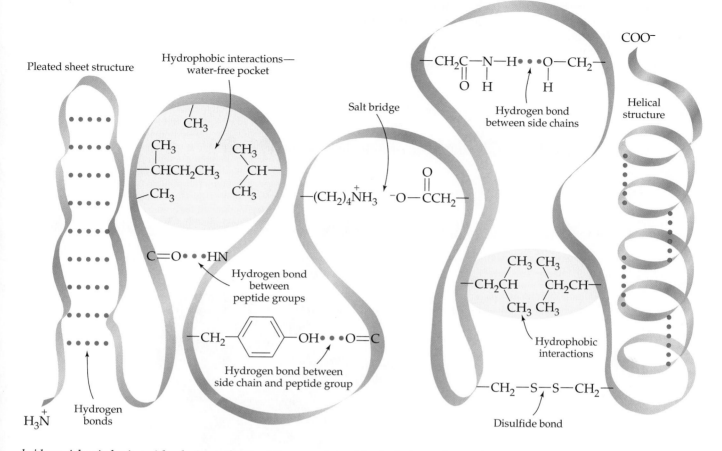

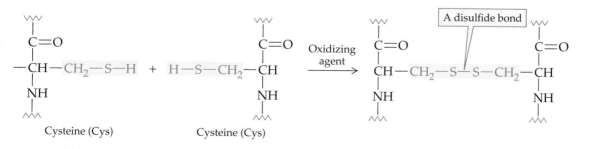

bridges. A basic lysine side chain and an acidic aspartate side chain have formed a salt bridge in the middle of the protein shown in Figure 18.4.

Hydrophobic Interactions Between R Groups

Hydrocarbon side chains are attracted to each other by the dispersion forces caused by the momentary uneven distribution of electrons (Section 8.11). The result is that these groups cluster together in the same way that oil molecules cluster on the surface of water (Section 9.2), so that these interactions are often referred to as *hydrophobic*. By clustering in this manner, the hydrophobic group shown in Figure 18.4 create a water-free pocket in the protein chain. Although the individual attractions are weak, their large number in proteins plays a major role in stabilizing the folded structures.

Covalent Sulfur-Sulfur Bonds

In addition to the noncovalent interactions, one type of covalent bond plays a role in determining protein shape. Cysteine amino acid residues have side chains containing thiol functional groups ($-SH$) that can react to form sulfur–sulfur bonds ($-S-S-$). (Section 14.10)

▲ **FIGURE 18.4 Interactions that determine protein shape.**   The regular pleated sheet (*left*) and helical structure (*right*) are created by hydrogen bonding between neighboring backbone atoms, while the other interactions involve side-chain groups that can be nearby or quite far apart in the protein chain.

If the cysteines are in different protein chains, the otherwise separate chains are linked together. If the cysteines are in the same chain, a loop is formed in the

Disulfide bond An S—S bond formed between two cysteine side chains; can join two peptide chains together or cause a loop in a peptide chain.

chain. Insulin provides a good example. It consists of two polypeptide chains connected by **disulfide bonds**, in two places. One of the chains also has a loop caused by a third disulfide bond.

Structure of insulin

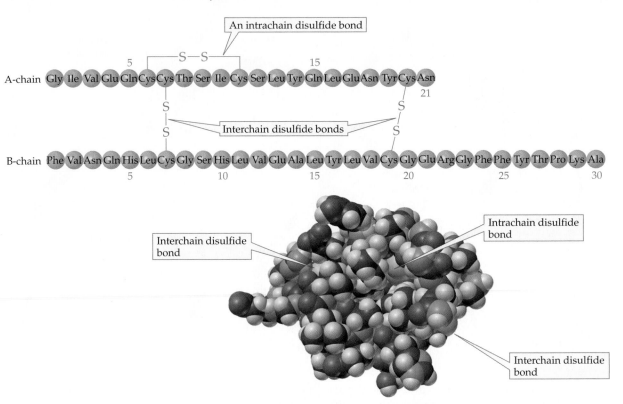

The structure and function of insulin are of intense interest because of its role in glucose metabolism and the need for supplementary insulin by individuals with diabetes (discussed in Section 23.9). Undoubtedly because of this need, studies of insulin have led the way in our still-developing ability to determine the structure of a biomolecule and prepare it synthetically.

In an historically important accomplishment, the amino acid sequence of insulin was determined in 1951—it was the *first* protein for which this was done. It took 15 years before the cross-linking and complete structure of the molecule were determined and a successful laboratory synthesis was carried out. With the advent of biotechnology in the 1980s, once again insulin was first. Until then, diabetic individuals had had to rely on insulin extracted from the pancreas of cows, and because of differences in three amino acids from human insulin, allergic reactions occasionally resulted. In 1982, human insulin became the first commercial product of genetic engineering to be licensed by the U.S. government for clinical use.

∞∞∞ LOOKING AHEAD

Insulin and angiotensin II (p. 516) are representative of a class of small polypeptides that are hormones—they are released when there is a chemical message that must be carried from one place to another. Hormones are discussed in Sections 20.2–20.5. ∞∞∞

■ WORKED EXAMPLE 18.5

What type of noncovalent interaction would occur between the threonine and glutamine side chains? Draw the structures of these amino acids to show the interaction.

Application Protein Analysis by Electrophoresis

Protein molecules in solution can be separated from each other by taking advantage of their overall positive or negative charges. In the electric field between two electrodes, a positively charged particle moves toward the negative electrode and a negatively charged particle moves toward the positive electrode. This movement, known as *electrophoresis*, varies with the strength of the electric field, the charge of the particle, the size and shape of the particle, and the nature of the medium in which the protein is moving.

The overall positive or negative charge of a protein is determined by how many of the acidic or basic side-chain functional groups in the protein are ionized, and this, like the charge of an amino acid (Section 18.4), depends on the pH. Thus, the mobility of a protein during electrophoresis depends on the pH of the medium. If the medium is at a pH equal to the isoelectric point of the protein, the protein does not move.

By varying the nature of the medium between the electrodes and other conditions, proteins can be separated in a variety of ways, including by their molecular weight. Once the separation is complete, the protein segments on the medium can be made visible by the addition of a dye.

Electrophoresis is routinely used in the clinical laboratory for determination of protein concentrations in blood. One application is in the diagnosis of sickle-cell anemia (p. 516). Normal adult hemoglobin (HbA) and hemoglobin showing the inherited sickle-cell trait (HbS) differ in their charges. Therefore, HbA and HbS move different distances during electrophoresis in a medium with constant pH. The accompanying diagram compares the results of electrophoresis of the material from broken-down red blood cells for a normal individual, one with sickle-cell anemia (two inherited sickle-cell genes), and one with sickle-cell trait (one normal and one inherited sickle-cell gene). With sickle-cell trait, an individual is likely to suffer symptoms of the disease only under conditions of severe oxygen deprivation.

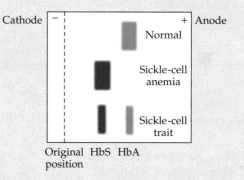

▲ **Gel electrophoresis of hemoglobin.** Hemoglobin in samples placed at the original position have moved left to right as shown. The normal individual has only HbA. The individual with sickle-cell anemia has no HbA, and the individual with sickle-cell trait has roughly equal amounts of HbA and HbS. HbA and HbS have negative charges of different magnitudes because HbS has two fewer Glu residues than HbA.

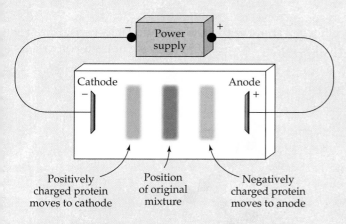

▲ **Movement of charged molecules in electrophoresis.**

See Additional Problem 18.77–18.78 at the end of the chapter.

ANALYSIS The side chains of threonine and glutamine contain an amide group and a hydroxyl group, respectively. These groups do not form salt bridges because they do not ionize. They are polar and therefore not hydrophobic. They can form a hydrogen bond between the oxygen of the amide carbonyl group and the hydrogen of the hydroxyl group.

SOLUTION The noncovalent, hydrogen bond interaction between threonine and glutamine is as follows:

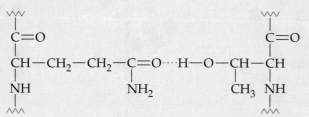

18.9 Secondary Protein Structure

The arrangement in space of the polypeptide backbones of proteins constitutes **secondary protein structure**. The secondary structure includes two kinds of repeating patterns known as the α-helix and the β-sheet. In both, hydrogen bonding between *backbone* atoms holds the polypeptide chain in place. The hydrogen bonding connects the carbonyl oxygen atom of one peptide unit with the amide hydrogen atom of another peptide unit ($-C=O\cdots H-N-$). In large protein molecules, regions of α-helix and β-sheet structure are connected by randomly arranged loops or coils that are a third type of secondary structure.

α-Helix

A single protein chain coiled in a spiral with a right-handed (clockwise) twist is known as an **α-helix** (Figure 18.5). The helix, which resembles a coiled tele-

Secondary protein structure Regular and repeating structural patterns (for example, α-helix, β-sheet) created by hydrogen bonding between backbone atoms in neighboring segments of protein chains.

Alpha- (α-) helix Secondary protein structure in which a protein chain forms a right-handed coil stabilized by hydrogen bonds between peptide groups along its backbone.

FIGURE 18.5 Alpha-helix ▶ secondary structure. The coil is held in place by hydrogen bonds (dotted red lines) between each carbonyl oxygen and the amide hydrogen four amino acid residues above it. The chain is a right-handed coil (shown separately on the right), and the hydrogen bonds lie parallel to the vertical axis.

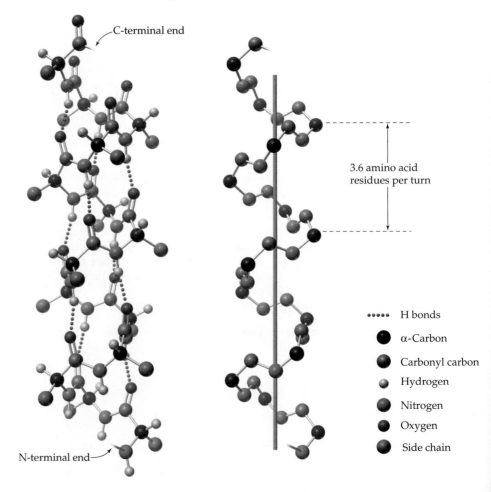

C-terminal end

3.6 amino acid residues per turn

N-terminal end

••••• H bonds

● α-Carbon

● Carbonyl carbon

● Hydrogen

● Nitrogen

● Oxygen

● Side chain

phone cord, is stabilized by hydrogen bonds between each backbone carbonyl oxygen and an amide hydrogen four amino acid residues farther along the backbone. The hydrogen bonds lie vertically along the helix, and the amino acid R groups extend to the outside of the coil. Although the strength of each individual hydrogen bond is small, the large number of bonds in the helix results in an extremely stable secondary structure.

β-Sheet

In the **β-sheet** structure, the polypeptide chains are held in place by hydrogen bonds between pairs of peptide units along neighboring backbone segments. The protein chains, which are extended to their full length, bend at each α carbon so that the sheet has a pleated contour, with the R groups extending above and below the sheet (Figure 18.6).

Beta- (β-) sheet   Secondary protein structure in which adjacent protein chains in the same or different molecules are held in place by hydrogen bonds along the backbones.

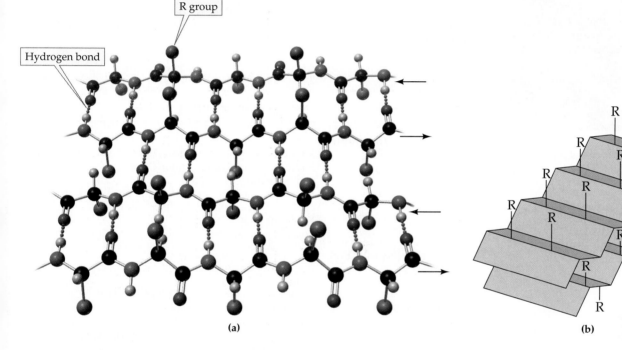

(a)

(b)

■ **PROBLEM 18.22**

Examine the α-helix in Figure 18.5 and determine how many backbone C and N atoms are included in the loop between an amide hydrogen atom and the carbonyl oxygen to which it is hydrogen-bonded.

▲   **FIGURE 18.6   Beta-sheet secondary structure.**   (a) The hydrogen bonds between neighboring protein chains. The protein chains usually lie side by side so that alternating chains run from the N-terminal end to the C-terminal end and from the C-terminal end to the N-terminal end (known as the *antiparallel* arrangement). (b) A pair of stacked pleated sheets illustrate how the R groups point above and below the sheets.

Secondary Structure in Fibrous and Globular Proteins

Proteins can be classified in several ways, one of which is to identify them as either *fibrous proteins* or *globular proteins*. In an example of the integration of molecular structure and function that is central to biochemistry, fibrous and globular proteins each have functions made possible by their distinctive structures.

Secondary structure is primarily responsible for the nature of **fibrous proteins**—tough, insoluble proteins in which the chains form long fibers or sheets. Wool, hair, and fingernails are made of fibrous proteins known as *α-keratins*, which are composed almost completely of α-helixes. In α-keratins, pairs of α-helixes are twisted together into small fibrils that are in turn twisted into larger and larger bundles. The hardness, flexibility, and stretchiness of the material varies with the number of disulfide bonds. In fingernails, for example, large numbers of disulfide bonds hold the bundles in place.

Fibrous protein   A tough, insoluble protein whose protein chains form fibers or sheets.

▲ Egg white is a globular protein and a spider web is a fibrous protein.

Globular protein A water-soluble protein whose chain is folded in a compact shape with hydrophilic groups on the outside.

Natural silk and spider webs are made of *fibroin*, a fibrous protein almost entirely composed of stacks of β-sheets. For such close stacking, the R groups must be relatively small (see Figure 18.6). Fibroin contains regions of alternating glycine (—H on α carbon) and alanine (—CH_3 on α carbon). The sheets stack so that sides with the smaller glycine hydrogens face each other and sides with the larger alanine methyl groups face each other.

Globular proteins are water-soluble proteins whose chains are folded into compact, globelike shapes. Their structures, which vary widely with their functions, are not regular like those of fibrous proteins. Where the protein chain folds back on itself, sections of α-helix and β-sheet are usually present, as illustrated in Figure 18.4. The presence of hydrophilic side chains on the outer surfaces of globular proteins accounts for their water solubility, allowing them to travel through the blood and other body fluids to sites where their activity is needed. The overall shapes of globular proteins represent another level of structure, tertiary structure, discussed in the next section.

Table 18.4 compares the occurrences and functions of some fibrous and globular proteins.

TABLE 18.4 Some Common Fibrous and Globular Proteins	
Name	**Occurrence and Function**
Fibrous proteins (insoluble)	
Keratins	Found in skin, wool, feathers, hooves, silk, fingernails
Collagens	Found in animal hide, tendons, bone, eye cornea, and other connective tissue
Elastins	Found in blood vessels and ligaments, where ability of the tissue to stretch is important
Myosins	Found in muscle tissue
Fibrin	Found in blood clots
Globular proteins (soluble)	
Insulin	Regulatory hormone for controlling glucose metabolism
Ribonuclease	Enzyme that catalyzes RNA hydrolysis
Immunoglobulins	Proteins involved in immune response
Hemoglobin	Protein involved in oxygen transport
Albumins	Proteins that perform many transport functions in blood; protein in egg white

18.10 Tertiary Protein Structure

Tertiary protein structure The way in which an entire protein chain is coiled and folded into its specific three-dimensional shape.

The overall three-dimensional shape that results from the folding of a protein chain is the protein's **tertiary structure**. In contrast to secondary structure, which depends mainly on attraction between backbone atoms, tertiary structure depends mainly on interactions of amino acid side chains that are far apart along the same backbone.

Although the bends and twists of the protein chain in a globular protein may appear irregular and the three-dimensional structure may appear random, this is not the case. Each protein molecule folds in a distinctive manner that is determined by its primary structure and results in its maximum stability. A protein with the shape in which it functions in living systems is known as a **native protein**.

Native protein A protein with the shape (secondary, tertiary, and quaternary structure) in which it exists naturally in living organisms.

The noncovalent interactions and disulfide covalent bonds described in Section 18.8 govern tertiary structure. An enzyme, *ribonuclease*, is drawn on the facing page in a style that shows the combination of α-helix and β-sheet regions, the loops connecting them, and four disulfide bonds:

Ribonuclease is representative of the tertiary structure of globular, water-soluble proteins. The hydrophobic, nonpolar side chains congregate in a hydro-

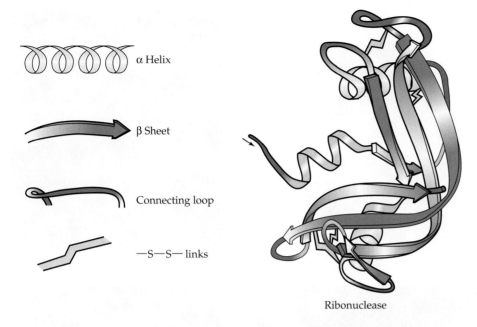

α Helix

β Sheet

Connecting loop

—S—S— links

Ribonuclease

carbonlike interior, and the hydrophilic side chains, which provide water solubility, congregate on the outside. Ribonuclease is classified as a **simple protein** because it is composed only of amino acid residues (124 of them). The drawing above shows ribonuclease in a style that clearly represents the combination of secondary structures in the overall tertiary structure of a globular protein.

Myoglobin is another example of a small globular protein. A relative of hemoglobin (described in the next section), myoglobin stores oxygen in skeletal muscles for use when there is an immediate need for energy. Structurally, the 153 amino acid residues of myoglobin are arranged in eight α-helical segments connected by short segments looped so that hydrophilic amino acid residues are on the exterior of the compact, spherical tertiary structure. Like many proteins, myoglobin is not a simple protein, but is a **conjugated protein**—a protein that is aided in its function by an associated non–amino acid nonprotein group. The oxygen-carrying portion of myoglobin has a heme group embedded within the polypeptide chain. In Figure 18.7 the myoglobin molecule is drawn in four different ways, each often used to illustrate the shapes of protein molecules. Some examples of other kinds of conjugated proteins are listed in Table 18.5.

Simple protein A protein composed of only amino acid residues.

Conjugated protein A protein that incorporates one or more non–amino acid units in its structure.

TABLE 18.5 Some Examples of Conjugated Proteins

Class of Protein	Nonprotein Part	Examples
Glycoproteins	Carbohydrates	Glycoproteins in cell membranes (Section 24.7)
Lipoproteins	Lipids	High- and low-density lipoproteins that transport cholesterol and other lipids through the body (Section 25.2)
Metalloproteins	Metal ions	The enzyme cytochrome oxidase, necessary for biological energy production, and many other enzymes
Phosphoproteins	Phosphate groups	Milk casein, which provides essential nutrients to infants
Hemoproteins	Heme	Hemoglobin and myoglobin, which transport and store oxygen, respectively
Nucleoproteins	RNA (ribonucleic acid)	Found in cell ribosomes, where they take part in protein synthesis

FIGURE 18.7 Myoglobin, ▶
drawn in four different styles.
(a) The sausagelike shape is often used alone to represent the helical portions of a globular protein. The red structure embedded in the protein is a molecule of heme, to which O_2 binds. (b) A protein *ribbon model* shows the helical portions as a ribbon. Here also, the heme molecule is red. (c) A ball-and-stick molecular model of myoglobin. (d) A space-filling model of myoglobin in which the hydrophobic residues are blue and the hydrophilic residues are purple.

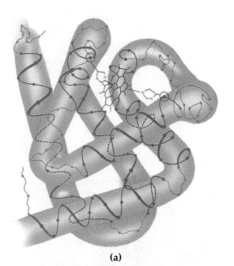

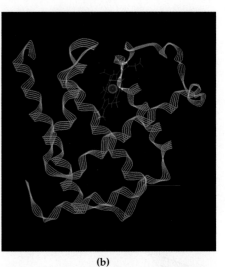

(a)

(b)

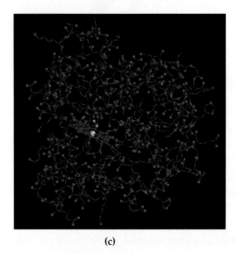

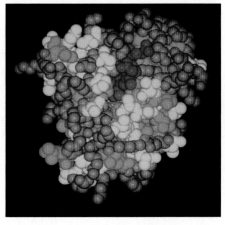

(c)

(d)

■ KEY CONCEPT PROBLEM 18.23

Hydrogen bonds are important in stabilizing both the secondary and tertiary structures of proteins. How do the groups that form hydrogen bonds in the secondary and tertiary structures differ?

18.11 Quaternary Protein Structure

The fourth and final level of protein structure, and the most complex, is **quaternary protein structure**—the way in which two or more polypeptide subunits associate to form a single three-dimensional protein unit. The individual polypeptides are held together by the same noncovalent forces responsible for tertiary structure. In some cases, there are also covalent bonds and the protein may incorporate a non–amino acid portion. *Hemoglobin* and *collagen* are both well-understood examples of proteins with quaternary structure essential to their function.

Hemoglobin

Hemoglobin (Figure 18.8b) is a conjugated quaternary protein composed of four polypeptide chains (two α chains and two β chains) held together primarily by the interaction of hydrophobic groups, and four heme groups. The polypeptides are similar in composition and tertiary structure to myoglobin (Figure 18.7). The α chains have 141 amino acids, and the β chains have 146 amino acids.

Quaternary protein structure
The way in which two or more protein chains aggregate to form large, ordered structures.

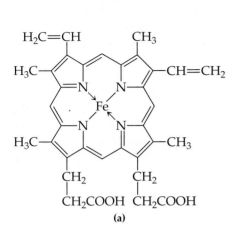

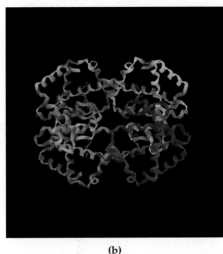

(a)

(b)

◀ **FIGURE 18.8   Heme and hemo-globin, a protein with quaternary structure.**   (a) A heme unit is present in each of the four polypetides in hemoglobin. (b) The polypeptides are shown in purple, green, blue, and yellow, with their heme units in red. Each polypeptide resembles myglobin in structure.

The hemes (Figure 18.8a), one in each of the four polypeptides, each contain an iron atom that is essential to their function. Hemoglobin is the oxygen carrier in red blood cells. In the lungs, O_2 bonds to the Fe, so that each hemoglobin can carry a maximum of four O_2 molecules. In cells in need of oxygen, the O_2 is released, and CO_2 (the product of respiration) is picked up and carried back to the lungs. (Oxygen transport is discussed further in Section 29.6.)

Collagen

Collagen is the most abundant of all proteins in mammals, making up 30% or more of the total. A fibrous protein, collagen is the major constituent of skin, tendons, bones, blood vessels, and other connective tissues. The basic structural unit of collagen (*tropocollagen*) consists of three intertwined chains of about 1000 amino acids each. Each stiff and rodlike chain is loosely coiled in a left-handed (counterclockwise) direction (Figure 18.9a). Three of these coiled chains wrap around each other (in a clockwise direction) to form a stiff, rodlike triple helix (Figure 18.9b) in which the chains are held together by hydrogen bonds.

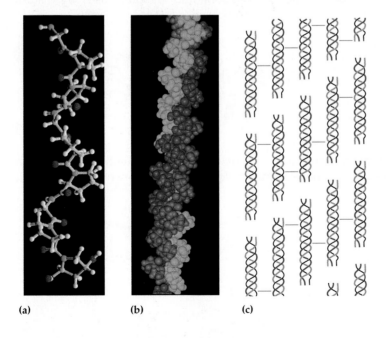

(a)   (b)   (c)

◀ **FIGURE 18.9   Collagen.**   (a) A single collagen helix (carbon, green; hydrogen, light blue; nitrogen, dark blue; oxygen, red). (b) The triple helix of tropocollagen. (c) The quaternary structure of a cross-linked collagen, showing the assemblage of tropocollagen molecules.

The various kinds of collagen have in common a glycine residue at every third position. Only glycine residues (with —H on the α carbon) can fit in the center of the tightly coiled triple helix. The larger side chains face the exterior of the helix. Proline is incorporated into the originally synthesized collagen molecules. A hydroxyl group is then added to some proline residues in a reaction that requires vitamin C (Section 19.10). Herein lies the explanation for the symptoms of scurvy, the disease that results from vitamin C deficiency. When vitamin C is in short supply, collagen is deficient in hydroxylated proline residues and, as a result, forms fibers poorly. The results are the skin lesions and fragile blood vessels that accompany scurvy.

The tropocollagen triple helixes are assembled into collagen in a quaternary structure formed by a great many strands overlapping lengthwise (Figure 18.9c). Depending on the exact purpose collagen serves in the body, further structural modifications occur. In connective tissue like tendons, covalent bonds between strands give collagen fibers a rigid, cross-linked structure. In teeth and bones, calcium hydroxyapatite [$Ca_5(PO_4)_3OH$] deposits in the gaps between chains to further harden the overall assembly.

Protein Structure Summary

- **Primary structure**—the sequence of amino acids connected by peptide bonds in the polypeptide chain, for example; Asp-Arg-Val-Tyr.

- **Secondary structure**—the arrangement in space of the polypeptide chain, which incudes the regular patterns of the α-helix and the β-sheet (held together by hydrogen bonds between amino acid residues along adjacent chains segments) plus the loops and coils that connect these segments.

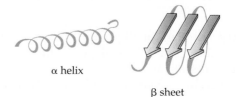

α helix

β sheet

- **Tertiary structure**—the folding of a protein molecule into a specific three-dimensional shape held together by noncovalent interactions that can be quite far apart along the backbone and, in some cases, by disulfide bonds between side chain thiol groups.

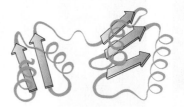

- **Quaternary structure**—two or more protein chains assembled in a larger three-dimensional structure held together by noncovalent interactions.

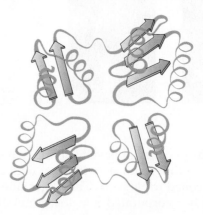

- **Classes of proteins**
  - *Fibrous proteins* are tough, insoluble, and composed of fibers and sheets; *globular proteins* are water-soluble and have chains folded into compact shapes.
  - *Simple proteins* contain only amino acid residues; *conjugated proteins* include one or more non–amino acid units.

■ **KEY CONCEPT PROBLEM 18.24**

Identify each of the following statements as descriptive of the secondary, tertiary, or quaternary structure of a protein. What type(s) of interaction(s) stabilize each type of structure?

(a) The polypeptide chain has a number of bends and twists resulting in a compact structure.

(b) The polypeptide backbone forms a right-handed coil.

(c) The four polypeptide chains are arranged in a spherical shape.

18.12 Chemical Properties of Proteins

Protein Hydrolysis

Just as a simple amide can be hydrolyzed to yield an amine and a carboxylic acid, a protein can be hydrolyzed. (∞ Section 17.6) In protein hydrolysis, the reverse of protein formation, peptide bonds are hydrolyzed to yield amino acids. In fact, digestion of proteins in the diet involves nothing more than hydrolyzing peptide bonds. For example:

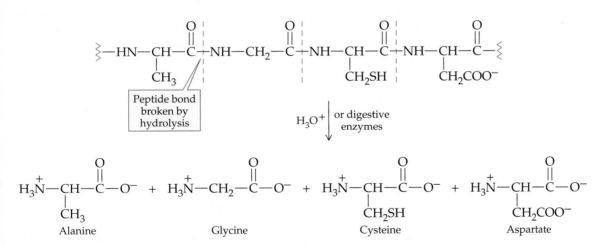

Alanine   Glycine   Cysteine   Aspartate

A chemist in the laboratory might choose to hydrolyze a protein by heating it with a solution of hydrochloric acid. Most digestion of proteins in the body takes place in the stomach and small intestine, where the process is catalyzed by enzymes (Section 28.1). Once formed, individual amino acids are absorbed through the wall of the intestine and transported in the bloodstream to wherever they are needed.

Protein Denaturation

Since the overall shape of a protein is determined by a delicate balance of noncovalent forces, it's not surprising that a change in protein shape often results when the balance is disturbed. Such a disruption in shape that does not affect the protein's primary structure is known as **denaturation**. When denaturation

Denaturation The loss of secondary, tertiary, or quaternary protein structure due to disruption of noncovalent interactions and/or disulfide bonds that leaves peptide bonds and primary structure intact.

of a globular protein occurs, for example, the structure unfolds from a well-defined globular shape to a randomly looped chain:

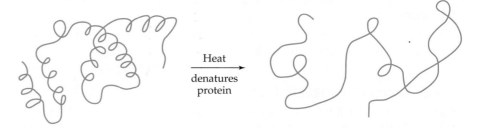

Denaturation is accompanied by changes in physical, chemical, and biological properties. Solubility is often decreased by denaturation, as occurs when egg white is cooked and the albumins coagulate into an insoluble white mass. Enzymes lose their catalytic activity and other proteins are no longer able to carry out their biological functions when their shapes are altered by denaturation.

Agents that cause denaturation include heat, mechanical agitation, detergents, organic solvents, extremely acidic or basic pH, and inorganic salts.

▲ Protein denaturation in action: The egg white denatures as the egg fries.

- **Heat** The weak side-chain attractions in globular proteins are easily disrupted by heating, in many cases only to temperatures above 50°C. Cooking meat converts some of the insoluble collagen into soluble gelatin, which can be used in glue and for thickening sauces.

- **Mechanical agitation** The most familiar example of denaturation by agitation is the foam produced by beating egg whites. Denaturation of proteins at the surface of the air bubbles stiffens the protein and causes the bubbles to be held in place.

- **Detergents** Even very low concentrations of detergents can cause denaturation by disrupting the association of hydrophobic side chains.

- **Organic compounds** Polar solvents such as acetone and ethanol interfere with hydrogen bonding by competing for bonding sites. The disinfectant action of ethanol, for example, results from its ability to denature bacterial protein.

- **pH change** Excess H^+ or OH^- ions react with the basic or acidic side chains in amino acid residues and disrupt salt bridges. One familiar example of denaturation by pH change is the protein coagulation that occurs when milk turns sour because it has become acidic.

- **Inorganic salts** Sufficiently high concentrations of ions can disturb salt bridges.

Most denaturation is irreversible: Hard-boiled eggs don't soften when their temperature is lowered. Many cases are known, however, in which unfolded proteins spontaneously undergo *renaturation*—a return to their native state when returned to a nondenaturing medium. Renaturation is accompanied by recovery of biological activity, indicating that the protein has completely returned to its stable secondary and tertiary structure. By spontaneously refolding into their native shapes, proteins demonstrate that all the information needed to determine these shapes is present in the primary structure.

Application Prions: Proteins that Cause Disease

At first, no one believed it. The existence of a protein that could duplicate itself and cause disease seemed impossible. Only bacteria, viruses, and other microorganisms that have DNA are able to reproduce and cause disease. That had been believed for a long time.

Even more unbelievable was the proposal that a form of protein could be responsible for disease that might be inherited, might be transmitted between individuals, and might arise spontaneously in the absence of inheritance or transmission. Stanley B. Prusiner, a neurologist, received the Nobel Prize for Physiology and Medicine in 1997 for his research demonstrating that indeed such proteins exist and can cause disease in all of these ways.

Prusiner named these proteins *prions* (pronounced "pree-ons"), for "proteinaceous infectious particles." Dr. Prusiner began his prion research in 1974. Some are still skeptical and there is much yet to be learned, but evidence is accumulating that all of the unbelievable premises are correct.

Prion-caused disease leaped into worldwide notice in the 1990s when individuals in Great Britain began to die from Creutzfeldt–Jakob disease (CJD) and it became apparent that the cause was eating beef from cows infected with mad-cow disease, known technically as bovine spongiform encephalopathy (BSE). The BSE name summarizes a major symptom of these and other prion diseases. Brain tissue develops open spaces, thereby becoming spongelike. Other diseases in the spongiform encephalopathy family include scrapie in sheep, a chronic wasting disease in elk and mule deer, and in humans, inherited CJD and kuru, which occurred among natives in New Guinea who honored their dead relatives by eating their brains but disappeared when the cannibalism ceased. There is no therapy for the spongiform encephalopathies. All are fatal and characterized by loss of muscular control and symptoms of dementia.

The following statements summarize some well-supported facts about prion proteins known in early 2001:

- Humans and all animals tested thus far have a gene for making a normal prion protein that resides in the brain and does not cause disease.

- An inherited genetic defect can cause one amino acid (proline) in normal prion protein to be replaced by another (leucine), resulting in a prion that is responsible for one of the inherited human prion diseases.

- The difference between normal and disease-causing prions lies in their secondary structure. Alpha helixes in the normal prion are replaced by β-sheets, resulting in a prion with a different shape.

- A misfolded, disease-causing prion can induce a normal prion to flip from the normal shape to disease-causing shape. It does not matter whether the disease-causing prion arises from a genetic defect, enters the body in

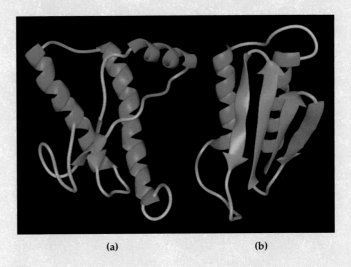

(a) (b)

▲ (a) A normal prion. (b) Proposed conformation for a disease-causing prion. (To interpret these drawings, see the picture of ribonuclease, p. 527.)

food or in some other manner, or is formed randomly and spontaneously. Exactly how the change in shape occurs is not yet understood, but the result is an accelerating spread of the disease-causing form.

Many unanswered questions lie behind the ongoing efforts to understand prions and their disease-causing potential. Are some individuals more susceptible to prion-caused diseases than others? To what extent can the disease move from one species to another? Can there be a diagnostic test to identify those who are susceptible or already exposed to abnormal prions? Looming over all is a most intriguing, major question: Could it be that other neurodegenerative diseases that create abnormal structures in the brain, including Alzheimer's disease and Parkinson's disease, are also prion diseases?

Meanwhile, the epidemic of mad-cow disease in Great Britain has been devastating. Millions of cattle with BSE have been slaughtered and more than 80 people have died from CJD.

An initial ban by on export of beef from Great Britain to other European countries was lifted in 1998, but their beef exports remain just a fraction of what they were and bans on import continue in some countries. Meanwhile, BSE has been identified in cattle in several other European countries and fears are growing that it can be transmitted by sheep and that it might spread outside of Europe.

In November, 2001 the U.S. Department of Agriculture released the results of a major scientific study that assessed the risk of mad cow disease in the United States. The principal conclusions of the study were that early actions had prevented the disease from entering the United States, but that vigilance and testing must continue.

See Additional Problems 18.79–18.80 at the end of the chapter.

It has often been said that an army marches on its stomach, and that's still true even for today's high-tech army. Distributing food to our far-flung military forces is obviously a huge task. Less obvious, but equally important, is making sure that this food is wholesome—and can stay wholesome long enough to be useful. That's a big part of Linnea Hallberg's job.

Lin works in a U.S. government laboratory as a food technologist. As she explains it, "In the food research lab we are continually exploring new methods of processing and packaging foods. Many of these products will be included as part of the combat rations provided to U.S. military personnel throughout the world, so long-term storage is an important consideration.

"Recently we had to develop a sandwich that wouldn't go stale or spoil over a long period of time. To do this, we needed to investigate what chemical processes were at work in the sandwich. One of the factors we examined was its water content. We knew that we could stabilize a food by reducing its water content or by reducing the mobility of the water. So we had to determine exactly how tightly bound the water was within the sandwich. We wanted to know if the water was going from the filling to the bread, or from the bread to the filling, and how quickly this process was occurring during storage.

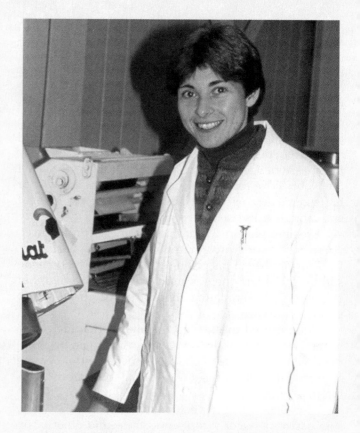

"First we used a dye that dissolved into the water in the sandwich, together with a specialized scanning microscope through which we could actually watch the movement of the water over a period of time. We were then able to excite electrons in the dyes to higher energy states with laser beams and then observe the light emitted by the electrons when they returned to lower energy levels (Chapter 3). We took photographs of the light over time and used them to gather information about the movement of the water within the food.

"There are other dyes that can be used that bond specifically to proteins. If we were studying a food high in proteins, for example, we could use this same method to determine whether the proteins were becoming soluble and moving or were a stable part of the structure.

"By tracking the migration of the water within the food, we can determine where its affinity lies and what affects it. For example, if the water from the filling is saturating the starch in the bread and making it soggy, we can prevent that by coating the starch with something water-resistant like silica or a fat. A food technologist also has to have a knowledge of nutrition, since it's important to know what effect these changes that we make may have on the nutrients within the food.

"There are many protective measures that food technologists and packaging specialists use to preserve foods. For example, we try to formulate a food with an optimal balance of preservatives that target oxidation, a type of chemical spoilage. In order to prevent fats from going rancid, we need a knowledge of chemistry to understand how we can combine synthetic phenolic antioxidants such as butylated hydroxytoluene (BHT), butylated hydroxyanisole (BHA), *tert*-butyl hydroquinone (TBHQ) with natural antioxidants like tocopherols (Vitamin E), plant phenols, spice extracts, amino acids, proteins, and ascorbic acid (Vitamin C) (Chapters 17, 21). Sequestrants or chelators like ethylenediaminetetraacetic acid (EDTA) are scavengers of catalytic metals like copper and iron (Chapter 19), and they can prevent oxidation reactions that could also lead to rancidity of fats (Chapters 14, 25).

"The work that I do every day as a food technologist is very firmly rooted in my knowledge of basic chemistry. Whether I'm working in the lab or sharing research findings at a conference with my colleagues, I am continually learning new information and constantly presented with fresh challenges. It is this ongoing process of learning and investigating that I find to be most exciting about my job. That's why I loved chemistry back in school, years ago—because it can explain so many mysteries in our lives, and we can use it to improve the quality of our lives."

Summary: Revisiting the Chapter Goals

1. What are the structural features of amino acids? *Amino acids* in body fluids have an ionized carboxylic acid group ($-COO^-$), an ionized amino group ($-NH_3^+$), and a side-chain R group bonded to a central carbon atom (the α carbon). Twenty different amino acids occur in *proteins* (Table 18.3), connected by *peptide bonds* (amide bonds) formed between the carboxyl group of one amino acid and the amino group of the next.

2. What are the properties of amino acids? Amino acid side chains have acidic or basic functional groups or neutral groups that are either polar or nonpolar. In glycine the "side chain" is a hydrogen atom. The dipolar ion in which the amino and carboxylic acid groups are both ionized is known as a *zwitterion*. For each amino acid, there is a distinctive *isoelectric point*—the pH at which the numbers of positive and negative charges in a solution are equal. At more acidic pH, some carboxylic acid groups are not ionized; at more basic pH, some amino groups are not ionized.

3. Why do amino acids have "handedness"? An object, including a molecule, has "handedness,"—that is, is *chiral*—when it has no plane of symmetry and thus has mirror images that cannot be superimposed on each other. A simple molecule can be identified as chiral if it contains a carbon atom bonded to four different groups. All α-amino acids except glycine meet this condition by having four different groups bonded to the α carbon.

4. What is the primary structure of a protein and what conventions are used for drawing and naming primary structures? Proteins are polymers of amino acids (*polypeptides*). Their *primary structure* is the linear sequence in which the amino acids are connected by peptide bonds. Using formulas or amino acid abbreviations, the primary structures are written with the amino-terminal end on the left (H_3^+N-) and the carboxyl-terminal end on the right ($-COO^-$). To name a peptide, the names of the amino acids are combined, starting at the amino-terminal end, with the endings of all but the carboxyl-terminal acid changed to *-yl*. Primary structures are often represented by combining three-letter abbreviations for the amino acids.

5. What types of interactons determine the overall shapes of proteins? Protein chains are drawn into their distinctive and biochemically active shapes by attractions between atoms along their backbones and between atoms in side-chain groups. Hydrogen bonding can occur between the backbone carbonyl groups and amide hydrogens of adjacent protein chains. *Noncovalent interactions* between side chains include hydrogen bonding between acidic and basic groups (*salt bridges*), and *hydrophobic interactions* among nonpolar groups. Covalent sulfur–sulfur bonds (*disulfide bonds*) can form bridges between the side chains in cysteine.

6. What are the secondary and tertiary structures of proteins? *Secondary structures* inclue the regular, repeating three-dimensional structures held in place by hydrogen bonding between backbone atoms within a chain or in adjacent chains. The α-*helix* is a coil with hydrogen bonding between carbonyl oxygens and amide hydrogens four amino acid residues farther along the same chain. The β-*sheet* is a pleated sheet with adjacent protein chain segments connected by hydrogen bonding between peptide groups. The adjacent chains in the β-sheet may be parts of the same protein chain or different protein chains. Secondary structure mainly determines the properties of *fibrous proteins*, which are tough and insoluble. *Tertiary structure* is the overall three-dimensional shape of a folded protein chain. Tertiary structure determines the properties of *globular proteins*, which are water-soluble, with hydrophilic groups on the outside and hydrophobic groups on the inside.

Key Words

Achiral, p. 512

Alpha- (α-) amino acid, p. 505

Alpha- (α-) helix, p. 524

Amino acid, p. 505

Amino-terminal (N-terminal) amino acid, p. 516

Beta- (β-) sheet, p. 525

Carboxyl-terminal (C-terminal) amino acid, p. 516

Chiral, p. 511

Chiral carbon atom, p. 513

Conjugated protein, p. 527

Denaturation, p. 531

Disulfide bond (in protein), p. 522

Enantiomers (optical isomers), p. 513

Fibrous protein, p. 525

Globular protein, p. 526

Hydrophilic, p. 508

Hydrophobic, p. 508

Isoelectric point (pI), p. 509

Native protein, p. 526

Noncovalent forces, p. 508

Peptide bond, p. 505

Primary protein structure, p. 515

Protein, p. 505

Quaternary protein structure, p. 528

Residue (amino acid), p. 516

Secondary protein structure, p. 524

Side chain (amino acid), p. 506

Simple protein, p. 527

Stereoisomers, p. 513

Tertiary protein structure, p. 526

Zwitterion, p. 509

Globular proteins often contain regions of α-helix and/or β-sheet secondary structures.

7. **What is quaternary protein structure?** Proteins that incorporate more than one peptide chain are said to have *quaternary structure*. In a quaternary structure, two or more folded protein subunits are united in a single structure by noncovalent interactions. Hemoglobin, for example, consists of two pairs of subunits, with a nonprotein heme molecule in each of the four subunits. Collagen is a fibrous protein composed of protein chains twisted together in triple helixes.

8. **What chemical properties do proteins have?** The peptide bonds are broken by *hydrolysis*, which may occur in acidic solution or during enzyme-catalyzed digestion of proteins in food. The end result of hydrolysis is production of the individual amino acids from the protein. *Denaturation* is the loss of overall structure by a protein while retaining its primary structure. Among the agents that cause denaturation are heat, mechanical agitation, pH change, and exposure to a variety of chemical agents, including detergents.

▪ Understanding Key Concepts

18.25 Draw the structure of each of the following tripeptides at low pH and high pH. At each pH, assume that all functional groups that might do so are ionized.
   (a) Val-Gly-Leu   (b) Arg-Lys-His   (c) Tyr-Pro-Ser
   (d) Glu-Asp-Phe   (e) Gln-Ala-Asn   (f) Met-Trp-Cys

18.26 Interactions of amino acids on the interior of proteins are key to the shapes of proteins. In group (a) below, which pairs of amino acids would form hydrophobic interactions? In group (b), which pairs would form ionic interactions? Which pairs in group (c) would form hydrogen bonds?

(a)	(b)	(c)
1. Pro...Leu	1. Arg...Asp	1. Gly...Ala
2. Arg...Thr	2. Cys...Met	2. Ala...Phe
3. Ile...Val	3. Gly...Ala	3. Cys...Cys
4. Tyr...Phe	4. His...Glu	4. Thr...Glu

18.27 Draw the hexapeptide Asp-Gly-Phe-Leu-Glu-Ala, and show (using dotted lines) the hydrogen bonding that would stabilize this structure if it were part of an α-helix.

18.28 The primary structure or sequence of peptides can be deduced by "overlap" of the sequence of partial hydrolysis products. For an octapeptide of the following composition, deduce the sequence from the partial hydrolysis products listed below.

2 Phe, 2 Glu, 2 His, Asp, Tyr

   (1) His-Tyr   (2) Glu-Phe   (3) Asp-His-Tyr
   (4) Tyr-Glu-His   (5) Phe-Glu-Phe   (6) Glu-Phe-Asp

18.29 Compare and contrast the characteristics of fibrous and globular proteins. Consider biological function, water solubility, amino acid composition, secondary structure, and tertiary structure. Give examples of three fibrous and three globular proteins.

▪ Additional Problems

Amino Acids

18.30 The amino acids in most biological systems are said to be α-L-acids. What does the prefix "α" mean?

18.31 What does the prefix "L" in α-L-acid mean?

18.32 What amino acids do the following abbreviations stand for? Draw the structure of each amino acid.
   (a) Ser   (b) Thr   (c) Pro

18.33 What amino acids do the following abbreviations stand for? Draw the structure of each amino acid.
   (a) Phe   (b) Lys   (c) Gln

18.34 Name and draw the structures of the amino acids that fit these descriptions:
   (a) Contains an isopropyl group
   (b) Contains a secondary alcohol group

18.35 Name and draw the structures of the amino acids that fit these descriptions:
   (a) Contains a thiol group
   (b) Contains a phenol group

18.36 At neutral pH, which of the following amino acids have a net positive charge, which have a net negative charge, and which are neutral?
   (a) Lysine   (b) Phenylalanine   (c) Leucine

18.37 At neutral pH, which of the following amino acids have a net positive charge, which have a net negative charge, and which are neutral?

(a) Glutamic acid   **(b)** Proline   **(c)** Alanine

18.38 Which of the following forms of aspartic acid would you expect to predominate at low pH, neutral pH, and high pH?

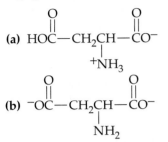

18.39 Which of the following forms of lysine would you expect to predominate at low pH, neutral pH, and high pH?

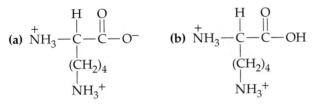

Handedness in Molecules

18.40 What does the term *chiral* mean? Give two examples.

18.41 What does the term *achiral* mean? Give two examples.

18.42 Which of the following objects are chiral?

(a) A shoe   **(b)** A bed   **(c)** A light bulb

18.43 Which of the following objects are achiral?

(a) A flowerpot   **(b)** A house key
(b) A pair of scissors

18.44 Draw the structures of the following compounds. Which of them is chiral? Mark each chiral carbon with an asterisk.

(a) 2-Bromo-2-chloropropane
(b) 2-Bromo-2-chlorobutane
(c) 2-Bromo-2-chloro-3-methylbutane

18.45 Draw the structures of the following compounds. Which of them is chiral? Mark each chiral carbon with an asterisk.

(a) 3-Chloropentane
(b) Cyclopentane
(c) 2-Methylpropanol

18.46 Which of the carbon atoms marked with arrows in the following compound are chiral?

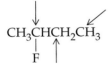

18.47 Which of the carbon atoms marked with arrows in the following compound are chiral?

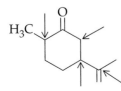

Peptides and Proteins

18.48 What is the difference between a simple protein and a conjugated protein?

18.49 What kinds of molecules are found in the following kinds of conjugated proteins in addition to the protein part?

(a) Nucleoproteins   **(b)** Lipoproteins
(b) Glycoproteins   **(d)** Hemoproteins

18.50 Name four different biological functions of proteins in the human body, and give an example of a protein with each function.

18.51 What is meant by each of the following terms as they apply to proteins, and what primary interactions stabilize the structure?

(a) Primary structure   **(b)** Secondary structure
(b) Tertiary structure   **(d)** Quaternary structure

18.52 Why is cysteine such an important amino acid for defining the tertiary structure of some proteins?

18.53 What conditions are required for disulfide bonds to form between cysteine residues in a protein?

18.54 How do the following noncovalent interactions help to stabilize the tertiary and quaternary structure of a protein? Give an example of a pair of amino acids that could give rise to each interaction.

(a) Hydrophobic interactions
(b) Salt bridges (ionic interactions)

18.55 How do the following noncovalent interactions help to stabilize the tertiary and quaternary structure of a protein? Give an example of a pair of amino acids that could give rise to each interaction.

(a) Side-chain hydrogen bonding
(b) Disulfide bonds

18.56 What kinds of changes take place in a protein when it is denatured?

18.57 Explain how a protein is denatured by the following:

(a) Heat   **(b)** Addition of a strong base
(c) Detergents

18.58 Use the three-letter abbreviations to name all tripeptides containing methionine, isoleucine, and lysine.

18.59 Write structural formulas for the two dipeptides containing phenylalanine and glutamate.

18.60 Which amino acid side chains are most likely to be found on the outside of a globular protein, and which on the inside? Explain each answer.

(a) (a) Valine   **(b)** Leucine
(b) (c) Aspartate   **(d)** Asparagine

18.61 The side chains of which of the following amino acids are most likely to be found on the outside of a globular protein? On the inside? Explain each answer.

(a) Phenylalanine   **(b)** Glutamate
(c) Lysine   **(d)** Tryptophan

18.62 Why do you suppose diabetics must receive insulin subcutaneously by injection rather than orally?

18.63 Individuals with phenylketonuria (PKU) are sensitive to phenylalanine in their diet. Why is a warning on foods containing aspartame (L-aspartyl-L-phenylalanine methyl ester) of concern for PKU individuals?

18.64 The *endorphins* are a group of naturally occurring neurotransmitters that act in a manner similar to morphine to control pain. Research has shown that the biologically active parts of the endorphin molecules are simple pentapeptides called *enkephalins*. Draw the structure of the methionine enkephalin with the sequence Tyr-Gly-Gly-Phe-Met. Identify the N-terminal and C-terminal amino acids.

18.65 Refer to Problem 18.64. Draw the structure of the leucine enkephalin with the sequence Tyr-Gly-Gly-Phe-Leu. Identify the N-terminal and C-terminal amino acids.

Properties and Reactions of Amino Acids and Proteins

18.66 Much of the chemistry of amino acids is the familiar chemistry of carboxylic acids and amine functional groups. What products would you expect to obtain from the following reactions of glycine?

(a) $\overset{+}{H_3N}-CH_2-\overset{\overset{O}{\|}}{C}O^- + HCl \longrightarrow$?

(b) $\overset{+}{H_3N}-CH_2-\overset{\overset{O}{\|}}{C}OH + CH_3OH \xrightarrow{H^+ \text{ catalyst}}$?

18.67 A scientist tried to prepare the simple dipeptide glycylglycine by the following reaction:

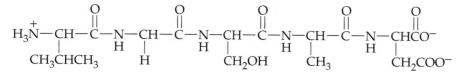

$$2\ \overset{+}{H_3N}CH_2\overset{\overset{O}{\|}}{C}O^- \longrightarrow \overset{+}{H_3N}CH_2\overset{\overset{O}{\|}}{C}NHCH_2\overset{\overset{O}{\|}}{C}O^-$$

An unexpected product formed during the reaction. This product was found to have the molecular formula $C_4H_6N_2O_2$ and to contain two peptide bonds. What happened?

18.68 (a) Identify the amino acids present in the peptide shown below.
(b) Identify the N-terminal and C-terminal amino acids of the peptide.
(c) Show the structures of the products that would be obtained on digestion of the peptide at physiological pH.

$$\overset{+}{H_3N}-CH-\overset{O}{\overset{\|}{C}}-\underset{H}{N}-CH-\overset{O}{\overset{\|}{C}}-\underset{H}{N}-CH-\overset{O}{\overset{\|}{C}}-\underset{H}{N}-CH-\overset{O}{\overset{\|}{C}}-\underset{H}{N}-CHCO^-$$

with substituents: CH_3CHCH_3, H, CH_2OH, CH_3, CH_2COO^-

18.69 (a) Identify the amino acids present in the peptide shown below.
(b) Identify the N-terminal and C-terminal amino acids of the peptide.
(c) Show the structures of the products that would be obtained on digestion of the peptide at physiological pH.

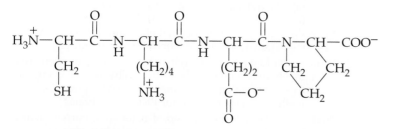

18.70 Which would you expect to be more soluble in water, a peptide rich in aspartate and lysine, or a peptide rich in valine and alanine? Explain.

18.71 Proteins are usually less soluble in water at their isoelectric points. Explain.

Applications

18.72 (a) Tyramine is said to be the "decarboxylation" product of the amino acid tyrosine. Write the reaction and explain what is meant by "decarboxylation."
(b) Phenelzine inhibits the "deamination" of tyramine. What key characteristic of phenelzine makes it similar enough to tyramine that it is able to block deamination of tyramine? [*Nutrition in Health and Disease*]

18.73 Why is it more important to have a daily source of protein than a daily source of fat or carbohydrates? [*Proteins in the Diet*]

18.74 What is an incomplete protein? [*Proteins in the Diet*]

18.75 In general, which is more likely to contain a complete (balanced) protein for human use—food from plant sources or food from animal sources? Explain. [*Proteins in the Diet*]

18.76 Two of the most complete (balanced) proteins (that is, proteins that have the best ratio of each of the amino acids for humans) are cow's milk protein (casein) and egg white protein. Explain why (not surprisingly) these are very balanced proteins for human growth and development. [*Proteins in the Diet*]

18.77 The proteins collagen, bovine insulin, and human hemoglobin have isoelectric points of 6.6, 5.4, and 7.1, respectively. If a sample containing these proteins were applied to an electrophoresis strip in a buffer at pH 6.6, describe the motion of each with respect to the positive and negative electrodes in the electrophoresis apparatus. [*Protein Analysis by Electrophoresis*]

[Problems continue on next page]

18.78 Three dipeptides were separated by electrophoresis at pH 6.4, and the results are shown below. If the three dipeptides were Lys-Trp, Glu-Tyr, and Ala-Cys, identify each spot with the appropriate dipeptide. [*Protein Analysis by Electrophoresis*]

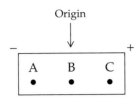

18.79 The change from a normal to a disease-causing prion results in a change from α-helices to β-sheets. How might this change alter the overall structure and intermolecular forces in the prion? [*Prions: Proteins that Cause Disease*]

18.80 List the properties of disease-causing prions that made their existence difficult to accept. [*Prions: Proteins that Cause Disease*]

General Questions and Problems

18.81 Fresh pineapple can't be used in gelatin desserts because it contains an enzyme that hydrolyzes the proteins in gelatin, destroying the gelling action. Canned pineapple can be added to gelatin with no problem. Why?

18.82 Both α-keratin and tropocollagen have helical secondary structure. How do they differ?

18.83 Bradykinin, a peptide that helps to regulate blood pressure, has the primary structure Arg-Pro-Pro-Gly-Phe-Ser-Pro-Phe-Arg.
  (a) Draw the complete structural formula of bradykinin.
  (b) Bradykinin has a very kinked secondary structure. Why?

18.84 For each amino acid listed, tell whether its influence on tertiary structure is largely through hydrophobic interactions, hydrogen bonding, formation of salt bridges, covalent bonding, or some combination of these effects.
  (a) Glutamate
  (b) Methionine
  (c) Glutamine
  (d) Threonine
  (e) Histidine
  (f) Phenylalanine
  (g) Cysteine
  (h) Valine

18.85 When subjected to oxidation, the chiral carbon in 2-pentanol becomes achiral. Why does this happen?

18.86 Why is hydrolysis of a protein not considered to be denaturation?

18.87 Oxytocin is a small peptide that can be used to induce labor by causing contractions in uterine walls. It has the primary structure Cys-Tyr-Ile-Gln-Asn-Cys-Pro-Leu-Gln. This peptide is held in a cyclic configuration by a disulfide bridge. Diagram oxytocin, showing the disulfide bridge.

18.88 Methionine has a sulfur atom in its formula. Why can't methionine be used for disulfide bridges?

18.89 List the amino acids that would be capable of hydrogen bonding if included in a peptide chain. Draw an example of two of these amino acids hydrogen-bonding to one another. For each one, draw a hydrogen bond to water in a separate sketch. Refer to Section 8.11 for help with drawing hydrogen bonds.

18.90 Four of the most abundant amino acids in proteins are leucine, alanine, glycine, and valine. What do these amino acids have in common? Would you expect these amino acids to be found on the interior or on the exterior of the protein?

18.91 Globular proteins are water-soluble, whereas fibrous proteins are insoluble in water. Indicate whether you would expect each of the following amino acids to be on the surface of a globular protein or on the surface of a fibrous protein.
  (a) Ala
  (b) Glu
  (c) Leu
  (d) Phe
  (e) Ser
  (f) Val

18.92 See Figure 18.7. The purple segments that connect successive regions of secondary structure are often referred to as "reverse turns" or "bends." The two most common amino acids in these looped regions are glycine and proline. From your knowledge of the structures of these two amino acids, can you speculate on why they might be found in these structures?

18.93 During sickle-cell anemia research to determine the modification involved in sickling, sequencing of the affected person's hemoglobin β-subunit revealed that the sixth amino acid was valine rather than glutamate; thus, the replacement of glutamate for valine was altering the structure of hemoglobin severely. Which amino acid, if it replaced the Glu, might cause the least disruption in hemoglobin structure? Why?

19

Enzymes and Vitamins

Leafy green vegetables provide several of the vitamins (riboflavin, folate, vitamins A and K) without which certain of our essential biochemical reactions could not occur.

Think of your body as a walking chemical laboratory. Although the analogy isn't perfect, there's a good deal of truth to it. In your body, as in the laboratory, chemical reactions are the major activity. In a laboratory, however, chemical reactions are carried out one at a time by mixing pure chemicals in individual containers. In your body, many thousands of reactions take place simultaneously in the same cells or body fluids.

Another difference between chemistry in a laboratory and chemistry in a living organism is control. In a laboratory, the speed of a reaction is controlled by adjusting experimental conditions such as temperature, solvent, and pH. In an organism these conditions can't be adjusted. The human body must maintain a temperature of 37°C, the solvent must be water, and the pH must be close to 7.4 in most body fluids.

How does an organism control its thousands of different reactions so that all occur to the proper extent? The answer is that all reactions in living organisms are governed by *enzymes*—powerful and highly selective biological catalysts. Enzymes are proteins, and many of them are in turn under the control of the chemical messengers classified as *hormones* and *neurotransmitters*.

In this chapter our focus will be on enzymes. We'll also look at *vitamins*, because they are essential to the function of certain enzymes. Chapter 20 is devoted to the role of hormones and neurotransmitters in keeping our biochemistry under control.

Here, we'll address the following questions:

1. What are enzymes?
The goal: Be able to describe the chemical nature of enzymes and their function in biochemical reactions.

2. How do enzymes work, and why are they so specific?
The goal: Be able to provide an overview of what happens as one or more substrates and an enzyme come together so that the catalyzed reaction can occur, and be able to list the properties of enzymes that make their specificity possible.

3. What effects do temperature, pH, enzyme concentration, and substrate concentration have on enzyme activity?
The goal: Be able to describe the changes in enzyme activity that result when temperature, pH, enzyme concentration, or substrate concentration change.

4. How is enzyme activity regulated?
The goal: Be able to define feedback, allosteric control, enzyme inhibition, zymogens, phosphorylation-dephosphorylation, and genetic control.

5. What are vitamins?
The goal: Be able to describe the two major classes of vitamins, the reasons vitamins are necessary in our diets, and the general results of excesses or deficiencies.

CONCEPTS TO REVIEW

Coordinate covalent bonds (Section 5.4)
Reaction rates (Section 7.4)
pH (Section 10.9)
Effects of conditions on reaction rates (Section 7.5)
Tertiary protein structure (Section 18.10)

19.1 Catalysis by Enzymes

Catalysts accelerate the rates of chemical reactions but at the end of the reaction have undergone no change. **Enzymes**, the catalysts for biochemical reactions, fit this definition. Like all catalysts, an enzyme doesn't affect the equilibrium point of a reaction and can't bring about a reaction that is energetically unfavorable. What an enzyme *does* do is decrease the time it takes to reach equilibrium by lowering the activation energy. (∞, p. 176)

Enzymes, with just a few exceptions, are water-soluble globular proteins. (∞, p. 526) As proteins, they are far larger and more complex molecules than simple inorganic catalysts. Because of their size and complexity, enzymes have available more ways in which to connect with reactants, speed up reactions, and be controlled by other molecules.

Enzyme A protein or other molecule that acts as a catalyst for a biological reaction.

 Functions of an Enzyme

Active site A pocket in an enzyme with the specific shape and chemical makeup necessary to bind a substrate.

Substrate A reactant in an enzyme-catalyzed reaction.

Specificity (enzyme) The limitation of the activity of an enzyme to a specific substrate, specific reaction, or specific type of reaction.

Within the folds of an enzyme's protein chain is the **active site**—the region where the reaction takes place. The active site has the specific shape and chemical reactivity needed to catalyze the reaction. One or more **substrates**, the reactants in an enzyme-catalyzed reaction, are held in place by attractions to groups that line the active site.

The extent to which an enzyme's activity is limited to a certain substrate and a certain type of reaction is referred to as the **specificity** of the enzyme. Enzymes differ greatly in their specificity. *Catalase*, for example, is almost completely specific for one reaction—the decomposition of hydrogen peroxide (Figure 19.1), a reaction needed to destroy the peroxide before it damages essential biomolecules by oxidizing them.

$$2\,H_2O_2 \underset{}{\overset{\text{Catalase}}{\rightleftharpoons}} 2\,H_2O + O_2(g)$$

FIGURE 19.1 Catalase in action. ▶ As ground beef liver is added to hydrogen peroxide solution in the beaker, catalase in the liver catalyzes rapid decomposition of the hydrogen peroxide, as shown by the formation of oxygen-filled bubbles.

Thrombin is specific for catalyzing hydrolysis of a peptide bond adjacent to arginine and does so primarily in a protein essential to blood clotting. With this bond broken, the product (fibrin) proceeds to polymerize into a blood clot (Section 29.5). *Carboxypeptidase A* is less limited—it removes many different C-terminal amino acid residues from protein chains during digestion. And the enzyme *papain* from papaya fruit catalyzes the hydrolysis of peptide bonds in many locations. (∞, p. 531) It's this ability to break down proteins that accounts for the use of papain in meat tenderizers, in contact-lens cleaners, and in cleansing dead or infected tissue from wounds (*debridement*).

Since the amino acids in enzymes are all L-amino acids, it should come as no surprise that enzymes are also specific with respect to stereochemistry. If a substrate is chiral, an enzyme usually catalyzes the reaction of only one of the pair of enantiomers because only one fits the active site in such a way that the reaction can occur. The enzyme lactate dehydrogenase, for example, catalyzes the removal of hydrogen from L-lactate but not from D-lactate:

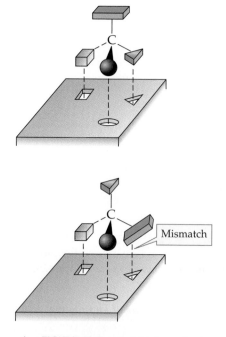

▲ **FIGURE 19.2 A chiral reactant and a chiral reaction site.** The enantiomer at the top fits the reaction site like a hand in a glove, but the enantiomer at the bottom doesn't fit and therefore can't be a substrate for this enzyme.

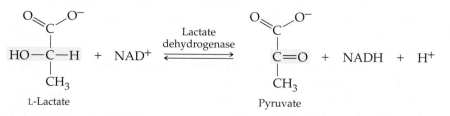

This is another example of the importance of molecular shape in biochemistry. The specificity of an enzyme for one of two enantiomers is a matter of fit. A left-handed enzyme can't fit with a right-handed substrate any more than a left-handed glove can fit on a right hand (Figure 19.2).

The catalytic activity of an enzyme is measured by its **turnover number**, the maximum number of substrate molecules acted upon by one molecule of enzyme per unit time (Table 19.1). Most enzymes turn over 10-1000 molecules per second, but some are much faster. Catalase, with its essential role in protecting against molecular damage, is one of the fastest—it can turn over 10 million molecules per second. This is the fastest reaction rate attainable in the body because it is the rate at which molecules collide.

Turnover number The maximum number of substrate molecules acted upon by one molecule of enzyme per unit time.

TABLE 19.1 Turnover Numbers for Some Enzymes

Enzyme	Reaction Catalyzed	Turnover Number (Maximum Number of Catalytic Events per Second)
Papain	Hydrolysis of peptide bonds	10
Ribonuclease	Hydrolysis of phosphate ester link in RNA	10^2
Kinases	Transfer of phosphoryl group between substrates	10^3
Acetylcholinesterase	Deactivation of the neurotransmitter acetylcholine	10^4
Carbonic anhydrase	Converts CO_2 to HCO_3^-	10^6
Catalase	Decomposition of H_2O_2 to $H_2O + O_2$	10^7

■ **PROBLEM 19.1**

Which of the enzymes listed in Table 19.1 catalyzes a maximum of 1000 reactions per second?

■ **PROBLEM 19.2**

Why is it essential to rinse contact lenses in saline solution after they have been cleansed by an enzyme solution and before they are placed in the eyes?

19.2 Enzyme Cofactors

Many enzymes are conjugated proteins that require nonprotein portions known as **cofactors**. Some cofactors are metal ions, others are nonprotein organic molecules called **coenzymes**. To be active, an enzyme may require a metal ion, a coenzyme, or both. Some enzyme cofactors are tightly held by noncovalent attractions or are covalently bound to their enzymes, while others are more loosely bound so that they can enter and leave the active site as needed. For example, NAD^+, written as a reactant in the equation for the dehydrogenation of L-lactate in the preceding section, is a coenzyme.

Why are cofactors necessary? The functional groups in proteins are limited to those of the amino acid side chains. By joining up with cofactors, enzymes acquire chemically reactive groups not available as side chains. For example, the NAD^+ reactant in the equation for the dehydrogenation of L-lactate shown in the preceding section is a coenzyme, and it is the oxidizing agent that makes the reaction possible. (Vitamins that function as cofactors are discussed in Section 19.10.)

The requirement that many enzymes have for metal ion cofactors explains our dietary need for trace minerals. The ions of iron, zinc, copper, manganese, molybdenum, cobalt, nickel, vanadium, and selenium all function as enzyme cofactors. These ions are able to form coordinate covalent bonds by accepting

Cofactor A nonprotein part of an enzyme that is essential to the enzyme's catalytic activity; a metal ion or a coenzyme.

Coenzyme An organic molecule that acts as an enzyme cofactor.

lone-pair electrons present on nitrogen or oxygen atoms in enzymes or substrates. (∞ Section 5.4) This bonding may anchor a substrate in the active site and may also allow the metal ion to participate in the catalyzed reaction. For example, every molecule of the digestive enzyme carboxypeptidase A contains one Zn^{2+} ion that is essential for its catalytic action. We say that the zinc ion is "coordinated" to two nitrogens in histidine side chains and one oxygen in a glutamate side chain. In this way the ion is held in place in the active site of the enzyme.

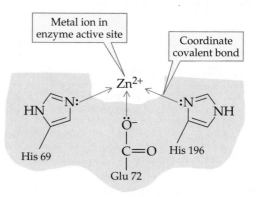

Like the trace minerals, certain vitamins (Section 19.10) are a dietary necessity for humans because they are function as building blocks for coenzymes and we cannot synthesize them.

■ **PROBLEM 19.3**

Check the label on a bottle of multivitamin/multimineral tablets and identify the metal ion cofactors listed above that are included in the supplement.

■ **KEY CONCEPT PROBLEM 19.4**

Coenzymes are often divided into two categories: vitamin-derived coenzymes and coenzymes not derived from vitamins. How might the sources of these kinds of enzymes differ?

19.3 Enzyme Classification

Enzymes are divided into six main classes according to the general kind of reaction they catalyze, and each main class is further subdivided (Table 19.2). Most of the names of the main classes are self-explanatory; they're listed below with examples:

- *Oxidoreductases* catalyze oxidation-reduction reactions of substrate molecules, most commonly addition or removal of oxygen or hydrogen. Because oxidation and reduction must occur together, these enzymes require coenzymes that are reduced or oxidized as the substrate is oxidized or reduced.

A(Reduced) + B(Oxidized) $\longrightarrow$ A'(Oxidized) + B'(Reduced)

$$\underset{\substack{\text{Reduced}\\\text{substrate}}}{CH_3-CH_2-OH} + \underset{\substack{\text{Oxidized}\\\text{coenzyme}}}{NAD^+} \underset{\text{dehydrogenase}}{\overset{\text{Alcohol}}{\rightleftarrows}} \underset{\substack{\text{Oxidized}\\\text{product}}}{CH_3-\overset{\displaystyle O}{\overset{\displaystyle \|}{C}}-H} + \underset{\substack{\text{Reduced}\\\text{coenzyme}}}{NADH} + H^+$$

Application Lead Poisoning and an Antidote

Lead is not known to be an essential metal and it can be toxic. It exerts its poisonous effect in two different ways. One is by bonding to the sulfur-containing side chains on cysteine residues in enzymes, thereby denaturing the enzymes. Targets for this effect include enzymes that function in the nervous system. At low levels, lead can thus cause decreased attention span and mental difficulties. These symptoms are noticed in children who develop the habit of ingesting flakes of lead-containing paint, which have a sweet taste. Primarily for this reason, lead-containing paint has not been used since the 1950s. It is, however, still present in older homes.

Another toxic effect of lead is displacement of an essential metal from the active site of an enzyme. Lead can, for example, displace zinc in an enzyme essential in the synthesis of heme, which is the oxygen-carrying part of hemoglobin. Moderate exposure to lead in this manner can cause anemia that requires medical attention.

Coordinate covalent bond formation is the key to the action of ethylenediamminetetraacetic acid (EDTA)

in treating lead poisoning. EDTA can form six coordinate covalent bonds to a lead ion. The resulting compound is water-soluble and carries lead away in urine. The effectiveness of the treatment can be followed by monitoring the lead concentration in urine.

See Additional Problem 19.62 at the end of the chapter.

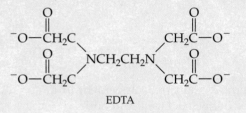

EDTA

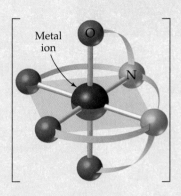

▲ **Coordinate covalent bond region at the center of EDTA.** EDTA is one of the few molecules able to form six coordinate covalent bonds. The bonds connect the metal ion to EDTA's two N atoms (blue) and the four negatively charged O atoms in the carboxylate groups (red). Because the coordination compound is water-soluble, it is used to clear lead ions and other toxic metal ions from the blood.

TABLE 19.2 Classification of Enzymes	
Main Class and Subclass Examples	**Examples of Types of Reactions Catalyzed**
Oxidoreductases	**Oxidation-reduction reactions**
Oxidases	Addition of O_2 to a substrate
Reductases	Reduction of a substrate
Dehydrogenases	Removal of 2 H's to form a double bond
Transferases	**Transfer of functional groups**
Transaminases	Transfer of amino group between substrates
Kinases	Transfer of a phosphoryl group between substrates
Hydrolases	**Hydrolysis reactions**
Lipases	Hydrolysis of ester groups in lipids
Proteases	Hydrolysis of peptide bonds in proteins
Nucleases	Hydrolysis of phosphate ester bonds in nucleic acids
Isomerases	**Isomerization of a substrate**
Lyases	**Group elimination to form double bond or addition to a double bond**
Dehydrases	Removal of H_2O from substrate to give double bond
Decarboxylases	Replacement of a carboxyl group by a hydrogen
Synthases	Addition of small molecule to a double bond
Ligases	**Bond formation coupled with ATP hydrolysis to provide energy**
Synthetases	Formation of bond between two substrates
Carboxylases	Formation of bond between substrate and CO_2 to add a carboxyl group ($-COO^-$)

▲ Adehylate kinase, a small transferase enzyme (194 amino acid residues), that adds a phosphate group to adenosine monophosphate.

- *Transferases* catalyze transfer of a group from one molecule to another. Kinases, for example, transfer a phosphate group from adenosine triphosphate (ATP) to give adenosine diphosphate (ADP) and a phosphorylated product

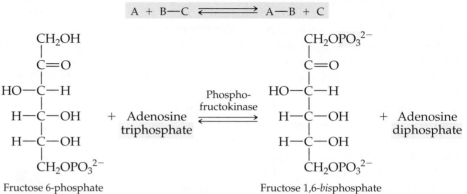

$$A + B{-}C \rightleftharpoons A{-}B + C$$

Fructose 6-phosphate   Fructose 1,6-*bis*phosphate

- *Hydrolases* catalyze the hydrolysis of substrates—the breaking of bonds with addition of water. The digestion of carbohydrates and proteins by hydrolysis requires these enzymes.

$$A{-}B + H_2O \longrightarrow A{-}OH + B{-}H$$

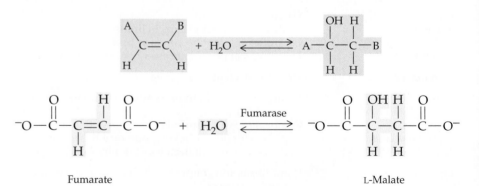

Polypeptide   Shortened polypeptide   Amino acid

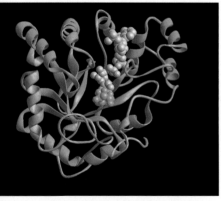

▲ Aldose reductase, an oxidoreductase enzyme that reduces a $C{=}O$ group in a sugar molecule to a $-C{-}OH$ group with the aid of the coenzyme NADH. The sugar glucose (orange) and the NADH (gray) are shown in the active site of the enzyme.

- *Isomerases* catalyze the isomerization (rearrangement of atoms) of a substrate in reactions that have but one substrate and one product.

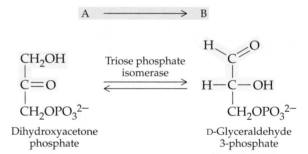

$$A \longrightarrow B$$

Dihydroxyacetone phosphate   D-Glyceraldehyde 3-phosphate

- *Lyases* (from the Greek *lein*, meaning "to break") catalyze the addition of a molecule such as H_2O, CO_2, or NH_3 to a double bond or the reverse reaction in which a molecule is eliminated to leave a double bond.

Fumarate   L-Malate

- *Ligases* (from the Latin *ligare*, meaning "to tie together") catalyze the bonding together of two substrate molecules. Because such reactions are generally not favorable, they require the simultaneous release of energy by a hydrolysis reaction, usually by the conversion of ATP to ADP (such energy release is discussed in Section 21.5).

$$A \ + \ B \ + \ \text{Adenosine triphosphate (ATP)} \ \longrightarrow \ A\text{–}B \ + \ \text{Adenosine diphosphate (ADP)} \ + \ HOPO_3^{2-} \ + \ H^+$$

$$CH_3-\overset{\overset{\displaystyle O}{\|}}{C}-CO^- \ + \ CO_2 \ + \ ATP \ \underset{\text{carboxylase}}{\overset{\text{Pyruvate}}{\rightleftharpoons}} \ {}^-O\overset{\overset{\displaystyle O}{\|}}{C}-CH_2-\overset{\overset{\displaystyle O}{\|}}{C}-CO^- \ + \ ADP \ + \ HOPO_3^{2-} \ + \ H^+$$

Pyruvate   Oxaloacetate

Note in the preceding examples that the enzymes have the family-name ending *-ase*. Exceptions to this rule occur for enzymes such as papain and trypsin, which are still referred to by older common names. The more informative modern systematic names always have two parts: The first identifies the substrate on which the enzyme operates, and the second part is an enzyme subclass name like those shown in Table 19.2. For example, *pyruvate carboxylase* is a ligase that acts on the substrate *pyruvate* to add a *carboxyl group*. Note also that enzymes are capable of catalyzing both forward and reverse reactions, and where both directions are of significance, the equations are often written with both arrows.

■ WORKED EXAMPLE 19.1

To what class does the enzyme that catalyzes the following reaction belong?

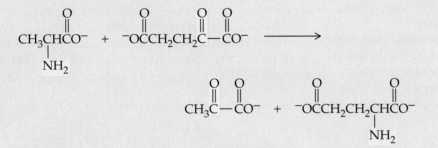

ANALYSIS   First, identify the type of reaction that has occurred. An amino group and a carbonyl group have changed places. Then, determine what class of enzyme catalyzes this type of reaction.

SOLUTION   The reaction is a transfer of an amino functional group, meaning that the enzyme is a transferase.

■ PROBLEM 19.5

Describe the reactions that you would expect these enzymes to catalyze.
(a) Retinal isomerase   (b) Squalene oxidase
(c) Glucose kinase   (d) Cellulose hydrolase

■ PROBLEM 19.6

To what class of enzymes does pyruvate decarboxylase belong? Describe in general the reaction it catalyzes.

■ PROBLEM 19.7

Identify and describe the chemical change in the isomerase-catalyzed reaction on p. 546.

Application Biocatalysis: Food and Chemicals

We've known for a very long time that enzymes can create desirable changes in food products. One step in the preparation of black tea, for example, relies on an enzyme present in the tea leaves. During a short period in which the dried, crushed tea leaves are exposed to the oxygen in air, a polyphenoloxidase breaks up polyphenols into tannins, smaller segments that also contain many phenolic hydroxyl groups. The tannins impart the darker color and characteristic flavors to black tea.

And then there's corn syrup, which is created by the action of enzymes on starch granules from corn. Starch (as you will see in Section 22.9) is a polymer of sugars. Digestion of the starch by yeast enzymes yields a mixture of individual sugar molecules that make corn syrup sweet and larger carbohydrate molecules that have no sweetness but thicken the syrup and make it sticky.

Creative application of an enzyme is also responsible for chocolate candies with liquid centers. When these candies are produced, the center is a moist fudge-like mixture of flavoring, sucrose (which is table sugar), and a yeast enzyme that breaks sucrose into the two smaller sugars it is made of. Because these sugars are more water-soluble than sucrose, the center gradually liquefies.

Enzymes have now moved far beyond food processing and into big-time chemistry. They are the focus of a rapidly expanding area christened *biocatalysis*. Industrial chemists are turning to enzymes as catalysts when the enzymes make otherwise difficult reactions possible, when they allow reactions that produce dangerous or wasteful by-products to be avoided, and especially in the production of chiral compounds desirable in the pharmaceutical industry for their application in drug manufacture.

A simple example of biocatalysis is the hydration of acrylonitrile to produce acrylamide, an important industrial chemical used in permanent press fabrics, dyes, adhesives, paper, and many other products:

$$CH_2{=}CHC{\equiv}N \xrightarrow{\text{Nitrile hydratase}} CH_2{=}CH{-}\overset{\displaystyle O}{\overset{\|}{C}}{-}NH_2$$

Acrylonitrile   Acrylamide

Biocatalysis has often been exploited in the synthesis of intermediates—chemicals that then become reactants in the synthesis of other substances. It is used, for example, in the production of *p*-hydroxyphenylglycine, which is needed in the synthesis of the antibiotic amoxicillin.

The production of agricultural chemicals and synthetic vitamins are other major areas of chemistry in which biocatalysis is an active area of study. Already on the horizon is biocatalysis by microorganisms that have been genetically altered to produce a desired chemical (Section 27.5).

See Additional Problems 19.63-64 at the end of the chapter.

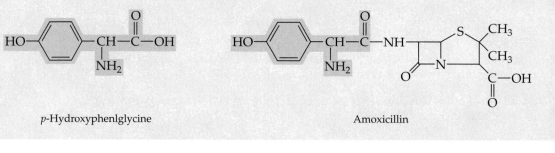

p-Hydroxyphenlglycine   Amoxicillin

■ PROBLEM 19.8

Which of the following reactions can be catalyzed by a decarboxylase and which cannot?

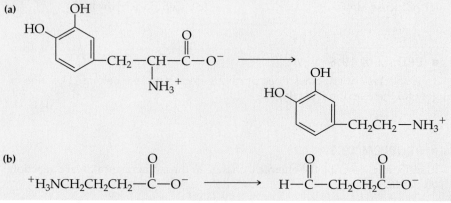

19.4 How Enzymes Work

Any theory of how enzymes work must explain why they are so specific and how they lower activation energies. The explanation for enzyme *specificity* is found in the active site. Exactly the right environment for the reaction is provided within the active site. It has side-chain groups that attract and hold the substrate or substrates in position by noncovalent interactions and sometimes by temporary covalent bonding. The active site also has the groups needed for catalysis of the reaction.

Two models are invoked to represent the interaction between substrates and enzymes. Historically, the *lock-and-key model* came first; it was proposed when the need for a spatial fit between substrates and enzymes was first recognized. The substrate is described as fitting into the active site as a key fits into a lock.

When it became possible to study enzyme-substrate interaction more closely, a new model was needed. Our modern understanding of molecular structure makes it clear that enzyme molecules are not totally rigid, like locks. The **induced-fit model** accounts for changes in the shape of the enzyme active site that accommodate the substrate and facilitate the reaction. As an enzyme and substrate come together, their interaction *induces* exactly the right fit for catalysis of the reaction.

Induced-fit model A model of enzyme action in which the enzyme has a flexible active site that changes shape to best fit the substrate and catalyze the reaction.

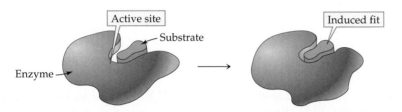

A well-studied example of induced fit, the interaction between glucose and hexokinase, is illustrated in Figure 19.3. The reaction, a common one, is a phosphorylation—the addition of a phosphoryl group to an —OH group, catalyzed by a kinase. The reaction is the first step in glucose (a hexose) metabolism (Section 23.2). Notice in Figure 19.3 how the enzyme closes in once the glucose molecule has entered the active site—this is the induced fit.

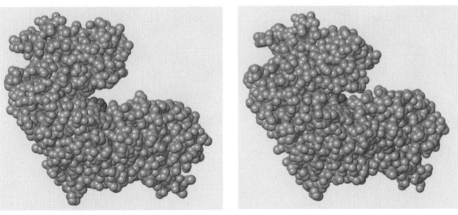

(a)   (b)

◀ **FIGURE 19.3   The induced fit of hexokinase (blue) and its substrate, glucose (red).** (a) The active site is a groove in the hexokinase molecule. (b) When glucose enters the active site, the enzyme changes shape, wrapping itself more snugly around the substrate.

Enzyme-catalyzed reactions begin with migration of the substrate or substrates into the active site to form an *enzyme-substrate complex*. The substrate is first drawn into position by the same kinds of noncovalent forces that govern the shapes of protein molecules. (∞ Figure 18.4, p. 521)

Before complex formation, the substrate molecule is in its most stable, lowest-energy shape. Within the complex, the molecule is forced into a less stable shape, and bonding electrons may be drawn away from some bonds in preparation for breaking them and forming new bonds. The result is to *lower the activation energy barrier* between substrate and product.

Within the enzyme-substrate complex, atoms that will form new bonds must connect with each other. The new bonds might be with a second substrate or temporary bonds with atoms in the enzyme. Also, groups needed for catalysis must be close to the necessary locations in the substrate. Many organic reactions, for example, require acidic, basic, or metal ion catalysts. An enzyme's active site can provide acidic and basic groups without disrupting the constant-pH environment in body fluids, while the necessary metal ions are present as cofactors. Once the chemical reaction is completed, enzyme and product molecules separate from each other and the enzyme, restored to its original condition, becomes available for another substrate.

The hydrolysis of a peptide bond by chymotrypsin, shown in Figure 19.4, illustrates how an enzyme functions. Chymotrypsin is one of the enzymes that participates in the digestion of proteins by breaking them down to smaller molecules. It cleaves polypeptide chains by breaking the peptide bond on the carbonyl side of amino acid residues that include an aromatic ring:

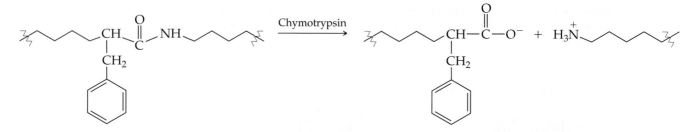

The enzyme-substrate complex forms (Figure 19.4a and b) by attraction of a hydrophobic side chain into a hydrophobic pocket in the active site and

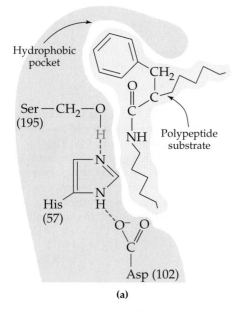

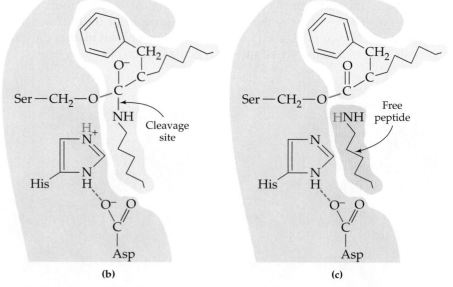

▲ **FIGURE 19.4** **Hydrolysis of a peptide bond by chymotrypsin.** (a) The polypeptide enters the active site with its hydrophobic side chain in the hydrophobic pocket and the peptide bond to be broken (red) opposite serine and histidine residues. (b) Hydrogen transfer from serine to histidine allows formation of a strained intermediate in which the serine side chain bonds to the peptide bond carbon (green). (c) The peptide bond is broken and the segment with the new terminal —NH₂ group leaves the active site. In subsequent steps (not shown) a water molecule enters the active site; its H atom restores the serine side chain and its —OH bonds to the other piece of the substrate protein to give a new terminal —COOH group so that this piece can leave the active site.

formation of a covalent bond (green) to the substrate. The result is to position the substrate with the peptide bond to be broken (red) next to the amino acid side chains that function as catalysts. The enzyme has not only joined up with the substrate (the **proximity effect**), but has done so in such a way as to bring the groups that must connect close to each other (the **orientation effect**). Aspartate, histidine, and serine provide side chains needed for catalysis within the active site (the **catalytic effect**). As an illustration of the critical nature of protein folding, note that in the 241-amino-acid primary structure of chymotrypsin, aspartate is number 102, histidine is number 57, and serine is number 195. These amino acids are distant from each other along the backbone, but are brought close together at exactly the positions needed in the active site.

With the peptide bond carbon temporarily bonded to serine in the active site, it is easier for the peptide bond to break because the activation energy barrier has been lowered (the **energy effect**). As the bond breaks, the nitrogen picks up a hydrogen (blue) from histidine to form the new terminal amino group and this portion of the substrate is set free (Figure 19.4c). Reaction with a water molecule then restores the hydrogen to serine and supplies an OH group to form the new terminal carboxyl group. This part of the substrate is set free and the enzyme is restored to its original state.

In summary, enzymes act as catalysts because of their ability to

- Bring substrate(s) and catalytic sites together (*proximity effect*)
- Hold substrate(s) at the exact distance and in the exact orientation necessary for reaction (*orientation effect*)
- Provide acidic, basic, or other types of groups required for catalysis (*catalytic effect*).
- Lower the energy barrier by inducing strain in bonds in the substrate molecule (*energy effect*)

■ **KEY CONCEPT PROBLEM 19.9**

The active sites of enzymes usually contain amino acids with acidic, basic, and polar side chains. Some enzymes also have amino acids with nonpolar side chains in their active sites. Which types of side chains would you expect to participate in holding the substrate in the active site and which types would you expect to be involved in the catalytic activity of the enzyme?

19.5 Effect of Concentration on Enzyme Activity

For a reaction to occur, the enzyme and substrate molecules must come together and form the enzyme-substrate complex. Therefore, variation in the reaction rate can be expected if the enzyme or substrate concentration changes.

Substrate Concentration

Consider the common situation in which the substrate concentration varies while the enzyme concentration remains unchanged. If the substrate concentration is low relative to that of the enzyme, not all the enzyme molecules are in use. The rate therefore increases with the concentration of substrate because more of the enzyme molecules are put to work. In this situation, shown at the far left of the curve in Figure 19.5, the rate increases as the available substrate increases. If the substrate concentration doubles, the rate doubles (a directly proportional relationship).

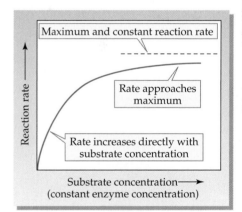

▲ **FIGURE 19.5 Change of reaction rate with substrate concentration when enzyme concentration is constant.** At low substrate concentration (at left), the reaction rate is directly proportional to the substrate concentration (at constant pH and temperature). With increasing substrate concentration, the rate drops off as more of the active sites are occupied. Eventually, with all active sites occupied, the rate reaches a maximum and constant rate.

As the substrate concentration continues to increase, however, the increase in the rate begins to level off as more and more of the active sites are occupied. (Think of people waiting on line to take their seats in a theater. The line moves more slowly as more seats fill up and it becomes more difficult to find an empty one.) Eventually, the substrate concentration reaches a point at which none of the available active sites are free. Since the reaction rate is now determined by how fast the enzyme-substrate complex is converted to product, the reaction rate becomes constant—the enzyme is saturated.

Once the enzyme is saturated, increasing substrate concentration has no effect on the rate. In the absence of a change in the concentration of the enzyme, the rate when the enzyme is saturated is determined by the efficiency of the enzyme, the pH, and the temperature.

Under most conditions, an enzyme is not likely to be saturated. Therefore, at a given pH and temperature, the reaction rate is controlled by the amount of substrate and the overall efficiency of the enzyme. If the enzyme-substrate complex is rapidly converted to product, the rate at which enzyme and substrate combine to form the complex becomes the limiting factor. Calculations show an upper limit to this rate: Enzyme and substrate molecules moving at random in solution can collide with each other no more often than about 10^8 collisions per mole per liter per second. Remarkably, a few enzymes actually operate with close to this efficiency—every one of the collisions results in the formation of product! One example of such an efficient enzyme is catalase, the enzyme that breaks down hydrogen peroxide at the rate of 10^7 catalytic events per second (∞ Table 19.1, p. 543).

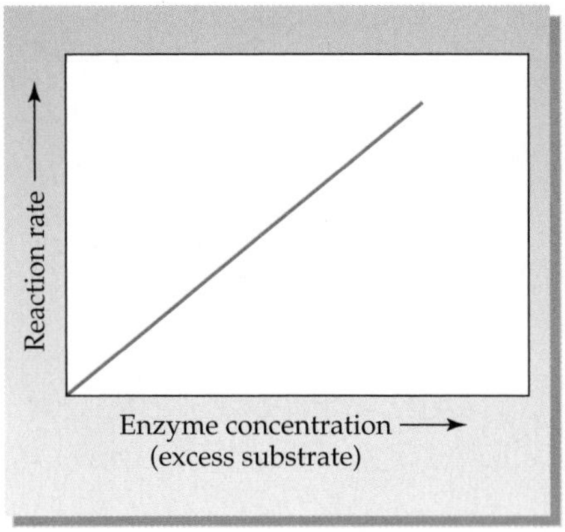

Enzyme Concentration

It is possible for the concentration of an active enzyme to vary according to our metabolic needs. So long as the concentration of substrate does not become a limitation, the reaction rate varies directly with the enzyme concentration (Figure 19.6). If the enzyme concentration doubles, the rate doubles; if the enzyme concentration triples, the rate triples; and so on.

▲ **FIGURE 19.6 Change of reaction rate with enzyme concentration in the presence of excess substrate.**

■ **KEY CONCEPT PROBLEM 19.10**

What do we mean when we say an enzyme is saturated with substrate? When an enzyme is saturated with substrate, how does adding more (a) substrate and (b) enzyme affect the rate of the reaction?

19.6 Effect of Temperature and pH on Enzyme Activity

Enzymes have been finely tuned through evolution so that their maximum catalytic activity is highly dependent on pH and temperature. As you might expect, optimum conditions vary slightly for each enzyme but are generally near normal body temperature and the pH of the body fluid in which the enzyme functions.

Effect of Temperature on Enzyme Activity

An increase in temperature increases the rate of most chemical reactions, and enzyme-catalyzed reactions are no exception. Unlike many simple reactions, however, the rates of enzyme-catalyzed reactions do not increase continuously with rising temperature. Instead, the rates reach a maximum and then begin to decrease, as shown in Figure 19.7a. This falloff in rate occurs because enzymes begin to denature when heated too strongly. (∞, p. 532) The noncovalent

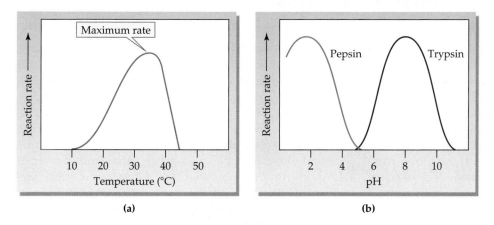

FIGURE 19.7   Effect of temperature (a) and pH (b) on reaction rate.   (a) The rate increases with increasing temperature until a temperature is reached at which the protein begins to denature; then the rate decreases rapidly. (b) The optimum activity for an enzyme occurs at the pH where it acts, as illustrated for two protein hydrolysis enzymes—pepsin, which acts in the highly acidic environment of the stomach, and trypsin, which acts in the small intestine.

attractions between protein side chains are disrupted, the delicately maintained three-dimensional shape of the enzyme begins to come apart, and as a result the active site needed for catalytic activity is destroyed.

Most enzymes denature and lose their catalytic activity above 50–60°C, a fact that explains why medical instruments and laboratory glassware can be sterilized by heating with steam in an autoclave. The high temperature of the steam denatures the enzymes of any bacteria present, thereby killing them.

A severe drop in body temperature creates the potentially fatal condition of hypothermia, which is accompanied by a slowdown in metabolic reactions. This effect is used to advantage by cooling the body during cardiac surgery.

Effect of pH on Enzyme Activity

The catalytic activity of many enzymes depends on pH and usually has a well-defined optimum point at the normal, buffered pH of the enzyme's environment. For example, pepsin, which initiates protein digestion in the highly acidic environment in the stomach, has its optimum activity at pH 2 (Figure 19.7b). By contrast, trypsin—like chymotrypsin, an enzyme that aids digestion of proteins in the small intestine—has optimum activity at pH 8. Most enzymes have their maximum activity at pH 5-9. Eventually, extremes of pH will denature a protein.

■ **PROBLEM 19.11**

Will the reaction catalyzed by the enzyme represented in Figure 19.7a have a higher rate of reaction at 20°C or at 30°C? Will it have a higher rate of reaction at 25°C or at 40°C?

■ **PROBLEM 19.12**

How will the rates of the reaction catalyzed by trypsin (Figure 19.7b) compare at pH 8 and pH 10?

19.7 Enzyme Regulation: Feedback and Allosteric Control

The control of biochemical reactions by enzymes is only part of the enzyme story. In the body, the concentrations of thousands of different compounds must vary continuously to meet changing conditions as we eat, sleep, exercise, or fall ill. Thus, enzymes must do more than merely speed up reactions. At a moment's notice, they must be able to turn some reactions off, slow some down a bit, or quickly accelerate others to their maximum possible rate. Clearly, then, the enzymes themselves must be regulated. How is this regulation achieved?

Activation (of an enzyme)   Any process that initiates or increases the action of an enzyme.

Inhibition (of an enzyme)   Any process that slows or stops the action of an enzyme.

Application Enzymes in Medical Diagnosis

In a healthy person, certain enzymes, such as those responsible for forming and dissolving blood clots, are normally present in high concentration in blood serum. Other enzymes function mainly within cells and are normally in low concentration in blood serum, which they enter only during normal degeneration of healthy cells. When tissue is injured, however, large quantities of enzymes are released into the blood from dying cells, with the distribution of enzymes dependent on the identity of the injured cells. Measurement of blood levels of enzymes is therefore a valuable diagnostic tool. For example, higher-than-normal levels of the enzymes included in a routine blood analysis indicate the following conditions:

Aspartate transaminase (AST)	Damage to heart or liver
Alanine transaminase (ALT)	Damage to heart or liver
Lactate dehydrogenase (LH)	Damage to heart, liver, or red blood cells
Alkaline phosphatase (ALP)	Damage to bone and liver cells
γ-Glutamyl transferase (GGT)	Damage to liver cells; alcoholism

Enzyme analysis relies on measuring the activity of an enzyme rather than its concentration. Because activity is influenced by pH, temperature, and substrate concentration, it is measured in international units at standard conditions. One international unit (U) is defined as the amount of an enzyme that converts 1 μmol of its substrate to product per minute under defined standard conditions of pH, temperature, and substrate concentration. The analytical results are thus reported in units per liter (U/L).

Among the most useful enzyme assays are those done to diagnose heart attacks (*myocardial infarctions, MI*). The levels of three enzymes are measured: creatine phosphokinase (CPK), aspartate transaminase (AST, formerly referred to as GOT), and lactate dehydrogenase (LDH). The CPK level rises almost immediately following an MI, reaching a sixfold increase over normal values after about 30 hours; the AST level triples after about 40 hours; and the LDH level doubles after about 4 days.

More specific verification of a heart attack is gained by analysis of the *isoenzymes* of LDH. Isoenzymes are structural variations of the same enzyme that catalyze the same reaction but in different tissues. Lactate dehydrogenase is a mixture of five isoenzymes, denoted LDH_1, LDH_2, LDH_3, LDH_4, and LDH_5. Of the five, heart tissue contains primarily LDH_1. A characteristic change following a heart attack is a flip in the levels of LDH_1 and LDH_2 together with a slight elevation of LDH_2, as illustrated in the accompanying bar graph.

Other changes in the LDH isoenzyme levels are indicative of other conditions. For example, elevation of LDH_4 and LDH_5 indicates acute hepatitis.

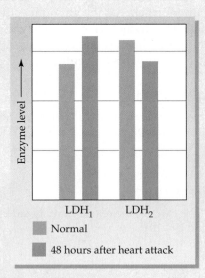

▲ Blood levels of lactate dehydrogenase (LDH) isoenzymes in a normal person (blue) and in a heart attack victim after 48 hours (red). Note the flip of LDH_1 and LDH_2 levels following a heart attack.

See Additional Problems 19.65–66 at the end of the chapter.

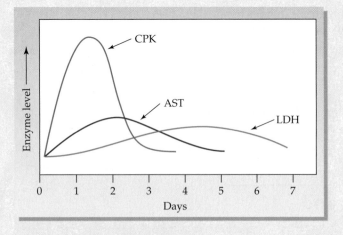

▲ Blood levels of creatine phosphokinase (CPK), aspartate transaminase (AST), and lactate dehydrogenase (LDH) in the days following a heart attack.

A variety of strategies are utilized to adjust the rates of enzyme-catalyzed reactions. Any process that starts or increases the action of an enzyme is **activation**. Any process that slows or stops the action of an enzyme is an **inhibition**. Although we'll describe the strategies of enzyme control one by one, keep in mind that several strategies are usually operating together. Considering that a cell contains thousands of proteins and hundreds of other kinds of biomolecules, all in the concentrations required to maintain constant conditions, the achievement of enzyme control is awe-inspiring.

Feedback

As you'll see in subsequent chapters, biochemical reaction pathways are dependent on series of consecutive reactions in which the product of one reaction is the reactant for the next. Such pathways are subject to feedback control. *Feedback* is a general term applied when the result of a process feeds information back to affect the beginning of the process. Any device that maintains a constant temperature, such as an oven, is regulated by feedback. An oven has a sensor that detects temperature and feeds back that information to turn the heating elements on or off.

Consider a biochemical pathway in which A is converted to B, then B is converted to C, and so on, with each reaction catalyzed by its own enzyme:

$$A \xrightarrow{\text{Enzyme 1}} B \xrightarrow{\text{Enzyme 2}} C \xrightarrow{\text{Enzyme 3}} D$$

What will happen if product D inhibits enzyme 1? This inhibition will cause the amount of A converted to B to decrease, and the synthesis of B and C will decrease in turn. The effect of this **feedback control** mechanism is to control the concentration of D. When more D than needed for other biochemical pathways is present, it inhibits enzyme 1 and its synthesis is slowed or stopped. No energy is then wasted making the unneeded intermediates B and C. When all of the available D is used up in other reactions, none is available for feedback. As a result, enzyme 1 is no longer inhibited and the production of D accelerates.

Suppose that the conversion of A to B, the reaction catalyzed by enzyme 1 in the pathway above, is at a point where control is critical. Perhaps the amount of product D must vary over a wide range, or perhaps it is needed only in an emergency. Feedback control is a common strategy at such key points in biochemical pathways.

Allosteric Control

Feedback is commonly exercised by what is known as *allosteric* control (from the Greek *allos*, meaning "other" and *steros*, meaning "space"). In **allosteric control** the binding of a molecule (an *allosteric regulator* or *effector*) at one site on a protein affects the binding of another molecule at a different site. Most **allosteric enzymes** have more than one protein chain and two kinds of binding sites—those for substrate and those for regulators (Figure 19.8). Binding of a regulator, usually by noncovalent interactions, changes the shape of the enzyme. This change alters the shape of the active site, thereby affecting the ability of the enzyme to bind its substrate and catalyze its reaction. An advantage of allosteric enzyme control is that the regulators need not be structurally similar to the substrate because they don't bind to the active site.

Allosteric control can be either positive or negative. Binding a positive regulator changes the active sites so that the enzyme becomes a better catalyst and the rate accelerates. Binding a negative regulator changes the active sites so that the enzyme is a less effective catalyst and the rate slows down. Because allosteric enzymes can have several substrate binding sites and several

Feedback control Regulation of an enzyme's activity by the product of a reaction later in a pathway.

Allosteric control An interaction in which the binding of a regulator at one site on a protein affects the protein's ability to bind another molecule at a different site.

Allosteric enzyme An enzyme whose activity is controlled by the binding of an activator or inhibitor at a location other than the active site.

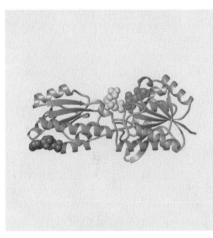

(a)

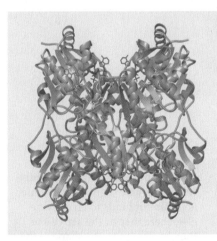

(b)

▲ **FIGURE 19.8 An allosteric enzyme.** (a) One of the four identical subunits in a phosphofructokinase, an enzyme that catalyzes transfer of a phosphoryl group from ATP to fructose 6-phosphate (see transferase reaction on p. 546). The subunit is shown with the diphosphate (yellow) and ADP (green) in the active site and the allosteric activator (red, also ADP) in the regulatory site. (b) The four subunits of the enzyme, two blue and two in purple.

regulator binding sites and because there may be interaction among them, very fine control is achieved.

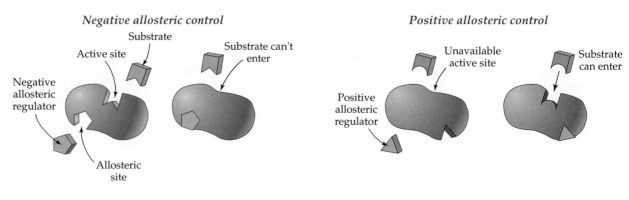

Negative allosteric control   *Positive allosteric control*

19.8 Enzyme Regulation: Inhibition

The inhibition of an enzyme can be *reversible* or *irreversible*. In reversible inhibition, the inhibitor can leave, restoring the enzyme to its uninhibited level of activity. In irreversible inhibition, the inhibitor remains permanently bound and the enzyme is permanently inhibited. The inhibition can also be *competitive* or *noncompetitive*, depending on whether the inhibitor binds to the active site or elsewhere.

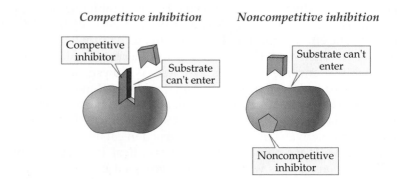

Competitive inhibition   *Noncompetitive inhibition*

Reversible Noncompetitive Inhibition

Noncompetitive (enzyme) inhibition Enzyme regulation in which an inhibitor binds to an enzyme elsewhere than at the active site, thereby changing the shape of the enzyme's active site and reducing its efficiency.

In **noncompetitive inhibition**, the inhibitor doesn't compete with the substrate for the active site. It binds to the enzyme in a different place. A noncompetitive inhibitor exerts allosteric control by changing the enzyme's shape so that the active site is less accessible or the reaction occurs less efficiently.

The rates without and with a noncompetitive inhibitor are compared in the top and bottom curves in Figure 19.9. With the inhibitor, the reaction rate increases with increasing substrate concentration more gradually than when no

FIGURE 19.9 Enzyme ▶
inhibition. The top curve and dashed line show the reaction rate and maximum rate with no inhibitor. With a competitive inhibitor (middle curve), the maximum rate is unchanged, but a higher substrate concentration is required to reach it. With a noncompetitive inhibitor (bottom curve) the maximum rate (bottom dashed line) is lowered.

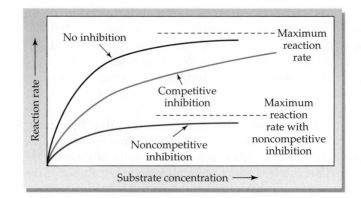

inhibitor is present. The maximum rate is lowered and once that rate is reached, no amount of substrate can increase it further. So long as the inhibitor is connected to the enzyme, this upper limit doesn't change.

Reversible Competitive Inhibition

What would happen if an enzyme were to encounter a molecule very much like its normal substrate in shape, size, and functional groups? The impostor molecule could enter the enzyme's active site, bind to it, and thereby prevent the usual substrate molecule from binding to the same site. As a result, the enzyme would be tied up and unavailable as a catalyst. This is **competitive inhibition**—the inhibitor *competes* with substrate for binding to the active site. A competitive inhibitor binds reversibly to an active site through noncovalent interactions, but undergoes no reaction. While it's there, it prevents the substrate from getting into the active site. Whether the substrate or the inhibitor occupies the active site depends on their relative concentrations.

Competitive (enzyme) inhibition Enzyme regulation in which an inhibitor competes with a substrate for binding to the enzyme active site.

Substrate + Enzyme $\rightleftharpoons$ Substrate–enzyme complex

Inhibitor + Enzyme $\rightleftharpoons$ Inhibitor–enzyme complex

 Energy and Enzymes

A substrate in relatively high concentration will occupy more of the active sites, so the reaction will be less inhibited. An inhibitor in relatively high concentration will occupy more of the active sites, so the reaction will be more inhibited.

The middle curve in Figure 19.9 shows that in the presence of a competitive inhibitor, the reaction rate increases more gradually with increasing substrate concentration than in its absence. Unlike noncompetitive inhibition, however, the maximum reaction rate is unchanged. Eventually, all of an enzyme's active sites can be occupied by substrate, but a higher substrate concentration is required to reach that condition.

The product of a reaction may be a competitive inhibitor for the enzyme that catalyzes that reaction. For example, glucose 6-phosphate is an allosteric competitive inhibitor for hexokinase, which catalyzes formation of this phosphate from glucose. Thus, when supplies of glucose 6-phosphate are ample, the glucose is available for other reactions.

A competitive inhibitor can sometimes be put to use in treating an unhealthy condition. For example, competitive inhibition is used to good advantage in the treatment of methanol poisoning. Although not harmful itself, methanol (wood alcohol) is oxidized in the body to formaldehyde, which is highly toxic ($CH_3OH \longrightarrow H_2C{=}O$). Because of its molecular similarity to methanol, ethanol acts as a competitive inhibitor of the methanol dehydrogenase enzyme. With the oxidation of methanol blocked by ethanol, the methanol is excreted without causing harm. Thus, the medical treatment of methanol poisoning includes administering ethanol.

Irreversible Inhibition

If an inhibitor forms a bond that is not easily broken with a group in an active site, the result is **irreversible inhibition**. The enzyme's reaction cannot occur because the substrate cannot appropriately connect with the active site. Many irreversible inhibitors are poisons as a result of their ability to completely shut down the active site. Heavy metal ions such as mercury (Hg^{2+}) or lead (Pb^{2+}) are irreversible inhibitors that form covalent bonds to the sulfur atoms in the —SH groups of cysteine residues. (∞, p. 521)

Organophosphorus insecticides such as parathion and malathion, and nerve gases like Sarin are irreversible inhibitors of the enzyme acetylcholinesterase, which breaks down a chemical messenger (*acetylcholine*) that

Irreversible (enzyme) inhibition Enzyme deactivation in which an inhibitor forms covalent bonds to the active site, permanently blocking it.

transmits nerve impulses (Section 20.7) The acetylcholinesterase inhibitors bond covalently to a serine residue in the enzyme's active site.

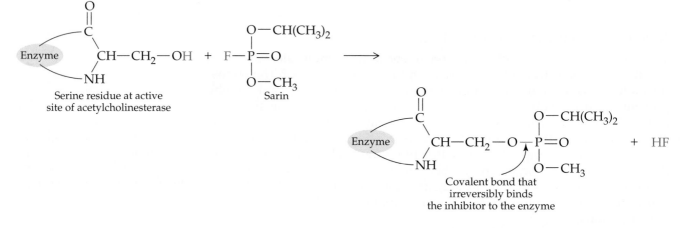

Normally, acetylcholinesterase breaks down acetylcholine immediately after that molecule has transmitted a nerve impulse. Without this enzyme activity, acetylcholine blocks transmission of further nerve impulses, resulting in paralysis of muscle fibers and death from respiratory failure. Sarin was the poison released by terrorists in the Tokyo subway system in 1995. The attack resulted in 12 deaths and varying degrees of injury to more than 5000 people.

■ **PROBLEM 19.13**

Could either of the molecules shown below be a competitive inhibitor for the enzyme that has *p*-aminobenzoate as its substrate? If so, why?

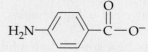

p-Aminobenzoate, the substrate

(a) $H_2NCH_2CH_3$ **(b)** H_2N—⟨ ⟩—$\overset{\overset{O}{\|}}{\underset{\underset{O}{\|}}{S}}$—$NH_2$

■ **PROBLEM 19.14**

What kind of reaction product might be a competitive inhibitor for the enzyme that catalyzes its formation?

19.9 Enzyme Regulation: Covalent Modification and Genetic Control

Covalent Modification

There are two general modes of enzyme regulation by covalent modification—removal of a covalently bonded portion of an enzyme, or addition of a group. Some enzymes are synthesized in inactive forms that differ from the active forms in composition. Activation of such enzymes, known as **zymogens** or

Zymogen A compound that becomes an active enzyme after undergoing a chemical change.

proenzymes, requires a chemical reaction that splits off part of the molecule. Blood clotting, for example, is initiated by activation of zymogens.

Three of the enzymes that digest proteins in the small intestine, for example, are produced in the pancreas as the zymogens *trypsinogen*, *chymotrypsinogen*, and *proelastase*. These enzymes must be inactive when they're synthesized so that they do not immediately digest the pancreas. Each zymogen has a polypeptide segment at one end that is not present in the active enzymes. The extra segments are snipped off to produce trypsin, chymotrypsin, and elastase when the zymogens reach the small intestine, where protein digestion occurs.

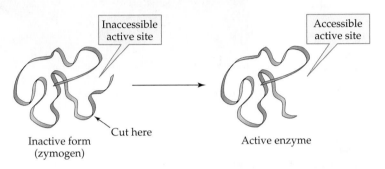

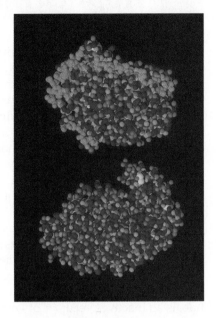

▲ Pepsinogen (a zymogen) at top and the active enzyme pepsin at bottom.

One danger of traumatic injury to the pancreas or the duct that leads to the small intestine is premature activation of these zymogens, resulting in acute pancreatitis, a painful and potentially fatal condition in which the activated enzymes attack the pancreas.

The reversible addition of phosphoryl groups ($-PO_3^{2-}$) to a serine or threonine residue is another mode of covalent modification. *Kinase* enzymes catalyze the addition of a phosphoryl group supplied by ATP (*phosphorylation*). *Phosphatase enzymes* catalyze the removal of the phosphoryl group (*dephosphorylation*). This control strategy swings into action, for example, when glycogen stored in muscles must be hydrolyzed to glucose that is needed for quick energy. Two serine residues in glycogen phosphorylase, the enzyme that initiates glycogen breakdown, are phosphorylated. Only with these phosphoryl groups in place is glycogen phosphorylase active. The groups are removed once the need to break down glycogen for quick energy has passed.

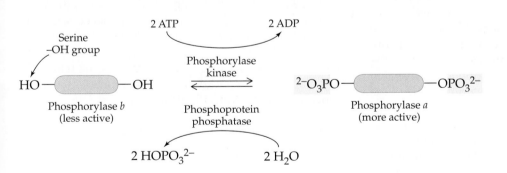

The curved arrows shown above are used frequently in biochemical equations in later chapters. Their focus is on changes in the major biomolecule reactant. The participation of other reactants needed to accomplish the chemical change is shown by the curved arrows adjacent to the main reaction arrow. Coenzymes and energy-providing molecules like ATP are often included in this manner. Here, the top curved arrow shows that the reaction in the forward direction requires ATP to supply the phosphoryl groups and produces ADP. The bottom curved arrow shows that water is needed for the reverse reaction, the hydrolysis that removes the phosphoryl groups as hydrogen phosphate anions.

Application Enzyme Inhibitors as Drugs

Consider a situation in which the chemical structures of a substrate and the active site to which it binds are known. A drug designer can create a molecule that is sufficiently similar to the substrate that it will bind to the active site and act as an inhibitor. Inhibiting a particular enzyme can be desirable for a number of reasons. We might want to increase the concentration of its substrate by using it up more slowly, or lower the concentration of its product by making less of it, or decrease the production of some other substance further along the same biochemical pathway.

The family of drugs known as ACE inhibitors are a good example of enzyme inhibitors as drugs. Angiotensin II, the octapeptide illustrated earlier (∞, p. 516), is a potent *pressor*—it elevates blood pressure, in part by causing contraction of blood vessels. Angiotensin I, a decapeptide, is an inactive precursor of angiotensin II. To become active, two amino acid residues—His and Leu—must be cut off the end of angiotensin I, a reaction catalyzed by angiotensin-converting enzyme (ACE).

Asp-Arg-Val-Tyr-Ile-His-Pro-Phe-His-Leu $\xrightarrow{\substack{\text{Angiotensin-}\\ \text{converting}\\ \text{enzyme}}}$

Angiotensin I

Asp-Arg-Val-Tyr-Ile-His-Pro-Phe + His-Leu

Angiotensin II

This reaction is part of a normal pathway for blood pressure control and is accelerated when blood pressure drops because of bleeding or dehydration.

The development of inhibitors for ACE was aided by knowing that a zinc(II) ion is present in the ACE active site. Another contributing bit of information was that the extract of venom from a South American pit viper is a mild ACE inhibitor and that this extract contains a pentapeptide with a proline residue at the carboxyl-terminal end. This information led to a search for proline-containing molecule that would bind to the zinc(II) ion.

The first ACE inhibitor on the market, *captopril*, was developed by experimenting with modifications of the proline structure. Success was achieved by introducing an —SH group that binds to the zinc ion in the active site:

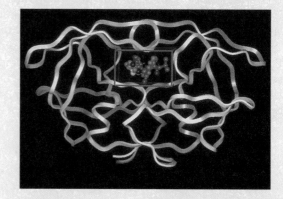

▲ Ritonovir, an enzyme inhibitor, in the active site of HIV protease.

Several other ACE inhibitors have subsequently been developed, and they are common medications for patients with high blood pressure.

The development of enzyme inhibitors is also playing a major role in the battle against *acquired immunodeficiency disease*, AIDS. The battle is far from won, but two important AIDS-fighting drugs are enzyme inhibitors. The first, known as AZT (*azidothymidine*, also called *zidovudine*), resembles in structure a molecule essential to reproduction of the AIDS-causing *human immunodeficiency virus (HIV)*. Because AZT is accepted by an HIV enzyme, it prevents the virus from producing duplicate copies of itself.

The most successful AIDS drug thus far inhibits a *protease*, an enzyme that cuts a long protein chain into smaller pieces needed by the HIV. *Protease inhibitors*, such as Ritonovir, cause dramatic decreases in the virus population and AIDS symptoms. The success is only achieved, however, by taking a "cocktail" of several drugs including AZT. The cocktail is expensive and requires precise adherence to a schedule of taking 20 pills a day. These conditions make it unavailable or too difficult for many individuals to use.

See Additional Problem 19.67 at the end of the chapter.

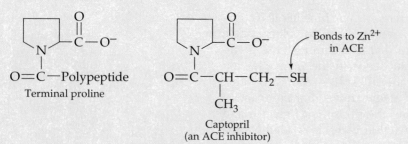

Captopril
(an ACE inhibitor)

Genetic Control

Yet another enzyme control strategy affects the supply of the enzyme itself. The synthesis of enzymes, like that of all proteins, is regulated by genes (Chapter 26). The **genetic control** strategy is especially useful for enzymes needed only at certain stages of development. Mechanisms controlled by hormones (Section 20.2) can accelerate or decelerate enzyme synthesis.

In summary, we have described the most important strategies that control the activity of enzymes. In healthy individuals several of these strategies are likely to be in use at once in any biochemical pathway.

Genetic (enzyme) control Regulation of enzyme activity by control of the synthesis of enzymes.

Mechanisms of Enzyme Control

- *Feedback*, which is exerted on an earlier reactant by a later product in a reaction pathway and is made possible by *allosteric control*, which is bonding with the feedback molecule in a way that alters the shape and therefore the efficiency of the enzyme.

- *Inhibition*, which may be *reversible* and occur away from the active site (*noncompetitive inhibition*) or at the active site (*competitive inhibition*), or may be *irreversible* because of covalent bonding of the inhibitor to the enzyme. Competitive inhibition is a strategy often utilized in medications, and irreversible inhibition is a mode of action of poisons.

- *Production of inactive enzymes (zymogens)*, which must be activated by cleaving a portion of the molecule.

- *Covalent modification of an enzyme by addition and removal of a phosphoryl group*, with the phosphoryl group supplied by ATP.

- *Genetic control*, whereby the quantity of enzyme is controlled by controlling the amount that is synthesized.

■ **PROBLEM 19.15**

Which type of enzyme regulation would probably be best for each of the following situations?

(a) An enzyme that must spring into action when a traumatic injury occurs
(b) An enzyme that becomes overactive during a disease
(c) An enzyme needed only during adolescence
(d) An enzyme needed only when there is low blood glucose

19.10 Vitamins

Vitamin An organic molecule, essential in trace amounts that must be obtained in the diet because it is not synthesized in the body.

Long before the reasons were understood, it was known that lime and other citrus juices cure scurvy, meat and milk cure pellagra, and cod-liver oil prevents rickets. Eventually, it was discovered that these diseases are caused by deficiencies of **vitamins**—organic molecules required in only trace amounts that must be obtained through the diet. Vitamins are a dietary necessity for humans because we do not have the ability to synthesize them.

The symptoms of scurvy—muscle weakness, swollen gums, and the easy bruising that result from defective collagen synthesis—are cured by consuming foods containing vitamin C, the coenzyme necessary for collagen synthesis. (∞, p. 529) Pellagra, with such varied symptoms as weight loss, dermatitis, depression, and dementia, results from a deficiency of niacin. And rickets, the occurrence of soft bones in children because of inadequate availability of calcium and phosphate, is due to a deficiency of vitamin D, which is essential to the incorporation of calcium in bones.

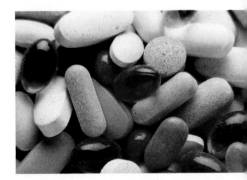

▲ **A source of vitamins.**

Water-Soluble Vitamins

Vitamins are grouped by solubility into two classes: water-soluble and fat-soluble. The water-soluble vitamins, listed in Table 19.3, are found in the aqueous environment inside cells, where most of them are needed as components of coenzymes. Over time, an assortment of names, letters, and numbers for designating the vitamins have accumulated. (One reason is that what was originally known as vitamin B turned out to be several different vitamins.) Among the water-soluble vitamins, three remain best known by letters rather than names—vitamins C, B_6, and B_{12}. Structurally, the water-soluble vitamins have in common the presence of —OH, —COOH, or other polar groups that impart their water solubility, but otherwise they range from simple molecules like vitamin C to quite large and complex structures like vitamin B_{12}.

TABLE 19.3   The Water-Soluble Vitamins*

Vitamin	Significance	Sources	Reference Daily Intake**	Effects of Deficiency	Effects of Excess
Thiamine (B_1)	In coenzyme for decarboxylation reactions	Milk, meat, bread, legumes	1.5 mg	Muscle weakness, and cardiovascular problems including heart disease, causing beriberi	Low blood pressure
Riboflavin (B_2)	In coenzymes FMN and FAD	Milk, meat	1.7 mg	Skin and mucous membrane deterioration	Itching, tingling sensations
Niacin (nicotinic acid, nicotinamide, B_3)	In coenzyme NAD^+	Meat, bread, potatoes	2.0 mg	Nervous system, gastrointestinal, skin, and mucous membrane deterioration, causing pellagra	Itching, burning sensations, blood vessel dilation, death after large dose
B_6 (pyridoxine)	In coenzyme for amino acid and lipid metabolism	Meat, legumes	2.0 mg	Retarded growth, anemia, convulsions, epithelial changes	Central nervous system alterations, perhaps fatal
Folic acid	In coenzyme for amino acid and nucleic acid metabolism	Vegetables, cereal, bread	0.4 mg	Retarded growth, anemia, gastrointestinal disorders; neural tube defects	Few noted except at massive doses
B_{12} (cobalamin)	In coenzyme for nucleic acid metabolism	Milk, meat	6 μg	Pernicious anemia	Excess red blood cells
Biotin	In coenzyme for carboxylation reactions	Eggs, meat, vegetables	0.3 mg	Fatigue, muscular pain, nausea, dermatitis	None reported
Pantothenic acid (B_5)	In coenzyme A	Milk, meat	10 mg	Retarded growth, central nervous system disturbances	None reported
C (ascorbic acid)	Coenzyme; delivers hydride ions; antioxidant	Citrus fruits, broccoli; greens	60 mg	Epithelial and mucosal deterioration, causing scurvy	Kidney stones

* Adapted in part from Frederic H. Martini, *Fundamentals of Anatomy and Physiology*, 4th edition (Prentice Hall, 1998).

** RDI values are the basis for information on the Nutrition Facts Label included on most packaged foods. The values are based on the Recommended Dietary Allowances of 1968.

Vitamin C is biologically active without any change in structure from the molecules present in foods. It is increasingly shown to be a valuable *antioxidant*, as described later in this section. *Biotin* is connected to enzymes by an amide bond at its carboxyl group but otherwise undergoes no structural change from dietary biotin.

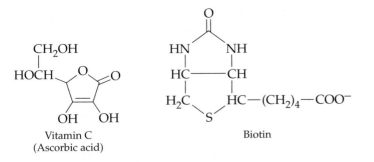

Vitamin C
(Ascorbic acid)

Biotin

The other water-soluble vitamins are incorporated into coenzymes. The vitamin-derived portions of two of the most important coenzymes, NAD^+ and coenzyme A, are illustrated in Figure 19.10. The functions, deficiency symptoms, and major dietary sources of the water-soluble vitamins are included in Table 19.3.

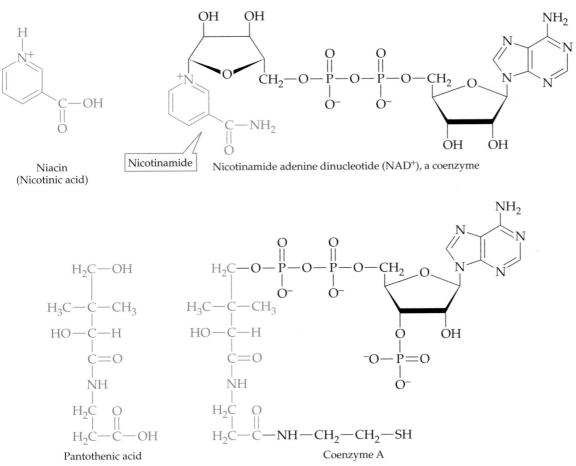

Niacin
(Nicotinic acid)

Nicotinamide

Nicotinamide adenine dinucleotide (NAD^+), a coenzyme

Pantothenic acid

Coenzyme A

▲ FIGURE 19.10 The enzyme-derived portions of NAD^+ and coenzyme A.

Fat-Soluble Vitamins

The fat-soluble vitamins A, D, E, and K can be stored in the body's fat deposits. Although the clinical effects of deficiencies of these vitamins are well documented, the molecular mechanisms by which they act are not nearly as well understood as those of the water-soluble vitamins. None has been identified as a coenzyme. Their functions, sources, and deficiency symptoms are summarized in Table 19.4. The hazards of overdosing on fat-soluble vitamins are greater than those of the water-soluble vitamins because of their ability to accumulate in body fats. Excesses of the water-soluble vitamins are more likely to be excreted in the urine.

TABLE 19.4   The Fat-Soluble Vitamins*

Vitamin	Significance	Sources	Reference Daily Intake**	Effects of Deficiency	Effects of Excess
A	Maintains epithelia; required for synthesis of visual pigments; antioxidant	Leafy green and yellow vegetables	1000 μg	Retarded growth, night blindness, deterioration of epithelial membranes	Liver damage, skin peeling, central nervous system effects (nausea, anorexia)
D	Required for normal bone growth, calcium and phosphorus absorption at gut, and retention at kidneys	Synthesized in skin exposed to sunlight	10 μg	Rickets, skeletal deterioration	Calcium deposits in many tissues, disrupting functions
E	Prevents breakdown of vitamin A and fatty acids; antioxidant	Meat, milk, vegetables	10 mg	Anemia; other problems suspected	None reported
K	Essential for liver synthesis of prothrombin and other clotting factors	Vegetables; production by intestinal bacteria	80 μg	Bleeding disorders	Liver dysfunction, jaundice

*Adapted in part from Frederic H. Martini, *Fundamentals of Anatomy and Physiology*, 4th edition (Prentice Hall, 1998).

**RDI values are the basis for information on the Nutrition Facts Label included on most packaged foods. The values are based on the Recommended Dietary Allowances of 1968. RDIs for fat-soluble vitamins are often reported in International Units (IU), which are defined differently for each vitamin. The values given here are approximate equivalents in mass units.

Vitamin A, which is essential for night vision, healthy eyes, and normal development of epithelial tissue, has three active forms: retinol, retinal, and retinoic acid. It is also produced in the body by cleavage of β-carotene, the molecule that gives an orange color to carrots and other vegetables.

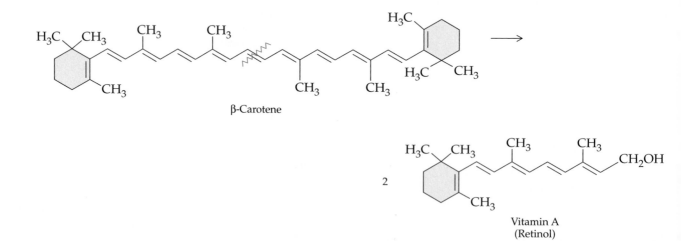

β-Carotene

Vitamin A
(Retinol)

Vitamin D, which is related in structure to cholesterol, is synthesized when ultraviolet light from the sun strikes a cholesterol derivative in the skin. In the kidney, vitamin D is converted to a hormone that regulates calcium absorption and bone formation. Vitamin D deficiencies are most likely to occur in malnourished individuals living where there is little sunlight. (It is interesting to note that a sunscreen of SPF factor 6-8 will completely block vitamin D synthesis.)

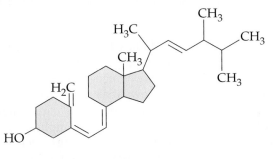

Vitamin D

Vitamin E comprises a group of structurally similar compounds called tocopherols, the most active of which is α-tocopherol. Like vitamin C, it is an antioxidant: It prevents the breakdown by oxidation of vitamin A and polyunsaturated fats. Vitamin E apparently is not toxic in overdosage as are the other fat-soluble vitamins. (Nevertheless, it is best to avoid excessively large doses of vitamin E.)

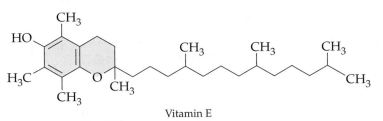

Vitamin E

Vitamin K also includes a number of structurally related compounds, in this case with hydrocarbon side chains of varying length. This vitamin is essential to the synthesis of several blood-clotting factors. It is produced by intestinal bacteria, so deficiencies are rare.

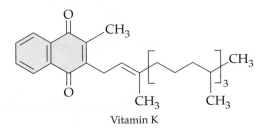

Vitamin K

■ **PROBLEM 19.16**

Compare the structures of vitamin A and vitamin C. What structural features does each have that make one water-soluble and the other fat-soluble?

■ **PROBLEM 19.17**

Based on the structure shown above for retinol (vitamin A) and the names of the two related forms of vitamin A, retinal and retinoic acid, what do you expect to be the structural differences among these three compounds?

Antioxidants

An **antioxidant** is a substance that prevents oxidation. Earlier we noted that antioxidants are used to combat air oxidation of unsaturated fats that cause deterioration of baked goods. (ooo, p. 404) In the body, we need similar protection against active oxidizing agents that are by-products of normal metabolism.

Antioxidant A substance that prevents oxidation by reacting with an oxidizing agent.

Application Vitamins, Minerals, and Food Labels

It's not uncommon to encounter incomplete or incorrect information about vitamins and minerals. We've been frightened by the possibility that aluminum causes Alzheimer's disease and tantalized by the possibility that vitamin C defeats the common cold. Sorting out fact from fiction or distinguishing preliminary research results from scientifically proven relationships is especially difficult in this area of nutrition. Much is yet to be learned about the functions of vitamins and minerals in the body, and new information is continuously being reported. It's tempting for health-conscious individuals to look for guaranteed routes to better health by taking vitamins or minerals, and it's tempting to profit-making organizations to take advantage of this motivation.

One consistent source of information on nutrition is the Food and Nutrition Board of the National Academy of Sciences-National Research Council. They periodically survey the latest nutritional information and publish Recommended Dietary Allowances (RDAs) that are "designed for the maintenance of good nutrition of the majority of healthy persons in the United States." Another source is the U.S. Food and Drug Administration (FDA), which has among its many responsibilities setting the rules for food labeling.

Since 1994, as mandated by the FDA, most packaged food products carry standardized *Nutrition Facts* labels. The nutritional value of a food serving of a specified size is reported as *% Daily Value*. For vitamins and minerals, these percentages are calculated from *Reference Daily Intake* values (RDIs). The RDIs are mostly derived from the 1968 RDAs and were designed to avoid deficiencies. The RDIs are averages for adults and children over 4 years of age. The values for vitamins are

included in Tables 19.3 and 19.4. For minerals, they're listed in the accompanying table.

In choosing which vitamins and minerals *must* be listed on the new labels, the government has focused on those of greatest importance in maintaining good health. The choices reflect a new emphasis on preventing disease rather than preventing deficiencies. The *mandatory* listings are for vitamin A, vitamin C, calcium, and iron. These recommendations are based on evidence for the benefits of high dietary levels of the antiooxidants vitamin A (or the related compound, β-carotene) and vitamin C. Calcium deficiencies are related to osteoporosis, and iron deficiencies are a special concern for women because of their menstrual blood loss. Thiamin, riboflavin, and niacin listings are no longer mandatory because deficiencies of these vitamins are no longer a public health problem in the United States.

Nutrition Facts

Serving Size 2 oz.
(56g/About 1/3 package)
(About 1 cup cooked)
Servings Per Container about 3

Amount Per Serving

Calories 190   Calories from Fat 15

	% Daily Value*
Total Fat 1.5g	2%
Saturated Fat 0g	0%
Cholesterol 0mg	0%
Sodium 730mg	30%
Total Carbohydrate 40g	13%
Dietary Fiber 1g	4%
Sugars 2g	
Protein 5g	

Vitamin A 0% • Vitamin C 0%

Calcuim 0% • Iron 4%

*Percent Daily Values are based on a 2,000 calorie diet. Your daily values may be higher or lower depending on your caloric needs:

	Calories:	2,000	2,500
Total Fat	Less than	65g	80g
Sat Fat	Less than	20g	25g
Cholesterol	Less than	300mg	300mg
Sodium	Less than	2,400mg	2,400mg
Total Carbohydrate		300g	375g
Dietary Fiber		25g	30g

Calories per gram:
Fat 9 • Carbohydrate 4 • Protein 4

Reference Daily Intake Values* for Minerals

Mineral	RDI	Mineral	RDI
Calcium	1.0 g	Selenium	70 μg
Iron	18 mg	Manganese	2 mg
Phosphorus	1.0 g	Fluoride	2.5 mg
Iodine	150 μg	Chromium	120 μg
Magnesium	400 mg	Molybdenum	75 μg
Zinc	15 mg	Chloride	3.4 g
Copper	2 mg		

*On Nutrition Facts Labels, calcium and iron must be listed; phosphorus, iodine, magnesium, zinc, and copper listings are optional; the others cannot be listed.

See Additional Problems 19.68-70 at the end of the chapter.

Our principal dietary antioxidants are vitamin C, vitamin E, β-carotene, and the mineral selenium. They work together to defuse the potentially harmful action of **free radicals**, highly reactive molecular fragments with unpaired electrons (for example, superoxide ion, $\cdot O_2^-$). Free radicals quickly gain stability by picking up electrons from nearby molecules, which are thereby damaged. (We'll have more to say about this in Section 21.10.)

Vitamin E is unique in having antioxidant activity as its principal biochemical role. It acts by giving up the hydrogen from its $-OH$ group to oxygen-containing free radicals. The hydrogen can then be restored by reaction with vitamin C. Selenium joins the list of important antioxidants because it is a cofactor in an enzyme that converts hydrogen peroxide (H_2O_2) to water before the peroxide can go on to produce free radicals.

Evidence for the benefits of antioxidants in disease prevention, especially in the prevention of cancer and heart disease, is accumulating. Laboratory experiments have demonstrated anticancer activity for vitamin C, vitamin E, β-carotene, and selenium, and the results are supported by a variety of studies in defined human populations. For example, vitamin E appears to reduce the risk of cancer among smokers, though not among nonsmokers. Low levels of serum selenium and vitamin E have been associated with a greater risk of breast cancer in a group of Finnish women.

In a study of over 100,000 people, those who took vitamin E supplements had fewer heart attacks. Vitamin C may slow the development of blocked arteries and may also prevent cancer by inhibiting the formation of carcinogens in the gastrointestinal tract.

Free radical   An atom or molecule with an unpaired electron.

■ **KEY CONCEPT PROBLEM 19.18**
Vitamins are a diverse group of compounds that must be present in the diet. List four functions of vitamins in the body.

Summary: Revisiting the Chapter Goals

1. **What are enzymes?** *Enzymes* are the catalysts for biochemical reactions. They are mostly water-soluble, globular proteins, and many incorporate *cofactors*, which are either metal ions or the nonprotein organic molecules known as *coenzymes*. One or more *substrate* molecules (the reactants) enter an *active site* lined by those protein side chains and cofactors necessary for catalyzing the reaction. There are six major classes of reactions catalyzed by enzymes and many subclasses (Table 19.2).

2. **How do enzymes work, and why are they so specific?** A substrate is drawn into the active site by noncovalent interactions. As the substrate enters the active site, the enzyme shape adjusts to best accommodate the substrate and catalyze the reaction (the *induced fit*). Within the *enzyme-substrate complex*, the substrate is held in the best orientation for reaction and in a strained condition that allows the activation energy to be lowered. When the reaction is complete, the product is released and the enzyme returns to its original condition. The *specificity* of each enzyme is determined by the presence within the active site of catalytically active groups, hydrophobic pockets, and ionic or polar groups that exactly fit the chemical makeup of the substrate.

3. **What effects do temperature, pH, enzyme concentration, and substrate concentration have on enzyme activity?** With increasing temperature, reaction rate increases to a maximum and then decreases as the enzyme protein denatures. Reaction rate is maximal at a pH that reflects the pH of the enzyme's site of action in the body. In the presence of excess substrate, reaction rate is directly proportional to enzyme concentration. With fixed

Key Words

Activation (of an enzyme), p. 553
Active site, p. 542
Allosteric control, p. 555
Allosteric enzyme, p. 555
Antioxidant, p. 565
Coenzyme, p. 543
Cofactor, p. 543
Competitive (enzyme) inhibition, p. 557
Enzyme, p. 541
Feedback control, p. 555
Free radical, p. 567
Genetic (enzyme) control, p. 561
Induced-fit model, p. 549
Inhibition (of an enzyme), p. 553
Irreversible (enzyme) inhibition, p. 557
Noncompetitive (enzyme) inhibition, p. 556
Specificity (enzyme), p. 542
Substrate, p. 542
Turnover number, p. 543
Vitamin, p. 561
Zymogen, p. 558

enzyme concentration, reaction rate first increases with increasing substrate concentration and then approaches a fixed maximum at which all active sites are occupied. (See Figures 19.5, 19.6, and 19.7.)

4. **How is enzyme activity regulated?** The effectiveness of enzymes is controlled by a variety of *activation* and *inhibition* strategies. A product of a later reaction can exercise *feedback control* over an enzyme for an earlier reaction in a pathway. Feedback acts through *allosteric control* of enzymes that have regulatory sites separate from their active sites. Binding a regulator induces a change of shape in the active site, increasing or decreasing the efficiency of the enzyme. *Allosteric inhibitors* are *noncompetitive* because they act away from the active site; they lower the maximum reaction rate. *Competitive inhibitors* typically resemble the substrate and reversibly block the active site; they slow the reaction rate but do not change the maximum rate. *Irreversible inhibitors* form covalent bonds to an enzyme that permanently inactivate it; most are poisons. Enzyme activity is also regulated by *reversible* phosphorylation and dephosphorylation, and by synthesis of inactive *zymogens* that are later activated by removal of part of the molecule. *Genetic control* is exercised by regulation of the synthesis of enzymes.

5. **What are vitamins?** *Vitamins* are organic molecules required in small amounts in the diet because our bodies cannot synthesize them. The water-soluble vitamins (Table 19.3) are coenzymes or parts of coenzymes. The fat-soluble vitamins (Table 19.4) have diverse and less well understood functions. In general, excesses of water-soluble vitamins are excreted and excesses of fat-soluble vitamins are stored in body fat, making excesses of the fat-soluble vitamins potentially more harmful. Vitamin C, β-carotene (a precursor of vitamin A), vitamin E, and selenium work together as *antioxidants* to protect biomolecules from damage by free radicals.

▪ Understanding Key Concepts

19.19 On the diagram shown below, indicate with dotted lines the bonding between the enzyme (a dipeptidase; select amino acid residues in black) and the substrate (in blue) that might occur to form the enzyme-substrate complex. What are the two types of bonding likely to occur?

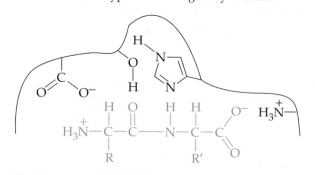

19.20 Answer questions (a)–(d) concerning the following reaction:

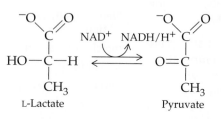

(a) The enzyme involved in this reaction belongs to what class of enzymes?

(b) Since hydrogens are removed, the enzyme belongs to what subclass of the enzyme class from part (a)?

(c) What is the substrate for the reaction as written?

(d) The enzyme name is derived from the substrate name and the subclass of the enzyme, and ends in the family-name ending for an enzyme. Name the enzyme.

19.21 In the reaction shown in Problem 19.20, will the enzyme likely use D-lactate as a substrate? Explain your answer. If D-lactate binds to the enzyme, how is it likely to affect the enzyme?

19.22 In the reaction shown in Problem 19.20, identify the coenzyme required for catalytic activity. Is the coenzyme an oxidizing agent or a reducing agent? Coenzymes that are modified during the reaction are called *stoichiometric coenzymes*, since they are needed in stoichiometric proportions to the substrate. What vitamin is a part of the coenzyme for this reaction?

19.23 Explain how each of the following changes will affect the rate of an enzyme-catalyzed reaction in the presence of a noncompetitive inhibitor: (a) increasing the substrate concentration at a constant inhibitor concentration, (b) decreasing the inhibitor concentration at a constant substrate concentration.

19.24 Explain how each of the following mechanisms regulate enzyme activity.

(a) Allosteric regulation
(b) Covalent modification
(c) Inhibition
(d) Genetic control

19.25 Acidic and basic groups are often found in the active sites of enzymes. Identify the acidic and basic amino acids in the active site in the diagram at the right.

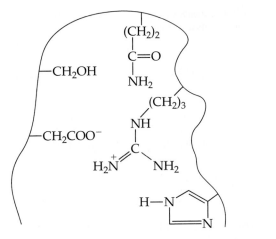

▪ Additional Problems

Structure and Classification of Enzymes

19.26 What general kinds of reactions do the following enzymes catalyze?

(a) Hydrolases   (b) Isomerases   (c) Lyases

19.27 What general kinds of reactions do the following enzymes catalyze?

(a) Oxidoreductases   (b) Ligases   (c) Transferases

19.28 What features of enzymes makes them so specific in their action?

19.29 Describe in general terms how enzymes act as catalysts.

19.30 What classes of enzymes would you expect to catalyze the following reactions?

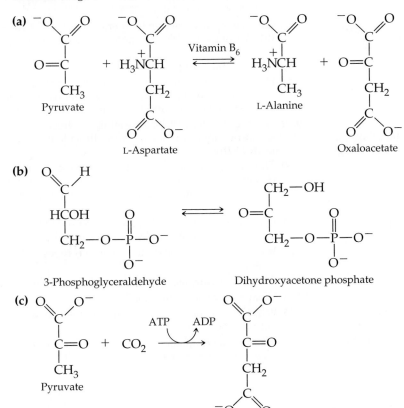

19.31 What classes of enzymes would you expect to catalyze the following reactions?

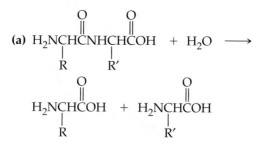

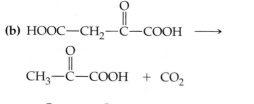

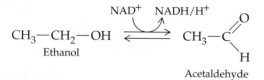

19.32 What kind of reaction does each of the following enzymes catalyze?

(a) A protease
(b) A DNA ligase
(c) A transmethylase

19.33 What kind of reaction does each of the following enzymes catalyze?

(a) A reductase
(b) A kinase
(c) A synthetase

19.34 The following reaction is catalyzed by the enzyme urease. To what class of enzymes does urease belong?

$$H_2N-\overset{\overset{\displaystyle O}{\|}}{C}-NH_2 \;+\; 2\,H_2O \;\xrightarrow{\text{Urease}}\; 2\,NH_3 \;+\; H_2CO_3$$
Urea

19.35 Alcohol dehydrogenase (ADH) catalyzes the following reaction. To what class of enzymes does ADH belong?

$$CH_3-CH_2-OH \;\underset{\text{Ethanol}}{\overset{NAD^+ \quad NADH/H^+}{\rightleftarrows}}\; CH_3-\overset{\overset{\displaystyle O}{\|}}{C}\diagdown H$$
Acetaldehyde

Enzyme Function and Regulation

19.36 What is the difference between the lock-and-key model of enzyme action and the induced-fit model?

19.37 Why is the induced-fit model a more likely model than the lock-and-key model?

19.38 Must the amino acid residues in the active site be near each other along the polypeptide chain? Explain.

19.39 The active site of an enzyme is a small portion of the enzyme molecule. What is the function of the rest of the huge molecule?

19.40 How do you explain the observation that pepsin, a digestive enzyme found in the stomach, has a high catalytic activity at pH 1.5, while trypsin, an enzyme of the small intestine, has no activity at pH 1.5?

19.41 Amino acid side chains in the active sites of enzymes can act as acids or bases during catalysis. List the amino acid side chains that can accept H^+ and those that can donate H^+ during enzyme-catalyzed reactions.

19.42 Draw an energy diagram for the exothermic enzyme-catalyzed hydrolysis of urea (Problem 19.34). Label the energy levels of reactants and products, the activation energy, and the overall energy difference of the reaction.

19.43 Discuss how the rate of reaction changes as enzyme concentration is increased.

19.44 Discuss how the rate of reaction changes as substrate concentration is increased.

19.45 Why don't increases in concentration of substrate or enzyme change reaction rates in exactly the same way?

19.46 What general effects would you expect the following changes to have on the rate of an enzyme-catalyzed reaction for an enzyme that has its maximum activity at body temperature (about 37°C)?

(a) Lowering the temperature from 37°C to 27°C
(b) Raising the pH from 7.5 to 10.5
(c) Adding a heavy metal salt such as $Hg(NO_3)_2$

19.47 What general effects would you expect the following changes to have on the rate of an enzyme-catalyzed reaction for an enzyme that has its maximum activity at body temperature (about 37°C)?

(a) Raising the reaction temperature from 37°C to 40°C
(b) Adding an oxidizing agent such as hydrogen peroxide
(c) Adding a drop of dilute $AgNO_3$

19.48 How do each of the three kinds of inhibitors mentioned in the text work? What kinds of bonds are formed between an enzyme and these three kinds of inhibitors?

19.49 Refer to Problem 19.48. Poisoning by which of the three kinds of inhibitors is the most difficult to treat medically? Why?

19.50 The meat tenderizer used in cooking is primarily papain, a protease enzyme isolated from the papaya tree. Why do you suppose papain is so effective at tenderizing meat?

19.51 Refer to Problem 19.50. Papain is also used to help relieve the pain of bee stings. Why do you suppose it works?

19.52 Why do allosteric enzymes have two types of binding sites?

19.53 Discuss the purpose of positive and negative regulation.

19.54 What is feedback inhibition?

19.55 What would be a reason to have "feed-forward activation"?

19.56 What is a zymogen? Why must some enzymes be secreted as zymogens?

19.57 Activation of a zymogen is by covalent modification. How might phosphorylation or dephosphorylation (also covalent modification) modify an enzyme to make it more active (or more inactive)?

Vitamins

19.58 What are the criteria for a compound to be called a vitamin?

19.59 What is the relationship between vitamins and enzymes?

19.60 Why is daily ingestion of vitamin C more critical than daily ingestion of vitamin A?

19.61 What are the four fat-soluble vitamins? Why is excess consumption of some of these vitamins of concern?

Applications

19.62 Lead exerts its poisonous effect on enzymes by two different mechanisms. Which of these mechanisms would you expect to be irreversible and which could be reversed by EDTA? [*Lead Poisoning and an Antidote*]

19.63 A pharmaceutical chemist needs to synthesize *p*-hydroxyphenylglycine, whose structure is shown below. What feature of enzymes makes biocatalysis particularly useful in this reaction? [*Biocatalysis: Food and Chemicals*]

19.64 Give two reasons why an industrial chemist might choose to use enzymes rather than nonbiological catalysts for a particular reaction. [*Biocatalysis: Food and Chemicals*]

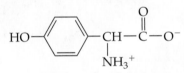

p-Hydroxyphenylglycine

19.65 What three enzymes show marked concentration increases in the blood after a heart attack? Why? [*Enzymes in Medical Diagnosis*]

19.66 Why must enzyme activity be monitored under standard conditions? [*Enzymes in Medical Diagnosis*]

19.67 The primary structure of angiotensin II has ...Pro-Phe at the C-terminal end of the octapeptide. The angiotensin-converting enzyme (ACE) inhibitor from the South American pit viper is a pentapeptide with a C-terminal proline and is a mild ACE inhibitor. Captopril has a modified proline structure and is also a mild ACE inhibitor. Why do you suppose that a mild ACE inhibitor is more valuable for the treatment of high blood pressure than a very potent ACE inhibitor? What structural modifications to the pit viper peptide might make it a more powerful ACE inhibitor? [*Enzyme Inhibitors as Drugs*]

19.68 Why are overdoses less common with water-soluble vitamins than with fat-soluble vitamins? [*Vitamins, Minerals, and Food Labels*]

19.69 Read the labels on foods that you eat for a day, or look the foods up in a nutrition table and determine what percent of your daily dosage of vitamins and minerals you get from each. Are you getting the recommended amounts from the food you eat, or should you be taking a vitamin or mineral supplement? [*Vitamins, Minerals, and Food Labels*]

19.70 For what reasons are listings for vitamin A, vitamin C, iron, and calcium mandatory on food labels? [*Vitamins, Minerals, and Food Labels*]

General Questions and Problems

19.71 Look up the structures of vitamin C and vitamin E in Section 19.10, and identify the functional groups in these vitamins.

19.72 The adult recommended daily allowance (RDA) of riboflavin is 1.6 mg. If one glass (100 mL) of apple juice contains 0.014 mg of riboflavin, how much apple juice would an adult have to consume to obtain the RDA?

19.73 Many vegetables are "blanched" (dropped into boiling water) for a few minutes before being frozen. Why is blanching necessary?

19.74 How can you distinguish between a competitive inhibitor and a noncompetitive inhibitor experimentally?

19.75 Trypsin is an enzyme that cleaves on the C-terminal side (that is, to the right of) all basic residues. Consider the peptide shown below. Predict the fragments that would formed by treatment of this peptide with trypsin: N = terminal end-Leu-Gly-Arg-Ile-Met-His-Tyr-Trp-Ala

19.76 The ability to change a selected amino acid residue to another amino acid is referred to as "point mutation" by biochemists. Referring to the reaction for peptide bond hydrolysis in Figure 19.4, speculate on the effects that the following point mutations might have on the mechanism of the chymotrypsin mechanism shown: serine to valine; aspartate to glutamate.

20

Chemical Messengers

Hormones, Neurotransmitters, and Drugs

A flood of chemical messengers is helping these soccer fans express their anger.

At this point, you've seen some of the many kinds of enzyme-catalyzed reactions that take place in cells. What hasn't been discussed yet is how the individual reactions are tied together. Clearly, the many thousands of reactions taking place in the billions of individual cells of our bodies don't occur randomly. There must be overall control mechanisms that coordinate these reactions, keeping us in chemical balance.

Two systems share major responsibility for regulating body chemistry—the *endocrine system* and the *nervous system*. The endocrine system depends on *hormones*, chemical messengers that circulate in the bloodstream. The nervous system relies primarily on a much faster means of communication—electrical impulses in nerve cells. But the nervous system also has its chemical messengers, the *neurotransmitters* that carry signals from one nerve cell to another and from nerve cells to their targets, the ultimate recipients of the messages.

Given the crucial role of hormones and neurotransmitters in the proper functioning of our bodies, it should not be surprising to find that many drugs act by mimicking, modifying, or opposing the action of chemical messengers.

In this chapter, we'll address the following questions about hormones, neurotransmitters, and the drugs that affect their actions:

1. What are hormones, and how do they function?
The goal: Be able to describe in general the origins, pathways, and actions of hormones.

2. What is the chemical nature of hormones?
The goal: Be able to list, with examples, the different chemical types of hormones.

3. **How does the hormone epinephrine deliver its message, and what is its major action?**

   The goal: Be able to outline the sequence of events in epinephrine's action as a hormone.

4. **What are neurotransmitters, and how do they function?**

   The goal: Be able to describe the general origins, pathways, and actions of neurotransmitters.

5. **How does acetylcholine deliver its message, and how do drugs alter its function?**

   The goal: Be able to outline the sequence of events in acetylcholine's action as a neurotransmitter and give examples of its agonists and antagonists.

6. **Which neurotransmitters and what kinds of drugs play roles in allergies, mental depression, drug addiction, and pain?**

   The goal: Be able to identify neurotransmitters and drugs active in these conditions.

7. **What are some of the methods used in drug discovery and design?**

   The goal: Be able to explain the general roles of ethnobotany, chemical synthesis, combinatorial chemistry, and computer-aided design in the development of new drugs.

CONCEPTS TO REVIEW

Amino acids (Sections 18.3, 18.4)
Shapes of proteins (Section 18.8)
Tertiary and quaternary protein structure (Sections 18.10, 18.11)
How enzymes work (Section 19.4)

20.1 Messenger Molecules

Coordination and control of our vital functions is accomplished by chemical messengers. Whether the messengers are hormones that arrive via the bloodstream or neurotransmitters released by nerve cells, such messengers ultimately connect with a *target*. The message is delivered by interaction between the chemical messenger and a **receptor** at the target. Many hormone receptors are proteins embedded in the membrane surrounding a cell. Other hormone receptors are inside the cell. The messengers of the nervous system connect with receptors on the next nerve cell in a pathway of nerve cells or at the tissue that must respond to the message—for example, a muscle that must contract.

Noncovalent attractions draw messengers and receptors together, much as a substrate is drawn into the active site of an enzyme. (⊙⊙⊙ Sections 18.8, 19.4) These attractions hold the messenger and receptor together long enough for the

> **Receptor** A molecule or portion of a molecule with which a hormone, neurotransmitter, or other biochemically active molecule connects to initiate a response in a target cell.

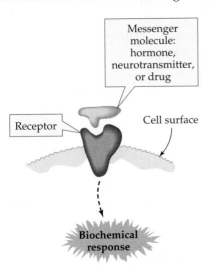

◀ A general representation of the interaction between a messenger molecule and a cellular receptor.

Messenger molecule: hormone, neurotransmitter, or drug

Receptor

Cell surface

Biochemical response

Hormone   A chemical messenger secreted by cells of the endocrine system and transported through the bloodstream to target cells with appropriate receptors where it elicits a response.

message to be delivered, but without any permanent chemical change to the messenger or the receptor. The results of the interaction are chemical changes within the target cell.

The chemical messengers of the endocrine system are the **hormones**. These molecules are produced by endocrine glands and tissues in various parts of the body. Because hormones travel through the bloodstream to their targets, the responses they produce require seconds to hours to begin. Their influence, however, may last a long time and can be wide-ranging. A single hormone will often affect many different tissues and organs—any cell with the appropriate receptors is a target. Insulin, for example, is a hormone secreted by the pancreas in response to elevated blood glucose levels. At target cells throughout the body, insulin accelerates uptake and utilization of glucose; in muscles it accelerates formation of glycogen, a glucose polymer that is metabolized when muscles need quick energy; and in fatty tissue it stimulates storage of fat molecules (lipids).

Neurotransmitter   A chemical messenger that travels between a neuron and a neighboring neuron or other target cell to transmit a nerve impulse.

The chemical messengers of the nervous system are **neurotransmitters**. The electrical signals of the nervous system travel along nerve fibers, taking only a fraction of a second to reach their highly specific destinations. Most nerve cells, however, do not make direct contact with the cells they stimulate. A neurotransmitter must carry the message across the tiny gap separating the nerve cell from its target. Because neurotransmitters are released in very short bursts and are quickly broken down or reabsorbed by the nerve cell, their effects are typically short-lived. The nervous system is organized so that nearly all of its vital switching, integrative, and information-processing functions depend on neurotransmitters.

In this chapter, we first discuss hormones and the endocrine system and include a detailed description of how one hormone—epinephrine (also known as adrenalin)—performs its functions. Then we'll cover neurotransmitters, using the action of acetylcholine to illustrate how neurotransmitters act. It is essential to recognize that hormones and neurotransmitters play a fundamental role in maintaining our health by their influence on metabolic. (See Application "Homeostasis.") Finally, we'll look briefly at the discovery and design of drugs as chemical messengers.

20.2 Hormones and the Endocrine System

Endocrine system   A system of specialized cells, tissues, and ductless glands that excretes hormones and shares with the nervous system the responsibility for maintaining constant internal body conditions and responding to changes in the environment.

The **endocrine system** includes all cells that secrete hormones into the bloodstream. Some of these cells are found in organs that also have nonendocrine functions (for example, the pancreas, which also produces digestive enzymes); others occur in glands devoted solely to hormonal control (for example, the thyroid gland). In no case do hormones carry out chemical reactions. Hormones alter the biochemistry of a cell by inhibiting or activating an existing enzyme, by initiating or altering the rate of synthesis of a specific protein, or in other ways.

The major endocrine glands are the pituitary gland in the brain, the thyroid gland, the adrenal gland, and the ovaries and testes. The hypothalamus, a section of the brain just above the pituitary gland, is in charge of the endocrine system. It communicates with other tissues in three ways:

- *Direct neural control*   A nervous system message from the hypothalamus initiates release of hormones by the adrenal gland. For example:

$$\text{Hypothalamus} \xrightarrow{\text{Nerve message}} \text{Adrenal gland} \longrightarrow \text{Epinephrine}$$

  Epinephrine is targeted to many cells; it increases heart rate, blood pressure, glucose availability.

- *Direct release of hormones*   Hormones move from the hypothalamus to the posterior pituitary gland, where they are stored until needed. For example:

$$\text{Hypothalamus} \longrightarrow \text{Antidiuretic hormone}$$

  Antidiuretic hormone is stored in the posterior pituitary gland; it is targeted to kidneys; causes retention of water and elevation of blood pressure.

Application Homeostasis

Homeostasis—the maintenance of a constant internal environment in the body—is as important to the study of living things as atomic structure is to the study of chemistry. The phrase "internal environment" is a general way to describe all the conditions within cells, organs, and body systems. Such conditions as body temperature, the availability of chemical compounds that supply energy, and the disposal of waste products must remain within specific limits for an organism to function properly. Throughout our bodies, sensors track the internal environment and send signals to restore proper balance if the environment changes. If oxygen is in short supply, for example, a signal is sent that makes us breathe harder. When we are cold, a signal is sent to constrict surface blood vessels and prevent further loss of heat.

At the chemical level, homeostasis requires that the concentrations of ions and many different organic compounds stay near normal levels. The predictability of the concentrations of such substances is the basis for

clinical chemistry—the chemical analysis of body tissues and fluids. In the clinical lab, various tests measure concentrations of significant ions and compounds in blood, urine, feces, spinal fluid, or other samples from a patient's body. Comparing the lab results with norms (average concentration ranges in a population of healthy individuals) shows which body systems are struggling, or possibly failing, to maintain homeostasis. To give just one example, urate is an anion that helps to carry waste nitrogen from the body. A urate concentration higher than the normal range of about 2.5–7.7 mg/dL in blood can signal kidney malfunction.

A copy of a clinical lab report for a routine blood analysis is shown in the Figure. (Fortunately, this individual has no significant variations from normal.) The metal names in the report refer to the various cations, and the heading "Phosphorus" refers to the anion.

See Additional Problems 20.64–65 at the end of this chapter.

Test	Result	Normal Range	Test	Result	Normal Range
Albumin	4.3 g/dL	3.5–5.3 g/dL	SGOT*	23 U/L	0–28 U/L
Alk. Phos.*	33 U/L	25–90 U/L	Total protein	5.9 g/dL	6.2–8.5 g/dL
BUN*	8 mg/dL	8–23 mg/dL	Triglycerides	75 mg/dL	36–165 mg/dL
Bilirubin T.*	0.1 mg/dL	0.2–1.6 mg/dL	Uric Acid	4.1 mg/dL	2.5–7.7 mg/dL
Calcium	8.6 mg/dL	8.5–10.5 mg/dL	GGT*	23 U/L	0–45 U/L
Cholesterol	227 mg/dL	120–250 mg/dL	Magnesium	1.7 mEq/L	1.3–2.5 mEq/L
Chol., HDL*	75 mg/dL	30–75 mg/dL	Phosphorus	2.6 mg/dL	2.5–4.8 mg/dL
Creatinine	0.6 mg/dL	0.7–1.5 mg/dL	SGPT*	13 U/L	0–26 U/L
Glucose	86 mg/dL	65–110 mg/dL	Sodium	137.7 mEq/L	135–155 mEq/L
Iron	101 mg/dL	35–140 mg/dL	Potassium	3.8 mEq/L	3.5–5.5 mEq/L
LDH*	48 U/L	50–166 U/L			

▲ A clinical lab report for routine blood analysis. The abbreviations marked with asterisks are for the following tests (alternative standard abbreviations are in parentheses): Alk. Phos., alkaline phosphatase (ALP); BUN, blood urea nitrogen; Bilirubin T., total bilirubin; Chol., HDL, cholesterol, high-density lipoproteins; LDH, lactate dehydrogenase; SGOT, serum glutamic oxaloacetic transaminase (AST); GGT, γ-glutamyl transferase; SGPT, serum glutamic pyruvic transaminase (ALT).

- **Indirect control through release of regulatory hormones** In the most common control mechanism, *regulatory hormones* from the hypothalamus stimulate or inhibit the release of hormones by the anterior pituitary gland. Many of these pituitary hormones in turn stimulate release of still other hormones by their own target tissues. For example:

$$\text{Hypothalamus} \xrightarrow{\text{Releasing factor}} \text{Pituitary gland} \longrightarrow$$
$$\text{Thyrotropin (a regulatory hormone)} \longrightarrow$$
$$\text{Thyroid gland} \longrightarrow \text{Thyroid hormones}$$

Thyroid hormones are targeted to cells throughout the body; they affect oxygen availability, blood pressure, other endocrine tissues.

Chemically, hormones are of three major types: (1) amino acid derivatives, small molecules containing amino groups; (2) polypeptides, which range from just a few amino acids to several hundred amino acids; (3) steroids, which are lipids with a distinctive molecular structure based on four connected rings.

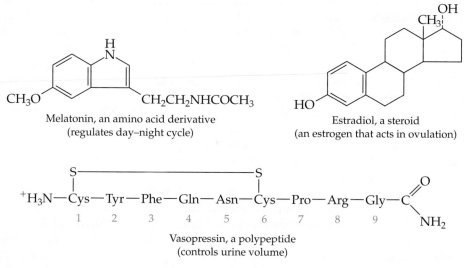

Melatonin, an amino acid derivative
(regulates day–night cycle)

Estradiol, a steroid
(an estrogen that acts in ovulation)

Vasopressin, a polypeptide
(controls urine volume)

Examples of the targets and actions of some hormones of each type are given in Table 20.1.

TABLE 20.1   Examples of Each Chemical Class of Hormones

Chemical Class	Hormone Examples	Source	Target	Major Action
Amino acid derivatives	Epinephrine and norepinephrine	Adrenal medulla	Most cells	Release glucose from storage; increase heart rate and blood pressure
	Thyroxine	Thyroid gland	Most cells	Influence energy use, oxygen consumption, growth, and development
Polypeptides (regulatory hormones)	Adrenocorticotropic hormone	Anterior pituitary	Adrenal cortex	Stimulate release of glucocorticoids (steroids), which control glucose metabolism
	Growth hormone	Anterior pituitary	Peripheral tissues	Stimulate growth of muscle and skeleton
	Follicle-stimulating hormone, luteinizing hormone	Anterior pituitary	Ovaries and testes	Stimulate release of steroid hormones
	Vasopressin	Posterior pituitary	Kidneys	Cause retention of water, elevation of blood volume and blood pressure
	Thyrotropin	Anterior pituitary	Thyroid gland	Stimulates release of thyroid hormones
Steroids	Cortisone and cortisol (glucocorticoids)	Adrenal cortex	Most cells	Counteract inflammation; control metabolism when glucose must be conserved
	Testosterone; estrogen, progesterone	Testes; ovaries	Most cells	Control development of secondary sexual characteristics, maturation of sperm and eggs

Upon arrival at its target cell, a hormone must deliver its signal to create a chemical response inside the cell. The signal enters the cell in one of two ways determined by the chemical nature of the hormone (Figure 20.1). Because the cell is surrounded by a membrane composed of hydrophobic molecules, only hydrophobic molecules can move across it on their own. The steroid hormones are of this type, so they can enter the cell directly. Within the cell's cytoplasm, a steroid hormone encounters a receptor that carries it to its target, DNA in the nucleus of the cell. The result is some variation in production of a protein governed by a particular gene.

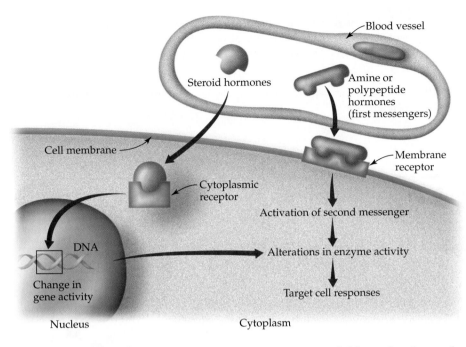

◄ **FIGURE 20.1   Interaction of hormones and receptors at the cellular level.**   Steroid hormones are hydrophobic and can cross the cell membrane to find receptors inside the cell. Amine and polypeptide hormones are hydrophilic and, because they cannot cross the cell membrane, act via second messengers.

The polypeptide and amine hormones are water-soluble molecules and so cannot cross the hydrophobic cell membranes. They act by noncovalent binding with receptors on cell surfaces. The result is release of a **second messenger** within the cell. There are several different second messengers, and the specific sequence of events varies. In general, three membrane-bound proteins participate in release of the second messenger: the receptor (1) and a *G protein* (2) that transfers the message to an enzyme (3). Connection of the hormone with its receptor causes a change in the receptor (not unlike the effect of an allosteric regulator on an enzyme; Section 19.7). The G protein then activates an enzyme that participates in release of the second messenger. The action of epinephrine by way of a second messenger is described in Section 20.3. Further examples of amino acid, polypeptide, and steroid hormones are given in Sections 20.4 and 20.5.

Second messenger   Chemical messenger released inside a cell when a hydrophilic hormone or neurotransmitter connects with a receptor on the cell surface.

⬤⬤⬤ LOOKING AHEAD

The cell membrane is hydrophobic because it is composed mainly of *lipids*—molecules that are not water-soluble and cling together like oil floating on the surface of a pond. Steroids are classified as lipids because they are not water-soluble, but are discussed here because they function as hormones. The chemical nature of other lipids and cell membranes is explored in Chapter 24.   ⬤⬤⬤

■ WORKED EXAMPLE 20.1

Classify each of the following hormones as an amino acid derivative, a polypeptide, or a steroid.

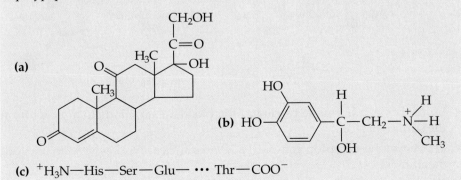

(c) ^+H_3N—His—Ser—Glu— ••• Thr—COO^-

20.3 How Hormones Work: Epinephrine and Fight-or-Flight

Epinephrine, also known as *adrenaline*, is often called the *fight-or-flight hormone* because it is released when we need an instant response to danger.

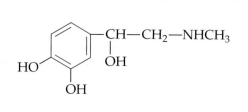

Epinephrine
(Adrenaline)

▼ FIGURE 20.2 Activation of cyclic-AMP as a second messenger. (a) The hormone receptor, inactive G protein, and inactive adenylate cyclase enzyme reside in the cell membrane. (b) On formation of the hormone–receptor complex the guanosine diphosphate (GDP) of the G protein is converted to guanosine triphosphate (GTP). (c) The active G protein–GTP complex activates adenylate cyclase, causing production of cyclic AMP inside the cell, where it will initiate the action called for by the hormone.

We've all felt the rush of epinephrine that accompanies a near-miss accident or a sudden loud noise. The main function of epinephrine in this type of situation is a dramatic increase in the availability of glucose as a source of energy to deal with whatever stress is at hand. The time elapsed from initial stimulus to glucose release into the bloodstream is a few only seconds.

Epinephrine acts via *cyclic adenosine monophosphate (cyclic AMP, or cAMP)*, an important second messenger. The sequence of events in this action, shown in Figure 20.2 and described below, illustrates one type of biochemical response to a change in an individual's external or internal environment.

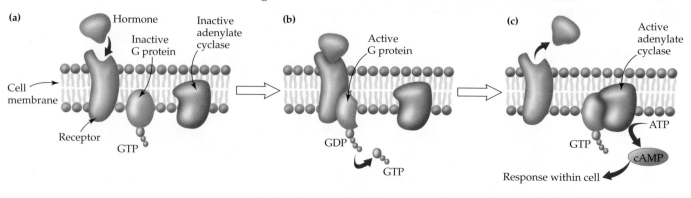

The Action of Non-Steroid Hormones

- Epinephrine, a hormone carried in the bloodstream, binds to a receptor on the surface of a cell.
- The hormone–receptor complex activates a nearby *G protein* in the cell membrane.

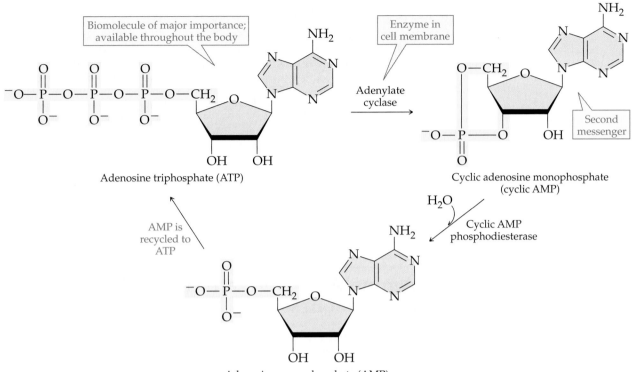

Adenosine triphosphate (ATP)

Cyclic adenosine monophosphate (cyclic AMP)

Adenosine monophosphate (AMP)

▲ **FIGURE 20.3 Production of cyclic AMP as a second messenger.** The reactions shown take place within the target cell after epinephrine or some other chemical messenger has interacted with a receptor on the cell surface. (The major role of ATP in providing energy for biochemical reactions is discussed in Section 21.5.)

- GDP (guanosine diphosphate) associated with the G protein is converted to GTP (guanosine triphosphate) by addition of a phosphate group.
- The G protein–GTP complex activates *adenylate cyclase*, an enzyme that also resides in the cell membrane.
- Adenylate cyclase catalyzes production within the cell of the second messenger—*cyclic AMP*—from ATP, as shown in Figure 20.3.
- Cyclic AMP initiates reactions that activate glycogen phosphorylase, the enzyme responsible for release of glucose from storage. (Other hormones result in initiation by cyclic AMP of other reactions.)
- When the emergency has passed, cyclic AMP is converted back to ATP.

In addition to making glucose available, epinephrine reacts with other receptors that increase blood pressure, heart rate, and respiratory rate; decrease blood flow to the digestive system (digestion isn't important during an emergency); and counteract spasms in the respiratory system. The resulting combined and rapid effects make epinephrine the most crucial drug for treatment of the serious medical emergency known as *anaphylactic shock*. Anaphylactic shock is the result of a severe allergic reaction, perhaps to a bee or wasp sting or to a drug. The major symptoms include a severe drop in blood pressure due to blood vessel dilation and difficulty breathing due to bronchial constriction. Epinephrine directly counters these symptoms. Individuals who know they are subject to this overwhelming and life-threatening allergic response carry epinephrine with them at all times.

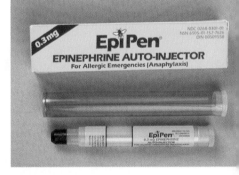

▲ An epinephrine autoinjection pen. Such devices are carried by individuals at risk of an anaphylactic reaction to an allergen.

■ **PROBLEM 20.1**

A phosphorus-containing anion is removed from ATP in its conversion to cyclic AMP, as shown in Figure 20.3. The anion is often abbreviated as PP_i. Which of the following anions is represented by PP_i?

(a) PO_4^{3-} **(b)** $P_2O_7^{4-}$ **(c)** $H_2PO_4^{-}$ **(d)** $P_3O_{10}^{5-}$

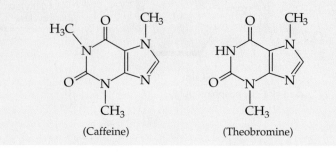

■ KEY CONCEPT PROBLEM 20.2

Caffeine and theobromine (from chocolate) act as stimulants. They work by altering the cAMP signal. Refer to Figure 20.3 and decide how these molecules might interact with an enzyme in the cAMP pathway to enhance the effect of cAMP.

(Caffeine)   (Theobromine)

20.4 Amino Acid Derivatives and Polypeptides as Hormones

Amino acid derivatives.   The biochemistry of the brain is an active area of research. As our understanding of chemical messages in the brain grows, the traditional distinctions between hormones and neurotransmitters have been breaking down. Several amino acid derivatives classified as hormones because of their roles in the endocrine system are also synthesized in neurons and function as neurotransmitters in the brain. (Because a barrier limits entry into the brain of chemicals traveling in the bloodstream, the brain cannot rely on a supply of chemical messengers synthesized elsewhere; see Section 29.3.) Epinephrine, the fight-or-flight hormone we described in Section 20.3, is one of the amino acid derivatives that is both a hormone and a neurotransmitter. The pathway for the synthesis of epinephrine is shown below in Figure 20.4, where you can see that several other chemical messengers are formed along the way.

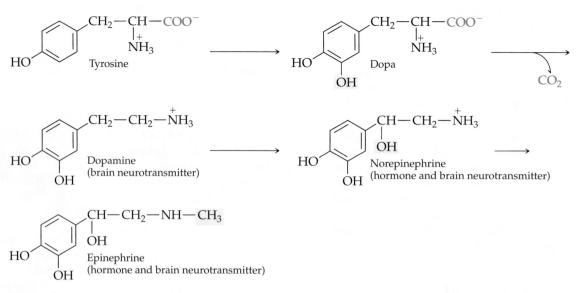

▲ **FIGURE 20.4 Synthesis of chemical messengers from tyrosine.** The changes in each step are highlighted.

Thyroxine, another amino acid derivative, is also a hormone. It is one of two iodine-containing hormones produced by the thyroid gland, and our need for dietary iodine is determined by these hormones. Unlike the other hormones derived from amino acids, thyroxine is a nonpolar compound that can cross the cell membrane and enter the cell, where it activates the synthesis of various enzymes. A greatly enlarged thyroid gland (a goiter) is a symptom of iodine deficiency. In developed countries where iodine is added to table salt, goiter is uncommon. In some regions of the world, however, iodine deficiency is a common and serious problem that results not only in goiter but in severe mental retardation in infants (*cretinism*).

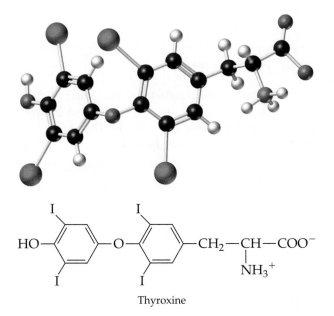

Thyroxine

Polypeptides. Polypeptides are the largest class of hormones. They range widely in molecular size and complexity as illustrated by two hormones that control the thyroid gland, *thyrotropin-releasing hormone (TRH)* and *thyroid-stimulating hormone (TSH)*. TRH, a modified tripeptide, is a regulatory hormone released by the hypothalamus. At the pituitary gland, TRH activates release of TSH, a protein that has 208 amino acid residues in two chains. The TSH in turn triggers release of amino acid derivative hormones from the thyroid gland.

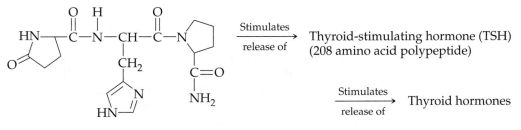

Thyrotropin-releasing hormone
(TRH)

Insulin, a protein with 51 amino acids (○○○, p. 522), is released by the pancreas in response to high concentrations of glucose in the blood. It stimulates cells to take up glucose and put it to use or into storage.

○○○ LOOKING AHEAD

Because of its importance in glucose metabolism and diabetes mellitus, the function of insulin as a hormone is described in Section 23.9.   ○○○

■ PROBLEM 20.3
Examine the TRH structure and identify the three amino acids from which it is derived. (The N-terminal amino acid has undergone ring formation, and the carboxyl group at the C-terminal end has been converted to an amide.)

■ **KEY CONCEPT PROBLEM 20.4**

Based on the structure of thyroxine, what might be the reason that thyroxine, although it is an amino acid derivative, is hydrophobic rather than hydrophilic?

Steroid A lipid whose structure is based on the following tetracyclic (four-ring) carbon skeleton:

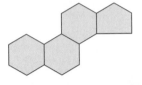

▲ A fellow with a good supply of steroid hormones.

The Action of Steroid Hormones, Part 1 and Part 2

20.5 Steroid Hormones

Steroids have in common a central structure composed of the four connected rings shown in the margin. Because they are soluble in hydrophobic solvents and not in water, steroids are classified as lipids. The steroid hormones can be divided according to function into three types:

- *Mineralocorticoids*, such as *aldosterone*, regulate the delicate cellular fluid balance between Na^+ and K^+ ions (hence the "mineral" in their name).

- *Glucocorticoids*, such as cortisol (also known as *hydrocortisone*) and its close relative cortisone, help to regulate glucose metabolism and inflammation. You have probably used an anti-inflammatory ointment containing hydrocortisone to reduce the swelling and itching of poison ivy or some other skin irritation.

- *Sex hormones*, including *testosterone* and *androsterone*, which are the two most important male sex hormones, or *androgens*. They are responsible for the development of male secondary sex characteristics during puberty and for promoting tissue and muscle growth.

Male sex hormones (androgens)

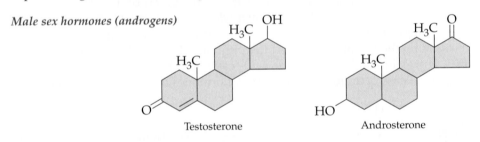

Estrone and estradiol, the female hormones known as *estrogens*, are synthesized from testosterone, mainly in the ovaries and also to a small extent in the adrenal cortex. Estrogens govern development of female secondary sex characteristics and participate in regulation of the menstrual cycle. The *progestins*, principally *progesterone*, are released by the ovaries during the second half of the menstrual cycle and prepare the uterus for implantation of a fertilized ovum should conception occur.

Female sex hormones

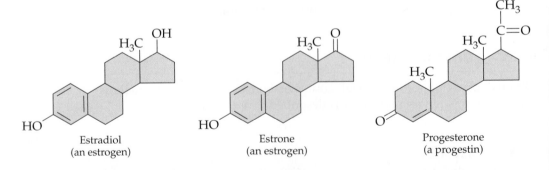

In addition to the several hundred known steroids isolated from plants and animals, a great many more have been synthesized in the laboratory in the

Application Plant Hormones

Yes, plants do have hormones. They are just as important to the health and development of plants as human hormones are to us. If you think about it for a moment, however, you'll realize that there must be some differences. Plants don't have endocrine systems, nor do they have fluids like blood that continuously circulate so that chemicals can be picked up where they are created and distributed to wherever they are needed.

Unlike animal hormones, which must be transported to their targets, plant hormones affect the cells in which they are synthesized. They may also reach nearby cells by diffusion or travel upward with water from the roots or downward with sugars made by photosynthesis in the leaves. The chemically most simple plant hormone is ethylene, $CH_2{=}CH_2$, a gas. At one time, citrus growers ripened oranges that had been picked green in rooms heated with kerosene stoves. Mysteriously, when the stoves were replaced with other means of heating, the oranges no longer ripened. It turns out that the ripening was hastened by ethylene from the kerosene, not by the warmth. Plants produce ethylene when it is time for fruit to ripen. Nowadays, bananas, tomatoes, and other fruits are picked hard and unripe, a condition in which they are easier to ship, and then ripened by exposure to ethylene. You can try this at home. Enclose some less than perfectly ripe pears or peaches in a brown paper bag along with, for example, a very ripe banana. Be careful, though. "One rotten apple can spoil the barrel." With too much exposure to ethylene, ripening can be overdone.

Just by watching our house plants, we know that plants turn toward the sun, a phenomenon known as *phototropism*. Charles Darwin was one of the first to wonder why this happens. He observed that covering

▲ An amaryllis leaning toward the sun.

the growing tips of the plants prevented phototropism. The explanation lies in the formation in the tip of an *auxin*, a hormone that travels downward and stimulates elongation of the stem. When light distribution is uneven, auxin concentrates on the shady side of the stem, causing it to grow faster so that the stem bends toward the sun. Auxin is produced in seed embryos, young leaves, and growing tips of plants. Interestingly, plants synthesize auxin from tryptophan, the starting compound in the synthesis of several mammalian chemical messengers (see Figure 20.6). As for ethylene, the effects of auxin can be overdone. An excessive concentration of auxin kills plants by over accelerating their growth. The most familiar synthetic auxin, 2,4-D, is widely used to kill broad-leaved weeds in this manner.

See Additional Problems 20.66–67 at the end of this chapter.

search for new drugs. Most birth control pills are a mixture of the synthetic estrogen *ethynyl estradiol* and the synthetic progestin *norethindrone*. These steroids function by tricking the body into a false pregnant state, making it temporarily infertile. The compound known as *RU-486*, or *mifepristone*, is effective as a "morning-after" pill. It prevents pregnancy by binding strongly to the progesterone receptor, thereby blocking implantation in the uterus of a fertilized egg cell.

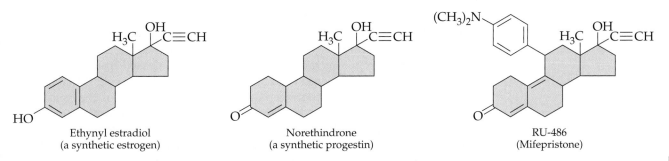

Ethynyl estradiol
(a synthetic estrogen)

Norethindrone
(a synthetic progestin)

RU-486
(Mifepristone)

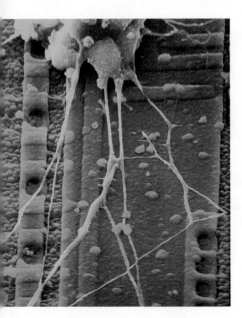

▲ Two ways to send messages. The picture shows a human neuron growing on the surface of a silicon-computer chip.

Synapse   The place where the tip of a neuron and its target cell lie adjacent to each other.

FIGURE 20.5   A nerve cell and ▶ transmission of a nerve signal by neurotransmitters. Transmission occurs between neurons when a neurotransmitter is released by the presynaptic neuron, crosses the synapse, and fits into a receptor on the postsynaptic neuron.

■ **PROBLEM 20.5**

Nandrolone is an anabolic, or tissue-building, steroid sometimes taken by athletes seeking to build muscle mass (though not recommended by medical experts). Among its effects are a high level of androgenic activity. Which of the androgens shown on p. 582 does it most closely resemble? How does it differ from that androgen?

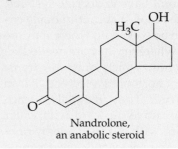

Nandrolone,
an anabolic steroid

20.6 Neurotransmitters

Neurotransmitters are the chemical messengers of the nervous system. They are released by nerve cells (*neurons*) and transmit signals to neighboring target cells, which might be other nerve cells, muscle cells, or endocrine cells. Structurally, nerve cells that rely on neurotransmitters typically have a bulb-like body connected to a long, thin stem called an *axon* (Figure 20.5). Short, tentacle-like appendages, the *dendrites*, protrude from the bulbous end of the neuron, while numerous filaments protrude from the axon at the opposite end. The filaments lie close to the target cell, separated only by a narrow gap—the **synapse**.

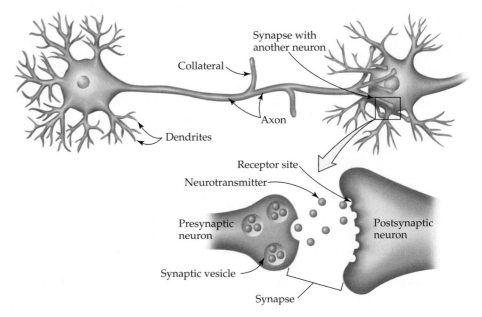

A nerve impulse is transmitted along a nerve cell by variations in electrical potential caused by the exchange of positive and negative ions across the cell membrane. Chemical transmission of the impulse between a nerve cell and its target occurs when neurotransmitter molecules are released from a *presynaptic neuron*, cross the synapse, and bind to receptors on the target cell. When the target is another nerve cell, the receptors lie on the dendrites of the next, *postsynaptic* neuron, as shown in Figure 20.5. Once neurotransmitter–receptor binding has occurred, the message has been delivered. The postsynaptic neuron then transmits the nerve impulse down its own axon until a neurotransmitter delivers the message to the next neuron or other target cell.

Neurotransmitter molecules are synthesized in the presynaptic neurons and stored there in small pockets, known as *vesicles*, from which they are released as needed. After a neurotransmitter has done its job, it must be *rapidly* removed so that the postsynaptic neuron is ready to receive another impulse. This is accomplished in one of two ways. Either a chemical change catalyzed by an enzyme available in the synaptic cleft inactivates the neurotransmitter, or alternatively, the neurotransmitter is returned to the presynaptic neuron and placed in storage until it is needed again.

 Nerves and Action Potential

Most neurotransmitters are amines, as you will see in Sections 20.8–20.10. The synthesis of dopamine, norepinephrine, and epinephrine from tyrosine was illustrated in Figure 20.4. The synthesis of serotonin and melatonin from tryptophan is illustrated in Figure 20.6. Some neurotransmitters act directly by causing changes in adjacent cells as soon as they connect with their receptors. Others rely on second messengers, often cyclic AMP, the same second messenger utilized by hormones. Increasingly, individual neurotransmitters are being associated with effects on emotions, drug addiction, pain relief, and other brain functions, as illustrated in the following sections.

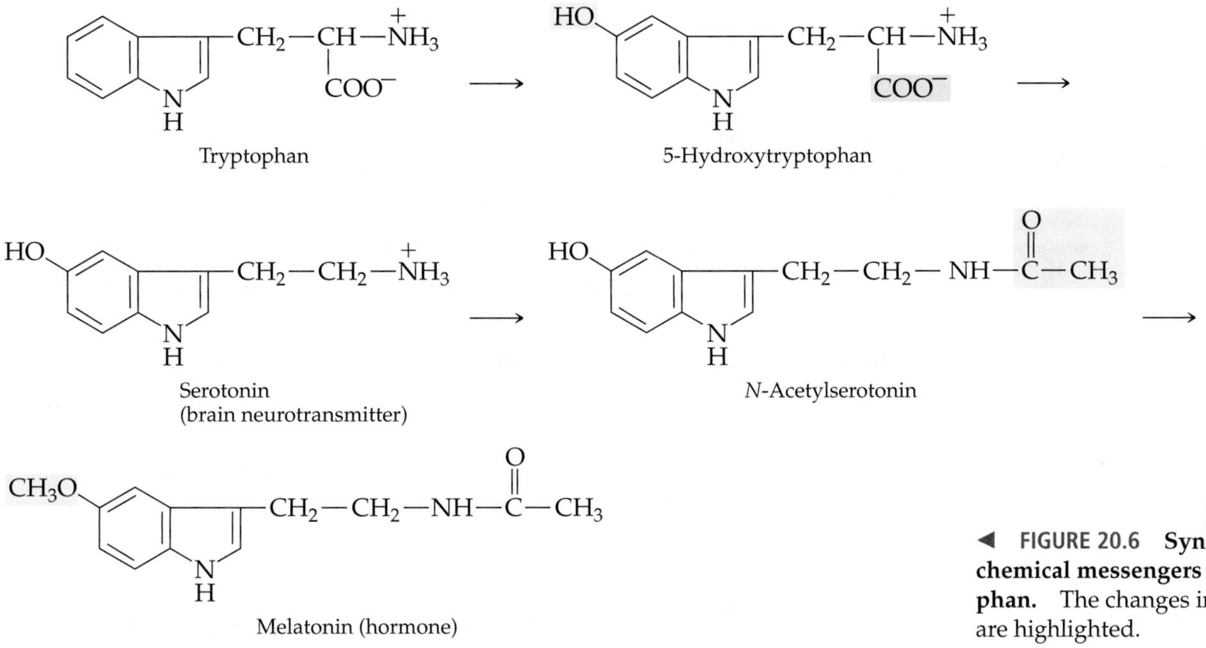

Tryptophan

5-Hydroxytryptophan

Serotonin
(brain neurotransmitter)

N-Acetylserotonin

Melatonin (hormone)

◀ **FIGURE 20.6 Synthesis of chemical messengers from tryptophan.** The changes in each step are highlighted.

■ **PROBLEM 20.6**

Which of the following transformations of amines in Figure 20.6 is (1) a methylation, (2) a decarboxylation, (3) an acetylation?

(a) 5-Hydroxytryptophan to serotonin
(b) Serotonin to N-acetylserotonin
(c) N-Acetylserotonin to melatonin

20.7 How Neurotransmitters Work: Acetylcholine, Its Agonists and Antagonists

Acetylcholine in Action

Acetylcholine (ACh) is a neurotransmitter responsible for the control of skeletal muscles. It is also widely distributed in the brain, where it may play a role in the sleep–wake cycle, learning and memory, and mood. Nerves that rely on ACh as their neurotransmitter are classified as *cholinergic* nerves.

FIGURE 20.7 Acetylcholine ▶ (ACh) release and re-uptake.
ACh is stored in vesicles in the presynaptic neuron. After it is released into the synapse and connects with its receptor, it is broken down by hydrolysis into acetate and choline in the reaction catalyzed by acetylcholinesterase. The choline is taken back into the synaptic knob and used to synthesize ACh, which is then stored in the vesicles until needed.

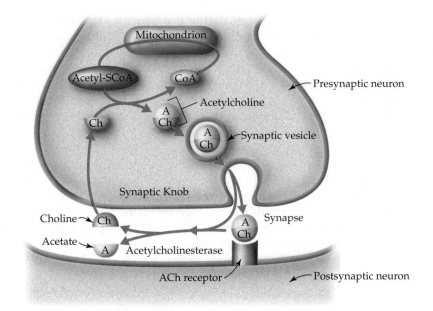

ACh is synthesized in presynaptic neurons and stored in their vesicles. The rapid sequence of events in the action of ACh in communicating between nerve cells, illustrated in Figure 20.7, is as follows:

- A nerve impulse arrives at the presynaptic neuron.
- The vesicles move to the cell membrane, fuse with it, and release their ACh molecules (several thousand molecules from each vesicle).
- ACh crosses the synapse and binds to receptors on the postsynaptic neuron.
- The resulting change in the permeability to ions of the postsynaptic neuron initiates the nerve impulse in that neuron.
- With the message delivered, acetylcholinesterase present in the synaptic cleft catalyzes the decomposition of acetylcholine:

Acetylcholine

$$CH_3 - \overset{\overset{\text{O}}{\|}}{C} - O - CH_2 - CH_2 - N^+(CH_3)_3 \xrightarrow[\text{H}_2\text{O}]{\text{Acetylcholinesterase}} HO - CH_2 - CH_2 - N^+(CH_3)_3 \ + \ CH_3COO^-$$

Acetylcholine (ACh)   Choline   Acetate

- Choline is absorbed back into the presynaptic neuron where new ACh is synthesized.

■ **PROBLEM 20.7**

Propranolol (trade name, Inderal) is an antagonist for certain epinephrine receptors and is a member of the class of drugs known as beta blockers (because they block what are known as beta receptors). Circle the functional groups in propranolol and name them. Compare the structure of propranolol with the structure of epinephrine and describe the differences.

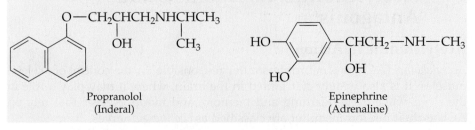

Propranolol
(Inderal)

Epinephrine
(Adrenaline)

Drugs and Acetylcholine

Many drugs act at acetylcholine synapses (∞, p. 557ff.), the places where the tip of a neuron that releases acetylcholine and its target cell lie adjacent to each other. A **drug** is any molecule that alters normal functions when it enters the body from an external source. The action is at the molecular level, and it can be either therapeutic or poisonous. To have an effect, many drugs must connect with a receptor just as a substrate must bind to an enzyme or as a hormone or neurotransmitter must bind to a receptor.

Pharmacologists classify some drugs as **agonists**—substances that act to produce or prolong the normal biochemical response of a receptor. Other drugs are classified as **antagonists**—substances that block or inhibit the normal response of a receptor. To illustrate the ways in which drugs can affect our biochemical activity, we next describe the action of a group of drugs. These drugs are all members of the same drug family in the sense that their biochemical activity occurs at acetylcholine synapses in the central nervous system. The locations of their actions can be seen in Figure 20.7.

Drug Any substance that alters body function when it is introduced from an external source.

▲ A Peruvian native inserting a poison-tipped dart into his blowgun.

Agonist A substance that interacts with a receptor to cause or prolong the receptor's normal biochemical response.

Antagonist A substance that blocks or inhibits the normal biochemical response of a receptor.

- *Botulinus toxin (an antagonist), blocks acetylcholine release and causes botulism.* The toxin, which is produced by bacterial growth in improperly canned food, binds irreversibly to the presynaptic neuron, where acetylcholine would be released. It prevents this release, frequently causing death due to muscle paralysis.

- *Black widow spider venom (an agonist), releases excess acetylcholine.* In the opposite reaction from that of botulism, the synapse is flooded with acetylcholine, resulting in muscle cramps and spasms.

- *Organophosphorus insecticides (antagonists), inhibit acetylcholinesterase.* All of the organophosphorus insecticides (a few examples are shown below) prevent the cholinesterase enzyme from breaking down acetylcholine within the synapse. As a result, the nerves are overstimulated, causing a variety of symptoms including muscle contraction and weakness, lack of coordination, and at high doses, convulsions.

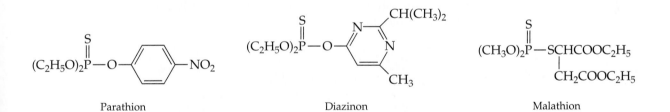

Parathion   Diazinon   Malathion

- *Nicotine binds to acetylcholine receptors.* *Nicotine* at low doses is a stimulant (an agonist) because it activates acetylcholine receptors. The sense of alertness and well-being produced by inhaling tobacco smoke is a result of this effect. At high doses delivered orally or through the skin, nicotine is an antagonist. It irreversibly blocks the acetylcholine receptors and can cause their degeneration. Nicotine has no therapeutic use other than in overcoming addiction to smoking.

- *Atropine (an antagonist), competes with acetylcholine at receptors.* Chemically atropine is an alkaloid (deadly nightshade; Section 15.6) that is a poison at high doses. At controlled doses, its therapeutic uses include acceleration of abnormally slow heart rate, paralysis of eye muscles during surgery, and relaxation of intestinal muscles in gastrointestinal disorders. Most importantly, it is a specific antidote for cholinesterase poisons such as organophosphorus insecticides. By blocking activation of the receptors, it counteracts the excess of acetylcholine created by cholinesterase inhibitors.

■ **PROBLEM 20.8**

The LD_{50} values (lethal dose in mg/kg, for rats; see p. 455) for the three organophosphorus insecticides listed above are parathion, 3–13 mg/kg; diazoninon, 250–285 mg/kg; and malathion, 1000–1375 mg/kg. Which would you choose for use in your garden and why?

■ **KEY CONCEPT PROBLEM 20.9**

Some drugs are classified as agonists while others are classified as antagonists. Explain the difference between agonists and antagonists and give an example of each.

▲ Ragweed. Exposure to its pollen can trigger release of histamine in your body.

20.8 Histamine and Antihistamines

Histamine is the neurotransmitter responsible for the symptoms of the allergic reaction so familiar to hay fever sufferers. It's also the chemical that causes an itchy bump when an insect bites you. In the body, histamine is produced by decarboxylation of the amino acid histidine:

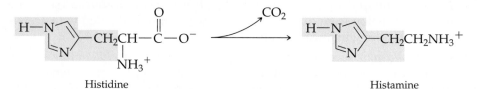

Histidine   Histamine

The *antihistamines* are a family of drugs that counteract the effect of histamine because they are histamine receptor antagonists. They competitively block the attachment of histamine to its receptors. Members of this family have in common a disubstituted ethylamine side chain, usually with two *N*-methyl groups. As illustrated by the examples below, the R' and R" groups at the other end of the molecule tend to be bulky.

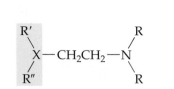

General antihistamine structure

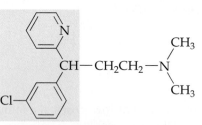

Chlorpheniramine
(an antihistamine)

Doxylamine
(an antihistamine)

Histamine also activates secretion of acid in the stomach. Development of an antagonist for this function of histamine succeeded after synthesis of about 200 different compounds with variations on the histamine structure. The result was *cimetidine*, widely publicized as a treatment for heartburn under its trade name of Tagamet.

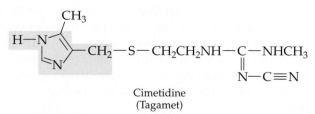

Cimetidine
(Tagamet)

Chemists have risen to a series of challenges and identified a chemical compound that may fill a major gap in our arsenal of painkillers. The final word will not be in for a few more years, as the compound works its way through tests for safety and efficacy. But the story is worth telling because it illustrates a not uncommon sequence of events in drug development.

The story began in a tropical rain forest in Ecuador more than 20 years ago. Dr. John W. Daly of the National Institutes of Health had traveled often to Central and South America to collect new species of frogs. In 1978, Daly returned to Ecuador to collect more of a poison frog (*Epipedobates tricolor*) in order to follow up on an exciting lead. Something in these frogs acted like morphine, but appeared to do so by a completely different mechanism.

Back home, he set out to find the structure of this compound by analyzing 60 mg of a mixture of frog skin extract. Step by step, with the skill needed to handle such tiny amounts, he isolated the pure compound, determined its molecular weight, broke it into two fragments, and identified the atoms in each fragment (one fragment with 6 C atoms, 10 H atoms, and 1 N atom; the other with 5 C atoms, 3 H atoms, 1 N atom, and 1 Cl atom). At this point he had only 0.5 mg of the compound left. Meanwhile, he had discovered that frogs grown in the laboratory do not produce the compound of interest. And no more frogs could be collected, because they had been put on the endangered species list.

Afraid to destroy his tiny sample without finishing the identification, Daly put the sample away and waited 13 years. By then, modern instrumentation had developed to the point where the job could be finished, and he succeeded in determining the structure of the compound he named *epibatidine*. But, of course, there wasn't enough of the compound to study its physiological effects. The next step was for someone to devise a way to synthesize the compound in the laboratory. This was accomplished by other chemists in 1996.

With an adequate supply, epibatidine's effect on living things could be studied. The bad news was that it is too toxic for use as a drug. The exciting news was that it does indeed act as a painkiller, but by a completely dif-

▲ *Epipedobates tricolor.* Identification of a chemical from the skin of this rare poison arrow frog may lead to a new class of painkilling drugs.

ferent mechanism than morphine and other opioids. Instead of acting at an opioid receptor, the compound acts at an acetylcholine receptor in the central nervous system. Nicotine, which is similar in structure to epibatidine, acts at the same receptor but is only a weak painkiller.

These discoveries revealed the possibility that a compound structurally similar to nicotine and epibatidine, but less poisonous, might be an effective painkiller but without the addictive properties and negative side effects of morphine and other opioids. Now the hunt was on. In 1998, after screening more than 500 similar compounds, the chemists at Abbot Laboratories zeroed in on a compound with the structure shown above as the best candidate for the long-sought completely new type of painkiller. Animal experiments indicate that the new compound, now known as ABT-594, does not create morphine-like addiction. The compound was soon started down the long road of drug evaluation experiments.

See Additional Problems 20.68–69 at the end of the chapter.

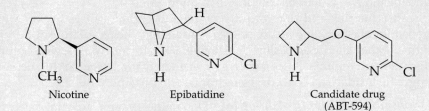

Nicotine   Epibatidine   Candidate drug (ABT-594)

20.9 Serotonin, Norepinephrine, and Dopamine

The Monoamines and Therapeutic Drugs   Serotonin, norepinephrine, and dopamine could be called the "big three" of neurotransmitters. They regularly make news as discoveries about them accumulate. Collectively, serotonin, norepinephrine, and dopamine are known as *monoamines*. (Their biochemical synthesis was shown in Figures 20.4 and 20.6.) All are active in the brain and all have been identified in various ways with mood, the experiences of fear and pleasure, mental illness, and drug addiction. Chemistry plays a central role in mental illness—that has become an inescapable conclusion.

One well-established relationship is the connection between major depression and a deficiency of serotonin, norepinephrine, and dopamine. The evidence came from the different modes of action of three families of drugs used to treat depression. Amitriptyline, phenelzine, and fluoxetine are representative of these three types of drugs. Each in its own way increases the concentration of the neurotransmitters at synapses.

Amitriptyline, a tricyclic antidepressant
(Elavil)

$CHCH_2CH_2N(CH_3)_2$

Phenelzine, an MAO inhibitor
(Nardil)

$CH_2CH_2NHNH_2$

Fluoxetine, an SSRI
(Prozac)

F_3C —⬡— $CHCH_2CH_2NHCH_3$

- Amitriptyline is representative of the *tricyclic antidepressants*, which were the first generation of these drugs. The tricyclics prevent the re-uptake of serotonin and norepinephrine from within the synapse.

- Phenelzine is a *monamine oxidase (MAO) inhibitor*, one of a group of medications that inhibit the enzyme that breaks down monoamine neurotransmitters. This inhibition of MAO allows the concentrations of monoamines at synapses to increase.

- Fluoxetine represents the newest class of antidepressants, the *selective serotonin re-uptake inhibitors (SSRI)*. They are more selective than the tricyclics because they inhibit only the re-uptake of serotonin. Fluoxetine (Prozac) has rapidly become the most widely prescribed drug for all but the most severe forms of depression. One major benefit of fluoxetine is that it does not cause the unpleasant side effects common to other antidepressants.

The relief of the symptoms of depression by these drugs is not, it should be noted, evidence that either the chemical basis of depression is fully understood or that increasing neurotransmitter concentration is the only action of these drugs. The brain still has many secrets. As one of the pharmacologists who developed fluoxetine put it, "If the human brain were simple enough for us to understand, we would be too simple to understand it."

The complex and not yet fully understood relationships between neurotransmitter activity and behavior are illustrated by the use of fluoxetine for conditions other than depression. It is being used to treat obsessive compulsive disorder and bulimia. Also, it is being considered as a treatment for alcoholism, migraine headaches, Tourette's syndrome, and obesity.

Dopamine and Drug Addiction   Dopamine plays a role in the brain in processes that control movement, emotional responses, and the experiences of pleasure and pain. It interacts with five different kinds of receptors in different parts of the brain. An oversupply of dopamine is associated with schizophrenia, while an undersupply results in the loss of fine motor control in Parkinson's disease (see Application "The Blood–Brain Barrier," Section 29.3). Dopamine also plays an important role in the brain's reward system. An ample supply of brain dopamine produces the pleasantly satisfied feeling that results from a rewarding experience—a natural high. Herein lies the role of dopamine in drug addiction. The more the dopamine receptors are stimulated, the greater the high.

Experiments have shown that cocaine blocks re-uptake of dopamine from the synapse, and amphetamines accelerate release of dopamine. Studies have linked increased brain levels of dopamine to alcohol and nicotine addiction as well. The higher-than-normal stimulation of dopamine receptors by drugs results in tolerance. In the drive to maintain constant conditions (see Application "Homeostasis"), the number of dopamine receptors decreases and the sensitivity of those that remain decreases. Brain cells thus require more and more of a drug for the same result, a condition that contributes to addiction.

Marijuana also creates an increase in dopamine levels, in the same brain areas where dopamine levels increase after administration of heroin and cocaine. The most active ingredient of the many in marijuana is tetrahydrocannabinol (THC).

Tetrahydrocannabinol (THC)

■ KEY CONCEPT PROBLEM 20.10

Identify the functional groups present in THC. Is the molecule likely to be hydrophilic or hydrophobic? Would you expect THC to build up in fatty tissues in the body, or would it be readily eliminated in the bloodstream?

■ KEY CONCEPT PROBLEM 20.11

The relationship between the structure of a molecule and its biochemical function is an essential area of study in biochemistry and the design of drugs. Based solely on what you learned in this section and Section 20.8, predict which of the following to compounds is an antihistamine and which is an antidepressant.

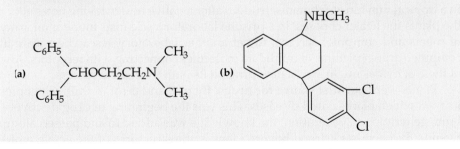

20.10 Neuropeptides and Pain Relief

Studies of morphine and other opium derivatives in the 1970s revealed that these addictive but effective pain-killing substances act via their own specific brain receptors. This raised some interesting questions: Why are there brain receptors for chemicals from a plant? Could it be that there are animal neurotransmitters that act at the same receptors?

The two pentapeptides *Met-enkephalin* and *Leu-enkephalin* (Met and Leu stand for the terminal amino acids) were discovered in the effort to answer these questions.

Met-enkephalin: Tyr-Gly-Gly-Phe-Met
Leu-enkephalin: Tyr-Gly-Gly-Phe-Leu

Both exert morphine-like suppression of pain when injected into the brains of experimental animals. The structural similarity between Met-enkephalin and morphine, highlighted below, supports the concept that both interact with the same receptors, which are located in regions of the brain and spinal cord that act in the perception of pain.

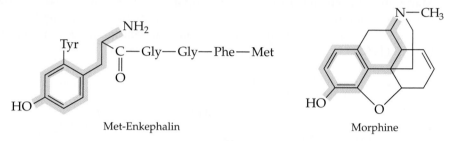

Met-Enkephalin   Morphine

Subsequently, about a dozen natural pain-killing polypeptides that act via the opiate receptors have been found. They are classified as *endorphins*. A 31-amino-acid polypeptide that ends with the same five-amino-acid sequence as Met-enkephalin is a more potent pain suppressor than morphine. Disappointingly, though, none of these compounds is the long-sought ideal, nonaddicting painkiller—all are addictive.

While there is much to be learned, enkephalins have been implicated in the runners' "high," the regulation of complex behavior states such as anger and sexual excitement, and the suppression of pain by acupuncture or during extreme stress—for example, in the competitive athlete who continues to play though injured. The term *endorphin* has entered the popular idiom to the extent that there is an endurance trial known as the Endorphin Fix Adventure Race and there is also an Endorphin Rush hot sauce.

▲ Endorphin Rush Hot Sauce.

20.11 Drug Discovery and Drug Design

In a tropical rain forest, a botanist trudges after a native healer, taking notes about the plants the healer chooses. In a pristine laboratory, scientists monitor an army of robots and computer screens. In yet another laboratory, researchers stare at computer-drawn pictures of candidate molecules connecting with receptors. Any of these activities might start a new drug on its path to medical success.

Plants were our first source for drugs. By trial and error, primitive peoples learned which plants cured diseases. This was the beginning of drug discovery. From generation to generation, the knowledge was added to and passed along. Eventually, chemists learned how to identify the structures of the active molecules and sometimes to improve upon them. This is how we got codeine for pain (from opium poppies), quinine for malaria (from fever tree), vinblastine for Hodgkins disease (from rosy periwinkle), scopolamine for motion sickness (from Jimson weed), and others. Estimates are that 25% of the prescriptions written each year in North America are for plant-derived drugs.

Today *ethnobotanists* work in remote regions of the world to learn what natives have discovered about the healing powers of plants. The botanists are pursuing drug discovery in a race against time. Forests and jungles are disappearing with the pressures of population expansion, and aging native healers are not finding apprentices to learn what they have to teach.

Probably the first *synthetic* chemicals used in medicine were diethyl ether and chloroform as anesthetics:

$$CH_3CH_2OCH_2CH_3 \quad CHCl_3$$

Diethyl ether   Chloroform

The technique of modifying a known structure to improve its biochemical activity was developed after cocaine was first used as a local anesthetic in 1884. The actual structure of cocaine was not known, but its hydrolysis products could be identified and showed that cocaine might be an ester of benzoic acid. Experiments with other benzoic acid esters in the early 1900s yielded benzocaine and procaine (Novocain), both still in use:

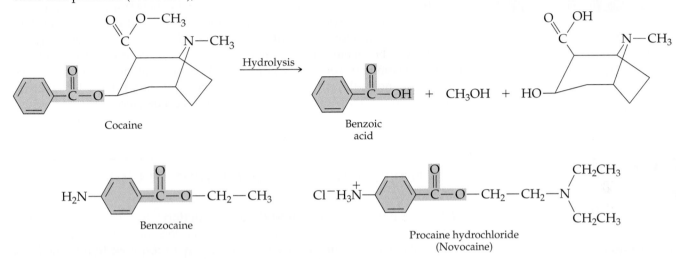

Cocaine   Benzoic acid

Benzocaine

Procaine hydrochloride
(Novocaine)

Also in the late 1800s, phenacetin was introduced as an analgesic. It had been preceded by the use of acetanilide, which was, however, soon withdrawn because of its toxicity. Derived as it was from the results of animal experiments with aniline, phenacetin was one of the first drugs designed with some knowledge of biochemistry. It remained on the market for many years until it was eventually withdrawn as evidence accumulated for its toxicity and possible carcinogenicity. Acetaminophen, introduced in 1893, is widely used today under such familiar trade names as Tylenol. It is now known to be produced in the body during metabolism of acetanilide and phenacetin.

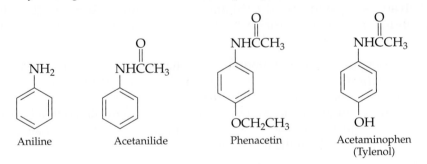

Aniline   Acetanilide   Phenacetin   Acetaminophen
(Tylenol)

Meanwhile, expanding knowledge of the structure of biochemically active molecules combined with advancing technology have opened the door to a new era. Drug *discovery* is merging with drug *design*.

One new technology, *combinatorial chemistry*, has been on the scene since 1991. It is mass production at the molecular level. Suppose it is believed that some combination of a defined set of molecular building blocks will yield an effective drug. The techniques of combinatorial chemistry allow the building blocks to react in every possible combination, not one reaction at a time but hundreds at a time. Reactions are carried out at a micro scale in tiny tubes or with molecules held down on solid supports. By combining reactants, dividing up the products, adding other reactants, and continuing this process, millions

▲ Stereo glasses provide a researcher with a three-dimensional view of a computer-generated model. Such 3-D views help scientists to visualize possible drug–receptor interactions.

of related compounds can be synthesized. Robots help with the mixing of chemicals. Computers keep track of the combinations and manage screening the products for some type of activity. Hope is high that the result will be a significant decrease in the average of 12 years and $359 million needed for development of a new drug.

In another rapidly developing technology, supercomputers and molecular graphics now allow an approach right to the heart of drug action—the drug–receptor connection. The ability to find the structure of proteins, once a tedious and lengthy activity, is advancing every day. Let's say that the complete tertiary structure of an enzyme has been found, the active site identified, and a search for an inhibitor for this enzyme is underway. The computer can consult a data base of quantitative information about drug–receptor interactions and other important properties such as hydrophobic versus hydrophilic solubilities. Once potential inhibitors are identified, pictures of such molecules entering the active site can be created on the computer screen. The pictures can be rotated and the fit examined from many angles. In this way, it is increasingly possible to design a molecule with just the right chemical and physical properties needed to connect with a biomolecule and produce a desired result.

Key Words

Agonist, p. 587

Antagonist, p. 587

Drug, p. 587

Endocrine system, p. 574

Hormone, p. 574

Neurotransmitter, p. 574

Receptor, p. 573

Second messenger, p. 577

Steroid, p. 582

Synapse, p. 584

Summary: Revisiting the Chapter Goals

1. **What are hormones, and how do they function?** *Hormones* are the chemical messengers of the *endocrine system*. Under control of the hypothalamus they are released from various locations, many in response to intermediate, regulatory hormones. Hormones travel in the bloodstream to target cells, where they connect with receptors that initiate chemical changes within cells.

2. **What is the chemical nature of hormones?** Hormones are *polypeptides, steroids,* or *amino acid derivatives.* Many are polypeptides, which range widely in size and include small molecules such as vasopressin and oxytocin, larger ones like insulin, and all of the regulatory hormones. Steroids have a distinctive four-ring structure and are classified as lipids because they are hydrophobic. All of the sex hormones are steroids. Hormones that are amino acid derivatives are synthesized from amino acids (Figures 20.5 and 20.6). Epinephrine and norepinephrine act as hormones throughout the body and also act as neurotransmitters in the brain.

3. **How does the hormone epinephrine deliver its message, and what is its major action?** Epinephrine, the fight-or-flight hormone, acts via a cell-surface receptor and a G protein that connects with an enzyme, both of which reside in the cell membrane. The enzyme adenylate cyclase transfers the message to a *second messenger,* a cyclic adenosine monophosphate (cyclic AMP), that acts within the target cell.

4. **What are neurotransmitters, and how do they function?** *Neurotransmitters* are synthesized in presynaptic neurons and stored there in vesicles from which they are released when needed. They travel across a *synaptic cleft* to *receptors* on adjacent target cells. Some act directly via their receptors; others utilize cyclic AMP or other second messengers. After their message is delivered, neurotransmitters must be quickly broken down or taken back into the presynaptic neuron so that the receptor is free to receive further messages.

5. **How does acetylcholine deliver its message, and how do drugs alter its function?** Acetylcholine is released from the vesicles of a presynaptic neuron and connects with receptors that initiate continuation of a nerve impulse

in the postsynaptic neuron. It is then broken down in the synaptic cleft by acetylcholinesterase to form choline that is returned to the presynaptic neuron where it is converted back to acetylcholine. *Agonists* such as nicotine at low doses activate acetylcholine receptors and are stimulants. *Antagonists* such as tubocurarine or atropine, which block activation of the receptors, are toxic in high doses, but at low doses can be useful as muscle relaxants.

6. **Which neurotransmitters and what kinds of drugs play roles in allergies, mental depression, drug addiction, and pain?** *Histamine*, an amino acid derivative, causes allergic symptoms. *Antihistamines* are antagonists with a general structure that resembles histamines, but with bulky groups at one end. Monoamines (serotonin, norepinephrine, and dopamine) are brain neurotransmitters whose deficiency is associated with mental depression. *Drugs* that increase their activity include *tricyclic antidepressants* (for example, amitriptyline), *monamine oxidase (MAO) inhibitors*, (for example, phenelzine), and *selective serotonin reuptake inhibitors (SSRI)* (for example, fluoxetine). An increase of dopamine activity in the brain is associated with the effects of most addictive substances. A group of neuropeptides act at opiate receptors to counteract pain.

7. **What are some of the methods used in drug discovery and design?** *Ethnobotanists* work to identify the medicinal products of plants known to native peoples. *Chemical synthesis* is used to improve on the medicinal properties of known compounds by creating similar structures. *Combinatorial chemistry* produces many related molecules for drug screening. *Computer design* is used to select the precise molecular structure to fit a given receptor.

■ Understanding Key Concepts

20.12 Diabetes occurs when there is a malfunction in the uptake of glucose from the bloodstream into the cells. Your friend's little brother was just diagnosed with type I diabetes, and she has asked you the following questions. How would you answer them?
  (a) What hormone is involved, and what class is it?
  (b) Where is the hormone released?
  (c) How is this hormone transported to the cells that need it to allow glucose to enter?
  (d) Would you expect the hormone to enter the cell to carry out its function? Explain.

20.13 In many species of animals, at the onset of pregnancy luteinizing hormone, is released; it promotes the synthesis of progesterone—a major hormone in maintaining the pregnancy.
  (a) Where is LH produced, and to what class of hormones does it belong?
  (b) Where is progesterone produced, and to what class of hormones does it belong?
  (c) Do progesterone-producing cells have LH receptors on their surface, or does LH enter the cell to carry out its function?
  (d) Does progesterone bind to a cell-surface receptor, or does it enter the cell to carry out its function? Explain.

20.14 The "rush of epinephrine" in response to danger causes the release of glucose in muscle cells so that those muscles can carry out either "fight or flight." Very small amounts of the hormone produced in the adrenal gland cause a powerful response. To get such a response, the original signal (epinephrine) must be amplified many times. At what step in the sequence of events (Section 20.3) would you predict that the signal is amplified? Explain. How might that amplification take place?

20.15 When an impulse arrives at the synapse, the synaptic vesicles open and release neurotransmitters into the cleft within a thousandth of a second. Within another ten-thousandth of a second, these molecules have diffused across the cleft and bound to receptor sites in the effector cell. In what two ways is transmission across a synapse terminated so that the neuron's signal is concluded?

20.16 Give two mechanisms by which neurotransmitters exert their effects.

20.17 What is the significance of dopamine in the addictive effects of cocaine, amphetamines, and alcohol?

▪ Additional Problems

Chemical Messengers

20.18 What is a hormone? What is the function of a hormone? How is a hormone detected by its target?

20.19 Describe what is meant by the terms chemical messenger, target tissue, and hormone receptor.

20.20 What is the difference between a hormone and a vitamin?

20.21 What is the difference between a hormone and a neurotransmitter?

20.22 How is hormone binding to its receptor more like an allosteric regulator binding to an enzyme than a substrate binding to an enzyme?

20.23 Is a hormone changed as a result of binding to a receptor? Is the receptor changed as a result of binding the hormone? What are the binding forces between hormone and receptor?

Hormones and the Endocrine System

20.24 List the three major classes of hormones.

20.25 Give two examples of each of the three major classes of hormones.

20.26 What is the purpose of the body's endocrine system?

20.27 Name as many endocrine glands as you can.

20.28 What is the structural difference between an enzyme and a hormone?

20.29 What is the relationship between enzyme specificity and tissue specificity for a hormone?

20.30 Describe in general terms how a peptide hormone works.

20.31 Describe in general terms how a steroid hormone works.

How Hormones Work: Epinephrine

20.32 In what gland is epinephrine produced and released? Under what circumstances is epinephrine released? What is its main function at the target tissues?

20.33 In order of their involvement, name the three membrane-bound proteins involved in transmitting the epinephrine message across the cell membrane. What is the "second messenger" inside the cell that results from the epinephrine message? Is the ratio of epinephrine molecules to second messenger less than 1:1, 1:1, or greater than 1:1? Explain.

20.34 What role does the second messenger play in a cell stimulated by epinephrine? What enzyme catalyzes hydrolysis of the second messenger to terminate the message? What is the product called?

20.35 Epinephrine is used clinically in the treatment of what life-threatening allergic response?

Hormones

20.36 Give an example of a polypeptide hormone. How many amino acids are in the hormone? Where is the hormone released? Where does the hormone function? What is the result of the hormone message?

20.37 Give an example of a steroid hormone. What is the structure of the hormone? Where is the hormone released? Where does the hormone function? What is the result of the hormone message?

20.38 What are the three major classes of steroid hormones? Give an example of each class.

20.39 Name the two primary male sex hormones and the three principal female sex hormones.

20.40 What characteristics in their mechanism of action does thyroxine share with the steroid hormones?

20.41 List two hormones that also function as neurotransmitters.

20.42 Identify the class of hormones to which each of the following belongs.

(a)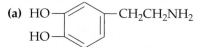

(b) Insulin

(c)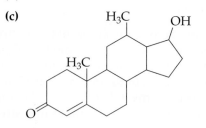

20.43 Identify the class of hormones to which each of the following belongs.

(a) Glucagon

(b)

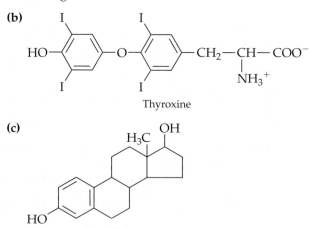

Thyroxine

(c)

Estradiol

Neurotransmitters

20.44 What is a synapse, and what role does it play in nerve transmission?

20.45 Describe in general terms how a nerve impulse is passed from one neuron to another.

20.46 List three cell types that might receive a message transmitted by a neurotransmitter.

20.47 List the three steps in chemical transmission of the impulse between a nerve cell and its target.

20.48 What are the two methods for removing the neurotransmitter once its job is done?

20.49 Write an equation for the reaction that is catalyzed by acetylcholinesterase.

20.50 Outline the six steps in cholinergic nerve transmission.

20.51 Why are enkephalins sometimes called "neurohormones"?

Chemical Messengers and Drugs

20.52 Differentiate between drugs that are agonists and those that are antagonists.

20.53 Give examples of two histamine antagonists that have very different tissue specificities and functions.

20.54 Name the "big three" monoamine neurotransmitters.

20.55 Name three families of drugs used to treat depression.

20.56 What is the impact and mode of action of cocaine on dopamine levels in the brain? What is the impact and mode of action of amphetamines on dopamine levels in the brain?

20.57 How is the tetrahydrocannabinol of marijuana similar in action to heroin and cocaine?

20.58 Why do we have brain receptors that respond to morphine and other opium derivatives from plants?

20.59 "Runner's high," sexual excitement, and other complex behaviors are believed to involve what neuropeptides?

20.60 What does an ethnobotanist do?

20.61 Combinatorial chemistry has added hundreds of drugs to the pharmaceutical market in recent years. What is the basis of the combinatorial approach to drug design? What advantages might the combinatorial approach have for the pharmaceutical industry?

20.62 In what ways are studies of the exact size and shape of biomolecules (such as enzymes, receptors, signal transducers, and so on.) leading to the development of new drugs to treat disease?

20.63 How are computers used in the development of new drugs to treat disease?

Applications

20.64 One of the responsibilities of the endocrine system is the maintenance of homeostasis in the body. Briefly, what is meant by the term homeostasis? [*Homeostasis*]

20.65 What is the goal of the measurements of clinical chemistry? [*Homeostasis*]

20.66 In animals, hormones are produced by the endocrine glands and tissues in various parts of the body. Why is it necessary for plants to synthesize the hormones in the cells where they are needed rather than in specialized cells? [*Plant Hormones*]

20.67 How does 2,4-D, a weed killer, take advantage of the function of a plant hormone? [*Plant Hormones*]

20.68 What is believed to be the mode of action of epibatidine? [*And from This Little Frog ...*]

20.69 Suggest a chemical modification to epibatidine that might be synthesized and tested as a painkiller. What is your reasoning for the suggested chemical modification? [*And from This Little Frog ...*]

General Questions and Problems

20.70 List and describe the functions of the three types of proteins involved in transmission of a hormone signal.

20.71 The cyclic AMP (second messenger) of signal transmission is very reactive and breaks down rapidly after synthesis. Why is this important to the signal transmission process?

20.72 We say that there is signal amplification in the transmission process. Explain how signal amplification occurs and what it means for transmission of the signal to the sites of cellular activity.

20.73 The phosphodiesterase that catalyzes hydrolysis of cyclic AMP is inhibited by caffeine. What overall effect would caffeine have on a signal that is mediated by cyclic AMP?

20.74 Compare the structures of the sex hormones testosterone and progesterone. What portions of the structures are the same? Where do they differ?

20.75 When you compare the structures of ethynyl estradiol to norethindrone, where do they differ? Where is ethynyl estradiol similar to estradiol? Where is norethindrone similar to progesterone?

20.76 Anandamides have been isolated from brain tissues and appear to be the natural ligand for the receptor that also binds tetrahydrocannabinol. Anandamides have also been discovered in chocolate and cocoa powder. How might the craving for chocolate be explained?

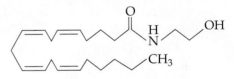

An anandamide structure

20.77 Identify the structural changes that occur in the first two steps in the conversion of tyrosine to epinephrine (Figure 20.4). To what main classes and subclasses of enzymes do the enzymes that catalyze these reactions belong?

20.78 Look at the structures of male and female sex hormones at the beginning of Section 20.5. Identify the type of functional-group change that can interconvert testosterone and androsterone, or estradiol and estrone. To what class of chemical reaction does this change belong?

21

The Generation of Biochemical Energy

One way to use biochemical energy.

All organisms must draw energy from their surroundings to stay alive. In animals, the energy comes from food and is released in the exquisitely interconnected reaction pathways of metabolism. We are powered by the oxidation of biomolecules made mainly of carbon, hydrogen, and oxygen. The end products are carbon dioxide, water, and energy.

$$\text{C, H, O (Food molecules)} + O_2 \longrightarrow CO_2 + H_2O + \text{Energy}$$

The principal food molecules— lipids, proteins, and carbohydrates— differ in structure and are broken down by distinctive pathways that are examined in later chapters. For the present, we're going to concentrate on the final, common pathways by which energy is released from all types of food molecules.

The major questions to be answered include the following:

1. **What is the source of our energy, and what is its fate in the body?**
   The goal: Be able to provide an overview of the sources of our energy and how we use it, identify the cellular location of energy generation, and explain the significance of exergonic and endergonic reactions in metabolism.
2. **How are the reactions that break down food molecules organized?**
   The goal: Be able to list the stages in catabolism and describe the role of each.
3. **What are the major strategies of metabolism?**
   The goal: Be able to explain and give examples of the roles of ATP, coupled reactions, and oxidized and reduced coenzymes in metabolic pathways.
4. **What is the citric acid cycle?**
   The goal: Be able to describe what happens in the citric acid cycle and explain its role in energy production.
5. **How is ATP generated in the final stage of catabolism?**
   The goal: Be able to describe in general the electron-transport chain, oxidative phosphorylation, and how they are coupled.
6. **What are the harmful by-products produced from oxygen, and what protects against them?**
   The goal: Be able to identify the highly reactive oxygen-containing products formed during metabolism and the enzymes and vitamins that counteract them.

CONCEPTS TO REVIEW

Oxidation-reduction reactions (Sections 6.12, 6.13)
Energy in chemical reactions (Sections 7.1– 7.3)
Enzymes (Sections 19.1, 19.4)

21.1 Energy and Life

Living things must do mechanical work— microorganisms engulf food, plants bend toward the sun, humans walk about. All organisms must also do the chemical work of synthesizing biomolecules needed for energy storage, growth, repair, and replacement. In addition, cells need energy for the work of moving molecules and ions across cell membranes. In humans, it is the energy released from food that allows these various kinds of work to be done.

Energy can be converted from one form to another but can be neither created nor destroyed. (⚬⚬⚬, Section 7.1) Ultimately, the energy used by all but a very few living things on earth comes from the sun (Figure 21.1). Plants convert

▼ **FIGURE 21.1   The flow of energy through the biosphere.**
Energy from the sun is ultimately stored in chemical bonds, used for work, used to maintain body temperature, or lost as heat.

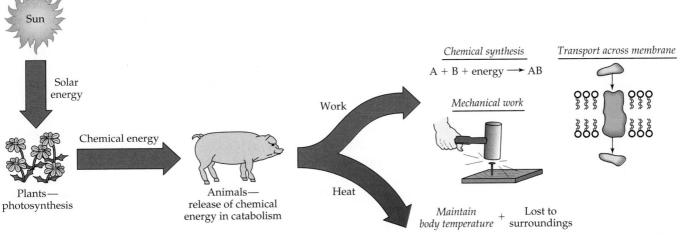

Chemical synthesis
A + B + energy ⟶ AB

Transport across membrane

Work

Mechanical work

Solar energy

Chemical energy

Heat

Plants— photosynthesis

Animals— release of chemical energy in catabolism

Maintain body temperature + Lost to surroundings

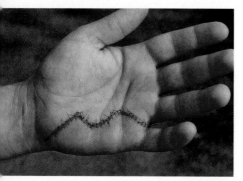

▲ Energy at work in the body. The biomolecules needed to heal this wound will be synthesized using energy from the catabolism of food molecules.

sunlight to potential energy stored mainly in the chemical bonds of carbohydrates. Plant-eating animals then utilize this energy, some of it for immediate needs and the rest to be stored for future needs, mainly in the chemical bonds of fats. Other animals, including humans, are able to eat plants or animals and use the chemical energy these organisms have stored.

Our bodies can't simply produce energy by burning up a steak all at once, however, because the large quantity of heat released would be harmful to us. Furthermore, it's difficult to capture energy for storage once it has been converted to heat. What we need is energy that can be stored and then released in the right amounts when and where it is needed, whether we're running away from an angry dog, studying for an exam, or sleeping peacefully. We therefore have some specific requirements for energy:

- Energy must be released from food gradually.
- Energy must be stored in readily accessible forms.
- The release of energy from storage must be finely controlled so that it is available exactly when and where it is needed.
- Just enough energy must be released as heat to maintain constant body temperature.
- Energy in a form other than heat must be available to drive chemical reactions that aren't favorable at body temperatures.

This chapter looks at some of the ways these requirements for energy management are met. We begin by reviewing basic concepts about energy. Then we take an overview of *metabolism* and the strategies it relies on. Next, we look at the *citric acid cycle* and *oxidative phosphorylation*, which together are the common pathway for the production of energy from all sources and for all needs.

21.2 Energy and Biochemical Reactions

Chemical reactions either release energy as they proceed or must absorb energy in order to proceed. Whether a reaction is favorable and can proceed on its own depends on the release or absorption of energy as heat (ΔH, the enthalpy change), together with the increase or decrease in disorder (ΔS, the entropy change) caused by the reaction. The net effect of these changes is given by the free-energy change of a reaction: $\Delta G = \Delta H - T\Delta S$. (∞, p. 171)

Reactions in living organisms are no different from reactions in a chemistry laboratory. Both follow the same laws, and both have the same energy requirements. Spontaneous reactions— that is, those that are *favorable* in the forward direction— release free energy and the energy released is available to do work. Such reactions, described as *exergonic*, are the source of our biochemical energy. (Remember the difference between the terms *exothermic* and *exergonic*. "Exergonic" applies to the release of free energy, represented by a negative ΔG. "Exothermic" applies only to the release of heat, represented by a negative value for the heat of reaction, ΔH. (∞, p. 166)

As shown by the energy diagram in Figure 21.2a, the products of a favorable, exergonic reaction are farther *downhill* on the energy scale than the reactants. That is, the products are more stable than the reactants, and as a result the free-energy change (ΔG) has a negative value. Oxidation reactions, for example, are usually downhill reactions that release energy. Oxidation of glucose, the principal source of energy for animals, produces 686 kcal of free energy per mole of glucose:

$$C_6H_{12}O_6 + 6\,O_2 \longrightarrow 6\,CO_2 + 6\,H_2O \qquad \Delta G = -686\ \text{kcal/mol}$$

The greater the amount of free energy released, the further a reaction proceeds toward product formation before reaching equilibrium.

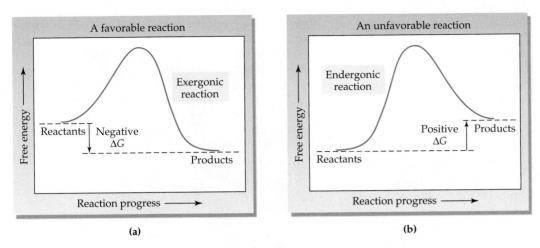

(a) A favorable reaction

(b) An unfavorable reaction

Reactions in which the products are higher in energy than the reactants can also take place, but such *unfavorable* reactions can't occur without the input of energy from an external source. In other words, energy has to be added to the reactants for an energetically *uphill* change to occur (Figure 21.2b). You might think of these as reactions that have to be *pushed* up the hill. Such reactions are described as *endergonic*. The larger the positive free-energy change, the greater the amount of energy that must be added to convert the reactants to products. (Remember the difference here also: "Endergonic" applies to reactions that require an input of free energy and have a positive value of ΔG. "Endothermic" refers to reactions that absorb heat from their surroundings and have a positive value for the heat of reaction, ΔH.)

Like the heat of reaction, the free-energy change switches sign for the reverse of a reaction, but the value doesn't change. Photosynthesis, the process whereby plants convert CO_2 and H_2O to glucose plus O_2, is the reverse of the oxidation of glucose. Its ΔG is therefore positive and of the same numerical value as that for the oxidation of glucose (see the Application "Plants and Photosynthesis," p. 623). The sun provides the necessary external energy for photosynthesis (686 kcal per mole of glucose formed).

$$6\,CO_2 + 6\,H_2O \underset{\text{Oxidation}}{\overset{\text{Photosynthesis}}{\rightleftharpoons}} C_6H_{12}O_6 + 6\,O_2$$

$\Delta G = +686$ kcal/mol (endergonic, energy required)

$\Delta G = -686$ kcal/mol (exergonic, energy released)

Living systems make constant use of this principle in the series of chemical reactions we know as the biochemical *pathways*. Energy is stored in the products of an overall endergonic reaction pathway. This stored energy is released as needed in an overall exergonic reaction pathway that regenerates the original reactants. It isn't necessary that every reaction in the pathways between the reactants and products be the same, so long as the pathways connect the same reactants and products

▲ **FIGURE 21.2 Energy diagrams for favorable and unfavorable reactions.** (a) In the favorable reaction, the products have less energy than the reactants. (b) In the unfavorable reaction, the products have more energy than the reactants.

▲ Our source of energy for photosynthesis.

■ PROBLEM 21.1

Each of the following reactions occurs in the citric acid cycle, an energy-producing sequence of reactions that we'll discuss later in this chapter. Which of the reactions listed is (are) exergonic? Which is (are) endergonic? Which will release the most energy? Write the complete equation for the reverse of reaction (c). (Recall that organic acids are usually referred to in biochemistry with the *-ate* ending because they exist as anions in body fluids.)

(a) Acetyl coenzyme A + Oxaloacetate + $H_2O \longrightarrow$ Citrate + Coenzyme A
$\Delta G = -9$ kcal/mol

(b) Citrate $\longrightarrow$ Isocitrate $\Delta G = +3$ kcal/mol

(c) Fumarate + $H_2O \longrightarrow$ L-Malate $\Delta G = -0.9$ kcal/mol

Application Life Without Sunlight

Before we had the equipment to descend deep into the ocean, no one imagined that life existed there. What could provide the food and energy? Textbooks firmly stated that all life depends on sunlight.

Not true. In 1977, hydrothermal vents—openings spewing water heated to 400°C deep within the earth—were found on the ocean floor. The hydrothermal vents were dubbed "black smokers" because the water was black with mineral sulfides precipitating from the hot, acidic water as it exited the vents. At 2200 meters below the ocean surface, there is no chance for the penetration of energy from sunlight. Therefore, the discovery of thriving clusters of tube worms, giant clams, mussels, and other creatures surrounding the black smokers was a great surprise.

Distinctive types of bacteria form the basis for the web of life in these locations. What replaces sunlight as their source of energy? The hot water is rich in dissolved inorganic substances that are reducing agents and therefore electron donors. Life-supporting energy is set free by their oxidation. Hydrogen sulfide, for example, is abundant in the hot seawater, which has passed through sulfur-bearing mineral deposits on its way to the surface. This is the same gas produced during anaerobic decomposition of organic matter in a swamp; it is also the gas that gives the awful odor to rotten eggs. As the hydrogen sulfide is converted to sulfate ions, the electrons set free in the oxidation move through an electron-transport chain that makes ATP formation possible.

Carbon dioxide dissolved in the seawater is the raw material used by the bacteria to make their own essential carbon-containing biomolecules. Experiments have shown that the tube worms, giant clams, and other creatures surrounding the black smokers do not eat the bacteria. Rather, the bacteria colonize their gastrointestinal tracts, where the waste products and dead bodies of the bacteria are the carbon source for biosynthesis by their hosts.

An opportunity to observe the colonization of a hot deep-ocean environment came in 1991 when scientists

▲ Tube worms at a hydrothermal vent in the Galapagos rift.

discovered a volcano erupting underneath the ocean. Initially, all life in the vicinity was wiped out, yet soon afterward, the area was thriving with bacteria. This discovery and others have raised some intriguing questions. The same black smoker bacteria have been found in the vicinity of the Mount St. Helens volcanic eruption, and hydrothermal vents with their communities of living things have been found in the fresh waters of the deepest lake on earth, Lake Baikal in Russia. Could it be that a thriving population of bacteria has been living in the hot interior of the earth ever since it formed? Were these anaerobic bacteria Earth's first inhabitants, and could they exist beneath the surface of other planets?

See Additional Problem 21.71 at the end of the chapter.

■ **KEY CONCEPT PROBLEM 21.2**

In a cell, sugar can be oxidized via metabolic pathways. Alternatively, you could burn sugar in the laboratory. Which of these methods would consume or produce more energy?

■ **KEY CONCEPT PROBLEM 21.3**

The overall equation in this section shows the cycle between photosynthesis and oxidation. Pathways operating in opposite directions cannot be exergonic in both directions:

(a) Which of the two pathways in this cycle is exergonic and which is endergonic?

(b) Where does the energy for the endergonic pathway come from?

21.3 Cells and Their Structure

Before we proceed with our overview of metabolism, it's important to see where the energy-generating reactions take place within the cells of living organisms. There are two main categories of cells: *prokaryotic cells*, usually found in single-celled organisms including bacteria and blue-green algae, and *eukaryotic cells*, found in some single-celled organisms and all plants and animals.

Eukaryotic cells are about 1000 times larger than bacterial cells, have a membrane-enclosed nucleus containing their DNA, and include several other kinds of internal structures known as *organelles*— small, functional units that perform specialized tasks. A diagram of a generalized eukaryotic cell is shown in Figure 21.3, along with a description of the functions of its major parts. Everything between the cell membrane and the nuclear membrane in a eukaryotic cell, including the various organelles, is referred to as the **cytoplasm**. The organelles are surrounded by the fluid part of the cytoplasm, the **cytosol**, which contains electrolytes, nutrients, and many enzymes, all in aqueous solution.

Cytoplasm   The region between the cell membrane and the nuclear membrane in a eukaryotic cell.

Cytosol   The fluid part of the cytoplasm surrounding the organelles within a cell.

◀ **FIGURE 21.3   A generalized eukaryotic cell.**   The table below lists the functions of the cell components most important for metabolism.

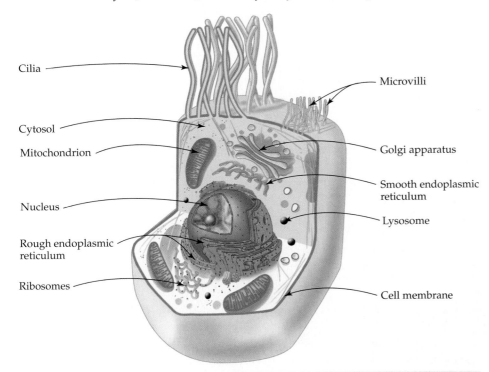

Cell Component	Principal Function
Cilia	Movement of materials; for example, mucus in lungs (not present in all cells)
Golgi apparatus	Synthesis and packaging of secretions and cell membrane
Mitochondrion	Synthesis of ATP from ADP
Rough endoplasmic reticulum	Protein synthesis and transport
Nucleus	Replication of DNA, which carries genetic information and governs protein synthesis
Ribosome	Protein synthesis
Microvilli	Absorption of extracellular substances; for example, in digestive tract (not present in all cells)
Cytosol	Intracellular fluid; contains dissolved proteins and nutrients
Lysosome	Removal of pathogens or damaged organelles
Smooth endoplasmic reticulum	Lipid and carbohydrate synthesis
Cell membrane	Composed of lipids plus proteins that govern entry and exit from cell and deliver signals to interior of cell

Mitochondrion (plural, **mitochondria**) An egg-shaped organelle where small molecules are broken down to provide the energy for an organism.

Mitochondrial matrix The space surrounded by the inner membrane of a mitochondrion.

Adenosine triphosphate (ATP) The principal energy-carrying molecule; removal of a phosphoryl group to give ADP releases free energy.

The **mitochondria** (singular, **mitochondrion**), often called the cell's "power plants," are the most important of the organelles for energy production. It is in the mitochondria that about 90% of the body's energy-carrying molecule, ATP, is produced.

A mitochondrion is a roughly egg-shaped structure composed of a smooth outer membrane and a folded inner membrane (Figure 21.4). The space enclosed by the inner membrane is the **mitochondrial matrix**. It is within the matrix that the citric acid cycle (Section 21.8) and the production of most of the body's **adenosine triphosphate (ATP)** take place. The coenzymes and proteins that manage the transfer of energy to the chemical bonds of ATP (Section 21.9) are embedded in the inner membrane of the mitochondrion.

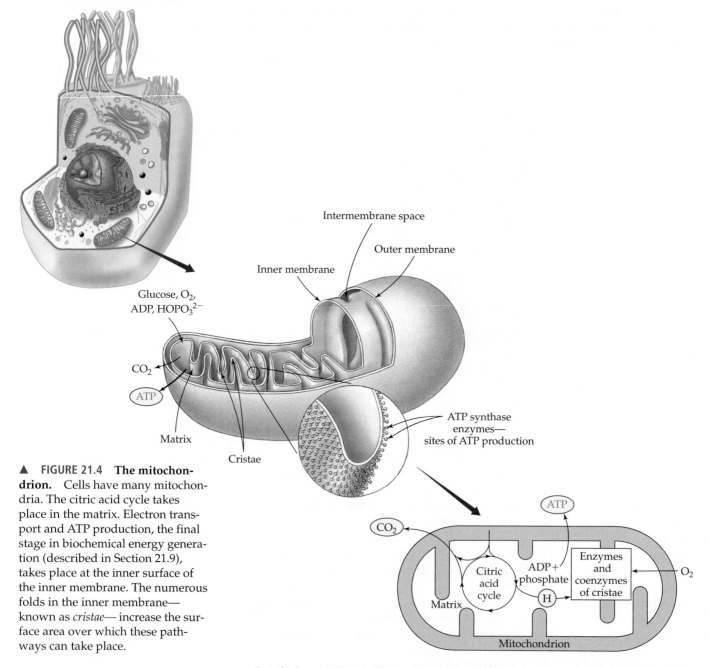

▲ **FIGURE 21.4 The mitochondrion.** Cells have many mitochondria. The citric acid cycle takes place in the matrix. Electron transport and ATP production, the final stage in biochemical energy generation (described in Section 21.9), takes place at the inner surface of the inner membrane. The numerous folds in the inner membrane—known as *cristae*—increase the surface area over which these pathways can take place.

It is believed that millions of years ago mitochondria were free-living bacteria that became trapped within single-celled plants and animals. As evidence for this, consider that mitochondria contain their own DNA, can synthesize some of their own proteins, and can multiply without outside assistance. The relationship of mitochondria to their host cells became a symbiotic one—the mitochondria produced energy needed by the eukaryotic cells. Thus, the mitochondria remained within the cells throughout evolution. The number of mitochondria is greatest in eye, brain, heart, and muscle cells, where the need for energy is greatest. The ability of mitochondria to reproduce is called upon in athletes who put heavy energy demands on their bodies—they develop an increased number of mitochondria to aid in energy production.

Interestingly, all mitochondria in our bodies develop from those in the egg that was fertilized, meaning that only our mothers contribute our inherited mitochondrial DNA. This condition is proving valuable in anthropological studies. The more similar the mitochondrial DNA found in ancient tissue, the more closely in time the beings from which the tissue came were alive.

21.4 An Overview of Metabolism and Energy Production

Together, all of the chemical reactions that take place in an organism constitute its metabolism. Most of these reactions occur in the reaction sequences of *metabolic pathways*. Such pathways may be linear (that is, the product of one reaction serves as the starting material for the next); cyclic (a series of reactions regenerates one of the first reactants); or spiral (the same set of enzymes progressively builds up or breaks down a molecule).

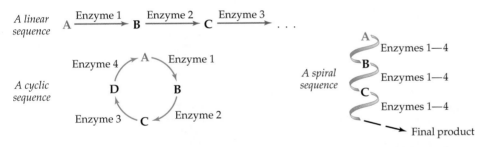

Those pathways that break molecules apart are known collectively as **catabolism**, whereas those that put building blocks back together to assemble larger molecules are known collectively as **anabolism**. The purpose of catabolism is to release energy from food, and the purpose of anabolism is to synthesize new biomolecules, including those that store energy.

Catabolism   Metabolic reaction pathways that break down food molecules and release biochemical energy.

Anabolism   Metabolic reactions that build larger biological molecules from smaller pieces.

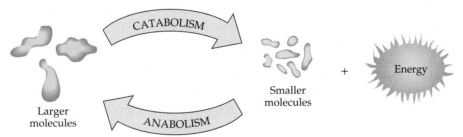

The overall picture of digestion, catabolism, and energy production is simple: Eating provides fuel, breathing provides oxygen, and our bodies oxidize the fuel to extract energy. The process can be roughly divided into the four stages described below and shown in Figure 21.5.

Stage 1: Digestion.   Enzymes in saliva, the stomach, and the small intestine convert the large molecules of lipids, carbohydrates, and proteins to smaller molecules. Carbohydrates are broken down to glucose and other sugars, proteins are broken down to amino acids, and triacylglycerols, the lipids commonly known as fats and oils, are broken down to glycerol plus long-chain

FIGURE 21.5 Pathways for the ▶ **digestion of food and the production of biochemical energy.** This diagram summarizes pathways covered in this chapter (the citric acid cycle and electron transport), and also the pathways discussed in Chapter 23 for carbohydrate metabolism, in Chapter 25 for lipid metabolism, and in Chapter 28 for protein metabolism.

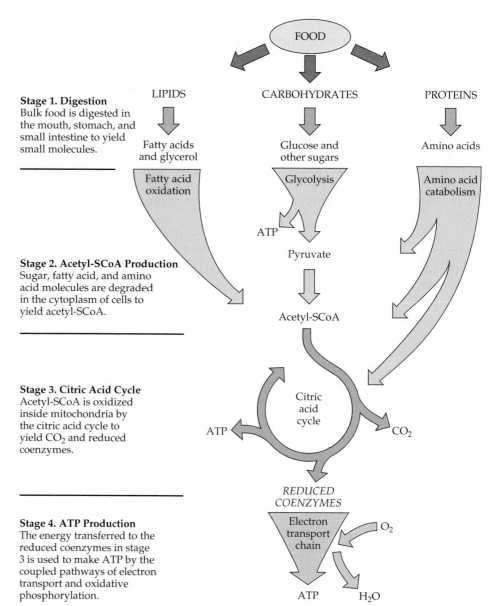

Stage 1. Digestion
Bulk food is digested in the mouth, stomach, and small intestine to yield small molecules.

Stage 2. Acetyl-SCoA Production
Sugar, fatty acid, and amino acid molecules are degraded in the cytoplasm of cells to yield acetyl-SCoA.

Stage 3. Citric Acid Cycle
Acetyl-SCoA is oxidized inside mitochondria by the citric acid cycle to yield CO_2 and reduced coenzymes.

Stage 4. ATP Production
The energy transferred to the reduced coenzymes in stage 3 is used to make ATP by the coupled pathways of electron transport and oxidative phosphorylation.

Acetyl-S-coenzyme A (acetyl-SCoA) Acetyl-substituted coenzyme A— the common intermediate that carries acetyl groups into the citric acid cycle.

Acetyl-S-coenzyme A

carboxylic acids, the fatty acids. These smaller molecules are transferred into the blood for transport to cells throughout the body.

Stage 2: Acetyl-S-coenzyme A production. The small molecules from digestion follow separate pathways that move their carbon atoms into two-carbon acetyl groups. The acetyl groups are attached to coenzyme A (⚬⚬⚬, Figure 19.9, p. 563) by a bond between the sulfur atom of the thiol (—SH) group at the end of the coenzyme A molecule and the carbonyl carbon atom of the acetyl group:

Attachment of acetyl group to coenzyme A

$$CH_3 - \underset{\underset{\textstyle O}{\|}}{C} - S - [\text{Coenzyme A}]$$

The resultant compound, **acetyl-S-coenzyme A**, which we abbreviate **acetyl-SCoA**, is an intermediate in the breakdown of *all* classes of food molecules. It carries the acetyl groups into the common pathways of catabolism— stage 3, the citric acid cycle (Section 21.8) and stage 4, electron transport and ATP production (Section 21.9).

Stage 3: Citric acid cycle. Within mitochondria, the acetyl-group carbon atoms are oxidized to the carbon dioxide that we exhale. Most of the energy released in the oxidation leaves the citric acid cycle in the chemical bonds of reduced coenzymes (NADH, FADH$_2$). Some energy also leaves the cycle stored in the chemical bonds of *adenosine triphosphate (ATP)* or a related triphosphate.

Stage 4: ATP production. Electrons from the reduced coenzymes are passed from molecule to molecule down an electron-transport chain. Along the way, their energy is harnessed to produce more ATP. At the end of the process these electrons, along with hydrogen ions from the reduced coenzymes, combine with oxygen that we breathe to produce water. Thus, the reduced coenzymes are in effect oxidized by atmospheric oxygen, while the energy that they carried is stored in the chemical bonds of ATP molecules.

∞∞∞ LOOKING AHEAD

Digestion and conversion of food molecules to acetyl-SCoA, stages 1 and 2 in Figure 21.5, occur by different metabolic pathways for carbohydrates, lipids, and proteins. Each of these pathways is discussed separately in later chapters: carbohydrate metabolism in Chapter 23, lipid metabolism in Chapter 25, and protein metabolism in Chapter 28.   ∞∞∞

■ **PROBLEM 21.4**

(a) Identify in Figure 21.5 the stages in the complete pathway for the conversion of the energy from carbohydrates to energy stored in ATP molecules.

(b) Identify in Figure 21.5 the three places at which the products of amino acid catabolism can join the central metabolism pathway.

21.5 Strategies of Metabolism: ATP and Energy Transfer

We have described ATP as the body's energy-transporting molecule. What exactly does that mean? Consider that the molecule has three —PO_3^- groups.

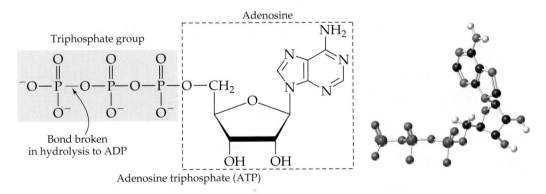

Adenosine triphosphate (ATP)

Removal of one of the —PO_3^{2-} groups from ATP by hydrolysis gives adenosine diphosphate (ADP). The ATP → ADP reaction is exergonic; it releases chemical energy that was held in the bond to the —PO_3^- group.

$$ATP + H_2O \longrightarrow ADP + HOPO_3^{2-} + H^+ \qquad \Delta G = -7.3 \text{ kcal/mol}$$

The reverse of ATP hydrolysis—a phosphorylation reaction—is, of course, endergonic (Section 21.2).

$$ADP + HOPO_3^{2-} + H^+ \longrightarrow ATP + H_2O \qquad \Delta G = +7.3 \text{ kcal/mol}$$

(In equations for biochemical reactions, we represent ATP and other energy-carrying molecules in red and their lower-energy equivalent molecules in blue.)

ATP is an energy transporter because its production from ADP requires an input of energy that is then released wherever the reverse reaction occurs. Biochemical energy is gathered from exergonic reactions that produce ATP. The ATP then travels to where energy is needed, and ATP hydrolysis releases the energy for whatever energy-requiring work must take place. *Biochemical energy production, transport, and use all depend upon the ATP $\rightleftharpoons$ ADP interconversion.*

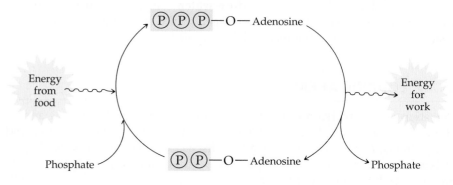

The hydrolysis of ATP to give ADP and its reverse, the phosphorylation of ADP, are reactions perfectly suited to their role in metabolism for two major reasons. One reason is the slow rate of ATP hydrolysis in the absence of a catalyst, so that the stored energy is released only in the presence of the appropriate enzymes.

The second reason is the intermediate value of the free energy of hydrolysis of ATP, as illustrated in Table 21.1. Since the primary metabolic function of ATP is to transport energy, it is often referred to as a "high-energy" molecule or as containing "high-energy" phosphorus–oxygen bonds. These terms are misleading because they promote the idea that ATP is somehow different from other compounds. The terms mean only that ATP is reactive and that a useful amount of energy is released when a phosphoryl group is removed from it by hydrolysis.

TABLE 21.1 Free Energies of Hydrolysis of Some Phosphates

$$R-O-\overset{\overset{\displaystyle O}{\|}}{\underset{\underset{\displaystyle O^-}{|}}{P}}-O^- \ + \ H_2O \ \rightleftharpoons \ ROH \ + \ HO-\overset{\overset{\displaystyle O}{\|}}{\underset{\underset{\displaystyle O^-}{|}}{P}}-O^-$$

Compound Name	Function	ΔG (kcal/mol)
Phosphoenol pyruvate	Final intermediate in conversion of glucose to pyruvate (glycolysis)— stage 2, Figure 21.5	−14.8
1, 3-Bisphosphoglycerate	Another intermediate in glycolysis	−11.8
Creatine phosphate	Energy storage in muscle cells	−10.3
ATP ($\longrightarrow$ ADP)	Principal energy carrier	−7.3
Glucose 1-phosphate	First intermediate in breakdown of carbohydrates stored as starch or glycogen	−5.0
Glucose 6-phosphate	First intermediate in glycolysis	−3.3
Fructose 6-phosphate	Second intermediate in glycolysis	−3.3

In fact, if removal of a phosphoryl group from ATP released *unusually* large amounts of energy, other reactions wouldn't be able to provide enough energy to convert ADP back to ATP. ATP is a convenient energy carrier in metabolism *because* its free energy of hydrolysis has an intermediate value. For this reason, the phosphorylation of ADP can be driven by coupling with a more exergonic reaction, as illustrated in the next section.

■ **PROBLEM 21.5**

Acetyl phosphate, whose structure is given below, is another compound with a relatively high free energy of hydrolysis.

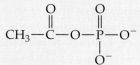

Using structural formulas, write the equation for the hydrolysis of this phosphate.

■ **PROBLEM 21.6**

A common metabolic strategy is the lack of reactivity—that is, the slowness to react—of compounds whose breakdown is exergonic. For example, hydrolysis of ATP to ADP or AMP is exergonic but does not take place without an appropriate enzyme present. Why would the cell use this metabolic strategy?

21.6 Strategies of Metabolism: Metabolic Pathways and Coupled Reactions

Now that you're acquainted with ATP, we'll explore how stored chemical energy is gradually released and how it can be used to drive endergonic (uphill) reactions. We've noted before that our bodies can't burn up a steak all at once. As shown in Figure 21.2a, however, the energy difference between a reactant (the steak) and the ultimate products of its catabolism (mainly carbon dioxide and water) is a fixed quantity. The same amount of energy is released no matter what pathway is taken between reactants and products. The metabolic pathways of catabolism take advantage of this fact by releasing energy bit by bit in a series of reactions, somewhat like the stepwise release of potential energy as water flows down an elaborate waterfall (Figure 21.6).

The overall reaction and the overall free-energy change for any series of reactions can be found by summing up the equations and the free-energy changes for the individual steps. For example, glucose is converted to pyruvate via the 10 reactions of the glycolysis pathway (part of stage 2, Figure 21.5). The overall free-energy change for glycolysis is about −8 kcal, showing that the pathway is exergonic— that is, downhill and favorable. The reactions of all metabolic pathways *add up* to favorable processes with negative free-energy changes.

Unlike the waterfall, however, not every individual step in every metabolic pathway is downhill. The metabolic strategy for dealing with what would be an energetically unfavorable reaction is to *couple* it with an energetically favorable reaction so that the overall energy change for the two reactions is favorable. For example, consider the reaction of glucose with hydrogen phosphate ion ($HOPO_3^{2-}$) to yield glucose 6-phosphate plus water, for which $\Delta G = +3.3$ kcal/mol. The reaction is unfavorable because the two products are 3.3 kcal higher in energy than the starting materials. This phosphorylation of glucose is, however, the essential first step toward all metabolic use of glucose.

▲ **FIGURE 21.6 Stepwise release of potential energy.** No matter what the pathway from the top to the bottom of this waterfall, the amount of potential energy released as the water falls from the top to the very bottom is the same.

To accomplish this reaction, it is coupled with the endergonic hydrolysis of ATP to give ADP:

(*Unfavorable*) $\quad$ Glucose + $HOPO_3^{2-}$ ⟶ Glucose 6-phosphate + H_2O $\qquad \Delta G = +3.3$ kcal/mol

(*Favorable*) $\quad$ ATP + H_2O ⟶ ADP + $HOPO_3^{2-}$ + H^+ $\qquad \Delta G = -7.3$ kcal/mol

(*Favorable*) $\quad$ Glucose + ATP ⟶ Glucose 6-phosphate + ADP $\qquad \Delta G = -4.0$ kcal/mol

The net energy change for these two coupled reactions is favorable: 4.0 kcal of free energy is released for each mole of glucose that is phosphorylated. Only by such coupling can the energy stored in one chemical compound be transferred to other compounds. Any excess energy is released as heat and contributes to maintaining body temperature (Figure 21.7).

Although we've written these reactions separately to show how their energies combine, coupled reactions don't take place separately. The net change occurs all at once as represented by the overall equation. The phosphoryl group is transferred directly from ATP to glucose without the intermediate formation of $HOPO_3^{2-}$. (Also, under physiological conditions, a reaction may be more or less exergonic than in the examples given here. We have stated the free-energy values at standard conditions.)

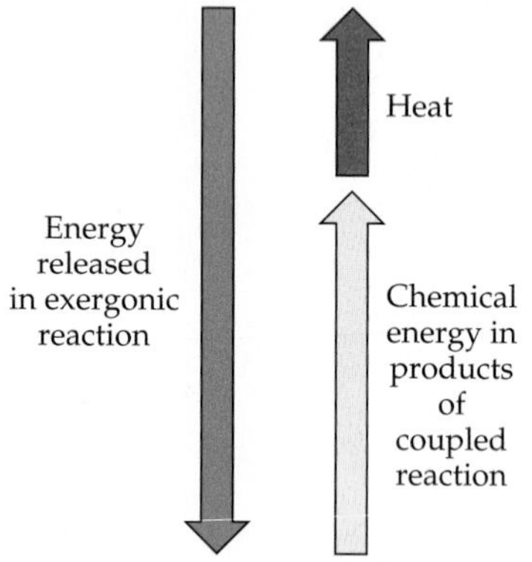

What about the endergonic synthesis of ATP from ADP, which has $\Delta G = +7.3$ kcal/mol? The same principle of coupling is put to use. For this endergonic reaction to occur, it must be coupled with a reaction that releases *more* than 7.3 kcal/mol. In a different step of glycolysis, for example, the formation of ATP is coupled with the hydrolysis of phosphoenolpyruvate, a phosphate of higher energy than ATP (Table 21.1). Here, the overall reaction is transfer of a phosphoryl group from phosphoenolpyruvate to ADP.

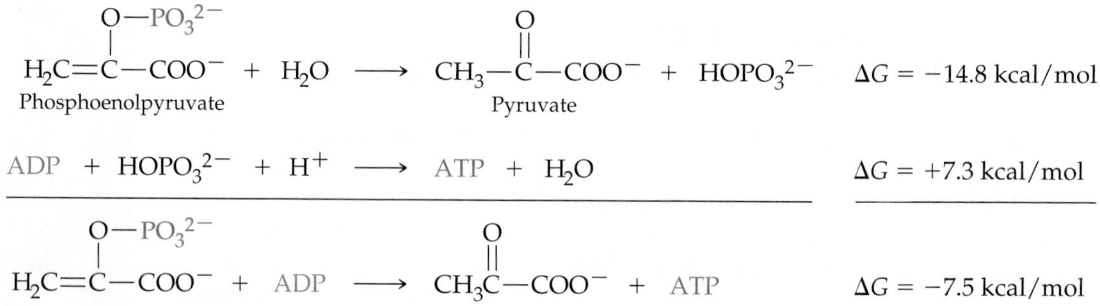

▲ **FIGURE 21.7 Energy exchange in coupled reactions.** The energy provided by an exergonic reaction is either released as heat or stored as chemical potential energy in the bonds of products of the coupled endergonic reaction.

$$\begin{array}{c}
\underset{\text{Phosphoenolpyruvate}}{H_2C{=}\overset{\displaystyle O{-}PO_3^{2-}}{\underset{|}{C}}{-}COO^-} + H_2O \longrightarrow \underset{\text{Pyruvate}}{CH_3{-}\overset{\displaystyle O}{\overset{||}{C}}{-}COO^-} + HOPO_3^{2-} \qquad \Delta G = -14.8 \text{ kcal/mol}
\end{array}$$

$$ADP + HOPO_3^{2-} + H^+ \longrightarrow ATP + H_2O \qquad\qquad \Delta G = +7.3 \text{ kcal/mol}$$

$$H_2C{=}\overset{\displaystyle O{-}PO_3^{2-}}{\underset{|}{C}}{-}COO^- + ADP \longrightarrow CH_3\overset{\displaystyle O}{\overset{||}{C}}{-}COO^- + ATP \qquad \Delta G = -7.5 \text{ kcal/mol}$$

Remember that in equations representing coupled reactions, a curved arrow often connects the reactants and products in one of the two chemical changes. (∞, p. 559) For example, the reaction of phosphenolpyruvate illustrated above can be written as follows:

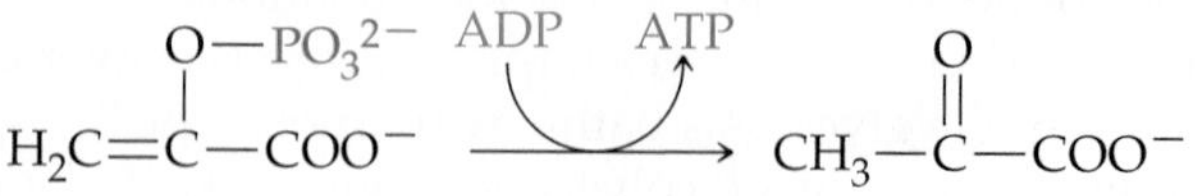

$$H_2C{=}\overset{\displaystyle O{-}PO_3^{2-}}{\underset{|}{C}}{-}COO^- \xrightarrow{\quad\overset{\displaystyle ADP \quad ATP}{\curvearrowright}\quad} CH_3{-}\overset{\displaystyle O}{\overset{||}{C}}{-}COO^-$$

■ **PROBLEM 21.7**

One of the steps in lipid metabolism is the reaction of glycerol (1,2,3-propanetriol, $HOCH_2CH(OH)CH_2OH$), with ATP to yield glycerol 1-phosphate. Write the equation for this reaction using the curved arrow symbolism.

■ **PROBLEM 21.8**

Why must a metabolic pathway that synthesizes a given molecule occur by a different series of reactions than a pathway that breaks down the same molecule?

Application Basal Metabolism

The minimum amount of energy expenditure required per unit of time to stay alive—to breathe, maintain body temperature, circulate blood, and keep all body systems functioning—is referred to as the *basal metabolic rate*. Ideally, it is measured in a person who is awake, is lying down at a comfortable temperature, has fasted and avoided strenuous exercise for 12 hours, and is not under the influence of any medications. The basal metabolic rate can be measured by monitoring respiration and finding the rate of oxygen consumption, which is proportional to the energy used.

An *average* basal metabolic rate is 70 kcal/hr, or about 1700 kcal/day. The rate varies with many factors, including sex, age, weight, and physical condition. A rule of thumb used by nutritionists to estimate basal energy needs per day is the requirement for 1 kcal/hr per kilogram of body weight by a male and 0.95 kcal/hr per kilogram of body weight by a female. For example, a 50 kg (110 lb) female has an estimated basal metabolic rate of $(50\,\text{kg})(0.95\,\text{kcal/kg hr}) = (48\,\text{kcal/hr})$ giving a daily requirement of approximately 1200 kcal.

The total calories a person needs each day is determined by his or her basal requirements plus the energy used in additional physical activities. The caloric consumption rates associated with some activities are listed in the accompanying table. A relatively inactive person requires about 30% above basal requirements per day, a lightly active person requires about 50% above basal, and a very active person such as an athlete or construction worker can use 100% above basal requirements in a day. Each day that you consume food with more calories than you use, the excess calories are stored as potential energy in the chemical bonds of fats in your body and your weight rises. Each day that you consume food with fewer calories than you burn, some chemical energy in your body is taken out of storage to make up the deficit. Fat is metabolized to CO_2 and H_2O, which the body gets rid of, and your weight drops.

Calories Used in Various Activities

Activity	Kilocalories (nutrition calories) used per minute
Sleeping	1.2
Reading	1.3
Listening to lecture	1.7
Weeding garden	5.6
Walking, 3.5 mph	5.6
Pick-and-shovel work	6.7
Recreational tennis	7.0
Soccer, basketball	9.0
Walking up stairs	10.0–18.0
Running, 12 min/mi (5 mph)	10.0
Running, 5 min/mi (12 mph)	25.0

▲ The cola drink contains 160 Calories (kcal) and the hamburger contains 500 Calories. How long would you have to play tennis to burn off these calories?

See Additional Problems 21.72– 74 at the end of the chapter.

■ **PROBLEM 21.9**

The hydrolysis of acetyl phosphate to give acetate and hydrogen phosphate ion has $\Delta G = -10.3$ kcal/mol. Combine the equations and ΔG values to determine whether coupling of this reaction with phosphorylation of ADP is favorable. (You need only give compound names or abbreviations in the equations.)

21.7 Strategies of Metabolism: Oxidized and Reduced Coenzymes

The net result of catabolism is the oxidation of food molecules to release energy. Many metabolic reactions are therefore oxidation–reduction reactions, which means that a steady supply of oxidizing and reducing agents must be available.

To deal with this requirement, a few coenzymes continuously cycle between their oxidized and reduced forms, just as adenosine continuously cycles between its triphosphate and diphosphate forms.

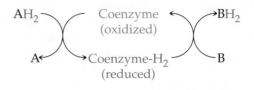

Coenzyme	As Oxidizing Agent	As Reducing Agent
Nicotinamide adenine dinucleotide	NAD^+	$NADH/H^+$
Nicotinamide adenine dinucleotide phosphate	$NADP^+$	$NADPH/H^+$
Flavin adenine dinucleotide	FAD	$FADH_2$
Flavin mononucleotide	FMN	$FMNH_2$

To review briefly, the important points to keep in mind about oxidation and reduction are the following:

- **Oxidation** can be loss of electrons, loss of hydrogen, or addition of oxygen.
- **Reduction** can be gain of electrons, gain of hydrogen, or loss of oxygen.
- **Oxidation and reduction** always occur together.

Each increase in the number of carbon–oxygen bonds is an oxidation, and each decrease in the number of carbon–hydrogen bonds is a reduction, as shown in Table 21.2.

Nicotinamide adenine dinucleotide and its phosphate are widespread, independent coenzymes that enter and leave enzyme active sites in which they are required for redox reactions. As oxidizing agents (NAD^+ and $NADP^+$) they remove hydrogen from a substrate and as reducing agents (NADH and NADPH) they provide hydrogen that adds to a substrate. The complete structure of NAD^+ is shown below with the change that converts it to NADH. The only difference between the structures of NAD^+/NADH and $NADP^+$/NADH is that the color-shaded —OH group here is instead an —OPO_3^{2-} group in $NADP^+$ and NADPH.

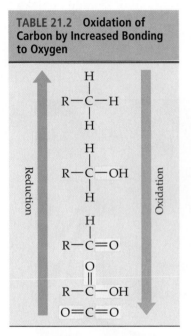

TABLE 21.2   Oxidation of Carbon by Increased Bonding to Oxygen

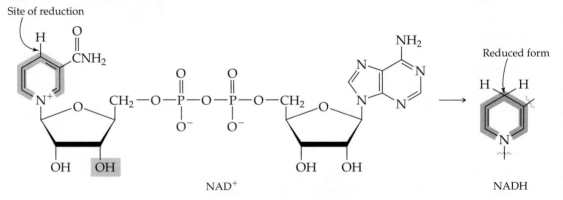

Site of reduction

NAD^+

Reduced form

NADH

As an example, consider a reaction in the citric acid cycle (step 8 in Figure 21.9, Section 21.8) from the oxidation–reduction, or redox, point of view:

The oxidation of malate to oxaloacetate requires the removal of two hydrogen atoms to convert a secondary alcohol to a ketone. (∞, p. 399) The oxidizing agent, which will be reduced, is NAD^+, a *coenzyme*, in this case for malate dehydrogenase. (Sometimes NAD^+ is written as a reactant or product to emphasize its role in a reaction. Keep in mind that although it is free to enter and leave the active site, it always functions as a coenzyme with the appropriate enzyme for the reaction.)

When considering enzyme-catalyzed redox reactions, it's important to recognize that a hydrogen atom is equivalent to a hydrogen *ion*, H^+, plus an electron, e^-. Thus, for the two hydrogen atoms removed in the oxidation of malate,

$$2 \text{ H atoms} = 2 \text{ H}^+ + 2 \text{ e}^-$$

When NAD^+ is reduced, both electrons accompany one of the hydrogens to give a hydride ion,

$$\text{H}^+ + 2 \text{ e}^- = \text{H:}^-$$

The reduction of NAD occurs by addition of H^- to the ring in the nicotinamide part of the structure, where the two electrons of H^- form a covalent bond:

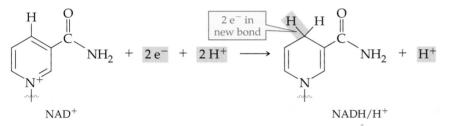

NAD$^+$ NADH/H$^+$

The second hydrogen removed from the oxidized substrate enters the surrounding aqueous solution as a hydrogen ion, H^+. The product of NAD^+ reduction is therefore often represented as $NADH/H^+$ to show that two hydrogen atoms have been removed from the reactant, one of which has bonded to NAD^+ and the other of which is a hydrogen ion in solution. ($NADP^+$ is reduced in the same way to form $NADPH/H^+$.)

Flavin adenine dinucleotide (FAD), another common oxidizing agent in catabolic reactions, is reduced by the formation of covalent bonds to two hydrogen atoms to give $FADH_2$. It participates in several reactions of the citric acid cycle, which is described in the next section.

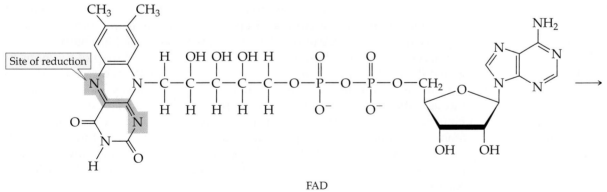

FAD

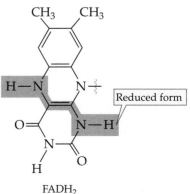

FADH$_2$

Because the reduced coenzymes, NADH and FADH$_2$, have picked up electrons (in their bonds to hydrogen) that are passed along in subsequent reactions, they are often referred to as *electron carriers*. As these coenzymes cycle through their oxidized and reduced forms, they also carry energy along from reaction to reaction. Ultimately, this energy is passed on to the bonds in ATP, as described in Section 21.9.

■ **PROBLEM 21.10**

Is adenosine diphosphate one of the building blocks of the FAD coenzyme?

■ **PROBLEM 21.11**

Look ahead to Figure 21.9 for the citric acid cycle. Draw the structures of the reactants in steps 3, 6, and 8, and indicate which hydrogen atoms are removed in these reactions.

Citric acid cycle The series of biochemical reactions that breaks down acetyl groups to produce energy carried by reduced coenzymes and carbon dioxide.

21.8 The Citric Acid Cycle

The carbon atoms from the first two stages of catabolism are carried into the third stage as acetyl groups bonded to coenzyme A. Like the phosphoryl groups in ATP molecules, the acetyl groups in acetyl coenzyme A molecules are readily removed in an energy-releasing hydrolysis reaction:

$$CH_3-\overset{\displaystyle O}{\overset{\displaystyle \|}{C}}-SCoA \ + \ H_2O \ \longrightarrow \ CH_3-\overset{\displaystyle O}{\overset{\displaystyle \|}{C}}-O^- \ + \ H-SCoA \ + \ H^+ \qquad \Delta G = -7.5 \text{ kcal/mol}$$

Acetyl-SCoA   Coenzyme A

Oxidation of two carbons to give CO$_2$ and transfer of energy to reduced coenzymes occurs in the **citric acid cycle**, also known as the *tricarboxylic acid cycle (TCA)* or *Krebs cycle* (after Hans Krebs, who unraveled its complexities in 1937). As its name implies, the citric acid *cycle* is a closed loop of reactions in which the product of the final step (step 8), oxaloacetate, which has four carbon atoms, is the reactant in the first step. The pathway of carbon atoms through the cycle and the significant products formed are summarized in Figure 21.8. The two carbon atoms of the acetyl group add to the four carbon atoms of oxaloacetate in step 1, and two carbon atoms are set free as carbon dioxide in steps 3 and 4. The cycle continues as four-carbon intermediates progress toward regeneration of oxaloacetate and production of additional reduced coenzymes.

The eight steps of the citric acid cycle and a brief description of each reaction are given in Figure 21.9. The names of the enzymes for each of the steps are listed in the table that accompanies the figure. The cycle takes place in mitochondria, where seven of the enzymes are dissolved in the matrix and one (for step 6) is embedded in the inner mitochondrial membrane (Section 21.3).

The cycle operates as long as (1) acetyl groups are available from acetyl-SCoA and (2) the oxidizing agent coenzymes NAD$^+$ and FAD are available. To meet condition 2, the reduced coenzymes NADH and FADH$_2$ must be reoxidized via the electron-transport chain in stage 4 of catabolism, (described in the next section). Because stage 4 relies on oxygen as the final electron acceptor, the cycle is also dependent on the availability of oxygen.

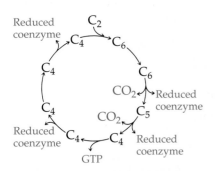

▲ **FIGURE 21.8 Significant outcomes of the citric acid cycle.** The eight steps of the cycle produce two molecules of carbon dioxide, four molecules of reduced coenzymes, and one energy-rich phosphate (GTP). The final step regenerates the reactant for step 1 of the next turn of the cycle.

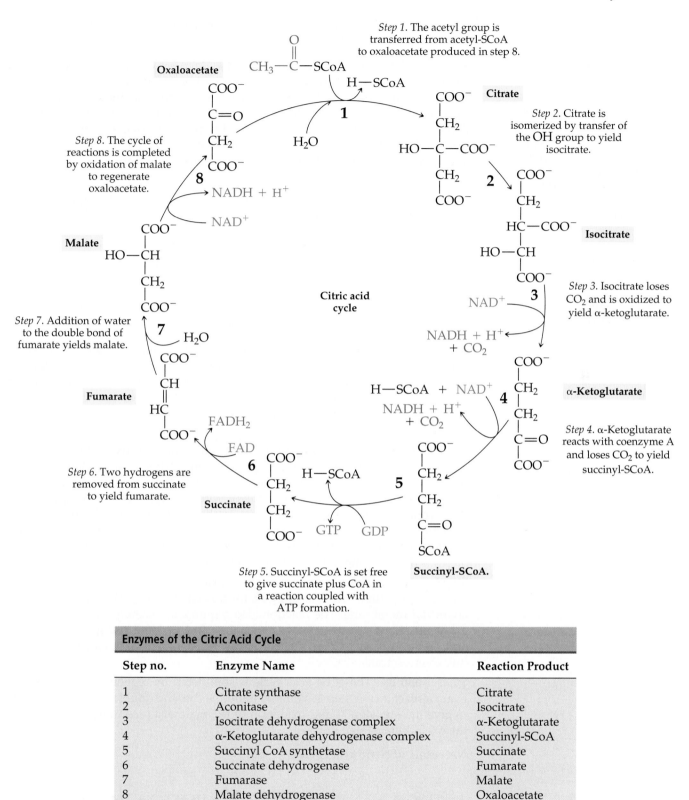

▲ **FIGURE 21.9   The citric acid cycle.**   The net effect of this eight-step cycle of reactions is the metabolic breakdown of acetyl groups (from acetyl-SCoA) into two molecules of carbon dioxide and energy carried by reduced coenzymes. Here and throughout this and the following chapters, energy-rich reactants or products (ATP, reduced coenzymes) are shown in red and their lower-energy counterparts (ADP, oxidized coenzymes) are shown in blue.

Enzymes of the Citric Acid Cycle

Step no.	Enzyme Name	Reaction Product
1	Citrate synthase	Citrate
2	Aconitase	Isocitrate
3	Isocitrate dehydrogenase complex	α-Ketoglutarate
4	α-Ketoglutarate dehydrogenase complex	Succinyl-SCoA
5	Succinyl CoA synthetase	Succinate
6	Succinate dehydrogenase	Fumarate
7	Fumarase	Malate
8	Malate dehydrogenase	Oxaloacetate

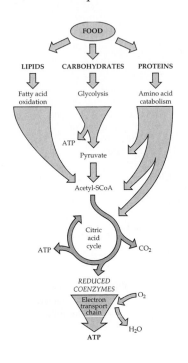

The steps of the citric acid cycle are summarized below, with an emphasis on what they accomplish.

Steps 1 and 2: The first two steps set the stage for oxidation. Acetyl groups enter the cycle at step 1 by addition to four-carbon oxaloacetate to give citrate, a six-carbon intermediate. Citrate is a tertiary alcohol and therefore cannot be oxidized; it is converted in step 2 to its isomer, isocitrate, a secondary alcohol that can be oxidized to a ketone in step 3. The two steps of the isomerization are catalyzed by the same enzyme, aconitase. Water is first removed and then added back to the intermediate, which remains in the active site, so that the —OH is on a different carbon atom.

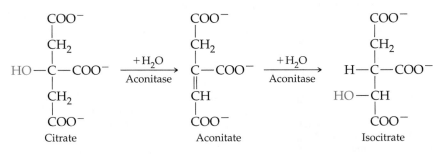

Steps 3 and 4: Both steps are oxidations that rely on NAD^+ as the oxidizing agent. One CO_2 leaves at step 3 as the —OH group of isocitrate is simultaneously oxidized to a keto group. A second CO_2 leaves at step 4, and the resulting succinyl group is added to coenzyme A. In both steps electrons and energy are transferred in the reduction of NAD^+. The succinyl-SCoA carries four carbon atoms along to the next step.

Step 5: With two carbon atoms now removed as carbon dioxide (though not the original two from the acetyl group), the four-carbon oxaloacetate must be restored for step 1 of the next cycle. In step 5, the exergonic conversion of succinyl-SCoA to succinate is coupled with phosphorylation of guanosine diphosphate (GDP) to give guanosine triphosphate (GTP). GTP is similar in structure to ATP and, like ATP, carries energy that can be released during transfer of one of its phosphoryl groups. In some cells, GTP is directly converted to ATP. Step 5 is the only step in the cycle that generates an energy-rich triphosphate.

Step 6: The succinate from step 5 is next oxidized by removal of two hydrogen atoms to give fumarate. The enzyme for this reaction is part of the inner mitochondrial membrane. The reaction also requires the coenzyme FAD, which is covalently bound to its enzyme rather than free to come and go. This enzyme and the FAD participate in stage 4 of catabolism by passing electrons directly into electron transport.

Steps 7 and 8: The citric acid cycle is completed by regeneration of oxaloacetate, a reactant for step 1. Water is added across the double bond of fumarate to give malate (step 7) and oxidation of the malate, a secondary alcohol, gives the oxaloacetate (step 8).

Net result of citric acid cycle

$$Acetyl\text{-}SCoA + 3\,NAD^+ + FAD + ADP + HOPO_3^{2-} + H_2O \longrightarrow$$
$$HSCoA + 3\,NADH + 3\,H^+ + FADH_2 + ATP + 2\,CO_2$$

- Production of four reduced coenzyme molecules (3 NADH, 1 $FADH_2$)
- Conversion of an acetyl group to two CO_2 molecules
- Production of one energy-rich molecule (GTP)

The rate of the citric acid cycle is controlled by the body's need for ATP and reduced coenzymes, and for the energy derived from them. For example, when energy is being used at a high rate, ADP accumulates and acts as an allosteric activator for the enzyme for step 3. When the body's supply of energy is abundant, NADH is present in excess and acts as an inhibitor of the enzyme for step 3. By such feedback mechanisms, as well as variations in the concentrations of necessary reactants, the cycle is activated when energy is needed and inhibited when energy is in good supply.

 Mitochondrial Electron Transport

■ **PROBLEM 21.12**

Which of the substances in the citric acid cycle are tricarboxylic acids (thus giving the cycle its alternative name)?

■ **PROBLEM 21.13**

In Figure 21.9, identify the steps at which reduced coenzymes are produced.

■ **PROBLEM 21.14**

Describe the reaction in the citric acid cycle that is catalyzed by succinate dehydrogenase.

■ **PROBLEM 21.15**

Identify the participants in the citric acid cycle that are α-keto acids.

■ **PROBLEM 21.16**

Which of the reactants in the citric acid cycle have two chiral carbon atoms?

■ **KEY CONCEPT PROBLEM 21.17**

The citric acid cycle can be divided into two stages. In one stage, carbon atoms are added and removed, and in the second stage, oxaloacetate is regenerated. Which steps of the citric acid cycle correspond to each stage?

Electron-transport chain The series of biochemical reactions that passes electrons from reduced coenzymes to oxygen and is coupled to ATP formation.

21.9 The Electron-Transport Chain and ATP Production

Keep in mind that in some ways catabolism is just like burning petroleum or natural gas. In both cases, the goal is to produce useful energy and the reaction products are water and carbon dioxide. The difference is that in catabolism the products are not released all at once and not all of the energy is released as heat.

At the conclusion of the citric acid cycle, the reduced coenzymes formed in the cycle are ready to donate their energy to making additional ATP. The energy is released in a series of oxidation–reduction reactions that move electrons from one electron carrier to the next as each carrier is reduced (gains an electron from the preceding carrier) and then oxidized (loses an electron by passing it along to the next carrier). Each reaction in the series is favorable, that is, it is exergonic. You can think of each reaction as a step along the way down our waterfall. The sequence of reactions that move the electrons along is known as the **electron-transport chain** (also the *respiratory chain*). The enzymes and coenzymes of the chain and ATP synthesis are embedded in the inner membrane of the mitochondrion (Figure 21.10). In the last step of the chain, the electrons

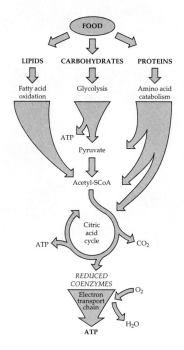

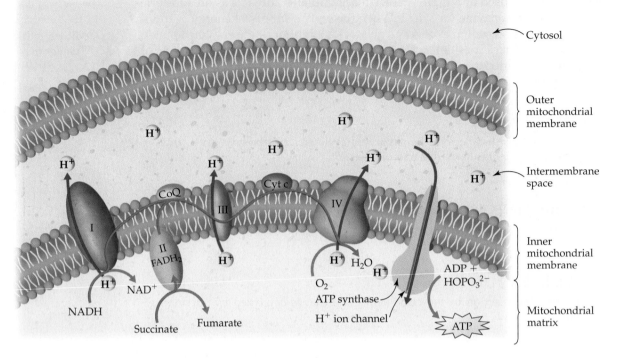

Cytosol

Outer mitochondrial membrane

Intermembrane space

Inner mitochondrial membrane

Mitochondrial matrix

▲ **FIGURE 21.10** **The mitochondrial electron-transport chain and ATP synthase.** The red line shows the path of electrons, and the green lines show the paths of hydrogen ions. The movement of hydrogen ions across the inner membrane at complexes I, III, and IV creates a higher concentration on the intermembrane side of the inner membrane than on the matrix side. The energy released by hydrogen ions returning to the matrix through ATP synthase provides the energy needed for ATP synthesis.

combine with the oxygen that we breathe and with hydrogen ions from their surroundings to produce water.

$$O_2 + 4\,e^- + 4\,H^+ \longrightarrow 2\,H_2O$$

This reaction is fundamentally the combination of hydrogen and oxygen gases. Carried out all at once with the gases themselves, the reaction is explosive. What happens to all that energy during electron transport?

As electrons move down the electron-transport pathway, the energy released is used to move hydrogen ions out of the mitochondrial matrix and into the intermembrane space. Because the inner membrane is otherwise impermeable to the H^+ ion, the result is a higher H^+ concentration in the intermembrane space than in the mitochondrial matrix. Moving ions from a region of lower concentration to one of higher concentration opposes the natural tendency for random motion to equalize concentrations throughout a mixture and therefore requires energy to make it happen. This energy is recaptured for ATP synthesis.

Electron Transport

Electron transport proceeds in four enzyme complexes held in fixed positions within the inner membrane of mitochondria and two electron carriers that move from one complex to another. The complexes and mobile electron carriers are organized in the sequence of their ability to pick up electrons, as illustrated in Figure 21.10. The four fixed complexes are very large assemblages of polypeptides and electron acceptors. The most important electron acceptors are of three types: (1) various cytochromes, which are proteins that contain heme groups (Figure 21.11a) in which the iron cycles between Fe^{2+} and Fe^{3+}; (2) proteins with

(a)

(b)

▲ Explosion of a balloon filled with a hydrogen-oxygen mixture. The amount of energy released is apparent.

Application A Molecular Motor in the Movies

The ultimate goal of biochemistry is to understand exactly how change occurs at the molecular level. For the synthesis of ATP this means knowing the molecular structure of the ATP synthase enzyme complex, identifying where the ADP substrate connects with the enzyme, and understanding how the enzyme changes shape as the reaction occurs and ATP is released.

In early 1997 an experiment provided dramatic proof of what had been proposed based on earlier experiments. Would you believe ATP synthase is a tiny molecular motor that spins active sites around on a drive shaft? Scientists attached the ATP-synthesizing segment of ATP synthase to a glass slide. At the top of the enzyme, they bonded a long filament of actin (a fibrous muscle protein) to the tip of the drive shaft. Because the actin is fluorescent, they were able to make a movie of the filament being spun around as the enzyme functions.

Three identical segments of the enzyme, which contain the active sites for ATP formation, are arranged around the central shaft-like portion of the molecule. (The active segments alternate with three segments that are not catalytically active.) Each of the three active segments changes shape in the following sequence as the central core of the enzyme rotates:

Loose stage—able to loosely bind ADP and $HOPO_3^-$

Tight stage—catalytically active stage in which ATP is formed

Open stage—ATP departs

The steps below show a portion of rotation in which one ATP is formed and another ATP departs.

Roughly a year after the molecular motor movie was announced, Paul D. Boyer (University of California at Los Angeles) and John E. Walker where chosen co-winners of the 1997 Nobel Prize in Chemistry for Chemistry for their demonstration of the structure of ATP synthase and the rotating mechanism by which it functions. Dr. Boyer began his search for an explanation of how ATP is synthesized in the 1950s, and Dr. Walker developed an understanding of the structure of ATP synthase beginning in the 1970s.

See Additional Problems 21.75– 21.76 at the end of the chapter.

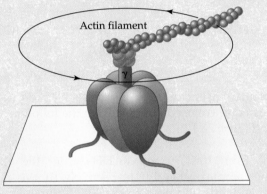

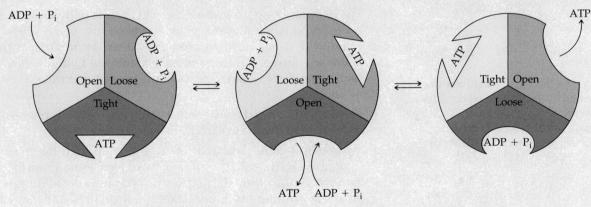

FIGURE 21.11 **A heme group** ▶
and a cytochrome (a) Heme
groups, in which the substituents at
the bonds marked in red vary, are
iron-containing coenzymes in the
cytochromes of the electron-
transport chain. They are also the
oxygen carriers in hemoglobin in
red blood cells. (b) In the
cytochrome shown here, the coiled
olive green ribbon is the amino acid
chain and the heme group is in red
with its central iron atom in gray.

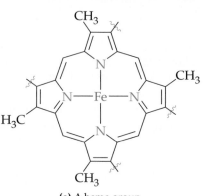

(a) A heme group

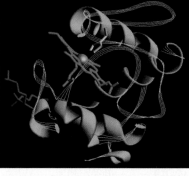

(b) A cytochrome

iron–sulfur groups in which the iron also cycles between Fe^{2+} and Fe^{3+}; and (3)
coenzyme Q, often known as *ubiquinone* because of its widespread occurrence
and because its ring structure with the two ketone groups is a *quinone*.

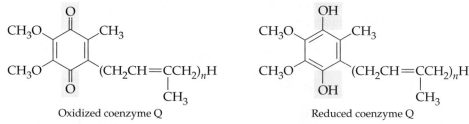

Oxidized coenzyme Q   Reduced coenzyme Q

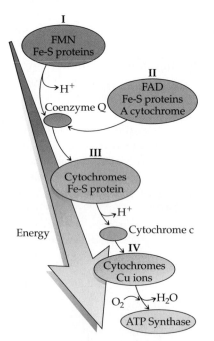

▲ **FIGURE 21.12 Pathway of electrons in electron transport.** Each
of the enzyme complexes I– IV contains several electron carriers.
(FMN in complex I is similar in
structure to FAD.)

The details of the reactions that move electrons in the electron-transport
chain are not important to us here. We need only focus on the following essential features of the pathway (Figure 21.12).

- Hydrogen and electrons from NADH and $FADH_2$ enter the electron-
transport chain at enzyme complexes I and II, respectively. (In this case, the
complexes function independently and not necessarily in numerical order.)
The enzyme for step 6 of the citric acid cycle is part of complex II, so $FADH_2$
is produced when that step of the cycle occurs. The $FADH_2$ does not leave
complex II. It is immediately oxidized there by reaction with coenzyme Q.
Following formation of the mobile coenzyme Q, hydrogen ions no longer
participate directly in the reductions. Instead, electrons are transferred one
by one.

- Electrons are passed from weaker to increasingly stronger oxidizing agents,
with energy released at each transfer.

- Hydrogen ions are released for transport through the inner membrane at
complexes I, III, and IV. Some of these ions come from the reduced coenzymes and some from the matrix— exactly how the hydrogen ions are transported to the intermembrane space is not yet fully understood.

- The H^+ concentration difference creates a potential energy difference across
the two sides of the inner membrane (like the energy difference between
water at the top and bottom of the waterfall). The maintenance of this concentration gradient across the membrane is *crucial*— it is the mechanism by
which energy for ATP formation is made available.

Oxidative phosphorylation The
synthesis of ATP from ADP using
energy released in the electron-
transport chain.

ATP synthase The enzyme complex in the inner mitochondrial
membrane at which hydrogen ions
cross the membrane and ATP is
synthesized from ADP.

ATP Synthesis

The reactions of the electron-transport chain are tightly coupled to **oxidative
phosphorylation**, the conversion of ADP to ATP by a reaction that is both an
oxidation and a phosphorylation. Hydrogen ions can return to the matrix only
by passing through a channel that is part of the **ATP synthase** enzyme complex (green pathway at the right in Figure 21.10). In doing so, they release the

potential energy they gained as they were moved against the concentration gradient at the enzyme complexes of the electron-transport chain. The energy they release drives the phosphorylation of ADP by reaction with hydrogen phosphate ion ($HOPO_3^{2-}$).

$$ADP + HOPO_3^{2-} \longrightarrow ATP + H_2O$$

ATP synthase has knob-tipped stalks that protrude into the matrix and are clearly visible in electron micrographs. ADP and $HOPO_3^{2-}$ are attracted into the knob portion. As hydrogen ions flow through the complex, ATP is produced and released back into the matrix. The reaction is facilitated by changes in the shape of the enzyme complex that are induced by the flow of hydrogen ions. (See Application "A Molecular Motor in the Movies.")

■ **PROBLEM 21.18**

Within the mitochondrion, is the pH higher in the intermembrane space or in the mitochondrial matrix?

■ **KEY CONCEPT PROBLEM 21.19**

The reduced coenzymes NADH and $FADH_2$ are oxidized in the electron-transport system. What is the final electron acceptor of the electron-transport system? What is the function of the H^+ ion in ATP synthesis?

21.10 Harmful Oxygen By-Products and Antioxidant Vitamins

More than 90% of the oxygen we breathe is used in the coupled electron-transport–ATP synthesis reactions. In these and other oxygen-consuming redox reactions the product may not be water, but one or more of three highly reactive species. Two are free radicals, which contain unpaired electrons (represented by single dots in the formulas). Like all free radicals, these two oxygen-containing species, the superoxide ion ($O_2^-\cdot$) and the hydroxyl free radical ($OH^-\cdot$), react as soon as possible to get rid of the unpaired electron. Often, they do this by grabbing an electron from a bond in another molecule, which results in breaking that bond. The third oxygen by-product is hydrogen peroxide, H_2O_2, which is a relatively strong oxidizing agent. The conditions that can enhance production of these three reactive oxygen species are represented in the drawing below. Some causes are environmental, such as exposure to smog or radiation. Others are physiological changes, including aging and inflammation. (Reperfusion is the return of blood to an area that had little blood circulation.)

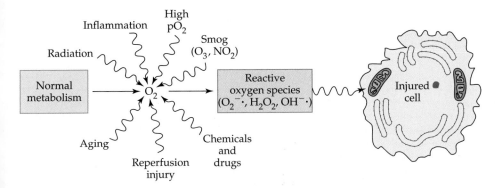

The reactivity of the superoxide free radical is beneficial in destroying infectious microorganisms. In what is known as a respiratory burst, the *phagocytes* (cells that engulf bacteria) produce superoxide ions that react destructively with bacteria:

$$2\,O_2 + \text{NADPH} \longrightarrow 2\,O_2^- \cdot + \text{NADP}^+ + H^+$$

Reactive oxygen species are also dangerous, however. They can break covalent bonds in enzymes and other proteins, DNA, and the lipids in cell membranes. Among the possible outcomes of such destruction are cancer, liver damage, rheumatoid arthritis, heart disease, immune system damage, and, according to some theories, the deterioration that normally accompanies aging. One of the hazards of breathing polluted air and cigarette smoke is breathing free radicals.

The potentially harmful oxygen species are constantly being generated in the body. Our protection against them is provided by superoxide dismutase and catalase, which as we have mentioned earlier, are among the fastest-acting enzymes. (∞, Section 19.5)

$$\underset{\substack{\\ \text{Superoxide ion}}}{2\,O_2^- \cdot + 2\,H^+} \xrightarrow{\text{Superoxide dismutase}} \underset{\substack{\\ \text{Hydrogen} \\ \text{peroxide}}}{H_2O_2 + O_2}$$

$$2\,H_2O_2 \xrightarrow{\text{Catalase}} 2\,H_2O + O_2$$

These and other enzymes are active inside cells where oxygen by-products are generated. Nevertheless, it is estimated that 1 in 50 of the harmful oxygen species escapes destruction inside a cell.

Protection is also provided by the antioxidant vitamins E, C, and A (or its precursor β-carotene), all of which can disarm free radicals by bonding with them. (∞, Section 19.10) Vitamin E is fat-soluble, and its major function is to protect cell membranes from potential damage initiated when a cell membrane lipid (RH) is converted to an oxygen-containing free radical, ROO. Vitamin E gives up electrons and hydrogen atoms to convert the lipid free radical to ROOH, a peroxide that is then enzymatically converted to ROH.

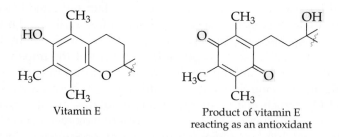

Vitamin E

Product of vitamin E
reacting as an antioxidant

Vitamin C, being water-soluble, is a free-radical scavenger in the blood. There are also many other natural antioxidants among the chemical compounds distributed in fruits and vegetables. We still have much to learn about them.

Application Plants and Photosynthesis

The principal biochemical difference between our-selves and plants is that plants can derive energy directly from sunlight and we cannot. In the process of *photosynthesis* plants use solar energy to synthesize oxy-gen and energy-rich carbohydrates from energy-poor reactants— CO_2 and water. Our own metabolism breaks down energy-rich reactants to extract the useful energy and produce energy-poor products— CO_2 and water. Is it surprising to discover that despite this difference in the direction of their reactions, plants rely on biochemi-cal pathways very much like our own?

The energy-capturing phase of photosynthesis takes place mainly in green leaves. Plant cells contain *chloroplasts*, which, though larger and more complex in structure, resemble mitochondria. Embedded in mem-branes within the chloroplasts, are large groups of *chlorophyll* molecules and the enzymes of an electron-transport chain. Chlorophyll is similar in structure to heme (Figure 21.12), but contains magnesium ions (Mg^{2+}) instead of iron ions (Fe^{2+}).

As solar energy is absorbed, chlorophyll molecules pass it along to specialized reaction centers, where it is used to boost the energy of electrons. The excited elec-trons then give up their extra energy as they pass down a pair of electron-transport chains. Some of this energy

▲ **Flowers along a roadside in North Carolina.** They are converting the potential energy of the sun into chemical potential energy stored in the bonds of carbohydrates.

is used to oxidize water, splitting it into oxygen, hydro-gen ions, and electrons (which replace those entering the electron-transport chain). At the end of the chain, the hydrogen ions, together with the electrons, are used to reduce $NADP^+$ to NADPH. Along the way, part of the energy of the electrons is used to pump hydrogen ions across a membrane to create a concentration gradi-ent. As in mitochondria, the hydrogen ions can only return across the membrane at enzyme complexes that convert ADP to ATP. Water needed for these *light-depen-dent reactions* enters the plant through the roots and leaves, and the oxygen that is formed is released through openings in the leaves.

The energy-carrying ATP and NADPH enter the fluid interior of the chloroplasts. Here their energy is used to drive the synthesis of carbohydrate molecules. So long as ATP and NADH are available, this part of photosynthesis is *light-independent*— it can proceed in the absence of sunlight.

Plants have mitochondria as well as chloroplasts, so they can also carry out the release of energy from stored carbohydrates. Because the breakdown of carbohy-drates continues in many harvested fruits and vegeta-bles, the goal in storage is to slow it down. Refrigeration is one measure that is taken, since (like most chemical reactions) the rate of respiration decreases at lower tem-peratures. Another is replacement of air over stored fruits and vegetables with carbon dioxide or nitrogen.

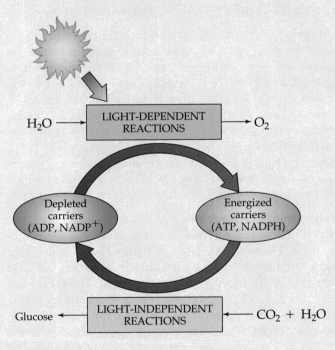

▲ The coupled reactions of photosynthesis.

See Additional Problems 21.77– 79 at the end of the chapter.

623

Key Words

Summary: Revisiting the Chapter Goals

1. What is the source of our energy, and what is its fate in the body? *Endergonic* reactions are unfavorable and require an external source of free energy to occur. *Exergonic* reactions are favorable, proceed spontaneously, and release free energy. We derive energy by oxidation of food molecules that contain energy captured by plants from sunlight. The energy is released gradually in exergonic reactions and is available to do work, to drive endergonic reactions, to provide heat, or to be stored until needed. Energy generation in eukaryotic cells takes place in *mitochondria*.

2. How are the reactions that break down food molecules organized? Food molecules undergo *catabolism* (are broken down) to provide energy in four stages (Figure 21.5): (1) digestion to form smaller molecules that can be absorbed into cells; (2) decomposition (by separate pathways for lipids, carbohydrates, and proteins) into two-carbon acetyl groups that are bonded to coenzyme A in *acetyl coenzyme A*; (3) reaction of the acetyl groups via the *citric acid cycle* to generate energy-rich reduced coenzymes and liberate carbon dioxide; and (4) *electron transport* and transfer of the energy of the reduced coenzymes from the citric acid cycle to our principal energy transporter, ATP.

3. What are the major strategies of metabolism? Using the energy from exergonic reactions, ADP is *phosphorylated* to give ATP. Where energy must be expended, it is released by removal of a phosphoryl group from ATP to give back ADP. An otherwise "uphill" reaction in a metabolic pathway is driven by coupling with an exergonic, "downhill" reaction that provides enough energy so that their combined outcome is exergonic and favorable. The oxidizing and reducing agents needed by the many redox reactions of metabolism are coenzymes that constantly cycle between their oxidized and reduced forms.

4. What is the citric acid cycle? The *citric acid cycle* (Figure 21.9) is a pathway of eight reactions, in which the product of the final reaction is the substrate for the first reaction. The reactions of the citric acid cycle (1) set the stage for oxidation of the acetyl group (steps 1 and 2); (2) remove two carboxyl groups as CO_2 molecules (oxidative decarboxylation) from the tricarboxylic acid isocitrate (steps 3 and 4); and (3) oxidize the four-carbon dicarboxylic acid succinate and regenerate oxaloacetate so that the cycle can start again (steps 5–8). Along the way, four reduced coenzyme molecules and one molecule of ATP are produced for each acetyl group oxidized. The reduced coenzymes carry energy for the subsequent production of additional ATP. The cycle is activated when energy is in short supply and inhibited when energy is in good supply.

5. How is ATP generated in the final stage of catabolism? ATP generation is accomplished by a series of enzyme complexes in the inner membranes of mitochondria (Figure 21.10). Electrons and hydrogen ions enter the first two complexes of the electron-transport chain from succinate (in the citric acid cycle) and NADH, where they are transferred to *coenzyme Q*. Then, the electrons and hydrogen ions proceed independently, the electrons gradually giving up their energy to the transport of hydrogen ions across the inner mitochondrial membrane to maintain different concentrations on opposite sides of the membrane. The hydrogen ions return to the matrix by passing through *ATP synthase*, where the energy they release is used to convert ADP to ATP.

6. What are the harmful by-products produced from oxygen, and what protects against them? Harmful by-products of oxygen-consuming reactions are the hydroxyl free radical, superoxide ion (also a free radical), and hydrogen peroxide. These reactive species damage other molecules by breaking bonds. Superoxide dismutase and catalase are enzymes that disarm these oxygen by-products. Vitamins E, C, and A (or its precursor β-carotene) are also antioxidants.

▪ Understanding Key Concepts

21.20 The following coupled reaction is the result of an exergonic reaction and an endergonic reaction:

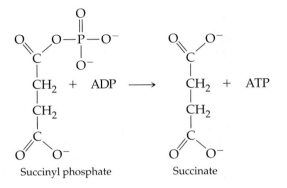

Succinyl phosphate   Succinate

    (a) Write the exergonic portion of the reaction.
    (b) Write the endergonic portion of the reaction.

21.21 Each of the reactions below is involved in one of the four stages of metabolism shown in Figure 21.5. Identify the stage of metabolism in which each reaction occurs.
    (a) Hydrolysis of starch to produce glucose
    (b) The oxidation of NADH coupled with the synthesis of ATP
    (c) The oxidation of acetyl-SCoA, in a series of reactions where NAD^+ is reduced and CO_2 is produced
    (d) Conversion of glucose to acetyl-SCoA

21.22 For the first step in fatty acid catabolism, we say that ATP is used to "drive" the reaction that links the fatty acid with coenzyme-A. Without the ATP hydrolysis, would you predict that the linking of fatty acid to coenzyme-A would be exergonic or endergonic? In the fatty acid SCoA synthesis, the hydrolysis of the ATP portion is based on what major strategy of metabolism?

21.23 Since no molecular oxygen participates in the citric acid cycle, the steps in which acetyl groups are oxidized to CO_2 involve removal of hydride ions and hydrogen ions. What is the acceptor of hydride ions? What is the acceptor of hydrogen ions?

21.24 The reaction below is catalyzed by isocitrate dehydrogenase and occurs in two steps, the first of which (step A) is formation of an unstable intermediate (shown in brackets).
    (a) In which step is a coenzyme needed? Identify the coenzyme.
    (b) In which step is CO_2 evolved and a hydrogen ion added?
    (c) Which of the structures shown can be described as a β-keto acid?

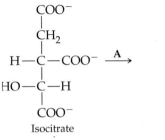

Isocitrate

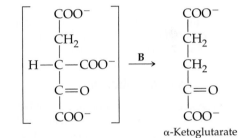

α-Ketoglutarate

21.25 For each of the reactions of the citric acid cycle, give the type of reaction occurring, name the enzyme involved, and indicate which of the six classes of enzymes it belongs to. (Some may have more than one kind of enzyme activity.)

21.26 The electron-transport chain uses several different metal ions, especially iron, copper, zinc, and manganese. Why are metals used frequently in these two pathways? What can metals do better than organic biomolecules?

▪ Additional Problems

Free Energy and Biochemical Reactions

21.27 What is the difference between an endergonic and an exergonic process?

21.28 What energy requirements must be met in order for a reaction to be favorable?

21.29 Why is ΔG a useful quantity for predicting the favorability of biochemical reactions?

21.30 Many biochemical reactions are catalyzed by enzymes. Do enzymes have an influence on the magnitude or sign of ΔG? Why or why not?

21.31 Each of the following reactions occurs during the catabolism of acetyl-SCoA. Which is exergonic? Which is endergonic? Which reaction produces a phosphate that can later yield energy by giving up a phosphate group?
    (a) Acetyl-SCoA + Oxaloacetate → Citrate + CoA-SH
        $\Delta G = -8$ kcal/mol

    (b) Succinyl-SCoA + GDP + Phosphate (P_i) →
                Succinate + CoA-SH + GTP + H_2O
        $\Delta G = -0.4$ kcal/mol
    (c) L-Malate + NAD^+ →
                 Oxaloacetate + NADH + H^+
        $\Delta G = +7$ kcal/mol

21.32 Each of the following reactions occurs during the catabolism of glucose. Which is exergonic? Which is endergonic? Which proceeds farthest toward products at equilibrium?
    (a) Phosphoenol pyruvate + H_2O →
                 Pyruvate + Phosphate (P_i)
        $\Delta G = -14.8$ kcal/mol
    (b) Glucose + P_i → Glucose 6-phosphate + H_2O
        $\Delta G = +3.3$ kcal/mol
    (c) 1,3-Bisphosphoglycerate + H_2O →
                 3-Phosphoglycerate + P_i
        $\Delta G = -11.8$ kcal/mol

Cells and Their Structure

21.33 List five differences between prokaryotic and eukaryotic cells.

21.34 What kinds of organisms have prokaryotic cells, and what kinds have eukaryotic cells?

21.35 What is an organelle?

21.36 What is the difference between the cytoplasm and the cytosol?

21.37 Describe in general terms the structural makeup of a mitochondrion.

21.38 Why are mitochondria called the "power plants" of the cell?

Metabolism

21.39 What is the difference between digestion and metabolism?

21.40 What is the difference between catabolism and anabolism?

21.41 What key metabolic substance is formed from the catabolism of all three major classes of foods: carbohydrates, lipids, and proteins?

21.42 Put the following events in the correct order of their occurrence: electron transport, digestion, oxidative phosphorylation, citric acid cycle.

Strategies of Metabolism

21.43 What is the full name of the substance formed during catabolism to store chemical energy?

21.44 Why is ATP sometimes called a high-energy molecule?

21.45 What is the chemical difference between ATP and ADP?

21.46 What general kind of chemical reaction does ATP carry out?

21.47 What does it mean when we say that two reactions are coupled?

21.48 Show why coupling the reaction for the hydrolysis of 1,3-bisphosphoglycerate to the phosphorylation of ADP is energetically favorable. Combine the equations and calculate ΔG for the coupled process. You need only give names or abbreviations, not chemical structures.

21.49 Write the reaction for the hydrolysis of 1,3-bisphosphoglycerate coupled to the phosphorylation of ADP using the curved arrow symbolism.

21.50 Would the hydrolysis of fructose 6-phosphate be favorable for phosphorylating ADP? Why or why not? (Refer to Table 21.1.)

21.51 NAD^+ is a coenzyme for dehydrogenation.
  (a) When a molecule is dehydrogenated, is NAD^+ oxidized or reduced?
  (b) Is NAD^+ an oxidizing agent or a reducing agent?
  (c) What type of substrate is NAD^+ associated with, and what type of product molecule is formed after dehydrogenation?
  (d) What is the form of NAD^+ after the dehydrogenation process?
  (e) Write a general equation for a reaction involving the operation of NAD^+ using the curved arrow symbolism.

21.52 FAD is a coenzyme for dehydrogenation.
  (a) When a molecule is dehydrogenated, is FAD oxidized or reduced?
  (b) Is FAD an oxidizing agent or a reducing agent?
  (c) What type of substrate is FAD associated with, and what is the type of product molecule after dehydrogenation?
  (d) What is the form of FAD after the dehydrogenation process?
  (e) Write a general equation for a reaction involving FAD using the curved arrow symbolism.

The Citric Acid Cycle

21.53 Where in the cell does the citric acid cycle take place?

21.54 By what other names is the citric acid cycle known?

21.55 What substance acts as the starting point of the citric acid cycle, reacting with acetyl-SCoA in the first step and being regenerated in the last step? Draw its structure.

21.56 What is the final fate of the carbons in acetyl-SCoA after several turns of the citric acid cycle?

21.57 Look at the eight steps of the citric acid cycle (Figure 21.9) and answer these questions:
  (a) Which steps involve oxidation reactions?
  (b) Which steps involve decarboxylation (loss of CO_2)?
  (c) Which step or steps involve a hydration reaction?

21.58 How many NADH and how many $FADH_2$ molecules are formed in the citric acid cycle?

The Electron-Transport Chain; Oxidative Phosphorylation

21.59 In what way are the processes of the citric acid cycle and the electron-transport chain interrelated?

21.60 What are the two primary functions of the electron-transport chain?

21.61 What two coenzymes are involved with initial events of the electron-transport chain?

21.62 What are the ultimate products of the electron-transport chain?

21.63 What do the following abbreviations stand for?
  (a) FAD
  (b) CoQ
  (c) $NADH/H^+$
  (d) Cyt c

21.64 What atom in the cytochromes undergoes oxidation and reduction in the electron-transport chain? What atoms in CoQ?

21.65 Put the following substances in the correct order of their action in the electron-transport chain: cytochrome c, coenzyme Q, NADH.

21.66 Fill in the missing substances in these coupled reactions:

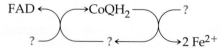

21.67 What would happen to the citric acid cycle if NADH and $FADH_2$ were not reoxidized?

21.68 Across what membrane is there a pH differential caused by the release of H^+ ions? On which side of the membrane are there more H^+ ions?

21.69 What does the term "oxidative phosphorylation" mean?

21.70 In oxidative phosphorylation, what is oxidized and what is phosphorylated?

Applications

21.71 Photosynthetic plants use a sunlight-driven electron-transport system to remove electrons from H_2O to produce O_2 and generate ATP and NADPH. In the bacteria surrounding black smokers, H_2S is the electron donor and corresponds to what component of the electron-transport system in photosynthetic plants? [*Life without Sunlight*]

21.72 How is basal metabolic rate defined? [*Basal Metabolism*]

21.73 Estimate your basal metabolic rate using the guidelines in the application. [*Basal Metabolism*]

21.74 Why do activities such as walking raise a body's needs above the basal metabolic rate? [*Basal Metabolism*]

21.75 Describe the three stages through which active sites of ATP synthase cycle as the enzyme rotates and the processes that occur in the active site in each stage. [*A Molecular Motor in the Movies*]

21.76 What provides the energy required to power the rotation of ATP synthase? [*A Molecular Motor in the Movies*]

21.77 Chlorophyll is similar in structure to heme in red blood cells but does not have an iron atom. What metal ion is present in chlorophyll? [*Plants and Photosynthesis*]

21.78 Photosynthesis consists of both light-dependent and light-independent reactions. What is the purpose of each type of reaction? [*Plants and Photosynthesis*]

21.79 One step of the cycle that incorporates CO_2 into glyceraldehyde in plants is the production of two 3-phosphoglycerates. This reaction has $\Delta G = -0.84$ kcal/mol. Is this process endergonic or exergonic? [*Plants and Photosynthesis*]

General Questions and Problems

21.80 Why must the breakdown of molecules for energy in the body occur in several steps, rather than in one step?

21.81 The first step in the citric acid cycle involves the reaction of acetyl-SCoA and oxaloacetate. Show the product of this reaction before hydrolysis to yield citrate.

21.82 The fumarate produced in step 6 of the citric acid cycle must have a trans double bond to continue on in the cycle. Suggest a reason why the corresponding cis double-bond isomer can't continue in the cycle.

21.83 With what class of enzymes are the coenzymes NAD^+ and FAD associated?

21.84 We talk of burning food in a combustion process, producing CO_2 and H_2O from food and O_2. Explain how O_2 is involved in the process although there is no O_2 directly involved in the citric acid cycle.

21.85 One of the steps that occurs when lipids are metabolized is shown below. Does this process require FAD or NAD as the coenzyme? What is the general class of enzyme that catalyzes this process?

21.86 Solutions of hydrogen peroxide can be kept for months in a brown closed bottle with only moderate decomposition. When used on a cut as an antiseptic, hydrogen peroxide begins to bubble rapidly. Give a possible explanation for this observation.

21.87 Identify the chiral intermediates in the TCA cycle.

21.88 If you use a flame to burn a pile of glucose completely to give carbon dioxide and water, the overall reaction is identical to the metabolic oxidation of glucose. Explain the differences in the fate of the energy released in each case.

21.89 What highly reactive oxygen species are by-products of oxygen consuming reactions in the body? Which enzymes and vitamins are used to destroy these reactive species?

22 Carbohydrates

An oat plant growing in a field of oats. The oats will provide polysaccharides that we need in our diet.

The word *carbohydrate* was used originally to describe glucose, the simplest and most readily available sugar. Because glucose has the formula $C_6H_{12}O_6$, it was once thought to be a "hydrate of carbon," $C_6(H_2O)_6$. Although this view was soon abandoned, the name "carbohydrate" persisted and is now used to refer to a large class of biomolecules with similar structures. Carbohydrates have in common many hydroxyl groups on adjacent carbons together with either an aldehyde or ketone group. Glucose, for example, has five —OH groups and one —CHO group:

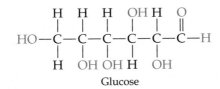

Glucose

Carbohydrates are synthesized by plants and stored in the form of starch, a polymer of glucose. When starch is eaten and digested, the freed glucose becomes a major source of the energy required by living organisms. Thus, carbohydrates are intermediaries by which energy from the sun is made available to animals.

In this chapter, we'll look at the answers to these questions about carbohydrates:

1. What are the different kinds of carbohydrates?
The goal: Be able to define monosaccharides, disaccharides, and polysaccharides, and to recognize examples.

2. Why are monosaccharides handed, and how does this influence the numbers and types of their isomers?
The goal: Be able to identify the chiral carbon atoms in monosaccharides, predict the number of isomers for different monosaccharides, and identify pairs of enantiomers.

3. What are the structures of monosaccharides, and how are they represented in written formulas?
The goal: Be able to explain relationships among the open-chain and cyclic monosaccharide structures, describe the isomers of monosaccharides, and show how they are represented by Fischer projections and cyclic structural formulas.

4. How do monosaccharides react with oxidizing agents and alcohols?
The goal: Be able to identify reducing sugars and the products of their oxidation, recognize acetals of monosaccharides, and describe glycosidic linkages in disaccharides.

5. What are the structures of some important disaccharides?
The goal: Be able to identify the monosaccharides combined in maltose, lactose, and sucrose, and describe the types of linkages between the monosaccharides.

6. What are the functions of some important carbohydrates that contain modified monosaccharide structures?
The goal: Be able to identify the functions of chitin, connective tissue polysaccharides, heparin, and glycoproteins.

7. What are the structures and functions of cellulose, starch, and glycogen?
The goal: Be able to describe the monosaccharides and linkages in these polysaccharides, and their fates in metabolism.

CONCEPTS TO REVIEW

Molecular shape (Section 5.7)
Chirality (Section 18.5)
Oxidation–reduction reactions (Sections 6.12, 6.13, 14.5, 16.5, 16.6)

22.1 An Introduction to Carbohydrates

Carbohydrates are a large class of naturally occurring polyhydroxy aldehydes and ketones. **Monosaccharides**, sometimes known as **simple sugars**, are the simplest carbohydrates. They have from three to seven carbon atoms, and each

Carbohydrate A member of a large class of naturally occurring polyhydroxy ketones and aldehydes.

Monosaccharide (simple sugar) A carbohydrate with 3–7 carbon atoms.

Aldose A monosaccharide that contains an aldehyde carbonyl group.

Ketose A monosaccharide that contains a ketone carbonyl group.

has one aldehyde or one ketone functional group. If it has an aldehyde group, the sugar is classified as an **aldose**. If it has a ketone group, the sugar is classified as a **ketose**. The aldehyde group is always at the end of the carbon chain, and the ketone group is always on the second carbon of the chain. In either case, there is a —CH$_2$OH group at the other end of the chain.

Monosaccharides

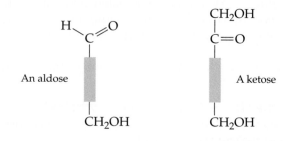

An aldose   A ketose

There are hydroxyl groups on all the carbon atoms between the carbonyl carbon atom and the —CH$_2$OH at the other end, and also on the end carbon next to a ketone group, as illustrated in the following three structures. The family-name ending *-ose* indicates a carbohydrate, and simple sugars are known by common names like *glucose*, *ribose*, and *fructose* rather than systematic names.

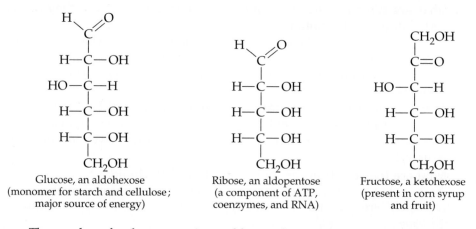

Glucose, an aldohexose
(monomer for starch and cellulose;
major source of energy)

Ribose, an aldopentose
(a component of ATP,
coenzymes, and RNA)

Fructose, a ketohexose
(present in corn syrup
and fruit)

The number of carbon atoms in an aldose or ketose can be specified by one of the prefixes *tri-*, *tetr-*, *pent-*, *hex-*, or *hept-*. Thus, glucose is an aldo*hex*ose (*aldo-* = aldehyde, *-hex* = six carbons; *-ose* = sugar); fructose is a keto*hex*ose (a six-carbon ketone sugar); and ribose is an aldo*pent*ose (a five-carbon aldehyde sugar). Most naturally occurring simple sugars are aldehydes with either five or six carbons.

Because of their many functional groups, monosaccharides undergo a variety of structural changes and chemical reactions. They react with each other to form **disaccharides** and **polysaccharides** (also known as **complex carbohydrates**), which are polymers of monosaccharides. Their functional groups also allow reactions with alcohols and with lipids and proteins to form biomolecules with specialized functions. These and other carbohydrates are introduced in later sections of this chapter. First, we are going to discuss two important aspects of carbohydrate structure:

Disaccharide A carbohydrate composed of two monosaccharides.

Polysaccharide (complex carbohydrate) A carbohydrate that is a polymer of monosaccharides

- Monosaccharides are chiral molecules (Sections 22.2, 22.3).
- Monosaccharides exist mainly in cyclic forms rather than the straight-chain forms written above (Section 22.4).

■ PROBLEM 22.1

Classify each of the following monosaccharides as an aldose or a ketose, and name each according to the number of carbon atoms:

(a) HOCH$_2$—CH—CH—CH—C—H
         | | | ||
         OH OH OH O

(b) HOCH$_2$—C—CH$_2$OH
          ||
          O

(c) HOCH$_2$—CH—CH—C—H
        | | ||
        OH OH O

■ **PROBLEM 22.2**

Draw the structures of an aldohexose and a ketotetrose.

22.2 Handedness of Carbohydrates

You have seen that amino acids are chiral (that is, not superimposible on their mirror images) because they contain carbon atoms bonded to four different groups. Glyceraldehyde, the three-carbon monosaccharide that is the simplest naturally occurring carbohydrate, has the structure shown below. Because there are four different groups bonded to the number 2 carbon atom (—CHO, —H, —OH, and —CH$_2$OH), this molecule is also chiral. (∞, p. 512)

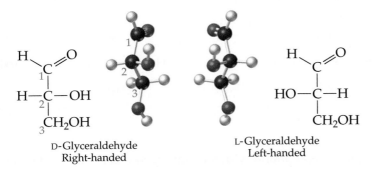

D-Glyceraldehyde
Right-handed

L-Glyceraldehyde
Left-handed

Chiral compounds lack a plane of symmetry and exist as a pair of enantiomers in either a "right-handed" D form or a "left-handed" L form. Like all enantiomers, the two forms of glyceraldehyde have the same physical properties except for the way in which they affect polarized light.

Light as we usually see it consists of electromagnetic waves oscillating in all planes at right angles to the direction of travel of the light beam. (∞, p. 64) When ordinary light is passed through a polarizer, only the waves in one plane get through, producing what is known as *plane-polarized light*. Solutions of *optically active* chemical compounds change the plane in which the light is polarized. The angle by which the plane is rotated can be measured in an instrument known as a *polarimeter*, which works on the principle diagrammed in Figure 22.1. Each enantiomer of a pair rotates the plane of the light by the

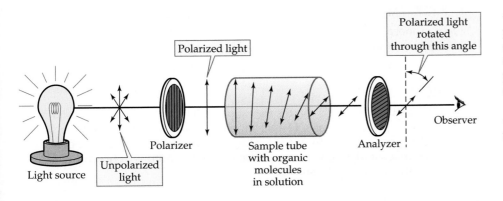

▲ **FIGURE 22.1   Principle of a polarimeter, used to determine optical activity.**   A solution of an optically active isomer rotates the plane of the polarized light by a characteristic amount.

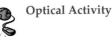 **Optical Activity**

same amount, but the directions of rotation are *opposite*. If one enantiomer rotates the plane of the light to the left, the other rotates it to the right.

Compounds like glyceraldehyde that have *one* chiral carbon atom can exist as two enantiomers. But what about compounds with more than one chiral carbon atom? How many isomers are there of compounds that have two, three, four, or more chiral carbons? Aldotetroses, for example, have two chiral carbon atoms and can exist in the four isomeric forms shown in Figure 22.2. These four aldotetrose stereoisomers consist of two mirror-image pairs of enantiomers, one pair named *erythrose* and one pair named *threose*. Because erythrose and threose are stereoisomers but are not mirror images of each other, they are described as **diastereomers**.

Diastereomers Stereoisomers that are not mirror images of each other.

FIGURE 22.2 Two pairs of ▶ enantiomers: The four isomeric aldotetroses (2,3,4-trihydroxybutanals). Carbon atoms 2 and 3 are chiral. Their —H atoms and —OH groups are written here to show their mirror-image relationship. Erythrose and threose exist as enantiomeric pairs.

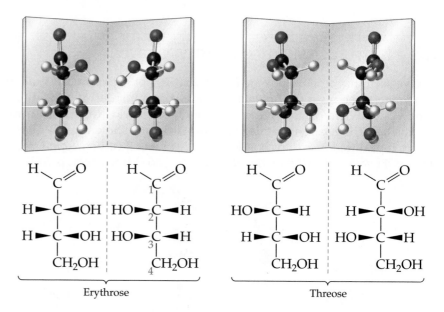

Erythrose   Threose

In general, a compound with n chiral carbon atoms has a maximum of 2^n possible stereoisomers and half that many pairs of enantiomers. The aldotetroses, for example, have $n = 2$ so that $2^n = 2^2 = 4$, meaning that four stereoisomers are possible. Glucose, an aldohexose, has four chiral carbon atoms and a total of $2^4 = 16$ possible stereoisomers (8 pairs of isomers). All 16 stereoisomers of glucose are known. (In some cases, fewer than the maximum predicted number of stereoisomers exist because some of the molecules have symmetry planes that make them identical to their mirror images.)

■ **PROBLEM 22.3**

Aldopentoses have three chiral carbon atoms. What is the maximum possible number of aldopentose stereoisomers?

■ **PROBLEM 22.4**

From monosaccharides (a)–(d) on the facing page, choose the one that is the enantiomer of the monosaccharide shown above (a)–(d).

■ **PROBLEM 22.5**

Notice in structures (a)–(d) on the facing page that the bottom carbon and its substituents are written as CH_2OH in every case. How does the C in this group differ in each case from the C atoms above it? Why must the locations of the H atoms and —OH groups attached to the carbons between this one and the carbonyl group be shown?

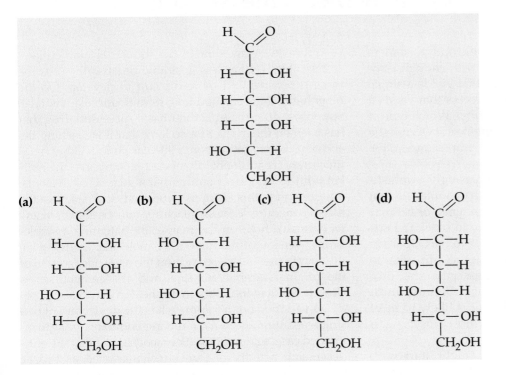

22.3 The D and L Families of Sugars: Drawing Sugar Molecules

A standard method of representation called a **Fischer projection** has been adopted for drawing stereoisomers on a flat page so that we can tell one from another. A chiral carbon atom is represented in a Fischer projection as the intersection of two crossed lines, and this carbon atom is considered to be on the printed page. Bonds that point up out of the page are shown as horizontal lines, and bonds that point behind the page are shown as vertical lines. Until now, we have used solid wedges and dashed lines to represent bonds above and behind the printed page, respectively, with ordinary solid lines for bonds in the plane of the page. (⚬⚬⚬, p. 115) The relationship between such a structure and a Fischer projection is illustrated below.

Fischer projection Structure that represents chiral carbon atoms as the intersections of two lines, with the horizontal lines representing bonds pointing out of the page and the vertical lines representing bonds pointing behind the page. For sugars, the aldehyde or ketone is at the top.

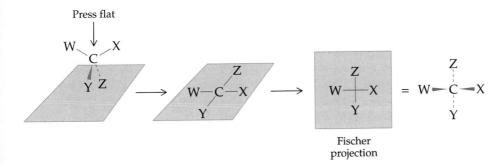

Fischer projection

In a Fischer projection, the aldehyde or ketone carbonyl group of a monosaccharide is always placed at the top. The result is that —H and —OH groups projecting above the page are on the left and right of the chiral carbons, and groups projecting behind the page are above and below the chiral carbons. The

Application Chirality Meets the Marketplace

Nature is better at synthesizing single optical isomers than chemists are. It is possible, with enough laboratory research, to devise a process that yields a single isomer, but it is easier to design a process that yields a mixture of the isomers (a *racemic mixture*). Which type of process to use in an industrial setting becomes an issue of scientific, medical, and commercial importance in the drug industry.

As drug design becomes increasingly sophisticated, the goal is often creation of a drug molecule that will bind with a specific hormone, enzyme, or cellular receptor. The desired result might be to block the target molecule so that it is not available to participate in a given reaction. Or it might be the exact opposite, to design a drug that will make the connection that produces a therapeutic chemical change. Because most biomolecules are chiral, a chiral drug is likely to meet the need most effectively by fitting most closely with the target.

The route to a chiral molecule might start with natural chiral reactants, might utilize a natural enzyme, or might rely on synthesizing the pair of enantiomers and then relying on a method of separating them. Also, note that it can be important to have the separate enantiomers available during the elaborate testing needed to prove a drug effective. Sometimes, marketing the mixture of isomers would be the wrong thing to do. Naproxen, for example, the active ingredient in the pain killer and anti-inflammatory Aleve, is sold as a single enantiomer. The other enantiomer causes liver damage.

From the viewpoint of the industrial marketplace, chiral syntheses are less desirable because they can be more expensive and more difficult to develop. On the other hand, there can be a long-term financial advantage. Suppose a drug manufacturer has a successful drug that has been on the market for so long that it is reaching the end of its patent protection, a 20-year period. Other manufacturers are undoubtedly poised to jump into the market with lower prices and perhaps generic equivalents. Suppose that this drug is produced and sold as the racemic mixture. The manufacturer may be able to renew its exclusive hold on the market by patenting a single-isomer version of the same drug molecule. By doing so, the enantiomer gets a free ride on the name recognition of the original racemic mixture, and the competition is slowed down in its ability to duplicate a synthesis.

A variation on this approach is the development of a single-enantiomer by a competing company. In a documented case, a competitor (Sepracor) found that the antidepressant activity of fluoxetine, marketed as Prozac (see Section 20.9), which is an enantomeric pair, resides in one of the enantiomers. The competitor patented the active isomer and began the clinical trials needed for its FDA approval as a new drug. Then, the original Prozac manufacturer (Lilly) bought a license on the single-isomer patent and will pay $90 million during the period needed to complete FDA approval of the isomer.

The size of the single-enantiomer drug market has been increasing every year since it began to develop about 10 years ago. With expanding research efforts, the ability to produce such drugs is growing easier. The top five classes of single-enantiomer drugs are cardiovascular drugs, antibiotics, hormones, cancer drugs, and those for central nervous system disorders. In 2000 the total worldwide market for these categories of drugs was about $170 billion, with about $85 billion, or 50% accounted for by single enantiomers.

See Additional Problems 22.73–74 at the end of the chapter.

Naproxen

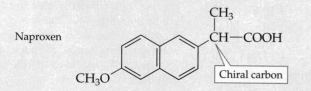

Fischer projection of one of the enantiomers of glyceraldehyde is therefore interpreted as follows:

Fischer projection of a glyceraldehyde enantiomer

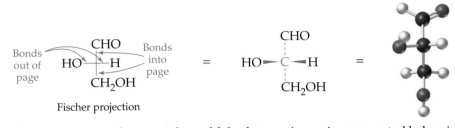

Fischer projection

For comparison, the same glyceraldehyde enantiomer is represented below in the conventional manner, showing the tetrahedron of bonds to the chiral carbon.

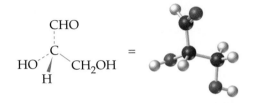

Monosaccharides are divided into two families—the D **sugars** and the L **sugars**—based on their structural relationships to glyceraldehyde. The value of consistently writing formulas for monosaccharides as Fischer projections is that these formulas allow us to identify the D and L forms at a glance. Look again at the structural formulas of the D and L forms of glyceraldehyde.

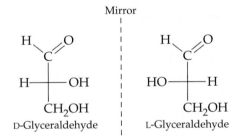

Mirror

D-Glyceraldehyde | L-Glyceraldehyde

In the D form, the —OH group on carbon 2 comes out of the plane of the paper and points to the *right*; in the L form, the —OH group at carbon 2 comes out of the plane of the paper and points to the *left*. If you mentally place a mirror plane between these Fischer projections, you can see that they're mirror images.

Nature has a strong preference for one type of handedness in carbohydrates, just as it does in amino acids and also in snail shells. It happens, however, that carbohydrates and amino acids have opposite handedness. Most naturally occurring α-amino acids belong to the L family, but most carbohydrates belong to the D family. Over the years, a variety of explanations for handedness in biochemical molecules have been proposed, with several focusing on physical influences on the origin of life. As yet, none of these theories has surfaced as more nearly correct than any of the others.

The designations D and L derive from the Latin *dextro* for "right" and *levo* for "left." *In all Fischer projections, the D form of a monosaccharide has the hydroxyl group on the chiral carbon atom farthest from the carbonyl group pointing toward the right, while the mirror-image L form has the hydroxyl group on this same carbon pointing toward the left.*

Fischer projections of molecules with more than one chiral carbon atom are written with the chiral carbons one above the other in a vertical line. To simplify visualizing the structures, we will often include the C's for the chiral carbons in the plane of the page. Otherwise, the structures are to be interpreted as Fischer projections. Two pairs of aldohexose enantiomers are represented

D **Sugar** Monosaccharide with the —OH group on the chiral carbon atom farthest from the carbonyl group pointing to the right in a Fischer projection.

L **Sugar** Monosaccharide with the —OH group on the chiral carbon atom farthest from the carbonyl group pointing to the left in a Fischer projection.

 D and L Notation for Carbohydrates

▲ Nature's preference. Not only molecules, but also snail shells have a preferred handedness. Most snail shells, like the one shown, are right-handed.

below in this manner. Given such a projection of one enantiomer, you can draw the other by reversing the substituents on the left and right of each chiral atom. Note that each pair of enantiomers has a different name.

Two pairs of aldohexose enantiomers

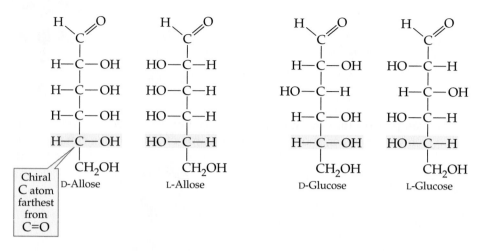

It would be easy to assume that the use of D and L, because they stand for *dextro* and *levo*, carries some meaning about the direction of rotation of plane-polarized light. Logical as it seems, this is not the case. The D and L relate directly only to the position of that —OH group on the bottom carbon in a Fischer projection. And the D and L isomers do indeed rotate plane-polarized light in opposite directions. But, and here's the point to remember, *the direction of rotation cannot be predicted*. There are D isomers that rotate polarized light to the left and L isomers that rotate it to the right.

■ WORKED EXAMPLE 22.1

Identify the following monosaccharides as (a) D-ribose or L-ribose, (b) D-mannose or L-mannose.

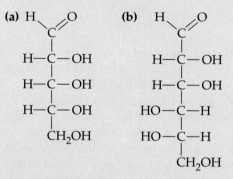

ANALYSIS To identfy D or L isomers, one must check the location of the —OH group on the carbon atom farthest from the carbonyl group. In a Fischer projection, this is the carbon atom above the bottom one. The —OH group points left in an L enantiomer and right in a D enantiomer.

SOLUTION In (a) the —OH group on the carbon above the bottom of the structure points right, so this is D-ribose. In (b), this —OH group points to the left, so this is L-mannose.

■ PROBLEM 22.6

Draw the enantiomer of each of the following monosaccharides, and in each pair identify which is the D sugar and which is the L sugar.

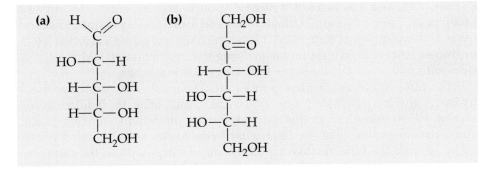

22.4 Structure of Glucose and Other Monosaccharides

D-Glucose, sometimes called *dextrose* or *blood sugar*, is the most widely occurring of all monosaccharides and has the most important function. In nearly all living organisms, D-glucose serves as a source of energy to fuel biochemical reactions. It is stored in polymeric form as starch in plants and as glycogen in animals (Section 22.9).

The discussion here of the structure of D-glucose illustrates a major point about the structure of monosaccharides. Although they can be written with the carbon atoms in a straight chain, monosaccharides with five or six carbon atoms exist mainly in their cyclic forms.

Look at the Fischer projection of D-glucose at the top left-hand corner of Figure 22.3 and notice the locations of the aldehyde group and the hydroxyl groups. You have seen that aldehydes and ketones react reversibly with alcohols to yield hemiacetals as shown below. (⊙⊙⊙, Section 16.7) (Remember that the key to recognizing the hemiacetal is a carbon atom bonded to both an —OH and an — OR group.)

$$R-\overset{\overset{\displaystyle O}{\|}}{C}-H \;+\; \overset{\overset{\displaystyle H}{|}}{O}-R' \;\rightleftharpoons\; R-\overset{\overset{\displaystyle O-H}{|}}{\underset{\underset{\displaystyle H}{|}}{C}}-O-R'$$

An aldehyde   An alcohol   A hemiacetal

▼ **FIGURE 22.3 The structure of D-glucose.** D-Glucose can exist as an open-chain hydroxy aldehyde or as a pair of cyclic hemiacetals. The cyclic forms differ only at C1, where the —OH group is either on the opposite side of the six-membered ring from the $CH_2OH(\alpha)$ or on the same side (β). To convert the Fischer projection into the six-membered ring formula, the Fischer projection is laid down with C1 to the right and the other end curled around at the back. Then the single bond between C4 and C5 is rotated so that the —CH_2OH group is vertical. Finally, the hemiacetal O—R bond is formed by connecting oxygen from the —OH group on C5 to C1, and the hemiacetal O—H group is placed on C1. (H's on carbons 2–5 are omitted here for clarity.)

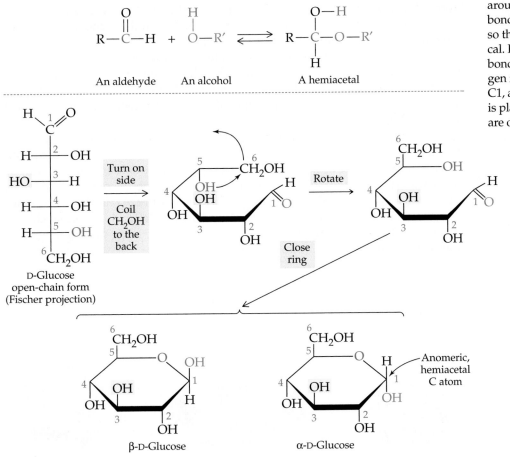

Since glucose has alcohol hydroxyl groups and an aldehyde carbonyl group in the same molecule, *internal* hemiacetal formation is possible. The aldehyde carbonyl group at carbon 1 (C1) and the hydroxyl group at carbon 5 (C5) in glucose react to form a six-membered ring that is a hemiacetal. Monosaccharides with five or six carbon atoms form rings in this manner.

The three structures at the top in Figure 22.3 show how to picture the 5-hydroxyl and the aldehyde group approaching each other for hemiacetal formation. When visualized in this manner, Fischer projections are converted to cyclic structures that (like the Fischer projections) can be interpreted consistently because the same relative arrangements of the groups on the chiral carbon atoms are maintained.

In the cyclic structures at the bottom of Figure 22.3, note how the —OH group on carbon 3, which is on the left in the Fischer projection, points *up* in the cyclic structure, and —OH groups that were on the right on carbons 2 and 4 point *down*. When cyclic structures (called *Haworth projections*) are drawn as illustrated in Figure 22.3, such relationships are always maintained. Note also that the —CH₂OH group in D sugars is always *above* the plane of the ring.

The hemiacetal carbon atom (C1) in the cyclic structures, like that in other hemiacetals, is bonded to two oxygen atoms (one in —OH and one in the ring). This carbon is chiral. As a result, there are two cyclic forms of glucose, known as the α and β forms. To see the difference, compare the locations of the hemiacetal —OH groups on C1 in the two bottom structures in Figure 22.3. In the β form, the hydroxyl at C1 points *up* and is on the same side of the ring as the —CH₂OH group at C5. In the α form, the hydroxyl at C1 points *down* and is on the opposite side of the ring from the —CH₂OH group. This relationship is maintained in cyclic monosaccharide structures drawn as demonstrated in Figure 22.3.

Cyclic monosaccharides that differ only in the positions of substituents at carbon 1 are known as **anomers**, and carbon 1 is said to be an **anomeric carbon atom**. It's the one that was the carbonyl carbon atom (C1 in an aldose and C2 in a hexose) and is now bonded to two O atoms. Note that the α and β anomers of a given sugar are not optical isomers because they aren't mirror images.

Although the structural difference between anomers might seem small, it has enormous biological consequences. You'll see that this one small change in structure accounts for the vast difference between the digestibility of starch, which we can digest, and that of cellulose, which we cannot digest (Section 22.9).

Ordinary crystalline glucose is entirely in the cyclic α form. Once dissolved in water, however, equilibrium is established among the open-chain form and the two anomers. The optical rotation of a freshly made solution of α-D-glucose gradually changes from its original value until it reaches a constant value that represents the optical activity of the equilibrium mixture. A solution of β-D-glucose or a mixture of the α and β forms would also undergo this gradual change in rotation, known as **mutarotation**, until the ring opening and closing reactions come to the following equilibrium:

Anomers Cyclic sugars that differ only in positions of substituents at the hemiacetal carbon (the anomeric carbon); the α form has the —OH on the opposite side from the —CH₂OH; the β form has the —OH on the same side as the —CH₂OH.

Anomeric carbon atom The hemiacetal C atom in a cyclic sugar; the C atom bonded to an —OH group and an O in the ring.

Mutarotation Change in rotation of plane-polarized light resulting from the equilibrium between cyclic anomers and the open-chain form of a sugar.

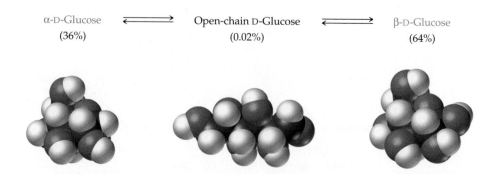

α-D-Glucose		Open-chain D-Glucose		β-D-Glucose
(36%)	⇌	(0.02%)	⇌	(64%)

All monosaccharides with five or six carbon atoms establish similar equilibria, but with different percentages of the different forms.

Monosaccharide Structures—Summary

- Monosaccharides are polyhydroxy aldehydes or ketones.
- Monosaccharides have three to seven carbon atoms, and a maximum of 2^n possible stereoisomers, where n is the number of chiral carbon atoms.
- D and L enantiomers differ in the orientation of the —OH group on the chiral carbon atom farthest from C1. In Fischer projections, D sugars have this —OH on the right and L sugars have this —OH on the left.
- D-Glucose (and other six-carbon aldoses) forms cyclic hemiacetals conventionally represented (as in Figure 22.3) so that —OH groups on chiral carbons on the left in Fischer projections point up and those on the right in Fischer projections point down.
- In glucose, the hemiacetal carbon (*the anomeric carbon*) is chiral, and α and β anomers differ in the orientation of the —OH groups on this carbon. The α anomer has the —OH on the opposite side from the —CH$_2$OH, and the β anomer has the —OH on the same side as the —CH$_2$OH.

Enantioners

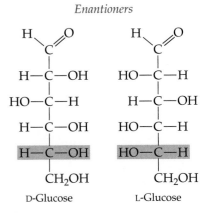

D-Glucose   L-Glucose

Anomers

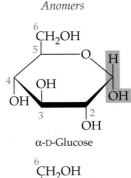

α-D-Glucose

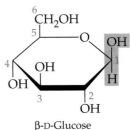

β-D-Glucose

■ WORKED EXAMPLE 22.2

The open-chain form of D-altrose, an aldohexose isomer of glucose, has the following structure. Draw D-altrose in its cyclic hemiacetal form:

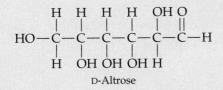

D-Altrose

SOLUTION First, coil D-altrose into a circular shape by mentally grasping the end farthest from the carbonyl group and bending it backward into the plane of the paper:

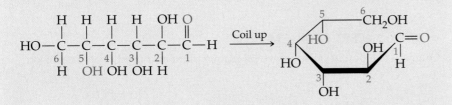

Next, rotate the bottom of the structure around the single bond between C4 and C5 so that the —CH$_2$OH group at the end of the chain is pointing up and the —OH group on C5 is pointing toward the aldehyde carbonyl group on the right:

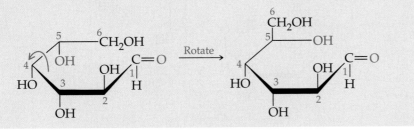

Finally, add the —OH group at C5 to the carbonyl C=O to form a hemiacetal ring. The new —OH group formed on C1 can be either up (β) or down (α):

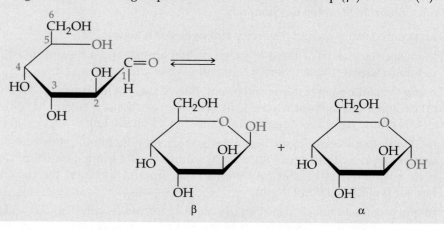

■ PROBLEM 22.7

D-Talose, a constituent of certain antibiotics, has the open-chain structure shown below. Draw D-talose in its cyclic hemiacetal form.

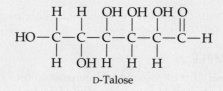

D-Talose

■ PROBLEM 22.8

The cyclic structure of D-mannose, an aldohexose, is drawn below. Convert this to the straight-chain Fischer projection structure.

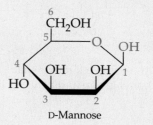

D-Mannose

■ KEY CONCEPT PROBLEM 22.9

Ouabain is a potent poison derived from a plant and used as a dart poison.

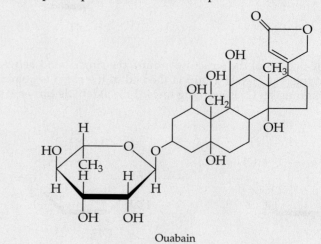

Ouabain

Application Carbohydrates in the Diet

The major monosaccharides in our diets are fructose and glucose from fruits and honey. The major disaccharides are sucrose, which is common table sugar, and lactose from milk. In addition, our diets contain large amounts of the digestible polysaccharide starch, present in grains such as wheat and rice, root vegetables such as potatoes, and legumes such as beans and peas. Nutritionists often refer to polysaccharides as *complex carbohydrates*.

The body's major use of digestible carbohydrates is to provide energy, 4 kcal per gram of carbohydrate. A small amount of any excess carbohydrate is converted to glycogen for storage in the liver and muscles, but most dietary carbohydrate in excess of our immediate needs for energy is converted into fat.

The Food Guide Pyramid (p. 510) reflects the emphasis on decreasing the amounts of meat and increasing the amounts of nonmeat products in our diet. At the base of the pyramid is the recommendation for 6–11 servings per day of bread, cereal, rice, and pasta, all foods high in complex carbohydrates. If your reaction is, I can't possibly eat that much, you should know that one serving, as defined for the pyramid, is quite small: 1 slice of bread: 1/2 cup of cooked rice, pasta, or cereal; or 1 ounce of cold cereal.

In terms of *total* carbohydrate, which includes both simple sugars and fiber, the *Nutrition Facts* labels on packaged foods (p. 566) give percentages based on a recommended 300 g per day of total carbohydrate and 25 g per day of dietary fiber (the nondigestible carbohydrates). These quantities provide 2000 Calories a day, with 60% of the calories from carbohydrates. The label also gives the total grams of sugars in the food, without a percentage because there is no recommended daily quantity of sugars. For purposes of the label, "sugars" are defined as all monosaccharides and disaccharides, whether naturally present or added.

▲ Some healthy dietary carbohydrates.

As an option, the label may also include grams of *soluble fiber* and *insoluble fiber*. Taken together, these are the types of polysaccharides that cannot be hydrolyzed to monosaccharides and absorbed into the bloodstream.

The U.S. Food and Drug Administration has the responsibility for reviewing the scientific basis for health claims for foods. Two of the allowed claims relate to carbohydrates. The first states that diets high in fiber might lower the risk of cancer and might lower the risk of heart disease in a diet that is also low in saturated fats and cholesterol. The second states that foods high in the soluble fiber from whole oats (oat bran) in reducing the risk of heart disease, again when the diet is also low in saturated fats and cholesterol. (For further information, see Application "Dietary Fiber" at the end of this chapter.)

See Additional Problems 22.75–76 at the end of the chapter.

The structure can be roughly divided into three sections: a monosaccharide ring, a four-ring system, and an oxygen-containing ring known as a lactone. (a) Identify the monosaccharide ring according to the number of carbons in the ring. Based on the location of the linkage between the monosaccharide ring and the larger ring system, is the monosaccharide the α or the β form? (b) The large ring system is similar to that in a class of molecules that you encountered in an earlier chapter. Identify this class of molecule. (c) Look closely at the "lactone" ring. A lactone is a cyclic version of what common organic functional group?

22.5 Some Important Monosaccharides

The monosaccharides, with their many opportunities for hydrogen bonding through their hydroxyl groups, are generally high-melting, white, crystalline solids that are soluble in water and insoluble in nonpolar solvents. Most monosaccharides and disaccharides are sweet-tasting (Table 22.1), digestible, and

TABLE 22.1   Relative Sweetness of Some Sugars and Sugar Substitutes

Name	Type	Sweetness
Lactose	Disaccharide	16
Galactose	Monosaccharide	30
Maltose	Disaccharide	33
Glucose	Monosaccharide	75
Sucrose	Disaccharide	100
Fructose	Monosaccharide	175
Cyclamate	Artificial	3,000
Aspartame	Artificial	15,000
Saccharin	Artificial	35,000

nontoxic (Figure 22.4). Except for glyceraldehyde (an aldotriose) and fructose (a ketohexose), the carbohydrates of interest in human biochemistry are all aldohexoses or aldopentoses. Most are in the D family.

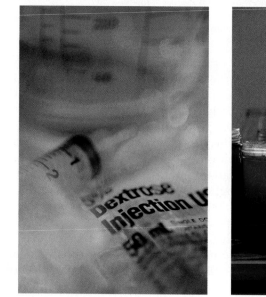

(a) (b) (c)

▲ **FIGURE 22.4 Monosaccharides**
(a) Intravenous fluid, which supplies glucose. (b) Jelly, with galactose in the pectin that stiffens it.
(c) Honey, which is high in fructose.

Glucose

Glucose is the most important simple carbohydrate in human metabolism. It is the final product of carbohydrate digestion and provides acetyl groups for entry into the citric acid cycle as acetyl-SCoA (∞∞ , Section 21.8) Maintenance of an appropriate blood glucose level is essential to human health. The hormones insulin and glucagon regulate its concentration in the blood. Because glucose is metabolized without further digestion, glucose solutions can be supplied intravenously to restore blood glucose levels.

∞∞ LOOKING AHEAD

In Chapter 23 we will describe the metabolic pathway (glycolysis) by which glucose is converted to pyruvate and then to acetyl-SCoA for entry into the citric acid cycle. The role of insulin in controlling blood glucose concentrations and the way in which those concentrations are affected by diabetes mellitus are also examined there.   ∞∞

Galactose

D-Galactose is widely distributed in plant gums and pectins, the sticky polysaccharides present in plant cells. It's also a component of the disaccharide lactose (milk sugar) and is produced from lactose during digestion. Like glucose, galactose is an aldohexose; it differs from glucose only in the spatial orientation of the —OH group at carbon 4. In the body, galactose is converted to glucose to provide energy and is synthesized from glucose to produce lactose for milk and compounds needed in brain tissue.

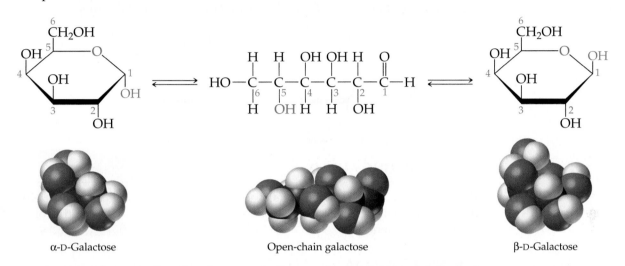

α-D-Galactose   Open-chain galactose   β-D-Galactose

A group of genetic disorders known as *galactosemias* result from an inherited deficiency of any of several enzymes needed to metabolize galactose. The result is a buildup of galactose or its 1-phosphate in blood and tissues. Early symptoms in infants include vomiting, an enlarged liver, and general failure to thrive. Other possible outcomes are liver failure, mental retardation, and development of cataracts when galactose in the eye is reduced to a polyhydroxy alcohol that accumulates. Treatment of galactosemia consists of a galactose-free diet.

Fructose

D-Fructose, often called *levulose* or *fruit sugar*, occurs in honey and many fruits. It is also one of the two monosaccharides combined in sucrose. Like glucose and galactose, fructose is a six-carbon sugar. It differs in being a ketohexose rather than an aldohexose. In solution, in addition to six-membered rings, it forms five-membered rings, as shown below:

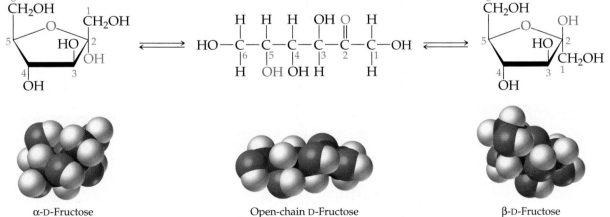

α-D-Fructose   Open-chain D-Fructose   β-D-Fructose

Fructose is sweeter than sucrose and is an ingredient in many sweetened beverages and prepared foods. As a phosphate, it is an intermediate in glucose metabolism.

Ribose and 2-Deoxyribose

Ribose and its relative 2-deoxyribose are both five-carbon aldehyde sugars. These two sugars are most important as parts of larger biomolecules. You have already seen ribose as a constituent of coenzyme A (Figure 19.10), in ATP and the second messenger cyclic AMP (Figure 20.3) and in oxidizing and reducing agent coenzymes (p. 612).

As its name shows, *2-deoxy*ribose differs from ribose by the absence of one oxygen atom, that in the —OH group at C2. Both ribose and 2-deoxyribose exist in the usual mixture of open-chain and cyclic hemiacetal forms.

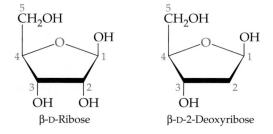

β-D-Ribose   β-D-2-Deoxyribose

⚬⚬⚬ LOOKING AHEAD

Ribose is part of RNA, ribonucleic acid, and deoxyribose is part of DNA, deoxyribonucleic acid. Chapter 26 is devoted to the roles of DNA in protein synthesis and heredity.   ⚬⚬⚬

■ **PROBLEM 22.10**

In the following monosaccharide hemiacetal, identify the anomeric carbon atom, number all the carbon atoms, and identify it as the α or β anomer.

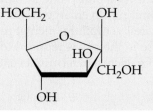

■ **PROBLEM 22.11**

Identify the chiral carbons in α-D-fructose, α-D-ribose, and and β-D-2-deoxyribose.

■ **PROBLEM 22.12**

Draw the structures of cyclic AMP and ATP (Figure 20.3), and identify the portion of the molecule from ribose.

■ **PROBLEM 22.13**

L-Fucose is one of the naturally occurring L monosaccharides. It is present in the short chains of monosaccharides by which blood groups are classified (see the Application "Cell-Surface Carbohydrates and Blood Type," p. 654). Compare the structure of L-fucose given below with the structures of α- and β-D-galactose and answer the following questions.

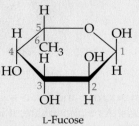

L-Fucose

(a) Is L-fucose an α or β anomer?

(b) Compared with galactose, on which carbon is L-fucose missing an oxygen?

(c) How do the positions of the —OH groups above and below the plane of the ring on carbons 2, 3, and 4 compare in D-galactose and L-fucose?

(d) "Fucose" is a common name. Is 6-deoxy-L-galactose a correct name for fucose?

22.6 Reactions of Monosaccharides

Reaction with Oxidizing Agents: Reducing Sugars

Aldehydes can be oxidized to carboxylic acids (RCHO $\rightarrow$ RCOOH), a reaction that applies to the open-chain form of aldose monosaccharides. (∞, p. 451) As the open-chain aldehyde is oxidized, its equilibrium with the cyclic form is displaced, in accord with Le Châtelier's principle, so that the open-chain form continues to be produced. (∞, p. 182) As a result, the aldehyde group of the monosaccharide is ultimately oxidized to a carboxylic acid group. For glucose, the reaction is

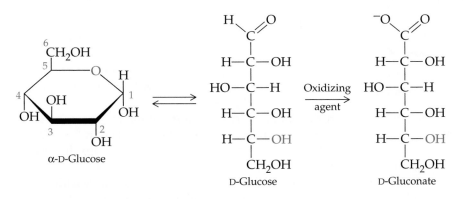

Carbohydrates that react with mild oxidizing agents are classified as **reducing sugars** (they reduce the oxidizing agent).

You probably would not predict it, but in basic solution, ketoses are also reducing sugars. The explanation is that, under these conditions, a ketone that has a hydrogen atom on the carbon adjacent to the carbonyl carbon undergoes a rearrangement. This hydrogen moves over to the carbonyl oxygen. The product is an *enediol*, "ene" for the double bond and "diol" for the two hydroxyl groups. The enediol rearranges to give an aldose, which is susceptible to oxidation.

Reducing sugar A carbohydrate that reacts in basic solution with a mild oxidizing agent.

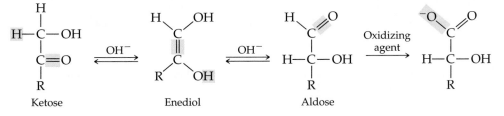

Here also, oxidation of the aldehyde to an acid drives the equilibria toward the right, and complete oxidation of the ketose occurs. Thus, *in basic solution, all monosaccharides, whether aldoses or ketoses, are reducing sugars.* This ability to act as reducing agents is the basis for most laboratory tests for the presence of monosaccharides.

The first equilibrium above—between the ketose and the enediol—is an example of *keto–enol tautomerism*, an equilibrium that results from a shift in position of a hydrogen atom and a double bond. Keto–enol tautomerism is possible whenever there is a hydrogen atom on a carbon adjacent to a carbonyl carbon.

Reaction with Alcohols: Glycoside and Disaccharide Formation

Hemiacetals react with alcohols with the loss of water to yield acetals, compounds with two — OR groups bonded to the same carbon. (∞, Section 16.7)

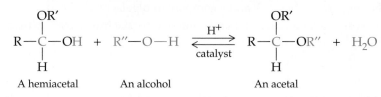

A hemiacetal   An alcohol   An acetal

Glycoside   A cyclic acetal formed by reaction of a monosaccharide with an alcohol, accompanied by loss of H_2O.

Because glucose and other monosaccharides are cyclic hemiacetals, they also react with alcohols to form acetals, which are called **glycosides**. In a glycoside, the —OH group on the anomeric carbon atom is replaced by an — OR group. For example, glucose reacts with methanol to produce methyl glucoside. (Note that a *gluc*oside is a cyclic acetal formed by glucose. A cyclic acetal derived from *any* sugar is a *gly*coside.)

Formation of a glycoside

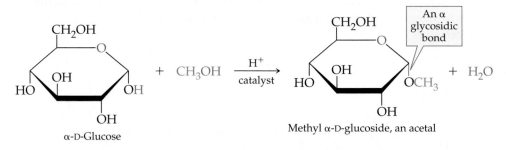

α-D-Glucose

Methyl α-D-glucoside, an acetal

Glycosidic bond   Bond between the anomeric carbon atom of a monosaccharide and an —OR group.

📀 **Common Carbohydrate Terms**

The bond between the anomeric carbon atom of the monosaccharide and the oxygen atom of the — OR group is called a **glycosidic bond**. Since glycosides like the one shown above do not contain hemiacetal groups that establish equilibria with open-chain forms, they are *not* reducing sugars.

In larger molecules, including disaccharides and polysaccharides, monosaccharides are connected to each other by glycosidic bonds. For example, a disaccharide forms by reaction of the anomeric carbon of one monosaccharide with an —OH group of a second monosaccharide.

Formation of a glycosidic bond between two monosaccharides

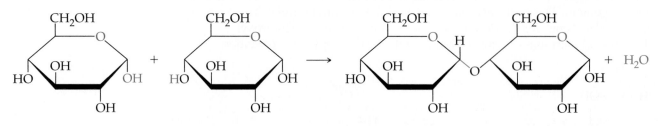

The reverse of this reaction is a *hydrolysis* and is the reaction that takes place during digestion of all carbohydrates.

Hydrolysis of a disaccharide

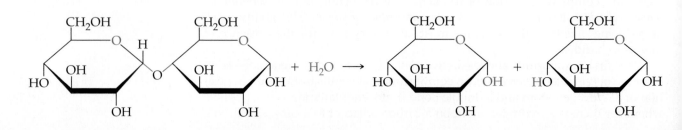

■ **PROBLEM 22.14**

Draw the structure of the α and β anomers that would result from the reaction of methanol and ribose (see p. 644) Are these compounds acetals or hemiacetals?

Formation of Phosphate Esters of Alcohols

Phosphate esters of alcohols contain a $-PO_3^{2-}$ group bonded to the oxygen atom of an $-OH$ group. The $-OH$ groups of sugars can add $-PO_3^{2-}$ groups to form phosphate esters in the same manner. The resulting phosphate esters of monosaccharides appear as reactants and products throughout the metabolism of carbohydrates. Glucose phosphate is the first to be formed and sets the stage for subsequent reactions. It is produced by the transfer of a $-PO_3^{2-}$ group from ATP to glucose in the first step of glycolysis, the metabolic pathway followed by glucose and other sugars, which is described in Chapter 23. Glycolysis leads to the ultimate conversion of glucose to the acetyl groups that are carried into the citric acid cycle.

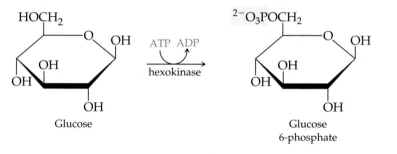

22.7 Disaccharides

Every day, you eat a disaccharide—sucrose. It's table sugar, which is made of two monosaccharides, one glucose and one fructose, bonded to each other. Sucrose is present in modest amounts, along with other mono- and disaccharides, in most fresh fruits and many fresh vegetables. But most sucrose in our diets has been added to something. Perhaps you add it to your coffee or tea. Or it is there in a ready-to-eat food product that you buy, maybe breakfast cereal, ice cream, or a "super sized" soda, or even bread. Excessive consumption of high-sucrose foods has been blamed for everything from criminal behavior to heart disease to hyperactivity in children, but without any widely accepted scientific proof. A proven connection with heart disease does exist, of course, but by way of the contribution of excess sugar calories to obesity.

Disaccharide Structure

The two monosaccharides in a disaccharide are connected by a glycosidic bond. The bond may be α or β as in cyclic monsaccharides: α points below the ring and β points above the ring (see Figure 22.3). The structures below show glycosidic bonds that create a **1,4 link**, that is, a link between C1 of one monosaccharide and C4 of the second monosaccharide.

1,4 Link A glycosidic link between the hemiacetal hydroxyl group at C1 of one sugar and the hydroxyl group at C4 of another sugar.

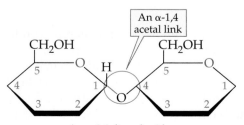

An α-1,4 disaccharide

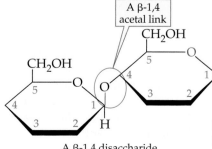

A β-1,4 disaccharide

 Identifying Glycosidic Linkages and Numbering Carbohydrate Rings

The three naturally occurring disaccharides discussed in the following sections are the most common ones. They illustrate the three different ways monosaccharides can be linked: by a glycosidic bond in the α orientation (maltose), a glycosidic bond in the β orientation (lactose), or a bond that connects two anomeric carbon atoms (sucrose).

Maltose

Maltose, often called malt sugar, is present in fermenting grains and can be prepared by enzyme-catalyzed degradation of starch. It is used in prepared foods as a sweetener. In the body, it is produced during starch digestion by α-amylase in the small intestine and then hydrolyzed to glucose by a second enzyme.

Two α-D-glucose molecules are joined in maltose by an α-1,4 link. A careful look at maltose shows that it is both an acetal (at C1 in the glucose on the left below) and a hemiacetal (at C1 in the glucose on the right below). Since the acetal ring on the left doesn't open and close spontaneously, it can't react with an oxidizing agent. The hemiacetal group on the right, however, establishes equilibrium with the aldehyde, making maltose a reducing sugar.

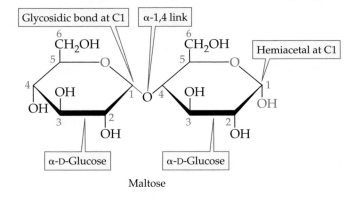

Maltose

Lactose

Lactose, or milk sugar, is the major carbohydrate in mammalian milk. Human milk, for example, is about 7% lactose. Structurally, lactose is a disaccharide composed of β-D-galactose and β-D-glucose. The two monosaccharides are connected by a β-1,4-link. Like maltose, lactose is a reducing sugar because the glucose ring (on the right in the following structure) is a hemiacetal at C1.

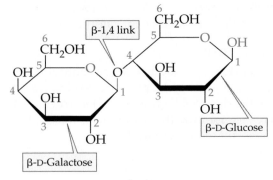

Lactose

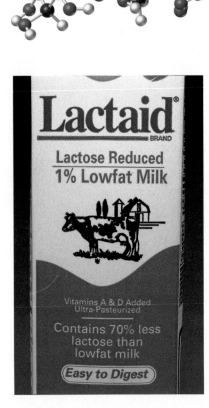

▲ Milk for lactose-intolerant individuals. The lactose content of the milk has been decreased by treating it with lactase.

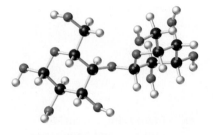

Lactose intolerance in adults is an unpleasant, though not life-threatening condition that is prevalent in all populations other than those of Northern European descent. In fact, it has been suggested that the *absence* of this condition in adults rather than its presence is the deviation from the usual. The activity of lactase, the enzyme that allows lactose digestion by infants, apparently gradually diminishes over the years. Because lactose remains in the intestines rather than being absorbed, it raises the osmolarity there, which draws in excess water. (∞, Section 9.12) Bacteria in the intestine also ferment the lactose to produce lactate, carbon dioxide, hydrogen, and methane. The result is bloating, cramps, flatulence, and diarrhea. The condition may be treated by a lactose-free diet, which extends to limitations on taking the many medications and artificial sweeteners in which lactose is an inactive ingredient. An alternative is the use of commercial enzyme preparations taken before milk products are consumed and Lactaid, milk that has been treated with lactase to reduce its lactose content.

Sucrose

Sucrose—plain table sugar—is probably the most common highly purified organic chemical used in the world. Sugar beets and sugarcane are the most common sources of sucrose. Hydrolysis of sucrose yields one molecule of D-glucose and one molecule of D-fructose. The 50:50 mixture of glucose and fructose that results, often referred to as *invert sugar*, is commonly used as a food additive because it's sweeter than sucrose.

Sucrose differs from maltose and lactose in that it has no hemiacetal group because a 1,2 link joins *both* anomeric carbon atoms. The absence of a hemiacetal group means that sucrose is not a reducing sugar. Sucrose is the only common disaccharide that is not a reducing sugar.

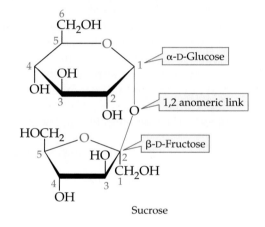

Sucrose

■ **WORKED EXAMPLE 22.3**

The disaccharide cellobiose can be obtained by enzyme-catalyzed hydrolysis of cellulose. Would you expect cellobiose to be a reducing or a nonreducing sugar?

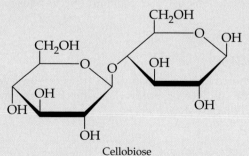

Cellobiose

ANALYSIS To be a reducing sugar, a disaccharide must contain a hemiacetal group, that is, a carbon bonded to one —OH group and one —OR group. The ring at the right in the structure above has such a group.

SOLUTION Cellobiose is a reducing sugar.

Application Frozen Fish?

Frozen fish are there in the supermarket. What about fish in the Antarctic? Do they suffer a frozen death when trapped beneath the polar ice? Ice crystals, after all, are like broken glass. When they form, living cells are cut to pieces, vital fluids and their dissolved biochemicals escape from cells, and cell functions cease.

The temperature beneath the sea ice in McMurdo Sound is −1.9°C. Yet there are fish swimming about. A first guess might be that the ions and molecules dissolved in body fluids provide a freezing point lowering (p. 253) sufficient to protect them, but that is wrong. Freezing point lowering can't do the job.

Biochemistry has provided some species of Antarctic fish with a most efficient antifreeze, reportedly 100–200 times more efficient than all other antifreezes. It's a glycoprotein. The repeating unit is a tripeptide (alanine-alanine-threonine) with a disaccharide (galactosyl-N-acetylgalactosamine) bonded to every threonine. The unit repeats from 17 to 50 times.

The glycoprotein antifreeze doesn't function by lowering the freezing point. Consider that the sugars have many polar —OH groups available to be attracted to water molecules. By binding with water molecules at the surface of tiny ice crystals, the glycoprotein slows down growth of the crystals. In other words, the glycoprotein gets in the way of more water molecules joining the ice crystal surface. When the blood containing the tiny growth-inhibited crystals circulates to the liver, the crystals melt because the liver

▲ Atlantic codfish, which can survive in frigid water.

remains warm enough to melt them before the blood flows back to colder parts.

Not surprisingly, there is widespread speculation about possible applications of antifreeze glycoprotein and proteins that serve a similar function in some species of fish. Could the large crystals that form in ice cream stored for a long time be avoided? Could genes for production of antifreeze be introduced into fruits that are damaged if the temperature drops below freezing? Could the preservation of organs for transplant or blood for transfusion be improved? Only time will tell.

See Additional Problems 22.77–78 at the end of the chapter.

■ **PROBLEM 22.15**

Refer to the cellobiose structure in the preceding Worked Example. How would you classify the link between the monosaccharides in cellobiose?

■ **PROBLEM 22.16**

Refer again to the cellobiose structure in the preceding Worked Example. Show the structures of the two monosaccharides that are formed on hydrolysis of cellobiose. What are their names?

■ **KEY CONCEPT PROBLEM 22.17**

Identify the following disaccharides. (a) The disaccharide contains two glucose units joined by an α-glycosidic linkage. (b) The disaccharide contains fructose and glucose. (c) The disaccharide contains galactose and glucose.

22.8 Variations on the Carbohydrate Theme

Monosaccharides with modified functional groups are components of a wide variety of biomolecules. Also, short chains of monosaccharides (known as *oligosaccharides*) enhance the functions of proteins and lipids to which they are bonded.

In this section we mention a few of the more interesting and important variations on the carbohydrate theme, several of which incorporate the modified glucose molecules shown below. Their distinctive functional groups are highlighted in yellow.

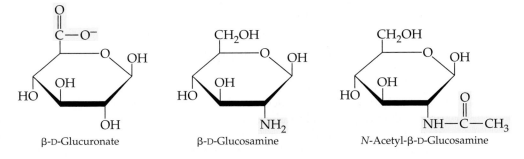

β-D-Glucuronate   β-D-Glucosamine   N-Acetyl-β-D-Glucosamine

Chitin

The shells of lobsters, beetles, and spiders are made of chitin, which is reported to be the polysaccharide second only to cellulose in its abundance in the natural world. Chitin is a hard, structural polymer. It is composed of N-acetyl-D-glucosamine rather than glucose but is otherwise identical to cellulose (p. 654).

▲ An Australian rock lobster with its chitin shell.

Connective Tissue and Polysaccharides

Connective tissues such as blood vessels, cartilage, and tendons are composed of protein fibers embedded in a syrupy matrix that contains unbranched polysaccharides (*mucopolysaccharides*). The gel-like mixtures of these polysaccharides with water serve as lubricants and shock absorbers around joints and in extracellular spaces. The repeating disaccharide units in two of these polysaccharides, hyaluronate and chondroitin, are illustrated below.

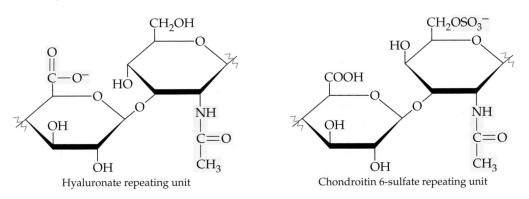

Hyaluronate repeating unit   Chondroitin 6-sulfate repeating unit

Hyaluronate molecules contain up to 25,000 disaccharide units and form a quite rigid, very viscous mixture with water molecules attracted to its negative charges. This mixture is the *synovial fluid* that lubricates joints. It is also present within the eye. *Chondroitin 6-sulfate* (also the 4-sulfate) is present in tendons and cartilage, where it is linked to proteins. It has been used in artificial skin. Chondroitin sulfates and glucosamine sulfate are available as dietary supplements in health food stores and are promoted as cures for osteoarthritis, in which there is a deterioration of cartilage at joints. They are prescribed by veterinarians for arthritic dogs, and there is anecdotal evidence for benefits in humans.

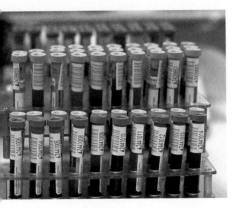

Heparin

Another of the polysaccharides associated with connective tissue, heparin is valuable medically as an *anticoagulant* (an agent that prevents or retards the clotting of blood). Heparin is composed of a variety of different monosaccharides, many of them containing sulfate groups.

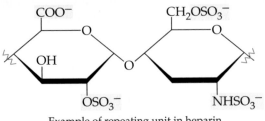

Example of repeating unit in heparin

In the structure above, notice the large number of negative charges. Heparin binds strongly to a blood-clotting factor and in this way presents clot formation. It is used clinically to prevent clotting after surgery or serious injury. Also, a coating of heparin is applied to any surfaces that will come into contact with blood that must not clot, such as the interiors of containers for blood samples collected for analysis or materials in prosthetic implants for the body.

Glycoproteins

Glycoprotein A protein that contains a short carbohydrate chain.

Proteins that contain short carbohydrate chains (*oligosaccharide* chains) are known as **glycoproteins**. (The prefix *glyco-* always refers to carbohydrates.) The carbohydrate is connected to the protein by a glycosidic bond between an anomeric carbon and a side chain of the protein. The bond is either a C—N glycosidic bond or a C—O glycosidic bond, as illustrated below.

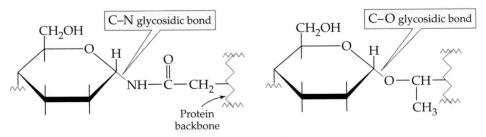

Glycoproteins have important functions at the surfaces of all cells. (You might say our cells are sugar-coated.) The protein portion of the molecule lies within the cell membrane, and the hydrophilic carbohydrate portion extends into the surrounding fluid. There, the oligosaccharide chains can function as receptors for molecular messengers, other cells, pathogenic microorganisms, or drugs. They are also responsible for the familiar A, B, O system of typing blood. (See Application "Cell-Surface Carbohydrates and Blood Type," p. 654.)

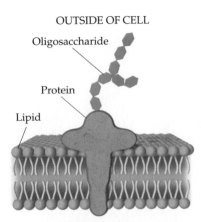

OUTSIDE OF CELL

Oligosaccharide

Protein

Lipid

INSIDE OF CELL

∞∞ LOOKING AHEAD

The basic components of cell membranes are lipid molecules. The wonderfully complex structure and function of the membrane are explored in Sections 24.5 and 24.6. Glycolipids—carbohydrates bonded to lipids—are, like glycoproteins, essential in cell membranes and are also discussed in Section 24.5. ∞∞

■ **PROBLEM 22.18**

In *N*-linked glycoproteins, the sugar is usually attached to the protein by a bond to the N atom in a side-chain amide. Which amino acids can form such a bond?

■ **PROBLEM 22.19**

Identify the type of glycosidic linkage in the repeating unit of heparin illustrated in this section.

22.9 Some Important Polysaccharides

Polysaccharides are polymers of tens, hundreds, or even many thousands of monosaccharides linked together through glycosidic bonds of the same type as in maltose and lactose. Three of the most important polysaccharides are *cellulose, starch,* and *glycogen.* Their repeating units are compared in the following structures.

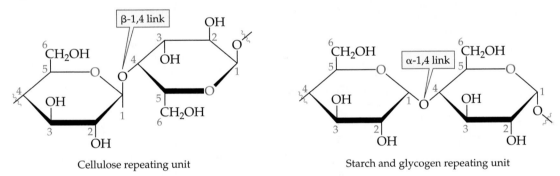

Cellulose repeating unit   Starch and glycogen repeating unit

Cellulose

Cellulose is the fibrous substance that provides structure in plants. Each huge cellulose molecule consists entirely of several thousand β-D-glucose units joined in a long straight chain by β-1,4 links. The bonding in cellulose is illustrated above by the flat hexagons we have used so far for monosaccharides. In reality, because of the tetrahedral bonding at each carbon atom, the carbohydrate rings are not flat but are bent up at one end and down at the other in what is known as the *chair conformation.* You may often see this shape in other books.

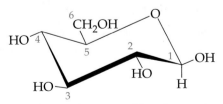

Chair conformation of β-D-glucose

By drawing the cellulose structure using chair conformations it is much easier to see how each glucose ring is reversed relative to the next. In the cellulose structure at the bottom of page 654, compare the locations of the O atoms in the rings. The hydrogen bonds within chains and between chains (shown in red) contribute to the rigidity and toughness of cellulose fibers.

Application

Cell-Surface Carbohydrates and Blood Type

It was discovered more than 90 years ago that human blood can be classified into four blood group types, called A, B, AB, and O. This classification results from the presence on red blood cell surfaces of three different oligosaccharide units, designated A, B, and O (see the diagram below). Individuals with type AB blood have both A and B oligosaccharides. (The structure of N-acetylglucosamine is shown on p. 651; L-fucose is shown on p. 644.)

Knowing the ABO type of blood is vitally important in choosing blood for transfusions. This is because a major component of the body's immune system (Chapter 28) is a collection of proteins called *antibodies* that recognize and attack foreign substances, such as

viruses, bacteria, and potentially harmful macromolecules. Among the targets of these antibodies are cell-surface molecules not present on the individual's own cells. For example, if you have type A blood, your plasma (the fluid portion of the blood) contains antibodies to the type B oligosaccharide shown above. Thus, if type B blood enters your body, its red blood cells will be recognized as foreign and your immune system will launch an attack on them. The result is clumping of the cells (agglutination), blockage of capillaries, and possibly death.

Because of the danger of such interactions, both the blood types that individuals can receive and the blood types of recipients to whom they can donate blood are

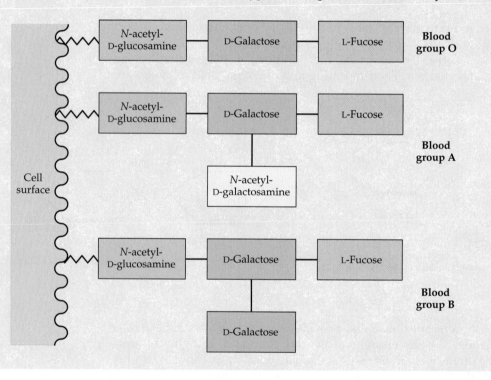

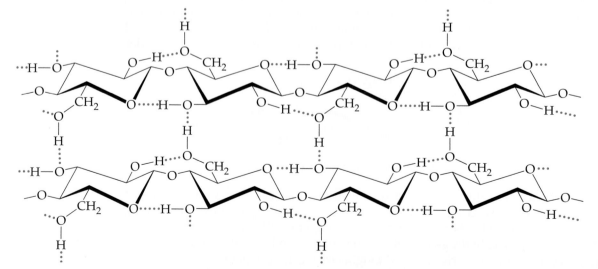

Cellulose

654

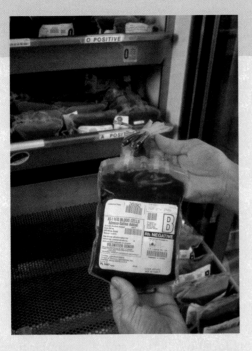

people with blood types A, B, and AB all lack antibodies to type O cells. Individuals with type O blood are therefore known as "universal donors"—in an emergency, their blood can safely be given to individuals of all blood types.

- Similarly, type AB individuals are known as "universal recipients." Because people with type AB blood have both A and B molecules on their red cells, their blood contains no antibodies to A, B, or O, and they can, if necessary, receive blood of all types.

- In theory, antibodies in the plasma of donated blood could also attack the red cells of the recipient. In practice, such reactions are unlikely to cause significant harm. Unless very large quantities of whole blood or plasma (the fluid portion of the blood) are being transfused, the donor's blood is quickly diluted by mixing with the much larger volume of the recipient's blood. Moreover, many transfusions today consist of packed red cells, with a minimum of the antibody-containing plasma. Nevertheless, exact matching of blood types is preferred whenever possible.

limited, as indicated in the accompanying table. A few features of the table deserve special mention:

- Note in the diagram on the opposite page that type O cell-surface oligosaccharides are similar in composition to those of both types A and B. Consequently,

(Another blood type classification of major clinical significance, which is more complex and less fully understood, is the presence or absence of Rh factor.)

Person with blood of type	Will have cell-surface oligosaccharides	Will have antibodies to cell-surface oligosaccharides	Can receive blood of type	Can donate blood to person with blood of type
O ("universal donor")	O	A and B	O	O, A, B, and AB (Red blood cells only)
A	A	B	O and A	A and AB
B	B	A	O and B	B and AB
AB	A and B	None	O, A, B, and AB (Red blood cells only)	AB

See Additional Problems 22. 79–80 at the end of the chapter.

Earlier we noted that the seemingly minor distinction between the α and β forms of cyclic sugars accounts for a vast difference between cellulose and starch. Cows and other grazing animals, termites, and moths are able to digest cellulose because microorganisms colonizing their digestive tracts produce enzymes that hydrolyze its β glycosidic bonds. Humans do not produce such enzymes nor harbor such organisms, and therefore can't hydrolyze cellulose, although some is broken down by bacteria in the large intestine. Cellulose is what grandma used to call "roughage," and we do need it in our diets in addition to starch.

Starch

Starch, like cellulose, is a polymer of glucose. In starch, individual glucose units are joined by α-1,4 links rather than by the β-1,4 links of cellulose. Starch is fully digestible and is an essential part of the human diet. It is present only in plant material; our major sources are beans, wheat, rice, and potatoes.

▲ **FIGURE 22.5 Helical structure of amylose.**

Unlike cellulose, which has only one form, there are two kinds of starch—amylose and amylopectin. *Amylose*, which accounts for about 20% of starch, is somewhat soluble in hot water and consists of several hundred to a thousand α-D-glucose units linked in long chains by the α-1,4 glycosidic bonds. Instead of lying side by side and flat as in cellulose, amylose tends to coil into helices (Figure 22.5). Amylose is what makes the cooking water cloudy when you boil potatoes.

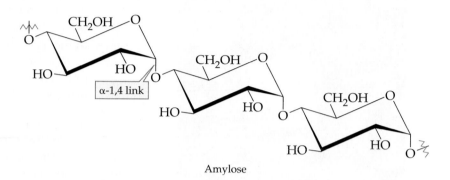

Amylose

Amylopectin, which accounts for about 80% of starch, is similar to amylose but has much larger molecules (up to 100,000 glucose units per molecule) and has α-1,6 branches approximately every 25 units along its chain. A glucose molecule at one of these branch points (shaded below) is linked to three other sugars. Amylopectin is not water-soluble.

Branch point in amylopectin (also glycogen)

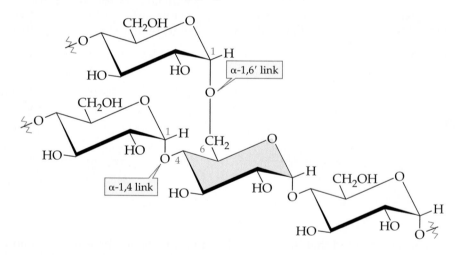

Starch molecules are digested mainly in the small intestine by α-amylase, which catalyzes hydrolysis of the α-1,4 links. As is usually the case in enzyme-catalyzed reactions, α-amylase is highly specific in its action. It hydrolyzes only α acetal links between glucose units (as in starch), while leaving β acetal links (as in cellulose) untouched.

■ **PROBLEM 22.20**

An individual starch molecule contains thousands of glucose units but has only a single hemiacetal group at the end of the long polymer chain. Would you expect starch to be a reducing carbohydrate? Explain.

Dietary fiber includes cellulose and all other indigestible polysaccharides in vegetables, both soluble and insoluble. The major categories of noncellulose fiber are hemicellulose, pectins and gums, and lignins.

Hemicellulose is a collective term for insoluble plant polysaccharides other than cellulose. These polysaccharides are composed of xylose, mannose, galactose, and modifications of these monosaccharides.

Pectins and vegetables gums, which contain galactose modified by the addition of carboxylic acid and *N*-acetyl groups, comprise the "soluble" portion of dietary fiber. Their outstanding characteristic is solubility in water or the formation of sticky or gelatinous mixtures with water. Pectins, which are present in fruits, are responsible for the "gel" in jelly. Because the texture of their dispersions in water is a desirable characteristic, pectins are often added to prepared foods to retain moisture, thicken sauces, or give a creamier texture.

Lignin, which like cellulose provides rigid structure in plants and especially in trees, is an insoluble dietary fiber. It is not a polysaccharide, however, but a polymer of complex structure that contains phenyl groups connected by carbon–carbon and carbon–oxygen bonds.

Foods high in insoluble fiber include wheat, bran cereals, and brown rice. Beans and peas have both soluble and insoluble fiber. These and other legumes are high in small polysaccharides containing galactose bonded to glucose residues. The polysaccharides are digested by bacteria in the gut, with the production of lactate, short-chain fatty acids, and gaseous by-products including hydrogen, carbon dioxide, and methane.

Fiber functions in the body to soften and add bulk to solid waste. Studies have shown that increased fiber in the diet may reduce the risk of colon and rectal cancer, hemorrhoids, diverticulosis, and cardiovascular disease. Cancer reduction may also occur because potentially carcinogenic substances are absorbed on fiber surfaces and eliminated before doing any harm. Pectin may also absorb and carry away bile acids,

▲ Beano, a product that contains α-galactosidase. Beano promises to diminish the production of gas in the large intestine.

causing an increase in their synthesis from cholesterol in the liver and a resulting decrease in blood cholesterol levels.

The U.S. Food and Drug Administration periodically reviews the Food Guide Pyramid and the related Dietary Guidelines for Americans. In releasing the fifth edition in 2000, the guidelines were described as "the gold standard when it comes to applying scientific research to what people should be eating." A significant change in the new edition was the recommendation to "Choose a variety of grains daily, especially whole grains." To accomplish this, one should choose foods with ingredients such as the following whole grains listed *first* on the ingredients label: brown rice, oatmeal, graham flour, pearl barley, or whole oats, wheat, or rye. (Note that "wheat flour" and "enriched flour" are not whole grains.)

See Additional Problems 22.81–82 at the end of the chapter.

Glycogen

Glycogen, sometimes called *animal starch*, serves the same energy storage role in animals that starch serves in plants. Some of the glucose from starches in our diet is used immediately as fuel, and some is stored as glycogen for later use. The largest amounts of glycogen are stored in the liver and muscles. In the liver, glycogen is a source of glucose, which is formed there when hormones signal a need for glucose in the blood. In muscles, glycogen is converted to glucose 6-phosphate for the synthesis of ATP.

Structurally, glycogen is similar to amylopectin in being a long polymer of α-D-glucose with the same type of branch points in its chain. Glycogen has many more branches than amylopectin, however, and is much larger—up to 1 million glucose units per molecule.

Comparison of branching in amylopectin and glycogen

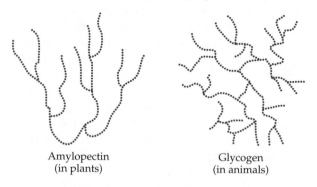

Amylopectin
(in plants)

Glycogen
(in animals)

Summary: Revisiting the Chapter Goals

1. **What are the different kinds of carbohydrates?** *Monosaccharides* are compounds with three to seven carbons, an aldehyde group on carbon 1 (an *aldose*) or a ketone group on carbon 2 (a *ketose*), and hydroxyl groups on all other carbons. *Disaccharides* consist of two monosaccharides; *polysaccharides* are polymers composed of up to thousands of monosaccharides.

2. **Why are monosaccharides handed, and how does this influence the numbers and types of their isomers?** Monosaccharides can contain several chiral carbon atoms, each bonded to one —H, one —OH, and two other carbon atoms in the carbon chain. A monosaccharide with n chiral carbon atoms may have 2^n stereoisomers and half that number of pairs of enantiomers. The members of different enantiomeric pairs are *diastereomers*—they are *not* mirror images of each other.

3. **What are the structures of monosaccharides, and how are they represented in written formulas?** *Fischer projection formulas* represent the open-chain structures of monosaccharides. They are interpreted as shown below, with the D and L enantiomers in a pair identified by having the —OH group on the chiral carbon farthest from the carbonyl group on the right (the D isomer) or the left (the L isomer).

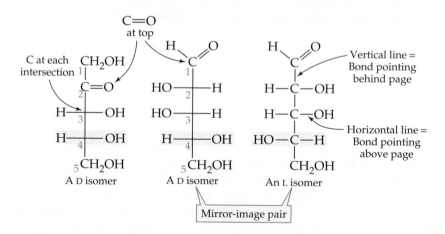

In solution, open-chain monosaccharides with five or six carbons establish equilibria with cyclic forms that are hemiacetals. The hemiacetal carbon (bonded to two O atoms) is referred to as the *anomeric carbon*, and this carbon is chiral. Two isomers of the cyclic form of a D or L monosaccharide, known as *anomers*, are possible because the —OH on the anomeric carbon may lie above or below the plane of the ring.

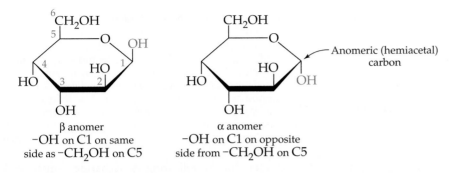

β anomer
−OH on C1 on same
side as −CH$_2$OH on C5

α anomer
−OH on C1 on opposite
side from −CH$_2$OH on C5

4. **How do monosaccharides react with oxidizing agents and alcohols?** Oxidation of a monosaccharide can result in a carboxyl group on the first carbon atom (C1 in the Fischer projection). Ketoses, as well as aldoses, are *reducing sugars* because the ketose is in equilibrium with an aldose form (via an enediol) that can be oxidized.

   Reaction of a hemiacetal with an alcohol produces an acetal. For a cyclic monosaccharide, reaction with an alcohol converts the —OH group on the anomeric carbon to an — OR group. The bond to the — OR group, known as a *glycosidic bond*, is α or β to the ring as was the —OH group. Disaccharides result from glycosidic bond formation between two monosaccharides.

5. **What are the structures of some important disaccharides?** In *maltose*, two D-glucose molecules are joined by a α-glycosidic bond that connects C1 (the anomeric carbon) of one molecule to C4 of the other—an *α-1,4 link*. In *lactose*, D-galactose and D-glucose are joined by a *β-1,4* link. In *sucrose*, D-fructose and D-glucose are joined by a glycosidic bond between the two anomeric carbons, a *1,2* link. Unlike maltose and lactose, sucrose is not a *reducing sugar* because it has no hemiacetal that can establish equilibrium with an aldehyde.

6. **What are the functions of some important carbohydrates that contain modified monosaccharide structures?** *Chitin* is a hard structural polysaccharide in the shells of lobsters and insects. Joints and intracellular spaces are lubricated by polysaccharides like *hyaluronate* and *chondroitin 6-sulfate*, which have ionic functional groups and form gel-like mixtures with water. *Heparin*, a polysaccharide with many ionized sulfate groups, bonds to a clotting factor and thus acts as an anticoagulant. *Glycoproteins* have short carbohydrate chains bonded to proteins; the carbohydrate segments (*oligosaccharides*) function as receptors at cell surfaces.

7. **What are the structures and functions of cellulose, starch, and glycogen?** *Cellulose* is a straight-chain polymer of β-D-glucose with β-1,4 links; it provides structure in plants. Cellulose is not digestible by humans, but is digestible by animals in which bacteria provide enzymes that hydrolyze the β-glycosidic bonds. *Starch* is a polymer of α-D-glucose connected by α-1,4 links in straight-chain (*amylose*) and branched-chain (*amylopectin*) forms. Starch is a storage form of glucose for plants and is digestible by humans. *Glycogen* is a storage form of glucose for animals, including humans. It is structurally similar to amylopectin, but is more highly branched. Glycogen from meat in the diet is also digestible.

▪ Understanding Key Concepts

22.21 During the digestion of starch from potatoes, the enzyme α-amylase catalyzes the hydrolysis of starch into maltose. Subsequently, the enzyme maltase catalyzes the hydrolysis of maltose into two glucose units. Classify each of these carbohydrates as a disaccharide, monosaccharide, or polysaccharide.

22.22 Identify each of the following as diastereomers, enantiomers, and/or anomers. (a) α-D-fructose and β-D-fructose (b) D-galactose and L-galactose (c) L-allose and D-glucose (both aldohexoses)

22.23 Consider the trisaccharide A, B, C at the bottom of the page.
  (a) Identify the hemiacetal and acetal linkages.
  (b) Identify the anomeric carbons, and indicate whether each is α or β.
  (c) What are the numbers of the carbons that form glycosidic linkages between monosaccharide A and monosaccharide B?
  (d) What are the numbers of the carbons that form glycosidic linkages between monosaccharide B and monosaccharide C?

22.24 Hydrolysis of both of the glycosidic bonds in the trisaccharide A, B, C at the bottom of the page would yield three monosaccharides.
  (a) Are any two of these monosaccharides the same?
  (b) Are any two of these monosaccharides enantiomers?
  (c) Draw the Fischer projections for the three monosaccharides.
  (d) Assign a name to each of the monosaccharides.

22.25 The trisaccharide shown at the bottom of the page has a specific sequence of monosaccharides. To determine this sequence, we could react the trisaccharide with an oxidizing agent. Since one of the monosaccharides in the trisaccharide is a reducing sugar, it would be oxidized from an aldehyde to a carboxylate. Which of the monosaccharides (A, B, or C) would be oxidized? Write the structure of the oxidized monosaccharide that would result after hydrolysis of the trisaccharide. How would this reaction assist in identifying the sequence of the trisaccharide?

22.26 Are any of the following disaccharides—maltose, lactose, cellobiose, sucrose—part of the trisaccharide shown at the bottom of this page? If so, identify which disaccharide and its location. (*Hint*: Look for an α-1,4 link, β-1,4 link or 1,2 link, and then determine if the correct monosaccharides are present.)

22.27 In this chapter, the polysaccharides of glucose that are considered include cellulose, amylose, amylopectin, and glycogen. The major criteria that distinguish these four polysaccharides include α-glycosidic links or β-glycosidic links, 1,4 links or both 1,4 and 1,6 links, and the degree of branching. Describe each polysaccharide using these five criteria.

22.28 In solution, glucose exists predominantly in the cyclic hemiacetal form, which does not contain an aldehyde group. How is it possible for mild oxidizing agents to oxidize glucose?

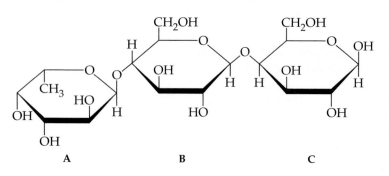

A   B   C

▪ Additional Problems

Classification and Structure of Carbohydrates

22.29 What is a carbohydrate?

22.30 What is the family-name ending for a sugar?

22.31 What is the structural difference between an aldose and a ketose?

22.32 Classify each of the following carbohydrates by indicating the nature of its carbonyl group and the number of carbon atoms present. For example, glucose is an aldohexose.

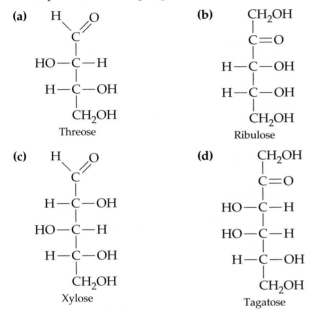

(a) Threose　(b) Ribulose　(c) Xylose　(d) Tagatose

22.33 Draw the open-chain structure of a ketoheptose.

22.34 Draw the open-chain structure of a four-carbon deoxy sugar.

22.35 Give the names of four important monosaccharides and tell where each occurs in nature.

22.36 "Dextrose" is an alternative name for what sugar?

Handedness in Carbohydrates

22.37 How are enantiomers related?

22.38 What is the structural relationship of L-glucose to D-glucose?

22.39 Only three stereoisomers are possible for 2,3-dibromo-2,3-dichlorobutane. Draw them, indicating which pair are enantiomers (optical isomers). Why does the other isomer not have an enantiomer?

22.40 You saw in Section 16.6 that aldehydes react with reducing agents to yield primary alcohols (RCH=O → RCH₂OH). The structures of two D-aldotetroses are shown. One of the two can be reduced to yield a chiral product, but the other yields an achiral product. Explain.

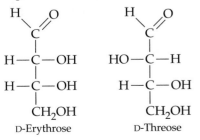

D-Erythrose　　D-Threose

22.41 What is an optically active compound?

22.42 What does a polarimeter measure?

22.43 Sucrose and D-glucose rotate plane-polarized light to the right; D-fructose rotates light to the left. When sucrose is hydrolyzed, the glucose–fructose mixture rotates light to the left.

　(a) What does this indicate about the relative degrees of rotation of light of glucose and fructose?

　(b) Why do you think that the mixture is called "invert sugar"?

22.44 What generalization can you make about the direction and degree of rotation of light by enantiomers?

Reactions of Carbohydrates

22.45 What does the term *reducing sugar* mean?

22.46 What is mutarotation, and what is its significance?

22.47 What are anomers, and how do anomers of a sugar differ?

22.48 What is the structural difference between the α hemiacetal form of a carbohydrate and the β form?

22.49 D-Gulose, an aldohexose isomer of glucose, has the following cyclic structure. Which is shown, the α form or the β form?

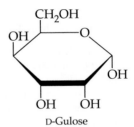

D-Gulose

22.50 D-Mannose, an aldohexose found in orange peels, has the following structure in open-chain form. Coil mannose around and draw it in cyclic hemiacetal α and β forms.

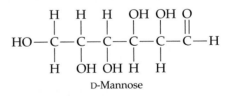

D-Mannose

22.51 Draw D-gulose (Problem 22.49) in its open-chain aldehyde form, both coiled and uncoiled.

22.52 D-Ribulose, a ketopentose related to ribose, has the following structure in open-chain form. Coil ribulose around and draw it in its five-membered cyclic β hemiacetal form.

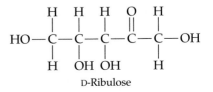

D-Ribulose

22.53 D-Allose, an aldohexose, is identical to D-glucose except that the hydroxyl group at C3 points down rather than up in the cyclic hemiacetal form. Draw the β form of this cyclic form of D-allose.

22.54 Draw D-allose (Problem 22.53) in its open-chain form.

22.55 Treatment of D-glucose with a reducing agent yields sorbitol, a substance used as a sugar substitute by diabetics. Draw the structure of sorbitol.

22.56 Reduction of D-fructose with a reducing agent yields a mixture of D-sorbitol along with a second, isomeric product. What is the structure of the second product?

22.57 Treatment of an aldose with an oxidizing agent like Tollens' reagent (Section 16.5) yields a carboxylic acid. Gluconic acid, the product of glucose oxidation, is used as its magnesium salt for the treatment of magnesium deficiency. Draw the structure of gluconic acid.

22.58 Oxidation of the aldehyde group of ribose would yield a carboxylic acid. Draw the structure of ribonic acid.

22.59 What is the structural difference between a hemiacetal and an acetal?

22.60 What are glycosides, and how can they be formed?

22.61 Look at the structure of D-mannose (Problem 22.50) and draw the two glycosidic products that you would expect to obtain by reacting D-mannose with methanol.

22.62 Draw a disaccharide of two mannose molecules attached by an α-1,4 glycosidic linkage. Explain why the glycosidic products in Problem 22.61 are *not* reducing sugars, but the product in this problem *is* a reducing sugar.

Disaccharides and Polysaccharides

22.63 Give the names of three important disaccharides. Tell where each occurs in nature. From which two monosaccharides is each made?

22.64 Lactose and maltose are reducing disaccharides, but sucrose is a nonreducing disaccharide. Explain.

22.65 Amylose (a form of starch) and cellulose are both polymers of glucose. What is the main structural difference between them? What roles do they serve in nature?

22.66 How are amylose and amylopectin similar, and how are they different?

22.67 *Gentiobiose*, a rare disaccharide found in saffron, has the following structure. What simple sugars would you obtain on hydrolysis of gentiobiose?

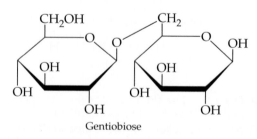

Gentiobiose

22.68 Look carefully at the structure of gentiobiose (Problem 22.67). Does gentiobiose have an acetal grouping? A hemiacetal grouping? Would you expect gentiobiose to be a reducing or nonreducing sugar? How would you classify the linkage (α or β and carbon numbers) between the two monosaccharides?

22.69 *Trehalose*, a disaccharide found in the blood of insects, has the following structure. What simple sugars would you obtain on hydrolysis of trehalose?

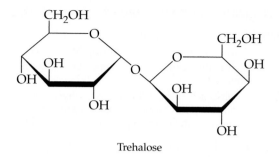

Trehalose

22.70 Does trehalose (Problem 22.69) have an acetal grouping? A hemiacetal grouping? Would you expect trehalose to be a reducing or nonreducing sugar? Classify the linkage between the two monosaccharides.

22.71 Amylopectin (a form of starch) and glycogen are both α-linked polymers of glucose. What is the structural difference between them? What roles do they serve in nature?

22.72 *Amygdalin* (Laetrile) is a glycoside isolated in 1830 from almond and apricot seeds. It is called a cyanogenic glycoside because hydrolysis with aqueous acid liberates hydrogen cyanide (HCN) along with benzaldehyde and two molecules of glucose. Structurally, amygdalin is a glycoside comprised of gentiobiose (Problem 22.67) and an alcohol, mandelonitrile. Draw the structure of amygdalin.

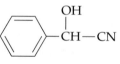

Mandelonitrile

Applications

22.73 Give some advantages and disadvantages of synthesizing and marketing a single-enantiomer drug. [*Chirality Meets the Marketplace*]

22.74 What is the advantage of using enzymes in the synthesis of single-enantiomer drugs? [*Chirality Meets the Marketplace*]

22.75 Carbohydrates provide 4 kcal per gram. If a person eats 400 g per day of digestible carbohydrates, what percentage of a 2000 kcal daily diet would be digestible carbohydrate? [*Carbohydrates in the Diet*]

22.76 What would be an example of a complex carbohydrate in the diet? What would be an example of a simple carbohydrate in the diet? Are soluble fiber and insoluble fiber complex or simple carbohydrates? [*Carbohydrates in the Diet*]

22.77 What type of intermolecular interaction occurs between water molecules and the carbohydrate groups on an antifreeze protein? [*Frozen Fish?*]

22.78 Antifreeze proteins use a different method of preventing freezing than the antifreeze we use in our cars. Explain the difference. [*Frozen Fish?*]

22.79 Look at the structures of the blood group determinants (p. 654). What groups do all blood types have in common? [*Cell-Surface Carbohydrates and Blood Type*]

22.80 People with type O blood can donate blood to anyone, but they cannot receive blood from everyone. From whom can they not receive blood? People with type AB blood can receive blood from anyone, but they cannot give blood to everyone. To whom can they give blood? Why? [*Cell-Surface Carbohydrates and Blood Type*]

22.81 Our bodies don't have the enzymes required to digest cellulose, yet it is a necessary addition to a healthy diet. Why? [*Dietary Fiber*]

22.82 Name two types of soluble fiber and their sources. [*Dietary Fiber*]

General Questions and Problems

22.83 What is the relationship between D-ribose and L-ribose? What generalizations can you make about D-ribose and L-ribose with respect to the following?
  (a) Melting point
  (b) Rotation of plane-polarized light
  (c) Density
  (d) Solubility in water
  (e) Chemical reactivity

22.84 What is the relationship between D-ribose and D-xylose (Problem 22.32c)? What generalizations can you make about D-ribose and D-xylose with respect to the following?
  (a) Melting point
  (b) Rotation of plane-polarized light
  (c) Density
  (d) Solubility in water
  (e) Chemical reactivity

22.85 L-Sorbose, which is used in the commercial production of vitamin C, differs from D-fructose only at carbon 5. Draw the open-chain structure of D-sorbose and the five-membered ring in the β form.

22.86 D-Fructose can form a six-membered cyclic hemiacetal as well as the more prevalent five-membered cyclic form. Draw the α isomer of D-fructose in the six-membered ring.

22.87 Are the α and β forms of monosaccharides enantiomers? Why or why not?

22.88 *Raffinose*, found in sugar beets, is the most prevalent trisaccharide. It is formed by an α-1,6 linkage of D-galactose to the glucose portion of sucrose. Draw the structure of raffinose.

22.89 Does raffinose (Problem 22.88) have a hemiacetal grouping? An acetal grouping? Is raffinose a reducing sugar?

22.90 When you chew a cracker for several minutes, it begins to taste sweet. What do you think the saliva in your mouth does to the starch in the cracker?

22.91 Write the open-chain structure of the only ketotriose. Name this compound and explain why it doesn't have any optical isomers.

22.92 What is the name of the group of disorders that result when the body lacks an enzyme necessary to digest galactose? What are the symptoms?

22.93 When a person cannot digest galactose, its reduced form, called dulcitol, often accumulates in the blood and tissues. Write the structure of the open-chain form of dulcitol. Does dulcitol have an enantiomer? Why or why not?

22.94 What is lactose intolerance, and what are its symptoms?

22.95 Many people who are lactose intolerant can eat yogurt, which is prepared from milk curdled by bacteria, with no problems. Give a reason why this is possible.

22.96 L-Fucose is also known as 6-deoxy-α-L-galactose. How many chiral carbons are present in L-fucose?

23 Carbohydrate Metabolism

The metabolism of carbohydrates by yeast makes this bread baking possible.

The story of carbohydrate metabolism is essentially the story of glucose: how it is converted to acetyl-SCoA for entrance into the citric acid cycle, how it is stored and then released, and how it is synthesized when carbohydrates are in short supply. Because of the importance of glucose, the body has several alternative strategies for maintaining the glucose concentration in blood and providing glucose to cells that depend on it.

In this chapter, we'll answer the following questions about carbohydrate metabolism:

1. What happens during digestion of carbohydrates?

The goal: Be able to describe where carbohydrates are digested and what the major products are.

2. **What are the major pathways in the metabolism of glucose?**

   The goal: Be able to identify the pathways available for the synthesis and breakdown of glucose and describe their interrelationships.

3. **What is glycolysis?**

   The goal: Be able to give an overview of the glycolysis pathway and its products, and identify where the major monosaccharides enter the pathway.

4. **What happens to pyruvate once it is formed?**

   The goal: Be able to describe each of the alternative pathways available to pyruvate and their respective outcomes.

5. **How is glucose metabolism regulated, and what are the influences of starvation and diabetes mellitus?**

   The goal: Be able to identify the hormones that influence glucose metabolism and describe the changes in metabolism during starvation and diabetes mellitus.

6. **What are glycogenesis and glycogenolysis?**

   The goal: Be able to define these pathways and their purpose.

7. **What is the role of gluconeogenesis in metabolism?**

   The goal: Be able to identify the functions, substrates, and products of this pathway.

CONCEPTS TO REVIEW

Phosphorylation (Section 17.8)
Function of ATP (Sections 21.5, 21.9)
Oxidized and reduced coenzymes (Section 21.7)
Carbohydrate structure (Chapter 22)

Digestion A general term for the breakdown of food into small molecules.

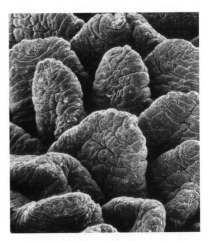

▲ A micrograph showing *villi*, the projections that line the small intestine. Each villus is covered with microvilli, where the digested food molecules are absorbed into the bloodstream.

23.1 Digestion of Carbohydrates

The first stage in catabolism is **digestion**, the breakdown of food into small molecules. Digestion entails the physical grinding, softening, and mixing of food, as well as the enzyme-catalyzed hydrolysis of carbohydrates, proteins, and fats. Digestion begins in the mouth, continues in the stomach, and concludes in the small intestine.

The products of digestion are mostly small molecules that are absorbed from the intestinal tract. The absorption happens through millions of tiny projections (the *villi*) that provide a total surface area as big as a football field. Once in the bloodstream, the small molecules are transported into target cells where many are further broken down for the purpose of releasing energy as their carbon atoms are converted to carbon dioxide. Others are excreted, and some are used as building blocks to synthesize new biomolecules.

The digestion of carbohydrates is summarized in Figure 23.1. The α-amylase in saliva catalyzes hydrolysis of the glycosidic bonds in carbohydrates (∞, p. 646). Starch from plants and glycogen from meat are hydrolyzed to give smaller polysaccharides and the disaccharide maltose. Salivary α-amylase continues to act on dietary polysaccharides in the stomach until, after an hour or so, it is inactivated by stomach acid. No further carbohydrate digestion takes place in the stomach.

α-Amylase is also secreted by the pancreas and enters the small intestine, where conversion of polysaccharides to maltose continues. Other enzymes from the mucous lining of the small intestine hydrolyze maltose and the dietary disaccharides sucrose and lactose to the monosaccharides glucose, fructose, and galactose, which are then transported across the intestinal wall into the bloodstream. Our focus in this chapter is on the metabolism of glucose; both

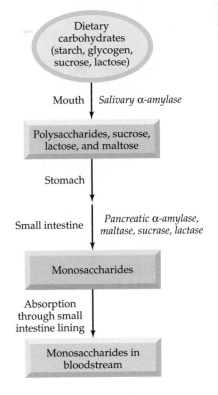

▲ **FIGURE 23.1   The digestion of carbohydrates.**

665

fructose and galactose can be converted to intermediates that enter the same metabolic pathway followed by glucose.

23.2 Glucose Metabolism: An Overview

Glucose is the major fuel for our bodies. It yields the energy carried by ATP. The initial metabolic fate of glucose is conversion into acetyl-SCoA, the common intermediate in the catabolism of all foods. The acetyl-SCoA proceeds down the central pathway of metabolism to the ultimate formation of ATP.

In this chapter, we go back to examine the central position of glucose in metabolism, as summarized in Figure 23.2. Glycolysis is the major pathway on the road to ATP synthesis. The reverse of glycolysis (gluconeogenesis) and the pathways leading to and from glycogen are discussed in the sections of this chapter noted in the figure. As you read this chapter, you will find it helpful to use Figure 23.2 and its accompanying table in sorting out the pathways that have such similar names.

When glucose enters a cell from the bloodstream, it is immediately converted to glucose 6-phosphate. Once this phosphate is formed, glucose is trapped within the cell because the phosphate cannot cross the cell membrane. Like the first step in many metabolic pathways, the formation of glucose-6-phosphate is highly exergonic and not reversible, thereby committing the initial substrate to subsequent reactions.

Several pathways are available to the glucose 6-phosphate:

- When energy is needed, glucose 6-phosphate moves down the central catabolic pathway shown in yellow in Figure 23.2, proceeding via the reactions of *glycolysis* to pyruvate and then to acetyl-S-coenzyme A, which enters the citric acid cycle. (⚭, Section 21.8)

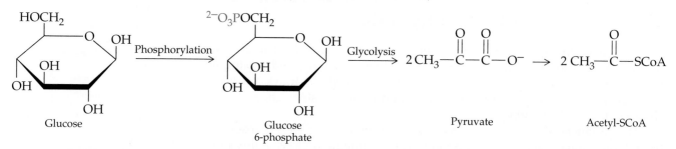

Glucose — Glucose 6-phosphate — Pyruvate — Acetyl-SCoA

- When cells are already well supplied with glucose, the excess glucose is converted to other forms for storage: to glycogen, the glucose storage polymer, by the *glycogenesis* pathway, or to fatty acids by entrance of acetyl-SCoA into the pathways of lipid metabolism (Chapter 25) rather than the citric acid cycle.

Pentose phosphate pathway
The biochemical pathway that produces ribose (a pentose), NADPH, and other sugar phosphates from glucose; an alternative to glycolysis.

- Glucose-6-phosphate can also enter the **pentose phosphate pathway**. This multistep pathway yields two products of importance to our metabolism. One is a supply of the coenzyme NADPH, a reducing agent that is essential for various biochemical reactions. The other is the 5-phosphate of the five-carbon sugar ribose, which is necessary for the synthesis of nucleic acids (DNA and RNA). Glucose-6-phosphate enters the pentose phosphate pathway when a cell's need for NADPH or ribose-6-phosphate exceeds its need for ATP.

■ PROBLEM 23.1

Name each of the following pathways:
- **(a)** pathway for release of glucose from glycogen
- **(b)** pathway for synthesis of glucose from lactate
- **(c)** pathway for synthesis of glycogen

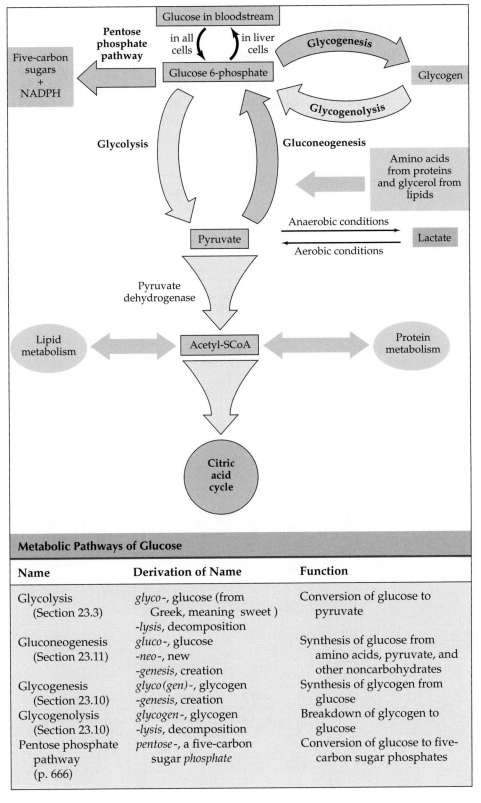

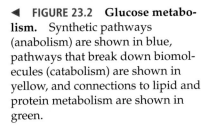

◀ **FIGURE 23.2 Glucose metabolism.** Synthetic pathways (anabolism) are shown in blue, pathways that break down biomolecules (catabolism) are shown in yellow, and connections to lipid and protein metabolism are shown in green.

Metabolic Pathways of Glucose

Name	Derivation of Name	Function
Glycolysis (Section 23.3)	*glyco-*, glucose (from Greek, meaning sweet) *-lysis*, decomposition	Conversion of glucose to pyruvate
Gluconeogenesis (Section 23.11)	*gluco-*, glucose *-neo-*, new *-genesis*, creation	Synthesis of glucose from amino acids, pyruvate, and other noncarbohydrates
Glycogenesis (Section 23.10)	*glyco(gen)-*, glycogen *-genesis*, creation	Synthesis of glycogen from glucose
Glycogenolysis (Section 23.10)	*glycogen-*, glycogen *-lysis*, decomposition	Breakdown of glycogen to glucose
Pentose phosphate pathway (p. 666)	*pentose-*, a five-carbon sugar *phosphate*	Conversion of glucose to five-carbon sugar phosphates

■ **PROBLEM 23.2**

Name the synthetic pathways that have glucose 6-phosphate as their first reactant.

23.3 Glycolysis

Glycolysis The biochemical pathway that breaks down a molecule of glucose into two molecules of pyruvate plus energy.

Glycolysis is a series of 10 enzyme-catalyzed reactions that breaks down each glucose molecule into two pyruvate molecules, and in the process yields two ATPs and two NADHs. The steps of glycolysis (also called the *Embden–Meyerhoff pathway* after its co-discoverers) are summarized in Figure 23.3, where the reactions and structures of intermediates should be looked at as you read the fol-

FIGURE 23.3   The glycolysis ▶ **pathway for converting glucose to pyruvate.**

 Glycolysis Part 1

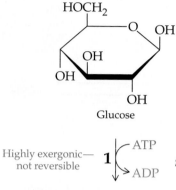

Glucose

Highly exergonic—not reversible   **1**   ⌐ ATP ↘ ADP

Step 1. Glucose undergoes reaction with ATP to yield glucose 6-phosphate plus ADP in a reaction catalyzed by *hexokinase*.

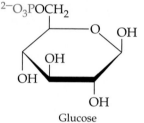

Glucose
6-phosphate

2

Step 2. Isomerization of glucose 6-phosphate yields fructose 6-phosphate. The reaction is catalyzed by the mutase enzyme, *glucose 6-phosphate isomerase*.

Fructose
6-phosphate

Highly exergonic—not reversible   **3**   ⌐ ATP ↘ ADP

Step 3. Fructose 6-phosphate reacts with a second molecule of ATP to yield fructose 1,6-bisphosphate plus ADP. *Phosphofructokinase*, the enzyme for step 3, provides a major control point in glycolysis.

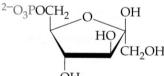

Fructose
1,6-bisphosphate

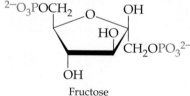

4

Step 4. The six-carbon chain of fructose 1,6-bisphosphate is cleaved into two three-carbon pieces by the enzyme *aldolase*. (Continued on next page.)

lowing paragraphs. Almost all organisms carry out glycolysis; in humans it occurs in the cytosol of all cells.

Steps 1–3 of Glycolysis: Phosphorylation Glucose is carried in blood to cells where it is transported across the cell membrane. As soon as it enters the cell, glucose is phosphorylated in *step 1* of glycolysis, which requires an energy investment from ATP. (Phosphorylation is addition of a $—PO_3^{2-}$ group. (∞, Section 17.8) This is the first of three highly exergonic and irreversible steps in glycolysis. From here on, all intermediates are sugar phosphates and are trapped within the cells because phosphates cannot cross cell membranes. The

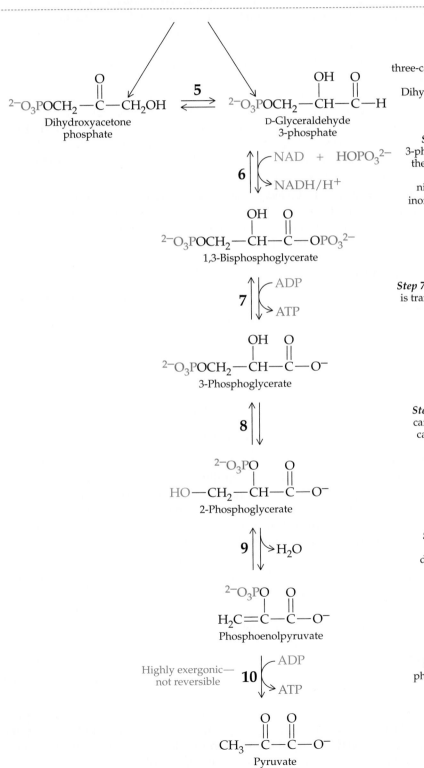

Step 5. The two products of step 4 are both three-carbon sugars, but only glyceraldehyde 3-phosphate can continue in the glycolysis pathway. Dihydroxyacetone phosphate must first be isomerized by the enzyme *triose phosphate isomerase*.

Step 6. Two reactions occur as glyceraldehyde 3-phosphate is first oxidized to a carboxylic acid and then phosphorylated by the enzyme *glyceraldehyde 3-phosphate dehydrogenase*. The coenzyme nicotinamide adenine dinucleotide (NAD) and inorganic phosphate ion ($HOPO_3^{2-}$) are required.

Step 7. A phosphate group from 1,3-bisphosphoglycerate is transferred to ADP, resulting in synthesis of ATP, and catalyzed by *phosphoglycerate kinase*.

Step 8. A phosphate group is next transferred from carbon 3 to carbon 2 of phosphoglycerate in a step catalyzed by the enzyme *phosphoglycerate mutase*.

Step 9. Loss of water from 2-phosphoglycerate produces phosphoenolpyruvate (PEP). The dehydration is catalyzed by the enzyme *enolase*.

Step 10. Transfer of the phosphate group from phosphoenolpyruvate to ADP yields pyruvate and generates ATP, catalyzed by *pyruvate kinase*.

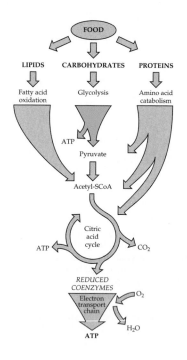

product of step 1, glucose 6-phosphate, is an allosteric inhibitor for the enzyme for this step (*hexokinase*), which thus plays an important role in the elaborate and delicate control of glucose metabolism. (◌◌◌, Figure 19.3, p. 549)

Step 2 is the isomerization of glucose 6-phosphate to fructose 6-phosphate. The enzyme (*glucose 6-phosphate isomerase*) acts by converting glucose 6-phosphate (an aldohexose) to fructose 6-phosphate (a ketohexose). The result is conversion of the six-membered glucose ring to a five-membered ring with a —CH$_2$OH group, which prepares the molecule for addition of another phosphoryl group in the next step.

Step 3 makes a second energy investment as fructose 6-phosphate is converted to fructose 1,6-bisphosphate by reaction with ATP in another exergonic reaction. (*Bis-* means "two"—that is, fructose with two phosphate groups. The "bis" prefix is used to distinguish between a molecule containing two phosphate groups in different locations and a "diphosphate"—a compound that contains a single diphosphate group, —OP$_2$O$_6{}^{4-}$) (◌◌◌, Section 17.8) Step 3 is another major control point for glycolysis. When the cell is short of energy, ADP and AMP (adenosine monophosphate) concentrations build up and activate the step 3 enzyme, *phosphofructokinase*. When energy is in good supply, ATP and citrate build up and allosterically inhibit the enzyme. The outcome of steps 1–3 is formation of a molecule ready to be split into the two-carbon intermediates that will ultimately become two molecules of pyruvate.

Steps 4 and 5 of Glycolysis: Cleavage and Isomerization *Step 4* converts the six-carbon bisphosphate from step 3 into two three-carbon monophosphates, one an aldose phosphate and one a ketose phosphate. The bond between carbons 3 and 4 in fructose 1,6-bisphosphate breaks, and a C=O group is formed.

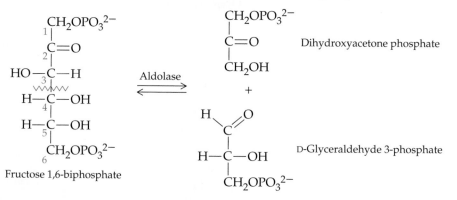

The two three-carbon sugar phosphates produced in step 4 are isomers that are interconvertible in an aldose–ketose equilibrium (*step 5* in Figure 23.3). Only glyceraldehyde 3-phosphate can continue on the glycolysis pathway, however. As the glyceraldehyde 3-phosphate reacts in step 6, the equilibrium of step 4 is shifted to the right. The overall result of steps 4 and 5 is therefore the production of *two* molecules of glyceraldehyde 3-phosphate.

Steps 1–5 are referred to as the *energy investment* part of glycolysis. So far, two ATPs have been invested and no income has been earned, but the stage has been set for a small profit. Note that since one glucose molecule gives two glyceraldehyde 3-phosphates that pass separately down the rest of the pathway, steps 6–10 of glycolysis each take place twice for every glucose molecule that enters at step 1.

Steps 6–10 of Glycolysis: Energy Generation The second half of glycolysis is devoted to generating molecules with phosphate groups that can be transferred to ATP.

Step 6 is the oxidation of glyceraldehyde 3-phosphate to 1,3-bisphosphoglycerate. NAD$^+$ is the oxidizing agent. Some of the energy from the exergonic oxidation is captured in NADH, and some is devoted to forming the phosphate. This is the first energy-generating step of glycolysis.

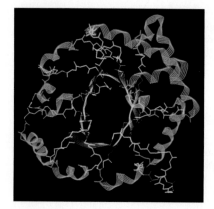

▲ **Triosephosphate isomerase, enzyme for step 5 of glycolysis.** β-Sheets are shown in blue, α-helixes in green, and random coil regions in yellow.

Step 7 generates the first ATP of glycolysis by transferring a phosphate group from 1,3-bisphosphoglycerate to ADP. Because this step occurs twice for each glucose molecule, the ATP energy balance sheet in glycolysis is even after step 7. Two ATPs were spent in steps 1–5, and now they've been replaced.

Steps 8 and 9—an isomerization followed by dehydration—generate phosphoenolpyruvate, the second energy-providing phosphate of glycolysis.

Step 10 is a highly exergonic, irreversible transfer of a phosphate group to ADP. The large amount of free energy released is accounted for by rearrangement of the less stable enol form of pyruvate to the more stable keto form. (∞, Section 22.6)

The two ATPs formed by the two occurrences of step 10 are pure profit, and the overall results of glycolysis are as follows:

Net result of glycolysis

$$C_6H_{12}O_6 + 2\,NAD^+ + 2\,HOPO_3^{2-} + 2\,ADP \longrightarrow 2\,CH_3-\underset{O}{\overset{O}{\underset{\|}{C}}}-\underset{O}{\overset{O}{\underset{\|}{C}}}-O^- + 2\,NADH + 2\,ATP + 2\,H_2O + 2\,H^+$$

Glucose   Pyruvate

- Conversion of glucose to two pyruvates
- Production of two ATPs
- Production of two molecules of reduced coenzyme NADH from NAD^+

■ **PROBLEM 23.3**
Identify the two pairs of steps in glycolysis in which phosphate intermediates are synthesized and their energy harvested as ATP.

■ **PROBLEM 23.4**
Identify each step in glycolysis that is an isomerization.

■ **PROBLEM 23.5**
Verify the isomerization that occurs in step 2 of glycolysis by drawing the open-chain forms of glucose 6-phosphate and fructose 6-phosphate.

■ **KEY CONCEPT PROBLEM 23.6**
In Figure 23.3 compare the starting compound (glucose) and the final product (pyruvate).
(a) Which is oxidized to a greater extent?
(b) Are there any steps in the glycolytic pathway in which an oxidation or reduction occurs? Identify the oxidizing or reducing agents that are involved in these steps.

23.4 Entry of Other Sugars into Glycolysis

The major monosaccharides from digestion other than glucose also eventually join the glycolysis pathway.

Fructose, from fruits or hydrolysis of the disaccharide sucrose, is converted to glycolysis intermediates in two ways: In muscle, it is phosphorylated to

Major dietary monosaccharides ▶ other than glucose.

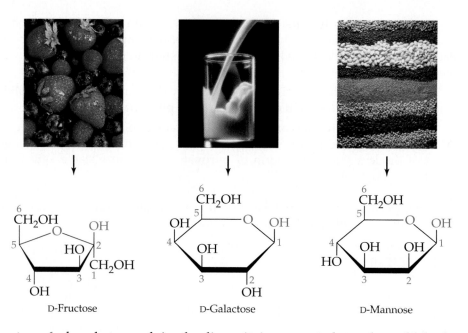

D-Fructose   D-Galactose   D-Mannose

fructose 6-phosphate, and in the liver, it is converted to glyceraldehyde 3-phosphate.

Galactose from hydrolysis of the disaccharide lactose is converted to glucose 6-phosphate by a five-step pathway. A hereditary defect affecting any of the enzymes in this pathway can be a cause of galactosemia. (∞∞, p. 643)

Mannose is a product of the hydrolysis of plant polysaccharides other than starch. It is converted by hexokinase to a 6-phosphate, which then undergoes a multistep, enzyme-catalyzed rearrangement and enters glycolysis as fructose 6-phosphate.

■ **PROBLEM 23.7**

Use curved arrows (like those in Figure 23.3) to write an equation for the conversion of fructose to fructose 6-phosphate by ATP. At what step does fructose 6-phosphate enter glycolysis?

■ **PROBLEM 23.8**

Compare glucose and galactose (see Section 22.5), and explain how their structures differ.

Aerobic In the presence of oxygen.

Anaerobic In the absence of oxygen.

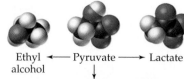

Ethyl ◀—— Pyruvate ——▶ Lactate
alcohol
            │
            ▼
      Acetyl-SCoA

▲ The biochemical transformations of pyruvate.

23.5 The Fate of Pyruvate

The conversion of glucose to pyruvate is a central metabolic pathway in most living systems. The further reactions of pyruvate, however, depend on metabolic conditions and on the nature of the organism. Under normal oxygen-rich (**aerobic**) conditions, pyruvate is converted to acetyl-SCoA. This pathway, however, is short-circuited in some tissues, especially when there's not enough oxygen (**anaerobic** conditions). Under anaerobic conditions, pyruvate is instead reduced to lactate. When sufficient oxygen again becomes available, the lactate is recycled to pyruvate. A third pathway for pyruvate is conversion back to glucose by *gluconeogenesis* (Section 23.11). This pathway is essential when the body is starved for glucose. The pyruvate for gluconeogenesis may come not only from glycolysis, but also from amino acids or glycerol from lipids.

Yeast is an organism with a different pathway for pyruvate, one that we put to use in a variety of ways. Yeast converts pyruvate to ethanol under anaerobic conditions.

Application Tooth Decay

Tooth decay is a complex interaction between food, bacteria, and a host organism. The clinical term for tooth decay is *dental caries*. It is recognized as an infectious microbial disease that results in the destruction of the calcified structures of the teeth.

The mouth is home to many different species of bacteria. A variety of habitats are provided by the diverse surfaces of the teeth, tongue, gums, and cheeks, and there are nutrients specific to each. Two permanent bacterial residents of the oral cavity, *Streptococcus sanguis* and *Streptococcus mutans*, compete for the same habitat on the biting surfaces of the teeth.

Dental plaque is defined as bacterial aggregations on the teeth that cannot be removed by the mechanical action of a strong water spray. Immediately after plaque has been removed by scrubbing with an abrasive paste, a coating of organic material composed of glycoproteins from the saliva begins to form. It completely covers the teeth within 2 hours after a visit to the dentist's office.

Bacteria then quickly colonize this newly formed film. They secrete a sticky matrix of an insoluble polysaccharide known as *dextran*, a branched polymer of the glucose that the bacteria have produced by hydrolysis of sucrose from food. The dextran allows the bacteria to stick firmly to the teeth so that the bacteria cannot be washed away by the saliva and swallowed. Bacteria that live comfortably in the mouth do not take well to the acidic environment of the stomach, so staying on the teeth is essential for their survival. The mass of bacteria, their sticky matrix, and the glycoprotein film together comprise dental plaque. Plaque is therefore not simply adherent food debris, but rather a community of microorganisms that forms through an orderly sequence of events

The bacteria resident in plaque release products consisting of proteins and carbohydrates. Some polysaccharides form intracellular granules that serve as energy storage depots for periods of low nutrient availability (between meals for the host). Other products are toxic to the gums and can promote periodontal disease. Carbohydrates, including structural components of the bacteria themselves, the storage granules, and the sticky matrix, constitute 20% of the dry weight of plaque.

What our dentists and parents told us—that eating candy would create cavities—is true! A diet high in sucrose favors the growth of *S. mutans* over that of *S. sanguis*. While both bacteria can cause tooth decay, *S. mutans* attacks teeth much more vigorously. It has an enzyme (a glucosyltransferase) that transfers glucose units from sucrose to the dextran polymer. The enzyme is specific to sucrose, and does not act on free glucose or the glucose from other carbohydrates. The mature plaque community then metabolizes fructose from the sucrose to lactate, and this acid causes the local pH in the area of the tooth to drop dramatically. If the pH stays low enough for a long enough time, the minerals in the teeth are dissolved away and the tooth begins to decay.

Cleaning teeth by brushing and flossing disrupts the bacterial plaque, removing many of the bacteria. However, enough bacteria always remain so that the colonization process can begin anew almost immediately. The disruption of plaque via oral hygiene and a diet low in sucrose will, however, favor the growth of *S. sanguis* over the evil *S. mutans*. To control the decay process, it is necessary to limit both the amount of sucrose in the diet and the frequency with which it is ingested.

The third factor in tooth decay is the host—ourselves. There are many variables that can prevent or promote tooth decay, including the composition of saliva, the shape of the teeth, and exposure to fluoride. As individuals, however, we have little control over these elements, leaving us reliant on proper oral hygiene habits, low sucrose diets, and preventive maintenance by dental professionals.

See Additional Problems 23.61–63 at the end of the chapter.

Aerobic Oxidation of Pyruvate to Acetyl-SCoA

For aerobic oxidation to proceed, the pyruvate first diffuses across the outer mitochondrial membrane from the cytosol where it was produced. Then it must be carried by a transporter protein across the otherwise impenetrable inner mitochondrial membrane. Once within the mitochondrial matrix, the pyruvate encounters the *pyruvate dehydrogenase complex* (Figure 23.4), a large multienzyme complex that catalyzes the conversion of pyruvate to acetyl-SCoA.

$$CH_3-\overset{O}{\overset{||}{C}}-\overset{O}{\overset{||}{C}}-O^- + HS-CoA \xrightarrow[\substack{\text{Pyruvate} \\ \text{dehydrogenase} \\ \text{complex}}]{NAD^+ \quad NADH/H^+} CH_3\overset{O}{\overset{||}{C}}-SCoA + CO_2$$

Pyruvate $\qquad\qquad\qquad\qquad\qquad\qquad\qquad\qquad\qquad$ Acetyl-SCoA

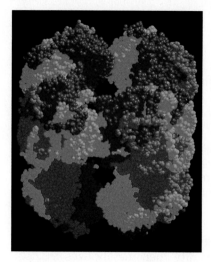

▲ **FIGURE 23.4 The central core of the pyruvate dehydrogenase complex.** The core is composed of 24 proteins arranged in three kinds of subunits shown in three different colors. The entire huge complex contains 60 subunits of three different enzymes. It requires NAD^+, CoA, FAD, and two other coenzymes. The enzyme subunits, which are adjacent to each other, swing into position one after the other as pyruvate loses CO_2 and is converted to an acetyl group that is then transferred to coenzyme A.

Fermentation The production of energy under anaerobic conditions.

Alcoholic fermentation The anaerobic breakdown of glucose to ethanol plus carbon dioxide by the action of yeast enzymes

Anaerobic Reduction to Lactate

Why does pyruvate take an alternative pathway when oxygen is in short supply? Since no oxygen has been needed in glucose catabolism thus far, what's the connection? The problem lies with NADH formed in step 6 of glycolysis (Figure 23.3). Under aerobic conditions, NADH is continually reoxidized to NAD^+ during electron transport (∞∞, Section 21.9), so NAD^+ is in good supply. If electron transport slows down because of insufficient oxygen, however, NADH concentration builds up, NAD^+ is in short supply, and glycolysis can't continue. An alternative way to reoxidize NADH is therefore essential because glycolysis *must* continue—it is the only available source of fresh ATP.

The reduction of pyruvate to lactate solves the problem. NADH serves as the reducing agent and is reoxidized to NAD^+, which is then available in the cytosol for glycolysis. Lactate formation serves no purpose other than NAD^+ production, and the lactate is reoxidized to pyruvate when oxygen is available.

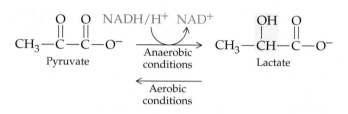

Red blood cells have no mitochondria and therefore always form lactate as the end product of glycolysis. Tissues where oxygen is in short supply also rely on anaerobic production of ATP by glycolysis. Examples are the cornea of the eye, where there is little blood circulation, and muscles during intense activity. The resulting buildup of lactate in the muscles causes fatigue and discomfort (see Application "The Biochemistry of Running," p. 682).

Alcoholic Fermentation

Microorganisms often must survive in the absence of oxygen and have evolved numerous anaerobic strategies for energy production, generally known as **fermentation**. When pyruvate undergoes fermentation by yeast, it is converted into ethanol plus carbon dioxide. This process, known as **alcoholic fermentation**, is used to produce beer, wine, and other alcoholic beverages and also to make bread. The carbon dioxide causes the bread to rise, and the alcohol evaporates during baking. The first leavened, or raised, bread was probably made by accident when airborne yeasts got into the dough. The tempting aroma of baking bread includes the aroma of alcohol vapors. Beer can be made by exposing the mash to outside air, where the airborne yeasts drift in from the surroundings.

■ **PROBLEM 23.9**

Complete oxidation of glucose produces six molecules of carbon dioxide. Describe the stage of catabolism at which each one is formed.

■ **KEY CONCEPT PROBLEM 23.10**

In alcoholic fermentation, each mole of pyruvate is converted to one mole of carbon dioxide and one mole of ethanol. In the process, about 50 kcal/mol of pyruvate is produced. Under the most favorable conditions, more than half of this energy is stored as ATP.

(a) What happens to the remaining energy produced in alcoholic fermentation?

(b) Give two reasons why it would be nearly impossible to reverse the reaction that converts pyruvate to ethanol and carbon dioxide.

23.6 Energy Output in Complete Catabolism of Glucose

The total energy output from oxidation of glucose is the combined result of glycolysis, conversion of pyruvate to acetyl-SCoA, conversion of two acetyl groups to four molecules of CO_2 in the citric acid cycle, and, finally, the passage of reduced coenzymes from each of these pathways through electron transport and the production of ATP by oxidative phosphorylation.

To determine the total number of ATPs generated from one glucose molecule, we can first add up the net equations for each of the pathways that precedes oxidative phosphorylation. Since each glucose yields two pyruvates and two acetyl-SCoAs, the net equations for pyruvate oxidation and the citric acid cycle must be multiplied by 2:

Net result of catabolism of one glucose molecule

Glycolysis (Section 23.3)

$$\text{Glucose} + 2\,\text{NAD}^+ + 2\,\text{HOPO}_3{}^{2-} + 2\,\text{ADP} \longrightarrow 2\,\text{Pyruvate} + 2\,\text{NADH} + 2\,\text{ATP} + 2\,\text{H}_2\text{O} + 2\,\text{H}^+$$

Pyruvate oxidation (Section 23.5)

$$2\,\text{Pyruvate} + 2\,\text{NAD}^+ + 2\,\text{HSCoA} \longrightarrow 2\,\text{Acetyl-SCoA} + 2\,\text{CO}_2 + 2\,\text{NADH} + 2\,\text{H}^+$$

Citric acid cycle (Section 21.8)

$$2\,\text{Acetyl-SCoA} + 6\,\text{NAD}^+ + 2\,\text{FAD} + 2\,\text{ADP} + 2\,\text{HOPO}_3{}^{2-} + 4\,\text{H}_2\text{O} \longrightarrow$$
$$2\,\text{HSCoA} + 6\,\text{NADH} + 6\,\text{H}^+ + 2\,\text{FADH}_2 + 2\,\text{ATP} + 4\,\text{CO}_2$$

$$\text{Glucose} + 10\,\text{NAD}^+ + 2\,\text{FAD} + 2\,\text{H}_2\text{O} + 4\,\text{ADP} + 4\,\text{HOPO}_3{}^{2-} \longrightarrow$$
$$10\,\text{NADH} + 10\,\text{H}^+ + 2\,\text{FADH}_2 + 4\,\text{ATP} + 6\,\text{CO}_2$$

The summation above shows a total of 4 ATPs per glucose molecule. The remainder of our ATP is generated via electron transport and oxidative phosphorylation. Thus, the total number of ATPs per glucose molecule is the 4 ATPs from glucose catabolism plus the number of ATPs produced for each reduced coenzyme that enters electron transport.

For a long time, based on the belief that 3 ATPs are generated per NADH and 2 ATPs per $FADH_2$, the maximum yield was taken as 38 ATPs, as calculated below:

$$10\,\text{NADH}\left(\frac{3\,\text{ATP}}{\text{NADH}}\right) + 2\,\text{FADH}_2\left(\frac{2\,\text{ATP}}{\text{FADH}_2}\right) + 4\,\text{ATP} = 38\,\text{ATP}$$

Our ever-expanding understanding of biochemical pathways has led to a revision in the potential number of ATPs per reduced coenzyme. The 38 ATPs per glucose molecule is viewed as a maximum yield of ATP, most likely possible in bacteria and other prokaryotes. In humans and other organisms more like us, the maximum is most likely 30–32 ATPs per glucose molecule.

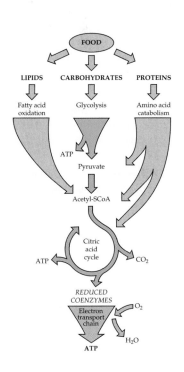

23.7 Regulation of Glucose Metabolism and Energy Production

Normal blood glucose concentration a few hours after a meal ranges roughly from 65 to 110 mg/dL. When departures from normal occur, we're in trouble (Figure 23.5). Low blood glucose (**hypoglycemia**) causes weakness, sweating,

Hypoglycemia Lower-than-normal blood glucose concentration.

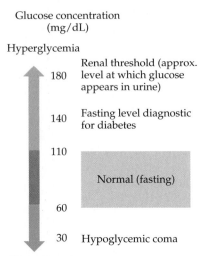

Glucose concentration (mg/dL)

Hyperglycemia

180 — Renal threshold (approx. level at which glucose appears in urine)

140 — Fasting level diagnostic for diabetes

110

Normal (fasting)

60

30 — Hypoglycemic coma

Hypoglycemia

▲ **FIGURE 23.5   Blood glucose.** The ranges for low blood glucose (in green; hypoglycemia), normal blood glucose (in purple), and high blood glucose (in orange; hyperglycemia) are indicated.

and rapid heartbeat, and in severe cases, low glucose in brain cells causes mental confusion, convulsions, coma, and eventually death. The brain can use *only* glucose as a source of energy. At a blood glucose level of 30 mg/dL, consciousness is impaired or lost, and prolonged hypoglycemia can cause permanent dementia. Excess blood glucose (**hyperglycemia**) causes increased urine flow as the normal osmolarity balance of fluids within the kidney is disturbed. Prolonged hyperglycemia can cause low blood pressure, coma, and death.

Two hormones from the pancreas have the major responsibility for blood glucose regulation. The first, insulin, is released when blood glucose concentration rises (Figure 23.6). Its role is to decrease blood glucose concentrations by accelerating the passage of glucose into cells where it is used for energy production, and by stimulating synthesis of glycogen, proteins, and lipids.

The second hormone, glucagon, is released when blood glucose concentration drops. In a reversal of insulin's effects, glucagon stimulates the break down of glycogen in the liver and release of glucose. Proteins and lipids are also broken down so that amino acids from proteins and glycerol from lipids can be converted to glucose in the liver by the gluconeogenesis pathways (see Section 23.11). Epinephrine (the "fight-or-flight" hormone) also accelerates the breakdown of glycogen, but primarily in muscle tissue, where the glucose is used to generate energy needed for quick action. (∞, Section 20.3)

Rising blood glucose concentration   *Falling blood glucose concentration*

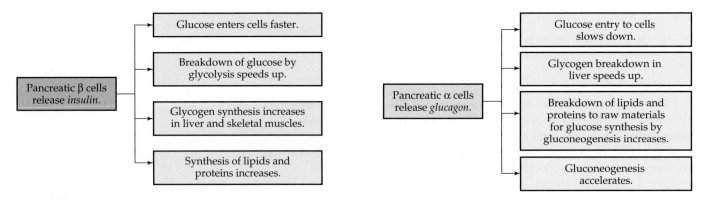

▲ **FIGURE 23.6   Regulation of glucose concentration by insulin and glucagon from the pancreas.**

Hyperglycemia   Higher-than-normal blood glucose concentration.

23.8 Metabolism in Fasting and Starvation

Imagine that you're lost in the woods. You've had no carbohydrates and very little else to eat for hours, and you're exhausted. The glycogen stored in your liver and muscles will soon be used up, but your brain must still rely on glucose to keep functioning. What happens next? Fortunately, your body's not yet ready to give up. It has mechanisms to assure that the limited glucose supplies will be delivered preferentially to your brain. In the liver, the gluconeogenesis pathway (Section 23.11) can make glucose from proteins. And if you're lost for a long time, there's a further backup system that extracts energy from compounds other than glucose.

The metabolic changes in the absence of food begin with a gradual decline in blood glucose concentration accompanied by an increased release of glucose from glycogen (Figure 23.7) (glycogenolysis, Section 23.10). All cells contain glycogen, but most is stored in liver cells (about 90 g in a 70 kg man) and muscle cells (about 350 g in a 70 kg man). Free glucose and glycogen represent less than 1% of our energy reserves and are used up in 15–20 hours of normal activity (3 hours in a marathon race).

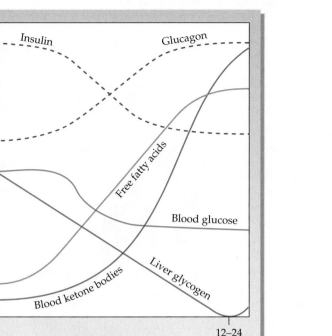

Fats are our largest energy reserve, but adjusting to dependence on them for energy takes time as there is no direct pathway to glucose for the fatty acids from fats (as shown in Figure 23.2). Energy from fatty tissue must be generated via acetyl-SCoA, the citric acid cycle, and electron transport.

As glucose and glycogen are exhausted, metabolism turns first to breakdown of proteins and glucose production from amino acids via gluconeogenesis in the liver. During the first few days of starvation, protein is used up at a rate as high as 75 g/day. Meanwhile, lipid catabolism is being mobilized, and acetyl-SCoA molecules derived from breakdown of lipids are accumulating. Eventually, the citric acid cycle is overloaded and can't degrade acetyl-SCoA as rapidly as it is produced. Acetyl-SCoA therefore builds up inside cells and begins to be removed by a new series of metabolic reactions that transform it into a group of compounds known as *ketone bodies*.

Ketone bodies

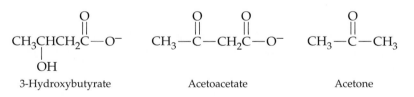

The ketone bodies enter the bloodstream and, as starvation continues, the brain and other tissues are able to switch over to producing up to 50% of their ATP from catabolizing ketone bodies instead of glucose. By the 40th day of starvation, metabolism has stabilized at the use of about 25 g of protein and 180 g of fat each day, a condition in which glucose and protein are conserved as much as possible. So long as adequate water is available, an average person can survive in this state for several months; those with more fat can survive longer.

⬭⬭⬭ LOOKING AHEAD

The breakdown of triacylglycerols from the fatty tissue produces not only ketone bodies but also glycerol, one of the compounds that can be converted to

Application Glucose in Blood and Urine

Glucose measurements are essential in the diagnosis of diabetes mellitus and in the management of diabetic patients, either in a clinical setting or on a day-to-day basis by the patients themselves. In the absence of therapeutic controls, glucose levels rise in blood and urine because the glucose is not entering cells and being metabolized as it should be.

Most tests for glucose in urine or blood rely on detecting a color change that accompanies the oxidation of glucose. Because glucose and its oxidation product, gluconate, are colorless, the oxidation must be tied chemically to the color change of a suitable indicator.

Benedict's reagent is an oxidizing agent that oxidizes aldehydes to carboxylic acids. (∞, Section 16.5) During the reaction the detectable color change is the reduction of blue copper(II) ion in aqueous solution to a brick-red precipitate of copper(I) oxide. Tablets containing the necessary Benedict's test chemicals in solid form (Clinitest tablets) are dropped into a diluted urine sample, and the color of the resulting mixture is compared to a standard chart to estimate glucose concentration. Unfortunately, Benedict's test is flawed because it is positive for any reducing sugar, such as lactose in the urine of a pregnant woman. Although Clinitest tablets are still available in drugstores for home use by diabetics, there are now better tests.

Modern methods for glucose detection rely on the action of an enzyme specific to glucose. In one such test, the enzyme is glucose oxidase, and the products of the oxidation are gluconate and hydrogen peroxide (H_2O_2). A second enzyme in the reaction mixture, a peroxidase, catalyzes the reaction of hydrogen peroxide with a dye that gives a detectable color change.

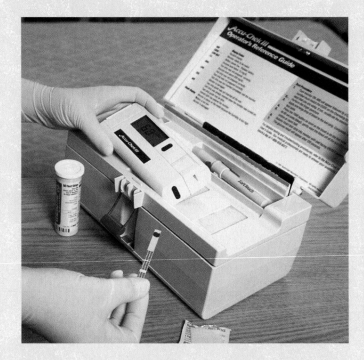

▲ Glucose blood test. A drop of blood has been placed on the strip, and the blood glucose level (indicated by the color of the strip) will be read by the glucose monitor being removed from the case.

The glucose oxidase test is available for urine and blood. Increasingly, diabetic individuals are monitoring their glucose levels in blood rather than urine, often with a modestly priced instrument that reads the color change electronically. The blood test is desirable because it is more specific and it detects rising glucose levels earlier than the urine test.

$$\text{Glucose} + O_2 \xrightarrow{\text{Glucose oxidase}} \text{Gluconate} + H_2O_2$$

$$\underset{\text{(Colorless)}}{H_2O_2 + \text{Reduced dye}} \xrightarrow{\text{Peroxidase}} \underset{\text{(Colored)}}{H_2O_2 + \text{Oxidized dye}}$$

See Additional Problems 22.64–65 at the end of the chapter.

---

glucose by gluconeogenesis. The production of glycerol and ketone bodies from triacylglycerols is described in Chapter 25, which is devoted to lipid metabolism. ∞

> ■ **PROBLEM 23.11**
> Refer to Figure 23.7 and summarize the changes in each substance shown during the starvation period represented in the figure.

Diabetes mellitus A chronic condition due to either insufficient insulin or failure of insulin to activate crossing of cell membranes, by glucose.

23.9 Metabolism in Diabetes Mellitus

Diabetes mellitus is one of the most common metabolic diseases. It is not a single disease but is classified into two major types, insulin-dependent and non-insulin-dependent. The insulin-dependent type, also called Type I or juvenile-onset diabetes (because it often appears in childhood), is caused by failure of the pancreatic cells to produce enough insulin. By contrast, in non-insulin-dependent diabetes, also called Type II or adult-onset diabetes (because it usually occurs in obese individuals over about 40 years of age), insulin is in good supply but fails to promote the passage of glucose across cell membranes. An estimated 6.6% of the U.S. population has Type II diabetes. Although often thought of only as a disease of glucose metabolism, diabetes affects protein and fat metabolism as well, and in some ways the metabolic response resembles starvation.

The symptoms by which diabetes is usually detected are excessive thirst accompanied by frequent urination, abnormally high glucose concentrations in urine and blood, and wasting of the body despite a good diet. These symptoms result when available glucose does not enter cells where it is needed. Glucose builds up in the blood, causing the symptoms of hyperglycemia and spilling over into the urine (glucosuria). In untreated diabetes, metabolism responds to the glucose shortage within cells by proceeding through the same stages as in starvation, from depletion of glycogen stores to breakdown of proteins and fats.

Adult-onset diabetes is thought to result when cell membrane receptors fail to recognize insulin. Drugs that increase insulin levels are an effective treatment because more of the undamaged receptors are put to work.

Juvenile-onset diabetes is classified as an autoimmune disease, a condition in which the body misidentifies some part of itself as an invader (Section 29.4). Gradually, the immune system wrongly identifies pancreatic beta cells as foreign matter, develops antibodies to them, and destroys them. To treat juvenile-onset diabetes, the missing insulin must be supplied by injection. Commercially available human insulin is now produced by bacteria modified by recombinant DNA techniques (Section 27.4) and was the first product of genetic engineering approved for use in humans.

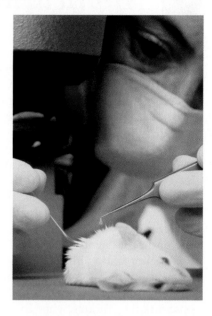

▲ **Insulin study.** The researcher is implanting human pancreatic islet cells in the mouse in order to study insulin production.

Diabetic individuals are subject to several serious conditions that result from elevated blood glucose levels. One reasonably well understood outcome is blindness due to cataracts. Increased glucose levels within the eye increase the quantity of glucose converted to sorbitol (in which the —CHO group of glucose is converted to —CH_2OH). Because sorbitol is not transported out of the cell, as is glucose, its rising concentration increases the osmolarity of fluid in the eye, causing increased pressure and cataracts. Elevated sorbitol is also associated with blood vessel lesions and gangrene in the legs, conditions that can accompany diabetes.

An insulin-dependent diabetic is at risk for two types of medical emergencies: ketoacidosis and hypoglycemia. Ketoacidosis results from the buildup of acidic ketones in the late stages of uncontrolled diabetes. It may lead to coma and diminished brain function but can be reversed by timely insulin administration. Hypoglycemia, or "insulin shock," by contrast, may be due to an overdose of insulin or failure to eat. If untreated, diabetic hypoglycemia can cause nerve damage or death.

The arrival at the emergency room of a diabetic patient in a coma requires quick determination of whether the condition is due to ketoacidosis or excess insulin. One indication of ketoacidosis is the aroma of acetone on the breath. Another is rapid respiration driven by the need to diminish acid concentration by eliminating carbon dioxide:

$$H^+ + HCO_3^- \longrightarrow H_2CO_3 \longrightarrow H_2O + CO_2 \text{ (Exhaled)}$$

An overdose of insulin does not cause rapid respiration.

Observations are backed up by bedside tests for glucose and ketones in blood and urine. A patient in insulin shock will, for example, have a very low blood glucose concentration.

■ **PROBLEM 23.12**

Sorbitol is the alcohol that accumulates in the eye and can cause cataracts. Draw the open-chain structure of sorbitol, which is identical to that of D-glucose except that the aldehyde group has been reduced to an alcohol group. Can sorbitol form a five- or six-membered cyclic hemiacetal?

Application The Glucose Tolerance Test

The glucose tolerance test is among the clinical laboratory tests usually done to pin down a diagnosis of diabetes mellitus. The patient must fast for 10–16 hours, avoid a diet high in carbohydrates prior to the fast, and refrain from taking any of a long list of drugs that can interfere with the test.

First, a blood sample is drawn to determine the fasting glucose concentration. The average normal fasting glucose level is 65–110 mg/dL. Then, the patient drinks a solution containing 75 g of glucose, and additional blood samples are taken at regular intervals thereafter. The accompanying figure compares the changes in diabetic and nondiabetic individuals. The diabetic patient has a higher fasting blood glucose level than the nondiabetic individual. In both, there is a rise in blood glucose concentration in the first hour. A difference becomes apparent after 2 hours, when the concentration in a nondiabetic individual has dropped to close to the fasting level but that in a diabetic individual remains high.

As listed below, a fasting blood glucose concentration of 140 mg/dL or higher and/or a glucose tolerance test concentration that remains above 200 mg/dL beyond 1 hour are considered diagnostic criteria for diabetes. For a firm diagnosis, the glucose tolerance test is usually given more than once.

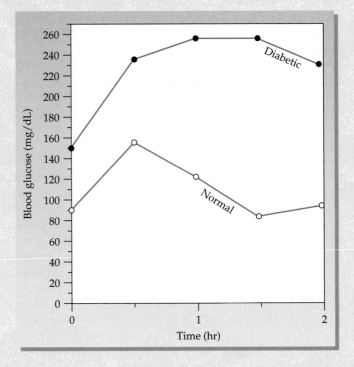

▲ Blood glucose concentration in glucose tolerance test for normal and diabetic individuals.

Key Diagnostic Features of Diabetes Mellitus

- Classic symptoms such as frequent urination, excessive thirst, rapid weight loss
- Random blood glucose concentration (without fasting) greater than 200 mg/dL
- Fasting blood glucose greater than 140 mg/dL

- Sustained blood glucose concentration greater than 200 mg/dL after glucose administration in glucose tolerance test

See Additional Problems 23.66–67 at the end of the chapter.

■ KEY CONCEPT PROBLEM 23.13

Ketoacidosis is relieved by rapid breathing, which converts bicarbonate ions and hydrogen ions in the blood to gaseous carbon dioxide and water, as shown in the equation on p. 679.

(a) Assuming that these reactions can go in either direction, how does a state of acidosis help to increase the generation of carbon dioxide?

(b) What principle describes the effect of added reactants and products on an equilibirum?

23.10 Glycogen Metabolism: Glycogenesis and Glycogenolysis

Glycogenesis The biochemical pathway for synthesis of glycogen.

Glycogen, the storage form of glucose in animals, is a branched polymer of glucose. (∞, p. 657) Glycogen synthesis, known as **glycogenesis**, occurs when glucose concentrations are high. It begins with glucose 6-phosphate and occurs via the three steps shown on the right in Figure 23.8. Glucose 6-phosphate is

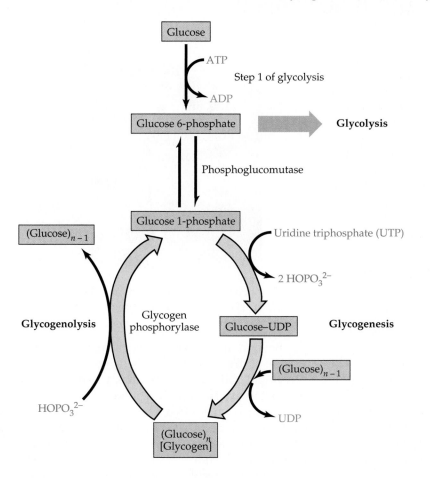

◀ **FIGURE 23.8 Glycogenolysis and glycogenesis.** Reading from the top down shows the pathway for glycogen synthesis from glucose (glycogenesis). Reading from the bottom up shows the pathway for release of glucose from glycogen (glycogenolysis).

first isomerized to glucose 1-phosphate. The glucose residue is then attached to uridine diphosphate (UDP):

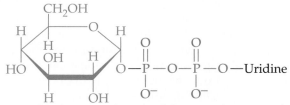

Glucose-UDP, the activated carrier of glucose in glycogen synthesis

The resulting glucose-UDP transfers glucose to a growing glycogen chain in an exergonic reaction.

As is usually true in metabolism, synthesis and breakdown are not accomplished by exactly reverse pathways. **Glycogenolysis** occurs in the two steps on the left in Figure 23.8. The first step is formation of glucose 1-phosphate by a glucose residue from glycogen. The glucose 1-phosphate is then converted to glucose 6-phosphate in the reverse of the reaction by which it is formed.

In muscle cells, glycogenolysis occurs when there is an immediate need for energy. The glucose 6-phosphate produced from glycogen goes directly into glycolysis. In the liver, glycogenolysis occurs when blood glucose is low (for example, during starvation; see Section 23.8). Because phosphates can't cross cell membranes, liver cells contain an enzyme that hydrolyzes glucose 6-phosphate to free glucose, which can then be released into the bloodstream.

Glycogenolysis The biochemical pathway for breakdown of glycogen to free glucose.

$$\text{Glucose 6-phosphate} + H_2O \xrightarrow{\text{Glucose 6-phosphate phosphatase}} \text{Glucose} + HOPO_3^{2-}$$

Application The Biochemistry of Running

A runner is poised tense and expectant, waiting for the sound of the starting gun. Long hours of training have prepared heart, lungs, and red blood cells to deliver the maximum amount of oxygen to the muscles, which have been conditioned to use it as efficiently as possible. In the moments before the race, mounting levels of epinephrine have readied the body for action. Now, everything depends on biochemistry: Chemical reactions in muscle cells will provide the energy to see the race through. How will that energy be produced?

The first source is the supply of immediately available ATP, but this is used up very quickly—probably within a matter of seconds. Additional ATP is then provided by the reaction of ADP with creatine phosphate, an amino acid phosphate in muscle cells that maintains the following equilibrium:

$$\text{ADP} + \text{Creatine phosphate} \rightleftharpoons \text{ATP} + \text{Creatine}$$

After about 30 seconds to a minute, stores of creatine phosphate are depleted, and glucose from glycogenolysis becomes the chief energy source. During maximum muscle exertion, oxygen can't enter muscle cells fast enough to keep the citric acid cycle and oxidative phosphorylation going. Under these anaerobic conditions, the pyruvate from glycolysis is converted to lactate rather than entering the citric acid cycle.

In a 100 m sprint, all the energy comes from available ATP, creatine phosphate (CP in the figure), and glycolysis of glucose from muscle glycogen. The anaerobic glycolysis suffices for only a minute or two of maximum exertion, because a buildup of lactate causes muscle fatigue.

▲ Glycogen stores sustained Antonio Pinto of Portugal long enough to set a European marathon record in 2000.

Beyond this, other pathways must come into action. As breathing and heart rate speed up and oxygen-carrying blood flows more quickly to muscles, the aerobic pathway is activated and ATP is once again generated by oxidative phosphorylation. The trick to avoiding muscle exhaustion in a long race is to run at a speed just under the "anaerobic threshold"—the rate of exertion at which oxygen is in short supply, ATP is supplied only by glycolysis, and lactate is produced.

Now the question is, which fuel will metabolism rely on during a long race—carbohydrate or fat? Burning fatty acids from fats is more efficient. More than twice as many calories are generated by burning a gram of fat than by burning a gram of carbohydrate. When we're sitting quietly, in fact, our muscle cells are burning mostly fat, and the fat in storage could support the exertion of marathon running for several days. By contrast, glycogen alone can provide enough glucose to fuel only 2–3 hours of such running under aerobic conditions.

The difficulty is that fatty acids can't be delivered to muscle cells fast enough to maintain the ATP level needed for running, so metabolism compromises and the glycogen stored in muscles remains the limiting factor for the marathon runner. Once glycogen is gone, extreme exhaustion and mental confusion set in—the condition known as "hitting the wall." Running speed is then limited to that sustainable by fats only. To delay this point as long as possible, a runner encourages glycogen synthesis by a diet high in carbohydrates prior to and during a race. In the hours just before the race, however, carbohydrates are avoided. Their effect of triggering insulin release is undesirable at this point because the resulting faster use of glucose will hasten depletion of glycogen.

See Additional Problems 23.68–70 at the end of the chapter.

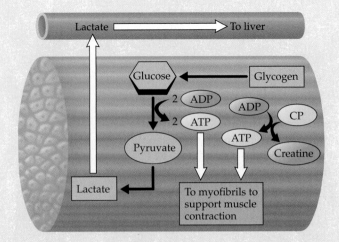

▲ At peak activity, ATP formation relies on creatine phosphate (CP) and glucose from muscle glycogen. Pyruvate is converted to lactate, which enters the bloodstream for transport to the liver, where it is recycled to pyruvate.

■ **PROBLEM 23.14**

The following overall reaction is a good example of the coupling of endergonic and exergonic reactions:

$$\text{UTP} + \text{Glucose 1-phosphate} + H_2O \longrightarrow \text{Glucose-UDP} + 2\,HOPO_3^{2-}$$

The coupled reactions are

$$\text{UTP} + \text{Glucose 1-phosphate} \longrightarrow \text{Glucose-UDP} + OP_2O_6^{4-}$$
$$\Delta G = 1.1\ \text{kcal/mol}$$

$$OP_2O_6^{4-} + H_2O \longrightarrow 2\,HOPO_3^{2-} \qquad \Delta G = -8.0\ \text{kcal/mol}$$

What is the common intermediate in these coupled reactions? What is ΔG for the coupled reactions? Based on these ΔG values, is the change favorable or unfavorable?

23.11 Gluconeogenesis: Glucose from Noncarbohydrates

Gluconeogenesis, which occurs mainly in the liver, is the pathway for making glucose from noncarbohydrate molecules—lactate, amino acids, and glycerol. This pathway becomes critical during fasting and the early stages of starvation. Failure of gluconeogenesis is usually fatal.

    Lactate is a normal product of glycolysis in red blood cells and of vigorous muscle activity. The body deals with the conditions during and after vigorous muscle activity—that is, high concentrations of lactate and depleted muscle glycogen—via the pathway shown in Figure 23.9. This pathway allows the liver to assume the burden of producing glucose by recycling it from lactate. Lactate absorbed from the blood is converted to pyruvate, the reactant for the first step of gluconeogenesis. The glucose synthesized in the liver is then returned to the muscles, where it can provide energy once more or be placed back into storage as muscle glycogen.

Gluconeogenesis   The biochemical pathway for the synthesis of glucose from noncarbohydrates, such as lactate, amino acids, or glycerol.

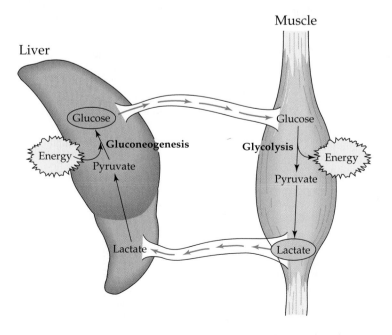

◀ **FIGURE 23.9   Glucose production during exercise (the Cori cycle).**   L-Lactate produced in muscles under anaerobic conditions during exercise is sent to the liver, where it is converted back to glucose. The glucose can then return via the bloodstream to the muscles, to be stored as glycogen or used for energy production. Gluconeogenesis requires energy, so shifting this pathway to the liver frees the muscles from the burden of having to produce even more energy.

We noted earlier that for metabolic pathways to be favorable, they must be exergonic. As a result, most are not reversible, because the amount of energy required by the reverse, endergonic pathway would be too large to be supplied by cellular metabolism. Glycolysis and gluconeogenesis provide another good example of this relationship and of the way around it.

Three reactions in glycolysis—steps 1, 3, and 10 in Figure 23.3—are too exergonic to be directly reversed. Gluconeogenesis therefore uses alternative reactions catalyzed by different enzymes that effectively reverse these steps. For example, gluconeogenesis begins with conversion of pyruvate to phospho-enolpyruvate, the reverse of the highly exergonic step 10 of glycolysis. In gluconeogenesis, two steps are required, utilizing two enzymes and the energy provided by two triphosphates, ATP and GTP. (Guanosine triphosphate, GTP, is similar in structure to ATP, and its phosphate groups can be removed in exergonic reactions. GDP is guanosine diphosphate.)

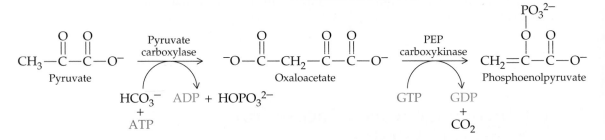

The seven other reactions of the 10-step glycolysis pathway are reversible because they operate at near-equilibrium conditions. They occur in reverse during gluconeogenesis, catalyzed by the same enzymes as in glycolysis. Glycerol from triacylglycerol catabolism (Section 25.3) is converted to dihydroxyacetone phosphate and enters the gluconeogenesis pathway at step 5 (Figure 23.3). The carbon atoms from certain amino acids (the glucogenic amino acids, Section 28.5) enter gluconeogenesis as either pyruvate or oxaloacetate.

■ **PROBLEM 23.15**

What two types of reactions will convert glycerol to dihydroxyacetone phosphate?

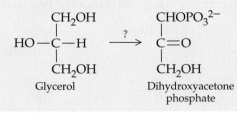

Application

Polysaccharides—What Are They Good For?

For a moment, forget about the metabolism of carbohydrates and think of them as raw materials—sources of connected carbon, hydrogen, and oxygen atoms. For years, we have taken advantage of the fiber-forming properties of cellulose. Cotton clothing is about 90% cellulose, and we make rayon (cellulose acetate) by converting the hydroxyl groups in cellulose to acetate ester groups. Many chemists are now hard at work exploring new ways to make useful substances from carbohydrates. This research is part of a movement to expand the use of *biomass*—raw materials from plants and animals. Biomass can be derived from fast-growing trees and grasses; agricultural leftovers such as wheat straw; and wood waste, which includes sawdust, tree prunings, waste paper, and yard waste.

One possible way to use cellulose is in production of a nonpolluting, energy-efficient fuel—hydrogen. This would be environmentally friendly. The only by-product of burning hydrogen is water, so replacement of fossil fuels with hydrogen is a tantalizing concept. Burning hydrogen from biomass would also be economical. It has been calculated that the cellulose from the huge quantity of U.S. waste newspaper alone could provide enough hydrogen to meet the energy needs of 1 million city residents. And as another benefit, biomass is renewable—we can keep growing it, so that it holds promise for the future as supplies of fossil fuels dwindle and become more difficult to recover.

The first step would be production of glucose by cooking cellulose with strong acid. Then, a bacterial enzyme (a glucose dehydrogenase) could be used to oxidize the glucose while converting $NADP^+$ to NADPH. Finally, another enzyme could be used to produce hydrogen from NADPH. Will we ever see this process in practical use? Right now, no one knows.

We are quite familiar with another use of renewable biomass—the production of ethyl alcohol in beverages by yeast fermentation of the glucose in grapes, corn, rye, and other plant materials. Did you know that, with some modification, automobile engines can run on straight ethyl alcohol? A 10% alcohol 90% gasoline mixture that can be used in existing engines is already commercially available. The search is on for a combination of chemical and enzymatic treatments that would permit large-scale production of fuel-grade ethyl alcohol from materials such as agricultural waste (straw, corn cobs, hulls, etc.), hardwood sawdust, and waste from pulp and paper production.

A variety of carbohydrate derivatives are widely used in commercial food products. Their major advantage is that they provide a texture and mouth feel like that of fats, but accompanied by zero calories or fewer calories than fat. The Calorie Control Council, a non-profit association that represents manufacturers of low-calorie foods and their ingredients, lists 12 carbohydrate-based fat replacers. They include zero-calorie food additives from cellulose and plant fibers. A finely ground cellulose, for example, can replace some of the fat in dairy products, frozen desserts, and salad dressings. A product from the insoluble fibers in oat, soybean and other plant hulls is heat stable and can be thus be used in baked goods, hot dogs, and hamburgers. There are soluble gums (guar gum, xanthan gum from bacteria, carrageenan from seaweed, and pectin, for example) that incorporate water molecules within their highly branched structures to produce thick, creamy textures. The gums, which also have just about zero calories, are used in salad dressings, desserts, and processed meats. Fat replacers with a few calories per gram include modified starches, dextrins (found in tapioca), and a group of polyols that can serve as both sweeteners with fewer calories than sugar and fat replacers.

You might try reading the ingredients list of prepared foods to see how many fat replacers you are consuming every day.

See Additional Problem 23.71 at the end of the chapter.

▲ This biomass could be put to practical use.

In the outpatient clinic of a 500-bed veteran's hospital in New England, a retired Army officer has just been told by his physician that he has diabetes. Now he is talking with Janice Makarchuk, a registered dietitian at the hospital. The information that Janice is explaining to him will be critically important in helping him understand his medical condition and what he can do to improve his overall health. Janice counsels many patients every day about the impact that diet can have on their lives.

She explains, "Although I've been a registered dietitian for the past 15 years, my interest in health and fitness goes back further than that. I began by studying physical education, and then went on to dietetics. Chemistry was very important in my education. I can recall being frustrated while I was trying to learn some of the abstract principles of basic chemistry, but as I progressed through organic chemistry and biochemistry, the principles that I had learned earlier began to fall into place."

Every day Janice must explain complex processes of body metabolism to patients in a manner that they can understand. "As a registered dietitian I have to assess patients' nutritional needs and provide proper dietary support. Most often there is a particular health concern that has brought the patient to me, such as cardiac disease, obesity, kidney disease, or diabetes. Frequently the patient has only a slight understanding of his condition and requires a great deal of education to grasp exactly what's going on in his body and how his diet can affect him."

"I rely on my chemistry and biochemistry background to help the patient understand the effect that diet can have on his body's altered functioning, and the problems that can occur as a result of his disease. For example, a knowledge of the metabolism of carbohydrates is essential to this man, who's learning how to manage his diabetic condition. He must become aware of the possibility of ketoacidosis, an acute, life-threatening complication of insulin-dependent diabetes. A number of compounds can affect the utilization of glucose. Hormones such as catecholamines and glucagon increase glucose production through glycogenolysis and gluconeogenesis, raising the risk of hyperglycemia.

"When working with sick people, it's very important to remember that the chemical reactions of a body are sometimes completely altered from what is typical. It is my role as a dietitian to constantly assess and monitor these changes and to make modifications in patients' diets as needed. Some patients who are critically ill may need tube feedings. Their situations require that I have specific knowledge of electrolyte balance,

hydration, osmolality of feeding formulas and energy/protein requirements.

"I also need to be aware of chemical interactions that can occur between specific foods and the patients' medications. For example, a patient taking certain high blood pressure medications may need to pay special attention to the amount of potassium in his diet [Chapter 28]. Another man who is on psychiatric medication may need to monitor the intake of a specific amino acid in his diet [Chapters 18 and 19]. Even certain vitamin supplements may be contraindicated for some cardiac patients on blood thinning medications."

Back in the outpatient clinic, Janice says goodbye to her new patient. "As a result of this visit, he understands how diabetes is affecting all his body systems. He's begun the process of learning about the effects of his new medications, and he's also gaining an awareness of the changes he'll need to make to help him manage his disease."

With Janice's continued support and guidance, it is likely that this man will be able to make the necessary alterations in his life to manage his diabetes and enjoy a healthier lifestyle. "Whenever a patient takes some control over his illness, makes changes, and begins to feel better, I realize how much I enjoy being a dietitian."

Summary: Revisiting the Chapter Goals

1. **What happens during digestion of carbohydrates?** Carbohydrate *digestion*, the hydrolysis of disaccharides and polysaccharides, begins in the mouth and continues in the stomach and small intestine. The products that enter the bloodstream from the small intestine are monosaccharides—mainly glucose, fructose, and galactose.

2. **What are the major pathways in the metabolism of glucose?** The major pathway for glucose, once inside a cell and converted to glucose 6-phosphate, is *glycolysis*. Pyruvate, the end product of glycolysis, is then fed into the citric acid cycle via acetyl-SCoA. One alternative pathway for glucose is *glycogenesis*, the synthesis of glycogen, which is stored mainly in the liver and muscles. Another alternative is the *pentose phosphate pathway*, which provides five-carbon sugars and NADPH needed for biosynthesis (see Figure 23.2).

3. **What is glycolysis?** Glycolysis (Figure 23.3) is a 10-step pathway that produces two molecules of pyruvate, two reduced coenzymes (NADH), and two ATPs for each molecule of glucose metabolized. Glycolysis begins with phosphorylation (steps 1–3) to form fructose 1,6-bisphosphate, followed by cleavage and isomerization that produce two molecules of glyceraldehyde 3-phosphate (steps 4–5). Each glyceraldehyde 3-phosphate then proceeds through the energy-generating steps (6–10) in which phosphates are alternately created and then donate their phosphate groups to ADP. Dietary monosaccharides other than glucose enter glycolysis at various points—fructose as fructose 6-phosphate or glyceraldehyde 3-phosphate, galactose as glucose 6-phosphate, and mannose as fructose 6-phosphate.

4. **What happens to pyruvate once it is formed?** When oxygen is in good supply, pyruvate is transported into mitochondria and converted to acetyl-SCoA for energy generation via the citric acid cycle and oxidative phosphorylation. When there is insufficient oxygen, pyruvate is reduced to L-lactate, with the production of the oxidized coenzyme NAD^+ that is essential to the continuation of glycolysis. This production of NAD^+ compensates for the shortage of NAD^+ created by the slowdown of electron transport when oxygen is in short supply. The lactate is transported in the bloodstream to the liver, where oxygen is available, and is oxidized back to pyruvate. In the presence of yeast, pyruvate undergoes *anaerobic fermentation* to yield ethyl alcohol.

5. **How is glucose metabolism regulated, and what are the influences of starvation and diabetes mellitus?** *Insulin*, produced when blood glucose concentration rises, accelerates glycolysis and glycogen synthesis. *Glucagon*, produced when blood glucose concentration drops, accelerates production of glucose in the liver from stored glycogen and from other precursors via the *gluconeogenesis* pathway. Adaptation to starvation begins with the effects of glucagon and energy production from protein, and then proceeds to reliance on ketone bodies from fatty tissue for energy generation. *Diabetes mellitus* may be insulin-dependent (the pancreas fails to produce insulin) or non-insulin-dependent (insulin receptors fail to recognize insulin). Among the serious outcomes of uncontrolled diabetes are cataracts, blood vessel lesions, ketoacidosis, and *hypoglycemia*.

6. **What are glycogenesis and glycogenolysis?** *Glycogenesis* (Figure 23.8), the synthesis of the polysaccharide glycogen, puts excess glucose into storage, mainly in muscles and the liver. *Glycogenolysis* is the release of stored glucose from glycogen. Glycogenolysis occurs in muscles when there is an immediate need for energy, producing glucose 6-phosphate for intracellular glycolysis. It occurs in the liver when blood glucose concentration is low and must be elevated. The liver has an enzyme to convert glucose 6-phosphate to glucose, which is released to the bloodstream.

Key Words

Aerobic, p. 672

Alcoholic fermentation, p. 674

Anaerobic, p. 672

Diabetes mellitus, p. 678

Digestion, p. 665

Fermentation, p. 674

Gluconeogenesis, p. 683

Glycogenesis, p. 680

Glycogenolysis, p. 681

Glycolysis, p. 668

Hyperglycemia, p. 676

Hypoglycemia, p. 675

Pentose phosphate pathway, p. 666

7. What is the role of gluconeogenesis in metabolism? *Gluconeogenesis* (Figure 23.9) maintains glucose levels by synthesizing it from lactate, from certain amino acids derived from proteins, and from glycerol derived from fatty tissue; it is critical during fasting and starvation. The gluconeogenesis pathway uses alternate enzymes for the reverse of the three highly exergonic steps of glycolysis, but otherwise utilizes the same enzymes for reactions that run in reverse of their direction in glycolysis.

▪ Understanding Key Concepts

23.16 What class of enzymes catalyzes the majority of the reactions involved in carbohydrate digestion?

23.17 Glucose 6-phosphate is in a pivotal position in metabolism. Depending on conditions, glucose 6-phosphate can follow one of several pathways. Under what conditions do each of the following occur?
(a) Pentose phosphate pathway
(b) Glycogenesis
(c) Hydrolysis to free glucose
(d) Glycolysis

23.18 What investments are made to get glycolysis started, and why are they made? What happens in the middle of the pathway to generate two three-carbon compounds? What are the outcomes of the reactions of these three-carbon compounds?

23.19 Outline the condition(s) that would direct pyruvate toward:
(a) Glucose synthesis (gluconeogenesis)
(b) Conversion to lactate
(c) Entry into the citric acid cycle
(d) Conversion to ethanol and CO_2
In what tissues or organisms is each pathway present?

23.20 Classify each one of the enzymes of glycolysis into one of the six classes of enzymes. What class of enzymes has the most representatives in glycolysis? Why is this consistent with the goals of glycolysis? Why are ligases *not* represented in glycolysis?

23.21 When blood glucose levels rise following a meal, what is the appropriate sequence of the events below?
(a) Glycolysis occurs to replenish the ATP supplies.
(b) Glucagon is secreted.
(c) Blood levels pass through normal to below normal (hypoglycemic).
(d) Insulin levels rise.
(e) The liver releases glucose into the bloodstream.
(f) Glucose is absorbed by cells.
(g) Glycogen synthesis (glycogenesis) occurs with excess glucose.

23.22 A friend was told that it is important to avoid air when making wine, so she added yeast to fresh grape juice and placed it in a bottle sealed to avoid air. Several days later, the lid exploded off the bottle. Explain what biochemistry was responsible for the exploding lid.

23.23 What are the sources of compounds for gluconeogenesis? Fatty acids from stored triacylglycerols (fat) are *not* available for gluconeogenesis. Speculate on why we do not have the enzymes to convert fatty acids into glucose. Plants (especially seeds) *do* have enzymes to convert fatty acids into carbohydrates. Why are they so lucky?

23.24 The pathway that converts glucose to acetyl-SCoA is often referred to as an "aerobic oxidation pathway." (a) Is molecular oxygen involved in any of the steps of glycolysis? (b) Thinking back to Chapter 21, where does molecular oxygen enter the picture?

▪ Additional Problems

Digestion and Metabolism

23.25 Where in the body does digestion occur, and what kinds of chemical reactions does it involve?

23.26 Complete the following word equation:

$$\text{Maltose} + H_2O \rightarrow ? + ?$$

Where in the digestion system does this process occur?

23.27 What are the products of digestion of proteins, triacylglycerols, sucrose, lactose, and starch?

23.28 What are the major monosaccharide products produced by digestion of carbohydrates?

23.29 What is meant by the words *aerobic* and *anaerobic*?

23.30 What three products are formed from pyruvate under aerobic, anaerobic, and fermentation conditions?

23.31 Differentiate between glycolysis and gluconeogenesis.

23.32 Differentiate between glycogenolysis and glycogenesis.

23.33 What is the major purpose of the pentose phosphate pathway?

23.34 Depending on the body's needs, into what type of compounds is glucose converted in the pentose phosphate pathway?

Glycolysis

23.35 What is the name and structure of the final product of glycolysis? Is there also a changed coenzyme product?

23.36 Glycolysis can occur under anaerobic conditions. What are two possible pathways to handle the NADH generated in this case? What are the products of the two pathways?

23.37 Where in the cell does glycolysis occur?

23.38 Where in the cell does gluconeogenesis occur?

23.39 Look at the 10 steps in glycolysis (Figure 23.3) and then answer the following questions:
(a) Which steps involve phosphorylation?
(b) Which step is an oxidation?
(c) Which step is a dehydration?

23.40 How many moles of ATP are produced by phosphorylation in the following?
   (a) Glycolysis of 1 mol of glucose
   (b) Aerobic conversion of 1 mol of pyruvate to 1 mol of acetyl-SCoA
   (c) Catabolism of 1 mol of acetyl-SCoA in the citric acid cycle

23.41 What is the purpose of the formation of lactate under anaerobic conditions?

23.42 Lactate can be converted into pyruvate by the enzyme lactate dehydrogenase and the coenzyme NAD^+. Write the reaction in the standard biochemical format, using a curved arrow to show the involvement of NAD^+.

23.43 How many moles of CO_2 are produced by the complete catabolism of 1 mol of sucrose?

23.44 How many moles of acetyl-SCoA are produced by the catabolism of 1 mol of sucrose?

Regulation of Glucose Metabolism/Metabolism in Diabetes Mellitus

23.45 Differentiate between blood sugar levels and resulting symptoms in hyperglycemia and hypoglycemia.

23.46 Differentiate between the effect of insulin and glucogon on blood sugar concentration.

23.47 From what molecules is glucose initially produced during starvation or fasting?

23.48 As starvation continues, to what is acetyl-SCoA converted to prevent buildup in the cells?

23.49 What are the symptoms of diabetes?

23.50 Many diabetics suffer blindness due to cataracts or undergo amputation of limbs. Why are these conditions associated with this disease?

23.51 Why is juvenile diabetes considered to be an autoimmune disease?

23.52 What is the difference between juvenile and adult-onset diabetes?

23.53 Where is most of the glycogen in the body stored?

23.54 What major site of glycogen storage is not able to add glucose to the bloodstream?

23.55 Of what use is UTP in the formation of glycogen from glucose?

23.56 Why is glycogenolysis not the exact reverse of the process of glycogenesis?

Glucose Anabolism

23.57 What is the name of the anabolic pathway for making glucose?

23.58 What two molecules serve as starting materials for glucose synthesis?

23.59 To what molecule is pyruvate initially converted in the anabolism of glucose? To what substance is that molecule in turn converted?

23.60 Why can't pyruvate be converted to glucose in an exact reverse of the glycolysis pathway?

Applications

23.61 What is the function of the insoluble polysaccharide known as dextran in formation of dental plaque? [*Tooth Decay*]

23.62 Name four of the major components of dental plaque. [*Tooth Decay*]

23.63 How is dental plaque associated with periodontal disease? [*Tooth Decay*]

23.64 Why is Benedict's test not infallible in testing for glucose? [*Glucose in Blood and Urine*]

23.65 Briefly describe the enzymatic process for determination of glucose. [*Glucose in Blood and Urine*]

23.66 How do fasting levels of glucose in a diabetic person compare to those in a nondiabetic person? [*The Glucose Tolerance Test*]

23.67 Discuss the differences in the response of a diabetic person compared to those of a nondiabetic person after drinking a glucose solution. [*The Glucose Tolerance Test*]

23.68 Why is creatine phosphate a better source of quick energy than glucose? [*The Biochemistry of Running*]

23.69 Why is it not possible for a person to sprint for miles? [*The Biochemistry of Running*]

23.70 Order the following sources of energy (from first used to last used) when muscles are called upon to do extensive work:
   (a) Glycogen
   (b) Glucose
   (c) ATP
   (d) Creatine phosphate
   (e) Fatty acids from triacylglycerols
   [*The Biochemistry of Running*]

23.71 Cellulose is the most abundant biomolecule in nature. Describe three products on the market that are derived from cellulose. What derivatives of cellulose are used to make these products? [*Polysaccharides—What Are They Good For?*]

General Questions and Problems

23.72 Why can pyruvate cross the mitochondrial membrane but no other molecule after step 1 in glycolysis can?

23.73 Look at the glycolysis pathway (Figure 23.3). With what type of process are kinase enzymes usually associated?

23.74 Is the same net production of ATP observed in the complete oxidation of fructose as is observed in the oxidation of glucose? Why or why not?

23.75 Explain why one more ATP is produced when glucose is obtained from glycogen rather than used directly from the blood.

24

Lipids

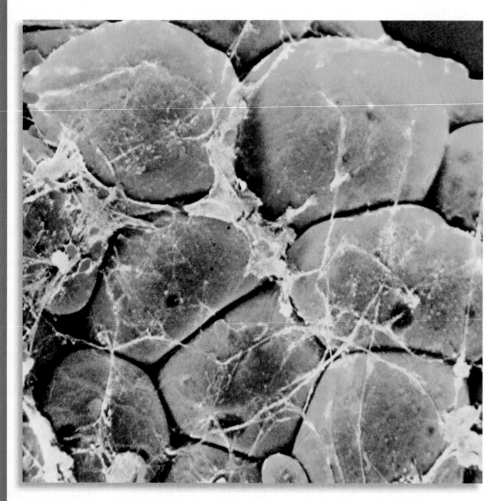

Where the fat goes. Each of these adipose tissue cells holds a globule of fat. (Magnified more than 500 times.)

Lipids are less well known than carbohydrates and proteins, yet they're just as essential to our diet and well-being. Lipids have three major roles in human biochemistry: (1) Within fat cells (*adipocytes*), they store energy left over from metabolism of food. (2) As part of all cell membranes, they help to keep separate the different chemical environments inside and outside the cells. (3) In the endocrine system and elsewhere, they serve as chemical messengers.

Chemically, lipids are defined as naturally occurring organic molecules that are nonpolar and therefore dissolve in nonpolar organic solvents but not in water. For example, if a sample of plant or animal tissue is placed in a kitchen blender, finely ground, and then extracted with ether, anything that dissolves in the ether is a lipid, while anything that does not dissolve in the ether (including carbohydrates, proteins, and inorganic salts) is not a lipid. In this chapter, we'll answer the following questions about lipids:

1. **What are the major classes of fatty acids and lipids?**
   The goal: Be able to describe the chemical structures and general properties of fatty acids, waxes, fats, and oils.

2. **What reactions do triacylglycerols undergo?**
   The goal: Be able to describe the results of hydrogenation and hydrolysis of triacylglycerols, and, given the reactants, predict the products.

3. **What are the membrane lipids?**
   The goal: Be able to identify the membrane lipids and describe their structures.

4. **What is the nature of a cell membrane?**
   The goal: Be able to describe the general structure of a cell membrane and its chemical composition.

5. **How do substances cross cell membranes?**
   The goal: Be able to distinguish between passive and active transport, and describe simple diffusion and facilitated diffusion.

6. **What are eicosanolds?**
   The goal: Be able to describe the general structure of prostaglandins and leukotrienes, and some of their functions.

CONCEPTS TO REVIEW

Intermolecular forces (Section 8.11)
Cis–trans isomerism (Section 13.3)
Esters and amides (Sections 17.4, 17.6)
Phosphoric acid derivatives (Section 17.8)

24.1 Structure and Classification of Lipids

Since **lipids** are defined by solubility in nonpolar solvents (a physical property) rather than by chemical structure, it's not surprising that there are a great many different kinds and that they serve a variety of functions in the body. Note in the following examples of lipid structures that the molecules contain large hydrocarbon portions and not many polar groups, which accounts for their solubility behavior.

Lipid A naturally occurring molecule from a plant or animal that is soluble in nonpolar organic solvents.

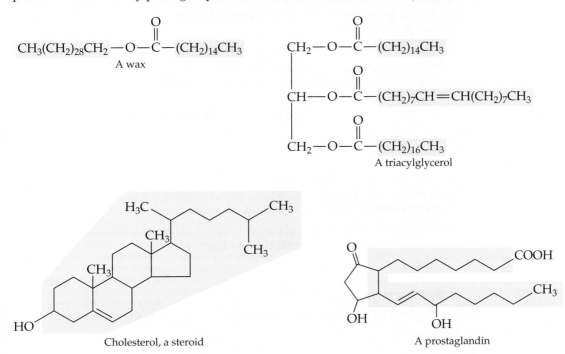

Cholesterol, a steroid

A prostaglandin

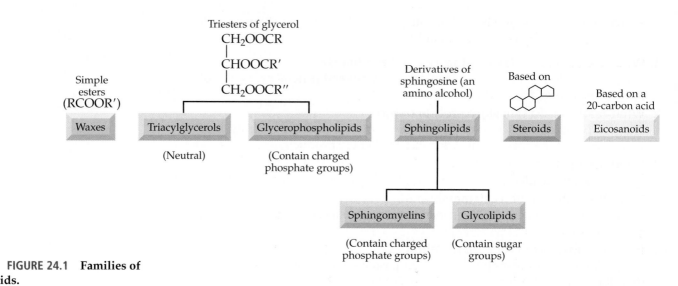

▲ **FIGURE 24.1** **Families of lipids.**

Fatty acid A long-chain carboxylic acid; those in animal fats and vegetable oils often have 12–22 carbon atoms.

Figure 24.1 organizes the classes of lipids discussed in this chapter according to their chemical structures. Many lipids are esters or amides of carboxylic acids with long, straight hydrocarbon chains, which are known as **fatty acids**.

Lipids that are esters or amides of fatty acids

- *Waxes* are carboxylic acid esters (RCOOR′) with long, straight hydrocarbon chains in both R groups; they are secreted by sebaceous glands in the skin of animals and perform mostly external protective functions (Section 24.2).

- *Triacylglycerols* are carboxylic acid triesters of glycerol, a three-carbon trialcohol. Triacylglycerols (Sections 24.2–24.3) make up the fats stored in our bodies and most dietary fats and oils. They are a major source of biochemical energy, a function described in Chapter 25.

- *Glycerophospholipids* (Section 24.5) are triesters of glycerol that contain charged phosphate diester groups and are abundant in cell membranes. Together with other lipids, they help to control the flow of molecules into and out of cells.

- *Sphingomyelins*, which are amides derived from an amino alcohol (*sphingosine*), also contain charged phosphate diester groups; they are essential to the structure of cell membranes (Section 24.5) and are abundant in nerve cell membranes.

- *Glycolipids*, which are also amides derived from *sphingosine*, contain polar carbohydrate groups; on cell surfaces the carbohydrate portion is recognized and connected with by intracellular messengers (Section 24.5).

There are also two groups of lipids that are not esters or amides: the *steroids* and the *eicosanoids*.

Other types of lipids

- *Steroids*. You have already seen the role of steroids as hormones. (∞∞, Section 20.5) In this chapter we focus on cholesterol (Section 24.6), which contributes to the structure of cell membranes.

- *Eicosanoids*. The eicosanoids (Section 24.9) are carboxylic acids that are a special type of intracellular chemical messenger.

■ PROBLEM 24.1

Use Figure 24.1 to identify the family of lipids to which the following molecules belong.

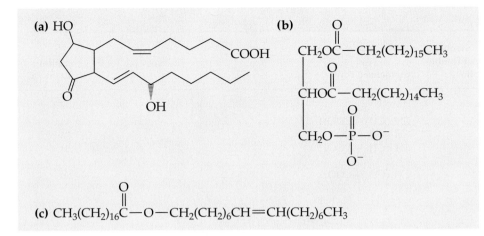

(a) [structure]

(b) [structure]

(c) $CH_3(CH_2)_{16}\overset{\displaystyle O}{\overset{\|}{C}}-O-CH_2(CH_2)_6CH=CH(CH_2)_6CH_3$

24.2 Fatty Acids and Their Esters

The naturally occurring fats and oils are triesters formed between glycerol and fatty acids. (Recall that an ester, RCOOR', is formed from a carboxylic acid and an alcohol. (∞, p. 483) Fatty acids are long, straight hydrocarbon chains with a carboxylic acid group at one end. Most have even numbers of carbon atoms. Fatty acids may or may not contain carbon–carbon double bonds. Those without double bonds are known as **saturated fatty acids**; those with double bonds are known as **unsaturated fatty acids**. If double bonds are present in naturally occurring fats and oils, the double bonds are usually cis rather than trans. (∞, p. 360)

Saturated fatty acid A long-chain carboxylic acid containing only carbon–carbon single bonds.

Unsaturated fatty acid A long-chain carboxylic acid containing one or more carbon–carbon double bonds.

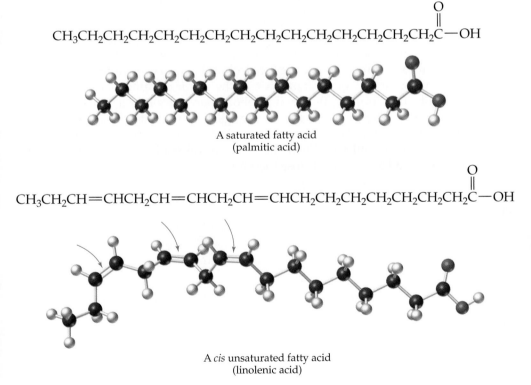

$CH_3CH_2CH_2CH_2CH_2CH_2CH_2CH_2CH_2CH_2CH_2CH_2CH_2CH_2CH_2\overset{\displaystyle O}{\overset{\|}{C}}-OH$

A saturated fatty acid
(palmitic acid)

$CH_3CH_2CH=CHCH_2CH=CHCH_2CH=CHCH_2CH_2CH_2CH_2CH_2CH_2CH_2\overset{\displaystyle O}{\overset{\|}{C}}-OH$

A *cis* unsaturated fatty acid
(linolenic acid)

Some of the common fatty acids are listed in Table 24.1. Palmitic acid (16 carbons) and stearic acid (18 carbons) are the most common saturated acids; oleic and linoleic acids (both with 18 carbons) are the most common unsaturated ones. Oleic acid is *monounsaturated*, that is, it has only one carbon–carbon double bond. The **polyunsaturated fatty acids** have more than one carbon–carbon double bond.

Polyunsaturated fatty acid A long-chain carboxylic acid that has two or more carbon–carbon double bonds.

TABLE 24.1 **Structures of Some Common Fatty Acids**

Name	Typical Source	Number of Carbons	Number of Double Bonds	Condensed Formula	Melting Point (°C)
Saturated					
Lauric	Coconut oil	12	0	$CH_3(CH_2)_{10}COOH$	44
Myristic	Butter fat	14	0	$CH_3(CH_2)_{12}COOH$	58
Palmitic	Most fats and oils	16	0	$CH_3(CH_2)_{14}COOH$	63
Stearic	Most fats and oils	18	0	$CH_3(CH_2)_{16}COOH$	70
Unsaturated					
Oleic	Olive oil	18	1	$CH_3(CH_2)_7CH{=}CH(CH_2)_7COOH$ (cis)	4
Linoleic	Vegetable oils	18	2	$CH_3(CH_2)_4CH{=}CHCH_2CH{=}CH(CH_2)_7COOH$ (all cis)	−5
Linolenic	Soybean and canola oils	18	3	$CH_3CH_2CH{=}CHCH_2CH{=}CHCH_2CH{=}CH(CH_2)_7COOH$ (all cis)	−11
Arachidonic	Lard	20	4	$CH_3(CH_2)_4(CH{=}CHCH_2)_4CH_2CH_2COOH$ (all cis)	−50

Two of the polyunsaturated fatty acids, linoleic and linolenic, are essential in the human diet because the body can't synthesize them, but they are needed for the synthesis of other lipids. Infants grow poorly and develop severe skin lesions if fed a diet lacking these acids. Adults usually have sufficient reserves in body fat to avoid such problems. A deficiency in adults can arise, however, after long-term intravenous feeding that contains inadequate essential fatty acids or among those surviving on limited and inadequate diets.

Waxes

Wax A mixture of monoesters of long-chain carboxylic acids with long-chain alcohols.

The simplest fatty acid esters in nature are the waxes. A **wax** is a mixture of fatty acid—long-chain alcohol esters. The acids usually have an even number of carbons from 16 through 36, while the alcohols have an even number of carbons from 24 through 36. For example, one of the major components in beeswax is the ester formed from a 30-carbon alcohol (triacontanol) and a 16-carbon acid (palmitic acid). The waxy protective coatings on most fruits, berries, leaves, and animal furs have similar structures. Aquatic birds have a water-repellent waxy coating on their feathers. When caught in an oil spill, the waxy coating dissolves in the oil and the birds lose their buoyancy.

Example of a wax

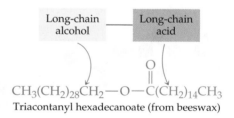

Triacontanyl hexadecanoate (from beeswax)

▲ **A duck afloat.** The water-repellant waxy coating on its feathers helps to prevent this fellow from sinking.

Triacylglycerol (triglyceride) A triester of glycerol with three fatty acids.

Triacylglycerols

Animal fats and vegetable oils are the most plentiful lipids in nature. Although they appear different—animal fats like butter and lard are solid, while vegetable oils like corn, olive, soybean, and peanut oil are liquid—their structures are closely related. All fats and oils are composed of triesters of glycerol (1,2,3-propanetriol, also known as glycerine) with three fatty acids. They are named chemically as **triacylglycerols**, but are often called **triglycerides**.

Application Lipids in the Diet

The major recognizable sources of fats and oils in our diet are butter and margarine, vegetable oils, the visible fat in meat, and chicken skin. In addition, there are "invisible" lipids in milk and milk products, meat, fish, poultry, nuts and seeds, whole-grain cereals, and prepared foods. The triacylglycerols in meat, poultry, fish, dairy products, and eggs are mostly saturated and are accompanied by small quantities of cholesterol. Vegetable oils have a higher unsaturated fatty acid content and contain no cholesterol. They never have, so the salad oil labels that proclaim. "No cholesterol" are stating a fact, not announcing something special the manufacturer has done for us.

Fats and oils are a popular component of our diet: They taste good, give a pleasant texture to food, and, because they are digested slowly, give a feeling of satisfaction after a meal. The percentage of calories from fats and oils in the average U.S. diet in the early 1980s was 40–45%, considerably more than needed. Excess dietary fats and oils are mostly stored as fat in adipose tissue.

Concern for the relationships among saturated fats, cholesterol levels, and various diseases—most notably heart disease and cancer (see Application "Lipids and Atherosclerosis," Chapter 25)—caused a modest decrease to 37% of the average calories from fats and oils in the U.S. diet by 1990. We significantly decreased our consumption of butter, eggs, beef, and whole milk (all containing relatively high proportions of saturated fat and cholesterol). This change did not coincide with a reduction in obesity, which is a weight 20% over the desirable weight for a person's height, sex, and activity level. In fact, concern has been developing in recent years over a rise in obesity in the U.S. population and its inevitable association with heart disease.

Several organizations, including the U.S. Food and Drug Administration, recommend a diet with not more than 30% of its calories from fats and oils. In a daily diet of 2200 Calories, which is about right for teenage girls, active women, and sedentary men, 30% from fats and oils is approximately 73 g, the amount in about 6 tablespoons of butter. Teenage boys and men, or women with very active lifestyles, require more daily calories, and therefore can include proportionately more fats in their diets.

The Nutrition Facts labels (illustrated in Application "Vitamins, Minerals, and Food Labels," Chapter 19) list the calories from fat, grams of fat (which includes all

▲ A selection of appealing but high-fat foods.

triacylglycerols), and grams of saturated fat in a single serving of a commercially prepared food. To check up on what you're eating, remember that you have to check the number of servings in the container. The label on a package of hot dogs, for example, says 12 g of fat. But checking shows that one hot dog = a single serving. If you eat two or three hot dogs, you'll be getting 24 g or 36 g of fat.

The FDA further recommends that not more than 10% of daily calories come from saturated fat and not more than 300 mg of cholesterol be included in the daily diet. For those with the goal of 2200 Calories per day, the limit is 24 g of saturated fats. (Those hot dogs contain 5 g of saturated fat per hot dog.) To reduce saturated fats requires choosing the low fat varieties of foods when possible. For example, 1% milk provides 1.6 g of saturated fat per cup compared with 5.1 g per cup in whole milk. The foods highest in cholesterol are high-fat dairy products, liver, and egg yolks (see Section 24.6, Cell Membrane Lipids: Cholesterol, p. 707).

See Additional Problems 24. 65–66 at the end of the chapter.

Triacylglycerols

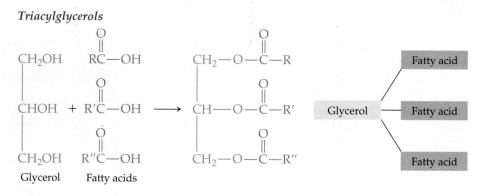

The three fatty acids of any specific triacylglycerol are not necessarily the same, as shown in the following example.

Example of a triacylglycerol

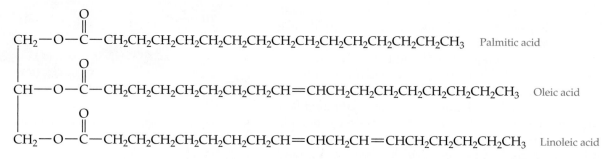

Furthermore, the fat or oil from a given natural source is a complex mixture of many different triacylglycerols. Table 24.2 lists the average composition of fats and oils from several different sources. Note particularly that vegetable oils consist almost entirely of unsaturated fatty acids, whereas animal fats contain a much larger percentage of saturated fatty acids. This difference in composition is the primary reason for the different melting points of fats and oils, as explained in the next section.

TABLE 24.2 Approximate Composition of Some Common Fats and Oils*

Source	Saturated Fatty Acids (%)				Unsaturated Fatty Acids (%)	
	C_{12} Lauric	C_{14} Myristic	C_{16} Palmitic	C_{18} Stearic	C_{18} Oleic	C_{18} Linoleic
Animal Fat						
Lard	—	1	25	15	50	6
Butter	2	10	25	10	25	5
Human fat	1	3	25	8	46	10
Whale blubber	—	8	12	3	35	10
Vegetable Oil						
Corn	—	1	8	4	46	42
Olive	—	1	5	5	83	7
Peanut	—	—	7	5	60	20
Soybean	—	—	7	4	34	53

*Where totals are less than 100%, small quantities of several other acids are present, with cholesterol also present in animal fats.

■ **PROBLEM 24.2**

One of the constituents of the carnauba wax used in floor and furniture polish is an ester of a C_{32} straight-chain alcohol with a C_{20} straight-chain carboxylic acid. Draw the structure of this ester. (Use subscripts to show the numbers of connected CH_2 groups, as in Table 24.1.)

■ **PROBLEM 24.3**

Draw the structure of a triacylglycerol whose components are glycerol and three oleic acid acyl groups.

■ **KEY CONCEPT PROBLEM 24.4**

(a) Which animal fat has the largest percentage of saturated fatty acids?
(b) Which vegetable oil has the largest percentage of polyunsaturated fatty acids?

24.3 Properties of Fats and Oils

The melting points listed in Table 24.1 show that the more double bonds a fatty acid has, the lower its melting point. For example, the saturated C_{18} acid (stearic) melts at 70°C, the monounsaturated C_{18} acid (oleic) melts at 4°C, and the diunsaturated C_{18} acid (linoleic) melts at −5°C. This same trend also holds true for triacylglycerols: The more highly unsaturated the acyl groups in a triacylglycerol, the lower its melting point. The difference in melting points between fats and oils is a consequence of this difference. Vegetable **oils** are lower melting because they generally have a higher proportion of unsaturated fatty acids than animal **fats**.

How do the double bonds make such a significant difference in properties? Compare the shapes of a saturated and an unsaturated fatty acid molecule:

Oil A mixture of triacylglycerols that is liquid because it contains a high proportion of unsaturated fatty acids.

Fat A mixture of triacylglycerols that is solid because it contains a high proportion of saturated fatty acids.

Stearic acid, an 18-carbon saturated fatty acid

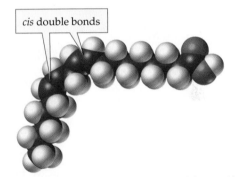

cis double bonds

Linoleic acid, an 18-carbon unsaturated fatty acid

The hydrocarbon chains in saturated acids are uniform in shape with identical angles at each carbon atom, and the chains are flexible, allowing them to nestle together. By contrast, the carbon chains in unsaturated acids have rigid kinks wherever they contain cis double bonds. The kinks make it difficult for such chains to fit next to each other in the orderly fashion necessary to form a solid. The more double bonds there are in a triacylglycerol, the harder it is for it to solidify. The shapes of the molecular models in Figure 24.2 further illustrate this concept.

The triacylglycerols are uncharged, nonpolar molecules that are hydrophobic. When stored in fatty tissue they coalesce, and the interior of an adipocyte is occupied by one large fat droplet with the nucleus pushed to one side. In addition to energy storage, adipose tissue serves to provide thermal insulation and protective padding. Most fatty tissue is located under the skin or in the abdominal cavity, where it cushions the organs.

We are accustomed to the characteristic yellow color and flavors of cooking oils, but these are contributed by natural materials carried along during

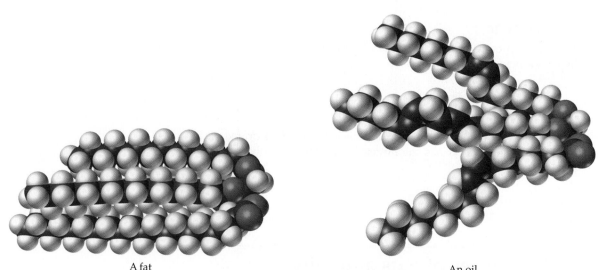

A fat   An oil

▲ **FIGURE 24.2 Triacylglycerols from a fat and an oil.**

production of the oils from plants; pure oils are colorless and odorless. Over-heating, or exposure to air or oxidizing agents, causes decomposition to products with unpleasant odors or flavors, creating what we call a *rancid oil*. Antioxidants are added to prepared foods to prevent oxidation of their oils. (See Application, "Phenols as Antioxidants," p. 404.)

Properties of the Triacylglycerols in Natural Fats and Oils

- Nonpolar and hydrophobic
- No ionic charges
- Solid triacylglycerols (fats)—high proportion of saturated fatty acid chains
- Liquid triacylglycerols (oils)—high proportion of unsaturated fatty acid chains

∞ LOOKING AHEAD

Triacylglycerols from plants and animals are a major component of our diet. In our bodies, they are the depots for energy storage. Therefore, in considering the metabolism of lipids, it is the metabolism of triacylglycerols that is of greatest interest. This topic is discussed in Chapter 25.   ∞

■ **PROBLEM 24.5**

Draw the complete structural formula of arachidonic acid (Table 24.1) in a way that shows the cis stereochemistry of its four double bonds.

■ **PROBLEM 24.6**

Which of the following two fatty acids has the higher melting point?

(a) $CH_3(CH_2)_4CH=CHCH_2CH=CHCH_2(CH_2)_6\overset{\overset{\displaystyle O}{\|}}{C}-OH$

(b) $CH_3(CH_2)_5CH=CHCH_2(CH_2)_6\overset{\overset{\displaystyle O}{\|}}{C}-OH$

■ **PROBLEM 24.7**

Can there be any chiral carbon atoms in triacylglycerols? If so, which one(s) can be chiral and what determines their chirality?

24.4 Chemical Reactions of Triacylglycerols

Hydrogenation

The carbon–carbon double bonds in vegetable oils can be hydrogenated to yield saturated fats in the same way that any alkene can react with hydrogen to yield an alkane. (∞, p. 365) Margarine and solid cooking fats like Crisco are produced commercially by hydrogenation of vegetable oils to give a product chemically similar to that found in animal fats:

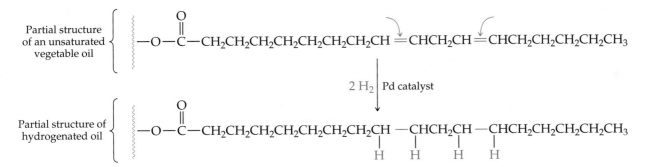

The extent of hydrogenation varies with the number of double bonds in the unsaturated acids and their locations. In general, the number of double bonds is reduced in stepwise fashion from three to two to one. By controlling the extent of hydrogenation and monitoring the composition of the product, it is possible to control consistency. In margarine, for example, only about two-thirds of the double bonds present in the starting vegetable oil are hydrogenated. The remaining double bonds, which vary in their locations, are left unhydrogenated so that the margarine has exactly the right consistency to keep soft in the refrigerator and melt on warm toast. (See Application "Butter and Its Substitutes," p. 708.)

Hydrolysis of Triacylglycerols; Soap

Triacylglycerols, like all esters, can be hydrolyzed, that is, they can react with water to form their carboxylic acids and alcohols. In the body, this hydrolysis is catalyzed by enzymes (hydrolases) and is the first reaction in the digestion of dietary fats and oils (Section 25.1).

In the laboratory and in commercial production of soap, hydrolysis of fats and oils is usually carried out by strong aqueous bases (NaOH or KOH) and is called *saponification*. (∞, p. 489) The initial products of saponification of a fat

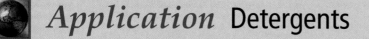

Application Detergents

S trictly speaking, anything that washes away dirt is a *detergent*. The term is usually applied, however, to synthetic materials made from petroleum chemicals. In the 1950s, synthetic detergents began to replace natural soaps. The goal was to overcome the problems caused when soap is used in hard water. Such water contains metal ions (mostly Ca^{2+} and Mg^{2+}) that have dissolved into it from minerals. When the metal ions in solution encounter fatty acid anions, they form what we call soap scum—precipitates of salts (for example, $[CH_3(CH_2)_{14}COO]_2Ca]$). The results are that soap is wasted, residues are left in washed clothing, and the hard-to-remove scum is left behind in bathtubs and washing machines.

Like soaps, synthetic detergent molecules have hydrophobic hydrocarbon tails and hydrophilic heads, and they cleanse by the same mechanism as soap—forming micelles around greasy dirt. All substances that function in this manner are described as surface-active agents, or *surfactants*. The hydrophilic heads may be anionic, cationic, or nonionic. Anionic surfactants are commonly used in home laundry products; cationic surfactants are used in fabric softeners and in disinfectant soaps; and nonionic surfactants are low-sudsing and are effective at low temperatures.

Note that the hydrocarbon chains in the representative surfactants shown below are unbranched. Some of the first detergents contained branched-chain hydrocarbons, but this was soon discovered to be a mistake. The bacteria in natural waters and sewage treatment plants are slow to consume branched-chain hydrocarbons, so detergents containing them were not decomposed and produced suds in streams and lakes.

See Additional Problems 24.67–69 at the end of the chapter.

$$Na^+ \; {}^-O-\underset{\underset{O}{\|}}{\overset{\overset{O}{\|}}{S}}-\!\!\!\left\langle\bigcirc\right\rangle\!\!\!-CH_2CH_2CH_2CH_2CH_2CH_2CH_2CH_2CH_2CH_2CH_2CH_3$$

Sodium dodecylbenzenesulfonate
(An ionic detergent)

$$CH_3(CH_2)_{10}CH_2OCH_2CH_2(OCH_2CH_2)_7OH$$

A polyether
(A nonionic detergent)

$$\left\langle\bigcirc\right\rangle\!\!-CH_2-\overset{\overset{\displaystyle CH_3}{|}}{\underset{\underset{\displaystyle CH_3}{|}}{N^+}}-R \; Cl^-$$

A benzalkonium chloride; $R=C_8H_{17}$ to $C_{18}H_{37}$
(A cationic detergent)

▲ Where are the lipids in this picture?

Soap The mixture of salts of fatty acids formed on saponification of animal fat.

or oil molecule are one molecule of glycerol and three molecules of fatty acid carboxylate salts:

Saponification

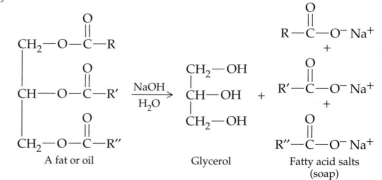

A fat or oil   Glycerol   Fatty acid salts
(soap)

The complex mixture of fatty acid salts produced by saponification of animal fat with NaOH or KOH is what we call **soap**. Crude soap curds, which contain glycerol and excess alkali as well as soap, are first purified by boiling with water and adding sodium chloride to precipitate the pure sodium carboxylate salts. The smooth soap that precipitates is then dried, perfumed, and pressed into bars for household use.

Soaps act as cleaning agents because the two ends of a soap molecule are so different. The sodium salt end is ionic and therefore hydrophilic (water-loving);

it tends to dissolve in water. The long hydrocarbon chain portion of the molecule, however, is nonpolar and therefore hydrophobic (water-fearing). Like an alkane, it tends to avoid water and to dissolve in nonpolar substances like grease, fat, and oil. Because of these opposing tendencies, soap molecules are attracted to both grease and water.

When soap is dispersed in water, the big organic anions cluster together so that their long, hydrophobic hydrocarbon tails are in contact. By doing so, they avoid disrupting the strong hydrogen bond interactions of water and instead create a nonpolar microenvironment. At the same time, their hydrophilic ionic heads on the surface of the cluster stick out into the water. The resulting spherical clusters are called **micelles** (Figure 24.3). Grease and dirt are suspended in water because they are coated by the nonpolar tails of the soap molecules and trapped in the center of the micelles. Once suspended within micelles, the grease and dirt can be washed away.

Micelle A spherical cluster formed by the aggregation of soap or detergent molecules so that their hydrophobic ends are in the center and their hydrophilic ends are on the surface.

◀ **FIGURE 24.3 Soap or detergent molecules in water.** The hydrophilic ionic ends remain in the water. At the surface of the water, a film forms with the hydrocarbon chains on the surface. Within the solution, the hydrocarbon chains cluster together at the centers of micelles. Greasy dirt is dissolved in the oily center and carried away. Lipids are transported in the bloodstream in similar micelles, as described in Section 25.2.

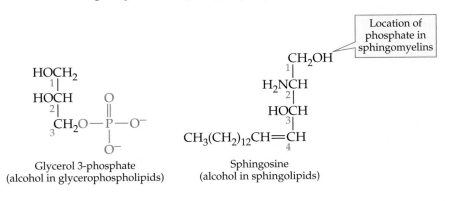

■ **PROBLEM 24.10**
Write the complete equation for the hydrolysis of a triacylglycerol in which the fatty acids are two molecules of stearic acid and one of oleic acid (see Table 24.1).

 Surfactant Molecules

24.5 Cell Membrane Lipids: Phospholipids and Glycolipids

Cell membranes separate the aqueous interior of cells from the aqueous environment surrounding the cells. To accomplish this, the membranes establish a hydrophobic barrier between the two watery environments. Lipids are ideal for this function.

The three major kinds of cell membrane lipids in animals are phospholipids, glycolipids, and cholesterol. The **phospholipids** contain a phosphate ester link. They are built up from either glycerol (to give *glycerophospholipids*) or from the alcohol sphingosine (to give *sphingomyelins*).

Phospholipid A lipid that has an ester link between phosphoric acid and an alcohol (either glycerol or sphingosine)

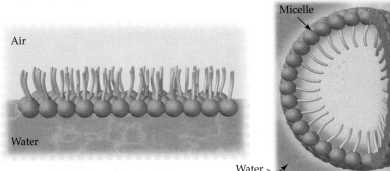

Glycerol 3-phosphate
(alcohol in glycerophospholipids)

Sphingosine
(alcohol in sphingolipids)

Glycolipid A lipid with a fatty acid bonded to the $C2-NH_2$ group and a sugar bonded to the $C1-OH$ group of sphingosine.

Sphingolipid A lipid derived from the amino alcohol sphingosine.

The **glycolipids** are also derived from sphingosine. They contain no phosphate group, but have an attached carbohydrate that is a monosaccharide or a short chain of monosaccharides. The general structures of these lipids and the relationships of their classification are shown at the top in Figure 24.4. Note the overlapping classes of membrane lipids. Glycolipids and sphingomyelins both contain sphingosine and are therefore classified as **sphingolipids**, whereas glycerophospholipids and sphingomyelins both contain phosphate groups and are therefore classified as phospholipids.

FIGURE 24.4 Membrane lipids. ▶ All have two hydrocarbon tails and polar, hydrophilic head groups. In the sphingolipids (sphingomyelins and glycolipids), one of the two hydrocarbon tails is part of the alcohol sphingosine (blue).

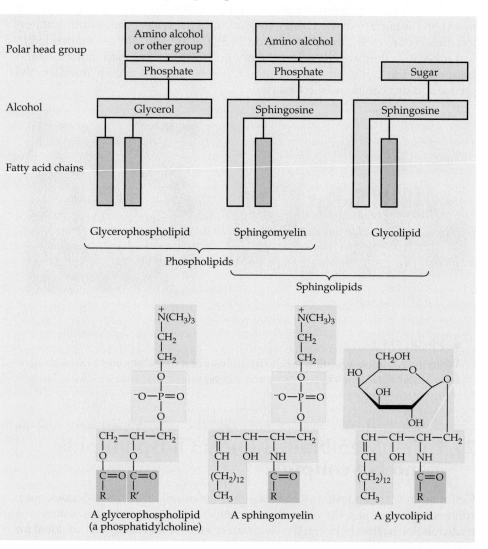

Cholesterol is a steroid, a class of biomolecules that are characterized by a system of four fused rings. We have discussed steroids in the context of their most significant role as hormones. (∞, Section 20.5) The presence of cholesterol in cell membranes is explored in Section 24.6, and its connection to heart disease is discussed in an Application "Lipids and Atherosclerosis," in the next chapter (p. 722).

Phospholipids

Because phospholipids have ionized phosphate groups at one end, they are similar to soap and detergent molecules in having ionic, hydrophilic heads and hydrophobic tails (see Figure 24.3). They differ, however, in having *two* tails instead of one.

Glycerophospholipid (phosphoglyceride) A lipid in which glycerol is linked by ester bonds to two fatty acids and one phosphate, which is in turn linked by another ester bond to an amino alcohol (or other alcohol).

Glycerophospholipids (also known as **phosphoglycerides**) are triesters of glycerol 3-phosphate, and are the most abundant membrane lipids. Two of the ester bonds are with fatty acids, which provide the two hydrophobic tails (purple in the general glycerophospholipid structure in Figure 24.4). The fatty acids may be any of the fatty acids normally present in fats or oils. The fatty acid acyl group (R—C=O) bonded to C1 of glycerol is usually saturated, while that at C2 is usually unsaturated. At the third position in glycerophospholipids there is a phosphate ester group (green in Figure 24.4). This phosphate has a second ester link to one of several different OH-containing compounds, often ethanolamine, choline, or serine (orange in Figure 24.4; see structures in Table 24.3).

TABLE 24.3   Some Glycerophospholipids

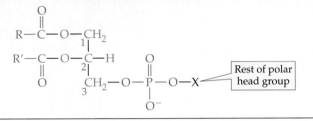

Precursor of X (HO—X)	Formula of X	Name of Resulting Glycerophospholipid Family	Function
Water	—H	Phosphatidate	Basic structure of glycerophospholipids
Choline	$-CH_2CH_2\overset{+}{N}(CH_3)_3$	Phosphatidylcholine	Basic structure of lecithins; most abundant membrane phospholipids
Ethanolamine	$-CH_2CH_2\overset{+}{N}H_3$	Phosphatidylethanolamine	Membrane lipids
Serine	$-CH_2-\underset{\underset{COO^-}{\mid}}{\overset{\overset{+}{N}H_3}{\overset{\mid}{CH}}}$	Phosphatidylserine	Present in most tissues; abundant in brain
myo-Inositol	(ring structure, Bond site)	Phosphatidylinositol	Relays chemical signals across cell membranes

The glycerophospholipids are named as derivatives of phosphatidic acids. In the example shown below, the second phosphate ester link is to the amino alcohol choline, $HOCH_2CH_2N^+(CH_3)_3$. Lipids of this type are known as *phosphatidylcholines*, or *lecithins*. (A substance referred to in the singular as lecithin, or phosphatidylserine, or any of the other classes of phospholipids, is usually a mixture of molecules with different R and R' tails.) Examples of some other classes of glycerophospholipids are included in Table 24.3.

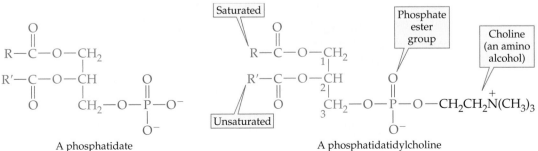

A phosphatidate

A phosphatidatidylcholine
(a glycerophospholipid that is a lecithin)

Application Colloids

What do blood plasma, Jell-O dessert, and mayonnaise have in common? All are *colloids*, or *colloidal dispersions*—mixtures of a dispersing medium (equivalent to a solvent) and dispersed particles larger than those in solutions. The dispersing medium and the dispersed particles can be any possible combination of substances in the solid, liquid, or gaseous state.

Several kinds of colloids are important in living systems. The mixture that remains after blood has clotted, the blood serum (Section 29.3), is a colloid in which the dispersing medium is a liquid (water) and the dispersed particles are protein molecules. Egg white and Jell-O dessert are also liquid/solid colloids. The Jell-O is a less mobile *gel* when it is cold because it contains the larger protein molecules of collagen that have been separated from the bone, skin, and connective tissue of animals. In gelatin at cool temperatures, the large molecules tangle together into a network that gives the mixture some solidity.

Milk is an *emulsion*, a colloidal dispersion of fat droplets in water. For such a colloid to exist, it must be stabilized by an emulsifying agent, which surrounds the fat droplets so that they do not coalesce and form separate fat and water layers. In milk the emulsifying agent is casein, a protein. If you have ever made mayonnaise, you know that the emulsifying agent there is egg yolk, which must be carefully beaten in to establish the colloidal dispersion. Soaps and detergents are emulsifying agents that help remove greasy dirt from clothing and household objects.

Solid colloidal particles pass through a normal filter but are too large to pass through a semipermeable membrane. Protein molecules, in particular, don't cross semipermeable membranes and thus play an essential role in determining the osmolarity of body fluids (Section 29.1). The distribution of water and solutes across the capillary walls that separate blood plasma from the

▲ Vigorous whipping blends the ingredients into the colloidal dispersion we know as mayonnaise.

fluid surrounding cells is controlled by the balance between blood pressure and osmotic pressure. The pressure of blood inside the capillary tends to push water out of the plasma (filtration), but the osmotic pressure of colloidal protein molecules tends to draw water into the plasma (reabsorption). The balance between the two processes varies with location in the body. At the arterial end of a capillary, where blood pumped from the heart has a higher pressure, filtration is favored. At the venous end, where blood pressure is lower, reabsorption is favored, causing waste products from metabolism to enter the bloodstream.

Types of Colloids

Dispersing Medium	Dispersed Substance	Colloid Type	Examples
Solid	Gas	Solid foam	Polyurethane foam, marshmallow
	Liquid	Solid emulsion or gel	Cheese, butter
	Solid	Solid sol	Ruby glass, certain alloys
Liquid	Gas	Foam*	Soapsuds, whipped cream
	Liquid	Emulsion*	Mayonnaise, milk, face cream
	Solid	Sols, gels	Gelatin, starch, blood serum, egg white
Gas	Liquid	Liquid aerosol	Fog, aerosol spray
	Solid	Solid aerosol	Smoke, airborne bacteria and viruses

*Require emulsifying agent.

See Additional Problems 24.70–24.71 at the end of the chapter.

Because of their combination of hydrophobic tails and hydrophilic head groups, the glycerophospholipids are *emulsifying agents*—substances that can surround droplets of nonpolar liquids and hold them in suspension in water (see micelle in Figure 24.3). You can find lecithin, usually obtained from soybean oil, listed as an ingredient in chocolate bars and other foods to which it is added to keep oils from separating out. It's the lecithin in egg yolk that emulsifies the oil droplets in mayonnaise. (See Application, "Colloids.")

In sphingolipids, the amino alcohol sphingosine provides one of the two hydrophobic hydrocarbon tails (blue below and in Figure 24.4). The second hydrocarbon tail is from a fatty acid acyl group connected by an amide link to the —NH_2 group in sphingosine (red below; purple in Figure 24.4).

▲ In most chocolates, lecithin is the emulsifying agent.

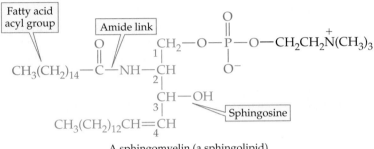

A sphingomyelin (a sphingolipid)

Sphingomyelins are sphingosine derivatives with a phosphate ester group at C1 of sphingosine. The sphingomyelins are major components of the coating around nerve fibers (the *myelin sheath*) and are present in large quantities in brain tissue. A diminished amount of sphingomyelins and phospholipids in brain myelin has been associated with multiple sclerosis. The orientation of the hydrophilic and hydrophobic regions of a sphingomyelin is shown in Figure 24.5, together with a general representation of this and other types of cell membrane lipids used in drawing cell membranes.

■ **PROBLEM 24.11**

For (a) the phosphatidylcholine on p. 703 and (b) the sphingomyelin in Figure 24.5, identify the products that would be formed by complete hydrolysis of all ester bonds.

Glycolipids

Glycolipids, like sphingomyelins, are derived from sphingosine. They differ in having a carbohydrate group at C1 (orange in the glycolipid in Figure 24.4) instead of a phosphate bonded to choline.

Glycolipids reside in cell membranes with their carbohydrate segments extending into the fluid surrounding the cells, just as do the carbohydrate segments of glycoproteins. (∞, p. 652) In this location they function as receptors that, as you saw in Chapter 20, are essential for recognizing chemical messengers, other cells, pathogens, and drugs.

The glycolipid molecule on the top of next page is classified as a *cerebroside*. Cerebrosides, which contain a single monosaccharide, are particularly abundant in nerve cell membranes in the brain, where the monosaccharide is D-galactose. They are also found in other cell membranes, where the sugar unit is D-glucose.

A sphingomyelin

▲ **FIGURE 24.5 A sphingomyelin, showing its polar, hydrophilic head group and its two hydrophobic tails.** The drawing on the right is the representation of phospholipids used in picturing cell membranes. It shows the relative positions of the hydrophilic head and the hydrophobic tails.

A glycolipid

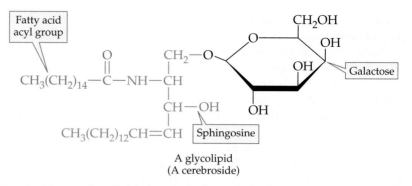

A glycolipid
(A cerebroside)

Gangliosides are glycolipids in which the carbohydrate is a small polysaccharide (an oligosaccharide) rather than a monosaccharide. Over 60 gangliosides are known. The oligosaccharides responsible for blood types are part of ganglioside molecules (see Application "Cell-Surface Carbohydrates and Blood Type" in Chapter 22).

Tay-Sachs disease, a genetic disorder, is the result of a defect in lipid metabolism that causes a greatly elevated concentration of a particular ganglioside in the brain. An infant born with this defect suffers mental retardation and liver enlargement, and often dies by age 3. Tay-Sachs is one of a group of sphingolipid storage diseases. These metabolic diseases result from deficiencies in the supply of enzymes that break down sphingolipids. The harmful consequences result from the *storage* of the excess sphingolipids.

■ **WORKED EXAMPLE 24.1**

A class of membrane lipids known as *plasmalogens* has the general structure shown here. Identify the component parts of this lipid and choose the terms that apply to it: phospholipid, glycerophospholipid, sphingolipid, glycolipid. Is it most similar to a phosphatidylethanolamine, a phosphatidylcholine, a cerebroside, or a ganglioside?

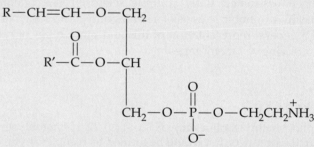

SOLUTION The molecule contains a phosphate group and thus is a phospholipid. The glycerol backbone of three carbon atoms bonded to three oxygen atoms is also present, so the compound is a glycerophospholipid, but one in which there is an ether linkage ($-CH_2-O-CH=CHR$) in place of one of the ester linkages. The phosphate group is bonded to ethanolamine ($HOCH_2CH_2NH_2$). The compound is not a sphingolipid or a glycolipid because it is not derived from sphingosine; for the same reason it is not a cerebroside or a ganglioside. Except for the ether group in place of an ester group, the compound has the same structure as a phosphatidylethanolamine.

■ **PROBLEM 24.12**

Draw the structure of the sphingomyelin that contains a myristic acid acyl group. Identify the hydrophilic head group and the hydrophobic tails in this molecule.

■ **PROBLEM 24.13**

Draw the structure of the glycerophospholipid that contains a stearic acid acyl group, an oleic acid acyl group, and a phosphate bonded to ethanolamine.

■ **PROBLEM 24.14**

Which of the following terms apply to the compound shown below?

(a) A phospholipid (b) A steroid (c) A sphingolipid

(d) A glycerophospholipid (e) A lipid (f) A phosphate ester

(g) A ketone

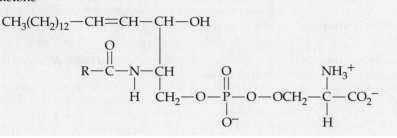

24.6 Cell Membrane Lipids: Cholesterol

Animal cell membranes also contain significant amounts of cholesterol. Cholesterol is a steroid, a member of the class of lipids that all contain the same four-ring system. (∞, Section 20.5) Steroids have many roles throughout both the plant and animal kingdoms. In human biochemistry, the major functions of steroids other than cholesterol are as hormones. and as the bile acids that are essential to the digestion of fats and oils in the diet (Section 25.1).

Cholesterol has the molecular structure and shape shown below:

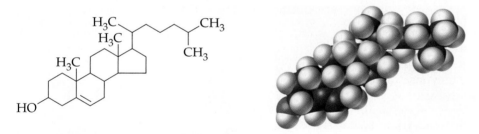

It is the most abundant animal steroid. The body of a 60 kg person contains about 175 g of cholesterol that serves two important functions: as a component of cell membranes and as the starting material for the synthesis of all other steroids. "Cholesterol" has become a household word because of its presence in the arterial plaque that contributes to heart disease (see Application "Lipids and Atherosclerosis," Chapter 25). Some cholesterol is obtained from the diet, but cholesterol is also synthesized in the liver. Even on a strict no-cholesterol diet, the body of an adult can manufacture approximately 800 mg of cholesterol per day.

The molecular model of cholesterol shown above reveals the nearly flat shape of the molecule. Except for its —OH group, cholesterol is hydrophobic. Within a cell membrane, cholesterol molecules are distributed among the hydrophobic tails of the phospholipids. Because the cholesterol molecules are more rigid than the hydrophobic tails, they help to maintain the structure of the membrane.

24.7 Structure of Cell Membranes

Phospholipids provide the basic structure of cell membranes, where they aggregate in a closed, sheet-like structure—the **lipid bilayer** (Figure 24.6). The bilayer is formed by two parallel layers of lipids oriented so that their ionic head groups protrude into the aqueous environments on either side of the

Lipid bilayer The basic structural unit of cell membranes; composed of two parallel sheets of membrane lipid molecules arranged tail to tail.

Application Butter and Its Substitutes

Sometimes the more scientific evidence we accumulate about the relationship between diet and health, the more difficult it is to choose what to eat. The choice between butter and margarine provides an excellent example.

It has become pretty well known that butter can contribute to elevated blood cholesterol, which is to be avoided because of its role in causing heart disease. (We'll have more to say about the cholesterol–heart disease connection in Application "Lipids and Atherosclerosis," Chapter 25.) Many individuals have switched from butter to margarine in response to this information. Margarine is made from vegetable oils; it contains no cholesterol and much less saturated fat than butter.

Gradually, however, information has been accumulating that margarine contains what *might* be an even more unhealthful ingredient—*trans fatty acids*. Oils are catalytically hydrogenated to give them a firmer consistency and also to lessen their tendency to become rancid as oxidation breaks the double bonds. (∞, p. 666) During the partial hydrogenation, some of the cis double bonds are inevitably converted to trans double bonds.

A variety of studies have linked the quantity of trans fatty acids in a person's diet to a greater risk for heart disease and cancer. One such study, widely reported in 1997, came from following the diet and health of 80,000 nurses for 14 years. Those with higher quantities of hydrogenated oils in their diets had a significantly higher risk of heart disease. One suggested explanation is that trans fats alter the metabolism of polyunsaturated fats, which are protective against heart disease.

Meat and dairy products contain a very small amount of trans fatty acids (about 2% in butter), but the quantities in foods containing hydrogenated oils are much higher—up to 40% in the stiffer margarines. If you choose to be serious about avoiding the trans fats, much more is involved than choosing a softer margarine, however. By reading the lists of ingredients on food labels, you will discover that almost all commercial baked goods, cookies, and crackers, as well as many other packaged food products, contain partially hydrogenated oils. It has been suggested that the quantity of trans fats present should be added to the items required to be listed on food labels.

For use in cooking and baking, and at the table, another choice became available in the United States a few years ago. The commercial spread Benecol results from the discovery in Finland that a substance from pine trees lowers cholesterol levels. The compound is a sitostannol, one of a group of plant steroids that differ slightly in structure from cholesterol, mainly in having an additional small hydrocarbon group, represented below by R.

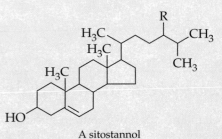

A sitostannol

Because of its similarity to cholesterol, a sitostannol blocks the uptake of cholesterol into the lipoproteins that transport it from the digestive tract into the bloodstream. In this way, as shown by extensive clinical trials, consumption of 3 g of the sitostanol per day can lower serum cholesterol by up to 10%.

▲ Is butter better?

See Additional Problems 24.72–24.73 at the end of the chapter.

Liposome A spherical structure in which a lipid bilayer surrounds a water droplet.

bilayer. Their nonpolar tails cluster together in the middle of the bilayer where they can interact and avoid water.

The bilayer is a favorable arrangement for phospholipids—it is highly ordered and stable, but still flexible. When phospholipids are shaken vigorously with water, they spontaneously form **liposomes**—small spherical vesicles with a lipid bilayer surrounding an aqueous center, as shown in Figure 24.6. Water-soluble substances can be trapped in the center of liposomes, and lipid-soluble substances can be incorporated into the bilayer. Liposomes have potential usefulness as media for drug delivery because they can fuse with cell membranes and empty their contents into the cell (see Application, "Liposomes for Health and Beauty," p. 710).

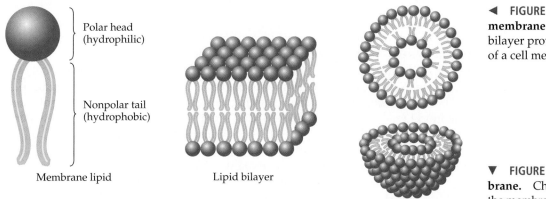

The overall structure of cell membranes is represented by the *fluid-mosaic model*. The membrane is described as *fluid* because it is not rigid and molecules can move around within it, and as a *mosaic* because it contains many kinds of molecules. The components of the cell membrane are represented in Figure 24.7.

▼ **FIGURE 24.7 The cell membrane.** Cholesterol forms part of the membrane, proteins are embedded in the lipid bilayer, and the carbohydrate chains of glycoproteins and glycolipids extend into the extracellular space, where they act as receptors. Integral proteins form channels to the outside of the cell and also participate in transporting large molecules across the membrane.

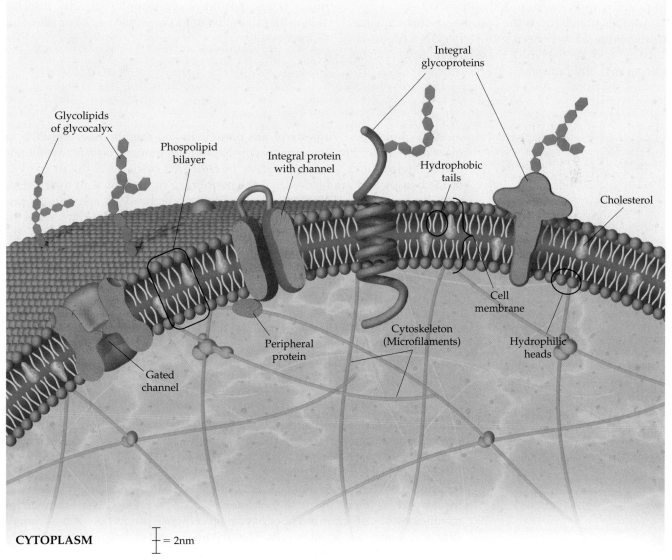

When phospholipids are shaken vigorously with water, they spontaneously form *liposomes*—small spherical vesicles with a lipid bilayer surrounding an aqueous center. Water-soluble substances can be trapped in the center of liposomes, and lipid-soluble substances can be incorporated into the lipid layer.

There is little doubt that liposomes have an exciting future in the marketplace. They provide a new way of delivering chemicals to the body. Research is under way throughout the industrial and academic communities to discover just how this potential can be put to work.

Liposomes are valuable for drug delivery because they can fuse with cell membranes and empty their contents into the cell. One focus is on cancer drugs. A major difficulty with cancer chemotherapy is that the drugs are toxic in varying degrees to both cancer cells and healthy cells. A properly designed liposome should be able to travel intact to the location of a tumor, carrying its water-soluble, tumor-destroying drug safely in its interior. Once it arrives at the cancer site, the liposome can fuse with the membrane of a cancerous cell and deliver the drug directly to where it can act most effectively.

Another possibility is the use of liposomes in gene therapy. It has already been shown that DNA (deoxyribonucleic acid, Section 26.1), the carrier of genes, can be incorporated into liposomes. The hope is that DNA carrying a normal gene can be inserted from the liposome into cells where it can replace the flawed gene.

Liposome drug delivery has already met the requirements of the Food and Drug Administration in an intravenous drug that targets systemic fungal infections. Individuals with compromised immune systems due to AIDS are especially susceptible to this kind of infection. The liposomes carry amphotericin B, an antibiotic that attacks the fungal cell membrane. By delivering amphotericin to the fungal cells, the liposomal drug diminishes the serious side effects of attack by this antibiotic on kidney cells and cells in other healthy organs.

Meanwhile, a broad array of liposome-based products has reached the marketplace as cosmetics. These products are expensive, but they promise what is obviously desirable to many individuals. The creams and lotions are described as creating healthier and more beautiful skin by "energizing" cells, preventing oxidative damage, and erasing wrinkles. Because they can merge with cell membranes, the liposomes deliver their moisturizers, perfumes, or vitamins directly into the skin rather than just onto the surface as a conventional lotion does.

See Additional Problems 24.74–24.75 at the end of the chapter.

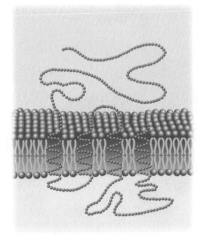

▲ **General structure for an integral membrane protein.** The green circles represent the amino acids. Many membrane proteins pass in and out of the membrane numerous times.

 Membrane Structure and Function

Glycolipids and cholesterol are present in cell membranes, and 20% or more of the weight of a membrane consists of protein molecules, many of them glycoproteins (∞, p. 652). *Peripheral proteins* are associated with just one face of the bilayer and are held within the membrane by noncovalent interactions with the hydrophobic lipid tails or the hydrophilic head groups. *Integral proteins* extend completely through the cell membrane and are anchored by hydrophobic regions that extend through the bilayer. In some cases, the hydrophobic amino acid chain may twist in and out of the membrane many times before ending on the exterior of the membrane with a hydrophilic sugar group. The carbohydrate parts of glycoproteins and glycolipids mediate the interactions of the cell with outside agents. Some integral proteins form channels to allow specific molecules or ions to enter or leave the cell.

Because the bilayer membrane is fluid rather than rigid, it is not easily ruptured. The lipids in the bilayer simply flow back together to repair any small hole or puncture. The effect is similar to what's observed in cooking when a thin film of oil or melted butter lies on top of the water. The film can be punctured and broken, but it immediately flows together when left alone. Still other consequences of bilayer fluidity are that small *nonpolar* molecules can easily enter the cell through the membrane and that some individual lipid or protein molecules can diffuse rapidly from place to place within the membrane.

The fluidity of the membrane varies with the relative amounts of saturated and unsaturated fatty acids in the glycerophospholipids. Such variation is put to use in the adaptation of organisms to their environment. In reindeer, for example, the membranes of cells near the hooves contain a higher proportion of unsaturated fatty acid chains than in other cells. These chains do not pack tightly together. The result is a membrane that remains fluid while the animals stand in the snow.

■ **KEY CONCEPT PROBLEM 24.15**

Integral membrane proteins are not water soluble. Why? How must these proteins differ from globular proteins?

▲ **Oil floating on water, an analogy for the fluid-mosaic cell membrane model.** When the oil layer is disturbed, it soon flows back together.

24.8 Transport Across Cell Membranes

The cell membrane must accommodate opposing needs in allowing the passage of molecules and ions into and out of a cell. On one hand, the membrane surrounding a living cell can't be impermeable, because nutrients must enter and waste products must leave the cell. On the other hand, the membrane can't be completely permeable, or substances would just move back and forth until their concentrations were equal on both sides—hardly what is required for homeostasis (see Application "Homeostasis," p. 575).

The problem is solved by two modes of passage across the membrane (Figure 24.8). In **passive transport**, substances move across the membrane freely by diffusion from regions of higher concentration to regions of lower concentration. In **active transport**, substances can cross the membrane only when energy is supplied because they must go in the reverse direction—from lower to higher concentration regions.

Passive Transport by Simple Diffusion

Some solutes enter and leave cells by **simple diffusion**—they simply wander by normal molecular motion into areas of lower concentration. Small molecules,

Passive transport Movement of a substance across a cell membrane without the use of energy, from a region of higher concentration to a region of lower concentration.

Active transport Movement of substances across a cell membrane with the assistance of energy (for example, from ATP).

Simple diffusion Passive transport by the random motion of diffusion through the cell membrane.

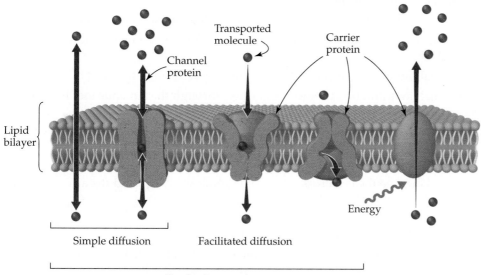

◀ **FIGURE 24.8 Modes of transport across cell membranes.**

Facilitated diffusion Passive transport across a cell membrane with the assistance of a protein that changes shape.

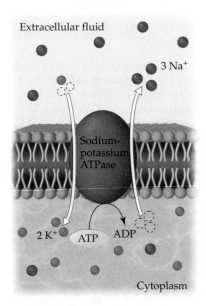

Extracellular fluid

3 Na$^+$

Sodium-potassium ATPase

2 K$^+$ ATP ADP

Cytoplasm

▲ **FIGURE 24.9 An example of active transport.** A protein known as sodium–potassium ATPase uses energy from ATP to move Na$^+$ and K$^+$ ions across cell membranes against their concentration gradients.

Concentration gradient A difference in concentration within the same system.

 Structure of a Typical Eukaryotic Plasma Membrane

such as CO_2 and O_2, and lipid-soluble substances, including steroid hormones, move through the hydrophobic lipid bilayer in this way. Hydrophilic substances similarly pass through the aqueous solutions inside channels formed by integral proteins. What can pass through the protein channels is limited by the size of the molecules relative to the size of the openings. The lipid bilayer is just about impermeable to ions and larger polar molecules, which aren't soluble in the nonpolar hydrocarbon region.

Passive Transport by Facilitated Diffusion

Like simple diffusion, **facilitated diffusion** is passive transport and requires no energy input. The difference is that in facilitated diffusion solutes are helped across the membrane by proteins. The interaction is similar to that between enzymes and substrates. The molecule to be transported binds to a membrane protein, which changes shape so that the transported molecule is released on the other side of the membrane. Glucose is transported into many cells in this fashion.

> ■ **PROBLEM 24.16**
>
> Would you predict that NO can cross a lipid bilayer by simple diffusion?

> ■ **PROBLEM 24.17**
>
> As noted earlier (Section 23.3), the first step in glycolysis, which occurs within cells, is phosphorylation of glucose to glucose 6-phosphate. Why does this step prevent passive diffusion of glucose back out of the cell?

Active Transport

It is essential to life that the concentrations of some solutes be different inside and outside cells. Such differences are contrary to the natural tendency of solutes to move about until the concentration equalizes. Therefore, maintaining **concentration gradients** (differences in concentration within the same system) requires the expenditure of energy. An important example of active transport is the continuous movement of sodium and potassium ions across cell membranes. Only by this means is it possible to maintain homeostasis, which requires low Na$^+$ concentrations within cells and higher Na$^+$ concentrations in extracellular fluids, with the opposite concentration ratio for K$^+$. Energy from the conversion of ATP to ADP is used to change the shape of an integral protein, simultaneously bringing two K$^+$ ions into the cell and moving three Na$^+$ ions out of the cell (Figure 24.9).

Some important points to remember about cell membranes are listed below:

Properties of cell membranes

- Cell membranes are composed of a fluid-like phospholipid bilayer.
- The bilayer incorporates cholesterol, proteins (including glycoproteins), and glycolipids.
- Small nonpolar molecules cross by diffusion through the lipid bilayer.
- Small ions and polar molecules diffuse through the aqueous medium in protein pores.
- Glucose and certain other substances (including amino acids) cross with the aid of proteins and without energy input (*facilitated diffusion*).
- Na$^+$, K$^+$, and other substances that maintain concentration gradients inside and outside the cell cross with expenditure of energy and the aid of proteins (*active transport*).

> ■ **KEY CONCEPT PROBLEM 24.18**
>
> The compositions of the inner and outer surfaces of the lipid bilayer are different. Why do these differences exist and how might they be of use to a living cell?

24.9 Eicosanoids: Prostaglandins and Leukotrienes

The **eicosanoids** are a group of compounds derived from 20-carbon unsaturated fatty acids (*eicosanoic acids*) and synthesized throughout the body. They function as short-lived chemical messengers that act near their points of synthesis ("local hormones").

The *prostaglandins* (named for their discovery in prostate cells) and the *leukotrienes* (named for their discovery in leukocytes) are two classes of eicosanoids that differ somewhat in their structure. The prostaglandins all contain a five-membered ring, which the leukotrienes lack.

Prostaglandins and leukotrienes are synthesized in the body from the 20-carbon unsaturated fatty acid arachidonic acid. Arachidonic acid, in turn, is synthesized from linolenic acid, helping to explain why linolenic is one of the two essential fatty acids. To illustrate the relationships among arachidonic acid, the prostaglandins, and the leukotrienes, we show arachidonic acid and a leukotriene so that they're "bent" into similar shapes:

Eicosanoid A lipid derived from a 20-carbon unsaturated carboxylic acid.

$$\overset{20}{C}H_3\overset{19}{C}H_2\overset{18}{C}H_2\overset{17}{C}H_2\overset{16}{C}H_2\overset{15}{C}H=\overset{14}{C}H\overset{13}{C}H\overset{12}{C}H_2\overset{11}{C}H=\overset{10}{C}H\overset{9}{C}H_2\overset{8}{C}H\overset{7}{=}\overset{6}{C}H\overset{5}{C}H_2\overset{4}{C}H=\overset{3}{C}H\overset{2}{C}H_2\overset{1}{C}H_2CH_2COOH$$

Arachidonic acid

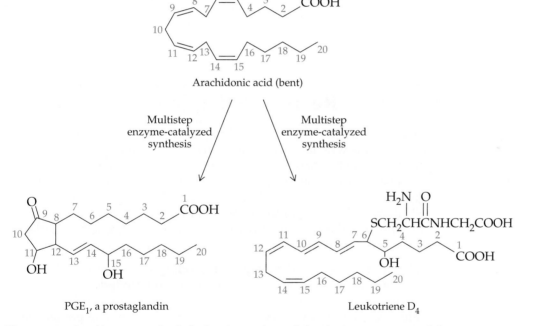

Arachidonic acid (bent)

Multistep enzyme-catalyzed synthesis

PGE$_1$, a prostaglandin

Leukotriene D$_4$

The prostaglandins are actively being investigated for their many potential therapeutic applications. The several dozen known prostaglandins have an extraordinary range of biological effects. They can lower blood pressure, influence platelet aggregation during blood clotting, stimulate uterine contractions, and lower the extent of gastric secretions. In addition, they are responsible for some of the pain and swelling that accompany inflammation. The first approved clinical uses of prostaglandins include stimulation of uterine contractions in therapeutic abortion (for example, of a dead fetus) and to halt persistent bleeding after delivery of a baby (postpartum bleeding).

It was discovered in 1971 that the anti-inflammatory and fever-reducing (*antipyretic*) action of aspirin results in part from its inhibition of prostaglandin synthesis. The aspirin transfers its acetyl group to a serine side chain in the enzyme that catalyzes the first step in conversion of arachidonic acid to prostaglandins, irreversibly inhibiting the enzyme. This inhibition is also thought to explain the effect of aspirin on combating heart attacks. The prostaglandin is a precursor for another type of eicosanoid (a *thromboxane*) that causes aggregation of blood platelets to occur.

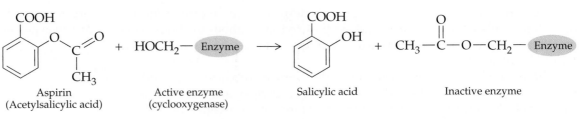

Aspirin   Active enzyme   Salicylic acid   Inactive enzyme
(Acetylsalicylic acid) (cyclooxygenase)

There is also great interest in the leukotrienes. Leukotriene release has been found to trigger the asthmatic response, severe allergic reactions, and inflammation. Asthma treatment with drugs that inhibit leukotriene synthesis is being studied.

■ **KEY CONCEPT PROBLEM 24.19**

In the eiconsanoid shown here identify all of the functional groups. Which groups would be capable of hydrogen bonding? Which would be most acidic? Would this molecule be overall nonpolar, overall polar, or something in between?

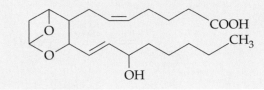

Key Words

Summary: Revisiting the Chapter Goals

1. **What are the major classes of fatty acids and lipids?** *Fatty acids* are carboxylic acids with long, straight hydrocarbon chains; they may be saturated or unsaturated. *Waxes* are esters of fatty acids and alcohols that have long, straight hydrocarbon chains. *Fats* and *oils* are *triacylglycerols*—triesters of glycerol with fatty acids. In fats, the fatty acid chains are mostly saturated; in oils there are varying proportions of unsaturated fatty acid chains. Fats are solid because the saturated hydrocarbon chains pack together neatly; oils are liquids because the kinks at the cis double bonds prevent such packing.

2. **What reactions do triacylglycerols undergo?** The principal reactions of triacylglycerols are catalytic *hydrogenation* and *hydrolysis*. Hydrogen adds to the double bonds of unsaturated hydrocarbon chains in oils, thereby thickening the consistency of the oils and raising their melting points. Treatment of a fat or oil with a strong base such as NaOH hydrolyzes the triacylglycerols to give glycerol and salts of fatty acids. Such *saponification* reactions produce soap, a mixture of fatty acid salts.

3. **What are the membrane lipids?** The membrane lipids include *phospholipids* and *glycolipids* (which have hydrophilic, polar head groups and a pair of hydrophobic tails) and cholesterol (a steroid). *Phospholipids*, which are either *glycerophospholipids* (derived from glycerol) or *sphingomyelins* (derived from the amino alcohol sphingosine), have charged phosphate diester groups in their hydrophilic heads. *Sphingolipids*, which are either sphingomyelins or *glycolipids*, are sphingosine. The glycolipids have carbohydrate head groups.

4. **What is the nature of a cell membrane?** The basic structure of cell membranes is a *bilayer of lipids*, with their hydrophilic heads in the aqueous environment outside and inside the cells, and their hydrophobic tails clustered

together in the center of the bilayer. *Cholesterol* molecules fit between the hydrophobic tails and help to maintain the membrane structure. The membrane also contains *glycoproteins* and *glycolipids* (with their carbohydrate segments at the cell surface, where they serve as receptors), as well as *proteins*. Some of the proteins extend through the membrane (*integral proteins*), and others are only partially embedded at one surface (*peripheral proteins*).

5. How do substances cross cell membranes? Small molecules and lipid-soluble substances can cross the lipid bilayer by simply diffusing through it. Ions and hydrophilic substances can move through aqueous fluid-filled channels in membrane proteins. Some substances cross the membrane by binding to an integral protein, which then releases them inside the cell. These modes of crossing are all *passive transport*—no energy is required because the substances move from regions of higher concentration to regions of lower concentration. *Active transport*, which requires energy and is carried out by certain integral proteins, moves substances against their *concentration gradients*.

6. What are eicosanoids? The *eicosanoids* are a group of compounds derived from 20-carbon unsaturated fatty acids. They are *local hormones*—that is, they act near their point of origin and are short-lived. *Prostaglandins*, which contain a five-membered ring, have a wide range of actions (such as stimulating uterine contractions and causing inflammation). *Leukotrienes*, which do not contain a five-membered ring, trigger the asthmatic response, severe allergic reactions, and inflammation.

▪ Understanding Key Concepts

24.20 The fatty acid composition of three triacylglycerols (A, B, and C) is reported below. Predict which one has the highest melting point. Which one would you expect to be liquid (oil) at room temperature? Explain.

	Palmitic acid	Stearic acid	Oleic acid	Linoleic acid
A	21.4%	27.8%	35.6%	11.9%
B	12.2%	16.7%	48.2%	22.6%
C	11.2%	8.3%	28.2%	48.6%

24.21 Complete hydrogenation of triacylglycerol B above would yield a triacylglycerol of what fatty acid composition? Would the hydrogenation product of triacylglycerol B be more like the hydrogenation product of triacylglycerol A or C? Explain.

24.22 A membrane lipid was isolated and completely hydrolyzed. The following products were detected: serine, phosphate, glycerol, palmitic acid, and oleic acid. Propose a structure for this membrane lipid, and name the family (Table 24.3) to which it belongs.

24.23 According to the fluid-mosaic model (Figure 24.7), the cell membrane is held together mostly by hydrophobic interactions. Considering the forces applied, why doesn't the cell membrane rupture as you move, press against objects, etc.?

24.24 Dipalmitoyl phosphatidylcholine (DPPC) is a surfactant on the surface of the alveoli in the lungs. What is the nature of its fatty acid groups? In what arrangement is it likely to exist at the lung surfaces?

▪ Additional Problems

Waxes, Fats, and Oils

24.25 What is a lipid?

24.26 Why are there so many different kinds of lipids?

24.27 What is a fatty acid?

24.28 Differentiate between saturated, monounsaturated, and polyunsaturated fatty acids.

24.29 What does it mean when we say that a fatty acid is a dietary essential fatty acid?

24.30 Name two dietary essential fatty acids. What are good sources of these fatty acids in the diet?

24.31 Are the double bonds in natural fatty acids primarily cis or trans?

24.32 Why does the presence of double bonds lower the melting temperature of a fatty acid? Do cis fatty acids have a higher or lower melting temperature than the corresponding trans fatty acids? Why?

24.33 What does it mean to say that fats and oils are triacylglycerols?

24.34 How do fats differ from oils in terms of physical properties and typical sources?

24.35 Draw the structure of glyceryl trilaurate, which is made from glycerol and three lauric acid molecules.

24.36 There are two isomeric triacylglycerol molecules whose components are glycerol, one palmitic acid unit, and two stearic acid units. Draw the structures of both, and explain how they differ.

24.37 What function do fats serve in an animal?

24.38 What function does a wax serve in a plant or animal?

24.39 *Spermaceti*, a fragrant substance isolated from sperm whales, was commonly used in cosmetics until it was banned in 1976 to protect the whales from extinction. Chemically, spermaceti is cetyl palmitate, the ester of palmitic acid with cetyl alcohol (the straight-chain C_{16} alcohol). Show the structure of spermaceti.

24.40 What kind of lipid is spermaceti—a fat, a wax, or a steroid?

Chemical Reactions of Lipids

24.41 How would you convert a vegetable oil like soybean oil into a solid cooking fat?

24.42 How would you convert a vegetable oil like corn oil into a soft margarine?

24.43 Draw the structures of all products you would obtain by saponification of the following lipid with aqueous KOH. What are the names of the products?

$$
\begin{array}{l}
\text{CH}_2\text{OC(CH}_2)_{16}\text{CH}_3 \\
\quad\quad\;\; \text{O} \quad\;\; \text{H}\;\; \text{H} \\
\text{CHOC(CH}_2)_7\text{C}=\text{C(CH}_2)_7\text{CH}_3 \\
\quad\quad\;\; \text{O} \quad \text{H}\;\text{H}\;\;\text{H}\;\text{H}\;\;\text{H}\;\text{H} \\
\text{CH}_2\text{OC(CH}_2)_7\text{C}=\text{CCH}_2\text{C}=\text{CCH}_2\text{C}=\text{CCH}_2\text{CH}_3
\end{array}
$$

24.44 Draw the structure of the product you would obtain on complete hydrogenation of the triacylglycerol in Problem 24.43. What is its name? Would it have a higher or lower melting temperature than the original triacylglycerol?

Phospholipids, Glycolipids, and Cell Membranes

24.45 What is the difference between a triacylglycerol and a phospholipid?

24.46 Why is it that glycerophospholipids, rather than triacylglycerols, are found in cell membranes?

24.47 How do sphingomyelms and cerebrosides differ structurally?

24.48 What are the names of the two different kinds of sphingosine-based lipids?

24.49 Why are glycerophospholipids more soluble in water than triacylglycerols?

24.50 How does a soap micelle differ from a membrane bilayer?

24.51 What constituents besides phospholipids are present in a cell membrane?

24.52 What would happen if cell membranes were freely permeable to all molecules?

24.53 Show the structure of a cerebroside made up of D-galactose, sphingosine, and myristic acid.

24.54 Draw the structure of a sphingomyelin that contains a stearic acid unit.

24.55 Draw the structure of a glycerophospholipid that contains palmitic acid, oleic acid, and the phosphate bonded to propanolamine.

24.56 *Cardiolipin*, a compound found in heart muscle, has the following structure. What products would be formed if all ester bonds in the molecule were saponified by treatment with aqueous NaOH?

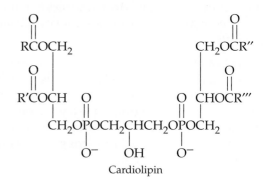

Cardiolipin

24.57 Which process requires energy—passive or active transport? Why is energy sometimes required to move solute across the cell membrane?

24.58 How does facilitated diffusion differ from simple diffusion?

Eicosanoids

24.59 Why are the eicosanoids often called "local hormones"?

24.60 Give an example of an eicosanoid serving as a local hormone.

24.61 Arachidonic acid is used to produce prostaglandins and leukotrienes. From what common fatty acid is arachidonic acid synthesized?

24.62 *Thromboxane A$_2$* is a lipid involved in the blood-clotting process. To what category of lipids does thromboxane A$_2$ belong?

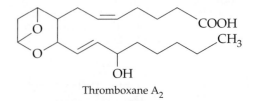

Thromboxane A$_2$

What fatty acid do you think serves as a biological precursor of thromboxane A$_2$?

24.63 Why is it desirable to inhibit the production of leukotrienes?

24.64 How does aspirin function to inhibit the formation of prostaglandins from arachidonic acid?

Applications

24.65 Fats and oils are major sources of triacylglycerols. List some other foods that are associated with high lipid content. [*Lipids in the Diet*]

24.66 According to the FDA, what is the maximum percentage of your daily calories that should come from fats and oils? [*Lipids in the Diet*]

24.67 Describe the mechanism by which soaps and detergents provide cleaning action. [*Detergents*]

24.68 Cationic detergents are rarely used for cleaning. For what purposes are they used? [*Detergents*]

24.69 Why are branched-chain hydrocarbons no longer used for detergents? [*Detergents*]

24.70 Which of the following are colloidal dispersions?
(a) A solution of sugar in water
(b) Fog
(c) Homogenized milk
(d) Smoke [*Colloids*]

24.71 In some types of kidney disease plasma proteins are eliminated in the urine. How will this affect the osmotic pressure of the blood? What effect will it have on the reabsorption of fluid from tissues? [*Colloids*]

24.72 Butter and an equally solid margarine both contain an abundance of saturated fatty acids. What lipid that has been identified as a health hazard is not present in margarine but is present in butter? [*Butter and Its Substitutes*]

24.73 It has been suggested recently that use of oils with more monounsaturated fatty acids (for example, oleic acid) is better for our health than those with polyunsaturated fatty acids or saturated fatty acids. What would be good sources for oils with predominantly monounsaturated fatty acids? (*Hint:* See Table 24.2.) [*Butter and Its Substitutes*]

24.74 A liposome designed to deliver a chemotherapy drug should be able to distinguish between the tumor cell and noncancerous cells. One way to accomplish this would be to have a molecule attached to the surface of the liposome that recognizes specific groups on the tumor cells. What kind(s) of groups might be recognized? [*Liposomes for Health and Beauty*]

24.75 Why does a liposome deliver a moisturizer to skin cells more efficiently than applying the moisturizer directly to the skin? [*Liposomes for Health and Beauty*]

General Questions and Problems

24.76 Which of the following are saponifiable lipids?
(a) Prostaglandin E$_1$
(b) A lecithin
(c) Progesterone
(d) A sphingomyelin
(e) A cerebroside
(f) Glyceryl trioleate

24.77 Identify the component parts of each saponifiable lipid listed in Problem 24.76.

24.78 Draw the structure of a triacylglycerol made from two molecules of linolenic acid and one molecule of myristic acid.

24.79 Would the triacylglycerol described in Problem 24.78 have a higher or lower melting temperature than the triacylglycerol made from one molecule each of linolenic, myristic, and stearic acids? Why?

24.80 Why is cholesterol not saponifiable?

24.81 Jojoba wax, used in candles and cosmetics, is partially composed of the ester of stearic acid and a straight-chain C$_{22}$ alcohol. Draw the structure of this wax component. Compare this to spermaceti in Problem 24.39. Do you think jojoba wax could replace spermaceti in the cosmetic industry?

24.82 What three types of lipids are particularly abundant in brain tissue?

24.83 What are two major roles of cholesterol?

24.84 What are some of the functions that prostaglandins serve in the body?

24.85 Lecithins are often used as food additives to provide emulsification. How do they accomplish this purpose?

24.86 If the average molar mass of a sample of soybean oil is 1500 amu, how many grams of NaOH are needed to saponify 5.0 g of the oil?

24.87 The concentration of cholesterol in the blood serum of a normal adult is approximately 200 mg/dL. How much total blood cholesterol does a person with a blood volume of 5.75 L have?

25

Lipid Metabolism

Lipid metabolism will be very important for this little harp seal pup.

Carbohydrate metabolism (discussed in Chapter 23) is one of our two major sources of energy. Lipid metabolism, the topic of this chapter, is the second. Of the various classes of lipids you saw in Chapter 24, the majority of the lipids in our diet are triacylglycerols. Our focus here, therefore, is on the metabolism of tri-acylglycerols, which as stored in fatty tissue constitute our chief energy reserve.

The main questions addressed in this chapter are the following:

1. What happens during the digestion of triacylglycerols?
The goal: Be able to list the sequence of events in the digestion of dietary tria-cylglycerols and their transport into the bloodstream.

2. What are the various roles of lipoproteins in lipid transport?
The goal: Be able to name the major classes of lipoproteins, specify the nature and function of the lipids they transport, and identify their destinations.

3. **What are the major pathways in the metabolism of triacylglycerols?**

The goal: Be able to name the major pathways for the synthesis and breakdown of triacylglycerols and fatty acids, and identify their connections to other metabolic pathways.

4. **How are triacylglycerols moved into and out of storage in adipose tissue?**

The goal: Be able to explain the reactions by which triacylglycerols are stored and mobilized, and how these reactions are regulated.

5. **How are fatty acids oxidized, and how much energy is produced by their oxidation?**

The goal: Be able to explain what happens to a fatty acid from its entry into a cell until its conversion to acetyl-SCoA.

6. **What is the function of ketogenesis?**

The goal: Be able to identify ketone bodies, describe their properties and synthesis, and explain their role in metabolism.

7. **How are fatty acids synthesized?**

The goal: Be able to compare the pathways for fatty acid synthesis and oxidation, and describe the reactions of the synthesis pathway.

CONCEPTS TO REVIEW

Types of lipids (Sections 24.1, 24.2)
Cell membranes (Sections 24.5–24.7)

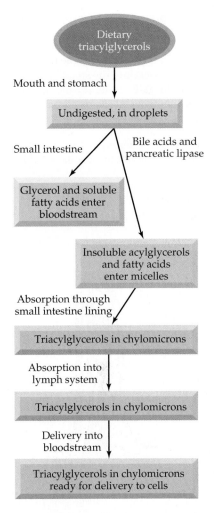

▲ **FIGURE 25.1 Digestion of triacylglycerols.**

25.1 Digestion of Triacylglycerols

When eaten, triacylglycerols (TAGs) pass through the mouth unchanged and enter the stomach (Figure 25.1). (Recall that an *acyl* group is the R—C=O portion of an ester. The acyl groups from fatty acids have relatively long-chain R groups.) The heat and churning action of the stomach break lipids into smaller droplets, a process that takes longer than the physical breakdown and digestion of other foods in the stomach. To be sure that there's time for this breakdown, the presence of lipids in consumed food slows down the rate at which the mixture of partially digested foods leaves the stomach. One of the reasons lipids are a pleasing part of the diet is that the stomach feels full for a longer time after a fatty meal.

The pathway of lipids from the diet through digestion to their ultimate biochemical fate in the body is not as straightforward as that of carbohydrates. The complications arise because dietary lipids—mainly triacylglycerols—are not water-soluble, but nevertheless must enter an aqueous environment. To be moved around within the body, they must therefore be dispersed and surrounded by a water-soluble coating, a process that must happen more than once as lipids travel along their metabolic pathways. During these travels, the lipids are packaged in various kinds of **lipoproteins**, which consist of droplets of hydrophobic lipids surrounded by phospholipids (∞, Section 24.5) and other molecules with their hydrophilic ends to the outside (Figure 25.2).

When partially digested food leaves the stomach, it enters the upper end of the small intestine (the *duodenum*), where its arrival triggers the release of *pancreatic lipases*—enzymes for the hydrolysis of lipids. The gallbladder simultaneously releases **bile**, a mixture that is manufactured in the liver and stored in the gallbladder until needed. Among other components, bile contains **bile acids** and cholesterol, which are steroids, and phospholipids.

Enzymes in the small intestine can't attack the lipids within their water-insoluble droplets, so it's the job of the bile acids and phospholipids to emulsify them by form-

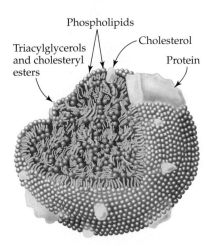

▲ **FIGURE 25.2 A lipoprotein.** A lipoprotein contains a core of neutral lipids, including triacylglycerols and cholesteryl esters. Surrounding the core is a layer of phospholipids in which varying proportions of proteins and cholesterol are embedded.

Lipoprotein A lipid–protein complex that transports lipids.

Bile Fluid secreted by the liver and released into the small intestine from the gallbladder during digestion; contains bile acids, bicarbonate ion, and other electrolytes.

Bile acids Steroid acids derived from cholesterol that are secreted in bile.

Lipid Digestion Part 3 and Part 4

ing micelles much like soap micelles. (∞, p. 701) You can see from the structure below of the anion of cholic acid, the major bile acid, that it resembles soaps and detergents because it also contains both hydrophilic and hydrophobic regions that allow it to act as an emulsifying agent. (see Application, "Colloids," p. 704).

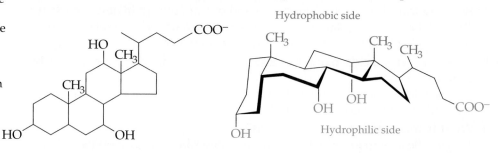

Pancreatic lipase partially hydrolyzes the emulsified triacylglycerols, producing mainly mono- and diacylglycerols, plus fatty acids and a small amount of glycerol.

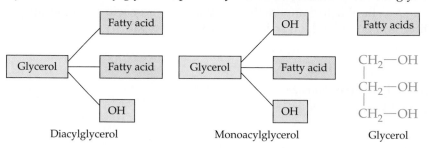

Diacylglycerol   Monoacylglycerol   Glycerol

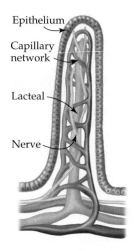

▲ **FIGURE 25.3 A villus, site of absorption in the intestinal lining.** A huge number of villi provide the surface at which lipids and other nutrients are absorbed. Small molecules enter the capillary network, and larger lipids enter the lacteals.

The smaller fatty acids and glycerol are water-soluble and are absorbed directly through the surface of the villi that line the small intestine. Within the villi (Figure 25.3), they enter the bloodstream through capillaries and are carried by the blood to the liver (via the hepatic portal vein).

The still-insoluble acylglycerols and larger fatty acids within the intestine are once again emulsified. Then, at the intestinal lining they are released from the micelles and absorbed. Because these lipids, and also cholesterol and partially hydrolyzed phospholipids, must next enter the aqueous bloodstream for transport, they are again packaged into water-soluble units, in this case the lipoproteins known as *chylomicrons*.

Chylomicrons are too large to enter the bloodstream through capillary walls. Instead, they are absorbed into the lymphatic system through lacteals within the villi (see Figure 25.2). The chylomicrons are then carried to the thoracic duct (just below the collarbone), where the lymphatic system empties into the bloodstream. At this point, these lipids are ready to be used for energy generation or put into storage. The pathways of lipids through the villi and into the transport systems of the bloodstream and the lymphatic system are summarized in Figure 25.4.

FIGURE 25.4 Pathways of ▶ lipids through the villi.

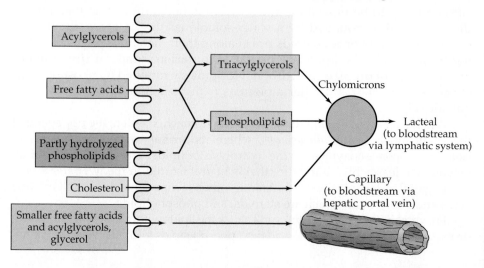

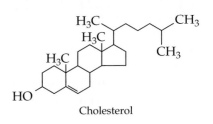

Cholesterol

■ **KEY CONCEPT PROBLEM 25.1**

Cholesterol (see structure in margin) and cholate (a bile acid anion, structure earlier in this section) are both steroids with very similar structures. However, the roles they play in the body are different: Cholate is an emulsifier and cholesterol plays an important role in membrane structure. Identify the small differences in their structures that make them well suited to their jobs in the body. Given their similar structures, could the roles of these molecules be reversed?

25.2 Lipoproteins for Lipid Transport

Lipids enter metabolism from three different sources: (1) the diet, (2) storage in adipose tissue, and (3) synthesis in the liver. Whatever their source these lipids must eventually be transported in blood, an aqueous medium, as summarized in Figure 25.5.

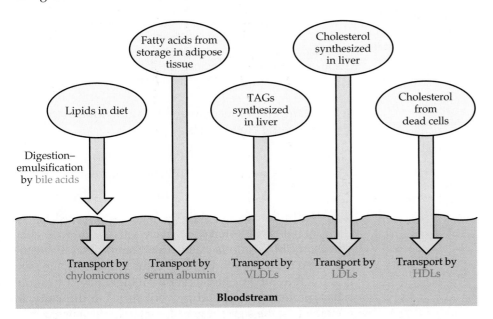

◀ **FIGURE 25.5   Transport of lipids.**   Fatty acids released from storage are carried by albumin, which is a large protein. All of the other lipids are carried packaged in various lipoproteins.

To become water-soluble, fatty acids released from adipose tissue associate with albumin, a very large protein that can bind up to 10 fatty acid molecules. All other lipids are carried by lipoproteins. (The role of lipoproteins in heart disease, where they are of great concern, is discussed in the Application, "Lipids and Atherosclerosis," next page).

Because lipids are less dense than proteins, the density of lipoproteins depends on their ratio of lipids to proteins. They can be somewhat arbitrarily divided into five major types distinguishable by their composition and densities. Chylomicrons, which are the only lipoproteins devoted to transport of lipids from the diet, carry triacylglycerols (TAGs) through the lymphatic system into the blood. These are the lowest density lipoproteins (less than 0.95 g/cm^3) because they carry the highest ratio of lipids to proteins. The four denser lipoprotein fractions have the following roles:

- **Very-low-density lipoproteins (VLDLs)** ($0.96–1.006 \text{ g/cm}^3$) carry TAGs from the liver (where they are synthesized) to peripheral tissues for storage or energy generation.

- **Intermediate-density lipoproteins (IDLs)** ($1.006–1.019 \text{ g/cm}^3$) carry remnants of the VLDLs from peripheral tissues back to the liver for use in synthesis.

Application Lipids and Atherosclerosis

According to the U.S. Food and Drug Administration (FDA), and in agreement with many other authorities, there is "strong, convincing and consistent evidence" for the connection between heart disease and diets high in saturated fats and cholesterol. (The FDA also found evidence that high dietary fat is one risk factor for certain types of cancer.)

Several points are clear:

- A diet rich in saturated animal fats leads to an increase in blood-serum cholesterol.

- A diet lower in saturated fat and higher in unsaturated fat can lower the serum cholesterol level.

- High levels of cholesterol are correlated with *atherosclerosis*, a condition in which yellowish deposits (*arterial plaque*) composed of cholesterol and other lipid-containing materials form within the larger arteries. The result of atherosclerosis is an increased risk of coronary artery disease and heart attack brought on by blockage of blood flow to heart muscles, or an increased risk of stroke due to blockage of blood flow to the brain.

Factors considered in an overall evaluation of an individual's risk of heart disease are the following:

Risk factors for heart disease

High blood levels of cholesterol and low levels of high-density lipoproteins (HDLs)

Cigarette smoking

High blood pressure

Diabetes

Obesity

Low level of physical activity

Family history of early heart disease

As discussed in Section 25.2, lipoproteins are complex assemblages of lipids and proteins that transport lipids throughout the body. If LDL (the so-called "bad" cholesterol) delivers more cholesterol than is needed to peripheral tissues, and if not enough HDL (the so-called "good" cholesterol) is present to remove it, the excess cholesterol is deposited in cells and arteries. Thus, the higher the HDL level, the less the likelihood of deposits and the lower the risk of heart disease. There is some evidence that a low HDL level (less than 35 mg/dL) may be the single best predictor of heart attack potential. Also, LDL has the negative potential to trigger inflammation and the buildup of plaque in artery walls. (Remember it this way—**low** LDL is good; **high** HDL is good.)

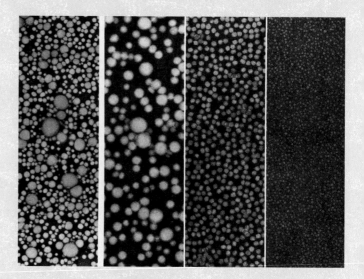

▲ **Lipoproteins.** The lipoproteins were separated from blood plasma in an ultracentrifuge and photographed with an electron microscope. From left to right, the lipoproteins shown are chylomicrons (magnified 43,000×), VLDLs (130,000×), LDLs (130,000×), and HDLs (130,000×).

Many groups recommend that individuals strive for the following blood levels:

Total cholesterol	200 mg/dL or lower
LDL	160 mg/dL or lower
HDL	60 mg/dL or higher

To further assess the risk level represented by an individual's cholesterol and HDL values, the total cholesterol/HDL ratio is calculated. The ideal ratio is considered to be 3.5. A ratio of 4.5 indicates an average risk, and a ratio of 5 or higher shows a high and potentially dangerous risk. The ratio overcomes the difficulty in evaluating the significance of, for example, a negatively high cholesterol level of 290 mg/dL combined with a positively high HDL value of 75 mg/dL. The resultant ratio of 3.9 indicates a low risk level.

Decreasing saturated fats and cholesterol in the diet, adopting an exercise program, and ceasing to smoke constitute the first line of defense for those at risk. For those at high risk or for whom the first-line defenses are inadequate, drugs are available that prevent or slow the progress of coronary artery disease by lowering serum cholesterol levels. Among the drugs are indigestible resins (*cholestyramine* and *colestipol*) that bind bile acids and accelerate their excretion, causing the liver to use up more cholesterol in bile acid synthesis. Another class of effective drugs are the statins (for example, lovastatin), which inhibit an enzyme crucial to the synthesis of cholesterol.

See Additional Problems 25.44–25.47 at the end of the chapter.

- **Low-density lipoproteins (LDLs)** $(1.019–1.063 \text{ g/cm}^3)$ transport cholesterol from the liver to peripheral tissues, where it may be used in cell membranes or for steroid synthesis (and is also available for formation of arterial plaque).
- **High-density lipoproteins (HDLs)** $(1.063–1.210 \text{ g/cm}^3)$ transport cholesterol *from* dead or dying cells back to the liver, where it is converted to bile acids. The bile acids are then available for use in digestion or are excreted when in excess.

25.3 Triacylglycerol Metabolism: An Overview

The metabolic pathways for triacylglycerols are summarized in Figure 25.6 and further explained in following sections of this chapter.

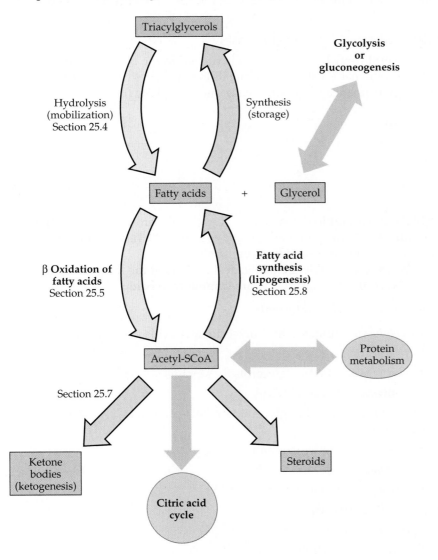

◄   **FIGURE 25.6   Metabolism of triacylglycerols.**   Pathways that break down molecules (catabolism) are shown in yellow, and synthetic pathways (anabolism) are shown in blue. Connections to other pathways or intermediates of metabolism are shown in green.

Dietary Triacylglycerols

For TAGs from the diet, hydrolysis occurs when chylomicrons in the bloodstream encounter lipoprotein lipase anchored in capillary walls. The resulting fatty acids then have two possible fates: (1) When energy is in good supply, they are converted back to TAGs for storage in adipose tissue. (2) When cells are in need of energy, the fatty acid carbon atoms are transferred to the acetyl groups of acetyl-SCoA.

The major pathway for acetyl-SCoA is the generation of energy via the citric acid cycle and oxidative phosphorylation. (⚬⚬⚬, Figure 21.5, p. 606). In resting muscle, fatty acids are the major energy source. Acetyl-SCoA serves as the starting material for the biosynthesis of fatty acids (*lipogenesis*) in the liver (Section 25.8). In addition, it can enter the *ketogenesis* pathway for production of ketone bodies, a source of energy called on when glucose is in short supply (Section 25.7). Acetyl-SCoA is also the starting material for the synthesis of cholesterol, from which all other steroids are made.

Triacylglycerols from Adipocytes

When stored TAGs are needed as an energy source, lipases within fat cells are activated by hormones (insulin and glucagon, Section 23.7). The stored TAGs are hydrolyzed to fatty acids, and the free fatty acids and glycerol are then released into the bloodstream. These fatty acids travel in association with *albumins* to cells where they are converted to acetyl-SCoA for energy generation.

Glycerol from Triacylglycerols

The glycerol from TAG hydrolysis is carried in the bloodstream to the liver or kidneys, where it is converted in the following series of reactions to glyceraldehyde 3-phosphate and dihydroxyacetone phosphate (DHAP).

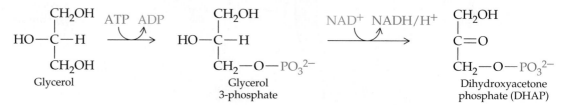

DHAP is a reactant in the synthesis of triacylglycerols (Section 25.4). It can also enter the glycolysis pathway (at step 6, Figure 23.4), and thus provides a link between carbohydrate and lipid metabolism.

The varied possible metabolic destinations of the fatty acids, glycerol, and acetyl-SCoA from dietary TAGs are summarized below.

Fate of Dietary Triacylglycerols

- *Triacylglycerols* undergo hydrolysis to fatty acids and glycerol.
- *Fatty acids* undergo
  - Resynthesis of triacylglycerols for storage
  - Conversion to acetyl-SCoA
- *Glycerol* is converted to glyceraldehyde 3-phosphate and DHAP, which participate in
  - *Glycolysis*—energy generation (⚬⚬⚬, Section 23.3)
  - *Gluconeogenesis*—glucose formation (⚬⚬⚬, Section 23.11)
  - *Triacylglycerol synthesis* (Section 25.4)
- *Acetyl-SCoA* participates in
  - *Triacylglycerol synthesis* (Section 25.4)
  - Ketone body synthesis (*ketogenesis*, Section 25.7)
  - Synthesis of steroids and other lipids
  - *Citric acid cycle and oxidative phosphorylation* (⚬⚬⚬, Sections 21.8, 21.9)

■ **PROBLEM 25.2**

Examine Figure 23.3 (p. 668) and explain how dihydroxyacetone phosphate can enter the glycolysis pathway and be converted to pyruvate

Nutritional Value of Medium-Chain Triacylglycerols

Failure of the complex processes needed for absorption of lipids into the bloodstream is referred to as *malabsorption*. Clinically, malabsorption is recognized by the presence of abnormally large quantities of fat in the feces (*steatorrhea*).

Among the many conditions that cause malabsorption are (1) diseases that decrease secretion of bile salts, resulting in poor micelle formation; (2) damage to the intestinal mucous lining that diminishes absorption (for example, by inflammatory diseases such as ulcerative colitis); and (3) failure of proper digestion.

Dietary fat can be provided to individuals suffering from malabsorption by feeding them synthetic medium-chain triacylglycerols (MCTs); these are synthetic triacylglycerols that contain fatty acids with 10 or fewer carbon atoms. Because they are more soluble, these smaller triacylglycerols bypass the need for pancreatic lipase and micelle formation. They are absorbed directly into cells lining the small intestine, where enzymes break them down to fatty acids and glycerol. The resulting medium-chain fatty acids need not be repackaged into chylomicrons to enter the lymph system. Instead, they are directly absorbed into the bloodstream and transported by the albumins. MCTs are also used with patients receiving *total parenteral nutrition*, in which the gastrointestinal tract is bypassed and all nutrients are supplied intravenously.

▲ The team of scientists that studied nutritional needs during their 1994 climb of Mt. Everest. Photo courtesy John W. Finley, University of Illinois.

A potential application of MCTs was uncovered by a team of scientists during a climb of Mt. Everest in 1994. The rapid absorption and conversion to energy of the MCTs make them an excellent addition to the diet during extremely demanding physical activity. Above 20,000 ft, the climbers needed up to 10,000 kcal per day, more than four times a normal daily caloric intake.

See Additional Problems 25.48–25.49 at the end of the chapter.

25.4 Storage and Mobilization of Triacylglycerols

We've noted that adipose tissue is the storage depot for TAGs and that TAGs are our primary energy storage form. TAGs don't just sit unused until needed, however. The passage of fatty acids in and out of storage in adipose tissue is a continuous process essential to maintaining homeostasis (see Application, "Homeostasis," p. 575).

TAG Synthesis

To see how storage and **mobilization** of TAGs are regulated, look back at Figure 23.6 (∞, p. 676), which shows the effects of the hormones insulin and glucagon on metabolism. After a meal, blood glucose levels are high, insulin levels rise, and glucagon levels drop. Glucose is entering cells and glycolysis is proceeding actively. Under these conditions, insulin activates the synthesis of TAGs for storage.

The reactants in TAG synthesis are glycerol 3-phosphate and fatty acid acyl groups carried by coenzyme A. TAG synthesis proceeds by transfer of

Mobilization (of triacylglycerols)
Hydrolysis of triacylglycerols in adipose tissue and release of fatty acids into the bloodstream.

first one and then another fatty acid acyl group from coenzyme A to glycerol 3-phosphate:

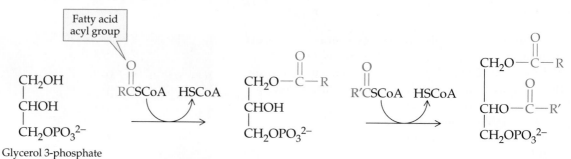

Glycerol 3-phosphate

Next, the phosphate group is removed and the third fatty acid group is added to give a triacylglycerol:

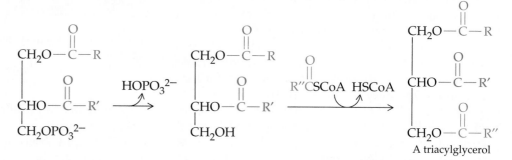

A triacylglycerol

Since adipocytes don't have the enzyme needed to convert glycerol to glycerol 3-phosphate, they can't make triacylglycerols unless some dihydroxyacetone phosphate is available from step 6 of glycolysis (Figure 23.3, p. 677). In other words, the metabolism of glucose is needed to supply dihydroxyacetone phosphate that will isomerize to give the necessary glycerol 3-phosphate.

TAG Mobilization

When digestion of a meal is finished, blood glucose levels are low, so that insulin levels drop and glucagon levels rise. The lower insulin level and higher glucagon level together activate *triacylglycerol lipase*, the enzyme within adipocytes that controls hydrolysis of stored TAGs. When glycerol 3-phosphate is in short supply—an indication that glycolysis is not producing sufficient energy—the fatty acids and glycerol produced by hydrolysis of the stored TAGs are released to the bloodstream for transport to energy-generating cells. Otherwise, the fatty acids and glycerol are cycled back into new TAGs for storage.

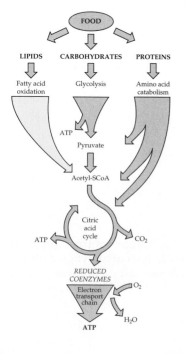

25.5 Oxidation of Fatty Acids

Once a fatty acid enters the cytosol of a cell that needs energy, three successive processes must occur.

1. *Activation* The fatty acid must be activated by conversion to fatty acyl-SCoA. This activation, which occurs in the cytosol, serves the same purpose as the first few steps in oxidation of glucose by glycolysis. Some energy from ATP must initially be invested in converting the fatty acid to fatty acyl-SCoA, a form that can be broken down more easily.

$$R-\overset{\overset{\displaystyle O}{\|}}{C}-O^- + HSCoA + ATP \longrightarrow R-\overset{\overset{\displaystyle O}{\|}}{C}-SCoA + AMP + P_2O_7^{4-}$$

Fatty acid   Fatty acyl-SCoA

2. *Transport*   The fatty acyl-SCoA must be transported into the mitochondrial matrix where energy generation will occur.

3. *Oxidation*   The fatty acyl-SCoA must be oxidized by enzymes in the mitochondrial matrix to produce acetyl-SCoA plus the reduced coenzymes to be used in ATP generation. The oxidation occurs by repetition of the series of four reactions shown in Figure 25.7, which make up the **β-oxidation pathway**. Each repetition of these reactions cleaves a two-carbon acetyl group from the end of a fatty acid acyl group and produces one acetyl-SCoA. The pathway is more of a spiral than a cycle because the long-chain fatty acyl group must continue to return to the pathway until each pair of carbon atoms is removed.

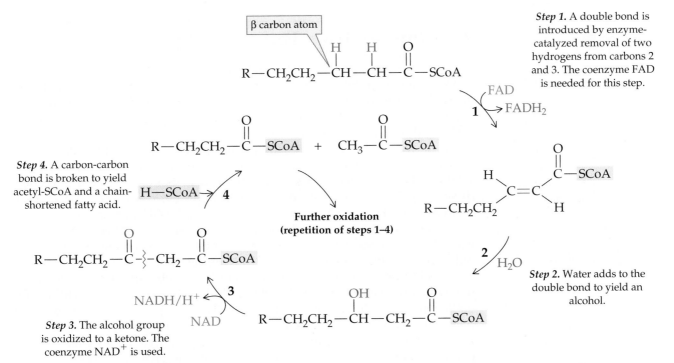

Step 1. A double bond is introduced by enzyme-catalyzed removal of two hydrogens from carbons 2 and 3. The coenzyme FAD is needed for this step.

Step 4. A carbon-carbon bond is broken to yield acetyl-SCoA and a chain-shortened fatty acid.

Further oxidation (repetition of steps 1–4)

Step 2. Water adds to the double bond to yield an alcohol.

Step 3. The alcohol group is oxidized to a ketone. The coenzyme NAD$^+$ is used.

▲ **FIGURE 25.7**   β **Oxidation of fatty acids.**   Passage of an acyl-SCoA through these four steps cleaves one acetyl group from the end of the fatty acid chain. In this manner, carbon atoms are removed from a fatty acid two at a time.

The β-oxidation pathway

The name "β oxidation" refers to the oxidation of the carbon atom β to the thioester linkage in two steps of the pathway.

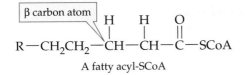

A fatty acyl-SCoA

Step 1: The first β oxidation   The oxidizing agent FAD removes hydrogen atoms from the carbon atoms α and β to the C=O group in the fatty acyl-SCoA, forming a carbon–carbon double bond. These hydrogen atoms and their electrons are passed directly from FADH$_2$ to coenzyme Q so that the electrons can enter the electron transport chain. (∞, Section 21.9)

Step 2: Hydration   A water molecule adds across the newly created double bond to give an alcohol with the —OH group on the β carbon.

Step 3: The second β oxidation   NAD$^+$ is the oxidizing agent for conversion of the β —OH group to a carbonyl group.

Step 4: Cleavage to remove an acetyl group   An acetyl group is split off and attached to a new coenzyme A molecule, leaving behind an acyl-SCoA that is two carbon atoms shorter.

β-Oxidation pathway   A repetitive series of biochemical reactions that degrades fatty acids to acetyl-SCoA by removing carbon atoms two at a time.

For a fatty acid with an even number of carbon atoms, all of the carbons are transferred to acetyl-SCoA molecules by an appropriate number of trips through the β-oxidation spiral. Additional steps are required for fatty acids with odd numbers of carbon atoms and those with double bonds. Ultimately, all fatty acid carbons are released for further oxidation in the citric acid cycle.

> ■ **KEY CONCEPT PROBLEM 25.3**
>
> In β oxidation, (a) identify the steps that are oxidations and describe the changes that occur; (b) identify the oxidizing agents; (c) identify the reaction that is an addition; (d) identify the reaction that is a substitution.

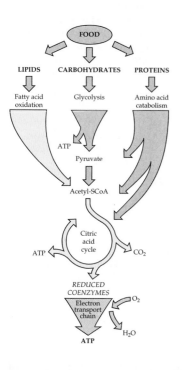

25.6 Energy from Fatty Acid Oxidation

The total energy output from fatty acid catabolism, like that from glucose catabolism, is measured by the total number of ATPs produced. In the case of fatty acids, this is the total number of ATPs from passage of acetyl-SCoA through the citric acid cycle plus those produced by the reduced coenzymes (NADH and $FADH_2$) during fatty acid oxidation.

First, we need to know the number of molecules of acetyl-SCoA from the fatty acids. Since two carbon atoms are removed with each acetyl-SCoA, the total number of acetyl-SCoAs for a fatty acid is its number of carbon atoms divided by 2. These acetyl-SCoAs proceed to the citric acid cycle, where each one yields 1 ATP and a total of 3 NADHs and 1 $FADH_2$. (∞, Section 21.8) Using the current best estimates of 2.5 ATPs for each NADH and 1.5 ATPs for each $FADH_2$ (∞, Section 23.6), each acetyl-SCoA generates 9 ATPs from reduced coenzymes, giving a total of 10 ATPs per acetyl-SCoA.

In addition, we must take into account the number of ATPs derived from the two reduced coenzymes (1 $FADH_2$ and 1 NADH) from each repetition of β oxidation. With 1.5 ATPs per $FADH_2$ and 2.5 ATPs per NADH, the total is 4 ATPs for each β oxidation. Note that the number of repetitions is always one less than the number of acetyl-SCoA molecules produced, because the last repetition of β oxidation cleaves a four-carbon chain to give 2 acetyl-SCoAs. Also, we must subtract the equivalent of 2 ATPs spent in activation of a fatty acid, the step that precedes the β oxidations.

As an example, the 12-carbon fatty acid, lauric acid ($CH_3(CH_2)_{10}COOH$), yields 78 ATPs:

From the citric acid cycle:

12 C atoms/2 = 6 acetyl-SCoAs

$$\frac{10 \text{ ATPs}}{\text{acetyl-SCoA}} \times 6 \text{ acetyl-SCoAs } = 60 \text{ ATPs}$$

Activation of the fatty acid: = −2 ATPs

From the 5 β-oxidations:

$$\frac{4 \text{ ATPs}}{\beta \text{ oxidation}} \times 5 \text{ β-oxidations } = 20 \text{ ATPs}$$

Total = 78 ATPs

Comparing the amount of ATP produced by fatty acid catabolism to the amount produced by glucose catabolism illustrates why our bodies use triacylglycerols rather than carbohydrates for long-term energy storage. We use lauric acid as our example because it has a molar mass close to that of glucose. Our best estimates show that 1 mol of glucose (180 g) generates 30–32 mol of ATP (∞, Section 23.6), whereas 1 mol of lauric acid (200 g) generates 78 mol of ATP.

Thus, fatty acids yield more than twice as much energy per gram as carbohydrates. In terms of nutritional Calories (that is, kilocalories), carbohydrates yield 4 Cal/g, whereas fats and oils yield 9 Cal/g.

In addition, stored fats have a greater "energy density" than stored carbohydrates. Because glycogen—the storage form of carbohydrates—is hydrophilic, about 2 g of water are held with each gram of glycogen. The hydrophobic fats do not hold water in this manner.

■ **PROBLEM 25.4**

How many molecules of acetyl-SCoA are produced by catabolism of the following fatty acids, and how many β oxidations are needed?
(a) Palmitic acid, $CH_3(CH_2)_{14}COOH$
(b) Eicosanoic acid, $CH_3(CH_2)_{18}COOH$

■ **PROBLEM 25.5**

Look back at the reactions of the citric acid cycle (Figure 21.9) and identify the three reactions that are similar to the first three reactions of the β oxidation of a fatty acid.

We have now looked at the metabolism of carbohydrates and lipids, which together provide most of our energy. Protein metabolism is utilized for energy production primarily when the carbohydrate and lipid supply is inadequate. Proteins are more important for their essential roles in providing structure and regulating function.

Figure 25.8 is a general summary of ATP production from the products of digestion and their cycling between the buildup (anabolism) and breakdown (catabolism) of biomolecules.

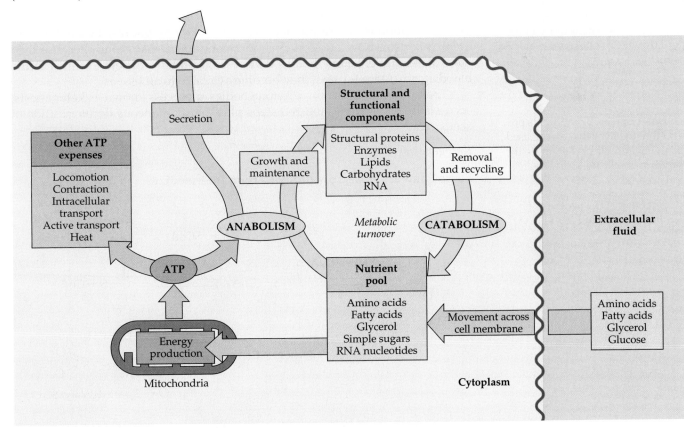

▲ **FIGURE 25.8 Summary of pathways of nutrients through anabolism and catabolism.**

☉☉☉ LOOKING AHEAD

The structure and function of proteins was discussed in Chapters 18 and 19, but we have not yet reviewed their metabolism, which is referred to in the general diagram in Figure 25.8. In Chapter 28 you will be introduced to the digestion of protein, the catabolism of amino acids, the fate of the nitrogen and carbon atoms from amino acids, and the biosynthesis of nonessential amino acids. Ribonucleic acids (RNAs), also shown in Figure 25.8, are covered in Chapter 26, as is the synthesis of proteins.   ☉☉☉

▲ Fat as a source of water. A camel's hump is almost entirely fat, which serves as a source of energy and also water. As reduced coenzymes from fatty acid oxidation pass through electron transport to generate ATP, large amounts of water are formed (about 8 H_2Os for each C atom in a fatty acid). This water sustains camels during long periods when no drinking water is available.

Ketone bodies   Compounds produced in the liver that can be used as fuel by muscle and brain tissue; 3-hydroxybutyrate, acetoacetate, and acetone.

Ketogenesis   The synthesis of ketone bodies from acetyl-SCoA.

25.7 Ketone Bodies and Ketoacidosis

What happens if lipid catabolism (or any other condition) produces more acetyl-SCoA than the citric acid cycle can handle? The energy is preserved by conversion of acetyl-SCoA in liver mitochondria to 3-hydroxybutyrate and acetoacetate. Because it is a β-keto acid and therefore somewhat unstable, acetoacetate undergoes spontaneous, nonenzymatic decomposition to form acetone.

Ketone bodies

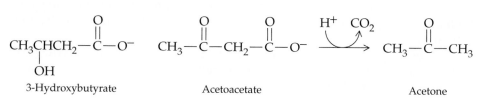

These compounds are traditionally known as **ketone bodies**, although one of them, 3-hydroxybutyrate, contains no ketone functional group. Because they are water-soluble, ketone bodies do not need protein carriers to travel in the bloodstream. Once formed, they become available to all tissues.

The formation of the three ketone bodies, a process known as **ketogenesis**, occurs in four enzyme-catalyzed steps plus the spontaneous decomposition of acetoacetate.

Ketogenesis

Steps 1 and 2:   Assembly of Six-Carbon Intermediate

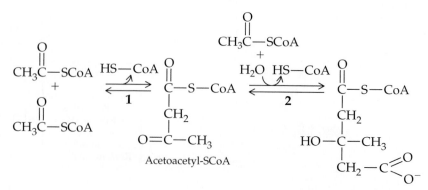

In *step 1*, two acetyl-SCoA molecules combine with each other in a reversible reaction. Then, in *step 2*, a third acetyl-SCoA and a water molecule react with acetoacetyl-SCoA to give 3-hydroxy-3-methylglutaryl-SCoA.

Steps 3 and 4 of Ketogenesis: Formation of the Ketone Bodies

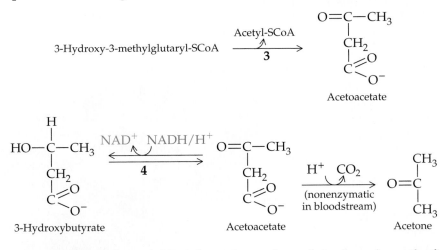

In step 3, removal of acetyl-SCoA from the product of *step 2* produces the first of the ketone bodies, *acetoacetate*. The first three steps add up to combination of the acetyl groups from two acetyl-SCoAs to give acetoacetate, the precursor of the other two ketone bodies. The acetoacetate from step 3 is reduced to 3-hydroxybutyrate in step 4. (Note that 3-hydroxybutyrate and acetoacetate are connected by a reversible reaction. In tissues that need energy, acetoacetate is produced by different enzymes than those used for ketogenesis. Acetyl-SCoA can then be produced from the acetoacetate.) As acetoacetate and 3-hydroxybutyrate are synthesized by ketogenesis in liver mitochondria, they are released to the bloodstream. Acetone is then formed in the bloodstream by the decomposition of acetoacetate.

Under well-fed, healthy conditions, skeletal muscles derive a small portion of their daily energy needs from acetoacetate, and heart muscles use it in preference to glucose. But consider the situation when energy production from glucose is inadequate due to starvation or because glucose is not being metabolized normally due to diabetes (⚬⚬⚬, Section 23.9). The body must respond by providing other energy sources in what can become a precarious balancing act. Under these conditions, the production of ketone bodies accelerates because acetoacetate and 3-hydroxybutyrate can be converted to acetyl-SCoA for oxidation in the citric acid cycle.

During the early stages of starvation, heart and muscle tissues burn larger quantities of acetoacetate, thereby preserving glucose for use in the brain. In prolonged starvation, even the brain can switch to ketone bodies to meet up to 75% of its energy needs.

The condition in which ketone bodies are produced faster than they are utilized (*ketosis*) occurs in diabetes. It is indicated by the odor of acetone (a highly volatile ketone) on the patient's breath and the presence of ketone bodies in the urine (*ketonuria*) and the blood (*ketonemia*).

Because two of the ketone bodies are carboxylic acids, continued ketosis such as might occur in untreated diabetes leads to the potentially serious condition known as **ketoacidosis**—acidosis resulting from increased concentrations of ketone bodies in the blood. The blood's buffers are overwhelmed and blood pH drops. An individual experiences dehydration due to increased urine flow, labored breathing because acidic blood is a poor oxygen carrier, and depression. Ultimately, if untreated, the condition leads to coma and death.

Ketoacidosis Lowered blood pH due to accumulation of ketone bodies.

■ **PROBLEM 25.6**

Which of the following classifications apply to the formation of 3-hydroxybutyrate from acetoacetate?

(a) Hydrolysis   **(b)** Oxidation   **(c)** Reduction   **(d)** Condensation

The Liver, Clearinghouse for Metabolism

The liver is the largest reservoir of blood in the body and also the largest internal organ, making up about 2.5% of the body's mass. Blood carrying the end products of digestion enters the liver through the hepatic portal vein before going into general circulation, so the liver is therefore ideally situated to regulate the concentrations of nutrients and other substances in the blood. The liver is important as the gateway for entry of drugs into the circulation and contains the enzymes needed to inactivate toxic substances.

Various functions of the liver have been described in scattered sections of this book, but it's only by taking an overview that the central role of the liver in metabolism can be appreciated. Among its many functions, the liver synthesizes glycogen from glucose, glucose from noncarbohydrates, triacylglycerols from mono- and diacylglycerols, and fatty acids from acetyl-SCoA. It is also the site of synthesis of cholesterol, bile acids, plasma proteins, and blood clotting factors. In addition, liver cells can catabolize glucose, fatty acids, and amino acids to yield carbon dioxide and energy stored in ATP. The *urea cycle*, by which nitrogen from amino acids is converted to urea for excretion, takes place in the liver (Section 28.4).

The liver stores reserves of glycogen, certain lipids and amino acids, iron, and fat-soluble vitamins, and releases them as needed to maintain homeostasis. In addition, only liver cells have the enzyme needed to convert glucose 6-phosphate from glycogenolysis and gluconeogenesis to glucose that can enter the bloodstream.

Given its central role in metabolism, the liver is subject to a number of pathologic conditions based on excessive accumulation of various metabolites. One example is *cirrhosis*, the development of fibrous tissue that is preceded by excessive triacylglycerol buildup.

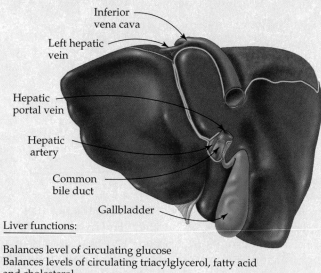

Liver functions:

Balances level of circulating glucose
Balances levels of circulating triacylglycerol, fatty acid and cholesterol
Removes excess amino acids from circulation; converts their nitrogen to urea for excretion
Stores reserves of fat-soluble vitamins and iron
Removes drugs from circulation and breaks them down

▲ **Anatomy of the liver.** Blood carrying metabolites from the digestive system enters the liver through the hepatic portal vein. The gallbladder is the site for storage of bile.

Cirrhosis occurs in alcoholism, uncontrolled diabetes, and metabolic conditions in which the synthesis of lipoproteins from triacylglycerols is blocked. Another example is *Wilson's disease*, a genetic defect in copper metabolism. In Wilson's disease, copper accumulates in the liver rather than being excreted or recycled for use in coenzymes. Chronic liver disease, as well as brain damage and anemia, are symptoms of Wilson's disease. The disease is treated by a low-copper diet and drugs that enhance the excretion of copper.

See Additional Problems 25.50–25.51 at the end of the chapter.

■ **PROBLEM 25.7**

Consider the reactions of ketogenesis.

(a) What role does acetyl-SCoA play?

(b) How many acetyl-SCoA molecules are used in the production of the ketone bodies?

(c) What is the essential role of ketone bodies during prolonged starvation?

25.8 Biosynthesis of Fatty Acids

The biosynthesis of fatty acids from acetyl-SCoA, a process known as **lipogenesis**, provides a link between carbohydrate, lipid, and protein metabolism. Because acetyl-SCoA is an end product of carbohydrate and amino acid catabolism, using it to make fatty acids allows the body to divert the energy of excess carbohydrates and amino acids into storage as triacylglycerols.

Fatty acid synthesis and catabolism are similar in that they both proceed two carbon atoms at a time. But as is usually the case, the biochemical pathway in one direction is not the exact reverse of the pathway in the other direction because the reverse of an energetically favorable pathway would be energetically unfavorable. This principle applies to β oxidation of fatty acids and its reverse, lipogenesis. The two pathways are compared in Table 25.1.

Lipogenesis The biochemical pathway for synthesis of fatty acids from acetyl-SCoA.

TABLE 25.1 Comparison of Fatty Acid Oxidation and Synthesis	
Oxidation	**Synthesis**
Occurs in mitochondria	Occurs in cytosol
Enzymes different from synthesis	Enzymes different from oxidation
Intermediates carried by coenzyme A	Intermediates carried by acyl carrier protein
Coenzymes: FAD, NAD$^+$	Coenzyme: NADPH
Carbon atoms removed two at a time	Carbon atoms added two at a time

The stage is set for lipogenesis by two reactions: (1) transfer of an acetyl group from acetyl-SCoA to an acyl carrier protein (ACP), and (2) conversion of acetyl-SCoA to malonyl-SCoA in a reaction that requires investment of energy from ATP. The malonyl group carries the carbon atoms that will be incorporated two at a time into the fatty acid. The malonyl-SCoA is then transferred to the acyl carrier protein (ACP), which is part of a multienzyme complex that contains all of the enzymes for lipogenesis.

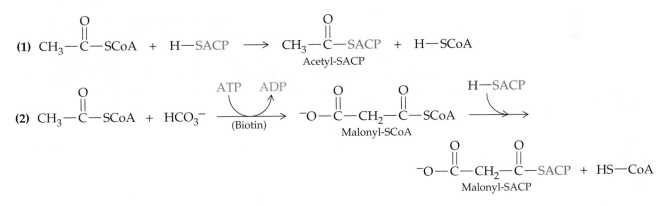

Once malonyl-SACP and acetyl-SACP have been generated, a series of four reactions lengthens the growing fatty acid chain by two carbon atoms with each repetition (Figure 25.9). Fatty acids with up to 16 carbon atoms (palmitic acid) are produced by this route.

Chain Elongation of Fatty Acid

Step 1: Condensation The malonyl group from malonyl-SACP transfers to acetyl-SACP with the loss of CO_2. The loss of the CO_2 that was added in the endothermic, ATP-driven formation of malonyl-SACP releases energy to drive the reaction.

Steps 2–4: Reduction, Dehydration, and Reduction These three reactions accomplish the reverse of steps 3, 2, and 1 in β oxidation of fatty acids (Figure 25.6). The

FIGURE 25.9   Chain elongation ▶ in the biosynthesis of fatty acids. The steps shown begin with acetyl acyl carrier protein (acetyl-SACP), the reactant in the first spiral of palmitic acid synthesis. Each new pair of carbon atoms is carried into the next spiral by a new malonyl-SACP. The growing chain remains attached to the carrier protein from the original acetyl-SACP.

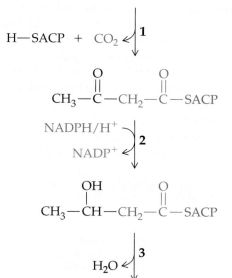

Step 1. Acetyl groups from acetyl-SACP and malonyl-SACP are joined by a C–C bond, with loss of the CO_2.

Step 2. In this reduction using the coenzyme NADPH, the carbonyl group of the original acetyl group is reduced to a hydroxyl group.

Step 3. Dehydration at the C atoms α and β to the remaining carbonyl group introduces a double bond.

Step 4. In another reduction, the double bond introduced in step 3 is converted to a single bond.

carbonyl group is reduced to an —OH group, dehydration yields a carbon–carbon double bond, and the double bond is reduced by addition of hydrogen.

The result of the first cycle in fatty acid synthesis is the addition of two carbon atoms to an acetyl group to give a four-carbon acyl group. The next cycle then adds two more carbon atoms to give a six-carbon acyl group:

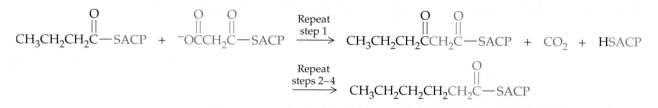

After seven trips through the elongation spiral, a 16-carbon palmitoyl group has been produced. Larger fatty acids are synthesized from palmitoyl-SCoA with the aid of specific enzymes.

Summary: Revisiting the Chapter Goals

1. **What happens during the digestion of triacylglycerols?** *Triacylglycerols (TAGs)* from the diet are broken into droplets in the stomach and enter the small intestine, where they are emulsified by *bile acids* and form micelles. Pancreatic lipases partially hydrolyze the TAGs in the micelles. Small fatty acids and glycerol from TAG hydrolysis are absorbed directly into the bloodstream at the intestinal surface. Insoluble hydrolysis products are carried to the lining in micelles, where they are absorbed and reassembled into TAGs. These TAGs are then assembled into *chylomicrons* (which are *lipoproteins*) and absorbed into the lymph system for transport to the bloodstream.

2. **What are the various roles of lipoproteins in lipid transport?** In addition to chylomicrons, which carry TAGs from the diet into the bloodstream, there are VLDLs (*very-low-density lipoproteins*), which carry TAGs synthesized in the liver to peripheral tissues for energy generation or storage; LDLs (*low-density lipoproteins*), which transport cholesterol from the liver to peripheral tissues for cell membranes or steroid synthesis; and HDLs (*high-density lipoproteins*), which transport cholesterol from peripheral tissues back to the liver for conversion to bile acids that are used in digestion or excreted.

3. **What are the major pathways in the metabolism of triacylglycerols?** Dietary TAGs carried by chylomicrons in the bloodstream undergo hydrolysis to fatty acids and glycerol by enzymes in capillary walls. TAGs in storage are similarly hydrolyzed within adipocytes. The fatty acids from either source undergo *β oxidation* to acetyl-SCoA or resynthesis into TAGs for storage. Acetyl-SCoA can participate in resynthesis of fatty acids (*lipogenesis*), formation of *ketone bodies (ketogenesis)*, steroid synthesis, or energy generation via the citric acid cycle and oxidative phosphorylation. Glycerol can participate in glycolysis, gluconeogenesis, or TAG synthesis.

4. **How are triacylglycerols moved into and out of storage in adipose tissue?** Synthesis of TAGs for storage is activated by insulin when glucose levels are high. The synthesis requires dihydroxyacetone phosphate (from glycolysis or glycerol) for conversion to glycerol 3-phosphate, to which fatty acyl groups are added one at a time to yield TAGs. Hydrolysis of TAGs stored in adipocytes is activated by glucagon when glucose levels drop and also by epinephrine.

5. **How are fatty acids oxidized, and how much energy is produced by their oxidation?** Fatty acids are activated (in the cytosol) by conversion to fatty acyl coenzyme A, a reaction that requires the equivalent of two ATPs in the conversion of ATP to AMP. The fatty acyl-SCoAs are transported into the mitochondrial matrix and are then oxidized two carbon atoms at a time to acetyl-SCoA by repeated trips through the β-oxidation spiral.

6. **What is the function of ketogenesis?** The ketone bodies are 3-hydroxybutyrate, acetoacetate, and acetone. They are produced from two acetyl-SCoA molecules. Their production is increased when energy generation from the citric acid cycle cannot keep pace with the quantity of acetyl-SCoA available. This occurs during the early stages of starvation and in unregulated diabetes. The ketone bodies are water-soluble and can travel unassisted in the bloodstream to tissues where acetyl-SCoA is produced from acetoacetate and 3-hydroxybutyrate. In this way, acetyl-SCoA is made available for energy generation when glucose is in short supply.

Key Words

β-oxidation pathway, p. 727

Bile, p. 720

Bile acids, p. 720

Ketoacidosis, p. 731

Ketogenesis; p. 730

Ketone bodies, p. 730

Lipoprotein, p. 720

Lipogenesis, p. 733

Mobilization (of triacylglycerols), p. 725

7. How are fatty acids synthesized? Fatty acid synthesis (lipogenesis), like β oxidation, proceeds two carbon atoms at a time in a four-step pathway. The pathways utilize different enzymes and coenzymes. In synthesis the initial four carbons are transferred from acetyl-SCoA to the malonyl carrier protein. Each additional pair of carbons is then added to the growing chain bonded to the carrier protein, with the final three steps of the four-step synthesis sequence the reverse of the first three steps in β oxidation.

■ Understanding Key Concepts

25.8 Oxygen is not a reactant in the β oxidation of fatty acids. Can β oxidation occur under anaerobic conditions? Explain your answer.

25.9 Identify each lipoprotein described below as either chylomicron, HDL, LDL, or VLDL.

   (a) Which lipoprotein has the highest ratio of protein to lipid?

   (b) Which lipoprotein has the lowest density? Why?

   (c) Which lipoprotein carries TAGs from the liver to peripheral tissues? How are TAGs used?

   (d) Which lipoprotein carries TAGs from the diet?

   (e) Which lipoprotein transports cholesterol from the liver to peripheral tissues?

   (f) Which lipoprotein removes cholesterol from circulation?

   (g) Which lipoprotein contains "bad cholesterol" from a vascular disease risk standpoint?

25.10 Lipid metabolism, especially TAG anabolism and catabolism, is closely associated with carbohydrate (glucose) metabolism. Insulin and glucagon levels in blood are regulated by the glucose levels in blood. Draw lines from the appropriate phrases in column A to appropriate phrases in columns B and C.

A	B	C
High blood glucose	High glucagon/ low insulin	Fatty acid and TAGs synthesis
Low blood glucose	High insulin/ low glucagon	TAG hydrolysis; fatty acid oxidation

25.11 One strategy used in many different biochemical pathways is an initial investment of energy early on, and a large payoff in energy at the end of the pathway. How is this strategy utilized in the catabolism of fats?

25.12 When oxaloacetate in liver tissue is being used for gluconeogenesis, what impact does this have on the citric acid cycle? Explain.

25.13 Why is it more efficient to store energy as TAGs rather than as glycogen?

■ Additional Problems

Digestion and Catabolism of Lipids

25.14 Why do lipids make you feel full for a long time after a meal?

25.15 Where does digestion of lipids occur?

25.16 What is the purpose of bile in lipid digestion?

25.17 Lipases break down triacylglycerols by catalyzing hydrolysis. What are the products of this hydrolysis?

25.18 What are chylomicrons, and how are they involved in lipid metabolism?

25.19 What is the origin of the triacylglycerols transported by very-low-density lipoproteins?

25.20 How are the fatty acids from adipose tissue transported?

25.21 The glycerol derived from lipolysis of triacylglycerols is converted into glyceraldehyde 3-phosphate, which then enters into step 6 of the glycolysis pathway. What further transformations are necessary to convert glyceraldehyde 3-phosphate into pyruvate?

25.22 If the conversion of glycerol to glyceraldehyde 3-phosphate releases one molecule of ATP, how many molecules of ATP are released during the conversion of glycerol to pyruvate?

25.23 How many molecules of ATP are released in the overall catabolism of glycerol to acetyl-SCoA? How many molecules of ATP are released in the complete catabolism of glycerol to CO_2 and H_2O?

25.24 What is an adipocyte?

25.25 What is the primary function of adipose tissues, and where in the body are they located?

25.26 What are the primary tissues that carry out fatty acid oxidation?

25.27 Where in the cell does β oxidation take place?

25.28 What initial chemical transformation takes place on a fatty acid to activate it for catabolism?

25.29 What must take place before an activated fatty acid undergoes β-oxidation?

25.30 Why is the stepwise oxidation of fatty acids called β oxidation?

25.31 Why is the sequence of reactions that catabolize fatty acids a *spiral* rather than a *cycle*?

25.32 How many moles of ATP are produced by one cycle of β oxidation?

25.33 Arrange these four molecules in order of their biological energy content (per mole):
  (a) Glucose
  (b) Capric acid, $CH_3(CH_2)_8COOH$
  (c) Sucrose
  (d) Myristic acid, $CH_3(CH_2)_{12}COOH$

25.34 Show the products of each step in the following fatty acid oxidation of hexanoic acid:

(a) $CH_3(CH_2)_4\overset{\displaystyle O}{\overset{\displaystyle \|}{C}}SCoA$ $\xrightarrow[\text{Acetyl-SCoA dehydrogenase}]{\text{FAD} \quad \text{FADH}_2}$?

(b) Product of (a) + H_2O $\xrightarrow[\text{hydratase}]{\text{Enoyl-SCoA}}$?

(c) Product of (b) $\xrightarrow[\beta\text{-Hydroxyacl-SCoA dehydrogenase}]{\text{NAD}^+ \quad \text{NADH/H}^+}$?

(d) Product of (c) + HSCoA $\xrightarrow[\text{transferase}]{\text{Acetyl-SCoA}}$?

25.35 Write the equation for the final step in the catabolism of any fatty acid with an even number of carbons.

25.36 How many molecules of acetyl-SCoA result from complete catabolism of the following compounds?
  (a) Caprylic acid, $CH_3(CH_2)_6COOH$
  (b) Myristic acid, $CH_3(CH_2)_{12}COOH$

25.37 How many cycles of β oxidation are necessary to completely catabolize caprylic and myristic acids?

Fatty Acid Anabolism

25.38 What is the name of the anabolic pathway for synthesizing fatty acids?

25.39 Why can't β oxidation proceed backward to produce triacylglycerols?

25.40 What is the starting material for fatty acid synthesis?

25.41 Why are fatty acids generally composed of an even number of carbons?

25.42 How many rounds of the lipogenesis cycle are needed to synthesize palmitic acid, $C_{15}H_{31}COOH$?

25.43 How many molecules of NADPH are needed to synthesize palmitic acid, $C_{15}H_{31}COOH$?

Applications

25.44 What are desirable goals for total cholesterol, HDL, and LDL values? [*Lipids and Atherosclerosis*]

25.45 What is atherosclerosis? [*Lipids and Atherosclerosis*]

25.46 What is the difference in the roles of LDL and HDL? [*Lipids and Atherosclerosis*]

25.47 Explain the significance of cholesterol/HDL ratios of 3.5, 4.5, and 5.5. [*Lipids and Atherosclerosis*]

25.48 What is malabsorption, and how is it detected? [*Nutritional Value of Medium-Chain Triacylglycerols*]

25.49 What are medium-chain triacylglycerols, and how do they address the problem of malabsorption? [*Nutritional Value of Medium-Chain Triacylglycerols*]

25.50 Give some reasons for why the liver is so vital to proper metabolic function. [*The Liver, Clearinghouse for Metabolism*]

25.51 What is cirrhosis of the liver, and what can trigger it? [*The Liver, Clearinghouse for Metabolism*]

General Questions and Problems

25.52 Consuming too many carbohydrates causes deposition of fats in adipose tissue. How can this happen? Why aren't the molecules stored in glycogen instead?

25.53 Are any of the intermediates in β oxidation chiral? Explain.

25.54 What three compounds are classified as ketone bodies? Why are they so designated? What process in the body produces them? Why do they form?

25.55 What is ketosis? What condition results from prolonged ketosis? Why is it dangerous?

25.56 How many molecules of acetyl-SCoA result from catabolism of glyceryl trimyristate?

25.57 Compare fats and carbohydrates as energy sources in terms of the amount of energy released per mole, and account for the observed energy difference.

25.58 Lipoproteins that transport lipids from the diet are described as exogenous. Those that transport lipids produced in metabolic pathways are described as endogenous. Which of the following lipoproteins is exogenous and which is endogenous?
  (a) high-density lipoprotein (HDL)
  (b) chylomicrons

25.59 High blood cholesterol levels are dangerous because of their correlation with atherosclerosis and consequent heart attacks and strokes. Is it possible to eliminate all cholesterol from the bloodstream by having a diet that includes no cholesterol? Would it be desirable to have no cholesterol at all in your body?

25.60 Individuals suffering from ketoacidosis have acidic urine. What effect do you expect ketones to have on pH? Why is pH lowered when ketone bodies are present?

25.61 In the synthesis of cholesterol, acetyl-SCoA is converted to 2-methyl-1,3-butadiene. Molecules of 2-methyl-1,3-butadiene are then joined to give the carbon skeleton of cholesterol. Draw the condensed structure of 2-methyl-1,3-butadiene. How many carbon atoms does cholesterol contain? What is the minimum number of 2-methyl-1,3-butadiene molecules that would be required to make one molecule of cholesterol?

26

Nucleic Acids and Protein Synthesis

A DNA model inspires thoughts about many aspects of life and science.

How does a seed know what kind of plant to become? How does a fertilized egg know how to grow into a human being? And how does a cell know what part of the body it's in so that it can function appropriately? The answers to these and a multitude of other questions about all living organisms reside in the biological molecules known as *nucleic acids*.

The nucleic acids are the chemical carriers of an organism's genetic information. Coded in an organism's *deoxyribonucleic acid (DNA)* is all the information that determines the nature of the organism, be it a dandelion, goldfish, or human being. What makes this possible is the storage in DNA of the blueprint for every protein in the body. As you saw in Chapters 18 and 19, proteins have a wide array of structures and functions, and nearly all reactions in the body are catalyzed by enzymes, which are proteins. It is the nature of its proteins that determines the nature of the organism.

In this chapter, we'll answer the following questions about nucleic acids:

1. **What is the composition of the nucleic acids, DNA and RNA?**
   The goal: Be able to describe and identify the components of nucleosides, nucleotides, DNA, and RNA.
2. **What is the structure of DNA?**
   The goal: Be able to describe the double helix and base pairing in DNA.
3. **How is DNA reproduced?**
   The goal: Be able to explain the process of DNA replication.
4. **What are the functions of RNA?**
   The goal: Be able to list the types of RNA, their locations in the cell, and their functions.
5. **How do organisms synthesize messenger RNA?**
   The goal: Be able to explain the process of transcription.
6. **How does RNA participate in protein synthesis?**
   The goal: Be able to explain the genetic code, and describe the initiation, elongation, and termination steps of translation.

CONCEPTS TO REVIEW

Hydrogen bonding (Section 8.11)
Phosphoric acid derivatives (Section 17.8)
Protein structure (Sections 18.7, 18.8)

26.1 DNA, Chromosomes, and Genes

The terms "chromosome" and "gene" were coined long before the chemical nature of these cell components was understood. A "chromosome" was a structure in the cell nucleus thought to be the carrier of genetic information—all the information needed by an organism to duplicate itself. A "gene" was the portion of a chromosome that controlled a specific inheritable trait such as brown eyes or red hair.

We've come a long way since these terms were first introduced. The molecular structure of chromosomes and genes is now understood. The sequence of events in which genetic information is reproduced when a cell divides can be described. And the relationship of genes to the synthesis of proteins is no longer a mystery. These concepts are described in this chapter. The story continues in the next chapter with an introduction to an achievement that opens the door to a revolution in science and technology—a complete map of the genetic information passed along during cell division.

When a cell is not actively dividing, its nucleus is occupied by *chromatin*, which is a compact tangle of **DNA (deoxyribonucleic acid)**, the carrier of genetic information, twisted around proteins (known as *histones*). During cell division, chromatin organizes itself into **chromosomes**. Each chromosome

DNA (deoxyribonucleic acid) The nucleic acid that stores genetic information; a polymer of deoxyribonucleotides.

Chromosome A complex of proteins and DNA; visible during cell division.

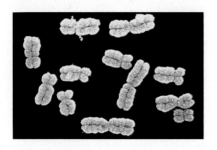

▲ Chromosomes during cell division.

contains a different DNA molecule, and the DNA is duplicated so that each new cell receives a complete copy.

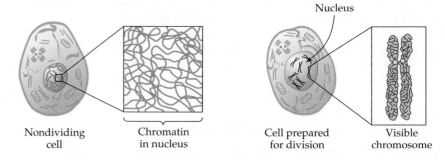

Nondividing cell — Chromatin in nucleus — Cell prepared for division — Visible chromosome

Nucleus

Each DNA molecule, in turn, is made up of many **genes**—individual segments of DNA that contain the instructions that direct the synthesis of a single polypeptide.

Organisms differ widely in their numbers of chromosomes. A horse, for example, has 64 chromosomes (32 pairs), a cat has 38 chromosomes (19 pairs), and a mosquito has 6 chromosomes (3 pairs). A human has 46 chromosomes (23 pairs).

Gene Segment of DNA that directs the synthesis of a single polypeptide.

26.2 Composition of Nucleic Acids

Like proteins and carbohydrates, nucleic acids are polymers. Proteins are polypeptides, carbohydrates are polysaccharides, and **nucleic acids** are *polynucleotides*. Each **nucleotide** has three parts: a five-membered ring monosaccharide, a nitrogen-containing cyclic compound that is a base, and a phosphate group ($-OPO_3^{2-}$).

Nucleic acid A polymer of nucleotides.

Nucleotide A five-carbon sugar bonded to a cyclic amine base and a phosphate group; monomer for nucleic acids.

A nucleotide

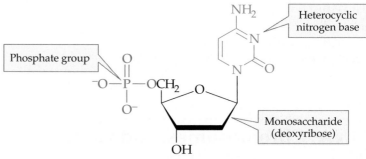

Phosphate group

Heterocyclic nitrogen base

Monosaccharide (deoxyribose)

RNA (ribonucleic acid) Nucleic acids responsible for putting the genetic information to use in protein synthesis; polymer of ribonucleotides. Includes messenger, transfer, and ribosomal RNA.

There are two types of nucleic acids, DNA and **ribonucleic acid (RNA)**. The function of RNA is to put the information stored in DNA to use. Before we discuss how the nucleic acids fulfill their functions, it's important to understand how their component parts are joined together and how DNA and RNA differ.

The Sugars

One difference between DNA and RNA is in the sugar parts of the molecules. In RNA, the sugar is ribose, as indicated by the name "*ribo*nucleic acid." In DNA, the sugar is 2-*deoxy*ribose, giving "*deoxy*ribonucleic acid." (The prefix "2-deoxy-" means that an oxygen atom is missing from the C2 position of ribose.)

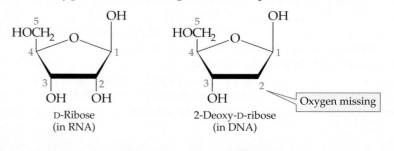

D-Ribose (in RNA)

2-Deoxy-D-ribose (in DNA)

Oxygen missing

The Bases

The five heterocyclic amine bases (screened in blue) in Table 26.1 are present in DNA and RNA. The first two are derivatives of purine, which contains two fused nitrogen-containing rings, and the other three are derivatives of pyrimidine, which has one nitrogen-containing ring. (∞, p. 427)

TABLE 26.1 Bases in DNA and RNA	

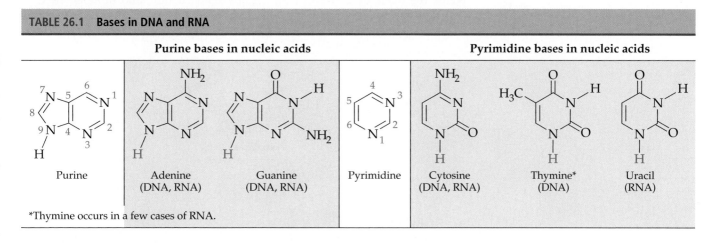

Purine bases in nucleic acids			Pyrimidine bases in nucleic acids			
Purine	Adenine (DNA, RNA)	Guanine (DNA, RNA)	Pyrimidine	Cytosine (DNA, RNA)	Thymine* (DNA)	Uracil (RNA)

*Thymine occurs in a few cases of RNA.

In addition to the difference in their sugars, RNA and DNA differ in the identity of one of their four bases.

- Thymine is present only in DNA molecules.
- Uracil is present only in RNA molecules.
- Adenine, guanine, and cytosine are present in both DNA and RNA.

Sugar + Base = Nucleoside

The combination of ribose or deoxyribose and one of the bases listed in Table 26.1 produces a **nucleoside**. The sugar and the base are connected by a bond between one of the nitrogen atoms in a ring and the sugar's anomeric carbon atom (the one bonded to two O atoms). The bond connecting the sugar and the base is a β-N-glycosidic bond. (∞, p. 652) For example, combination of ribose and adenine gives the nucleoside known as adenosine, which you should recognize as the parent molecule of adenosine triphosphate (ATP). (∞, p. 607)

Nucleoside A five-carbon sugar bonded to a cyclic amine base; like a nucleotide but with no phosphate group

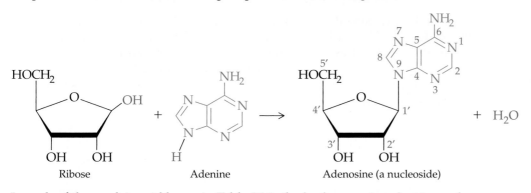

Ribose + Adenine → Adenosine (a nucleoside) + H_2O

In each of the nucleic acid bases in Table 26.1, the hydrogen atom lost in nucleoside formation is shown in red.

Nucleosides are named with the base name modified by the ending *-osine* for the purine bases and *-idine* for the pyrimidine bases. No prefix is used for nucleosides containing ribose, but the prefix *deoxy-* is added for those that contain deoxyribose. The four *ribonucleosides* are thus adenosine, guanosine, cytidine, and uridine. To distinguish between atoms in the two rings of a nucleoside, numbers without primes are used for atoms in the nitrogen base, and numbers with primes are used for atoms in the sugar ring, as illustrated above for adenosine.

■ **KEY CONCEPT PROBLEM 26.1**

Name the nucleoside shown below. Copy the structure, and number the C and N atoms.

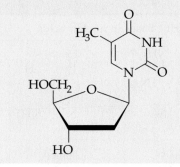

■ **PROBLEM 26.2**

Write the molecular formulas for the sugars D-ribose and 2-deoxy-D-ribose. Exactly how do they differ in composition?

Nucleoside + Phosphate = Nucleotide

Nucleotides are the building blocks of nucleic acids; they are the monomers of the DNA and RNA polymers. Each nucleotide is a 5′-monophosphate ester of a nucleoside.

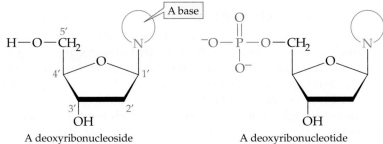

A deoxyribonucleoside   A deoxyribonucleotide

Nucleotides are named by adding 5′-*monophosphate* at the end of the name of the nucleoside. The nucleotides corresponding, for example, to adenosine and deoxycytidine are thus adenosine 5′-monophosphate (AMP) and deoxycytidine 5′-monophosphate (dCMP). Nucleotides containing D-ribose are classified as **ribonucleotides** those containing 2-deoxy-D-ribose are known as **deoxyribonucleotides**. For example,

Ribonucleotide A nucleotide containing D-ribose.

Deoxyribonucleotide A nucleotide containing 2-deoxy-D-ribose.

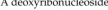

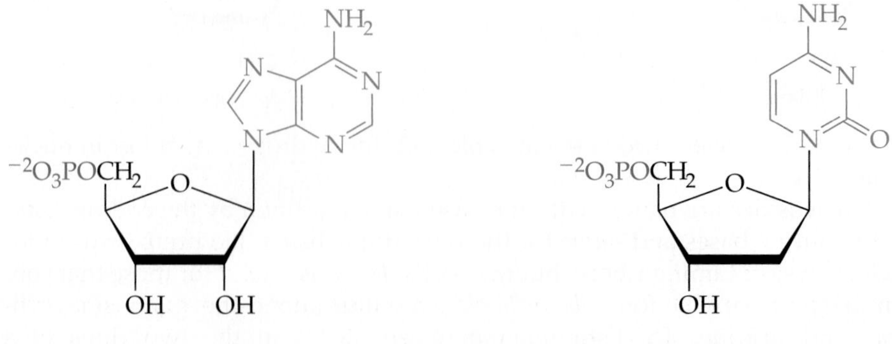

Adenosine 5′-monophosphate (AMP)
(a ribonucleotide)

Deoxycytidine 5′-monophosphate (dCMP)
(a deoxyribonucleotide)

Any of the nucleotides can add additional phosphate groups to form diphosphate or triphosphate esters. As illustrated by *adenosine triphosphate*, these esters are named with the nucleoside name plus *diphosphate* or *triphosphate*. Throughout the preceding chapters, you have seen that adenosine triphosphate (ATP) plays an essential role as a source of biochemical energy, which is released during its conversion to adenosine diphosphate (ADP)

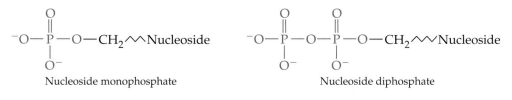

Nucleoside monophosphate   Nucleoside diphosphate

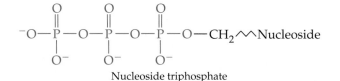

Nucleoside triphosphate

The names of the bases, nucleosides, and nucleotides are summarized in Table 26.2 together with their abbreviations, which are commonly used in writing about biochemistry.

TABLE 26.2 Names of Bases, Nucleosides, and Nucleotides in DNA and RNA

Bases	Nucleosides	Nucleotides*
DNA		
	Deoxyribonucleosides	**Deoxyribonucleotides**
Adenine (A)	Deoxyadenosine	Deoxyadenosine 5′-monophosphate (dAMP)
Guanine (G)	Deoxyguanosine	Deoxyguanosine 5′-monophosphate (dGMP)
Cytosine (C)	Deoxycytidine	Deoxycytidine 5′-monophosphate (dCMP)
Thymine (T)	Deoxythymidine	Deoxythymidine 5′-monophosphate (dTMP)
RNA		
	Ribonucleosides	**Ribonucleotides**
Adenine (A)	Adenosine	Adenosine 5′-monophosphate (AMP)
Guanine (G)	Guanosine	Guanosine 5′-monophosphate (GMP)
Cytosine (C)	Cytidine	Cytidine 5′-monophosphate (CMP)
Uracil (U)	Uridine	Uridine 5′-monophosphate (UMP)

*The nucleotides are also named as, for example, deoxyadenylate and adenylate.

▲ These chickens are very different, but each has 39 chromosomes.

Summary—Nucleoside, Nucleotide, and Nucleic Acid Composition

Nucleoside

- A sugar and a base

Nucleotide

- A sugar, a base, and a phosphate group ($-OPO_3^{2-}$)

DNA (deoxyribonucleic acid)

- A polymer of deoxyribonucleotides
- The sugar is 2-deoxy-D-ribose
- The bases are adenine, guanine, cytosine, and *thymine*

RNA (ribonucleic acid)

- A polymer of ribonucleotides
- The sugar is D-ribose
- The bases are adenine, guanine, cytosine, and *uracil*

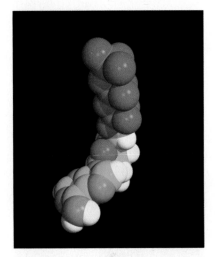

▲ ATP is the triphosphate of the adenosine nucleotide (Section 21.5).

■ **WORKED EXAMPLE 26.1**

Classify the following structure as that of a nucleoside or a nucleotide, identify its sugar and base components, and name it.

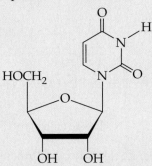

SOLUTION   The compound contains a sugar, recognizable by the O atom in the ring and the —OH groups. It also contains a nitrogen base, recognizable by the nitrogen-containing ring. There is, however, no phosphate group, so the compound is a nucleoside. The sugar has an —OH in the 2' position and is therefore ribose. Checking the base structures in Table 26.1 shows that this one is uracil, a pyrimidine base. Using the *-idine* ending for pyrimidine bases, the nucleoside is therefore named uridine.

■ **WORKED EXAMPLE 26.2**

Draw the structure of the nucleotide represented by dTMP.

SOLUTION   Consulting Table 26.2 shows that dTMP is deoxythymidine 5'-monophosphate. Therefore, the nitrogen base in this nucleotide is thymine, whose structure is shown in Table 26.1. This base must be bonded (by replacing the H that is red in Table 26.1) to deoxyribose, and there must be a phosphate group in the 5' position of the deoxyribose. The structure is

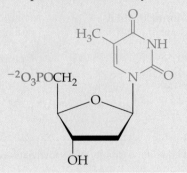

■ **KEY CONCEPT PROBLEM 26.3**

Draw the structure of 2′-deoxyadenosine 5′-monophosphate; include all of the numbers in the structure.

■ **PROBLEM 26.4**

Draw the structure of the triphosphate of guanosine, a triphosphate that, like ATP, provides energy for certain reactions.

■ **PROBLEM 26.5**

Write the full names of dCMP, CMP, UDP, AMP, and ATP.

26.3 The Structure of Nucleic Acid Chains

Keep in mind that nucleic acids are polymers of nucleotides. The nucleotides are connected in DNA and RNA by phosphate diester linkages between the —OH group on C3′ of the sugar ring of one nucleotide and the phosphate group on C5′ of the next nucleotide:

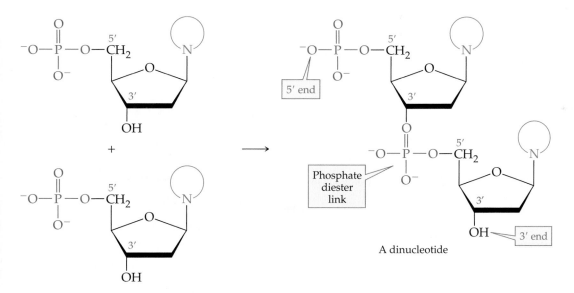

A dinucleotide

A nucleotide chain commonly has a free phosphate group on a 5′ carbon at one end (known as the 5′ *end*) and a free —OH group on a 3′ carbon at the other end (the 3′ *end*), as illustrated in the dinucleotide just above and in the trinucleotide in Figure 26.1. Additional nucleotides join by formation of additional phosphate diester linkages between these groups until the polynucleotide chain of a DNA molecule is formed.

Just as the structure and function of a protein depend on the sequence in which its individual amino acids are connected (ⓄⓄⓄ, Section 18.7), the structure and function of a nucleic acid depend on the sequence in which its individual nucleotides are connected. To carry the analogy even further, note that both proteins and nucleic acids have backbones that do not vary in composition. The differences between different proteins and between different nucleic acids are provided by the order of the groups bonded to the backbone—amino acid side chains in proteins and bases in nucleic acids.

Comparison of protein and nucleic acid backbones and side chains

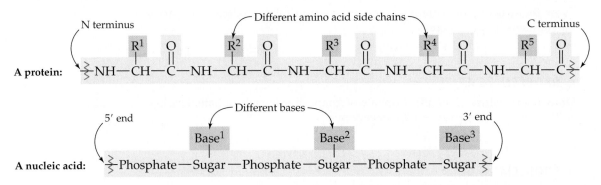

The sequence of nucleotides in a nucleic acid chain is described by starting at the 5′ end and identifying the bases in order of occurrence. Rather than write the full name of each nucleotide or each base, however, it's more convenient and customary to use the simple one-letter abbreviations of the bases to designate the order in which they are attached to the sugar–phosphate backbone: A for adenine, G for guanine, C for cytosine, T for thymine, and U for uracil in RNA. The trinucleotide in Figure 26.1, for example, would be represented by T-A-G or TAG.

■ **PROBLEM 26.6**
Identify the bases in their order in the G-G-T-A tetranucleotide.

FIGURE 26.1   A trinucleotide. ▶
In all polynucleotides, as illustrated here, there is a phosphate group at the 5′ end; there is a sugar —OH group at the 3′ end; and the nucleotides are connected by 3′,5′-phosphate diester links.

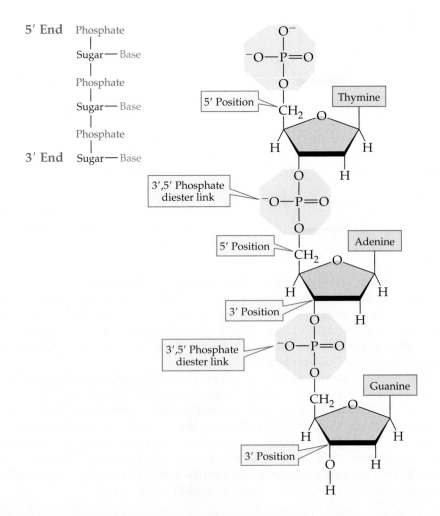

■ **PROBLEM 26.7**

Draw the full structure of the DNA dinucleotide C-T. Identify the 5′ and 3′ ends of this dinucleotide.

26.4 Base Pairing in DNA: The Watson–Crick Model

DNA Structure and Analysis

DNA samples from different cells of the same species have the same proportions of the four heterocyclic bases, but samples from different species can have quite different proportions of bases. For example, human DNA contains 30% each of adenine and thymine, and 20% each of guanine and cytosine. The bacterium *Escherichia coli* contains 24% each of adenine and thymine, and 26% each of guanine and cytosine. Note that in both cases, A and T are present in equal amounts, as are G and C. The bases occur in pairs. Why should this be?

In 1953, James Watson and Francis Crick proposed a structure for DNA that not only accounts for the pairing of bases but also accounts for the storage and transfer of genetic information. According to the Watson–Crick model, a DNA molecule consists of *two* polynucleotide strands coiled around each other in a helical, screw-like fashion. The sugar–phosphate backbone is on the *outside* of this right-handed **double helix**, and the heterocyclic bases are on the *inside*, so that a base on one strand points directly toward a base on the second strand. The double helix resembles a twisted ladder, with the sugar–phosphate backbone making up the sides and the hydrogen-bonded base pairs, the rungs.

Double helix   Two strands coiled around each other in a screwlike fashion; in most organisms the two polynucleotides of DNA form a double helix.

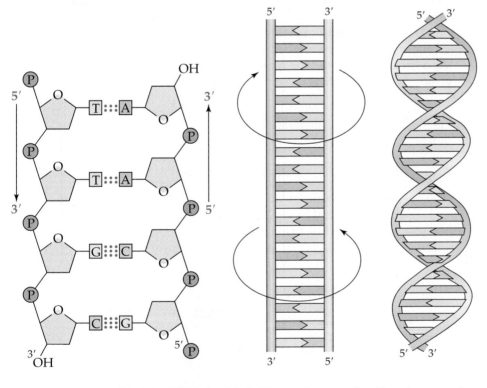

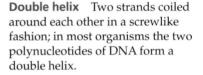

The two strands of the DNA double helix run in opposite directions–one in the 5′ to 3′ direction, the other in the 3′ to 5′ direction. The stacking of the hydrophobic bases in the interior and the alignment of the hydrophilic sugars and phosphate groups on the exterior provide stability to the structure. Hydrogen bonding also enhances DNA stability. Each pair of bases in the center of the double helix is connected by hydrogen bonding. As illustrated in Figure 26.2, adenine and thymine (A-T) form two hydrogen bonds to each other, and cytosine and guanine (C-G) form three hydrogen bonds to each other. Although individual hydrogen bonds are not especially strong, the several thousand along a DNA chain collectively contribute to stability.

FIGURE 26.2 Base pairing in DNA. Hydrogen bonds of similar lengths connect the pairs of bases; thymine and adenine, cytosine and guanine.

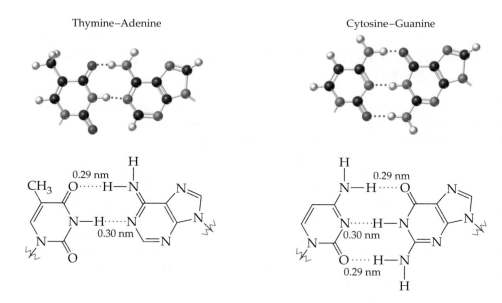

Base pairing The pairing of bases connected by hydrogen bonding (G-C and A-T), as in the DNA double helix.

The pairing of the bases strung like beads along the two polynucleotide strands of the DNA double helix is described as *complementary*. Wherever a T base occurs in one strand, an A base falls opposite it in the other strand; wherever a C base occurs in one strand, a G base falls opposite it on the other strand. This **base pairing** explains why A and T and why C and G occur in equal amounts in double-stranded DNA.

As a trick for remembering how the bases pair up, note that if the symbols are arranged in alphabetical order, the first and last ones pair, and the two middle ones pair.

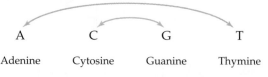

The strength and shape of the double helix, illustrated in Figure 26.3, is dependent upon the fit and hydrogen bonding of the bases. The base pairing is also, as you will see, the key to understanding how DNA functions.

FIGURE 26.3 A segment of DNA (a) In this model, notice that the base pairs are nearly perpendicular to the sugar–phosphate backbones. (b) A space-filling model of the same DNA segment. (c) An abstract representation of the DNA double helix and base pairing.

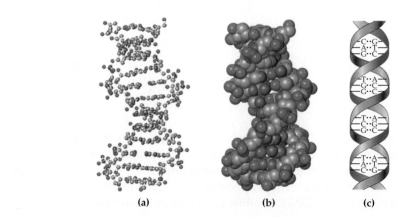

■ **WORKED EXAMPLE**

What sequence of bases of one strand of DNA (reading in the 3′ to 5′ direction) is complementary to the sequence 5′ T-A-T-G-C-A-G 3′ on the other strand?

SOLUTION Remembering that A always bonds to T and C always bonds to G, go through the original 5′ to 3′ sequence replacing each A by T, each T by A, each C by G, and each G by C. Keep in mind that when a 5′ to 3′ strand is matched in this manner to its complementary strand, the complementary strand will read from left to right in the 3′ to 5′ direction. [Where the direction in which a base sequence is written is *not* specified, you can assume it follows the customary 5′ to 3′ direction.]

Original strand	5′ T-A-T-G-C-A-G 3′
Complementary strand	3′ A-T-A-C-G-T-C 5′

■ **PROBLEM 26.8**

What sequence of bases on one DNA strand is complementary to each of the following sequences on another strand?
(a) 5′ G-C-C-T-A-G-T 3′   **(b)** 5′ A-A-T-G-G-C-T-C-A 3′

■ **PROBLEM 26.9**

Draw the structures of adenine and uracil (which replaces thymine in RNA), and show the hydrogen bonding that can occur between them.

■ **PROBLEM 26.10**

Is a DNA molecule neutral, negatively charged, or positively charged?

■ **KEY CONCEPT PROBLEM 26.11**

(a) DNA and RNA, like proteins, can be denatured to produce unfolded or uncoiled strands. Heating DNA to what is referred to as its "melting temperature" denatures it. Why does a longer strand of DNA have a higher melting temperature than a shorter one? (b) The DNA melting temperature also varies with base composition. Would you expect a DNA with a high percentage of G/C base pairs to have a higher or lower melting point than one with a high percentage of A/T base pairs? How can you account for your choice?

26.5 Nucleic Acids and Heredity

Your heredity is determined by the DNA in the fertilized egg from which you grew. A sperm cell carrying DNA from your father united with an egg cell carrying DNA from your mother. Their combination produced the full complement of chromosomes and genes that you carry through life. Each of your 23 pairs of chromosomes contains one DNA molecule copied from that of your father and one DNA molecule copied from that of your mother. Most cells in your body contain copies of these originals. (The exceptions are red blood cells, which have no nuclei and no DNA, and egg or sperm cells, which have 23 single DNA molecules, rather than pairs.)

Cell division is an ongoing process—no cell lives as long as we do. Therefore, every time a cell divides, its DNA must be copied. The double helix of DNA and complementary base pairing make this duplication possible. Because of how the bases pair, each strand of the double helix is a blueprint for the other

▲ Heredity visualized—four generations of the same family. Their total complement of DNA is similar but not the same.

Application Viruses and AIDS

Viruses are submicroscopic infectious agents consisting of a piece of nucleic acid, typically wrapped in a protective coat of protein. They are not cellular and can multiply only inside a living cell. The viral nucleic acid may be either DNA or RNA, it may be either single-stranded or double-stranded, and it may consist of either a single piece or several pieces. Many hundreds of viruses are known, each of which can infect a particular plant or animal cell.

Viruses occupy the gray area between living and nonliving. By itself, a virus has none of the cellular machinery necessary for replication. But once it enters a living cell, a virus takes over the host cell and forces it to produce virus copies. Some of the infected cells may eventually die, but others continue to produce copies of the virus, which then leave the host cell and spread the infection to other cells.

Viral infection begins when a virus enters a host cell, loses its protein coat, and releases its nucleic acid. What happens next depends on whether the virus is based on DNA or RNA. If it is a DNA virus, the host cell first replicates the viral DNA, producing many copies, and then decodes it in the normal way. The viral DNA is transcribed to produce RNA, and the RNA is translated to synthesize viral coat proteins. Copies of the viral DNA are packaged in newly synthesized protein envelopes, producing new virus particles that are then released from the cell.

If the infectious agent is an RNA virus, however, a problem exists. Either the cell must transcribe and produce proteins directly from the viral RNA template, or else it must first produce DNA from the viral RNA by

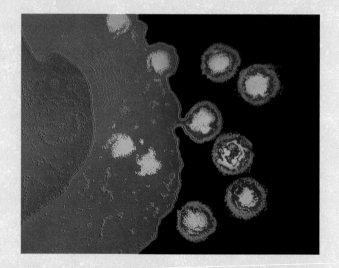

▲ HIV viruses emerging from an infected lymphocyte.

reverse transcription. Viruses that follow the reverse transcription route are known as *retroviruses*.

The *human immunodeficiency virus (HIV-1)* responsible for most cases of AIDS in the United States is a retrovirus. As shown in the diagram, *reverse transcriptase* first produces single DNA strands complementary to the HIV RNA and then produces double-stranded DNA that enters a host cell chromosome. There, the cell's normal replication and translation produce RNA and proteins that are assembled into new virus particles.

The HIV virus attacks mainly *T lymphocytes*, which are part of the immune system that defends the body against foreign invaders (Section 29.4). As T lymphocytes

strand. The copying process is an awesome aspect of how DNA functions—a process that we'll discuss in Section 26.6.

But first, we want to answer two related questions: How do nucleic acids carry the information that determines our inherited traits? And how is that stored information interpreted and put into action?

Our genetic information is carried in the sequence of bases along the DNA strands (Section 26.9). Every time a cell divides, the information is passed along to the daughter cells. Within our cells, the genetic information encoded in the DNA is put into action by the synthesis of proteins.

Three fundamental processes known as *replication*, *transcription*, and *translation* take place in the duplication, transfer, and use of genetic information:

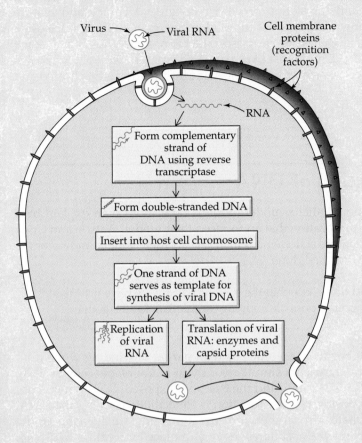

Virus → Viral RNA

Cell membrane proteins (recognition factors)

RNA

Form complementary strand of DNA using reverse transcriptase

Form double-stranded DNA

Insert into host cell chromosome

One strand of DNA serves as template for synthesis of viral DNA

Replication of viral RNA

Translation of viral RNA: enzymes and capsid proteins

die off, the body is open to attack. Most AIDS victims succumb to infection by bacteria, other viruses, or fungi that thrive in the absence of defense by T lymphocytes.

Viral infections, whether HIV infections or the common cold, are difficult to treat with chemical agents. The challenge is to design a drug that can act on viruses within cells without damaging the cells and their genetic machinery. Development of drugs for the treatment of AIDS is especially challenging because HIV has the highest mutation rate of any known virus. Drugs active against one strain of HIV may soon encounter a mutant they can't combat.

The best success with AIDS drugs thus far has been with a three-drug therapy, the first therapy that allows HIV-infected individuals to survive. Two of the drugs (AZT, azidothymidine; 3TC, lamivudine) are false nucleosides. The viral reverse transcriptase incorporates them into the viral DNA, and they then slow down production of new viral RNA. The third drug (saquinavir) inhibits an enzyme (protease) that is necessary for production of proteins coded for by the viral genes. Taken together, these three drugs can reduce the amount of HIV in a patient's body to undetectable levels.

The success of the three-drug therapy is accompanied by some problems. It requires carefully following a challenging schedule for taking a large number of pills each day. It is not effective for everyone. It is very expensive, so expensive that it is beyond the reach of the poor populations in countries where AIDS is rampant. Also, there is the fear that the virus will eventually develop resistance to the protease inhibitor that is greatly responsible for the success of the therapy. Thus, the search for other kinds of therapies continues. One thus-far unattained goal is development of a vaccine that would prevent the disease.

See Additional Problems 26.66–68 at the end of the chapter.

- **Replication** (Section 26.6) is the process by which a replica, or identical copy, of DNA is made when a cell divides, so that each of the two daughter cells has the same DNA.

- **Transcription** (Section 26.8) is the process by which the genetic messages contained in DNA are read and copied. The products of transcription are individual ribonucleic acids, which carry the instructions stored in DNA out of the nucleus to the sites of protein synthesis.

- **Translation** (Section 26.10) is the process by which the genetic messages carried by RNA are decoded and used to build proteins.

Replication The process by which copies of DNA are made when a cell divides.

Transcription The process by which the information in DNA is read and used to synthesize RNA.

Translation The process by which RNA directs protein synthesis.

In the following sections we'll look at these important processes. Replication, transcription, and translation proceed with great accuracy and require participation by many other types of molecules. Energy-supplying nucleoside triphosphates and many enzymes play essential roles, some understood, some not yet fully understood, and most likely some not yet discovered. Our goal is simply to present an overview of how the genetic information is duplicated and put to work.

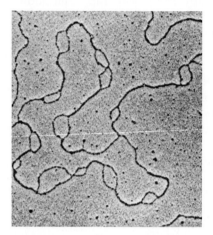

▲ **FIGURE 26.4** **Formation of numerous replication sites as DNA unwinds.**

 DNA Replication

26.6 Replication of DNA

The Watson–Crick double-helix model of DNA does more than explain base pairing. In the short publication that announced their model, Watson and Crick made the following simple statement, which reveals their understanding of the significance of what they had done:

> It has not escaped our notice that the specific pairing we have postulated immediately suggests a possible copying mechanism for the genetic material.

DNA replication begins in the nucleus with partial unwinding of the double helix. The unwinding occurs simultaneously in many places (Figure 26.4). As the two DNA strands separate and the bases are exposed, DNA polymerase enzymes move into position—their function is to facilitate copying the DNA. Nucleoside triphosphates carrying each of the four bases are available in the vicinity. One by one, the triphosphates move into place by forming hydrogen bonds to the bases exposed on the DNA strand being copied (the *template* strand). The hydrogen bond formation requires that A can pair only with T, and G can pair only with C. DNA polymerase then catalyzes bond formation between the 5′ phosphate group of the arriving nucleoside triphosphate and the 3′—OH of the growing polynucleotide strand, as the two extra phosphate groups are removed:

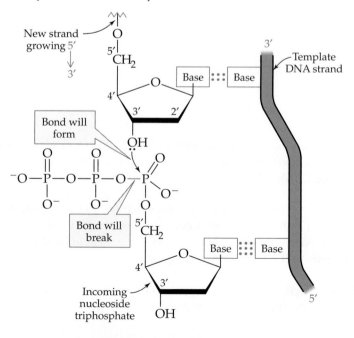

Bond formation in DNA replication

DNA polymerase catalyzes the reaction between the 5′ phosphate on an incoming nucleotide and the free 3′—OH on the growing polynucleotide. Therefore, the template strand can only be copied in the 3′ to 5′ direction, and the new DNA strand can grow only in the 5′ to 3′ direction. This limitation means that strand growth must begin at the 3′ end of the strand that is being copied, the template strand.

Since each new strand is complementary to its template strand, two identical copies of the DNA double helix are produced during replication. In each new double helix, one strand is the template and the other is the newly synthesized strand. The result is described by saying that the replication is *semiconservative* (one of the two strands is conserved).

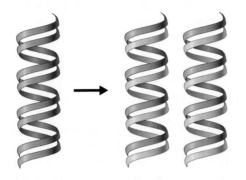

◄ Semiconservative replication produces a pair of DNA double helixes (green and orange) in which one strand (green) is the original strand and the other (orange) is the strand that has been copied from the original.

Note in Figure 26.5 that the incoming nucleoside triphosphate is adding to the 3′ end of the new strand. Only one new strand can grow continuously in this manner as the point of replication (the *replication fork*) moves along, because the original DNA strands are complementary. The other strand would have to grow continuously by addition of nucleotides at the 5′ end, but DNA polymerase does not carry out that reaction. The directions of growth are illustrated in Figure 26.5, where the right-hand strand (orange) is the continuously growing strand of the new DNA. The other strand must grow in the opposite direction. To accomplish growth in this direction, short pieces of RNA known

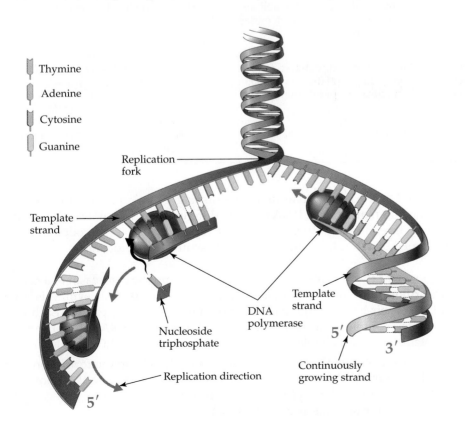

Thymine
Adenine
Cytosine
Guanine

Replication fork

Template strand

Nucleoside triphosphate

DNA polymerase

Template strand

5′

3′

Replication direction

Continuously growing strand

5′

◄ **FIGURE 26.5 DNA replication.** Because the polynucleotide must grow in the 5′ to 3′ direction, one strand (shown here at the right, in orange) grows continuously toward the replication fork and the other (at the left, in blue) grows in segments as the fork moves. The segments are later joined by a ligase enzyme.

as primers are produced at the replication fork. The primers are complementary to the DNA where they are formed and show where replication should begin. They are then extended at their 3′ ends to produce a series of segments of this strand (blue) each growing 3′ to 5′ as the fork moves along. To complete formation of this new strand, the sections are joined by the action of a *DNA ligase enzyme*.

Consider the magnitude of the job in replication. The total number of base pairs in a human cell—the human **genome**—is 3 billion base pairs. Yet the base sequences of the huge DNA molecules are faithfully copied during replication, and a random error occurs only about once in each 10 billion to 100 billion bases. The complete copying process in human cells takes several hours. To replicate such huge molecules as human DNA at this speed requires not one, but many replication forks, producing many segments of DNA strands that are ultimately united.

Genome All of the genetic material in the chromosomes of an organism; its size is given as the number of base pairs.

26.7 Structure and Function of RNA

RNA is similar to DNA—both are sugar–phosphate polymers and both have nitrogen-containing bases attached—but there are differences (Table 26.3). For one, as we have already explained (Section 26.2), RNA and DNA differ in composition: The sugar in RNA is ribose rather than the deoxyribose in DNA, and the base in ribose is uracil instead of thymine. They also differ in size and structure—RNA molecules are smaller than DNA molecules. The RNAs are single-stranded molecules (as distinct from the double-stranded DNA) with complex folds, sometimes folding back on themselves to form double helixes in some regions.

TABLE 26.3	**Comparison of DNA and RNA**			
	Sugar	**Bases**	**Shape and Size**	**Function**
DNA	Deoxyribose	Adenine Guanine Cytosine Thymine	Paired strands in double helix; 50 million or more nucleotides per strand	Stores genetic information
RNA	Ribose	Adenine Guanine Cytosine Uracil	Single-stranded with folded regions; <100 to about 50,000 nucleotides per RNA	**mRNA**—Encodes copy of genetic information ("blueprints" for protein synthesis) **tRNA**—Carries amino acids for incorporation into protein **rRNA**—Component of ribosomes (sites of protein synthesis)

Also, DNA has only one function—storing genetic information. By contrast, the different kinds of RNA perform different functions. Working together, however, the following three RNAs make it possible for the encoded information carried by DNA to be put to use in the synthesis of proteins.

Ribosome The structure in the cell where protein synthesis occurs; composed of protein and rRNA.

Ribosomal RNA (rRNA) The RNA that is complexed with proteins in ribosomes.

Messenger RNA (mRNA) The RNA that carries the code transcribed from DNA and directs protein synthesis.

• **Ribosomal RNAs** Outside the nucleus but within the cytoplasm of a cell are the **ribosomes**—small granular organelles where protein synthesis takes place. (Their location in the cell is shown in Figure 21.3, p. 603). Each ribosome is a complex consisting of about 60% **ribosomal RNA (rRNA)** and 40% protein, with a total molecular weight of approximately 5,000,000 amu.

• **Messenger RNAs** The **messenger RNAs (mRNA)** carry information transcribed from DNA in the cell nucleus out of the nucleus to the ribosomes, where proteins will be synthesized. They are polynucleotides that carry the same code for proteins as does the DNA.

• **Transfer RNAs** The **transfer RNAs (tRNA)** are smaller RNAs that deliver amino acids one by one to protein chains growing at ribosomes. Each tRNA recognizes and carries only one amino acid.

Transfer RNA (tRNA) The RNA that transports amino acids into position for protein synthesis.

26.8 Transcription: RNA Synthesis

Ribonucleic acids are synthesized in the cell nucleus. Before leaving the nucleus, all types of RNA molecules are modified in the various ways needed for their different functions. We focus here on messenger RNA (mRNA; in eukaryotes) because its synthesis (transcription) is the first step in transferring the information carried by DNA into protein synthesis.

In transcription, as in replication, a small section of the DNA double helix unwinds, the bases on the two strands are exposed, and one by one the complementary nucleotides are attached. rRNA, tRNA, and mRNA are all synthesized in essentially the same manner. Only one of the two DNA strands is transcribed during RNA synthesis. The DNA strand that is transcribed is the *template strand*; its complement in the original helix is the *informational strand*. The mRNA molecule is complementary to the template strand. Thus, the messenger RNA molecule produced is a duplicate of the DNA informational strand, but with the U base wherever the DNA has a T base. The relationships are illustrated by the following short DNA and mRNA segments:

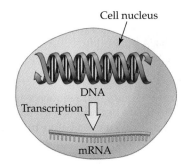

DNA informational strand	5′	ATG	CCA	GTA	GGC	CAC	TTG	TCA	3′
DNA template strand	3′	TAC	GGT	CAT	CCG	GTG	AAC	AGT	5′
mRNA	5′	AUG	CCA	GUA	GGC	CAC	UUG	UCA	3′

The transcription process, illustrated in Figure 26.6, begins when RNA polymerase recognizes a control segment in DNA that precedes the nucleotides to be transcribed. *The genetic code*, which we'll discuss in Section 26.9, relies on

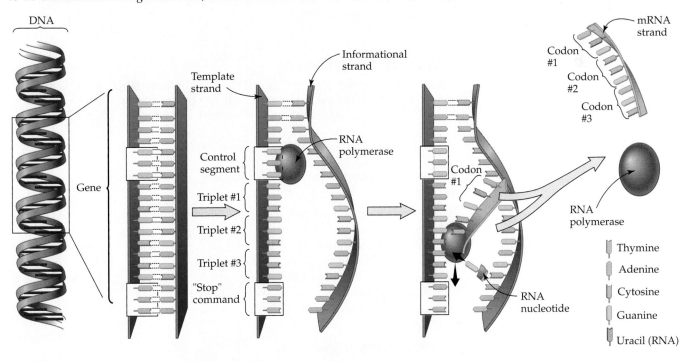

▲ **FIGURE 26.6 Transcription of DNA to produce mRNA.** The transcription shown here produces a hypothetical three-codon mRNA. From left to right, the DNA unwinds, the RNA polymerase connects with the control segment on the template strand, the mRNA is assembled as the polymerase moves along the template strand, and transcription ends when the polymerase reaches the stop command.

Initiation of Transcription in
E. Coli

Exon A nucleotide sequence in a
gene that codes for part of a protein.

Intron A nucleotide sequence in
mRNA that does not code for part
of a protein; removed before mRNA
proceeds to protein synthesis.

triplets of consecutive bases—the *codons*. The triplets carried by mRNA code for
amino acids to be assembled into proteins (Section 26.10). The RNA polymerase
moves down the DNA segment to be transcribed, adding complementary
nucleotides one by one to the growing RNA strand as it goes. Transcription
ends when the RNA polymerase reaches a codon triplet that signals the end of
the sequence to be copied.

At the end of transcription, the mRNA molecule contains a base for every
base that was on the informational DNA strand, from the start to the stop
codon. Further steps are necessary before the mRNA is ready to direct protein
synthesis. It turns out that genes occupy only about 10% of the base pairs in
DNA. A gene will begin in one small section of DNA called an **exon** (it is
*ex*pressed), be interrupted by a section called an **intron** (it *in*tervenes) that does
not code for any part of the protein to be synthesized, and then continue farther down the chain in another exon. In the final mRNA molecule released
from the nucleus, the intron sections have been cut out and the remaining
pieces spliced together.

The possible functions of introns, which are noncoding DNA, are the subject of great speculation and study. No clear-cut function has been assigned to
introns. One intriguing possibility is that they are leftovers from an earlier stage
in evolution. An interesting observation is that the exon portions of different
genes often code for protein segments of similar structure and function. It has
been suggested that introns might separate exons that can be mixed and
matched to create new genes.

■ WORKED EXAMPLE 26.4

The nucleotide sequence in a segment of a DNA informational strand is given
below. What is the nucleotide sequence in the complementary DNA template
strand? What is the sequence transcribed from this template into mRNA?

5' ATC GCA ACT GTC 3'

SOLUTION To provide the requested sequences it is important to remember
several things:

1. In the informational and template strands the base pairs are A–T and
   C–G.
2. Matching base pairs along the informational strand gives the template
   strand written in the 3' to 5' direction.
3. The mRNA strand is identical to the DNA informational strand except
   that it has a U wherever the informational strand has a T.
4. Matching base pairs along the template strand gives the mRNA strand
   written in the 5' to 3' direction.

Applying these principles gives

DNA informational strand	5' ATC GCA ACT GTC 3'
DNA template strand	3' TAG CGT TGA CAG 5'
mRNA	5' AUC GCA ACU GUC 3'

■ **PROBLEM 26.12**

What mRNA base sequences are complementary to each of the following DNA template sequences? Be sure to label the 5′ and 3′ ends of the complementary sequences.

(a) 5′ GAT TAC CGT 3'   **(b)** 3′ TAT GGC TAG GCA 5'

26.9 The Genetic Code

The ribonucleotide sequence in an mRNA chain is like a coded sentence that specifies the order in which amino acid residues should be joined to form a protein. Each "word," or **codon**, in the mRNA sentence is a series of three ribonucleotides that code for a specific amino acid. For example, the series uracil-uracil-guanine (UUG) on an mRNA chain is a codon directing incorporation of the amino acid leucine into a growing protein chain. Similarly, the sequence guanine-adenine-uracil (GAU) codes for aspartate.

Of the 64 possible three-base combinations in RNA, 61 code for specific amino acids and 3 code for chain termination (the *stop codons*). The meaning of each codon—the **genetic code** universal to all but a few living organisms—is shown in Table 26.4. Note that most amino acids are specified by more than one codon and that codons are always written in the 5′ to 3′ direction.

Codon   A sequence of three ribonucleotides in the messenger RNA chain that codes for a specific amino acid; also a three nucleotide sequence that is a stop codon and stops translation.

Genetic code   The sequence of nucleotides, coded in triplets (codons) in mRNA, that determines the sequence of amino acids in protein synthesis.

TABLE 26.4   Codon Assignments of Base Triplets in mRNA

First Base (5′ end)	Second Base	Third Base (3′ end)			
		U	C	A	G
U	U	Phe	Phe	Leu	Leu
	C	Ser	Ser	Ser	Ser
	A	Tyr	Tyr	*Stop*	*Stop*
	G	Cys	Cys	*Stop*	Trp
C	U	Leu	Leu	Leu	Leu
	C	Pro	Pro	Pro	Pro
	A	His	His	Gln	Gln
	G	Arg	Arg	Arg	Arg
A	U	Ile	Ile	Ile	Met
	C	Thr	Thr	Thr	Thr
	A	Asn	Asn	Lys	Lys
	G	Ser	Ser	Arg	Arg
G	U	Val	Val	Val	Val
	C	Ala	Ala	Ala	Ala
	A	Asp	Asp	Glu	Glu
	G	Gly	Gly	Gly	Gly

The relationships among the DNA informational and template strand segments illustrated earlier are repeated below, along with the protein segment for which they code:

DNA informational strand	5′	ATG CCA GTA GGC CAC TTG TCA	3'
DNA template strand	3′	TAC GGT CAT CCG GTG AAC AGT	5'
mRNA	5′	AUG CCA GUA GGC CAC UUG UCA	3'
Protein		Met Pro Val Gly His Leu Ser	

■ **WORKED EXAMPLE 26.5**

In Worked Example 26.4 we derived the mRNA sequence of nucleotides shown below. What is the sequence of proteins coded for by the mRNA sequence?

5′ AUC GCA ACU GUC 3′

SOLUTION The codons must be identified by consulting Table 26.4. They are

5′ AUC GCA ACU GUC 3′
Ile Ala Thr Val

Written out in full, the protein sequence is

isoleucine-alanine-threonine-valine

■ **PROBLEM 26.13**

List possible codon sequences for the following amino acids.
(a) Ala   **(b)** Phe   **(c)** Leu   **(d)** Val   **(e)** Tyr

■ **PROBLEM 26.14**

Name the base represented by the codon GUC and identify the amino acid this codon codes for.

■ **PROBLEM 26.15**

What amino acids do the following sequences code for?
(a) AUU   **(b)** GCG   **(c)** CGA   **(d)** AAC

■ **PROBLEM 26.16**

A hypothetical tripeptide Ile-Ile-Ile could be synthesized by the cell. What three different base triplets in mRNA could be combined to code for this tripeptide?

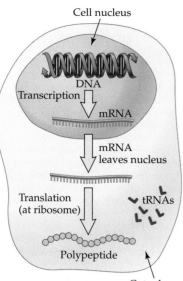

Cell nucleus

DNA
Transcription
mRNA

mRNA leaves nucleus

Translation (at ribosome)

tRNAs

Polypeptide

Cytoplasm

26.10 Translation: Transfer RNA and Protein Synthesis

Now we come to translation of the messages carried by mRNA and the synthesis of proteins. The synthesis occurs at ribosomes, which are outside the nucleus and within the cytoplasm of cells. The mRNA connects with the ribosome, and the amino acids, which are available in the cytosol, are delivered one by one attached to transfer RNA (tRNA) molecules.

First, let's look at the structure of the tRNAs. Every cell contains 20 or more different tRNAs, each designed to carry a specific amino acid, though they are similar in overall structure. A tRNA molecule is a single polynucleotide chain held together by regions of base pairing in a partially helical structure something like a cloverleaf (Figure 26.7a). In three dimensions, a tRNA molecule is L-shaped, as shown in Figures 26.7b and c.

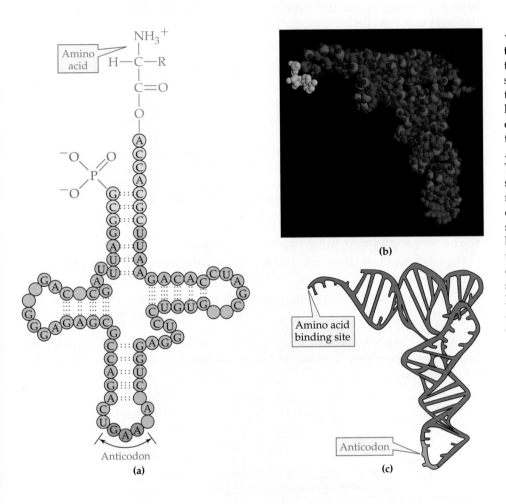

◀ **FIGURE 26.7 Structure of tRNA.** (a) Schematic, flattened tRNA molecule. The cloverleaf-shaped tRNA contains an anticodon triplet on one "leaf" and a covalently bonded amino acid at its 3' end. (The example shown is a yeast tRNA that codes for phenylalanine. All tRNAs have similar structures. The nucleotides not identified are slightly altered analogs of the four normal ribonucleotides.) (b) A computer-generated model of the serine tRNA molecule. The serine binding site is shown in yellow and the anticodon in red. (c) The three-dimensional shape (the tertiary structure) of a tRNA molecule. Note how the anticodon is at one end and the amino acid is at the other end.

Each amino acid is bonded to its specific tRNA by an ester linkage between the —COOH of the amino acid and an —OH group on a ribose at the 3' end of the tRNA chain (which is at one end of the "L"). Individual synthetase enzymes are responsible for connecting each of the amino acids with their partner tRNAs.

At the other end of the L-shaped tRNA molecule is a sequence of three nucleotides called an **anticodon** (Figure 26.7). The anticodon of each tRNA is complementary to an mRNA codon—*always the one designating the particular amino acid that the tRNA carries*. For example, the tRNA carrying the amino acid leucine, which is coded for by 5' CUG 3' in mRNA, has the complementary sequence 3' GAC 5' as its anticodon. This is how the genetic message of nucleotide triplets, the codons, is translated into the sequence of amino acids in a protein. When the tRNA codon pairs off with its complementary mRNA codon, leucine is delivered to its proper place in the growing protein chain.

The three stages in protein synthesis are *initiation*, *elongation*, and *termination*. As you read the descriptions, follow along in the diagram of translation in Figure 26.8.

Anticodon A sequence of three ribonucleotides on tRNA that recognizes the complementary sequence (the codon) on mRNA.

Protein Synthesis

(1) Initiation

Protein synthesis begins with the coming together of an mRNA, the first tRNA, and the *small* subunit of a ribosome. The first codon on the 5′ end of mRNA, an AUG, acts as a "start" signal for the translation machinery and codes for a methionine-carrying tRNA. Initiation is completed when the *large* ribosomal subunit has joined the small one and the methionine-bearing tRNA occupies one of the two binding sites on the united ribosome. (Not all proteins should have methionine at one end—if it's not needed, the methionine from chain initiation is removed before the new protein goes to work.)

(2) Elongation

Next to the first site on the ribosome is a second binding site where the next codon on mRNA is exposed and the tRNA carrying the next amino acid will be attached. All available tRNA molecules can approach and try to fit, but only one with an appropriate anticodon sequence can bind. Once the tRNA with amino acid 2 has arrived, an enzyme in the large subunit catalyzes formation of the new peptide bond and breaks the bond of amino acid 1 with its tRNA. The first tRNA then leaves the ribosome, and the entire ribosome shifts three positions (one codon) along the mRNA chain. As a result, the second binding site is opened up to accept the tRNA carrying the next amino acid.

The three elongation steps now repeat:

- The next appropriate tRNA binds to the ribosome.
- Peptide bond formation attaches the newly arrived amino acid to the growing chain.
- Ribosome position shifts takes place to free the second site for the next tRNA.

A single mRNA can be "read" simultaneously by many ribosomes. The growing polypeptides increase in length as the ribosomes move down the mRNA strand.

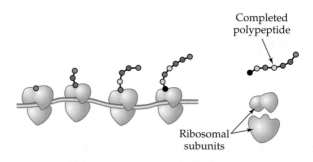

Completed polypeptide

Ribosomal subunits

(3) Termination

When synthesis of the protein is completed, a "stop" codon signals the end of translation. An enzyme called a *releasing factor* than catalyzes cleavage of the polypeptide chain from the last tRNA, the tRNA and mRNA molecules are released from the ribosome, and the two ribosome subunits again separate.

In our discussion in this and the preceding sections, we've left many questions about replication, transcription, and translation unanswered. What tells a cell when to start replication? Are there mechanisms to repair damaged DNA or correct random errors made during replication? (There are.) Keep in mind that synthesis of mRNA is the beginning of synthesis of a protein. Since each cell contains the entire genome and since cells differ widely in their function, what keeps genes for unneeded proteins turned off? What determines just

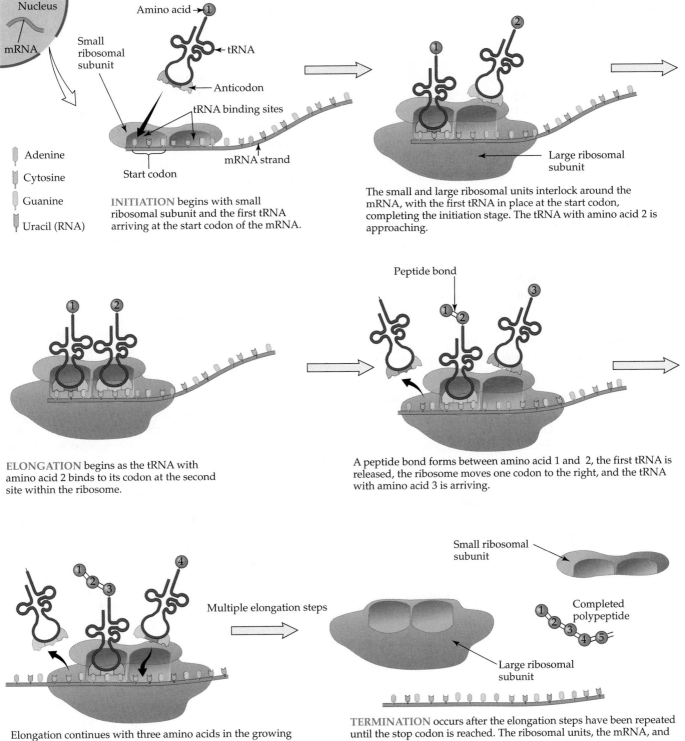

INITIATION begins with small ribosomal subunit and the first tRNA arriving at the start codon of the mRNA.

The small and large ribosomal units interlock around the mRNA, with the first tRNA in place at the start codon, completing the initiation stage. The tRNA with amino acid 2 is approaching.

ELONGATION begins as the tRNA with amino acid 2 binds to its codon at the second site within the ribosome.

A peptide bond forms between amino acid 1 and 2, the first tRNA is released, the ribosome moves one codon to the right, and the tRNA with amino acid 3 is arriving.

Elongation continues with three amino acids in the growing chain and the fourth one arriving with its tRNA

TERMINATION occurs after the elongation steps have been repeated until the stop codon is reached. The ribosomal units, the mRNA, and the polypeptide separate.

▲ **FIGURE 26.8** **Translation: The Initiation, Elongation, and Termination Stages in Protein Synthesis.**

Before the 1980s, studying DNA meant facing the frustration of working with very small, hard-to-obtain samples. Everyone wished for a way to copy DNA, to make millions of copies whenever they needed them. The wish was granted one evening by Kary B. Mullis, a young biochemist who was mentally attacking a different problem—how to identify the sequence of nucleotides in DNA. In his own words, here is what happened:

> Sometimes a good idea comes to you when you are not looking for it. Through an improbable combination of coincidences, naiveté, and lucky mistakes, such a revelation came to me one Friday night in April 1983 as I gripped the steering wheel of my car and snaked along a moonlit road into northern California's redwood country .... I liked night driving; every weekend I went north to my cabin and sat still for three hours in the car, my hands occupied, my mind free. On that particular night I was thinking about my proposed DNA sequencing experiment.*

Mullis drove along, trying out in his mind and then rejecting various ways to approach the experiment, which required combining DNA polymerase, natural DNA, nucleoside triphosphates, and short synthetic nucleotide chains (*oligonucleotides*) in just the right way. Then,

> I was suddenly jolted by a realization: The strands of DNA in the target and the extended oligonucleotides would have the same base sequences. In effect, the mock reaction would

have doubled the number of DNA targets in the sample! ... Excited, I started running powers of two in my head: 2, 4, 8, 16, 32—I remembered vaguely that 2 to the tenth power was about 1,000 and that therefore 2 to the twentieth was around a million.*

The quotations above are from "The Unusual Origin of the Polymerase Chain Reaction," Kary B. Mullis, Scientific American, April 1990, p. 56, which gives an extended account of the thought processes that led to the discovery.

The outcome of this night drive was development of the *polymerase chain reaction (PCR)*, now carried out automatically by instruments in every molecular biology lab. In 1993 Mullis shared the Nobel Prize in chemistry for this work.

The goal of the PCR is to produce a large quantity of a specific segment of DNA. The DNA might be part of a genome study, it might be from a crime scene or a fossil, or it might be from a specimen preserved as a medical record. The raw materials for the reaction are the DNA that contains the nucleotide sequence to be amplified, *primers* (synthetic oligonucleotides with bases complementary to the base sequences on either side of the sequence of interest), the deoxyribonucleoside triphosphates that carry the four DNA bases, and DNA polymerase that will copy the DNA between the primers.

The reaction is carried out in three steps:

1. *Heating of a DNA solution to cause the helix to unravel into single strands.*

when a particular gene in a particular cell is transcribed? What indicates the spot on DNA where transcription should begin? How are transcription and the resulting protein synthesis regulated? You have seen that steroid hormones function by directly entering the nucleus to activate enzyme synthesis. (∞, Section 20.5) This is just one mode of gene regulation. It can also occur by modification of DNA, during transcription, and during translation. Much is known beyond what we have covered. There are also many questions that cannot yet be fully answered.

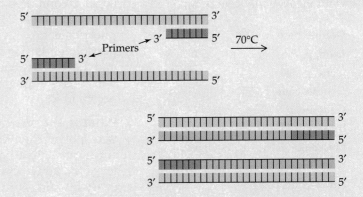

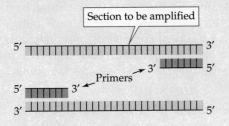

5′ ⊥⊥⊥⊥⊥⊥⊥⊥⊥⊥⊥⊥⊥⊥⊥⊥⊥⊥⊥⊥⊥⊥⊥⊥⊥ 3′   90°C →

3′ ⊥⊥⊥⊥⊥⊥⊥⊥⊥⊥⊥⊥⊥⊥⊥⊥⊥⊥⊥⊥⊥⊥⊥⊥⊥ 5′

5′ ⊥⊥⊥⊥⊥⊥⊥⊥⊥⊥⊥⊥⊥⊥⊥⊥⊥⊥⊥⊥⊥⊥⊥⊥ 3′

3′ ⊥⊥⊥⊥⊥⊥⊥⊥⊥⊥⊥⊥⊥⊥⊥⊥⊥⊥⊥⊥⊥⊥⊥⊥ 5′

2. *Addition of primers complementary to the DNA on either side of the single-stranded DNA sequence to be amplified.* Because DNA polymerase copies DNA at the point where the double helix is unwinding, it is necessary to create double-stranded DNA at the point where copying is to start. The primers indicate this starting point.

5′ ⊥⊥⊥⊥⊥⊥⊥⊥⊥⊥⊥⊥⊥⊥⊥⊥⊥⊥⊥⊥⊥⊥⊥⊥ 3′

    50°C →

3′ ⊥⊥⊥⊥⊥⊥⊥⊥⊥⊥⊥⊥⊥⊥⊥⊥⊥⊥⊥⊥⊥⊥⊥⊥ 5′

Section to be amplified

5′ ⊥⊥⊥⊥⊥⊥⊥⊥⊥⊥⊥⊥⊥⊥⊥⊥⊥⊥⊥⊥⊥⊥⊥⊥ 3′
                          3′ ⊥⊥⊥⊥⊥⊥ 5′
                     Primers
5′ ⊥⊥⊥⊥⊥ 3′
3′ ⊥⊥⊥⊥⊥⊥⊥⊥⊥⊥⊥⊥⊥⊥⊥⊥⊥⊥⊥⊥⊥⊥⊥⊥ 5′

3. *Extension of the primers by DNA polymerase to create double-stranded DNA identical to the original.* The DNA polymerase starts copying at the ends of the primers so that the new DNA segment incorporates the primers.

5′ ⊥⊥⊥⊥⊥⊥⊥⊥⊥⊥⊥⊥⊥⊥⊥⊥⊥⊥⊥⊥⊥⊥⊥⊥ 3′
                       3′ ⊥⊥⊥⊥⊥⊥ 5′   70°C →
           Primers
5′ ⊥⊥⊥⊥⊥ 3′
3′ ⊥⊥⊥⊥⊥⊥⊥⊥⊥⊥⊥⊥⊥⊥⊥⊥⊥⊥⊥⊥⊥⊥⊥⊥ 5′

5′ ⊥⊥⊥⊥⊥⊥⊥⊥⊥⊥⊥⊥⊥⊥⊥⊥⊥⊥⊥⊥⊥⊥⊥⊥ 3′
3′ ⊥⊥⊥⊥⊥⊥⊥⊥⊥⊥⊥⊥⊥⊥⊥⊥⊥⊥⊥⊥⊥⊥⊥⊥ 5′

5′ ⊥⊥⊥⊥⊥⊥⊥⊥⊥⊥⊥⊥⊥⊥⊥⊥⊥⊥⊥⊥⊥⊥⊥⊥ 3′
3′ ⊥⊥⊥⊥⊥⊥⊥⊥⊥⊥⊥⊥⊥⊥⊥⊥⊥⊥⊥⊥⊥⊥⊥⊥ 5′

The reactants are combined in a closed container and the temperature cycled from about 90°C for step 1, to about 50°C for step 2, and to about 70°C for step 3. The temperature cycle requires only a few minutes and can be repeated over and over again for the same mixture. The first cycle produces two molecules of DNA; the second produces four molecules; and so on, with doubling at each cycle. Twenty-five amplification cycles yield over 30 million copies of the original DNA segment.

Automation of the PCR was made possible by the discovery of a heat-stable polymerase (*Taq polymerase*) isolated from a bacterium that lives in hot springs. Because the enzyme survives the temperature needed for separating the DNA strands, it is not necessary to add fresh enzyme for each three-step cycle.

See Additional Problems 26.69–70 at the end of the chapter.

■ **PROBLEM 26.17**

What amino acid sequence is coded for by the following mRNA base sequence? CUU-AUG-GCU-UGG-CCC-UAA

■ **PROBLEM 26.18**

What anticodon sequences of tRNAs match the mRNA codons in Problem 26.17?

Key Words

Summary: Revisiting the Chapter Goals

1. What is the composition of the nucleic acids, DNA, and RNA? *Nucleic acids* are polymers of *nucleotides*. Each nucleotide contains a sugar, a base, and a phosphate group. The sugar is D-ribose in *ribonucleic acids (RNAs)* and 2-deoxy-D-ribose in *deoxyribonucleic acids (DNAs)*. The C5 —OH of the sugar is bonded to the phosphate group, and the anomeric carbon of the sugar is connected by an *N*-glycosidic bond to one of five heterocyclic nitrogen bases (Table 26.1). A *nucleoside* contains a sugar and a base, but not the phosphate group. In DNA and RNA, the nucleotides are connected by phosphate diester linkages between the 3′ —OH group of one nucleotide and the 5′ phosphate group of the next nucleotide. DNA and RNA both contain adenine, guanine, and cytosine; thymine occurs in DNA and uracil occurs in RNA.

2. What is the structure of DNA? The DNA in each *chromosome* consists of two polynucleotide strands twisted together in a *double helix*. The sugar–phosphate backbones are on the outside, and the bases are in the center of the helix. The bases on the two strands are complementary—opposite every thymine is an adenine, opposite every guanine is a cytosine. The base pairs are connected by hydrogen bonds (two between T and A; three between G and C). Because of the *base pairing*, one DNA strand runs in the 5′ to 3′ direction and its complementary partner in the 3′ to 5′ direction.

3. How is DNA reproduced? *Replication* (Figure 26.5) requires DNA polymerase enzymes and deoxyribonucleoside triphosphates. The DNA helix partially unwinds and the enzymes move along incorporating nucleotides with bases complementary to those on the unwound DNA strand being copied. The enzymes copy only in the 3′ to 5′ direction of the template strand, so that one strand is copied continuously and the other strand is copied in segments as the replication fork moves along. In each resulting double helix, one strand is the original template strand and the other is the new copy.

4. What are the functions of RNA? *Messenger RNA (mRNA)* carries the genetic information out of the nucleus to the *ribosomes* in the cytosol, where protein synthesis occurs. *Transfer RNAs (tRNAs)* circulate in the cytosol, where they bond to amino acids that they will then deliver to protein synthesis. *Ribosomal RNAs (rRNAs)* are incorporated into ribosomes.

5. How is messenger RNA synthesized? In *transcription* (Figure 26.6), one DNA strand serves as the template and the other, the informational strand, is not copied. Nucleotides carrying bases complementary to the template bases between a control segment and a stop *codon* are connected one by one to form mRNA. The mRNA is identical to the matching segment of the informational strand, but with uracil replacing thymine. *Introns*, which are base sequences that do not code for amino acids in the protein, are cut out before the mRNA leaves the nucleus.

6. How does RNA participate in protein synthesis? The genetic information is read as a sequence of codons—triplets of bases in DNA that give the sequence of amino acids in a protein. Of the 64 possible codons (Table 26.4), 61 specify amino acids and 3 are stop codons. Each tRNA has at one end an *anticodon* consisting of three bases complementary to those of the codon that specifies the amino acid it carries. Initiation of *translation* (Figure 26.8) is the coming together of the large and small subunits of the ribosome, an mRNA, and the first amino acid–bearing tRNA connected at the first of the two binding sites in the ribosome. *Elongation* proceeds as the next tRNA arrives at the second binding site, its amino acid is bonded to the first one, the first tRNA leaves, and the ribosome moves along so that once again there is a vacant second site. These steps repeat until the stop codon is reached. The termination step consists of separation of the two ribosome subunits, the mRNA, and the protein.

▪ Understanding Key Concepts

26.19 Combine the structures below to create a ribonucleotide. Show where water is removed to form an *N*-glycosidic linkage and where water is removed to form a phosphate ester. Draw the resulting ribonucleotide structure, and name it.

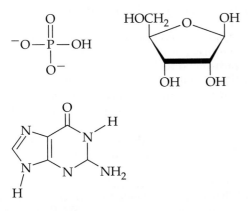

26.20 Copy the diagram shown below and indicate with dotted lines where the hydrogen bonding would occur between complementary strands of DNA. What would be the sequence of each strand of DNA drawn (remember that the sequence is written from 5′ to 3′ end)?

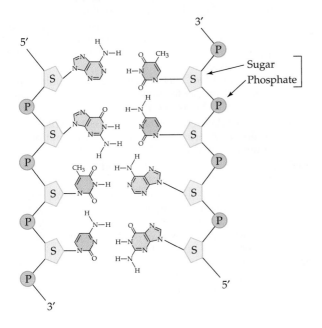

26.21 Copy the simplified drawing of a DNA replication fork below:

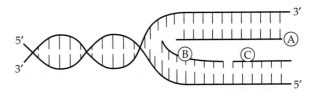

(a) On the drawing, indicate the direction of synthesis of the new strand labeled A and the location of DNA polymerase on the strand.

(b) On the drawing, indicate the direction of synthesis of the new strand labeled B and the location of DNA polymerase on the strand.

(c) How will strand C and strand B be connected?

26.22 What groups are found on the exterior of the DNA double helix? In the nucleus, DNA strands are wrapped around the proteins called histones. Would you expect the histones to be neutral, positively charged or negatively charged? Which amino acids would you expect to be abundant in histones?

26.23 In addition to RNA polymerase, transcription of DNA for the synthesis of RNA requires (a) a control segment of DNA (also called an initiation sequence), (b) an informational sequence, (c) a template strand of DNA, and (d) an end of the sequence (termination sequence). Determine the direction of RNA synthesis on the RNA strand in the diagram shown below. Draw in the locations of elements (a)–(d).

26.24 Tyr-Gly-Gly-Phe is the sequence of the first four amino acids in enkephalin. If we were searching for enkephalin genes, we would need to know what sequence of bases in DNA we should be looking for. Use the boxes below to indicate answers to parts (a)–(d).

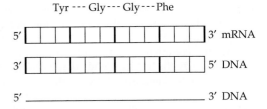

(a) What RNA sequence could code for these four amino acids?

(b) What is a double-stranded DNA sequence (gene) that could code for these amino acids?

(c) Which strand of DNA is the template strand, and which is the informational strand?

(d) How many possible DNA sequences are there?

■ Additional Problems

Structure and Function of Nucleic Acids

26.25 What is a nucleotide, and what three kinds of components does a nucleotide contain?

26.26 What are the names of the sugars in DNA and RNA, and how do they differ?

26.27 What are the names of the four major heterocyclic bases in DNA? In RNA?

26.28 What are the two types of bases in DNA and RNA? Which bases correspond to each type?

26.29 What are the three main kinds of RNA, and what are their functions?

26.30 Rank the following kinds of nucleic acids in order of size: tRNA, DNA, mRNA.

26.31 (a) What is meant by the term *base pairing*?
   (b) Which bases pair with which other bases?
   (c) How many hydrogen bonds does each base pair have?

26.32 How are replication, transcription, and translation similar? How are they different?

26.33 What is the difference between a gene and a chromosome?

26.34 What are the two major components of chromatin?

26.35 What genetic information does a single gene contain?

26.36 How many chromosomes are present in a human cell?

26.37 What kind of intermolecular attraction holds the DNA double helix together?

26.38 What does it mean to speak of bases as being "complementary"?

26.39 The DNA from sea urchins contains about 32% A and about 18% G. What percentages of T and C would you expect in sea urchin DNA? Explain.

26.40 If a double-stranded DNA molecule was 28% T, what would be the percentage of A, C, and G? Explain.

26.41 What is the difference between the 3′ end and the 5′ end of a polynucleotide?

26.42 When polynucleotides are written, are they written 3′ to 5′, or 5′ to 3′?

26.43 Show by drawing structures how the phosphate and sugar components of a nucleic acid are joined. What type of linkage is formed between the sugar and the phosphate?

26.44 Show by drawing structures how the sugar and heterocyclic base components of a nucleic acid are joined. What small molecule is formed?

26.45 Draw the complete structure of deoxycytidine 5′-phosphate, one of the four major deoxyribonucleotides.

26.46 Draw the complete structure of the RNA dinucleotide U-C. Identify the 5′ and 3′ ends of the dinucleotide.

26.47 One part of the synthesis of RNA is the splicing out of the introns. The cell uses this strategy often: Make a biomolecule that is more elaborate than its biologically active form and convert it into its active form when it is needed. You have also seen this strategy with proinsulin and with zymogens. What is the advantage of this approach?

Nucleic Acids and Heredity

26.48 Transcribed RNA is complementary to which strand of DNA?

26.49 Why are numerous replication forks needed when human DNA is duplicated?

26.50 Why do we say that DNA replication is semiconservative?

26.51 What is a codon, and on what kind of nucleic acid is it found?

26.52 What is an anticodon, and on what kind of nucleic acid is it found?

26.53 What is the general shape and structure of a tRNA molecule?

26.54 There are different tRNAs for each amino acid. What is one major way to differentiate among the tRNAs for each amino acid?

26.55 What are exons and introns? Which of the two carries genetic information for coding protein structure?

26.56 Look at Table 26.4 and find codons for the following amino acids:
   (a) Pro   (b) Lys   (c) Met

26.57 What amino acids are specified by the following codons?
   (a) A-C-U   (b) G-G-A   (c) C-U-U

26.58 What anticodon sequences are complementary to the codons listed in Problem 26.57? (Remember that the anticodons are opposite in direction to the codons, so label the 3′ and 5′ ends!)

26.59 What anticodon sequences are complementary to the codons for the amino acids given in Problem 26.56? (Remember that the anticodons are opposite in direction to the codons, so label the 3′ and 5′ ends!)

26.60 If the sequence T-A-C-C-G-A appeared on the information strand of DNA, what sequence would appear opposite it on the template strand? Label your answer with 3′ and 5′ ends.

26.61 Refer to Problem 26.60. What sequence would appear on the mRNA molecule transcribed from the DNA sequence T-A-C-C-G-A? Label your answer with 3′ and 5′ ends.

26.62 Refer to Problems 26.60 and 26.61. What dipeptide would be synthesized from the informational DNA sequence T-A-C-C-G-A?

26.63 What tetrapeptide would be synthesized from the informational DNA sequence C-C-T-G-A-C-G-C-G-G-T-T?

26.64 Metenkephalin is a small peptide found in animal brains that has morphine-like properties. Give an mRNA sequence that could code for the synthesis of metenkephalin: Tyr-Gly-Gly-Phe-Met. Label your answer with 3′ and 5′ ends.

26.65 Refer to Problem 26.64. Give a double-stranded DNA sequence that could code for metenkephalin. Label your answer with 3′ and 5′ ends.

Applications

26.66 How do viruses differ from living organisms? [*Viruses and AIDS*]

26.67 Explain the process of reverse transcription. What is the name given to viruses that use this process? [*Viruses and AIDS*]

26.68 How do vaccines work? Why is it so difficult to design drugs effective against AIDS? [*Viruses and AIDS*]

26.69 Briefly describe how a polymerase chain reaction works. [*Serendipity and the Polymerase Chain Reaction*]

26.70 What is the purpose of a polymerase chain reaction? [*Serendipity and the Polymerase Chain Reaction*]

General Questions and Problems

26.71 Normal hemoglobin has a glutamic acid at the same site in which sickle-cell hemoglobin has a valine. List all the possible mRNA codons that could be present for each type of hemoglobin. Is it possible that one base change could result in a change from Glu to Val in hemoglobin?

26.72 Insulin is synthesized as preproinsulin, which has 81 amino acids. How many nitrogen bases must be present in the DNA to code for preproinsulin?

26.73 Human and horse insulin are both composed of two polypeptide chains with one chain containing 21 amino acids and the other containing 30 amino acids. Human and horse insulin differ at two amino acids: the ninth position in one chain (human has serine and horse has glycine) and the thirtieth position on the other chain (human has threonine and horse has alanine). How must the DNA differ to account for this? Identify the 5' and 3' ends of the four trinucleotide complementary DNA sequences.

26.74 If the initiation codon for proteins is AUG, how can you account for the case of a protein that does not include methionine as its first amino acid?

26.75 Suppose that 22% of the nucleotides of a DNA molecule are deoxyadenosine, and during replication the relative amounts of available deoxynucleoside triphosphates are 22% dATP, 22% dCTP, 28% dGTP, and 28% dTTP. What deoxynucleoside triphosphate would be limiting to the replication? Explain.

27

Genomics

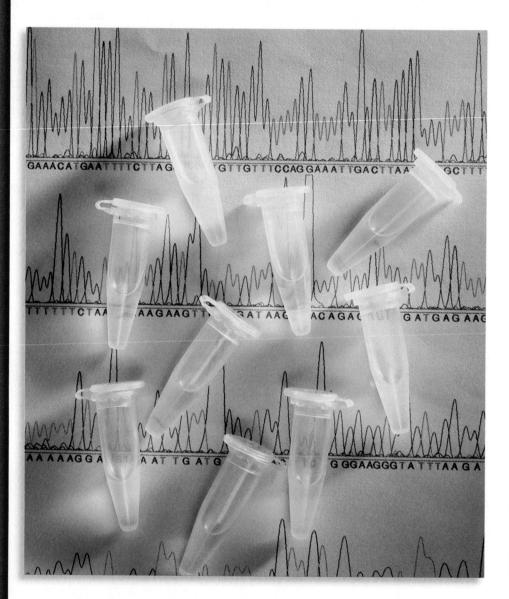

A genetic blueprint. *The vials hold DNA samples dissolved in ethanol. Behind the vials is the printout from an automatic DNA sequencing machine.*

In Chapter 26 we described the fundamentals of DNA structure and function, the synthesis of DNA, and its role in making proteins from amino acids. With the arrival of the 21st century, we stand at the brink of a scientific revolution centered on DNA. Before many years pass, a complete and accurate map of human DNA will be available, thanks to the power of computers, creative instrument design, and the widespread efforts of many individuals. The map will catalog every base in every gene. Creation of this map has been compared to landmark achievements such as harnessing nuclear power and flight into outer space. In significance for individual human beings, there has never been anything like it.

This chapter opens with a description of the event at which mapping the human genome hit the news worldwide. We then explore variations in the content of the DNA in each chromosome and a technique for manipulating DNA. Finally, we look briefly at ways in which genomic information can be put to use. Along the way, we'll address the following questions:

1. **What is the working draft of the human genome and the circumstances of its creation?**
   The goal: Be able to describe the genome mapping projects and the major accomplishments of their working drafts.

2. **What are the various segments along the length of the DNA in a chromosome?**
   The goal: Be able to describe the nature of telomeres, centromeres, exons and genes, and noncoding DNA.

3. **What are mutations?**
   The goal: Be able to define mutations, identify what can cause them, and also identify their possible results.

4. **What are polymorphisms and single nucleotide polymorphisms (SNPs) and how can identifying them be useful?**
   The goal: Be able to define polymorphisms and SNPs, and explain the significance of knowing the locations of SNPs.

5. **What is recombinant DNA?**
   The goal: Be able to define recombinant DNA and explain how it is used for production of proteins by bacteria.

6. **What does the future hold for uses of genomic information?**
   The goal: Be able to provide an overview of the current and possible future applications of the human genome map.

CONCEPTS TO REVIEW

Structure, synthesis, and function of DNA (Chapter 26)

27.1 Mapping the Human Genome

The Genome in the News

Genetic Code of Human Life Is Cracked by Scientists

Front page headline in the New York Times, June 27, 2000

When science makes the front page of a major newspaper, you can be sure it is big news. The events that inspired the headline were major achievements by two vigorously competing yet complementary projects with the same goal: to map the human genome by identifying the sequence on the 23 human chromosomes of 3 billion base pairs.

The major participants in the press conference at which the achievement was announced were Bill Clinton, then President of the United States, and Tony Blair, Prime Minister of England, both representing countries with major participants in the international consortium known as the Human Genome Project; Dr. Francis S. Collins leader of the Human Genome Project; and Dr. S. Craig Venter, then president and a founder of Celera Genomics, a commercial company that is also sequencing the genome. Each of these individuals in their own way, as shown in the nearby box, expressed the significance of what was

▲ William Clinton, Dr. Francis S. Collins (right), and S. Craig Venter (left) at the press conference announcing completion of the working drafts of the human genome.

"In genetic terms all human beings, regardless of race, are more than 99.9 percent the same. What that means is that modern science has confirmed what we first learned from ancient faiths. The most important fact of life on this earth is our common humanity."

U.S. President William Clinton

"Nothing better demonstrates the way technology and science are driving us—fast-forwarding us—all into the future."

Prime Minister of England, Anthony Blair

"Science is a voyage of exploration into the unknown. We are here today to celebrate a milestone along a truly unprecedented voyage. This one into ourselves."

Dr. Francis S. Collins, Director of the National Human Genome Research Institute and leader of the international Human Genome Project

"The basic knowledge that we are providing the world will have a profound impact on the human condition and the treatments for disease and our view of our place in the biological continuum."

Dr. J. Craig Venter, President, Celera Genomics

described as a "working draft" of the human genome, the sequencing of about 90% of the gene-rich segments of DNA.

The 20 groups around the world that comprise the Human Genome Project reside at not-for-profit institutes and universities. They make their progress in genome studies freely available to other scientists as it develops. To do this, they post their results on the World Wide Web each day. Although the Human Genome Project and Celera Genomics are equally aware of the genome's intrinsic value to humanity, Celera Genomics seeks to make a profit from activities that utilize their genomic information.

Despite their different approaches, both groups proceeded successfully to the working draft stage. Their competition, in fact, inspired acceleration of the Human Genome Project, which completed 90% of its working draft of the sequence in 15 months instead of the originally anticipated four years. Furthermore, the results of the two groups are to a great extent in agreement and supportive of each other.

Mapping the human genome cannot be done by starting at one end of the DNA in each chromosome and proceeding base by base. As a small demonstration of the mapping challenge, consider that the nucleotides that code for proteins (the *exons*) are interspersed with noncoding nucleotides (the *introns*). There is no spacing between "words" in the genetic code, nor is there any "punctuation". In the English language, one would have to start with something like this:

sfdggmaddrydkdkdkrrrsjfljhadxccctmctmaqqqoumlittgklejagkjghjoailambrsslj

to uncover this:

sfdgg**ma**dd**ry**dkdkdkrrrsjflj**had**xccctmctm**a**qqqoum**litt**gk**le**jagkjghjoai**lambr**sslj

The string of C's, G's, T's, and A's of the human genome would, it has been estimated, fill 75,490 pages of standard type in a newspaper like the *New York Times*.

How to Map a Genome

The two groups used different approaches to taking DNA apart, analyzing its base sequences, and reassembling the information. The Human Genome Project proceeded through a series of maps of finer and finer resolution (think of progressing from a satellite photo of the United States, to a map of your state, to a map of the city where you live). Celera followed a random "shotgun" approach in which they fragmented DNA and then relied on instrumental and computer-driven techniques to establish the sequence.

The strategy of the Human Genome Project for generating the complete map is illustrated in Figure 27.1. Pictured at the top is a familiar type of chromosome drawing for chromosome 21 (the smallest one, with 37 million base pairs), which shows the location of banding visible in electron micrographs.

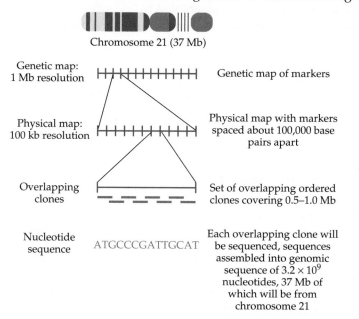

Chromosome 21 (37 Mb)

Genetic map:
1 Mb resolution — Genetic map of markers

Physical map:
100 kb resolution — Physical map with markers spaced about 100,000 base pairs apart

Overlapping clones — Set of overlapping ordered clones covering 0.5–1.0 Mb

Nucleotide sequence — ATGCCCGATTGCAT — Each overlapping clone will be sequenced, sequences assembled into genomic sequence of 3.2×10^9 nucleotides, 37 Mb of which will be from chromosome 21

◄ **FIGURE 27.1   Human Genome Project mapping strategy.**

▲   A sample of DNA ready for analysis.

Next is a *genetic map*, which shows the physical location of *markers*, identifiable physical locations on either introns or exons that are known to be inherited. The markers are an average of one million nucleotides apart. This is known as a genetic map because the order and locations of the markers are established by genetic studies of inheritance in related individuals.

The next map, the *physical map*, refines the distance between markers to about 100,000 base pairs. It includes markers identified by a variety of experimental methods.

To proceed to a map of finer resolution, a chromosome is cut into large segments and multiple copies of the segments are produced. The segments are **clones**, a term that refers to identical copies of organisms, cells, or in this case DNA segments.

The overlapping clones, which cover the entire length of the chromosome, are organized in sequence to produce the next level of map (see Figure 27.1). The clones are fragmented into 500 base pieces, each of which ends in a C, G, T, or A and in which each C,G,T, or A fluoresces in a different color. These fragments are separated by electrophoresis (∞, p. 523). In the final step, the sequences are assembled into a completed nucleotide map of the chromosome.

Clones   Identical copies of organisms, cells, or DNA segments from a single ancestor.

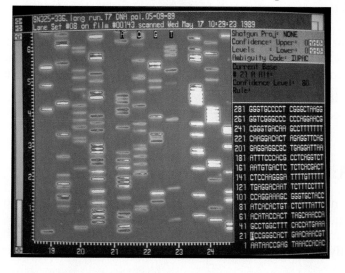

◄   Sequencing gel and computer readout showing the result of DNA sequencing.

In their shotgun approach, Celera broke the entire genome into fragments without any identification of their origin. The fragments were copied many times and modified so that ultimately they were cut into 500-base pieces with fluorescently labeled bases that could be sequenced by high-speed machines. The resulting sequences were reassembled by identifying overlapping ends. At Celera, this task was carried out in the world's largest nongovernmental supercomputing center.

> ■ **PROBLEM 27.1**
>
> Decode the following sequence of letters to find an English phrase.
>
> uuiioouoppagfdtttrrroncetrtrnnnaedigopuponsldjflsjfxxxbvgfaqqqeutimeab-
> rrxoknuw

A Genomic Status Report

The working drafts announced in 2000 comprised not-yet-complete surveys of the base sequences in the DNA of all the human chromosomes. At that time there were both major accomplishments and major uncertainties:

- One group reported 3.12 billion base pairs in the human genome; the other reported 3.15 billion base pairs. (Both groups skipped sequencing long stretches of repetitive nucleotide patterns that do not code for proteins.)

- The genome contains 30,000–40,000 genes, a *major* surprise. The standard estimate had long been 100,000. The function of many of the identified genes is as yet unknown, but the function of others has been discovered by using information from the map.

- Through reorganization of exons during messenger RNA synthesis, a gene can produce not one, but many different proteins.

- About 200 of the human genes are identical to genes in bacteria, from which they may have originated.

- Roughly 50% of the genome is devoted to repetitive base sequences of unknown function or perhaps with no function at all. Studies of this noncoding DNA and its possible origins may provide clues to the evolution of humans, however.

- Both working draft maps were acknowledged to contain interruptions and possibly misoriented segments.

- The next steps are refinement of the working drafts and completion of the sequencing. The Human Genome Project states that its goal is 99.99% accuracy. Sequencing the repetitive noncoding DNA may, it is thought, reveal that it has a function. Annotation of the map with the proteins it codes for and the functions of the proteins is also an ongoing study.

In what was surely an intentional parallel to Watson and Crick's announcement of the double helix structure of DNA (∞, p. 752), the scientific paper from the Human Genome Project ends with the following statement:

> Finally, it has not escaped our notice that the more we learn about the human genome, the more there is to explore.

27.2 A Trip Along a Chromosome

Information about the human genome is flooding forth from universities, research institutes, and biotechnology companies at a rapid rate. There are dozens of informative World Wide Web sites to explore. Some of them change almost daily as new information is uncovered and old information is displaced.

Application Whose Genome Is It?

One might wonder whose genome provided the standard against which those of all other human beings will be evaluated. What individual might symbolize all of us? A star athlete? A brilliant scientist? A truly average person?

It shouldn't take more than a moment's thought to realize that selection of a single individual is a bad idea. No person would want their DNA profile distributed around the world, nor would they wish to run the risk of being identified. Suppose their DNA harbors a major but not-yet-known harmful mutation. Also, some sort of average is needed, since no two individuals other than identical twins, have exactly the same base sequences in their genomes.

▲ How DNA varies: The parents and their children have many similarities in their DNA, but between the parent-child pairs there are wide variations in DNA.

The obvious choice, recognized by both genome mapping projects, was DNA from a group of anonymous individuals. The Human Genome Project collected blood from 5 to 10 more volunteers than would ultimately be used for DNA sequencing. All identifying labels were removed before the samples to be used were chosen. The ultimate map comes from a composite of these samples. Where variations occurred among the samples, the most common variation was chosen. The Celera project also relied on anonymous donors. Their genome is based on DNA from five individuals: two Caucasian men, and three women, one black, one Hispanic, and one of Chinese descent.

▲ How DNA varies: Twins have identical DNA.

See Additional Problems 27.48–49 at the end of the chapter.

The ongoing activity is fascinating, but much of it is beyond what most of us need to know. In this section we take a trip along the major regions and structural variations in the DNA folded into each chromosome. Understanding this much of the complexities of genomic information will help to put the biotechnology revolution that lies ahead into context.

Telomeres and Centromeres

The DNA in a chromosome begins with the **telomeres** that occur at both ends of every chromosome. The telomeres in human DNA are long series of a repeating group of nucleotides, $(TTAGGG)_n$. These repeats do not code for anything. They protect the ends of the chromosome from accidental changes that might alter the DNA's coding sequences. They also prevent the DNA from bonding to the DNA in other chromosomes or DNA fragments. Telomeres were not sequenced in the mapping projects described in Section 27.1, nor were the **centromeres** of chromosomes, the constrictions that determine the shapes of chromosomes during cell division. The centromere, like the telomeres, contains large repetitive base sequences that do not code for proteins.

Each new cell starts life with a long stretch of telomeric DNA on each of its chromosomes, 1000 or more of the repeating group. Some of it is lost with each cell division. Ultimately, a much shorter telomere is associated with the stage at which the cell stops dividing (known as *senescence*). Continuation of shortening beyond this stage is associated with DNA instability and cell death.

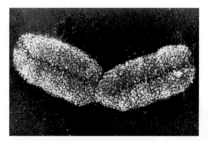

▲ Color-enhanced electron microscope image of a single chromosome with its centromere in place during metaphase.

Telomeres The ends of chromosomes; in humans contain long series of repeating groups of nucleotides.

Centromeres The central regions of chromosomes.

Telomerase is the enzyme responsible for adding telomeres to DNA. Under normal, healthy conditions, telomerase is active in young cells destined to become specialized. It is not active in the others (the *somatic* cells in adults). There is widespread speculation that telomere shortening plays a role in the natural progression of human aging. Some support for this concept comes from experiments with mice whose telomerase activity has been knocked out. These mice age prematurely, and if they become pregnant, their embryos do not survive.

What would happen if telomerase remains active in a cell rather than declining in activity with age? With the length of its telomeres constantly being replenished by telomerase, the cell would not age and instead would continue to divide. Consider that such continuing division is characteristic of cancer cells. In fact, the majority of cancer cells are known to contain active telomerase. Where this activity stands in relation to the presence of cancer-causing genes and environmental factors is not yet understood. As you might suspect, however, there are ongoing experiments on the inactivation of telomerase.

Noncoding DNA

In addition to the noncoding telomeres, centromeres, and introns along a chromosome, there are promoter sequences, which are noncoding but regulatory regions of DNA that determine which of its genes are turned on. Remember that all of your cells (except red blood cells) contain all of your genes, but only the genes needed by the individual cells go to work.

The human genome, it turns out, has much more noncoding DNA than the mapped genomes known for other animals. Perhaps the segments of repetitive but noncoding DNA are needed to accommodate the folding of DNA within the nucleus. Perhaps they played a role in evolution.

Genes

In learning about transcription (⊂∞⊃, Section 26.8), you saw that the nucleotides of a single gene are not consecutive along a stretch of DNA. Instead, the coding segments (the *exons*) are interspersed with noncoding segments (the *introns*) (p. 756). As a small example of what must be dealt with in mapping the genome, we have reproduced below an illustration of a one million base sequence of chromosome 22.

A one million base segment of chromosome 22. The red bars are interspersed regions of noncoding nucleotides. The blue bars are exons of known genes. (This illustration appeared in the February 15, 2001 publication of the Human Genome Project working draft. *Nature* **409**.)

Chromosome 22 is one of the smaller chromosomes and was one of the first to have all of its nonrepetitive DNA mapped. The map identified 34 million bases that are attributed to 545 genes. This chromosome carries genes known to be associated with the immune system, congenital heart disease, schizophrenia, leukemia, various cancers, and many other genetically-related conditions. The map also revealed several hundred previously unknown genes.

27.3 Mutations and Polymorphisms

The base-pairing mechanism of DNA replication and RNA transcription provides an extremely efficient and accurate method for preserving and using genetic information, but it's not perfect. Occasionally an error occurs, resulting in the incorporation of an incorrect base at some point.

An occasional error during the transcription of a messenger RNA molecule may not create a serious problem. After all, large numbers of mRNA molecules

▲ An error in nucleic acid composition that occurs once in 3–4 million lobsters is responsible for the beautiful color of this crustacean.

are continually being produced and an error that occurs perhaps one out of a million times would hardly be noticed in the presence of many correct mRNAs. If an error occurs during the replication of a DNA molecule, however, the consequences can be far more damaging. Each chromosome in a cell contains only *one* kind of DNA, and if it is miscopied during replication, then the error is passed on when the cell divides.

An error in base sequence that is carried along during DNA replication is called a **mutation**. The term commonly refers to variations in DNA sequence that are present in a very small number of individuals of a species. Some mutations result from spontaneous and random events. Others are induced by exposure to a **mutagen**—an external agent that can cause a mutation. Viruses, chemicals, and ionizing radiation can all be mutagenic. Some types of mutations are listed in Table 27.1.

Mutation   An error in base sequence that is carried along in DNA replication.

Mutagen   A substance that causes mutations.

TABLE 27.1   Types of Mutations

Type	Description
Point mutations	A single base change
Silent	A change that specifies the same amino acid; for example, GUU → GUC, gives Val → Val
Missense	A change that specifies a different amino acid; for example, GUU → GCU gives Val → Ala
Nonsense	A change that produces a stop codon, for example, CGA → UGA gives Arg → Stop
Frameshift	The number of inserted or deleted bases is not a multiple of 3, so that all triplets following the mutation are read differently
Insertion	Addition of one or more bases
Deletion	Loss of one or more bases

The biological effects of incorporating an incorrect amino acid into a protein range from negligible to catastrophic, depending on the location of the change as well as its nature. The effect might be a hereditary disease like sickle-cell anemia or a birth defect. There are thousands of known human hereditary diseases. Some of the more common ones are listed in Table 27.2. Mutations, sometimes the combination of several mutations, can also produce vulnerability to certain diseases, which may or may not develop in an individual.

TABLE 27.2   Some Common Hereditary Diseases, Their Causes, and Their Prevalence

Name	Nature and Cause of Defect	Prevalence in Population
Phenylketonuria (PKU)	Brain damage in infants caused by the defective enzyme phenylalanine hydroxylase	1 in 40,000
Albinism	Absence of skin pigment caused by the defective enzyme tyrosinase	1 in 20,000
Tay-Sachs disease	Mental retardation caused by a defect in production of the enzyme hexosaminidase A	1 in 6000 (Ashkenazi Jews) 1 in 100,000 (General population)
Cystic fibrosis	Bronchopulmonary, liver, and pancreatic obstructions by thickened mucus; defective gene and protein identified	1 in 3000
Sickle-cell anemia	Anemia and obstruction of blood flow caused by a defect in hemoglobin	1 in 185 (African-Americans)

Polymorphism A variation in DNA sequence within a population.

Polymorphisms are also variations in the nucleotide sequence of DNA, but here the reference is to variations that are common within a given population. The location of polymorphisms responsible for some inherited human diseases are shown in Figure 27.2.

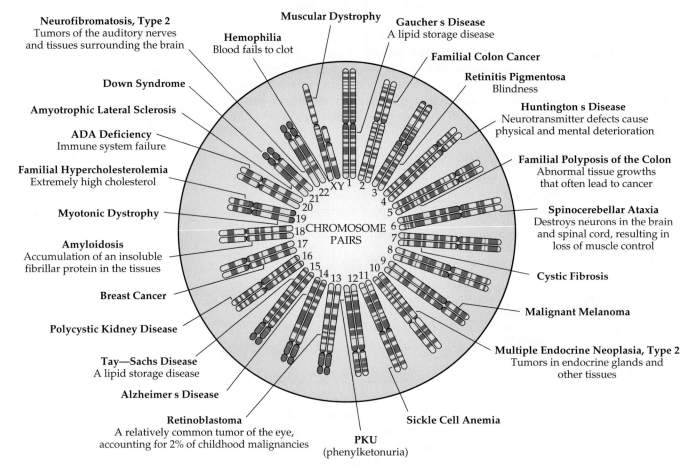

Neurofibromatosis, Type 2
Tumors of the auditory nerves and tissues surrounding the brain

Muscular Dystrophy

Hemophilia
Blood fails to clot

Gaucher s Disease
A lipid storage disease

Familial Colon Cancer

Retinitis Pigmentosa
Blindness

Down Syndrome

Amyotrophic Lateral Sclerosis

ADA Deficiency
Immune system failure

Familial Hypercholesterolemia
Extremely high cholesterol

Myotonic Dystrophy

Amyloidosis
Accumulation of an insoluble fibrillar protein in the tissues

Breast Cancer

Polycystic Kidney Disease

Tay—Sachs Disease
A lipid storage disease

Alzheimer s Disease

Retinoblastoma
A relatively common tumor of the eye, accounting for 2% of childhood malignancies

PKU
(phenylketonuria)

Sickle Cell Anemia

Multiple Endocrine Neoplasia, Type 2
Tumors in endocrine glands and other tissues

Malignant Melanoma

Cystic Fibrosis

Spinocerebellar Ataxia
Destroys neurons in the brain and spinal cord, resulting in loss of muscle control

Familial Polyposis of the Colon
Abnormal tissue growths that often lead to cancer

Huntington s Disease
Neurotransmitter defects cause physical and mental deterioration

CHROMOSOME PAIRS

▲ **FIGURE 27.2   A human chromosome map.** Regions on each chromosome that have been identified as responsible for inherited diseases are indicated.

The replacement of one nucleotide by another in the same location along the DNA sequence is known as a **single-nucleotide polymorphism** (**SNP**, pronounced "snip"). In other words, two different nucleotides at the same position along two defined stretches of DNA are SNPs. A SNP is expected to occur in at least 1% of a specific population and therefore provides a link to a characteristic of that population.

Depending on their location as well as their nature, the biological effects of SNPs range from negligible, to normal variations such as those in eye or hair color, to genetic diseases. *SNPs are the most common source of variations between individual human beings.* Most genes carry one or more SNPs, and in different individuals most SNPs occur in the same location.

Imagine that the sequence A-T-G on the informational strand of DNA is miscopied to give messenger RNA with the codon sequence A-C-G rather than the correct sequence A-U-G. Because A-C-G codes for threonine, whereas A-U-G codes for methionine, threonine will be inserted into the corresponding protein during translation. Furthermore, every copy of the protein will have the same variation. The outcome will depend on the function of the protein and the effect of the amino acid change on its structure and reactivity.

In addition to producing a change in the identity of an amino acid, a SNP might specify the same amino acid (for example, changing GUU to GUC, both of which code for valine) or it might terminate protein synthesis by introducing a stop codon (for example, changing CGA to UGA).

Concurrent with the work of the Human Genome Project, an international team of industrial and academic scientists is compiling a catalog of SNPs. As of early 2001, the exact locations in the human genome of 1.4 million SNPs had

Single-nucleotide polymorphism (SNP) Common single-base-pair variation in DNA.

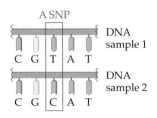

A SNP

DNA sample 1

C G T A T

DNA sample 2

C G C A T

been recorded. Their frequency is roughly one SNP for about every 2000 bases, with 60,000 of them in coding regions. These are considered but a fraction of the total that will eventually be identified.

We have described the single amino acid change that results in sickle-cell anemia (∞, p. 516). It took years of research to identify the SNP responsible for that disease. With a computerized catalog of SNPs it might have been found in a few hours. Another known SNP is associated with the risk of developing Alzheimer's disease. But not all SNPs create susceptibility to diseases. There is also one that imparts a resistance to HIV and AIDS.

The SNP catalog, while far from complete, has been of value from the start. Early in its development it was used to locate SNPs responsible for 30 abnormal conditions, including total color blindness, one type of epilepsy, and susceptibility to the development of breast cancer.

An investigator studying the relationship between testosterone and prostate cancer used the catalog to find 15 SNPs in a gene that affects testosterone levels. Examination of the DNA from prostate cancer patients showed that these SNPs occurred in four combinations in these people. The next step is to hunt down the role of each of those four genetic variations in the disease. Perhaps that information will inspire the development of a new drug.

The search for disease-related SNPs is not easy, although it is expected to become so as the SNP catalog expands and the related proteins are understood. Consider that one research group set out to identify SNPs associated with diabetes. Eventually they sequenced a 66,000 nucleotide stretch of DNA in an area implicated in diabetes. It contained 180 SNPs. Comparing these sequences for 100 diabetics and 100 nondiabetic controls turned up a combination of three SNPs that signaled a susceptibility to diabetes.

The cataloging of SNPs is expected to usher in the era of genetic medicine. It will be possible to predict for an individual the potential age at which inherited diseases will become active, their severity, and their reactions to various types of treatment. The therapeutic course will be designed to meet the distinctive genomic profile of the person.

■ WORKED EXAMPLE 27.1

The result of a mutation that changes a single amino acid in a DNA sequence depends on the type of amino acid replaced and the nature of the new amino acid. (a) What kind of change would have little effect on the protein containing the alternative amino acid? (b) What kind of change could have a major effect on the protein containing the alternative amino acid? Give an example of each type of mutation.

ANALYSIS The result of exchanging one amino acid for another depends on the change in the nature of the amino acid side chains. To speculate on the result of such a change requires considering the structure of the side chains, which are all shown in Table 18.3 The question to consider is whether the mutation introduces an amino acid with such a different side chain that it is likely to alter the function of the resulting protein.

SOLUTION (a) Exchange of an amino acid with a small nonpolar side chain for another with the same type of side chain (for example, glycine for alanine) or exchange of amino acids with very similar side chains (for example, serine for threonine) might have little effect.
(b) Conversion of an amino acid with a nonpolar side chain to one with a polar, acidic, or basic side chain could have a major effect because of a possible change in the side chain interactions that affect protein folding (see Figure 18.4). Some examples of this type include exchanging threonine, glutamate, or lysine for isoleucine.

■ KEY CONCEPT PROBLEM 27.2

Consider that a SNP alters the base sequence in an mRNA codon by changing UGU to UGG. Speculate on the significance of this change.

Acrime scene does not always yield fingerprints. It may, instead, yield samples of blood or semen or bits of hair. DNA analysis of such samples provides a new kind of "fingerprinting" for identifying criminals or proving suspects innocent.

DNA fingerprinting relies on finding variations in the DNA isolated from a crime scene sample and exam-

ining the match of these variations to those of a suspect or victim. The naturally occurring variability of the base sequence in DNA is like a fingerprint. It is the same in all cells from a given individual and is sufficiently different from that of other individuals that it can be used for identification.

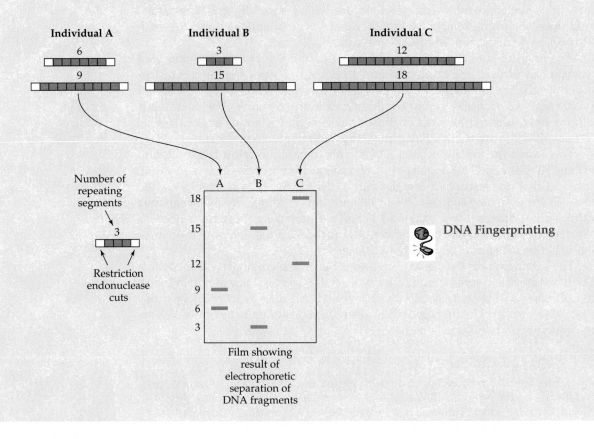

Film showing result of electrophoretic separation of DNA fragments

DNA Fingerprinting

27.4 Recombinant DNA

In this section, we describe a technique that makes it possible to manipulate, alter, and reproduce pieces of DNA. The technique requires the creation of **recombinant DNA**—DNA that contains two or more DNA segments not found together in nature. Progress in all aspects of genomics has built upon information gained in the application of recombinant DNA. The two other techniques that play major roles in DNA studies are the polymerase chain reaction (PCR) and electrophoresis. PCR allows for the synthesis of large quantities of identical pieces of DNA (∞, Application, p. 762). Electrophoresis, which can be carried out simultaneously on large numbers of samples, separates proteins or DNA fragments according to their size (∞, Application p. 523).

Using recombinant DNA technology, it is possible to cut a gene out of one organism and recombine it into the genetic machinery of a second organism. Bacteria provide excellent hosts for recombinant DNA. Bacterial cells, unlike the cells of higher organisms, contain part of their DNA in circular pieces called *plasmids*, each of which carries just a few genes. Plasmids are extremely easy to isolate, several copies of each plasmid may be present in a cell, and each plasmid replicates through the normal base-pairing pathway. The ease of isolating

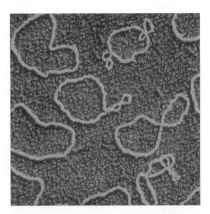

▲ Plasmids from *Escherichia coli* bacteria, hosts for recombinant DNA.

▲ Examination of an autoradiogram.

Throughout a DNA sample there are regions of non-coding DNA that contain repeating nucleotide sequences. The repetitive patterns used in DNA fingerprinting are known as *variable number tandem repeats (VNTRS)*. They are from 15 to 100 base pairs long, lie within or between genes, and consist of repetitive base pair patterns. For statistical significance, the known VNTRS on several genes are examined to create a DNA fingerprint. The probability of a fingerprint match with someone other than the correct individual is estimated at 1 in 1.5 billion.

Fingerprinting relies on a restriction endonuclease that cuts the DNA at either end of the repeating sequence. The procedure, in general, is as follows:

- Digest the DNA sample with the restriction endonuclease.
- Separate the resulting DNA fragments according to their size by gel electrophoresis. (∞, p. 523)
- Transfer the fragments to a nylon membrane (a *blotting* technique).
- Treat the blot with a radioactive DNA probe complementary to the sticky ends of the DNA fragments, so that the probe binds to its matching DNA fragments.
- Identify the locations of the now-radioactive fragments by exposing an X-ray film to the blot. The film result of this procedure is known as an *autoradiogram*.

An autoradiogram resembles a bar code, with dark bands arrayed in the order of the increasing molecular size of the DNA fragments. To compare the DNA of different individuals, the DNA samples are run in parallel columns on the same electrophoresis gel. In this way, the comparison is validated by having been run under identical conditions. The accompanying diagram illustrates what might be the DNA fingerprint patterns of three unrelated individuals with repeating segments of varying length on the same pair of chromosomes.

Only identical twins have identical DNA fingerprints. Offspring have patterns with some similarities to those of their parents, making such fingerprints valuable in proving or disproving paternity.

See Additional Problems 27.50–27.51 at the end of the chapter.

---

and manipulating plasmids plus the rapid replication of bacteria creates ideal conditions for production of recombinant DNA and the proteins whose synthesis it directs in bacteria.

To prepare a plasmid for insertion of a foreign gene, the plasmid is cut open with a a bacterial enzyme that cleaves a DNA molecule between the same two nucleotides wherever that pair of nucleotides appears together. For example, the restriction endonuclease *Eco*RI cuts a plasmid between G and A in the sequence G-A-A-T-T-C. This restriction enzyme makes its cut at the same spot in the sequence of both strands of the double-stranded DNA read in the same 5' to 3' direction. As a result, the cut is offset so that both DNA strands are left with a few unpaired bases on each end. These groups of unpaired bases are known as *sticky ends* because they are available to match up with complementary base sequences.

Recombinant DNA DNA that contains two or more DNA segments not found together in nature.

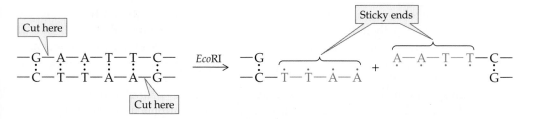

Recombinant DNA Technology

Recombinant DNA is produced by cutting the two DNA segments to be combined with the same restriction endonuclease. The result is DNA fragments with sticky ends that are complementary to each other.

Consider a gene fragment that has been cut from human DNA and is to be inserted into a plasmid. The gene and the plasmid are both cut with the same enzyme, one that produces sticky ends. The sticky ends on the gene fragment are complementary to the sticky ends on the opened plasmid. The two are mixed in the presence of a DNA ligase enzyme that joins them together by re-forming their phosphodiester bonds and reconstitutes the now-altered plasmid.

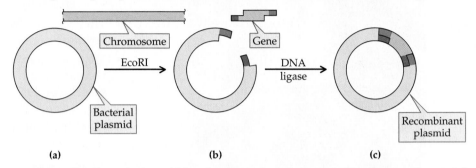

Once the altered plasmid has been made, it is inserted back into a bacterial cell where the normal processes of transcription and translation take place to synthesize the protein encoded by the inserted gene. Since bacteria multiply rapidly, there are soon a large number of them, all containing the recombinant DNA and all manufacturing the protein encoded by the recombinant DNA. Huge numbers of the bacteria can be put to work as a protein factory.

Proteins manufactured in this manner have already reached the marketplace, and many more are on the way. Human insulin was the first such protein to become available. Others now include human growth hormone used for children who would otherwise be abnormally small, and blood clotting factors for hemophiliacs. A major advantage of these proteins is that they cannot carry diseases from the host organisms that were their previous sources, human pituitary glands for growth hormone and human blood for blood clotting factors.

■ **PROBLEM 27.3**

A restriction enzyme known as BgIII cuts DNA in the place marked below.

5′ -A-//-G-A-T-C-T- 3′

Draw the complementary 3′ to 5′ strand and show where it is cut by the same enzyme.

■ **PROBLEM 27.4**

Are the following base sequences "sticky" or are they not?
(a) A-C-G-G-A and T-G-C-C-T
(b) G-T-G-A-C and -C-A-T-G-G
(c) G-T-A-T-A and -A-C-G-C-G

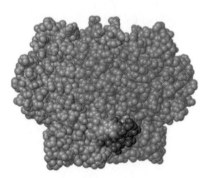

▲ *Eco*RI restriction endonuclease (purple and yellow) bound to a DNA fragment (blue and green strands).

Genomics The study of whole sets of genes and their functions.

27.5 Genomics: Using What We Know

Genomics has a simple and straightforward definition. It is the study of whole sets of genes and their functions. Genomics is inspiring studies that reach into all aspects of plant and animal life. Bacterial genomics will lead to better understanding of how bacteria cause disease and the development of new therapies. Plant genomics will enhance the value and utility of agricultural crops. The genomic study of farm animals will improve their health and availability. Humans will benefit from all of these studies, as well as those that contribute to their own health.

To provide a glimpse of where genomics is headed, we have provided descriptions of some of its applications in Table 27.3. These descriptions are not quite definitions. Many of the fields of study are so new that their territory is viewed differently by different individuals. At the opening of this chapter we noted that we stand at the beginning of a revolution. You may well encounter some of the endeavors listed in Table 27.3 as the revolution proceeds.

TABLE 27.3   **Genomics-Related Fields of Study**

Biotechnology
A collective term for the application of biological and biochemical research to the development of products that improve the health of humans, other animals, and plants.

Bioinformatics
The use of computers to manage and interpret genomic information, and to make predictions about biological systems. Applications of bioinformatics include studies of individual genes and their functions, drug design, and drug development.

Functional genomics
Use of genome sequences to solve biological problems.

Comparative genomics
Comparison of the genome sequences of different organisms to discover regions with similar functions and perhaps similar evolutionary origins.

Proteomics
Study of the complete set of proteins coded for by a genome or synthesized within a given type of cell, including the quest for an understanding of the role of each protein in healthy or diseased conditions. This understanding has potential application in drug design and is being pursued by more than one commercial organization.

Pharmacogenomics
The genetic basis of responses to drug treatment. Goals include the design of more effective drugs and an understanding of why certain drugs work in some patients but not in others.

Pharmacogenetics
The matching of drugs to individuals based on the content of their personal genome in order to avoid administration of drugs that are ineffective or toxic and focus on drugs that are most effective for that individual.

Toxicogenomics
A newly developing application that combines genomics and bioinformatics in studying how toxic agents affect genes and in screening possibly harmful agents.

Genetic engineering
Alteration of the genetic material of a cell or an organism The goals may be to make the organism produce new substances or perform new functions. Examples are introduction of a gene that causes bacteria to produce a desired protein or allows a crop plant to withstand the effects of a pesticide that will repel harmful insects.

Gene therapy
Alteration of an individual's genetic makeup with the goal of curing or preventing a disease.

Bioethics
The ethical implications of how knowledge of the human genome is used.

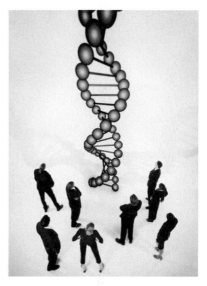

Genetically Modified Plants and Animals

The development of new varieties of plants and animals has been proceeding for centuries as the result of natural accidents and occasional success in the hybridization of known varieties. The techniques for mapping, studying, and modifying human genes apply equally as well to plants and animals. Their application can greatly accelerate our ability to generate crop plants and farm animals with desirable characteristics.

Some genetically modified crops have already been planted in large quantities in the United States. Each year millions of tons of corn are destroyed by a caterpillar (the European corn borer) that does its damage deep inside the corn stalk and out of reach of pesticides. To solve this problem, a bacterial gene (from *Bacillus thuringiensis*, Bt) has been transplanted into corn. The gene causes the corn to produce a toxin that kills the caterpillars. In 2000, one-quarter of all corn planted in the United States was Bt corn. Soybeans genetically modified to withstand herbicides are also widely grown. The soybean crop remains unharmed when the surrounding weeds are killed by the herbicide.

▲ The yellow rice has been genetically modified to provide vitamin A.

Tests are underway with genetically modified coffee beans that are caffeine-free, potatoes that absorb less fat when they are fried, and a yellow rice that provides the vitamin A desperately needed in poor populations where insufficient vitamin A causes death and blindness.

Fish farming is an expanding industry as natural populations of fish diminish. There are genetically engineered salmon that can grow to 7–10 pounds, a marketable size, in up to one-half the time of their unmodified cousins. Similar genetic modifications are anticipated for other varieties of fish. And there is the prospect of cloning leaner pigs.

Will genetically modified plants and animals intermingle with natural varieties and cause harm to them? Should food labels state whether the food contains genetically modified ingredients? Might unrecognized harmful substances enter the food supply? These are vigorously debated questions.

Gene Therapy

Gene therapy is based on the premise that replacement of a disease-causing gene with the healthy gene will cure or prevent a disease. The most clear-cut expectations for gene therapy lie in treating monogenic diseases, those that result from flawed DNA in a single gene.

The focus has been on using viruses as *vectors*, the agents that deliver therapeutic quantities of DNA directly into cell nuclei. The expectation was that this method could result in lifelong elimination of an inherited disease, and many studies have been undertaken. Unfortunately, expectations remain greater than achievements thus far. Investigations into the direct injection of naked DNA have begun, with one early report of success in encouraging blood vessel growth in patients with inadequate blood supply to their hearts.

A Personal Genomic Survey

Suppose that during the examination prior to diagnosis and treatment for a health problem a patient's entire genome could be surveyed. What benefits might result? One possibility is that the choice of drugs could be directed toward those that will be most effective for that individual. It is no secret that not everyone reacts in the same manner to a given medication. Perhaps the patient lacks an enzyme needed for a drug's metabolism. It is known, for example, that codeine is ineffective as a pain killer in those who lack the enzyme that converts codeine into morphine, which is the active analgesic (∞, p. 434). Perhaps the patient has a *monogenic defect*, a flaw in a single gene that is the direct cause of the disease. Such a patient might, at some time in the future, be a candidate for gene therapy.

In cancer therapy there may be advantages in understanding the genetic differences between normal cells and tumor cells. Such knowledge could assist in chemotherapy, where the goal is use of an agent that kills the tumor cells but does the least possible amount of harm to noncancerous cells.

Another application of human genetic information could arise from genetic screening of infants. The immediate use of gene therapy might eliminate the threat of a monogenically based disease. Or perhaps a lifestyle adjustment might be in order for an individual with one or more SNPs that predict a susceptibility to heart disease, diabetes, or some other disease that results from combinations of genetic and environmental influences. And consider that once done, an individual's genetic map would be available for the rest of their life. Perhaps someday we may even carry a wallet card encoded with our genetic information.

Snips and Chips

Our understanding of SNPs is already at work in screening implemented by DNA chips. A chip can be used, for example, to screen for the polymorphism that wipes out the analgesic effect of codeine. Or consider that a gene with several polymorphisms codes for an enzyme that metabolizes cardiovascular agents, antipsychotics, and many other drugs. Genomic screening can determine whether an

Application Tomatoes for the Desert

Does sufficient irrigation make it possible for agriculture to thrive in a desert? It often can. But after several seasons of watering, the land can become infertile if the water contains even a low concentration of sodium chloride. With each watering, the salt remains in the soil as the water evaporates. When the salt concentration gets high enough, traditional crop plants are stunted in their growth or die.

One-fourth or more of all irrigated lands worldwide are estimated to suffer from salt buildup. Obviously a cash crop that thrives in salt-laden soil would be a benefit to those struggling to farm such lands. Genetic engineering has opened the door to creation of a candidate crop of this type. In a sort of gene therapy for plants, a salt-tolerant tomato has been created.

Working together, scientists at the University of California at Davis (Eduardo Blumwald) and the University of Toronto (Hong-Xia Zhang) identified a single gene in the weed, *Arabidopsis thaliana*, that makes that weed salt-tolerant. (*Arabidopsis thaliana* is one of the organisms whose complete genome has been mapped.) The gene codes for a protein that transports sodium ions into sacks (*vacuoles*) inside plant cells. This isolation of the ions prevents them from damaging the cells and killing the plant.

Inserting this gene into tomato plants has yielded a tomato that can grow with its roots in a water solution (that is, hydroponically) with sodium chloride concentrations as high as 200 millimolar. A salt concentration of 50 millimolar is usually high enough to begin killing off crop plants. (For comparison, seawater has a salt concentration of 530 millimolar.) The next step will be to explore how this tomato grows in soil rather than water.

The plant accumulates the salt in sacs within its leaves, not in the fruit, which is reported to taste like any good fresh tomato. Perhaps growing such a plant for several seasons will remove enough salt from the soil to allow other crops to grow there.

See Additional Problem 27.52 at the end of the chapter.

---

individual with polymorphisms in that gene will get no effect from a drug, the expected effect, or perhaps have a greater-than-normal response to the drug. Eventually, screening tests for polymorphisms of this enzyme will be a diagnostic test carried out by a DNA chip in a doctor's office. The results will aid in choosing the right drug and dosage.

DNA chip screening has already revealed the genetic variations responsible for two types of pediatric leukemia, a distinction that could not be made by examining diseased cells under the microscope. Because the two leukemias require quite different therapies, use of the chip to identify the types is a valuable development.

A DNA chip is a solid support bearing large numbers of short, single-stranded bits of DNA of known composition. The DNA is organized on the chip in whatever manner is best for a particular type of screening, for example, to identify the presence or absence of polymorphisms. A sample to be screened is labeled with a fluorescent tag and applied to the chip. During an incubation period, the sample will bond to the DNA segments with complementary nucleic acid sequences. Then, the fluorescence is read to discover where the bonding has occurred and what DNA variations are present.

Bioethics

We can mention only briefly an area of major concern that arises from the revolution in genomics. It is not chemical, nor is it directly related to curing and preventing disease. The existence of this concern is recognized in the ELSI program of the National Human Genome Research Institute. ELSI deals with the Ethical, Legal, and Social Implications of human genetic research. The scope of ELSI is broad and thought-provoking. It deals with many questions such as the following

- Who should have access to personal genetic information and how will it be used?
- Who should own and control genetic information?
- Should genetic testing be performed when no treatment is available?
- Should we re-engineer the genes we pass on to our children?

[If you are interested in the ELSI program, their web page is an excellent resource (http://www.ornl.gov/hgmis/elsi/elsi.html).]

"I take my job very, very seriously. In forensic science, your analysis could contribute to someone being put to death or spending the rest of their life in jail. I try to be as careful and thorough as possible," says Amy Price.

For the past five years, Amy has worked as a forensic scientist in a state forensic laboratory. "Because of the popularity of certain television shows, a lot of people have misconceptions about what it really means to be a forensic scientist. The work that a crime scene investigator does on a TV show is actually done by about ten different people. For example, I don't question suspects or arrest anybody. I don't go to the crime scene or collect the evidence. The evidence is submitted to me in the lab for analysis."

Amy works in the Forensic Biology section, and her job is to identify body fluids that are found as evidence in criminal cases. Her job starts when she receives a piece of evidence such as a knife, a swab from a body, a piece of clothing, or even a door or a car. Amy visually examines the evidence to look for suspected body fluids. She can also use an "alternate light source," which is an ultraviolet light, sort of like a black light, that emits light at a wavelength of 450 nanometers. Almost every body fluid, except blood, will fluoresce and glow under this light.

She then uses balances and pH meters to prepare reagents for "presumptive" chemical tests. ("Presumptive" tests are screening tests used to narrow down the many thousands of possibilities so that more specific "confirmatory" tests can be used later to confirm a substance.) With these reagents, specific chemical reactions occur in the presence of blood or seminal fluid to produce a color change. For example, one commonly used presumptive test for blood uses a solution of phenolphthalein and hydrogen peroxide on a piece of filter paper; when blood is present, the sample turns pink. If these tests reveal the presence of body fluids, then Amy can proceed to the next step—DNA analysis.

To start the DNA analysis, Amy uses a number of different chemical solvents, reagents, and probes to extract DNA from the sample, precipitate it, and quantitate it. She then uses PCR (polymerase chain reaction, see p. 762), gel electrophoresis (see p. 523), fluorescent labels, lasers, and computers to get a DNA "profile" or "type" (see "DNA Fingerprinting," p. 778). With the DNA analysis technology currently used in Amy's lab, the probability of finding two individuals with the same DNA type is less than one in 200 quadrillion (the population of the world is approximate 6 billion).

Amy describes the various components of her analysis clearly, simply, and confidently, thus demonstrating another important aspect of her job. Forensic scientists must be able to explain and defend their procedures and results to a jury. "I've testified in court a

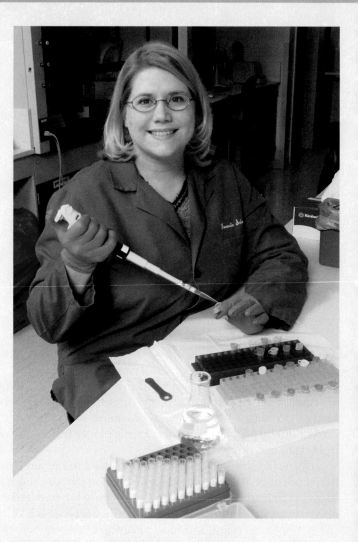

dozen times in the last three and a half years," reports Amy. "Recently there's been a decline in the number of times I've needed to testify, because DNA analysis is becoming more accepted and understood by juries and the general public. They've seen it on TV and heard about it on the news, so it's not as mysterious as in the past. The attorneys and the judge don't feel it needs to be explained in as much detail."

How did she get interested in forensic science? "I always knew I really liked science. But in high school, I thought that to be a scientist meant I'd have to do research. That didn't appeal to me. Did I really want to spend 40 years of my life investigating something and perhaps never finding an answer? Then when my high school chemistry class took a field trip to the Maryland State Crime Lab, I got hooked. It seemed like every single case was a mini-research project and once you finished working the case, you got the answer."

Amy chose a college where she could get a degree in forensic science. "Biology came really easy for me, and chemistry was a lot harder for me in high school. I knew I needed to become stronger in chemistry, so I decided to major in forensic chemistry and toxicology, with a minor in biology. I planned to become an expert in drug chemistry, but when I got a chance to learn about DNA analysis on an internship after my sophomore year, I switched my area of interest."

In her job, Amy interacts with a wide range of people, from police officers and detectives to attorneys for the prosecution and defense. She also works with the medical examiner and with many of the other experts in her forensic lab. Other forensic scientists in her lab specialize in areas such as toxicology, latent fingerprints, firearms, controlled substances, bloodstain pattern analysis, and questioned documents. Some forensic specialties require extensive training in chemistry; some require less. All forensic scientists, however, find themselves using chemistry on their job.

When she discusses forensic science with students who are considering going into the field, Amy stresses that forensic science involves many different disciplines. "Very few people do it all," she points out. "If you think you might want to work in forensic science, try to do an internship in a lab in the specific field that most interests you."

What does she like most about her job?

"In my job, I'm posed a question by an investigator or an attorney, such as 'Is this the suspect's blood or is it not?' When I can find something that can help answer that question—one single hair or a tiny stain that no one expected to find—that's the most exhilarating part of my job. It's very satisfying, because I know that if I weren't there, that information might never have been brought to light—either to prosecute or exonerate."

■ **PROBLEM 27.5**

Classify each of the following activities according to the fields of study listed in Table 27.3.

(a) Identification of genes that perform identical functions in mice and humans.

(b) Creation of a variety of wheat that will not be harmed by an herbicide that kills weeds that threaten wheat crops.

(c) Screening of individual's genome to choose the most appropriate pain-killing medication for that person.

(d) Computer analysis of base sequence information from groups of people with and without a given disease to discover where the disease-causing polymorphism lies.

Summary: Revisiting the Chapter Goals

1. **What is the working draft of the human genome and the circumstances of its creation?** In June, 2000, the Human Genome Project, an international consortium of not-for-profit institutions, and Celera Genomics, a for-profit company, jointly announced completion of working drafts of the human genome. With the exception of large areas of repetitive DNA, the DNA base sequences of all chromosomes had been examined. The Human Genome Project utilized successive maps of DNA segments of known location. Celera began by fragmenting all of the DNA. In both groups the fragments were cloned, labeled, ordered, and the individual maps assembled by computers. The results of the two projects are generally supportive of each other. There are about 3 billion base pairs and 30,000–40,000 genes, each able to direct the synthesis of more than one protein. Roughly 50% of the genome consists of noncoding, repetitive sequences. About 200 of the human genes are identical to those in bacteria.

2. **What are the various segments along the length of the DNA in a chromosome?** Telomeres, which fall at the ends of chromosomes, are regions of noncoding, repetitive DNA that protect the ends from accidental changes. At each cell division, the telomeres are shortened, with significant shortening associated with senescence and death of the cell. In cancer cells, telomerase

Key Words

Clones, p. 771

Centromeres, p. 773

Genomics, p. 780

Mutagen, p. 775

Mutation, p. 775

Polymorphism, p. 776

Recombinant DNA, p. 779

Single-nucleotide-polymorphism (SNP), p. 776

Telomeres, p. 773

remains active in telomere synthesis. Centromeres are the constricted regions of chromosomes that form during cell division and also carry noncoding DNA. Exons are the protein coding regions of DNA and together make up the genes that direct protein synthesis. The repetitive noncoding segments of DNA are of either no function or unknown function.

3. **What are mutations?** A *mutation* is an error in the base sequence of DNA that will be passed along during replication. Mutations arise by random error during replication or may be caused by ionizing radiation, viruses, or chemical agents (*mutagens*). Mutations cause inherited diseases and the tendency to acquire others.

4. **What are polymorphisms and single nucleotide polymorphisms (SNPs) and how can identifying them be useful?** A polymorphism is a variation in DNA that is linked to a trait within a population. A SNP is the replacement of one nucleotide by another. The result might be the replacement of one amino acid by another in a protein, no change because the new codon specifies the same amino acid, or the introduction of a stop codon. Many inherited diseases are known to be caused by SNPs; they can also be beneficial. Understanding the location and effect of SNPs is expected to lead to new therapies.

5. **What is recombinant DNA?** Recombinant DNA is produced by combining DNA segments that do not normally occur together. A gene from one organism is inserted into the DNA of another organism. Recombinant DNA is conveniently created in plasmids (circular DNA) from bacteria. Bacteria carrying these plasmids then serve as factories for the synthesis of large quantities of the encoded protein.

6. **What does the future hold for use of genomic information?** Mapping the human genome holds major promise for applications in health and medicine. Drugs can be precisely chosen based on a patient's own DNA, thereby avoiding drugs that are ineffective or toxic for that individual. Perhaps one day inherited diseases will be prevented or cured by gene therapy. By genetic modification of crop plants and farm animals, the productivity, marketability, and health benefits of these products can be enhanced. Progress in each of these areas is bound to be accompanied by controversy and ethical dilemmas.

▪ Understanding Key Concepts

27.6 What information is contained in the working draft of the human genome? What areas will be refined as work on the project continues?

27.7 Clearly, all humans have variations in their DNA sequences. How is it possible to sequence the human genome if every individual is unique?

27.8 List the types of noncoding DNA. Give the function of each, if it is known.

27.9 What are the similarities and differences between mutations and polymorphisms?

27.10 What is recombinant DNA? How it can be used to produce human proteins in bacteria?

27.11 Identify some major potential benefits of the applications of genomics and some major negative outcomes

---

▪ Additional Problems

The Human Genome Map

27.12 Identify the group of nonprofit organizations and the private corporation that combined to announce working drafts of the human genome.

27.13 How did the competition that developed between the groups developing the human genome map benefit the Human Genome Project?

27.14 Approximately what portion of the human genome is composed of repeat sequences?

27.15 Approximately how many base pairs were identified in the human genome working drafts?

27.16 Among results of the genome working drafts, (a) were any human genes found to be identical to genes in bacteria and (b) what was learned about the number of proteins produced by a given gene?

27.17 What is the most surprising result found thus far in the human genome studies?

27.18 You may have heard of Dolly, the cloned sheep that grew from an embryo created in a laboratory. In the context of DNA mapping, what are clones and what essential role do they play?

Chromosomes, Mutations, and Polymorphisms

27.19 What is thought to be the primary purpose of telomeres?

27.20 How is the age of a cell predicted by its telomeric sequences?

27.21 What is the role of the enzyme telomerase? In what cell type of cell is it normally most active and most inactive?

27.22 What is the centromere?

27.23 What is a mutagen?

27.24 Identify three types of mutagens.

27.25 Why is a mutation of a base in a DNA sequence much more serious than a mutation in a transcribed mRNA sequence?

27.26 What are the two general and common ways that mutations can occur in a DNA sequence?

27.27 What is a SNP?

27.28 How are SNPs linked to traits in individual human beings?

27.29 List some potential biological effects of SNPs.

27.30 What would be a medical advantage of having a catalog of SNPs?

27.31 Does a single base pair substitution in a strand of DNA necessarily result in a new amino acid in the protein coded for by that gene?

27.32 What determines the significance of a change in the identity of an amino acid in a protein?

27.33 Compare the severity of DNA mutations that produce the following changes in mRNA codons:
  (a) UUU to UUC   **(b)** CUU to CGU

27.34 Compare the severity of DNA mutations that produce the following changes in mRNA codons:
  (a) GUC to GCC   **(b)** CCC to CAC

Recombinant DNA

27.35 Why are bacterial plasmids the preferred host for DNA sequences that are to be cloned?

27.36 Name two advantages of using recombinant DNA to make proteins such as insulin, human growth hormone, or blood clotting factors.

27.37 In the formation of recombinant DNA, a restriction endonuclease cuts a bacterial plasmid to give sticky ends. The DNA segments that are to be added to the plasmid are cleaved with the same restriction endonuclease. What are sticky ends and why is it important that the target DNA and the plasmid it will be incorporated in have complementary sticky ends?

27.38 In addition to recombinant DNA, what two other techniques discussed in earlier chapters play major roles in DNA studies?

27.39 Give the sequence of unpaired bases that would be sticky with the following sequences.
  (a) CGTAA   **(b)** TGACC   **(c)** AGTCT

27.40 Are the following base sequences sticky or not sticky?
  (a) TTAGC and AAACG   **(b)** CGTACG and GTACGT

Genomics

27.41 How might the work of a person who practices pharmacogenomics differ from that of a pharmacogeneticist?

27.42 What is proteomics and how might it benefit health care?

27.43 Genetic engineering and gene therapy are similar fields within genomics. What do they have in common and what distinguishes them?

27.44 Provide two examples of genetically engineered crops that are improvements over their predecessors.

27.45 Imagine that you become a parent in an age when a full genetic workup is available for every baby. What advantages and disadvantages might there be to having this information?

27.46 What type of technology might be used to diagnose inherited diseases in the doctor's office of the future?

27.47 What is a DNA chip?

Applications

27.48 Whose DNA was used by scientists working on the two projects that sequenced the human genome? [*Whose Genome Is It?*]

27.49 How were adjustments made in the standard genome for DNA variations among the DNA samples being used? [*Whose Genome Is It?*]

27.50 Briefly state the five basic steps of DNA fingerprinting. Why is it so important that the DNA fingerprinting techniques be standardized? [*DNA Fingerprinting*]

27.51 Why is it possible in DNA fingerprinting to compare segments of DNA from different tissues (for example, semen and blood samples)? [*DNA Fingerprinting*]

27.52 In what other directions might study of genetically engineered salt-tolerant plants proceed? [*Tomatoes for the Desert*]

General Questions and Problems

27.53 What is a monogenic disease?

27.54 What is the role of a vector in gene therapy?

27.55 If the DNA sequence A-T-T-G-G-C-C-T-A on an informational strand mutated and became A-C-T-G-G-C-C-T-A, what effect would the mutation have on the sequence of the protein produced?

27.56 Write the base sequence that would be sticky with the sequence T-A-T-G-A-C-T.

27.57 In general terms, what is the cause of hereditary diseases?

27.58 In the DNA of what kind of cell must a mutation occur for the genetic change to be passed down to future generations?

28

Protein and Amino Acid Metabolism

A large source of amino acids. In this chapter, you'll learn about the metabolic fate of these amino acids and their nitrogen atoms.

Before we discuss the topics at hand in this chapter, a review of what has already been covered on proteins and amino acids is in order. In general, for each major class of biomolecules, our goal has been to describe their structure, their sources in the diet or from biochemical synthesis, their function, and their fate via the reactions of catabolism. The structures of the amino acids and the various levels of protein structure were covered in Chapter 18. The hydrolysis of dietary protein is our major source for amino acids. The essential function of proteins as enzymes was explored in Chapter 19, and their function within cell membranes was described in Section 24.8. In addition, the role of amino acids and proteins as hormones or precursors to neurotransmitter synthesis was examined in Chapter 20, and we covered the biosynthesis of proteins as directed by the genetic code in Chapter 26. The remaining major aspect of protein biochemistry, to which we now turn, is the metabolic fate of amino acids.

1. What happens during the digestion of proteins, and what is the destination of the amino acids?

The goal: Be able to list the sequence of events in the digestion of proteins, and describe the nature of the amino acid pool.

2. What are the major strategies in the catabolism of amino acids?

The goal: Be able to identify the major reactions and products of amino acid catabolism and the fate of the products.

3. What is the urea cycle?

The goal: Be able to list the major reactants and products of the urea cycle.

4. What are the essential and nonessential amino acids, and how, in general, are amino acids synthesized?

The goal: Be able to define essential and nonessential amino acids, and describe the general strategy of amino acid synthesis.

CONCEPTS TO REVIEW

Amino acids (Sections 18.3, 18.4)
Primary protein structure (Section 18.7)
Overview of metabolism (Section 21.4)

28.1 Digestion of Protein

The end result of protein digestion is simple—the hydrolysis of all peptide bonds to produce a collection of amino acids:

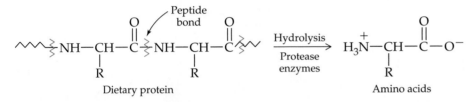

Dietary protein   Amino acids

The digestion of dietary proteins, as summarized in Figure 28.1, begins with their denaturation in the strongly acidic environment of the stomach (pH 1–2). In addition to hydrochloric acid, gastric secretions include pepsinogen, a zymogen that is activated by acid to give the enzyme pepsin. Unlike most proteins, pepsin is stable and active at pH 1–2. Protein digestion begins with hydrolysis by pepsin of some of the peptide bonds in the denatured proteins.

The polypeptides produced by pepsin then enter the small intestine, where the pH is about 7–8. Pepsin is inactivated in the less acidic environment, and a group of pancreatic zymogens is secreted. The activated enzymes (proteases) then take over further hydrolysis of peptide bonds in the partially digested proteins.

The combined action of the pancreatic proteases in the small intestine and other proteases in the cells of the intestinal lining frees the amino acids from dietary proteins. After active transport across cell membranes lining the intestine, the amino acids are absorbed directly into the bloodstream.

The active transport of amino acids into cells is managed by several transport systems devoted to different groups of amino acids. For this reason, an excess of one amino acid in the diet can dominate the transport and produce a deficiency of others. This condition usually arises only in individuals taking large quantities of a single amino acid dietary supplement, of the type often sold in health food stores.

28.2 Amino Acid Metabolism: An Overview

The entire collection of free amino acids throughout the body—the **amino acid pool**—occupies a central position in protein and amino acid metabolism (Figure 28.2, p. 790). All tissues and biomolecules in the body are constantly being degraded, repaired, and replaced. Cells throughout the body have the enzymes for hydrolysis of waste protein, and a healthy adult turns over about 300 g of protein every day. Thus, amino acids are continuously entering the pool, not only from digestion but also from breakdown of old protein, and are continuously being withdrawn for synthesis of new nitrogen-containing biomolecules.

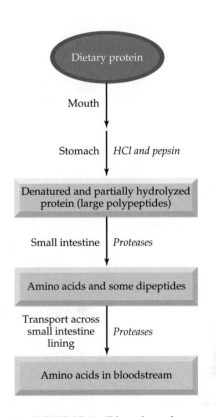

▲ **FIGURE 28.1 Digestion of proteins.**

Amino acid pool The entire collection of free amino acids in the body.

789

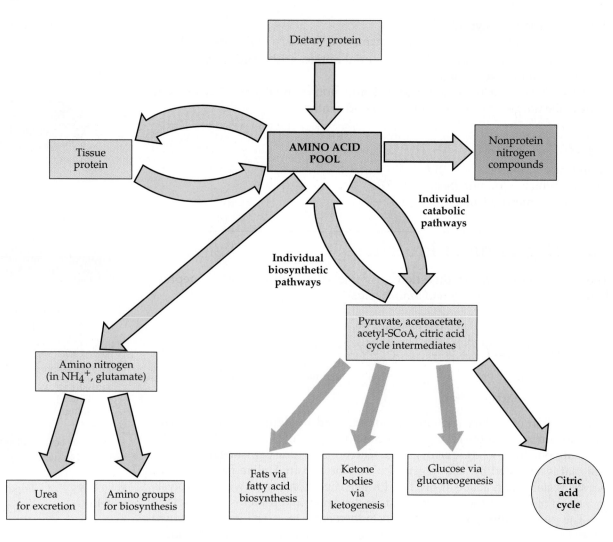

▲ **FIGURE 28.2   Protein and amino acid metabolism.**   Amino acids move in and out of the amino acid pool, as shown by blue arrows for catabolic pathways and yellow arrows for synthetic pathways. Connections to fat and glucose metabolism are shown by green arrows.

Each of the 20 protein amino acids is degraded via its own unique pathway. The important point to remember is that the general scheme is the same for each one.

Amino acid catabolism

- Removal of the amino group
- Use of nitrogen in synthesis of new nitrogen compounds
- Passage of nitrogen into the urea cycle
- Incorporation of the carbon atoms into compounds that can enter the citric acid cycle

Our bodies don't store nitrogen-containing compounds. Therefore, the amino nitrogen from dietary protein has just two possible fates. It may be used in the synthesis of new nitrogen-containing compounds, including the following types of biomolecules:

Some nitrogen-containing biomolecules synthesized from amino acids

Nitric oxide (NO, a chemical messenger, p. 428)

Hormones (Section 20.4)

Neurotransmitters (Section 20.6)

Nicotinamide (in NAD^+ and $NADP^+$; Section 21.7)

Heme (in red blood cells; p. 620)

Creatine phosphate (for quick energy in muscles; p. 651)

Purine and pyrimidine bases (for nucleic acids; Section 26.2)

Or the nitrogen must be incorporated into urea and excreted.

Once the carbon atoms of amino acids are converted to compounds that can enter the citric acid cycle, they are available for several alternative pathways. They can continue through the cycle in the body's main energy-generating pathway to give CO_2 and energy stored in ATP. About 10–20 % of our energy is normally produced in this way from amino acids. If not needed immediately for energy, the carbon-carrying intermediates produced from amino acids can enter storage as triacylglycerols (via lipogenesis, Section 25.8) or glycogen (via gluconeogenesis and glycogen synthesis, Sections 23.10, 23.11). They can also be converted to ketone bodies (Section 25.7).

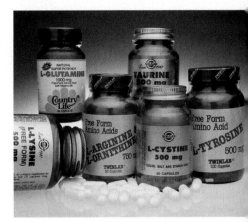

▲ Individual amino acids are promoted for a variety of unproven health benefits. Because amino acids are classified as foods, they need not undergo the testing for purity, safety, and efficacy required for FDA approval.

■ **PROBLEM 28.1**

Decide whether each of the following statements is true or false. If false, explain why.

(a) The amino acid pool is found mainly in the liver.

(b) Nitrogen-containing compounds can be stored in fatty tissue.

(c) Some hormones and neurotransmitters are synthesized from amino acids.

(d) Amino groups can be stored in fatty tissue.

(e) Glycine is an essential amino acid because it is present in every protein.

■ **KEY CONCEPT PROBLEM 28.2**

Serotonin is a monoamine neurotransmitter. It is formed in the body from the amino acid tryptophan (Figure 20.6, p 585)". What class of enzyme would catalyze each of the two steps that converts tryptophan to serotonin?

28.3 Amino Acid Catabolism: The Amino Group

The first step in amino acid catabolism is removal of the amino group. In this process, known as **transamination,** the amino group of the amino acid and the keto group of an α-keto acid change places:

Transamination The interchange of the amino group of an amino acid and the keto group of an α-keto acid.

$$R'-\underset{\underset{NH_3^+}{|}}{CH}-COO^- \ + \ R''-\overset{\overset{O}{\|}}{C}-COO^- \ \underset{}{\overset{\alpha\text{-Transaminase}}{\rightleftharpoons}} \ R'-\overset{\overset{O}{\|}}{C}-COO^- \ + \ R''-\underset{\underset{NH_3^+}{|}}{CH}-COO^-$$

Amino acid 1　　　　　α-Keto acid 1　　　　　　　　　α-Keto acid 2　　　　Amino acid 2

There are a number of transaminase enzymes. Most are specific for α-ketoglutarate as the amino group acceptor and work with several amino acids. The α-ketoglutarate is converted to glutamate, and the amino acid is converted to an α-keto acid:

$$R'-\underset{\underset{NH_3^+}{|}}{CH}-COO^- \ + \ ^-OOC-CH_2CH_2-\overset{\overset{O}{\|}}{C}-COO^- \ \rightleftharpoons \ R'-\overset{\overset{O}{\|}}{C}-COO^- \ + \ ^-OOC-CH_2CH_2\underset{\underset{NH_3^+}{|}}{CH}-COO^-$$

Amino acid 1　　　　　α-Ketoglutarate　　　　　　　　　　α-Keto acid 2　　　　Glutamate
　　　　　　　　　(amino group acceptor)　　　　　　　　　　　　　　　　　(amino acid 2)

For example, alanine is converted to pyruvate by transamination. The enzyme for this conversion, alanine aminotransferase (ALT), is especially abundant in the liver, and above-normal ALT concentrations in the blood are taken as an indication of liver damage that has allowed ALT to leak into the bloodstream.

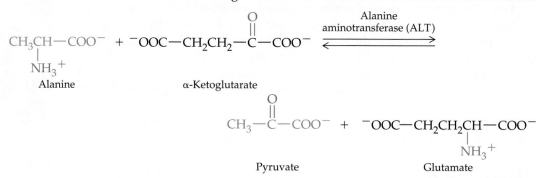

Transamination is a key reaction in many biochemical pathways, where it interconverts amino acid amino groups and carbonyl groups as necessary. The transamination reactions are reversible and go easily in either direction, depending on the concentrations of the reactants. In this way, amino acid concentrations are regulated by keeping synthesis and breakdown in balance. For example, the reaction of pyruvate with glutamate (the reverse of the preceding reaction) is the main synthetic route for alanine.

The glutamate from transamination serves as an amino group carrier. It provides amino groups for the synthesis of new amino acids, but most of it is recycled to fresh α-ketoglutarate. In the process, known as **oxidative deamination**, the glutamate amino group is removed as ammonium ion and replaced by a carbonyl group to give back α-ketoglutarate.

Oxidative deamination
Conversion of an amino acid —NH$_2$ group to an α-keto group, with removal of NH$_4^+$.

$$^-OOC-CH_2CH_2CH-COO^- \;+\; H_2O \;\xrightarrow[\substack{\text{Glutamate} \\ \text{dehydrogenase}}]{\substack{NAD^+(NADP^+)\;\;NADH(NADPH)}}\; NH_4^+ \;+\; ^-OOC-CH_2CH_2C-COO^-$$

Glutamate (with NH$_3^+$) α-Ketoglutarate

The ionized ammonia formed in this reaction proceeds to the urea cycle for conversion to urea that is eliminated in the urine. The pathway of nitrogen from an amino acid to urea is summarized in Figure 28.3.

■ WORKED EXAMPLE 28.1

The blood serum concentration of the transaminase from heart muscle, aspartate aminotransferase (AST), is used in the diagnosis of heart disease because the enzyme escapes into the serum from damaged heart cells. AST catalyzes transamination of aspartate with α-ketoglutarate. What are the products of this reaction?

SOLUTION   The reaction is the interchange of an amino group from aspartate with the keto group from α-ketoglutarate. We know that α-ketoglutarate always gives glutamate in transamination, so one product is glutamate. The product from the amino acid will have a keto group instead of the amino group. Consulting Table 18.3 shows that the structure of aspartate (aspartic acid) is

$$^-OOC-CH_2\overset{\alpha}{CH}-COO^-$$
$$| $$
$$NH_3^+$$

Aspartate

Removing the $-NH_3^+$ and $-H$ groups bonded to the α carbon and replacing them by a $=O$ gives the desired α-keto acid, oxaloacetate, the starting compound for the citric acid cycle:

$$^-OOC-CH_2-\overset{\overset{\displaystyle O}{\|}}{C}-COO^-$$
Oxaloacetate

The reaction is

$$\text{Aspartate} + \alpha\text{-Ketoglutarate} \longrightarrow \text{Oxaloacetate} + \text{Glutamate}$$

■ **PROBLEM 28.3**

What are the structure and name of the α-keto acid formed by transamination of the amino acid leucine? (Refer to Table 18.3.)

■ **PROBLEM 28.4**

What is the product of the following reaction?

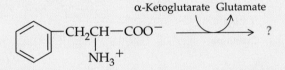

■ **PROBLEM 28.5**

Explain how the conversion of glutamate to α-ketoglutarate can be identified as an oxidation reaction.

■ **PROBLEM 28.6**

Unlike most amino acids, branched-chain amino acids are broken down in tissues other then the liver. Identify the three amino acids with branched-chain R groups (◁◁◁, Table 18.3, p. 507). For one of these amino acids, write the equation for its transamination.

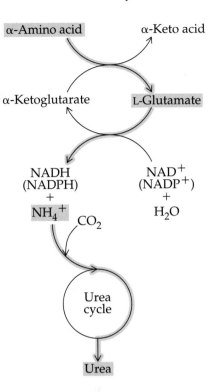

▲ **FIGURE 28.3 Pathway of nitrogen from an amino acid to urea.** The nitrogen-bearing compounds and their pathway are highlighted in yellow.

Urea cycle   The cyclic biochemical pathway that produces urea for excretion.

28.4 The Urea Cycle

Ammonia is highly toxic to living things and must be eliminated in a way that does no harm. Fish are able to excrete ammonia through their gills directly into their watery surroundings where it is immediately diluted, but mammals must first convert ammonia, in solution as ammonium ion, to nontoxic urea via the **urea cycle.** By so doing, they avoid elimination of ammonia in the urine, which could only be done safely if it were dissolved in such a large volume of water that it would cause dehydration.

The conversion of ammonium ion to urea takes place in the liver. From there, the urea is transported to the kidneys and transferred to urine for excretion. Like many other biochemical pathways, urea formation begins with an energy investment. Ammonium ion from oxidative deamination of amino acids, carbon dioxide from the citric acid cycle (in solution as bicarbonate ion), and ATP combine to form carbamoyl phosphate. This reaction, like the citric acid cycle, takes place in the mitochondrial matrix. Two ATPs are

▲ Fish don't need to convert ammonia to urea for elimination because it is quickly diluted in the surrounding water. Birds and reptiles eliminate nitrogen as solid uric acid.

invested and one phosphate is transferred to form the carbamoyl phosphate (an energy-rich phosphate like ATP):

$$NH_4^+ + HCO_3^- \xrightarrow[\substack{\text{Carbamoyl} \\ \text{phosphate synthetase I}}]{\text{2 ATP} \quad \text{2 ADP}} \overset{+}{H_3}N-\overset{\displaystyle \overset{O}{\|}}{C}-O-PO_3^{2-} + HOPO_3^{2-} + H_2O$$

Carbamoyl phosphate next reacts in the first step of the four-step urea cycle, shown in Figure 28.4.

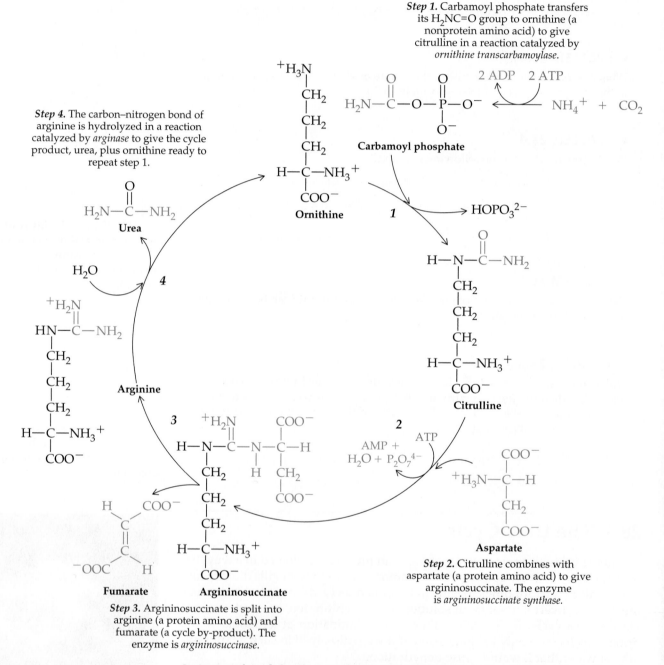

Step 1. Carbamoyl phosphate transfers its $H_2NC{=}O$ group to ornithine (a nonprotein amino acid) to give citrulline in a reaction catalyzed by *ornithine transcarbamoylase.*

Step 4. The carbon–nitrogen bond of arginine is hydrolyzed in a reaction catalyzed by *arginase* to give the cycle product, urea, plus ornithine ready to repeat step 1.

Step 2. Citrulline combines with aspartate (a protein amino acid) to give argininosuccinate. The enzyme is *argininosuccinate synthase.*

Step 3. Argininosuccinate is split into arginine (a protein amino acid) and fumarate (a cycle by-product). The enzyme is *argininosuccinase.*

▲ **FIGURE 28.4   The urea cycle.** The formation of carbamoyl phosphate and step 1, the formation of citrulline, take place in the mitochondrial matrix. Steps 2–4 take place in the cytosol.

Steps 1 and 2 of the Urea Cycle: Building Up a Reactive Intermediate The first step of the urea cycle transfers the carbamoyl group, $H_2NC{=}O$, from carbamoyl phosphate to ornithine, an amino acid not found in proteins, to give citrulline, another nonprotein amino acid. This exergonic reaction introduces the first nitrogen of the urea end product into the cycle.

Next, a molecule of aspartate, a standard protein amino acid, combines with citrulline in a reaction driven by conversion of ATP to AMP and pyrophos-

Application Pathways to Gout

As explained in an earlier applications, gout is a severely painful condition caused by the precipitation of sodium urate crystals in joints. (∞, p. 149) A small amount of our waste nitrogen is excreted in urine and feces as urate rather than urea. Because the urate salt is highly insoluble, any excess of the urate anion causes precipitation of sodium urate. The pain of gout results from a cascade of inflammatory responses to these crystals in the affected tissue.

That the symptoms of gout are caused by urate crystals has been known for a very long time. Understanding the many possible causes of the crystal formation is taking much longer and is far from complete. Looking at a few of the pathways to gout illustrates some of the many ways that the delicate balance of our biochemistry can be disrupted.

Uric acid is an end product of the breakdown of purine nucleosides, and loss of its acidic H (red) gives urate ion. For example, for adenosine,

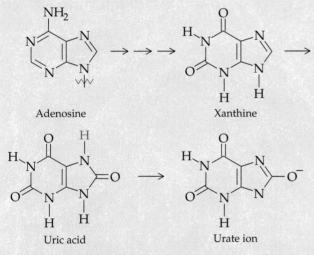

Adenosine → → → Xanthine →

Uric acid → Urate ion

Anything that increases the production of uric acid or inhibits its excretion in the urine is a possible cause of gout. For example, several known hereditary enzyme defects increase the quantity of purines and therefore of uric acid. Sometimes, gouty attacks follow injury or severe muscle exertion. Complicating matters is the observation that the presence of crystals in a joint is not always accompanied by inflammation and pain.

One significant cause of increased uric acid production is accelerated breakdown of ATP, ADP, or AMP. For example, alcohol abuse generates acetaldehyde that must be metabolized in the kidney by a pathway that requires ATP and produces excess AMP. Inherited fructose intolerance, glycogen storage diseases, and circulation of poorly oxygenated blood also accelerate uric acid production by this route. With low oxygen, ATP is not efficiently regenerated from ADP in mitochondria, leaving the ADP to be disposed of.

Conditions that diminish excretion of uric acid include kidney disease, dehydration, hypertension, lead poisoning, and competition for excretion from anions produced by ketoacidosis.

One treatment for gout relies on allopurinol, a structural analog of hypoxanthine, which is a precursor of xanthine in the formation of urate. Allopurinol inhibits the enzyme for conversion of hypoxanthine and xanthine to urate. Because hypoxanthine and xanthine are more soluble than sodium urate, they are more easily eliminated.

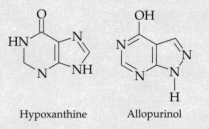

Hypoxanthine   Allopurinol

See Additional Problems 28.40–41 at the end of the chapter.

phate ($P_2O_7^{4-}$), followed by the additional exergonic hydrolysis of pyrophosphate. Both nitrogen atoms destined for urea are now bonded to the same carbon atom in argininosuccinate (red C in Figure 28.4).

Urea

Steps 3 and 4 of the Urea Cycle: Cleavage and Hydrolysis of the Step 2 Product
Step 3 cleaves argininosuccinate into two pieces: arginine, an amino acid, and fumarate. Now all that remains, in *step 4*, is hydrolysis of arginine to give urea and restore ornithine, the reactant in step 1 of the cycle.

Net Result of the Urea Cycle

$$HCO_3^- + NH_4^+ + 3\,ATP + {}^-OOC-CH_2-CH-COO^- + 2\,H_2O \longrightarrow$$
$$\underset{\text{Aspartate}}{\overset{|}{NH_3^+}}$$

$$\underset{\text{Urea}}{H_2N-\overset{\overset{\displaystyle O}{\|}}{C}-NH_2} + 2\,ADP + AMP + 4\,HOPO_3^{2-} + \underset{\text{Fumarate}}{{}^-OOC-CH=CH-COO^-}$$

- Elimination as urea of carbon from CO_2, nitrogen from NH_4^+, and nitrogen from the amino acid aspartate
- Breaking of four phosphate bonds to provide energy
- Production of the citric acid cycle intermediate, fumarate

Hereditary diseases associated with defects in the enzymes for each step in the urea cycle have been identified. The resulting abnormally high levels of ammonia in the blood (*hyperammonemia*) cause vomiting in infancy, lethargy, irregular muscle coordination (*ataxia*), and mental retardation. Immediate treatment consists of transfusions, blood dialysis (*hemodialysis*), and use of chemical agents to remove ammonia. Long-term treatment requires a low-protein diet and frequent small meals to avoid protein overload.

■ **KEY CONCEPT PROBLEM 28.7**

Fumarate from step 3 of the urea cycle may be recycled into aspartate for use in step 2 of the cycle. The sequence of reactions for this recycling is shown below. Which of the following classifications apply to each reaction?

(i) Oxidation (ii) Reduction (iii) Transamination (iv) Elimination (v) Addition

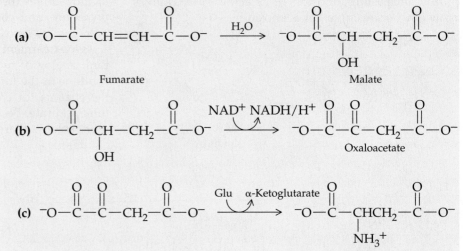

▼ **FIGURE 28.5 Fate of amino acid carbon atoms.** The carbon atoms of the amino acids are converted to the seven compounds shown in red and blue type in the figure, each of which is either an intermediate in the citric acid cycle or a precursor to citrate. The amino acids in the blue boxes are glucogenic—they can form glucose via the entry of oxaloacetate into gluconeogenesis. Those in the pink boxes are ketogenic—they are available for ketogenesis.

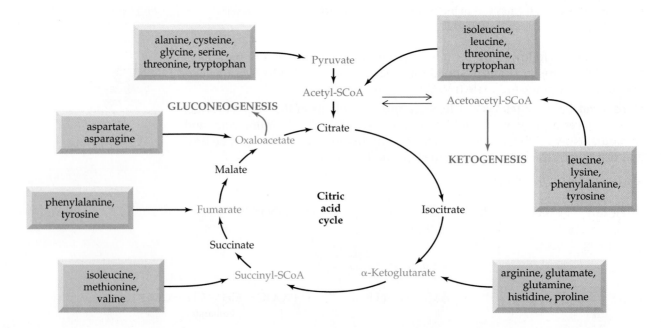

28.5 Amino Acid Catabolism: The Carbon Atoms

The carbon atoms of each protein amino acid arrive by distinctive pathways at pyruvate, acetyl-SCoA, or one of the citric acid cycle intermediates shown in blue type in Figure 28.5. Eventually, all of the amino acid carbon skeletons can be used to generate energy, either by passing through the citric acid cycle and into the gluconeogenesis pathway to form glucose or by entering the ketogenesis pathway to form ketone bodies.

Those amino acids that are converted to acetoacetyl-SCoA or acetyl-SCoA (pink boxes in Figure 28.5) can enter the ketogenesis pathway. They are therefore referred to as *ketogenic amino acids.* (⊙⊙⊙, Section 25.7)

Those amino acids that can proceed by way of oxaloacetate to the gluconeogenesis pathway (⊙⊙⊙, Section 23.11) are known as *glucogenic amino acids* (Table 28.1). Both ketogenic and glucogenic amino acids are able to enter fatty acid biosynthesis via acetyl-SCoA. (⊙⊙⊙, Section 25.8)

> ### ■ PROBLEM 28.8
> Arginine is converted to ornithine in the last step of the urea cycle (Figure 28.4). To carry arginine carbon atoms to the citric acid cycle, ornithine undergoes transamination at its terminal amino group to give an aldehyde, followed by oxidation to glutamate and conversion to α-ketoglutarate. Write the structures of the five molecules in the pathway beginning with arginine and ending with α-ketoglutarate. Circle the region of structural change in each.

28.6 Biosynthesis of Nonessential Amino Acids

Humans are able to synthesize about half of the 20 amino acids found in proteins. These are known as the **nonessential amino acids** because they don't have to be supplied by our diet. The remaining amino acids—the **essential amino acids** (Table 28.2)—are synthesized only by plants and microorganisms. Humans must obtain the essential amino acids from food (see Application "Proteins in the Diet," p. 519). Meats contain all of the essential amino acids. The foods that do not have all of them are described as having *incomplete amino acids.* Food combinations that together contain all of the amino acids are *complementary* sources of protein. It is interesting to note that we synthesize the nonessential amino acids in pathways with 1–3 steps, whereas synthesis of the essential amino acids by other organisms requires many more steps and a big energy investment. It seems that it's easier to let the others do it.

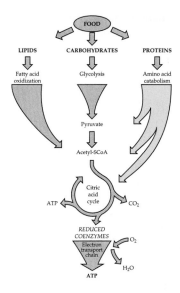

TABLE 28.1   Glucogenic and Ketogenic Amino Acids

Glucogenic

Alanine	Glycine
Arginine	Histidine
Aparagine	Methionine
Aspartate	Proline
Cysteine	Serine
Glutamate	Threonine
Glutamine	Valine

Both glucogenic and ketogenic
Isoleucine
Lysine
Phenylalanine
Tryptophan
Tyrosine

Ketogenic
Leucine

Nonessential amino acid   One of 11 amino acids that are synthesized in the body and are therefore not necessary in the diet.

Essential amino acid   An amino acid that cannot be synthesized by the body and thus must be obtained in the diet.

TABLE 28.2   Essential Amino Acids

Amino acids essential for adults

Histidine	Lysine	Threonine
Isoleucine	Methionine	Tryptophan
Leucine	Phenylalanine	Valine

Some foods with incomplete amino acids
Grains, nuts, and seeds: High in methionine, low in lysine
Legumes: High in lysine, low in methionine
Corn: High in methionine, low in lysine and tryptophan

Some examples of complementary sources of protein

Peanut butter on bread	Nuts and soybeans
Rice and beans	Black-eyed peas and corn bread
Beans and corn	

Reductive deamination
Conversion of an α-keto acid to an amino acid by reaction with NH_4^+.

All of the nonessential amino acids derive their amino groups from glutamate. As you've seen, this is also the molecule that picks up ammonia in amino acid catabolism and carries it into the urea cycle. Glutamate can also be made from NH_4^+ and α-ketoglutarate by **reductive deamination**, the reverse of oxidative deamination (Section 28.3). The same glutamate dehydrogenate enzyme carries out the reaction:

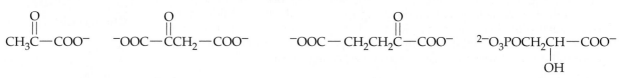

Glutamate also provides nitrogen for the synthesis of other nitrogen-containing compounds, including the purines and pyrimidines that are part of DNA. (∞, Section 26.2)

The following four common metabolic intermediates, which you've seen by now in many roles, are the precursors for synthesis of the nonessential amino acids:

Precursors in synthesis of nonessential amino acids

$$CH_3\overset{O}{\overset{\|}{C}}-COO^- \qquad {}^-OOC-\overset{O}{\overset{\|}{C}}CH_2-COO^- \qquad {}^-OOC-CH_2CH_2\overset{O}{\overset{\|}{C}}-COO^- \qquad {}^{2-}O_3POCH_2\underset{OH}{CH}-COO^-$$

Pyruvate   Oxaloacetate   α-Ketoglutarate   3-Phosphoglycerate

Glutamine, for example, is made from glutamate and asparagine is made by reaction of glutamine with aspartate:

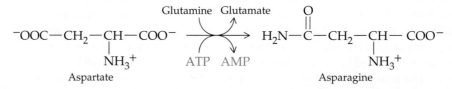

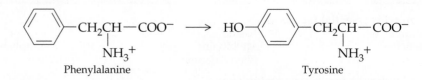

The amino acid tyrosine is sometimes classified as nonessential because we can synthesize it from phenylalanine, an essential amino acid:

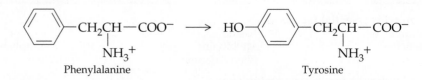

Whatever the classification, we have a high nutritional requirement for phenylalanine, and several metabolic diseases are associated with defects in the enzymes needed to convert it to tyrosine and other metabolites. The best known of these diseases is phenylketonuria (PKU), the first inborn error of metabolism for which the biochemical cause was recognized. In 1947 it was found that failure of the conversion of phenylalanine to tyrosine is the cause of PKU.

Application Xenobiotics

ow do our bodies tackle a chemical compound they've never encountered before? The compound might be a drug, a pesticide, or an environmental pollutant. It could be addictive, it could be a carcinogen, or it could be a harmless compound that needs only to be excreted. In medical terminology, such compounds are known as *xenobiotics*—chemical compounds foreign to the body.

A xenobiotic is likely to be met with a two-pronged attack when it enters the body. First, it undergoes chemical reactions that make it hydrophilic and water-soluble, so that it doesn't accumulate in fatty tissues. Second, enzymes further modify its structure, converting it to a compound that is even more water-soluble and is likely to be excreted.

In the first reaction for many xenobiotics, a hydrogen atom is replaced by a hydroxyl group in a hydroxylation reaction carried out by a monooxygenase enzyme complex known as a *cytochrome P-450*. (The enzyme name derives from the absorption of light with a wavelength of 450 nm. There is a large family of P-450 enzymes.) For example, a xenobiotic (RH), oxygen, and NADPH react as follows:

$$RH + O_2 \xrightarrow{\quad NADPH \quad NADP^+ \quad} R\text{---}OH + H_2O$$

About half of all drugs are metabolized by cytochrome P-450. As you might expect, this protective action takes place mainly in the liver as blood passes through on its way to general circulation.

The second phase of xenobiotic metabolism is usually bonding through an oxygen, nitrogen, or sulfur atom to an even more polar group—for example, the tripeptide glutathione:

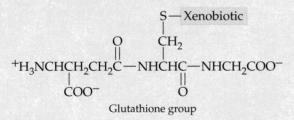

Glutathione group

The cytochrome P-450 enzymes have both good and bad effects: good in detoxifying some poisons; bad in converting some compounds, especially polycyclic aromatic hydrocarbons like those in cigarette smoke, to carcinogenic compounds that react with and damage DNA. Like many biomolecules, cytochromes are *inducible*, meaning that when more is needed, more is

▲ Computer-generated model of a cytochrome P-450, an enzyme that catalyzes structural changes in xenobiotics. The red structure is a heme group.

made. It's well known that smokers have more cytochrome P-450 than nonsmokers. In fact, the wide variability of xenobiotic enzyme activity with age, sex, genetic makeup, and previous exposure to xenobiotics is thought to play a role in the wide variability in individual responses to carcinogens.

▲ Individuals who breathe cigarette smoke usually have a higher concentration of cytochrome P-450 in their blood.

See Additional Problems 28.43–45 at the end of the chapter.

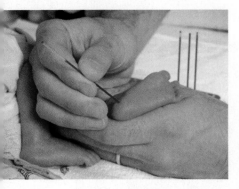

▲ Drawing blood from a newborn infant for the PKU test.

PKU results in elevated blood serum and urine concentrations of phenylalanine, phenylpyruvate, and several other metabolites produced when the body diverts phenylalanine to metabolism by other pathways. Undetected PKU causes mental retardation by the second month of life. Estimates are that, prior to the 1960s, 1% of those institutionalized for mental retardation were PKU victims. The only defense against PKU and simular treatable metabolic disorders that take their toll early in life is widespread screening of newborn infants. In the 1960s a test for PKU was introduced, and virtually all hospitals in the United States now routinely screen for it. Treatment consists of a diet low in phenylalanine, which is maintained in infants with special formulas and in older individuals by eliminating meat and using low-protein grain products. Individuals with PKU must be alert for foods sweentened with aspartame (Nutrasweet), which is a derivative of phenylalanine.

■ KEY CONCEPT PROBLEM 28.9

In the pathway for synthesis of serine shown below, identify which of the reaction is:

(a) A transamination **(b)** A hydrolysis **(c)** An oxidation

$$^-OOCCHCH_2OPO_3^{2-} \longrightarrow \; ^-OOCCCH_2OPO_3^{2-} \longrightarrow$$

$$\underset{\text{OH}}{|} \qquad\qquad \overset{\displaystyle O}{\overset{\displaystyle \|}{}}$$

3-Phosphoglycerate $\qquad$ 3-Phosphohydroxypyruvate

$$^-OOCCHCH_2OPO_3^{2-} \longrightarrow \; ^-OOCCHCH_2OH$$

$$\underset{\text{NH}_3^+}{|} \qquad\qquad\qquad \underset{\text{NH}_3^+}{|}$$

3-Phosphoserine $\qquad\qquad$ Serine

Summary: Revisiting the Chapter Goals

1. **What happens during the digestion of proteins, and what is the destination of the amino acids?** Protein digestion begins in the stomach and continues in the small intestine. The result is virtually complete hydrolysis to yield free amino acids. The amino acids enter the bloodstream after active transport into cells lining the intestine. The body does not store nitrogen compounds, but amino acids are constantly entering the amino acid pool from dietary protein or broken down body protein and being withdrawn from the pool for biosynthesis or further catabolism.

2. **What are the major strategies in the catabolism of amino acids?** Each amino acid is catabolized by a distinctive pathway, but in most of them the amino group is removed by *transamination* (the transfer of an amino group from an amino acid to a keto acid), usually to form glutamate. Then, the amino group of glutamate is removed as ammonium ion by *oxidative deamination*. The ammonium ion is destined for the *urea cycle*. The carbon atoms from amino acids are incorporated into compounds that can enter the *citric acid cycle*. These carbon compounds are also available for conversion to fatty acids or glycogen for storage, or for synthesis of ketone bodies.

3. **What is the urea cycle?** Ammonium ion (from amino acid catabolism) and bicarbonate ion (from carbon dioxide) react to produce carbamoyl phosphate, which enters the urea cycle. The first two steps of the urea cycle produce a reactive intermediate in which both of the nitrogens that will be part of the urea end product are bonded to the same carbon atom. Then arginine is formed and split by hydrolysis to yield urea, which will be excreted. The net result of the urea cycle is reaction of ammonium ion with aspartate to give urea and fumarate.

4. **What are the essential and nonessential amino acids, and how, in general, are amino acids synthesized?** *Essential amino acids* must be obtained in the diet because our bodies do not synthesize them. They are made only by plants and microorganisms, and their synthetic pathways are complex. Our bodies do synthesize the so-called *nonessential amino acids*. Their synthetic pathways are quite simple and generally begin with pyruvate, oxaloacetate, α-ketoglutarate, or 3-phosphoglycerate. The nitrogen is commonly supplied by glutamate.

Key Words

Amino acid pool, p. 791

Essential amino acids, p. 799

Nonessential amino acid, p. 799

Oxidative deamination, p. 794

Reductive deamination, p. 800

Transamination, p. 793

Urea cycle, p. 795

■ Understanding Key Concepts

28.10 In the following diagram, fill in the sources for the amino acid pool.

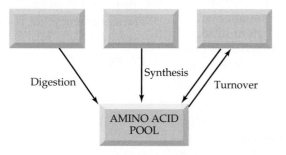

28.11 What are the fates of the carbon and nitrogen atoms a catabolized amino acid?

28.12 A treatment for hyperammonemia (excess NH_4^+ in the blood) is to administer pyruvate. What two enzymes would be necessary to detoxify the ammonium ion in the presence of pyruvate? What would be the product?

28.13 In the liver, the relative activity of ornithine transcarbamylase is high, that of argininosuccinate synthetase is low, and that of arginase is high. Why is it important that ornithine transcarbamylase activity be high? What might be the consequence if arginase activity were low or defective?

28.14 We say that some amino acids are "essential" and others are "nonessential." What do we really mean when we use these two terms?

28.15 Three metabolites that can result from the breakdown of the carbon skeleton of amino acids are ketone bodies, acetyl-SCoA, and glucose. Briefly describe how each of these metabolites can be produced from amino acid catabolism.

■ Additional Problems

Amino Acid Pool

28.16 In what part of the digestive tract does the digestion of proteins begin?

28.17 Where is the body's amino acid pool?

28.18 What glycolytic intermediates are precursors to amino acids?

28.19 What citric acid cycle intermediates are precursors to amino acids?

Amino Acid Catabolism

28.20 What is meant by transamination?

28.21 Pyruvate and oxaloacetate can be acceptors for the amino group in transamination. Write the structures for the products formed from transamination of these two compounds.

28.22 What is the structure of the α-keto acid formed from transamination of each of the following amino acids?

    (a) Isoleucine   **(b)** Cysteine

28.23 What is the structure of the α-keto acid formed from transamination of each of the following amino acids?

    (a) Phenylalanine
    (b) Serine

28.24 What is meant by an oxidative deamination?

28.25 What coenzymes are associated with oxidative deamination?

28.26 Write the structure of the α-keto acid produced by oxidative deamination of each of the following:

    (a) Phenylalanine
    (b) Tryptophan

28.27 What is the other product formed in oxidative deamination besides an α-keto acid?

28.28 What is a glucogenic amino acid?

28.29 What is a ketogenic amino acid?

Urea Cycle

28.30 Why does the body convert NH_4^+ to urea for excretion?

28.31 Why might the urea cycle be called the arginine cycle?

28.32 What is the source of carbon in the formation of urea?

28.33 What are the sources of the two nitrogens in the formation of urea?

Amino Acid Biosynthesis

28.34 How do "essential" and "nonessential" amino acids differ in the number of steps required for their synthesis in organisms that synthesize both?

28.35 Which amino acid serves as the source of nitrogen for synthesis of the other amino acids?

28.36 What is the name of the process by which amino acids are made from common non-nitrogen metabolites? This is the reverse of what process from amino acid catabolism?

28.37 How is tyrosine biosynthesized in the body? What disease prevents this biosynthesis, thereby making tyrosine an essential amino acid for those who have this condition?

28.38 PKU is an abbreviation for what disorder? What are the symptoms of PKU? How can PKU be treated for a nearly normal life?

28.39 Diet soft drinks that are sweetened with aspartame carry a warning label for phenylketonurics. Why?

28.40 Which of the following biomolecules contain nitrogen from an amino acid?

    (a) glycogen
    (b) nitric oxide
    (c) collagen
    (d) epinephrine
    (e) stearic acid
    (f) fructose

Applications

28.41 Your grandfather is complaining of pain in his swollen and inflamed big toe, and the doctor indicates that it is caused by gout.

   (a) How would you explain to him what gout is and its biochemical cause?

   (b) What can you suggest to him to prevent these gouty attacks? [*Pathways to Gout*]

28.42 Allopurinol is a drug often used to assist in the control of gout. At what step(s) in the catabolism of purines is allopurinol effective? What is its effect? Compare the structure of allopurinol with hypoxanthine and xanthine. Where does allopurinol differ in structure from hypoxanthine? Is this the site on the molecule that corresponds to the site where hypoxanthine or xanthine is oxidized? [*Pathways to Gout*]

28.43 What does the term *xenobiotic* mean? [*Xenobiotics*]

28.44 In what two ways does a body respond to a xenobiotic agent? [*Xenobiotics*]

28.45 Many of the cytochrome P-450 enzymes are inducible. What does this mean? [*Xenobiotics*]

General Questions and Problems

28.46 What form of energy is used in the formation of urea?

28.47 Write the equation for the transamination reaction that occurs between valine and pyruvate.

28.48 Name the four products (carbon skeletons) of amino acid catabolism that can enter the citric acid cycle, and show where in the cycle they enter.

28.49 Can an amino acid be both glucogenic and ketogenic? Explain why or why not.

28.50 Briefly explain how the carbons from amino acids can end up in fatty acids located in adipose tissues.

28.51 Consider all of the metabolic processes we have studied. Why do we say that tissue biochemistry is dynamic? Describe some examples of these dynamic relationships.

28.52 Two major differences between the amino acid pool and the fat and carbohydrate pools in the body center on storage and on energy. Discuss these major differences.

28.53 When some of the carbons of glutamate are converted to glycogen, what is the order of the following compounds in that pathway?

   (a) Glucose

   (b) Glutamate

   (c) Glycogen

   (d) Oxaloacetate

   (e) α-Ketoglutarate

   (f) Phosphoenolpyruvate

28.54 The pancreatic proteases are synthesized and stored as zymogens. They are activated after the pancreatic juices enter the small intestine. Why is it essential that these enzymes be synthesized and stored in their inactive forms?

28.55 Why might an excess of a particular amino acid in the diet cause a deficiency of other amino acids?

28.56 The net reaction for the urea cycle-shows that 3 ATPs are hydrolyzed; however the total ATP "cost" is 4 ATPs. Explain why this is true.

28.57 What are the two possible fates of the amino nitrogen from dietary proteins in animals?

29

Body Fluids

CONTENTS

While on the way to further treatment, body fluid balance must be maintained.

Just about every aspect of chemistry you've studied so far applies to the subject of this chapter—body fluids. Electrolytes, nutrients and waste products, metabolism intermediates, and chemical messengers flow through the body in blood, and wastes exit in the urine. The chemical compositions of blood and urine therefore mirror chemical reactions throughout the body. Fortunately, samples of these fluids are easily collected. Many advances in understanding biological chemistry have been based on information from blood and urine analysis. In addition, studies of blood and urine chemistry provide information essential for the diagnosis and treatment of disease.

The questions answered in this chapter demonstrate the central role of chemistry in understanding physiology:

1. How are body fluids classified?
The goal: Be able to describe the major categories of body fluids, their general composition, and the exchange of solutes between them.

2. What are the roles of blood in maintaining homeostasis?
The goal: Be able to explain the composition and functions of blood.

3. How do blood components participate in the body's defense mechanisms?
The goal: Be able to identify and describe the roles of blood components that participate in inflammation, the immune response, and blood clotting.

4. How do red blood cells participate in the transport of blood gases?

The goal: Be able to explain the relationships among O_2 and CO_2 transport, and acid–base balance.

5. How is the composition of urine controlled?

The goal: Be able to describe the transfer of water and solutes during urine formation, and give an overview of the composition of urine.

CONCEPTS TO REVIEW

Solutions (Sections 9.1, 9.2, 9.10)
pH (Sections 10.9, 10.10)

29.1 Body Water and Its Solutes

Water is the solvent in all body fluids, and the water content of the human body averages about 60% (by weight). Physiologists describe body water as occupying two different "compartments"—the *intracellular* and the *extracellular* compartments. Thus far we have looked mainly at chemical reactions occurring in the **intracellular fluid** the fluid inside cells, which includes about two-thirds of all body water (Figure 29.1). The remaining one-third of body water is **extracellular fluid**, which includes mainly **blood plasma** (the fluid portion of blood) and **interstitial fluid** (the fluid that fills the spaces between cells).

To be soluble in water, a substance must be an ion, a gas, a small polar molecule, or a large molecule with many polar or ionic groups on its surface. All four types of solutes are present in body fluids. The majority are inorganic ions and ionized biomolecules (mainly proteins), as shown in the comparison of blood plasma, interstitial fluid, and intracellular fluid in Figure 29.2. Although these fluids have different compositions, their osmolarities are the same; that is, they

Intracellular fluid Fluid inside cells.

Extracellular fluid Fluid outside cells.

Blood plasma Liquid portion of the blood: an extracellular fluid.

Interstitial fluid Fluid surrounding cells: an extracellular fluid.

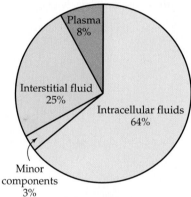

▲ **FIGURE 29.1 Distribution of body water.** About two-thirds of body water is intracellular—within cells. The extracellular fluids include blood plasma, fluids surrounding cells (interstitial), and such minor components as lymph, cerebrospinal fluid, and the fluid that lubricates joints (synovial fluid).

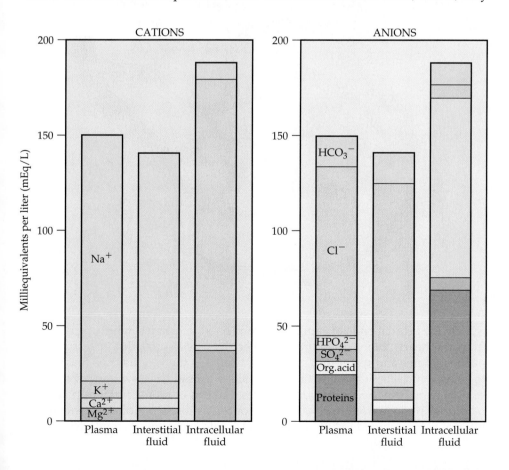

◄ **FIGURE 29.2 The distribution of cations and anions in body fluids.** Outside cells, Na^+ is the major cation and Cl^- is the major anion. Inside cells, K^+ is the major cation and HPO_4^{2-} is the major anion. Note that at physiological pH, proteins are negatively charged.

have the same number of moles of dissolved solute particles (ions or molecules) per liter. The osmolarity is kept in balance by the passage of water across cell membranes by osmosis, which occurs in response to osmolarity differences. (In osmosis, water moves from the more dilute to the more concentrated solution.)

Inorganic ions, known collectively as *electrolytes* (Section 9.9), have many roles in the reactions of metabolism, are major contributors to the osmolarity of body fluids, and move about as necessary to maintain charge balance. Water-soluble proteins make up a large proportion of the solutes in blood plasma and intracellular fluid; 100 mL of blood contains about 7 g of protein. Blood proteins transport lipids and other molecules, and they play essential roles in the protection afforded by blood clotting (Section 29.5) and the immune response (Section 29.4). The blood gases oxygen and carbon dioxide, along with glucose, amino acids, and the nitrogen-containing by-products of protein catabolism, are the major small molecules in body fluids.

Blood travels through peripheral tissue in a network of tiny, hair-like capillaries that connect the arterial and venous parts of the circulatory system (Figure 29.3). It is at capillaries that nutrients and end products of metabolism are exchanged between blood and interstitial fluid. Capillary walls consist of a single layer of loosely spaced cells. Water and many small solutes move freely through the walls in response to differences in fluid pressure and concentration.

FIGURE 29.3 The capillary network. ▶ Solute exchange between blood and interstitial fluid occurs across capillary walls.

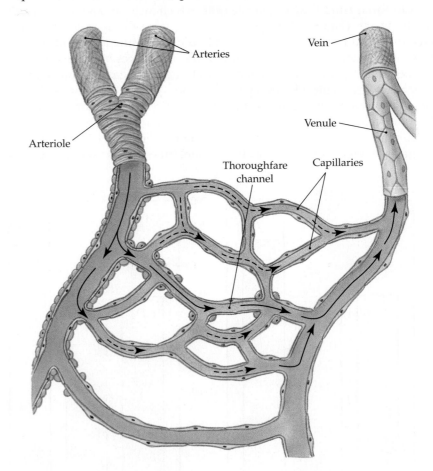

On the arterial ends of capillaries (connected to arteries, which carry blood from the heart), blood pressure is higher than interstitial fluid pressure and pushes solutes and water into interstitial fluid. On the venous ends (connected to veins, which carry blood to the heart), blood pressure is lower, and water and solutes reenter the blood plasma. Solutes that can cross membranes passively by diffusion move from regions of high concentration to regions of low concentration. The combined result of water and solute exchange at capillaries is that, except for proteins, blood plasma and interstitial fluid are similar in composition (see Figure 29.2).

In addition to blood capillaries, peripheral tissue is networked with lymph capillaries that terminate in blind pockets (Figure 29.4). The lymphatic system collects excess interstitial fluid, debris from cellular breakdown, and proteins and lipid droplets too large to pass through capillary walls. Interstitial fluid and the substances that accompany it into the lymphatic system are referred to as *lymph*, and the walls of lymph capillaries are constructed so that lymph cannot return to the surrounding tissue. Ultimately lymph enters the bloodstream at the thoracic duct.

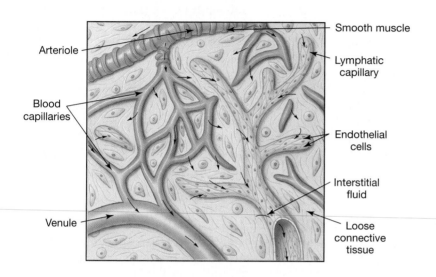

◄ **FIGURE 29.4 Blood and lymph capillaries.** The arrows show the flow of fluids.

Solutes in the interstitial fluid and the intracellular fluid are exchanged by crossing cell membranes. Here, major differences in concentration are maintained by active transport against concentration gradients (from regions of low concentration to regions of high concentration) and by the impermeability of cell membranes to certain solutes, notably the sodium ion (Figure 29.5). Sodium ion concentration is high in extracellular fluids and low in intracellular fluids (as shown in Figure 29.2). Potassium ion concentrations are just the reverse: high inside cells and low outside cells.

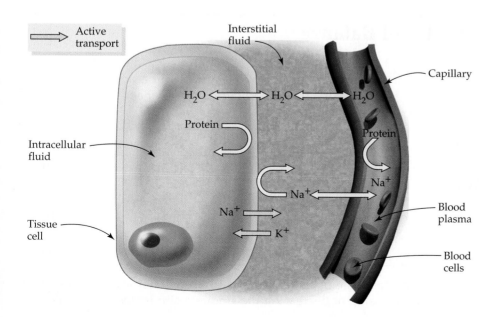

◄ **FIGURE 29.5 Exchange among body fluids.** Water exchanges freely in most tissues, with the result that the osmolarities of blood plasma, interstitial fluid, and intracellular fluid are the same. Large proteins cross neither capillary walls nor cell membranes, leaving the interstitial fluid protein concentration low. Concentration differences between interstitial fluid and intracellular fluid are maintained by active transport.

The skin plays a role in fluid balance by providing a surface for the cooling evaporation of water. Preventing skin from losing too much moisture is a concern for individuals with dry or aging skin. What are we buying with the more than $2 billion spent on moisturizers each year? A good moisturizer should smooth and soften the skin, and should help it retain the 10–30% of water that is normally part of its outermost layer (the *epidermis*).

There are two kinds of moisturizers. The first kind prevents evaporation of water from the skin surface. Moisturizers of this kind contain animal fats such as lanolin (from sheep oil glands and wool) and vegetable oils such as olive oil, safflower oil, palm kernel oil, apricot oil, or sesame seed oil (all used in moisturizing products). The body's own moisturizer of this type is the waxy *sebum* excreted from *sebaceous glands* in the skin. Because these moisturizers are nonpolar substances, water in the skin's outer layer does not pass through them. Petroleum jelly (for example, Vaseline) is a fine moisturizer of this type, but like some of the others it is not popular because it feels greasy and is messy to use.

The second type of moisturizer, known as a *humectant*, holds moisture in the skin by attracting it from the skin and the air. From a list of some common humectants—glycerol, propylene glycol, urea, lactic acid, and lecithin (a membrane phospholipid)—you can see that these are all compounds capable of forming hydrogen bonds with water. (Look up the structures if you're not sure.)

Some moisturizers—usually the more expensive ones—claim the ability to beautify skin and reverse the aging process by restoring missing proteins such as collagen and elastin or by supplying hormones, vitamins, and even DNA. Do you think these molecules can cross the outer layer of skin?

See Additional Problem 29.52–53 at the end of the chapter.

■ **KEY CONCEPT PROBLEM 29.1**

The drug cisplatin is used to treat various forms of cancer in humans. As with many drugs, the difficult part of their design is ensuring transport into the cell. The reaction shown below takes place in the body when cisplatin is administered.

$$Pt(NH_3)_2Cl_2(aq) + H_2O(l) \rightleftharpoons [Pt(NH_3)_2(H_2O)Cl]^+(aq) + Cl^-(aq)$$

Which form of cisplatin would you expect to exist in the cell (where chloride concentrations are small)? Which form of cisplatin would you expect to exist outside the cell (where chloride concentrations are high)? Which form of cisplatin would most readily enter the cell? Why?

29.2 Fluid Balance

The body is kept in overall fluid balance by roughly equal daily intake and output of water as summarized in the following table. In a hot environment or when doing strenuous work, the intake of drinking water and loss in sweat and exhaled gases both increase greatly.

Water intake (mL/day)		Water Output (mL/day)	
Drinking water	1200	Urine	1400
Water from food	1000	Skin	400
Water from metabolic		Lungs	400
oxidation of food	300	Sweat	100
		Feces	200
Total	2500		2500

One important role of the kidneys is to keep water and electrolytes in balance by increasing or decreasing the amounts eliminated. The intake and output of water are, in turn, controlled by hormones. Receptors in the hypothalamus monitor the concentration of solutes in blood plasma, and as little as a 2% change in osmolarity can cause an adjustment in hormone secretion (see Figure 9.1). For example, when a rise in blood osmolarity indicates an increased concentration of solutes and therefore a shortage of water, secretion of *antidiuretic hormone* (also known as *vasopressin*) increases. In the kidney, antidiuretic hormone causes a decrease in the water content of the urine. At the same time, the thirst mechanism is activated to cause increased water intake.

A long list of abnormal conditions cause what physicians refer to as the *syndrome of inappropriate antidiuretic hormone secretion (SIADH)*, the result of excess secretion of the hormone. When this occurs, the kidney excretes too little water, the water content of body compartments increases, and serum concentrations of electrolytes drop to dangerously low levels. Among the causes of SIADH are regional low blood volume due to decreased return of blood to the heart (caused by, for example, asthma, pneumonia, pulmonary obstruction, or heart failure) and misinterpretation by the hypothalamus of osmolarity (due, for example, to central-nervous-system disorders, barbiturates, or morphine).

The reverse problem, inadequate secretion of antidiuretic hormone, is often a result of injury to the hypothalamus, and causes *diabetes insipidus*. In this condition (unrelated to diabetes mellitus), up to 15 L of dilute urine is excreted each day. Administration of synthetic hormone can control the problem.

▲ Intravenous fluids ready to be used for restoring fluid balance, electrolyte balance, and glucose supply.

29.3 Blood

Blood flows through the body in the circulatory system, which in the absence of trauma or disease is an essentially closed system. About 55% of blood is plasma, which contains the proteins and other solutes shown in Figure 29.6, and the remaining 45% is a mixture of red blood cells (**erythrocytes**), platelets, and white blood cells.

Erythrocytes   Red blood cells; transporters of blood gases.

The plasma and cells together make up **whole blood**, which is usually collected for clinical laboratory analysis directly into evacuated tubes. An anticoagulant must be added to whole blood to prevent clotting, which otherwise will occur within 20–60 minutes at room temperature. Often, the anticoagulant is supplied in the collection tube. Heparin, a natural polysaccharide (Section 22.8), interferes with the action of enzymes needed for clotting. Other anticoagulants, such as citrate ion and oxalate ion, form precipitates with calcium ion, which is also needed for blood clotting, thereby removing it from solution. Plasma is separated from blood cells by spinning the sample in a centrifuge, which causes the blood cells to clump together at the bottom of the tube and leaves the plasma at the top.

Whole blood   Blood plasma plus blood cells.

Many laboratory analyses are performed on **blood serum**, the fluid remaining after blood has completely clotted. When a serum sample is desired, whole blood is collected in the presence of an agent that hastens clotting. Thrombin, a natural component of the clotting system (Section 29.5), is often used for this purpose. Centrifugation separates the clot and cells to leave behind the serum.

Blood serum   Fluid portion of blood remaining after clotting has occurred.

Major Components of Blood

- **Whole blood**
    Blood plasma—fluid part of blood containing water-soluble solutes
    Blood cells—red blood cells (carry gases)
        —white blood cells (part of immune system)
- **Blood serum**—fluid portion of plasma left after blood has clotted

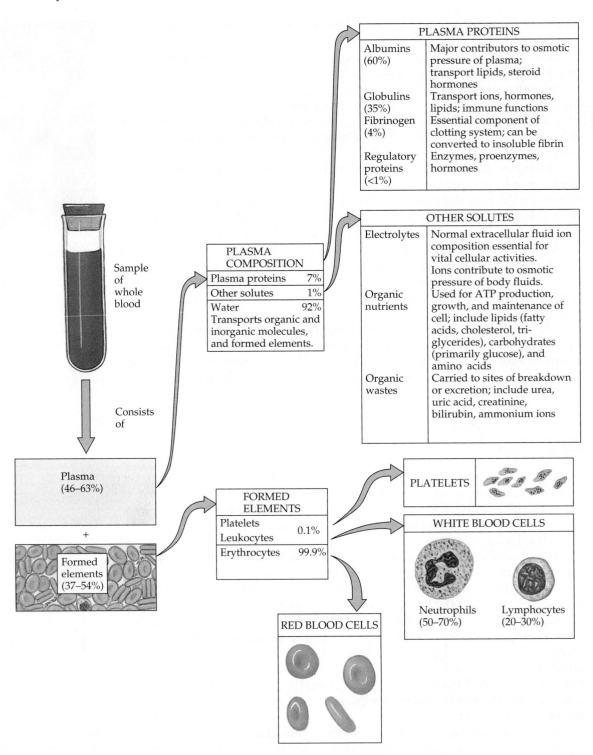

▲ FIGURE 29.6   **The composition of whole blood.**

The functions of the major protein and cellular components of blood are summarized in Table 29.1. These functions fall into three categories:

- **Transport** The circulatory system is the body's equivalent of the interstate highway network, transporting materials from where they enter the system to where they are used or disposed of. Oxygen and carbon dioxide are carried to and from red blood cells. Nutrients are carried from the intestine to the sites of their catabolism. Waste products of metabolism are carried to the kidneys. Hormones from endocrine glands are delivered to their target tissues.

TABLE 29.1 Protein and Cellular Components of Blood	
Blood Component	**Function**
Proteins	
Albumins	Transport lipids, hormones, drugs; major contributor to plasma osmolarity
Globulins	
Immunoglobulins (γ-globulins, antibodies)	Identify antigens (microorganisms and other foreign invaders) and initiate their destruction
Transport globulins	Transport lipids and metal ions
Fibrinogen	Forms fibrin, the basis of blood clots
Blood cells	
Red blood cells (erythrocytes)	Transport O_2, CO_2, H^+
White blood cells	
Lymphocytes (T cells and B cells)	Defend against specific pathogens and foreign substances
Neutrophils, eosinophils, and monocytes	Carry out phagocytosis—engulf foreign invaders
Basophils	Release histamine during inflammatory response of injured tissue
Platelets	Help to initiate blood clotting

▲   **Result of separating blood serum from blood clot.**   Analysis of the serum is an essential part of medical diagnosis.

- **Regulation**   Blood redistributes body heat as it flows along, thereby participating in the regulation of body temperature. It also picks up or delivers water and electrolytes as they are needed. In addition, blood buffers are essential to the maintenance of acid–base balance.

- **Defense**   Blood carries the molecules and cells needed for two major defense mechanisms: (1) the immune response, which destroys foreign invaders; and (2) blood clotting, which prevents loss of blood and begins healing of wounds.

In the following two sections, we take an overview of the defense functions of blood—the immune response and blood clotting. The transport of blood gases is then covered in Section 29.6. (Lipid transport was discussed in Chapter 25.)

■ **PROBLEM 29.2**

The following terms refer to types of body fluid. Match each term with its definition below. (a) Interstitial fluid (b) Whole blood (c) Blood serum (d) Intracellular fluid (e) Blood plasma

(i)   Fluid that remains when blood cells are removed.

(ii)   Fluid, solutes, and cells that together flow through veins and arteries.

(iii) Fluid that fills spaces between cells.

(iv) Fluid that remains when blood clotting agents are removed from plasma.

(v)   Fluid within cells.

29.4 Plasma Proteins, White Blood Cells, and Immunity

A foreign invader, known as an **antigen**, is any molecule or portion of a molecule recognized by the body as not part of itself. The antigen might be a foreign molecule or a recognizable molecular segment on the surface of a

Antigen   A substance foreign to the body that triggers the immune response.

microorganism. Antigens can also be small molecules, known as *haptens*, that are recognized as antigens only after they have bonded to carrier proteins. Haptens include some antibiotics, environmental pollutants, and allergens from plants and animals.

The recognition of an antigen can initiate three different responses. The first, the **inflammatory response**, is a general localized response that is not specific to a given antigen. Unlike inflammation, the two types of **immune response** depend on recognition of *specific* invaders, which might be viruses, bacteria, toxic substances, or infected cells (Figure 29.7). At the molecular level, the invading antigen is detected by an interaction very much like that between an enzyme and its substrate. Non-covalent attraction allows a spatial fit between the antigen and a defender specific to that antigen. The *cell-mediated immune response* depends on white blood cells known as *T cells*. The *antibody-mediated immune response* depends on **antibodies (or immunoglobulins)** produced by the white blood cells known as *B cells*.

Inflammatory response A nonspecific defense mechanism triggered by antigens or tissue damage.

Immune response Defense mechanism of the immune system dependent on the recognition of specific antigens, including viruses, bacteria, toxic substances, and infected cells; either cell-mediated or antibody-mediated.

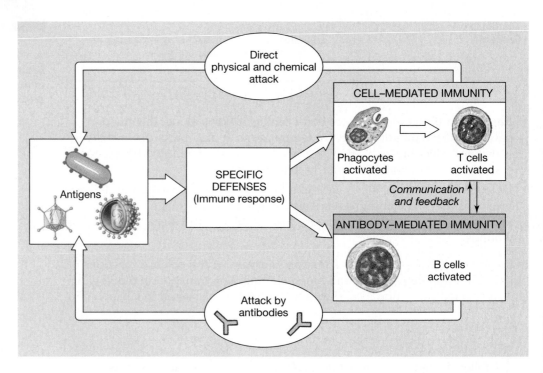

▲ **FIGURE 29.7 The immune response.** The attack on antigens occurs by cell-mediated and antibody-mediated immune responses.

Inflammation and the immune responses require normal numbers of white blood cells to be effective. If the white blood cell count falls below 1000 per milliliter of blood, any infection can be life-threatening. The devastating results of white blood cell destruction in AIDS is an example of this condition (see Application "Viruses and AIDS," p. 750.)

Antibody (immunoglobulin) Glycoprotein molecule that identifies antigens.

Inflammation Result of the inflammatory response: includes swelling, redness, warmth, and pain.

Inflammatory Response

Cell damage due to infection or injury initiates **inflammation**, a nonspecific defense mechanism that produces swelling, redness, warmth, and pain. For example, the swollen, painful, red bump that develops around a splinter in your finger is an inflammation. Chemical messengers released at the site of the injury direct the inflammatory response. One such messenger is histamine, which is synthesized from the amino acid histidine and is stored in cells throughout the body. Histamine release is also triggered by an allergic response.

Application The Blood–Brain Barrier

Nowhere in human beings is the maintenance of a constant internal environment more important than in the brain. If the brain were exposed to the fluctuations in concentrations of hormones, amino acids, neurotransmitters, and potassium that occur elsewhere in the body, inappropriate nervous activity would result. Therefore, the brain must be rigorously isolated from variations in blood composition.

How can the brain receive nutrients from the blood in capillaries and yet be protected? The answer lies in the unique structure of the *endothelial cells* that form the walls of brain capillaries. Unlike the cells in most other capillaries, those in brain capillaries form a series of continuous tight junctions so that nothing can pass between them. To reach the brain, therefore, a substance must cross this blood–brain barrier by crossing the endothelial cell membranes.

The brain, of course, can't be completely isolated or it would die from lack of nourishment. Glucose, the main source of energy for brain cells, and certain amino acids the brain can't manufacture are recognized and brought across the cell membranes by transporters specific to each nutrient. Similar specific transporters move surplus substances out of the brain.

An asymmetric (one-way) transport system exists for glycine, a small amino acid that is a potent neurotransmitter. Glycine inhibits rather than activates transmission of nerve signals, and its concentration must be held at a lower level in the brain than in the blood. To accomplish this, there is a glycine transport system in the cell membrane closest to the brain, but no matching transport system on the other side. Thus, glycine can be transported out of the brain but not into it.

The brain is also protected by the "metabolic" blood–brain barrier. In this case, a compound that gets into an endothelial cell is converted within the cell to a metabolite that is unable to enter the brain. A striking demonstration of the metabolic brain barrier is provided by *dopamine*, a neurotransmitter, and L-*dopa*, a metabolic precursor of dopamine.

L-Dopa can both enter and leave the brain because it is recognized by an amino acid transporter. However, the brain is protected from an entering excess of L-dopa by its conversion to dopamine within the endothelial cells. Like glycine, dopamine, which is also produced from L-dopa within the brain, can leave the brain but cannot enter it. The dopamine deficiency that occurs in Parkinson's disease can therefore be treated by administration of L-dopa.

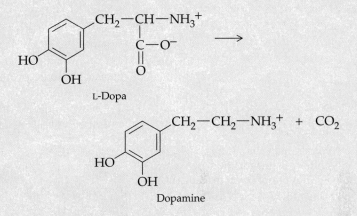

L-Dopa

Dopamine

Since crossing the endothelial cell membrane is the route into the brain, substances soluble in the membrane lipids readily breach the blood–brain barrier. Among such substances are nicotine, caffeine, codeine, diazepam (Valium, an antidepressant), and heroin. Heroin differs from morphine in having two nonpolar acetyl groups where the morphine has polar hydroxyl groups. The resulting difference in lipid solubility allows heroin to enter the brain much more efficiently than morphine. Once heroin is inside the brain, enzymes remove the acetyl groups to give morphine, in essence trapping it in the brain. When developing therapeutic drugs, it is possible to take advantage of such trapping so that a carefully controlled dosage can have a potent effect.

See Additional Problems 29.54–56 at the end of the chapter.

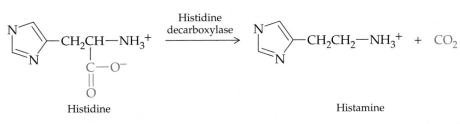

Histidine

Histamine

Histamine sets off dilation of capillaries and increases the permeability of capillary walls. The resulting increased blood flow into the damaged area reddens and warms the skin, and swelling occurs as plasma carrying blood-clotting factors and defensive proteins enters the intercellular space. At the same time, white blood cells cross capillary walls to attack invaders.

Bacteria or other antigens at the site of inflammation are destroyed by white blood cells known as *phagocytes*, which surround invading cells and destroy them by enzyme-catalyzed hydrolysis reactions. Phagocytes also emit chemical messengers that help to direct the inflammatory response. An inflammation caused by a wound will heal completely only after all infectious agents have been removed, with dead cells and other debris absorbed into the lymph system.

Cell-Mediated Immune Response

Cell-Mediated Immune Response

The cell-mediated immune response is under the control of several kinds of *T lymphocytes*, or *T cells*. It guards principally against abnormal cells and bacteria or viruses that have entered normal cells. Cell-mediated immunity also guards against some cancer cells and causes the rejection of transplanted organs.

A complex series of events begins when a T cell recognizes an antigen that is presented to it on the surface of a phagocyte or other cell. The result is production of *cytotoxic*, or *killer*, T cells that can destroy the invader (for example, by releasing a toxic protein that kills by perforating cell membranes) and *helper* T cells that enhance defenses against the invader in a variety of ways. Thousands of *memory* T cells are also produced; they remain on guard and will immediately generate the appropriate killer T cells if the same pathogen reappears.

White blood cells. (a) A ▶ lymphocyte phagocytizing a yeast cell. (b) A lymphocyte reaches out to snare several *E. coli* bacteria.

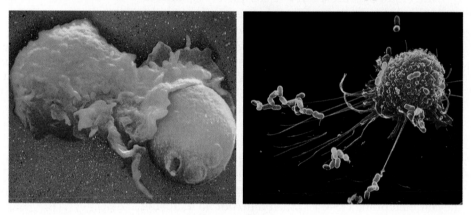

Antibody-Mediated Immunity

Antibody-Mediated Immune Response

The white blood cells known as *B lymphocytes* or *B cells*, with the assistance of T cells, are responsible for the antibody-mediated immune response. Unlike T cells, which identify only antigenic cells, B cells identify antigens adrift in body fluids. A B cell is activated when it first bonds to an antigen and then encounters a helper T cell that recognizes the same antigen. This activation can take place anywhere in the body but often occurs in lymph nodes, tonsils, or the spleen, which have large concentrations of lymphocytes.

Once activated, B cells divide to form plasma cells that secrete antibodies specific to the antigen. The antibodies are immunoglobulins. The body contains up to 10,000 different immunoglobulins at any given time, and we have the capacity to make more than 100 million others. The immunoglobulins are glycoproteins composed of two "heavy" polypeptide chains and two "light" polypeptide chains joined by disulfide bonds, as shown in Figure 29.8a. The variable regions are sequences of amino acids that will bind a specific antigen. Once synthesized, antibodies spread out to find their antigens.

Formation of an antigen–antibody complex (Figure 29.9) inactivates the antigen by one of several methods. The complex may, for example, attract phagocytes, or it may block the mechanism by which the invader connects with a target cell.

Activated B-cell division also yields memory cells that remain on guard and quickly produce more plasma cells if the same antigen reappears. The long-lived B and T memory cells are responsible for long-term immunity to diseases after the first illness or after a vaccination.

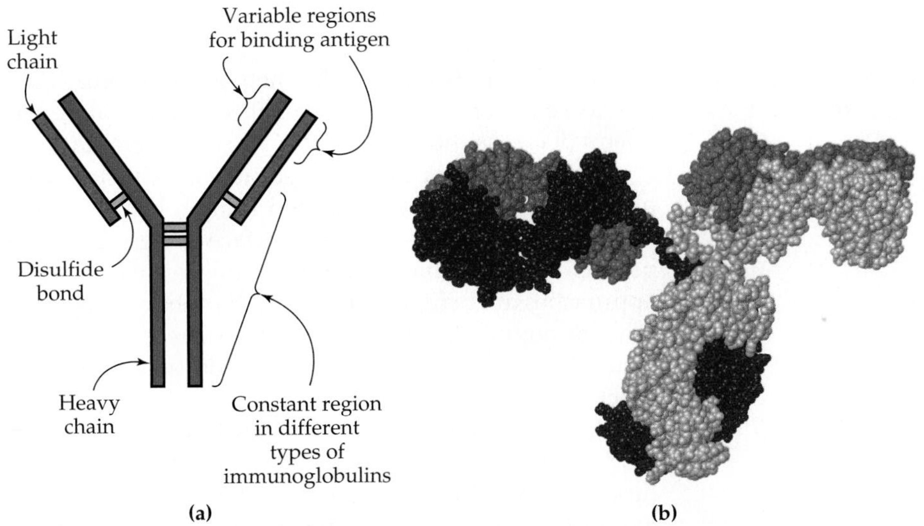

◀ **FIGURE 29.8   Structure of an immunoglobulin, which is an antibody.** (a) The regions of an immunoblobulin. (b) Molecular model of an immunoglobulin; the heavy chains are gray and blue and both light chains are red.

Several different kinds of immunoglobins have been identified. *Immunoglobulin G antibodies* (known as *gamma globulins*), for example, protect against viruses and bacteria. Allergies and asthma are caused by a oversupply of *immunoglobulin E*. Numerous disorders result from the mistaken identification of normal body constituents as foreign and the overproduction of antibodies to combat them. These **autoimmune diseases** include attack on connective tissue at joints in rheumatoid arthritis, attack on pancreatic islet cells in some forms of diabetes mellitus, and a generalized attack on nucleic acids and blood components in systemic lupus erythematosus.

Autoimmune disease   Disorder in which the immune system identifies normal body components as antigens and produces antibodies to them.

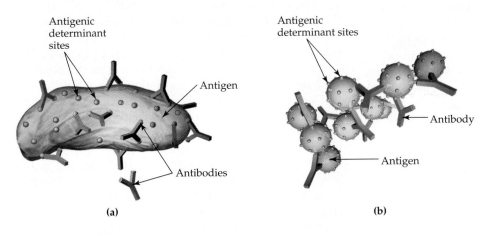

(a)   (b)

◀ **FIGURE 29.9   Antigen–antibody complexes.** (a) Antigens bond to antigenic-determinant sites on the surface of, for example, a bacterium. (b) Because each antibody has two binding sites, the interaction of many antigens and antibodies creates a large immune complex.

29.5 Blood Clotting

A blood clot consists of blood cells trapped in a mesh of the insoluble fibrous protein known as **fibrin**. Clot formation is a multiple-step process requiring participation of 12 clotting factors. The calcium ion is one of the clotting factors. Others, most of which are glycoproteins, are synthesized in the liver by pathways that require vitamin K as a coenzyme. Therefore, a deficiency of vitamin K, the presence of a competitive inhibitor of vitamin K, or a deficiency of a clotting factor can cause excessive bleeding, sometimes from even minor tissue damage. Hemophilia is a disorder caused by an inherited genetic defect that results in the absence of one or another of the clotting factors. Hemophilia occurs in 1 in 10,000 individuals, and 80–90% of hemophiliacs are males.

The body's mechanism for halting blood loss from even the tiniest capillary is referred to as **hemostasis**. The first events in hemostasis are (1) constriction of surrounding blood vessels and (2) formation of a plug composed of the blood cells known as *platelets* at the site of tissue damage.

Fibrin   Insoluble protein that forms the fiber framework of a blood clot.

Hemostasis   The stopping of bleeding.

Application Blood Substitutes

At present, we rely on donated human blood for transfusions, but there are quite a few reasons why an alternative would be desirable:

- Population and the need for blood transfusions are both growing rapidly.
- The number of blood donors is declining.
- There is fear of contaminated blood.
- Errors in blood typing can be fatal.
- Blood has a short shelf-life (42 days).
- Blood must be stored at low temperature (4°C), making transport and storage difficult, especially in remote regions.

Looking at the number of components in blood (Figure 29.6) makes creation of a substitute seem impossible. However, by focusing on just the most vital function of blood—oxygen transport—researchers have been able to develop a number of potential blood substitutes.

There are two major possibilities for oxygen-carrying blood substitutes. One relies on *modified hemoglobin*, which chemically binds oxygen as does regular hemoglobin. The other uses *perfluorocarbons*—compounds in which all hydrogens have been replaced by fluorine, making them extremely stable. These substances do not bind oxygen; they dissolve it.

Hemoglobin consists of four protein subunits, two identical α subunits and two identical β subunits, firmly held together in α,β pairs. Each subunit contains a heme molecule with an iron atom to which the oxygen binds. Because of allosteric effects, binding of each oxygen molecule increases the affinity of the hemoglobin

for the next one. (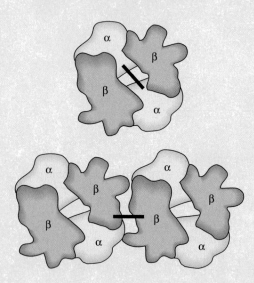, p. 555) Then, when it comes time to release the oxygen in peripheral cells, other allosteric effects decrease the oxygen affinity of hemoglobin.

Hemoglobin for blood substitutes is extracted from blood cells from either outdated human blood or cow blood. In the process, the cell-surface antigens are left behind, making blood typing unnecessary. Several approaches are being used to bind the subunits together more strongly, a change needed to prevent the loosely connected pairs in free hemoglobin from separating and quickly being cleared from the body.

Six of the ten blood substitutes in clinical trials in 1998 contain modified hemoglobin isolated from human or cow blood. Another uses cross-linked hemoglobin produced in *E. coli* by recombinant DNA techniques. (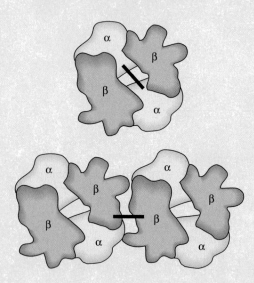, Section 27.4) The modified hemoglobins have a shelf-life of up to 2 years, some without refrigeration. They are administered intravenously after being mixed with a standard blood-volume extender. Because free hemoglobin has a shorter lifetime in the body than transfused blood cells, its use is limited mainly to surgery and trauma. The first FDA approcal of a hemoglobin-based blood substitute went to a preparation for use in dogs.

The Teflon-like perfluorocarbons (PFCs) account for the rest of the blood substitutes being tested in humans. Because PFCs are not miscible with water, they are mixed with an emulsifying agent such as lecithin. The amount of oxygen dissolved by PFCs is proportional to the amount a patient breathes, making it possible for the PFCs to deliver a higher concentration of oxygen than hemoglobin can. This possibility is accompanied by the need to protect against delivery of too much oxygen, which could cause damage.

One advantage of the PFCs is that they are much less expensive than other blood substitutes. Another advantage is that the PFCs are so resistant to chemical change that they are not converted by any biochemical reactions to harmful substances. Instead, because they are so volatile, they are simply exhaled.

Several safety issues remain to be fully proven, however, such as the freedom of hemoglobins from disease transmission and the absence of toxicity in perfluorocarbons. Nevertheless, given the predictions of blood shortages in years to come and government interest in blood substitutes that can more easily be made available in battlefield situations, one or more of the preparations currently being tested is likely to be approved.

See Additional Problems 29.57–58 at the end of the chapter.

Next, **blood clotting** swings into action. It is be triggered by two different pathways: (1) The *intrinsic pathway* begins when blood makes contact with the negatively charged surface of the fibrous protein collagen exposed at the site of tissue damage. Glass is also negatively charged, and clotting is activated in exactly the same manner when blood is placed in a glass tube. (2) The *extrinsic pathway* begins when damaged tissue releases an integral membrane glycoprotein known as *tissue factor*.

The result in either case is a cascade of reactions in which several inactive clotting factors that are zymogens (Section 19.9) are activated one after another by cleavage of polypeptides, often producing enzymes that then catalyze activation of the next factor in the cascade. The two pathways merge and, in the final step of the common pathway, the enzyme *thrombin* catalyzes cleavage of small polypeptides from the soluble plasma protein fibrinogen. Negatively charged groups in these polypeptides make fibrinogen soluble and keep the molecules apart. Once these polypeptides are removed, the resulting insoluble fibrin molecules immediately associate with each other by noncovalent interactions. Then they are bound into fibers by formation of amide cross-links between lysine and glutamine side chains in a reaction catalyzed by another of the clotting factors:

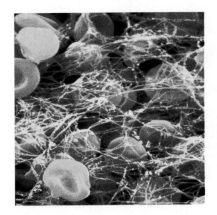

▲   A blood clot (false-color electron micrograph). Red blood cells can be seen enmeshed in the network of fibrin threads.

$$\underset{\displaystyle \text{Gln}}{} - CH_2CH_2 - \overset{\displaystyle O}{\overset{\|}{C}} - NH_2 \;+\; {}^+H_3NCH_2CH_2CH_2CH_2 - \underset{\displaystyle \text{Lys}}{} \longrightarrow$$

Protein chain

$$\underset{\displaystyle \text{Gln}}{} - CH_2CH_2 - \overset{\displaystyle O}{\overset{\|}{C}} - NHCH_2CH_2CH_2CH_2 - \underset{\displaystyle \text{Lys}}{} \;+\; NH_4{}^+$$

Cross-link between protein chains

Once the clot has done its job of preventing blood loss and binding together damaged surfaces as they heal, the clot is broken down by hydrolysis of its peptide bonds.

Blood clot   A network of fibrin fibers and trapped blood cells that forms at the site of blood loss.

29.6   Red Blood Cells and Blood Gases

Red blood cells, or erythrocytes, have one major purpose: to transport blood gases. Erythrocytes in mammals have no nuclei or ribosomes and cannot replicate themselves. In addition, they have no mitochondria or glycogen and must obtain glucose from the surrounding plasma. Their enormous number—about 250 million in a single drop of blood—and their large surface area provide for rapid exchange of gases throughout the body. Because they are small and flexible, erythrocytes can squeeze through the tiniest capillaries one at a time.

Of the protein in an erythrocyte, 95% is hemoglobin, the transporter of oxygen and carbon dioxide. Hemoglobin (Hb) is composed of four protein chains with the quaternary structure shown earlier in Section 18.11. Each protein chain has a central heme molecule in a crevice in its nonpolar interior, and each of the four hemes can combine with one O_2 molecule.

Oxygen Transport

The iron(II) ion, Fe^{2+}, in the center of each heme molecule is the site of the action in transporting oxygen, which is held to the iron by bonding through an unshared electron pair. In contrast to the cytochromes of the respiratory chain, where iron cycles between Fe^{2+} and Fe^{3+}, heme iron must remain in the reduced Fe^{2+} state to maintain its oxygen-carrying ability. Hemoglobin carrying four oxygens (oxyhemoglobin) is bright red. Hemoglobin that has lost one or more oxygens (deoxyhemoglobin) is dark red-purple, which accounts for the darker color of venous blood. Dried blood is brown, because exposure to atmospheric oxygen has oxidized the iron. The color of arterial blood carrying oxygen is

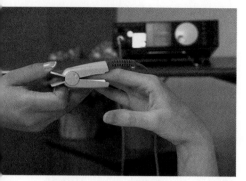

▲ **FIGURE 29.10** **A pulse oxi-metry sensor for continuous monitoring of blood oxygen.** One side of the sensor contains two light-emitting diodes (LEDs), one that emits in the visible red range (better absorbed by dark-red deoxygenated blood) and one that emits in the infrared range (better absorbed by bright-red oxygenated blood). On the opposite side of the sensor, a photodetector measures the light that passes through and sends the signal to an instrument that computes the percent oxygen saturation of the blood and also records the pulse. Normal oxygen saturation is 95–100%. Below 85% tissues are at risk, and below 70% is typically life-threatening.

used in a clinically valuable method for monitoring oxygenation (known as *pulse oximetry*, Figure 29.10).

At normal physiological conditions, the percentage of heme molecules that carry oxygen, known as the *percent saturation,* is dependent on the partial pressure of oxygen in surrounding tissues, as shown in Figure 29.11. The relation is not a simple equilibrium, as demonstrated by the shape of the curve, but is influenced by allosteric interaction. (∞, Section 19.7) Binding each O_2 causes changes in hemoglobin quaternary structure that enhance binding of the next O_2, and releasing each oxygen enhances release of the next. As a result, oxygen is more readily released to tissue where the partial pressure of oxygen is low. The average oxygen partial pressure in peripheral tissue is 40 mm Hg, a pressure at which Hb remains 75% saturated by oxygen, leaving a large amount of O_2 in reserve for emergencies. Note, however, the rapid drop in the curve between 40 mm Hg and 20 mm Hg, the oxygen pressure in tissue where metabolism is occurring rapidly.

Carbon Dioxide Transport, Acidosis, and Alkalosis

Oxygen and carbon dioxide are the "blood gases" transported by erythrocytes. By way of the bicarbonate ion/carbon dioxide buffer, the intimate relationships among H^+ and HCO_3^- concentrations and O_2 and CO_2 partial pressures are essential to maintaining electrolyte and acid–base balance:

$$CO_2(aq) + H_2O(l) \rightleftharpoons H_2CO_3(aq) \rightleftharpoons HCO_3^-(aq) + H^+(aq)$$
Controlled by the lungs   Controlled by the kidneys

In a clinical setting, therefore, monitoring "blood gases" usually refers to measuring the pH of blood as well as the gas concentrations. Carbon dioxide from metabolism in peripheral cells diffuses into interstitial fluid and then into capillaries, where it is transported in blood three ways: (1) dissolved, $CO_2(aq)$, (2) bonded to Hb, or (3) as HCO_3^- in solution. About 7% of the CO_2 dissolves in blood plasma. The rest enters erythrocytes, where some of it bonds to hemoglobin, not in the heme portion of the molecule but by reaction with nonionized amino acid $-NH_2$ groups in the protein part of hemoglobin:

$$Hb-NH_2 + CO_2 \rightleftharpoons Hb-NHCOO^- + H^+$$

Most of the CO_2 is rapidly converted to bicarbonate ion within erythrocytes, which contain a large concentration of carbonic anhydrase. The resulting water-soluble HCO_3^- ion can leave the erythrocyte and travel in the blood to the lungs, where it will be converted back to CO_2 for exhalation. To maintain electrolyte balance, a Cl^- ion enters the erythrocyte for every HCO_3^- ion that leaves, and the process is reversed when the blood reaches the lungs. A cell-membrane protein controls the ion exchange. The exchange is passive because the ions move down their natural concentration gradients, from higher to lower concentrations.

FIGURE 29.11 **Oxygen saturation of hemoglobin at normal physiological conditions.** ▶ Oxygen pressure is about 100 mm Hg in arteries and 20 mm Hg in active muscles. Note the big release of oxygen between 40 mm Hg and 20 mm Hg.

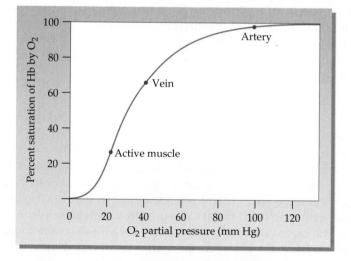

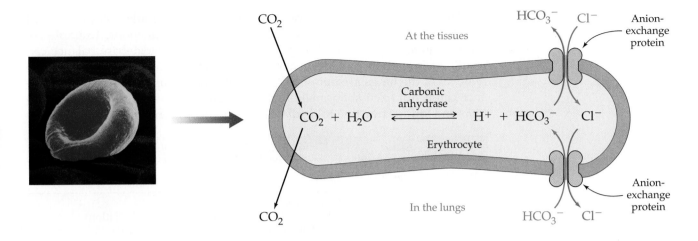

Without some compensating change, the result of hemoglobin binding of CO_2 and the action of carbonic anhydrase would be an unacceptably large increase in acidity. To cope with this, hemoglobin responds by reversibly binding hydrogen ions:

$$Hb \cdot 4\,O_2 + 2\,H^+ \rightleftharpoons Hb \cdot 2\,H^+ + 4\,O_2$$

The release of oxygen is enhanced by allosteric effects when the hydrogen ion binding increases, and oxygen is held more firmly when the hydrogen ion binding decreases.

The changes in the oxygen saturation curve with CO_2 and H^+ concentrations and with temperature are shown in Figure 29.12. The curve shifts to the right, indicating decreased affinity of Hb for O_2, when the H^+ and CO_2 concentrations increase and when the temperature increases. These are exactly the conditions in muscles that are working hard and need more oxygen. The curve shifts to the left, indicating increased affinity of Hb for oxygen, under the opposite conditions of decreased H^+ and CO_2 concentrations and lower temperature.

Homeostasis requires a blood pH between 7.35 and 7.45. A pH outside this range creates either **acidosis** or **alkalosis**.

Acidosis The abnormal condition associated with a blood plasma pH below 7.35; may be respiratory or metabolic.

Alkalosis The abnormal condition associated with a blood plasma pH above 7.45; may be respiratory or metabolic.

Acidosis	Normal	Alkalosis
Below 7.35	7.35–7.45	Above 7.45

◀ **FIGURE 29.12 Changes in oxygen affinity of hemoglobin with changing conditions.** The normal curve of Figure 29.11 is shown in red.

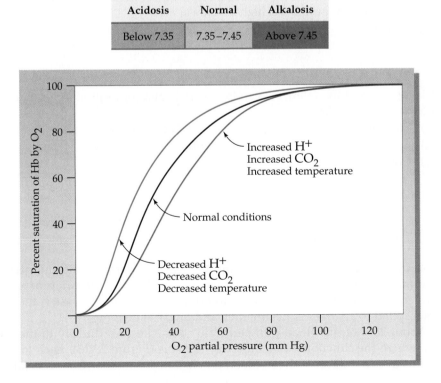

The wide variety of conditions that cause acidosis or alkalosis are divided between respiratory malfunctions and metabolic malfunctions. Examples of each are given in Table 29.2. *Respiratory* disruption of acid–base balance can result when carbon dioxide generation by metabolism and carbon dioxide removal at the lungs are out of balance. *Metabolic* disruption of acid–base balance can result from abnormally high acid generation or failure of buffer systems and kidney function to regulate bicarbonate concentration.

TABLE 29.2   Causes of Acidosis and Alkalosis

Type of Imbalance	Causes

Respiratory acidosis	CO_2 buildup due to:
	    Decreased respiratory activity (hypoventilation)
	    Cardiac insufficiency (for example, congestive failure,
	        cardiac arrest)
	    Deterioration of pulmonary function (for example, asthma,
	        emphysema, pulmonary obstruction, pneumonia)

Respiratory alkalosis	Loss of CO_2 due to:
	    Excessive respiratory activity (hyperventilation, due, for
	        example, to high fever, nervous condition)

Metabolic acidosis	Increased production of metabolic acids due to:
	    Fasting or starvation
	    Diabetes
	    Excessive exercise
	    Decreased acid excretion in urine due to:
	    Poisoning
	    Renal failure
	    Decreased plasma bicarbonate concentration due to:
	    Diarrhea

Metabolic alkalosis	Elevated plasma bicarbonate concentration due to:
	    Vomiting
	    Diuretics
	    Antacid overdose

■ **KEY CONCEPT PROBLEM 29.3**

Carbon dioxide dissolved in body fluids has a pronounced effect on pH.

(a) Does pH go up or down when carbon dioxide dissolves in these fluids? Does this change indicate higher or lower acidity?

(b) What does a blood gas analysis measure?

■ **PROBLEM 29.4**

Classify each of the following conditions as a cause of respiratory or metabolic acidosis or alkalosis.

(a) Pneumonia   **(b)** Kidney failure

(c) Overdose of an antacid   **(d)** Hyperventilation

(e) Congestive heart failure

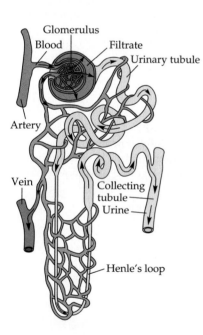

▲ **FIGURE 29.13   Structure of a nephron.** Water moves out of the urinary tubule and the collecting tubule. The concentration of solutes in urine is established as they move both in and out along the tubules.

Labels: Glomerulus, Blood, Filtrate, Urinary tubule, Artery, Vein, Collecting tubule, Urine, Henle's loop

29.7 The Kidney and Urine Formation

The kidneys bear the major responsibility for maintaining a constant internal environment in the body. By managing the elimination of appropriate amounts of water, electrolytes, hydrogen ions, and nitrogen-containing wastes, the kidneys respond to changes in health, diet, and physical activity.

About 25% of the blood pumped from the heart goes directly to the kidneys, where the functional units are the *nephrons* (Figure 29.13). Each kidney contains over a million of them. Blood enters a nephron at a *glomerulus* (at the top in Figure 29.13), a tangle of capillaries surrounded by a fluid-filled space. **Filtration**, the first of three essential kidney functions, occurs here. The pressure of blood pumped into the glomerulus directly from the heart is high enough to push plasma and all its solutes except large proteins across the capillary membrane into the surrounding fluid, the **glomerular filtrate**. The filtrate flows from the capsule into the tubule that makes up the rest of the nephron, and the blood enters the network of capillaries intertwined with the tubule.

About 125 mL of filtrate per minute enters the kidneys, and they produce 180 L of filtrate per day. This filtrate contains not only waste products but also many solutes the body can't afford to lose, such as glucose and electrolytes. Since we excrete only about 1.4 L of urine each day, you can see that another important function of the kidneys is **reabsorption**—the recapture of water and essential solutes by moving them out of the tubule.

Reabsorption alone, however, is not sufficient to provide the kind of control over urine composition that is needed. More of certain solutes must be excreted than is present in the filtrate. This situation is dealt with by **secretion**—the transfer of solutes *into* the kidney tubule.

Reabsorption and secretion require the transfer of solutes and water among the filtrate, the interstitial fluid surrounding the tubule, and blood in the capillaries. Some of the substances reabsorbed and secreted are listed in Table 29.3. Solutes cross the tubule and capillary membranes by passive diffusion in response to concentration or ionic charge differences, or by active transport. Water moves in response to differences in the osmolarity of the fluids on the two sides of the membranes. Solute and water movement is also controlled by hormone-directed variations in the permeability of the tubule membrane.

29.8 Urine Composition and Function

Urine contains the products of glomerular filtration, minus the substances reabsorbed in the tubules, plus the substances secreted in the tubules. The actual concentrations of these substances in urine at any time are determined by the amount of water being excreted, which can vary significantly with water intake, exercise, temperature, and state of health. (For identical quantities of solutes, concentration *decreases* when the quantity of solvent water *increases* and concentration *increases* when the quantity of water *decreases*.)

About 50 g of solids in solution are excreted every day—about 20 g of electrolytes and 30 g of nitrogen-containing wastes (urea and ammonia from amino acid catabolism, creatinine from breakdown of creatine phosphate in muscles, and uric acid from purine catabolism). Normal urine composition is usually reported as the quantity of each solute excreted per day, and laboratory urinalysis often requires collection of all urine excreted during a 24-hr period.

The following paragraphs briefly describe a few of the mechanisms that control the composition of urine.

Acid–Base Balance

Respiration, buffers, and excretion of hydrogen ions in urine combine to maintain acid–base balance. Metabolism normally produces an excess of hydrogen ions, 50–100 mEq of which must be excreted each day to prevent acidosis. Very little free hydrogen ion exists in blood plasma, and therefore very little enters the glomerular filtrate. Instead, the H^+ to be eliminated is produced by the reaction of CO_2 with water in the cells lining the tubules of the nephrons.

$$CO_2 + H_2O \xrightarrow{\text{Carbonic anhydrase}} H^+ + HCO_3^-$$

To bloodstream
To filtrate

Filtration (kidney) Filtration of blood plasma through a glomerulus and into a kidney nephron.

Glomerular filtrate Fluid that enters the nephron from the glomerulus; filtered blood plasma.

Reabsorption (kidney) Movement of solutes out of filtrate in a kidney tubule.

Secretion (kidney) Movement of solutes into filtrate in a kidney tubule.

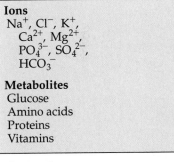

TABLE 29.3 Reabsorption and Secretion in Kidney Tubules

Reabsorbed

Ions
Na^+, Cl^-, K^+,
Ca^{2+}, Mg^{2+},
PO_4^{3-}, SO_4^{2-},
HCO_3^-

Metabolites
Glucose
Amino acids
Proteins
Vitamins

Secreted

Ions
K^+, H^+, Ca^{2+}

Wastes
Creatinine
Ammonia
Organic acids and bases

Miscellaneous
Neurotransmitters
Histamine
Drugs (penicillin, atropine, morphine, numerous others)

 Kidney and Renal Function

The HCO_3^- ions return to the bloodstream, and the H^+ ions enter the filtrate. Thus, the more hydrogen ions there are to be excreted, the more bicarbonate ions are returned to the bloodstream.

The urine must carry away the necessary quantity of H^+ without becoming excessively acid. To accomplish this, the H^+ is tied up by reaction with HPO_4^{2-} absorbed at the glomerulus, or by reaction with NH_3 produced in the tubule cells by deamination of glutamate:

$$H^+ + HPO_4^{2-} \longrightarrow H_2PO_4^-$$

$$H^+ + NH_3 \longrightarrow NH_4^+$$

When acidosis occurs, the kidney responds by synthesizing more ammonia, thereby increasing the quantity of H^+ eliminated.

A further outcome of H^+ production in tubule cells is the net reabsorption of HCO_3^- that entered the filtrate at the glomerulus. The body cannot afford to lose HCO_3^-. The result would be production of additional acid from carbon dioxide by reaction with water. Instead, H^+ secreted into the filtrate combines with HCO_3^- in the filtrate to produce CO_2 and water. Upon returning to the bloodstream, the CO_2 is reconverted to HCO_3^-.

$$\underset{\text{In the filtrate}}{H^+ \quad + \quad HCO_3^-} \quad \longrightarrow \quad \underset{\searrow \text{ To bloodstream}}{CO_2 \quad + \quad H_2O}$$

In summary, acid–base reactions in the kidneys have the following results:

- Secreted H^+ is eliminated in the urine as NH_4^+ or $H_2PO_4^-$.
- Secreted H^+ combines with filtered HCO_3^-, producing CO_2 that returns to the bloodstream and again is converted to HCO_3^-.

Fluid and Na^+ Balance

The amount of water reabsorbed is dependent on the osmolarity of the fluid passing through the kidneys, the antidiuretic hormone–controlled permeability of the collecting duct membrane, and the amount of Na^+ actively reabsorbed. Increased sodium reabsorption means higher interstitial osmolarity, greater water reabsorption, and decreased urine volume. In the opposite condition of decreased sodium reabsorption, less water is reabsorbed and urine volume increases. Diuretic drugs such as furosemide (Endural or Lasix), which is used in treating hypertension and congestive heart failure, act by inhibiting the active transport of Na^+ out of the loop of Henle. Caffeine acts as a diuretic in a similar way.

The reabsorption of Na^+ is normally under the control of the steroid hormone aldosterone. The arrival of chemical messengers signaling a decrease in total blood plasma volume accelerates the secretion of aldosterone. The result is increased Na^+ reabsorption in the kidney tubules accompanied by increased water reabsorption.

Application Automated Clinical Laboratory Analysis

What happens when a physician orders chemical tests of blood, urine, or spinal fluid? The sample goes to a clinical chemistry laboratory, often in a hospital, where most tests are done by automated clinical chemistry analyzers. There are basically two types of chemical analysis, one for the quantity of a chemical (a natural biochemical, a drug, or a toxic substance) and the other for the quantity of an enzyme with a specific metabolic activity.

Many chemical components are measured by mixing a reagent with the sample—the *analyte*—and determining the quantity of a colored product with a photometer, an instrument that measures the absorption of light of a wavelength specific to the product. For each test specified, a portion of the sample is mixed with the appropriate reagent and the photometer is adjusted to the exact wavelength necessary.

When it's not possible to directly from a product that can be seen by a photometer, other reaction sequences that end with a detectable product are devised. Many analytes are the substrates for enzyme-catalyzed reactions. Analysis of the substrate concentration is therefore often made possible by treating the analyte with appropriate enzymes. Glucose is determined in this manner by a pair of enzyme-catalyzed reactions: The glucose is converted to glucose 6-phosphate by the hexokinase-catalyzed reaction with ATP; the phosphate is then oxidized by $NADP^+$; and the quantity of NADPH produced is measured photometrically.

The second type of analysis, determination of the quantity of a specific enzyme or the ratio of two or more enzymes, is valuable in detecting organ damage that allows enzymes to leak into body fluids. For example, elevation of both ALT (alanine aminotransferase) and AST (aspartate aminotransferase) with an AST/ALT ratio greater than 1.0 is characteristic of liver disease. If, however, the AST is greatly elevated and the AST/ALT ratio is higher than 1.5, a myocardial infarction (heart attack) is likely to have occurred. When the analyte is an enzyme, advantage is often taken of its action on a substrate. ALT, for example, is determined by photometrically monitoring the disappearance of NADH in the following pair of coupled reactions (where LD = lactate dehydrogenase):

▲ Blood samples being loaded into an automated analysis instrument. The bar codes on the tubes specify what analyses are to be performed.

$$\text{L-Alanine} + \alpha\text{-Ketoglutarate} \xrightarrow{\text{ALT}} \text{Pyruvate} + \text{L-Glutamate}$$

$$\text{Pyruvate} + \text{NADH/H}^+ \xrightarrow{\text{LD}} \text{Lactate} + \text{NAD}^+$$

Automated analyzers rely on premixed reagents and automatic division of a fluid sample into small portions for each test. A low-volume analyzer that can provide rapid results for a few tests accepts a bar-coded serum or plasma sample cartridge (about the size of a small cassette tape cartridge) followed by bar-coded reagent cartridges. The instrument software reads the bar codes and directs an automatic pipettor (which removes small samples of precisely measured volumes) to transfer the appropriate volume of sample to each test cartridge. The instrument then moves the test cartridge along as the sample and reagents are mixed, the reaction takes place for a measured amount of time, and the photometer reading is taken and converted to the test result.

A high-volume analyzer with more complex software can randomly access 40 or more tests and can run over 400 tests per hour at a cost of less than 10 cents per test. The end result is a printed report on each sample listing the types of tests, the sample values, and a normal range for each test.

See Additional Problems 29.59–61 at the end of the chapter.

Imagine being a medical technologist in a busy hospital laboratory, performing a variety of tests: blood chemistries, hematology, urinalyses, and other assays. Patients are being treated for diseases ranging from diabetes and cancer to thyroid imbalances and rare hereditary metabolic disorders such as phosphofructokinase deficiency. You're busy but you love your work, and when you're called in periodically to draw blood, you're always surprised that some of the patients are so nice and compliant, while others are so angered by the procedure and uncooperative, they act like animals.

Well, don't be surprised, because if you're Jonathan Harris, your patients are animals!

Jonathan is laboratory supervisor at a modern animal hospital in a large metropolitan area. This is no ordinary local veterinary clinic. There are 140 employees, with more than 25 doctors including board-certified specialists and interns. The hospital is staffed around the clock seven days a week.

This wasn't the career Jonathan thought he'd have when he got out of school; then, he worked as a generalist at a human hospital. He switched to a veterinary facility because it was the only job available at the time, and found that he enjoyed it. He liked the fact that the medical technologists worked closely with the doctors, getting a feel for the cases and for what was going on with each animal. So when Jonathan's lab began doing less clinical work and more paperwork, he moved on to his current position.

He is now laboratory supervisor and he loves it. He observes, "Although I didn't care for the introductory chemistry I had early on, almost everything that I learned in the chemistry that I took for the med tech courses I find very useful. It enables me to understand the tests we do here."

For example, take kidney function. An animal with urinary tract blockage will exhibit a variety of symptoms: weakness, lack of appetite, and (because it will have stopped drinking) dehydration. Since the kidneys filter the blood, kidney function can be monitored by measuring blood levels of creatinine and blood urea nitrogen (BUN), protein byproducts. BUN increases in dehydration, regardless of the cause, and this will show up on the admitting blood tests. Rising creatinine levels, on the other hand, typically indicate a more serious underlying kidney problem. Creatinine is not as strongly affected by dehydration, so elevated levels are commonly due to urinary blockage or kidney failure. Once the blockage—usually stones—is removed, creatinine is monitored daily in-house until the animal has recovered.

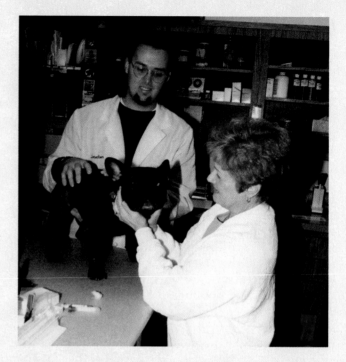

The hospital is also a referring facility for cases that the smaller clinics in the area cannot handle. Jonathan is often called in to draw blood for diagnosis. He says, "I was hired to help the doctors make better diagnoses. Before I came they sent out everything except a few tests that were easy to do on the spot. Now I run blood chemistries, and I do almost all the urinalyses in house. I also do a lot of microscope work—looking at slides of blood smears, skin scrapings, ear smears, cytologies."

The blood smears are examined for a deficiency of platelets or for regenerative anemia: Is that anemic cat making any red cells? The laboratory staff look carefully at red cell morphology. They also look at the white blood cells because, as Jonathan explains, "If you examine a smear and see some strange white cells, cells that looks a little bit immature, and it's an anemic cat, you need to consider the possibility of feline leukemia."

Diabetic dogs and cats are treated routinely. Many animals come in just for blood work. The owner will bring the cat or dog into the lab, Jonathan will draw blood for a blood glucose level and report the results to the doctor.

Jonathan finds his work gives him a lot of satisfaction. He says, "In a different part of the country where there might not be access to a place like this, many animals would be euthanized. But when they come in here, we can do so much for them. Often we can make it possible for them to live relatively normal, happy lives for many, many years."

Summary: Revisiting the Chapter Goals

1. **How are body fluids classified?** Body fluids are either intracellular or extracellular. *Extracellular fluid* includes *blood plasma* (the fluid part of blood) and *interstitial fluid*. *Blood serum* is the fluid remaining after blood has clotted. Solutes in body fluids include blood gases, electrolytes, metabolites, and proteins. Solutes are carried throughout the body in blood and lymph. Exchange of solutes between blood and interstitial fluid occurs at the network of capillaries in peripheral tissues. Exchange of solutes between interstitial fluid and intracellular fluid occurs by passage across cell membranes.

2. **What are the roles of blood in maintaining homeostasis?** The principal functions of blood are (1) transport of solutes and blood gases, (2) regulation, including regulation of heat and acid–base balance, and (3) defense, which includes the *immune response* and *blood clotting*. In addition to plasma and proteins, blood is composed of red blood cells (*erythrocytes*), which transport oxygen; white blood cells (for defense functions); and *platelets*, which participate in blood clotting (Table 29.1).

3. **How do blood components participate in the body's defense mechanisms?** The presence of an *antigen* (a substance foreign to the body) initiates (1) the inflammatory response, (2) the cell-mediated immune response, and (3) the antibody-mediated immune response. The *inflammatory response* is initiated by histamine and accompanied by the destruction of invaders by *phagocytes*. The *cell-mediated response* is effected by *T cells* that can, for example, release a toxic protein that kills invaders. The *antibody-mediated response* is effected by *B cells*, which generates *antibodies (immunoglobins)*, proteins that complex with antigens and destroy them. Blood clotting occurs in a cascade of reactions in which a series of zymogens are activated, ultimately resulting in the formation of a clot composed of *fibrin* and platelets.

4. **How do red blood cells participate in the transport of blood gases?** Oxygen is transported bonded to Fe^{2+} ions in hemoglobin. The percent saturation of hemoglobin with oxygen (Figure 29.12) is governed by the partial pressure of oxygen in surrounding tissues and allosteric variations in hemoglobin structure. Carbon dioxide is transported in blood as a solute, bonded to hemoglobin, or in solution as bicarbonate ion. In peripheral tissues, carbon dioxide diffuses into red blood cells, where it is converted to bicarbonate ion. Acid–base balance is controlled as hydrogen ions generated by bicarbonate formation are bound by hemoglobin. At the lungs, oxygen enters the cells, and bicarbonate and hydrogen ions leave. A blood pH outside the normal range of 7.35–7.45 can be caused by respiratory or metabolic imbalance, resulting in the potentially serious conditions of *acidosis* or *alkalosis*.

5. **How is the composition of urine controlled?** The first essential kidney function is *filtration*, in which plasma and most of its solute cross capillary membranes and enter the *glomerular filtrate*. Water and essential solutes are then reabsorbed, while additional solutes for elimination are secreted into the filtrate. Urine is thus composed of the products of filtration, minus the substances reabsorbed, plus the secreted substances. It is composed of water, nitrogen-containing wastes, and electrolytes (including $H_2PO_4^-$ and NH_4^+) that are excreted to help to maintain acid–base balance. The balance between water and Na^+ excreted or absorbed is governed by the osmolarity of fluid in the kidney, the hormone aldosterone, and various chemical messengers.

Key Words

Acidosis, 821

Alkalosis, 821

Antibody (immunoglobulin), 814

Antigen, 813

Autoimmune disease, 818

Blood clot, 819

Blood plasma, 807

Blood serum, 811

Erythrocytes, 811

Extracellular fluid, 807

Fibrin, 818

Filtration (kidney), 823

Glomerular filtrate, 823

Hemostasis, 818

Immune response, 814

Inflammation, 814

Inflammatory response, 814

Interstitial fluid, 807

Intracellular fluid, 807

Reabsorption (kidney), 823

Secretion (kidney), 823

Whole blood, 811

▪ Understanding Key Concepts

29.5 Body fluids occupy two different compartments, either inside the cells or outside the cells.
  (a) What are the body fluids found inside the cell called?
  (b) What are the body fluids found outside the cell called?
  (c) What are the two major subclasses of fluids found outside the cells?
  (d) What are the major electrolytes found inside the cells?
  (e) What are the major electrolytes found outside the cells?

29.6 Fill in the blanks in the following diagram with the names of the principal components of whole blood.

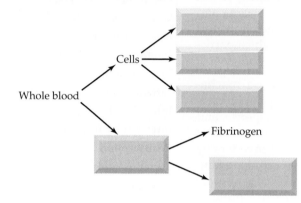

29.7 Fill in the blanks to identify some of the major functions of blood:
  (a) Blood carries _____ from the tissues to lungs.
  (b) Blood carries _____ from lungs to tissues.
  (c) Blood transports _____ from the digestive system to the tissues.
  (d) Blood carries _____ from the tissues to the site of excretion.
  (e) Blood transports _____ from the endocrine glands to their site of binding.
  (f) Blood transports defensive agents such as _____ to destroy foreign material and _____ to prevent blood loss.

29.8 List four symptoms of inflammation. Explain how the chemical messenger histamine is biosynthesized, and how it can elicit each of the symptoms of inflammation.

29.9 Differentiate between cell-mediated immune response and antibody-mediated immune response.

▪ Additional Problems

Body Fluids

29.10 What are the three principal body fluids?

29.11 Which body fluid contains the majority of the body's water?

29.12 What characteristics are needed for a substance to be soluble in body fluids?

29.13 How are components that are not soluble in the blood carried in blood?

29.14 How does blood pressure compare to interstitial fluid pressure in arterial capillaries? In venous capillaries?

29.15 What effects do the differences in pressure between arterial capillaries, interstitial fluids, and venous capillaries have on solutes crossing cell membranes?

29.16 What is the purpose of the lymphatic system?

29.17 Where in the body does the lymph enter the bloodstream?

29.18 What is another name for antidiuretic hormone?

29.19 What is the purpose of antidiuretic hormone?

29.20 What is blood plasma?

29.21 What is blood serum?

29.22 What are the three main types of blood cells?

29.23 What is the major function of each of the three types of blood cells?

29.24 What solutes in body fluids are referred to as electrolytes?

29.25 What are the major electrolytes inside cells and outside cells?

29.26 What is an antigen, and what are the three types of responses the body makes upon exposure to an antigen?

29.27 Antihistamines are often prescribed to counteract the effects of allergies. Explain how these drugs work. (*Hint*: see also Section 9.9.)

29.28 How are specific immune responses similar to the enzyme–substrate interaction?

29.29 What is the name of the class of plasma proteins that are involved in the antibody-mediated immune response?

29.30 What types of cells are associated with the antibody-directed immune response, and how do they work?

29.31 What are memory cells, and what do they "remember"?

29.32 T lymphocytes or T cells have three major functions. Name them.

29.33 T cells are often discussed in conjunction with the disease AIDS, in which a virus destroys these cells. How do T cells work to combat disease?

29.34 What is a blood clot?

29.35 What vitamin and what mineral are specifically associated with the clotting process?

29.36 What two pathways trigger blood clotting?

29.37 Why do you suppose that many of the enzymes involved in blood clotting are secreted by the body as zymogens?

29.38 How many O_2 molecules can be bound by each hemoglobin tetramer?

29.39 What must be the charge of the iron in hemoglobin for it to perform its function?

29.40 How do deoxyhemoglobin and oxyhemoglobin differ in color?

29.41 How does the degree of saturation of hemoglobin vary with the partial pressure of O_2 in the tissues?

29.42 Oxygen has an allosteric interaction with hemoglobin. What are the results of this interaction as oxygen is bonded and as it is released?

29.43 Use Figure 29.11 to estimate the partial pressure of O_2 at which hemoglobin is 50% saturated with oxygen under normal conditions. Dry air at sea level is about 21% oxygen. What is the percentage saturation of your hemoglobin under these conditions?

29.44 What are the three ways of transporting CO_2 in the body?

29.45 Explain two ways in which the CO_2 released by oxidation of glucose tends to raise the H^+ concentration of blood.

29.46 Does each of the following effects cause hemoglobin to release more O_2 to the tissues or to absorb more O_2?

(a) Raising the temperature
(b) Production of CO_2
(c) Increase of the H^+ concentration

29.47 Does a blood serum pH of 7.6 indicate a condition of acidosis or alkalosis?

29.48 Ketoacidosis is a condition that can arise in an individual with diabetes due to excessive production of ketone bodies. Is this condition classified as metabolic acidosis or respiratory acidosis? Explain.

29.49 Is acidosis caused by cardiac insufficiency due to congestive heart failure classified as metabolic acidosis or respiratory acidosis? Explain.

29.50 Kidneys are often referred to as filters that purify the blood. What other two essential functions do the kidneys perform to help maintain homeostasis?

29.51 Write the reactions by which HPO_4^{2-} and HCO_3^- absorb excess H^+ from the urine before elimination.

Applications

29.52 Emollients help keep the skin moist by providing an impenetrable barrier that keeps water from escaping the skin surface. What types of substances are used as emollients? [*Keeping Skin Hydrated*]

29.53 What is a humectant? List some substances that can be used as humectants. [*Keeping Skin Hydrated*]

29.54 How do endothelial cells in brain capillaries differ from those in other capillary systems? [*The Blood–Brain Barrier*]

29.55 What is meant by an asymmetric transport system? Give one specific example of such a system. [*The Blood–Brain Barrier*]

29.56 What type of substance is likely to breach the blood–brain barrier? Would ethanol be likely to cross this barrier? Why or why not? [*The Blood–Brain Barrier*]

29.57 What are four reasons why blood substitutes would be desirable? What two types of blood substitutes are discussed in the application? [*Blood Substitutes*]

29.58 What are the advantages and disadvantages of each of the two blood substitutes discussed? [*Blood Substitutes*]

29.59 How are photometers used in automated analysis? [*Automated Clinical Laboratory Analysis*]

29.60 Why is automated analysis useful to test for enzyme levels in body fluids? [*Automated Clinical Laboratory Analysis*]

29.61 What are some advantages of using automated analyzers compared to using technicians? [*Automated Clinical Laboratory Analysis*]

General Questions and Problems

29.62 Why is ethanol soluble in blood?

29.63 Nursing mothers are able to impart some immunity to their infants. Why do you think this is so?

29.64 Many people find they retain water after eating salty food, evidenced by swollen fingers and ankles. Explain this phenomenon in terms of how the kidneys operate.

29.65 How does active transport differ from osmosis?

29.66 When is active transport necessary to move substances through cell membranes?

29.67 Discuss the importance of the CO_2/HCO_3^- equilibrium in blood and in urine.

29.68 We have discussed homeostasis throughout this text. But what is *hemostasis*? Is it related to homeostasis?

29.69 When people panic, cry, or have a high fever, they often begin to hyperventilate. Hyperventilation is abnormally fast or deep respiration, which results in the loss of carbon dioxide from the blood. Explain how hyperventilation changes the blood chemistry. Why can breathing into a paper bag alleviate hyperventilation?

Appendix A
Scientific Notation

What Is Scientific Notation?

The numbers that you encounter in chemistry are often either very large or very small. For example, there are about 33,000,000,000,000,000,000,000 H_2O molecules in 1.0 mL of water, and the distance between the H and O atoms in an H_2O molecule is 0.000 000 000 095 7 m. These quantities are more conveniently written in *scientific notation* as 3.3×10^{22} molecules and 9.57×10^{-11} m, respectively. In scientific notation (also known as *exponential notation*), a quantity is represented as a number between 1 and 10 multiplied by a power of 10. In this kind of expression, the small raised number to the right of the 10 is the exponent.

Number	Exponential Form	Exponent
1,000,000	1×10^6	6
100,000	1×10^5	5
10,000	1×10^4	4
1,000	1×10^3	3
100	1×10^2	2
10	1×10^1	1
1		
0.1	1×10^{-1}	-1
0.01	1×10^{-2}	-2
0.001	1×10^{-3}	-3
0.0001	1×10^{-4}	-4
0.00001	1×10^{-5}	-5
0.000001	1×10^{-6}	-6
0.0000001	1×10^{-7}	-7

Numbers greater than 1 have *positive* exponents, which tell how many times a number must be *multiplied* by 10 to obtain the correct value. For example, the expression 5.2×10^3 means that 5.2 must be multiplied by 10 three times:

$$5.2 \times 10^3 = 5.2 \times 10 \times 10 \times 10 = 5.2 \times 1000 = 5200$$

Note that doing this means moving the decimal point three places to the right:

5200.

123

The value of a positive exponent indicates *how many places to the right the decimal point must be moved* to give the correct number in ordinary decimal notation.

Numbers less than 1 have *negative exponents*, which tell how many times a number must be *divided* by 10 (or multiplied by one-tenth) to obtain the correct value. Thus, the expression 3.7×10^{-2} means that 3.7 must be divided by 10 two times:

$$3.7 \times 10^{-2} = \frac{3.7}{10 \times 10} = \frac{3.7}{100} = 0.037$$

Note that doing this means moving the decimal point two places to the left:

$$0.037$$
$$21$$

The value of a negative exponent indicates *how may places to the left the decimal point must be moved* to give the correct number in ordinary decimal notation.

Representing Numbers in Scientific Notation

How do you convert a number from ordinary notation to scientific notation? If the number is greater than or equal to 10, shift the decimal point to the *left* by n places until you obtain a number between 1 and 10. Then, multiply the result by 10^n. For example, the number 8137.6 is written in scientific notation as 8.1376×10^3:

Number of places decimal point was shifted to the left

$$8137.6 = 8.1376 \times 10^3$$

Shift decimal point to the left by 3 places to get a number between 1 and 10

When you shift the decimal point to the left by three places, you are in effect dividing the number by $10 \times 10 \times 10 = 1000 = 10^3$. Therefore, you must multiply the result by 10^3 so that the value of the number is unchanged.

To convert a number less than 1 to scientific notation, shift the decimal point to the *right* by n places until you obtain a number between 1 and 10. Then, multiply the result by 10^{-n}. For example, the number 0.012 is written in scientific notation as 1.2×10^{-2}:

Number of places decimal point was shifted to the right

$$0.012 = 1.2 \times 10^{-2}$$

Shift decimal point to the right by 2 places to get a number between 1 and 10

When you shift the decimal point to the right by two places, you are in effect multiplying the number by $10 \times 10 = 100 = 10^2$. Therefore, you must multiply the result by 10^{-2} so that the value of the number is unchanged. ($10^2 \times 10^{-2} = 10^0 = 1$.)

The following table gives some additional examples. To convert from scientific notation to ordinary notation, simply reverse the preceding process. Thus, to write the number 5.84×10^4 in ordinary notation, drop the factor of 10^4 and move the decimal point 4 places to the *right* ($5.84 \times 10^4 = 58,400$). To write the number 3.5×10^{-1} in ordinary notation, drop the factor of 10^{-1} and move the decimal point 1 place to the *left* ($3.5 \times 10^{-1} = 0.35$). Note that you don't need scientific notation for numbers between 1 and 10 because $10^0 = 1$.

Number	Scientific notation
58,400	5.84×10^4
0.35	3.5×10^{-1}
7.296	$7.296 \times 10^0 = 7.296 \times 1$

Mathematical Operations with Scientific Notation

Addition and Subtraction in Scientific Notation

To add or subtract two numbers expressed in scientific notation, both numbers must have the same exponent. Thus, to add 7.16×10^3 and 1.32×10^2, first write the latter number as 1.32×10^3 and then add:

$$
\begin{array}{r}
7.16 \ \times 10^3 \\
+0.132 \times 10^3 \\
\hline
7.29 \ \times 10^3
\end{array}
$$

The answer has three significant figures. (Significant figures are discussed in Section 2.4.) Alternatively, you can write the first number as 71.6×10^2 and then add:

$$
\begin{array}{r}
71.6 \ \times 10^2 \\
+ \ 1.32 \times 10^2 \\
\hline
72.9 \ \times 10^2
\end{array} = 7.29 \times 10^3
$$

Subtraction of these two numbers is carried out in the same manner.

$$
\begin{array}{r}
7.16 \ \times 10^3 \\
+0.132 \times 10^3 \\
\hline
7.03 \ \times 10^3
\end{array} \quad \text{or} \quad
\begin{array}{r}
71.6 \ \times 10^2 \\
- \ 1.32 \times 10^2 \\
\hline
70.3 \ \times 10^2
\end{array} = 7.03 \times 10^3
$$

Multiplication in Scientific Notation

To multiply two numbers expressed in scientific notation, multiply the factors in front of the powers of 10 and then add the exponents. For example,

$$(2.5 \times 10^4)(4.7 \times 10^7) = (2.5)(4.7) \times 10^{4+7} = 12 \times 10^{11} = 1.2 \times 10^{12}$$
$$(3.46 \times 10^5)(2.2 \times 10^{-2}) = (3.46)(2.2) \times 10^{5+(-2)} = 7.6 \times 10^3$$

Both answers have two significant figures.

Division in Scientific Notation

To divide two numbers expressed in scientific notation, divide the factors in front of the powers of 10 and then subtract the exponent in the denominator from the exponent in the numerator. For example,

$$\frac{3 \times 10^6}{7.2 \times 10^2} = \frac{3}{7.2} \times 10^{6-2} = 0.4 \times 10^4 = 4 \times 10^3 \quad \text{(1 significant figure)}$$

$$\frac{7.50 \times 10^{-5}}{2.5 \times 10^{-7}} = \frac{7.50}{2.5} \times 10^{-5-(-7)} = 3.0 \times 10^2 \quad \text{(2 significant figures)}$$

Scientific Notation and Electronic Calculators

With a scientific calculator you can carry out calculations in scientific notation. You should consult the instruction manual for your particular calculator to learn how to enter and manipulate numbers expressed in an exponential format. On most calculators, you enter the number $A \times 10^n$ by (i) entering the number A, (ii) pressing a key labeled EXP or EE, and (iii) entering the exponent n. If the exponent is negative, you press a key labeled $+/-$ before entering the value of n. (Note that you do not enter the number 10.) The calculator displays the number $A \times 10^n$ with the number A on the left followed by some space and then the exponent n. For example,

$$4.625 \times 10^2 \quad \text{is displayed as} \quad 4.625\ 02$$

To add, subtract, multiply, or divide exponential numbers, use the same sequence of keystrokes as you would in working with ordinary numbers. When you add or subtract on a calculator, the numbers need not have the same exponent; the calculator automatically takes account of the different exponents. Remember, though, that the calculator often gives more digits in the answer than the allowed number of significant figures. It's sometimes helpful to outline the calculation on paper, as in the preceding examples, to keep track of the number of significant figures.

■ **PROBLEM A.1**

Perform the following calculations, expressing the results in scientific notation with the correct number of significant figures. (You don't need a calculator for these.)

(a) $(1.50 \times 10^4) + (5.04 \times 10^3)$ **(b)** $(2.5 \times 10^{-2}) - (5.0 \times 10^{-3})$
(c) $(6.3 \times 10^{15}) \times (10.1 \times 10^3)$ **(d)** $(2.5 \times 10^{-3}) \times (3.2 \times 10^{-4})$
(e) $(8.4 \times 10^4) \div (3.0 \times 10^6)$ **(f)** $(5.530 \times 10^{-2}) \div (2.5 \times 10^{-5})$

■ **ANSWERS:**

(a) 2.00×10^4 **(b)** 2.0×10^{-2} **(c)** 6.4×10^{19}
(d) 8.0×10^{-7} **(e)** 2.8×10^{-2} **(f)** 2.2×10^3

■ **PROBLEM A.2**

Perform the following calculations, expressing the results in scientific notation with the correct number of significant figures. (Use a calculator for these.)

(a) $(9.72 \times 10^{-1}) + (3.4823 \times 10^2)$
(b) $(3.772 \times 10^3) - (2.891 \times 10^4)$
(c) $(1.956 \times 10^3) \div (6.02 \times 10^{23})$
(d) $3.2811 \times (9.45 \times 10^{21})$
(e) $(1.0015 \times 10^3) \div (5.202 \times 10^{-9})$
(f) $(6.56 \times 10^{-6}) \times (9.238 \times 10^{-4})$

■ **ANSWERS:**

(a) 3.4920×10^2 **(b)** -2.514×10^4 **(c)** 3.25×10^{-21}
(d) 3.10×10^{22} **(e)** 1.925×10^{11} **(f)** 6.06×10^{-9}

Appendix B
Conversion Factors

Length SI Unit: Meter (m)

    1 meter $= 0.001$ kilometer (km)

            $= 100$ centimeters (cm)

            $= 1.0936$ yards (yd)

    1 centimeter $= 10$ millimeters (mm)

                 $= 0.3937$ inch (in.)

    1 nanometer $= 1 \times 10^{-9}$ meter

    1 Angstrom (Å) $= 1 \times 10^{-10}$ meter

    1 inch $= 2.54$ centimeters

    1 mile $= 1.6094$ kilometers

Volume SI Unit: Cubic meter (m^3)

    1 cubic meter $= 1000$ liters (L)

    1 liter $= 1000$ cubic centimeters (cm^3)

            $= 1000$ milliliters (mL)

            $= 1.056710$ quarts (qt)

    1 cubic inch $= 16.4$ cubic centimeters

Temperature SI Unit: Kelvin (K)

    $0\ K = -273.15°C$

        $= -459.67°F$

    $°F = (9/5)°C + 32°; (1.8 \times °C) + 32°$

    $°C = (5/9)(°F - 32°); \dfrac{(°F - 32°)}{1.8}$

    $K = °C + 273.15°$

Mass SI Unit: Kilogram (kg)

    1 kilogram $= 100$ grams (g)

               $= 2.205$ pounds (lb)

    1 gram $= 1000$ milligrams (mg)

            $= 0.03527$ ounce (oz)

    1 pound $= 453.6$ grams

    1 atomic mass unit $= 1.66054 \times 10^{-24}$ gram

Pressure SI Unit: Pascal (Pa)

    1 pascal $= 9.869 \times 10^{-7}$ atmosphere

    1 atmosphere $= 101,325$ pascals

                $= 760$ mm Hg (Torr)

                $= 14.70\ lb/in^2$

Energy SI Unit: Joule (J)

    1 joule $= 0.23901$ calorie (cal)

    1 calorie $= 4.184$ joules

Appendix C
The Henderson–Hasselbalch Equation and Amino Acids

The Henderson–Hasselbalch equation relates the pH of a weak-acid solution to the pK_a of the acid (the negative logarithm of the acid dissociation constant, K_a, discussed in Section 10.7) and to the concentrations of the acid and its anion. This equation can be obtained by taking the negative logarithms of both sides of the equation derived in Section 10.12:

$$[H_3O^+] = K_a \frac{[HA]}{[A^-]}$$

The Henderson–Hasselbalch Equation
$$-\log [H_3O^+] = -\log\left(K_a \frac{[HA]}{[A^-]} \right)$$

$$pH = pK_a + \log \frac{[A^-]}{[HA]}$$

The Henderson–Hasselbalch equation is particularly valuable for estimating $[A^-]/[HA]$ for amino acids at varying pHs. For example, when $pH = pK_a$, then $\log [A^-]/[HA] = 0$ and $[A^-]/[HA] = 1$. Thus, half of a weak acid is present as HA and half as A^-:

If $\qquad\qquad\qquad\qquad\qquad pH = pK_a$

then $\qquad\qquad\qquad\qquad \log [A^-]/[HA] = 0$

so $\qquad\qquad [A^-]/[HA] = 1/1 \quad$ and $\quad [A^-] = [HA]$

At the point where $pH = pK_a - 2$, then $[A^-]/[HA] = 1/100$ and the acid is only about 1% ionized:

If $\qquad\qquad\qquad\qquad\qquad pH = pK_a - 2$

then $\qquad\qquad\qquad\qquad \log [A^-]/[HA] = -2$

so $\qquad\qquad\qquad\qquad [A^-]/[HA] = 1/100$

The Henderson–Hasselbalch equation thus gives the following relationships:

$$pH = pK_a + 2 \quad [A^-]/[HA] = 100/1$$
$$pH = pK_a + 1 \quad [A^-]/[HA] = 10/1$$
$$pH = pK_a \qquad\quad [A^-]/[HA] = 1/1$$
$$pH = pK_a - 1 \quad [A^-]/[HA] = 1/10$$
$$pH = pK_a - 2 \quad [A^-]/[HA] = 1/100$$

The pK_a of any acid can be experimentally determined by a titration in which a base is slowly added to the acid solution and the change in pH is recorded. The pH at which one-half the acid present has reacted is equal to the pK_a. If the acid has two acidic hydrogens, however, the two pK_a's are measured. Take the amino acid alanine, for example (Section 18.3). The accompanying *titration curve* shows the results of starting at low pH with fully protonated alanine (structure at the bottom) and titrating with 2 equivalents of NaOH until the alanine is fully deprotonated (structure at the top).

The two "legs" of the alanine titration curve—the first leg from pH 1 to 6 and the second leg from pH 6 to 11—show first the titration of the more acidic —COOH group and second the titration of the $-NH_3^+$ group. When 0.5 equivalent of NaOH is added, the first deprotonation is half complete and the pH equals the first pK_a; when 1.0 equivalent of NaOH is added, the first deprotonation is fully complete and the isoelectric point is reached; when 1.5 equivalents of NaOH is added, the second deprotonation is half complete and the pH equals the second pK_a; and when 2.0 equivalents of NaOH is added, the second deprotonation is complete. Of course, the procedure can also be run in reverse. That is, once the pK_a's of an amino acid are known, a titration curve can be calculated with the Henderson–Hasselbalch equation.

The titration curve shows that the isoelectric point of an amino acid with two acidic groups is halfway between its two pK_a's. The value of pI, the pH at the isoelectric point, is therefore one-half the sum of the pK_a's, as illustrated here for alanine:

$$pI = \frac{pK_{a1} + pK_{a2}}{2} = \frac{2.3 + 9.8}{2} = 6.1$$

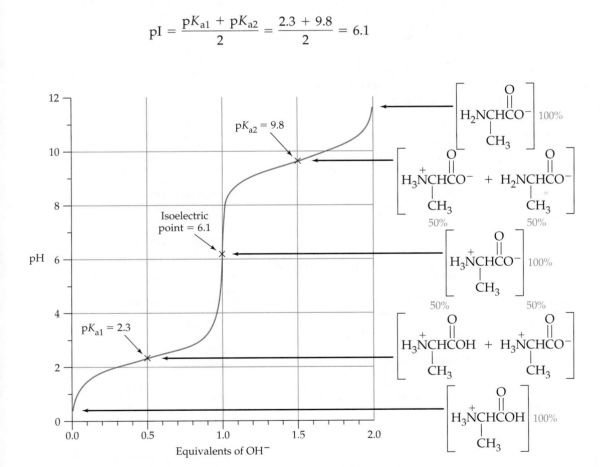

▲ **A titration curve for alanine.** At pH < 1, alanine is entirely protonated; at pH 2.3, alanine is a 50:50 mix of protonated and neutral forms; at pH 6.1, alanine is entirely neutral; at pH 9.8, alanine is a 50:50 mix of neutral and deprotonated forms; at pH > 11, alanine is entirely deprotonated.

Glossary

1,4 Link A glycosidic link between the hemiacetal hydroxyl group at C1 of one sugar and the hydroxyl group at C4 of another sugar.

Acetal A compound that has two ether-like —OR groups bonded to the same carbon atom.

Acetyl coenzyme A (acetyl-SCoA) Acetyl-substituted coenzyme A—the common intermediate that carries acetyl groups into the citric acid cycle.

Acetyl group A $CH_3C\text{=}O$ group.

Achiral The opposite of chiral; having no right- or left-handedness and no non-superimposable mirror images.

Acid A substance that provides H^+ ions in water.

Acid dissociation constant (K_a) The equilibrium constant for the dissociation of an acid (HA), equal to $[H^+][A^-]/[HA]$

Acidosis The abnormal condition associated with a blood plasma pH below 7.35; may be respiratory or metabolic.

Acid–base indicator A dye that changes color depending on the pH of a solution.

Activation (of an enzyme) Any process that initiates or increases the action of an enzyme.

Activation energy (E_{act}) The amount of energy necessary for reactants to surmount the energy barrier to reaction; determines reaction rate.

Active site A pocket in an enzyme with the specific shape and chemical makeup necessary to bind a substrate.

Active transport Movement of substances across a cell membrane with the assistance of energy (for example, from ATP).

Acyl group An $RC\text{=}O$ group.

Addition reaction A general reaction type in which a substance X—Y adds to the multiple bond of an unsaturated reactant to yield a saturated product that has only single bonds.

Addition reaction, aldehydes and ketones Addition of an alcohol or other compound to the carbon–oxygen double bond to give a carbon–oxygen single bond.

Adenosine triphosphate (ATP) The principal energy-carrying molecule; removal of a phosphoryl group to give ADP releases free energy.

Aerobic In the presence of oxygen.

Agonist A substance that interacts with a receptor to cause or prolong the receptor's normal biochemical response.

Alcohol A compound that has an —OH group bonded to a saturated, alkane-like carbon atom, R—OH.

Alcoholic fermentation The anaerobic breakdown of glucose to ethanol plus carbon dioxide by the action of yeast enzymes.

Aldehyde A compound that has a carbonyl group bonded to one carbon and one hydrogen, RCHO.

Aldol reaction A base-catalyzed addition reaction in which two aldehydes or two ketones combine to form a hydroxy aldehyde or a hydroxy ketone, with the —OH group on the second carbon atom away from the $C\text{=}O$ group.

Aldose A monosaccharide that contains an aldehyde carbonyl group.

Alkali metal An element in group 1A of the periodic table.

Alkaline earth metal An element in group 2A of the periodic table.

Alkaloid A naturally occurring nitrogen-containing compound isolated from a plant; usually basic, bitter, and poisonous.

Alkalosis The abnormal condition associated with a blood plasma pH above 7.45; may be respiratory or metabolic.

Alkane A hydrocarbon that has only single bonds.

Alkene A hydrocarbon that contains a carbon–carbon double bond.

Alkoxide ion The anion resulting from deprotonation of an alcohol, RO^-.

Alkoxy group An —OR group.

Alkyl group The part of an alkane that remains when a hydrogen atom is removed.

Alkyl halide A compound that has an alkyl group bonded to a halogen atom, R—X.

Alkyne A hydrocarbon that contains a carbon–carbon triple bond.

Allosteric control An interaction in which the binding of a regulator at one site on a protein affects the protein's ability to bind another molecule at a different site.

Allosteric enzyme An enzyme whose activity is controlled by the binding of an activator or inhibitor at a location other than the active site.

Alpha (α) particle A helium nucleus (He^{2+}), emitted as α- radiation.

Alpha- (α-) amino acid An amino acid in which the amino group is bonded to the carbon atom next to the —COOH group.

Alpha- (α-) helix Secondary protein structure in which a protein chain forms a right-handed coil stabilized by hydrogen bonds between peptide groups along its backbone.

Amide A compound that has a carbonyl group bonded to a carbon atom and a nitrogen atom group, $RCONR'_2$, where the R' groups may be alkyl groups or hydrogen atoms.

Amine A compound that has one or more organic groups bonded to nitrogen; primary, RNH_2; secondary, R_2NH; or tertiary, R_3N.

Amino acid A molecule that contains both an amino group and a carboxylic acid functional group.

Amino acid pool The entire collection of free amino acids in the body.

Amino group The —NH_2 functional group.

Amino-terminal (N-terminal) amino acid The amino acid with the free —NH_3^+ group at the end of a protein.

Ammonium ion A positive ion formed by addition of hydrogen to ammonia or an amine (may be primary, secondary, or tertiary).

Ammonium salt An ionic compound composed of an ammonium cation and an anion; an amine salt.

Amorphous solid A solid whose particles do not have an orderly arrangement.

Amphoteric A substance that can react as either an acid or a base.

Anabolism Metabolic reactions that build larger biological molecules from smaller pieces.

Anaerobic In the absence of oxygen.

Anion A negatively charged ion.

Anomeric carbon atom The hemiacetal C atom in a cyclic sugar; the C atom bonded to an —OH group and an O in the ring.

Anomers Cyclic sugars that differ only in positions of substituents at the hemiacetal carbon (the anomeric carbon); the α form has the —OH on the opposite side from the —CH_2OH; the β form has the —OH on the same side as the —CH_2OH.

Antagonist A substance that blocks or inhibits the normal biochemical response of a receptor.

Antibody (immunoglobulin) Glycoprotein molecule that identifies antigens.

Anticodon A sequence of three ribonucleotides on tRNA that recognizes the complementary sequence (the codon) on mRNA.

Antigen A substance foreign to the body that triggers the immune response.

Antioxidant A substance that prevents oxidation by reacting with an oxidizing agent.

Aqueous solution A solution in which water is the solvent.

Aromatic The class of compounds containing benzene-like rings.

Artificial transmutation The change of one atom into another brought about by a nuclear bombardment reaction.

Atom The smallest and simplest particle of an element.

Atomic mass unit (amu) A convenient unit for describing the mass of an atom; 1 amu = 1/12 the mass of a carbon-12 atom.

Atomic number (Z) The number of protons in an atom.

Atomic weight The weighted average mass of an element's atoms.

ATP synthase The enzyme complex in the inner mitochondrial membrane at which hydrogen ions cross the membrane and ATP is synthesized from ADP.

Autoimmune disease Disorder in which the immune system identifies normal body components as antigens and produces antibodies to them.

Avogadro's law Equal volumes of gases at the same temperature and pressure contain equal numbers of molecules (V/n = constant, or $V_1/n_1 = V_2/n_2$).

Avogadro's number (N_A) The number of units in a mole; 6.02×10^{23}.

Beta- (β-) Oxidation pathway A repetitive series of biochemical reactions that degrades fatty acids to acetyl-SCoA by removing carbon atoms two at a time.

Balanced equation Describing a chemical equation in which the numbers and kinds of atoms are the same on both sides of the reaction arrow.

Base A substance that provides OH^- ions in water.

Base pairing The pairing of bases connected by hydrogen bonding (G-C and A-T), as in the DNA double helix.

Beta (β) particle An electron (e^-), emitted as β radiation.

Beta-(β-) Sheet Secondary protein structure in which adjacent protein chains in the same or different molecules are held in place by hydrogen bonds along the backbones.

Bile acids Steroid acids derived from cholesterol that are secreted in bile.

Bile Fluid secreted by the liver and released into the small intestine during digestion; contains bile acids, bicarbonate ion, and other electrolytes.

Binary compound A compound formed by combination of two different elements.

Biotechnology The application of recombinant DNA and other new bioprocessing techniques to the development of products that improve human health, animal health, and agriculture.

Blood clot A network of fibrin fibers and trapped blood cells that forms at the site of blood loss.

Blood plasma Liquid portion of the blood: an extracellular fluid.

Blood serum Fluid portion of blood remaining after clotting has occurred.

Boiling point (bp) The temperature at which liquid and gas are in equilibrium.

Bond angle The angle formed by three adjacent atoms in a molecule.

Bond dissociation energy The amount of energy that must be supplied to break a bond and separate the atoms in an isolated gaseous molecule.

Bond length The optimum distance between nuclei in a covalent bond.

Boyle's law The pressure of a gas at constant temperature is inversely proportional to its volume (PV = constant, or $P_1V_1 = P_2V_2$).

Branched-chain alkane An alkane that has a branching connection of carbons.

Brønsted–Lowry acid A substance that can donate a hydrogen ion, H^+, to another molecule or ion.

Brønsted–Lowry base A substance that can accept H^+ from an acid.

Buffer A combination of substances that act together to prevent a drastic change in pH; usually a weak acid and its conjugate base.

Carbocation An ion that contains a positively charged carbon atom, R_3C^+.

Carbohydrate A member of a large class of naturally occurring polyhydroxy ketones and aldehydes.

Carbonyl compound Any compound that contains a carbonyl group C=O.

Carbonyl group A functional group that has a carbon atom joined to an oxygen atom by a double bond, C=O.

Carbonyl-group substitution reaction A reaction in which a new group replaces (substitutes for) a group attached to a carbonyl-group carbon in an acyl group.

Carboxyl group The —COOH functional group.

Carboxyl-terminal (C-terminal) amino acid The amino acid with the free —COO⁻ group at the end of a protein.

Carboxylate anion The anion that results from ionization of a carboxylic acid, $RCOO^-$.

Carboxylic acid A compound that has a carbonyl group bonded to a carbon atom and an —OH group, RCOOH.

Carboxylic acid salt An ionic compound containing a carboxylic anion and a cation.

Catabolism Metabolic reaction pathways that break down food molecules and release biochemical energy.

Catalyst A substance that speeds up the rate of a chemical reaction but is itself unchanged.

Cation A positively charged ion.

Chain reaction A reaction that, once started, is self-sustaining.

Change of state The conversion of a substance from one state to another—for example, from a liquid to a gas.

Charles's law The volume of a gas at constant pressure is directly proportional to its Kelvin temperature (V/T = constant, or $V_1/T_1 = V_2/T_2$).

Chemical change A change in the chemical makeup of a substance.

Chemical compound A pure substance that can be broken down into simpler substances by chemical reactions.

Chemical energy The potential energy stored in a chemical compound.

Chemical equation An expression in which symbols and formulas are used to represent a chemical reaction.

Chemical equilibrium A state in which the rates of forward and reverse reactions are the same.

Chemical formula A notation for a chemical compound using element symbols and subscripts to show how many atoms of each element are present.

Chemical property A property that involves change in the chemical nature of a substance.

Chemical reaction A process in which the identity and composition of one or more substances are changed.

Chemistry The study of the nature, properties, and transformations of matter.

Chiral carbon atom (chirality center) A carbon atom bonded to four different groups.

Chiral Having right- or left-handedness; able to have two different mirror-image forms.

Chromosome A complex of proteins and DNA; visible during cell division.

Cis–trans isomers Alkenes that have the same connections between atoms but differ in their three-dimensional structures because of the way that groups are attached to different sides of the double bond. The cis isomer has hydrogen atoms on the same side of the double bond; the trans isomer has them on opposite sides.

Citric acid cycle The series of biochemical reactions that breaks down acetyl groups to produce energy carried by reduced coenzymes and carbon dioxide.

Codon A sequence of three ribonucleotides in the messenger RNA chain that codes for a specific amino acid; also the three nucleotide sequence (a stop codon) that stops translation

Coefficient A number placed in front of a formula to balance a chemical equation.

Coenzyme An organic molecule that acts as an enzyme cofactor.

Cofactor A nonprotein part of an enzyme that is essential to the enzyme's catalytic activity; a metal ion or a coenzyme.

Colligative property A property of a solution that depends only on the number of dissolved particles, not on their chemical identity.

Colloid A homogeneous mixture that contains particles in the range 2–500 nm diameter.

Combined gas law The product of the pressure and volume of a gas is inversely proportional to its temperature (PV/T = constant, or $P_1V_1/T_1 = P_2V_2/T_2$).

Combustion A chemical reaction that produces a flame, usually because of burning with oxygen.

Competititve (enzyme) inhibition Enzyme regulation in which an inhibitor competes with a substrate for binding to the enzyme active site.

Concentration The quantity of a substance dissolved in a given volume.

Concentration gradient A difference in concentration within the same system.

Condensed structure A shorthand way of drawing structures in which C—C and C—H bonds are understood rather than shown.

Conformation The specific three-dimensional arrangement of atoms in a molecule at a given instant.

Conjugate acid The substance formed by addition of H^+ to a base.

Conjugate acid–base pair Two substances whose formulas differ by only a hydrogen ion, H^+.

Conjugate base The substance formed by loss of H^+ from an acid.

Conjugated protein A protein that incorporates one or more non-amino acid units in its structure.

Constitutional isomers Compounds with the same molecular formula but different connections among their atoms.

Conversion factor An expression of the relationship between two units.

Coordinate covalent bond The covalent bond that forms when both electrons are donated by the same atom.

Cosmic rays A mixture of high-energy particles—primarily of protons and various atomic nuclei—that shower the earth from outer space.

Covalent bond A bond formed by sharing electrons between atoms.

Critical mass The minimum amount of radioactive material needed to sustain a nuclear chain reaction.

Crystalline solid A solid whose atoms, molecules, or ions are rigidly held in an ordered arrangement.

Cycloalkane An alkane that contains a ring of carbon atoms.

Cycloalkene A cyclic hydrocarbon that contains a double bond.

Cytoplasm The region between the cell membrane and the nuclear membrane in a eukaryotic cell.

Cytosol The fluid part of the cytoplasm surrounding the organelles within a cell.

d-Block element A transition metal element that results from the filling of d orbitals.

D-Sugar Monosaccharide with the —OH group on the chiral carbon atom farthest from the carbonyl group pointing to the right in a Fischer projection.

Dalton's law The total pressure exerted by a mixture of gases is equal to the sum of the partial pressures exerted by each individual gas.

Decay series A sequential series of nuclear disintegrations leading from a heavy radioisotope to a nonradioactive product.

Dehydration The loss of water from an alcohol to yield an alkene.

Denaturation The loss of secondary, tertiary, or quaternary protein structure due to disruption of noncovalent interactions and/or disulfide bonds that leaves peptide bond and primary structure intact.

Density The physical property that relates the mass of an object to its volume; mass per unit volume.

Deoxyribonucleotide A nucleotide containing 2-deoxy-D-ribose.

Diabetes mellitus A chronic condition due to either insufficient insulin or failure of insulin to activate crossing of cell membranes, by glucose.

Diastereomers Stereoisomers that are not mirror images of each other.

Digestion A general term for the breakdown of food into small molecules.

Dilution factor The ratio of the initial and final solution volumes (V_1/V_2).

Dipole–dipole force The attractive force between positive and negative ends of polar molecules.

Disaccharide A carbohydrate composed of two monosaccharides.

Dissociation The splitting apart of an acid in water to give H^+ and an anion.

Disulfide A compound that contains a sulfur–sulfur bond, RS–SR.

Disulfide bond (in protein) An S–S bond formed between two cysteine side chains; can join two peptide chains together or cause a loop in a peptide chain.

DNA (deoxyribonucleic acid) The nucleic acid that stores genetic information; a polymer of deoxyribonucleotides.

Double bond A covalent bond formed by sharing two electron pairs.

Double helix Two strands coiled around each other in a screwlike fashion; in most organisms the two polynucleotides of DNA form a double helix.

Drug Any substance that alters body function when it is introduced from an external source.

Eicosanoid A lipid derived from a 20-carbon unsaturated carboxylic acid.

Electrolyte A substance that produces ions and therefore conducts electricity when dissolved in water.

Electron A negatively charged subatomic particle.

Electron affinity The energy released on adding an electron to a single atom in the gaseous state.

Electron capture A process in which the nucleus captures an inner-shell electron from the surrounding electron cloud, thereby converting a proton into a neutron.

Electron configuration The specific arrangement of electrons in an atom's shells and subshells.

Electron shell A grouping of electrons in an atom according to energy.

Electron subshell A grouping of electrons in a shell according to the shape of the region of space they occupy.

Electron-dot symbol An atomic symbol with dots placed around it to indicate the number of valence electrons.

Electron-transport chain The series of biochemical reactions that passes electrons from reduced coenzymes to oxygen and is coupled to ATP formation.

Electronegativity The ability of an atom to attract electrons in a covalent bond.

Element A fundamental substance that can't be broken down chemically into any simpler substance.

Elimination reaction A general reaction type in which a saturated reactant yields an unsaturated product by losing groups from two adjacent carbon atoms.

Enantiomers, optical isomers The two mirror-image forms of a chiral molecule.

Endergonic A nonspontaneous reaction or process that absorbs free energy and has a positive ΔG.

Endocrine system A system of specialized cells, tissues, and ductless glands that excretes hormones and shares with the nervous system the responsibility for maintaining constant internal body conditions and responding to changes in the environment.

Endothermic A process or reaction that absorbs heat and has a positive ΔH.

Energy The capacity to do work or supply heat.

Enthalpy change (ΔH) An alternative name for heat of reaction.

Entropy (S) The amount of disorder in a system.

Enzyme A protein or other molecule that acts as a catalyst for a biological reaction.

Equilibrium constant (K) Value of the equilibrium constant expression for a given reaction.

Equivalent (Eq), ion The mass in grams of ions that contains Avogadro's number of charges.

Erythrocytes Red blood cells; transporters of blood gases.

Essential amino acid An amino acid that cannot be synthesized by the body and thus must be obtained in the diet.

Ester A compound that has a carbonyl group bonded to a carbon atom and an —OR group, RCOOR′.

Esterification The reaction between an alcohol and a carboxylic acid to yield an ester plus water.

Ether A compound that has an oxygen atom bonded to two organic groups, R—O—R.

Ethyl group The —CH_2CH_3 alkyl group.

Exergonic A spontaneous reaction or process that releases free energy and has a negative ΔG.

Exon A nucleotide sequence in DNA that is part of a gene and codes for part of a protein.

Exothermic A process or reaction that releases heat and has a negative ΔH.

Extracellular fluid Fluid outside cells.

f-Block element An inner transition metal element that results from the filling of f orbitals.

Facilitated diffusion Passive transport across a cell membrane with the assistance of a protein that changes shape.

Factor-label method A problem-solving procedure in which equations are set up so that unwanted units cancel and only the desired units remain.

Fat A mixture of triacylglycerols that is solid because it contains a high proportion of saturated fatty acids.

Fatty acid A long-chain carboxylic acid; those in animal fats and vegetable oils often have 12–22 carbon atoms.

Feedback control Regulation of an enzyme's activity by the product of a reaction later in a pathway.

Fermentation The production of energy under anaerobic conditions.

Fibrin Insoluble protein that forms the fiber framework of a blood clot.

Fibrous protein A tough, insoluble protein whose protein chains form fibers or sheets.

Filtration (kidney) Filtration of blood plasma through a glomerulus and into a kidney nephron.

Fischer projection Structure that represents chiral carbon atoms as the intersections of two lines, with the horizontal lines representing bonds pointing out of the page and the vertical lines representing bonds pointing behind the page. For sugars, the aldehyde or ketone is at the top.

Formula unit The formula that identifies the smallest neutral unit of a compound.

Formula weight The sum of the atomic weights of the atoms in one formula unit of any compound.

Free radical An atom or molecule with an unpaired electron.

Free-energy change (ΔG) The criterion for spontaneous change (negative ΔG; $\Delta G = \Delta H - T\Delta S$).

Functional group An atom or group of atoms within a molecule that has a characteristic structure and chemical behavior.

Gamma (γ) radiation Radioactivity consisting of high-energy light waves.

Gas A substance that has neither a definite volume nor a definite shape.

Gas laws A series of laws that predict the influence of pressure (P), volume (V), and temperature (T) on any gas or mixture of gases.

Gay-Lussac's law For a fixed amount of gas at a constant voume, pressure is directly proportional to the Kelvin temperature ($P/T = $ constant, or $P_1/T_1 = P_2/T_2$).

Gene Segment of DNA that directs the synthesis of a single polypeptide.

Gene therapy Treatment of disease by replacing, manipulating, or supplementing nonfunctional genes.

Genetic (enzyme) control Regulation of enzyme activity by control of the synthesis of enzymes.

Genetic code The sequence of nucleotides, coded in triplets (codons) in mRNA, that determines the sequence of amino acids in protein synthesis.

Genetic engineering The alteration of the genetic makeup of cells or individuals by the removal, insertion, or modification of genes.

Genome All of the genetic material in the chromosomes of an organism; its size is given as the number of base pairs.

Globular protein A water-soluble protein whose chain is folded in a compact shape with hydrophilic groups on the outside.

Glomerular filtrate Fluid that enters the nephron from the glomerulus; filtered blood plasma.

Gluconeogenesis The biochemical pathway for the synthesis of glucose from non-carbohydrates, such as lactate, amino acids, or glycerol.

Glycerophospholipid (phosphoglyceride) A lipid in which glycerol is linked by ester bonds to two fatty acids and one phosphate, which is in turn linked by another ester bond to an amino alcohol (or other alcohol).

Glycogenesis The biochemical pathway for synthesis of glycogen.

Glycogenolysis The biochemical pathway for breakdown of glycogen to free glucose.

Glycol A dialcohol, or diol.

Glycolipid A lipid with a fatty acid bonded to the $C2-NH_2$ and a sugar bonded to the $C1-OH$ group of sphingosine.

Glycolysis The biochemical pathway that breaks down a molecule of glucose into two molecules of pyruvate plus energy.

Glycoprotein A protein that contains a short carbohydrate chain.

Glycoside A cyclic acetal formed by reaction of a monosaccharide with an alcohol, accompanied by loss of H_2O.

Glycosidic bond Bond between the anomeric carbon atom of a monosaccharide and an $-OR$ group.

Group One of the 18 vertical columns of elements in the periodic table.

Half-life ($t_{1/2}$) The amount of time required for one-half of a radioactive sample to decay.

Halogen An element in group 7A of the periodic table.

Halogenation (alkene) The addition of Cl_2 or Br_2 to a multiple bond to give a 1,2-dihalide product.

Halogenation (aromatic) The substitution of a halogen group ($-X$) for a hydrogen on an aromatic ring.

Heat The kinetic energy transferred from a hotter object to a colder object when the two are in contact.

Heat of fusion The quantity of heat required to completely melt a substance once it has reached its melting point.

Heat of reaction (ΔH) The amount of heat absorbed or released in a reaction.

Heat of vaporization The quantity of heat needed to completely vaporize a liquid once it has reached its boiling point.

Hemiacetal A compound with both an alcohol-like $-OH$ group and an ether-like $-OR$ group bonded to the same carbon atom.

Hemostasis The stopping of bleeding.

Henry's law The solubility of a gas in a liquid is directly proportional to its partial pressure over the liquid at constant temperature.

Heterocycle A ring that contains nitrogen or some other atom in addition to carbon.

Heterogeneous mixture A nonuniform mixture that has regions of different composition.

Homogeneous mixture A uniform mixture that has the same composition throughout.

Hormone A chemical messenger secreted by cells of the endocrine system and transported through the bloodstream to target cells with appropriate receptors where it elicits a response.

Hydration The addition of water to a multiple bond to give an alcohol product.

Hydrocarbon An organic compound that contains only carbon and hydrogen.

Hydrogen bond The attraction between a hydrogen atom bonded to an electronegative O, N, or F atom and another nearby electronegative O, N, or F atom.

Hydrogenation The addition of H_2 to a multiple bond to give a saturated product.

Hydrohalogenation The addition of HCl or HBr to a multiple bond to give an alkyl halide product.

Hydrolysis A reaction in which a bond or bonds are broken and the $H-$ and $-OH$ of water add to the atoms of the broken bond or bonds.

Hydronium ion The H_3O^+ ion, formed when an acid reacts with water.

Hydrophilic Water-loving; a hydrophilic substance dissolves in water.

Hydrophobic Water-fearing; a hydrophobic substance does not dissolve in water.

Hygroscopic Having the ability to pull water molecules from the surrounding atmosphere.

Hyperglycemia Higher-than-normal blood glucose concentration.

Hypoglycemia Lower-than-normal blood glucose concentration.

Ideal gas A gas that obeys all the assumptions of the kinetic–molecular theory.

Ideal gas law A general expression relating pressure, volume, temperature, and amount for an ideal gas: $PV = nRT$.

Immune response Defense mechanism of the immune system dependent on the recognition of specific antigens, including viruses, bacteria, toxic substances, and infected cells; either cell-mediated or antibody-mediated.

Induced-fit model A model of enzyme action in which the enzyme has a flexible active site that changes shape to best fit the substrate and catalyze the reaction.

Inflammation Result of the inflammatory response: includes swelling, redness, warmth, and pain.

Inflammatory response A nonspecific defense mechanism triggered by antigens or tissue damage.

Inhibition (of an enzyme) Any process that slows or stops the action of an enzyme.

Inner transition metal element An element in one of the 14 groups shown separately at the bottom of the periodic table.

Intermolecular force A force that acts between molecules and holds molecules close to one another in liquids and solids.

Interstitial fluid Fluid surrounding cells: an extracellular fluid.

Intracellular fluid Fluid inside cells.

Intron A portion of DNA between coding regions of a gene (exons); is transcribed and then removed from final messenger RNA.

Ion An electrically charged atom or group of atoms.

Ion-product constant for water (K_w) The product of the H_3O^+ and OH^- molar concentrations in water or any aqueous solution ($K_w = [H_3O^+][OH^-] = 1.00 \times 10^{-14}$).

Ionic bond The electrical attractions between ions of opposite charge in a crystal.

Ionic compound A compound that contains ionic bonds.

Ionic equation An equation in which ions are explicitly shown.

Ionic solid A crystalline solid held together by ionic bonds.

Ionization energy The energy required to remove one electron from a single atom in the gaseous state.

Ionizing radiation A general name for high-energy radiation of all kinds.

Irreversible (enzyme) inhibition Enzyme deactivation in which an inhibitor forms covalent bonds to the active site, permanently blocking it.

Isoelectric point (pI) The pH at which a sample of an amino acid has equal number of $+$ and $-$ charges.

Isomers Compounds with the same molecular formula but different structures.

Isopropyl group The branched-chain alkyl group $-CH(CH_3)_2$.

Isotonic Having the same osmolarity.

Isotopes Atoms with identical atomic numbers but different mass numbers.

Ketoacidosis Lowered blood pH due to accumulation of ketone bodies.

Ketogenesis The synthesis of ketone bodies from acetyl-SCoA.

Ketone A compound that has a carbonyl group bonded to two carbons in organic groups that can be the same or different, $R_2C{=}O$, RCOR'.

Ketone bodies Compounds produced in the liver that can be used as fuel by muscle and brain tissue; 3-hydroxybutyrate, acetoacetate, and acetone.

Ketose A monosaccharide that contains a ketone carbonyl group.

Kinetic energy The energy of an object in motion.

Kinetic–molecular theory of gases A group of assumptions that explain the behavior of gases.

L-Sugar Monosaccharide with the —OH group on the chiral carbon atom farthest from the carbonyl group pointing to the left in a Fischer projection.

Law of conservation of energy Energy can be neither created nor destroyed in any physical or chemical change.

Law of conservation of mass Matter can be neither created nor destroyed in any physical or chemical change.

Le Châtelier's principle When a stress is applied to a system in equilibrium, the equilibrium shifts to relieve the stress.

Lewis structure A molecular representation that shows both the connections among atoms and the locations of lone-pair valence electrons.

Line structure A shorthand way of drawing structures in which atoms aren't shown; instead, a carbon atom is understood to be at every intersection of lines, and hydrogens are filled in mentally.

Lipid A naturally occurring molecule from a plant or animal that is soluble in nonpolar organic solvents.

Lipid bilayer The basic structural unit of cell membranes; composed of two parallel sheets of membrane lipid molecules arranged tail to tail.

Lipogenesis The biochemical pathway for synthesis of fatty acids from acetyl-SCoA.

Lipoprotein A lipid–protein complex that transports lipids.

Liposome A spherical structure in which a lipid bilayer surrounds a water droplet.

Liquid A substance that has a definite volume but that changes shape to fill its container.

London dispersion force The short-lived attractive force due to the constant motion of electrons within molecules.

Lone pair A pair of electrons that is not used for bonding.

Main group element An element in one of the two groups on the left or the six groups on the right of the periodic table.

Markovnikov's rule In the addition of HX to an alkene, the H becomes attached to the carbon that already has the most H's, and the X becomes attached to the carbon that has fewer H's.

Mass A measure of the amount of matter in an object.

Mass number (A) The total number of protons and neutrons in an atom.

Matter The physical material that makes up the universe; anything that has mass and occupies space.

Melting point (mp) The temperature at which solid and liquid are in equilibrium.

Messenger RNA (mRNA) The RNA that carries the code transcribed from DNA and directs protein synthesis.

Metabolism The overall sum of the many reactions taking place in an organism.

Metal A malleable element with a lustrous appearance that is a good conductor of heat and electricity.

Metalloid An element whose properties are intermediate between those of a metal and a nonmetal.

Methyl group The —CH_3 alkyl group.

Micelle A spherical cluster formed by the aggregation of soap or detergent molecules so that their hydrophobic ends are in the center and their hydrophilic ends are on the surface.

Miscible Mutually soluble in all proportions.

Mitochondrial matrix The space surrounded by the inner membrane of a mitochondrion.

Mitochondrion (plural, mitochondria) An egg-shaped organelle where small molecules are broken down to provide the energy for an organism.

Mixture A blend of two or more substances, each of which retains its chemical identity.

Mobilization (of triacylglycerols) Hydrolysis of triacylglycerols in adipose tissue and release of fatty acids into the bloodstream.

Molar mass The mass in grams of one mole of a substance, numerically equal to the molecular weight.

Molarity (M) Concentration expressed as the number of moles of solute per liter of solution.

Mole The amount of a substance corresponding to 6.02×10^{23} units.

Molecular compound A compound that consists of molecules rather than ions.

Molecular formula A formula that shows the numbers and kinds of atoms in one molecule of a compound.

Molecular weight The sum of the atomic weights of the atoms in a molecule.

Molecule A group of atoms held together by covalent bonds.

Monomer A small molecule that is used to prepare a polymer.

Monosaccharide (simple sugar) A carbohydrate with 3–7 carbon atoms.

Mutagen A substance that causes mutations.

Mutarotation Change in rotation of plane-polarized light resulting from the equilibrium between cyclic anomers and the open-chain form of a sugar.

Mutation An error in base sequence that is carried along in DNA replication.

n-propyl group The straight-chain alkyl group —$CH_2CH_2CH_3$.

Native protein A protein with the shape (secondary, tertiary, and quaternary structure) in which it exists naturally in living organisms.

Net ionic equation An equation that does not include spectator ions.

Neurotransmitter A chemical messenger that travels between a neuron and a neighboring neuron or other target cell to transmit a nerve impulse.

Neutralization reaction The reaction of an acid with a base.

Neutron An electrically neutral subatomic particle.

Nitration The substitution of a nitro group (—NO_2) for a hydrogen on an aromatic ring.

Noble gas An element in group 8A of the periodic table.

Noncompetitive (enzyme) inhibition Enzyme regulation in which an inhibitor binds to an enzyme elsewhere than at the active site, thereby changing the shape of the enzyme's active site and reducing its efficiency.

Nonelectrolyte A substance that does not produce ions when dissolved in water.

Nonessential amino acid One of eleven amino acids that are synthesized in the body and are therefore not necessary in the diet.

Nonmetal An element that is a poor conductor of heat and electricity.

Normal boiling point The boiling point at a pressure of exactly 1 atmosphere.

Normality (N) A measure of acid (or base) concentration expressed as the number of acid (or base) equivalents per liter of solution.

Nuclear decay The spontaneous emission of a particle from an unstable nucleus.

Nuclear fission The fragmenting of heavy nuclei.

Nuclear fusion The joining together of light nuclei.

Nuclear reaction A reaction that changes an atomic nucleus, usually causing the change of one element into another.

Nucleic acid A polymer of nucleotides.

Nucleon A general term for both protons and neutrons.

Nucleoside A five-carbon sugar bonded to a cyclic amine base; like a nucleotide but missing the phosphate group.

Nucleotide A five-carbon sugar bonded to a cyclic amine base and one phosphate group (a nucleoside monophosphate); monomer for nucleic acids.

Nucleus The dense, central core of an atom that contains protons and neutrons.

Nuclide The nucleus of a specific isotope of an element.

Octet rule Main-group elements tend to undergo reactions that leave them with 8 valence electrons.

Oil A mixture of triacylglycerols that is liquid because it contains a high proportion of unsaturated fatty acids.

Oligonucleotide A small, linear sequence of nucleotides.

Orbital A region of space within an atom where an electron in a given subshell can be found.

Organic chemistry The study of carbon compounds.

Osmolarity (osmol) The sum of the molarities of all dissolved particles in a solution.

Osmosis The passage of solvent through a semipermeable membrane separating two solutions of different concentration.

Osmotic pressure The amount of external pressure applied to the more concentrated solution to halt the passage of solvent molecules across a semipermeable membrane.

Oxidation The loss of one or more electrons by an atom.

Oxidation number A number that indicates whether an atom is neutral, electron-rich, or electron-poor.

Oxidation–Reduction, or Redox, reaction A reaction in which electrons are transferred from one atom to another.

Oxidative deamination Conversion of an amino acid —NH_2 group to an α-keto group, with removal of NH_4^+.

Oxidative phosphorylation The synthesis of ATP from ADP using energy released in the electron-transport chain.

Oxidizing agent A reactant that causes an oxidation by taking electrons from or decreasing the oxidation number of another reactant.

p-**Block element** A main group element that results from the filling of *p* orbitals.

Partial pressure The pressure exerted by a gas in a mixture.

Parts per billion (ppb) Number of parts per one billion (10^9) parts.

Parts per million (ppm) Number of parts per one million (10^6) parts.

Passive transport Movement of a substance across a cell membrane without the use of energy, from a region of higher concentration to a region of lower concentration.

Pentose phosphate pathway The biochemical pathway that produces ribose (a pentose), NADPH, and other sugar phosphates from glucose; an alternative to glycolysis.

Peptide bond An amide bond that links two amino acids together.

Percent yield The percent of the theoretical yield actually obtained from a chemical reaction.

Period One of the 7 horizontal rows of elements in the periodic table.

Periodic table A table of the elements in order of increasing atomic number and grouped according to their chemical similarities.

pH A measure of the acid strength of a solution; the negative common logarithm of the H_3O^+ concentration.

Phagocytosis Engulfing and digestion of a particle by a cell.

Phenol A compound that has an —OH group bonded directly to an aromatic, benzene-like ring, Ar—OH.

Phenyl The C_6H_5— group.

Phosphate ester A compound formed by reaction of an alcohol with phosphoric acid; may be a monoester, $ROPO_3H_2$; a diester, $(RO)_2PO_3H$; or a triester, $(RO)_3PO$; also may be a di- or triphosphate.

Phospholipid A lipid that has an ester link between phosphoric acid and an alcohol (glycerol or sphingosine).

Phosphoryl group The —PO_3^{2-} group in organic phosphates.

Phosphorylation Transfer of a phosphoryl group, —PO_3^{2-}, between organic molecules.

Physical change A change that does not affect the chemical makeup of a substance or object.

Physical property A property that can be determined without changing the chemical nature of a substance or object.

Physical quantity A physical property that can be measured.

Polar covalent bond A bond in which the electrons are attracted more strongly by one atom than by the other.

Polyatomic ion An ion that is composed of more than one atom.

Polymer A large molecule formed by the repetitive bonding together of many smaller molecules.

Polysaccharide (complex carbohydrate) A carbohydrate that is a polymer of monosaccharides.

Polyunsaturated fatty acid A long-chain fatty acid that has two or more carbon–carbon double bonds.

Positron A "positive electron," which has the same mass as an electron but a positive charge.

Potential energy Energy that is stored because of position, composition, or shape.

Precipitate An insoluble solid that forms in solution during a chemical reaction.

Pressure The force per unit area pushing against a surface.

Primary carbon atom A carbon atom with one other carbon attached to it.

Primary protein structure The sequence in which amino acids are linked by peptide bonds in a protein.

Product A substance that is formed in a chemical reaction and is written on the right side of the reaction arrow in a chemical equation.

Property A characteristic useful for identifying a substance or object.

Protein A large biological molecule made of many amino acids linked together through amide (peptide) bonds.

Proton A positively charged subatomic particle.

Quaternary ammonium ion A positive ion with four organic groups bonded to the nitrogen atom.

Quaternary ammonium salt An ionic compound composed of a quaternary ammonium ion and an anion.

Quaternary carbon atom A carbon atom with four other carbons attached to it.

Quaternary protein structure The way in which two or more protein chains aggregate to form large, ordered structures.

Radioactivity The spontaneous emission of radiation from a nucleus.

Radioisotope A radioactive isotope.

Radionuclide The nucleus of a radioactive isotope.

Reabsorption (kidney) Movement of solutes out of filtrate in a kidney tubule.

Reactant A substance that undergoes change in a chemical reaction and is written on the left side of the reaction arrow in a chemical equation.

Reaction mechanism A description of the individual steps by which old bonds are broken and new bonds are formed in a reaction.

Reaction rate A measure of how rapidly a reaction occurs; determined by E_{act}.

Rearrangement reaction A general reaction type in which a molecule undergoes bond reorganization to yield an isomer.

Receptor A molecule or portion of a molecule with which a hormone, neurotransmitter, or other biochemically active molecule connects to initiate a response in a target cell.

Recombinant DNA DNA that contains segments from two different species.

Reducing agent A reactant that causes a reduction by giving up electrons or increasing the oxidation number of another reactant.

Reducing sugar A carbohydrate that reacts in basic solution with a mild oxidizing agent.

Reduction The gain of one or more electrons by an atom.

Reductive deamination Conversion of an α-keto acid to an amino acid by reaction with NH_4^+.

Regular tetrahedron A geometric figure with four identical triangular faces.

Replication The process by which copies of DNA are made when a cell divides.

Residue (amino acid) An amino acid unit in a polypeptide.

Resonance The phenomenon where the true structure of a molecule is an average among two or more conventional structures.

Reversible reaction A reaction that can go in either the forward direction or the reverse direction, from products to reactants or reactants to products.

Ribonucleotide A nucleotide containing D-ribose.

Ribosomal RNA (rRNA) The RNA that is complexed with proteins in ribosomes.

Ribosome The structure in the cell where protein synthesis occurs; composed of protein and rRNA.

RNA (ribonucleic acids) The nucleic acids (messenger, transfer, and ribosomal) responsible for putting the genetic information to use in protein synthesis; polymers of ribonucleotides.

Rounding off A procedure used for deleting nonsignificant figures.

s-**Block element** A main group element that results from the filling of an *s* orbital.

Salt An ionic compound formed from reaction of an acid with a base.

Saponification The reaction of an ester with aqueous hydroxide ion to yield an alcohol and the metal salt of a carboxylic acid.

Saturated A molecule whose carbon atoms bond to the maximum number of hydrogen atoms.

Saturated fatty acid A long-chain carboxylic acid containing only carbon–carbon single bonds.

Saturated solution A solution that contains the maximum amount of dissolved solute at equilibrium.

Scientific notation A number expressed as the product of a number between 1 and 10, times the number 10 raised to a power.

Second messenger Chemical messenger released inside a cell when a hydrophilic hormone or neurotransmitter connects with a receptor on the cell surface.

Secondary carbon atom A carbon atom with two other carbons attached to it.

Secondary protein structure Regular and repeating structural patterns (for example, α-helix, β-sheet) created by hydrogen bonding between backbone atoms in neighboring segments of protein chains.

Secretion (kidney) Movement of solutes into filtrate in a kidney tubule.

SI units Units of measurement defined by the International System of Units.

Side chain (amino acid) Group bonded to thecarbon next to the carboxyl group in an amino acid; different in different amino acids

Significant figures The number of meaningful digits used to express a value.

Simple diffusion Passive transport by the random motion of diffusion through the cell membrane.

Simple protein A protein composed of only amino acid residues.

Single bond A covalent bond formed by sharing one electron pair.

Soap The mixture of salts of fatty acids formed on saponification of animal fat.

Solid A substance that has a definite shape and volume.

Solubility The maximum amount of a substance that will dissolve in a given amount of solvent at a specified temperature.

Solute A substance dissolved in a liquid.

Solution A homogeneous mixture that contains particles the size of a typical ion or small molecule.

Solvation The clustering of solvent molecules around a dissolved solute molecule or ion.

Solvent The liquid in which another substance is dissolved.

Specific gravity The density of a substance divided by the density of water at the same temperature.

Specific heat The amount of heat that will raise the temperature of 1 g of a substance by 1°C.

Specificity (enzyme) The limitation of the activity of an enzyme to a specific substrate, specific reaction, or specific type of reaction.

Spectator ion An ion that appears unchanged on both sides of a reaction arrow.

Sphingolipid A lipid derived from the amino alcohol sphingosine.

Spontaneous process A process or reaction that, once started, proceeds on its own without any external influence.

Standard molar volume The volume of one mole of a gas at standard temperature and pressure (22.4 L).

Standard temperature and pressure (STP) Standard conditions for a gas, defined as 0°C (273 K) and 1 atm (760 mm Hg) pressure.

State of matter The physical state of a substance as a solid, a liquid, or a gas.

Stereoisomers Isomers that have the same molecular and structural formulas, but different spatial arrangements of their atoms.

Steroid A lipid whose structure is based on the following tetracyclic (four-ring) carbon skeleton:

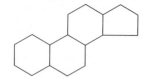

Straight-chain alkane An alkane that has all its carbons connected in a row.

Strong acid An acid that gives up H⁺ easily and is essentially 100% dissociated in water.

Strong base A base that has a high affinity for H⁺ and holds it tightly.

Strong electrolyte A substance that ionizes completely when dissolved in water.

Structural formula A molecular representation that shows the connections among atoms by using lines to represent covalent bonds.

Subatomic particles Three kinds of fundamental particles from which atoms are made: protons, neutrons, and electrons.

Substituent An atom or group of atoms attached to a parent compound.

Substitution reaction A general reaction type in which an atom or group of atoms in a molecule is replaced by another atom or group of atoms.

Substrate A reactant in an enzyme-catalyzed reaction.

Sulfonation The substitution of a sulfonic acid group ($-SO_3H$) for a hydrogen on an aromatic ring.

Supersaturated solution A solution that contains more than the maximum amount of dissolved solute; a nonequilibrium situation.

Synapse The place where the tip of a neuron and its target cell lie adjacent to each other.

Synaptic cleft The gap between a neuron and its target cell that is crossed by a neurotransmitter.

Temperature The measure of how hot or cold an object is.

Tertiary carbon atom A carbon atom with three other carbons attached to it.

Tertiary protein structure The way in which an entire protein chain is coiled and folded into its specific three-dimensional shape.

Thiol A compound that contains an $-SH$ group, $R-SH$.

Titration A procedure for determining the total acid or base concentration of a solution.

Transamination The interchange of the amino group of an amino acid and the keto group of an α-keto acid.

Transcription The process by which the information in DNA is read and used to synthesize RNA.

Transfer RNA (tRNA) The RNA that transports amino acids into position for protein synthesis.

Transition metal element An element in one of the 10 smaller groups near the middle of the periodic table.

Translation The process by which RNA directs protein synthesis.

Transmutation The change of one element into another.

Triacylglycerol (triglyceride) A triester of glycerol with three fatty acids.

Triple bond A covalent bond formed by sharing three electron pairs.

Turnover number The maximum number of substrate molecules acted upon by one molecule of enzyme per unit time.

Unit A defined quantity used as a standard of measurement.

Unsaturated A molecule that contains a carbon–carbon multiple bond, to which more hydrogen atoms can be added.

Unsaturated fatty acid A long-chain carboxylic acid containing one or more carbon–carbon double bonds.

Urea cycle The cyclic biochemical pathway that produces urea for excretion.

Valence electron An electron in the outermost, or valence, shell of an atom.

Valence shell The outermost electron shell of an atom.

Valence-shell electron-pair repulsion (VSEPR) model A method for predicting molecular shape by noting how many electron charge clouds surround atoms and assuming that the clouds orient as far away from one another as possible.

Vapor The gas molecules in equilibrium with a liquid.

Vapor pressure The partial pressure of gas molecules in equilibrium with a liquid.

Vitamin An organic molecule, essential in trace amounts that must be obtained in the diet because it is not synthesized in the body.

Volume/volume percent concentration [(v/v)%] Concentration expressed as the number of milliliters of solute dissolved in 100 mL of solution.

Wax A mixture of esters of long-chain carboxylic acids with long-chain alcohols.

Weak acid An acid that gives up H⁺ with difficulty and is less than 100% dissociated in water.

Weak base A base that has only a slight affinity for H⁺ and holds it weakly.

Weak electrolyte A substance that is only partly ionized in water.

Weight A measure of the gravitational force that the earth or other large body exerts on an object.

Weight/volume percent concentration [(w/v)%] Concentration expressed as the number of grams of solute per 100 mL of solution.

Whole blood Blood plasma plus blood cells.

X rays Electromagnetic radiation with an energy somewhat less than that of γ rays.

Yield The amount of product actually formed in a reaction.

Zwitterion A neutral dipolar ion that has one + charge and one − charge.

Zymogen A compound that becomes an active enzyme after undergoing a chemical change.

Answers to Selected Problems

Short answers are given for in-chapter problems, *Understanding Key Concepts* problems, and even-numbered end-of-chapter problems. Explanations and full answers for all problems are provided in the accompanying *Study Guide and Solutions Manual*.

Chapter 1

1.1 all **1.2** physical: **(a)**, **(d)**; chemical: **(b)**, **(c)** **1.3** gas **1.4** solid **1.5** mixture: **(a)**, **(d)**; pure: **(b)**, **(c)** **1.6** physical: **(a)**, **(c)**; chemical: **(b)** **1.7** **(a)** U **(b)** Ti **(c)** W **1.8** **(a)** sodium **(b)** calcium **(c)** palladium **(d)** potassium **(e)** strontium **(f)** tin **1.9** **(a)** 1 nitrogen atom, 3 hydrogen atoms **(b)** 1 sodium atom, 1 hydrogen atom, 1 carbon atom, 3 oxygen atoms **(c)** 8 carbon atoms, 18 hydrogen atoms **(d)** 6 carbon atoms, 8 hydrogen atoms, 6 oxygen atoms **1.11** Metalloids are at the boundary between metals and nonmetals. **1.12** helium (He), neon (Ne), argon (Ar), krypton (Kr), xenon (Xe), radon (Rn) **1.13** copper (Cu), silver (Ag), gold (Au) **1.14** red: yttrium, metal; green: arsenic, metalloid; blue: argon, nonmetal **1.15** americium, a metal **1.16** Chemistry is the study of matter. **1.18** physical: **(a)**, **(c)**, **(e)**; chemical **(b)**, d) **1.20** A gas has no definite shape or volume; a liquid has no definite shape but has a definite volume; a solid has a definite volume and a definite shape. **1.22** gas **1.24** mixture: **(a)**, **(b)**, **(d)**, **(f)**; pure: **(c)**, **(e)** **1.26** An element is a pure substance that can't be broken down chemically. A compound is a pure substance that can be broken down chemically to yield elements. **1.28** element: **(a)**; compound: **(b)**, **(c)**; mixture: **(d)**, **(e)**, **(f)** **1.30** **(a)** reactant: hydrogen peroxide; products: water, oxygen **(b)** compounds: hydrogen peroxide, water; element: oxygen **1.32** 114 elements; 90 occur naturally **1.34** Metals: lustrous, malleable, conductors of heat and electricity; nonmetals: gases or brittle solids, nonconductors; metalloids: properties intermediate between metals and nonmetals. **1.36** **(a)** Gd **(b)** Ge **(c)** Tc **(d)** As **(e)** Cd **1.38** **(a)** nitrogen **(b)** potassium **(c)** chlorine **(d)** calcium **(e)** phosphorus **(f)** manganese **1.40** Only the first letter of a chemical symbol is capitalized. **1.42** **(a)** "2" is a subscript, H_2O **(b)** Water is composed of hydrogen and oxygen. **1.44** **(a)** magnesium, sulfur, oxygen **(b)** iron, bromine **(c)** cobalt, phosphorus **(d)** arsenic, hydrogen **(e)** calcium, chromium, oxygen **1.46** Carbon, hydrogen, nitrogen, oxygen; 10 atoms **1.48** $C_{13}H_{18}O_2$ **1.50** **(a)** metal **(b)** nonmetal **1.54** **(a)** A physical change doesn't alter chemical makeup; a chemical change alters a substance's chemical makeup. **(b)** Melting point is the temperature at which a change of state from solid to liquid occurs; boiling point is the temperature at which a change of state from liquid to gas occurs. **(c)** A reactant is a substance that undergoes change in a chemical reaction; a product is a substance formed as a result of a chemical reaction. **(d)** A metal is a lustrous malleable element that is a good conductor of heat and electricity; a nonmetal is an element that is a poor conductor. **1.56** compounds: **(a)**, **(d)**, **(e)**; elements: **(b)**, **(c)** **1.58** mixture **1.60** **(a)** Fe **(b)** Cu **(c)** Co **(d)** Mo **(e)** Cr **(f)** F **(g)** S

Chapter 2

2.1 **(a)** milliliter **(b)** kilogram **(c)** centimeter **(d)** kilometer **(e)** microgram **2.2** **(a)** L **(b)** μL **(c)** nm **(d)** Mm **2.3** **(a)** 0.000 000 001 m **(b)** 0.1 g **(c)** 1000 m **(d)** 0.001 L **(e)** 0.000 000 001 g **2.4** **(a)** 3 **(b)** 4 **(c)** 5 **(d)** exact **2.5** 32.6°C; three significant figures **2.6** **(a)**

5.8 × 10^1 g **(b)** 4.6792 × 10^4 m **(c)** 6.720 × 10^{-4} cm **(d)** 3.453 × 10^2 kg **2.7** **(a)** 48,850 mg **(b)** 0.000 008 3 m **(c)** 0.0400 m **2.8** **(a)** 6.0 × 10^5 **(b)** 1.300 × 10^3 **(c)** 7.942 × 10^{11} **2.9** 2.78 × 10^{-10} m = 2.78 × 10^2 pm **2.10** **(a)** 2.30 g **(b)** 188.38 mL **(c)** 0.009 L **(d)** 1.000 kg **2.11** **(a)** 50.9 mL **(b)** 0.078 g **(c)** 11.9 m **(d)** 51 m g **(e)** 103 **2 . 12** **(a)** 1 L = 1000 mL; 1 mL = 0.001 L **(b)** 1 g = 0.03527 oz; 1 oz = 28.35 g **(c)** 1 L = 1.057 qt; 1 qt = 0.9461 L **2.13** **(a)** 454 g **(b)** 2.5 L **(c)** 105 qt **2.14** 795 mL **2.15** 7.36 m/s **2.16** **(a)** 3.4 kg **(b)** 120 mL **2.17** **(a)** 10.6 mg/kg **(b)** 36 mg/kg **2.18** 57.8°C **2.19** −38.0°F; 234.3 K **2.20** 7,700 cal **2.21** 0.21 cal/g·°C **2.22** float: ice, human fat, cork, balsa wood; sink: gold, table sugar, earth **2.23** 8.392 mL **2.24** 2.21 g/cm^3 **2.25** more dense **2.26** **(a)** 34 mL **(b)** 2.7 cm **2.27** **(a)** 0.977 **(b)** three **(c)** less dense **2.28** The smaller cylinder is more accurate because the gradations are smaller. **2.29** higher in chloroform **2.30** A physical quantity consists of a number and a unit. **2.32** mass (g); volume (m^3); length (m); temperature (K) **2.34** They are the same. **2.36** **(a)** centiliter **(b)** decimeter **(c)** millimeter **(d)** nanoliter **(e)** milligram **(f)** cubic meter **(g)** cubic centimeter **2.38** 10^9 pg, 3.5 × 10^4 pg **2.40** **(a)** 9.457 × 10^3 **(b)** 7 × 10^{-5} **(c)** 2.000 × 10^{10} **(d)** 1.2345 × 10^{-2} **(e)** 6.5238 × 10^2 **2.42** **(a)** 6 **(b)** 3 **(c)** 3 **(d)** 4 **(e)** 1–5 **(f)** 2–3 **2.44** **(a)** 7,926 mi, 7,900 mi, 7,926.38 mi **(b)** 7.926 381 × 10^3 mi **2.46** **(a)** 12.1 g **(b)** 96.19 cm **(c)** 263 mL **(d)** 20.9 mg **2.48** **(a)** 0.3614 cg **(b)** 0.0120 ML **(c)** 0.0144 mm **(d)** 60.3 ng **2.50** **(a)** 97.8 kg **(b)** 0.133 mL **(c)** 0.46 ng **2.52** **(a)** 62.1 mi/hr **(b)** 91.1 ft/s **2.54** 4 × 10^3 cells/in. **2.56** 10 g **2.58** 6 × 10^{10} cells **2.60** Potential energy is stored energy; kinetic energy is the energy of motion. **2.62** 37.0°C, 310.2 K **2.64** 537 cal = 0.537 kcal **2.66** 39°C **2.68** 0.179 cm^3 **2.70** 11.4 g/cm^3 **2.72** 0.7856 g/mL; sp gr = 0.7856 **2.74** **(a)** 1 mg = 0.0154 gr **(b)** 1 fluidram = 0.125 fluid ounce **(c)** 1 mL = 0.269 fluidram **(d)** 1 mL = 16 minim **(e)** 1 minim = 2.08 × 10^{-3} fluid ounce **2.76** BMI is a person's weight in kilograms, divided by the square of height in meters. **2.78** shock absorber, insulator, energy storehouse **2.80** 8.88 g **2.82** 8.5 × 10^{-3} g/kg; 1.2 tablets **2.84** **(a)** 3.9 × 10^4 g **(b)** 86 lb **2.86** 300 mL **2.88** 4.4 g; 0.0097 lb **2.90** 2200 mL **2.92** 2.2 tablespoons **2.94** 2.33 × 10^{10} L **2.96** **(a)** 2 × 10^3 mg/L **(b)** 2 × 10^3 μg/mL **(c)** 2 g/L **(d)** 2 × 10^3 ng/μL **2.98** 34.0°C **2.100** float

Chapter 3

3.1 14.0 amu **3.2** 3.27 × 10^{-17} g **3.3** 6.02 × 10^{23} atoms in all cases **3.4** When the mass in grams is numerically equal to the mass in amu, there are 6.02 × 10^{23} atoms. **3.5** **(a)** Re **(b)** Li **(c)** Te **3.6** 92 protons, 92 electrons, 143 neutrons **3.7** chromium **3.8** $^{35}_{17}Cl$; $^{37}_{17}Cl$ **3.9** **(a)** $^{11}_5B$ **(b)** $^{56}_{26}Fe$ **3.10** group 3A, period 3 **3.11** silver, calcium **3.12** nitrogen (1), phosphorus (2), arsenic (3), antimony (4), bismuth (5) **3.13** metals: titanium, scandium; nonmetals: selenium, argon; metalloids: tellurium, astatine **3.14** **(a)** nonmetal, main group, noble gas **(b)** metal, main group **(c)** nonmetal, main group **(d)** metal, transition element **3.15** zinc, metal. period 4, group 2B **3.16** 6, 2, 6 **3.17** 10, neon **3.18** 12, m a g n e s i u m **3.19** **(a)** $1s^2 2s^2 2p^2$ **(b)** $1s^2 2s^2 2p^6 3s^1$ **(c)** $1s^2 2s^2 2p^6 3s^2 3p^5$ **(d)** $1s^2 2s^2 2p^6 3s^2 3p^6 4s^2$ **3.20** $1s^2 2s^2 2p^6 3s^2 3p^2$; $1s^2 2s^2 2p^6 3s^2 3p^6 4s^2 3d^{10} 4p^6$ **3.21** $4p^3$, all are unpaired **3.22** gallium

3.23 group 2A   **3.24** group 7A, $1s^2\,2s^2\,2p^6\,3s^2\,3p^5$ **3.25** group 6A, $ns^2\,np^4$
3.26 (a) p orbital **(b)** s orbital
3.27

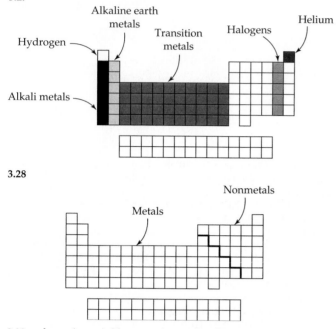

3.28

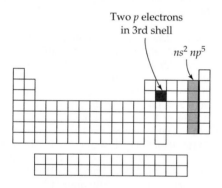

3.29 red: gas (argon); blue: atomic number 42 (molybdenum); green: (cesium); lithium, sodium, potassium, and rubidium are similar.
3.30

Two p electrons in 3rd shell

$ns^2\,np^5$

3.31 nickel   **3.32** Matter is composed of atoms. Atoms of different elements differ. Compounds consist of different atoms combined in specific proportions. Atoms do not change in chemical reactions.   **3.34 (a)** 3.4702×10^{-22} g **(b)** 2.1801×10^{-22} g **(c)** 6.6465×10^{-24} g   **3.36** 14.01 g   **3.38** 6.022×10^{23} **3.40** protons (+ charge, 1 amu); neutrons (no charge, 1 amu); electrons (− charge, 0.0005 amu).   **3.42 (a)** potassium **(b)** tin **(c)** zinc   **3.44** 18, 20, 22   **3.46 (a)** and **(c)**   **3.48 (a)** F-19 **(b)** Ne-19 **(c)** F-21 **(d)** Mg-21   **3.50 (a)** $^{14}_{6}\text{C}$ **(b)** $^{39}_{19}\text{K}$ **(c)** $^{20}_{10}\text{Ne}$ **3.52** $^{12}_{6}\text{C}$—six neutrons $^{13}_{6}\text{C}$—seven neutrons $^{14}_{6}\text{C}$—eight neutrons   **3.54** 63.55 amu   **3.56** Eight electrons are needed to fill the $3s$ and $3p$ subshells.   **3.58** Am, metal   **3.60 (a, b)** transition metals **(c)** $3d$ **3.62** selenium   **3.64** sodium, potassium, rubidium, cesium, francium   **3.66** 2   **3.68** 2, 8, 18   **3.70** 10, neon   **3.72 (a)** two paired, two unpaired **(b)** four paired, one unpaired **(c)** two unpaired   **3.74** 2   **3.76** beryllium, $2s$; arsenic, $4p$ **3.78 (a)** 8 **(b)** 4 **(c)** 2 **(d)** 1 **(e)** 3 **(f)** 7   **3.80** A scanning tunneling microscope has much higher resolution.   **3.82** H, He   **3.84 (a)** ultraviolet **(b)** gamma waves **(c)** X rays   **3.86** He, Ne, Ar, Kr, Xe, Rn   **3.88** Tellurium atoms have more neutrons than iodine atoms.   **3.90** 1 (2 e), 2 (8 e), 3 (18 e), 4 (32 e), 5 (18 e), 6 (4 e)   **3.92** 79.90 amu   **3.94** Carbon weighs more.   **3.96** 12 g   **3.98** Sr, metal, group 2A, period 5, 38 protons   **3.100** 2, 8, 18, 10, 2; metal   **3.102 (a)** The $4s$ subshell fills before $3d$. **(b)** The $2s$ subshell fills before $2p$. **(c)** Silicon has 14 electrons: $1s^2\,2s^2\,2p^6\,3s^2\,3p^2$. **(d)** The $3s$ electrons have opposite spins.   **3.104** $7p$

Chapter 4
4.1 Mg^{2+} is a cation.   **4.2** O^{2-} is an anion.   **4.3** less than Kr, but higher than most other elements   **4.4 (a)** B **(b)** Ca **(c)** Sc   **4.5 (a)** H **(b)** S **(c)** Na   **4.6** Potassium ($1s^2\,2s^2\,2p^6\,3s^2\,3p^6\,4s^1$) can gain the argon configuration by losing 1 electron.   **4.7** Aluminum can lose 3 electrons to form Al^{3+}.   **4.8** Oxygen can gain the neon configuration ($1s^2\,2s^2\,2p^6$) by adding two electrons.
4.9 $\cdot\ddot{\text{X}}\cdot$ **4.10** $:\ddot{\text{R}}\text{n}:$ $\cdot\dot{\text{P}}\text{b}\cdot$ $:\ddot{\text{X}}\text{e}:$ $\cdot\text{Ra}\cdot$
4.11 (a) $\text{Se} + 2\,e^- \rightarrow \text{Se}^{2-}$ **(b)** $\text{Ba} \rightarrow \text{Ba}^{2+} + 2\,e^-$ **(c)** $\text{Br} + e^- \rightarrow \text{Br}^-$ **4.12** cation   **4.13 (b)**   **4.14 (c)**   **4.15 (a)** copper(II) ion **(b)** fluoride ion **(c)** magnesium ion **(d)** sulfide ion   **4.16 (a)** Ag^+ **(b)** Fe^{2+} **(c)** Cu^+ **(d)** Te^{2-} **4.17** Na^+, sodium ion; K^+, potassium ion; Ca^{2+}, calcium ion; Cl^-, chloride ion   **4.18 (a)** nitrate ion **(b)** cyanide ion **(c)** hydroxide ion **(d)** hydrogen phosphate ion   **4.19 (a)** HCO_3^- **(b)** NH_4^+ **(c)** PO_4^{3-} **(d)** MnO_4^- **4.20 (a)** AgI **(b)** Ag_2O **(c)** Ag_3PO_4 **4.21 (a)** Na_2SO_4 **(b)** $FeSO_4$ **(c)** $Cr_2(SO_4)_3$ **4.22** $(NH_4)_2CO_3$ **4.23** $Al_2(SO_4)_3$, $Al(CH_3CO_2)_3$ **4.24** blue: K_2S; red: $BaBr_2$ **4.25** $BaSO_4$ **4.26** silver(I) sulfide   **4.27 (a)** copper(II) oxide **(b)** calcium cyanide **(c)** sodium nitrate **(d)** copper(I) sulfate **(e)** lithium phosphate **(f)** ammonium chloride   **4.28 (a)** $Ba(OH)_2$ **(b)** $CuCO_3$ **(c)** $Mg(HCO_3)_2$ **(d)** CuF **(e)** $Fe_2(SO_4)_3$ **(f)** $Fe(NO_3)_2$ **4.29** acids: **(a)**, **(d)**; bases **(b)**, **(c)**   **4.30 (a)** HCl **(b)** H_2SO_4
4.31

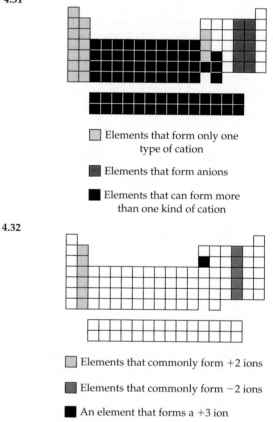

☐ Elements that form only one type of cation

☐ Elements that form anions

■ Elements that can form more than one kind of cation

4.32

☐ Elements that commonly form +2 ions

☐ Elements that commonly form −2 ions

■ An element that forms a +3 ion

4.33 (a) F^- **(b)** Na **(c)** Ca^{2+} **4.34 (a)** sodium atom (larger) **(b)** Na^+ ion (smaller)   **4.35 (a)** chlorine atom (smaller) **(b)** Cl^- anion (larger)   **4.36** red: SrF_2; blue: Li_2S; green: $AlBr_3$ **4.37 (a)** PbO **(b)** $ZnBr_2$ **(c)** $CrCl_3$
4.38 (a) $\cdot\text{Be}\cdot$ **(b)** $:\ddot{\text{N}}\text{e}:$ **(c)** $\cdot\text{Sr}\cdot$ **(d)** $\cdot\dot{\text{Al}}\cdot$
4.40 (a) $\text{Ca} \rightarrow \text{Ca}^{2+} + 2\,e^-$ **(b)** $\text{Au} \rightarrow \text{Au}^+ + e^-$ **(c)** $\text{F} + e^- \rightarrow \text{F}^-$ **(d)** $\text{Cr} \rightarrow \text{Cr}^{3+} + 3\,e^-$ **4.42** true: **(d)**; false: **(a)**, **(b)**, **(c)**   **4.44** Main group atoms undergo reactions that leave them with a noble gas electron configuration.   **4.46** −2   **4.48 (a)** Sr **(b)** Br   **4.50 (a)** $1s^2\,2s^2\,2p^6\,3s^2\,3p^6\,4s^2\,3d^{10}\,4p^6$ **(b)** $1s^2\,2s^2\,2p^6\,3s^2\,3p^6\,4s^2\,3d^{10}\,4p^6$ **(c)** $1s^2\,2s^2\,2p^6\,3s^2\,3p^6$ **(d)** $1s^2\,2s^2\,2p^6\,3s^2\,3p^6\,4s^2\,3d^{10}\,4p^6\,5s^2\,4d^{10}\,5p^6$ **(e)** $1s^2\,2s^2\,2p^6$ **4.52 (a)** O **(b)** Li **(c)** Zn **(d)** N   **4.54** none   **4.56** Cr^{2+}: $1s^2\,2s^2\,2p^6\,3s^2\,3p^6\,3d^4$; Cr^{3+}: $1s^2\,2s^2\,2p^6\,3s^2\,3p^6\,3d^3$ **4.58** greater

4.60 (a) sulfide ion **(b)** tin(II) ion **(c)** strontium ion **(d)** magnesium ion **(e)** gold(I) ion **4.62 (a)** Se^{2-} **(b)** O^{2-} **(c)** Ag^+ **4.64 (a)** OH^- **(b)** SO_4^{2-} **(c)** $CH_3CO_2^-$ **(d)** MnO_4^- **(e)** OCl^- **(f)** NO_3^- **(g)** HCO_3^- **4.66 (a)** $Al(NO_3)_3$ **(b)** $AgNO_3$ **(c)** $Zn(NO_3)_2$ **(d)** $Ba(NO_3)_2$ **4.68 (a)** $NaHCO_3$ **(b)** KNO_3 **(c)** $CaCO_3$
4.70

	S^{2-}	Cl^-	PO_4^{3-}	CO_3^{2-}
copper(II)	CuS	$CuCl_2$	$Cu_3(PO_4)_2$	$CuCO_3$
Ca^{2+}	CaS	$CaCl_2$	$Ca_3(PO_4)_2$	$CaCO_3$
NH_4^+	$(NH_4)_2S$	NH_4Cl	$(NH_4)_3PO_4$	$(NH_4)_2CO_3$
ferric ion	Fe_2S_3	$FeCl_3$	$FePO_4$	$Fe_2(CO_3)_3$

4.72 (a) magnesium carbonate **(b)** calcium acetate **(c)** silver(I) cyanide **(d)** sodium dichromate **4.74** $Ca_3(PO_4)_2$ **4.76** An acid gives H^+ ions in water; a base gives OH^- ions. **4.78 (a)** $HNO_2 \rightarrow H^+ + NO_2^-$ **(b)** $HCN \rightarrow H^+ + CN^-$ **(c)** $Ca(OH)_2 \rightarrow Ca^{2+} + 2\,OH^-$ **(d)** $CH_3CO_2H \rightarrow H^+ + CH_3CO_2^-$ **4.80** To a geologist, a mineral is a naturally occurring crystalline compound. To a nutritionist, a mineral is a metal ion essential for human health. **4.82** slight **4.84** Sodium protects against fluid loss and is necessary for muscle contraction and transmission of nerve impulses. **4.86** $10\,Ca^{2+}, 6\,PO_4^{3-}, 2\,OH^-$ **4.88** H^- has the helium configuration, $1s^2$ **4.90 (a)** CrO_3 **(b)** VCl_5 **(c)** MnO_2 **(d)** MoS_2 **4.92 (a)** -1 **(b)** 3 gluconate ions per iron(III) **4.94 (a)** $Co(CN)_2$ **(b)** UO_3 **(c)** $SnSO_4$ **(d)** K_3PO_4 **(e)** Ca_3P_2 **(f)** $LiHSO_4$ **4.96 (a)** metal **(b)** nonmetal **(c)** X_2Y_3 **(d)** X: group 3A; Y: group 6A

Chapter 5

5.1 $:\ddot{I}:\ddot{I}:$; xenon **5.2 (a)** P 3, H 1 **(b)** Se 2, H 1 **(c)** H 1, Cl 1 **(d)** Si 4, F 1 **5.3** $PbCl_4$ **5.4 (a)** CH_2Cl_2 **(b)** BH_3 **(c)** NI_3 **(d)** $SiCl_4$
5.5

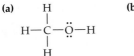

5.6

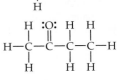

5.7

5.8

5.9

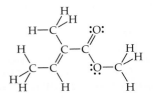

5.10
(a) **(b)** **(c)**

5.11 (a) $C_6H_{10}O_2$
(b)

5.12
(a) **(b)** **(c)** **(d)**

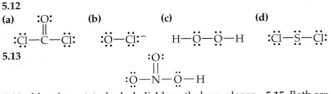

5.13

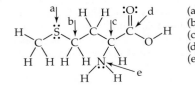

5.14 chloroform: tetrahedral; dichloroethylene: planar **5.15** Both are tetrahedral. **5.16** Both are bent.
5.17

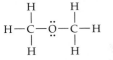

(a) bent
(b) tetrahedral
(c) tetrahedral
(d) planar triangular
(e) pyramidal

5.18 $H < P < S < N < O$ **5.19 (a)** polar covalent **(b)** ionic **(c)** covalent **(d)** polar covalent **5.20 (a)** $^{\delta^+}S-F^{\delta^-}$ **(b)** $^{\delta^+}P-O^{\delta^-}$ **(c)** $^{\delta^+}As-Cl^{\delta^-}$
5.21

5.22 The carbons are tetrahedral; the oxygen is bent.

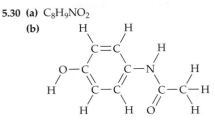

5.23

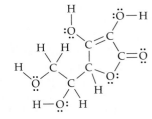

5.24 (a) disulfur dichloride **(b)** iodine chloride **(c)** iodine trichloride **5.25 (a)** SeF_4 **(b)** P_2O_5 **(c)** BrF_3 **5.26 (a)** ionic **(b)** covalent **5.27 (a)** **5.28 (a)** tetrahedral **(b)** pyramidal **(c)** planar triangular **5.29 (c)** is square planar

5.30 (a) $C_8H_9NO_2$
(b)

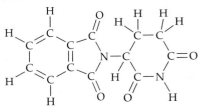

(c) All carbons are trigonal planar except the $-CH_3$ carbon. Nitrogen is pyramidal.
5.31

5.32 $C_{13}H_{10}N_2O_4$

5.33

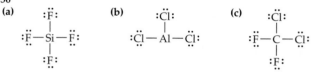

5.34 In a covalent bond, atoms share an electron pair. In an ionic bond, one atom donates an electron to another. **5.36 (a)** i, iv **(b)** iii **(c)** ii **(d)** i, iv **(e)** i, ii **(f)** iii **5.38** two bonds **5.40 (b)**, **(c)** **5.42** $SnCl_4$ **5.44** the N–O bond **5.46 (a)** A molecular formula shows the numbers and kinds of atoms; a structural formula shows how the atoms are bonded to one another. **(b)** A structural formula shows the bonds between atoms; a condensed structure shows atoms but not bonds. **(c)** A lone pair of valence electrons is not shared in a bond; a shared pair of electrons is shared between two atoms. **5.48 (a)** 10 **(b)** 10 **(c)** 24 **(d)** 20 **5.50** too many hydrogens

5.52

(a) $H-\ddot{\underset{..}{O}}-\ddot{N}=\ddot{\underset{..}{O}}$ **(b)** $H-\underset{\underset{H}{|}}{\overset{\overset{H}{|}}{C}}-C\equiv N:$ **(c)** $H-\ddot{\underset{..}{F}}:$

5.54 (a) $CH_3CH_2CH_3$ **(b)** $H_2C=CHCH_3$ **(c)** CH_3CH_2Cl

5.56

(a) $:\ddot{F}-\underset{\underset{\overset{|}{\ddot{F}}:}{|}}{\overset{\overset{:\ddot{F}:}{|}}{Si}}-\ddot{F}:$ **(b)** $:\ddot{\underset{..}{Cl}}-\overset{\overset{:\ddot{Cl}:}{|}}{Al}-\ddot{\underset{..}{Cl}}:$ **(c)** $:\ddot{F}-\underset{\underset{\overset{|}{\ddot{F}}:}{|}}{\overset{\overset{:\ddot{Cl}:}{|}}{C}}-\ddot{\underset{..}{Cl}}:$

(d) $:\ddot{\underset{..}{O}}-\overset{\overset{:O:}{||}}{S}-\ddot{\underset{..}{O}}:$ **(e)** $:\ddot{Br}-\overset{\overset{:\ddot{Br}:}{|}}{B}-\ddot{Br}:$ **(f)** $:\ddot{F}-\overset{\overset{:\ddot{F}:}{|}}{N}-\ddot{F}:$

5.58 CH_3CH_2OH **5.60** NH_2NH_2

5.62

$\ddot{S}=C=\ddot{S}$

5.64 The cyanide ion has one more electron than protons.

$:C\equiv N:^-$

5.66 (a) planar triangular **(b)** bent

5.68

(a) $H'\overset{\overset{Cl}{|}}{\underset{\underset{Cl}{}}{C}}-Cl$ **(b)** 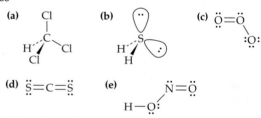   **(c)** $\ddot{\underset{..}{O}}=\ddot{\underset{..}{O}}$ $:\ddot{\underset{..}{O}}:$

(d) $\ddot{S}=C=\ddot{S}$ **(e)** $\overset{\overset{}{\ddot{N}}=\ddot{\underset{..}{O}}}{\underset{H-\ddot{\underset{..}{O}}:}{}}$

5.70 The C=O carbon is planar triangular; other carbons are tetrahedral. **5.72** most electronegative: upper right; least electronegative: left **5.74** K < Li < C < Br < Cl **5.76 (a)** $\overset{\delta+}{I}-\overset{\delta-}{Br}$ **(b)** $\overset{\delta-}{O}-\overset{\delta+}{H}$ **(c)** $\overset{\delta+}{C}-\overset{\delta-}{F}$ **(d)** $\overset{\delta-}{N}-\overset{\delta+}{C}$ **(e)** nonpolar **5.78** ionic: **(a)**; polar covalent: **(b)**, **(c)**, **(d)** **5.80** No; bond polarities cancel. **5.82 (a)** phosphorus triiodide **(b)** arsenic trichloride **(c)** tetraphosphorus trisulfide **(d)** dialuminum hexafluoride **(e)** nitrogen triiodide **(f)** iodine heptafluoride **5.84 (a)** NO_2 **(b)** SF_6 **(c)** BrI_5 **(d)** N_2O_3 **(e)** N_2O_4 **(f)** $AsCl_5$ **5.86** Carbon monoxide serves as a stimulus to production of cyclic GMP, which regulates many cellular functions. **5.88** A polymer is a long-chain molecule formed from repeating units. **5.90** Each of 20 million names must be unique and allow chemical identification. **5.92 (a)** Because they have 8 valence electrons.

(b) $:\ddot{F}-\overset{\overset{:\ddot{F}:}{|}}{\underset{\underset{:\ddot{F}:}{|}}{Xe}}-\ddot{F}:$

5.94 (a) C forms 4 bonds **(b)** N forms 3 bonds **(c)** S forms 2 bonds **(d)** could be correct **5.96 (b)** tetrahedral **(c)** contains a coordinate covalent bond **(d)** has 19 p and 18 e⁻

5.98

$\left[H-\overset{\overset{H}{|}}{\underset{\underset{H}{|}}{P}}-H \right]^+$

$:\ddot{\underset{..}{Cl}}-\underset{\underset{\overset{|}{\ddot{Cl}}:}{|}}{\overset{\overset{:\ddot{Cl}:}{|}}{C}}-\overset{\overset{:\ddot{O}-H}{|}}{\underset{\underset{H}{|}}{C}}-\ddot{\underset{..}{O}}-H$

5.100

$H-\ddot{\underset{..}{O}}-\overset{\overset{:O:}{||}}{\underset{\underset{:O:}{||}}{C}}-C-\ddot{\underset{..}{O}}-H$

5.102 (a) $:\ddot{\underset{..}{Cl}}-\overset{\overset{:O:}{||}}{C}-\ddot{\underset{..}{O}}-\underset{\underset{H}{|}}{\overset{\overset{H}{|}}{C}}-H$ **(b)** $H-\underset{\underset{H}{|}}{\overset{\overset{H}{|}}{C}}-C\equiv C-H$

Chapter 6

6.1 (a) Solid cobalt(II) chloride plus gaseous hydrogen fluoride gives solid cobalt(II) fluoride plus gaseous hydrogen chloride. **(b)** Aqueous lead(II) nitrate plus aqueous potassium iodide gives solid lead(II) iodide plus aqueous potassium nitrate. **6.2** balanced: **(a)**, **(c)** **6.3** $3 O_2 \rightarrow 2 O_3$ **6.4 (a)** $Ca(OH)_2 + 2 HCl \rightarrow CaCl_2 + 2 H_2O$ **(b)** $4 Al + 3 O_2 \rightarrow 2 Al_2O_3$ **(c)** $2 CH_3CH_3 + 7 O_2 \rightarrow 4 CO_2 + 6 H_2O$ **(d)** $2 AgNO_3 + MgCl_2 \rightarrow 2 AgCl + Mg(NO_3)_2$ **6.5** $2 A + B_2 \rightarrow A_2B_2$ **6.6 (a)** 206.0 amu **(b)** 232.0 amu **6.7** 1.71×10^{21} **6.8** 0.15 g **6.9** 111.0 amu **6.10** 0.217 mol; 4.6 g **6.11** 5.00 g weighs more **6.12 (a)** $Ni + 2 HCl \rightarrow NiCl_2 + H_2$; 4.90 mol **(b)** 3.00 mol **(c)** 6.00 mol **6.13** $6 CO_2 + 6 H_2O \rightarrow C_6H_{12}O_6 + 6 O_2$; 90.0 mol CO_2 **6.14 (a)** 39.6 mol **(b)** 13.8 g **6.15** 6.31 g WO_3; 0.165 g H_2 **6.16** 44.7 g; 57.0% **6.17** 47.3 g **6.18 (a)** precipitation **(b)** redox **(c)** acid–base neutralization **6.19** soluble: **(b)**, **(d)**; insoluble: **(a)**, **(c)**, **(e)** **6.20** precipitation: **(a)**, **(b)** **6.21 (a)** $2 CsOH(aq) + H_2SO_4(aq) \rightarrow Cs_2SO_4(aq) + 2 H_2O(l)$ **(b)** $Ca(OH)_2(aq) + 2 CH_3CO_2H(aq) \rightarrow Ca(CH_3CO_2)_2(aq) + 2 H_2O(l)$ **(c)** $NaHCO_3(aq) + HBr(aq) \rightarrow NaBr(aq) + CO_2(g) + H_2O(l)$ **6.22 (a)** $2 Li(s) + Pb^{2+}(aq) \rightarrow 2 Li^+(aq) + Pb(s)$ **(b)** $OH^-(aq) + H^+(aq) \rightarrow H_2O(l)$ **(c)** $CuS(s) + 2 H^+(aq) \rightarrow Cu^{2+}(aq) + H_2S(g)$ **6.23 (a)** oxidizing: Cu; reducing: Fe **(b)** oxidizing: Cl; reducing: Mg **(c)** oxidizing: Cr_2O_3; reducing: Al **6.24** $2 K + Br_2 \rightarrow 2 KBr$; oxidizing: Br_2; reducing: K **6.25 (a)** V(III) **(b)** Sn(IV) **(c)** Cr(VI) **(d)** Cu(II) **(e)** Ni(II) **6.26 (a)** N, +4; O, −2 **(b)** H, +1; Cl, +7; O, −2 **(c)** Mn, +7; O, −2 **(d)** H, +1; S, +6; O, −2 **6.27 (a)** not redox **(b)** Na oxidized from 0 to +1; H reduced from +1 to 0 **(c)** C oxidized from 0 to +4; O reduced from 0 to −2 **(d)** not redox **(e)** S oxidized from +4 to +6; Mn reduced from +7 to +2 **6.28 (d)** **6.29 (c)** **6.30** reactants: **(d)**; products: **(c)** **6.31** $C_5H_{11}NO_2S$; MW = 149.1 amu **6.32 (a)** $A_2 + 3 B \rightarrow 2 AB_3$ **(b)** 2 mol AB_3; 0.67 mol AB_3 **6.33 (a)** box 1 **(b)** box 2 **(c)** box 3 **6.34** Ag_2CO_3 and $Ag_2Cr_2O_4$ are possible **6.35** 22 g, 31 g **6.36** In a balanced equation, the numbers and kinds of atoms are the same on both sides of the reaction arrow. **6.38 (a)** $SO_2(g) + H_2O(g) \rightarrow H_2SO_3(aq)$ **(b)** $2 K(s) + Br_2(l) \rightarrow 2 KBr(s)$ **(c)** $C_3H_8(g) + 5 O_2(g) \rightarrow 3 CO_2(g) + 4 H_2O(l)$ **6.40 (a)** $2 C_2H_6(g) + 7 O_2(g) \rightarrow 4 CO_2(g) + 6 H_2O(g)$ **(b)** balanced **(c)** $2 Mg(s) + O_2(g) \rightarrow 2 MgO(s)$ **(d)** $2 K(s) + 2 H_2O(l) \rightarrow 2 KOH(aq) + H_2(g)$ **6.42** $2 NaHCO_3(aq) + H_2SO_4(aq) \rightarrow 2 CO_2(g) + Na_2SO_4(aq) + 2 H_2O(l)$ **6.44** One mole of a substance is an amount equal to the formula weight of the substance in grams. 6.02×10^{23} molecules **6.46** 6.02×10^{23} ions; 1.20×10^{24} ions **6.48** 2.43×10^{23} atoms **6.50** 6.44×10^{-4} mol **6.52** 284.5 g **6.54 (a)** 0.0641 mol **(b)** 0.0595 mol **(c)** 0.0418 mol **(d)** 0.0143 mol **6.56** 0.27 g **6.58 (a)** $N_2(g) + O_2(g) \rightarrow 2 NO(g)$ **(b)** 7.50 mol **(c)** 7.62 mol **(d)** 0.125 mol **6.60 (a)** $N_2(g) + 3 H_2(g) \rightarrow 2 NH_3(g)$ **(b)** 0.471 mol **(c)** 16.1 g **6.62 (a)** $Fe_2O_3(s) + 3 CO(g) \rightarrow 2 Fe(s) + 3 CO_2(g)$ **(b)** 1.59 g **(c)** 141 g **6.64** 158 kg **6.66 (a)** 11.4 g **(b)** 83.8% **6.68 (a)**

$CH_4(g) + 2 Cl_2(g) \rightarrow CH_2Cl_2(l) + 2 HCl(g)$ **(b)** 444 g **(c)** 202 g
6.70 (a), **(b)** $H^+(aq) + OH^-(aq) \rightarrow H_2O(l)$ **6.72** All are insoluble.
6.74 precipitation: **(b)** **6.76 (a)** $Mg(s) + Cu^{2+}(aq) \rightarrow Mg^{2+}(aq) +$
$Cu(s)$ **(b)** $2 Cl^-(aq) + Pb^{2+}(aq) \rightarrow PbCl_2(s)$ **(c)** $2 Cr^{3+}(aq) +$
$3 S^{2-}(aq) \rightarrow Cr_2S_3(s)$ **6.78** reducing agents: metals on left side; oxidizing agents: groups 6A and 7A **6.80 (a)** gains **(b)** loses **(c)** loses
(d) gains **6.82 (a)** N, +4; O, −2 **(b)** S, +6; O, −2 **(c)** C, +4; O,
−2; Cl, −1 **(d)** C, 0; H, +1; Cl, −1 **6.84 (a)** oxidized: Si; reduced Cl
(b) oxidized: Br; reduced: Cl **(c)** oxidized: Sb; reduced: Cl **6.86** area
of oil; thickness of oil; shape of oil molecules **6.88** zinc **6.90 (a)**
0.459 g **(b)** H^+ is reduced; Zn is oxidized. **6.92 (a)** $2 Al(s) +$
$Fe_2O_3(s) \rightarrow Al_2O_3(l) + 2 Fe(l)$ **(b)** $2 NH_4NO_3(s) \rightarrow 2 N_2(g) +$
$O_2(g) + 4 H_2O(g)$ **6.94** $3 Cu^{2+}(aq) + 2 PO_4^{3-}(aq) \rightarrow Cu_3(PO_4)_2(s)$
6.96 (a) $Cu(s) + 4 H^+(aq) + 2 NO_3^-(aq) \rightarrow Cu^{2+}(aq) + 2 NO(g) +$
$2 H_2O(l)$ **(b)** yes **6.98** 31.1 qt

Chapter 7

7.1 (a) endothermic **(b)** $\Delta H = +678$ kcal **(c)** $C_6H_{12}O_6(aq) +$
$6 O_2(g) \rightarrow 6 CO_2(g) + 6 H_2O(l) + 678$ kcal **7.2 (a)** e n d o t h e r
mic **(b)** 200 kcal **(c)** 74.1 kcal **7.3** 104 kcal **7.4 (a)** increase **(b)**
increase **(c)** decrease **7.5 (a)** no **(b)** increases **(c)** yes **7.6 (a)**
+0.06 kcal/mol; nonspontaneous **(b)** 0.00 kcal/mol; equilibrium **(c)**
−0.05 kcal/mol; spontaneous **7.7 (a)** positive **(b)** spontaneous at all
temperatures
7.8

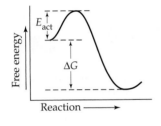

7.9

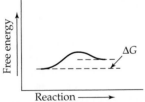

7.10 (a) rate increases **(b)** rate decreases **(c)** rate decreases

7.11 (a) $K = \dfrac{[NO_2]^2}{[N_2O_4]}$ **(b)** $K = \dfrac{[CH_3Cl][HCl]}{[CH_4][Cl_2]}$ **(c)** $K = \dfrac{[Br_2][F_2]^5}{[BrF_5]^2}$

7.12 (a) products favored **(b)** reactants favored **(c)** product favored
7.13 $K = 29.1$ **7.14** The reaction forming CD has larger K. **7.15**
reaction favored by high pressure and low temperature **7.16 (a)**
favors reactants **(b)** favors product **(c)** favors product **7.17** ΔH is
positive; ΔS is positive; ΔG is negative **7.18** ΔH is negative; ΔS
is negative; ΔG is negative **7.19 (a)** $2 A_2 + B \rightarrow 2 A_2B$ **(b)** ΔH is
negative; ΔS is negative; ΔG is negative **7.20 (a)** The blue curve represents faster reaction. **(b)** red curve is spontaneous **7.21** red curve
represents catalyzed reaction
7.22

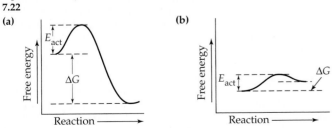

7.23 (a) positive **(b)** nonspontaneous at low temperature; spontaneous at high temperature **7.24** lower enthalpy for reactants **7.26**
(a) positive **(b)** 48 kcal **(c)** 3.5 kcal **7.28 (a)** $C_6H_{12}O_6 + 6 O_2 \rightarrow$

$6 CO_2 + 6 H_2O$ **(b)** 1.0×10^3 kcal **(c)** 57 kcal **7.30** increased disorder: **(a)**; decreased disorder: **(b)**, **(c)** **7.32** A spontaneous process,
once started, continues without external influence. **7.34** release or
absorption of heat, and increase or decrease in entropy **7.36** ΔH is
usually larger than $T\Delta S$ **7.38 (a)** endothermic **(b)** increases **(c)**
$T\Delta S$ is larger than ΔH **7.40 (a)** $H_2(g) + Br_2(l) \rightarrow 2 HBr(g)$ **(b)** increases **(c)** yes, because ΔH is negative and ΔS is positive **(d)**
$\Delta G = -25.6$ kcal/mol **7.42** the amount of energy needed for reactants to surmount the barrier to reaction
7.44

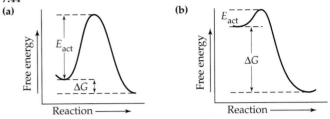

(a) **(b)**

7.46 Collisions increase in frequency and occur with more energy.
7.48 A catalyst lowers the activation energy. **7.50 (a)** yes **(b)**
reaction rate is slow **7.52** At equilibrium, the rates of forward and
reverse reactions are equal. Amounts of reactants and products need
not be equal.

7.54 (a) $K = \dfrac{[CO_2]^2}{[CO]^2[O_2]}$ **(b)** $K = \dfrac{[HCl]^2[C_2H_4Cl_2]}{[Cl_2]^2[C_2H_6]}$

(c) $K = \dfrac{[H_3O^+][F^-]}{[HF][H_2O]}$ **(d)** $K = \dfrac{[O_3]^2}{[O_2]^3}$

7.56 $K = 7.2 \times 10^{-3}$ **7.58 (a)** 2.0 mol/L **(b)** 0.029 mol/L **7.60**
more reactant **7.62 (a)** endothermic **(b)** reactants are favored **(c) (1)**
favors ozone; **(2)** favors ozone; **(3)** favors O_2; no effect; **(4)** favors ozone
7.64 (a) decrease **(b)** no effect **(c)** increase **7.66** decrease **7.68 (a)**
increase **(b)** decrease **(c)** no effect **(d)** decrease **7.70** fat **7.72**
Dilation cools the body by increasing blood flow to the surface. **7.74**
(a) exothermic **(b)** −5.5 kcal **7.76 (a)** Breathing CO favors formation of $Hb(CO)_4$, decreasing HbO_2. **(b)** Pure O_2 shifts equilibrium to
right, increasing HbO_2. **7.78 (a)** 5.40 kcal **(b)** 5.40 kcal **7.80 (a)**
$2 CH_3OH + 3 O_2 \rightarrow 2 CO_2 + 4 H_2O$ **(b)** 272 kcal **7.82** −31.2 kcal

Chapter 8

8.1 (a) disfavored by ΔH; favored by ΔS **(b)** 0 kcal/mol **(c)**
$\Delta H = -9.72$ kcal/mol; $\Delta S = -26.1$ cal/(mol·K) **8.2** 0.29 atm; 4.3
psi; 29,000 Pa **8.3** 1000 mm Hg **8.4** 450 L **8.5** 1.2 atm, 62 atm
8.6 0.38 L; 0.85 L **8.7** 33 psi **8.8** 231 L **8.9** 165 mm Hg **8.10**
4460 mol; 7.14×10^4 g CH_4; 1.96×10^5 g CO_2 **8.11** 5.0 atm **8.12**
1,100 mol; 4,400 g
8.13 (a) **(b)**

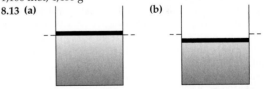

8.14 9.3 atm He; 0.19 atm O_2 **8.15** 75.4% N_2, 13.2% O_2, 5.3% CO_2,
6.2% H_2O **8.16** 90.4 mm Hg **8.17 (c)** **8.18 (a)** decrease **(b)** increase **8.19 (a)**, **(c)** **8.20 (a)** London forces **(b)** hydrogen bonds,
dipole–dipole forces, London forces **(c)** dipole–dipole forces, London
forces **8.21** 1.93×10^3 cal, 1.43×10^4 cal
8.22

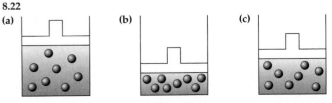

(a) **(b)** **(c)**

(a) volume increases by 50% **(b)** volume decreases by 50% **(c)**
volume unchanged **8.23 (c)** **8.24 (c)**

8.25

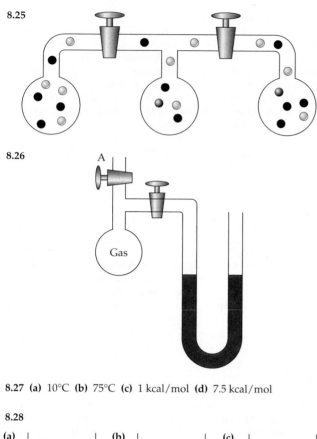

8.26

8.27 (a) 10°C **(b)** 75°C **(c)** 1 kcal/mol **(d)** 7.5 kcal/mol

8.28

(a)   **(b)**   **(c)**

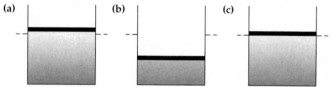

8.29 red = 300 mm Hg; yellow = 100 mm Hg; green = 200 mm Hg **8.30** One atmosphere is equal to exactly 760 mm Hg. **8.32** (1) A gas consists of tiny particles moving at random with no forces between them. (2) The amount of space occupied by the gas particles is small. (3) The average kinetic energy of the gas particles is proportional to the Kelvin temperature. (4) Collisions between particles are elastic. **8.34 (a)** 760 mm Hg **(b)** 190 mm Hg **(c)** 5.7×10^3 mm Hg **(d)** 711 mm Hg **(e)** 0.314 mm Hg **8.36** 930 mm Hg **8.38** V varies inversely with P when n and T are constant. **8.40** 101 mL **8.42** 7500 L **8.44** V varies directly with T when n and P are constant. **8.46** 364 K = 91°C **8.48** 130 mL **8.50** P varies directly with T when n and V are constant. **8.52** 1.3 atm **8.54** 493 K = 220°C **8.56** 49.5 mL **8.58 (a)** P increases by factor of 4. **(b)** P decreases by factor of 4. **8.60** 484 mL **8.62** Because gas particles are so far apart and have no interactions, their chemical identity does not matter. **8.64** 22.4 L **8.66** 63 mL **8.68** 15 kg **8.70** $PV = nRT$ **8.72** Cl_2 has fewer molecules but weighs more. **8.74** 2.2×10^4 mm Hg **8.76** 22 L **8.78** the pressure contribution of one component in a mixture of gases **8.80** 93 mm Hg **8.82** the partial pressure of the vapor above the liquid **8.84** Increased pressure raises a liquid's boiling point; decreased pressure lowers it. **8.86 (a)** all molecules **(b)** molecules with polar covalent bonds **(c)** molecules with —OH or —NH bonds **8.88** Ethanol has hydrogen bonds. **8.90 (a)** 29.2 kcal **(b)** 138 kcal **8.92** Atoms in a crystalline solid have a regular, orderly arrangement. **8.94** 4.57 kcal **8.96** Systolic pressure is the maximum pressure just after contraction; diastolic pressure is the minimum pressure at the end of the heart cycle. **8.98** by trapping solar radiation **8.100** The supercritical state is intermediate in properties between liquid and gas. **8.102** As temperature increases, molecular collisions become more violent. **8.104** 0.13 mol; 4.0 L **8.106** 590 g/day **8.108 (a)** 0.714 g/L **(b)** 1.96 g/L **(c)** 1.43 g/L

8.110 (a)

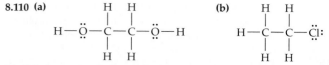

(b)

(c) Ethylene glycol forms hydrogen bonds. **8.112 (a)** 492°R **(b)** $R = 0.0455$ (L · atm)/(mol · °R)

Chapter 9
9.1 (a) heterogeneous mixture **(b)** homogeneous solution **(c)** homogeneous colloid **(d)** homogeneous solution **9.2 (c), (d) 9.3** $Na_2SO_4 \cdot 10H_2O$ **9.4** 322 g **9.5** 80 g/(100 mL) **9.6** 5.6 g/(100 mL) **9.7** 6.8×10^{-5} g/(100 mL) **9.8** 0.925 M **9.9 (a)** 0.031 mol **(b)** 0.68 mol **9.10** 0.48 g **9.11** 0.38 g **9.12** 0.0086% (w/v) **9.13** 6.6% (w/v) **9.14 (a)** 12 g **(b)** 1.5 g **9.15** Place 38 mL acetic acid in flask and dilute to 500.0 mL. **9.16 (a)** 22 mL **(b)** 18 mL **9.17** 1.6 ppm **9.18** Pb: 0.015 ppm, 0.0015 mg; Cu: 1.3 ppm, 0.13 mg **9.19** 2.40 M **9.20** 39.1 mL **9.21 (a)** 39.1 g **(b)** 79.9 g **(c)** 12.2 g **(d)** 48.0 g **9.22 (a)** 39.1 mg **(b)** 79.9 mg **(c)** 12.2 mg **(d)** 48.0 mg **9.23** 9.0 mg **9.24** 100.76°C **9.25** weak electrolyte **9.26 (a)** red curve is pure solvent; green curve is solution **(b)** solvent bp = 62°Cl solution bp = 69°C **(c)** 2 M **9.27** −1.9°C **9.28** 3 ions/mol **9.29 (a)** 0.70 Osmol **(b)** 0.30 Osmol **9.30** 0.33 Osmol
9.31

Before equilibrium   At equilibrium

9.32 HCl dissociates into ions; acetic acid does not. **9.33** HBr is dissociated; HF is not. **9.34** upper red line: liquid; lower green line: gas **9.35 (a) 9.36 (d) 9.37** green curve is solvent; red curve is solution **9.38** homogeneous: mixing is uniform; heterogeneous: mixing in nonuniform **9.40** polarity **9.42 (b), (d) 9.44** 17.0 g/100 mL **9.46** Concentrated solution can be saturated or not; saturated solutions can be concentrated or not. **9.48** Molarity is the number of moles of solute per liter of solution. **9.50** Dissolve 25.0 mL of ethyl alcohol in water and dilute to 500.0 mL. **9.52** Dissolve 1.5 g NaCl in water to a final volume of 250 mL. **9.54 (a)** 6.7% (w/v) **(b)** 4.3% (w/v) **9.56 (a)** 4.0 g **(b)** 15 g **9.58** 230 mL, 1600 mL **9.60** 10 ppm **9.62 (a)** 0.425 M **(b)** 1.53 M **(c)** 1.02 M **9.64** 5.3 mL **9.66** 37 g **9.68** 400 mL **9.70** 600 mL **9.72** a substance that conducts electricity when dissolved in water **9.74** Ca^{2+} concentration is 0.0015 M. **9.76** 40 mEq **9.78** 0.355 g **9.80** $Ba(OH)_2$ **9.82** 861 g **9.84** The insides of the cell has higher osmolarity than water, so water passes in and increases pressure. **9.86 (a)** 0.20 M Na_2SO_4 **(b)** 3% (w/v) NaOH **9.88** The body manufactures more hemoglobin. **9.90** Sports drinks contain electrolytes, carbohydrate, and vitamins. **9.92 (a)** 0.0067% (w/v) **(b)** 67 ppm **(c)** 0.000 40 M **9.94** 9.4 mL **9.96** 0.57 M **9.98** NaCl: 0.147 M; KCl: 0.0040 M; $CaCl_2$: 0.0030 M **9.100** 4.0 mL **9.102 (a)** $CoCl_2 + 6H_2O \rightarrow CoCl_2 \cdot 6H_2O$ **(b)** 1.13 g **9.104** 36% dissociated

Chapter 10
10.1 (a), (b) 10.2 (a), (c) 10.3 (a) H_2S **(b)** HPO_4^{2-} **(c)** HCO_3^- **(d)** NH_3 **10.4** acids: HF, H_2S; bases: HS^-, F^-; conjugate acid–base pairs: H_2S and HS^-, HF and F^- **10.5 (a)** base **(b)** acid **(c)** base **10.6** $Al(OH)_3 + 3HCl \rightarrow AlCl_3 + 3H_2O$; $Mg(OH)_2 + 2HCl \rightarrow MgCl_2 + 2H_2O$ **10.7 (a)** $2KHCO_3 + H_2SO_4 \rightarrow 2H_2O + 2CO_2 + K_2SO_4$ **(b)** $MgCO_3 + 2HNO_3 \rightarrow H_2O + CO_2 + Mg(NO_3)_2$ **10.8** $H_2SO_4 + 2NH_3 \rightarrow (NH_4)_2SO_4$ **10.9** $CH_3CH_2NH_2 + HCl \rightarrow CH_3CH_2NH_3^+Cl^-$ **10.10 (a)** NH_4^+ **(b)** H_2SO_4 **(c)** H_2CO_3 **10.11 (a)** F^- **(b)** OH^- **10.12** $HPO_4^{2-} + OH^- \rightleftharpoons PO_4^{3-} + H_2O$; favored in forward direction **10.13** The —NH_3 hydrogens are most acidic. **10.14** citric acid **10.15 (a)** acidic, $[OH^-] = 3.1 \times 10^{-10}$ M **(b)** basic, $[OH^-] = 3.2 \times 10^{-3}$ M **10.16** pH = 5 has higher $[H^+]$; pH 9 has higher $[OH^-]$ **10.17 (a)** 5 **(b)** 5 **10.18 (a)** 1×10^{-13} M **(b)** 1×10^{-3} M **(c)** 1×10^{-8} M; **(b)** is most acidic; **(a)** is most basic **10.19** pH = 4 **10.20 (a)** acidic, 3×10^{-7} M **(b)** most basic, 1×10^{-8} M **(c)** acidic 2×10^{-4} M **(d)**

most acidic, 3×10^{-4} M **10.21 (a)** 8.28 **(b)** 5.05 **10.22** 2.60 **10.23** 3.28 **10.24** 3.82 **10.25** Chloride ion is too weak a base to neutralize any added acid. **10.26 (a)** 0.079 Eq **(b)** 0.338 Eq **(c)** 0.14 Eq **10.27 (a)** 0.26 N **(b)** 1.13 N **(c)** 0.47 N **10.28** 0.730 M **10.29** 133 mL **10.30** 0.225 M **10.31 (a)** neutral **(b)** basic **(c)** basic **(d)** acidic **10.32**

(a)

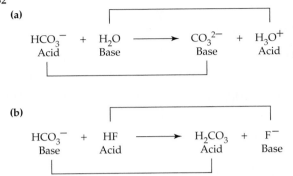

$$HCO_3^- + H_2O \longrightarrow CO_3^{2-} + H_3O^+$$
Acid   Base   Base   Acid

(b)

$$HCO_3^- + HF \longrightarrow H_2CO_3 + F^-$$
Base   Acid   Acid   Base

10.33 (a) box 2 **(b)** box 3 **(c)** box 1 **10.34** The O—H hydrogen in each is most acidic; acetic acid **10.35 (a)** box 1 **(b)** box 2 **(c)** box 1 **10.36 (a)** box 3 **(b)** box 1 **10.37** 0.67 M **10.38** HBr dissociates into ions. **10.40** KOH dissociates into ions. **10.42** A monoprotic acid can donate one proton; a diprotic acid can donate two. **10.44 (a), (e) 10.46 (a)** acid **(b)** base **(c)** neither **(d)** acid **(e)** neither **(f)** acid **10.48 (a)** CH_2ClCO_2H **(b)** $C_5H_5NH^+$ **(c)** $HSeO_4^-$ **(d)** $(CH_3)_3NH^+$ **10.50 (a)** $HCO_3^- + HCl \rightarrow H_2O + CO_2 + Cl^-$; $HCO_3^- + NaOH \rightarrow H_2O + Na^+ + CO_3^{2-}$ **(b)** $H_2PO_4^- + HCl \rightarrow H_3PO_4 + Cl^-$; $H_2PO_4^- + NaOH \rightarrow H_2O + Na^+ + HPO_4^{2-}$ **1 0 . 5 2** $2 HCl + CaCO_3 \rightarrow H_2O + CO_2 + CaCl_2$

10.54 $K_a = \dfrac{[H_3O^+][A^-]}{[HA]}$

10.56 $K_w = [H_3O^+][OH^-] = 1.0 \times 10^{-14}$

10.58 $K_a = \dfrac{[H_3O^+][NH_3]}{[NH_4^+]}$

10.60 (a) HF **(b)** HSO_4^- **(c)** $H_2PO_4^-$ **(d)** CH_3CO_2H **10.62** basic **10.64** 1×10^{-2} M **10.66** 1.00 **10.68 (a)** 7.60 **(b)** 3.30 **(c)** 11.64 **1 0 . 7 0 (a)** 1×10^{-10} M **(b)** 1×10^{-3} M **(c)** 1×10^{-14} M **(d)** 2.4×10^{-13} M **(e)** 9.1×10^{-7} M **10.72** A buffer contains a weak acid and its anion. The acid neutralizes any added base, and the anion neutralizes any added acid. **10.74** $CH_3CO_2^-Na^+ + H^+ \rightarrow CH_3CO_2H + Na^+$; $CH_3CO_2H + OH^- \rightarrow CH_3CO_2^- + H_2O$ **10.76** 9.19 **10.78** An equivalent is the formula weight in grams divided by the number of H_3O^+ or OH^- ions produced. **10.80** 0.25 Eq; 0.75 Eq **10.82** 25 mL; 25 mL **10.84 (a)** 0.50 Eq **(b)** 0.084 Eq **(c)** 0.25 Eq **10.86** 0.17 M; 0.34 N **10.88** 0.075 M **10.90 (a)** $NaAl(OH)_2CO_3 + 4 HCl \rightarrow CO_2 + 3 H_2O + NaCl + AlCl_3$ **(b)** 51.5 mg **10.92** 2×10^{-6} M **10.94** Citric acid reacts with sodium bicarbonate to release CO_2. **10.96** Both have the same amount of acid; HCl has higher $[H_3O^+]$ and lower pH. **10.98** 0.35 M **10.100 (a)** NH_4^+, acid; OH^-, base; NH_3, conjugate base; H_2O, conjugate acid **(b)** 5.56 g **10.102 (a)** $Na_2O(aq) + H_2O(l) \rightarrow 2 NaOH(aq)$ **(b)** 13.0 **(c)** 5000 mL

Chapter 11
11.1 $^{218}_{84}PO$ **11.2** $^{226}_{88}Ra$ **11.3** $^{14}_{6}C \rightarrow ^{0}_{-1}e + ^{14}_{7}N$ **11.4 (a)** $^{3}_{1}H \rightarrow ^{0}_{-1}e + ^{3}_{2}He$ **(b)** $^{210}_{82}Pb \rightarrow ^{0}_{-1}e + ^{210}_{83}Bi$ **(c)** $^{20}_{9}F \rightarrow ^{0}_{-1}e + ^{20}_{10}Ne$ **11.5 (a)** $^{38}_{20}Ca \rightarrow ^{0}_{1}e + ^{38}_{19}K$ **(b)** $^{118}_{54}Xe \rightarrow ^{0}_{1}e + ^{118}_{53}I$ **(c)** $^{79}_{37}Rb \rightarrow ^{0}_{1}e + ^{79}_{36}Kr$ **1 1 . 6 (a)** $^{62}_{30}Zn + ^{0}_{-1}e \rightarrow ^{62}_{29}Cu$ **(b)** $^{110}_{54}Sn + ^{0}_{-1}e \rightarrow ^{110}_{49}In$ **(c)** $^{81}_{36}Kr + ^{0}_{-1}e \rightarrow ^{81}_{35}Br$ **11.7** $^{120}_{49}In \rightarrow ^{0}_{-1}e + ^{120}_{50}Sn$ **11.8** 12% **11.9** 3 days **11.10** 13 m **11.11** 2% **11.12** 4.0 mL **11.13** $^{237}_{93}Np$ **11.14** $^{241}_{95}Am + ^{4}_{2}He \rightarrow 2 ^{1}_{0}n + ^{243}_{97}Bk$ **11.15** $^{40}_{18}Ar + ^{1}_{1}H \rightarrow ^{1}_{0}n + ^{40}_{19}K$ **11.16** $^{235}_{92}U + ^{1}_{0}n \rightarrow 2 ^{1}_{0}n + ^{137}_{52}Te + ^{97}_{40}Zr$ **11.17** 2 **11.18**

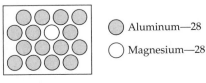

Aluminum—28
Magnesium—28

11.19 $^{14}_{6}C$ **11.20** The shorter arrow represents β emission; longer arrows represent α emission. **11.21** $^{241}_{94}Pu \rightarrow ^{241}_{95}Am \rightarrow ^{237}_{93}Np \rightarrow ^{233}_{91}Pa \rightarrow ^{233}_{92}U$ **11.22** $^{148}_{69}Tm \rightarrow ^{0}_{1}e + ^{148}_{68}Er$ or $^{148}_{69}Tm + ^{0}_{-1}e \rightarrow ^{148}_{68}Er$ **11.23** 3.5 years **11.24** It emits radiation. **11.26** A nuclear reaction changes the identity of the atom, is unaffected by temperature or catalysts, and often releases a large amount of energy. A chemical reaction does not change the identity of the atoms, is affected by temperature and catalysts, and involves relatively small energy changes. **11.28** $^{4}_{2}He$ **11.30** Gamma is highest and alpha is lowest. **11.32** by breaking bonds in DNA **11.34** A neutron decays. **11.36** The number of nucleons and the number of charges in the same on both sides. **11.38** α emission: Z decreases by 2 and A decreases by 4; β emission: Z increases by 1 and A is unchanged **11.40** In fission, a nucleus fragments to smaller pieces. **11.42 (a)** $^{35}_{17}Cl$ **(b)** $^{24}_{11}Na$ **(c)** $^{90}_{39}Y$ **11.44 (a)** $^{140}_{55}Cs$ **(b)** $^{246}_{96}Cm$ **11.46 (a)** $4 ^{1}_{0}n$ **(b)** $^{146}_{57}La$ **11.48** $^{198}_{80}Hg + ^{1}_{0}n \rightarrow ^{198}_{79}Au + ^{1}_{1}H$; a proton **11.50** Half of a sample decays in that time. **11.52** 0.006 g **11.54** 1 ng; 5×10^{-3} ng **11.56** The inside walls of a Geiger counter tube are negatively charged, and a wire in the center is positively charged. Radiation ionizes argon gas inside the tube, which creates a conducting path for current between the wall and the wire. **11.58** Rems indicate the amount of tissue damage and allow comparisons between equivalent doses of different types of radiation. **11.60** (1) c **(2)** b **(3)** d **(4)** a **11.62** 6.9 m **11.64** Ionizing radiation selectively destroys rapidly growing cancer cells. **11.66** Irradiation kills harmful microorganisms by destroying their DNA. **11.68** They yield more data, including three-dimensional images. **11.70** The leather is too new. **11.72** Nuclear decay is an intrinsic property of a nucleus and is not affected by external conditions. **11.74 (a)** β emission **(b)** Mo-98 **11.76 (a)** U-234 **(b)** for radiation shielding **11.78** Their cells divide rapidly. **11.80 (a)** $^{162}_{75}Re \rightarrow ^{4}_{2}He + ^{158}_{73}Ta$ **(b)** $^{188}_{74}W \rightarrow ^{0}_{-1}e + ^{188}_{75}Re$ **11.82** $^{238}_{92}U + ^{1}_{0}n \rightarrow 2 ^{0}_{-1}e + ^{239}_{94}Pu$ **11.84** 6 α particles; 4 β particles

Chapter 12
12.1 (a) alcohol, carboxylic acid **(b)** double bond, ester **(c)** aromatic ring, amine, carboxylic acid **12.2 (a)** CH_3CHO **(b)** $CH_3CH_2CO_2H$ **12.3** $CH_3CH_2CH_2CH_2CH_2CH_2CH_3$

12.4

$$\underset{}{CH_3CH_2CH_2CH_2\overset{\overset{\displaystyle CH_3}{|}}{C}HCH_3} \qquad \underset{}{CH_3CH_2CH_2\overset{\overset{\displaystyle CH_3}{|}}{C}HCH_2CH_3}$$

$$\underset{\underset{\displaystyle CH_3}{|}}{CH_3CH_2CH_2\overset{\overset{\displaystyle CH_3}{|}}{C}CH_3} \qquad \underset{\underset{\displaystyle CH_3}{|}}{CH_3CH_2\overset{\overset{\displaystyle CH_3}{|}}{C}HCHCH_3} \qquad \underset{\underset{\displaystyle CH_3}{|}}{CH_3CH_2\overset{\overset{\displaystyle CH_3}{|}}{C}CH_2CH_3}$$

$$\underset{\underset{\displaystyle CH_3}{|}\ \underset{\displaystyle CH_3}{|}}{CH_3CHCH_2CHCH_3} \qquad \underset{\underset{\displaystyle CH_2CH_3}{|}}{CH_3CH_2CHCH_2CH_3} \qquad \underset{\underset{\displaystyle H_3C}{|}}{\overset{\overset{\displaystyle H_3C\ \ CH_3}{|\ \ \ |}}{CH_3C-CHCH_3}}$$

12.5

(a) $CH_3CH_2CH_2CH_2CH_3$ **(b)** $CH_3\overset{\overset{\displaystyle CH_3}{|}}{C}HCH_2CH_3$ **(c)** $CH_3\overset{\overset{\displaystyle CH_3}{|}}{\underset{\underset{\displaystyle CH_3}{|}}{C}}CH_3$

12.6 Structures **(a)** and **(c)** are identical, and are isomers of **(b)**.

12.7

$$CH_3CH_2CH_2CH_2CH_2CH_3 \qquad CH_3CH_2CH_2\overset{\overset{\displaystyle CH_3}{|}}{C}HCH_3$$

$$CH_3CH_2\overset{\overset{\displaystyle CH_3}{|}}{C}HCH_2CH_3 \qquad CH_3\overset{\overset{\displaystyle CH_3}{|}}{\underset{\underset{\displaystyle CH_3}{|}}{C}}HCHCH_3 \qquad CH_3CH_2\overset{\overset{\displaystyle CH_3}{|}}{\underset{\underset{\displaystyle CH_3}{|}}{C}}CH_3$$

12.8 (a) 2,6-dimethyloctane **(b)** 3,3-diethylheptane

12.9 (a)

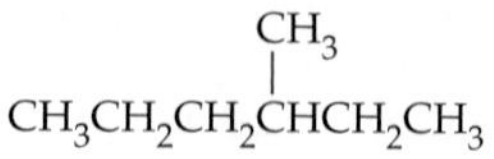

$$CH_3CH_2CH_2\underset{\underset{CH_3}{|}}{C}HCH_2CH_3$$

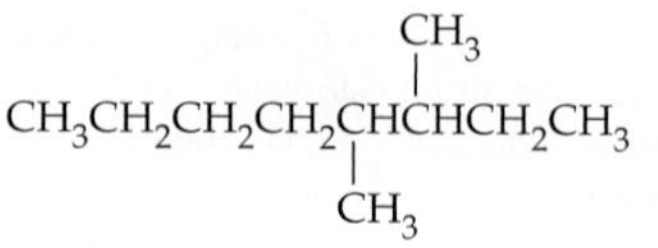

(b)

$$CH_3CH_2CH_2CH_2\underset{\underset{CH_3}{|}}{C}H\underset{\underset{CH_3}{|}}{C}HCH_2CH_3$$

(c)

$$CH_3\underset{\underset{CH_3}{|}}{\overset{\overset{CH_3}{|}}{C}}HCH_2\underset{\underset{CH_3}{|}}{\overset{\overset{CH_3}{|}}{C}}CH_3$$

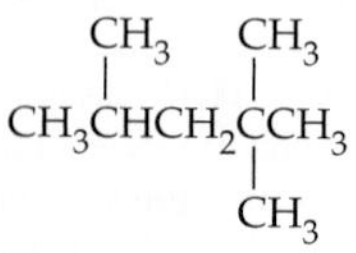

12.10 CH₃—primary, CH₂—secondary, CH—tertiary, C—quaternary

12.11 There are many possible answers; for example:

(a)

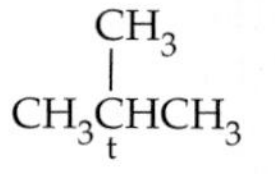

$$CH_3\underset{t}{\overset{\overset{CH_3}{|}}{C}}HCH_3$$

2-Methylpropane

(b)

$$CH_3\underset{t}{C}HCH_2CH_2\underset{q}{\overset{\overset{CH_3}{|}}{C}}CH_3$$

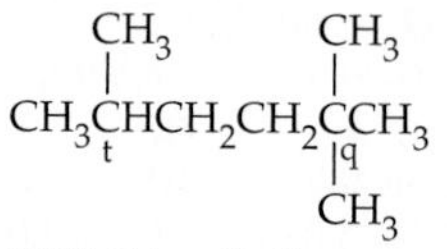

2,2,5-Trimethylhexane

12.12 (a) 2,2-dimethylpentane **(b)** 2,3,3-trimethylpentane **12.13** $2\,C_2H_6 + 7\,O_2 \rightarrow 4\,CO_2 + 6\,H_2O$

12.14

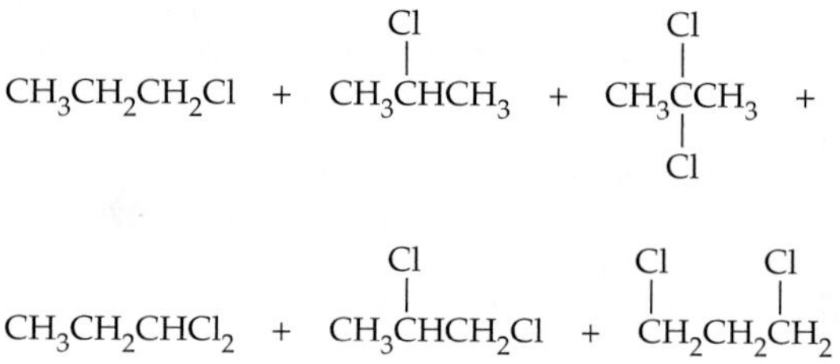

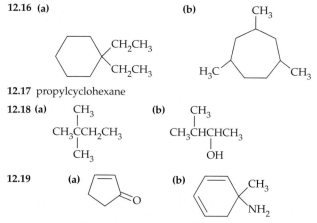

12.15 (a) 1-ethyl-4-methylcyclohexane **(b)** 1-ethyl-3-isopropylcyclopentane

12.16 (a)

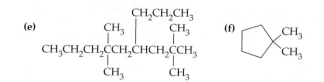

(b)

12.17 propylcyclohexane

12.18 (a)

$$CH_3\underset{\underset{CH_3}{|}}{\overset{\overset{CH_3}{|}}{C}}CH_2CH_3$$

(b)

$$CH_3\underset{\underset{OH}{|}}{\overset{\overset{CH_3}{|}}{C}}HCHCH_3$$

12.19   (a) 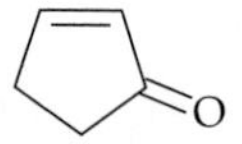   **(b)**

12.20 (a) double bond, ketone, ether **(b)** double bond, amine, carboxylic acid **12.21 (a)** 2,3-dimethylpentane **(b)** 2,5-dimethylhexane **12.22 (a)** 1,1-dimethylcyclopentane **(b)** isopropylcyclobutane **12.24** Carbon can form four strong bonds to itself and to many other elements. **12.26** Organic compounds are nonpolar. **12.28** A polar covalent bond is a bond in which electrons are shared unequally. **12.30 (a)** alcohol **(b)** aromatic ring, carboxylic acid, ester

12.32 (a)

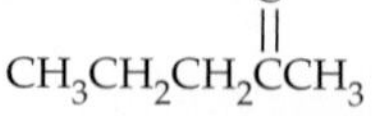

$$CH_3CH_2CH_2\overset{\overset{O}{||}}{C}CH_3$$
Ketone

(b)

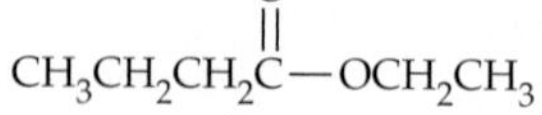

$$CH_3CH_2CH_2\overset{\overset{O}{||}}{C}-OCH_2CH_3$$
Ester

(c)

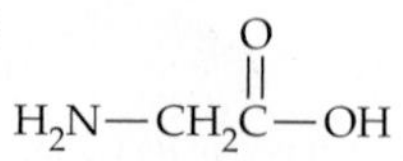

$$H_2N-CH_2\overset{\overset{O}{||}}{C}-OH$$
Amine
carboxylic acid

12.34 They must have the same formula but different structures. **12.36** A primary carbon is bonded to one other carbon; a secondary carbon is bonded to two other carbons; a tertiary carbon is bonded to three other carbons; and a quaternary carbon is bonded to four other carbons. **12.38 (a)** 2,3-dimethylbutane **(b)** cyclopentane

12.40

$$CH_3CH_2CH_2OH \qquad CH_3\underset{\underset{OH}{|}}{C}HCH_3 \qquad CH_3CH_2-O-CH_3$$

12.42

(a)

$$CH_3CH_2CH_2CH_2OH \qquad CH_3CH_2\underset{\underset{OH}{|}}{C}HCH_3 \qquad CH_3\underset{\underset{CH_3}{|}}{C}HCH_2OH$$

$$CH_3\underset{\underset{CH_3}{|}}{\overset{\overset{OH}{|}}{C}}CH_3$$

(b)

$$CH_3CH_2CH_2NH_2 \qquad CH_3\underset{\underset{NH_2}{|}}{C}HCH_3 \qquad CH_3CH_2\underset{\underset{H}{|}}{N}CH_3 \qquad CH_3\underset{\underset{CH_3}{|}}{N}CH_3$$

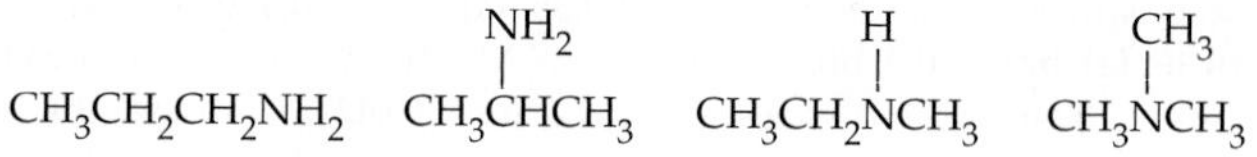

(c)

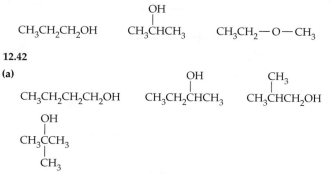

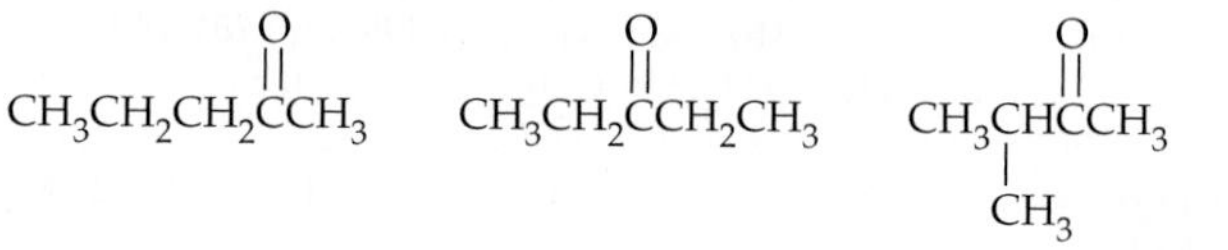

12.44 identical: **(a)**; isomers: **(b)**, **(d)**, **(e)**; unrelated: **(c)** **12.46** All have a carbon with five bonds. **12.48 (a)** 4-ethyl-3-methyloctane **(b)** 5-isopropyl-3-methyloctane **(c)** 2,2,6-trimethylheptane **(d)** 4-isopropyl-4-methyloctane **(e)** 2,2,4,4-tetramethylpentane **(f)** 4,4-diethyl-2-methylhexane **(g)** 2,2-dimethyldecane

12.50   (a)

$$CH_3CH_2CH_2\underset{\underset{CH_2CH_3}{|}}{C}HCH_2CH_3$$

(b)

$$CH_3CH_2\underset{\underset{CH_3}{|}}{\overset{\overset{H_3C\;\;CH_3}{|\;\;\;\;|}}{C}HC}CH_3$$

(c)

$$CH_3CH_2CH_2\underset{\underset{CH_3}{|}}{\overset{\overset{H_3C\;\;CH_2CH_3}{|\;\;\;\;\;\;\;|}}{CHC}}CH_2CH_3$$

(d)

$$CH_3CH_2CH_2\underset{\underset{CH_3CHCH_2}{|}}{C}HCH_2CH_2\underset{\underset{CH_3}{|}}{C}HCH_3$$

(e)

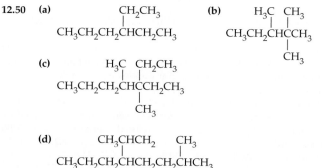

(f)

12.52 (a) 1-isopropyl-1-methylcyclopentane **(b)** 1,1,3,3-tetramethylcyclopentane **(c)** propylcyclohexane **(d)** 4-butyl-1,1-dimethylcyclohexane **12.54 (a)** 2,2-dimethylpentane **(b)** 3,5-dimethylheptane **(c)** *sec*-butylcyclobutane **12.56** heptane, 2-methylhexane, 3-methylhexane, 2,2-dimethylpentane, 2,3-dimethylpentane, 2,4-dimethylpentane, 3,3-dimethylpentane, 3-ethylpentane, 2,2,3-trimethylbutane **12.58** $C_3H_8 + 5\,O_2 \rightarrow 3\,CO_2 + 4\,H_2O$

12.60

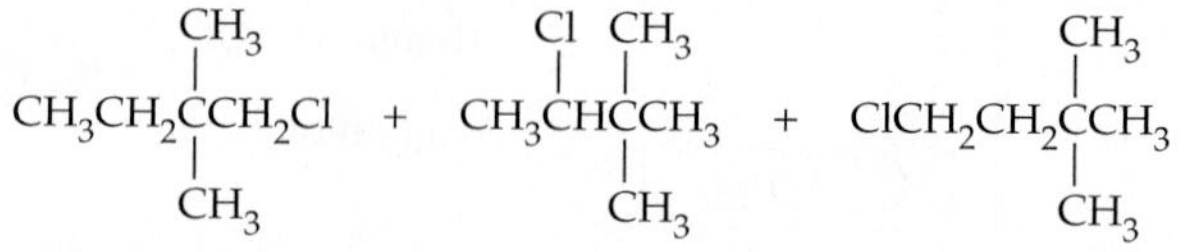

$$CH_3CH_2\underset{\underset{CH_3}{|}}{\overset{\overset{CH_3}{|}}{C}}CH_2Cl \;+\; CH_3\underset{\underset{CH_3}{|}}{C}H\overset{\overset{Cl}{|}}{C}CH_3 \;+\; ClCH_2CH_2\underset{\underset{CH_3}{|}}{\overset{\overset{CH_3}{|}}{C}}CH_3$$

12.62 They are identical. **12.64** Natural gas consists of C_1—C_4 alkanes; petroleum consists of larger alkanes. **12.66 (a)** ketone, alcohol, double bond **(b)** carboxylic acid, amine, amide, ester, aromatic ring **12.68** too many hydrogens

12.70 (a)

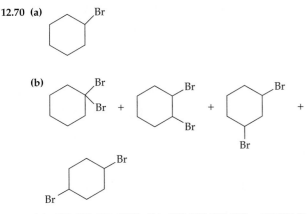

(b)

12.72 (a) $CH_3CH_2CH_2CHO$ **(b)** $CH_3CH_2CH_2CH{=}CHCH_2Br$ **(c)** cyclohexane **(d)** $CH_3CH{=}CHCH{=}CH_2$

Chapter 13
13.1 (a) 2-methyl-3-heptene **(b)** 2-methyl-1,5-hexadiene
13.2
(a)   **(b)**

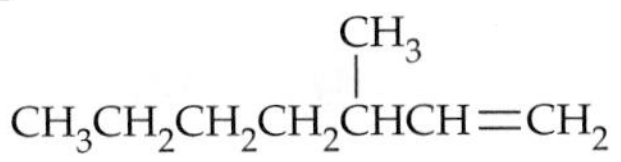

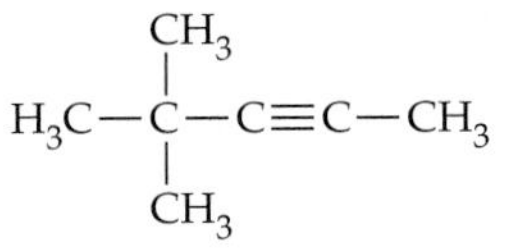

$CH_3CH_2CH_2CH_2CHCH{=}CH_2$ $H_3C{-}\overset{CH_3}{\underset{CH_3}{C}}{-}C{\equiv}C{-}CH_3$

(c) 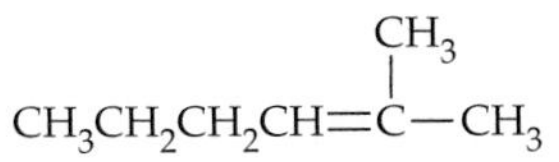   **(d)**

$CH_3CH_2CH_2CH{=}\overset{CH_3}{C}{-}CH_3$ $CH_3CH_2CH{=}\overset{CH_3CH_2}{C}{-}\overset{CH_3}{\underset{CH_3}{C}}{-}CH_3$

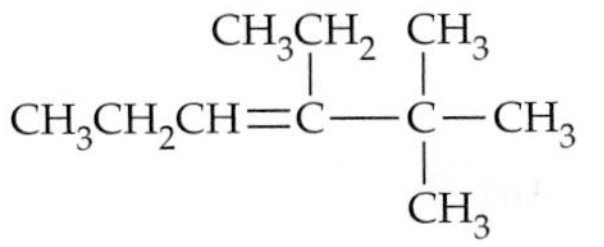

13.3 (a) 2,3-dimethyl-1-pentene **(b)** 2,3-dimethyl-2-hexene **13.4** **(a), (c)**
13.5

$\overset{CH_3CH_2}{\underset{H_3C}{}}C{=}C\overset{CH_2CH_3}{\underset{CH_3}{}}$ $\overset{CH_3CH_2}{\underset{H_3C}{}}C{=}C\overset{CH_3}{\underset{CH_2CH_3}{}}$

cis-3,4-Dimethyl-3-hexene   *trans*-3,4-Dimethyl-3-hexene

13.6 (a) *cis*-4-methyl-2-hexene **(b)** *trans*-5,6-dimethyl-3-heptene **13.7 (a)** substitution **(b)** addition **(c)** elimination **13.8 (a), (b), (c)** $CH_3CH_2CH_2CH_3$ **(d)** methylcyclohexane **13.9 (a)** 1,2-dibromo-2-methylpropane **(b)** 1,2-dichloropentane **(c)** 4,5-dichloro-2,4-dimethylheptane **13.10 (a)** bromocyclohexane **(b)** 2-chlorohexane **(c)** 2-chloro-3-methylbutane **13.11 (a)** 3-ethyl-2-pentene **(b)** 2,3-dimethyl-1-butene or 2,3-dimethyl-2-butene **13.12** 2-bromo-2,4-dimethylhexane **13.13 (a),**
(b)

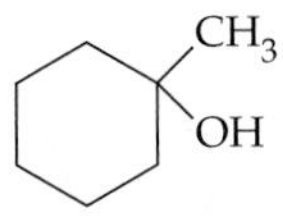

13.14 2-ethyl-1-butene or 3-methyl-2-pentene **13.15** $(CH_3)_3C^+$
13.16

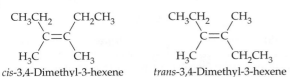

13.17

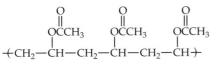

13.18 (a) *o*-bromochlorobenzene **(b)** butylbenzene **(c)** *o*-bromomethylbenzene or *o*-bromotoluene

13.19 (a)   **(b)**

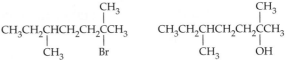

(c)   **(d)**

13.20 (a) *o*-isopropylbenzaldehyde **(b)** *p*-bromoaniline

13.21 (a)   **(b)**

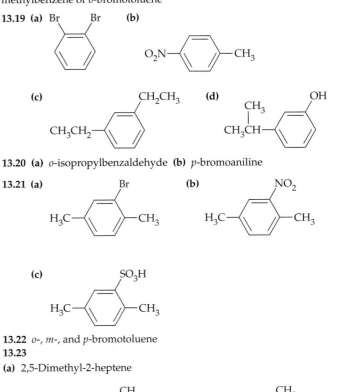

(c)

13.22 *o*-, *m*-, and *p*-bromotoluene
13.23
(a) 2,5-Dimethyl-2-heptene

$CH_3CH_2CHCH_2CH_2\overset{CH_3}{\underset{Br}{C}}CH_3$ $CH_3CH_2CHCH_2CH_2\overset{CH_3}{\underset{OH}{C}}CH_3$

(b) 3,3-Dimethylcyclopentene

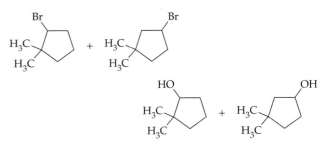

13.24 (a) 4,4-dimethyl-1-hexyne **(b)** 2,7-dimethyl-4-octyne **13.25 (a)** *m*-isopropylphenol **(b)** *o*-bromobenzoic acid

13.26 (a)

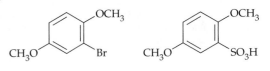

(b)

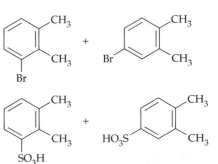

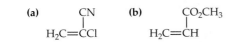

13.27

(a)

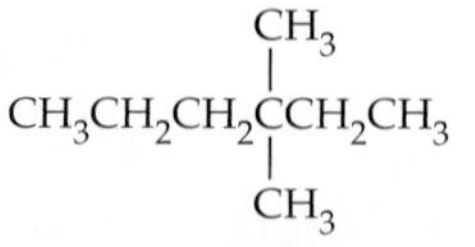

3,3-Dimethylhexane

(b)

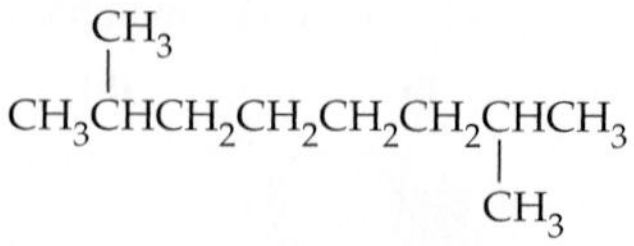

2,7-Dimethyloctane

13.28

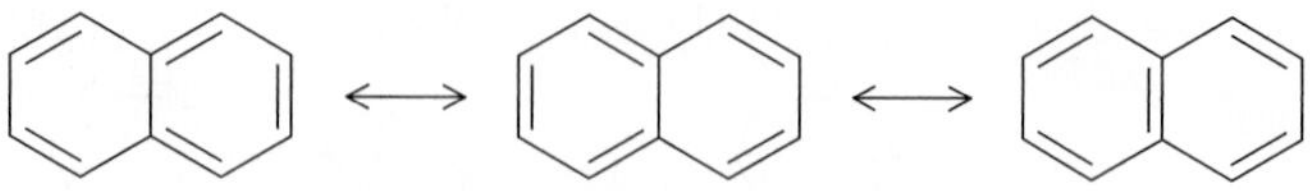

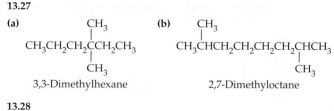

13.29 2-methyl-2-pentene **13.30** They have C—C multiple bonds and can add hydrogen. **13.32** alkene: *-ene;* alkyne: *-yne;* aromatic: *-benzene*

13.34 **(a)** $CH_3CH_2CH_2CH{=}CH_2$ 1-Pentene **(b)** $CH_3CH_2C{\equiv}CH$ 1-Butyne

(c)

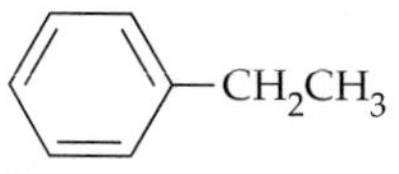

Ethylbenzene

13.36 **(a)** 1-pentene **(b)** 5-methyl-2-hexyne **(c)** 2,3-dimethyl-2-butene **(d)** 2-ethyl-3-methyl-1,3-pentadiene **(e)** 4-ethyl-3,5-dimethylcyclohexene **(f)** 3,3-diethylcyclobutene

13.38 **(a)**

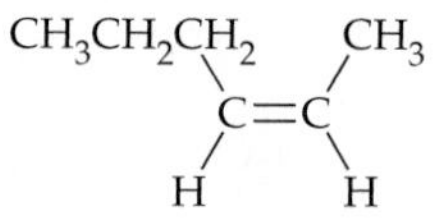

(b)

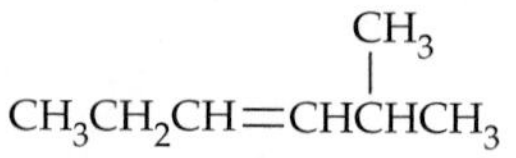

(c)

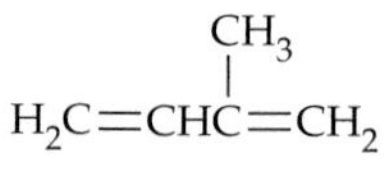

(d)

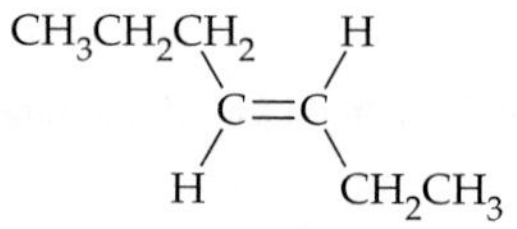

(e)

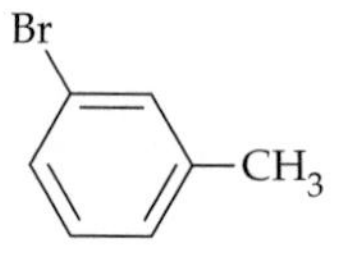

(f)

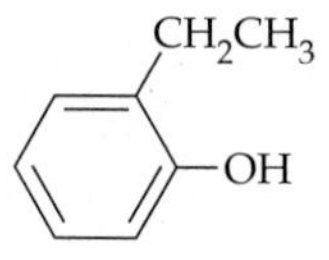

(g)

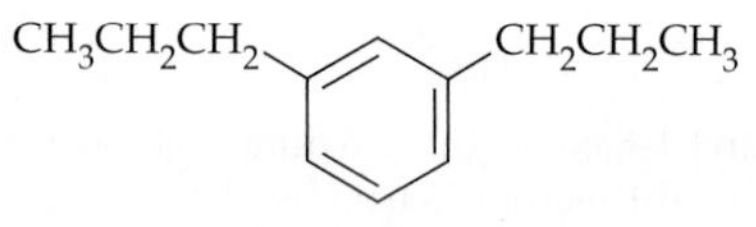

13.40 1-pentyne, 2-pentyne, 3-methyl-1-butyne **13.42** 1-pentene, 2-pentene, 2-methyl-1-butene, 3-methyl-1-butene, 2-methyl-2-butene **13.44** Each double bond carbon must be bonded to two different groups. **13.46** 2-pentene

13.48

(a)

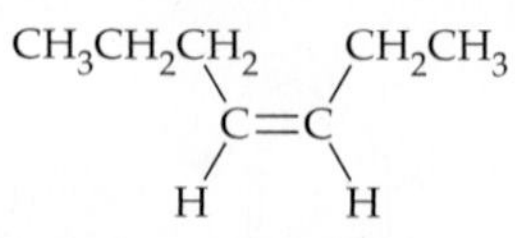

(b)

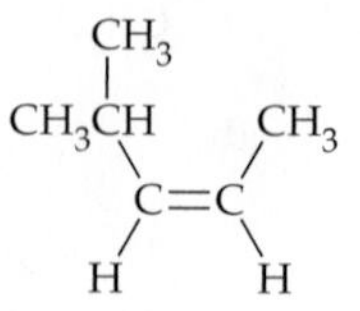

(c)

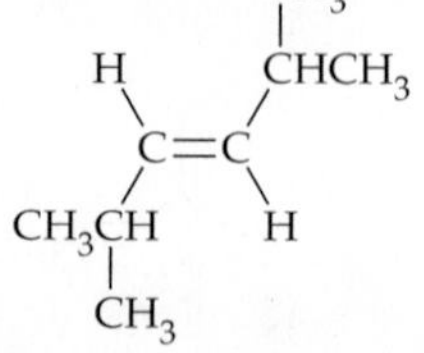

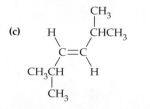

13.50 **(a)** identical **(b)** identical **13.52** substitution: two reactants exchange parts to give two products; addition: two reactants add to give one product **13.54** rearrangement **13.56** **(a)** substitution **(b)** substitution

13.58 **(a)**

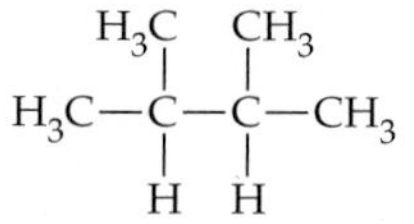

(b)

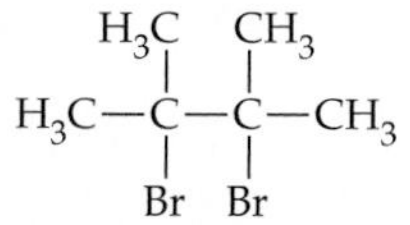

(c)

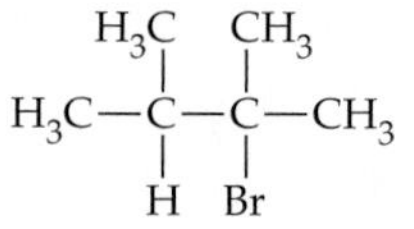

(d)

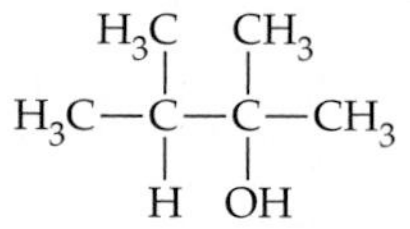

13.60

(a)

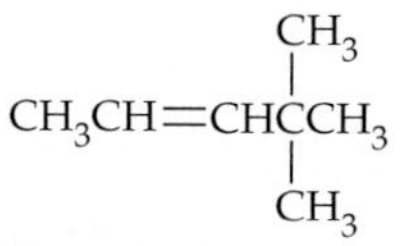

+ Cl_2

(b) $CH_3CH{=}CH_2$ + H_2

(c) $CH_3CH{=}CHCH_3$ or $H_2C{=}CHCH_2CH_3$ + HBr

(d)

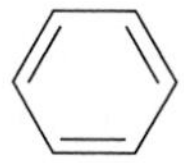

+ HCl

(e)

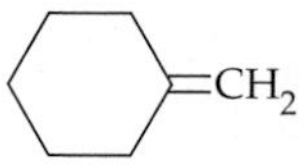

+ Cl_2

13.62

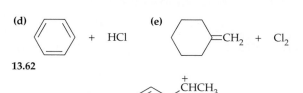

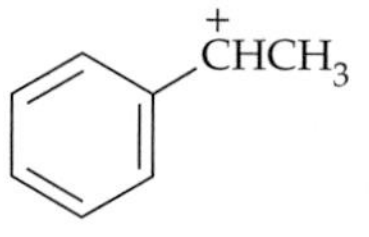

13.64

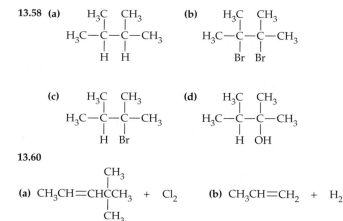

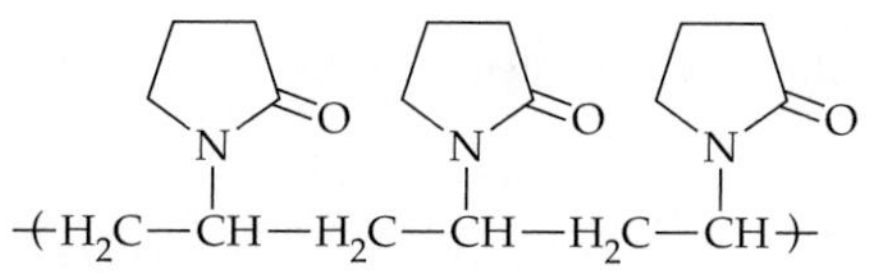

13.66 **(b)** benzene + Br_2 → bromobenzene + HBr **13.68** cyclohexane **13.70** Rod cells are responsible for vision in dim light; cone cells are responsible for color vision. **13.72** a compound that has two or more benzene-like rings joined together to share a common bond **13.74** ultraviolet **13.76** A trans double bond is too strained to exist in a small ring; large rings are more flexible.

13.78

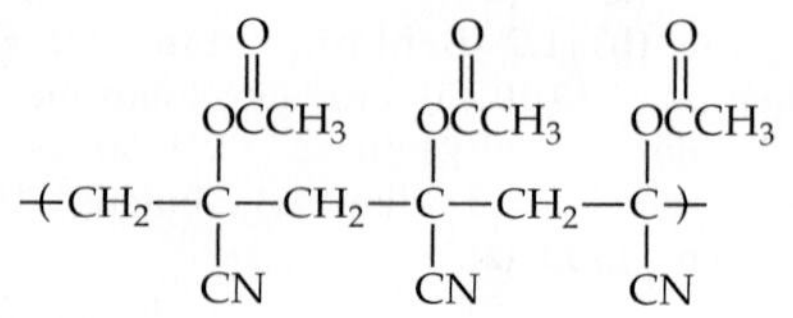

13.80 Cyclohexene reacts with Br_2.

13.82

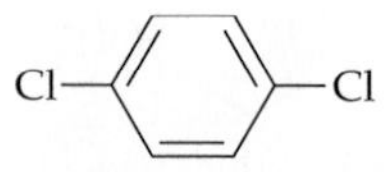

13.84 $C_6H_5CH_2CH_2CHO$ **13.86** 2-bromopentane and 3-bromopentane; the double bond has the same substitution on both carbons.

13.88 **(a)**

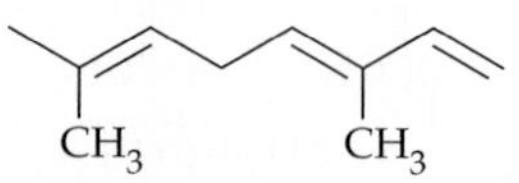

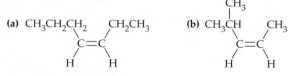

(b)

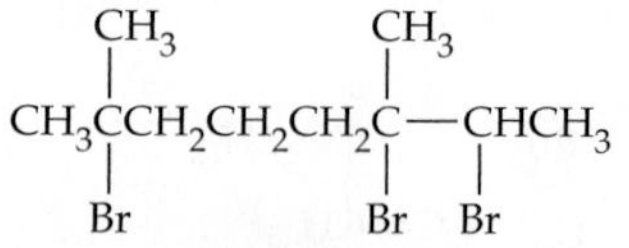

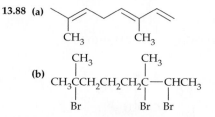

13.90 (a) and **(c)** have cis–trans isomers.

Chapter 14
14.1 (a) alcohol **(b)** alcohol **(c)** phenol **(d)** alcohol **(e)** ether **(f)** ether **14.2** A hydroxyl group is a part of a larger molecule.
14.3 (a)

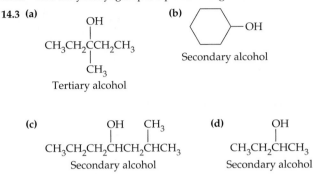

14.22

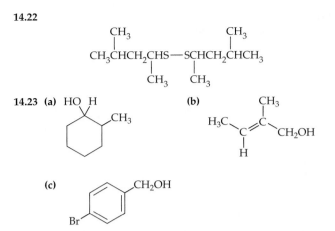

14.23 (a) HO H **(b)**

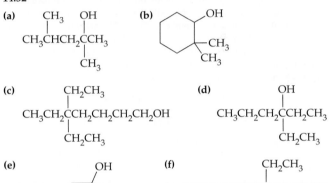

(c)

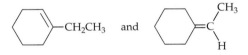

14.4 (a) 3-pentanol, secondary **(b)** 2-ethyl-1-pentanol, primary **(c)** 5-bromo-2-ethyl-1-pentanol, primary **(d)** 4,4-dimethylcyclohexanol, secondary **14.5** See 14.3 and 14.4. **14.6 (a) 14.7 (b) 14.8 (a)** propene **(b)** cyclohexene **(c)** 4-methyl-1-pentene (minor) and 4-methyl-2-pentene (major) **14.9 (a)** 2,3-dimethyl-2-butanol **(b)** 1-butanol or 2-butanol
14.10

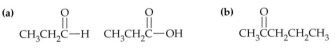

14.11

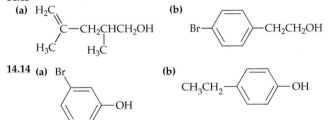

14.24 Alcohols have an —OH group bonded to an alkane-like carbon atom; ethers have an oxygen atom bonded to two carbon atoms; and phenols have an —OH group bonded to a carbon of an aromatic ring. **14.26** Alcohols form hydrogen bonds. **14.28** ketone, double bond, alcohol **14.30 (a)** 2-ethyl-1-pentanol **(b)** 3-methyl-1-butanol **(c)** 1,2,4-butanetriol **(d)** 2-phenyl-2-propanol **(e)** 2-methylcyclopentanol **(f)** 2-ethyl-2-methyl-1-butanol
14.32

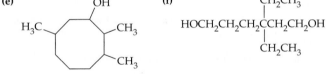

14.12 (a) 2-propanol **(b)** cycloheptanol **(c)** 3-methyl-1-butanol
14.13

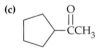

14.34 (a) primary **(b)** primary **(c)** primary, secondary **(d)** tertiary **(e)** secondary **(f)** primary **14.36 (a)** < **(c)** < **(d)** **14.38** a ketone **14.40** aldehyde or carboxylic acid **14.42** Phenols dissolve in aqueous NaOH; alcohols don't.
14.44

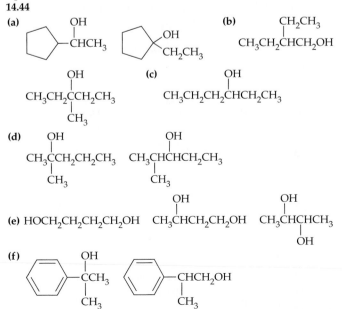

14.14

14.15 (a) p-chlorophenol **(b)** m-methylphenol **14.16 (a)** methyl propyl ether **(b)** diisopropyl ether **(c)** methyl phenyl ether **14.17 (a)** CH₃CH₂CH₂S–SCH₂CH₂CH₃ **(b)** (CH₃)₂CHCH₂CH₂S–SCH₂CH₂CH(CH₃)₂ **14.18 (a)** 1-chloro-1-ethylcyclopentane **(b)** 3-bromo-5-methylheptane **14.19 (a)** 5-methyl-3-hexanol **(b)** m-methoxytoluene **(c)** 3-methylcyclohexanol

14.20

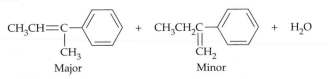

14.21 (CH₃)₂CHCH₂CH₂CHO, (CH₃)₂CHCH₂CH₂CO₂H

14.46

(a) 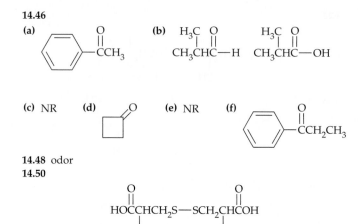 **(b)** H₃C O
CH₃CHC—H   CH₃CHC—OH

(c) NR **(d)** [cyclobutanone structure] **(e)** NR **(f)** [phenyl propyl ketone structure]

14.48 odor

14.50

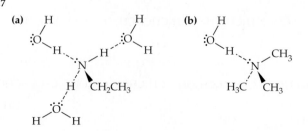

14.52 Alcohols can form hydrogen bonds; thiols and alkyl chlorides can't. **14.54** depressant **14.56** The liver is the site of alcohol metabolism. **14.58** a reactive species that contains an unpaired electron **14.60** diethyl ether **14.62** The ozone layer shields the earth from intense solar radiation. **14.64** alcohols: 1-butanol, 2-butanol, 2-methyl-1-propanol, 2-methyl-2-propanol; ethers: diethyl ether, methyl propyl ether, isopropyl methyl ether **14.66** Both are largely nonpolar hydrocarbons. **14.68** An antiseptic kills microorganisms on living tissue. **14.70** (a) *m*-dichlorobenzene (b) 1,2-dibromo-1-butene (c) *m*-propylphenol (d) 1,1-dibromocyclopentane (e) 5-chloro-3,5-dimethyl-3-hexanol (f) 4-methyl-2,5-heptanediol (g) 4-bromo-6,6-dimethyl-2-heptyne (h) 1-bromo-2-chlorocyclopentane **14.72** 3,7-Dimethyl-2,6-octadiene-1-ol

CH₃   CH₃ O
CH₃C=CHCH₂CH₂C=CHC—H

14.74 $C_2H_6O + 3 O_2 \rightarrow 2 CO_2 + 3 H_2O$

Chapter 15
15.1 (a) primary (b) secondary (c) primary (d) secondary (e) tertiary
15.2 (a) propylamine (b) dimethylamine (c) *N*-pentylaniline
15.3

(a) CH₃CH₂CH₂CH₂NH₂ **(b)** CH₃CH₂NHCH₃

(c) [phenyl-N(CH₃)₂ structure] **(d)** CH₃CH₂CHCH₂OH
                                          NH₂

15.4 The ion has one less electron than the neutral atoms.

CH₃
H₃C—N⁺—CH₃
CH₃

15.5 CH₃CH₂CH₂CH₂NHCH₂CH₃ *N*-Ethylbutylamine **15.6** Compound (a) is lowest boiling; (b) is highest boiling (strongest hydrogen bonds).
15.7

(a) [hydrogen bonding diagram] **(b)** [hydrogen bonding diagram]

15.8 (a) methylamine, ethylamine, dimethylamine, trimethylamine (b) pyridine (c) aniline **15.9** (a) piperidine: $C_5H_{11}N$ (b) purine: $C_5H_4N_4$ **15.10** (a) and (d)

15.11

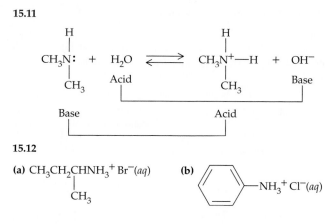

15.12

(a) CH₃CH₂CHNH₃⁺ Br⁻(aq) **(b)**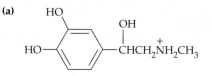
                CH₃

(c) CH₃CH₂NH₃⁺ CH₃COO⁻(aq) **(d)** CH₃NH₂ + H₂O(l) + NaCl(aq)

15.13 (a) *sec*-butylammonium bromide (b) anilinium chloride (c) ethylammonium acetate (d) methylamine **15.14** (a) ethylamine (b) triethylamine
15.15

(a) 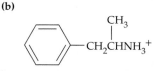 HO
HO—[ring]—CHCH₂NH₂CH₃ OH

(b)

CH₃
[phenyl]—CH₂CHNH₃⁺

15.16–15.17

(a)

CH₃
CH₃CH₂CH₂CH₂CH₂CH₂NH⁺ Cl⁻
CH₃

Hexyldimethylammonium chloride
or N,N-Dimethylhexylammonium chloride
Salt of a tertiary amine

(b)   CH₃CH₂NH₃⁺ Br⁻

Ethylammonium bromide
Salt of a primary amine

15.18
CH₃CH₂CH₂CH₂NH₃⁺ Cl⁻(aq) + NaOH(aq) →
CH₃CH₂CH₂CH₂NH₂ + H₂O(l) + NaCl(aq)
15.19 Benadryl has the general structure. In Benadryl, R = —CH₃, and R′ = R″ = C_6H_5—.
15.20

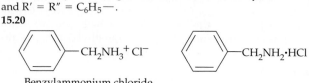

Benzylammonium chloride

15.21
(a) H₃C   CH₃   p = primary
       \t   s | s = secondary
        NCH₂CH₂NHCHCH₂—[ring]—ᵖNH₂   t = tertiary
       /
   H₃C

(b) All groups can form hydrogen bonds.   **15.22 (a)** All groups can participate in hydrogen bonding. **(b)** Arginine is water-soluble because it can form hydrogen bonds with water.

15.23

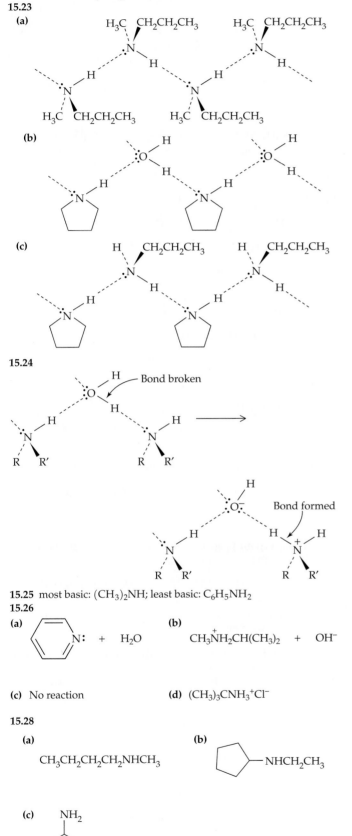

(a)

(b)

(c)

15.24

Bond broken

Bond formed

15.25 most basic: $(CH_3)_2NH$; least basic: $C_6H_5NH_2$

15.26

(a)

  + H_2O

(b) $CH_3\overset{+}{N}H_2CH(CH_3)_2$ + OH^-

(c) No reaction

(d) $(CH_3)_3CNH_3^+Cl^-$

15.28

(a)

$CH_3CH_2CH_2CH_2NHCH_3$

(b)

—$NHCH_2CH_3$

(c)

NH_2

$CH_2CH_2CH_3$

15.30 (a) *N*-methylcyclobutylamine (secondary) **(b)** *N*-cyclopentyl-*N*-methylaniline (tertiary)   **15.32** Diethylamine
15.34 (a) *N*-methylcyclopentylammonium nitrate (salt of a secondary amine)

(b)

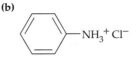

(Salt of a primary amine)

(c)

CH_3CHCH_3

$CH_3CH_2CH_2CH_2CH_2CH_2\overset{+}{N}H^+ Cl^-$

CH_2CH_3

(Salt of a tertiary amine)

15.36

Tertiary amine

Ester

Aromatic ring

Ester

Cocaine

15.38

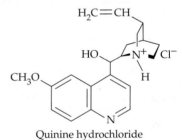

Quinine hydrochloride

15.40 (a)

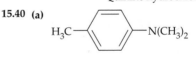

+ H_2O + NaBr

(b)

$CH_3CHNH_3^+$ + OH^-

CH_3

(c)

$\overset{+}{N}$—H Cl^-

15.42 Choline doesn't react with HCl because its nitrogen isn't basic.
15.44 (1) lowering of blood pressure   (2) memory enhancement   (3) reduction of sickling in sickle-cell anemia   (4) destruction of malaria parasites   (5) destruction of the tuberculosis bacterium
15.46

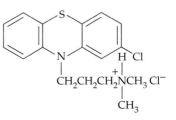

15.48 (a) chemical toxicology, forensic toxicology, environmental toxicology **(b)** the structure of the toxin, its mode of action, a mechanism to reverse its effects   **15.50** Its large hydrocarbon region is water-insoluble.

15.52

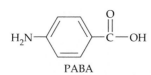

PABA

15.54

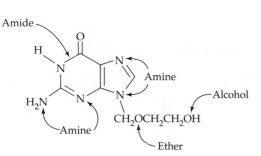

Acyclovir—related to purine

15.56

Amines	Alcohols
(a) foul smelling	pleasant smelling
(b) somewhat basic	not basic
(c) lower boiling, due to weaker hydrogen bonds, but higher boiling than ethers or alkyl halides	higher boiling, due to strong hydrogen bonds

15.58 (a) 6-methyl-2-heptene **(b)** *p*-isopropylphenol **(c)** dibutylamine **15.60** Molecules of hexylamine can form hydrogen bonds to each other, but molecules of triethylamine can't.   **15.62** Psilocin is related to indole.   **15.64** Pyridine forms H-bonds with water; benzene doesn't form H-bonds.

Chapter 16
16.1

(a)

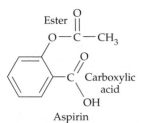

Aspirin

(b)

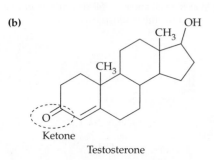

Testosterone

(c)

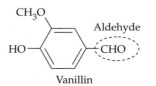

Vanillin

(d)   **(e)**   **(f)**

Ketone   Aldehyde   Ester

16.2

(d)

(e)

16.3

(a)   **(b)**

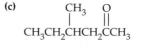

(c)   **(d)**

16.4 (a) pentanal **(b)** 3-pentanone **(c)** 4-methylhexanal **(d)** 4-heptanone
16.5 (a)   **(b)**

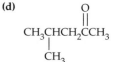

$C_6H_{12}O$ $C_7H_{14}O$
4-Methylpentanal   5-Methy-3-hexanone
An aldehyde   A ketone

16.6 (a) polar **(b)** flammable **(c)** liquid **(d)** bp of 100°C   **16.7** Alcohols form hydrogen bonds, which raise their boiling points. Aldehydes and ketones have higher boiling points than alkanes because they are polar.
16.8

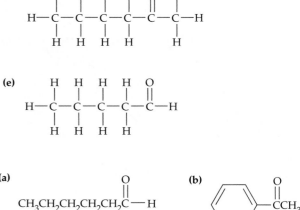

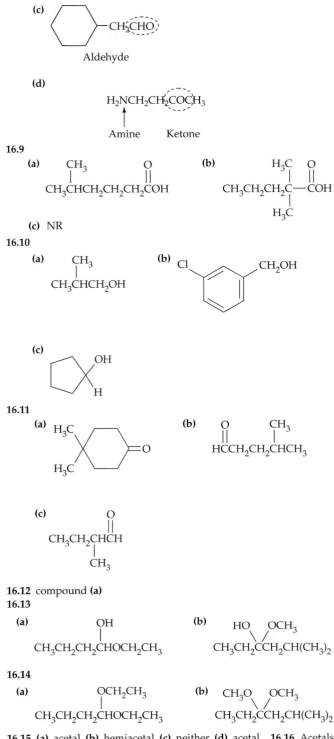

(c)

—CH₂CHO

Aldehyde

(d)

H₂NCH₂CH₂COCH₃

Amine   Ketone

16.9

(a)   CH₃

CH₃CHCH₂CH₂CH₂COH

(b)   H₃C   O

CH₃CH₂CH₂C—COH
H₃C

(c) NR

16.10

(a)   CH₃

CH₃CHCH₂OH

(b)   Cl —〈〉— CH₂OH

(c)   OH
H

16.11

(a) H₃C 〈 〉 =O
H₃C

(b)   O   CH₃

HCCH₂CH₂CHCH₃

(c)   O

CH₃CH₂CHCH
CH₃

16.12 compound **(a)**

16.13

(a)   OH

CH₃CH₂CH₂CHOCH₂CH₃

(b)   HO   OCH₃

CH₃CH₂CCH₂CH(CH₃)₂

16.14

(a)   OCH₂CH₃

CH₃CH₂CH₂CHOCH₂CH₃

(b)   CH₃O   OCH₃

CH₃CH₂CCH₂CH(CH₃)₂

16.15 (a) acetal **(b)** hemiacetal **(c)** neither **(d)** acetal **16.16** Acetals **(b)** and **(d)** were formed from aldehydes. Acetal **(a)** was formed from a ketone.

16.17 (a)

〈 〉—CH₂CCH₂CH₃   +   2 CH₃OH

(b) CH₃CH₂CHO   +   2 CH₃CH₂CH₂OH

16.18 (a) Hydride adds to the carbonyl carbon. **(b)** The arrow to the right represents reduction, and the arrow to the left represents oxidation. **16.19** Aldehydes can be oxidized to carboxylic acids. Tollens' reagent differentiates an aldehyde from a ketone.

16.20

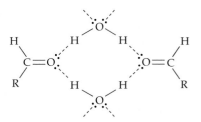

16.21

(a) Under acidic conditions, an alcohol adds to the carbonyl group of an aldehyde to form a hemiacetal, which is unstable and further reacts to form an acetal.

(b)

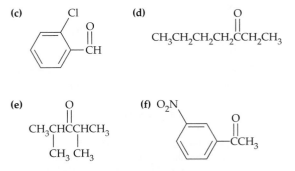

16.22 In solution, glucose exists as a cyclic hemiacetal because this structure is more stable. **16.23** One equivalent of an alcohol adds to the carbonyl group of an aldehyde or ketone to form a hemiacetal. Two equivalents of alcohol add to the carbonyl group to yield an acetal.

16.24 (a)   O

CH₃CCH₃

(b)   CH₃   O

CH₃CH₂CHCH₂CH

(c)   O   O

CH₃CCH₂CH

(d)   O

HOCH₂CCH₃

16.26 Structure **(a)** is an aldehyde, and structure **(f)** is a ketone.

16.28 (a)   O

CH₃CH₂CH₂CH₂CH₂CH₂CH

(b)   CH₃   O

CH₃CCH₂CH₂CH
CH₃

(c)   Cl
〈 〉 O
CH

(d)   O

CH₃CH₂CH₂CH₂CCH₂CH₃

(e)   O

CH₃CHCCHCH₃
CH₃ CH₃

(f) O₂N
〈 〉 O
CCH₃

16.30 (a) 2-methylbutanal **(b)** 2-methylpentanal **(c)** 2,2-dimethylpropanal **(d)** 2-butanone **(e)** 5-methyl-2-hexanone **16.32** For **(a)** and **(b)**, a ketone can't occur at the end of a carbon chain. For **(c)**, numbering must start at the end of the carbon chain closer to the carbonyl group. **16.34** A hemiacetal is produced.

16.36 (a) NR; cyclopentanol

(b)   O

CH₃CH₂CH₂CH₂CH₂COH ;   CH₃CH₂CH₂CH₂CH₂CH₂OH

(c)   COOH

CH₃CH₂CH₂CHCH₂CH₃ ;   CH₃CH₂CH₂CHCH₂CH₃
                                          CH₂OH

16.38

(a)

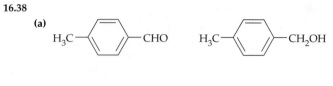

(b)

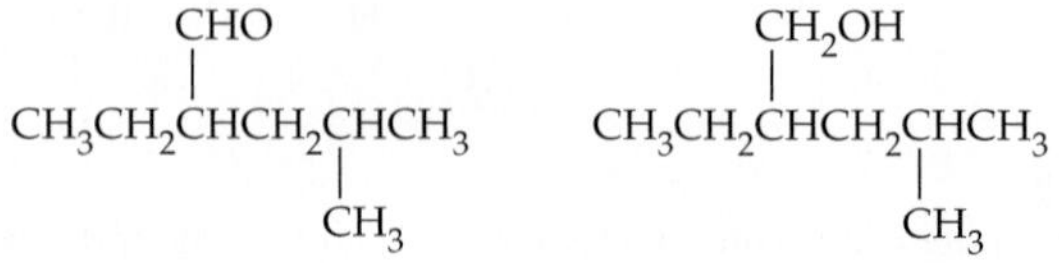

CHO
|
CH₃CH₂CHCH₂CHCH₃
|
CH₃

CH₂OH
|
CH₃CH₂CHCH₂CHCH₃
|
CH₃

(c) CH₃CH=CHCHO   CH₃CH=CHCH₂OH

16.40

(a)

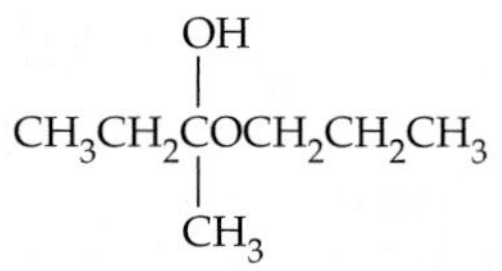

OH
|
CH₃CH₂COCH₂CH₂CH₃
|
CH₃

(b)

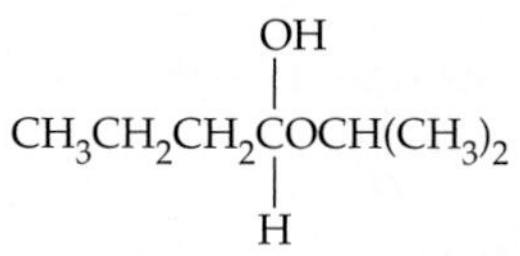

OH
|
CH₃CH₂CH₂COCH(CH₃)₂
|
H

(c)

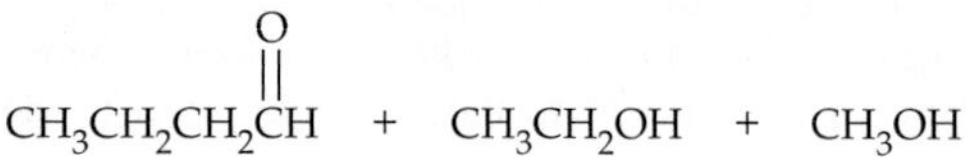

O
‖
CH₃CH₂CH₂CH + CH₃CH₂OH + CH₃OH

(d)

H₃C
      \
       C=O + HOCH₂CH₂OH
      /
H₃C

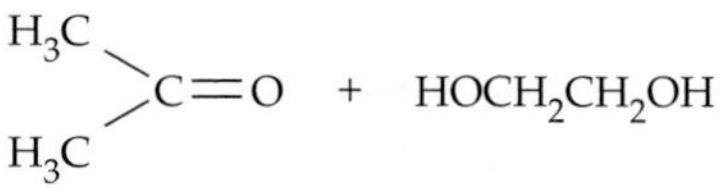

16.42 HOCH₂CH₂CH₂CH₂CHO   **16.44** HOCH₂CH₂CH₂OH   and CH₂O (formaldehyde)

16.46

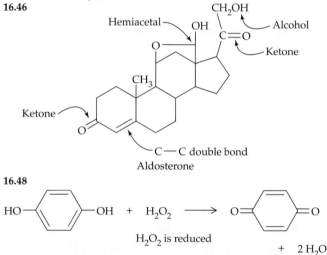

Aldosterone

16.48

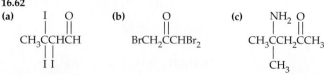

H₂O₂ is reduced

+ 2 H₂O

16.50 ethanol **16.52** **(a)** Advantages: inexpensive, no need to sacrifice animals. **(b)** Disadvantage: results of tests on cultured cells may not be reliable for more complex organisms. **16.54** *p*-methoxybenzaldehyde **16.56** Aldehydes are easily oxidized.

16.58

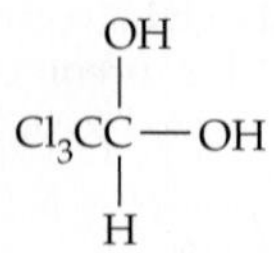

OH
|
Cl₃CC—OH
|
H

16.60 (a) *o*-isopropylmethoxybenzene **(b)** 5,5-dimethyl-3-hexyne **(c)** *N*-ethylcyclopentylammonium bromide **(d)** *N,N*-diethylhexylamine

16.62

(a)

I O
| ‖
CH₃CCHCH
| |
I I

(b)

O
‖
BrCH₂CCHBr₂

(c)

NH₂ O
| ‖
CH₃CCH₂CCH₃
|
CH₃

16.64 (a)

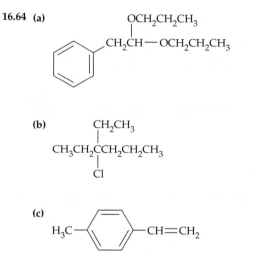

(b)

CH₂CH₃
|
CH₃CH₂CCH₂CH₂CH₃
|
Cl

(c)

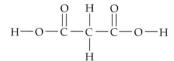

16.66 Highest boiling = 1-butanol (strongest hydrogen bonds)

Chapter 17

17.1 carboxylic acid: **(c)**; amides: **(a) (f) (h)**; ester: **(d)**; none: **(b) (e) (g)**

17.2 (a)

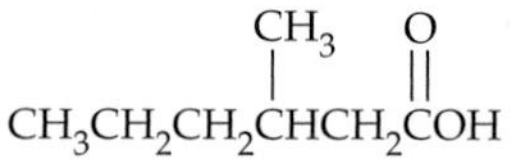

CH₃ O
| ‖
CH₃CH₂CH₂CHCH₂COH

(b)

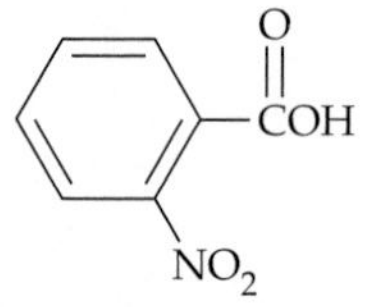

17.3

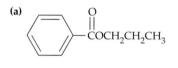

17.4 propanoic acid

17.5

(a)

O
‖
COCH₂CH₂CH₃ (benzene ring)

(b)

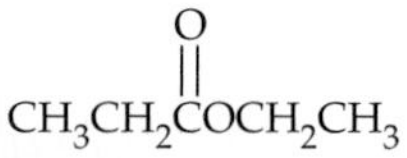

O
‖
CH₃CH₂COCH₂CH₃

(c)

O
‖
H₂C=CHCOCH₃

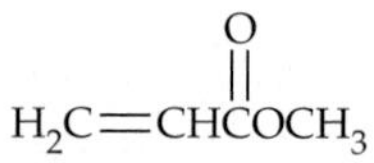

17.6 CH₃COOH is highest boiling (most H-bonding). CH₃CH₂CH₃ is lowest boiling (nonpolar). **17.7 (a)** C₃H₇COOH is more soluble (smaller —R group). **(b)** (CH₃)₂CHCOOH is more soluble (carboxylic acid).

17.8

O
‖
CH₃CH₂COH
Propanoic acid

O
‖
CH₃CH₂COCH₃
Methyl propanoate

O
‖
CH₃CH₂CNH₂
Propanamide

O
‖
CH₃CH₂CNHCH₃
N-Methylpropanamide

O
‖
CH₃CH₂CNCH₃
         |
         CH₃
N,N-Dimethylpropanamide

17.9 (a) 4-methylpentanoic acid **(b)** *N*-methyl-*p*-chlorobenzamide **17.10 (a)** (CH₃)₂CHCH₂CH₂CONH₂ **(b)** CH₃CH₂CH₂CONHCH₃ **17.11 (a)** (ii) **(b)** (i) **(c)** (iv) **(d)** (iii) **17.12 (a)** (ii) **(b)** (i) **(c)** (iii) **(d)** (i) **(e)** (i) **(f)** (iii)

17.13 **(a)**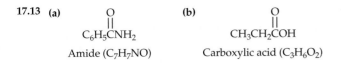

C₆H₅CNH₂
Amide (C₇H₇NO)

(b)

CH₃CH₂COH
Carboxylic acid (C₃H₆O₂)

(c)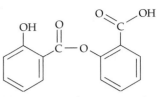

CH₃COCH₂CH₃
Ester (C₄H₈O₂)

17.14 **(a)**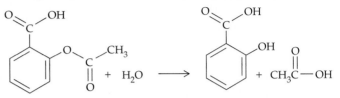

CH₃CH₂CH₂C—O⁻ K⁺ + H₂O

(b)

[CH₃CH₂CH₂CHC—O⁻]₂ Ba²⁺ + 2 H₂O
CH₃

17.15 **(a)** [HCOO⁻]₂ Ca²⁺ **(b)** H₂C = CHCOO⁻Na⁺ **17.16** CH₃COO⁻
⁻OOCCH₂CH₂CH₂COO⁻ Na⁺ **17.17** HCOOCH₂CH(CH₃)₂
17.18 **(a)**

⬡—OH + HOCCH₂CH₂CH(CH₃)₂

(b)

CH₃CH₂CH₂CH₂COH + HOCH(CH₃)₂

17.19
(a)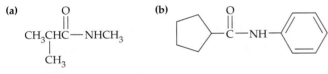

CH₃CHC—NHCH₃
CH₃

(b)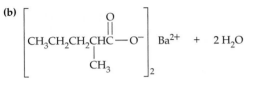

17.20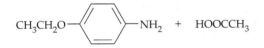

CH₃CH₂O—⬡—NH₂ + HOOCCH₃

17.21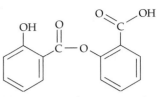

17.22 Aspirin is acidic (—COOH), lidocaine is basic (amine), benzocaine is weakly basic (aromatic amine), acetaminophen is weakly
acidic (phenol). **17.23** Moisture in the air hydrolyzes the ester bond.

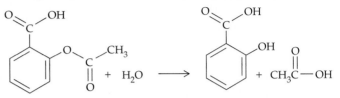

+ H₂O ⟶ + CH₃C—OH

17.24 **(a)** 2-methylpropanoic acid + 2-propanol **(b)** 2-butenoic acid +
ethanol **(c)** p-bromobenzoic acid + 1-propanol **17.25** **(a)** 2-butenoic
acid + methylamine **(b)** p-chlorobenzoic acid + diethylamine
17.26

(a)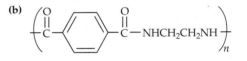

⟮CCH₂CH₂C—OCH₂CH₂O⟯ₙ

(b)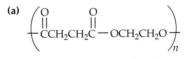

⟮C—⬡—C—NHCH₂CH₂NH⟯ₙ

17.27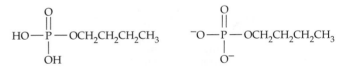

HO—P—OCH₂CH₂CH₂CH₃   ⁻O—P—OCH₂CH₂CH₂CH₃
OH   O⁻

17.28 **(a)** amide + H₂O → CH₃COOH + NH₃
(b) phosphate monoester + H₂O → CH₃CH₂OH + HOPO₃²⁻
(c) carboxylic acid ester + H₂O → CH₃CH₂COOH + HOCH₃
17.29

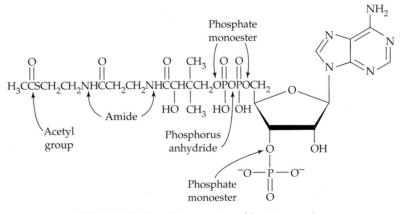

17.30 **(a)** At pH = 7.4, pyruvate and lactate are anions.
(b)

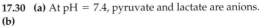

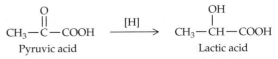

O   OH
CH₃—C—COOH ⟶[H] CH₃—CH—COOH
Pyruvic acid   Lactic acid

(c) Pyruvate and lactate have similar solubilities in water.
(d)

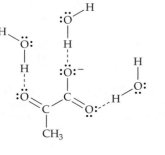

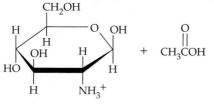

17.31 (a) H_2O + acid or base

(b)

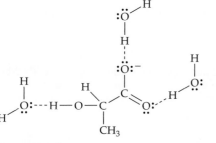

$+$ CH_3COH

17.32 (a) a phosphate ester linkage

(b)

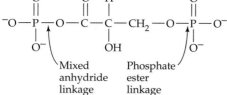

Mixed anhydride linkage   Phosphate ester linkage

17.33

$^-OOCCOO^-$ $^-OOCCH_2COO^-$
Oxalate   Malonate

$^-OOCCH_2CH_2COO^-$ $^-OOCCH_2CH_2CH_2COO^-$
Succinate   Glutarate

17.34

(a)

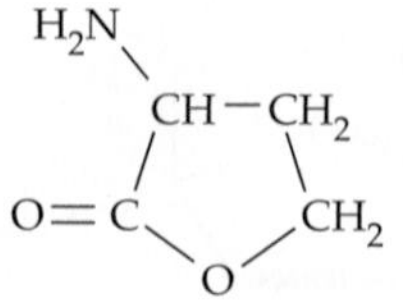

(b)

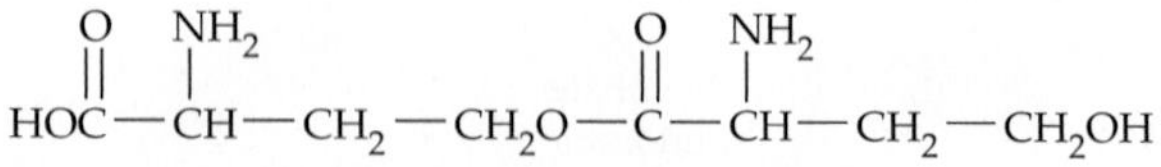

$$\underset{HOC}{\overset{O}{\|}} - \underset{CH}{\overset{NH_2}{|}} - CH_2 - CH_2O - \overset{O}{\overset{\|}{C}} - \underset{CH}{\overset{NH_2}{|}} - CH_2 - CH_2OH$$

(c)

$$\underset{HOC}{\overset{O}{\|}} - \underset{C}{\overset{H}{|}} - \underset{N}{\overset{H}{|}} - \overset{O}{\overset{\|}{C}} - \underset{CH}{\overset{NH_2}{|}} - CH_2 - CH_2OH$$
$$\underset{|}{\overset{}{}}$$
$$CH_2CH_2OH$$

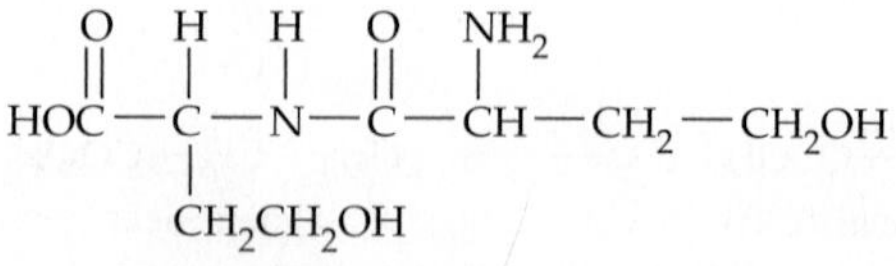

17.35

(a) (i)   **(ii)**   **(iii)**

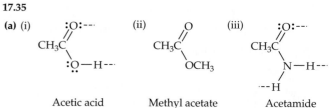

Acetic acid   Methyl acetate   Acetamide

(b) Methyl acetate is lowest boiling (no hydrogen bonds); acetamide is highest boiling.   **17.36 (a)** ethyl benzoate **(b)** 2,3-dimethylpentanoic acid **(c)** 2-methylpropanamide **(d)** *N,N*-dimethylformamide. Compound **(b)** is acidic; all other compounds are neutral.

17.37

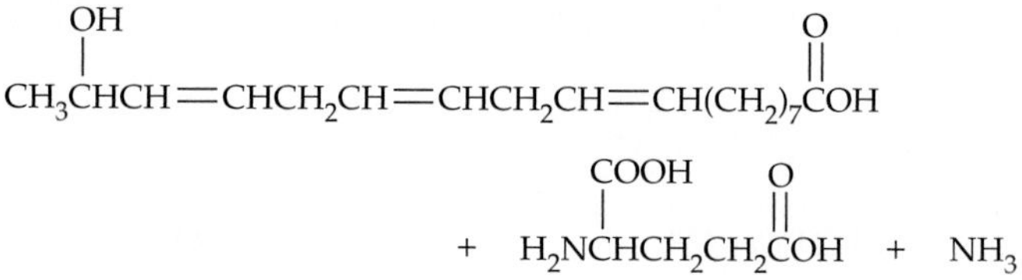

$$\underset{CH_3CHCH}{\overset{OH}{|}} = CHCH_2CH = CHCH_2CH = CH(CH_2)_7\overset{O}{\overset{\|}{C}}OH$$

$$+ \quad \underset{H_2NCHCH_2CH_2\overset{O}{\overset{\|}{C}}OH}{\overset{COOH}{|}} \quad + \quad NH_3$$

17.38

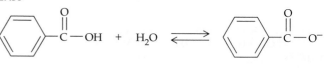

$+ \quad H_3O^+$

17.40

$CH_3CH_2CH_2CH_2COOH$ $\underset{CH_3CH_2CHCOOH}{\overset{CH_3}{|}}$
Pentanoic acid   2-Methylbutanoic acid

$\underset{CH_3CHCH_2COOH}{\overset{CH_3}{|}}$ $(CH_3)_3CCOOH$
3-Methylbutanoic acid   2,2-Dimethylpropanoic acid

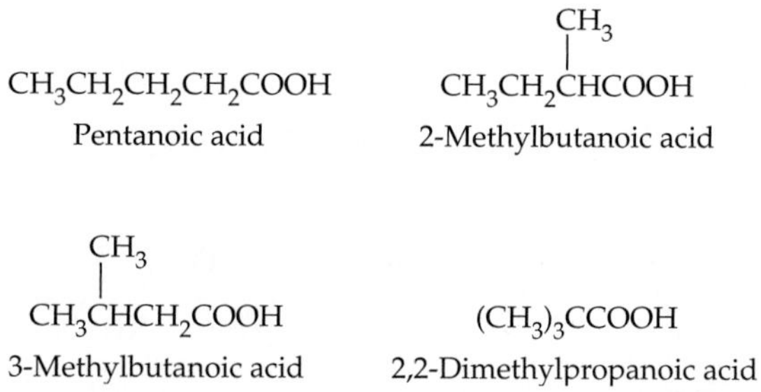

17.42 (a) hexanoic acid **(b)** 2-methylpentanoic acid **(c)** 2-ethylbutanoic acid **(d)** 3-cyclopropylpropanoic acid   **17.44 (a)** potassium 3-ethylpentanoate **(b)** ammonium benzoate **(c)** calcium propanoate

17.46

(a) $\underset{\underset{CH_3}{|}}{\overset{CH_3}{\overset{|}{CH_3CHCHCH_2COOH}}}$

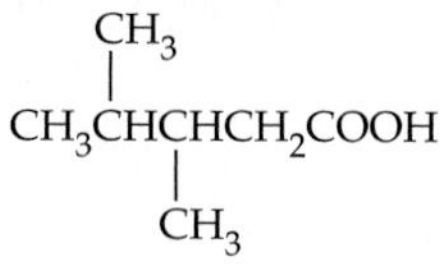

(b) $(C_6H_5)_3CCOOH$

(c) CH_3CH_2

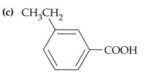

$-COOH$

(d) $CH_3CH_2CH_2COO^-$ $^+NH_3CH_3$

17.48

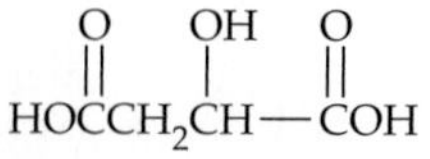

$$\overset{O}{\overset{\|}{HOCCH_2}}\underset{CH}{\overset{OH}{|}} - \overset{O}{\overset{\|}{COH}}$$

17.50

$$Na^+ {}^-\overset{O}{\overset{\|}{OCCH_2}}\underset{CH}{\overset{OH}{|}} - \overset{O}{\overset{\|}{CO}}{}^- Na^+$$

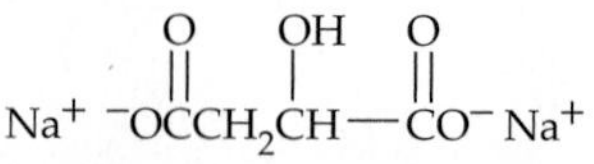

17.52 **(a)** CH₃CH₂CH₂CH₂CONH₂   CH₃CH₂CONHCH₂CH₃
Pentanamide   N-Ethylpropanamide

HCON(CH₂CH₃)₂
N,N-Diethylformamide

(b) CH₃CH₂CH₂CH₂COOCH₃   CH₃CH₂COOCH₂CH₂CH₃
Methyl pentanoate   Propyl propanoate

HCOOCH₂CH₂CH₂CH₂CH₃
Pentyl formate

17.54 **(a)** 3-methylbutyl acetate **(b)** methyl 4-methylpentanoate

(c)
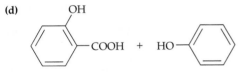

(d)
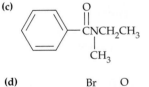

17.56 **(a)** CH₃COOH + HOCH₂CH₂CH(CH₃)₂
(b) (CH₃)₂CHCH₂CH₂COOH + HOCH₃

(c)

CH₃COOH   +   HO—⬡

(d)

⬡(OH)—COOH   +   HO—⬡

17.58 **(a)** 2-ethylbutanamide **(b)** N-phenylbenzamide

(c)

⬡—CNCH₂CH₃ (with O double bond and CH₃)

(d)

CH₃CH₂CH₂CHCHCNH₂ (with Br, Br, and O)

17.60 **(a)** 2-ethylbutanoic acid + ammonia **(b)** benzoic acid + aniline
(c) benzoic acid + N-methylethylamine **(d)** 2,3-dibromohexanoic acid + ammonia

17.62

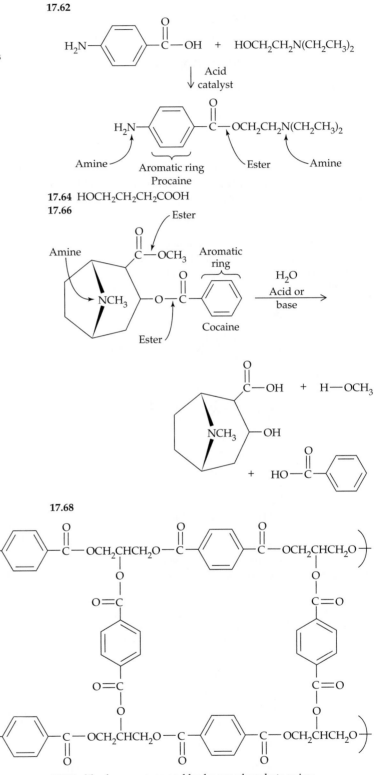

17.64 HOCH₂CH₂CH₂COOH

17.66

17.68

17.70 dihydroxyacetone and hydrogen phosphate anion

17.72

17.74 A cyclic phosphate diester is formed when a phosphate group forms an ester with two hydroxyl groups in the same molecule.

17.76 Trichloroacetic acid: used for chemical peeling of the skin. Lactic acid: used for wrinkle removal and moisturizing.

17.78

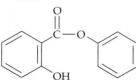

17.80 strong acids and bases

17.82

17.84 pH; Tollens' test **17.86 (a)** 2-chloro-3,4-dimethyl-3-hexene **(b)** *N*-methyl-*N*-phenylpropanamide **(c)** phenyl 2,2-diethylbutanoate **(d)** *N*-ethyl-*o*-nitrobenzamide

Chapter 18

18.1 Aromatic ring: phenylalanine, tyrosine, tryptophan
   Contain sulfur: cysteine, methionine
   Alcohols: serine, threonine, tyrosine (phenol)
   Alkyl side chain: alanine, valine, leucine, isoleucine

18.2

18.3

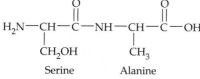

Serine   Alanine

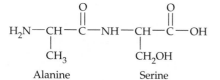

Alanine   Serine

18.4 α-amino acids: **(a)**, **(d)**   **18.5 (b)** Phe, Ser **(c)** Thr, Tyr
18.6

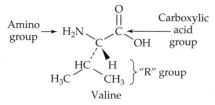

Valine

18.7

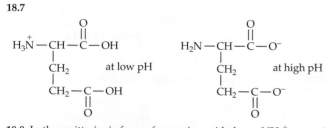

18.8 In the zwitterionic form of an amino acid, the —NH₃⁺ group is an acid, and the —COO⁻ group is a base. **18.9** chiral: **(a)**, **(b)**, **(d)**, **(e)**, **(g)** **18.10** Handed: wrench, corkscrew, jar lid. Not handed: thumbtack, pencil, straw **18.11** 2-Butanol has a carbon with 4 different groups bonded to it. **18.12** chiral: **(b)**, **(c)**

18.13

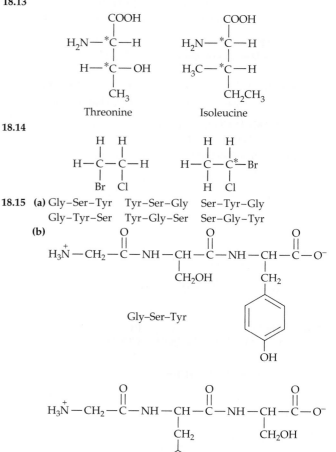

Threonine   Isoleucine

18.14

18.15 (a) Gly–Ser–Tyr   Tyr–Ser–Gly   Ser–Tyr–Gly
      Gly–Tyr–Ser   Tyr–Gly–Ser   Ser–Gly–Tyr
   (b)

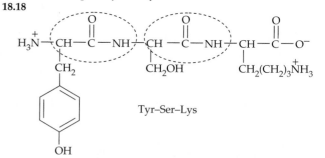

Gly–Ser–Tyr

Gly–Tyr–Ser

18.16 Val–His–Ala   His–Ala–Val   Ala–His–Val
      Val–Ala–His   His–Val–Ala   Ala–Val–His
18.17 (a) Leu-Asp **(b)** Tyr-Ser-Lys
18.18

Tyr–Ser–Lys

18.19 Asp–Tyr + Phe + Glu–Asn–Cys–Pro–Lys–Gly **18.20 (a)** hydrogen bond **(b)** hydrophobic interaction **(c)** salt bridge **(d)** hydrophobic interaction **18.21 (a)** Tyr, Asp, Ser **(b)** Ala, Ile, Val, Leu **18.22** eleven backbone atoms **18.23** Secondary structure: stabilized by hydrogen bonds between amide nitrogens and carbonyl oxygens of polypeptide backbone. Tertiary structure: stabilized by hydrogen bonds between amino acid side-chain groups. **18.24 (a)** tertiary; **(b)** secondary; **(c)** quaternary **18.25** At low pH, the groups at the end of the polypeptide chain exist as —NH₃⁺ and —COOH. At high pH, they exist as —NH₂ and —COO⁻. In addition, side chain functional groups may be ionized as follows: **(a)** no change **(b)** Lys, His, Arg positively charged at low pH; Lys, His neutral at high pH: **(c)** Tyr neutral at low pH, negatively charged at high pH: **(d)** Glu, Asp neutral at low pH, negatively charged at high pH: **(e)** no change: **(f)** Cys neutral at low pH, negatively charged at high pH. **18.26 (a)** 1, 3, 4 **(b)** 1, 4 **(c)** 4

18.27

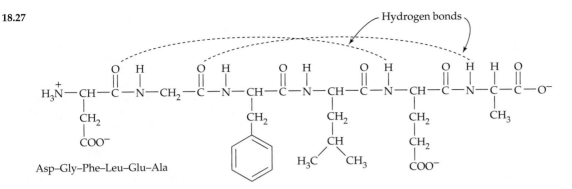

Asp–Gly–Phe–Leu–Glu–Ala

18.28 Phe–Glu–Phe–Asp–His–Tyr–Glu–His   **18.29** *Fibrous Proteins*: structural proteins, water-insoluble, contain many Gly and Pro residues, contain large regions of α-helix or β-sheet, few side-chain interactions. Examples: collagen, α-keratin, fibroin. *Globular Proteins*: enzymes and hormones, usually water-soluble, contain most amino acids, contain smaller regions of α-helix and β-sheet, complex tertiary structure. Examples: ribonuclease, hemoglobin, insulin.   **18.30** The prefix "α" means that —NH_2 and —COOH are bonded to the same carbon.

18.32

(a)   **(b)**

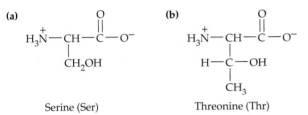

Serine (Ser)   Threonine (Thr)

(c)

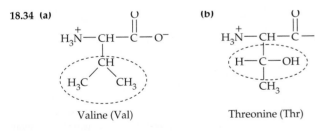

Proline (Pro)

18.34 (a)   **(b)**

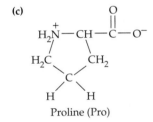

Valine (Val)   Threonine (Thr)

18.36 Lys**(a)** is positively charged. Phe**(b)** and Leu**(c)** are neutral.   **18.38 (a)** low pH **(b)** high pH   **18.40** A chiral object is handed. Examples: glove, car.   **18.42 (a), (c)**

18.44

(a)   **(b)**

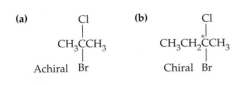

Achiral Br   Chiral Br

(c)

$(CH_3)_2CH\overset{*}{C}CH_3$

Chiral Br

18.46

Chiral ——   —— Achiral

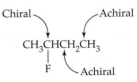

$CH_3CHCH_2CH_3$

|
F   —— Achiral

18.48 A simple protein is composed only of amino acids. A conjugated protein consists of a protein associated with one or more non-protein molecules.

18.50

Type of protein	Function	Example
Enzymes:	Catalyze biochemical reactions	Ribonuclease
Hormones:	Regulate body functions	Insulin
Storage proteins:	Store essential substances	Myoglobin
Transport proteins:	Transport substances through body fluids	Serum albumin
Structural proteins:	Provide shape and support	Collagen
Protective proteins:	Defend the body against foreign matter	Immunoglobulins
Contractile proteins:	Do mechanical work	Myosin and actin

18.52 Disulfide bonds stabilize tertiary structure.   **18.54** In *hydrophobic interactions*, hydrocarbon side chains cluster in the center of proteins and make proteins spherical. Examples: Phe, Ile. *Salt bridges* bring together distant parts of a polypeptide chain. Examples: Lys, Asp.   **18.56** When a protein is denatured, its nonprimary structure is disrupted, and it can no longer catalyze reactions.   **18.58** Met–Ile–Lys Met–Lys–Ile Ile–Met–Lys Ile–Lys–Met Lys–Met–Ile Lys–Ile–Met   **18.60** *Outside*: Asp, Asn (They can form H-bonds.) *Inside*: Val, Leu (They have hydrophobic interactions.)   **18.62** Enzymes would hydrolyze insulin.

18.64

N-terminal   C-terminal

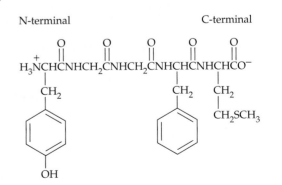

18.66 (a) $H_3\overset{+}{N}CH_2COOH$ **(b)** $H_3\overset{+}{N}CH_2COOCH_3$ **18.68** N-terminal: Val–Gly–Ser–Ala–Asp C-terminal   **18.70** A peptide rich in Asp and Lys is more soluble, because its side chains are more polar and can form hydrogen bonds with water.

18.72

(a)

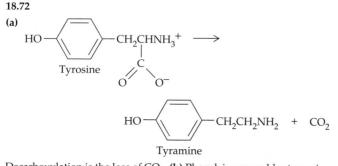

Tyrosine → Tyramine + CO_2

Decarboxylation is the loss of CO_2. **(b)** Phenelzine resembles tyramine and blocks the enzyme that catalyzes deamination of tyramine. **18.74** An incomplete protein lacks one or more essential amino acids. **18.76** They must provide complete nutrition to developing organisms. **18.78** A = Lys–Trp, B = Ala–Cys, C = Glu–Tyr **18.80** It was hard to accept that a protein might duplicate itself, cause disease, be responsible for inherited disease, be transmitted, and might arise spontaneously. **18.82** In α-keratin, pairs of α-helices twist together into small fibrils that are twisted into larger bundles. In tropocollagen, 3 coiled chains wrap into a triple helix.
18.84 Hydrophobic interactions: **(b) (f) (h)**
  Hydrogen bonding: **(a) (c) (d) (e)**
  Salt bridges: **(a) (e)**
  Covalent bonding: **(g)**
18.86 Denaturation disrupts shape without breaking peptide bonds; hydrolysis breaks bonds and destroys primary structure. **18.88** Met has a $-SCH_3$ group, which can't form disulfide bridges. **18.90** nonpolar side chains: interior of a protein **18.92** Glycine is small; proline is rigid and bent.

Chapter 19

19.1 kinases **19.2** (1) The enzyme might catalyze reactions within the eye. (2) Saline is sterile and isotonic. **19.3** iron, copper, manganese, molybdenum, vanadium, cobalt, nickel, chromium **19.4** Vitamin-derived coenzymes must be provided in the diet. Other coenzymes are synthesized in the body. **19.5** **(a)** isomerization of retinal **(b)** oxidation of squalene **(c)** phosphate transfer to glucose **(d)** hydrolysis of cellulose **19.6** A lyase. It catalyzes the elimination of CO_2 from pyruvate. **19.7** A terminal $-OH$ group is converted to an aldehyde, and the adjacent ketone is converted to an $-OH$ group. **19.8** Reaction **(a)** is catalyzed by an decarboxylase. **19.9** Acidic, basic and polar side chains take part in catalytic activity. Nonpolar side chains hold the enzyme in the active site. **19.10** Substrate molecules are bound to all of the active sites. **(a)** no effect **(b)** increases the rate. **19.11** higher at 30°. Somewhat higher at 40°. **19.12** The rate is much greater at pH = 8. **19.13** Molecule **(b)**, because it resembles the substrate. **19.14** a product that resembles the substrate **19.15** **(a)** covalent modification **(b)** competitive inhibition **(c)** genetic control **(d)** feedback control or covalent modification **19.16** Vitamin A—long hydrocarbon chain. Vitamin C—polar hydroxyl groups. **19.17** Retinal—aldehyde. Retinoic acid—carboxylic acid. **19.18** enzyme cofactors; antioxidants; aid in absorption of calcium and phosphate ions; aid in synthesis of visual pigments and blood clotting factors
19.19

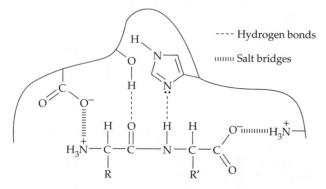

---- Hydrogen bonds

|||||| Salt bridges

19.20 **(a)** oxidoreductase **(b)** dehydrogenase **(c)** L-lactate **(d)** L-lactate dehydrogenase **19.21** No. An enzyme usually catalyzes the reaction of only one enantiomer. D-Lactate might be a competitive inhibitor. **19.22** NAD^+ is an oxidizing agent and includes the vitamin niacin. **19.23** **(a)** Rate increases when [substrate] is low, but max. rate is soon reached; max. rate is always lower than max. rate of uninhibited reaction. **(b)** Rate increases. **19.24** **(a)** Binding of the regulator at a site away from the catalytic site changes the shape of the enzyme. **(b)** Addition or removal of a covalently bonded group changes the activity of an enzyme **(c)** Noncompetitive inhibition is a type of allosteric regulation (see part **(a)**). Competitive inhibition occurs when an inhibitor reversibly occupies an enzyme's active site. Irreversible inhibition results when an inhibitor covalently binds to an enzyme and destroys its ability to catalyze a reaction. **(d)** Hormones control the synthesis of enzymes. **19.25** From left to right: aspartate (acidic), serine, glutamine, arginine (basic), histidine (basic). **19.26** **(a)** hydrolysis (bond breaking by addition of water) of a substrate **(b)** isomerization of a substrate **(c)** addition of a small molecule to a double bond, or removal of a small molecule to form a double bond **19.28** An enzyme is a large three-dimensional molecule with a catalytic site into which a substrate can fit. Enzymes are specific in their action because only one or a few molecules have the appropriate shape and functional groups to fit into the catalytic site. **19.30** **(a)** transferase **(b)** isomerase **(c)** ligase **19.32** **(a)** hydrolysis of peptide bonds **(b)** bond formation between two DNA chains **(c)** transfer of a methyl group **19.34** hydrolase **19.36** Lock-and-key: An enzyme is rigid (lock) and only one specific substrate (key) can fit in the active site. Induced fit: An enzyme can change its shape to accommodate the substrate and to catalyze the reaction. **19.38** No. Protein folding can bring the residues close to each other. **19.40** In the stomach, an enzyme must be active at an acidic pH. In the intestine, an enzyme needs to be active at a higher pH and need not be active at pH = 1.5.
19.42

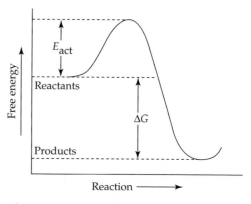

19.44 As substrate concentration is increased, the rate increases until all enzyme active sites are occupied. At this point, adding more substrate doesn't increase the rate. **19.46** **(a) (b)** The rate decreases. **(c)** The reaction stops (or its rate greatly decreases). **19.48** *Noncompetitive inhibition:* Inhibitor binds reversibly and noncovalently away from the active site and changes the shape of the site to make it difficult for the enzyme to catalyze reactions. *Competitive inhibition:* Inhibitor binds reversibly and noncovalently at the active site and keeps the substrate from entering. *Irreversible inhibition:* Inhibitor irreversibly forms a covalent bond at the active site and destroys the catalytic ability of the enzyme. **19.50** Papain hydrolyzes peptide bonds in meat. **19.52** One site is for catalysis, and one site is for regulation. **19.54** Feedback inhibition occurs when a product of a series of enzyme-catalyzed reactions is an inhibitor for an earlier step of the reaction sequence. **19.56** A zymogen is an enzyme synthesized in a form different from its active form. Some enzymes are secreted as zymogens because they might be harmful to the body if they were secreted in their active form. **19.58** A vitamin is a small essential organic molecule that must be supplied in the diet. **19.60** Vitamin C (water-soluble) is excreted in urine and must be replaced. Vitamin A (fat-soluble) is stored in fatty tissue. **19.62** (1) bonding to cysteine—irreversible; (2) displacing a metal from an enzyme's active site—reversed by

EDTA. **19.64** (1) Enzymes produce only one enantiomer of a product. (2) Enzymes allow difficult reactions to take place (3) Use of enzymes avoids reactions with hazardous byproducts. **19.66** Enzyme activity must be monitored under standard conditions because activity is affected by pH, temperature and substrate concentration. **19.68** Water-soluble vitamins are excreted, but fat-soluble vitamins accumulate in tissue. **19.70** They are considered to be most important for maintaining good health. **19.72** 11 L apple juice **19.74** In competitive inhibition, adding a large amount of substrate will cause the rate to return to its usual value. In noncompetitive inhibition, no amount of substrate will cause the rate to return to its usual value. **19.76** Ser → Val: enzyme would be inactivated. Asp → Glu: effect would probably be minor.

Chapter 20

20.1 (b) **20.2** The molecules resemble the heterocyclic part of cAMP, and they might act as inhibitors to the enzyme that inactivates cAMP. **20.3** Glu–His–Pro **20.4** The hydrophobic part of the structure is larger than the polar, hydrophilic part. **20.5** Testosterone has a —CH_3 group between the first two rings; nandrolone doesn't. Otherwise, their structures are identical. **20.6 (a)** 2 **(b)** 3 **(c)** 1 **20.7** Similarities: both structures have aromatic rings, secondary amine groups, alcohol groups. Differences: propanolol has an ether group and a naphthalene ring system; epinephrine has two phenol hydroxyl groups; the compounds have different side chain carbon skeletons. **20.8** Malathion: It's the least toxic. **20.9** An agonist (black widow spider venom) prolongs the biochemical response of a receptor. An antagonist (botulinus toxin) blocks or inhibits the normal response of a receptor. **20.10** phenol hydroxyl group, ether, carbon-carbon double bond, aromatic ring. THC is hydrophobic and is likely to accumulate in fatty tissue. **20.11 (a)** antihistamine **(b)** antidepressant **20.12 (a)** insulin (polypeptide hormone) **(b)** pancreas **(c)** in the bloodstream **(d)** Insulin doesn't enter cells directly because it can't pass through cell membranes. Instead, it binds with a cell surface receptor. **20.13 (a)** polypeptide hormone (produced in the anterior pituitary gland) **(b)** steroid hormone (produced in ovaries) **(c)** Progesterone-producing cells have LH receptors. **(d)** Progesterone is lipid-soluble and can enter cells. **20.14** (1) Adenylate cyclase can produce a great many molecules of cAMP. (2) A great amount of glucose is released when glycogen phosphorylase is activated. **20.15** enzymatic inactivation; reuptake by presynaptic neuron **20.16** Neurotransmitters can act by binding to receptors or by activating second messengers. They may be enzymatically deactivated or they may be returned to the presynaptic neuron. **20.17** These substances increase dopamine levels in the brain. The brain responds by decreasing the number and sensitivity of dopamine receptors. Thus more of the substance is needed to elevate dopamine levels, leading to addiction. **20.18** A hormone is a molecule that travels through the bloodstream to its target tissue, where it binds to a receptor and regulates biochemical reactions. **20.20** A vitamin is usually an enzyme cofactor, whereas a hormone regulates enzyme activity. **20.22** Hormone binding is noncovalent and controls the rate of a reaction, rather than taking part in a reaction. **20.24** polypeptide hormones, steroid hormones, amino acid derivatives **20.26** The endocrine system manufactures and secretes hormones. **20.28** Enzymes are proteins; hormones may be polypeptides, proteins, steroids or amino acid derivatives. **20.30** Polypeptide hormones travel through the bloodstream and bind to cell receptors, which are on the outside of a cell. The receptors cause production within cells of "second messengers" that activate enzymes. **20.32** Epinephrine is produced in the adrenal medulla, in response to a signal from the hypothalamus. It stimulates the production of glucose. **20.34** It initiates reactions that release glucose from storage. Termination occurs when phosphodiesterase converts cAMP to AMP. **20.36** Insulin contains 51 amino acids, is released from the pancreas, and acts at cells, causing them to take up glucose. **20.38** mineralocorticoids (aldosterone), glucocorticoids (cortisone), sex hormones (testosterone, estrone) **20.40** Both have large nonpolar regions and can cross cell membranes. **20.42 (a)** amino acid derivative **(b)** polypeptide hormone **(c)** steroid hormone **20.44** A synapse is the gap between two nerve cells that neurotransmitters cross to transmit

their message. **20.46** nerve cell, muscle cell, endocrine cell **20.48** enzymatic deactivation, reuptake **20.50** (1) An impulse arrives at the presynaptic neuron. (2) Vesicles release ACh (3) ACh crosses the cleft and binds to receptors on the postsynaptic neuron. (4) A nerve impulse is initiated. (5) Acetylcholinesterase catalyzes breakdown of ACh. (6) Choline is reabsorbed, and new ACh is synthesized. **20.52** Agonists prolong the response of a receptor. Antagonists block the response of a receptor. **20.54** serotonin, norepinephrine, dopamine **20.56** Cocaine increases dopamine levels by blocking reuptake. Amphetamines accelerate release of dopamine. **20.58** The brain receptors for neurotransmitters from animals also, coincidentally, serve as receptors for neurotransmitters from plants. **20.60** An ethnobotanist discovers what indigenous people have learned about the healing power of plants. **20.62** Scientists who know about the exact size and shape of enzymes and receptors can design drugs that interact with the active sites of these biomolecules. **20.64** Homeostasis is the maintenance of a constant internal environment. **20.66** Plants don't have endocrine systems or a circulatory fluid like blood. **20.68** Epibatidine acts as a painkiller at acetylcholine receptors in the central nervous system. **20.70** The *hormone receptor* recognizes the hormone, sets in motion the stimulation process, and interacts with *G-protein*. This protein binds GTP and activates *adenylate cyclase*, which catalyzes the formation of cAMP. **20.72** Signal amplification is the process by which a small signal induces a response much larger than the original signal. In the case of hormones, a small amount of hormone can bring about a very large response. **20.74** Testosterone has an —OH group in the 5-membered ring; progesterone has an acetyl group in that position. Otherwise, the two molecules are identical. **20.76** Chocolate acts at dopamine receptors and produces feelings of satisfaction. **20.78** Testosterone is converted to androsterone by reductions (C=O, C=C) in the first ring, and an oxidation of —OH in the five-membered ring. Estradiol is converted to estrone by oxidation of —OH in the five-membered ring. These conversions are oxidations and reductions.

Chapter 21

21.1 exergonic: **(a)**, **(c)**; endergonic: **(b)**; releases the most energy: **(a)** **21.2** Both pathways produce the same amount of energy. **21.3 (a)** exergonic—oxidation of glucose; endergonic—photosynthesis **(b)** sunlight

21.4 (a)

Carbohydrates $\xrightarrow{\text{Digestion}}$ Glucose, sugars $\xrightarrow{\text{Glycolysis}}$

Pyruvate $\longrightarrow$ Acetyl-SCoA $\xrightarrow{\text{Citric acid}}$
$\qquad\qquad\qquad\qquad\qquad$ cycle

Reduced coenzymes $\xrightarrow[\text{transport}]{\text{Electron}}$ ATP

(b) pyruvate, acetyl-SCoA, citric acid cycle intermediates

21.5

$$H_3C-\overset{\overset{\displaystyle O}{\|}}{C}-O-\overset{\overset{\displaystyle O}{\|}}{\underset{\underset{\displaystyle O^-}{|}}{P}}-O^- \;+\; H_2O \;\longrightarrow$$

$$H_3C-\overset{\overset{\displaystyle O}{\|}}{C}-O^- \;+\; {}^-O-\overset{\overset{\displaystyle O}{\|}}{\underset{\underset{\displaystyle OH}{|}}{P}}-O^- \;+\; H^+$$

21.6 Energy is produced only when it is needed.
21.7

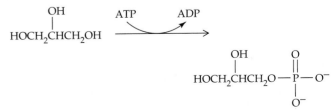

21.8 If a process is exergonic, its exact reverse is endergonic and can't occur unless it is coupled with an exergonic reaction in a different pathway. **21.9** favorable ($\Delta G = -3.0$ kcal/mol) **21.10** yes **21.11**

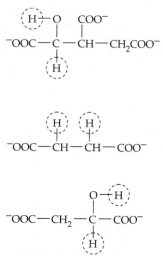

21.12 citric acid, isocitric acid **21.13** steps 3, 4, 6, 8 **21.14** Succinic dehydrogenase catalyzes the removal of two hydrogens from succinate to yield fumarate. **21.15** α-ketoglutarate, oxaloacetate **21.16** isocitrate **21.17** Steps 1–4 correspond to the first stage, and steps 5–8 correspond to the second stage. **21.18** mitochondrial matrix **21.19** O_2. Movement of H^+ from a region of high $[H^+]$ to a region of low $[H^+]$ releases energy that is used in ATP synthesis.
21.20
(a) Succinyl phosphate + $H_2O \longrightarrow$ Succinate + $HOPO_3^{2-}$ + H^+
(b) ADP + $HOPO_3^{2-}$ + $H^+ \longrightarrow$ ATP + H_2O $\Delta G = +7.3$ kcal/mol
21.21 **(a)** Stage 1 (digestion) **(b)** Stage 4 (ATP synthesis) **(c)** Stage 3 (citric acid cycle) **(d)** Stage 2 (glycolysis). **21.22** endergonic; coupling **21.23** NAD^+ accepts hydride ions; hydrogen ions are released to the mitochondrial matrix, and ultimately combine with reduced O_2 to form H_2O. **21.24** **(a)** Step A (NAD^+) **(b)** Step B **(c)** oxidoreductase **(d)** product of A **21.25** Step 1: lyase; Step 2: isomerase; Step 3: oxidoreductase; Step 4: oxidoreductase, lyase; Step 5: ligase; Step 6: oxidoreductase; Step 7: lyase; Step 8: oxidoreductase **21.26** Metals are better oxidizing and reducing agents. Also, they can accept and donate electrons in one-electron increments. **21.28** ΔG must be negative. **21.30** Enzymes affect only the rate of a reaction, not the size or sign of ΔG. **21.32** Exergonic: **(a)**, **(c)**; endergonic: **(b)**. Reaction **(a)** proceeds farthest toward products. **21.34** Prokaryotic cells: bacteria and algae. All other organisms have eukaryotic cells. **21.36** The cytoplasm consists of everything between the cell membrane and the nuclear membrane; the cytosol is the medium that fills the interior of the cell and contains electrolytes, nutrients and many enzymes, in aqueous solution. **21.38** 90% of the body's ATP is synthesized in mitochondria. **21.40** Catabolism is the breakdown of large molecules. Anabolism is the assembly of large molecules from smaller molecules. **21.42** *First*: digestion, citric acid cycle, electron transport, oxidative phosphorylation *Last* **21.44** Energy is released when ATP transfers a phosphoryl group. **21.46** ATP transfers a phosphoryl group. **21.48** $\Delta G = -4.5$ kcal/mol **21.50** Not favorable because overall ΔG is positive ($+4.0$ kcal/mol) **21.52** **(a)** FAD is reduced. **(b)** FAD is an oxidizing agent. **(c)** FAD oxidizes $-CH_2CH_2-$ to $-CH=CH-$. **(d)** $FADH_2$
(e)

$$-CH_2CH_2- \xrightarrow[\text{}]{\text{FAD} \quad \text{FADH}_2} -CH=CH-$$

21.54 krebs cycle, tricarboxylic acid cycle **21.56** Both carbons are oxidized to CO_2. **21.58** 3 NADH, 1 $FADH_2$. **21.60** ATP formation and oxidation of NADH and $FADH_2$ **21.62** H_2O, ATP, oxidized coenzymes **21.64** Fe^{2+}/Fe^{3+}; oxygen atoms in the CoQ ring

21.66

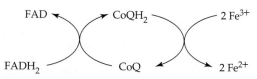

21.68 The pH differential occurs across the inner mitochondrial membrane. The intermembrane space has more H^+ ions. **21.70** Reduced coenzymes are oxidized, and ADP is phosphorylated. **21.72** Basal metabolic rate is the minimum amount of energy per unit time needed for breathing, maintaining body temperature, circulating blood and keeping all body organs functioning. **21.74** Daily activities use energy, and the body requires a larger caloric intake. **21.76** The energy is provided by the movement of hydrogen ions from a region of high concentration to a region of lower concentration as they pass through a channel in the ATP synthase complex. **21.78** The light reaction produces O_2, NADPH, and ATP. The dark reaction produces carbohydrates from water and CO_2. **21.80** Stepwise breakdown avoids production of large amounts of heat, allows for energy storage, and controls the rate of metabolism. **21.82** The cis double-bond isomer can't act as a substrate in the next step of the cycle. **21.84** Electrons from electron transport reduce O_2, which combines with H^+ to yield H_2O. **21.86** Catalase catalyzes the decomposition of H_2O_2 to water and O_2, which bubbles from the wound. **21.88** Energy from combustion is wasted as heat. Energy from metabolic oxidation is stored and used as needed.

Chapter 22
22.1 **(a)** aldopentose **(b)** ketotriose **(c)** aldotetrose
22.2

$$HOCH_2 \overset{OH}{\underset{|}{-}} CH \overset{OH}{\underset{|}{-}} CH \overset{OH}{\underset{|}{-}} CH \overset{OH}{\underset{|}{-}} CH \overset{O}{\overset{||}{-}} CH$$

An aldohexose

$$HOCH_2 \overset{OH}{\underset{|}{-}} CH \overset{O}{\overset{||}{-}} C - CH_2OH$$

A ketotetrose

22.3 eight stereoisomers **22.4** **(d)** **22.5** The bottom carbon is not chiral. The orientations of the hydroxyl groups bonded to the chiral carbons must be shown in order to indicate which stereoisomer is pictured.
22.6

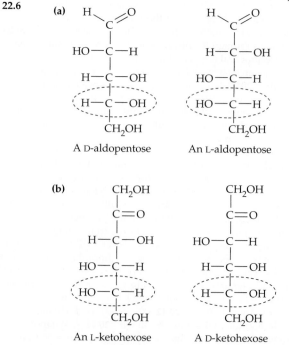

22.7

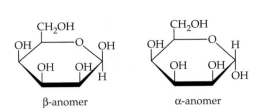

β-anomer   α-anomer

22.8

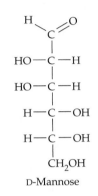

D-Mannose

22.9 (a) a hexose **(b)** a steroid **(c)** an ester

22.10

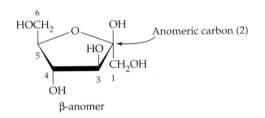

β-anomer

Anomeric carbon (2)

22.11

(a) 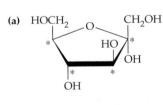 **(b)**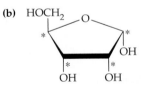

(c)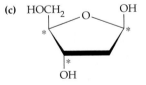

22.12

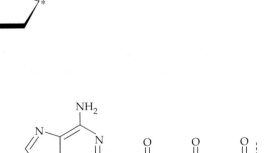

Cyclic AMP   from ribose   ATP

22.13 (a) an α anomer **(b)** carbon 6 **(c)** Groups that are below the plane of the ring in D-galactose are above the plane of the ring in L-fucose. Groups that are above the plane of the ring in D-galactose are below the plane of the ring in L-fucose. **(d)** yes

22.14

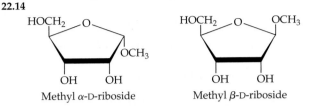

Methyl α-D-riboside   Methyl β-D-riboside

22.15 a β-1,4 glycosidic link **22.16** β-D-Glucose + β-D-Glucose **22.17 (a)** Maltose **(b)** Sucrose **(c)** Lactose **22.18** glutamine, asparagine **22.19** an α-1,4 glycosidic link **22.20** No. There are too few hemiacetal units to give a detectable result. **22.21**

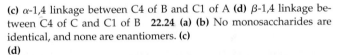

Starch polysaccharide —Amylase→ Maltose disaccharide —Maltase→ Glucose monosaccharide

22.22 (a) diastereomers, anomers **(b)** enantiomers **(c)** diastereomers
22.23 (a) (b)

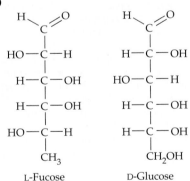

Acetal linkage

A   B   C
α-anomer   β-anomer   β-anomer

(c) α-1,4 linkage between C4 of B and C1 of A **(d)** β-1,4 linkage between C4 of C and C1 of B **22.24 (a) (b)** No monosaccharides are identical, and none are enantiomers. **(c)**
(d)

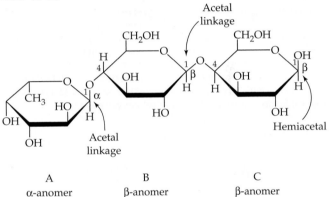

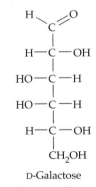

L-Fucose   D-Glucose   D-Galactose

22.25 Monosaccharide C is oxidized. Identification of the carboxylic acid also identifies the terminal monosaccharide.

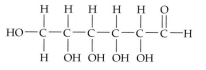

22.26 No
22.27

Polysaccharide	Linkage	Branching?
Cellulose	β-1,4	no
Amylose	α-1,4	no
Amylopectin	α-1,4	yes: α-1,6 branches occur ~ every 25 units
Glycogen	α-1,4	yes: even more α-1,6 branches than in amylopectin

22.28 Glucose is in equilibrium with its open-chain aldehyde form, which reacts with an oxidizing agent. **22.30** -ose **22.32** (a) aldotetrose (b) ketopentose (c) aldopentose (d) ketohexose
22.34

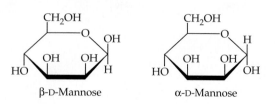

A four-carbon deoxy sugar

22.36 glucose **22.38** They are mirror images. **22.40** The reduction product of D-erythrose is achiral. **22.42** A polarimeter measures the degree of rotation of plane-polarized light by a solution of an optically active compound. **22.44** Equimolar solutions of enantiomers rotate light to the same degree but in opposite directions. **22.46** Mutarotation occurs when either a pure anomer or a mixture of anomers is dissolved in water. In either case, if the rotation of plane-polarized light is measured, the degree of rotation changes until it reaches a constant value. At this point, an equilibrium mixture of both anomers is present in the solution. **22.48** In the β form of a hemiacetal, the —OH at C1 is on the same side of the ring as the —CH₂OH group. In the α-form, the two groups are on opposite sides.
22.50

CH₂OH   CH₂OH

β-D-Mannose   α-D-Mannose

22.52

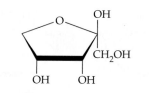

22.54

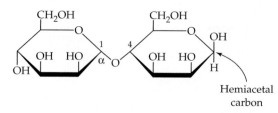

22.56

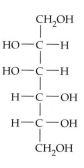

22.58

22.60 A glycoside is an acetal that is formed when the hemiacetal —OH group of a carbohydrate reacts with an alcohol.
22.62

Hemiacetal carbon

The hemiacetal carbon in this problem is in equilibrium with an open-chain aldehyde that is a reducing sugar. **22.64** Sucrose has no hemiacetal group. **22.66** Amylose and amylopectin both consist of long polymers of α-D-glucose linked by α-1,4 glycosidic bonds. Amylopectin is much larger and has α-1,6 branches every 25 units or so along the chain. **22.68** Gentiobiose contains both an acetal grouping and a hemiacetal grouping. Gentiobiose is a reducing sugar. A β-1,6 linkage connects the two monosaccharides. **22.70** Trehalose is a nonreducing sugar because it contains no hemiacetal linkages. The two D–glucose monosaccharides are connected by an α-1,1 acetal link.

22.72

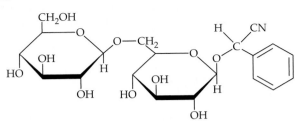

22.74 Enzyme-catalyzed reactions usually produce only one enantiomer. **22.76** Starch is a complex carbohydrate. Glucose is a simple carbohydrate. Soluble and insoluble fiber are complex carbohydrates. **22.78** Antifreeze in cars lowers the freezing point of water. Antifreeze proteins slow the formation of ice crystals by forming hydrogen bonds with water molecules at the surface of ice crystals that are starting to form. **22.80** People with type O blood can receive blood only from other donors that have type O blood. People with type AB blood can give blood only to other people with type AB blood. **22.82** pectin and vegetable gum: found in fruits, barley, oats and beans. **22.84** D-Ribose and D-xylose are diastereomers and differ in all properties listed.

22.86

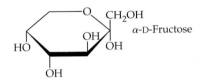

α-D-Fructose

22.88

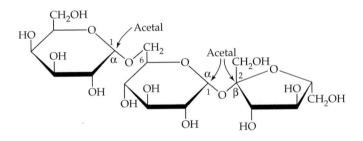

22.90 The starch is enzymatically hydrolyzed to glucose. **22.92** symptoms of galactosemia: vomiting, liver failure, mental retardation, cataracts **22.94** Lactose intolerance is an inability to digest lactose. Symptoms include bloating, cramps and diarrhea. **22.96** 4 chiral carbons

Chapter 23
23.1 (a) glycogenolysis (b) gluconeogenesis (c) glycogenesis **23.2** glycogenesis, pentose phosphate pathway, glycolysis **23.3** (a) steps 6 and 7 (b) steps 9 and 10 **23.4** isomerizations: steps 2, 5, 8
23.5

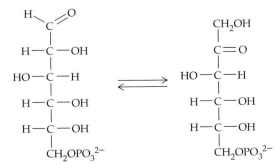

23.6 (a) pyruvate (b) Step 6: Glyceraldehyde 3-phosphate is oxidized; NAD^+ is the oxidizing agent. **23.7** Fructose 6-phosphate enters glycolysis at step 3.

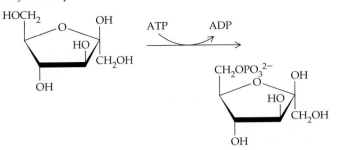

23.8 Glucose and galactose differ in configuration at C4. **23.9** Pyruvate → Acetyl-SCoA: 1 molecule CO_2 Isocitrate → α-Ketoglutarate: 1 molecule CO_2 α-Ketoglutarate → Succinyl-SCoA; 1 molecule CO_2 Since each molecule of glucose provides 2 pyruvates, 6 molecules of CO_2 are formed. **23.10** (a) The energy is lost as heat (b) The reverse of fermentation is very endothermic; loss of CO_2 drives the reaction to completion in the forward direction. **23.11** Insulin decreases; blood glucose decreases, the level of glucagon increases. Glucagon causes the breakdown of liver glycogen and the release of glucose. As glycogen is used up, the level of free fatty acids and ketone bodies increases. **23.12** Sorbitol can't form a cyclic acetal because it doesn't have a carbonyl group.

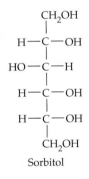

Sorbitol

23.13 (a) The increase in $[H^+]$ drives the equilibrium shown in Section 23.9 to the right, causing the production of CO_2. (b) Le Châtelier's principle. **23.14** The common intermediate is pyrophosphate $(P_2O_7^{4-})$. $\Delta G = -6.9$ kcal/ mol, indicating that the reaction is favorable. **23.15** phosphorylation, oxidation **23.16** hydrolases **23.17** (a) when ribose 5-phosphate or NADPH are needed (b) when glucose supply is adequate, and the body doesn't need to use glucose for energy production (c) when the body needs free glucose (d) when the supply of glucose is adequate and the body needs energy **23.18** Phosphorylations of glucose and fructose 6-phosphate produce important intermediates that repay the initial energy investment. Fructose 1,6-bisphosphate is cleaved into two three-carbon compounds, which are converted to pyruvate. **23.19** (a) when the body needs free glucose; in the liver (b) under anaerobic conditions; in muscle, red blood cells (c) when the body needs energy; in mitochondria (d) under anaerobic conditions, in yeast
23.20 Step 1: transferase
Step 2: isomerase
Step 3: transferase
Step 4: lyase
Step 5: isomerase
Step 6: oxidoreductase, transferase
Step 7: transferase
Step 8: isomerase
Step 9: lyase
Step 10: transferase; transferases (because many reactions involve phosphate transfers). Ligases are associated with reactions that synthesize molecules, not with reactions that break down molecules.

23.21 (d), (f), (a), (g), (c), (b), (e) **23.22** Yeast fermented the glucose in the wine to ethanol and CO_2. **23.23** Sources of compounds for gluconeogenesis: pyruvate, lactate, citric acid cycle intermediates, many amino acids. Germinating seeds need to synthesize carbohydrates from fats; humans obtain carbohydrates from food. **23.24 (a)** No **(b)** Molecular oxygen appears in the last step of the electron transport chain, where it combines with water, H^+ and electrons (from electron transport) to form H_2O. **23.26** 2 glucose; in the lining of the small intestine **23.28** glucose, fructose, galactose **23.30** acetyl-SCoA, lactate, ethanol + CO_2 **23.32** glycogenolysis: breakdown of glycogen to form glucose; glycogenesis: synthesis of glycogen from glucose **23.34** ribose 5-phosphate, glycolysis intermediates **23.36** (1) NADH can be reoxidized to NAD^+ under anaerobic conditions that also produce lactate. (2) In yeast, NADH is used in the conversion of pyruvate to yeast. Under aerobic conditions, NADH can enter the electron transport chain, where it is reoxidized to NAD^+ and ATP is produced. **23.38** in the cytosol of liver cells **23.40 (a)** 2 mol ATP **(b)** 0 **(c)** 1 mol ATP (most of the ATP in the citric acid cycle is produced in the electron transport chain)
23.42

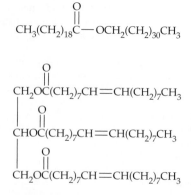

23.44 4 mol acetyl-SCoA **23.46** When blood glucose levels are high, insulin is released and glucose is metabolized. When blood glucose levels are low, glucagon is released and glycogen is metabolized to produce glucose. **23.48** ketone bodies **23.50** Excess glucose is converted to sorbitol, which can't be transported out of cells. This buildup changes osmolarity and causes cataracts and blindness. **23.52** *Juvenile diabetes* is caused by insufficient production of insulin in the pancreas. *Adult-onset diabetes* is caused by the failure of cell membrane receptors to recognize insulin. **23.54** muscle cells **23.56** The exact reverse of an energetically favorable pathway must occur by an alternate route in order to be favorable. **23.58** pyruvate, lactate **23.60** Several steps in the reverse of glycolysis are energetically unfavorable. **23.62** glycoproteins, bacteria, dextran, polysaccharide storage granules **23.64** Benedict's test detects all reducing sugars, not just glucose. **23.66** 140 g/dL vs 90 g/dL **23.68** Creatine phosphate provides ATP in one step. **23.70** first used: ATP, creatine phosphate, glucose, glycogen, fatty acids: last used **23.72** Pyruvate isn't a phosphate. **23.74** Yes. Fructose 6-phosphate enters glycolysis as a glycolysis intermediate.

Chapter 24
24.1 (a) eicosanoid **(b)** glycerophospholipid **(c)** wax
24.2

24.3

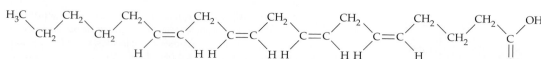

24.4 (a) butter **(b)** soybean oil
24.5

24.6 (b) **24.7** When two different fatty acids are bonded to C1 and C3 of glycerol, C2 is chiral. **24.8** London forces; weak; hydrogen bonds between water molecules are stronger than London forces. **24.9** The acyl groups are from stearic acid.

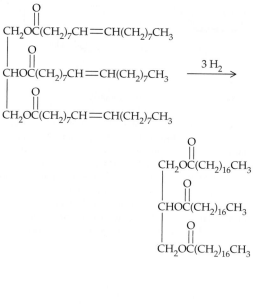

24.10

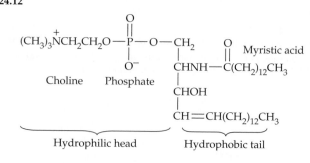

24.11 (a) glycerol, phosphate ion, choline, $RCOO^-\ Na^+$, $R'\ COO^-\ Na^+$, **(b)** sphingosine, phosphate ion, choline, sodium palmitate
24.12

24.13

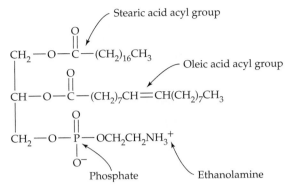

Stearic acid acyl group

$$CH_2-O-\overset{\displaystyle O}{\overset{\displaystyle \|}{C}}-(CH_2)_{16}CH_3$$

Oleic acid acyl group

$$CH-O-\overset{\displaystyle O}{\overset{\displaystyle \|}{C}}-(CH_2)_7CH=CH(CH_2)_7CH_3$$

$$CH_2-O-\overset{\displaystyle O}{\overset{\displaystyle \|}{\underset{\displaystyle \underset{O^-}{|}}{P}}}-OCH_2CH_2NH_3^+$$

Phosphate   Ethanolamine

24.14 (a), (c), (e), (f) **24.15** They must be hydrophobic, contain many amino acids with nonpolar side chains, and must be folded so that the hydrophobic regions face outward. **24.16** yes **24.17** Glucose 6-phosphate has a charged phosphate group and can't pass through the hydrophobic lipid bilayer. **24.18** The surfaces are in different environments and serve different functions. **24.19** carboxylic acid (most acidic), alcohol, C–C double bonds, ethers. The molecule has both polar and nonpolar regions. **24.20** A has the highest melting point. B and C are probably liquids at room temperature. **24.21** 12.2% palmitic acid, 87.5% stearic acid; more like C
24.22

$$CH_2O\overset{\displaystyle O}{\overset{\displaystyle \|}{C}}(CH_2)_{14}CH_3$$

$$CHO\overset{\displaystyle O}{\overset{\displaystyle \|}{C}}(CH_2)_7CH=CH(CH_2)_7CH_3$$

$$CH_2O-\overset{\displaystyle O}{\overset{\displaystyle \|}{\underset{\displaystyle \underset{O^-}{|}}{P}}}-OCH_2\underset{\displaystyle \underset{NH_3^+}{|}}{CH}COO^-$$

A glycerophospholipid

24.23 Because the membrane is fluid, it can flow together after an injury. **24.24** C_{16} saturated fatty acids. The polar head lies in lung tissue, and the hydrocarbon tails protrude into the alveoli. **24.26** Many different kinds of biomolecules dissolve in nonpolar solvents. **24.28** Saturated fatty acids have no double bonds. Monosaturated fatty acids have one double bond. Polyunsaturated fatty acids have 2 or more double bonds. **24.30** linoleic acid, linolenic acid; vegetable oils and nuts **24.32** Kinks from double bonds make crystal formation more difficult. Cis acids melt lower because they are kinkier. **24.34** Fats: saturated and unsaturated fatty acids, solids, from animal sources. Oils: mostly unsaturated, liquids, from plant sources.
24.36

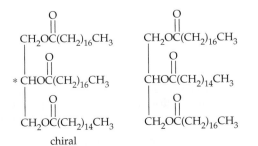

$$CH_2O\overset{\displaystyle O}{\overset{\displaystyle \|}{C}}(CH_2)_{16}CH_3$$

$$^*CHO\overset{\displaystyle O}{\overset{\displaystyle \|}{C}}(CH_2)_{16}CH_3$$

$$CH_2O\overset{\displaystyle O}{\overset{\displaystyle \|}{C}}(CH_2)_{14}CH_3$$

chiral

$$CH_2O\overset{\displaystyle O}{\overset{\displaystyle \|}{C}}(CH_2)_{16}CH_3$$

$$CHO\overset{\displaystyle O}{\overset{\displaystyle \|}{C}}(CH_2)_{14}CH_3$$

$$CH_2O\overset{\displaystyle O}{\overset{\displaystyle \|}{C}}(CH_2)_{16}CH_3$$

24.38 a protective coating **24.40** a wax **24.42** hydrogenate some of the double bonds **24.44** glyceryl tristearate; higher-melting

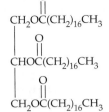

$$CH_2O\overset{\displaystyle O}{\overset{\displaystyle \|}{C}}(CH_2)_{16}CH_3$$

$$CHO\overset{\displaystyle O}{\overset{\displaystyle \|}{C}}(CH_2)_{16}CH_3$$

$$CH_2O\overset{\displaystyle O}{\overset{\displaystyle \|}{C}}(CH_2)_{16}CH_3$$

24.46 Triacylglycerols don't have an ionic head. **24.48** sphingomyelins, glycolipids **24.50** In a soap micelle, the polar hydrophilic heads are on the exterior, and the hydrophobic tails cluster in the center. In a membrane bilayer, hydrophilic heads are on both the exterior and interior surfaces of the membrane, and the region between the two surfaces is occupied by hydrophobic tails. **24.52** Concentrations of all substances would be the same on both sides of the membrane.
24.54

$$CH_2-O-\overset{\displaystyle O}{\overset{\displaystyle \|}{\underset{\displaystyle \underset{O^-}{|}}{P}}}-OCH_2CH_2\overset{+}{N}(CH_3)_3$$

$$CH-NH-\overset{\displaystyle O}{\overset{\displaystyle \|}{C}}-(CH_2)_{16}CH_3$$

$$CHOH$$

$$CH=CH(CH_2)_{12}CH_3$$

24.56 3 glycerols, $RCOO^-\,Na^+$, $R'\,COO^-\,Na^+$, $R'''COO^-\,Na^+$, 2 phosphates **24.58** Facilitated diffusion requires carrier proteins. **24.60** a prostaglandin that stimulates uterine contractions **24.62** an eicosanoid; arachidonic acid **24.64** It transfers its acetyl group to an enzyme that converts arachidonic acid to prostaglandins. **24.66** 30% **24.68** fabric softeners, disinfecting soaps **24.70** fog, homogenized milk, smoke **24.72** cholesterol **24.74** The liposome might carry a group that recognizes glycolipids or proteins on the exterior surface of the tumor cell. **24.76 (b), (d), (e), (f)**
24.78

$$CH_2-O-\overset{\displaystyle O}{\overset{\displaystyle \|}{C}}-(CH_2)_7CH=CHCH_2CH=CHCH_2CH=CHCH_2CH_3$$

$$CH-O-\overset{\displaystyle O}{\overset{\displaystyle \|}{C}}-(CH_2)_7CH=CHCH_2CH=CHCH_2CH=CHCH_2CH_3$$

$$CH_2-O-\overset{\displaystyle O}{\overset{\displaystyle \|}{C}}-(CH_2)_{12}CH_3$$

or the isomer

24.80 It has no ester linkages. **24.82** sphingomyelins, cerebrosides, gangliosides **24.84** lower blood pressure, assist in blood clotting, stimulate uterine contractions, lower gastric secretions, cause swelling **24.86** 0.4g NaOH

Chapter 25
25.1 Cholate has 4 polar groups on its hydrophilic side that allow it to interact with an aqueous environment; its hydrophobic side interacts with TAGs. Cholate and cholesterol can't change roles. **25.2** Dihydroxyacetone phosphate is isomerized to glyceraldehyde 3-phosphate, which enters glycolysis. **25.3 (a), (b)** *Step 1*; a C=C double bond is introduced; FAD is the oxidizing agent. *Step 3*; an alcohol is oxidized to a ketone; NAD^+ is the oxidizing agent. **(c)** *Step 2*; water is added to a carbon-carbon double bond. **(d)** *Step 4*; HSCoA displaces acetyl-SCoA, producing a chain-shortened acyl-SCoA fatty acid. **25.4 (a)** 8 acetyl-SCoA, 7 β oxidations **(b)** 10 acetyl-SCoA, 9 β oxidations **25.5** step 6, step 7, step 8 **25.6** (iii) **25.7 (a)** Acetyl-SCoA provides the acetyl groups used in synthesis of ketone bodies. **(b)** 3 **(c)** The body uses ketone bodies as an energy source during starvation.

25.8 Oxygen is needed to reoxidize reduced coenzymes, formed in β oxidation, that enter the electron transport chain. **25.9 (a)** HDL **(b)** chylomicrons; because they have the greatest ratio of lipid to protein **(c)** VLDL; used for storage or energy production **(d)** chylomicrons **(e)** LDL **(f)** HDL **(g)** LDL **25.10** high blood glucose → high insulin/low glucagon → fatty acid and TAG synthesis: low blood glucose → low insulin/high glucagon → TAG hydrolysis; fatty acid oxidation **25.11** yes. Formation of a fatty acyl-SCoA is coupled with conversion of ATP to AMP and pyrophosphate. **25.12** Less acetyl-SCoA can be catabolized in the citric acid cycle, and acetyl-SCoA is diverted to ketogenesis. **25.13** Catabolism of fat provides more calories/gram than does catabolism of glycogen, and, thus fats are a more efficient way to store calories. **25.14** They slow the rate of movement of food through the stomach. **25.16** Bile emulsifies lipid droplets. **25.18** Chylomicrons are lipoproteins that transport dietary lipids. **25.20** complexed with albumins **25.22** 5.5 molecules of ATP **25.24** a fat cell, where TAGs are stored and mobilized. **25.26** heart, liver, and muscle cells **25.28** conversion to a fatty-acyl-SCoA **25.30** The carbon β to the -SCoA group is oxidized. **25.32** 14 mol ATP
25.34

(a)

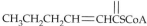

(b)

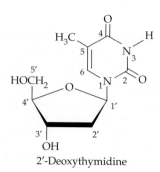

(c)

(d)

25.36 (a) 4 molecules of acetyl-SCoA **(b)** 7 molecules of acetyl-SCoA **25.38** lipogenesis **25.40** acetyl-SCoA **25.42** 7 rounds **25.44** Total cholesterol: 200 mg/dL or lower; LDL: 160 mg/dL or lower; HDL: 60 mg/dL or higher **25.46** LDL carries cholesterol from the liver to tissues; HDL carries cholesterol from tissues to the liver, where it is excreted. **25.48** an inability to absorb lipids in the bloodstream; fat in the feces **25.50** The liver synthesizes many important biomolecules, it catabolizes glucose, fatty acids and amino acids, it stores many substances, and it inactivates toxic substances. **25.52** The excess acetyl-SCoA from catabolism of carbohydrates is stored as fat. The body can't resynthesize carbohydrate from acetyl-SCoA. **25.54** acetoacetate, acetone and 3-hydroxybutyrate. They are ketones or derived from ketones; ketogenesis. They are formed during lipid catabolism when more acetyl-SCoA is formed than the citric acid cycle can handle. **25.56** 22 molecules of acetyl-SCoA **25.58** HDL is endogenous. Chylomicrons are exogenous. **25.60** Ketones have little effect on pH, but the two other ketone bodies are acidic, and they lower the pH of urine.

Chapter 26
26.1

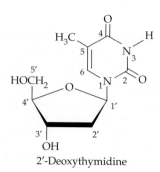

2′-Deoxythymidine

26.2 D-Ribose ($C_5H_{10}O_5$) has one more oxygen atom than 2-deoxy-D-ribose ($C_5H_{10}O_4$).

26.3

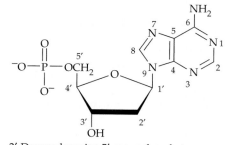

2′-Deoxyadenosine 5′-monophosphate

26.4

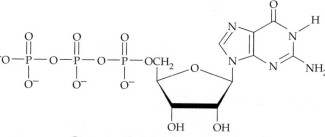

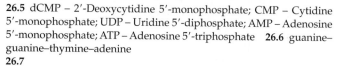

Guanosine 5′-triphosphate (GTP)

26.5 dCMP – 2′-Deoxycytidine 5′-monophosphate; CMP – Cytidine 5′-monophosphate; UDP – Uridine 5′-diphosphate; AMP – Adenosine 5′-monophosphate; ATP – Adenosine 5′-triphosphate **26.6** guanine–guanine–thymine–adenine
26.7

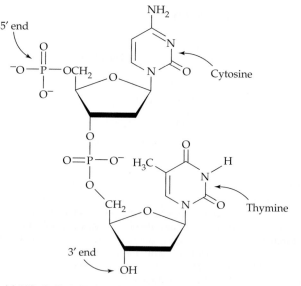

26.8 (a) 3′C–G–G–A–T–C–A 5′ **(b)** 3′ T–T–A–C–C–G–A–G–T 5′
26.9

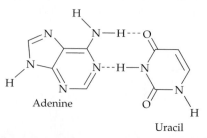

26.10 negatively charged **26.11 (a)** A longer strand has more hydrogen bonds. **(b)** A chain with a higher percent of G/C pairs has a higher melting point, because it has more hydrogen bonds. **26.12 (a)** 3′ C–U–A–A–U–G–G–C–A 5′ **(b)** 3′ A–U–A–C–C–G–A–U–C–C–G–U 5′ **26.13 (a)** G–C–U, G–C–C, G–C–A, G–C–G **(b)** U–U–U, U–U–C **(c)**

U–U–A, U–U–G, C–U–U, C–U–C, C–U–A, C–U–G **(d)** G–U–U, G–U–C, G–U–A, G–U–G **(e)** U–A–U, U–A–C **26.14** The sequence guanine, uracil, cytosine codes for valine. **26.15 (a)** Ile **(b)** Ala **(c)** Arg **(d)** Asn **26.16** Six DNA informational strands can code for Ile:

5′ ATT ATC ATA 3′ 5′ ATC ATT ATA 3′ 5′ ATA ATC ATT 3′
5′ ATT ATA ATC 3′ 5′ ATC ATA ATT 3′ 5′ ATA ATT ATC 3′

26.17–26.18

mRNA sequence:	5′ CUU—AUG—GCU—UGG—CCC—UAA 3′
Amino-acid sequence:	Leu—Met—Ala—Trp—Pro—Stop
tRNA anticodons:	3′ GAA UAC CGA ACC GGG 5′

26.19

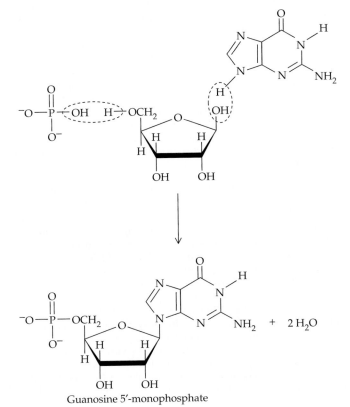

Guanosine 5′-monophosphate

26.20

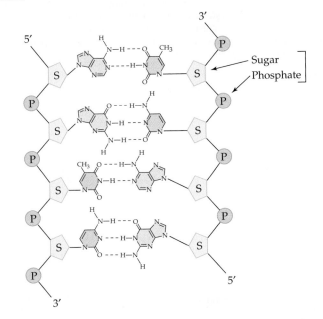

Sequence of the left chain: 5′ A–G–T–C 3′
Sequence of the right chain: 5′ G–A–C–T 3′

26.21

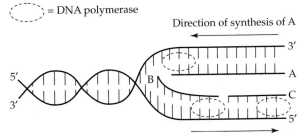

= DNA polymerase

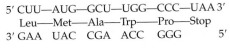

Segments B and C are joined by the action of a DNA ligase. **26.22** The sugar-phosphate backbone is found on the outside of the DNA double helix. Histones are positively charged; they contain groups such as Lys, Arg and His.

26.23

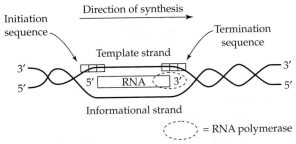

= RNA polymerase

26.24 More than one codon can code for each amino acid. Only one possibility is shown.

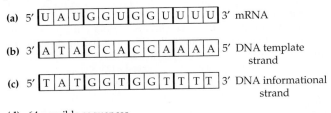

(a) 5′ | U | A | U | G | G | U | G | G | U | U | U | 3′ mRNA

(b) 3′ | A | T | A | C | C | A | C | C | A | A | A | 5′ DNA template strand

(c) 5′ | T | A | T | G | G | T | G | G | T | T | T | 3′ DNA informational strand

(d) 64 possible sequences

26.26 2-deoxyribose (DNA); ribose (RNA). 2-Deoxyribose is missing an —OH group at C2. **26.28** The purine bases (two fused heterocyclic rings) are adenine and guanine. The pyrimidine bases (one heterocyclic ring) are cytosine, thymine (in DNA) and uracil (in RNA). **26.30** DNA is largest; tRNA is smallest. **26.32** *Similarities*: All are polymerizations; all use a nucleic acid as a template; all use hydrogen bonding to bring the subunits into position. *Differences*: In replication, DNA makes a copy of itself. In transcription, DNA is used as a template for the synthesis of mRNA. In translation, mRNA is used as a template for the synthesis of proteins. Replication and transcription take place in the nucleus of cells, and translation takes place in ribosomes. **26.34** DNA, protein **26.36** 46 chromosomes (23 pairs) **26.38** They always occur in pairs: they always H-bond with each other. **26.40** 28% T, 28% A, 22% C, 22% G. (%T = %A; %C = %G: %T + %A + %C + %G = 100%) **26.42** 5′ to 3′

26.44

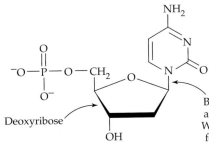

Deoxyribose

Bond between cytosine and C1 of deoxyribose. Water is removed in the formation of the bond.

26.46

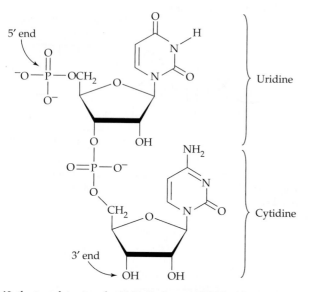

5′ end

Uridine

Cytidine

3′ end

26.48 the template strand   **26.50** Each new DNA has one newly synthesized strand and one old strand.   **26.52** An anticodon is a 3 nucleotide tRNA sequence that is complementary to an mRNA codon for a specific amino acid.   **26.54** tRNAs for each amino acid differ in their anticodon sequences.   **26.56 (a)** C–C–U, C–C–C, C–C–A, C–C–G **(b)** A–A–A, A–A–G **(c)** A–U–G   **26.58** (3′ → 5′) **(a)** U–G–A **(b)** C–C–U **(c)** G–A–A   **26.60** (3′ → 5′) A–T–G–G–C–T   **26.62** Tyr–Arg   **26.64** (5′ → 3′) UAU–GGU–GGU–UUU–AUG–UAA Other sequences are possible.   **26.66** Viruses consist of a strand of nucleic acid wrapped in a protein coat; viruses can't replicate or manufacture protein independent of a host cell.   **26.68** To be effective, a drug must be powerful enough to act on viruses within cells without damaging the cells and their genetic material.   **26.70** A polymerase chain reaction is used to produce a large number of copies of a specific DNA chain.   **26.72** 249 bases   **26.74** Met is removed after synthesis is complete.

Chapter 27

27.1 "once upon a time"   **27.2** As a result of the SNP, the base sequence codes for Trp, instead of Cys. This change would probably affect the functioning of the protein.   **27.3** 3′ –T–C–T–A–G– // –A– 5′   **27.4 (a)** sticky **(b) (c)** not sticky   **27.5 (a)** comparative genomics **(b)** genetic engineering **(c)** pharmacogenetics **(d)** bioinformatics   **27.6** The working draft is the sequence of the nucleotides in about 90% of genes. Further work: increasing accuracy to 99.9%, sequencing the repetitive noncoding DNA, and identifying the proteins coded for by the genes.   **27.7** The variations are only a small part of the genome; the rest is identical among humans.   **27.8** *telomeres* (protect the chromosome from damage, involved with aging), *centromeres* (involved with cell division), *promoter sequences* (determine which genes will be replicated), *introns* (function unknown)   **27.9** Similarities: both are variations in base sequences. Differences: A mutation is an error that is transferred during replication and affects only a few people; a polymorphism is a variation in sequence that is common within a population.   **27.10** Recombinant DNA contains two or more DNA segments that do not occur together in nature. The DNA that codes for a specific human protein can be incorporated into a bacterial plasmid using recombinant DNA technology. The plasmid is then reinserted into a bacterial cell, where its protein-synthesizing machinery makes the desired protein.   **27.11** Major benefits of genomics: creation of disease-resistant and nutrient-rich crops, gene therapy, genetic screening. Major negative outcomes: misuse of an individual's genetic information, prediction of a genetic disease for which there is no cure.   **27.12** Non-profit group: Human Genome Project. Private corporation: Celera Genomics.   **27.14** 50%   **27.16 (a)** Approx. 200 genes are shared between bacteria and humans. **(b)** A single gene may produce several proteins.   **27.18** The clones used in DNA mapping are identical copies of DNA segments from a single individual. In mapping, it is essential to have a sample large enough for experimental manipulation.   **27.20** The youngest cells have long telomeres, and the oldest cells have short telomeres.   **27.22** It is the constriction that determines the shape of a chromosome during cell division.   **27.24** ionizing radiation, mutagens, certain viruses   **27.26** random and spontaneous events, exposure to a mutagen   **27.28** A SNP can result in the change in identity of an amino acid inserted into a protein a particular location in a polypeptide chain. The effect of a SNP depends on the function of the protein and the nature of the SNP.   **27.30** A physician could predict the age at which inherited diseases might become active, their severity, and the response to various types of treatment.   **27.32** a change in the type of side chain   **27.34 (a)** Substitution of Ala for Val may have minor effects **(b)** Substitution of His for Pro is more serious because the amino acids have very different side chains.   **27.36 (a)** Proteins can be produced in large quantities. **(b)** Proteins are disease-free.   **27.38** electrophoresis, polymerase chain reaction   **27.40 (a) (b)** not sticky   **27.42** Proteomics, the study of the complete set of proteins coded for by a genome or synthesized by a given type of cell, might provide information about the role of a protein in both healthy and diseased cells.   **27.44** corn, soybeans   **27.46** a DNA chip   **27.48** a group of anonymous individuals   **27.50** (1) digestion with a restriction endonuclease (2) separation of fragments by electrophoresis (3) fragments transferred to a nylon membrane (4) treatment of the blot with a radioactive DNA probe (5) identification of fragments by exposure to X-ray film. All samples must be analyzed under the same conditions in order to be compared.   **27.52** insertion of the gene into other plants; development of plants that could be irrigated with sea water   **27.54** A vector is the agent used to carry therapeutic quantities of DNA directly into cell nuclei.   **27.56** ATACTGA   **27.58** a germ cell (sperm or egg)

Chapter 28

28.1 (a) false **(b)** true **(c)** true **(d)** false **(e)** false   **28.2** oxidoreductase; lyase

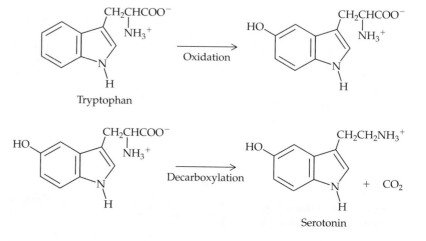

Tryptophan   Oxidation

Decarboxylation   Serotonin

28.3

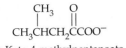

$$CH_3CHCH_2CCOO^-$$

α-Keto-4-methylpentanoate

28.4

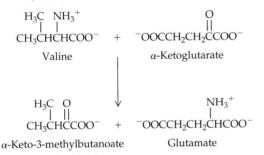

$$\text{—}CH_2C\text{—}COO^-$$

28.5 by the loss of two hydrogens to either NAD^+ or $NADP^+$ **28.6** valine, leucine, isoleucine

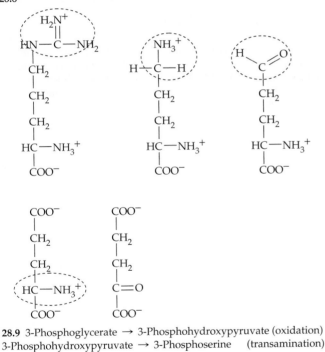

$$CH_3CHCHCOO^- \ + \ {}^-OOCCH_2CH_2CCOO^-$$

Valine   α-Ketoglutarate

$$CH_3CHCCOO^- \ + \ {}^-OOCCH_2CH_2CHCOO^-$$

α-Keto-3-methylbutanoate   Glutamate

28.7 (a) v **(b)** i **(c)** iii
28.8

28.9 3-Phosphoglycerate → 3-Phosphohydroxypyruvate (oxidation) 3-Phosphohydroxypyruvate → 3-Phosphoserine (transamination) 3-Phosphoserine → Serine (hydrolysis)
28.10

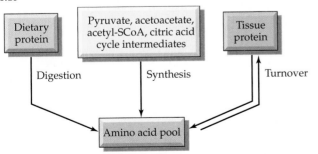

28.11 (1) Catabolism of an amino acid begins with a transamination reaction that removes the amino nitrogen (2) The resulting α-keto acid, which contains the carbon atoms, is converted to a common metabolic intermediate. (3) The amino group of glutamate (from the amino acid) is removed by oxidative deamination. (4) The amino nitrogen is transformed to urea in the urea cycle and is excreted.   **28.12** glutamate dehydrogenase; alanine aminotransferase. Alanine is the product.   **28.13** to quickly remove ammonia from the body; buildup of urea and shortage of ornithine   **28.14** All amino acids are necessary for protein synthesis. The body can synthesize only some of them; the others must be provided by food and are thus essential in the diet.   **28.15** The carbon atoms from ketogenic amino acids can be converted to ketone bodies or to acetyl-SCoA. The carbon atoms from glucogenic amino acids can be converted to compounds that can enter gluconeogenesis and can form glucose, which can enter glycolysis and also yield acetyl-SCoA.   **28.16** stomach   **28.18** pyruvate, 3-phosphoglycerate   **28.20** A keto group of an α-keto acid and an amino group of an amino acid change places.
28.22

(a)

$$CH_3CH_2CH\text{—}C\text{—}COO^-$$

(b)

$$HSCH_2\text{—}C\text{—}COO^-$$

28.24 An $\text{—}NH_3^+$ group of an amino acid is replaced by a carbonyl group, and ammonium ion is elimiated.
28.26

(a)

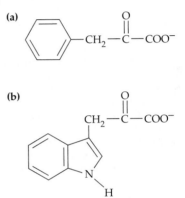

(b)

28.28 A glucogenic amino acid is catabolized to pyruvate or citric acid cycle intermediates.   **28.30** Ammonia is toxic.   **28.32** CO_2 from the citric acid cycle   **28.34** Nonessential amino acids are synthesized in humans in 1–3 steps. Essential amino acids are synthesized in microorganisms in 7–10 steps.   **28.36** reductive amination; the reverse of oxidative deamination   **28.38** phenylketonuria; mental retardation; restriction of phenylalanine in the diet   **28.40 (b) (c) (d)**   **28.42** Oxidized allopurinol inhibits the enzyme that converts xanthine to uric acid. The more soluble intermediates are excreted. The nitrogen at position 7 of hypoxanthine is at position 8 in allopurinol, where it blocks oxidation of xanthine.   **28.44** (1) Hydroxylation makes xenobiotic compounds more water-soluble. (2) Bonding to polar molecules makes xenobiotics easier to excrete.   **28.46** Three ATPs are used.   **28.48** α-ketoglutarate (step 4), succinyl-SCoA (step 5), fumarate (step 7), oxaloacetate (step 1)   **28.50** Amino acid carbons can be catabolized to acetyl-SCoA, which can be used in lipogenesis. The fatty acids can form TAGs, be transported to adipose tissue, where they can be hydrolyzed to fatty acids.   **28.52** *Storage*: Unlike fats and carbohydrates, amino acids aren't stored in the body. Instead, amino acid nitrogen is excreted and the resulting compounds are converted to either fats or carbohydrates. *Energy*: Fats and carbohydrates that are not stored are catabolized. Surplus amino acids must be converted to either fats or carbohydrates in order to be an energy source.   **28.54** The active forms of the proteases would hydrolyze the proteins in the lining of the pancreas.   **28.56** One of the ATPs in the urea cycle is hydrolyzed to AMP, which is the equivalent of spending two ATPs.

Chapter 29
29.1 In the cell: the charged form. Outside the cell: the uncharged form. The uncharged form enters the cell more readily.   **29.2 (a)** iii **(b)** ii **(c)** iv **(d)** v **(e)** i   **29.3 (a)** pH goes down **(b)** $[O_2]$, $[CO_2]$, pH   **29.4 (a)** respiratory acidosis **(b)** metabolic acidosis **(c)** metabolic alkalosis **(d)** respiratory alkalosis **(e)** respiratory alkalosis   **29.5 (a)** intracellular

fluid **(b)** extracellular fluid **(c)** blood plasma, interstitial fluid **(d)** K^+, Mg^{2+}, HPO_4^{2-} **(e)** Na^+, Cl^-
29.6

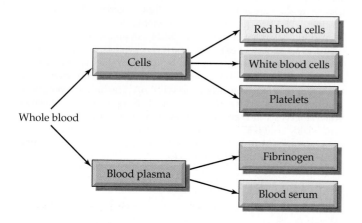

29.7 (a) CO_2 **(b)** O_2 **(c)** nutrients **(d)** waste products **(e)** hormones **(f)** white blood cells, platelets   **29.8** swelling, redness, warmth, pain; enzymatic decarboxylation of histidine; Histamine dilates capillaries, increasing blood flow that reddens and warms the skin. Blood-clotting factors and defensive proteins cause pain and swelling.   **29.9** *Cell-mediated immune response*: under control of T cells; arises when abnormal cells, bacteria or viruses enter cells; invaders killed by T cells. *Antibody-mediated immune response*: under control of B cells, assisted by T cells; occurs when antigens enter cells; B cells divide to produce plasma cells, which form antibodies; an antibody-antigen complex inactivates the antigen.   **29.10** intracellular fluid, extracellular fluid, interstitial fluid   **29.12** It must be an ion, a gas, a small molecule, or a molecule with many polar groups.   **29.14** Blood pressure in arterial capillaries is higher than interstitial fluid pressure, and blood pressure in venous capillaries is lower than interstitial fluid pressure.   **29.16** It collects interstitial fluid, cellular debris, proteins, and lipid droplets and returns them to the bloodstream.   **29.18** vasopressin   **29.20** the fluid part of blood containing water-soluble solutes   **29.22** red blood cells, white blood cells and platelets   **29.24** a substance that produces ions   **29.26** a foreign substance; inflammation, cell-mediated immune response, antibody-mediated immune response   **29.28** They are both noncovalent and specific interactions.   **29.30** B lymphocytes identify antigens and divide into plasma cells, which produce antibodies to the antigen.   **29.32** Killer T cells destroy the invader; helper T cells enhance defenses; memory T cells can produce new killer T cells if needed.   **29.34** a mass of blood cells trapped in a fibrin mesh   **29.36** intrinsic pathway; extrinsic pathway   **29.38** one O_2 molecule   **29.40** Oxyhemoglobin is bright red; deoxyhemoglobin is dark red-purple.   **29.42** Uptake of the first oxygen facilitates uptake of the remaining 3 oxygens. Release of the first oxygen facilitates release of the remaining 3 oxygens.   **29.44** a dissolved gas, bound to hemoglobin, bicarbonate ion   **29.46** All cause Hb to release more O_2.   **29.48** Metabolic acidosis (result of a metabolic process).   **29.50** reabsorption, secretion   **29.52** fats such as lanolin, vegetable oils or petroleum jelly   **29.54** They form tight junctions, so that no substances can pass between them.   **29.56** substances that are soluble in membrane lipids (such as ethanol)   **29.58** *Modified hemoglobin*—Advantages: long shelf life, blood typing unnecessary; Disadvantages: cost, short lifetime in the body, possible biochemical modification in the body. *Perfluorocarbons*—Advantages: cost, unreactivity; Disadvantages: possibly toxic, hard to regulate O_2 delivery.   **29.60** Automated analysis can reproducibly detect changes in enzyme levels that might indicate organ damage. **29.62** It is a small, polar molecule. **29.64** When $[Na^+]$ is high, secretion of ADH increases and causes the amount of water retained by the body to increase, causing swelling. **29.66** Active transport is necessary when a cell needs a substance that has a higher concentration inside the cell than outside, or when a cell needs to secrete a substance that has a higher concentration outside the cell than inside. **29.68** Hemostasis is the body's mechanism for preventing blood loss, and might be considered to be a part of homeostasis.

Photo Credits

Frontmatter: iv, AP/Wide World Photos; **v,** Richard Megna/Fundamental Photographs; **vi,** John McMurry; **vii,** Nubar Alexanian/Stock Boston; **viii,** SuperStock, Inc.; **ix,** Robert Mathena/Fundamental Photographs; **x,** John W. Finley; **xi,** Huiying Li and Thomas L. Poulos, University of California, Irvine.

Chapter 1: Opener, Kent Knudson/ Stock Boston; **Fig. 1.2,** Richard Megna/Fundamental Photographs; **4,** E.R. Degginger/Color-Pic, Inc.; **5,** Richard Megna/Fundamental Photographs; **9 (top left),** DPA/HMA/The Image Works; **9 (top middle),** L.S. Stepanowicz/Visuals Unlimited; **9 (top right),** Erich Schrempp/Photo Researchers, Inc.; **9 (bottom left),** Richard Megna/Fundamental Photographs; **9 (bottom middle),** Bill Pierce/Rainbow; **9 (bottom right),** Lester V. Bergman/CORBIS; **10 (left),** Lester V. Bergman/CORBIS; **10 (right),** Photo courtesy of Texas Instruments Incorporated; **11,** Stock Montage/Hulton | Archive/Getty Images Inc.

Chapter 2: Opener, Alan Kearney/The Viesti Collection, Inc.; **18,** National Institute for Biological Standards and Control (U.K.)/Science Photo Library/Photo Researchers, Inc.; **Fig. 2.1 (left),** Ohaus Corporation; **Fig. 2.1 (right),** Science VU/Visuals Unlimited; **20,** McCracken Photographers/Pearson Education/PH College; **21,** McCracken Photographers/Pearson Education/PH College; **22,** The Image Works; **23,** Doug Martin/Photo Researchers, Inc.; **24,** Phil Degginger/Color-Pic, Inc.; **26 (top left and top middle),** Dr. Tony Brain and David Parker/Science Photo Library/Photo Researchers, Inc.; **27 (top right),** CNRI/Science Photo Library/Photo Researchers, Inc.; **26 (bottom),** Frank LaBua/Pearson Education/PH College; **28,** Duomo Photography Incorporated; **30,** Dennis Kunkel/Phototake NYC; **31,** Michal Heron/Pearson Education/PH College; **32,** Tony Di Zinno/Tony Di Zinno Photographs; **Fig. 2.5,** Carey B. Van Loon; **36,** Richard Megna/Fundamental Photographs; **37,** William W. Bacon III/Photo Researchers, Inc.; **38,** Richard T. Nowitz/Science Source/Photo Researchers, Inc.; **39,** AP/Wide World Photos.

Chapter 3: Opener, Art Wolfe/ Art Wolfe, Inc.; **45 (top),** Paul Silverman/Fundamental Photographs; **45 (bottom),** Ken Karp/Omni-Photo Communications, Inc.; **46,** AP/Wide World Photos; **50,** Richard Megna/Fundamental Photographs; **51,** Colin Cuthbert/Science Photo Library/Photo Researchers, Inc.; **53 (top and bottom),** Richard Megna/Fundamental Photographs; **54 (top),** Stephen Frisch/Stock Boston; **54 (bottom),** Richard Cummins/The Viesti Collection, Inc.; **55,** Allan Morton & Dennis Milon/Science Photo Library/Photo Researchers, Inc.; **56,** L. Steinmark/Custom Medical Stock Photo, Inc.; **64,** SuperStock, Inc.; **65,** Randall Wedin/Wedin Communications.

Chapter 4: Opener, SuperStock, Inc.; **71,** Richard Megna/Fundamental Photographs; **Fig. 4.2 (left),** Richard Megna/Fundamental Photographs; **Fig. 4.2 (right),** Carey Van Loon; **Fig. 4.3,** Ed Degginger/Color-Pic, Inc.; **76,** Richard Megna/Fundamental Photographs; **77 (left),** Chip Clark; **76 (middle),** Jeffrey A. Scovil; **76 (right),** Jeffrey A. Scovil; **82,** Photo courtesy of James S. Aber, Emporia State University; **84,** Karl Hartmann/Traudel Sachs/Phototake NYC; **88,** CNRI/Science Photo Library/Photo Researchers, Inc.; **91,** P. Motta/Dept. Anatomy, U. La Sapienza/Science Photo Library/Photo Researchers, Inc.; **93,** David Kaplan/Pearson Education/PH College.

Chapter 5: Opener, Piotr Malecki/Newsmakers/ Getty Images, Inc.; **100,** UPI/Corbis-Bettmann; **105,** Liane Enkelis/Stock Boston; **107 (top),** Mark Richards/PhotoEdit; **107 (bottom),** Farmland Industries, Inc.; **117,** Mark S. Skalny/Visulas Unlimited; **123,** Alan & Linda Detrick/Photo Researchers, Inc.

Chapter 6: Opener, SuperStock, Inc.; **134,** Jeff Greenberg/Stock Boston; **135,** Jean Marc Barey/Agence Vandystadt/Photo Researchers, Inc.; **Fig. 6.1(a),** Phil Degginger/Color-Pic, Inc.; **137,** Tom Bochsler/Pearson Education/PH College; **139 (left),** Library of Congress; **139**

(right), Science Photo Library/Photo Researchers, Inc.; **144,** Richard Megna/Fundamental Photographs; **146,** Carey Van Loon; **147,** McCracken Photographers/Pearson Education/PH College; **149,** Etching by James Gillray, 1799/The Granger Collection; **152,** McCracken Photographers/Pearson Education/PH College; **153,** McCracken Photographers/Pearson Education/PH College; **154,** Richard Megna/Fundamental Photographs; **155 (top),** Willie Hill, Jr./The Image Works; **155 (bottom),** Tony Freeman/ PhotoEdit; **158,** Richard Megna/Fundamental Photographs.

Chapter 7: Opener, Alan Kearney/The Viesti Collection, Inc.; **166,** E.R. Degginger/Color-Pic, Inc.; **167,** Ken Graham/Getty Images Inc.; **169 (top),** SuperStock, Inc.; **169 (bottom),** Getty Images, Inc./PhotoDisc, Inc.; **Fig. 7.1,** Tom Bochsler/Pearson Education/PH College; **174,** Adam Hart-Davis/Science Photo Library/Photo Researchers, Inc.; **177,** AC/General Motors/Peter Arnold, Inc.; **178,** Novosti/Science Photo Library/Photo Researchers, Inc.; **179,** Martin Rogers/Stock Boston; **175,** Tom Pantages.

Chapter 8: Opener, Jeff & Alexa Henry/Peter Arnold, Inc.; **Fig. 8.3 (bkgd),** NASA Headquarters; **200,** Richard Megna/Fundmental Photographs; **201,** Jim Corwin/Photo Researchers, Inc.; **208,** Galen Rowell/CORBIS; **215,** Richard Megna/Fundamental Photographs; **216 (top),** Ulrich Sapountsis/Okapia/Photo Researchers, Inc.; **216 (bottom),** Herman Eisenbeiss/Photo Researchers, Inc.; **Fig. 8.21,** Tom Pantages; **217 (bottom),** John McMurry; **218 (left),** Jeffrey A. Scovil; **218 (right),** Herve Berthoule/Jacana Scientific Control/Photo Researchers, Inc.; **219,** Y. Lefevre/Bios/Peter Arnold, Inc.; **221,** Bob Kramer/Stock Boston; **223,** Wesley Bocxe/Photo Researchers, Inc.; **224,** David Kaplan/Pearson Education/PH College.

Chapter 9: Opener, Science VU/NOAA/Office of Oceanic and Atmospheric Research/National Undersea Research Program/Visuals Unlimited; **232 (left),** Japack/CORBIS; **232 (middle),** Japack/CORBIS; **232 (right),** Yoav Levy/Phototake NYC; **233,** Michael Baytoff/Black Star; **235,** Jonathan Blair/CORBIS; **Fig. 9.3,** Richard Megna/Fundamental Photographs; **238,** SuperStock; **240,** AP/Wide World Photos; **Fig. 9.5,** Richard Megna/Fundamental Photographs; **246,** Phil Degginger/Color-Pic, Inc.; **Fig. 9.7,** Richard Megna/Fundamental Photographs; **251,** Rod Kaye Photography/Aristock, Inc.; **253,** Craig Newbauer/Peter Arnold, Inc.; **Fig. 9.11,** Dennis Kunkel/Dennis Kunkel Microscopy, Inc.; **257,** Martin Dohrn/Science Photo Library/Photo Researchers, Inc.

Chapter 10: Opener, Horticultural Photography; **264,** Tony Freeman/PhotoEdit; **265,** Gail Mooney/CORBIS; **Fig. 10.1,** Carey B. Van Loon; **270 (left),** Dr. E. Walker/Science Photo Library/Photo Researchers, Inc.; **258 (right),** A.B. Dowsett/Science Photo Library/Photo Researchers, Inc.; **277,** Tom McCarthy/PhotoEdit; **Fig. 10.4 (left),** Richard Megna/Fundamental Photographs; **Fig. 10.4 (right),** Tom Bochsler/Pearson Education/PH College; **Fig. 10.5,** Tom Bochsler/Pearson Education/PH College; **Fig. 10.8,** Ed Degginger/Color-Pic, Inc.; **288,** William E. Ferguson.

Chapter 11: Opener, Paul Almasy/CORBIS; **305 (right),** Martin Dohrn/Science Photo Library/Photo Researchers, Inc.; **308,** Will & Deni McIntyre/Photo Researchers, Inc.; **310 (top),** Yoav Levy/Phototake NYC; **310 (bottom),** Kevin Schafer/Peter Arnold, Inc.; **311,** Lowell Georgia/Photo Researchers, Inc.; **Fig. 11.6,** Rennie Van Munchow/Phototake; **314,** Roger Tully/Getty Images Inc.; **315,** Stephen Frisch/Stock Boston; **316,** Todd Gipstein/CORBIS; **317,** Colorfoto Hans Hinz, Basel, Switzerland; **318,** David Kaplan/Pearson Education/PH College.

Chapter 12: Opener, Vince Streano/CORBIS; **327,** Andy Levin/Photo Researchers, Inc.; **333,** Paul Silverman/Fundamental Photographs; **337,** John McMurry; **343,** Tony Freeman/PhotoEdit; **345,** David Halpern/Photo Researchers, Inc.

Chapter 13: Opener, Joe Viesti/The Viesti Collection, Inc.; **356,** Michael Newman/PhotoEdit; **363,** Omikron/Photo Researchers, Inc.; **Fig. 13.1,** Joel Gordon Photography; **373,** Richard Megna/Fundamental Photographs; **374 (top),** Michal Heron/Pearson Education/PH College; **374 (bottom),** Fotopic/Omni-Photo Communications, Inc.; **376,** Robert Brenner/PhotoEdit; **379,** Fujifotos/The Image Works.

Chapter 14: Opener, Jaime Abecasis/SuperStock, Inc.; **391,** Fred Lyon; **392,** David Young-Wolff/PhotoEdit; **401,** Simon Ridgway/Newsmakers/Getty Images Inc.; **402,** John Shaw/Tom Stack & Associates; **404,** SuperStock, Inc.; **406,** Rod Planck/Photo Researchers, Inc.; **407,** John Greim/Science Photo Library/Photo Researchers, Inc.; **408,** D. Robert Franz/CORBIS; **409,** Brian Parker/Tom Stack & Associates; **410,** NASA Goddard Laboratory for Atmospheres; **412,** Jay Riggs/Wedin Communications.

Chapter 15: Opener, Gerard Lacz/Peter Arnold, Inc.; **425,** Nubar Alexanian/Stock Boston; **428,** Albert Normandin/Masterfile Corporation; **431,** Donald Clegg and Roxy Wilson/Simon & Schuster/PH College; **432,** SuperStock, Inc.; **434 (top),** Runk/Schoenberger/Grant Heilman Photography, Inc.; **434 (bottom),** Allan Rosenberg/Getty Images, Inc./PhotoDisc, Inc.; **435 (left),** Jonathan Blair/CORBIS; **435 (right),** James Marshall/CORBIS.

Chapter 16: Opener, Michael Freeman/CORBIS; **445 (photo),** Thomas Eisner and Daniel Aneshansley, Cornell University; **445 (art),** Adapted from *Introduction to Ecological Biochemistry* 2/e by J.B. Harbone with permission of Academic Press, Inc., San Diego.; **447,** Royalty Free/CORBIS; **448,** Reinhard Eisele/CORBIS; **449,** AP/Wide World Photos; **450,** Paul Silverman/Fundamental Photographs; **Fig. 16.2,** Joel Gordon Photography; **454,** Thomas Kitchin/Tom Stack & Associates; **455,** Scott Camazine/Photo Researchers, Inc.; **461,** David Kaplan/Pearson Education/PH College.

Chapter 17: Opener, Randy Faris/CORBIS; **474,** G. Biss/Masterfile Corporation; **478,** Michelle Garrett/CORBIS; **481,** Michael Dalton/Fundamental Photographs; **482,** Pearson Education Corporate Digital Archive; **483,** Brian Sytnyk/Masterfile Corporation; **Fig. 17.1,** The Granger Collection; **487 (top),** Reprinted with permission from *Chemical & Engineering News*, June 29, 1998, 76(26), p. 14. Copyright 1998 American Chemical Society; **487 (bottom),** GABR/Publiphoto/Photo Researchers, Inc.; **491,** Ed Degginger/Color-Pic, Inc.; **492,** Photo courtesy of DuPont-Kevlar(R).

Chapter 18: Opener, Larry Lefever/Grant Heilman Photography, Inc.; **503,** Will & Deni McIntyre/Photo Researchers, Inc.; **513 (top),** Grant Heilman/Grant Heilman Photography, Inc.; **513 (bottom),** Runk/Schoenberger/Grant Heilman Photography, Inc.; **514,** Richard Megna/Fundamental Photographs; **517,** Stanley Flegler/Visuals Unlimited; **519,** Tony Freeman/PhotoEdit; **526 (top),** SuperStock, Inc.; **526 (bottom),** Steven Fuller/Peter Arnold, Inc.; **Fig. 18.7(b),** Kim M. Gernert/Pearson Education/PH College; **Fig. 18.7(c),** Ken Eward/Photo Researchers, Inc.; **Fig. 18.7(d),** Kim M. Gernert/Pearson Education/PH College; **Fig. 18.8(b),** Coordinates by G. Fermi and M. F. Perutz; image by Laboratory for Molecular Modeling, School of Pharmacy, University of North Carolina at Chapel Hill. Pearson Education/PH College; **Fig. 18.9(a),** R. J. Feldmann; **Fig. 18.9(b),** Coordinates by Barbara Brodsky and Cynthia G. Long; image by Molecular Graphics and Modelling, Duke University. Pearson Education/PH College; **532,** Stephen Marks/Getty Images Inc.; **534,** David Kaplan/Pearson Education/PH College.

Chapter 19: Opener, Hans Reinhard/Okapia/Photo Researchers, Inc.; **Fig. 19.1,** Richard Megna/Fundamental Photographs; **546 (top),** Manuel C. Peitsch/CORBIS; **546 (bottom),** Manuel C. Peitsch/CORBIS; **559,** Ken Eward/Science Source/Photo Researchers, Inc.; **560,** Abbott Laboratories; **561,** Kent Knudson/Stock Boston; **565,** Mark Gibson/CORBIS.

Chapter 20: Opener, Phillipe Crochet/Getty Images, Inc.; **579,** Michal Heron/Pearson Education/PH College; **582,** Jon Freeman/Online USA, Inc./Getty Images, Inc.; **583,** Grant Heilamn/Grant Heilman Photography, Inc.; **584,** Don W. Fawcett/Science Source/Photo Researchers, Inc.; **587,** Jack Fields/Photo Researchers, Inc.; **588,** Scott Camazine & Sue Trainor/Photo Researchers, Inc.; **589,** Tom McHugh/Photo Researchers, Inc.; **592,** Garden Row Foods; **594,** Geoff Tompkinson/Science Photo Library/Photo Researchers, Inc.

Chapter 21: Opener, Joseph Nettis/Photo Researchers, Inc.; **600,** Barry L. Runk/Grant Heilman Photography, Inc.; **601,** Royalty Free/CORBIS; **602,** Al Giddings/Al Giddings Images, Inc.; **Fig. 21.6,** Alan Majchrowicz/Peter Arnold, Inc.; **611,** Richard Megna/Fundamental Photographs; **618,** Donald Clegg/Pearson Education/PH College; **Fig. 21.11(b),** Manuel C. Peitsch/CORBIS; **623,** Tim Barnwell/Stock Boston.

Chapter 22: Opener, Grant Heilman/Grant Heilman Photography, Inc.; **635,** James H. Carmichael, Jr./Photo Researchers, Inc.; **641,** Robert Mathena/Fundamental Photographs; **Fig. 22.4(a),** Randy Faris/CORBIS; **Fig. 22.4(b),** Peter Marbach, Grant Heilman Photography, Inc.; **Fig. 22.4(c),** John Lei/Stock Boston; **648,** Michael Dalton/Fundamental Photographs; **650,** Doug Allan/Oxford Scientific Films/Animals Animals/Earth Scenes; **651,** Kelvin Aitken/Peter Arnold, Inc.; **652,** David Pollack/Corbis/Stock Market; **655,** Larry Mulvehill/Science Source/Photo Researchers, Inc.; **626,** GlaxoSmithKline plc.

Chapter 23: Opener, Ramon Manent/CORBIS; **665,** Photo Lennart Nilsson/Albert Bonniers Forlag; **670,** Coordinates by T. Alber, G.A. Petsko, and E. Lolis; image by Molecular Graphics and Modelling, Duke University. Pearson Education/PH College; **672 (left),** SuperStock, Inc.; **672 (middle),** Michael Neveux/CORBIS; **672 (right),** John Heseltine/Science Photo Library/Photo Researchers, Inc.; **Fig. 23.4,** Andrea Mattevi and Wim G.J. Hol/Pearson Education/PH College; **678,** Michal Heron/Pearson Education/PH College; **679,** Roger Ressmeyer/CORBIS; **682,** Glenn Campbell/Getty Images, Inc.; **685,** Benelux/Photo Researchers, Inc.; **686,** David Kaplan/Pearson Education/PH College.

Chapter 24: Opener, Frank Lane Picture Agency/CORBIS; **694,** David Welling/Animals Animals/Earth Scenes; **695,** Charles D. Winters/Photo Researchers, Inc.; **700,** Michelle Garrett/CORBIS; **704,** Richard Megna/Fundamental Photographs; **705,** C Squared Studios/PhotoDisc, Inc.; **708,** Royalty Free/CORBIS; **711,** Kristen Brochmann/Fundamental Photographs.

Chapter 25: Opener, Robert Winslow/The Viesti Collection, Inc.; **722,** Robert L. Hamilton, Jr., University of California, San Francisco; **725,** John W. Finley; **730,** George Holton/Photo Researchers, Inc.

Chapter 26: Opener, James A. Sugar/CORBIS; **740,** Biophoto Associates/Science Source/Photo Researchers, Inc.; **743,** Larry Lefever/Grant Heilam Photography, Inc.; **744,** Prof. K. Seddon & Dr. T. Evans, Queen's University, Belfast/Photo Researchers, Inc.; **749,** Barros & Barros/Getty Images, Inc.; **750,** Chris Bjornberg/Photo Researchers, Inc.; **Fig. 26.4,** Reproduced by permission from H.J. Kreigstein and D.S. Hogness, *Proceedings of the National Academy of Sciences* 71:136 (1974), p. 137, Fig. 2.; **Fig. 26.7(b),** Ken Eward/Science Source/Photo Researchers, Inc.

Chapter 27: Opener, James King-Holmes/Photo Researchers, Inc.; **769,** Dennis Brack/Stockphoto.com; **771 (top),** BSIP/Ermakoff/Photo Researchers, Inc.; **771 (bottom),** Peter Menzel/Stock Boston; **773 (top),** SuperStock, Inc.; **773 (middle),** Pat Miller/Stockphoto.com; **773 (bottom),** Biophoto Associates/Photo Researchers, Inc.; **774,** Bob Luckey, Jr.; **778,** K.G. Murti/Visuals Unlimited; **779,** Sinclair Stammers/Photo Researchers, Inc.; **781,** Corbis RF; **782,** Peter Beyer, University of Freiburg, Germany; **784,** David M. Abell/Wedin Communications.

Chapter 28: Opener, C.J. Tollen/Bruce Coleman Inc.; **793,** Michael Dalton/Fundamental Photographs; **795,** Royalty Free/Corbis Digital Stock; **801 (top),** Huiying Li and Thomas L. Poulos, University of California, Irvine; **801 (bottom),** Richard Hutchings/PhotoEdit; **802,** Dan McCoy/Rainbow.

Chapter 29: Opener, Jim Olive/Peter Arnold, Inc.; **811,** Bryan F. Peterson/Corbis/Stock Market; **813,** Mark Burnett/Photo Researchers, Inc.; **816 (left),** Biology Media/Science Source/Photo Researchers, Inc.; **816 (right),** Photo Lennart Nilsson/Albert Bonniers Forlag; **819,** CNRI/Science Photo Library/Photo Researchers, Inc.; **Fig. 29.10,** Michal Heron/Pearson Education/PH College; **821,** Bill Longcore/Photo Researchers, Inc.; **825,** BSIP/Lenee/Photo Researchers, Inc.; **826,** David Kaplan/Pearson Education/PH College.

Index

Functional Groups of Importance in Biochemical Molecules

Functional Group	Structure	Type of Biomolecule
Amino group	$-NH_3^+, \ -NH_2$	Amino acids and proteins (Sections 18.3, 18.7)
Hydroxyl group	$-OH$	Monosaccharides (carbohydrates) and glycerol: a component of triacylglycerols (lipids) (Sections 22.4, 24.2)
Carbonyl group		Monosaccharides (carbohydrates); in acetyl group (CH_3CO) used to transfer carbon atoms during catabolism (Sections 22.4, 21.4, 21.8)
Carboxyl group		Amino acids, proteins, and fatty acids (lipids) (Sections 18.3, 18.7, 24.2)
Amide group		Links amino acids in proteins; formed by reaction of amino group and carboxyl group (Section 18.7)
Carboxylic acid ester		Triacylglycerols (and other lipids); formed by reaction of carboxyl group and hydroxyl group (Section 24.2)
Phosphates: mono-, di-, tri-		ATP and many metabolism intermediates (Sections 17.8, 21.5, and throughout metabolism sections)
Hemiacetal group		Cyclic forms of monosaccharides; formed by a reaction of carbonyl group with hydroxyl group (Section 16.7, 22.4)
Acetal group		Connects monosaccharides in disaccharides and larger carbohydrates; formed by reaction of carbonyl group with hydroxyl group (Sections 16.7, 22.7, 22.9)